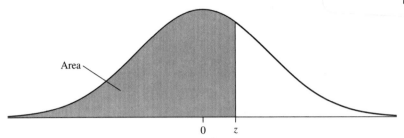

Area

z	0.00	0.01	0.02	0.03	0.04	0.05	0.06	0.07	0.08	0.09
0.0	.5000	.5040	.5080	.5120	.5160	.5199	.5239	.5279	.5319	.5359
0.1	.5398	.5438	.5478	.5517	.5557	.5596	.5636	.5675	.5714	.5753
0.2	.5793	.5832	.5871	.5910	.5948	.5987	.6026	.6064	.6103	.6141
0.3	.6179	.6217	.6255	.6293	.6331	.6368	.6406	.6443	.6480	.6517
0.4	.6554	.6591	.6628	.6664	.6700	.6736	.6772	.6808	.6844	.6879
0.5	.6915	.6950	.6985	.7019	.7054	.7088	.7123	.7157	.7190	.7224
0.6	.7257	.7291	.7324	.7357	.7389	.7422	.7454	.7486	.7517	.7549
0.7	.7580	.7611	.7642	.7673	.7704	.7734	.7764	.7794	.7823	.7852
0.8	.7881	.7910	.7939	.7967	.7995	.8023	.8051	.8078	.8106	.8133
0.9	.8159	.8186	.8212	.8238	.8264	.8289	.8315	.8340	.8365	.8389
1.0	.8413	.8438	.8461	.8485	.8508	.8531	.8554	.8577	.8599	.8621
1.1	.8643	.8665	.8686	.8708	.8729	.8749	.8770	.8790	.8810	.8830
1.2	.8849	.8869	.8888	.8907	.8925	.8944	.8962	.8980	.8997	.9015
1.3	.9032	.9049	.9066	.9082	.9099	.9115	.9131	.9147	.9162	.9177
1.4	.9192	.9207	.9222	.9236	.9251	.9265	.9279	.9292	.9306	.9319
1.5	.9332	.9345	.9357	.9370	.9382	.9394	.9406	.9418	.9429	.9441
1.6	.9452	.9463	.9474	.9484	.9495	.9505	.9515	.9525	.9535	.9545
1.7	.9554	.9564	.9573	.9582	.9591	.9599	.9608	.9616	.9625	.9633
1.8	.9641	.9649	.9656	.9664	.9671	.9678	.9686	.9693	.9699	.9706
1.9	.9713	.9719	.9726	.9732	.9738	.9744	.9750	.9756	.9761	.9767
2.0	.9772	.9778	.9783	.9788	.9793	.9798	.9803	.9808	.9812	.9817
2.1	.9821	.9826	.9830	.9834	.9838	.9842	.9846	.9850	.9854	.9857
2.2	.9861	.9864	.9868	.9871	.9875	.9878	.9881	.9884	.9887	.9890
2.3	.9893	.9896	.9898	.9901	.9904	.9906	.9909	.9911	.9913	.9916
2.4	.9918	.9920	.9922	.9925	.9927	.9929	.9931	.9932	.9934	.9936
2.5	.9938	.9940	.9941	.9943	.9945	.9946	.9948	.9949	.9951	.9952
2.6	.9953	.9955	.9956	.9957	.9959	.9960	.9961	.9962	.9963	.9964
2.7	.9965	.9966	.9967	.9968	.9969	.9970	.9971	.9972	.9973	.9974
2.8	.9974	.9975	.9976	.9977	.9977	.9978	.9979	.9979	.9980	.9981
2.9	.9981	.9982	.9982	.9983	.9984	.9984	.9985	.9985	.9986	.9986
3.0	.9987	.9987	.9987	.9988	.9988	.9989	.9989	.9989	.9990	.9990
3.1	.9990	.9991	.9991	.9991	.9992	.9992	.9992	.9992	.9993	.9993
3.2	.9993	.9993	.9994	.9994	.9994	.9994	.9994	.9995	.9995	.9995
3.3	.9995	.9995	.9995	.9996	.9996	.9996	.9996	.9996	.9996	.9997
3.4	.9997	.9997	.9997	.9997	.9997	.9997	.9997	.9997	.9997	.9998
3.5	.9998	.9998	.9998	.9998	.9998	.9998	.9998	.9998	.9998	.9998
3.6	.9998	.9998	.9999	.9999	.9999	.9999	.9999	.9999	.9999	.9999
3.7 or more	.9999									

Table A.3 Critical Values for the Student's *t* Distribution

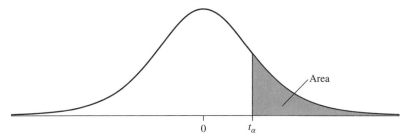

Degrees of Freedom	Area in Right Tail									
	0.40	0.25	0.10	0.05	0.025	0.01	0.005	0.0025	0.001	0.0005
1	0.325	1.000	3.078	6.314	12.706	31.821	63.657	127.321	318.309	636.619
2	0.289	0.816	1.886	2.920	4.303	6.965	9.925	14.089	22.327	31.599
3	0.277	0.765	1.638	2.353	3.182	4.541	5.841	7.453	10.215	12.924
4	0.271	0.741	1.533	2.132	2.776	3.747	4.604	5.598	7.173	8.610
5	0.267	0.727	1.476	2.015	2.571	3.365	4.032	4.773	5.893	6.869
6	0.265	0.718	1.440	1.943	2.447	3.143	3.707	4.317	5.208	5.959
7	0.263	0.711	1.415	1.895	2.365	2.998	3.499	4.029	4.785	5.408
8	0.262	0.706	1.397	1.860	2.306	2.896	3.355	3.833	4.501	5.041
9	0.261	0.703	1.383	1.833	2.262	2.821	3.250	3.690	4.297	4.781
10	0.260	0.700	1.372	1.812	2.228	2.764	3.169	3.581	4.144	4.587
11	0.260	0.697	1.363	1.796	2.201	2.718	3.106	3.497	4.025	4.437
12	0.259	0.695	1.356	1.782	2.179	2.681	3.055	3.428	3.930	4.318
13	0.259	0.694	1.350	1.771	2.160	2.650	3.012	3.372	3.852	4.221
14	0.258	0.692	1.345	1.761	2.145	2.624	2.977	3.326	3.787	4.140
15	0.258	0.691	1.341	1.753	2.131	2.602	2.947	3.286	3.733	4.073
16	0.258	0.690	1.337	1.746	2.120	2.583	2.921	3.252	3.686	4.015
17	0.257	0.689	1.333	1.740	2.110	2.567	2.898	3.222	3.646	3.965
18	0.257	0.688	1.330	1.734	2.101	2.552	2.878	3.197	3.610	3.922
19	0.257	0.688	1.328	1.729	2.093	2.539	2.861	3.174	3.579	3.883
20	0.257	0.687	1.325	1.725	2.086	2.528	2.845	3.153	3.552	3.850
21	0.257	0.686	1.323	1.721	2.080	2.518	2.831	3.135	3.527	3.819
22	0.256	0.686	1.321	1.717	2.074	2.508	2.819	3.119	3.505	3.792
23	0.256	0.685	1.319	1.714	2.069	2.500	2.807	3.104	3.485	3.768
24	0.256	0.685	1.318	1.711	2.064	2.492	2.797	3.091	3.467	3.745
25	0.256	0.684	1.316	1.708	2.060	2.485	2.787	3.078	3.450	3.725
26	0.256	0.684	1.315	1.706	2.056	2.479	2.779	3.067	3.435	3.707
27	0.256	0.684	1.314	1.703	2.052	2.473	2.771	3.057	3.421	3.690
28	0.256	0.683	1.313	1.701	2.048	2.467	2.763	3.047	3.408	3.674
29	0.256	0.683	1.311	1.699	2.045	2.462	2.756	3.038	3.396	3.659
30	0.256	0.683	1.310	1.697	2.042	2.457	2.750	3.030	3.385	3.646
31	0.256	0.682	1.309	1.696	2.040	2.453	2.744	3.022	3.375	3.633
32	0.255	0.682	1.309	1.694	2.037	2.449	2.738	3.015	3.365	3.622
33	0.255	0.682	1.308	1.692	2.035	2.445	2.733	3.008	3.356	3.611
34	0.255	0.682	1.307	1.691	2.032	2.441	2.728	3.002	3.348	3.601
35	0.255	0.682	1.306	1.690	2.030	2.438	2.724	2.996	3.340	3.591
36	0.255	0.681	1.306	1.688	2.028	2.434	2.719	2.990	3.333	3.582
37	0.255	0.681	1.305	1.687	2.026	2.431	2.715	2.985	3.326	3.574
38	0.255	0.681	1.304	1.686	2.024	2.429	2.712	2.980	3.319	3.566
39	0.255	0.681	1.304	1.685	2.023	2.426	2.708	2.976	3.313	3.558
40	0.255	0.681	1.303	1.684	2.021	2.423	2.704	2.971	3.307	3.551
50	0.255	0.679	1.299	1.676	2.009	2.403	2.678	2.937	3.261	3.496
60	0.254	0.679	1.296	1.671	2.000	2.390	2.660	2.915	3.232	3.460
80	0.254	0.678	1.292	1.664	1.990	2.374	2.639	2.887	3.195	3.416
100	0.254	0.677	1.290	1.660	1.984	2.364	2.626	2.871	3.174	3.390
200	0.254	0.676	1.286	1.653	1.972	2.345	2.601	2.839	3.131	3.340
z	0.253	0.674	1.282	1.645	1.960	2.326	2.576	2.807	3.090	3.291
	20%	50%	80%	90%	95%	98%	99%	99.5%	99.8%	99.9%
					Confidence Level					

Elementary STATISTICS

Second Edition

William Navidi
Colorado School of Mines

Barry Monk
Middle Georgia State College

McGraw Hill Education

ELEMENTARY STATISTICS, SECOND EDITION

Published by McGraw-Hill Education, 2 Penn Plaza, New York, NY 10121. Copyright © 2016 by McGraw-Hill Education. All rights reserved. Printed in the United States of America. Previous editions © 2013. No part of this publication may be reproduced or distributed in any form or by any means, or stored in a database or retrieval system, without the prior written consent of McGraw-Hill Education, including, but not limited to, in any network or other electronic storage or transmission, or broadcast for distance learning.

Some ancillaries, including electronic and print components, may not be available to customers outside the United States.

This book is printed on acid-free paper.

1 2 3 4 5 6 7 8 9 0 DOW/DOW 1 0 9 8 7 6 5

ISBN 978-0-07-783635-1
MHID 0-07-783635-9

ISBN 978-1-259-29248-4 (Annotated Instructor's Edition)
MHID 1-259-29248-7

Senior Vice President, Products & Markets: *Kurt L. Strand*
Vice President, General Manager, Products & Markets: *Marty Lange*
Vice President, Content Design & Delivery: *Kimberly Meriwether David*
Managing Director: *Ryan Blankenship*
Director, Product Development: *Rose Koos*
Product Developer: *Ashley Zellmer McFadden*
Marketing Manager: *Alex Gay*
Marketing Specialist: *Cherie Harshman*
Director of Digital Content Development: *Robert Brieler*
Digital Product Analyst: *Michael Lemke*
Digital Product Developer: *Adam Fischer*
Director, Content Design & Delivery: *Linda Avenarius*
Program Manager: *Rachel Townsend*
Content Project Managers: *Brent dela Cruz,Peggy J. Selle*
Buyer: *Laura M. Fuller*
Design: *Matt Backhaus*
Content Licensing Specialist: *Lori Hancock*
Cover Image: © *E+/Getty images; back cover image: © Score by Aflo/Getty images*
Compositor: *Laserwords Private Limited*
Typeface: *10/12 Times LT Std Roman*
Printer: *R. R. Donnelley*

All credits appearing on page or at the end of the book are considered to be an extension of the copyright page.

Library of Congress Cataloging-in-Publication Data
Navidi, William Cyrus.
  Elementary statistics / William Navidi, Colorado School of Mines, Barry Monk, Middle Georgia State College.—Second edition.
    pages cm
  ISBN 978-0-07-783635-1 (alk. paper)
  1. Mathematical statistics—Textbooks. I. Monk, Barry (Barry J.) II. Title.
  QA276.12.N385 2016
  519.5—dc23

                                                                      2014020908

The Internet addresses listed in the text were accurate at the time of publication. The inclusion of a website does not indicate an endorsement by the authors or McGraw-Hill Education, and McGraw-Hill Education does not guarantee the accuracy of the information presented at these sites.

www.mhhe.com

To Catherine, Sarah, and Thomas
—William Navidi

To Shaun and Dawn
—Barry Monk

About the Authors

William Navidi   is a professor of Applied Mathematics and Statistics at the Colorado School of Mines in Golden, Colorado. He received a Bachelor's degree in Mathematics from New College, a Master's degree in Mathematics from Michigan State University, and a Ph.D. in Statistics from the University of California at Berkeley. Bill began his teaching career at the County College of Morris, a two-year college in Dover, New Jersey. He has taught mathematics and statistics at all levels, from developmental through the graduate level. Bill has written two Engineering Statistics textbooks for McGraw-Hill. In his spare time, he likes to play racquetball.

Barry Monk   is a Professor of Mathematics and serves as the Department Chair of Mathematics at Middle Georgia State College in Macon, Georgia. Barry received a Bachelor of Science in Mathematical Statistics, a Master of Arts in Mathematics specializing in Optimization and Statistics, and a Ph.D. in Applied Mathematics, all from the University of Alabama. Barry has been teaching Introductory Statistics since 1992 in the classroom and online environments. Barry has a minor in Creative Writing and is a skilled jazz pianist.

Brief Contents

Contents

CHAPTER **1**   **Basic Ideas   1**

CHAPTER **2**   **Graphical Summaries of Data   35**

CHAPTER **3**   **Numerical Summaries of Data   91**

Preface

This book is designed for an introductory course in statistics. The mathematical prerequisite is basic algebra. In addition to presenting the mechanics of the subject, we have endeavored to explain the concepts behind them, in a writing style as straightforward, clear, and engaging as we could make it. As practicing statisticians, we have done everything possible to ensure that the material presented is accurate and correct. We believe that this book will enable instructors to explore statistical concepts in depth yet remain easy for students to read and understand.

To achieve this goal, we have incorporated a number of useful pedagogical features:

Features

- **Check Your Understanding Exercises:** After each concept is explained, one or more exercises are immediately provided for students to be sure they are following the material. These exercises provide students with confidence that they are ready to go on, or alert them to the need to review the material just covered.

- **Explain It Again:** Many important concepts are reinforced with additional explanation in these marginal notes.

- **Real Data:** Statistics instructors universally agree that the use of real data engages students and convinces them of the usefulness of the subject. A great many of the examples and exercises use real data. Some data sets explore topics in health or social sciences, while others are based in popular culture such as movies, contemporary music, or video games.

- **Integration of Technology:** Many examples contain screenshots from the TI-84 Plus calculator, MINITAB, and Excel. Each section contains detailed, step-by-step instructions, where applicable, explaining how to use these forms of technology to carry out the procedures explained in the text.

- **Interpreting Technology:** Many exercises present output from technology and require the student to interpret the results.

- **Write About It:** These exercises, found at the end of each chapter, require students to explain statistical concepts in their own words.

- **Case Studies:** Each chapter begins with a discussion of a real problem. At the end of the chapter, a case study demonstrates applications of chapter concepts to the problem.

Flexibility

We have endeavored to make our book flexible enough to work effectively with a wide variety of instructor styles and preferences. We cover both the *P*-value and critical value approaches to hypothesis testing, so instructors can choose to cover either or both of these methods. The material on two-sample inference is divided into two chapters—Chapter 10 on two-sample confidence intervals, and Chapter 11 on two-sample hypothesis tests. This gives instructors the option of covering all the material on confidence intervals before starting hypothesis testing, by covering Chapter 10 immediately after Chapter 8.

We have placed the material on descriptive statistics for bivariate data immediately following descriptive statistics for univariate data. Those who wish to cover bivariate description and inference together may postpone Chapter 4 until sometime before covering Chapter 13.

Instructors differ widely in their preferences regarding the depth of coverage of probability. A light treatment of the subject may be obtained by covering Section 5.1 and skipping the rest of the chapter. More depth can be obtained by covering Sections 5.2 and 5.3. Section 5.4 on counting can be included for an even more comprehensive treatment.

Supplements

Supplements, including online homework, videos, guided student notes, and PowerPoint presentations, play an increasingly important role in the educational process. As authors, we have adopted a hands-on approach to the development of our supplements, to make sure that they are consistent with the style of the text and that they work effectively with a variety of instructor preferences. In particular, our online homework package offers instructors the flexibility to choose whether the solutions that students view are based on tables or technology, where applicable.

New in This Edition

The second edition of the book is intended to extend the strengths of the first. Some of the changes are:

- A new chapter on Nonparametric Statistics has been added.
- A new section on Multiple Regression has been added.
- The material on percentiles and quartiles has been rewritten to make it easier for those who wish to cover quartiles without covering percentiles.
- A large number of new exercises have been included, many of which involve real data from recent sources.
- Several of the case studies have been updated.
- The exposition has been improved in a number of places.

William Navidi
Barry Monk

Acknowledgments

We are indebted to many people for contributions at every stage of development. Colleagues and students who reviewed the evolving manuscript provided many valuable suggestions. In particular, John Trimboli, Don Brown, and Duane Day contributed to the supplements, and Mary Wolfe helped create the video presentations. Ashlyn Munson contributed a number of exercises, and Jackie Miller and Tim Chappell played an important role in the development of our digital content.

The staff at McGraw-Hill has been extremely capable and supportive. Content Project Manager Peggy Selle was always patient and helpful. Robert Brieler was superb in directing the development of our digital content. We appreciate the guidance of Ryan Blankenship and Ashley Zellmer McFadden, whose input has considerably improved the final product.

William Navidi
Barry Monk

Reviewers

Over 150 instructors reviewed the first and second edition manuscript to give feedback on content, design, pedagogy, and organization that was used to guide the direction of the text.

Thomas Achatz, *Washtenaw Community College*
Andrea Adlman, *Ventura College*
Raid Amin, *University of West Florida*
Billie Anderson, *Bryant University*
Diana Asmus, *Greenville Technical College*
John Avioli, *Christopher Newport University*
Charla Baker, *Middle Georgia State College*
Robert Bass, *Gardner-Webb University*
Robbin Bates-Yelverton, *Park University*
Lynn Beckett-Lemus, *El Camino College*
Diane Benner, *Harrisburg Area Community College*
Abraham Biggs, *Broward College*
Gregory Bloxom, *Pensacola State College*
Donna Brouilette, *Georgia Perimeter College*
Allen Brown, *Wabash Valley College*
Denise Brown, *Collin County Community College*
Don Brown, *Middle Georgia State College*
Mary Brown, *Harrisburg Area Community College*
Jennifer Bryan, *Oklahoma Christian University*
William Burgin, *Gaston College*
Gerald Busald, *San Antonio College*
David Busekist, *Southeastern Louisiana University*
Lynn Cade, *Pensacola State College*
Elizabeth Carrico, *Illinois Central College*
Connie Carroll, *Guilford Technical Community College*
Joseph Castillo, *Broward College*
Linda Chan, *Mount San Antonio College & Pasadena City College*

Ayona Chatterjee, *University of West Georgia*
Chand Chauhan, *Indiana University Purdue University Fort Wayne*
Pinyuen Chen, *Syracuse University*
Askar Choudhury, *Illinois State University*
Lee Clendenning, *State College Gainesville*
James Condor, *State College of Florida-Manatee*
John Curran, *Eastern Michigan University*
John Daniels, *Central Michigan University*
Larry Dumais, *American River College*
Christina Dwyer, *State College of Florida-Manatee*
Wayne Ehler, *Anne Arundel Community College*
Mark Ellis, *Central Piedmont Community College*
Franco Fedele, *University of West Florida*
Art Fortgang, *Southern Oregon University*
Thomas Fox, *Cleveland State Community College*
Kristin Good, *Washtenaw Community College*
Carrie Grant, *Flagler College*
Delbert Greear, *Gainesville State College*
Jason Greshman, *Nova Southeastern University*
David Gurney, *Southeastern Louisiana University*
Chris Hail, *Union University-Jackson*
Phillip Harris, *Illinois Central College*
Matthew He, *Nova Southeastern University*
Mary Beth Headlee, *State College of Florida-Manatee*
James Helmreich, *Marist College*
Todd Hendricks, *Georgia Perimeter College-Newton Campus*

Jada Hill, *Richland College*

Mary Hill, *College of DuPage*

Patricia Humphrey, *Georgia Southern University*

Nancy Johnson, *State College of Florida-Manatee*

June Jones, *Middle Georgia State College*

Maryann Justinger, *Erie Community College-South Campus*

Joseph Karnowski, *Norwalk Community College*

Joseph Kazimir, *East Los Angeles College*

Esmarie Kennedy, *San Antonio College*

Lynette Kenyon, *Collin College*

Gary Kersting, *North Central Michigan College*

Raja Khoury, *Collin College*

Heidi Kiley, *Suffolk County Community College-Selden*

Daniel Kim, *Southern Oregon University*

Ann Kirkpatrick, *Southeastern Louisiana University*

John Klages, *County College of Morris*

Karon Klipple, *San Diego City College*

Alex Kolesnik, *Ventura College*

Erica Kwiatkowski-Egizio, *Joliet Junior College*

William Langston, *Finger Lakes Community College*

Jiawei Liu, *Georgia State University*

Fujia Lu, *Endicott College*

Timothy Maharry, *Northwestern Oklahoma State University*

Aldo Maldonado, *Park University*

Kenneth Mann, *Catawba Valley Community College*

James Martin, *Christopher Newport University*

Erin Martin-Wilding, *Parkland College*

Amina Mathias, *Cecil College*

Karen McKarnin, *Allen Community College*

Penny Morris, *Polk Community College*

B. K. Mudunuri, *Long Beach City College-CalPoly Pomona*

Linda Myers, *Harrisburg Area Community College*

Miroslaw Mystkowski, *Gardner-Webb University*

Shai Neumann, *Brevard Community College*

Karen Orr, *Roane State Community College*

Richard Owens, *Park University*

Irene Palacios, *Grossmont College*

Luca Petrelli, *Mount Saint Mary's University*

Blanche Presley, *Middle Georgia State College*

Robert Prince, *Berry College*

Richard Puscas, *Georgia Perimeter College*

Ramaswamy Radhakrishnan, *Illinois State University*

Leela Rakesh, *Central Michigan University*

Gina Reed, *Gainesville State College*

Andrea Reese, *Daytona State College-Daytona Beach*

Lisa Rombes, *Washtenaw Community College*

Jason Rosenberry, *Harrisburg Area Community College*

Yolanda Rush, *Illinois Central College*

Loula Rytikova, *George Mason University*

Vicki Schell, *Pensacola State College*

Carol Schoen, *University of Wisconsin-Eau Claire*

Pali Sen, *University of North Florida*

Rosa Seyfried, *Harrisburg Area Community College*

Larry Shrewsbury, *Southern Oregon University*

Abdallah Shuaibi, *Truman College*

Rick Silvey, *University of Saint Mary*

Russell Simmons, *Brookhaven College*

Peggy Slavik, *University of Saint Mary*

Karen Smith, *University of West Georgia*

Victor Swaim, *Southeastern Louisiana University*

Scott Sykes, *University of West Georgia*

John Trimboli, *Middle Georgia State College*

Barbara Tucker, *Tarrant County College South East*

Christopher Turner, *Pensacola State College*

Dave Vinson, *Pellissippi State Community College*

David Vogel, *Middle Georgia State College*

Suzanne Walker, *Bryant University*

Joseph Walker, *Georgia State University*

Barbara Ward, *Belmont University*

Xiaohong Wang, *Central Michigan University*

Jason Willis, *Gardner-Webb University*

Joseph Wolkan, *Washtenaw Community College*

Fuzhen Zhang, *Nova Southeastern University*

Yichuan Zhao, *Georgia State University*

Deborah Ziegler, *Hannibal LaGrange College*

Bashar Zogheib, *Nova Southeastern University*

Stephanie Zwyghuizen, *Jamestown Community College*

Supplements

McGraw-Hill conducted in-depth research to create a new learning experience that meets the needs of students and instructors today. The result is a reinvented learning experience rich in information, visually engaging, and easily accessible to both instructors and students.

- McGraw-Hill's Connect is a web-based assignment and assessment platform that helps students connect to their coursework and prepares them to succeed in and beyond the course.

- Connect enables math and statistics instructors to create and share courses and assignments with colleagues and adjuncts with only a few clicks of the mouse. All exercises, learning objectives, and activities are vetted and developed by math instructors to ensure consistency between the textbook and the online tools.

- Connect also links students to an interactive eBook with access to a variety of media assets and a place to study, highlight and keep track of class notes.

To learn more contact your sales rep or visit www.mcgraw-hillconnect.com.

ALEKS ALEKS® is a web-based program that uses artificial intelligence to assess a student's knowledge and provide personalized instruction on the exact topics the student is most ready to learn. By providing individualized assessment and learning, ALEKS helps students to master course content quickly and easily. ALEKS allows students to easily move between explanations and practice, and provides intuitive feedback to help students correct and analyze errors. ALEKS also includes a powerful instructor module that simplifies course management so instructors spend less time with administrative tasks and more time directing student learning.

ALEKS 360® is a new cost-effective total course solution: fully integrated, interactive eBook, including lecture and exercise videos tied to the textbook, combined with ALEKS personalized assessment and learning.

To learn more about ALEKS and ALEKS 360 contact your sales rep or visit www. aleks.com.

LEARNSMART **Give students the tools they need. Right when they need them.** McGraw-Hill LearnSmart® adaptively assesses students' skill levels to determine which topics students have mastered and which require further practice. Then it delivers customized learning content based on their strengths and weaknesses. The result: students learn faster and more efficiently because they get the help they need, right when they need it - instead of getting stuck on lessons, or being continually frustrated with stalled progress.

SMARTBOOK **SmartBook™** is the first and only adaptive reading experience available for the higher education market. Powered by the intelligent and adaptive LearnSmart engine, SmartBook facilitates the reading process by identifying what content a student knows and doesn't know. As a student reads, the material continuously adapts to ensure the student is focused on the content he or she needs the most to close specific knowledge gaps.

With **McGraw-Hill Create**™, you can easily rearrange chapters, combine material from other content sources, and quickly upload content you have written like your course syllabus or teaching notes. Find the content you need in Create by searching through thousands of leading McGraw-Hill textbooks. Arrange your book to fit your teaching style. Create even allows you to personalize your book's appearance by selecting the cover and adding your name, school, and course information. Assemble a Create book and you'll receive a complimentary print review copy in 3–5 business days or a complimentary electronic review copy (eComp) via email in minutes. Go to www.mcgrawhillcreate.com today and experience how McGraw-Hill Create™ empowers you to teach *your* students *your* way.

Videos

Author-produced lecture videos introduce concepts, definitions, formulas, and problem-solving procedures to help students better comprehend the topic at hand. Exercise videos illustrate the authors working through selected exercises, following the solution methodology employed in the text. These videos are closed-captioned for the hearing-impaired, are subtitled in Spanish, and meet the Americans with Disabilities Act Standards for Accessible Design.

Instructor's Testing and Resource Online

This computerized test bank, available online to adopting instructors, utilizes TestGen® cross-platform test generation software to quickly and easily create customized exams. Using hundreds of test items taken directly from the text, TestGen allows rapid test creation and flexibility for instructors to create their own questions from scratch with ability to randomize number values. Powerful search and sort functions help quickly locate questions and arrange them in any order, and built-in mathematical templates let instructors insert stylized text, symbols, graphics, and equations directly into questions without need for a separate equation editor. A separate CD version is available upon request.

MegaStat®

MegaStat® is a statistical add-in for Microsoft Excel, handcrafted by J. B. Orris of Butler University. When MegaStat is installed, it appears as a menu item on the Excel menu bar and allows you to perform statistical analysis on data in an Excel workbook.

Instructor's Solutions Manual

Derived from author-approved Connect solutions, this manual contains detailed solutions to all of the problems in the text.

TI-84 Plus Graphing Calculator Manual

This friendly, author-influenced manual teaches students to learn about statistics and solve problems by using this calculator while following each text chapter.

Excel Manual

This workbook, specially designed to accompany the text by the authors, provides additional practice in applying the chapter concepts while using Excel.

MINITAB 14 Manual

With guidance from the authors, this manual includes material from the book to provide seamless use from one to the other, providing additional practice in applying the chapter concepts while using the MINITAB program.

Guided Student Notes

Guided notes provide instructors with the framework of day-by-day class activities for each section in the book. Each lecture guide can help instructors make more efficient use of class time and can help keep students focused on active learning. Students who use the lecture guides have the framework of well-organized notes that can be completed with the instructor in class.

Data Sets

from selected exercises have been pre-populated into MINITAB, TI-Graph Link, Excel, SPSS, and comma-delimited ASCII formats for student and instructor use. These files are available on the text's website as well as on the Data CD included with the text.

Annotated Instructor's Edition (instructors only)

The Annotated Instructor's Edition contains answers to all exercises. The answers to most questions are printed in blue next to each problem. Answers not appearing on the page can be found in the Answer Appendix at the end of the book.

Student's Solutions Manual

Derived from author-approved Connect solutions, this manual contains detailed solutions to all odd-numbered text problems and answers to all Quizzes, Reviews, and Case Study problems found at the end of each chapter.

Index of Applications

Basic Ideas

Introduction

How does air pollution affect your health? Over the past several decades, scientists have become increasingly convinced that air pollution is a serious health hazard. The World Health Organization has estimated that air pollution causes 2.4 million deaths each year. The health effects of air pollution have been investigated by measuring air pollution levels and rates of disease, then using statistical methods to determine whether higher levels of pollution lead to higher rates of disease.

Many air pollution studies have been conducted in the United States. For example, the town of Libby, Montana, was the focus of a recent study of the effect of particulate matter — air pollution that consists of microscopic particles — on the respiratory health of children. As part of this study, parents were asked to fill out a questionnaire about their children's respiratory symptoms. It turned out that children exposed to higher levels of particulate pollution were more likely to exhibit symptoms of wheezing, as shown in the following table.

Level of Exposure	Percentage with Symptoms
High	8.89%
Low	4.56%

The rate of symptoms was almost twice as high among those exposed to higher levels of pollution. At first, it might seem easy to conclude that higher levels of pollution cause symptoms of wheezing. However, drawing accurate conclusions from information like this is rarely that simple. The case study at the end of this chapter will present more complete information and will show that additional factors must be considered.

Sampling

Objectives

1. Construct a simple random sample
2. Determine when samples of convenience are acceptable
3. Describe stratified sampling, cluster sampling, systematic sampling, and voluntary response sampling
4. Distinguish between statistics and parameters

In the months leading up to an election, polls often tell us the percentages of voters that prefer each of the candidates. How do pollsters obtain this information? The ideal poll would be one in which every registered voter were asked his or her opinion. Of course, it is impossible to conduct such an ideal poll, because it is impossible to contact every voter. Instead, pollsters contact a relatively small number of voters, usually no more than a couple of thousand, and use the information from these voters to predict the preferences of the entire group of voters.

The process of polling requires two major steps. First, the voters to be polled must be selected and interviewed. In this way the pollsters collect information. In the second step, the pollsters analyze the information to make predictions about the upcoming election. Both the collection and the analysis of the information must be done properly for the results to be reliable. The field of statistics provides appropriate methods for the collection, description, and analysis of information.

> **DEFINITION**
>
> **Statistics** is the study of procedures for collecting, describing, and drawing conclusions from information.

The polling problem is typical of a problem in statistics. We want some information about a large group of individuals, but we are able to collect information on only a small part of that group. In statistical terminology, the large group is called a *population*, and the part of the group on which we collect information is called a *sample*.

Explain It Again

Why do we draw samples?
It's usually impossible to examine every member of a large population. So we select a group of a manageable size to examine. This group is called a sample.

> **DEFINITION**
>
> - A **population** is the entire collection of individuals about which information is sought.
> - A **sample** is a subset of a population, containing the individuals that are actually observed.

Ideally, we would like our sample to represent the population as closely as possible. For example, in a political poll, we would like the proportions of voters preferring each of the candidates to be the same in the sample as in the population. Unfortunately, there are no methods that can guarantee that a sample will represent the population well. The best we can do is to use a method that makes it very likely that the sample will be similar to the population. The best sampling methods all involve some kind of random selection. The most basic, and in many cases the best, sampling method is the method of **simple random sampling**.

Simple Random Sampling

Objective 1 Construct a simple random sample

To understand the nature of a simple random sample, think of a lottery. Imagine that 10,000 lottery tickets have been sold, and that 5 winners are to be chosen. What is the fairest way to choose the winners? The fairest way is to put the 10,000 tickets in a drum, mix them thoroughly, then reach in and draw 5 tickets out one by one. These 5 winning tickets are a simple random sample from the population of 10,000 lottery tickets. Each ticket is equally likely to be one of the 5 tickets drawn. More importantly, each collection of 5 tickets that can be formed from the 10,000 is equally likely to comprise the group of 5 that is drawn.

> **DEFINITION**
>
> A **simple random sample** of size n is a sample chosen by a method in which each collection of n population items is equally likely to make up the sample, just as in a lottery.

Since a simple random sample is analogous to a lottery, it can often be drawn by the same method now used in many lotteries: with a computer random number generator. Suppose there are N items in the population. We number the items 1 through N. Then we generate a list of random integers between 1 and N, and choose the corresponding population items to comprise the simple random sample.

EXAMPLE 1.1

Choosing a simple random sample

There are 300 employees in a certain company. The Human Resources department wants to draw a simple random sample of 20 employees to fill out a questionnaire about their attitudes toward their jobs. Describe how technology can be used to draw this sample.

Solution

Step 1: Make a list of all 300 employees, and number them from 1 to 300.
Step 2: Use a random number generator on a computer or a calculator to generate 20 random numbers between 1 and 300. The employees who correspond to these numbers comprise the sample.

EXAMPLE 1.2

Determining whether a sample is a simple random sample

A physical education professor wants to study the physical fitness levels of students at her university. There are 20,000 students enrolled at the university, and she wants to draw a sample of size 100 to take a physical fitness test. She obtains a list of all 20,000 students, numbered from 1 to 20,000. She uses a computer random number generator to generate 100 random integers between 1 and 20,000, then invites the 100 students corresponding to those numbers to participate in the study. Is this a simple random sample?

Solution

Yes, this is a simple random sample because any group of 100 students would have been equally likely to have been chosen.

EXAMPLE 1.3

Determining whether a sample is a simple random sample

The professor in Example 1.2 now wants to draw a sample of 50 students to fill out a questionnaire about which sports they play. The professor's 10:00 A.M. class has 50 students. She uses the first 20 minutes of class to have the students fill out the questionnaire. Is this a simple random sample?

Solution

No. A simple random sample is like a lottery, in which each student in the population has an equal chance to be part of the sample. In this case, only the students in a particular class had a chance to be in the sample.

EXAMPLE 1.4

In a simple random sample, all samples are equally likely

To play the Colorado Lottery Lotto game, you must select six numbers from 1 to 42. Then lottery officials draw a simple random sample of six numbers from 1 to 42. If your six numbers match the ones in the simple random sample, you win the jackpot. Sally plays the lottery and chooses the numbers 1, 2, 3, 4, 5, 6. Her friend George says that this isn't a good choice, since it is very unlikely that a random sample will turn up the first six numbers. Is he right?

Solution

No. It is true that the combination 1, 2, 3, 4, 5, 6 is unlikely, but every other combination is equally unlikely. In a simple random sample of size 6, every collection of six numbers is equally likely (or equally unlikely) to come up. So Sally has the same chance as anyone to win the jackpot.

EXAMPLE 1.5

Using technology to draw a simple random sample

Use technology to draw a simple random sample of five employees from the following list.

1. Dan Aaron	11. Johnny Gaines	21. Jorge Ibarra	31. Edward Shingleton
2. Annie Bienh	12. Carlos Garcia	22. Maurice Jones	32. Michael Speciale
3. Oscar Bolivar	13. Julio Gonzalez	23. Jared Kerns	33. Andrew Steele
4. Dominique Bonnaud	14. Jacqueline Gordon	24. Kevin King	34. Neil Swain
5. Paul Campbell	15. James Graves	25. Frank Lipka	35. Sherry Thomas
6. Jeffrey Carnahan	16. Ronald Harrison	26. Carl Luther	36. Shequiea Thompson
7. Joel Chae	17. Andrew Huang	27. Laverne Mitchell	37. Barbara Tilford
8. Dustin Chen	18. Anthony Hunter	28. Zachary Quesada	38. Jermaine Tryon
9. Steven Coleman	19. Jonathan Jackson	29. Donnell Romaine	39. Lizbet Valdez
10. Richard Davis	20. Bruce Johnson	30. Gary Sanders	40. Katelyn Yu

Solution

We will use the TI-84 Plus graphing calculator. The step-by-step procedure is presented in the Using Technology section on page 9. We begin by choosing a **seed**, which is a number that the calculator uses to get the random number generator started. Display (a) shows the seed being set to 21. (The seed can be chosen in almost any way; this number was chosen by looking at the seconds display on a digital watch.) Display (b) presents the five numbers in the sample.

(a)

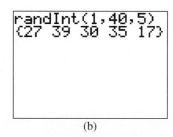

(b)

The simple random sample consists of the employees with numbers 27, 39, 30, 35, and 17. These are Laverne Mitchell, Lizbet Valdez, Gary Sanders, Sherry Thomas, and Andrew Huang.

CAUTION

If you use a different type of calculator, a different statistical package, or a different seed, you will get a different random sample. This is perfectly all right. So long as the sample is drawn by using a correct procedure, it is a valid random sample.

Check Your Understanding

1. A pollster wants to estimate the proportion of voters in a certain town who are Democrats. He goes to a large shopping mall and approaches people to ask whether they are Democrats. Is this a simple random sample? Explain.

2. A telephone company wants to estimate the proportion of customers who are satisfied with their service. They use a computer to generate a list of random phone numbers and call those people to ask them whether they are satisfied. Is this a simple random sample? Explain.

Answers are on page 12.

Samples of Convenience

In some cases, it is difficult or impossible to draw a sample in a truly random way. In these cases, the best one can do is to sample items by some convenient method. A sample obtained in such a way is called a *sample of convenience*.

DEFINITION

A **sample of convenience** is a sample that is not drawn by a well-defined random method.

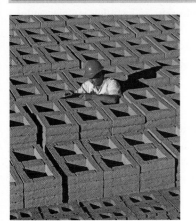

EXAMPLE 1.6

Drawing a sample of convenience

A construction engineer has just received a shipment of 1000 concrete blocks, each weighing approximately 50 pounds. The blocks have been delivered in a large pile. The engineer wishes to investigate the crushing strength of the blocks by measuring the strengths in a sample of 10 blocks. Explain why it might be difficult to draw a simple random sample of blocks. Describe how the engineer might draw a sample of convenience.

Solution

To draw a simple random sample would require removing blocks from the center and bottom of the pile, which might be quite difficult. One way to draw a sample of convenience would be to simply take 10 blocks off the top of the pile.

Problems with samples of convenience

The big problem with samples of convenience is that they may differ systematically in some way from the population. For this reason, samples of convenience should not be used, except in some situations where it is not feasible to draw a random sample. When it is necessary to draw a sample of convenience, it is important to think carefully about all the ways in which the sample might differ systematically from the population. If it is reasonable to believe that no important systematic difference exists, then it may be acceptable to treat the sample of convenience as if it were a simple random sample. With regard to the concrete blocks, if the engineer is confident that the blocks on the top of the pile do not differ systematically in any important way from the rest, then he can treat the sample of convenience as a simple random sample. If, however, it is possible that blocks in different parts of the pile may have been made from different batches of mix, or may have different curing times or temperatures, a sample of convenience could give misleading results.

SUMMARY

- A sample of convenience may be acceptable when it is reasonable to believe that there is no systematic difference between the sample and the population.
- A sample of convenience is not acceptable when it is possible that there is a systematic difference between the sample and the population.

Some Other Sampling Methods
Stratified sampling

In **stratified sampling**, the population is divided into groups, called **strata**, where the members of each stratum are similar in some way. Then a simple random sample is drawn from each stratum. Stratified sampling is useful when the strata differ from one another, but the individuals within a stratum tend to be alike.

EXAMPLE 1.7

Drawing a stratified sample

A company has 1000 employees, of whom 800 are full-time and 200 are part-time. The company wants to survey 50 employees about their opinions regarding benefits. Attitudes toward benefits may differ considerably between full-time and part-time employees. Why might it be a good idea to draw a stratified sample? Describe how one might be drawn.

Solution

If a simple random sample is drawn from the entire population of 1000 employees, it is possible that the sample will contain only a few part-time employees, and their attitudes will not be well represented. For this reason, it might be advantageous to draw a stratified sample. To draw a stratified sample, one would use two strata. One stratum would consist of the full-time employees, and the other would consist of the part-time employees. A simple random sample would be drawn from the full-time employees, and another simple random sample would be drawn from the part-time employees. This method guarantees that both full-time and part-time employees will be well represented.

Explain It Again

Example of a cluster sample: Imagine drawing a simple random sample of households, and interviewing every member of each household. This would be a cluster sample, with the households as the clusters.

Cluster sampling

In **cluster sampling**, items are drawn from the population in groups, or clusters. Cluster sampling is useful when the population is too large and spread out for simple random sampling to be feasible. Cluster sampling is used extensively by U.S. government agencies in sampling the U.S. population to measure sociological factors such as income and unemployment.

EXAMPLE 1.8

Drawing a cluster sample

To estimate the unemployment rate in a county, a government agency draws a simple random sample of households in the county. Someone visits each household and asks how many adults live in the household, and how many of them are unemployed. What are the clusters? Why is this a cluster sample?

Solution

The clusters are the groups of adults in each of the households in the county. This a cluster sample because a simple random sample of clusters is selected, and every individual in each selected cluster is part of the sample.

Explain It Again

The difference between cluster sampling and stratified sampling: In both cluster sampling and stratified sampling, the population is divided into groups. In stratified sampling, a simple random sample is chosen from each group. In cluster sampling, a random sample of groups is chosen, and every member of the chosen groups is sampled.

Systematic sampling

Imagine walking alongside a line of people and choosing every third one. That would produce a **systematic sample**. In a systematic sample, the population items are ordered. It is decided how frequently to sample items; for example, one could sample every third item, or every fifth item, or every hundredth item. Let k represent the sampling frequency. To begin the sampling, choose a starting place at random. Select the item in the starting place, along with every kth item after that.

Systematic sampling is sometimes used to sample products as they come off an assembly line, in order to check that they meet quality standards.

EXAMPLE 1.9

Describe a systematic sample

Automobiles are coming off an assembly line. It is decided to draw a systematic sample for a detailed check of the steering system. The starting point will be the third car, then every fifth car after that will be sampled. Which cars will be sampled?

Solution

We start with the third car, then count by fives to determine which cars will be sampled. The sample will consist of cars numbered 3, 8, 13, 18, and so on.

Voluntary response sampling

Voluntary response samples are often used by the media to try to engage the audience. For example, a news commentator will invite people to tweet an opinion, or a radio announcer will invite people to call the station to say what they think. How reliable are voluntary response samples? To put it simply, *voluntary response samples are never reliable*. People who go to the trouble to volunteer an opinion tend to have stronger opinions than is typical of the population. In addition, people with negative opinions are often more likely to volunteer their responses than those with positive opinions.

Figures 1.1–1.4 illustrate several valid methods of sampling.

Figure 1.1 Simple random sampling

Figure 1.2 Systematic sampling

Figure 1.3 Stratified sampling

Figure 1.4 Cluster sampling

Check Your Understanding

3. A radio talk-show host invites listeners to send an email to express their opinions on an upcoming election. More than 10,000 emails are received. What kind of sample is this?

4. Every 10 years, the U.S. Census Bureau attempts to count every person living in the United States. To check the accuracy of their count in a certain city, they draw a sample of census districts (roughly equivalent to a city block) and recount everyone in the sampled districts. What kind of sample is formed by the people who are recounted?

5. A public health researcher is designing a study of the effect of diet on heart disease. The researcher knows that the diets of men and women tend to differ, and that men are more susceptible to heart disease. To be sure that both men and women are well represented, the study comprises a simple random sample of 100 men and another simple random sample of 100 women. What kind of sample do these 200 people represent?

6. A college basketball team held a promotion at one of its games in which every twentieth person who entered the arena won a free basketball. What kind of sample do the winners represent?

Answers are on page 12.

Simple random sampling is the most basic method

Simple random sampling is not the only valid method of random sampling. But it is the most basic, and we will focus most of our attention on this method. From now on, unless otherwise stated, the terms *sample* and *random sample* will be taken to mean *simple random sample*.

Objective 4 Distinguish between statistics and parameters

Statistics and Parameters

We often use numbers to describe, or summarize, a sample or a population. For example, suppose that a pollster draws a sample of 500 likely voters in an upcoming election, and 68% of them say that the state of the economy is the most important issue for them. The quantity "68%" describes the sample. A number that describes a sample is called a *statistic*.

Explain It Again

Statistic and parameter: An easy way to remember these terms is that "statistic" and "sample" both begin with "s," and "parameter" and "population" both begin with "p."

> **DEFINITION**
>
> A **statistic** is a number that describes a sample.

Now imagine that the election takes place, and that one of the items on the ballot is a proposition to raise the sales tax to pay for the development of a new park downtown. Let's say that 53% of the voters vote in favor of the proposition. The quantity "53%" describes the population of voters who voted in the election. A number that describes a population is called a *parameter*.

> **DEFINITION**
>
> A **parameter** is a number that describes a population.

EXAMPLE 1.10   Distinguishing between a statistic and a parameter

Which of the following is a statistic and which is a parameter?

   a. 57% of the teachers at Central High School are female.

   b. In a sample of 100 surgery patients who were given a new pain reliever, 78% of them reported significant pain relief.

Solution

   a. The number 57% is a parameter, because it describes the entire population of teachers in the school.

   b. The number 78% is a statistic, because it describes a sample.

USING TECHNOLOGY

We use Example 1.5 to illustrate the technology steps.

TI-84 PLUS

Drawing a simple random sample

Step 1. Enter any nonzero number on the HOME screen as the seed.

Step 2. Press **STO >**

Step 3. Press **MATH**, select **PRB**, then **1: rand**, and then press **ENTER**. This enters the seed into the calculator memory. See Figure A, which uses the number 21 as the seed.

Step 4. Press **MATH**, select **PRB**, then **5: randInt**. Then enter **1, N, n**, where **N** is the population size and **n** is the desired sample size. In Example 1.5, we use $N = 40$ and $n = 5$ (Figure B).

Step 5. Press **ENTER**. The five values in the random sample for Example 1.5 are **27, 39, 30, 35, 17** (Figure C).

Figure A

Figure B

Figure C

MINITAB

Drawing a simple random sample

Step 1. Click **Calc**, then **Random Data**, then **Integer...**

Step 2. In the **Generate _____ rows of data** field, enter twice the desired sample size. For example, if the desired sample size is 10, enter 20. The reason for this is that some sample items may be repeated, and these will need to be deleted.

Step 3. In the **Store in column(s)** field, enter **C1**.

Step 4. Enter **1** for the **Minimum value** and the population size **N** for the **Maximum value**. We use **Maximum value** = 40 for Example 1.5. Click **OK**.

Step 5. **Column C1** of the worksheet will contain a list of randomly selected numbers between **1** and **N**. If any number appears more than once in **Column C1**, delete the replicates so that the number appears only once. For Example 1.5, our random sample begins with **16, 14, 30, 28, 17, ...** (Figure D).

↓	C1
1	16
2	14
3	30
4	28
5	17
6	13
7	4
8	8
9	6
10	15
11	35
12	5

Figure D

EXCEL

Drawing a simple random sample

Step 1. In **Column A**, enter the values **1** through the population size **N**. For Example 1.5, **N = 40**.

Step 2. In **Column B**, next to each value in **Column A**, enter the command **=rand( )**. This results in a randomly generated number between 0 and 1 in each cell in **Column B**.

Step 3. Select all values in **Columns A** and **B** and then click on the **Data** menu and select **Sort**.

Step 4. In the **Sort by** field, enter **Column B** and select **Smallest to Largest** in the **Order** field. Press **OK**. **Column A** now contains the random sample. Our random sample begins with **17, 12, 28, 20, 6, ...** (Figure E).

	A	B
1	17	0.919798
2	12	0.83333
3	28	0.105403
4	20	0.863496
5	6	0.820379
6	11	0.514735
7	36	0.051148
8	19	0.756971
9	39	0.826555
10	26	0.315564
11	8	0.119579
12	30	0.329793

Figure E

SECTION 1.1 Exercises

Exercises 1–6 are the Check Your Understanding exercises located within the section.

Understanding the Concepts

In Exercises 7–12, fill in each blank with the appropriate word or phrase.

7. The entire collection of individuals about which information is sought is called a _____ .

8. A _____ is a subset of a population.

9. A _____ is a type of sample that is analogous to a lottery.

10. A sample that is not drawn by a well-defined random method is called a _____ .

11. A _____ sample is one in which the population is divided into groups and a random sample of groups is drawn.

12. A _____ sample is one in which the population is divided into groups and a random sample is drawn from each group.

In Exercises 13–16, determine whether the statement is true or false. If the statement is false, rewrite it as a true statement.

13. A sample of convenience is never acceptable.

14. In a cluster sample, the population is divided into groups, and a random sample from each group is drawn.

15. Both stratified sampling and cluster sampling divide the population into groups.

16. One reason that voluntary response sampling is unreliable is that people with stronger views tend to express them more readily.

Practicing the Skills

In Exercises 17–20, determine whether the number described is a statistic or a parameter.

17. In a recent poll, 57% of the respondents supported a school bond issue.

18. The average age of the employees in a certain company is 35 years.

19. Of the students enrolled in a certain college, 80% are full-time.

20. In a survey of 500 high-school students, 60% of them said that they intended to go to college.

Exercises 21–24 refer to the population of animals in the following table. The population is divided into four groups: mammals, birds, reptiles, and fish.

Mammals		Birds	
1. Aardvark	6. Lion	11. Flamingo	16. Hawk
2. Buffalo	7. Zebra	12. Swan	17. Owl
3. Elephant	8. Pig	13. Sparrow	18. Chicken
4. Squirrel	9. Dog	14. Parrot	19. Duck
5. Rabbit	10. Horse	15. Pelican	20. Turkey
Reptiles		**Fish**	
21. Gecko	26. Python	31. Catfish	36. Shark
22. Iguana	27. Turtle	32. Tuna	37. Trout
23. Chameleon	28. Tortoise	33. Cod	38. Perch
24. Rattlesnake	29. Alligator	34. Salmon	39. Guppy
25. Boa constrictor	30. Crocodile	35. Goldfish	40. Minnow

21. **Simple random sample:** Draw a simple random sample of eight animals from the list of 40 animals in the table.

22. Another sample: Draw a sample of eight animals by drawing a simple random sample of two animals from each group. What kind of sample is this?

23. Another sample: Draw a simple random sample of two groups of animals from the four groups, and construct a sample of 20 animals by including all the animals in the sampled groups. What kind of sample is this?

24. Another sample: Choose a random number between 1 and 5. Include the animal with that number in your sample, along with every fifth animal thereafter, to construct a sample of eight animals. What kind of sample is this?

In Exercises 25–36, identify the kind of sample that is described.

25. Parking on campus: A college faculty consists of 400 men and 250 women. The college administration wants to draw a sample of 65 faculty members to ask their opinion about a new parking fee. They draw a simple random sample of 40 men and another simple random sample of 25 women.

26. Cruising the mall: A pollster walks around a busy shopping mall, and approaches people passing by to ask them how often they shop at the mall.

27. What's on TV? A pollster obtains a list of all the residential addresses in a certain town, and uses a computer random number generator to choose 150 of them. The pollster visits each of the 150 households and interviews all the adults in each household about their television viewing habits.

28. Don't drink and drive: Police at a sobriety checkpoint pull over every fifth car to determine whether the driver is sober.

29. Tell us your opinion: A television newscaster invites viewers to tweet their opinions on a proposed bill on immigration policy. More than 50,000 people express their opinions in this way.

30. Reading program: The superintendent of a large school district wants to test the effectiveness of a new program designed to improve reading skills among elementary school children. There are 30 elementary schools in the district. The superintendent chooses a simple random sample of five schools, and institutes the new reading program in those schools. A total of 4700 children attend these five schools.

31. Customer survey: All the customers who entered a store on a particular day were given a survey to fill out concerning their opinions of the service at the store.

32. Raffle: Five hundred people attend a charity event, and each buys a raffle ticket. The 500 ticket stubs are put in a drum and thoroughly mixed, and 10 of them are drawn. The 10 people whose tickets are drawn win a prize.

33. Hospital survey: The director of a hospital pharmacy chooses at random 100 people age 60 or older from each of three surrounding counties to ask their opinions of a new prescription drug program.

34. Bus schedule: Officials at a metropolitan transit authority want to get input from people who use a certain bus route about a possible change in the schedule. They randomly select 5 buses during a certain week and poll all riders on those buses about the change.

35. How much did you spend? A retailer samples 25 receipts from the past week by numbering all the receipts, generating 25 random numbers, and sampling the receipts that correspond to these numbers.

36. Phone features: A cell phone company wants to draw a sample of 600 customers to gather opinions about potential new features on upcoming phone models. The company draws a random sample of 200 from customers with iPhones, a random sample of 100 from customers with LG phones, a random sample of 100 from customers with Samsung phones, and a random sample of 200 from customers with other phones.

37. Computer network: Every third day, a computer network administrator analyzes the company's network logs to check for signs of computer viruses.

38. Smartphone apps: A smartphone app produces a message requesting customers to click on a link to rate the app.

Working with the Concepts

39. You're giving me a headache: A pharmaceutical company wants to test a new drug that is designed to provide superior relief from headaches. They want to select a sample of headache sufferers to try the drug. Do you think that it is feasible to draw a simple random sample of headache sufferers, or will it be necessary to use a sample of convenience? Explain your reasoning.

40. Pay more for recreation? The director of the recreation center at a large university wants to sample 100 students to ask them whether they would support an increase in their recreation fees in order to expand the hours that the center is open. Do you think that it is feasible to draw a simple random sample of students, or will it be necessary to use a sample of convenience? Explain your reasoning.

41. Voter preferences: A pollster wants to sample 500 voters in a town to ask them who they plan to vote for in an upcoming election. Describe a sampling method that would be appropriate in this situation. Explain your reasoning.

42. Quality control: Products come off an assembly line at the rate of several hundred per hour. It is desired to sample 10% of them to check whether they meet quality standards. Describe a sampling method that would be appropriate in this situation. Explain your reasoning.

43. On-site day care: A large company wants to sample 200 employees to ask their opinions about providing a day care center for the employees' children. They want to be sure to sample equal numbers of men and women. Describe a sampling method that would be appropriate in this situation. Explain your reasoning.

44. The tax man cometh: The Internal Revenue Service wants to sample 1000 tax returns that were submitted last year to determine the percentage of returns that had a refund. Describe a sampling method that would be appropriate in this situation. Explain your reasoning.

Extending the Concepts

45. Draw a sample: Imagine that you are asked to determine students' opinions at your school about a potential change in library hours. Describe how you could go about getting a sample of each of the following types: simple random sample, sample of convenience, voluntary response sample, stratified sample, cluster sample, systematic sample.

46. A systematic sample is a cluster sample: Explain how a systematic sample is actually a type of cluster sample.

Answers to Check Your Understanding Exercises for Section 1.1

1. No; this sample consists only of people in the town who visit the mall.

2. Yes; every group of n customers, where n is the sample size, is equally likely to be chosen.

3. Voluntary response sample

4. Cluster sample

5. Stratified sample

6. Systematic sample

SECTION 1.2   Types of Data

Objectives

1. Understand the structure of a typical data set
2. Distinguish between qualitative and quantitative variables
3. Distinguish between ordinal and nominal variables
4. Distinguish between discrete and continuous variables

Objective 1 Understand the structure of a typical data set

Data Sets

In Section 1.1, we described various methods of collecting information by sampling. Once the information has been collected, the collection is called a **data set**. A simple example of a data set is presented in Table 1.1, which shows the major, final exam score, and grade for several students in a certain statistics class.

Table 1.1 Major, Final Exam Score, and Grade for Several Students

Student	Major	Exam Score	Grade
1	Psychology	92	A
2	Business	75	B
3	Communications	82	B
4	Psychology	72	C
5	Art	85	B

Table 1.1 illustrates some basic features that are found in most data sets. Information is collected on **individuals**. In this example, the individuals are students. In many cases, individuals are people; in other cases, they can be animals, plants, or things. The characteristics of the individuals about which we collect information are called **variables**. In this example, the variables are major, exam score, and grade. Finally, the values of the variables that we obtain are the **data**. So, for example, the data for individual #1 are Major = Psychology, Exam score = 92, and Grade = A.

Check Your Understanding

1. A pollster asks a group of six voters about their political affiliation (Republican, Democrat, or Independent), their age, and whether they voted in the last election. The results are shown in the following table.

Voter	Political Affiliation	Age	Voted in Last Election?
1	Republican	34	Yes
2	Democrat	56	Yes
3	Democrat	21	No
4	Independent	28	Yes
5	Republican	61	No
6	Independent	46	Yes

a. How many individuals are there?
b. Identify the variables.
c. What are the data for individual #3?

Answers are on page 18.

Objective 2 Distinguish between qualitative and quantitative variables

Qualitative and Quantitative Variables

Variables can be divided into two types: qualitative and quantitative. **Qualitative variables**, also called **categorical variables**, classify individuals into categories. For example, college major and gender are qualitative variables. **Quantitative variables** are numerical and tell how much or how many of something there is. Height and score on an exam are examples of quantitative variables.

Explain It Again

Another way to distinguish qualitative from quantitative variables: Quantitative variables are counts or measurements, whereas qualitative variables are descriptions.

SUMMARY
- Qualitative variables classify individuals into categories.
- Quantitative variables tell how much or how many of something there is.

EXAMPLE 1.11

Distinguishing between qualitative and quantitative variables

Which of the following variables are qualitative and which are quantitative?

a. A person's age
b. A person's gender
c. The mileage (in miles per gallon) of a car
d. The color of a car

Solution

a. Age is quantitative. It tells how much time has elapsed since the person was born.
b. Gender is qualitative. It consists of the categories "male" and "female."
c. Mileage is quantitative. It tells how many miles a car will go on a gallon of gasoline.
d. Color is qualitative.

Objective 3 Distinguish between ordinal and nominal variables

Ordinal and Nominal Variables

Qualitative variables come in two types: **ordinal variables** and **nominal variables**. An ordinal variable is one whose categories have a natural ordering. The letter grade received in a class, such as A, B, C, D, or F, is an ordinal variable. A nominal variable is one whose

Figure 1.5 Qualitative variables come in two types: ordinal variables and nominal variables.

categories have no natural ordering. Gender is an example of a nominal variable. Figure 1.5 illustrates how qualitative variables are divided into nominal and ordinal variables.

SUMMARY

- Ordinal variables are qualitative variables whose categories have a natural ordering.
- Nominal variables are qualitative variables whose categories have no natural ordering.

EXAMPLE 1.12   Distinguishing between ordinal and nominal variables

Which of the following variables are ordinal and which are nominal?

  a. State of residence
  b. Gender
  c. Letter grade in a statistics class (A, B, C, D, or F)
  d. Size of soft drink ordered at a fast-food restaurant (small, medium, or large)

Solution

  a. State of residence is nominal. There is no natural ordering to the states.
  b. Gender is nominal.
  c. Letter grade in a statistics class is ordinal. The order, from high to low, is A, B, C, D, F.
  d. Size of soft drink is ordinal.

Objective 4 Distinguish between discrete and continuous variables

Discrete and Continuous Variables

Quantitative variables can be either *discrete* or *continuous*. **Discrete variables** are those whose possible values can be listed. Often, discrete variables result from counting something, so the possible values of the variable are 0, 1, 2, and so forth. **Continuous variables** can, in principle, take on any value within some interval. For example, height is a continuous variable because someone's height can be 68, or 68.1, or 68.1452389 inches. The possible values for height are not restricted to a list. Figure 1.6 illustrates how quantitative variables are divided into discrete and continuous variables.

Figure 1.6 Quantitative variables come in two types: discrete variables and continuous variables.

SUMMARY

- Discrete variables are quantitative variables whose possible values can be listed. The list may be infinite — for example, the list of all whole numbers.
- Continuous variables are quantitative variables that can take on any value in some interval. The possible values of a continuous variable are not restricted to any list.

EXAMPLE 1.13   Distinguishing between discrete and continuous variables

Which of the following variables are discrete and which are continuous?

  a. The age of a person at his or her last birthday
  b. The height of a person
  c. The number of siblings a person has
  d. The distance a person commutes to work

Solution

a. Age at a person's last birthday is discrete. The possible values are 0, 1, 2, and so forth.

b. Height is continuous. A person's height is not restricted to any list of values.

c. Number of siblings is discrete. The possible values are 0, 1, 2, and so forth.

d. Distance commuted to work is continuous. It is not restricted to any list of values.

Check Your Understanding

2. Which are qualitative and which are quantitative?
   a. The number of patients admitted to a hospital on a given day
   b. The model of car last sold by a particular car dealer
   c. The name of your favorite song
   d. The seating capacity of an auditorium

3. Which are nominal and which are ordinal?
   a. The names of the streets in a town
   b. The movie ratings G, PG, PG-13, R, and NC-17
   c. The winners of the gold, silver, and bronze medals in an Olympic swimming competition

4. Which are discrete and which are continuous?
   a. The number of female members of the U.S. House of Representatives
   b. The amount of water used by a household during a given month
   c. The number of stories in an apartment building
   d. A person's body temperature

Answers are on page 18.

SECTION 1.2   Exercises

Exercises 1–4 are the Check Your Understanding exercises located within the section.

Understanding the Concepts

In Exercises 5–10, fill in each blank with the appropriate word or phrase.

5. The characteristics of individuals about which we collect information are called _____.

6. Variables that classify individuals into categories are called _____.

7. _____ variables are always numerical.

8. Qualitative variables can be divided into two types: _____ and _____.

9. A _____ variable is a quantitative variable whose possible values can be listed.

10. _____ variables can take on any value in some interval.

In Exercises 11–14, determine whether the statement is true or false. If the statement is false, rewrite it as a true statement.

11. Qualitative variables describe how much or how many of something there is.

12. A nominal variable is a qualitative variable with no natural ordering.

13. A discrete variable is one whose possible values can be listed.

14. A person's height is an example of a continuous variable.

Practicing the Skills

In Exercises 15–24, determine whether the data described are qualitative or quantitative.

15. Your best friend's name

16. Your best friend's age

17. The number of touchdowns in a football game

18. The title of your statistics book

19. The number of files on a computer

20. The waist size of a pair of jeans

21. The ingredients in a recipe

22. Your school colors

23. The makes of cars sold by a particular car dealer

24. The number of cars sold by a car dealer last month

In Exercises 25–32, determine whether the data described are nominal or ordinal.

25. The categories Strongly disagree, Disagree, Neutral, Agree, and Strongly agree on a survey

26. The names of the counties in a state

27. The shirt sizes of Small, Medium, Large, and X-Large

28. I got an A in statistics, a B in biology, and C's in history and English.

29. This semester, I am taking statistics, biology, history, and English.

30. I ordered a pizza with pepperoni, mushrooms, olives, and onions.

31. In the track meet, I competed in the high jump and the pole vault.

32. I finished first in the high jump and third in the pole vault.

In Exercises 33–40, determine whether the data described are discrete or continuous.

33. The amount of caffeine in a cup of Starbucks coffee

34. The distance from a student's home to his school

35. The number of steps in a stairway

36. The number of students enrolled at a college

37. The amount of charge left in the battery of a cell phone

38. The number of patients who reported that a new drug had relieved their pain

39. The number of electrical outlets in a coffee shop

40. The time it takes for a text message to be delivered

Working with the Concepts

41. Ringtones: Following are the ten top-selling ringtones for the year 2012:

1. Alicia Keys — Girl on Fire
2. PSY — Gangam Style
3. Florida Georgia Line — Cruise
4. Taylor Swift — I Knew You Were Trouble
5. 2 Chainz — I'm Different
6. Rihanna — Diamonds
7. Gary Allan — Every Storm
8. Little Big Town — Pontoon
9. The Band Perry — Better Dig Two
10. Lil Wayne — No Worries

Source: www.billboard.com

Are these data nominal or ordinal?

42. More Ringtones: The following table presents the number of weeks that each of the ringtones in Exercise 41 spent on the top-ten chart as of February 2012.

Ringtone	Weeks in top ten
1. Alicia Keys — Girl on Fire	20
2. PSY — Gangam Style	19
3. Florida Georgia Line — Cruise	23
4. Taylor Swift — I Knew You Were Trouble	8
5. 2 Chainz — I'm Different	6
6. Rihanna — Diamonds	15
7. Gary Allan — Every Storm	8
8. Little Big Town — Pontoon	34
9. The Band Perry — Better Dig Two	8
10. Lil Wayne — No Worries	15

Source: www.billboard.com

Are these data discrete or continuous?

43. How's the economy? A poll conducted by the American Research Group asked individuals their views on how the economy will be a year from now. Respondents were given four choices: Better than today, Same as today, Worse than today, and Undecided. Are these choices nominal or ordinal?

44. Global warming: A Pew poll conducted in October 2012 asked people between the ages of 18 and 29 how serious a problem global warming is. Of those who responded, 43% thought it was very serious, 24% thought it was somewhat serious, 15% thought it was not too serious, and 17% thought it was not a problem. Are these percentages qualitative or quantitative?

45. Read any good books lately? According to *Time* magazine, some of the best fiction books in 2012 were:

This Is How You Lose Her by Junot Diaz
Where'd You Go, Bernadette? by Maria Semple

NW by Zadie Smith
Building Stories by Chris Ware
The Casual Vacancy by J.K. Rowling

Are these data nominal or ordinal?

46. Watch your language: According to MerriamWebster Online, the top ten Funny Sounding and Interesting words are:

1. Bumfuzzle	6. Snickersnee
2. Cattywampus	7. Widdershins
3. Gardyloo	8. Collywobbles
4. Taradiddle	9. Gubbins
5. Billingsgate	10. Diphthong

Are these data nominal or ordinal?

47. Top ten PC games: In February 2012, Nielsen Media published the following data about the top ten PC games:

Game Title	Publisher	Percentage of Gaming Audience	Average Minutes Played per Week
World of Warcraft	Blizzard Entertainment	7.154	500
League of Legends	Riot Games	3.776	402
Hanging Gardens of Babylon	Big Fish Games, Inc	6.894	267
Lord of the Rings Online: The Shadows of Angmar	Turbine	1.249	527
Half-Life 2	Vivendi Games	2.631	265
Warcraft III: Reign of Chaos	Blizzard Entertainment	1.062	364
The Sims 3	The Electronic Arts Inc.	1.308	227
Zuma's Revenge!–Adventure	PopCap Games	1.444	148
Dungeons & Dragons Online: Stormreach	Atari	0.388	445
Grand Theft Auto: San Andreas	Rockstar Games	0.752	188

a. Which of the columns represent qualitative variables?
b. Which of the columns represent quantitative variables?
c. Which of the columns represent nominal variables?
d. Which of the columns represent ordinal variables?

48. At the movies: The following table provides information about the top-grossing movies for the years 1997–2012.

Year	Movie Title	Creative Type	MPAA Rating	Ticket Sales (millions)	Tickets Sold (millions)
1997	*Men in Black*	Science fiction	PG-13	250.1	54.5
1998	*Titanic*	Dramatization	PG-13	443.3	94.5
1999	*Star Wars Ep. I: The Phantom Menace*	Science fiction	PG	430.4	84.7
2000	*How the Grinch Stole Christmas*	Kids fiction	PG	253.4	47.0
2001	*Harry Potter and the Sorcerer's Stone*	Fantasy	PG	291.6	51.5
2002	*Spider-Man*	Super hero	PG-13	403.7	69.5
2003	*Finding Nemo*	Kids fiction	G	339.7	56.3
2004	*Shrek 2*	Kids fiction	PG	436.5	70.3
2005	*Star Wars Ep. III: Revenge of the Sith*	Science fiction	PG-13	380.3	59.3
2006	*Pirates of the Caribbean: Dead Man's Chest*	Fantasy	PG-13	423.3	64.6
2007	*Spider-Man 3*	Super hero	PG-13	336.5	48.9
2008	*The Dark Knight*	Super hero	PG-13	531.0	74.0
2009	*Transformers: Revenge of the Fallen*	Science fiction	PG-13	402.1	53.6
2010	*Toy Story 3*	Kids fiction	G	415.0	52.6
2011	*Harry Potter and the Deathly Hallows: Part II*	Fantasy	PG-13	381.0	48.0
2012	*Marvel's The Avengers*	Super hero	PG-13	623.3	79.6

Source: http://www.the-numbers.com/

a. Which of the columns represent qualitative variables?
b. Which of the columns represent quantitative variables?
c. Which of the columns represent nominal variables?
d. Which of the columns represent ordinal variables?

Extending the Concepts

49. What do the numbers mean? A survey is administered by a marketing firm. Two of the people surveyed are Brenda and Jason. Three of the questions are as follows:

    i. Do you favor the construction of a new shopping mall?
      (1) Strongly oppose   (2) Somewhat oppose   (3) Neutral   (4) Somewhat favor   (5) Strongly favor
    ii. How many cars do you own?
    iii. What is your marital status?
      (1) Married   (2) Single   (3) Divorced   (4) Domestically partnered   (5) Other

  a. Are the responses for question (i) nominal or ordinal?

  b. On question (i), Brenda answers (2) and Jason answers (4). Jason's answer (4) is greater than Brenda's answer (2). Does Jason's answer reflect more of something?

  c. Jason's answer to question (i) is twice as large as Brenda's answer. Does Jason's answer reflect twice as much of something? Explain.

  d. Are the responses for question (ii) qualitative or quantitative?

  e. On question (ii), Brenda answers 2 and Jason answers 1. Does Brenda's answer reflect more of something? Does Brenda's answer reflect twice as much of something? Explain.

  f. Are the responses for question (iii) nominal or ordinal?

  g. On question (iii), Brenda answers (4) and Jason answers (2). Does Brenda's answer reflect more of something? Does Brenda's answer reflect twice as much of something? Explain.

Answers to Check Your Understanding Exercises for Section 1.2

1. a. 6   **b.** Political affiliation, Age, and Voted in last election
   c. Political affiliation = Democrat, Age = 21, Voted in last election = no

2. a. Quantitative   **b.** Qualitative   **c.** Qualitative
   d. Quantitative

3. a. Nominal   **b.** Ordinal   **c.** Ordinal

4. a. Discrete   **b.** Continuous   **c.** Discrete
   d. Continuous

SECTION 1.3   Design of Experiments

Objectives

1. Distinguish between a randomized experiment and an observational study
2. Understand the advantages of randomized experiments
3. Understand how confounding can affect the results of an observational study
4. Describe various types of observational studies

Objective 1 Distinguish between a randomized experiment and an observational study

Experiments and Observational Studies

Will a new drug help prevent heart attacks? Does one type of seed produce a larger wheat crop than another? Does exercise lower blood pressure? To illustrate how scientists address questions like these, we describe how a study might be conducted to determine which of three types of seed will result in the largest wheat yield.

- Prepare three identically sized plots of land, with similar soil types.
- Plant each type of seed on a different plot, choosing the plots at random.
- Water and fertilize the plots in the same way.
- Harvest the wheat, and measure the amount grown on each plot.
- If one type of seed produces substantially more (or less) wheat than the others, then scientists will conclude that it is better (or worse) than the others.

The following terminology is used for studies like this.

DEFINITION

The **experimental units** are the individuals that are studied. These can be people, animals, plants, or things. When the experimental units are people, they are sometimes called **subjects**.

In the wheat study just described, the experimental units are the three plots of land.

> **DEFINITION**
>
> The **outcome**, or **response**, is what is measured on each experimental unit.

In the wheat study, the outcome is the amount of wheat produced.

> **DEFINITION**
>
> The **treatments** are the procedures applied to each experimental unit. There are always two or more treatments. The purpose is to determine whether the choice of treatment affects the outcome.

In the wheat study, the treatments are the three types of seed.

In general, studies fall into two categories: *randomized experiments* and *observational studies*.

> **DEFINITION**
>
> A **randomized experiment** is a study in which the investigator assigns the treatments to the experimental units at random.

The wheat study described above is a randomized experiment. In some situations, randomized experiments cannot be performed, because it isn't possible to randomly assign the treatments. For example, in studies to determine how smoking affects health, people cannot be assigned to smoke. Instead, people choose for themselves whether to smoke, and scientists observe differences in health outcomes between groups of smokers and nonsmokers. Studies like this are called *observational studies*.

> **DEFINITION**
>
> An **observational study** is one in which the assignment to treatment groups is not made by the investigator.

When possible, it is better to assign treatments at random and perform a randomized experiment. As we will see, the results of randomized experiments are generally easier to interpret than the results of observational studies.

Objective 2 Understand the advantages of randomized experiments

Randomized Experiments

In July 2008, an article in *The New England Journal of Medicine* (359:339–354) reported the results of a study to determine whether a new drug called raltegravir is effective in reducing levels of virus in patients with human immunodeficiency virus (HIV). A total of 699 patients participated in the experiment. These patients were divided into two groups. One group was given raltegravir. The other group was given a placebo. (A placebo is a harmless tablet, such as a sugar tablet, that looks like the drug but has no medical effect.) Thus there were two treatments in this experiment, raltegravir and placebo.

The experimenters had decided to give raltegravir to about two-thirds of the subjects and the placebo to the others. To determine which patients would be assigned to which group, a simple random sample consisting of 462 of the 699 patients was drawn; this sample constituted the raltegravir group. The remaining 237 patients were assigned to the placebo group.

It was decided to examine subjects after 16 weeks and measure the levels of virus in their blood. Thus the outcome for this experiment was the number of copies of virus per milliliter of blood. Patients were considered to have a successful outcome if they had fewer than 50 copies of the virus per milliliter of blood. In the raltegravir group, 62% of the subjects had a successful outcome, but only 35% of the placebo group did. The conclusion was that raltegravir was effective in lowering the concentration of virus in HIV patients. We will examine this study, and determine why it was reasonable to reach this conclusion.

The raltegravir study was a randomized experiment, because the treatments were assigned to the patients at random. What are the advantages of randomized experiments? In a perfect study, the treatment groups would not differ from each other in any important way except that they receive different treatments. Then, if the outcomes differ among the groups, we may be confident that the differences in outcome must have been caused by differences in treatment. In practice, it is impossible to construct treatment groups that are exactly alike. But randomization does the next best thing. In a randomized experiment, any differences between the groups are likely to be small. In addition, the differences are due only to chance.

Because the raltegravir study was a randomized experiment, it is reasonable to conclude that the higher success rate in the raltegravir group was actually due to raltegravir.

SUMMARY

In a randomized experiment, if there are large differences in outcomes among the treatment groups, we can conclude that the differences are due to the treatments.

EXAMPLE 1.14   Identifying a randomized experiment

To assess the effectiveness of a new method for teaching arithmetic to elementary school children, a simple random sample of 30 first-graders was taught with the new method, and another simple random sample of 30 first-graders was taught with the currently used method. At the end of eight weeks, the children were given a test to assess their knowledge. What are the treatments in this study? Explain why this is a randomized experiment.

Solution

The treatments are the two methods of teaching. This is a randomized experiment because children were assigned to the treatment groups at random.

Double-blind experiments

We have described the advantages of assigning treatments at random. It is a further advantage if the assignment can be done in such a way that neither the experimenters nor the subjects know which treatment has been assigned to which subject. Experiments like this are called *double-blind* experiments. The raltegravir experiment was a double-blind experiment, because neither the patients nor the doctors treating them knew which patients were receiving the drug and which were receiving the placebo.

DEFINITION

An experiment is **double-blind** if neither the investigators nor the subjects know who has been assigned to which treatment.

Experiments should be run double-blind whenever possible, because when investigators or subjects know which treatment is being given, they may tend to report the results differently. For example, in an experiment to test the effectiveness of a new pain reliever, patients who know they are getting the drug may report their pain levels differently than those who know they are taking a placebo. Doctors can be affected as well; a doctor's diagnosis may be influenced by a knowledge of which treatment a patient received.

In some situations, it is not possible to run a double-blind experiment. For example, in an experiment that compares a treatment that involves taking medication to a treatment that involves surgery, both patients and doctors will know who got which treatment.

EXAMPLE 1.15   Determining whether an experiment is double-blind

Is the experiment described in Example 1.14 a double-blind experiment? Explain.

Solution

This experiment is not double-blind, because the teachers know whether they are using the new method or the old method.

Randomized block experiments

The type of randomized experiment we have discussed is sometimes called a **completely randomized experiment**, because there is no restriction on which subjects may be assigned which treatment. In some situations, it is desirable to restrict the randomization a bit. For example, imagine that two reading programs are to be tested in an elementary school that has children in grades 1 through 4. If children are assigned at random to the programs, it is possible that one of the programs will end up with more fourth-graders while the other one will end up with more first-graders. Since fourth-graders tend to be better readers, this will give an advantage to the program that happens to end up with more of them. This possibility can be avoided by randomizing the students within each grade separately. In other words, we randomly assign exactly half of the students within each grade to each reading program.

This type of experiment is called a **randomized block experiment**. In the example just discussed, each grade constitutes a block. In a randomized block experiment, the subjects are divided into blocks in such a way that the subjects in each block are the same or similar with regard to a variable that is related to the outcome. Age and gender are commonly used blocking variables. Then the subjects within each block are randomly assigned a treatment.

Observational Studies

Recall that an observational study is one in which the investigators do not assign the treatments. In most observational studies, the subjects choose their own treatments. Observational studies are less reliable than randomized experiments. To see why, imagine a study that is intended to determine whether smoking increases the risk of heart attack. Imagine that a group of smokers and a group of nonsmokers are observed for several years, and during that time a higher percentage of the smoking group experiences a heart attack. Does this prove that smoking increases the risk of heart attack? No. The problem is that the smoking group will differ from the nonsmoking group in many ways other than smoking, and these other differences may be responsible for differences in the rate of heart attacks. For example, smoking is more prevalent among men than among women. Therefore, the smoking group will contain a higher percentage of men than the nonsmoking group. It is known that men have a higher risk of heart attack than women. So the higher rate of heart attacks in the smoking group could be due to the fact that there are more men in the smoking group, and not to the smoking itself.

Objective 3 Understand how confounding can affect the results of an observational study

Confounding

The preceding example illustrates the major problem with observational studies. It is difficult to tell whether a difference in the outcome is due to the treatment or to some other difference between the treatment and control groups. This is known as **confounding**. In the preceding example, gender was a *confounder*. Gender is related to smoking (men are more likely to smoke) and to heart attacks (men are more likely to have heart attacks). For this reason, it is difficult to determine whether the difference in heart attack rates is due to differences in smoking (the treatment) or differences in gender (the confounder).

Explain It Again

Another way to describe a confounder: A confounder is something other than the treatment that can cause the treatment groups to have different outcomes.

> **SUMMARY**
>
> A **confounder** is a variable that is related to both the treatment and the outcome. When a confounder is present, it is difficult to determine whether differences in the outcome are due to the treatment or to the confounder.

How can we prevent confounding? One way is to design a study so that the confounder isn't a factor. For example, to determine whether smoking increases the risk of heart attack, we could compare a group of male smokers to a group of male nonsmokers, and a group of female smokers to a group of female nonsmokers. Gender wouldn't be a confounder here, because there would be no differences in gender between the smoking and nonsmoking groups. Of course, there are other possible confounders. Smoking rates vary among ethnic groups, and rates of heart attacks do, too. If people in ethnic groups that are more susceptible to heart attacks are also more likely to smoke, then ethnicity becomes a confounder. This can be dealt with by comparing smokers of the same gender and ethnic group to nonsmokers of that gender and ethnic group.

Designing observational studies that are relatively free of confounding is difficult. In practice, many studies must be conducted over a long period of time. In the case of smoking, this has been done, and we can be confident that smoking does indeed increase the risk of heart attack, along with other diseases. If you don't smoke, you have a much better chance to live a long and healthy life.

SUMMARY

In an observational study, when there are differences in the outcomes among the treatment groups, it is often difficult to determine whether the differences are due to the treatments or to confounding.

EXAMPLE 1.16

Determining the effect of confounding

In a study of the effects of blood pressure on health, a large group of people of all ages were given regular blood pressure checkups for a period of one year. It was found that people with high blood pressure were more likely to develop cancer than people with lower blood pressure. Explain how this result might be due to confounding.

Solution

Age is a likely confounder. Older people tend to have higher blood pressure than younger people, and older people are more likely to get cancer than younger people. Therefore people with high blood pressure may have higher cancer rates than younger people, even though high blood pressure does not cause cancer.

Check Your Understanding

1. To study the effect of air pollution on respiratory health, a group of people in a city with high levels of air pollution and another group in a rural area with low levels of pollution are examined to determine their lung capacity. Is this a randomized experiment or an observational study?

2. It is known that drinking alcohol increases the risk of contracting liver cancer. Assume that in an observational study, a group of smokers has a higher rate of liver cancer than a group of nonsmokers. Explain how this result might be due to confounding.

Answers are on page 26.

Objective 4 Describe various types of observational studies

Types of Observational Studies

There are two main types of observational studies: cohort studies and case-control studies. Cohort studies can be further divided into prospective, cross-sectional, and retrospective studies.

Cohort studies

In a **cohort study**, a group of subjects (the cohort) is studied to determine whether various factors of interest are associated with an outcome.

In a **prospective** cohort study, the subjects are followed over time. One of the most famous prospective cohort studies is the Framingham Heart Study. This study began in 1948 with 5209 men and women from the town of Framingham, Massachusetts. Every two years, these subjects are given physical exams and lifestyle interviews, which are studied to discover factors that increase the risk of heart disease. Much of what is known about the effects of diet and exercise on heart disease is based on this study.

Prospective studies are among the best observational studies. Because subjects are repeatedly examined, the quality of the data is often quite good. Information on potential confounders can be collected as well. Results from prospective studies are generally more reliable than those from other observational studies. The disadvantages of prospective studies are that they are expensive to run, and that it takes a long time to develop results.

In a **cross-sectional** study, measurements are taken at one point in time. For example, in a study published in the *Journal of the American Medical Association* (300:1303–1310), I. Lang and colleagues studied the health effects of bisphenol A, a chemical found in the linings of food and beverage containers. They measured the levels of bisphenol A in urine samples from 1455 adults. They found that people with higher levels of bisphenol A were more likely to have heart disease and diabetes.

Cross-sectional studies are relatively inexpensive, and results can be obtained quickly. The main disadvantage is that the exposure is measured at only one point in time, so there is little information about how past exposures may have contributed to the outcome. Another disadvantage is that because measurements are made at only one time, it is impossible to determine a time sequence of events. For example, in the bisphenol A study just described, it is possible that higher levels of bisphenol A cause heart disease and diabetes. But it is also possible that the onset of heart disease or diabetes causes levels of bisphenol A to increase. There is no way to determine which happened first.

In a **retrospective** cohort study, subjects are sampled after the outcome has occurred. The investigators then look back over time to determine whether certain factors are related to the outcome. For example, in a study published in *The New England Journal of Medicine* (357:753–761), T. Adams and colleagues sampled 9949 people who had undergone gastric bypass surgery between 5 and 15 years previously, along with 9668 obese patients who had not had bypass surgery. They looked back in time to see which patients were still alive. They found that the survival rates for the surgery patients were greater than for those who had not undergone surgery.

Retrospective cohort studies are less expensive than prospective cohort studies, and results can be obtained quickly. A disadvantage is that it is often impossible to obtain data on potential confounders.

One serious limitation of all cohort studies is that they cannot be used to study rare diseases. Even in a large cohort, very few people will contract a particular rare disease. To study rare diseases, case-control studies must be used.

Case-control studies

In a **case-control** study, two samples are drawn. One sample consists of people who have the disease of interest (the cases), and the other consists of people who do not have the disease (the controls). The investigators look back in time to determine whether a particular factor of interest differs between the two groups. For example, S. S. Nielsen and colleagues conducted a case-control study to determine whether exposure to pesticides is related to brain cancer in children (*Environmental Health Perspectives*, 118:144–149). They sampled 201 children under the age of 10 who had been diagnosed with brain cancer, and 285 children who did not have brain cancer. They interviewed the parents of the children to estimate the extent to which the children had been exposed to pesticides. They did not find a clear relationship between pesticide exposure and brain cancer. This study could not have been conducted as a cohort study, because even in a large cohort of children, very few will get brain cancer.

Case-control studies are always retrospective, because the outcome (case or control) has occurred before the sampling is done. Case-control studies have the same advantages and disadvantages as retrospective cohort studies. In addition, case-control studies have the advantage that they can be used to study rare diseases.

Check Your Understanding

3. In a study conducted at the University of Southern California, J. Peters and colleagues studied elementary school students in 12 California communities. Each year for 10 years, they measured the respiratory function of the children and the levels of air pollution in the communities.
   a. Was this a cohort study or a case-control study?
   b. Was the study prospective, cross-sectional, or retrospective?

4. In a study conducted at the University of Colorado, J. Ruttenber and colleagues studied people who had worked at the Rocky Flats nuclear weapons production facility near Denver, Colorado. They studied a group of workers who had contracted lung cancer, and another group who had not contracted lung cancer. They looked back at plant records to determine the amount of radiation exposure for each worker. The purpose of the study was to determine whether the people with lung cancer had been exposed to higher levels of radiation than those who had not gotten lung cancer.
   a. Was this a cohort study or a case-control study?
   b. Was the study prospective, cross-sectional, or retrospective?

Answers are on page 26.

SECTION 1.3   Exercises

Exercises 1–4 are the Check Your Understanding exercises located within the section.

Understanding the Concepts

In Exercises 5–10, fill in each blank with the appropriate word or phrase.

5. In a _____ experiment, subjects do not decide for themselves which treatment they will get.

6. In a _____ study, neither the investigators nor the subjects know who is getting which treatment.

7. A study in which the assignment to treatment groups is not made by the investigator is called _____.

8. A _____ is a variable related to both the treatment and the outcome.

9. In a _____ study, the subjects are followed over time.

10. In a _____ study, a group of subjects is studied to determine whether various factors of interest are associated with an outcome.

In Exercises 11–16, determine whether the statement is true or false. If the statement is false, rewrite it as a true statement.

11. In a randomized experiment, the treatment groups do not differ in any systematic way except that they receive different treatments.

12. A confounder makes it easier to draw conclusions from a study.

13. In an observational study, subjects are assigned to treatment groups at random.

14. Observational studies are generally more reliable than randomized experiments.

15. In a case-control study, the outcome has occurred before the subjects are sampled.

16. In a cross-sectional study, measurements are made at only one point in time.

Practicing the Skills

17. To determine the effectiveness of a new pain reliever, a randomly chosen group of pain sufferers is assigned to take the new drug, and another randomly chosen group is assigned to take a placebo.
    a. Is this a randomized experiment or an observational study?
    b. The subjects taking the new drug experienced substantially more pain relief than those taking the placebo. The research team concluded that the new drug is effective in relieving pain. Is this conclusion well justified? Explain.

18. A medical researcher wants to determine whether exercising can lower blood pressure. At a health fair, he measures the blood pressure of 100 individuals, and interviews them about their exercise habits. He divides the individuals into two categories: those whose typical level of exercise is low, and those whose level of exercise is high.
    a. Is this a randomized experiment or an observational study?

b. The subjects in the low-exercise group had considerably higher blood pressure, on the average, than subjects in the high-exercise group. The researcher concluded that exercise decreases blood pressure. Is this conclusion well justified? Explain.

19. A medical researcher wants to determine whether exercising can lower blood pressure. She recruits 100 people with high blood pressure to participate in the study. She assigns a random sample of 50 of them to pursue an exercise program that includes daily swimming and jogging. She assigns the other 50 to refrain from vigorous activity. She measures the blood pressure of each of the 100 individuals both before and after the study.
   a. Is this a randomized experiment or an observational study?
   b. On the average, the subjects in the exercise group substantially reduced their blood pressure, while the subjects in the no-exercise group did not experience a reduction. The researcher concluded that exercise decreases blood pressure. Is this conclusion well justified? Explain.

20. An agricultural scientist wants to determine the effect of fertilizer type on the yield of tomatoes. There are four types of fertilizer under consideration. She plants tomatoes on four plots of land. Each plot is treated identically except for receiving a different type of fertilizer.
   a. What are the treatments?
   b. Is this a randomized experiment or an observational study?
   c. The yields differ substantially among the four plots. Can you conclude that the differences in yield are due to the differences in fertilizer? Explain.

Working with the Concepts

21. **Air pollution and colds:** A scientist wants to determine whether people who live in places with high levels of air pollution get more colds than people in areas with little air pollution. Do you think it is possible to design a randomized experiment to study this question, or will an observational study be necessary? Explain.

22. **Cold medications:** A scientist wants to determine whether a new cold medicine relieves symptoms more effectively than a currently used medicine. Do you think it is possible to design a randomized experiment to study this question, or will an observational study be necessary? Explain.

23. **Taxicabs and crime:** A sociologist discovered that regions that have more taxicabs tend to have higher crime rates. Does increasing the number of taxicabs cause the crime rate to increase, or could the result be due to confounding? Explain.

24. **Recovering from heart attacks:** In a study of people who had suffered heart attacks, it was found that those who lived in smaller houses were more likely to recover than those who lived in larger houses. Does living in a smaller house increase the likelihood of recovery from a heart attack, or could the result be due to confounding? Explain.

25. **Eat your vegetables:** In an observational study, people who ate four or more servings of fresh fruits and vegetables each day were less likely to develop colon cancer than people who ate little fruit or vegetables. True or false:
   a. The results of the study show that eating more fruits and vegetables reduces your risk of contracting colon cancer.
   b. The results of the study may be due to confounding, since the lifestyles of people who eat large amounts of fruits and vegetables may differ in many ways from those of people who do not.

26. **Vocabulary and height:** A vocabulary test was given to students at an elementary school. The students' ages ranged from 5 to 11 years old. It was found that the students with larger vocabularies tended to be taller than the students with smaller vocabularies. Explain how this result might be due to confounding.

27. **Secondhand smoke:** A recent study compared the heart rates of 19 infants born to nonsmoking mothers with those of 17 infants born to mothers who smoked an average of 15 cigarettes a day while pregnant and after giving birth. The heart rates of the infants at one year of age were 20% slower on the average for the smoking mothers.
   a. What is the outcome variable?
   b. What is the treatment variable?
   c. Was this a cohort study or a case-control study?
   d. Was the study prospective, cross-sectional, or retrospective?
   e. Could the results be due to confounding? Explain.
   Source: *Environmental Health Perspectives* 118:a158−a159

28. **Pollution in China:** In a recent study, Z. Zhao and colleagues measured the levels of formaldehyde in the air in 34 classrooms in the schools in the city of Taiyuan, China. On the same day, they gave questionnaires to 1993 students aged 11−15 in those schools, asking them whether they had experienced respiratory problems (such as asthma attacks, wheezing, or shortness of breath). They found that the students in the classrooms with higher levels of formaldehyde reported more respiratory problems.
   a. What is the outcome variable?
   b. What is the treatment variable?
   c. Was this a cohort study or a case-control study?
   d. Was the study prospective, cross-sectional, or retrospective?
   e. Could the results be due to confounding? Explain.
   Source: *Environmental Health Perspectives* 116:90−97

Extending the Concepts

29. The Salk Vaccine Trial: In 1954, the first vaccine against polio, known as the Salk vaccine, was tested in a large randomized double-blind study. Approximately 750,000 children were asked to enroll in the study. Of these, approximately 350,000 did not participate, because their parents refused permission. The children who did participate were randomly divided into two groups of about 200,000 each. One group, the treatment group, got the vaccine, while the other group, the control group, got a placebo. The rate of polio in the treatment group was less than half of that in the control group.

  a. Is it reasonable to conclude that the Salk vaccine was effective in reducing the rate of polio?

  b. Polio is sometimes difficult to diagnose, as its early symptoms are similar to those of the flu. Explain why it was important for the doctors in the study not to know which children were getting the vaccine.

  c. Perhaps surprisingly, polio was more common among upper-income and middle-income children than among lower-income children. The reason is that lower-income children tended to live in less hygienic surroundings. They would contract mild cases of polio in infancy while still protected by their mother's antibodies, and thereby develop a resistance to the disease. The children who did not participate in the study were more likely to come from lower-income families. The rate of polio in this group was substantially lower than the rate in the placebo group. Does this prove that the placebo caused polio, or could this be due to confounding? Explain.

30. Another Salk Vaccine Trial: Another study of the Salk vaccine, conducted at the same time as the trial described in Exercise 29, used a different design. In this study, approximately 350,000 second graders were invited to participate. About 225,000 did so, and the other 125,000 refused. All of the participating second graders received the vaccine. The control group consisted of approximately 725,000 first and third graders. They were not given any placebo, so no consent was necessary.

  a. Was this a randomized experiment?

  b. Was it double-blind?

  c. The treatment group consisted of children who had consent to participate. The control group consisted of all first and third graders. It turned out that the results of this study seriously underestimated the effectiveness of the vaccine. Use the information provided in Exercise 29(c) to explain why.

Answers to Check Your Understanding Exercises for Section 1.3

1. Observational study

2. People who smoke may be more likely to drink alcohol than people who do not smoke. Therefore, it might be possible for smokers to have higher rates of liver cancer without it being caused by smoking.

3. a. Cohort study   **b.** Prospective

4. a. Case-control study   **b.** Retrospective

SECTION 1.4   Bias in Studies

Objectives

1. Define bias
2. Identify sources of bias

Objective 1   Define bias

Defining Bias

No study is perfect, and even a properly conducted study will generally not give results that are exactly correct. For example, imagine that you were to draw a simple random sample of students at a certain college to estimate the percentage of students who are Democrats. Your sample would probably contain a somewhat larger or smaller percentage of Democrats than the entire population of students, just by chance. However, imagine drawing many simple random samples. Some would have a greater percentage of Democrats than in the population, and some would have a smaller percentage of Democrats than in the population. But on the average, the percentage of Democrats in a simple random sample will be the same as the percentage in the population. A study conducted by a procedure that produces the correct result on the average is said to be *unbiased*.

Now imagine that you tried to estimate the percentage of Democrats in the population by selecting students who attended a speech made by a Democratic politician. On the average, studies conducted in this way would overestimate the percentage of Democrats in the population. Studies conducted with methods that tend to overestimate or underestimate a population value are said to be *biased*.

DEFINITION

Bias is the degree to which a procedure systematically overestimates or underestimates a population value.

- A study conducted by a procedure that tends to overestimate or underestimate a population value is said to be **biased**.
- A study conducted by a procedure that produces the correct result on the average is said to be **unbiased**.

Objective 2 Identify sources of bias

Identifying Sources of Bias

In practice, it is important to design studies to have as little bias as possible. Unfortunately, some studies are highly biased, and the conclusions drawn from them are not reliable. Here are some common types of bias.

Voluntary response bias

Recall that a voluntary response survey is one in which people are invited to log on to a website, send a text message, or call a phone number, in order to express their opinions on an issue. In many cases, the opinions of the people who choose to participate in such surveys do not reflect those of the population as a whole. In particular, people with strong opinions are more likely to participate. In general, voluntary response surveys are highly biased.

Self-interest bias

Many advertisements contain data that claim to show that the product being advertised is superior to its competitors. Of course, the advertiser will not report any data that tend to show that the product is inferior. Even more seriously, many people are concerned about a trend for companies to pay scientists to conduct studies involving their products. In particular, physicians are sometimes paid by drug companies to test their drugs and to publish the results of these tests in medical journals. People who have an interest in the outcome of an experiment have an incentive to use biased methods.

Social acceptability bias

People are reluctant to admit to behavior that may reflect negatively on them. This characteristic of human nature affects many surveys. For example, in political polls it is important for the pollster to determine whether the person being interviewed is likely to vote. A good way to determine whether someone is likely to vote in the next election is to find out whether they voted in the last election. It might seem reasonable to ask the following question:

> "Did you vote in the last presidential election?"

The problem with this direct approach is that people are reluctant to answer "No," because they are concerned that not voting is socially less acceptable than voting. Here is how the Pew Research Center asked the question in a 2010 poll:

> "In the 2008 presidential election between Barack Obama and John McCain, did things come up that kept you from voting, or did you happen to vote?"

People are more likely to answer this version of the question truthfully.

Leading question bias

Sometimes questions are worded in a way that suggests a particular response. For example, a political group that supports lowering taxes sent out a survey that included the following question:

> "Do you favor decreasing the heavy tax burden on middle-class families?"

The words "heavy" and "burden" suggest that taxes are too high, and encourage a "Yes" response. A better way to ask this question is to present it as a multiple choice:

> "What is your opinion on decreasing taxes for middle-class families?
> Choices: Strongly disagree, Somewhat disagree, Neither agree nor disagree, Somewhat agree, Strongly agree."

Nonresponse bias

People cannot be forced to answer questions or to participate in a study. In any study, a certain proportion of people who are asked to participate refuse to do so. These people are called **nonresponders**. In many cases, the opinions of nonresponders tend to differ from the opinions of those who do respond. As a result, surveys with many nonresponders are often biased.

Sampling bias

Sampling bias occurs when some members of the population are more likely to be included in the sample than others. For example, samples of convenience almost always have sampling bias, because people who are easy to sample are more likely to be included. It is almost impossible to avoid sampling bias completely, but modern survey organizations work hard to keep it at a minimum.

A Big Sample Size Doesn't Make Up for Bias

A sample is useful only if it is drawn by a method that is likely to represent the population well. If you use a biased method to draw a sample, then drawing a big sample doesn't help; a big nonrepresentative sample does not describe a population any better than a small nonrepresentative sample. In particular, voluntary response surveys often draw several hundred thousand people to participate. Although the sample is large, it is unlikely to represent the population well, so the results are meaningless.

Check Your Understanding

1. Eighty thousand people attending a professional football game filled out surveys asking their opinions on using tax money to upgrade the football stadium. Seventy percent said that they supported the use of tax money. Then a pollster surveyed a simple random sample of 500 voters, and only 30% of the voters in this sample supported the use of tax money. The owner of the football team claims that the survey done at the football stadium is more reliable, because the sample size was much larger. Is he right? Explain.

2. A polling organization placed telephone calls to 1000 people in a certain city to ask them whether they favor a tax increase to build a new school. Two hundred people answered the phone, and 150 of them opposed the tax. Can you conclude that a majority of people in the city oppose the tax, or is it likely that this result is due to bias? Explain.

Answers are on page 29.

SECTION 1.4   Exercises

Exercises 1 and 2 are the Check Your Understanding exercises located within the section.

Understanding the Concepts

In Exercises 3–5, fill in each blank with the appropriate word or phrase.

3. _____ are highly unreliable in part because people who have strong opinions are more likely to participate.

4. People who are asked to participate in a study but refuse to do so are called _____ .

5. A large sample is useful only if it is drawn by a method that is likely to represent the _____ well.

In Exercises 6–8, determine whether the statement is true or false. If the statement is false, rewrite it as a true statement.

6. The way that a question in a survey is worded rarely has an effect on the responses.

7. Surveys with many nonresponders often provide misleading results.

8. Large samples usually give reasonably accurate results, no matter how they are drawn.

Practicing the Skills

In Exercises 9–16, specify the type of bias involved.

9. A bank sent out questionnaires to a simple random sample of 500 customers asking whether they would like the bank to extend its hours. Eighty percent of those returning the questionnaire said they would like the bank to extend its hours. Of the 500 questionnaires, 20 were returned.

10. To determine his constituents' feelings about election reform, a politician sends a survey to people who have subscribed to his newsletter. More than 1000 responses are received.

11. An e-store that sells cell phone accessories reports that 98% of its customers are satisfied with the speed of delivery.

12. A sign in a restaurant claims that 95% of their customers believe them to have the best food in the world.

13. A television newscaster invites viewers to email their opinions about whether the U.S. Congress is doing a good job in handling the economy. More than 100,000 people send in an opinion.

14. A police department conducted a survey in which police officers interviewed members of their community to ask their opinions on the effectiveness of the police department. The police chief reported that 90% of the people interviewed said that they were satisfied with the performance of the police department.

15. In a study of the effectiveness of wearing seat belts, a group of people who had survived car accidents in which they had not worn seat belts reported that seat belts would not have helped them.

16. To estimate the prevalence of illegal drug use in a certain high school, the principal interviewed a simple random sample of 100 students and asked them about their drug use. Five percent of the students acknowledged using illegal drugs.

Working with the Concepts

17. **Nuclear power, anyone?** In a survey conducted by representatives of the nuclear power industry, people were asked the question: "Do you favor the construction of nuclear power plants in order to reduce our dependence on foreign oil?" A group opposed to the use of nuclear power conducted a survey with the question: "Do you favor the construction of nuclear power plants that can kill thousands of people in an accident?"
    a. Do you think that the percentage of people favoring the construction of nuclear power plants would be about the same in both surveys?
    b. Would either of the two surveys produce reliable results? Explain.

18. **Who's calling, please?** Random-digit dialing is a sampling method in which a computer generates phone numbers at random to call. In recent years, caller ID has become popular. Do you think that caller ID increases the bias in random digit dialing? Explain.

19. **Who's calling, please?** Many polls are conducted over the telephone. Some polling organizations choose a sample of phone numbers to call from lists that include landline phone numbers only, and do not include cell phones. Do you think this increases the bias in phone polls? Explain.

20. **Order of choices:** When multiple-choice questions are asked, the order of the choices is usually changed each time the question is asked. For example, in the 2012 presidential election, a pollster would ask one person "Who do you prefer for president, Barack Obama or Mitt Romney?" For the next person, the order of the names would be reversed: "Mitt Romney or Barack Obama?" If the choices were given in the same order each time, do you think that might introduce bias? Explain.

Extending the Concepts

21. *Literary Digest* **poll:** In the 1936 presidential election, Republican candidate Alf Landon challenged President Franklin Roosevelt. The *Literary Digest* magazine conducted a poll in which they mailed questionnaires to more than 10 million voters. The people who received the questionnaires were drawn from lists of automobile owners and people with telephones. The magazine received 2.3 million responses, and predicted that Landon would win the election in a landslide with 57% of the vote. In fact, Roosevelt won in a landslide with 62% of the vote. Soon afterward, the *Literary Digest* folded.
    a. In 1936 most people did not own automobiles, and many did not have telephones. Explain how this could have caused the results of the poll to be mistaken.
    b. What can be said about the response rate? Explain how this could have caused the results of the poll to be mistaken.
    c. The *Literary Digest* believed that its poll would be accurate, because it received 2.3 million responses, which is a very large number. Explain how the poll could be wrong, even with such a large sample.

Answers to Check Your Understanding Exercises for Section 1.4

1. No. The sample taken at the football stadium is biased, because football fans are more likely to be sampled than others. The fact that the sample is big doesn't make it any better.

2. No. There is a high degree of nonresponse bias in this sample.

Chapter 1 Summary

Section 1.1: Most populations are too large to allow us to study each member, so we draw samples and study those. Samples must be drawn by an appropriate method. Simple random sampling, stratified sampling, cluster sampling, and systematic sampling are all valid methods. When none of these methods are feasible, a sample of convenience may be used, so long as it is reasonable to believe that there is no systematic difference between the sample and the population.

Section 1.2: Data sets contain values of variables. Qualitative variables place items in categories, whereas quantitative variables are counts or measurements. Qualitative variables can be either ordinal or nominal. An ordinal variable is one for which the categories have a natural ordering. For nominal variables, the categories have no natural ordering. Quantitative variables can be discrete or continuous. Discrete variables are ones whose possible values can be listed, whereas continuous variables can take on values anywhere within an interval.

Section 1.3: Scientists conduct studies to determine whether different treatments produce different outcomes. The most reliable studies are randomized experiments, in which subjects are assigned to treatments at random. When randomized experiments are not feasible, observational studies may be performed. Results of observational studies may be hard to interpret, because of the potential for confounding.

Section 1.4: Some studies produce more reliable results than others. A study that is conducted by a method that tends to produce an incorrect result is said to be biased. Some of the most common forms of bias are voluntary response bias, self-interest bias, social acceptability bias, leading question bias, nonresponse bias, and sampling bias.

Vocabulary and Notation

bias   27
biased   27
case-control study   23
categorical variable   13
cluster sampling   6
cohort study   23
completely randomized experiment   21
confounder   21
confounding   21
continuous variable   14
cross-sectional study   23
data   12
data set   12
discrete variable   14
double-blind   20
experimental unit   18
individual   12
leading question bias   27

nominal variable   13
nonresponders   28
nonresponse bias   28
observational study   19
ordinal variable   13
outcome   19
parameter   8
population   2
prospective study   23
qualitative variable   13
quantitative variable   13
randomized block experiment   21
randomized experiment   19
response   19
retrospective study   23
sample   2
sample of convenience   5

sampling bias   28
seed   4
self-interest bias   27
simple random sample   3
simple random sampling   2
social acceptability bias   27
statistic   8
statistics   2
strata   5
stratified sampling   5
subject   18
systematic sample   6
treatment   19
unbiased   27
variable   12
voluntary response bias   27
voluntary response sample   7

Chapter Quiz

1. Provide an example of a qualitative variable and an example of a quantitative variable.

2. Is the name of your favorite author a qualitative variable or a quantitative variable?

3. True or false: Nominal variables do not have a natural ordering.

4. _____ variables are quantitative variables that can take on any value in some interval.

5. True or false: Ideally, a sample should represent the population as little as possible.

6. A utility company sends surveys to 200 of its customers in such a way that 100 surveys are sent to customers who pay their bills on time, 50 surveys are sent to customers whose bills are less than 30 days late, and 50 surveys are sent to customers whose bills are more than 30 days late. Which type of sample does this represent?

7. A sample of convenience is _____ when it is reasonable to believe that there is no systematic difference between the sample and the population. (*Choices:* acceptable, not acceptable)

8. The manager of a restaurant walks around and asks selected customers about the service they have received. Which type of sample does this represent?

9. True or false: An experiment where neither the investigators nor the subjects know who has been assigned to which treatment is called a double-blind experiment.

10. A poll is conducted of 3500 households close to major national airports, and another 2000 that are not close to an airport, in order to study whether living in a noisier environment results in health effects. Is this a randomized experiment or an observational study?

11. In a study, 200 patients with skin cancer are randomly divided into two groups. One group receives an experimental drug and the other group receives a placebo. Is this a randomized experiment or an observational study?

12. In a randomized experiment, if there are large differences in outcomes among treatment groups, we can conclude that the differences are due to the _____ .

13. In analyzing the course grades of students in an elementary statistics course, a professor notices that students who are seniors performed better than students who are sophomores. The professor is tempted to conclude that older students perform better than younger ones. Describe a possible confounder in this situation.

14. True or false: The way that questions are worded on a survey may have an effect on the responses.

15. A radio talk show host invites listeners to call the show to express their opinions about a political issue. How reliable is this survey? Explain.

Review Exercises

1. Qualitative or quantitative? Is the number of points scored in a football game qualitative or quantitative?

2. Nominal or ordinal? Is the color of an MP3 player nominal or ordinal?

3. Discrete or continuous? Is the area of a college campus discrete or continuous?

4. Which type of variable is it? A theater concession stand sells soft drink and popcorn combos that come in sizes small, medium, large, and jumbo. True or false:
   a. Size is a qualitative variable.
   b. Size is an ordinal variable.
   c. Size is a continuous variable.

In Exercises 5–8, identify the kind of sample that is described.

5. Website ratings: A popular website is interested in conducting a survey of 400 visitors to the site in such a way that 200 of them will be under age 30, 150 will be aged 30–55, and 50 will be over 55.

6. Favorite performer: Viewers of a television show are asked to vote for their favorite performer by sending a text message to the show.

7. School days: A researcher selects 4 of 12 high schools in a certain region and surveys all of the administrative staff members in each school about a potential change in the ordering of supplies. Which type of sample does this represent?

8. Political polling: A pollster obtains a list of registered voters and uses a computer random number generator to choose 100 of them to ask which candidate they prefer in an upcoming election.

9. Fluoride and tooth decay: Researchers examine the association between the fluoridation of water and the prevention of tooth decay by comparing the prevalence of tooth decay in countries that have fluoridated water with the prevalence in countries that do not.
   a. Is this a randomized experiment or an observational study?
   b. Assume that tooth decay was seen to be less common in countries with fluoridated water. Could this result be due to confounding? Explain.

10. Better gas mileage: A taxi company in a large city put a new type of tire with a special tread on a random sample of 50 cars, and the regular type of tire on another random sample of 50 cars. After a month, the gas mileage of each car was measured.
   a. Is this a randomized experiment or an observational study?
   b. Assume that one of the samples had noticeably better gas mileage than the other. Could this result be due to confounding? Explain.

11. Cell phones and driving: To determine whether using a cell phone while driving increases the risk of an accident, a researcher examines accident reports to obtain data about the number of accidents in which a driver was talking on a cell phone.
   a. Is this a randomized experiment or an observational study?
   b. Assume that the accident reports show that people were more likely to have an accident while talking on a cell phone. Could this result be due to confounding? Explain.

12. Turn in your homework: The English department at a local college is considering using electronic-based assignment submission in its English composition classes. To study its effects, each section of the class is divided into two groups at random. In one group, assignments are submitted by turning them in to the professor on paper. In the other group, assignments are submitted electronically.
   a. Is this a randomized experiment or an observational study?
   b. Assume that the electronically submitted assignments had many fewer typographical errors, on average, than the ones submitted on paper. Could this result be due to confounding? Explain.

In Exercises 13–15, explain why the results of the studies described are unreliable.

13. **Which TV station do you watch?** The TV columnist for a local newspaper invites readers to log on to a website to vote for their favorite TV newscaster.

14. **Longevity:** A life insurance company wants to study the life expectancy of people born in 1950. The company's actuaries examine death certificates of people born in that year to determine how long they lived.

15. **Political opinion:** A congressman sent out questionnaires to 10,000 constituents to ask their opinions on a new health-care proposal. A total of 200 questionnaires were returned, and 70% of those responding supported the proposal.

Write About It

1. Describe the difference between a stratified sample and a cluster sample.

2. Explain why it is better, when possible, to draw a simple random sample rather than a sample of convenience.

3. Describe circumstances under which each of the following samples could be used: simple random sample, a sample of convenience, a stratified sample, a cluster sample, a systematic sample.

4. Suppose that you were asked to collect some information about students in your class for a statistics project. Give some examples of variables you might collect that are ordinal, nominal, discrete, and continuous.

5. Quantitative variables are numerical. Are some qualitative variables numerical as well? If not, explain why not. If so, provide an example.

6. What are the primary differences between a randomized experiment and an observational study?

7. What are the advantages of a double-blind study? Are there any disadvantages?

8. Provide an example of a study, either real or hypothetical, that is conducted by people who have an interest in the outcome. Explain how the results might possibly be misleading.

9. Explain why each of the following questions is leading. Provide a more appropriate wording.
   a. Should Americans save more money or continue their wasteful spending?
   b. Do you support more funding for reputable organizations like the Red Cross?

Case Study: Air Pollution And Respiratory Symptoms

Air pollution is a serious problem in many places. One form of air pollution that is suspected to cause respiratory illness is particulate matter (PM), which consists of tiny particles in the air. Particulate matter can come from many sources, most commonly ash from burning, but also from other sources such as tiny particles of rubber that wear off of automobile and truck tires.

The town of Libby, Montana, was recently the focus of a study on the effect of PM on the respiratory health of children. Many houses in Libby are heated by wood stoves, which produce a lot of particulate pollution. The level of PM is greatest in the winter when more stoves are being used, and declines as the weather becomes warmer. The study attempted to determine whether higher levels of PM affect the respiratory health of children. In one part of the study, schoolchildren were given a questionnaire to bring home to their parents. Among other things, the questionnaire asked whether the child had experienced symptoms of wheezing during the past 60 days. Most parents returned the questionnaire within a couple of weeks. Parents who did not respond promptly were sent another copy of the questionnaire through the mail. Many of these parents responded to this mailed version.

Table 1.2 presents, for each day, the number of questionnaires that were returned by parents of children who wheezed, the number returned by those who did not wheeze, the average concentration of particulate matter in the atmosphere during the past 60 days (in units of micrograms per cubic meter), and whether the questionnaires were delivered in school or through the mail.

We will consider a PM level of 17 or more to be high exposure, and a PM level of less than 17 to be low exposure.

1. How many people had high exposure to PM?

2. How many of the high-exposure people had wheeze symptoms?

3. What percentage of the high-exposure people had wheeze symptoms?

4. How many people had low exposure to PM?

5. How many of the low-exposure people had wheeze symptoms?

6. What percentage of the low-exposure people had wheeze symptoms?

7. Is there a large difference between the percentage of high-exposure people with wheeze symptoms and the percentage of low-exposure people with wheeze symptoms?

8. Explain why the percentage of high-exposure people with wheeze symptoms is the same as the percentage of school-return people with wheeze symptoms.

Table 1.2

Date	PM Level	Number of People Returning Questionnaires	Number Who Wheezed	School/Mail
March 5	19.815	3	0	School
March 6	19.885	72	9	School
March 7	20.006	69	5	School
March 8	19.758	30	1	School
March 9	19.827	44	7	School
March 10	19.686	31	1	School
March 11	19.823	38	3	School
March 12	19.697	66	5	School
March 13	19.505	42	4	School
March 14	19.359	31	1	School
March 15	19.348	19	4	School
March 16	19.318	3	1	School
March 17	19.124	2	0	School
April 12	14.422	10	1	Mail
April 13	14.418	9	1	Mail
April 14	14.405	8	0	Mail
April 15	14.141	3	0	Mail
April 16	13.910	4	0	Mail
April 17	13.951	2	0	Mail
April 18	13.545	2	0	Mail
April 20	13.326	3	0	Mail
April 22	13.154	2	0	Mail

 9. Explain why the percentage of low-exposure people with wheeze symptoms is the same as the percentage of mail-return people with wheeze symptoms.

10. As the weather gets warmer, PM goes down because wood stoves are used less. Explain how this causes the mode of response (school or mail) to be related to PM.

11. It is generally the case in epidemiologic studies that people who have symptoms are often eager to participate, while those who are unaffected are less interested. Explain how this may cause the mode of response (school or mail) to be related to the outcome.

12. Rather than send out questionnaires, the investigators could have telephoned a random sample of people over a period of days. Explain how this might have reduced the confounding.

Graphical Summaries of Data

Introduction

Are cars becoming more fuel efficient? Increasing prices of gasoline, along with concerns about the environment, have made fuel efficiency an important concern. To determine whether recently built cars are getting better mileage than older cars, we will compare the U.S. Environmental Protection Agency highway mileage ratings for model year 2013 cars with the mileages for model year 2000 cars. The following tables present the results, in miles per gallon.

Highway Mileage Ratings for 2000 Compact Cars							
28	21	30	33	24	29	30	38
27	23	27	34	21	27	38	28
29	21	31	31	32	23	36	28
27	32	31	30	26	23	40	28
27	27	34	28	34	28	38	27
29	27	33	24	31	33	27	28
26	26	31	16	28	30	30	24
37	35	30	33	31	32		

Source: www.fueleconomy.gov

Highway Mileage Ratings for 2013 Compact Cars						
39	39	33	30	33	29	40
29	34	36	34	32	33	35
35	33	35	33	27	37	31
37	30	29	25	37	37	30
40	38	36	39	40	38	34
30	36	40	36	32	35	36
32	38	29	37	31	32	22
30	32	31	29	37	36	29

Source: www.fueleconomy.gov

It is hard to tell from the lists of numbers whether the mileages have changed much between 2000 and 2013. What is needed are methods to summarize the data, so that their most important features stand out. One way to do this is by constructing graphs that allow us to visualize the important features of the data. In this chapter, we will learn how to construct many of the most commonly used graphical summaries. In the case study at the end of the chapter, you will be asked to use graphical methods to compare the mileages between 2000 cars and 2013 cars.

SECTION 2.1 | Graphical Summaries for Qualitative Data

Objectives

1. Construct frequency distributions for qualitative data
2. Construct bar graphs
3. Construct pie charts

Objective 1 Construct frequency distributions for qualitative data

Frequency Distributions for Qualitative Data

How do retailers analyze their sales data to determine which methods of payment are most popular? Table 2.1 presents a list compiled by a retailer. The retailer accepts four types of credit cards: Visa, MasterCard, American Express, and Discover. The list contains the types of credit cards used by the last 50 customers.

Table 2.1 Types of Credit Cards Used

Discover	Visa	Visa	Am. Express	Visa
Visa	Visa	Am. Express	MasterCard	Visa
Am. Express	MasterCard	Visa	Visa	Visa
Visa	Am. Express	Am. Express	MasterCard	Visa
MasterCard	Visa	Discover	Am. Express	Discover
Visa	Am. Express	Discover	Visa	MasterCard
Visa	Visa	Visa	Visa	MasterCard
MasterCard	Am. Express	Visa	MasterCard	Visa
MasterCard	Discover	MasterCard	Visa	Visa
MasterCard	Discover	Am. Express	Discover	Visa

Table 2.1 is typical of data in raw form. It is a big list, and it's hard to gather much information simply by looking at it. To make the important features of the data stand out, we construct summaries. The starting point for many summaries is a frequency distribution.

DEFINITION

- The **frequency** of a category is the number of times it occurs in the data set.
- A **frequency distribution** is a table that presents the frequency for each category.

EXAMPLE 2.1   Construct a frequency distribution

Construct a frequency distribution for the data in Table 2.1.

Solution

To construct a frequency distribution, we begin by tallying the number of observations in each category and recording the totals in a table. Table 2.2 presents a frequency distribution for the credit card data. We have included the tally marks in this table, but in practice it is permissible to omit them.

Table 2.2 Frequency Distribution for Credit Cards

Credit Card	Tally	Frequency
MasterCard	⊪⊪ ⊪⊪ |	11
Visa	⊪⊪ ⊪⊪ ⊪⊪ ⊪⊪ |||	23
Am. Express	⊪⊪ ||||	9
Discover	⊪⊪ ||	7

It's a good idea to perform a check by adding the frequencies, to be sure that they add up to the total number of observations. In Table 2.2, the frequencies add up to 50, as they should.

Relative frequency distributions

A frequency distribution tells us exactly how many observations are in each category. Sometimes we are interested in the proportion of observations in each category. The proportion of observations in a category is called the *relative frequency* of the category.

> **DEFINITION**
>
> The **relative frequency** of a category is the frequency of the category divided by the sum of all the frequencies.
>
> $$\text{Relative frequency} = \frac{\text{Frequency}}{\text{Sum of all frequencies}}$$

We can add a column of relative frequencies to the frequency distribution. The resulting table is called a *relative frequency distribution*.

> **DEFINITION**
>
> A **relative frequency distribution** is a table that presents the relative frequency of each category. Often the frequency is presented as well.

CAUTION

When constructing a frequency distribution, be sure to check that the sum of all the frequencies is equal to the total number of observations.

Explain It Again

Difference between frequency and relative frequency: The frequency of a category is the number of items in the category. The relative frequency is the proportion of items in the category.

EXAMPLE 2.2   Constructing a relative frequency distribution

Construct a relative frequency distribution for the data in Table 2.2.

Solution

We compute the relative frequencies for each type of credit card in Table 2.2 by using the following steps.

Step 1: Find the total number of observations by summing the frequencies.

$$\text{Sum of frequencies} = 11 + 23 + 9 + 7 = 50$$

Step 2: Find the relative frequency for the first category, MasterCard.

$$\text{Relative frequency for MasterCard} = \frac{11}{50} = 0.22$$

Step 3: Find the relative frequencies for the remaining categories.

$$\text{Relative frequency for Visa} = \frac{23}{50} = 0.46$$

$$\text{Relative frequency for Am. Express} = \frac{9}{50} = 0.18$$

$$\text{Relative frequency for Discover} = \frac{7}{50} = 0.14$$

Table 2.3 on page 38 presents a relative frequency distribution for the data in Table 2.2.

Table 2.3 Relative Frequency Distribution for Credit Cards

Credit Card	Frequency	Relative Frequency
MasterCard	11	0.22
Visa	23	0.46
Am. Express	9	0.18
Discover	7	0.14

Check Your Understanding

1. The following table lists the types of aircraft for the landings that occurred during a day at a small airport. ("Single" refers to single-engine and "Twin" refers to twin-engine.)

Types of Aircraft Landing at an Airport

Twin	Single	Helicopter	Turboprop	Twin	Single
Turboprop	Jet	Jet	Turboprop	Turboprop	Single
Jet	Single	Single	Twin	Twin	Turboprop
Helicopter	Single	Single	Single	Twin	Single
Jet	Twin	Twin	Single	Twin	Twin

a. Construct a frequency distribution for these data.
b. Construct a relative frequency distribution for these data.

Answers are on page 48.

Objective 2 Construct bar graphs

Bar Graphs

A **bar graph** is a graphical representation of a frequency distribution. A bar graph consists of rectangles of equal width, with one rectangle for each category. The heights of the rectangles represent the frequencies or relative frequencies of the categories. Example 2.3 shows how to construct a bar graph for the credit card data in Table 2.3.

EXAMPLE 2.3 Constructing bar graphs

Construct a frequency bar graph and a relative frequency bar graph for the credit card data in Table 2.3.

Solution

Step 1: Construct a horizontal axis. Place the category names along this axis, evenly spaced.
Step 2: Construct a vertical axis to represent the frequency or the relative frequency.
Step 3: Construct a bar for each category, with the heights of the bars equal to the frequencies or relative frequencies of their categories. The bars should not touch and should all be of the same width.

Figure 2.1(a) presents a frequency bar graph, and Figure 2.1(b) presents a relative frequency bar graph. The graphs are identical except for the scale on the vertical axis.

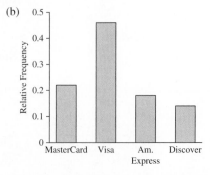

Figure 2.1 (a) Frequency bar graph. (b) Relative frequency bar graph.

Pareto charts

Sometimes it is desirable to construct a bar graph in which the categories are presented in order of frequency or relative frequency, with the largest frequency or relative frequency on the left and the smallest one on the right. Such a graph is called a **Pareto chart**. Pareto charts are useful when it is important to see clearly which are the most frequently occurring categories.

EXAMPLE 2.4 Constructing a Pareto chart

Construct a relative frequency Pareto chart for the data in Table 2.3.

Solution

Figure 2.2 presents the result. It is just like Figure 2.1(b) except that the bars are ordered from tallest to shortest.

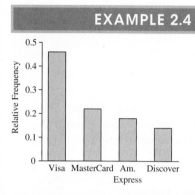

Figure 2.2 Pareto chart for the credit card data

Horizontal bars

The bars in a bar graph can be either horizontal or vertical. Horizontal bars are sometimes more convenient when the categories have long names.

EXAMPLE 2.5 Constructing bar graphs with horizontal bars

The following relative frequency distribution categorizes employed U.S. residents by type of employment in a recent year. Construct a relative frequency bar graph.

Type of Employment	Relative Frequency
Farming, forestry, fishing	0.007
Manufacturing, extraction, transportation, and crafts	0.203
Managerial, professional, and technical	0.373
Sales and office	0.242
Other services	0.176

Source: CIA — *The World Factbook*

Solution

The bar graph follows. We use horizontal bars, because the category names are long.

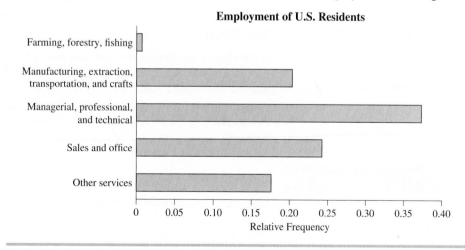

Side-by-side bar graphs

Sometimes we want to compare two bar graphs that have the same categories. The best way to do this is to construct both bar graphs on the same axes, putting bars that correspond to the same category next to each other. The result is called a **side-by-side bar graph**.

As an illustration, Table 2.4 presents the number of visitors, in millions, to several popular websites in December 2011 and in December 2012.

Table 2.4 Number of Visitors, in Millions

Website	December 2011	December 2012
Google	157	173
Facebook	147	166
Yahoo	146	164
YouTube	149	163
Wikipedia	93	92

Source: mostpopularwebsites.net

We would like to visualize the changes in the numbers of visitors between 2011 and 2012. Figure 2.3 presents the side-by-side bar graph. The bar graph clearly shows that the traffic for Google, Facebook, Yahoo, and YouTube grew somewhat from 2011 to 2012, while Wikipedia's traffic barely changed.

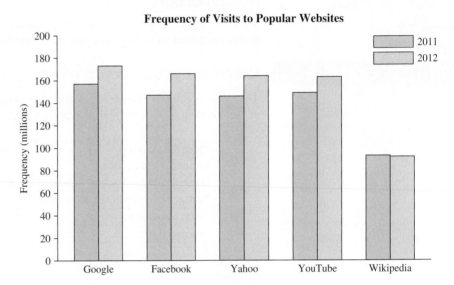

Figure 2.3 Side-by-side bar graph for the number of visitors to various websites

Objective 3 Construct pie charts

Pie Charts

A **pie chart** is an alternative to the bar graph for displaying relative frequency information. A pie chart is a circle. The circle is divided into sectors, one for each category. The relative sizes of the sectors match the relative frequencies of the categories. For example, if a category has a relative frequency of 0.25, then its sector takes up 25% of the circle. It is customary to label each sector with its relative frequency, expressed as a percentage. Example 2.6 illustrates the method for constructing a pie chart.

EXAMPLE 2.6   Constructing a pie chart

Construct a pie chart for the credit card data in Table 2.3.

Solution

For each category, we must determine how large the sector for that category must be. Since there are 360 degrees in a circle, we multiply the relative frequency of the category by 360 to determine the number of degrees in the sector for that category. For example, the relative frequency for the MasterCard category is 0.22. Therefore, the size of the sector for this category is $0.22 \cdot 360° = 79°$. Table 2.5 presents the results for all the categories.

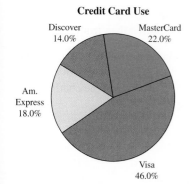

Credit Card Use

Figure 2.4 Pie chart for the credit card data in Table 2.5

Table 2.5 Sizes of Sectors for Pie Chart of Credit Card Data

Credit Card	Frequency	Relative Frequency	Size of Sector
MasterCard	11	0.22	79°
Visa	23	0.46	166°
Am. Express	9	0.18	65°
Discover	7	0.14	50°

Figure 2.4 presents the pie chart.

Constructing pie charts by hand is tedious. However, many software packages, such as MINITAB and Excel, can draw them. Step-by-step instructions for constructing a pie chart in MINITAB and Excel are presented in the Using Technology section on page 42.

Check Your Understanding

2. The following table presents a frequency distribution for the number of cars and light trucks sold in the month of June 2013.

Type of Vehicle	Frequency
Small car	276,200
Midsize car	333,515
Luxury car	98,414
Minivan	81,355
SUV	112,328
Pickup truck	191,664
Cross-over truck	300,442

Source: *The Wall Street Journal*

a. Construct a bar graph.
b. Construct a relative frequency distribution.
c. Construct a relative frequency bar graph.
d. Construct a pie chart.

3. The following bar graph presents the number of refugees admitted to the United States between October 2011 and October 2012 from each of the top six countries.

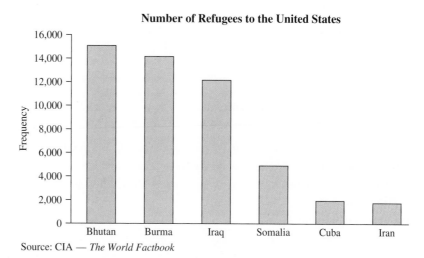

Number of Refugees to the United States

Source: CIA — *The World Factbook*

a. Which country provided the greatest number of refugees?
b. Someone says that Iraq provided more refugees than Somalia and Cuba together. Is this correct? Explain how you can tell.

c. Approximately how many refugees came from Burma?

d. Approximately how many more refugees came from Burma than from Iraq?

4. In January 2013, The CBS News/New York Times poll asked a sample of people the following question: Do you think things in the United States five years from now will be better, worse, or about the same as they are today? The following pie chart presents the percentages of people who gave each response.

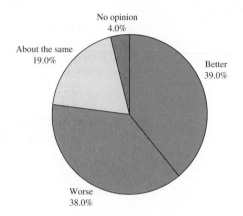

a. Which was the most common response?

b. What percentage of people said that things would be the same or worse in five years?

c. True or false: More than half of the people surveyed said that things would be the same or better in five years.

Answers are on page 48.

USING TECHNOLOGY

We use the data in Table 2.6 to illustrate the technology steps. Table 2.6 lists 20 responses to a survey question about the reason for visiting a local library.

Table 2.6

Study	Meet someone	Check out books	Meet someone	Check out books
Study	Study	Meet someone	Check out books	Check out books
Meet someone	Check out books	Study	Study	Check out books
Check out books	Study	Study	Meet someone	Study

MINITAB

Constructing a frequency distribution

Step 1. Name your variable *Reason*, and enter the data into **Column C1**.

Step 2. Click on **Stat**, then **Tables**, then **Tally Individual Variables...**

Step 3. Double-click on the *Reason* variable and check the **Counts** and **Percents** boxes.

Step 4. Press **OK** (Figure A).

```
Tally for Discrete Variables: Reason

           Reason  Count  Percent
  Check Out Books      7    35.00
    Meet Someone       5    25.00
           Study       8    40.00
             N=        20
```

Figure A

Constructing a bar graph

Step 1. Name your variable *Reason*, and enter the data into **Column C1**.

Step 2. Click on **Graph**, then **Bar Chart**. If given raw data as in Table 2.6, select **Bars Represent: Counts of Unique Values**. For the **Bar Chart type**, select **Simple**. Click **OK**. (If given a frequency distribution, select **Bars Represent: Values from a Table**.)

Step 3. Double-click on the *Reason* variable and click on any of the options desired.

Step 4. Press **OK** (Figure B).

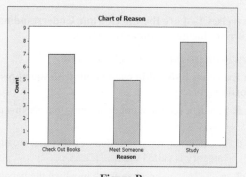

Figure B

Constructing a pie chart

Step 1. Name your variable *Reason*, and enter the data into **Column C1**.

Step 2. Click on **Graph**, then **Pie Chart**. If given raw data as in Table 2.6, select **Chart Raw Data**, and click **OK**. (If given a frequency distribution, select **Chart Values from a Table**.)

Step 3. Double-click on the *Reason* variable and click on any of the options desired.

Step 4. Press **OK** (Figure C).

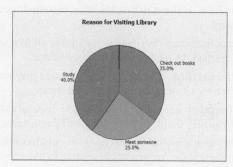

Figure C

EXCEL

Constructing a frequency distribution

Step 1. Enter the data in **Column A** with the label *Reason* in the topmost cell.

Step 2. Select **Insert**, then **Pivot Table**. Enter the range of cells that contain the data in the **Table/Range** field and click **OK**.

Step 3. In **Choose fields to add to report**, check Reason.

Step 4. Click on *Reason* and drag to the **Values** box. The result is shown in Figure D.

Row Labels	Count of Reason
Check out books	7
Meet someone	5
Study	8
Grand Total	**20**

Figure D

Constructing bar graphs and pie charts

Step 1. Enter the categories in **Column A** and the frequencies or relative frequencies in **Column B**.

Step 2. Highlight the values in **Column A** and **Column B**, and select **Insert**. For a bar graph, select **Column**. For a pie chart, select **Pie**.

SECTION 2.1 Exercises

Exercises 1–4 are the Check Your Understanding exercises located within the section.

Understanding the Concepts

In Exercises 5–8, fill in each blank with the appropriate word or phrase.

5. In a data set, the number of items that are in a particular category is called the _____ .

6. In a data set, the proportion of items that are in a particular category is called the _____ .

7. A _____ is a bar graph in which the bars are ordered by size.

8. A _____ is represented by a circle in which the sizes of the sectors match the relative frequencies of the categories.

In Exercises 9–12, determine whether the statement is true or false. If the statement is false, rewrite it as a true statement.

9. In a frequency distribution, the sum of all frequencies is less than the total number of observations.

10. In a pie chart, if a category has a relative frequency of 30%, then its sector takes up 30% of the circle.

11. The relative frequency of a category is equal to the frequency divided by the sum of all frequencies.

12. In bar graphs and Pareto charts, the widths of the bars represent the frequencies or relative frequencies.

Practicing the Skills

13. The following bar graph presents the average amount a U.S. family spent, in dollars, on various food categories in a recent year.

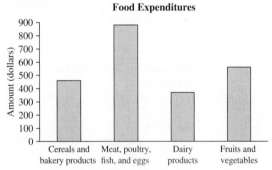

Food Expenditures

Source: Consumer Expenditure Survey

a. On which food category was the most money spent?

b. True or false: On the average, families spent more on cereals and bakery products than on fruits and vegetables.

c. True or false: Families spent more on animal products (meat, poultry, fish, eggs, and dairy products) than on plant products (cereals, bakery products, fruits, and vegetables).

14. The most common blood typing system divides human blood into four groups: A, B, O, and AB. The following bar graph presents the frequencies of these types in a sample of 150 blood donors.

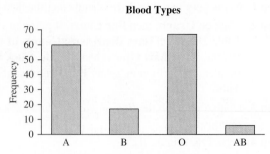

Blood Types

a. Which is the most frequent type?

b. True or false: More than half of the individuals in the sample had type O blood.

c. True or false: More than twice as many people had type A blood as had type B blood.

15. Following is a pie chart that presents the percentages of video games sold in each of four rating categories.

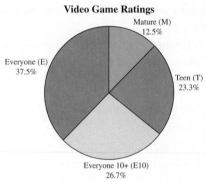

Video Game Ratings

Mature (M) 12.5%
Everyone (E) 37.5%
Teen (T) 23.3%
Everyone 10+ (E10) 26.7%

Source: Entertainment Software Association

a. Construct a relative frequency bar graph for these data.

b. Construct a relative frequency Pareto chart for these data.

c. In which rating category are the most games sold?

d. True or false: More than twice as many T-rated games are sold as M-rated games.

e. True or false: Fewer than one in five games sold is an M-rated game.

16. **Student expenses:** The following pie chart presents the percentages of a student's budget spent on average in various categories.

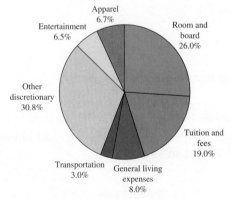

Apparel 6.7%
Entertainment 6.5%
Room and board 26.0%
Other discretionary 30.8%
Tuition and fees 19.0%
Transportation 3.0%
General living expenses 8.0%

Source: O'Donnell & Associates, LLC

a. Construct a relative frequency bar graph for these data.

b. Construct a relative frequency Pareto chart for these data.

c. In which category was the largest amount spent?

d. General living expenses, Room and board, and Transportation generally do not change drastically from month to month. What percentage of a student's budget do these categories make up?

17. Food sources: The following side-by-side bar graph presents the percentages of the sources of funds for food expenditures in the U.S. in 1940 and 2012.

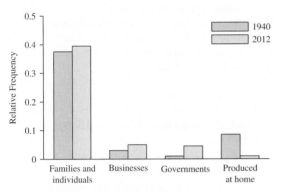

Source: U.S. Department of Agriculture

a. Which sources of funds for food expenditures increased as a proportion of the total from 1940 to 2012?

b. Which source of funds for food expenditures decreased as a proportion of the total from 1940 to 2012?

c. In 1940, the smallest source of funds for food expenditures was governments. Is this the smallest source of funds in 2012?

d. In 1940, the largest source of funds for food expenditures was families and individuals. Is this the largest source of funds in 2012?

18. Super Bowl: The following side-by-side bar graph presents the results of a survey in which men and women were asked to name their favorite thing about watching the Super Bowl.

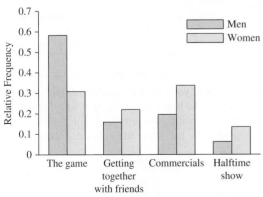

Source: BIGInsight

a. Which part of the Super Bowl do a greater proportion of men than women have as their favorite?

b. True or false: For both men and women, the smallest proportion have the halftime show as their favorite.

c. True or false: About twice as many men as women have the commercials as their favorite.

d. True or false: The proportion of men for whom the game or the half time show is their favorite is about the same as the proportion of women for whom the game or the commercials is their favorite.

Working with the Concepts

19. iPod sales: The following frequency distribution presents the number of iPods sold (in thousands) in each quarter of each year from 2009 through 2012.

Quarter	Number Sold (in thousands)
Jan.–Mar. 2009	11,013
Apr.–Jun. 2009	10,215
Jul.–Sep. 2009	10,177
Oct.–Dec. 2009	20,970
Jan.–Mar. 2010	10,885
Apr.–Jun. 2010	9,410
Jul.–Sep. 2010	9,050
Oct.–Dec. 2010	19,446
Jan.–Mar. 2011	9,020
Apr.–Jun. 2011	7,540
Jul.–Sep. 2011	6,622
Oct.–Dec. 2011	15,397
Jan.–Mar. 2012	7,673
Apr.–Jun. 2012	6,751
Jul.–Sep. 2012	5,334
Oct.–Dec. 2012	12,700

Source: Apple

a. Construct a frequency bar graph.

b. Construct a relative frequency distribution.

c. Construct a relative frequency bar graph.

d. True or false: The four quarters with the largest sales were all October to December.

20. Popular video games: The following frequency distribution presents the numbers of copies sold in 2012 for each of the ten best-selling video games.

Game	Platform	Sales (millions)
Call of Duty: Black Ops II	Xbox 360	6.1
Halo 4	Xbox 360	4.7
Call of Duty: Black Ops II	PS3	3.5
Kinect Adventures!	Xbox 360	2.9
Just Dance 4	Wii	2.5
Assassin's Creed III	Xbox 360	2.1
Madden NFL 13	Xbox 360	1.9
Assassin's Creed III	PS3	1.9
Pokemon Black/White Version 2	DS	1.7
New Super Mario Bros. 2	3DS	1.6
Mass Effect 3	Xbox 360	1.6
NBA 2K13 3	Xbox 360	1.5

Source: http://www.vgchartz.com

a. Construct a frequency bar graph.

b. Construct a relative frequency distribution.

c. Construct a relative frequency bar graph.
d. True or false: More than 20% of the games sold were Call of Duty: Black Ops II.

21. More iPods: Using the data in Exercise 19:
   a. Construct a frequency distribution for the total number of iPods sold in each of the four quarters Jan.–Mar., Apr.–Jun., Jul.–Sep., and Oct.–Dec.
   b. Construct a frequency bar graph.
   c. Construct a relative frequency distribution.
   d. Construct a relative frequency bar graph.
   e. Construct a pie chart.
   f. True or false: More than half of iPods were sold between October and December.

22. Popular platforms: Using the data in Exercise 20:
   a. Construct a frequency distribution that presents the total sales for each of the platforms among the top ten games.
   b. Construct a frequency bar graph.
   c. Construct a relative frequency distribution.
   d. Construct a relative frequency bar graph.
   e. Construct a pie chart.
   f. True or false: More than half the games sold among the top ten were Xbox 360 games.

23. Hospital admissions: The following frequency distribution presents the five most frequent reasons for hospital admissions in U.S. community hospitals in a recent year.

Reason	Frequency (in thousands)
Congestive heart failure	990
Coronary atherosclerosis	1400
Heart attack	744
Infant birth	3800
Pneumonia	1200

Source: Agency for Health Care Policy and Research

   a. Construct a frequency bar graph.
   b. Construct a relative frequency distribution.
   c. Construct a relative frequency bar graph.
   d. Construct a relative frequency Pareto chart.
   e. Construct a pie chart.
   f. The categories coronary atherosclerosis, congestive heart failure, and heart attack refer to diseases of the circulatory system. True or false: There were more hospital admissions for infant birth than for diseases of the circulatory system.

24. World population: Following are the populations of the continents of the world (not including Antarctica) in the year 2012.

Continent	Population (in millions)
Africa	1072
Asia	4260
Oceania	37
Europe	740
North America	349
South America	397

Source: U.S. Census Bureau

   a. Construct a frequency bar graph.
   b. Construct a relative frequency distribution.
   c. Construct a relative frequency bar graph.
   d. Construct a relative frequency Pareto chart.
   e. Construct a pie chart.
   f. True or false: In the year 2012, more than half of the people in the world lived in Asia.
   g. True or false: In the year 2012, there were more people in Europe than in North and South America combined.

25. Ages of video gamers: The Nielsen Company estimated the numbers of people in various gender and age categories who used a video game console. The results are presented in the following frequency distribution.

Gender and Age Group	Frequency (in millions)
Males 2–11	13.0
Females 2–11	10.1
Males 12–17	9.6
Females 12–17	6.2
Males 18–34	16.1
Females 18–34	11.6
Males 35–49	10.4
Females 35–49	9.3
Males 50+	3.5
Females 50+	3.9

Source: The Nielsen Company

   a. Construct a frequency bar graph.
   b. Construct a relative frequency distribution.
   c. Construct a relative frequency bar graph.
   d. Construct a pie chart.
   e. True or false: More than half of video gamers are male.
   f. True or false: More than 40% of video gamers are female.
   g. What proportion of video gamers are 35 or over?

26. How secure is your job? In a survey, employed adults were asked how likely they thought it was that they would lose their jobs within the next year. The results are presented in the following frequency distribution.

Response	Frequency
Very likely	741
Fairly likely	859
Not too likely	3789
Not likely	9773

Source: General Social Survey

a. Construct a frequency bar graph.
b. Construct a relative frequency distribution.
c. Construct a relative frequency bar graph.
d. Construct a pie chart.
e. True or false: More than half of the people surveyed said that it was not likely that they would lose their job.
f. What proportion of the people in the survey said that it was very likely or fairly likely that they would lose their job?

27. **Back up your data:** In a survey commissioned by the Maxtor Corporation, U.S. computer users were asked how often they backed up their computer's hard drive. The following frequency distribution presents the results.

Response	Frequency
More than once per month	338
Once every 1–3 months	424
Once every 4–6 months	212
Once every 7–11 months	127
Once per year or less	311
Never	620

Source: The Maxtor Corporation

a. Construct a frequency bar graph.
b. Construct a relative frequency distribution.
c. Construct a relative frequency bar graph.
d. Construct a pie chart.
e. True or false: More than 30% of the survey respondents never back up their data.
f. True or false: Less than 50% of the survey respondents back up their data more than once per year.

28. **Education levels:** The following frequency distribution categorizes U.S. adults aged 18 and over by educational attainment in the year 2012.

Educational Attainment	Frequency (in thousands)
None	834
1–4 years	1,764
5–6 years	3,618
7–8 years	4,575
9 years	4,068
10 years	4,814
11 years	11,429
High school graduate	70,441
Some college but no degree	45,685
Associate's degree (occupational)	9,380
Associate's degree (academic)	12,100
Bachelor's degree	43,277
Master's degree	16,625
Professional degree	3,099
Doctoral degree	3,191

Source: U.S. Census Bureau

a. Construct a frequency bar graph.
b. Construct a relative frequency distribution.
c. Construct a relative frequency bar graph.
d. Construct a frequency distribution with the following categories: 8 years or less, 9–11 years, High school graduate, Some college but no degree, College degree (Associate's or Bachelor's), Graduate degree (Master's, Professional, or Doctoral).

e. Construct a pie chart for the frequency distribution in part (d).
f. What proportion of people did not graduate from high school?

29. **Music sales:** The following frequency distribution presents the number of units sold for categories of physical and digital music in the years 2010 and 2011. The Mobile category refers to ringtones and other music downloaded to a mobile device. The category "Other" includes CD singles, cassettes, DVDs, and download albums.

Type of Music	Sales (in millions)	
	2010	2011
CDs	253.0	240.8
Download single	1177.4	1306.2
Mobile	188.5	115.9
Other	95.3	113.3

Source: Recording Industry Association of America

a. Construct a relative frequency distribution for the 2010 sales.
b. Construct a relative frequency distribution for the 2011 sales.
c. Construct a side-by-side relative frequency bar graph to compare the sales in 2010 and 2011.
d. True or false: More than half of all sales in 2011 were download singles.

30. **Bought a new car lately?** The following table presents the number of vehicles sold in the United States by several manufacturers in February 2012 and February 2013.

Manufacturer	2012	2013
General Motors	209,306	224,314
Ford	178,644	195,310
Chrysler LLC	133,521	139,015
Toyota	159,462	166,377
Honda	110,157	107,987
Nissan	106,731	99,636

Source: *The Wall Street Journal*

a. Construct a relative frequency distribution for the 2012 sales.
b. Construct a relative frequency distribution for the 2013 sales.
c. Construct a side-by-side relative frequency bar graph to compare the sales in 2012 and 2013.
d. True or false: For every manufacturer, sales were higher in 2013 than in 2012.

31. **Instagram followers:** The following frequency distribution presents the number of Instagram followers in January 2013 for each of five name brands.

Brand	Followers (thousands)
Victoria's Secret	1364
Starbucks	1062
Nike	838
Burberry	652
Redbull	536

Source: www.socialfresh.com

a. Construct a frequency bar graph.
b. Construct a relative frequency distribution.

c. Construct a relative frequency bar graph.

d. Construct a pie chart.

e. What proportion are following Starbucks?

32. Smartphones: The following table presents the number of global shipments of smartphones for several vendors in 2012.

Vendor	Shipments (millions)
Samsung	213.0
Nokia	35.0
Apple	135.8
Others	316.3

Source: Strategy Analytics

a. Construct a frequency bar graph.

b. Construct a relative frequency distribution.

c. Construct a relative frequency bar graph.

d. Construct a pie chart.

e. What proportion of global shipments in 2012 were for Samsung phones?

33. Smartphone sales: The following table presents the number of worldwide smartphone sales (in millions) for the first quarter of 2013.

Company	Sales (millions)
Samsung	64.74
Apple	38.33
LG Electronics	10.08
Huawei Technologies	9.33
ZTE	7.88
Others	79.68

Source: Gartner Research

a. Construct a frequency bar graph.

b. Construct a relative frequency distribution.

c. Construct a relative frequency bar graph.

d. Construct a pie chart.

e. True or false: More smartphones were sold by Samsung than by Apple, LG, Huawei, and ZTE combined.

34. Happy Halloween: The following table presents proportions of people who get ideas for Halloween costumes from various sources. Is this a valid relative frequency distribution? Why or why not?

Source	Proportion
Twitter	0.048
Facebook	0.152
Friends and Family	0.237
Retail Stores	0.357
Magazines	0.193
Pinterest	0.071

Source: National Retail Federation

Extending the Concepts

35. Native languages: The following frequency distribution presents the number of households (in thousands) categorized by the language spoken at home, for the cities of New York and Los Angeles in the year 2009. The Total column presents the numbers of households in both cities combined.

Language	New York	Los Angeles	Total
English	4098	1339	5437
Spanish	1870	1555	3425
Other Indo-European	1037	237	1274
Asian and Pacific Island	618	301	919

Source: U.S. Census Bureau

a. Construct a frequency bar graph for each city.

b. Construct a frequency bar graph for the total.

c. Construct a relative frequency bar graph for each city.

d. Construct a relative frequency bar graph for the total.

e. Explain why the heights of the bars for the frequency bar graph for the total are equal to the sums of the heights for the individual cities.

f. Explain why the heights of the bars for the relative frequency bar graph for the total are not equal to the sums of the heights for the individual cities.

Answers to Check Your Understanding Exercises for Section 2.1

1. a.

Aircraft	Frequency
Twin	9
Single	10
Helicopter	2
Turboprop	5
Jet	4

b.

Aircraft	Frequency	Relative Frequency
Twin	9	0.300
Single	10	0.333
Helicopter	2	0.067
Turboprop	5	0.167
Jet	4	0.133

2. a.

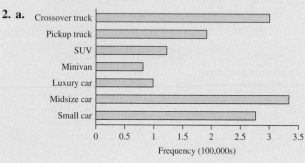

b.

Type of Vehicle	Frequency	Relative Frequency
Small car	276,200	0.198
Midsize car	333,515	0.239
Luxury car	98,414	0.071
Minivan	81,355	0.058
SUV	112,328	0.081
Pickup truck	191,664	0.138
Crossover truck	300,442	0.216

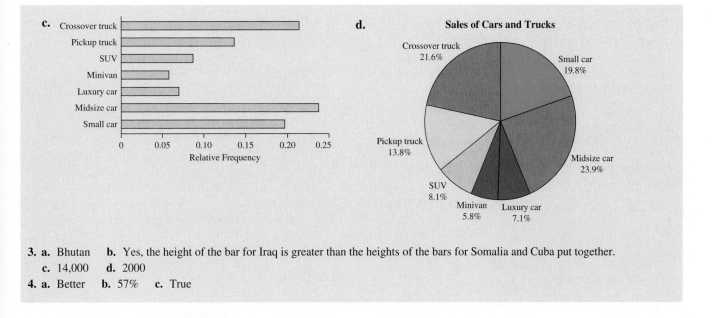

3. a. Bhutan   **b.** Yes, the height of the bar for Iraq is greater than the heights of the bars for Somalia and Cuba put together.
c. 14,000   **d.** 2000
4. a. Better   **b.** 57%   **c.** True

Frequency Distributions and Their Graphs

Objectives

1. Construct frequency distributions for quantitative data
2. Construct histograms
3. Determine the shape of a distribution from a histogram
4. Construct frequency polygons and ogives

Objective 1 Construct frequency distributions for quantitative data

Frequency Distributions for Quantitative Data

How much air pollution is caused by motor vehicles? This question was addressed in a study by Dr. Janet Yanowitz at the Colorado School of Mines. She studied the emissions of particulate matter, a form of pollution consisting of tiny particles, that has been associated with respiratory disease. The emissions for 65 vehicles, in units of grams of particles per gallon of fuel, are presented in Table 2.7.

Table 2.7 Particulate Emissions for 65 Vehicles

1.50	0.87	1.12	1.25	3.46	1.11	1.12	0.88	1.29	0.94	0.64	1.31	2.49
1.48	1.06	1.11	2.15	0.86	1.81	1.47	1.24	1.63	2.14	6.64	4.04	2.48
1.40	1.37	1.81	1.14	1.63	3.67	0.55	2.67	2.63	3.03	1.23	1.04	1.63
3.12	2.37	2.12	2.68	1.17	3.34	3.79	1.28	2.10	6.55	1.18	3.06	0.48
0.25	0.53	3.36	3.47	2.74	1.88	5.94	4.24	3.52	3.59	3.10	3.33	4.58

To summarize these data, we will construct a frequency distribution. Since these data are quantitative, there are no natural categories. We therefore divide the data into **classes**. The classes are intervals of equal width that cover all the values that are observed. For example, for the data in Table 2.7, we could choose the classes to be 0.00–0.99, 1.00–1.99, and so forth. We then count the number of observations that fall into each class, to obtain the class frequencies.

EXAMPLE 2.7

Construct a frequency distribution

Construct a frequency distribution for the data in Table 2.7, using the classes 0.00–0.99, 1.00–1.99, and so on.

Solution

First we list the classes. We begin by noting that the smallest value in the data set is 0.25 and the largest is 6.64. We list classes until we get to the class that contains the largest value. The classes are 0.00–0.99, 1.00–1.99, 2.00–2.99, 3.00–3.99, 4.00–4.99, 5.00–5.99, and 6.00–6.99. Since the largest number in the data set is 6.64, these are enough classes.

  Now we count the number of observations that fall into each class. The first class is 0.00–0.99. We count nine observations between 0.00 and 0.99 in Table 2.7. The next class is 1.00–1.99. We count 26 observations in this class. We repeat this procedure with classes 2.00–2.99, 3.00–3.99, 4.00–4.99, 5.00–5.99, and 6.00–6.99. The results are presented in Table 2.8. This is a frequency distribution for the data in Table 2.7.

Explain It Again

Frequency distributions for quantitative and qualitative data: Frequency distributions for quantitative data are just like those for qualitative data, except that the data are divided into classes rather than categories.

Table 2.8 Frequency Distribution for Particulate Data

Class	Frequency
0.00–0.99	9
1.00–1.99	26
2.00–2.99	11
3.00–3.99	13
4.00–4.99	3
5.00–5.99	1
6.00–6.99	2

We can also construct a relative frequency distribution. As with qualitative data, the relative frequency of a class is the frequency of that class, divided by the sum of all the frequencies.

DEFINITION

The **relative frequency** of a class is given by

$$\text{Relative frequency} = \frac{\text{Frequency}}{\text{Sum of all frequencies}}$$

EXAMPLE 2.8

Construct a relative frequency distribution

Construct a relative frequency distribution for the data in Table 2.7, using the classes 0.00–0.99, 1.00–1.99, and so on.

Solution

The frequency distribution is presented in Table 2.8. We compute the sum of all the frequencies:

$$\text{Sum of all frequencies} = 9 + 26 + 11 + 13 + 3 + 1 + 2 = 65$$

We can now compute the relative frequency for each class. For the class 0.00–0.99, the frequency is 9. The relative frequency is therefore

$$\text{Relative frequency} = \frac{\text{Frequency}}{\text{Sum of all frequencies}} = \frac{9}{65} = 0.138$$

Table 2.9 is a relative frequency distribution for the data in Table 2.7. The frequencies are shown as well.

Table 2.9 Relative Frequency Distribution for Particulate Data

Class	Frequency	Relative Frequency
0.00–0.99	9	0.138
1.00–1.99	26	0.400
2.00–2.99	11	0.169
3.00–3.99	13	0.200
4.00–4.99	3	0.046
5.00–5.99	1	0.015
6.00–6.99	2	0.031

Choosing the classes

In Examples 2.7 and 2.8, we chose the classes to be 0.00–0.99, 1.00–1.99, and so on. There are many other choices we could have made. For example, we could have chosen the classes to be 0.00–1.99, 2.00–3.99, 4.00–5.99, and 6.00–7.99. As another example, we could have chosen them to be 0.00–0.49, 0.50–0.99, and so on, up to 6.50–6.99. We now define some of the terminology that we will use when discussing classes.

DEFINITION

- The **lower class limit** of a class is the smallest value that can appear in that class.
- The **upper class limit** of a class is the largest value that can appear in that class.
- The **class width** is the difference between consecutive lower class limits.

CAUTION

The class width is the difference between the lower limit and the lower limit of the next class, not the difference between the lower limit and the upper limit.

Class limits should be expressed with the same number of decimal places as the data. The data in Table 2.7 are rounded to two decimal places, so the class limits for these data are expressed with two decimal places as well.

EXAMPLE 2.9

Find the class limits and widths

Find the lower class limits, the upper class limits, and the class widths for the relative frequency distribution in Table 2.9.

Solution

The classes are 0.00–0.99, 1.00–1.99, and so on, up to 6.00–6.99. The lower class limits are therefore 0.00, 1.00, 2.00, 3.00, 4.00, 5.00, and 6.00. The upper class limits are 0.99, 1.99, 2.99, 3.99, 4.99, 5.99, and 6.99.

We find the class width for the first class by subtracting consecutive lower limits:

Class width = Lower limit for second class − Lower limit for first class
$$= 1.00 - 0.00$$
$$= 1.00$$

Similarly, we find that all the classes have a width of 1.

When constructing a frequency distribution, there is no one right way to choose the classes. However, there are some requirements that must be satisfied:

Requirements for Choosing Classes

- Every observation must fall into one of the classes.
- The classes must not overlap.
- The classes must be of equal width.
- There must be no gaps between classes. Even if there are no observations in a class, it must be included in the frequency distribution.

The following procedure will produce a frequency distribution whose classes meet these requirements.

Procedure for Constructing a Frequency Distribution for Quantitative Data

Step 1: Choose a class width.

Step 2: Choose a lower class limit for the first class. This should be a convenient number that is slightly less than the minimum data value.

Step 3: Compute the lower limit for the second class by adding the class width to the lower limit for the first class:

Lower limit for second class = Lower limit for first class + Class width

Step 4: Compute the lower limits for each of the remaining classes, by adding the class width to the lower limit of the preceding class. Stop when the largest data value is included in a class.

Step 5: Count the number of observations in each class, and construct the frequency distribution.

EXAMPLE 2.10

Constructing a frequency distribution

Construct a frequency distribution for the data in Table 2.7, using a class width of 1.50.

Solution

Step 1: The class width is given to be 1.50.

Step 2: The smallest value in the data is 0.25. A convenient number that is smaller than 0.25 is 0.00. We will choose 0.00 to be the lower limit for the first class.

Step 3: The lower class limit for the second class is $0.00 + 1.50 = 1.50$.

Step 4: Continuing, the lower limits for the remaining classes are

$$1.50 + 1.50 = 3.00$$
$$3.00 + 1.50 = 4.50$$
$$4.50 + 1.50 = 6.00$$
$$6.00 + 1.50 = 7.50$$

Since the largest data value is 6.64, every data value is now contained in a class.

Step 5: We count the number of observations in each class to obtain the following frequency distribution.

Frequency Distribution for Particulate Data Using a Class Width of 1.5

Class	Frequency
0.00–1.49	28
1.50–2.99	18
3.00–4.49	15
4.50–5.99	2
6.00–7.49	2

1. Using the data in Table 2.7, construct a frequency distribution with classes of width 0.5.

Answer is on page 67.

Computing the class width for a given number of classes

In Example 2.10, the first step in computing the frequency distribution was to choose a class width. Sometimes we begin by choosing an approximate number of classes instead. In these cases, we compute the class width as follows:

Step 1: Decide approximately how many classes to have.

Step 2: Compute the class width as follows:

$$\text{Class width} = \frac{\text{Largest data value} - \text{Smallest data value}}{\text{Number of classes}}$$

Step 3: Round the class width to a convenient value. It is usually better to round up.

Once the class width is determined, we proceed just as in the case where the class width is given. We choose a lower limit for the first class by choosing a convenient number that is slightly less than the minimum data value. We then compute the lower limits for the remaining classes, count the number of observations in each class, and construct the frequency distribution. Note that the actual number of classes may differ somewhat from the chosen number, because the class width is rounded and because the lower limit of the first class will generally be less than the smallest data value.

EXAMPLE 2.11

Computing the class width

Find the class width for a frequency distribution for the data in Table 2.7, if it is desired to have approximately seven classes.

Solution

Step 1: We will have approximately seven classes.

Step 2: The smallest data value is 0.25 and the largest is 6.64. We compute the class width:

$$\text{Class width} = \frac{6.64 - 0.25}{7} = 0.91$$

Step 3: We round 0.91 up to 1, since this is the nearest convenient number. We will use a class width of 1.

A reasonable choice for the lower limit of the first class is 0. This choice will give us the frequency distribution in Table 2.8.

Objective 2 Construct histograms

Histograms

Once we have a frequency distribution or a relative frequency distribution, we can put the information in graphical form by constructing a **histogram**. Histograms based on frequency distributions are called **frequency histograms**, and histograms based on relative frequency distributions are called **relative frequency histograms**. Histograms are related to bar graphs and are appropriate for quantitative data. A histogram is constructed by drawing a rectangle for each class. The heights of the rectangles are equal to the frequencies or the relative frequencies, and the widths are equal to the class width. The left edge of each rectangle corresponds to the lower class limit, and the right edge touches the left edge of the next rectangle.

EXAMPLE 2.12

Construct a histogram

Table 2.10 (page 54) presents a frequency distribtion and the relative frequency distribution for the particulate emissions data.

Construct a frequency histogram based on the frequency distribution in Table 2.10. Construct a relative frequency histogram based on the relative frequency distribution in Table 2.10.

Table 2.10 Frequency and Relative Frequency Distributions for Particulate Data

Class	Frequency	Relative Frequency
0.00–0.99	9	0.138
1.00–1.99	26	0.400
2.00–2.99	11	0.169
3.00–3.99	13	0.200
4.00–4.99	3	0.046
5.00–5.99	1	0.015
6.00–6.99	2	0.031

Solution

We construct a rectangle for each class. The first rectangle has its left edge at the lower limit of the first class, which is 0.00, and its right edge at the lower limit of the next class, which is 1.00. The second rectangle has its left edge at 1.00 and its right edge at the lower limit of the next class, which is 2.00, and so on.

For the frequency histogram, the heights of the rectangles are equal to the frequencies. For the relative frequency histogram, the heights of the rectangles are equal to the relative frequencies.

Figure 2.5 presents a frequency histogram, and Figure 2.6 presents a relative frequency histogram, for the data in Table 2.10. Note that the two histograms have the same shape. The only difference is the scale on the vertical axis.

Figure 2.5 Frequency histogram for the frequency distribution in Table 2.10

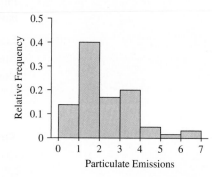

Figure 2.6 Relative frequency histogram for the relative frequency distribution in Table 2.10

How should I choose the number of classes for a histogram?

There are no hard-and-fast rules for choosing the number of classes. In general, it is good to have more classes rather than fewer, but it is also good to have reasonably large frequencies in some of the classes. The following two principles can guide the choice:

- Too many classes produce a histogram with too much detail, so that the main features of the data are obscured.
- Too few classes produce a histogram lacking in detail.

Figures 2.7 and 2.8 illustrate these principles. Figure 2.7 presents a histogram for the particulate data where the class width is 0.1. This narrow class width results in a large number of classes. The histogram has a jagged appearance, which distracts from the overall shape of the data. On the other extreme, Figure 2.8 presents a histogram for these data with

Figure 2.7 The class width is too narrow. The jagged appearance distracts from the overall shape of the data.

Figure 2.8 The class width is too wide. Only the most basic features of the data are visible.

a class width of 2.0. The number of classes is too small, so only the most basic features of the data are visible in this overly simple histogram.

Choosing a large number of classes will produce a narrow class width, and choosing a smaller number will produce a wider class width. It is appropriate to experiment with various choices for the number of classes, in order to find a good balance. The following guidelines are helpful.

Guidelines for Selecting the Number of Classes

- For many data sets, the number of classes should be at least 5 but no more than 20.
- For very large data sets, a larger number of classes may be appropriate.

EXAMPLE 2.13

Constructing a histogram with technology

Use technology to construct a frequency histogram for the emissions data in Table 2.7 on page 49.

Solution

The following figure shows the histogram constructed in MINITAB. Note that MINITAB has chosen a class width of 0.5. With this class width, there are two empty classes. These show up as a gap that separates the last two rectangles on the right from the rest of the histogram.

Step-by-step instructions for constructing histograms with the TI-84 Plus and with MINITAB are given in the Using Technology section on pages 61 and 62.

Check Your Understanding

2. Following is a frequency distribution that presents the number of live births to women aged 15–44 in the state of Wyoming in a recent year.

Distribution of Births by Age of Mother

Age	Frequency
15–19	795
20–24	2410
25–29	2190
30–34	1208
35–39	499
40–44	109

Source: Wyoming Department of Health

    a. List the lower class limits.
    b. What is the class width?
    c. Construct a frequency histogram.
    d. Construct a relative frequency distribution.
    e. Construct a relative frequency histogram.

Answers are on page 67.

Open-ended classes

It is sometimes necessary for the first class to have no lower limit or for the last class to have no upper limit. Such a class is called **open-ended**. Table 2.11 presents a frequency distribution for the number of deaths in the United States due to pneumonia in a recent year for various age groups. Note that the last age group is "85 and older," an open-ended class.

Table 2.11 Deaths Due to Pneumonia

Age	Number of Deaths
5–14	69
15–24	178
25–34	299
35–44	875
45–54	1872
55–64	3099
65–74	6283
75–84	17,775
85 and older	27,758

Source: U.S. Census Bureau

When a frequency distribution contains an open-ended class, a histogram cannot be drawn.

Histograms for discrete data

When data are discrete, we can construct a frequency distribution in which each possible value of the variable forms a class. Then we can draw a histogram in which each rectangle represents one possible value of the variable. Table 2.12 presents the results of a hypothetical survey in which 1000 adult women were asked how many children they had. Number of children is a discrete variable, and in this data set, the values of this variable are 0 through 8. To construct a histogram, we draw rectangles of equal width, centered at the values of the variables. The rectangles should be just wide enough to touch. Figure 2.9 presents a histogram.

Table 2.12 Women with a Given Number of Children

Number of Children	Frequency
0	435
1	175
2	222
3	112
4	38
5	9
6	7
7	0
8	2

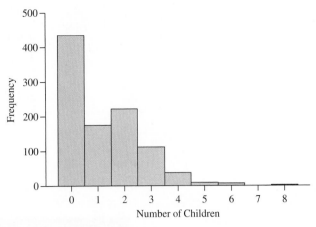

Figure 2.9 Histogram for data in Table 2.12

Objective 3 Determine the shape of a distribution from a histogram

Shapes of Histograms

The purpose of a histogram is to give a visual impression of the "shape" of a data set. Statisticians have developed terminology to describe some of the commonly observed shapes. A histogram is **symmetric** if its right half is a mirror image of its left half. Very few histograms are perfectly symmetric, but many are approximately symmetric. A histogram is **skewed** if one side, or tail, is longer than the other. A histogram with a long right-hand tail is said to be **skewed to the right**, or **positively skewed**. A histogram with a long left-hand tail is said to be **skewed to the left**, or **negatively skewed**. These terms apply to both frequency histograms and relative frequency histograms. Figure 2.10 presents some histograms for hypothetical samples. As another example, the histogram for particulate emissions, shown in Figure 2.5, is skewed to the right.

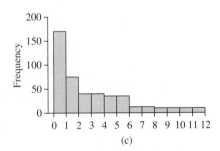

Figure 2.10   (a) A histogram skewed to the left. (b) An approximately symmetric histogram. (c) A histogram skewed to the right.

The examples in Figure 2.10 are straightforward to categorize. In real life, the classification is not always clear-cut, and people may sometimes disagree on how to describe the shape of a particular histogram.

Modes

A peak, or high point, of a histogram is referred to as a **mode**. A histogram is **unimodal** if it has only one mode, and **bimodal** if it has two clearly distinct modes. In principle, a histogram can have more than two modes, but this does not happen often in practice. The histograms in Figure 2.10 are all unimodal. Figure 2.11 (page 58) presents a bimodal histogram for a hypothetical sample.

As another example, it is reasonable to classify the histogram for particulate emissions, shown in Figure 2.5, as unimodal, with the rectangle above the class 1–2 as the only mode. While some might say that the rectangle above the class 3–4 is another mode, most would agree that it is too small a peak to count as a second mode.

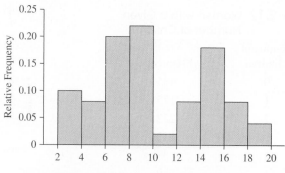

Figure 2.11 A bimodal histogram

Check Your Understanding

3. Classify each of the following histograms as skewed to the left, skewed to the right, or approximately symmetric.

a.

b.

c.

4. Classify each of the following histograms as unimodal or bimodal.

a.

b.

c.

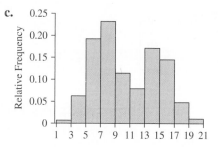

Answers are on page 67.

Objective 4 Construct frequency polygons and ogives

Frequency Polygons and Ogives

Histograms are the most commonly used graphs for representing a frequency or relative frequency distribution. Here we discuss two others. First is the **frequency polygon**. To construct a frequency polygon, we must compute the **class midpoints**, which we now define.

DEFINITION

The **midpoint** of a class is the average of its lower class limit and the lower class limit of the next class.

$$\text{Class midpoint} = \frac{\text{Lower limit} + \text{Lower limit of next class}}{2}$$

We construct a frequency polygon by plotting a point for each class. The x-coordinate of the point is the class midpoint, and the y-coordinate is the frequency of the class. We then connect all the points with straight lines.

EXAMPLE 2.14   Construct a frequency polygon

Table 2.13 presents frequencies and relative frequencies for the particulate emissions data. These were first given in Table 2.9. Construct a frequency polygon for the frequencies in Table 2.13.

Table 2.13 Relative Frequency Distribution for Particulate Data

Class	Frequency	Relative Frequency
0.00–0.99	9	0.138
1.00–1.99	26	0.400
2.00–2.99	11	0.169
3.00–3.99	13	0.200
4.00–4.99	3	0.046
5.00–5.99	1	0.015
6.00–6.99	2	0.031

Solution

We first compute the midpoints of the classes. For example, to compute the midpoint of the first class, we find its lower class limit, 0.00, and the lower class limit of the next class, which is 1.00.

$$\text{Class midpoint of first class} = \frac{0.00 + 1.00}{2} = 0.50$$

Table 2.14 presents the class midpoints for the data in Table 2.13, and shows how the first two and the last of them are computed. Note that to compute the midpoint for the last class, we pretend that there is a class following it whose lower limit is 7.00.

Table 2.14 Relative Frequency Distribution for Particulate Data

Class	Midpoint	Frequency	Relative Frequency
0.00–0.99	$\frac{0+1}{2} = 0.5$	9	0.138
1.00–1.99	$\frac{1+2}{2} = 1.5$	26	0.400
2.00–2.99	2.5	11	0.169
3.00–3.99	3.5	13	0.200
4.00–4.99	4.5	3	0.046
5.00–5.99	5.5	1	0.015
6.00–6.99	$\frac{6+7}{2} = 6.5$	2	0.031

We now plot the points whose x-coordinates are the class midpoints and whose y-coordinates are the frequencies. We then connect the points with straight lines to obtain the frequency polygon. The result is shown in Figure 2.12 (page 60).

We can also construct a **relative frequency polygon**. This is the same as a frequency polygon, except that the frequencies are replaced by relative frequencies. Figure 2.13 presents a relative frequency polygon for the data in Table 2.13.

Figure 2.12 Frequency polygon for the particulate emissions data in Table 2.13

Figure 2.13 Relative frequency polygon for the particulate emissions data in Table 2.13

Ogives

An **ogive** (pronounced "oh jive") is another type of polygon. A **frequency ogive** plots *cumulative frequencies*, and a **relative frequency ogive** plots *cumulative relative frequencies*. We now define these terms.

DEFINITION

The **cumulative frequency** of a class is the sum of the frequencies of that class and all previous classes.

The **cumulative relative frequency** of a class is the cumulative frequency divided by the sum of all the frequencies.

$$\text{Cumulative relative frequency} = \frac{\text{Cumulative frequency}}{\text{Sum of all frequencies}}$$

Table 2.15 presents the cumulative frequencies and cumulative relative frequencies for the data in Table 2.13, and shows how they are calculated. Note that the sum of all the frequencies is 65, which is the largest cumulative frequency.

Table 2.15

Class	Frequency	Relative Frequency	Cumulative Frequency	Cumulative Relative Frequency
0.00–0.99	9	0.138	9	9/65 = 0.138
1.00–1.99	26	0.400	9 + 26 = 35	35/65 = 0.538
2.00–2.99	11	0.169	9 + 26 + 11 = 46	46/65 = 0.708
3.00–3.99	13	0.200	59	59/65 = 0.908
4.00–4.99	3	0.046	62	62/65 = 0.954
5.00–5.99	1	0.015	63	63/65 = 0.969
6.00–6.99	2	0.031	65	65/65 = 1.000

To construct an ogive, plot the cumulative frequencies or cumulative relative frequencies on the *y*-axis, and the upper class limits on the *x*-axis. Then connect the points with straight lines.

EXAMPLE 2.15   Construct an ogive

Construct a frequency ogive and a relative frequency ogive for the data in Table 2.15.

Solution

The cumulative frequencies, cumulative relative frequencies, and upper class limits are shown in Table 2.15. Figures 2.14 and 2.15 present the ogives.

Figure 2.14 Frequency ogive for the data in Table 2.15

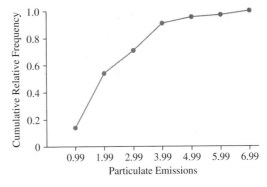

Figure 2.15 Relative frequency ogive for the data in Table 2.15

USING TECHNOLOGY

We use the data in Table 2.7 to illustrate the technology steps.

TI-84 PLUS

Entering Data

Step 1. We will enter the data into **L1** in the data editor. To clear out any data that may be in the list, press **STAT**, then **4: ClrList**, then enter **L1** by pressing **2nd, L1** (Figure A). Then press **ENTER**.

Step 2. Enter the data into **L1** in the data editor by pressing **STAT** then **1: Edit...** For the data in Table 2.7, we begin with **1.5, .87, 1.12, 1.25, 3.46, ...** (Figure B).

Figure A

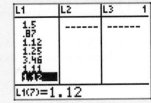

Figure B

Constructing a Histogram

Step 1. Press **2nd, Y=** to access the STAT PLOTS menu and select **Plot1** by pressing **1**.

Step 2. Select **On** and the histogram icon (Figure C).

Step 3. Press **WINDOW** and:

- Set **Xmin** to the lower class limit of the first class. We use 0 for our example.
- Set **Xmax** to the lower class limit of the class following the one containing the largest data value. We use 7.
- Set **Xscl** to the class width. We use 1.
- Set **Ymin** to 0.
- Set **Ymax** to a value greater than the largest frequency of all classes. We use 30.

Step 4. Press **GRAPH** to view the histogram (Figure D).

Figure C

Figure D

MINITAB

Constructing a Histogram

Step 1. Name your variable *Particulate Emissions* and enter the data from Table 2.7 into Column **C1**.

Step 2. Click on **Graph**. Select **Histogram**. Choose the **Simple** option. Press **OK**.

Step 3. Double-click on the *Particulate Emissions* variable and press **OK** (Figure E).

Figure E

EXCEL

Constructing a Histogram

Step 1. Enter the *Particulate Emissions* data from Table 2.7 in **Column A**.

Step 2. Press **Data**, then **Data Analysis**. Select **Histogram** and click **OK**.

Step 3. Enter the range of cells that contain the data in the **Input Range** field and check the **Chart Output** box.

Step 4. Click **OK**.

SECTION 2.2   Exercises

Exercises 1–4 are the Check Your Understanding exercises located within the section.

Understanding the Concepts

In Exercises 5–8, fill in each blank with the appropriate word or phrase.

5. When the right half of a histogram is a mirror image of the left half, the histogram is _____ .

6. A histogram is skewed to the left if its _____ tail is longer than its _____ tail.

7. A histogram is _____ if it has two clearly distinct modes.

8. The _____ of a class is the sum of the frequencies of that class and all previous classes.

In Exercises 9–12, determine whether the statement is true or false. If the statement is false, rewrite it as a true statement.

9. In a frequency distribution, the class width is the difference between the upper and lower class limits.

10. The number of classes used has little effect on the shape of the histogram.

11. There is no one right way to choose the classes for a frequency distribution.

12. A mode occurs at the peak of a histogram.

Practicing the Skills

In Exercises 13–16, classify the histogram as skewed to the left, skewed to the right, or approximately symmetric.

13.

14.

15.

16.

In Exercises 17 and 18, classify the histogram as unimodal or bimodal.

17.

18.

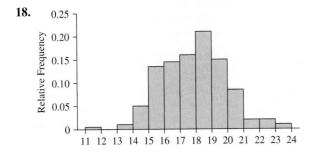

Working with the Concepts

19. Student heights: The following frequency histogram presents the heights, in inches, of a random sample of 100 male college students.

a. How many classes are there?
b. What is the class width?
c. Which class has the highest frequency?

d. What percentage of students are more than 72 inches tall?
e. Is the histogram most accurately described as skewed to the right, skewed to the left, or approximately symmetric?

20. Trained rats: Forty rats were trained to run a maze. The following frequency histogram presents the numbers of trials it took each rat to learn the maze.

a. What is the most frequent number of trials?
b. How many rats learned the maze in three trials or less?
c. How many rats took nine trials or more to learn the maze?
d. Is the histogram most accurately described as skewed to the right, skewed to the left, or approximately symmetric?

21. Interpret histogram: The following histogram shows the distribution of serum cholesterol level (in milligrams per deciliter) for a sample of men. Use the histogram to answer the following questions:

a. Is the percentage of men with cholesterol levels above 240 closest to 30%, 50%, or 70%?
b. In which interval are there more men: 240–260 or 280–340?

22. Interpret histogram: The following histogram shows the distribution of systolic blood pressure (in millimeters) for a sample of women. Use the histogram to answer the following questions:

a. Is the percentage of women with blood pressures above 120 closest to 25%, 50%, or 75%?
b. In which interval are there more women: 130–135 or 140–150?

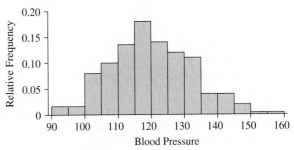

23. **Skewed which way?** For which of the following data sets would you expect a histogram to be skewed to the right? For which would it be skewed to the left?
   a. The lengths of the words in a book
   b. Dates of coins in circulation
   c. Scores of students on an easy exam

24. **Skewed which way?** For which of the following data sets would you expect a histogram to be skewed to the right? For which would it be skewed to the left?
   a. Annual incomes for residents of a town
   b. Amounts of time taken by students on a one-hour exam
   c. Ages of residents of a town

25. **Batting average:** The following frequency distribution presents the batting averages of Major League Baseball players who had 300 or more plate appearances during the 2012 season.

Batting Average	Frequency
0.180–0.199	4
0.200–0.219	14
0.220–0.239	41
0.240–0.259	59
0.260–0.279	58
0.280–0.299	51
0.300–0.319	29
0.320–0.339	10
0.340–0.359	1

Source: sports.espn.go.com

   a. How many classes are there?
   b. What is the class width?
   c. What are the class limits?
   d. Construct a frequency histogram.
   e. Construct a relative frequency distribution.
   f. Construct a relative frequency histogram.
   g. What percentage of players had batting averages of 0.300 or more?
   h. What percentage of players had batting averages less than 0.220?

26. **Batting average:** The following frequency distribution presents the batting averages of Major League Baseball players in both the American League and the National League who had 300 or more plate appearances during the 2012 season.

Batting Average	American League Frequency	National League Frequency
0.180–0.199	2	2
0.200–0.219	7	7
0.220–0.239	21	20
0.240–0.259	30	29
0.260–0.279	26	32
0.280–0.299	21	30
0.300–0.319	12	17
0.320–0.339	5	5
0.340–0.359	0	1

Source: sports.espn.go.com

   a. Construct a frequency histogram for the American League.
   b. Construct a frequency histogram for the National League.
   c. Construct a relative frequency distribution for the American League.

   d. Construct a relative frequency distribution for the National League.
   e. Construct a relative frequency histogram for the American League.
   f. Construct a relative frequency histogram for the National League.
   g. What percentage of American League players had batting averages of 0.300 or more?
   h. What percentage of National League players had batting averages of 0.300 or more?
   i. Compare the relative frequency histograms. What is the main difference between the distributions of batting averages in the two leagues?

27. **Time spent playing video games:** A sample of 200 college freshmen was asked how many hours per week they spent playing video games. The following frequency distribution presents the results.

Number of Hours	Frequency
1.0–3.9	25
4.0–6.9	34
7.0–9.9	48
10.0–12.9	29
13.0–15.9	23
16.0–18.9	17
19.0–21.9	13
22.0–24.9	7
25.0–27.9	3
28.0–30.9	1

   a. How many classes are there?
   b. What is the class width?
   c. What are the class limits?
   d. Construct a frequency histogram.
   e. Construct a relative frequency distribution.
   f. Construct a relative frequency histogram.
   g. What percentage of students play video games less than 10 hours per week?
   h. What percentage of students play video games 19 or more hours per week?

28. **Murder, she wrote:** The following frequency distribution presents the number of murders (including negligent manslaughter) per 100,000 population for each U.S. city with population over 250,000 in the year 2011.

Murder Rate	Frequency
0.0–4.9	21
5.0–9.9	23
10.0–14.9	12
15.0–19.9	6
20.0–24.9	5
25.0–29.9	0
30.0–34.9	2
35.0–39.9	2
40.0–44.9	0
45.0–49.9	0
50.0–54.9	2

Source: Federal Bureau of Investigation

   a. How many classes are there?
   b. What is the class width?
   c. What are the class limits?
   d. Construct a frequency histogram.
   e. Construct a relative frequency distribution.

f. Construct a relative frequency histogram.

g. What percentage of cities had murder rates less than 10 per 100,000 population?

h. What percentage of cities had murder rates of 30 or more per 100,000 population?

29. BMW prices: The following table presents the manufacturer's suggested retail price (in $1000s) for 2013 base models and styles of BMW automobiles.

50.1	89.8	55.2	90.5	30.8	62.7	38.9
70.4	48.0	89.2	47.5	86.2	53.4	90.2
55.2	93.5	39.3	73.6	60.1	140.7	31.2
64.2	44.1	80.6	38.6	68.8	32.5	64.2
56.7	96.7	36.9	65.0	59.8	114.7	43.3
74.9	57.7	108.4	47.4	82.4	44.0	77.6
55.7	93.7	47.8	86.8			

Source: autos.yahoo.com

a. Construct a frequency distribution using a class width of 10, and using 30 as the lower class limit for the first class.

b. Construct a frequency histogram from the frequency distribution in part (a).

c. Construct a relative frequency distribution using the same class width and lower limit for the first class.

d. Construct a relative frequency histogram.

e. Are the histograms unimodal or bimodal?

f. Repeat parts (a)–(d), using a class width of 20, and using 30 as the lower class limit for the first class.

g. Do you think that class widths of 10 and 20 are both reasonably good choices for these data, or do you think that one choice is much better than the other? Explain your reasoning.

30. Geysers: The geyser Old Faithful in Yellowstone National Park alternates periods of eruption, which typically last from 1.5 to 4 minutes, with periods of dormancy, which are considerably longer. The following table presents the durations, in minutes, of 60 dormancy periods that occurred during a recent year.

91	99	99	83	99	85	90	96	88	93
88	88	92	116	59	101	90	71	103	97
82	91	89	89	94	94	61	96	66	105
90	93	88	92	86	93	95	83	90	99
89	94	90	95	93	105	96	92	101	91
94	92	94	86	88	99	90	99	84	92

a. Construct a frequency distribution using a class width of 5, and using 55 as the lower class limit for the first class.

b. Construct a frequency histogram from the frequency distribution in part (a).

c. Construct a relative frequency distribution using the same class width and lower limit for the first class.

d. Construct a relative frequency histogram.

e. Are the histograms skewed to the left, skewed to the right, or approximately symmetric?

f. Repeat parts (a)–(d), using a class width of 10, and using 50 as the lower class limit for the first class.

g. Do you think that class widths of 5 and 10 are both reasonably good choices for these data, or do you think that one choice is much better than the other? Explain your reasoning.

31. Hail to the chief: There have been 57 presidential inaugurations in U.S. history. At each one, the president has made an inaugural address. Following are the number of words spoken in each of these addresses.

1431	135	2321	1730	2166	1177	1211	3375
4472	2915	1128	1176	3843	8460	4809	1090
3336	2831	3637	700	1127	1339	2486	2979
1686	4392	2015	3968	2218	984	5434	1704
1526	3329	4055	3672	1880	1808	1359	559
2273	2459	1658	1366	1507	2128	1803	1229
2427	2561	2320	1598	2155	1592	2071	2395
2096							

Source: The American Presidency Project

a. Construct a frequency distribution with approximately five classes.

b. Construct a frequency histogram from the frequency distribution in part (a).

c. Construct a relative frequency distribution using the same classes as in part (a).

d. Construct a relative frequency histogram from this relative frequency distribution.

e. Are the histograms skewed to the left, skewed to the right, or approximately symmetric?

f. Construct a frequency distribution with approximately nine classes.

g. Repeat parts (b)–(d), using the frequency distribution constructed in part (f).

h. Do you think that five and nine classes are both reasonably good choices for these data, or do you think that one choice is much better than the other? Explain your reasoning.

32. Internet radio: The following table presents the number of hours a sample of 40 subscribers listened to Pandora Radio in a given week.

52	18	2	20	9	9	11	6	18	6
4	12	9	16	10	37	15	18	8	23
4	3	17	19	12	20	11	14	10	37
21	36	17	3	23	28	19	20	29	12

a. Construct a frequency distribution with approximately eleven classes.

b. Construct a frequency histogram from this frequency distribution.

c. Construct a relative frequency distribution the same classes.

d. Construct a relative frequency histogram from this relative frequency distribution.

e. Are the histograms skewed to the left, skewed to the right, or approximately symmetric?

f. Construct a frequency distribution with approximately four classes.

g. Repeat parts (b)–(d), using the frequency distribution constructed in part (f).

h. Do you think that four and eleven classes are both reasonably good choices for these data, or do you think that one choice is much better than the other? Explain your reasoning.

33. Brothers and sisters: Thirty students in a first-grade class were asked how many siblings they have. Following are the results.

1	1	2	1	2	3	7	1	1	5
1	1	3	0	1	1	1	2	5	0
0	1	2	2	4	2	2	3	3	4

a. Construct a frequency histogram.
b. Construct a relative frequency histogram.
c. Are the histograms skewed to the left, skewed to the right, or approximately symmetric?

34. Cough, cough: The following table presents the number of days a sample of patients reported a cough lasting from an acute cough illness.

16	20	21	20	19	17	18	12	22
17	16	24	15	13	21	16	21	20
20	19	18	16	19	20	21	21	14
16	16	14	17	18	14	15	20	20
13	15	18	20	19	21	19	18	20

Based on data from *Annals of Family Medicine*

a. Construct a frequency histogram.
b. Construct a relative frequency histogram.
c. Are the histograms skewed to the left, skewed to the right, or approximately symmetric?

35. Frequency polygon: Using the data in Exercise 25:
a. Construct a frequency polygon for the frequency distribution.
b. Construct a relative frequency polygon, using the same classes.

36. Frequency polygon: Using the data in Exercise 26:
a. Construct a frequency polygon for the American League frequency distribution.
b. Construct a relative frequency polygon for the American League, using the same classes.
c. Construct a frequency polygon for the National League frequency distribution.
d. Construct a relative frequency polygon for the National League, using the same classes.

37. Frequency polygon: Using the data in Exercise 27:
a. Construct a frequency polygon for the frequency distribution.
b. Construct a relative frequency polygon, using the same classes.

38. Frequency polygon: Using the data in Exercise 28:
a. Construct a frequency polygon for the frequency distribution.
b. Construct a relative frequency polygon, using the same classes.

39. Ogive: Using the data in Exercise 25:
a. Compute the cumulative frequencies for the classes in the frequency distribution.
b. Construct a frequency ogive for the frequency distribution.
c. Compute the cumulative relative frequencies for the classes in the frequency distribution.
d. Construct a relative frequency ogive, using the same classes.

40. Ogive: Using the data in Exercise 26:
a. Compute the cumulative frequencies for the classes in the American League frequency distribution.
b. Construct a frequency ogive for the American League frequency distribution.
c. Compute the cumulative relative frequencies for the classes in the American League frequency distribution.

d. Construct a relative frequency ogive for the American League, using the same classes.
e. Compute the cumulative frequencies for the classes in the National League frequency distribution.
f. Construct a frequency ogive for the National League frequency distribution.
g. Compute the cumulative relative frequencies for the classes in the National League frequency distribution.
h. Construct a relative frequency ogive for the National League, using the same classes.

41. Ogive: Using the data in Exercise 27:
a. Compute the cumulative frequencies for the classes in the frequency distribution.
b. Construct a frequency ogive for the frequency distribution.
c. Compute the cumulative relative frequencies for the classes in the frequency distribution.
d. Construct a relative frequency ogive, using the same classes.

42. Ogive: Using the data in Exercise 28:
a. Compute the cumulative frequencies for the classes in the frequency distribution.
b. Construct a frequency ogive for the frequency distribution.
c. Compute the cumulative relative frequencies for the classes in the frequency distribution.
d. Construct a relative frequency ogive, using the same classes.

43. No histogram possible: A company surveyed 100 employees to find out how far they travel in their commute to work. The results are presented in the following frequency distribution.

Distance in Miles	Frequency
0.0–4.9	18
5.0–9.9	26
10.0–14.9	15
15.0–19.9	13
20.0–24.9	12
25.0–29.9	9
30 or more	7

Explain why it is not possible to construct a histogram for this data set.

44. Histogram possible? Refer to Exercise 43: Suppose you found out that none of the employees traveled more than 34 miles. Would it be possible to construct a histogram? If so, construct a histogram. If not, explain why not.

Extending the Concepts

45. Silver ore: The following histogram presents the amounts of silver (in parts per million) found in a sample of rocks. One rectangle from the histogram is missing. What is its height?

46. Classes of differing widths: Consider the following relative frequency distribution for the data in Table 2.7, in which the classes have differing widths.

Class	Frequency	Relative Frequency
0.00–0.99	9	0.138
1.00–1.49	19	0.292
1.50–1.99	7	0.108
2.00–2.99	11	0.169
3.00–3.99	13	0.200
4.00–6.99	6	0.092

a. Compute the class width for each of the classes.
b. Construct a relative frequency histogram. Compare it to the relative frequency histogram in Figure 2.6, in which the classes all have the same width. Explain why using differing widths gives a distorted picture of the data.
c. The *density* of a class is the relative frequency divided by the class width. For each class, divide the relative frequency by the class width to obtain the density.
d. Construct a histogram in which the height of each rectangle is equal to the density of the class. This is called a *density histogram*.
e. Compare the density histogram to the relative frequency histogram in Figure 2.6, in which the classes all have the same width. Explain why differing class widths in a density histogram do not distort the data.

47. Detecting skewness from ogives: Which of the following two ogives represents a skewed distribution and which represents a distribution that is approximately symmetric?

48. Skewed which way? Refer to Exercise 47. For the ogive that represents the skewed distribution, is it skewed to the right or to the left?

Answers to Check Your Understanding Exercises for Section 2.2

1.

Class	Frequency
0.00–0.49	2
0.50–0.99	7
1.00–1.49	19
1.50–1.99	7
2.00–2.49	7
2.50–2.99	4
3.00–3.49	9
3.50–3.99	4
4.00–4.49	2
4.50–4.99	1
5.00–5.49	0
5.50–5.99	1
6.00–6.49	0
6.50–6.99	2

2. a. 15, 20, 25, 30, 35, 40 **b.** 5 **c.**

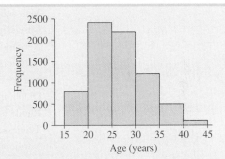

d.

Age	Frequency	Relative Frequency
15–19	795	0.110
20–24	2410	0.334
25–29	2190	0.304
30–34	1208	0.168
35–39	499	0.069
40–44	109	0.015

e.

3. a. Skewed to the right **b.** Skewed to the left **c.** Approximately symmetric
4. a. Unimodal **b.** Unimodal **c.** Bimodal

| SECTION 2.3 | More Graphs for Quantitative Data |

Objectives

1. Construct stem-and-leaf plots
2. Construct dotplots
3. Construct time-series plots

Histograms and other graphs that are based on frequency distributions can be used to summarize both small and large data sets. For small data sets, it is sometimes useful to have a summary that is more detailed than a histogram. In this section, we describe some commonly used graphs that provide more detailed summaries of smaller data sets. These graphs illustrate the shape of the data set, while allowing every value in the data set to be seen.

Objective 1 Construct stem-and-leaf plots

Stem-and-Leaf Plots

Stem-and-leaf plots are a simple way to display small data sets. For example, Table 2.16 presents the U.S. Census Bureau data for the percentage of the population aged 65 and over for each state and the District of Columbia.

Table 2.16 Percentage of Population Aged 65 and Over, by State

Alabama	14.1	Alaska	8.1	Arizona	13.9
Arkansas	14.3	California	11.5	Colorado	10.7
Connecticut	14.4	Delaware	14.1	District of Columbia	11.5
Florida	17.8	Georgia	10.2	Hawaii	14.3
Idaho	12.0	Illinois	12.4	Indiana	12.7
Iowa	14.9	Kansas	13.4	Kentucky	13.1
Louisiana	12.6	Maine	15.6	Maryland	12.2
Massachusetts	13.7	Michigan	12.8	Minnesota	12.4
Mississippi	12.8	Missouri	13.9	Montana	15.0
Nebraska	13.8	Nevada	12.3	New Hampshire	12.6
New Jersey	13.7	New Mexico	14.1	New York	13.6
North Carolina	12.4	North Dakota	15.3	Ohio	13.7
Oklahoma	13.8	Oregon	13.0	Pennsylvania	15.5
Rhode Island	14.1	South Carolina	13.6	South Dakota	14.6
Tennessee	13.3	Texas	10.5	Utah	9.0
Vermont	14.3	Virginia	12.4	Washington	12.2
West Virginia	16.0	Wisconsin	13.5	Wyoming	14.0

Source: U.S. Census Bureau

In a stem-and-leaf plot, the rightmost digit is the leaf, and the remaining digits form the stem. For example, the stem for Alabama is 14, and the leaf is 1. We construct a stem-and-leaf plot for the data in Table 2.16 by using the following three-step process:

Step 1: Make a vertical list of all the stems in increasing order, and draw a vertical line to the right of this list. The smallest stem in Table 2.16 is 8, belonging to Alaska, and the largest is 17, belonging to Florida. The list of stems is shown in Figure 2.16(a).

Step 2: Go through the data set, and for each value, write its leaf next to its stem. For example, the first value is 14.1, for Alabama. We write a "1" next to the stem 14. The next value is 8.1 for Alaska, so we write a "1" next to the stem 8. When we are finished, we have the result shown in Figure 2.16(b).

Step 3: For each stem, arrange its leaves in increasing order. The result is the stem-and-leaf plot, shown in Figure 2.16(c).

```
 8        8 | 1               8 | 1
 9        9 | 0               9 | 0
10       10 | 7 2 5          10 | 2 5 7
11       11 | 5 5            11 | 5 5
12       12 | 0 4 7 6 2 8 4 8 3 6 4 4 2   12 | 0 2 2 3 4 4 4 4 6 6 7 8 8
13       13 | 9 4 1 7 9 8 7 6 7 8 0 6 3 5  13 | 0 1 3 4 5 6 6 7 7 7 8 8 9 9
14       14 | 1 3 4 1 3 9 1 1 6 3 0       14 | 0 1 1 1 1 3 3 3 4 6 9
15       15 | 6 0 3 5        15 | 0 3 5 6
16       16 | 0              16 | 0
17       17 | 8              17 | 8
  (a)           (b)                  (c)
```

Figure 2.16 Steps in the construction of a stem-and-leaf plot

Rounding data for a stem-and-leaf plot

Table 2.17 presents the particulate emissions for 65 vehicles. The first digits range from 0 to 6, and we would like to construct a stem-and-leaf plot with these digits as the stems. The problem is that this leaves two digits for the leaf, but the leaf must consist of only one digit. The solution to this problem is to round the data so that there will be only one digit for the leaf. Table 2.18 presents the particulate data rounded to two digits.

We now follow the three-step process to obtain the stem-and-leaf plot. The result is shown in Figure 2.17.

Table 2.17 Particulate Emissions for 65 Vehicles

1.50	0.87	1.12	1.25	3.46	1.11	1.12	0.88	1.29	0.94	0.64	1.31	2.49
1.48	1.06	1.11	2.15	0.86	1.81	1.47	1.24	1.63	2.14	6.64	4.04	2.48
1.40	1.37	1.81	1.14	1.63	3.67	0.55	2.67	2.63	3.03	1.23	1.04	1.63
3.12	2.37	2.12	2.68	1.17	3.34	3.79	1.28	2.10	6.55	1.18	3.06	0.48
0.25	0.53	3.36	3.47	2.74	1.88	5.94	4.24	3.52	3.59	3.10	3.33	4.58

Table 2.18 Particulate Emissions for 65 Vehicles, Rounded to Two Digits

1.5	0.9	1.1	1.3	3.5	1.1	1.1	0.9	1.3	0.9	0.6	1.3	2.5
1.5	1.1	1.1	2.2	0.9	1.8	1.5	1.2	1.6	2.1	6.6	4.0	2.5
1.4	1.4	1.8	1.1	1.6	3.7	0.6	2.7	2.6	3.0	1.2	1.0	1.6
3.1	2.4	2.1	2.7	1.2	3.3	3.8	1.3	2.1	6.6	1.2	3.1	0.5
0.3	0.5	3.4	3.5	2.7	1.9	5.9	4.2	3.5	3.6	3.1	3.3	4.6

```
0 | 3 5 5 6 6 9 9 9 9
1 | 0 1 1 1 1 1 1 2 2 2 2 3 3 3 3 3 4 4 5 5 5 6 6 6 8 8 9
2 | 1 1 1 2 4 5 5 6 7 7 7
3 | 0 1 1 1 3 3 4 5 5 5 6 7 8
4 | 0 2 6
5 | 9
6 | 6 6
```

Figure 2.17 Stem-and-leaf plot for the data in Table 2.18

Split stems

Sometimes one or two stems contain most of the leaves. When this happens, we often use two or more lines for each stem. The plot is then called a **split stem-and-leaf plot**. We will use the data in Table 2.19 (page 70) to illustrate the method. These data consist of scores on a final examination in a statistics class, arranged in order.

Table 2.19 Scores on a Final Examination

58	66	68	70	70	71	71	72	73	73
75	76	78	78	79	80	80	80	81	82
82	82	82	83	84	86	86	86	87	88
89	89	89	90	92	93	95	97		

Figure 2.18 presents a stem-and-leaf plot for these data, using the stems 5, 6, 7, 8, and 9.

```
5 | 8
6 | 68
7 | 001123356889
8 | 0001222234666678999
9 | 02357
```

Figure 2.18 Stem-and-leaf plot for the data in Table 2.19

Most of the leaves are on two stems, 7 and 8. For this reason, the stem-and-leaf plot does not reveal much detail about the data. To remedy this situation, we will assign each stem two lines on the plot instead of one. Leaves with values 0–4 will go on the first line, and leaves with values 5–9 will go on the second line. So, for example, we will do the following with the stem 7:

```
7 | 001123356889     will become     7 | 0011233
                                      7 | 56889
```

The split stem-and-leaf plot is shown in Figure 2.19. Note that every stem is given two lines, even those that have only a few leaves. Each stem in a split stem-and-leaf plot must receive the same number of lines.

CAUTION

In a split stem-and-leaf plot, each stem must be given the same number of lines.

```
5 |
5 | 8
6 |
6 | 68
7 | 0011233
7 | 56889
8 | 0001222234
8 | 66678999
9 | 023
9 | 57
```

Figure 2.19 Split stem-and-leaf plot for the data in Table 2.19

Check Your Understanding

1. **Weights of college students:** The following table presents weights in pounds for a group of male college freshmen.

136	163	157	195	150	149	151	155	163	145
124	124	156	148	195	192	133	129	160	158
166	155	171	157	182	124	160	172	161	143

a. List the stems for a stem-and-leaf plot.
b. For each item in the data set, write its leaf next to its stem.
c. Rearrange the leaves in numerical order to create a stem-and-leaf plot.

Answers are on page 78.

Back-to-back stem-and-leaf plots

When two data sets have values similar enough so that the same stems can be used, we can compare their shapes with a **back-to-back stem-and-leaf plot**. In a back-to-back stem-and-leaf plot, the stems go down the middle. The leaves for one of the data sets go off to the right, and the leaves for the other go off to the left.

EXAMPLE 2.16   Constructing a back-to-back stem-and-leaf plot

In Table 2.18, we presented particulate emissions for 65 vehicles. In a related experiment carried out at the Colorado School of Mines, particulate emissions were measured for 35 vehicles driven at high altitude. Table 2.20 presents the results. Construct a back-to-back stem-and-leaf plot to compare the emission levels of vehicles driven at high altitude with those of vehicles driven at sea level.

Table 2.20 Particulate Emissions for 35 Vehicles Driven at High Altitude

8.9	4.4	3.6	4.4	3.8	2.4	3.8	5.3	5.8	2.9	4.7	1.9	9.1
8.7	9.5	2.7	9.2	7.3	2.1	6.3	6.5	6.3	2.0	5.9	5.6	5.6
1.5	6.5	5.3	5.6	2.1	1.1	3.3	1.8	7.6				

Solution

Figure 2.20 presents the results. It is clear that vehicles driven at high altitude tend to produce higher emissions.

High Altitude		Sea Level
	0	355669999
9851	1	0111111222233333344555666889
974110	2	11124556777
8863	3	0111334555678
744	4	026
9866633	5	9
5533	6	66
63	7	
97	8	
521	9	

Figure 2.20   Back-to-back stem-and-leaf plots comparing the emissions in vehicles driven at high altitude with emissions from vehicles driven at sea level

Dotplots

Objective 2 Construct dotplots

A **dotplot** is a graph that can be used to give a rough impression of the shape of a data set. It is useful when the data set is not too large, and when there are some repeated values. As an example, Table 2.21 presents the number of children of each of the presidents of the United States and their wives.

Table 2.21 Numbers of Children of U.S. Presidents and Their Wives

0	2	10	2	5	3	6	2	2	4	1
5	4	15	3	4	5	3	2	3	4	2
6	0	0	0	8	3	3	6	2	4	2
0	4	6	4	7	2	0	1	2	6	

Figure 2.21 (page 72) presents a dotplot for the data in Table 2.21. For each value in the data set, a vertical column of dots is drawn, with the number of dots in the column equal to the number of times the value appears in the data set. The dotplot gives a good indication of where the values are concentrated, and where the gaps are. For example, it is

immediately apparent from Figure 2.21 that the most frequent number of children is 2, and only four presidents had more than 6. (John Tyler holds the record with 15.)

Figure 2.21 Dotplot for the data in Table 2.21

EXAMPLE 2.17   Constructing a dotplot with technology

Use technology to construct a dotplot for the exam score data in Table 2.19 on page 70.

Solution

The following figure shows the dotplot constructed in MINITAB. Step-by-step instructions for constructing dotplots with MINITAB are given in the Using Technology section on page 74.

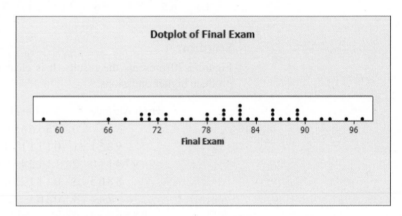

Objective 3 Construct time-series plots

Time-Series Plots

A **time-series plot** may be used when the data consist of values of a variable measured at different points in time. As an example, we consider the Dow Jones Industrial Average, which reflects the prices of 30 large stocks. Table 2.22 presents the closing value of the Dow Jones Industrial Average at the end of each year from 2000 to 2012.

Table 2.22 Dow Jones Industrial Average

Year	Average
2000	10,786.85
2001	10,021.50
2002	8,341.63
2003	10,453.92
2004	10,783.01
2005	10,717.50
2006	12,463.15
2007	13,264.82
2008	8,776.39
2009	10,428.05
2010	11,557.51
2011	12,217.56
2012	13,104.14

Source: Yahoo Finance

In a time-series plot, the horizontal axis represents time, and the vertical axis represents the value of the variable we are measuring. We plot the values of the variable at each of the times, then connect the points with straight lines. Example 2.18 shows how.

EXAMPLE 2.18

Constructing a time-series plot

Construct a time-series plot for the data in Table 2.22.

Solution

Step 1: Label the horizontal axis with the times at which measurements were made.

Step 2: Plot the value of the Dow Jones Industrial Average for each year.

Step 3: Connect the points with straight lines.

The result is shown in Figure 2.22. It is clear that the average declined from 2000 to 2002, generally increased from 2002 to 2007, dropped sharply in 2008, and increased from 2008 to 2012.

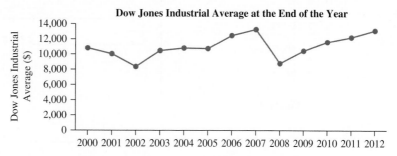

Figure 2.22 Time-series plot for the Dow Jones Industrial Average

Check Your Understanding

2. The National Institute on Drug Abuse surveyed U.S. high school seniors every two years to determine the percentage who said they had used marijuana one or more times. The following time-series plot presents the results.

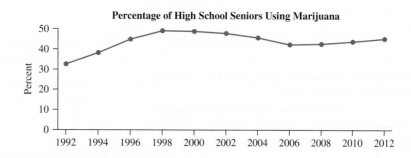

a. During what year was marijuana use among high school students the highest?

b. True or false: At one time, more than half of all high school students reported using marijuana.

c. During what periods of time was marijuana use increasing?

d. During what periods of time was marijuana use decreasing?

e. True or false: Marijuana use among high school students in 2012 was lower than it had been since 1992.

Answers are on page 78.

USING TECHNOLOGY

We use the data in Table 2.19 to illustrate the technology steps.

MINITAB

Constructing a stem-and-leaf plot and dotplot

Step 1. Name your variable *Final Exam* and enter the data from Table 2.19 into **Column C1**.

Step 2. Click on **Graph**. Select **Stem-and-Leaf** or **Dotplot**. For **Dotplot**, choose the **Simple** option. Press **OK**.

Step 3. Double-click on the *Final Exam* variable and press **OK**. (See Figures A and B.)

```
5   8
6
6   68
7   0011233
7   56889
8   0001222234
8   66678999
9   023
9   57
```

Figure A

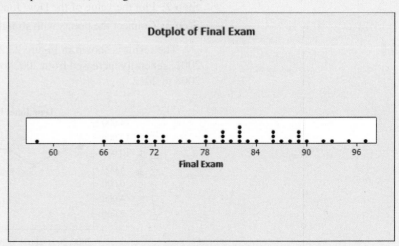

Figure B

SECTION 2.3   Exercises

Exercises 1 and 2 are the Check Your Understanding exercises located within the section.

Understanding the Concepts

In Exercises 3–6, fill in each blank with the appropriate word or phrase.

3. In a stem-and-leaf plot, the rightmost digit of each data value is the _____ .

4. In a back-to-back stem-and-leaf plot, each of the two data sets plotted must have the same _____ .

5. A _____ is useful when the data consist of values measured at different points in time.

6. In a time-series plot, the horizontal axis represents _____ .

In Exercises 7–10, determine whether the statement is true or false. If the statement is false, rewrite it as a true statement.

7. Stem-and-leaf plots and dotplots provide a simple way to display data for small data sets.

8. In a stem-and-leaf plot, each stem must be a single digit.

9. In a dotplot, the number of dots in a vertical column represents the number of times a certain value appears in a data set.

10. In a time-series plot, the vertical axis represents time.

Practicing the Skills

11. Construct a stem-and-leaf plot for the following data.

57	20	27	16	11	12	29	39	45	52	58	15
46	27	22	21	15	50	16	45	20	55	12	31

12. Construct a stem-and-leaf plot for the following data, in which the leaf represents the hundredths digit.

5.03	4.99	4.95	5.01	4.99	5.03	4.91	5.25	4.80
5.24	4.94	5.04	5.17	4.81	5.22	4.92	5.05	4.89
5.19	5.17	5.25	5.14	5.10	4.94	5.19	4.99	

13. List the data in the following stem-and-leaf plot. The leaf represents the ones digit.

```
3 | 0012
3 | 56779
4 | 234
4 | 567777889
5 | 011122224
5 | 67889
6 | 13
6 |
```

14. List the data in the following stem-and-leaf plot. The leaf represents the tenths digit.

```
14 | 4 6 8 9
15 | 1 2 2 4 5 7 7 8
16 | 0 1 1 1 2 3 7 7 9
17 |
18 | 2 3 8
```

15. Construct a dotplot for the data in Exercise 11.

16. Construct a dotplot for the data in Exercise 12.

Working with the Concepts

17. BMW prices: The following table presents the manufacturer's suggested retail price (in $1000s) for 2013 base models and styles of BMW automobiles.

50.1	89.8	55.2	90.5	30.8	62.7	38.9
70.4	48.0	89.2	47.5	86.2	53.4	90.2
55.2	93.5	39.3	73.6	60.1	140.7	31.2
64.2	44.1	80.6	38.6	68.8	32.5	64.2
56.7	96.7	36.9	65.0	59.8	114.7	43.3
74.9	57.7	108.4	47.4	82.4	44.0	77.6
55.7	93.7	47.8	86.8			

Source: autos.yahoo.com

a. Round the data to the nearest whole number (round .5 up) and construct a stem-and-leaf plot, using the numbers 3 through 14 as the stems.

b. Repeat part (a), but split the stems, using two lines for each stem.

c. Which stem-and-leaf plot do you think is more appropriate for these data, the one in part (a) or the one in part (b)? Why?

18. How's the weather? The following table presents the daily high temperatures for the city of Macon, Georgia, in degrees Fahrenheit, for the winter months of January and February, 2013.

70	67	48	57	53	56	56	63	67	69	74	78
77	77	76	75	65	58	62	68	66	53	58	68
52	61	61	66	76	73	56	52	59	64	55	71
73	52	64	63	67	66	56	60	53	66	51	51
58	65	61	67	56	48	71	51	68	63	58	

a. Construct a stem-and-leaf plot, using the digits 4, 5, 6, and 7 as the stems.

b. Repeat part (a), but split the stems, using two lines for each stem.

c. Which stem-and-leaf plot do you think is more appropriate for these data, the one in part (a) or the one in part (b)? Why?

19. Air pollution: The following table presents amounts of particulate emissions for 65 vehicles. These data also appear in Table 2.18.

1.5	0.9	1.1	1.3	3.5	1.1	1.1	0.9	1.3	0.9	0.6	1.3	2.5
1.5	1.1	1.1	2.2	0.9	1.8	1.5	1.2	1.6	2.1	6.6	4.0	2.5
1.4	1.4	1.8	1.1	1.6	3.7	0.6	2.7	2.6	3.0	1.2	1.0	1.6
3.1	2.4	2.1	2.7	1.2	3.3	3.8	1.3	2.1	6.6	1.2	3.1	0.5
0.3	0.5	3.4	3.5	2.7	1.9	5.9	4.2	3.5	3.6	3.1	3.3	4.6

a. Construct a split stem-and-leaf plot in which each stem appears twice, once for leaves 0–4 and again for leaves 5–9.

b. Compare the split stem-and-leaf plot to the plot in Figure 2.17. Comment on the advantages and disadvantages of the split stem-and-leaf plot for these data.

20. Technology salaries: The following table presents the annual salaries for the employees of a small technology firm. Round each number to the nearest thousand, and then construct a stem-and-leaf plot.

91,808	118,625	131,092	60,763
36,463	37,187	45,870	50,594
98,302	123,973	182,255	59,186
44,889	164,861	71,082	69,695
28,098	157,110	50,461	98,132
49,742	25,339	24,164	107,878
136,690	129,514	99,254	57,468

21. Tennis and golf: Following are the ages of the winners of the men's Wimbledon tennis championship and the Master's golf championship for the years 1968 through 2013.

Ages of Wimbledon Winners											
29	30	26	27	25	27	21	31	20	21	22	23
24	22	29	24	25	17	18	22	22	21	24	22
22	21	22	23	26	25	26	27	28	31	21	21
22	23	24	25	22	27	24	24	30	26		

Ages of Master's Winners											
39	29	38	33	32	36	38	35	33	27	42	27
23	31	28	26	32	27	46	28	30	31	32	33
32	35	28	43	38	23	41	33	37	25	26	32
31	29	33	31	28	39	39	26	33	32		

a. Construct back-to-back split stem-and-leaf plots for these data sets.

b. How do the ages of Wimbledon champions differ from the ages of Master's champions?

22. Pass the popcorn: Following are the running times (in minutes) for the 15 top-grossing movies rated PG or PG-13, and the 15 top-grossing movies rated R, for the weekend of March 15–17, 2013.

Movies Rated PG or PG-13	
Oz The Great and Powerful	127
Escape From Planet Earth	89
Life of Pi	126
Wreck-It Ralph	108
The Incredible Burt Wonderstone	100
Jack the Giant Slayer	114
Snitch	112
Safe Haven	115
The Last Exorcism Part II	88
Quartet	98
Emperor	98
Warm Bodies	97
Dark Skies	97
Lincoln	145
The Hobbit: An Unexpected Journey	166

Source: Box Office Mojo

Movies Rated R

The Call	95
Identity Thief	111
21 and Over	93
Silver Linings Playbook	122
Dead Man Down	110
A Good Day to Die Hard	97
Argo	120
Zero Dark Thirty	157
Django Unchained	165
Spring Breakers	94
No	118
Hansel and Gretel: Witch Hunters	88
Parker	118
Phantom	97

Source: Box Office Mojo

a. Construct back-to-back stem-and-leaf plots for these data sets.
b. Do the running times of R-rated movies differ greatly from the running times of movies rated PG or PG-13, or are they roughly similar?

23. **More weather:** Construct a dotplot for the data in Exercise 18. Are there any gaps in the data?

24. **Safety first:** Following are the numbers of hospitals in each of the 50 U.S. states plus the District of Columbia that won Patient Safety Excellence Awards. Construct a dotplot for these data and describe its shape.

2	0	9	3	24	6	1	0	1	14	3
0	2	10	10	11	3	1	4	0	5	12
5	12	0	3	4	0	0	0	5	1	7
11	2	15	3	5	20	1	2	1	5	16
2	0	8	6	0	8	0				

25. **Looking for a job:** The following table presents the U.S. unemployment rate for each of the years 1989 through 2012.

Year	Unemployment	Year	Unemployment
1989	5.3	2001	4.7
1990	5.6	2002	5.8
1991	6.8	2003	6.0
1992	7.5	2004	5.5
1993	6.9	2005	5.1
1994	6.1	2006	4.6
1995	5.6	2007	4.6
1996	5.4	2008	5.8
1997	4.9	2009	9.3
1998	4.5	2010	9.6
1999	4.2	2011	8.9
2000	4.0	2012	8.1

Source: National Bureau of Labor Statistics

a. Construct a time-series plot of the unemployment rate.
b. For which periods of time was the unemployment rate increasing? For which periods was it decreasing?

26. **Vacant apartments:** The following table presents the percentage of U.S. residential rental units that were vacant during each quarter from 2009 through 2012.

Quarter	Vacancy Rate	Quarter	Vacancy Rate
Mar. 2009	10.1	Mar. 2011	9.7
Jun. 2009	10.6	Jun. 2011	9.2
Sep. 2009	11.1	Sep. 2011	9.8
Dec. 2009	10.7	Dec. 2011	9.4
Mar. 2010	10.6	Mar. 2012	8.8
Jun. 2010	10.6	Jun. 2012	8.6
Sep. 2010	10.3	Sep. 2012	8.6
Dec. 2010	9.4	Dec. 2012	8.7

Source: Current Population Survey

a. Construct a time-series plot for these data.
b. From December 2009 through December 2012, the U.S. economy was recovering from a recession. What was the trend in the vacancy rate during this time period?

27. **Military spending:** The following table presents the amount spent, in billions of dollars, on national defense by the U.S. government every other year for the years 1951 through 2013. The amounts are adjusted for inflation, and represent 2013 dollars.

Year	Spending	Year	Spending
1951	477.5	1983	539.4
1953	504.9	1985	577.4
1955	391.0	1987	562.0
1957	415.9	1989	540.0
1959	410.4	1991	531.6
1961	415.4	1993	450.5
1963	447.0	1995	408.0
1965	424.4	1997	389.0
1967	540.4	1999	395.7
1969	543.5	2001	420.1
1971	452.6	2003	559.1
1973	414.7	2005	601.2
1975	383.5	2007	681.8
1977	406.2	2009	719.9
1979	402.8	2011	689.1
1981	451.0	2013	614.7

Source: Department of Defense

a. Construct a time-series plot for these data.
b. The plot covers six decades, from the 1950s through the decade of 2000–2009. During which of these decades did national defense spending increase, and during which decades did it decrease?
c. The United States fought in the Korean War, which ended in 1953. What effect did the end of the war have on military spending after 1953?
d. During the period 1965–1968, the United States steadily increased the number of troops in Vietnam from 23,000 at the beginning of 1965 to 537,000 at the end of 1968. Beginning in 1969, the number of Americans in Vietnam was steadily reduced, with the last of them leaving in 1975. How is this reflected in the national defense spending from 1965 to 1975?

28. **College students:** The following table presents the numbers of male and female students (in thousands) enrolled in college in the United States as undergraduates for each of the years 1993 through 2012.

Year	Male	Female	Year	Male	Female
1993	5484	6840	2003	6227	8253
1994	5422	6840	2004	6340	8441
1995	5401	6831	2005	6409	8555
1996	5421	6906	2006	6514	8671
1997	5469	6982	2007	6728	8876
1998	5446	6991	2008	7123	9372
1999	5559	7122	2009	7595	9970
2000	5778	7377	2010	7835	10244
2001	6004	7711	2011	7979	10347
2002	6192	8065	2012	8038	10489

Source: National Center for Educational Statistics

a. Construct a time-series plot for the male enrollment; then on the same axes, construct a time-series plot for the female enrollment.

b. Which is growing faster, male enrollment or female enrollment?

29. Let's eat: The following time-series plot presents the amount spent, in billions of dollars, on food by U.S. residents for the years 1995 through 2011.

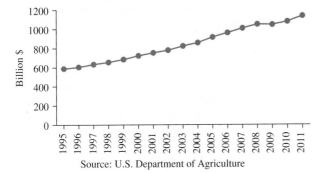

Source: U.S. Department of Agriculture

a. Estimate the amount spent on food in 2003.

b. Estimate the amount by which food expenditures increased between 2000 and 2008.

c. True or false: The amount spent in 2011 is approximately twice as much as the amount spent in 1995.

d. True or false: The amount spent increased every year from 1995 through 2011.

30. Going for gold: The following time-series plot presents the number of Summer Olympic events in which the United States won a gold medal in each Olympic year from 1952 through 2012.

Source: Wikipedia

a. In one year, the United States did not participate in Summer games that were held in Moscow, in protest of the invasion of Afghanistan by the Soviet Union. Which year was this?

b. In 1984, the Soviet Union did not participate in the Summer games held in Los Angeles, citing "undisguised threats" against their athletes. Estimate the number of gold medals won by the United States in that year.

c. Other than 1980 and 1984, has the number of gold medals won by the United States been generally increasing, generally decreasing, or staying about the same?

31. Let's go skiing: The following time-series plot presents the number of inches of snow falling in Denver each year from 1882 through 2012.

Source: National Weather Service

a. Estimate the largest annual snowfall ever recorded in Denver.

b. Was the year of the largest annual snowfall closest to 1900, 1910, or 1920?

c. Was the amount of snowfall in the years 2000–2012 greater than, less than, or about equal to the snowfall in most other years?

d. True or false: The year with the least snowfall ever recorded in Denver was in the 1800s.

e. True or false: It usually snows more than 80 inches per year in Denver.

32. Baseball salaries: The following time-series plot presents the average salary of Major League Baseball players for the years 1993–2012.

a. What is the first year that the average salary was more than 1.5 million dollars?

b. In 1994, the players went on strike, and the last two months of the season were canceled. In 2003, a new labor agreement was reached that penalized teams that paid unusually high salaries. How did these events affect the average salary?

33. Christmas shopping: The following table presents the average amount spent by each shopper over Thanksgiving weekend, which is regarded as the start of the Christmas shopping season in the U.S. for the years 2005 through 2012.

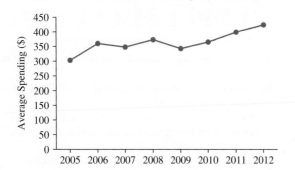

a. True or false: The number of years in which the amount decreased is the same as the number of years in which it increased.

b. The amount increased every year from 2010 through 2012.

c. True or false: The amount spent in 2009 was less than the amount spent in any previous year.

d. The amount spent in 2012 was greater than $400.

34. Arctic ice sheet: The following table presents the extent of ice coverage (in millions of square kilometers) in the Arctic region on January 1st of each year from 1989 through 2013.

a. What was the first year that the coverage dropped below 14 million square kilometers?

b. What was the first year that the coverage dropped below 13 million square kilometers?

c. True or false: The coverage has been less than 14 million square kilometers in every year since 2000.

d. True or false: The coverage has decreased in every year since 2000.

Extending the Concepts

35. Elections: In U.S. presidential elections, each of the 50 states casts a number of electoral votes equal to its number of senators (2) plus its number of members of the House of Representatives. In addition, the District of Columbia casts three electoral votes. Following are the numbers of electoral votes cast for president for each of the 50 states and the District of Columbia in the election of 2012.

9	3	11	6	55	9	7	3	29	16	4
4	20	11	6	6	8	8	4	10	11	16
10	6	10	3	5	6	4	14	5	29	15
3	18	7	7	20	4	9	3	11	38	6
3	13	12	5	10	3	3				

a. Construct a split stem-and-leaf plot for these data, using two lines for each stem.

b. Construct a frequency histogram, with the classes chosen so that there are two classes for each stem.

c. Explain why the stem-and-leaf plot and the histogram have the same shape.

Answers to Check Your Understanding Exercises for Section 2.3

1. a.

12
13
14
15
16
17
18
19

b.

12	4494
13	63
14	9583
15	70156857
16	330601
17	12
18	2
19	552

c.

12	4449
13	36
14	3589
15	01556778
16	001336
17	12
18	2
19	255

2. a. 1998   **b.** False   **c.** 1992–1998; 2006–2012   **d.** 1998–2006   **e.** False

Graphs Can Be Misleading

Objectives

1. Understand how improper positioning of the vertical scale can be misleading
2. Understand the area principle for constructing statistical graphs
3. Understand how three-dimensional graphs can be misleading

Statistical graphs, when properly used, are powerful forms of communication. Unfortunately, when graphs are improperly used, they can misrepresent the data and lead people to draw incorrect conclusions. We discuss here three of the most common forms of misrepresentation: incorrect position of the vertical scale, incorrect sizing of graphical images, and misleading perspective for three-dimensional graphs.

Objective 1 Understand how improper positioning of the vertical scale can be misleading

Positioning the Vertical Scale

Table 2.23 is a distribution of the number of passengers, in millions, at Denver International Airport in each year from 2006 through 2012.

Table 2.23 Passenger Traffic at Denver International Airport

Year	Number of Passengers (in millions)
2006	47.3
2007	49.9
2008	51.2
2009	50.2
2010	52.2
2011	52.7
2012	53.1

Source: Wikipedia

In order to get a better picture of the data, we can make a bar graph. Figures 2.23 and 2.24 present two different bar graphs of the same data. Figure 2.23 presents a clear picture of the data. We can see that the number of passengers has been fairly steady, with just a slight increase from 2006 to 2012. Now imagine that someone was eager to persuade us that passenger traffic had increased greatly since 2006. If they were to show us Figure 2.23, we wouldn't be convinced. So they might show us a misleading picture like Figure 2.24 instead. Figure 2.24 gives the impression of a truly dramatic increase.

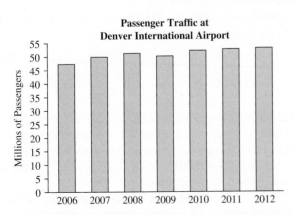

Figure 2.23 The bottom of the bars is at zero. This bar graph gives a correct impression of the data.

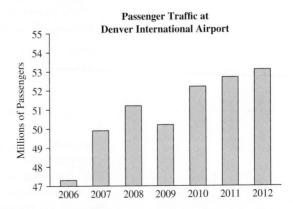

Figure 2.24 The bottom of the bars is not at zero. This bar graph exaggerates the differences between the bars.

Figures 2.23 and 2.24 are based on the same data. Why do they give such different impressions? The reason is that the baseline (the value corresponding to the bottom of the bars) is at zero in Figure 2.23, but not at zero in Figure 2.24. This exaggerates the differences between the bars. For example, in Figure 2.24, the bar for the year 2012 is more than 10 times as long as the bar for the year 2006, but the actual increase in passenger traffic is much less than that.

This sort of misleading information can be created with time-series plots as well. Figures 2.25 and 2.26 present two different time-series plots of the data. In Figure 2.25, the baseline is at zero, so an accurate impression is given. In Figure 2.26, the baseline is larger than zero, so the rate of increase is exaggerated.

> When a graph or plot represents how much or how many of something, check the baseline. If it isn't at zero, the graph may be misleading.

Figure 2.25   The baseline is at zero. This plot gives an accurate picture of the data.

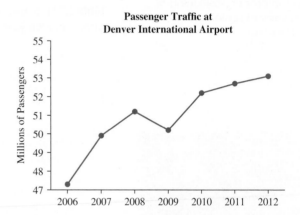

Figure 2.26   The baseline is not at zero. This plot exaggerates the rate of increase.

Objective 2 Understand the area principle for constructing statistical graphs

The Area Principle

We often use images to compare amounts. Larger images correspond to greater amounts. To use images properly in this way, we must follow a rule known as the **area principle**.

> ### The Area Principle
>
> When amounts are compared by constructing an image for each amount, the *areas* of the images must be proportional to the amounts. For example, if one amount is twice as much as another, its image should have twice as much area as the other image.
>
> When the area principle is violated, the images give a misleading impression of the data.

Bar graphs, when constructed properly, follow the area principle. The reason is that all the bars have the same width; only their height varies. Therefore, the areas of the bars are proportional to the amounts. For example, Figure 2.27 presents a bar graph that illustrates a comparison of the cost of jet fuel in 2003 and 2013. In 2003, the cost of jet fuel was $0.83 per gallon, and in 2008 it had risen to $2.87 per gallon.

The bars in the bar graph differ in only one dimension — their height. The widths are the same. For this reason, the bar graph presents an accurate comparison of the two prices. The price in 2013 is about 3.5 times as much as the price in 2003, and the area of the bar for 2013 is about 3.5 times as large as the area of the bar for 2003.

Cost of Jet Fuel

Source: www.airlines.org

Figure 2.27   Price per gallon of jet fuel in 2003 and 2013. The bar graph accurately represents the difference.

Unfortunately, people often mistakenly vary both dimensions of an image when making a comparison. This exaggerates the difference. Following is a comparison of the cost of jet fuel in the years 2003 and 2013 that uses a picture of an airplane to illustrate the difference.

Cost of Jet Fuel

2003   2013

The pictures of the planes make the difference appear much larger than the correctly drawn bar graph does. The reason is that both the height and the width of the airplane have been increased by a factor of 3.5. Thus, the area of the larger plane is more than 12 times the area of the smaller plane. This graph violates the area principle, and gives a misleading impression of the comparison.

Check Your Understanding

1. The population of country A is twice as large as the population of country B. True or false: If images are used to represent the populations, both the height and width of the image for country A should be twice as large as the height and width of the image for country B.

2. If the baseline of a bar graph or time-series plot is not at zero, then the differences may appear to be _____ than they actually are. (*Choices: larger, smaller*)

Answers are on page 85.

Objective 3 Understand how three-dimensional graphs can be misleading

Three-Dimensional Graphs and Perspective

The bar graph in Figure 2.27 presents an accurate picture of the prices of jet fuel in the years 2003 and 2013. Newspapers and magazines often prefer to present three-dimensional bar graphs, because they are visually more impressive. Unfortunately, in order to make the tops of the bars visible, these graphs are often drawn as though the reader is looking down on them. This can make the bars look shorter than they really are.

Figure 2.28 (page 82) presents a three-dimensional bar graph of the sort often seen in publications. The data are the same as in Figure 2.27: The price in 2003 is $0.83, and the price in 2013 is $2.87. However, because you are looking down on the bars, they appear shorter than they really are.

Beware of three-dimensional bar graphs. If you can see the tops of the bars, they may look shorter than they really are.

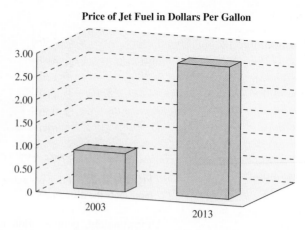

Figure 2.28   Price per gallon of jet fuel in 2003 and 2013. The bars appear shorter than they really are, because you are looking down at them.

SECTION 2.4   Exercises

Exercises 1 and 2 are the Check Your Understanding exercises located within the section.

Understanding the Concepts

In Exercises 3 and 4, fill in each blank with the appropriate word or phrase.

3. A plot that represents how much of something there is may be misleading if the baseline is not at _____ .

4. The area principle says that when images are used to compare amounts, the areas of the images should be _____ to the amounts.

Working with the Concepts

5. **CD sales decline:** Sales of CDs have been declining for several years as more music is downloaded over the Internet. Following are two bar graphs that illustrate the decline in CD sales. (Source: Recording Industry Association of America)

Choose one of the following options, and explain why it is correct:
(i) Graph A presents an accurate picture, and graph B exaggerates the decline.
(ii) Graph B presents an accurate picture, and graph A understates the decline.

6. **Xbox sales:** The following bar graph and time-series plot both present the sales of Xbox video game consoles for the years 2009–2012. Which of the graphs presents a more accurate picture? Why? (Source: Microsoft)

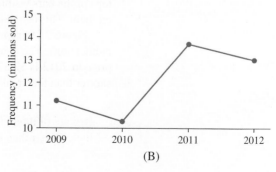

7. Stock market crash: The Dow Jones Industrial Average reached its highest level ever on October 9, 2007, when it closed at $14,164.53. One year later, on October 9, 2008, the average had dropped almost 40%, to $8,579.19. Which of the following graphs accurately represents the magnitude of the drop? Which one exaggerates it?

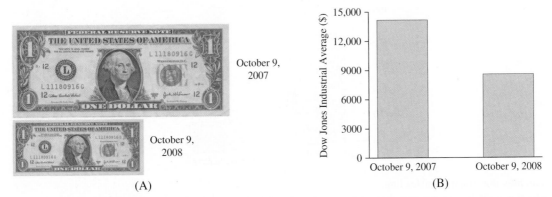

(A)   (B)

8. Save your money: In December 2011, U.S. residents saved a total of $392 billion. In December 2012, that amount approximately doubled, to $804 billion. Which of the following graphs compares these totals more accurately, and why? (Source: Bureau of Economic Analysis)

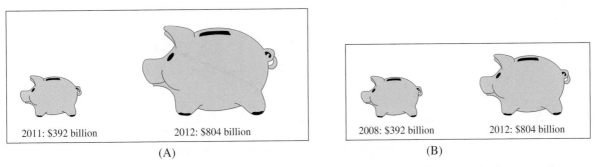

(A)   (B)

9. Tying the knot: Data compiled by the U.S. Census Bureau suggests that the percentage of women who have never been married has increased over time. The following bar graph presents the percentages of women who had never married in the years 1970 and 2012. Does the bar graph present an accurate picture of the increase, or is it misleading? Explain.

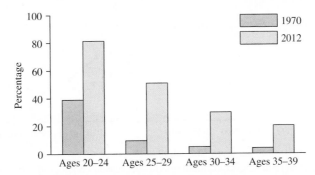

10. Time-shifting: Data from the Nielsen Company suggests that the number of hours per month spent watching time-shifted television using devices such as digital video recorders has increased over time. The following bar graph presents the number of hours per month people in several age groups spent watching time-shifted television in 2009 and in 2012. Does the bar graph present an accurate picture of the increase, or is it misleading? Explain.

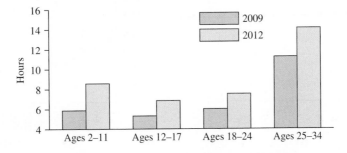

11. Female senators: Of the 100 members of the United States Senate in 2013, 80 were men and 20 were women. The following three-dimensional bar graph attempts to present this information.

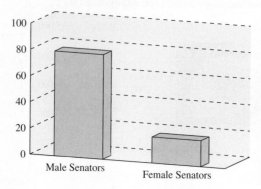

a. Explain how this graph is misleading.

b. Construct a graph (not necessarily three-dimensional) that presents this information accurately.

12. Age at marriage: Data compiled by the U.S. Census Bureau suggests that the age at which women first marry has increased over time. The following time-series plot presents the average age at which women first marry for the years 1950–2010. Does the plot present an accurate picture of the increase, or is it misleading? Explain.

13. College degrees: Both of the following time-series plots present the percentage of U.S. Bachelor degrees that were in science or mathematics during the years 1970 through 2010. (Source: U.S. Department of Education)

Which of the following statements is more accurate, and why?

(i) The percentage of degrees that were in science or mathematics decreased considerably between 1970 and 2010.

(ii) The percentage of degrees that were in science or mathematics decreased somewhat between 1970 and 2010.

14. Food expenditures: Both of the following time-series plots present the percentage of income spent on food by U.S. residents for the years 1995 through 2011. (Source: U.S. Department of Agriculture)

(A)   (B)

Which of the following statements is more accurate, and why?
 (i) The percentage of income spent on food decreased considerably between 1992 and 2008.
 (ii) The percentage of income spent on food decreased slightly between 1992 and 2008.

Extending the Concepts

15. Manipulating the *y*-axis: For the data in Table 2.23:
 a. Construct a bar graph in which the *y*-axis is labeled from 0 to 100.
 b. Compare this bar graph with the bar graphs in Figures 2.23 and 2.24. Does this bar graph tend to make the difference seem smaller than the other bar graphs do?
 c. Which of the three bar graphs do you think presents the most accurate picture of the data? Why?

Answers to Check Your Understanding Exercises for Section 2.4

1. False   **2.** Larger

Chapter 2 Summary

Section 2.1: The first step in summarizing qualitative data is to construct a frequency distribution or relative frequency distribution. Then a bar graph or pie chart can be constructed. Bar graphs can illustrate either frequencies or relative frequencies. Side-by-side bar graphs can be used to compare two qualitative data sets that have the same categories.

Section 2.2: Frequency distributions and relative frequency distributions are also used to summarize quantitative data. Histograms are graphical summaries that illustrate frequency distributions and relative frequency distributions, allowing us to visualize the shape of a data set. Histograms can show us whether a data set is skewed or symmetric, unimodal or bimodal. Frequency polygons can also be used to illustrate frequency distributions and relative frequency distributions. Ogives illustrate cumulative relative frequency distributions.

Section 2.3: Stem-and-leaf plots and dotplots are useful summaries for small data sets. They have an advantage over histograms: They allow every point in the data set to be seen. Back-to-back stem-and-leaf plots can be used to compare the shapes of two data sets. Time-series plots illustrate how the value of a variable has changed over time.

Section 2.4: To avoid constructing a misleading graph, be sure to start the vertical scale at zero. When images are used to compare amounts, the area principle should be followed. This principle states that the areas of the images should be proportional to the amounts. Three-dimensional bar graphs are often misleading, because the bars look shorter than they really are.

Vocabulary and Notation

area principle 80	frequency ogive 60	relative frequency histogram 53
back-to-back stem-and-leaf plot 71	frequency polygon 58	relative frequency ogive 60
bar graph 38	histogram 53	relative frequency polygon 60
bimodal 57	lower class limit 51	side-by-side bar graph 39
class 49	mode 57	skewed 57
class midpoint 58	negatively skewed 57	skewed to the left 57
class width 51	ogive 60	skewed to the right 57
cumulative frequency 60	open-ended class 56	split stem-and-leaf plot 69
cumulative relative frequency 60	Pareto chart 39	stem-and-leaf plot 68
dotplot 71	pie chart 40	symmetric 57
frequency 36	positively skewed 57	time-series plot 72
frequency distribution 36	relative frequency 37	unimodal 57
frequency histogram 53	relative frequency distribution 37	upper class limit 51

Chapter Quiz

1. Following is the list of letter grades for students in an algebra class: A, B, F, A, C, C, A, B, D, F, D, A, A, B, C, F, B, D, C, A, A, A, F, B, C, A, C. Construct a frequency distribution for these data.

2. Construct a relative frequency distribution for the data in Exercise 1.

3. Construct a frequency bar graph for the data in Exercise 1.

4. Construct a pie chart for the data in Exercise 1.

5. The first class in a relative frequency distribution is 2.0–4.9, and there are six classes. Find the remaining five classes. What is the class width?

6. True or false: A histogram can have more than one mode.

7. A sample of 100 students was asked how many hours per week they spent studying. The following frequency distribution shows the results.

Number of Hours	Frequency
1.0–4.9	14
5.0–8.9	34
9.0–12.9	29
13.0–16.9	15
17.0–20.9	8

    a. Construct a frequency histogram for these data.

    b. Construct a relative frequency histogram for these data.

8. Construct a frequency polygon for the data in Exercise 7.

9. Construct a relative frequency ogive for the data in Exercise 7.

10. List the data in the following stem-and-leaf plot. The leaf represents the ones digit.

```
1 | 1155999
2 | 223578
3 | 008
4 | 4578
5 | 0133568
```

11. Following are the prices (in dollars) for a sample of coffee makers.

    19   22   29   68   35   37   28   22   41   39   28

    Construct a stem-and-leaf plot for these data.

12. Following are the prices (in dollars) for a sample of espresso makers.

    99   50   31   65   50   99   70   40   25   56   30   77

    Construct a back-to-back stem-and-leaf plot for these data and the data in Exercise 11.

13. Construct a dotplot for the data in Exercise 11.

14. The following table presents the percentage of Americans who use a cell phone exclusively, with no landline phone, for the years 2008–2012. Construct a time-series plot for these data.

Time Period	Percent
January–June 2008	16.1
July–December 2008	18.4
January–June 2009	21.1
July–December 2009	22.9
January–June 2010	24.9
July–December 2010	27.8
January–June 2011	30.2
July–December 2011	32.3
January–June 2012	34.0

Source: National Health Interview Survey

15. According to the area principle, if one amount is twice as much as another, its image should have _____ as much area as the other image.

Review Exercises

1. Trust your doctor: The General Social Survey recently surveyed people to ask, "How much would you trust your doctor to put your health above costs?" The following relative frequency bar graph presents the results.

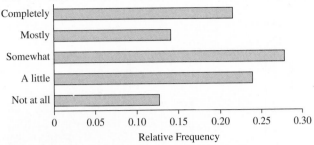

Source: General Social Survey

  a. Which was the most frequently given answer?

  b. True or false: Less than one-fourth of the respondents said that they trusted their doctor completely.

  c. True or false: More than half of the respondents said that they trusted their doctor either a little or not at all.

  d. A total of 2719 people responded to this question. True or false: More than 500 of them said that they completely trusted or mostly trusted their doctor.

2. Internet browsers: The following relative frequency distribution presents the useage in percent of various Internet browsers worldwide in 2013.

Browser	Usage
Internet Explorer	55.99
Firefox	20.63
Chrome	15.74
Safari	5.46
Opera	1.77
Others	0.41

Source: www.netmarketshare.com

  a. Construct a relative frequency bar graph.

  b. Construct a pie chart.

  c. True or false: The Chrome and Safari browsers account for more than 30% of usage.

3. Poverty rates: The following table presents the percentage of children who lived in poverty in several Colorado counties in 2008 and in 2013.

County	Percent in 2008	Percent in 2013
Adams	18.7	22.7
Arapahoe	13.7	16.4
Boulder	9.3	13.0
Denver	27.3	26.2
El Paso	15.7	18.0
Jefferson	11.2	12.2

Source: Colorado Children's Campaign

  a. Construct a side-by-side bar graph for these data.

  b. True or false: The poverty rate was higher in 2013 than in 2008 for each of the counties.

  c. Which county had the greatest increase?

4. Do your homework: The National Survey of Student Engagement asked a sample of college freshmen how often they came to class without completing their assignments. Following are the results:

Response	Percent
Never	35
Sometimes	48
Often	12
Very often	5

a. Construct a relative frequency bar graph.
b. Construct a pie chart.
c. True or false: More than half of the students reported that they sometimes come to class without completing their assignments.

5. **Quiz scores:** The following frequency histogram presents the scores on a recent statistics quiz in a class of 50 students.

a. What is the most frequent score?
b. How many students scored less than 6?
c. What percentage of students scored 10?
d. Is the histogram more accurately described as unimodal or as bimodal?

6. **House freshmen:** Newly elected members of the U.S. House of Representatives are referred to as "freshmen." The following frequency distribution presents the number of freshmen elected in each of the past 51 elections, from 1912 to 2012.

Number of Freshmen	Frequency
20–39	2
40–59	13
60–79	10
80–99	14
100–119	7
120–139	3
140–159	1
160–179	1

Source: Library of Congress

a. How many classes are there?
b. What is the class width?
c. What are the class limits?
d. Construct a frequency histogram.
e. Construct a relative frequency distribution.
f. Construct a relative frequency histogram.
g. In what percentage of elections were 100 or more freshmen elected?
h. In what percentage of elections were fewer than 60 freshmen elected?

7. **More freshmen:** For the data in Exercise 6:
a. Construct a frequency polygon.
b. Construct a relative frequency polygon.
c. Construct a frequency ogive.
d. Construct a relative frequency ogive.

8. **Royalty:** Following are the ages at death for all English and British monarchs since 1066.

59	40	67	58	56	28	41	49	65	68
43	64	33	46	35	49	40	12	32	52
55	15	42	69	58	48	54	67	51	49
67	76	81	67	71	81	68	70	77	56

a. Construct a frequency distribution with approximately eight classes.
b. Construct a frequency histogram based on this frequency distribution.
c. Construct a relative frequency distribution with approximately eight classes.
d. Construct a relative frequency histogram based on this frequency distribution.

9. **More royalty:** Construct a stem-and-leaf plot for the data in Exercise 8.

10. Presidents: Following are the ages at deaths for all U.S. presidents.

67	83	90	73	85	68	78	80	53	65
71	79	56	77	64	74	66	49	63	57
70	67	58	71	60	57	67	72	60	63
46	90	78	88	64	81	93	93		

  a. Construct a frequency distribution with a class width of 5 and a lower limit of 45 for the first class.
  b. Construct a frequency histogram based on this frequency distribution.
  c. Construct a relative frequency distribution with a class width of 5 and a lower limit of 45 for the first class.
  d. Construct a relative frequency histogram based on this frequency distribution.

11. Royalty and presidents: For the data in Exercises 8 and 10:
  a. Construct a back-to-back stem-and-leaf plot.
  b. Construct a back-to-back stem-and-leaf plot with split stems.
  c. Which plot do you think is more appropriate for these data?

12. Dotplot: Construct a dotplot for the data in Exercise 10.

13. Music sales: In recent years, sales of digital music (downloads) have been increasing, while sales of physical units (CDs, cassettes, and others) have been declining. The following table presents the number of units sold (in millions) for both digital and physical music formats in each of the years 2007 through 2011.

Year	Digital	Physical
2007	1230.4	543.9
2008	1450.7	401.8
2009	1542.6	309.2
2010	1471.9	267.7
2011	1569.2	255.7

Source: Recording Industry Association of America

  a. Construct a time-series plot for digital sales. On the same axes, construct a time-series plot for physical sales.
  b. Describe the trends in sales for digital and physical music formats.

14. More music: Refer to Exercise 13. Although physical formats sell fewer units than digital formats, their retail value is higher — CDs typically sell for $15 or more, while a download single typically costs a dollar or less. The following table presents the total number of music units, in millions, sold in each year from 2007 through 2011, and the total retail value, in millions of dollars, of the sold units.

Year	Units Sold	Retail Value
2007	1774.3	10,372
2008	1852.5	8,480
2009	1851.8	7,684
2010	1739.6	6,995
2011	1824.9	7,133

Source: Recording Industry Association of America

  a. Construct a time-series plot for units sold.
  b. Construct a time-series plot for total retail value.
  c. Explain why the total retail value has been decreasing while the total units sold has been increasing.

15. Falling birth rate: The following time-series plots both present estimates for the number of births per 1000 people worldwide for the years 1975–2010. (Source: The United Nations)

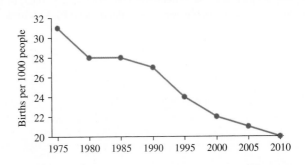

Which of the following statements is more accurate? Explain your reasoning.
(i) The birth rate decreased somewhat between 1975 and 2010.
(ii) The birth rate decreased dramatically between 1975 and 2010.

Write About It

1. Explain why the frequency bar graph and the relative frequency bar graph for a data set have a similar appearance.

2. In what ways do frequency distributions for qualitative data differ from those for quantitative data?

3. Provide an example of a data set whose histogram you would expect to be skewed to the right. Explain why you would expect the histogram to be skewed to the right.

4. Time-series data are discrete when observations are made at regularly spaced time intervals. The time-series data sets in this chapter are all discrete. Time-series data are continuous when there are observations at extremely closely spaced intervals that are connected to provide values at every instant of time. An example of continuous time-series data is an electrocardiogram. Provide some examples of time-series data that are discrete and some that are continuous.

5. Find examples of graphs in newspapers, magazines, or on the Internet that are misleading in some way. Explain how they are misleading. Then find some that present accurate comparisons and explain why you believe they are accurate.

Case Study: Do Late-Model Cars Get Better Gas Mileage?

In the chapter introduction, we presented gas mileage data for 2000 and 2013 model year compact cars. We will use histograms and back-to-back stem-and-leaf plots to compare the mileages between these two groups of cars. The following tables present the mileages, in miles per gallon.

Highway Mileage Ratings for 2000 Compact Cars							
28	21	30	33	24	29	30	38
27	23	27	34	21	27	38	28
29	21	31	31	32	23	36	28
27	32	31	30	26	23	40	28
27	27	34	28	34	28	38	27
29	27	33	24	31	33	27	28
26	26	31	16	28	30	30	24
37	35	30	33	31	32		

Source: www.fueleconomy.gov

Highway Mileage Ratings for 2013 Compact Cars						
39	39	33	30	33	29	40
29	34	36	34	32	33	35
35	33	35	33	27	37	31
37	30	29	25	37	37	30
40	38	36	39	40	38	34
30	36	40	36	32	35	36
32	38	29	37	31	32	22
30	32	31	29	37	36	29

Source: www.fueleconomy.gov

1. Construct a frequency distribution for the 2000 cars with a class width of 1.
2. Explain why a class width of 1 is too narrow for these data.
3. Construct a relative frequency distribution for the 2000 cars with a class width of 2, where the first class has a lower limit of 15.
4. Construct a histogram based on this relative frequency distribution. Is the histogram unimodal or bimodal? Describe the skewness, if any, in these data.
5. Construct a frequency distribution for the 2013 cars with an appropriate class width.
6. Using this class width, construct a relative frequency distribution for the 2013 cars.
7. Construct a histogram based on this relative frequency distribution. Is the histogram unimodal or bimodal? Describe the skewness, if any, in these data.
8. Compare the histogram for the 2000 cars with the histogram for the 2013 cars. Which cars tend to have higher gas mileage?
9. Construct a back-to-back stem-and-leaf plot for these data, using two lines for each stem. Which do you think illustrates the comparison better, the histograms or the back-to-back stem-and-leaf plot? Why?

Numerical Summaries of Data

Introduction

How do manufacturers increase quality and reduce costs? Companies continually consider new ideas to produce higher-quality, lower-cost products. To determine whether a new idea can lead to higher quality or lower cost, data must be collected and analyzed.

The following tables on page 92 present data produced by a manufacturer of computer chips, as described in the book *Statistical Case Studies for Industrial Process Improvement* by V. Czitrom and P. Spagon. Computer chips contain electronic circuits and are sealed with a thin layer of silicon dioxide. For the manufacturing process to work, the thickness of the layer must be carefully controlled. The manufacturer considered using recycled silicon wafers rather than new ones. Recycled wafers are much cheaper, so if the idea were feasible, it would lead to a reduction in cost. It must be determined whether the thicknesses of the oxide layers for recycled wafers are similar to those for the new wafers. The following tables present thickness measurements (in tenths of a nanometer) from some test runs.

New								
90.0	92.2	94.9	92.7	91.6	88.2	92.0	98.2	96.0
91.1	89.8	91.5	91.5	90.6	93.1	88.9	92.5	92.4
96.7	93.7	93.9	87.9	90.4	92.0	90.5	95.2	94.3
92.0	94.6	93.7	94.0	89.3	90.1	91.3	92.7	94.5

Recycled								
91.8	94.5	93.9	77.3*	92.0	89.9	87.9	92.8	93.3
92.6	90.3	92.8	91.6	92.7	91.7	89.3	95.5	93.6
92.4	91.7	91.6	91.1	88.0	92.4	88.7	92.9	92.6
91.7	97.4	95.1	96.7	77.5*	91.4	90.5	95.2	93.1

*Measurement is in error due to a defective gauge.

It is difficult to determine by looking at the tables whether the thicknesses tend to differ between new and recycled wafers. To interpret these data sets, we need to summarize them in ways that will reveal the important features. Histograms, stem-and-leaf plots, and dotplots are graphical summaries of data sets. While graphs are excellent tools for visualizing the important features of a data set, they have limitations. In particular, graphs often cannot measure a feature precisely; for precise descriptions, we need to use numbers.

In this chapter, we will learn about several of the most commonly used numerical summaries of data. Some of these describe the center of the data; these are called **measures of center**. Others describe how spread out the data values are; these are called **measures of spread**. Still others, called **measures of position**, specify the proportion of the data that is less than a given value.

In the case study at the end of the chapter, you will be asked to use some of the summaries introduced in the chapter to help determine which type of wafer will produce better results.

SECTION 3.1   Measures of Center

Objectives

1. Compute the mean of a data set
2. Compute the median of a data set
3. Compare the properties of the mean and median
4. Find the mode of a data set
5. Approximate the mean with grouped data

Objective 1 Compute the mean of a data set

The Mean

How do instructors determine your final grade? It's the end of the semester, and you have just finished your statistics class. During the semester, you took five exams, and your scores were 78, 83, 92, 68, and 85. Your instructor must find a single number to give a summary of your performance. The quantity he or she is most likely to use is the **arithmetic mean**, which is often simply called the **mean**. To find the mean of a list of numbers, add the numbers, then divide by how many numbers there are.

EXAMPLE 3.1   Computing the mean

Find the mean of the exam scores 78, 83, 92, 68, and 85.

Solution

Step 1: Add the numbers.
$$78 + 83 + 92 + 68 + 85 = 406$$

Step 2: Divide the sum by the number of observations. There were five observations. Therefore, the mean is
$$\text{Mean} = \frac{406}{5} = 81.2$$

Explain It Again

The mean and the average: Some people refer to the mean as the "average." In fact, there are many kinds of averages; the mean is just one of them.

In Example 3.1, we rounded the mean to one more decimal place than the data. We will follow this practice in general.

SUMMARY

We will round the mean to one more decimal place than the data.

Notation for the mean

Computing a mean involves adding a list of numbers. It is useful to have some notation that will allow us to discuss lists of numbers in general.

When we wish to write down a list of n numbers without specifying what the numbers are, we often write $x_1, x_2, ..., x_n$. To indicate that we are adding these numbers, we write $\sum x$. (The symbol $\sum$ is the uppercase Greek letter sigma.)

NOTATION

- A list of n numbers is denoted $x_1, x_2, ..., x_n$.
- $\sum x$ represents the sum of these numbers: $\sum x = x_1 + x_2 + \cdots + x_n$

Sample means and population means

Recall that a population consists of an entire collection of individuals about which information is sought, and a sample consists of a smaller group drawn from the population. The method for calculating the mean is the same for both samples and populations, except for the notation. If $x_1, x_2, ..., x_n$ is a sample, then the mean is called the *sample mean* and is denoted with the symbol $\bar{x}$. The mean of a population is called the *population mean* and is denoted by μ (the Greek letter mu).

> **Explain It Again**
>
> **Sample size and population size:** In general, we will use a lowercase n to denote a sample size and an uppercase N to denote a population size.

DEFINITION

If $x_1, ..., x_n$ is a sample, the **sample mean** is

$$\bar{x} = \frac{x_1 + x_2 + \cdots + x_n}{n} = \frac{\sum x}{n}$$

If $x_1, ..., x_N$ is a population, the **population mean** is

$$\mu = \frac{x_1 + x_2 + \cdots + x_N}{N} = \frac{\sum x}{N}$$

How the mean measures the center of the data

The mean is a measure of center. Figure 3.1 presents the exam scores in Example 3.1 on a number line, and shows the position of the mean. If we imagine each data value to be a weight, then the mean is the point at which the data set would balance.

Figure 3.1 The mean is the point where the data set would balance, if each data value were represented by an equal weight.

A misconception about the mean

> **Explain It Again**
>
> **The mean may not be a value in the data set:** The mean is not necessarily a typical value for the data. In fact, the mean may be a value that could not possibly appear in the data set.

Some people believe that the mean represents a "typical" data value. In fact, this is not necessarily so. This is shown in Example 3.1, where we computed the mean of five exam scores and obtained a result of 81.2. If, like most exams, the scores are always whole numbers, then 81.2 is certainly not a "typical" data value; in fact, it could not possibly be a data value.

Objective 2 Compute the median of a data set

The Median

The basic idea behind the **median** is simple: We try to find a number that splits the data set in half, so that half of the data values are less than the median and half of the data values are greater than the median. The procedure for computing the median differs, depending on whether the number of observations in the data set is even or odd.

Explain It Again

Finding the middle number: For large data sets, it may not be easy to find the middle number just by looking. If n is odd, the middle number is in position $\frac{n+1}{2}$, and if n is even, the two middle numbers are in positions $\frac{n}{2}$ and $\frac{n}{2}+1$.

Procedure for Computing the Median

Step 1: Arrange the data values in increasing order.
Step 2: Determine the number of data values, n.
Step 3: *If n is odd:* The median is the middle number.

     If n is even: The median is the average of the middle two numbers.

In Example 3.1, we found the mean of five exam scores. In Example 3.2, we will find the median.

EXAMPLE 3.2

Computing the median

Find the median of the exam scores 78, 83, 92, 68, and 85.

Solution

Step 1: We arrange the data values in increasing order to obtain

$$68 \quad 78 \quad 83 \quad 85 \quad 92$$

Step 2: There are $n = 5$ values in the data set, so n is odd.
Step 3: The middle number is 83, so the median is 83.

EXAMPLE 3.3

Computing the median

One of the goals of medical research is to develop treatments that reduce the time spent in recovery. Eight patients undergo a new surgical procedure, and the number of days spent in recovery for each is as follows.

$$20 \quad 15 \quad 12 \quad 27 \quad 13 \quad 19 \quad 13 \quad 21$$

Find the median time spent in recovery.

Solution

Step 1: We arrange the numbers in increasing order to obtain

$$12 \quad 13 \quad 13 \quad 15 \quad 19 \quad 20 \quad 21 \quad 27$$

Step 2: There are $n = 8$ numbers in the data set, so n is even.
Step 3: The middle two numbers are 15 and 19. The median is the average of these two numbers.

$$\text{Median} = \frac{15 + 19}{2} = 17$$

The median time spent in recovery is 17 days.

Using technology to compute the mean and median

In practice, technology is often used to compute means and medians, as Example 3.4 shows.

EXAMPLE 3.4

Using technology to compute the mean and median

Use technology to compute the mean and median of the recovery times in Example 3.3.

Solution

Enter the data into **L1**, then use the **1–Var Stats** command. Figure 3.2 presents the TI-84 Plus display. The mean is $\bar{x} = 17.5$ and the median (denoted "Med") is 17. Step-by-step instructions for computing the mean and median with the TI-84 Plus are presented in the Using Technology section on page 100.

	A	B
1	*Descriptive Statistics*	
2		
3	Mean	17.5
4	Standard Error	1.832251
5	Median	17
6	Mode	13
7	Standard Deviation	5.182388
8	Sample Variance	26.85714
9	Kurtosis	-0.09277
10	Skewness	0.759528
11	Range	15
12	Minimum	12
13	Maximum	27
14	Sum	140
15	Count	8

Figure 3.3

Figure 3.2 TI-84 Plus display showing the mean and median for the data in Example 3.3

Figure 3.3 presents Excel output and Figure 3.4 presents MINITAB output. The mean and median are highlighted in red. Step-by-step instructions for computing the mean and median in Excel and MINITAB are presented in the Using Technology section on page 100.

Variable	N	Mean	SE Mean	StDev	Minimum	Q1	Median	Q3	Maximum
Time	8	17.5	1.832	5.182	12.00	13.00	17.00	20.25	27.00

Figure 3.4

Objective 3 Compare the properties of the mean and median

Comparing the Properties of the Mean and Median

Both the mean and the median are frequently used as measures of center. It is important to know how their properties differ.

The mean is more influenced by extreme values than the median is

One important difference between the mean and the median is that the formula for the mean uses every value in the data set, but the formula for the median depends only on the middle number or the middle two numbers. This is particularly important for data sets in which one or more numbers are unusually large or unusually small. In most cases, these extreme values will have a large influence on the mean, but little or no influence on the median. Example 3.5 illustrates this principle.

EXAMPLE 3.5

Determining that the mean is more influenced by extreme values than the median is

Five families, named Smith, Jones, Gonzales, Brown, and Jackson, live in an apartment building. Their annual incomes, in dollars, are 25,000, 31,000, 34,000, 44,000, and 56,000. The Smith family, whose income is 25,000, wins a million-dollar lottery, so their income increases to 1,025,000. Find the mean and median income both before and after the Smiths win the lottery. Which measure of center is more influenced by the large number, the mean or the median?

Solution

We compute the mean and median before the lottery win. The mean income is

$$\text{Mean} = \frac{25,000 + 31,000 + 34,000 + 44,000 + 56,000}{5} = 38,000$$

The median is the middle number:

$$\text{Median} = 34,000$$

After the lottery win, the mean is

$$\text{Mean} = \frac{1{,}025{,}000 + 31{,}000 + 34{,}000 + 44{,}000 + 56{,}000}{5} = 238{,}000$$

To find the median, we arrange the numbers in order, obtaining

$$31{,}000 \quad 34{,}000 \quad 44{,}000 \quad 56{,}000 \quad 1{,}025{,}000$$

The median is the middle number:

$$\text{Median} = 44{,}000$$

The extreme value of 1,025,000 has influenced the mean quite a lot, increasing it from 38,000 to 238,000. In comparison, the median has been influenced much less, increasing only from 34,000 to 44,000.

Because the median is not much influenced by extreme values, we say that the median is *resistant*.

DEFINITION

A statistic is **resistant** if its value is not affected much by extreme values (large or small) in the data set.

We can summarize the results of Example 3.5 as follows.

SUMMARY

The median is resistant, but the mean is not.

The mean and median can help describe the shape of a data set

The mean and median measure the center of a data set in different ways. The mean is the point at which a data set balances (see Figure 3.1). The median is the middle number, so that half of the data values are less than the median and half are greater. It turns out that when a data set is symmetric, the mean and median are equal.

When a data set is skewed, however, the mean and median are often quite different. When a data set is skewed to the right, there are some large values in the right tail. Because the median is resistant while the mean is not, the mean is generally more affected by these large values than the median is. Therefore, for a data set that is skewed to the right, the mean is often greater than the median.

Figure 3.5 illustrates the idea. For most data sets that are skewed to the left, the mean will be to the left of, or less than, the median. For most data sets that are skewed to the right, the mean will be to the right of, or greater than, the median. When a data set is approximately symmetric, the balancing point is near the middle of the data, so the mean and the median will be approximately equal.

> **CAUTION**
>
> The relationship between the mean and median and the shape of the data set holds for most data sets, but not all.

Median Mean
(a)

Mean = Median
(b)

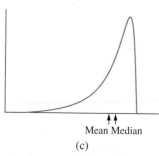
Mean Median
(c)

Figure 3.5 (a) When a data set is skewed to the right, the mean is generally greater than the median. (b) When a data set is approximately symmetric, the mean and median will be approximately equal. (c) When a data set is skewed to the left, the mean is generally less than the median.

SUMMARY

In most cases, the shape of a histogram reflects the relationship between the mean and median as follows:

Shape	Relationship Between Mean and Median
Skewed to the right	Mean is noticeably greater than median
Approximately symmetric	Mean is approximately equal to median
Skewed to the left	Mean is noticeably less than median

For an exception to this rule, see Exercise 74.

Which is a better measure of center, the mean or the median?

The short answer is that neither one is better than the other. They both measure the center in different, but appropriate, ways. When the data are highly skewed or contain extreme values, some people prefer to use the median, because the median is more representative of a typical value. However, the mean is still an appropriate measure of center, and is sometimes preferable, even when the data are highly skewed (see Exercises 63 and 64).

The following table summarizes the features of the mean and median.

	Advantages	Disadvantages
Mean	Takes every value into account	Highly influenced by extreme values: not resistant
Median	Not much influenced by extreme values: resistant	Depends only on middle value or middle two values

Check Your Understanding

1. Compute the mean and median of the following sample:
74   87   36   97   60   58   46

2. Compute the mean and median of the following sample:
69   17   75   96   74   80

3. Someone surveys the families in a certain town and reports that the mean number of children in a family is 2.1. Someone else says that this must be wrong, because it is impossible for a family to have 2.1 children. Comment.

4. A data set has a mean of 6 and a median of 4. Would you expect this data set to be skewed to the right or skewed to the left?

5. A data set has a mean of 5 and a median of 7. Would you expect this data set to be skewed to the right or skewed to the left?

Answers are on page 109.

Critical thinking about the mean and median

We can compute the mean and median for any list of numbers. However, they do not always produce meaningful results. The mean and median are useful for numbers that measure or count something. They are not useful for numbers that are used simply as labels. Example 3.6 illustrates the idea.

EXAMPLE 3.6   Determining whether the mean and median make sense

Following is information about the five starting players on a certain college basketball team:

Their heights, in inches, are 74, 76, 79, 80, and 82.

Their uniform numbers are 15, 32, 4, 43, and 26.

Will we obtain meaningful information by computing the mean and median height? How about the mean and median uniform number? Explain.

Solution

The mean and median height are meaningful, because heights are measurements.

The mean and median uniform numbers are not meaningful, because these numbers are just labels. They don't measure or count anything.

Objective 4 Find the mode of a data set

The Mode

In Section 2.2, we defined a mode to be the highest point of a histogram. There is another definition of this term. The **mode** of a data set is the value that appears most frequently.

EXAMPLE 3.7

Finding the mode

Ten students were asked how many siblings they had. The results, arranged in order, were

$$0 \quad 1 \quad 1 \quad 1 \quad 1 \quad 2 \quad 2 \quad 3 \quad 3 \quad 6$$

Find the mode of this data set.

Solution

The value that appears most frequently is 1. Therefore, the mode of this data set is 1.

When two or more values are tied for the most frequent, they are all considered to be modes. If the values all have the same frequency, we say that the data set has no mode.

> ### Explain It Again
>
> **The mode isn't really a measure of center:** The mode is sometimes classified as a measure of center. However, this isn't really accurate. The mode can be the largest value in a data set, or the smallest, or anywhere in between.

SUMMARY

- The mode of a data set is the value that appears most frequently.
- If two or more values are tied for the most frequent, they are all considered to be modes.
- If the values all have the same frequency, we say that the data set has no mode.

Computing the mode for qualitative data

The mean and median can be computed only for quantitative data. The mode, on the other hand, can be computed for qualitative data as well. For qualitative data, the mode is the most frequently appearing category.

EXAMPLE 3.8

Finding the mode for qualitative data

Following is a list of the makes of all the cars rented by an automobile rental company on a particular day. Which make of car is the mode?

Honda	Toyota	Toyota	Honda	Ford
Chevrolet	Nissan	Ford	Chevrolet	Chevrolet
Honda	Dodge	Ford	Ford	Toyota
Chevrolet	Toyota	Toyota	Toyota	Nissan

Solution

The most frequent category is "Toyota," which appears six times. Therefore, the mode is "Toyota."

Check Your Understanding

6. Find the mode or modes, if they exist:
   a. The sample is 3, 6, 0, 1, 1, 8, 0, 1, 1.
   b. The sample is 4, 7, 4, 1, 6, 5, 6.
   c. The sample is 4, 8, 5, 9, 6, 3.

Answers are on page 109.

Objective 5 Approximate the mean with grouped data

Approximating the Mean with Grouped Data

Sometimes we don't have access to the raw data in a data set, but we are given a frequency distribution. In these cases we can approximate the mean. We use Table 3.1 to illustrate the method. This table presents the number of text messages sent via cell phone by a sample of 50 high school students. We will approximate the mean number of messages sent.

Table 3.1 Number of Text Messages Sent by High School Students

Number of Text Messages Sent	Frequency
0–49	10
50–99	5
100–149	13
150–199	11
200–249	7
250–299	4

We present the method for approximating the mean with grouped data.

Procedure for Approximating the Mean with Grouped Data

Step 1: Compute the midpoint of each class. The midpoint of a class is found by taking the average of the lower class limit and the lower limit of the next larger class. For the last class, there is no next larger class, but we use the lower limit that the next larger class would have.

Step 2: For each class, multiply the class midpoint by the class frequency.

Step 3: Add the products Midpoint × Frequency over all classes.

Step 4: Divide the sum obtained in Step 3 by the sum of the frequencies.

EXAMPLE 3.9   Approximating the mean with grouped data

Compute the approximate mean number of messages sent, using Table 3.1.

Solution

The calculations are summarized in Table 3.2.

Step 1: Compute the midpoints: For the first class, the lower class limit is 0. The lower limit of the next class is 50. The midpoint is therefore

$$\frac{0 + 50}{2} = 25$$

We continue in this manner to compute the midpoint of each class. Note that for the last class, we average the lower limit of 250 with 300, which is the lower limit that the next class would have.

Step 2: Multiply the midpoints by the frequencies as shown in the column in Table 3.2 labeled "Midpoint × Frequency."

Step 3: Add the products Midpoint × Frequency, to obtain 6850.

Step 4: The sum of the frequencies is 50. The mean is approximated by 6850/50 = 137.

Table 3.2 Calculating the Mean Number of Text Messages Sent by High School Students

Class	Midpoint	Frequency	Midpoint × Frequency
0–49	25	10	25 × 10 = 250
50–99	75	5	75 × 5 = 375
100–149	125	13	125 × 13 = 1625
150–199	175	11	175 × 11 = 1925
200–249	225	7	225 × 7 = 1575
250–299	275	4	275 × 4 = 1100
	Sum = 50		Sum = 6850

$$\text{Mean} \approx \frac{6850}{50} = 137$$

USING TECHNOLOGY

We use Example 3.1 to illustrate the technology steps.

TI-84 PLUS

Computing the mean and median

Step 1. Enter the data into **L1** in the data editor by pressing **STAT** then **1: Edit...** For Example 3.1, we use **78, 83, 92, 68, 85** (Figure A).

Step 2. Press **STAT** and highlight the **CALC** menu.

Step 3. Select **1–Var Stats** and press **ENTER**. The **1–Var Stats** command is now shown on the home screen.

Step 4. Enter the list name **L1** next to the **1–Var Stats** command by pressing **2nd**, then **L1** (Figure B).

Step 5. Press **ENTER**. The descriptive statistics are displayed on the screen (Figures C and D).

Using the TI-84 PLUS Stat Wizards (see Appendix B for more information)

Step 1. Enter the data into **L1** in the data editor by pressing **STAT** then **1: Edit...** For Example 3,1, we use **78, 83, 92, 68, 85** (Figure A).

Step 2. Press **STAT** and highlight the **CALC** menu.

Step 3. Select **1–Var Stats** and press **ENTER**. Enter **L1** next to the **List** field. Keep the **FreqList** field blank.

Step 4. Select **Calculate** and press **ENTER**. The descriptive statistics are displayed on the screen (Figures C and D).

Figure A

Figure B

Figure C

Figure D

MINITAB

Computing the mean and median

Step 1. Enter the data in **Column C1**. For Example 3.1, we use **78, 83, 92, 68, 85**.

Step 2. Click on **Stat**, then **Basic Statistics**, then **Display Descriptive Statistics...** .

Step 3. Enter **C1** in the **Variables** field.

Step 4. Click **Statistics** and select the desired statistics. Press **OK**.

Step 5. Press **OK** (Figure E).

Descriptive Statistics: C1

Variable	Mean	SE Mean	StDev	Minimum	Q1	Median	Q3	Maximum
C1	81.20	3.99	8.93	68.00	73.00	83.00	88.50	92.00

Figure E

EXCEL

Computing the mean and median

Step 1. Enter the data in **Column A**. For Example 3.1, we use **78, 83, 92, 68, 85**.

Step 2. Select **Data**, then **Data Analysis**. Highlight **Descriptive Statistics** and press **OK**.

Step 3. Enter the range of cells that contain the data in the **Input Range field** and check the **Summary Statistics** box.

Step 4. Press **OK** (Figure F).

	A	B
1	*Descriptive Statistics*	
2		
3	Mean	81.2
4	Standard Error	3.992493
5	Median	83
6	Mode	#N/A
7	Standard Deviation	8.927486
8	Sample Variance	79.7
9	Kurtosis	0.715308
10	Skewness	-0.59281
11	Range	24
12	Minimum	68
13	Maximum	92
14	Sum	406
15	Count	5

Figure F

SECTION 3.1   Exercises

Exercises 1–6 are the Check Your Understanding exercises located within the section.

Understanding the Concepts

In Exercises 7–10, fill in each blank with the appropriate word or phrase.

7. The _____ is calculated by summing all data values and dividing by how many there are.

8. The _____ is a number that splits the data set in half.

9. The median is resistant because it is not affected much by _____.

10. The _____ is the value in the data set that appears most frequently.

In Exercises 11–14, determine whether the statement is true or false. If the statement is false, rewrite it as a true statement.

11. Every data set contains at least one mode.

12. The mean is resistant.

13. For most data sets that are skewed to the right, the mean is less than the median.

14. A mode is always a value that is in the data set.

Practicing the Skills

15. Find the mean, median, and mode for the following data set:

$$12 \quad 27 \quad 26 \quad 27 \quad 25$$

16. Find the mean, median, and mode for the following data set:

$$-20 \quad 15 \quad 21 \quad -20 \quad 19$$

17. Find the mean, median, and mode for the following data set:

$$28 \quad -31 \quad 28 \quad 0 \quad 31 \quad -23$$

18. Find the mean, median, and mode for the following data set:

$$83 \quad 98 \quad 22 \quad 89 \quad 99 \quad 98$$

In Exercises 19–22, use the given frequency distribution to approximate the mean.

19.

Class	Frequency
0–9	13
10–19	7
20–29	10
30–39	9
40–49	11

20.

Class	Frequency
0–15	2
16–31	14
32–47	6
48–63	13
64–79	15

21.

Class	Frequency
0–49	17
50–99	26
100–149	14
150–199	34
200–249	26
250–299	8

22.

Class	Frequency
0–19	18
20–39	11
40–59	6
60–79	6
80–99	10
100–119	5

23. Use the properties of the mean and median to determine which are the correct mean and median for the following histogram.

(i) Mean is 4.6; median is 5.0
(ii) Mean is 4.5; median is 4.2
(iii) Mean is 3.5; median is 4.3
(iv) Mean is 6.1; median is 5.6

24. Use the properties of the mean and median to determine which are the correct mean and median for the following histogram.

(i) Mean is 3.6; median is 4.8
(ii) Mean is 4.8; median is 3.6
(iii) Mean is 3.5; median is 3.1
(iv) Mean is 4.2; median is 4.1

25. Use the properties of the mean and median to determine which are the correct mean and median for the following histogram.

(i) Mean is 8.0; median is 8.1
(ii) Mean is 8.5; median is 6.5
(iii) Mean is 7.0; median is 8.3
(iv) Mean is 6.2; median is 6.1

26. Use the properties of the mean and median to determine which are the correct mean and median for the following histogram.

(i) Mean is 3.0; median is 4.1
(ii) Mean is 4.2; median is 3.2

(iii) Mean is 4.4; median is 5.0
(iv) Mean is 4.3; median is 4.1

27. Find the mean, median, and mode of the data in the following stem-and-leaf plot. The leaf represents the ones digit.

1	2
2	0779
3	78
4	13

28. Find the mean, median, and mode of the data in the following stem-and-leaf plot. The leaf represents the ones digit.

0	8
1	16
2	3559
3	0

Working with the Concepts

29. Facebook friends: In a study of Facebook users conducted in 2012 by the Pew Research Center, the mean number of Facebook friends per user was 245 and the median was 111. If a histogram were constructed for the numbers of Facebook friends for all Facebook users, would you expect it to be skewed to the right, skewed to the left, or approximately symmetric? Explain.

30. Mean and median height: The National Center for Health Statistics reported in a recent year that the mean height for U.S. women aged 20–29 was 64.3 inches, and the median was 64.2 inches. If a histogram were constructed for the heights of all U.S. women aged 20–29, would you expect it to be skewed to the right, skewed to the left, or approximately symmetric? Explain.

31. Hamburgers: An ABC News story reported the number of calories in hamburgers from six fast-food restaurants: McDonald's, Burger King, Wendy's, Hardee's, Sonic, and Dairy Queen. The results are

250   290   230   310   310   350

a. Find the mean number of calories.
b. Find the median number of calories.

32. Great swimmer: In the 2008 Olympic Games, Michael Phelps won eight gold medals in swimming. Following are the events he won, along with his margin of victory, in seconds.

Event	Margin of Victory
100m Butterfly	0.01
200m Freestyle	1.89
200m Butterfly	0.67
200m Individual medley	2.29
400m Individual medley	2.32
4 × 100m Freestyle	0.08
4 × 200m Freestyle	4.74
4 × 100m Medley	0.30

a. What was the mean margin of victory?
b. What was the median margin of victory?
c. Assume the margin of victory in the 200m freestyle was incorrectly entered as 18.9. Which would increase more, the mean or the median? Explain why.

33. Take your best shot: The following table presents the number of shots that can be taken on a fully charged battery for a sample of 19 digital cameras.

Camera	Number of Shots	Camera	Number of Shots
Panasonic Lumix DMC-3D1	200	Sony Cyber-shot DSC-W690	220
Canon ELPH 115 IS	200	Canon PowerShot A2300	210
Nikon Coolpix P330	200	Panasonic Lumix DMC-SZ1	250
Fujifilm FinePix JX500	170	Nikon Coolpix S3300	210
Sony Cyber-shot DSC-TX200V	220	Fujifilm FinePix JZ250	180
GE E1410SW	150	Fujifilm FinePix T400	160
Fujifilm FinePix JX580	170	Nikon Coolpix S3500	220
Panasonic Lumix DMC-FH8	260	Sony Cyber-shot DSC-W710	220
Nikon Coolpix S01	190	Canon PowerShot A4000 IS	175
Sony Cyber-shot DSC-W650	220		

Source: *Consumer Reports*

a. Find the mean number of shots.
b. Find the median number of shots.
c. Find the mode or modes, if any.
d. Would you expect the numbers of shots to be skewed to the left, skewed to the right, or approximately symmetric? Explain.

34. **Facebook applications:** The following table presents the number of monthly users for the 25 most popular Facebook applications as of January 2013.

Application	Monthly Users (millions)	Application	Monthly Users (millions)
Instagram	44.8	Samsung Mobile	21.5
FarmVille 2	43.5	ChefVille	20.9
TripAdvisor™	34.3	Pinterest	20.5
Texas HoldEm Poker	34.0	Bubble Safari	19.4
Microsoft Live	26.0	Bubble Safari Ocean	19.1
I want to add your birthday (K-factor Media)	25.5	Scribd	17.8
Spotify	24.6	Dragon City	17.6
Candy Crush Saga	23.5	Skype	17.2
Bing	23.3	Bubble Witch Saga	16.7
schoolFeed	23.1	Nokia	16.7
I want to add your birthday (My Calendar)	22.8	CityVille 2	15.9
Yahoo! Social Bar	22.3	Static Iframe Tab	15.4
Diamond Dash	21.6		

Source: insidefacebook

a. Find the mean number of monthly users.
b. Find the median number of monthly users.
c. Would you expect these data to be skewed to the right, skewed to the left, or approximately symmetric? Explain.

35. **What's your favorite TV show?** The following tables present the ratings for the top 20 prime-time shows for the 2007–2008 and 2012–2013 seasons. The rating is the percentage of households with TV sets that watched the program.

Top Rated TV Programs: 2007–2008

Program	Rating
American Idol — Tuesday	16.1
American Idol — Wednesday	15.9
Dancing With the Stars — Monday (2007)	14.0
Dancing With the Stars — Tuesday (2007)	12.7
Dancing With the Stars — Monday (2008)	12.6
Dancing With the Stars — Tuesday (2008)	11.8
Desperate Housewives	11.6
CSI	10.6
House — Tuesday	10.5
Grey's Anatomy	10.4
Sunday Night Football	9.7
CSI: Miami	9.2
House — Monday	9.2
NCIS	9.2
Survivor: China	9.0
Without a Trace	8.8
The Moment of Truth	8.8
Two and a Half Men	8.5
60 Minutes	8.4
Criminal Minds	8.2

Source: Nielsen Media Research

Top Rated TV Programs: 2012–2013

Program	Rating
NBC Sunday Night Football	7.9
Big Bang Theory	6.2
The Voice	5.1
Modern Family	4.9
The Voice — Tuesday	4.6
American Idol — Wednesday	4.6
American Idol — Thursday	4.3
The Following	4.3
Two and a Half Men	4.1
Grey's Anatomy	4.1
NCIS	4.0
Football Night in America Part 3	4.0
Revolution	3.9
2 Broke Girls	3.7
How I Met Your Mother	3.7
Family Guy	3.6
Once Upon a Time	3.6
Survivor: Philippines	3.5
X-Factor — Wednesday	3.5
NCIS: LA	3.4

Source: Nielsen Media Research

a. Find the mean and median ratings for 2007–2008.
b. Find the mean and median ratings for 2012–2013.
c. Since 2008, video streaming through companies such as Netflix has become more popular. Some media experts believe that this has resulted in a reduction in the ratings. Do the results of parts (a) and (b) support this claim? Explain.

36. **Beer:** The following table presents the number of active breweries for samples of states located east and west of the Mississippi River.

East		West	
State	**Number of Breweries**	**State**	**Number of Breweries**
Connecticut	18	Alaska	17
Delaware	10	Arizona	31
Florida	47	California	305
Georgia	22	Colorado	111
Illinois	52	Iowa	21
Kentucky	13	Louisiana	6
Maine	38	Minnesota	41
Maryland	23	Montana	30
Massachusetts	40	South Dakota	5
New Hampshire	16	Texas	37
New Jersey	20	Utah	15
New York	76		
North Carolina	46		
South Carolina	14		
Tennessee	19		
Vermont	20		

Source: http://www.beerinstitute.org/

a. Find the mean number of breweries for states east of the Mississippi.
b. Find the mean number of breweries for states west of the Mississippi.
c. Find the median number of breweries for states east of the Mississippi.
d. Find the median number of breweries for states west of the Mississippi.
e. Does one region have a lot more breweries per state than the other, or are they about the same?
f. The sample of western states happens to include California. Remove California from the sample of western states, and compute the mean and median for the remaining western states.
g. How does the removal of California affect the comparison of the number of breweries between eastern and western states?

37. **Gas prices:** The following table presents the average price, in U.S. dollars per gallon, of unleaded regular gasoline in several countries in the years 2003 and 2013.

Country	2003	2013
Belgium	1.22	3.39
France	1.14	3.33
Germany	1.20	3.42
Italy	1.40	3.57
Netherlands	1.38	3.38
United Kingdom	1.09	3.18
United States	1.26	3.25

Source: U.S. Energy Information Administration

a. Find the mean and median gas price for 2003.
b. Find the mean and median gas price for 2013.
c. Which increased more between 2003 and 2013, the mean or the median?

38. **House prices:** The following table presents prices, in thousands of dollars, of single-family homes for some of the largest metropolitan areas in the United States for the first quarter of 2012 and the first quarter of 2013.

Metro Area	2012	2013	Metro Area	2012	2013
Atlanta, GA	87.8	115.1	New York, NY	363.8	368.2
Baltimore, MD	218.1	226.5	Philadelphia, PA	193.5	197.7
Boston, MA	311.5	332.2	Phoenix, AZ	129.9	169.0
Chicago, IL	157.2	159.4	Portland, OR	208.6	246.5
Cincinnati, OH	112.5	121.0	Riverside, CA	174.3	216.7
Cleveland, OH	84.9	101.0	Saint Louis, MO	103.7	111.0
Dallas, TX	148.2	160.4	San Diego, CA	359.5	412.3
Denver, CO	226.4	261.2	San Francisco, CA	448.0	593.9
Houston, TX	152.1	163.7	Seattle, WA	265.4	312.6
Las Vegas, NV	122.1	155.1	Tampa, FL	131.9	141.8
Miami, FL	182.0	219.9	Washington, DC	311.6	348.7
Minneapolis, MN	147.3	170.6			

Source: National Realtors Association

a. Find the mean and median price for 2012.
b. Find the mean and median price for 2013.
c. In general, house prices increased from 2012 to 2013. Which increased more, the mean or the median?

39. Heavy football players: Following are the weights, in pounds, for offensive and defensive linemen on the New York Giants National Football League team at the beginning of a recent year.

Offense:	315	253	255	319	252	300	327	303	299	310	317	302
Defense:	304	309	306	305	265	270	264	317	255	274	261	296

a. Find the mean and median weight for the offensive linemen.
b. Find the mean and median weight for the defensive linemen.
c. Do offensive or defensive linemen tend to be heavier, or are they about the same?

40. Stock prices: Following are the closing prices of Google stock for each trading day in May and June 2013.

May					
880.37	877.07	873.65	866.20	869.79	829.61
880.93	884.74	900.68	900.62	886.25	820.43
875.04	877.00	871.98	879.81	890.22	
879.73	864.64	859.70	859.10	867.63	

June				
871.22	870.76	868.31	881.27	873.32
882.79	889.42	906.97	908.53	909.18
903.87	915.89	887.10	877.53	880.23
871.48	873.63	857.23	861.55	845.72

a. Find the mean and median price in May.
b. Find the mean and median price in June.
c. Does there appear to be a substantial difference in price between May and June, or are the prices about the same?

41. Flu season: The following tables present the number of specimens that tested positive for Type A and Type B influenza in the United States during the first 15 weeks of the 2012–13 flu season.

Type A				
36	99	177	200	258
384	584	999	1539	2748
3764	4841	5346	5405	5795

Source: Centers for Disease Control

Type B				
56	98	121	139	151
230	263	359	495	674
821	1096	1162	1106	1156

Source: Centers for Disease Control

a. Find the mean and median number of Type A cases in the first fifteen weeks of the flu season.
b. Find the mean and median number of Type B cases in the first fifteen weeks of the flu season.
c. A public health official says that there are more than twice as many cases of Type A influenza as Type B. Do these data support this claim?

42. News flash: The following table presents the circulation (in thousands) for the top 25 U.S. daily newspapers in both print and digital editions for March 2013.

Print				
1480.7	731.4	1424.4	432.9	360.5
300.0	431.1	184.8	213.8	368.1
190.6	265.8	231.2	159.4	180.3
241.0	216.1	184.8	227.7	285.9
125.7	126.3	192.8	172.0	149.5

Source: Alliance for Audited Media

Digital				
898.1	1133.9	249.9	177.7	155.7
200.6	42.3	77.7	192.8	46.8
65.9	112.0	102.3	15.5	160.5
17.1	95.5	68.0	73.7	7.0
69.0	16.0	21.6	73.5	6.7

Source: Alliance for Audited Media

a. Find the mean and median circulation for print editions.
b. Find the mean and median circulation for digital editions.
c. The editor of an Internet news source says that digital circulation is more than half of print circulation. Do the data support this claim?

43. Commercial break: Following are the amounts spent (in millions of dollars) on media advertising in the United States by a sample of ten companies in the first quarter of 2013.

295.9	722.5	280.3	286.9	362.9	331.2	394.6	257.0	463.5	340.1

a. Find the mean amount spent on advertising.
b. Find the median amount spent on advertising.
c. If the amount of 722.5 was incorrectly listed as 7225, how would this affect the mean? How would it affect the median?

44. Magazines: The following data represent the annual costs in dollars, of a sample of 22 popular magazine subscriptions.

14.98	9.97	20.99	7.97	9.99	11.97	116.99	14.95	11.97	16.00	9.97
17.98	38.95	14.98	14.98	16.00	24.99	14.98	10.99	18.00	12.99	11.99

Source: http://magazines.com

    a. Find the mean annual subscription cost.
    b. Find the median annual subscription cost.
    c. What is the mode?
    d. Which value in this data set is most accurately described as an extreme value?
    e. How would the mean, median, and mode be affected if the extreme value were removed from the list?

45. Don't drink and drive: The Insurance Institute for Highway Safety reported that there were 5037 fatalities among drivers in auto accidents in 2011. Following is a frequency distribution of their ages.

Age	Number of Fatalities
11–20	485
21–30	1736
31–40	988
41–50	868
51–60	585
61–70	249
71–80	126

Source: Insurance Institute for Highway Safety

    a. Approximate the mean age.
    b. The first class consists of ages 11 through 20, but most drivers are at least 16 years old. Does this tend to make the approximate mean too large or too small? Explain.

46. Age distribution: The ages of residents of Banks City, Oregon, are given in the following frequency distribution.

Age	Frequency
0–9	283
10–19	203
20–29	217
30–39	256
40–49	176
50–59	92
60–69	21
70–79	23
80–89	12
90–99	3

Source: U.S. Census Bureau

    a. Approximate the mean age.
    b. Assume all three people aged 90–99 were in fact 90 years old. Would this tend to make the approximate mean too large or too small? Explain.

47. Be my Valentine: The following frequency distribution presents the amounts, in dollars, spent for Valentine's Day gifts in a survey of 123 U.S. adults in 2013.

Amount	Frequency
0.00–19.99	19
20.00–39.99	13
40.00–59.99	21
60.00–79.99	19
80.00–99.99	12
100.00–119.99	10
120.00–139.99	7
140.00–159.99	8
160.00–179.99	7
180.00–199.99	1
200.00–219.99	3
220.00–239.99	2
240.00–259.99	1

Source: Based on data from the National Retail Federation

    a. Approximate the mean amount spent on Valentine's Day gifts.
    b. If the majority of the people in the category 0.00–19.99 actually didn't spend any money, would this tend to make the approximate mean too large or too small?

48. Get your degree: The following frequency distribution presents the number of U.S. adults (in thousands) ages 25 to 74 who have received a Bachelor's degree in 2012.

Age	Frequency
25–29	5501
30–34	4643
35–39	4322
40–44	4564
45–49	4464
50–54	4262
55–59	3838
60–64	3261
65–69	2195
70–74	1335

Source: U.S. Census

a. Approximate the mean age.
b. If the majority of the people ages 70–74 were in fact 74, would this tend to make the approximate mean too large or too small?

49. Income: The personal income per capita of a state is the total income of all adults in the state, divided by the number of adults. The following table presents the personal income per capita (in thousands of dollars) for each of the 50 states and the District of Columbia.

32	40	33	30	42	41	54	41	61	38	33	39	31	40	34	35	37
31	35	34	46	49	35	41	29	34	32	36	40	42	49	31	47	34
35	35	34	35	39	39	31	34	33	37	31	37	41	38	30	36	43

Source: U.S. Bureau of Economic Analysis

a. What is the mean state income?
b. What is the median state income?
c. Based on the mean and median, would you expect the data to be skewed to the left, skewed to the right, or approximately symmetric? Explain.
d. Construct a frequency histogram. Do the results agree with your expectation?

50. Take in a show: The following table presents the weekly attendance, in thousands, at Broadway shows during the 2012–2013 season.

240.9	241.0	240.2	245.9	251.0	260.2	252.1	256.0	269.4	222.4	213.6	196.6	192.2
191.3	185.5	162.5	180.9	183.0	196.4	187.7	235.2	292.4	240.9	231.6	232.1	220.6
246.5	236.8	219.7	149.4	214.9	228.3	214.0	205.3	184.2	183.1	189.9	181.6	190.4
197.0	208.5	218.2	238.5	235.8	239.3	240.4	234.9	244.1	254.9	269.2	265.6	257.5

Source: Broadway League

a. Find the mean and median weekly attendance.
b. Based on the mean and median, would you expect that a histogram would be skewed to the left, skewed to the right, or approximately symmetric? Explain.
c. Construct a frequency histogram. Do the results agree with your expectation?

51. Read any good books lately? The following data represent the responses of 24 library patrons when asked about their favorite type of book. Which type of book is the mode?

Biography	Fiction	Biography	Historical	Fiction	Biography
Fiction	Nonfiction	Fiction	Fiction	Historical	Biography
How-to guide	Nonfiction	How-to guide	Nonfiction	Historical	Nonfiction
Fiction	Fiction	How-to guide	How-to guide	Biography	How-to guide

52. Sources of news: A sample of 32 U.S. adults was surveyed and asked, "Do you get most of your information about current events from newspapers, magazines, the Internet, television, radio, or some other source?" The results are shown below. What is the mode?

Television	Internet	Television	Other	Newspapers	Internet	Television	Radio
Magazines	Newspapers	Other	Radio	Radio	Internet	Other	Television
Internet	Magazines	Other	Television	Internet	Newspapers	Newspapers	Internet
Newspapers	Other	Television	Newspapers	Television	Magazines	Television	Television

Source: General Social Survey

53. Find the mean: The National Center for Health Statistics sampled 5844 American women over the age of 20 and found that their median weight was 157.2 pounds. A histogram of the data set was skewed to the right. Which of the following is a possible value for the mean weight: 150 pounds, 155 pounds, or 160 pounds? Explain.

54. **Find the median:** According to the 2011 Current Population Survey, the mean personal income for American adults was $38,337. A histogram for incomes is skewed to the right. Which of the following is a possible value for the median income, $26,000, $40,000, or $50,000?

55. **Find the median:** In 2012, approximately 1.7 million people took the mathematics section of the SAT. Of these, 42% scored less than 490, 45% scored less than 500, 49% scored less than 510, 52% scored less than 520, and 55% scored less than 530. In what interval is the median score?
   i. Between 490 and 500
   ii. Between 500 and 510
   iii. Between 510 and 520
   iv. Between 520 and 530

56. **Find the median:** The National Health and Nutrition Examination Survey measured the heights, in inches, of a large number of American adult men. Of these heights, 33% were less than 68, 42% were less than 69, 59% were less than 70, and 71% were less than 71. In what interval is the median height?
   i. Between 68 and 69
   ii. Between 69 and 70
   iii. Between 70 and 71
   iv. Between 71 and 72

57. **Heights:** There are 2500 women and 2000 men enrolled in a certain college. The mean height for the women is 64.4 inches, and the mean height for the men is 69.8 inches.
   a. Find the sum of the heights of the women.
   b. Find the sum of the heights of the men.
   c. Find the mean height for all 4500 people.
   d. The average of the means for men and women is $(64.4 + 69.8)/2 = 67.1$. Explain why this is not equal to the mean height for all the people.

58. **Exam scores:** There are two sections of an introductory statistics course. Section A has 25 students and section B has 30 students. All students took the same final exam. The mean score in section A was 82 and the mean score in section B was 78.
   a. Find the sum of the scores in section A.
   b. Find the sum of the scores in section B.
   c. Find the mean score for all 55 students.
   d. The average of the means in the two sections is $(82 + 78)/2 = 80$. Explain why this is not equal to the mean score for all the students.

59. **How many numbers?** A data set has a median of 17, and six of the numbers in the data set are less than 17. The data set contains a total of n numbers.
   a. If n is odd, and exactly one number in the data set is equal to 17, what is the value of n?
   b. If n is even, and none of the numbers in the data set are equal to 17, what is the value of n?

60. **How many numbers?** A data set has a median of 10, and eight of the numbers in the data set are less than 10. The data set contains a total of n numbers.
   a. If n is odd, and exactly one of the numbers in the data set is equal to 10, what is the value of n?
   b. If n is even, and two of the numbers in the data set are equal to 10, what is the value of n?

61. **What's the score?** Jermaine has entered a bowling tournament. To prepare, he bowls five games each day and writes down the score of each game, along with the mean of the five scores. He is looking at the scores from one day last week and finds that one of the numbers has accidentally been erased. The four remaining scores are 201, 193, 221, and 187. The mean score is 202. What is the missing score?

62. **What's your grade?** Addison has been told that her average on six homework assignments in her history class is 85. She can find only five of the six assignments, which have scores of 91, 72, 96, 88, and 75. What is the score on the lost homework assignment?

63. **Mean or median?** The Smith family in Example 3.5 had the good fortune to win a million-dollar prize in a lottery. Their annual income for each of the five years leading up to their lottery win are as follows:

   15,000   18,000   20,000   25,000   1,025,000

   a. Compute the mean annual income.
   b. Compute the median annual income.
   c. Which provides a more appropriate description of the Smiths' financial position, the mean or the median? Explain.

64. **Mean or median?** The incomes in a certain town of 1000 households are strongly skewed to the right. The mean income is $60,000, and the median income is only $40,000. The town is going to impose a 1% income tax, and the town council wants to estimate how much revenue will be generated. Which is the more relevant measure of center for the town council, the mean income or the median income? Explain.

65. **Properties of the mean:** Make up a data set in which the mean is equal to one of the numbers in the data set.

66. **Properties of the median:** Make up a data set in which the median is equal to one of the numbers in the data set.

67. **Properties of the mean:** Make up a data set in which the mean is *not* equal to one of the numbers in the data set.

68. **Properties of the median:** Make up a data set in which the median is *not* equal to one of the numbers in the data set.

69. **The midrange:** The **midrange** is a measure of center that is computed by averaging the largest and smallest values in a data set. In other words,

$$\text{Midrange} = \frac{\text{Largest value} + \text{Smallest value}}{2}$$

Is the midrange resistant? Explain.

70. **Mean, median, and midrange:** A data set contains only two values. Are the mean, median, and midrange all equal? Explain.

Extending the Concepts

71. **Changing units:** A sample of five college students have heights, in inches, of 65, 72, 68, 67, and 70.
    a. Compute the sample mean.
    b. Compute the sample median.
    c. Convert each of the heights to units of feet, by dividing by 12.
    d. Compute the sample mean of the heights in feet. Is this equal to the sample mean in inches divided by 12?
    e. Compute the sample median of the heights in feet. Is this equal to the sample median in inches divided by 12?

72. **Effect on the mean and median:** Four employees in an office have annual salaries of $30,000, $35,000, $45,000, and $70,000.
    a. Compute the mean salary.
    b. Compute the median salary.
    c. Each employee gets a $1000 raise. Compute the new mean. Does the mean increase by $1000?
    d. Each employee gets a 5% raise. Compute the new mean. Does the mean increase by 5%?

73. **Nonresistant median:** Consider the following data set:

        0   0   1   1   1   1   1   1   2   2   8   8   9   9   9   9   9   9   9   10   10

    a. Show that the mean and median are both equal to 5.
    b. Suppose that a value of 26 is added to this data set. Which is affected more, the mean or the median?
    c. Suppose that a value of 100 is added to this data set. Which is affected more, the mean or the median?
    d. It is possible for an extreme value to affect the median more than the mean, but if the value is extreme enough, the mean will be affected more than the median. Explain.

74. **Exception to the skewness rule:** Consider the following data set:

        0   0   0   0   0   0   0   0   0   1   1   1   1   1   1   1   2   2   2   3

    a. Compute the mean and median.
    b. Based on the mean and median, would you expect the data set to be skewed to the left, skewed to the right, or approximately symmetric? Explain.
    c. Construct a frequency histogram. Does the histogram have the shape you expected?

Answers to Check Your Understanding Exercises for Section 3.1

1. Mean is 65.4; median is 60.

2. Mean is 68.5; median is 74.5.

3. The mean does not have to be a value that could possibly appear in the data set.

4. Skewed to the right

5. Skewed to the left

6. **a.** 1   **b.** 4 and 6   **c.** No mode

SECTION 3.2   Measures of Spread

Objectives

1. Compute the range of a data set
2. Compute the variance of a population and a sample
3. Compute the standard deviation of a population and a sample
4. Approximate the standard deviation with grouped data
5. Use the Empirical Rule to summarize data that are unimodal and approximately symmetric
6. Use Chebyshev's Inequality to describe a data set
7. Compute the coefficient of variation

Would you rather live in San Francisco or St. Louis? If you had to choose between these two cities, one factor you might consider is the weather. Table 3.3 presents the average monthly temperatures, in degrees Fahrenheit, for both cities.

Table 3.3 Temperatures in San Francisco and St. Louis

	Jan	Feb	Mar	Apr	May	Jun	Jul	Aug	Sep	Oct	Nov	Dec
San Francisco	51	54	55	56	58	60	60	61	63	62	58	52
St. Louis	30	35	44	57	66	75	79	78	70	59	45	35

Source: National Weather Service

To compare the temperatures, we will compute their means.

$$\text{Mean for San Francisco} = \frac{51+54+55+56+58+60+60+61+63+62+58+52}{12}$$
$$= 57.5$$

$$\text{Mean for St. Louis} = \frac{30+35+44+57+66+75+79+78+70+59+45+35}{12}$$
$$= 56.1$$

The means are similar: 57.5° for San Francisco and 56.1° for St. Louis. Does this mean that the temperatures are similar in both cities? Definitely not. St. Louis has a cold winter and a hot summer, while the temperature in San Francisco is much the same all year round. Another way to say this is that the temperatures in St. Louis are more spread out than the temperatures in San Francisco. The dotplots in Figure 3.6 illustrate the difference in spread.

Figure 3.6   The monthly temperatures for St. Louis are more spread out than those for San Francisco.

The mean does not tell us anything about how spread out the data are; it only gives us a measure of the center. It is clear that the mean by itself is not adequate to describe a data set. We must also have a way to describe the amount of spread. Dotplots allow us to visualize the spread, but we need a numerical summary to measure it precisely.

Objective 1 Compute the range of a data set

The Range

The simplest measure of the spread of a data set is the *range*.

DEFINITION

The **range** of a data set is the difference between the largest value and the smallest value.

$$\text{Range} = \text{Largest value} - \text{Smallest value}$$

EXAMPLE 3.10   Compute the range of a data set

Compute the range of the temperature data for San Francisco and for St. Louis, and interpret the results.

Solution

The largest value for San Francisco is 63 and the smallest is 51. The range for San Francisco is $63 - 51 = 12$.

The largest value for St. Louis is 79 and the smallest is 30. The range for St. Louis is $79 - 30 = 49$.

The range is much larger for St. Louis, which indicates that the spread in the temperatures is much greater there.

Although the range is easy to compute, it is not often used in practice. The reason is that the range involves only two values from the data set — the largest and the smallest. The measures of spread that are most often used are the variance and the standard deviation, which use every value in the data set.

Objective 2 Compute the variance of a population and a sample

The Variance

When a data set has a small amount of spread, like the San Francisco temperatures, most of the values will be close to the mean. When a data set has a larger amount of spread, more of the data values will be far from the mean. The **variance** is a measure of how far the values in a data set are from the mean, on the average. We will describe how to compute the variance of a population.

The difference between a population value, x, and the population mean, μ, is $x - \mu$. This difference is called a **deviation**. Values less than the mean will have negative deviations, and values greater than the mean will have positive deviations. If we were simply to add the deviations, the positive and the negative ones would cancel out. So we square the deviations to make them all positive. Data sets with a lot of spread will have many large squared deviations, while those with less spread will have smaller squared deviations. The average of the squared deviations is the *population variance*.

Explain It Again

Another formula for the population variance: An alternate formula for the population variance is

$$\sigma^2 = \frac{\sum x^2 - N\mu^2}{N}$$

This formula always gives the same result as the one in the definition.

CAUTION

The population variance will *never* be negative. It will be equal to zero if all the values in a population are the same. Otherwise, the population variance will be positive.

DEFINITION

Let $x_1, ..., x_N$ denote the values in a population of size N. Let μ denote the population mean. The **population variance**, denoted by σ^2, is

$$\sigma^2 = \frac{\sum(x - \mu)^2}{N}$$

We present the procedure for computing the population variance.

Procedure for Computing the Population Variance

Step 1: Compute the population mean μ.

Step 2: For each population value x, compute the deviation $x - \mu$.

Step 3: Square the deviations, to obtain quantities $(x - \mu)^2$.

Step 4: Sum the squared deviations, obtaining $\sum(x - \mu)^2$.

Step 5: Divide the sum obtained in Step 4 by the population size N to obtain the population variance σ^2.

In practice, variances are usually calculated with technology. It is a good idea to compute a few by hand, however, to get a feel for the procedure.

EXAMPLE 3.11   Computing the population variance

Compute the population variance for the San Francisco temperatures.

Solution

The calculations are shown in Table 3.4 (page 112).

Step 1: Compute the population mean:

$$\mu = \frac{51 + 54 + 55 + 56 + 58 + 60 + 60 + 61 + 63 + 62 + 58 + 52}{12} = 57.5$$

Step 2: Subtract μ from each value to obtain the deviations $x - \mu$. These calculations are shown in the second column of Table 3.4.

Table 3.4 Calculations for the Population Variance in Example 3.11

x	$x - \mu$	$(x - \mu)^2$
51	-6.5	$(-6.5)^2 = 42.25$
54	-3.5	$(-3.5)^2 = 12.25$
55	-2.5	$(-2.5)^2 = 6.25$
56	-1.5	$(-1.5)^2 = 2.25$
58	0.5	$0.5^2 = 0.25$
60	2.5	$2.5^2 = 6.25$
60	2.5	$2.5^2 = 6.25$
61	3.5	$3.5^2 = 12.25$
63	5.5	$5.5^2 = 30.25$
62	4.5	$4.5^2 = 20.25$
58	0.5	$0.5^2 = 0.25$
52	-5.5	$(-5.5)^2 = 30.25$

$\mu = 57.5$ $\qquad\qquad\qquad$ $\sum(x - \mu)^2 = 169$

$$\sigma^2 = \frac{169}{12} = 14.083$$

Step 3: Square the deviations. These calculations are shown in the third column of Table 3.4.

Step 4: Sum the squared deviations to obtain

$$\sum(x - \mu)^2 = 169$$

Step 5: The population size is $N = 12$. Divide the sum obtained in Step 4 by N to obtain the population variance σ^2.

$$\sigma^2 = \frac{\sum(x - \mu)^2}{N} = \frac{169}{12} = 14.083$$

In Example 3.11, note how important it is to make all the deviations positive, which we do by squaring them. If we simply add the deviations without squaring them, the positive and negative ones will cancel each other out, leaving 0.

Check Your Understanding

1. Compute the population variance for the St. Louis temperatures. Compare the result with the variance for the San Francisco temperatures, and interpret the result.

Answer is on page 128.

The sample variance

When the data values come from a sample rather than a population, the variance is called the *sample variance*. The procedure for computing the sample variance is a bit different from the one used to compute a population variance.

CAUTION

The sample variance will *never* be negative. It will be equal to zero if all the values in a sample are the same. Otherwise, the sample variance will be positive.

DEFINITION

Let $x_1, \ldots, x_n$ denote the values in a sample of size n. The **sample variance**, denoted by s^2, is

$$s^2 = \frac{\sum(x - \bar{x})^2}{n - 1}$$

The formula is the same as for the population variance, except that we replace the population mean μ with the sample mean $\bar{x}$, and we divide by $n - 1$ rather than N.

We present the procedure for computing the sample variance.

Explain It Again

Another formula for the sample variance: An alternate formula for the sample variance is

$$s^2 = \frac{\sum x^2 - n\bar{x}^2}{n-1}$$

This formula will always give the same result as the one in the definition.

Procedure for Computing the Sample Variance

Step 1: Compute the sample mean $\bar{x}$.

Step 2: For each sample value x, compute the difference $x - \bar{x}$. This quantity is called a deviation.

Step 3: Square the deviations, to obtain quantities $(x - \bar{x})^2$.

Step 4: Sum the squared deviations, obtaining $\sum(x - \bar{x})^2$.

Step 5: Divide the sum obtained in Step 4 by $n - 1$ to obtain the sample variance s^2.

EXAMPLE 3.12

Computing the sample variance

A company that manufactures batteries is testing a new type of battery designed for laptop computers. They measure the lifetimes, in hours, of six batteries, and the results are 3, 4, 6, 5, 4, and 2. Find the sample variance of the lifetimes.

Solution

The calculations are shown in Table 3.5.

Step 1: Compute the sample mean:

$$\bar{x} = \frac{3 + 4 + 6 + 5 + 4 + 2}{6} = 4$$

Step 2: Subtract $\bar{x}$ from each value to obtain the deviations $x - \bar{x}$. These calculations are shown in the second column of Table 3.5.

CAUTION

Don't round off the value of $\bar{x}$ when computing the sample variance.

Table 3.5 Calculations for the Sample Variance in Example 3.12

x	$x - \bar{x}$	$(x - \bar{x})^2$
3	-1	$(-1)^2 = 1$
4	0	$0^2 = 0$
6	2	$2^2 = 4$
5	1	$1^2 = 1$
4	0	$0^2 = 0$
2	-2	$(-2)^2 = 4$

$$\bar{x} = 4 \qquad \sum(x - \bar{x})^2 = 10$$

$$s^2 = \frac{10}{6-1} = 2$$

Step 3: Square the deviations. These calculations are shown in the third column of Table 3.5.

Step 4: Sum the squared deviations to obtain

$$\sum(x - \bar{x})^2 = 10$$

Step 5: The sample size is $n = 6$. Divide the sum obtained in Step 4 by $n - 1$ to obtain the sample variance s^2.

$$s^2 = \frac{\sum(x - \bar{x})^2}{n - 1} = \frac{10}{6 - 1} = 2$$

Why do we divide by $n - 1$ rather than n?

It is natural to wonder why we divide by $n - 1$ rather than n when computing the sample variance. When computing the sample variance, we use the sample mean to compute the deviations $x - \bar{x}$. For the population variance, we use the population mean for the deviations $x - \mu$. Now it turns out that the deviations using the sample mean tend to be a bit smaller than the deviations using the population mean. If we were to divide by n when computing a sample variance, the value would tend to be a bit smaller than the population variance.

It can be shown mathematically that the appropriate correction is to divide the sum of the squared deviations by $n - 1$ rather than n.

The quantity $n - 1$ is sometimes called the **degrees of freedom** for the sample standard deviation. The reason is that the deviations $x - \bar{x}$ will always sum to 0. Thus, if we know the first $n - 1$ deviations, we can compute the nth one. For example, if our sample consists of the four numbers 2, 4, 9, and 13, the sample mean is

$$\bar{x} = \frac{2 + 4 + 9 + 13}{4} = 7$$

The first three deviations are

$$2 - 7 = -5, \quad 4 - 7 = -3, \quad 9 - 7 = 2$$

The sum of the first three deviations is

$$-5 + (-3) + 2 = -6$$

We can now determine that the last deviation must be 6, in order to make the sum of all four deviations equal to 0. When we know the first three deviations, we can determine the fourth one. Thus, for a sample of size 4, there are 3 degrees of freedom. In general, for a sample of size n, there will be $n - 1$ degrees of freedom.

Objective 3 Compute the standard deviation of a population and a sample

The Standard Deviation

There is a problem with using the variance as a measure of spread. Because the variance is computed using squared deviations, the units of the variance are the squared units of the data. For example, in Example 3.12, the units of the data are hours, and the units of variance are squared hours. In most situations, it is better to use a measure of spread that has the same units as the data. We do this simply by taking the square root of the variance. The quantity thus obtained is called the **standard deviation**. The standard deviation of a sample is denoted s, and the standard deviation of a population is denoted σ.

> **DEFINITION**
>
> - The **sample standard deviation** s is the square root of the sample variance s^2.
> $$s = \sqrt{s^2}$$
> - The **population standard deviation** σ is the square root of the population variance σ^2.
> $$\sigma = \sqrt{\sigma^2}$$

EXAMPLE 3.13   Computing the standard deviation

The lifetimes, in hours, of six batteries (first presented in Example 3.12) were 3, 4, 6, 5, 4, and 2. Find the standard deviation of the battery lifetimes.

Solution

In Example 3.12, we computed the sample variance to be $s^2 = 2$. The sample standard deviation is therefore

$$s = \sqrt{s^2} = \sqrt{2} = 1.414$$

EXAMPLE 3.14   Computing standard deviations with technology

Compute the standard deviation for the data in Example 3.13.

Solution

Following is the display from the TI-84 Plus.

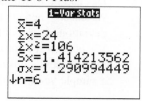

```
        1-Var Stats
x̄=4
Σx=24
Σx²=106
Sx=1.414213562
σx=1.290994449
↓n=6
```

The TI-84 Plus calculator denotes the sample standard deviation by Sx. The display shows that the sample standard deviation is equal to 1.414213562. The calculator does not know whether the data set represents a sample or an entire population. Therefore, the display also presents the population standard deviation, which is denoted σx.

Check Your Understanding

2. Find the sample variance and standard deviation for the following samples.
   a. 2   2   1   5   4   5   0   1
   b. 22   27   13   53   47   50

3. Find the population variance and standard deviation for the population values 7, 3, 2, 5, 4, 9, 6, 3.

Answers are on page 128.

Recall: A statistic is resistant if its value is not affected much by extreme data values.

The standard deviation is not resistant

Example 3.15 shows that the standard deviation is *not* resistant.

EXAMPLE 3.15

Show that the standard deviation is not resistant

In Example 3.14, we found that the sample standard deviation for six battery lifetimes was $s = 1.414$. The six lifetimes were 3, 4, 6, 5, 4, and 2. Assume that the battery with a lifetime of 6 actually had a lifetime of 20, so that the lifetimes are 3, 4, 20, 5, 4, and 2. Compute the standard deviation.

Solution
Following is the display from the TI-84 Plus.

```
    1-Var Stats
x̄=6.333333333
Σx=38
Σx²=470
Sx=6.772493386
σx=6.18241233
↓n=6
```

Including the extreme value of 20 in the data set has increased the sample standard deviation from 1.414 to 6.772. Clearly, the standard deviation is not resistant.

Approximating the Standard Deviation with Grouped Data

Objective 4 Approximate the standard deviation with grouped data

Sometimes we don't have access to the raw data in a data set, but we are given a frequency distribution. In Section 3.1, we learned how to approximate the sample mean from a frequency distribution. We now show how to approximate the standard deviation.

We use Table 3.6 to illustrate the method. Table 3.6 presents the number of text messages sent via cell phone by a sample of 50 high school students. In Section 3.1, we computed the approximate sample mean to be $\bar{x} = 137$. We will now approximate the standard deviation.

Table 3.6 Number of Text Messages Sent
by High School Students

Number of Text Messages Sent	Frequency
0–49	10
50–99	5
100–149	13
150–199	11
200–249	7
250–299	4

We present the procedure for approximating the standard deviation from grouped data.

> ### Procedure for Approximating the Standard Deviation with Grouped Data
>
> **Step 1:** Compute the midpoint of each class. The midpoint of a class is found by taking the average of the lower class limit and the lower limit of the next larger class. Then compute the mean as described in Section 3.1.
>
> **Step 2:** For each class, subtract the mean from the class midpoint to obtain Midpoint − Mean.
>
> **Step 3:** For each class, square the difference obtained in Step 2 to obtain (Midpoint − Mean)^2, and multiply by the frequency to obtain (Midpoint − Mean)$^2 \times$ Frequency.
>
> **Step 4:** Add the products (Midpoint − Mean)$^2 \times$ Frequency over all classes.
>
> **Step 5:** Compute the sum of the frequencies n. To compute the *population* variance, divide the sum obtained in Step 4 by n. To compute the *sample* variance, divide the sum obtained in Step 4 by $n - 1$.
>
> **Step 6:** Take the square root of the variance obtained in Step 5. The result is the standard deviation.

EXAMPLE 3.16 | Computing the standard deviation for grouped data

Compute the approximate sample standard deviation of the number of messages sent, using the data given in Table 3.6.

Solution

The calculations are summarized in Table 3.7.

Table 3.7 Calculating the Variance and Standard Deviation of the Number of Text Messages

Class	Midpoint	Frequency	Mean	Midpoint − Mean	(Midpoint − Mean)$^2 \times$ Frequency
0–49	25	10	137	−112	$12544 \times 10 = 125{,}440$
50–99	75	5	137	−62	$3844 \times 5 = 19{,}220$
100–149	125	13	137	−12	$144 \times 13 = 1{,}872$
150–199	175	11	137	38	$1444 \times 11 = 15{,}884$
200–249	225	7	137	88	$7744 \times 7 = 54{,}208$
250–299	275	4	137	138	$19044 \times 4 = 76{,}176$
		Sum = 50			Sum = 292,800

$$\text{Variance} = \frac{292{,}800}{50 - 1} = 5975.51020 \qquad \text{Standard deviation} = \sqrt{5975.51020} = 77.3014$$

Step 1: Compute the midpoints: For the first class, the lower class limit is 0. The lower limit of the next class is 50. The midpoint is therefore

$$\frac{0 + 50}{2} = 25$$

We continue in this manner to compute the midpoints of each of the classes. Note that for the last class, we average the lower limit of 250 with 300, which is the lower limit that the next class would have. We computed the sample mean in Example 3.9 in Section 3.1. The sample mean is $\bar{x} = 137$.

Step 2: For each class, subtract the mean from the class midpoint as shown in the column labeled "Midpoint − Mean."

Step 3: For each class, square the difference obtained in Step 2 and multiply by the frequency as shown in the column labeled "(Midpoint − Mean)$^2 \times$ Frequency."

Step 4: Add the products (Midpoint − Mean)$^2 \times$ Frequency over all classes, to obtain the sum 292,800.

Step 5: The sum of the frequencies is 50. Since we are considering the data to be a sample, we subtract 1 from this sum to obtain 49. The sample variance is 292,800/49 = 5975.51020.

Step 6: The sample standard deviation is $\sqrt{5975.51020} = 77.3014$.

Objective 5 Use the Empirical Rule to summarize data that are unimodal and approximately symmetric

The Empirical Rule

Many histograms have a single mode near the center of the data, and are approximately symmetric. Such histograms are often referred to as *bell-shaped*. Other histograms are strongly skewed; these are not bell-shaped. When a data set has a bell-shaped histogram, it is often possible to use the standard deviation to provide an approximate description of the data using a rule known as the **Empirical Rule**.

The Empirical Rule

When a population has a histogram that is approximately bell-shaped, then

- Approximately 68% of the data will be within one standard deviation of the mean. In other words, approximately 68% of the data will be between $\mu - \sigma$ and $\mu + \sigma$.
- Approximately 95% of the data will be within two standard deviations of the mean. In other words, approximately 95% of the data will be between $\mu - 2\sigma$ and $\mu + 2\sigma$.
- All, or almost all, of the data will be within three standard deviations of the mean. In other words, all, or almost all, of the data will be between $\mu - 3\sigma$ and $\mu + 3\sigma$.

CAUTION

The Empirical Rule should not be used for data sets that are not approximately bell-shaped.

The Empirical Rule holds for many bell-shaped data sets. Figure 3.7 illustrates the Empirical Rule.

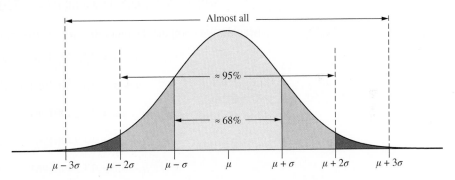

Figure 3.7 The Empirical Rule. Approximately 68% of the data values are between $\mu - \sigma$ and $\mu + \sigma$, approximately 95% are between $\mu - 2\sigma$ and $\mu + 2\sigma$, and almost all are between $\mu - 3\sigma$ and $\mu + 3\sigma$.

EXAMPLE 3.17 Using the Empirical Rule to describe a data set

Table 3.8 presents the percentage of the population aged 65 and over in each state and the District of Columbia. Figure 3.8 (page 118) presents a histogram of these data. Compute the mean and standard deviation, and use the Empirical Rule to describe the data.

Table 3.8 Percentage of People Aged 65 and Over in Each of the 50 States and District of Columbia

14.1	8.1	13.9	14.3	11.5	10.7	14.4	14.1	11.5	17.8	14.0
10.2	14.3	12.0	12.4	12.7	14.9	13.4	13.1	12.6	15.6	
12.2	13.7	12.8	12.4	12.8	13.9	15.0	13.8	12.3	12.6	
13.7	14.1	13.6	12.4	15.3	13.7	13.8	13.0	15.5	14.1	
13.6	14.6	13.3	10.5	9.0	14.3	12.4	12.2	16.0	13.5	

Figure 3.8 Histogram for the data in Table 3.8

Solution

Step 1: Figure 3.8 shows that the histogram is approximately bell-shaped, so we may use the Empirical Rule.

Step 2: We use the TI-84 Plus to compute the mean and standard deviation. The display is shown here:

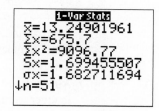

Note that the 51 entries (corresponding to 50 states plus the District of Columbia) are an entire population. Therefore we will interpret the quantity $\bar{x} = 13.24901961$ produced by the TI-84 Plus as the population mean μ, and we will use $\sigma = 1.682711694$ for the standard deviation.

Step 3: We compute the quantities $\mu - \sigma$, $\mu + \sigma$, $\mu - 2\sigma$, $\mu + 2\sigma$, $\mu - 3\sigma$, and $\mu + 3\sigma$.

$$\mu - \sigma = 13.24901961 - 1.682711694 = 11.57$$

$$\mu + \sigma = 13.24901961 + 1.682711694 = 14.93$$

$$\mu - 2\sigma = 13.24901961 - 2(1.682711694) = 9.88$$

$$\mu + 2\sigma = 13.24901961 + 2(1.682711694) = 16.61$$

$$\mu - 3\sigma = 13.24901961 - 3(1.682711694) = 8.20$$

$$\mu + 3\sigma = 13.24901961 + 3(1.682711694) = 18.30$$

We conclude that the percentage of the population aged 65 and over is between 11.57 and 14.93 in approximately 68% of the states, between 9.88 and 16.61 in approximately 95% of the states, and between 8.20 and 18.30 in almost all the states.

The Empirical Rule can be used for samples as well as populations. When we work with a sample, we use $\bar{x}$ in place of μ and s in place of σ.

EXAMPLE 3.18   **Using the Empirical Rule to describe a data set**

A sample of size 200 has sample mean $\bar{x} = 50$ and sample standard deviation $s = 10$. The histogram is approximately bell-shaped.

   a. Find an interval that is likely to contain approximately 68% of the data values.

   b. Approximately what percentage of the data values will be between 30 and 70?

Solution

a. We use the Empirical Rule. Approximately 68% of the data will be between $\bar{x} - s$ and $\bar{x} + s$. We compute

$$\bar{x} - s = 50 - 10 = 40 \qquad \bar{x} + s = 50 + 10 = 60$$

It is likely that approximately 68% of the data values are between 40 and 60.

b. The value 30 is two standard deviations below the mean, since

$$\bar{x} - 2s = 50 - 20 = 30$$

and the value 70 is two standard deviations above the mean, since

$$\bar{x} + 2s = 50 + 20 = 70$$

Therefore, it is likely that approximately 95% of the data values are between 30 and 70.

EXAMPLE 3.19

Determining whether the Empirical Rule is appropriate

Following is a histogram for a data set. Should the Empirical Rule be used?

Solution

No. The distribution is skewed, rather than bell-shaped. Therefore, the Empirical Rule should not be used.

Check Your Understanding

4. A data set has a mean of 20 and a standard deviation of 3. A histogram is shown here. Is it appropriate to use the Empirical Rule to approximate the proportion of the data between 14 and 26? If so, find the approximation. If not, explain why not.

5. A data set has a mean of 50 and a standard deviation of 8. A histogram is shown here. Is it appropriate to use the Empirical Rule to approximate the proportion of the data between 42 and 58? If so, find the approximation. If not, explain why not.

Answers are on page 128.

Objective 6 Use Chebyshev's Inequality to describe a data set

Chebyshev's Inequality

When a distribution is bell-shaped, the Empirical Rule gives us an approximation to the proportion of data that will be within one or two standard deviations of the mean. **Chebyshev's Inequality** is a rule that holds for any data set.

Explain It Again

Chebyshev's Inequality is always valid: Chebyshev's Inequality can be used with any data set, whether or not it is bell-shaped.

Chebyshev's Inequality

In any data set, the proportion of the data that will be within K standard deviations of the mean is at least $1 - 1/K^2$. Specifically, by setting $K = 2$ or $K = 3$, we obtain the following results:

- At least 3/4 (75%) of the data will be within two standard deviations of the mean.
- At least 8/9 (88.9%) of the data will be within three standard deviations of the mean.

EXAMPLE 3.20   Using Chebyshev's Inequality

As part of a public health study, systolic blood pressure was measured for a large group of people. The mean was $\bar{x} = 120$ and the standard deviation was $s = 10$. What information does Chebyshev's Inequality provide about these data?

Solution

We compute:

$$\bar{x} - 2s = 120 - 2(10) = 100 \qquad \bar{x} + 2s = 120 + 2(10) = 140$$
$$\bar{x} - 3s = 120 - 3(10) = 90 \qquad \bar{x} + 3s = 120 + 3(10) = 150$$

We conclude:

- At least 75% of the people had systolic blood pressures between 100 and 140.
- At least 88.9% of the people had systolic blood pressures between 90 and 150.

Comparing Chebyshev's Inequality to the Empirical Rule

Both Chebyshev's Inequality and the Empirical Rule provide information about the proportion of a data set that is within a given number of standard deviations of the mean. An advantage of Chebyshev's Inequality is that it applies to any data set, whereas the Empirical Rule applies only to data sets that are approximately bell-shaped. A disadvantage of Chebyshev's Inequality is that for most data sets, it provides only a very rough approximation. Chebyshev's Inequality produces a minimum value for the proportion of the data that will be within a given number of standard deviations of the mean. For most data sets, the actual proportions are much larger than the values given by Chebyshev's Inequality.

Check Your Understanding

6. A group of elementary school students took a standardized reading test. The mean score was 70 and the standard deviation was 10. Someone says that only 50% of the students scored between 50 and 90. Is this possible? Explain.

7. A certain type of bolt used in an aircraft must have a length between 122 and 128 millimeters in order to be acceptable. The manufacturing process produces bolts whose mean length is 125 millimeters with a standard deviation of 1 millimeter. Can you be sure that more than 85% of the bolts are acceptable? Explain.

Answers are on page 128.

Objective 7 Compute the coefficient of variation

The Coefficient of Variation

The coefficient of variation (CV for short) tells how large the standard deviation is relative to the mean. It can be used to compare the spreads of data sets whose values have different units.

DEFINITION

The **coefficient of variation** is found by dividing the standard deviation by the mean.

$$CV = \frac{\sigma}{\mu}$$

EXAMPLE 3.21   Computing the coefficient of variation

National Weather Service records show that over a 30-year period, the annual precipitation in Atlanta, Georgia, had a mean of 49.8 inches with a standard deviation of 7.6 inches, and the annual temperature had a mean of 62.2 degrees Fahrenheit with a standard deviation of 1.3 degrees. Compute the coefficient of variation for precipitation and for temperature. Which has greater spread relative to its mean?

Solution

The coefficient of variation for precipitation is

$$CV \text{ for precipitation} = \frac{\text{Standard deviation of precipitation}}{\text{Mean precipitation}} = \frac{7.6}{49.8} = 0.153$$

The coefficient of variation for temperature is

$$CV \text{ for temperature} = \frac{\text{Standard deviation of temperature}}{\text{Mean temperature}} = \frac{1.3}{62.2} = 0.021$$

The CV for precipitation is larger than the CV for temperature. Therefore, precipitation has a greater spread relative to its mean.

Note that we cannot compare the standard deviations of precipitation and temperature because they have different units. It does not make sense to ask whether 7.6 inches is greater than 1.3 degrees. The CV is unitless, however, so we can compare the CVs.

Check Your Understanding

8. Lengths of newborn babies have a mean of 20.1 inches with a standard deviation of 1.9 inches. Find the coefficient of variation of newborn lengths.

Answer is on page 128.

USING TECHNOLOGY

TI-84 Plus

Computing the sample standard deviation

The TI-84 PLUS procedure to compute the mean and median, described on page 100, will also compute the standard deviation.

MINITAB

Computing the sample standard deviation

The MINITAB procedure to compute the mean and median, described on page 100, will also compute the standard deviation.

EXCEL

Computing the sample standard deviation

The EXCEL procedure to compute the mean and median, described on page 100, will also compute the standard deviation.

SECTION 3.2 Exercises

Exercises 1–8 are the Check Your Understanding exercises located within the section.

Understanding the Concepts

In Exercises 9–12, fill in each blank with the appropriate word or phrase.

9. If all values in a data set are the same, then the sample variance is equal to _____.

10. The standard deviation is the square root of the _____.

11. For a bell-shaped data set, approximately _____ of the data will be in the interval $\mu - \sigma$ to $\mu + \sigma$.

12. Chebyshev's Inequality states that for any data set, the proportion of data within K standard deviations of the mean is at least _____.

In Exercises 13–16, determine whether the statement is true or false. If the statement is false, rewrite it as a true statement.

13. The variance and standard deviation are measures of center.

14. The range of a data set is the difference between the largest value and the smallest value.

15. In a bell-shaped data set with $\mu = 15$ and $\sigma = 5$, approximately 95% of the data will be between 10 and 20.

16. For some data sets, Chebyshev's Inequality may be used but the Empirical Rule should not be.

Practicing the Skills

17. Find the sample variance and standard deviation for the following sample:

    17 40 24 18 16

18. Find the sample variance and standard deviation for the following sample:

    59 25 12 29 16 8 26 30 17

19. Find the sample variance and standard deviation for the following sample:

    15 9 5 12 9 21 4 24 18

20. Find the population variance and standard deviation for the following population:

    16 6 18 3 25 22

21. Find the population variance and standard deviation for the following population:

    20 8 11 23 27 29 62 4

22. Find the population variance and standard deviation for the following population:

    26 25 29 23 14 20 12 18 24 31 22 32

23. Approximate the sample variance and standard deviation given the following frequency distribution:

Class	Frequency
0–9	13
10–19	7
20–29	10
30–39	9
40–49	11

24. Approximate the sample variance and standard deviation given the following frequency distribution:

Class	Frequency
0–15	2
16–31	14
32–47	6
48–63	13
64–79	15

25. Approximate the population variance and standard deviation given the following frequency distribution:

Class	Frequency
0–49	17
50–99	26
100–149	14
150–199	34
200–249	26
250–299	8

26. Approximate the population variance and standard deviation given the following frequency distribution:

Class	Frequency
0–19	18
20–39	11
40–59	6
60–79	6
80–99	10
100–119	5

27. The following TI-84 Plus display presents some population parameters.

    ```
    1-Var Stats
    x̄=32
    Σx=480
    Σx²=15900
    Sx=6.210590034
    σx=6
    ↓n=15
    ```

    a. Assume the population is bell-shaped. Approximately what percentage of the population values are between 26 and 38?

    b. Assume the population is bell-shaped. Between what two values will approximately 95% of the population be?

    c. If we do not assume that the population is bell-shaped, at least what percentage of the population will be between 20 and 44?

28. The following TI-84 Plus display presents some population parameters.

    ```
    1-Var Stats
    x̄=134
    Σx=2680
    Σx²=359620
    Sx=5.12989176
    σx=5
    ↓n=20
    ```

    a. Assume the population is bell-shaped. Approximately what percentage of the population values are between 124 and 144?

b. Assume the population is bell-shaped. Between what two values will approximately 68% of the population be?

c. If we do not assume that the population is bell-shaped, at least what percentage of the population will be between 119 and 149?

29. The following TI-84 Plus display presents some sample statistics.

a. Assume that a histogram of the sample is bell-shaped. Approximately what percentage of the sample values are between 104 and 248?

b. Assume that a histogram for the sample is bell-shaped. Between what two values will approximately 68% of the sample be?

c. If we do not assume that the histogram is bell-shaped, at least what percentage of the sample values will be between 68 and 284?

30. The following TI-84 Plus display presents some sample statistics.

a. Assume that a histogram of the sample is bell-shaped. Approximately what percentage of the sample values are between 72 and 86?

b. Assume that a histogram for the sample is bell-shaped. Between what two values will approximately 95% of the sample be?

c. If we do not assume that the histogram is bell-shaped, at least what percentage of the sample values will be between 65 and 93?

Working with the Concepts

31. eCommerce: The following table presents the amount spent (in billions of dollars) on online purchases in several categories for the years 2011 and 2013.

Category	2011	2013
Computer and Consumer Electronics	42.0	56.8
Apparel and Accessories	37.6	54.2
Auto Parts	21.6	25.5
Books/Music/Video	17.4	24.4
Furniture and Home Furnishings	15.1	20.2
Toys and Hobbies	9.5	13.1
Health and Personal Care	9.3	12.5
Office Equipment and Supplies	6.5	8.0
Food and Beverages	4.3	5.8

Source: www.emarketer.com

a. Find the sample standard deviation of the online purchases for 2011.

b. Find the sample standard deviation of the online purchases for 2013.

c. The amount of online purchases increased in every category between 2011 and 2013. Did the spread in online purchases increase as well?

32. Sports car or convertible? The following table presents the fuel efficiency, in miles per gallon, for a sample of convertibles and a sample of sports cars.

Convertible Model	MPG	Sports Model	MPG
Volkswagen Eos	25	BMW 135i	23
Mini Cooper	25	Mazda3 Mazdaspeed	24
Saab 9-3	24	Subaru Impreza WRX STi	21
BMW 328i	21	Mazda RX-8	18
Toyota Camry Solara	21	Mitsubishi Lancer Evolution	21
Volvo C70	21	Volkswagen GTI	25
Ford Mustang V6	20	Honda Civic Si	27

Source: *Consumer Reports*

a. Find the sample standard deviation of the mileage for the sample of convertibles.

b. Find the sample standard deviation of the mileage for the sample of sports cars.

c. Which sample has greater spread?

33. Heavy football players: Following are the weights, in pounds, for samples of offensive and defensive linemen in the National Football League.

Offense:	335	301	307	252	260	307	325	310	305	305	264	325
Defense:	284	290	286	355	305	295	297	325	310	297	314	348

a. Find the sample standard deviation for the weights for the offensive linemen.
b. Find the sample standard deviation for the weights for the defensive linemen.
c. Is there greater spread in the weights of the offensive or the defensive linemen?

34. **Beer:** The following table presents the number of active breweries for samples of states located east and west of the Mississippi River.

East		West	
State	**Number of Breweries**	**State**	**Number of Breweries**
Connecticut	18	Alaska	17
Delaware	10	Arizona	31
Florida	47	California	305
Georgia	22	Colorado	111
Illinois	52	Iowa	21
Kentucky	13	Louisiana	6
Maine	38	Minnesota	41
Maryland	23	Montana	30
Massachusetts	40	South Dakota	5
New Hampshire	16	Texas	37
New Jersey	20	Utah	15
New York	76		
North Carolina	46		
South Carolina	14		
Tennessee	19		
Vermont	20		

Source: http://www.beerinstitute.org/

a. Compute the sample standard deviation for the number of breweries east of the Mississippi River.
b. Compute the sample standard deviation for the number of breweries west of the Mississippi River.
c. Compute the range for each data set.
d. Based on the standard deviations, which region has the greater spread in the number of breweries?
e. Based on the ranges, which region has the greater spread in the number of breweries?
f. The sample of western states happens to include California. Remove California from the sample of western states, and compute the sample standard deviation for the remaining western states. Does the result show that the standard deviation is not resistant? Explain.
g. Compute the range for the western states with California removed. Is the range resistant? Explain.

35. **What's your favorite TV show?** The following tables present the ratings for the top 20 prime-time shows for the 2007–2008 season and for the 2012–2013 season. The rating is the percentage of households with TV sets that watched the program.

Top Rated TV Programs: 2007–2008		Top Rated TV Programs: 2012–2013	
Program	**Rating**	**Program**	**Rating**
American Idol — Tuesday	16.1	*NBC Sunday Night Football*	7.9
American Idol — Wednesday	15.9	*Big Bang Theory*	6.2
Dancing With the Stars — Monday (2007)	14.0	*The Voice*	5.1
Dancing With the Stars — Tuesday (2007)	12.7	*Modern Family*	4.9
Dancing With the Stars — Monday (2008)	12.6	*The Voice* — Tuesday	4.6
Dancing With the Stars — Tuesday (2008)	11.8	*American Idol* — Wednesday	4.6
Desperate Housewives	11.6	*American Idol* — Thursday	4.3
CSI	10.6	*The Following*	4.3
House — Tuesday	10.5	*Two and a Half Men*	4.1
Grey's Anatomy	10.4	*Grey's Anatomy*	4.1
Sunday Night Football	9.7	*NCIS*	4.0
CSI: Miami	9.2	*Football Night in America Part 3*	4.0
House — Monday	9.2	*Revolution*	3.9
NCIS	9.2	*2 Broke Girls*	3.7
Survivor: China	9.0	*How I Met Your Mother*	3.7
Without a Trace	8.8	*Family Guy*	3.6
The Moment of Truth	8.8	*Once Upon a Time*	3.6
Two and a Half Men	8.5	*Survivor: Philippines*	3.5
60 Minutes	8.4	*X-Factor* — Wednesday	3.5
Criminal Minds	8.2	*NCIS: LA*	3.4

Source: Nielsen Media Research   Source: Nielsen Media Research

a. Find the population standard deviation of the ratings for 2007–2008.
b. Find the population standard deviation of the ratings for 2012–2013.
c. Compute the range for the ratings for both seasons.
d. Based on the standard deviations, did the spread in ratings increase or decrease over the two seasons?
e. Based on the ranges, did the spread in ratings increase or decrease over the two seasons?

36. **House prices:** The following table presents prices, in thousands of dollars, of single-family homes for some of the largest metropolitan areas in the United States for the first quarter of 2012 and the first quarter of 2013.

Metro Area	2012	2013	Metro Area	2012	2013
Atlanta, GA	87.8	115.1	New York, NY	363.8	368.2
Baltimore, MD	218.1	226.5	Philadelphia, PA	193.5	197.7
Boston, MA	311.5	332.2	Phoenix, AZ	129.9	169.0
Chicago, IL	157.2	159.4	Portland, OR	208.6	246.5
Cincinnati, OH	112.5	121.0	Riverside, CA	174.3	216.7
Cleveland, OH	84.9	101.0	Saint Louis, MO	103.7	111.0
Dallas, TX	148.2	160.4	San Diego, CA	359.5	412.3
Denver, CO	226.4	261.2	San Francisco, CA	448.0	593.9
Houston, TX	152.1	163.7	Seattle, WA	265.4	312.6
Las Vegas, NV	122.1	155.1	Tampa, FL	131.9	141.8
Miami, FL	182.0	219.9	Washington, DC	311.6	348.7
Minneapolis, MN	147.3	170.6			

Source: National Realtors Association

a. Find the population standard deviation for 2012.
b. Find the population standard deviation for 2013.
c. In general, house prices increased from 2012 to 2013. Did the spread in house prices decrease as well, or did it increase?

37. **Stock prices:** Following are the closing prices of Google stock for each trading day in May and June 2013.

May					
880.37	877.07	873.65	866.20	869.79	829.61
880.93	884.74	900.68	900.62	886.25	820.43
875.04	877.00	871.98	879.81	890.22	
879.73	864.64	859.70	859.10	867.63	

June				
871.22	870.76	868.31	881.27	873.32
882.79	889.42	906.97	908.53	909.18
903.87	915.89	887.10	877.53	880.23
871.48	873.63	857.23	861.55	845.72

a. Find the population standard deviation for the prices in May.
b. Find the population standard deviation for the prices in June.
c. Financial analysts use the word *volatility* to refer to the variation in prices of assets such as stocks. In which month was the price of Google stock more volatile?

38. **Stocks or bonds?** Following are the annual percentage returns for the years 1993–2012 for three categories of investment: stocks, Treasury bills, and Treasury bonds. Stocks are represented by the Dow Jones Industrial Average.

Year	Stocks	Bills	Bonds	Year	Stocks	Bills	Bonds
1993	13.72	2.98	14.21	2003	25.32	1.03	0.38
1994	2.14	3.99	−8.04	2004	3.15	1.23	4.49
1995	33.45	5.52	23.48	2005	−0.61	3.01	2.87
1996	26.01	5.02	1.43	2006	16.29	4.68	1.96
1997	22.64	5.05	9.94	2007	6.43	4.64	10.21
1998	16.10	4.73	14.92	2008	−33.84	1.59	20.10
1999	25.22	4.51	−8.25	2009	18.82	0.14	−11.12
2000	−6.18	5.76	16.66	2010	11.02	0.13	8.46
2001	−7.10	3.67	5.57	2011	5.53	0.03	16.04
2002	−16.76	1.66	15.12	2012	7.26	0.05	2.97

Source: Federal Reserve

a. The standard deviation of the return is a measure of the risk of an investment. Compute the population standard deviation for each type of investment. Which is the riskiest? Which is least risky?
b. Treasury bills are short-term (1 year or less) loans to the U.S. government. Treasury bonds are long-term (30-year) loans to the government. Finance theory states that long-term loans are riskier than short-term loans. Do the results agree with the theory?
c. Finance theory states that the more risk an investment has, the higher its mean return must be. Compute the mean return for each class of investment. Do the results follow the theory?

39. Time to review: The following table presents the time taken to review articles that were submitted for publication to the journal *Technometrics* during a recent year. A few articles took longer than 9 months to review; these are omitted from the table. Consider the data to be a population.

Time (Months)	Number of Articles
0.0–0.9	45
1.0–1.9	17
2.0–2.9	18
3.0–3.9	19
4.0–4.9	12
5.0–5.9	14
6.0–6.9	13
7.0–7.9	22
8.0–8.9	11

a. Approximate the variance of the times.
b. Approximate the standard deviation of the times.

40. Age distribution: The ages of residents of Banks City, Oregon, are given in the following frequency distribution. Consider these data to be a population.

Age	Frequency
0–9	283
10–19	203
20–29	217
30–39	256
40–49	176
50–59	92
60–69	21
70–79	23
80–89	12
90–99	3

Source: U.S. Census Bureau

a. Approximate the variance of the ages.
b. Approximate the standard deviation of the ages.

41. Lunch break: In a recent survey of 655 working Americans ages 25–34, the average weekly amount spent on lunch was $44.60 with standard deviation $2.81. The weekly amounts are approximately bell-shaped.
a. Estimate the percentage of amounts that are between $36.17 and $53.03.
b. Estimate the percentage of amounts that are between $41.79 and $47.41.
c. Between what two values will approximately 95% of the amounts be?

42. Pay your bills: In a large sample of customer accounts, a utility company determined that the average number of days between when a bill was sent out and when the payment was made is 32 with a standard deviation of 7 days. Assume the data to be approximately bell-shaped.
a. Between what two values will approximately 68% of the numbers of days be?
b. Estimate the percentage of customer accounts for which the number of days is between 18 and 46.
c. Estimate the percentage of customer accounts for which the number of days is between 11 and 53.

43. Newborn babies: A study conducted by the Center for Population Economics at the University of Chicago studied the birth weights of 621 babies born in New York. The mean weight was 3234 grams with a standard deviation of 871 grams. Assume that birth weight data are approximately bell-shaped. Estimate the number of newborns who weighed between
a. 2363 grams and 4105 grams
b. 1492 grams and 4976 grams

44. Internet providers: In a survey of 600 homeowners with high-speed Internet, the average monthly cost of a high-speed Internet plan was $64.20 with standard deviation $11.77. Assume the plan costs to be approximately bell-shaped. Estimate the number of plans that cost between
a. $40.66 and $87.74
b. $52.43 and $75.97

45. Lunch break: For the data in Exercise 41, estimate the percentage of amounts that were
a. less than $41.79
b. greater than $50.22
c. between $44.60 and $47.41

46. Pay your bills: For the data in Exercise 42, estimate the percentage of bills for which the number of days between when a bill was sent and when payment was made was
   a. greater than 39
   b. less than 18
   c. between 18 and 32

47. Newborn babies: For the data in Exercise 43, estimate the number of newborns whose weight was
   a. less than 4105 grams
   b. greater than 1492 grams
   c. between 3234 and 4976 grams

48. Internet providers: For the data in Exercise 44, estimate the number of Internet plans whose cost is
   a. greater than $52.43
   b. less than $87.74
   c. between $52.43 and $64.20

49. Empirical Rule OK? The following histogram presents a data set with a mean of 4.5 and a standard deviation of 2. Is it appropriate to use the Empirical Rule to approximate the proportion of the data between 0.5 and 8.5? If so, find the approximation. If not, explain why not.

50. Empirical Rule OK? The following histogram presents a data set with a mean of 62 and a standard deviation of 17. Is it appropriate to use the Empirical Rule to approximate the proportion of the data between 45 and 79? If so, find the approximation. If not, explain why not.

51. Empirical Rule OK? The following histogram presents a data set with a mean of 35 and a standard deviation of 9. Is it appropriate to use the Empirical Rule to approximate the proportion of the data between 26 and 44? If so, find the approximation. If not, explain why not.

52. Empirical Rule OK? The following histogram presents a data set with a mean of 16 and a standard deviation of 2. Is it appropriate to use the Empirical Rule to approximate the proportion of the data between 12 and 20? If so, find the approximation. If not, explain why not.

53. **What's the temperature?** The temperature in a certain location was recorded each day for two months. The mean temperature was 62.4°F with a standard deviation of 3.1°F. What can you determine about these data by using Chebyshev's Inequality with $K = 2$?

54. **Find the standard deviation:** The National Center for Health Statistics sampled a number of women aged 20–29 and found that the mean height was 64.2 inches and the histogram for the data set was approximately bell-shaped. Assume the heights in the data set ranged from 55 to 74 inches. Is the standard deviation of the data closest to 2, 3, or 4? Explain.

55. **Find the standard deviation:** The National Center for Health Statistics sampled a number of men aged 20–29 and found that the mean height was 69.4 inches and the histogram for the data set was approximately bell-shaped. Assume the heights in the data set ranged from 62 to 78 inches. Is the standard deviation of the data closest to 2.5, 3.5, or 4.5? Explain.

56. **Price of electricity:** The Energy Information Administration records the price of electricity in the United States each month. In July 2013, the average price of electricity was 11.92 cents per kilowatt-hour. Suppose that the standard deviation is 2.1 cents per kilowatt-hour. What can you determine about these data by using Chebyshev's Inequality with $K = 3$?

57. **Possible or impossible?** A data set has a mean of 20 and a standard deviation of 5. Which of the following might possibly be true, and which are impossible?
    a. Less than 50% of the data values are between 10 and 30.
    b. Only 1% of the data values are greater than 35.
    c. More than 15% of the data values are less than 5.
    d. More than 90% of the data values are between 5 and 35.

58. **Possible or impossible?** A data set has a mean of 50 and a standard deviation of 10. Which of the following might possibly be true, and which are impossible?
    a. More than 10% of the data values are negative.
    b. Only 5% of the data values are greater than 70.
    c. More than 20% of the data values are less than 30.
    d. Less than 75% of the data values are between 30 and 70.

59. **Standard deviation and mean:** For a list of positive numbers, is it possible for the sample standard deviation to be greater than the mean? If so, give an example. If not, explain why not.

60. **Standard deviation equal to 0?** Is it possible for the sample standard deviation of a list of numbers to equal 0? If so, give an example. If not, explain why not.

61. **Height and weight:** A National Center for Health Statistics study states that the mean height for adult men in the United States is 69.4 inches with a standard deviation of 3.1 inches, and the mean weight is 194.7 pounds with a standard deviation of 68.3 pounds.
    a. Compute the coefficient of variation for height.
    b. Compute the coefficient of variation for weight.
    c. Which has greater spread relative to its mean, height or weight?

62. **Test scores:** Scores on a statistics exam had a mean of 75 with a standard deviation of 10. Scores on a calculus exam had a mean of 60 with a standard deviation of 9.
    a. Compute the coefficient of variation for statistics exam scores.
    b. Compute the coefficient of variation for calculus exam scores.
    c. Which has greater spread relative to their mean, statistics scores or calculus scores?

Extending the Concepts

63. **Mean absolute deviation:** A measure of spread that is an alternative to the standard deviation (SD) is the **mean absolute deviation** (MAD). For a data set containing values $x_1, ..., x_n$, the mean absolute deviation is given by

$$\text{Mean absolute deviation} = \frac{\sum |x - \bar{x}|}{n}$$

   a. Compute the mean $\bar{x}$ for the data set 1, 3, 4, 7, 9.
   b. Construct a table like Table 3.5 that contains an additional column for the values $|x - \bar{x}|$.
   c. Use the table to compute the SD and the MAD.
   d. Now consider the data set 1, 3, 4, 7, 9, 30. Compute the SD and the MAD for this data set.
   e. Which measure of spread is more resistant, the SD or the MAD? Explain.

Answers to Check Your Understanding Exercises for Section 3.2

1. The variance of the St. Louis temperatures is 291.9. This is greater than the variance of the San Francisco temperatures, which indicates that there is greater spread in the St. Louis temperatures.

2. a. Variance is 3.7143; standard deviation is 1.9272.
   b. Variance is 281.8667; standard deviation is 16.7889.

3. Variance is 4.8594; standard deviation is 2.2044.

4. Approximately 95% of the data values are between 14 and 26.

5. The Empirical Rule should not be used because the data are skewed.

6. No. The interval between 50 and 90 is the interval within two standard deviations of the mean. At least 75% of the data must be between 50 and 90.

7. Yes. The interval between 122 and 128 is the interval within three standard deviations of the mean. At least 8/9 (88.9%) of the data must be between 122 and 128.

8. 0.0945

| SECTION 3.3 | Measures of Position |

Objectives

1. Compute and interpret z-scores
2. Compute the quartiles of a data set
3. Compute the percentiles of a data set
4. Compute the five-number summary for a data set
5. Understand the effects of outliers
6. Construct boxplots to visualize the five-number summary and outliers

Objective 1 Compute and interpret z-scores

The z-Score

Who is taller, a man 73 inches tall or a woman 68 inches tall? The obvious answer is that the man is taller. However, men are taller than women on the average. Let's ask the question this way: Who is taller relative to their gender, a man 73 inches tall or a woman 68 inches tall? One way to answer this question is with a *z-score*.

The z-score of an individual data value tells how many standard deviations that value is from its population mean. So, for example, a value one standard deviation above the mean has a z-score of 1. A value two standard deviations below the mean has a z-score of -2.

DEFINITION

Let x be a value from a population with mean μ and standard deviation σ. The **z-score** for x is

$$z = \frac{x - \mu}{\sigma}$$

EXAMPLE 3.22

Computing and interpreting z-scores

A National Center for Health Statistics study states that the mean height for adult men in the United States is $\mu = 69.4$ inches, with a standard deviation of $\sigma = 3.1$ inches. The mean height for adult women is $\mu = 63.8$ inches, with a standard deviation of $\sigma = 2.8$ inches. Who is taller relative to their gender, a man 73 inches tall, or a woman 68 inches tall?

Solution

We compute the z-scores for the two heights:

$$z\text{-score for man's height} = \frac{x - \mu}{\sigma} = \frac{73 - 69.4}{3.1} = 1.16$$

$$z\text{-score for woman's height} = \frac{x - \mu}{\sigma} = \frac{68 - 63.8}{2.8} = 1.50$$

The height of the 73-inch man is 1.16 standard deviations above the mean height for men. The height of the 68-inch woman is 1.50 standard deviations above the mean height for women. Therefore, the woman is taller, relative to the population of women, than the man is, relative to the population of men.

z-scores and the Empirical Rule

z-scores work best for populations whose histograms are approximately bell-shaped — that is, for populations for which we can use the Empirical Rule. Recall that the Empirical Rule says that for a bell-shaped population, approximately 68% of the data will be within one standard deviation of the mean, approximately 95% will be within two standard deviations, and almost all will be within three standard deviations. Since the z-score is the number of standard deviations from the mean, we can easily interpret the z-score for bell-shaped populations.

---

z-Scores and the Empirical Rule

When a population has a histogram that is approximately bell-shaped, then

- Approximately 68% of the data will have z-scores between -1 and 1.
- Approximately 95% of the data will have z-scores between -2 and 2.
- All, or almost all, of the data will have z-scores between -3 and 3.

The z-score is less useful for populations that are not bell-shaped. For example, in some skewed populations there will be no values with z-scores greater than 1, while in others, values with z-scores greater than 1 occur frequently. We can't be sure how to interpret z-scores when the population is skewed. It is best, therefore, to use z-scores only for populations that are approximately bell-shaped. See Exercise 45 for an illustration.

Objective 2 Compute the quartiles of a data set

Quartiles

The weather in Los Angeles is dry most of the time, but it can be quite rainy in the winter. The rainiest month of the year is February. Table 3.9 presents the annual rainfall in Los Angeles, in inches, for each February from 1969 to 2013.

Table 3.9 Annual Rainfall in Los Angeles During February

Year	Rainfall	Year	Rainfall	Year	Rainfall	Year	Rainfall	Year	Rainfall
1969	8.03	1978	8.91	1987	1.22	1996	4.94	2005	11.02
1970	2.58	1979	3.06	1988	1.72	1997	0.08	2006	2.37
1971	0.67	1980	12.75	1989	1.90	1998	13.68	2007	0.92
1972	0.13	1981	1.48	1990	3.12	1999	0.56	2008	1.64
1973	7.89	1982	0.70	1991	4.13	2000	5.54	2009	3.57
1974	0.14	1983	4.37	1992	7.96	2001	8.87	2010	4.27
1975	3.54	1984	0.00	1993	6.61	2002	0.29	2011	3.29
1976	3.71	1985	2.84	1994	3.21	2003	4.64	2012	0.16
1977	0.17	1986	6.10	1995	1.30	2004	4.89	2013	0.20

There is a lot of spread in the amount of rainfall in Los Angeles in February. For example, in 1984 there was no measurable rain at all, while in 1998 it rained more than 13 inches.

In Section 3.1 we learned how to compute the mean and median of a data set, which describe the center of a distribution. For data sets like the Los Angeles rainfall data, which exhibit a lot of spread, it is useful to compute measures of positions other than the center, to get a more detailed description of the distribution. **Quartiles** provide a way to do this. Quartiles divide a data set into four approximately equal pieces.

DEFINITION

Every data set has three quartiles:

- The **first quartile**, denoted Q_1, separates the lowest 25% of the data from the highest 75%.
- The **second quartile**, denoted Q_2, separates the lower 50% of the data from the upper 50%. Q_2 is the same as the median.
- The **third quartile**, denoted Q_3, separates the lowest 75% of the data from the highest 25%.

Explain It Again

The second quartile is the same as the median: The second quartile, Q_2, divides the data in half. Therefore Q_2 is the same as the median.

There are several methods for computing quartiles, all of which give similar results. We present a fairly straightforward method here.

Procedure for Computing the Quartiles of a Data Set

Step 1: Arrange the data in increasing order.

Step 2: Let n be the number of values in the data set. To compute the second quartile, simply compute the median. For the first or third quartiles, proceed as follows.

For the first quartile, compute $L = 0.25n$.
For the third quartile, compute $L = 0.75n$.

Step 3: *If L is a whole number*, the quartile is the average of the number in position L and the number in position $L + 1$.

If L is not a whole number, round it *up* to the next higher whole number. The quartile is the number in the position corresponding to the rounded-up value.

EXAMPLE 3.23

Computing quartiles

Compute the first and third quartiles of the Los Angeles rainfall data.

Solution
Step 1: Table 3.10 presents the data in increasing order.

Step 2: There are $n = 45$ values in the data set. For the first quartile we compute

$$L = 0.25(45) = 11.25$$

Step 3: Since $L = 11.25$ is not a whole number, we round it *up* to 12. The first quartile, Q_1, is the number in the 12th position. From Table 3.10 we can see that the first quartile is $Q_1 = 0.92$.

Step 4: For the third quartile we compute

$$L = 0.75(45) = 33.75$$

Step 5: Since $L = 33.75$ is not a whole number, we round it *up* to 34. The third quartile, Q_3, is the number in the 34th position. From Table 3.10 we can see that the third quartile is $Q_3 = 4.94$.

> **CAUTION**
>
> Always round *L up*. Do not round down.

Table 3.10 Annual Rainfall in Los Angeles During February, in Increasing Order

Year	Rainfall	Year	Rainfall	Year	Rainfall	Year	Rainfall	Year	Rainfall
1984	0.00	1971	0.67	2006	2.37	1976	3.71	1993	6.61
1997	0.08	1982	0.70	1970	2.58	1991	4.13	1973	7.89
1972	0.13	2007	0.92	1985	2.84	2010	4.27	1992	7.96
1974	0.14	1987	1.22	1979	3.06	1983	4.37	1969	8.03
2012	0.16	1995	1.30	1990	3.12	2003	4.64	2001	8.87
1977	0.17	1981	1.48	1994	3.21	2004	4.89	1978	8.91
2013	0.20	2008	1.64	2011	3.29	1996	4.94	2005	11.02
2002	0.29	1988	1.72	1975	3.54	2000	5.54	1980	12.75
1999	0.56	1989	1.90	2009	3.57	1986	6.10	1998	13.68

Figure 3.9 (page 132) presents a dotplot of the Los Angeles rainfall data set with the quartiles indicated. The quartiles divide the data set into four parts, with approximately 25% of the data in each part. Recall that the median is the same as the second quartile. Since there are 45 values in this data set, the median is the 23rd value when the data are arranged in order. From Table 3.10, we can see that the median is 3.12.

Figure 3.9 The quartiles of the Los Angeles rainfall data set

EXAMPLE 3.24 Using technology to compute quartiles

Use technology to compute the first and third quartiles for the Los Angeles rainfall data presented in Table 3.9.

Solution
Figure 3.10 presents MINITAB output. The quartiles are highlighted in red.

Variable	N	Mean	SE Mean	StDev	Minimum	Q1	Median	Q3	Maximum
Rainfall	42	3.749	0.559	3.624	0.000	0.810	3.120	5.240	13.680

Figure 3.10

The **1–Var Stats** command for the TI-84 Plus calculator described in Section 3.1 also computes the quartiles of a data set. Figure 3.11 presents the results for the Los Angeles rainfall data.

Figure 3.11

Note that the values produced by both MINITAB and the TI-84 Plus differ slightly from the results obtained in Example 3.23, because they use a slightly different procedure than the one we describe here. Step-by-step instructions are presented in the Using Technology section on page 140.

Objective 3 Compute the percentiles of a data set

Percentiles

Quartiles describe the shape of a distribution by dividing it into fourths. Sometimes it is useful to divide a data set into a greater number of pieces to get a more detailed description of the distribution. *Percentiles* provide a way to do this. Percentiles divide a data set into hundredths.

DEFINITION

For a number p between 1 and 99, the pth **percentile** separates the lowest $p\%$ of the data from the highest $(100 - p)\%$.

There are several methods for computing percentiles, all of which give similar results. We present a fairly straightforward method here.

Procedure for Computing the Data Value Corresponding to a Given Percentile

Step 1: Arrange the data in increasing order.

Step 2: Let n be the number of values in the data set. For the pth percentile, compute the value

$$L = \frac{p}{100} \cdot n$$

Step 3: *If L is a whole number*, then the pth percentile is the average of the number in position L and the number in position $L + 1$.

If L is not a whole number, round it *up* to the next higher whole number. The pth percentile is the number in the position corresponding to the rounded-up value.

EXAMPLE 3.25

Computing a percentile

Compute the 60th percentile of the Los Angeles rainfall data.

Solution

Step 1: Table 3.10 presents the data in increasing order.

Step 2: There are $n = 45$ values in the data set. For the 60th percentile, we take $p = 60$ and compute

$$L = \frac{60}{100} \cdot 45 = 27$$

CAUTION

Always round *L* up. Do not round down.

Step 3: Since $L = 27$ is a whole number, the 60th percentile is the average of the numbers in the 27th and 28th positions. From Table 3.10 we can see that the 60th percentile is $\dfrac{3.57 + 3.71}{2} = 3.64$.

Computing the percentile corresponding to a given data value

Sometimes we are given a value from a data set, and wish to compute the percentile corresponding to that value. Following is a simple procedure for doing this.

Procedure for Computing the Percentile Corresponding to a Given Data Value

Step 1: Arrange the data in increasing order.

Step 2: Let x be the data value whose percentile is to be computed. Use the following formula to compute the percentile:

$$\text{Percentile} = 100 \cdot \frac{(\text{Number of values less than } x) + 0.5}{\text{Number of values in the data set}}$$

Round the result to the nearest whole number.

EXAMPLE 3.26

Computing the percentile corresponding to a given data value

In 1989, the rainfall in Los Angeles during the month of February was 1.90. What percentile does this correspond to?

Solution

Step 1: Table 3.10 presents the data in increasing order.

Step 2: There are 45 values in the data set. There are 17 values less than 1.90. Therefore,

$$\text{Percentile} = 100 \cdot \frac{17 + 0.5}{45} = 38.9$$

We round the result to 39. The value 1.90 corresponds to the 39th percentile.

Check Your Understanding

1. Following are final exam scores, arranged in increasing order, for 28 students in an introductory statistics course.

58	59	62	64	67	68	69	71	73	74	74	75	76	76
76	77	78	78	78	82	82	84	86	87	87	88	91	97

   a. Find the first quartile of the scores.
   b. Find the third quartile of the scores.
   c. Fred got a 73 on the exam. On what percentile is this score?
   d. Students whose scores are on the 80th percentile or above will get a grade of A. Louisa got an 88 on the exam. Will she get an A?

2. For the years 1869–2007, the 90th percentile of annual snowfall in Central Park in New York City was 50.1 inches. Approximately what percentage of years had snowfall less than 50.1 inches?

Answers are on page 145.

Objective 4 Compute the five-number summary for a data set

The Five-Number Summary

The five-number summary of a data set consists of the median, the first quartile, the third quartile, the minimum value, and the maximum value. These values are generally arranged in order.

DEFINITION

The **five-number summary** of a data set consists of the following quantities:

Minimum	First quartile	Median	Third quartile	Maximum

EXAMPLE 3.27   ## Constructing a five-number summary

Table 3.11 presents the number of students absent in a middle school in northwestern Montana for each school day in January of a recent year. Construct the five-number summary.

Table 3.11 Number of Absences

Date	Number Absent	Date	Number Absent	Date	Number Absent
Jan. 2	65	Jan. 14	59	Jan. 23	42
Jan. 3	67	Jan. 15	49	Jan. 24	45
Jan. 4	71	Jan. 16	42	Jan. 25	46
Jan. 7	57	Jan. 17	56	Jan. 28	100
Jan. 8	51	Jan. 18	45	Jan. 29	59
Jan. 9	49	Jan. 21	77	Jan. 30	53
Jan. 10	44	Jan. 22	44	Jan. 31	51
Jan. 11	41				

Solution

Step 1: We arrange the numbers in increasing order. The ordered numbers are:

41 42 42 44 44 45 45 46 49 49 51 51 53 56 57 59 59 65 67 71 77 100

Step 2: The minimum is 41 and the maximum is 100.

Step 3: We use the methods described in Example 3.23 to compute the first and third quartiles. The first quartile is $Q_1 = 45$, and the third quartile is $Q_3 = 59$.

Step 4: We use the method described in Section 3.1 to compute the median. The median is 51.

Step 5: The five-number summary is

$$41 \quad 45 \quad 51 \quad 59 \quad 100$$

Objective 5 Understand the effects of outliers

Outliers

An **outlier** is a value that is considerably larger or considerably smaller than most of the values in a data set. Some outliers result from errors; for example a misplaced decimal point may cause a number to be much larger or smaller than the other values in a data set. Some outliers are correct values, and simply reflect the fact that the population contains some extreme values.

When it is certain that an outlier resulted from an error, the value should be corrected or deleted. However, if it is possible that the value of an outlier is correct, it should remain in the data set. Deleting an outlier that is not an error will produce misleading results.

> **CAUTION**
>
> Do not delete an outlier unless it is certain that it is an error.

EXAMPLE 3.28

Determining whether an outlier should be deleted

The temperature in a downtown location in a certain city is measured for eight consecutive days during the summer. The readings, in degrees Fahrenheit, are 81.2, 85.6, 89.3, 91.0, 83.2, 8.45, 79.5, and 87.8. Which reading is an outlier? Is it certain that the outlier is an error, or is it possible that it is correct? Should the outlier be deleted?

Solution

The outlier is 8.45, which is much smaller than the rest of the data. This outlier is certainly an error; it is likely that a decimal point was misplaced. The outlier should be corrected if possible, or deleted.

EXAMPLE 3.29

Determining whether an outlier should be deleted

The following table presents the populations, as of July 2009, of the five largest cities in the United States.

City	Population in millions
New York	8.4
Los Angeles	3.9
Chicago	2.7
Houston	2.2
Philadelphia	1.5

Source: U.S. Census Bureau

Which value is an outlier? Is it certain that the outlier is an error, or is it possible that it is correct? Should the outlier be deleted?

Solution

The population of New York, 8.4 million, is an outlier because it is much larger than the other values. This outlier is not an error. It should not be deleted. If it were deleted, the data would indicate that the largest city in the United States is Los Angeles, which would be incorrect.

The interquartile range

The *interquartile range* (IQR for short) is a measure of spread that is often used to detect outliers. The IQR is the difference between the first and third quartiles.

> **DEFINITION**
>
> The **interquartile range** is found by subtracting the first quartile from the third quartile.
>
> $$\text{IQR} = Q_3 - Q_1$$

The IQR method for finding outliers

In Examples 3.28 and 3.29, we determined the outlier just by looking at the data and finding an extreme value. In many cases, this is a good way to find outliers. There are some formal methods for finding outliers as well. The most frequently used method is the **IQR method**.

> **The IQR Method for Finding Outliers**
>
> **Step 1:** Find the first quartile, Q_1, and the third quartile, Q_3, of the data set.
> **Step 2:** Compute the interquartile range.
>
> $$\text{IQR} = Q_3 - Q_1$$
>
> **Step 3:** Compute the **outlier boundaries**. These boundaries are the cutoff points for determining outliers.
>
> $$\text{Lower outlier boundary} = Q_1 - 1.5\,\text{IQR}$$
> $$\text{Upper outlier boundary} = Q_3 + 1.5\,\text{IQR}$$
>
> **Step 4:** Any data value that is less than the lower outlier boundary or greater than the upper outlier boundary is considered to be an outlier.

EXAMPLE 3.30   Detecting outliers

Use the IQR method to determine which values, if any, in the absence data in Table 3.11 are outliers.

Solution

Step 1: In Example 3.27, we computed the first and third quartiles: $Q_1 = 45$ and $Q_3 = 59$.
Step 2: $\text{IQR} = Q_3 - Q_1 = 59 - 45 = 14$
Step 3: The outlier boundaries are:

$$\text{Lower outlier boundary} = 45 - 1.5(14) = 24$$
$$\text{Upper outlier boundary} = 59 + 1.5(14) = 80$$

Step 4: There are no values in the data set less than the lower boundary of 24. There is one value, 100, that is greater than the upper boundary of 80. Thus there is one outlier, 100.

Check Your Understanding

3. Table 3.12 presents the 2012 payrolls (in millions of dollars) for each of the 30 Major League Baseball teams.

Table 3.12 2012 Payrolls for Major League Baseball Teams ($ millions)

Team	Payroll	Team	Payroll	Team	Payroll
New York Yankees	198	Chicago White Sox	97	Cleveland Indians	78
Philadelphia Phillies	175	Los Angeles Dodgers	95	Colorado Rockies	78
Boston Red Sox	173	Minnesota Twins	94	Toronto Blue Jays	75
Los Angeles Angels	154	New York Mets	93	Arizona Diamondbacks	74
Detroit Tigers	132	Chicago Cubs	88	Tampa Bay Rays	64
Texas Rangers	121	Atlanta Braves	83	Pittsburgh Pirates	63
Miami Marlins	118	Cincinnati Reds	82	Kansas City Royals	61
San Francisco Giants	118	Seattle Mariners	82	Houston Astros	61
St. Louis Cardinals	110	Baltimore Orioles	81	Oakland Athletics	55
Milwaukee Brewers	98	Washington Nationals	81	San Diego Padres	55

Source: *USA Today*

    a. Construct the five-number summary.
    b. Find the IQR.
    c. Find the upper and lower outlier boundaries.
    d. Which values, if any, are outliers?

Answers are on page 145.

Objective 6 Construct boxplots to visualize the five-number summary and outliers

Boxplots

A **boxplot** is a graph that presents the five-number summary along with some additional information about a data set. There are several kinds of boxplots. The one we describe here is sometimes called a **modified boxplot**.

Explain It Again

Another name for boxplots: Boxplots are sometimes called *box-and-whisker diagrams.*

Procedure for Constructing a Boxplot

Step 1: Compute the first quartile, the median, and the third quartile.

Step 2: Draw vertical lines at the first quartile, the median, and the third quartile. Draw horizontal lines between the first and third quartiles to complete the box.

Step 3: Compute the lower and upper outlier boundaries.

Step 4: Find the largest data value that is less than the upper outlier boundary. Draw a horizontal line from the third quartile to this value. This horizontal line is called a **whisker**.

Step 5: Find the smallest data value that is greater than the lower outlier boundary. Draw a horizontal line (whisker) from the first quartile to this value.

Step 6: Determine which values, if any, are outliers. Plot each outlier separately.

EXAMPLE 3.31 Constructing a boxplot

Construct a boxplot for the absence data in Table 3.11.

Solution

Step 1: In Example 3.27, we computed the median to be 51 and the first and third quartiles to be $Q_1 = 45$ and $Q_3 = 59$.

Step 2: We draw vertical lines at 45, 51, and 59, then horizontal lines to complete the box, as follows:

Number of Absences

Step 3: We compute the outlier boundaries as shown in Example 3.30:

$$\text{Lower outlier boundary} = 45 - 1.5(14) = 24$$

$$\text{Upper outlier boundary} = 59 + 1.5(14) = 80$$

Step 4: The largest data value that is less than the upper boundary is 77. We draw a horizontal line from 59 up to 77, as follows:

Step 5: The smallest data value that is greater than the lower boundary is 41. We draw a horizontal line from 45 down to 41, as follows:

Step 6: We determine, as shown in Example 3.30, that the value 100 is the only outlier. We plot this point separately, to produce the boxplot shown in Figure 3.12.

Figure 3.12 Boxplot for the absence data in Table 3.11

Check Your Understanding

4. Construct a boxplot for the payroll data in Table 3.12.

Answer is on page 145.

Determining the shape of a data set from a boxplot

In Section 2.2, we learned how to determine from a histogram whether a data set is symmetric or skewed. In many cases, a boxplot can give us the same information. For example, in the boxplot for the absence data (Figure 3.12), the median is closer to the first quartile than to the third quartile, and the upper whisker is longer than the lower one. This indicates that the data are skewed to the right.

Figure 3.13 presents a histogram of the absence data. The skewness is clearly apparent.

Figure 3.13 Histogram for the absence data in Table 3.11

Determining Skewness from a Boxplot

- If the median is closer to the first quartile than to the third quartile, or the upper whisker is longer than the lower whisker, the data are skewed to the right.
- If the median is closer to the third quartile than to the first quartile, or the lower whisker is longer than the upper whisker, the data are skewed to the left.
- If the median is approximately halfway between the first and third quartiles, and the two whiskers are approximately equal in length, the data are approximately symmetric.

Figures 3.14–3.16 illustrate the way in which boxplots reflect skewness and symmetry.

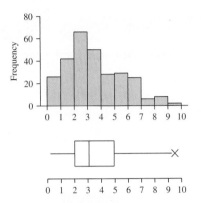

Figure 3.14 Skewed to the right

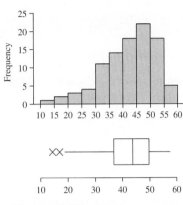

Figure 3.15 Skewed to the left

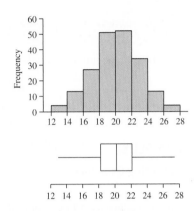

Figure 3.16 Approximately symmetric

Comparative boxplots

Boxplots do not provide as much detail as histograms do regarding the shape of a data set. However, they provide an excellent method for comparing data sets. We can plot two or more boxplots, one above another, to provide an easy visual comparison.

As an example, Table 3.13 presents the rainfall in Los Angeles each February for the years 1924–1968. We would like to compare these data with the February rainfall for the years 1969–2013, which was presented in Table 3.9.

Table 3.13 Annual Rainfall in Los Angeles During February: 1924–1968

Year	Rainfall	Year	Rainfall	Year	Rainfall	Year	Rainfall	Year	Rainfall
1924	0.03	1933	0.00	1942	1.05	1951	1.48	1960	2.26
1925	0.53	1934	2.04	1943	3.07	1952	0.63	1961	0.15
1926	2.70	1935	2.23	1944	8.65	1953	0.33	1962	11.57
1927	9.03	1936	7.25	1945	3.34	1954	2.98	1963	2.88
1928	1.43	1937	7.87	1946	1.52	1955	0.68	1964	0.00
1929	2.15	1938	9.81	1947	0.86	1956	0.59	1965	0.23
1930	0.45	1939	1.13	1948	1.29	1957	1.47	1966	1.51
1931	3.25	1940	5.43	1949	1.41	1958	6.46	1967	0.11
1932	5.33	1941	12.42	1950	1.67	1959	3.32	1968	0.49

Figure 3.17 (page 140) presents **comparative boxplots** for February rainfall during the years 1924–1968 and 1969–2013.

We can see that the boxplot for 1969–2013 extends farther to the right than the boxplot for 1924–1968. This tells us that 1969–2013 was, on the whole, rainier than 1924–1968. We can also see the boxplot for 1969–2013 is longer than the one for 1924–1968. This tells us that the rainfall was more variable during 1969–2013. Finally, the upper whisker is longer than the lower whisker in both boxplots. This tells us that both rainfall data sets are skewed to the right. Finally, we note that there were more outliers in 1924–1968 than in 1969–2013.

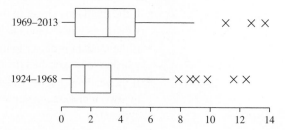

Figure 3.17 Comparative boxplots for February rainfall in Los Angeles, 1924–1968 and 1969–2013

USING TECHNOLOGY

Table 3.14 lists the number of calories in 11 fast-food restaurant hamburgers. We use these data to illustrate the technology steps.

Table 3.14 Number of Calories in Fast-Food Hamburgers

840	1090	680	950	1070	860
940	1285	900	1120	720	

TI-84 PLUS

Computing Quartiles

The procedure used to compute the mean and median, presented on page 100, will also compute the quartiles. The quartiles for the data in Table 3.14 are shown in Figure A.

Figure A

Drawing Boxplots

Step 1. Enter the data from Table 3.14 into **L1** in the data editor.

Step 2. Press **2nd, Y=** to access the STAT PLOTS menu and select **Plot1** by pressing **1**.

Step 3. Select **On** and the boxplot icon in the lower left. Press **ENTER** (Figure B).

Step 4. Press **ZOOM** and then **9: ZoomStat** (Figure C).

Figure B

Figure C

MINITAB

Computing Quartiles

The MINITAB procedure used to compute the mean and median, described on page 100, will also compute the quartiles. The quartiles for the data in Table 3.14 are shown in Figure D.

Descriptive Statistics: C1

Variable	Mean	SE Mean	StDev	Minimum	Q1	Median	Q3	Maximum
C1	950.5	54.2	179.9	680.0	840.0	940.0	1090.0	1285.0

Figure D

Drawing Boxplots

Step 1. Enter the data from Table 3.14 into **Column C1**.

Step 2. Click on **Graph** and select **Boxplot**. Choose the **One Y, Simple option** and press **OK**.

Step 3. Double-click on **C1** and press **OK**.

EXCEL

Computing Quartiles

Step 1. Enter the data from Table 3.14 in Column **A**.

Step 2. Select the **Insert Function** icon f_x and highlight **Statistical** in the category field.

Step 3. Highlight **QUARTILE.EXC** and press **OK**. Enter the range of cells that contain the data from Table 3.14 in the **Array** field. In the **Quart** field, enter **1** for Q_1, **2** for Q_2, or **3** for Q_3.

Step 4. Click **OK**.

SECTION 3.3   Exercises

Exercises 1–4 are the Check Your Understanding exercises located within the section.

Understanding the Concepts

In Exercises 5–8, fill in each blank with the appropriate word or phrase.

5. _____ divide the data set approximately into quarters.

6. The median is the same as the _____ quartile.

7. The quantity $Q_3 - Q_1$ is known as the _____.

8. A value that is considerably larger or smaller than most of the values in a data set is called an _____.

In Exercises 9–12, determine whether the statement is true or false. If the statement is false, rewrite it as a true statement.

9. The third quartile, Q_3, separates the lowest 25% of the data from the highest 75%.

10. The 25th percentile is the same as the first quartile.

11. The five-number summary consists of the minimum, the first quartile, the mode, the third quartile, and the maximum.

12. In a boxplot, if the lower whisker is much longer than the upper whisker, then the data are skewed to the left.

Practicing the Skills

13. A population has mean $\mu = 7$ and standard deviation $\sigma = 2$.
   a. Find the z-score for a population value of 5.
   b. Find the z-score for a population value of 10.
   c. What number has a z-score of 2?

14. A population has mean $\mu = 25$ and standard deviation $\sigma = 4$.
   a. Find the z-score for a population value of 16.
   b. Find the z-score for a population value of 31.
   c. What number has a z-score of 2.5?

In Exercises 15 and 16, identify the outlier. Then tell whether the outlier seems certain to be due to an error, or whether it could conceivably be correct.

15. A rock is weighed five times. The readings in grams are 48.5, 47.2, 4.91, 49.5, and 46.3.

16. A sociologist samples five families in a certain town and records their annual income. The incomes are $34,000, $57,000, $13,000, $1,200,000, and $62,000.

17. For the data set

| 37 | 82 | 20 | 25 | 31 | 10 | 41 | 44 | 4 | 36 | 68 |

   a. Find the first and third quartiles.
   b. Find the IQR.
   c. Find the upper and lower outlier boundaries.
   d. List all the values, if any, that are classified as outliers.

18. For the data set

| 15 | 7 | 2 | 4 | 4 | 3 | 4 | 3 | 4 | 25 | 4 | 9 | 3 |
| 12 | 2 | 8 | 3 | 2 | 2 | 6 | 7 | 3 | 10 | 4 | 5 | 4 |

   a. Find the first and third quartiles.
   b. Find the IQR.
   c. Find the upper and lower outlier boundaries.
   d. List all the values, if any, that are classified as outliers.

19. For the data set

2	2	2	2	5	7	8	8	9	9	14	14
14	16	19	20	21	22	22	24	24	27	32	33
33	33	34	34	35	35	35	37	38	38	38	40
40	40	41	42	46	47	48	48	48	48	48	49

   a. Find the 58th percentile.
   b. Find the 22nd percentile.
   c. Find the 78th percentile.
   d. Find the 15th percentile.

20. For the data set

1	5	8	8	8	11	13	14	15	15	16	17
20	23	24	25	25	26	26	29	31	34	35	35
38	44	45	47	47	51	53	53	54	55	55	57
57	59	60	62	65	69	70	75	75	76	78	79
81	83	83	84	89	91	92	93	93	96	96	99

a. Find the 80th percentile.
b. Find the 43rd percentile.
c. Find the 18th percentile.
d. Find the 65th percentile.

Working with the Concepts

21. Standardized tests: In 2012, the mean score on the ACT test was 21.1 and the standard deviation was 5.2. The mean score on the SAT mathematics test was 514 and the standard deviation was 117. The distributions of both scores were approximately bell-shaped.
  a. Find the z-score for an ACT score of 27.
  b. Find the z-score for an SAT math score of 650.
  c. Which score is higher, relative to its population of scores?
  d. Jose's ACT score had a z-score of 0.75. What was his ACT score?
  e. Emma's SAT score had a z-score of -2.0. What was her SAT score?

22. A fish story: The mean length of one-year-old spotted flounder, in millimeters, is 126 with standard deviation of 18, and the mean length of two-year-old spotted flounder is 162 with a standard deviation of 28. The distribution of flounder lengths is approximately bell-shaped.
  a. Anna caught a one-year-old flounder that was 150 millimeters in length. What is the z-score for this length?
  b. Luis caught a two-year-old flounder that was 190 millimeters in length. What is the z-score for this length?
  c. Whose fish is longer, relative to fish the same age?
  d. Joe caught a one-year-old flounder whose length had a z-score of 1.2. How long was this fish?
  e. Terry caught a two-year-old flounder whose length had a z-score of -0.5. How long was this fish?

  Source: *Turkish Journal of Veterinary and Animal Science*, 29:1013–1018

23. Blood pressure in men: The three quartiles for systolic blood pressure in a sample of 3179 men were $Q_1 = 108$, $Q_2 = 116$, and $Q_3 = 127$.
  a. Find the IQR.
  b. Find the upper and lower outlier boundaries.
  c. A systolic blood pressure greater than 140 is considered high. Would a blood pressure of 140 be an outlier?

  Source: *Journal of Human Hypertension*, 16:305–312

24. Blood pressure in women: The article referred to in Exercise 23 reported that the three quartiles for systolic blood pressure in a sample of 1213 women between the ages of 20 and 29 were $Q_1 = 100$, $Q_2 = 108$, and $Q_3 = 115$.
  a. Find the IQR.
  b. Find the upper and lower outlier boundaries.
  c. A systolic blood pressure greater than 140 is considered high. Would a blood pressure of 140 be an outlier?

25. Hazardous waste: Following is a list of the number of hazardous waste sites in each of the 50 states of the United States in a recent year. The list has been sorted into numerical order.

0	1	2	2	3	5	6	9	9	9
9	9	11	12	12	12	12	12	13	13
14	14	14	14	15	15	15	16	19	19
20	21	25	26	29	30	32	32	32	38
40	48	49	49	52	67	86	97	97	116

Source: U.S. Environmental Protection Agency

a. Find the first and third quartiles of these data.
b. Find the median of these data.
c. Find the upper and lower outlier boundaries.
d. Are there any outliers? If so, list them.
e. Construct a boxplot for these data.
f. Describe the shape of this distribution.
g. What is the 30th percentile?
h. What is the 85th percentile?
i. The state of Georgia has 16 hazardous waste sites. What percentile is this?

26. Cholesterol levels: The National Health and Nutrition Examination Survey (NHANES) measured the serum HDL cholesterol levels in a large number of women. Following is a sample of 40 HDL levels (in milligrams per deciliter) that are based on the results of that survey. They have been sorted into numerical order.

27	28	30	32	34	36	37	37	37	37
37	40	45	47	48	49	53	53	54	56
57	58	61	62	63	63	64	64	64	65
66	70	72	73	73	74	80	80	81	84

Source: NHANES

a. Find the first and third quartiles of these data.
b. Find the median of these data.
c. Find the upper and lower outlier boundaries.
d. Are there any outliers? If so, list them.
e. Construct a boxplot for these data.
f. Describe the shape of this distribution.
g. What is the 20th percentile?
h. What is the 67th percentile?
i. One woman had a cholesterol level of 58. What percentile is this?

27. Commuting to work: Jamie drives to work every weekday morning. She keeps track of the time it takes, in minutes, for 35 days. The results follow.

15	17	17	17	17	18	19
19	19	19	19	19	20	20
20	20	20	21	21	21	21
21	21	21	21	21	22	23
23	24	26	31	36	38	39

a. Find the first and third quartiles of these data.
b. Find the median of these data.
c. Find the upper and lower outlier boundaries.
d. Are there any outliers? If so, list them.
e. Construct a boxplot for these data.
f. Describe the shape of this distribution.
g. What is the 14th percentile?
h. What is the 87th percentile?
i. One day, the commute time was 31 minutes. What percentile is this?

28. Windy city by the bay: Following are wind speeds (in mph) for 29 randomly selected days in San Francisco.

13.4	23.1	27.2	31.8	36.3	40.3	14.1	24.6
27.7	32.6	38.1	40.9	18.3	25.2	29.7	33.6
38.7	44.2	20.8	26.8	30.1	34.5	40.2	46.8
22.9	26.9	30.8	35.8	40.3			

a. Find the first and third quartiles of these data.
b. Find the median of these data.
c. Find the upper and lower outlier boundaries.
d. Are there any outliers? If so, list them.
e. Construct a boxplot for these data.
f. Describe the shape of this distribution.
g. What is the 40th percentile?
h. What is the 10th percentile?
i. One day, the wind speed was 30.1 mph. What percentile is this?

29. Caffeine: Following are the number of grams of carbohydrates in 12-ounce espresso beverages offered at Starbucks.

14	43	38	44	31	27	39	59	9	10	54
14	25	26	9	46	30	24	41	26	27	14

Source: www.starbucks.com

a. Find the first and third quartiles of these data.
b. Find the median of these data.
c. Find the upper and lower outlier boundaries.
d. The beverage with the most carbohydrates is a Peppermint White Chocolate Mocha, with 59 grams. Is this an outlier?
e. The beverages with the least carbohydrates are an Iced Skinny Flavored Latte, and a Cappuccino, each with 9 grams. Are these outliers?
f. Construct a boxplot for these data.
g. Describe the shape of this distribution.
h. What is the 31st percentile?
i. What is the 71st percentile?
j. There are 38 grams of carbohydrates in an Iced Dark Cherry Mocha. What percentile is this?

30. Nuclear power: The following table presents the number of nuclear reactors as of August 2, 2013, in each country that had one or more reactors.

Country	Number of Reactors	Country	Number of Reactors
Argentina	2	South Korea	23
Armenia	1	Mexico	2
Belgium	7	Netherlands	1
Brazil	2	Pakistan	3
Bulgaria	2	Romania	2
Canada	19	Russia	33
China	18	Slovakia	4
Czech Republic	6	Slovenia	1
Finland	4	South Africa	2
France	58	Spain	8
Germany	9	Sweden	10
Hungary	4	Switzerland	5
India	20	Ukraine	15
Iran	1	United Kingdom	16
Japan	50	United States	100

Source: International Atomic Energy Agency

a. Find the first and third quartiles of these data.
b. Find the median of these data.
c. Find the upper and lower outlier boundaries.
d. Which countries are outliers?
e. Construct a boxplot for these data.
f. Describe the shape of this distribution.
g. What is the 45th percentile?
h. What is the 88th percentile?
i. India has 20 nuclear reactors. What percentile is this?

31. Place your bets: As of July, 2013, 28 states in the U.S. had one or more tribal gambling casinos. The following table presents the number of tribal casinos in each of those states.

3	2	26	70	2	2	8
7	3	4	3	22	39	3
14	7	3	21	8	2	11
114	8	14	1	34	31	4

Source: American Gaming Association

a. Find the first and third quartiles of these data.
b. Find the median of these data.
c. Find the upper and lower outlier boundaries.
d. Which values, if any, are outliers?
e. Construct a boxplot for these data.
f. Describe the shape of this distribution.
g. What is the 40th percentile?
h. What is the 83rd percentile?
i. North Dakota has 11 tribal casinos. What percentile is this?

32. Hail to the chief: There have been 57 presidential inaugurations in U.S. history. At each one, the president has made an inaugural address. Following are the number of words spoken in each of these addresses.

1431	135	2321	1730	2166	1177	1211	3375
4472	2915	1128	1176	3843	8460	4809	1090
3336	2831	3637	700	1127	1339	2486	2979
1686	4392	2015	3968	2218	984	5434	1704
1526	3329	4055	3672	1880	1808	1359	559
2273	2459	1658	1366	1507	2128	1803	1229
2427	2561	2320	1598	2155	1592	2071	2395
2096							

Source: The American Presidency Project

a. Find the first and third quartiles of these data.
b. Find the median of these data.
c. Find the upper and lower outlier boundaries.
d. The two shortest speeches were 135 words, by George Washington in 1793, and 559 words, by Franklin Roosevelt in 1945. Are either of these outliers?
e. The two longest speeches were 8460 words, by William Henry Harrison in 1841, and 5434 words, by William Howard Taft in 1909. Are either of these outliers?
f. Construct a boxplot for these data.
g. Describe the shape of this distribution.
h. What is the 15th percentile?

i. What is the 65th percentile?

j. Barack Obama used 2395 words in his inauguration speech in 2009. What percentile is this?

33. Bragging rights: After learning his score on a recent statistics exam, Ed bragged to his friends: "My score is the first quartile of the class." Did Ed have a good reason to brag? Explain.

34. Who scored the highest? On a final exam in a large statistics class, Tom's score was the tenth percentile, Dick's was the median, and Harry's was the third quartile. Which of the three scores was the highest? Which was the lowest?

35. Baseball salaries: In 2012, the San Francisco Giants defeated the Detroit Tigers to become the champions of Major League Baseball. Following are the salaries, in millions of dollars, of the players on each of these teams.

Giants							
19.00	18.25	16.17	10.00	8.50	6.00	6.00	5.00
4.85	4.25	3.20	3.00	2.20	1.58	1.30	1.25
1.00	0.75	0.63	0.62	0.56	0.48	0.48	0.48
0.48	0.48	0.48	0.48	0.48			

Tigers							
23.00	21.00	20.10	13.00	9.00	6.73	5.50	5.50
5.50	3.75	3.10	3.00	2.10	2.10	1.10	1.00
0.90	0.51	0.51	0.50	0.50	0.50	0.49	0.49
0.49	0.48	0.48	0.48	0.48			

a. Find the median, the first quartile, and the third quartile of the Giants' salaries.

b. Find the median, the first quartile, and the third quartile of the Tigers' salaries.

c. Find the upper and lower outlier bounds for the Giants' salaries.

d. Find the upper and lower outlier bounds for the Tigers' salaries.

e. Construct comparative boxplots for the two data sets. What conclusions can you draw?

36. Automotive emissions: Following are levels of particulate emissions for 65 vehicles driven at sea level, and for 35 vehicles driven at high altitude.

Sea Level									
1.5	0.9	1.1	1.3	3.5	1.1	1.1	0.9	1.3	0.9
0.6	1.3	2.5	1.5	1.1	1.1	2.2	0.9	1.8	1.5
1.2	1.6	2.1	6.6	4.0	2.5	1.4	1.4	1.8	1.1
1.6	3.7	0.6	2.7	2.6	3.0	1.2	1.0	1.6	3.1
2.4	2.1	2.7	1.2	3.3	3.8	1.3	2.1	6.6	1.2
3.1	0.5	0.3	0.5	3.4	3.5	2.7	1.9	5.9	4.2
3.5	3.6	3.1	3.3	4.6					

High Altitude						
8.9	4.4	3.6	4.4	3.8	2.4	3.8
5.3	5.8	2.9	4.7	1.9	9.1	8.7
9.5	2.7	9.2	7.3	2.1	6.3	6.5
6.3	2.0	5.9	5.6	5.6	1.5	6.5
5.3	5.6	2.1	1.1	3.3	1.8	7.6

a. Find the median, the first quartile, and the third quartile of the sea-level emissions.

b. Find the median, the first quartile, and the third quartile of the high-altitude emissions.

c. Find the upper and lower outlier bounds for the sea-level emissions.

d. Find the upper and lower outlier bounds for the high-altitude emissions.

e. Construct comparative boxplots for the two data sets. What conclusions can you draw?

37. Comparative boxplots: Following are boxplots of the level of fine particle air pollution, in micrograms per cubic meter, in the cities of Denver and Greeley, Colorado, during a recent winter.

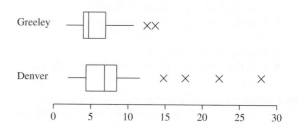

a. In which city is the pollution level generally higher?

b. Approximately what percentage of the values for Greeley are greater than the median value for Denver? Is it closest to 25%, 50%, 75%, or 90%?

c. In which city is there more spread in the pollution levels?

d. Are the pollution levels for Greeley skewed right, skewed left, or approximately symmetric?

e. Are the pollution levels for Denver skewed right, skewed left, or approximately symmetric?

38. Comparative boxplots: Following are boxplots of diastolic blood pressure, in millimeters, for a sample of people aged 18–39 years, and a sample over 60 years old. The data are consistent with results reported by the National Health Statistics Reports.

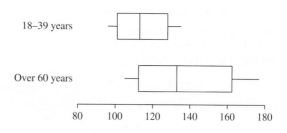

a. Which group generally has higher blood pressure?

b. Approximately what percentage of people over 60 have blood pressure higher than the median for people aged 18–39? Is it closest to 25%, 50%, 75%, or 90%?

c. For which group are the blood pressures more spread out?

d. Are the blood pressures for people 18–39 skewed right, skewed left, or approximately symmetric?

e. Are the blood pressures for people over 60 skewed right, skewed left, or approximately symmetric?

39. Boxplot possible? The most recent United States Census reported the per capita incomes for each of the 50 states. The five-number summary, in thousands of dollars, is
Minimum: 20.0, Q_1: 24.1, Median: 25.7, Q_3: 29.0, Maximum: 34.8

  a. Does the five-number summary provide enough information to construct a boxplot? If so, construct the boxplot. If not, explain why not.

  b. Are these data skewed to the right, skewed to the left, or approximately symmetric?

40. Boxplot possible? Following is the five-number summary for the populations, in millions, for the 50 states of the United States.
Minimum: 0.58, Q_1: 1.86, Median: 4.49, Q_3: 6.90, Maximum: 38.04

  a. Does the five-number summary provide enough information to construct a boxplot? If so, construct the boxplot. If not, explain why not.

  b. Are these data skewed to the right, skewed to the left, or approximately symmetric?

41. Unusual boxplot: Ten residents of a town were asked how many children they had. The responses were as follows.

      0   0   3   1   0   0   4   0   7   0

  a. Explain why the median and first quartile are the same.

  b. Construct a boxplot for these data.

  c. Explain why the boxplot has no left whisker.

42. Clean-shaven boxplot: Make up a list of numbers whose boxplot has no whiskers.

Extending the Concepts

43. The vanishing outlier: Seven families live on a small street in a certain town. Their annual incomes (in $1000s) are 15, 20, 30, 35, 50, 60, and 150.

  a. Find the first and third quartiles, and the IQR.

b. Show that 150 is an outlier.
A big new house is built on the street, and the income (in $1000s) of the family that moves in is 200.

  c. Find the first and third quartiles, and the IQR of the eight incomes.

  d. Are there any outliers now?

  e. Explain how adding the value 200 to the data set eliminated the outliers.

44. Beyond quartiles and percentiles: If we divide a data set into four approximately equal parts, the three dividing points are called quartiles. If we divide a data set into 100 approximately equal parts, the 99 dividing points are called percentiles. In general, if we divide a data set into k approximately equal parts, we can call the dividing points k-tiles. How would you find the ith k-tile of a data set of size n?

45. z-scores and skewed data: Table 3.9 presents the February rainfalls in Los Angeles for the period 1969–2013.

  a. Show that the mean of these data is $\mu = 3.759$ and the standard deviation is $\sigma = 3.489$.

  b. Show that the z-score for a rainfall of 0 (rounded to two decimal places) is $z = -1.08$.

  c. Show that the z-score for a rainfall of 7.52 (rounded to two decimal places) is $z = 1.08$.

  d. What percentage of the years had rainfalls of 0?

  e. What percentage of the years had rainfalls of 7.52 or more?

  f. The z-scores indicate that a rainfall of 0 and a rainfall of 7.52 are about equally extreme. Is a rainfall of 7.52 really as extreme as a rainfall of 0, or is it less extreme?

  g. These data are skewed to the right. Explain how skewness causes the z-score to give misleading results.

Answers to Check Your Understanding Exercises for Section 3.3

1. a. 70   **b.** 83   **c.** 30th   **d.** Yes

2. 90%

3. a. 55   75   85.5   118   198   **b.** 43

  c. Lower outlier bound is −10.5; upper bound is 182.5.

  d. 198 is the only outlier.

4.

Chapter 3 Summary

Section 3.1: We can describe the center of a data set with the mean or the median. When a data set is skewed to the left, the mean is generally less than the median, and when a data set is skewed to the right, the mean is generally greater than the median. The mode of a data set is the most frequently occurring value.

Section 3.2: The spread of a data set is most often measured with the standard deviation. For data sets that are unimodal and approximately symmetric, the Empirical Rule can be used to approximate the proportion of the data that lies within a given number of standard deviations of the mean. Chebyshev's Inequality, which is valid for all data sets, provides a lower bound for the proportion of the data that lies within a given number of standard deviations of the mean. The coefficient of variation (CV) measures the spread of a data set relative to its mean. The CV provides a way to compare spreads of data sets whose values are in different units.

Section 3.3: For bell-shaped data sets, the z-score gives a good description of the position of a value in a data set. Quartiles and percentiles can be used to describe the positions for any data set. Quartiles are used to compute the five-number summary, which consists of the minimum value, the first quartile, the median, the third quartile, and the maximum value. Outliers are values that are considerably larger or smaller than most of the values in a data set. Boxplots are graphs that allow us to visualize the five-number summary, along with any outliers. Comparative boxplots allow us to visually compare the shapes of two or more data sets.

Vocabulary and Notation

$\sum x = x_1 + \cdots + x_n$ 93
arithmetic mean   92
boxplot   137
Chebyshev's Inequality   119
coefficient of variation (CV)   121
comparative boxplots   139
degrees of freedom   114
deviation   111
Empirical Rule   117
first quartile Q_1 130
five-number summary   134
interquartile range (IQR)   136
IQR method   136

mean   92
mean absolute deviation (MAD)   128
measure of center   92
measure of position   92
measure of spread   92
median   94
mode   98
modified boxplot   137
outlier   135
outlier boundaries   136
percentile   132
population mean μ 93
population standard deviation σ 114

population variance σ^2 111
quartile   130
range   110
resistant   96
sample mean $\bar{x}$ 93
sample standard deviation s 114
sample variance s^2 112
second quartile Q_2 130
standard deviation   114
third quartile Q_3 130
variance   111
whisker   137
z-score   129

Important Formulas

Sample mean:
$$\bar{x} = \frac{\sum x}{n}$$

Population mean:
$$\mu = \frac{\sum x}{N}$$

Range:
Range = largest value − smallest value

Population variance:
$$\sigma^2 = \frac{\sum(x-\mu)^2}{N}$$

Sample variance:
$$s^2 = \frac{\sum(x-\bar{x})^2}{n-1}$$

Coefficient of variation:
$$CV = \frac{\sigma}{\mu}$$

z-score:
$$z = \frac{x-\mu}{\sigma}$$

Interquartile range:
$$IQR = Q_3 - Q_1 = \text{third quartile} - \text{first quartile}$$

Lower outlier boundary:
$$Q_1 - 1.5\,IQR$$

Upper outlier boundary:
$$Q_3 + 1.5\,IQR$$

Chapter Quiz

1. Of the mean, median, and mode, which must be a value that actually appears in the data set?
2. The prices (in dollars) for a sample of personal computers are: 550, 700, 420, 580, 550, 450, 690, 390, 350. Calculate the mean, median, and mode for this sample.
3. If a computer with a price of $2000 were added to the list in Exercise 2, which would be affected more, the mean or the median?
4. In general, a histogram is skewed to the left if the _____ is noticeably less than the _____.
5. A sample of 100 students was asked how many hours per week they spent studying. The following frequency table shows the results:

Number of Hours	Frequency
1.0–4.9	14
5.0–8.9	34
9.0–12.9	29
13.0–16.9	15
17.0–20.9	8

   a. Approximate the mean time this sample of students spent studying.
   b. Approximate the standard deviation of the time this sample of students spent studying.
6. A sample has a variance of 16. What is the standard deviation?

7. Each of the following histograms represents a data set with mean 20. One has a standard deviation of 3.96 and the other has a standard deviation of 2.28. Which is which? Fill in the blanks: Histogram I has a standard deviation of _____ and histogram II has a standard deviation of _____.

 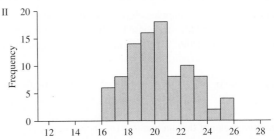

In Exercises 8–11, suppose that the mean starting salary of social workers in a specific region is $37,480 with a standard deviation of $1,400.

8. Assume that the histogram of starting salaries is approximately bell-shaped. Approximately what percentage of the salaries will be between $34,680 and $40,280?

9. Assume it is not known whether the histogram of starting salaries is bell-shaped. Fill in the blank: At least _____ percent of the salaries will be between $34,680 and $40,280.

10. John's starting salary is $38,180. What is the z-score of his salary?

11. Find the coefficient of variation of the salaries.

12. True or false: If a student's exam grade is on the 55th percentile, then approximately 45% of the scores are below his or her grade.

13. The five-number summary for a sample is 7, 18, 35, 62, 85. What is the IQR?

14. The prices (in dollars) for a sample of coffee makers are:

    19   22   29   68   35   37   28   22   41   39   28

    a. Find the first and third quartiles.
    b. Find the upper and lower outlier boundaries.
    c. Are there any outliers? Of so, list them.

15. Construct a boxplot for the data in Exercise 14.

Review Exercises

1. **Support your local artist:** Following are the annual amounts of federal support (in millions of dollars) for National Endowment for the Arts programs for the years 2006 through 2012.

Year	Amount
2006	124.4
2007	124.6
2008	144.7
2009	155.0
2010	167.5
2011	154.7
2012	146.0

   Source: National Endowment for the Arts

   a. Find the mean annual amount of federal aid from 2006 through 2012.
   b. Find the median annual amount of federal aid from 2006 through 2012.

2. **Corporate profits:** The following table presents the profit, in billions of dollars, for the year 2012 for each of the 15 largest U.S. corporations in terms of revenue.

Corporation	Profit	Corporation	Profit
Exxon Mobil	41.1	Ford Motor	20.2
Walmart	15.7	Hewlett-Packard	7.1
Chevron	26.9	AT&T	3.9
ConocoPhillips	12.4	Valero Energy	2.1
General Motors	9.2	Bank of America Corp.	1.5
General Electric	14.2	McKesson	1.2
Berkshire Hathaway	10.2	Verizon Communications	2.4
Fannie Mae	−16.9		

   Source: CNNMoney

   a. Find the mean profit.
   b. Find the median profit.
   c. Are these data skewed to the right, skewed to the left, or approximately symmetric? Explain.

3. **Computer chips:** A computer chip is a wafer made of silicon that contains complex electronic circuitry made up of microscopic components. The wafers are coated with a very thin coating of silicon dioxide. It is important that the coating be of uniform thickness over the wafer. To check this, engineers measured the thickness of the coating, in millionths of a meter, for samples of wafers made with two different processes.

Process 1:	90.0	92.2	94.9	92.7	91.6	88.2	92.0	98.2	96.0
Process 2:	76.1	90.2	96.8	84.6	93.3	95.7	90.9	100.3	95.2

   a. Find the mean of the thicknesses for each process.
   b. Find the median of the thicknesses for each process.
   c. If it is desired to obtain as thin a coating as possible, is one process much better than the other? Or are they about the same?

4. **More computer chips:** Using the data in Exercise 3:
   a. Find the sample variance of the thicknesses for each process.
   b. Find the sample standard deviation of the thicknesses for each process.
   c. Which process appears to be better in producing a uniform thickness? Explain.

5. **Stock prices:** Following are the closing prices of Microsoft stock for each trading day in May and June 2013.

May						June				
34.54	34.62	34.35	33.67	33.72		34.90	35.03	34.88	35.02	34.27
33.27	33.49	34.59	34.98	35.00		34.15	34.61	34.85	35.08	34.87
34.40	34.72	35.00	34.84	35.47		34.08	33.85	33.53	33.03	32.69
35.67	34.96	34.78	34.99	35.59		32.66	32.99	33.31	33.75	33.49
33.16	32.72									

   a. Find the mean and median price in May.
   b. Find the mean and median price in June.
   c. Does there appear to be a substantial difference in price between May and June? Or are the prices about the same?

6. **More stock prices:** Using the data in Exercise 5:
   a. Find the population standard deviation of the prices in May.
   b. Find the population standard deviation of the prices in June.
   c. Financial analysts use the word *volatility* to refer to the variation in stock prices. Was the volatility for the price of Microsoft stock greater in May or June?

7. **Measure that ball:** Each of 16 students measured the circumference of a tennis ball by two different methods:
   A: Estimate the circumference by eye.
   B: Measure the circumference by rolling the ball along a ruler.
   The results (in centimeters) are given below, in increasing order for each method:

A:	18.0	18.0	18.0	20.0	22.0	22.0	22.5	23.0	24.0	24.0	25.0	25.0	25.0	25.0	26.0	26.4
B:	20.0	20.0	20.0	20.0	20.2	20.5	20.5	20.7	20.7	20.7	21.0	21.1	21.5	21.6	22.1	22.3

   a. Compute the sample standard deviation of the measurements for each method.
   b. For which method is the sample standard deviation larger? Why should one expect this method to have the larger standard deviation?
   c. Other things being equal, is it better for a measurement method to have a smaller standard deviation or a larger standard deviation? Or doesn't it matter? Explain.

8. **Time in surgery:** Records at a hospital show that a certain surgical procedure takes an average of 162.8 minutes with a standard deviation of 4.9 minutes. If the data are approximately bell-shaped, between what two values will about 95% of the data fall?

9. **Rivets:** A machine makes rivets that are used in the manufacture of airplanes. To be acceptable, the length of a rivet must be between 0.9 centimeter and 1.1 centimeters. The mean length of a rivet is 1.0 centimeter, with a standard deviation of 0.05 centimeter. What is the maximum possible percentage of rivets that are unacceptable?

10. **How long can you talk?** A manufacturer of cell phone batteries determines that the average length of talk time for one of its batteries is 470 minutes. Suppose that the standard deviation is known to be 32 minutes and that the data are approximately bell-shaped. Estimate the percentage of batteries that have z-scores between -1 and 1.

11. **Paying rent:** The monthly rents for apartments in a certain town have a mean of $800 with a standard deviation of $150. What can you determine about these data by using Chebyshev's Inequality with $K = 3$?

12. Advertising costs: The amounts spent (in billions) on media advertising in the United States for a sample of categories in 2012 are presented in the following table.

Advertising Category	Amount Spent
Retail	16.35
Automotive	14.84
Local Services	8.98
Telecom	8.66
Financial Services	7.89
Personal Care Products	6.84
Food & Candy	6.57
Direct Response	6.34
Restaurants	6.19
Insurance	4.86

Source: Kantar Media

a. Find the mean amount spent on advertising.
b. Find the median amount spent on advertising.
c. Find the sample variance of the advertising amounts.
d. Find the sample standard deviation of the advertising amounts.
e. Find the first quartile of the advertising amounts.
f. Find the third quartile of the advertising amounts.
g. Find the 40th percentile of the advertising amounts.
h. Find the 65th percentile of the advertising amounts.

13. Matching: Match each histogram to the boxplot that represents the same data set.

a.

b.

c.

d.

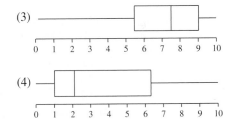

14. Weights of soap: As part of a quality control study aimed at improving a production line, the weights (in ounces) of 50 bars of soap are measured. The results are shown below, sorted from smallest to largest.

11.6	12.6	12.7	12.8	13.1	13.3	13.6	13.7
13.8	14.1	14.3	14.3	14.6	14.8	15.1	15.2
15.6	15.6	15.7	15.8	15.8	15.9	15.9	16.1
16.2	16.2	16.3	16.4	16.5	16.5	16.5	16.6
17.0	17.1	17.3	17.3	17.4	17.4	17.4	17.6
17.7	18.1	18.3	18.3	18.3	18.5	18.5	18.8
19.2	20.3						

a. Find the first and third quartiles of these data.
b. Find the median of these data.
c. Find the upper and lower outlier boundaries.
d. Are there any outliers? If so, list them.
e. Construct a boxplot for these data.

15. More corporate profits: Using the data in Exercise 2:
  a. Find the first and third quartiles of the profit.
  b. Find the median profit.
  c. Find the upper and lower outlier boundaries.
  d. Are there any outliers? If so, list them.
  e. Construct a boxplot for these data.

Write About It

1. The U.S. Department of Labor annually publishes an Occupational Outlook Handbook, which reports the job outlook, working conditions, and earnings for thousands of different occupations. The handbook reports both the mean and median annual earnings. For most occupations, which is larger, the mean or the median? Why do you think so?

2. Explain why the Empirical Rule is more useful than Chebyshev's Inequality for bell-shaped distributions. Explain why Chebyshev's Inequality is more useful for distributions that are not bell-shaped.

3. Does Chebyshev's Inequality provide useful information when $K = 1$? Explain why or why not.

4. Is Chebyshev's Inequality true when $K < 1$? Is it useful? Explain why or why not.

5. Percentiles are values that divide a data set into hundredths. The values that divide a data set into tenths are called deciles, denoted $D_1, D_2, ..., D_9$. Describe the relationship between percentiles and deciles.

Case Study: Can Recycled Materials Be Used In Electrical Devices?

Electronic devices contain electric circuits etched into wafers made of silicon. These silicon wafers are sealed with an ultrathin layer of silicon dioxide, in a process known as oxidation. This can be done with either new or recycled wafers.

   In a study described in the book *Statistical Case Studies for Industrial Process Improvement* by V. Czitrom and P. Spagon, both new and recycled wafers were oxidized, and the thicknesses of the layers were measured to determine whether they tended to differ between the two types of wafers. Recycled wafers are cheaper than new wafers, so the hope was that they would perform at least as well as the new wafers. Following are 36 thickness measurements (in tenths of a nanometer) for both new and recycled wafers.

New								
90.0	92.2	94.9	92.7	91.6	88.2	92.0	98.2	96.0
91.1	89.8	91.5	91.5	90.6	93.1	88.9	92.5	92.4
96.7	93.7	93.9	87.9	90.4	92.0	90.5	95.2	94.3
92.0	94.6	93.7	94.0	89.3	90.1	91.3	92.7	94.5

Recycled								
91.8	94.5	93.9	77.3*	92.0	89.9	87.9	92.8	93.3
92.6	90.3	92.8	91.6	92.7	91.7	89.3	95.5	93.6
92.4	91.7	91.6	91.1	88.0	92.4	88.7	92.9	92.6
91.7	97.4	95.1	96.7	77.5*	91.4	90.5	95.2	93.1

*Measurement is in error due to a defective gauge.

1. Construct comparative boxplots for the thicknesses of new wafers and recycled wafers.

2. Identify all outliers.

3. Should any of the outliers be deleted? If so, delete them and redraw the boxplots.

4. Identify any outliers in the redrawn boxplots. Should any of these be deleted? Explain.

5. Are the distributions of thicknesses skewed, or approximately symmetric?

6. Delete outliers as appropriate, and compute the mean thickness for new and for recycled wafers.

7. Delete outliers as appropriate, and compute the median thickness for new and for recycled wafers.

8. Delete outliers as appropriate, and compute the standard deviation of the thicknesses for new and for recycled wafers.

9. Suppose that it is desired to use the type of wafer whose distribution has less spread. Write a brief paragraph that explains which type of wafer to use and why. Which measure is more useful for spread in this case, the standard deviation or the interquartile range? Explain.

Summarizing Bivariate Data

Introduction

Inflation and unemployment are measures of the health of the economy. Inflation is the percentage increase in prices over the course of a year, and unemployment is the percentage of the labor force that is out of work. The following table presents levels of inflation and unemployment, as reported by the Bureau of Labor Statistics, for the years 1985 through 2012.

Year	Inflation	Unemployment	Year	Inflation	Unemployment
1985	3.8	7.2	1999	2.7	4.2
1986	1.1	7.0	2000	3.4	4.0
1987	4.4	6.2	2001	1.6	4.7
1988	4.4	5.5	2002	2.4	5.8
1989	4.6	5.3	2003	1.9	6.0
1990	6.1	5.6	2004	3.3	5.5
1991	3.1	6.8	2005	3.4	5.1
1992	2.9	7.5	2006	2.5	4.6
1993	2.7	6.9	2007	4.1	4.6
1994	2.7	6.1	2008	0.1	5.8
1995	2.5	5.6	2009	2.7	9.3
1996	3.3	5.4	2010	1.5	9.6
1997	1.7	4.9	2011	3.0	8.9
1998	1.6	4.5	2012	1.7	8.1

Source: Bureau of Labor Statistics

Economists have long studied the relationship between inflation and unemployment. One theory states that inflation and unemployment follow a pattern called "the Phillips curve," in which higher inflation leads to lower unemployment, while lower inflation leads to higher unemployment. This theory is now widely regarded as too simple, and economists continue to study data, looking for more complex relationships. In the case study at the end of this chapter, we will examine some methods for predicting unemployment.

Questions about relationships between variables arise frequently in science, business, public policy, and other areas where informed decisions need to be made. For example, how does a person's level of education affect his or her income? How does the amount of time spent studying for an exam affect an exam score? Data used to study questions like these consist of **ordered pairs**. An ordered pair consists of values of two variables for each individual in the data set. In the preceding table, the ordered pairs are (inflation rate, unemployment rate). To study the relationship between education and income, the ordered pair might be (number of years of education, annual income).

Data that consist of ordered pairs are called **bivariate data**. The basic graphical tool used to study bivariate data is the *scatterplot*, in which each ordered pair is plotted as a point. In many cases, the points on a scatterplot tend to cluster around a straight line. In these cases, the summary statistic most often used to measure the closeness of the relationship between the two variables is the *correlation coefficient*, which we will study in Section 4.1. When two variables are closely related to each other, it is often of interest to try to predict the value of one of them when given the value of the other. This is done with the equation of the *least-squares regression line*, which we will study in Sections 4.2 and 4.3.

SECTION 4.1 | Correlation

Objectives

1. Construct scatterplots for bivariate data
2. Compute the correlation coefficient
3. Interpret the correlation coefficient
4. Understand that correlation is not the same as causation

Objective 1 Construct scatterplots for bivariate data

Scatterplots

A real estate agent wants to study the relationship between the size of a house and its selling price. Table 4.1 presents the size in square feet and the selling price in thousands of dollars, for a sample of houses in a suburban Denver neighborhood.

Table 4.1 Size and Selling Price for a Sample of Houses

Size (square feet)	Selling Price ($1000s)
2521	400
2555	426
2735	428
2846	435
3028	469
3049	475
3198	488
3198	455

Source: Sue Bays Realty

It is reasonable to suspect that the selling price is related to the size of the house. Specifically, we expect that houses with larger sizes are more likely to have higher selling prices. A good way to visualize a relationship like this is with a **scatterplot**. In a scatterplot, each individual in the data set contributes an ordered pair of numbers, and each ordered pair is plotted on a set of axes.

EXAMPLE 4.1   Construct a scatterplot

Construct a scatterplot for the data in Table 4.1.

Solution

Each of the eight houses in Table 4.1 contributes an ordered pair of numbers. We will take the size as the first number and selling price as the second number. So the ordered pairs to be plotted are $(2521, 400)$, $(2555, 426)$, and so on. Figure 4.1 presents the scatterplot. We can see that houses with larger sizes tend to have larger selling prices, and houses with smaller sizes tend to have smaller (lower) selling prices.

Figure 4.1 Scatterplot of selling price versus size for a sample of houses

Let's take a closer look at Figure 4.1. We have observed that larger sizes tend to be associated with larger prices, and smaller sizes tend to be associated with smaller prices. We refer to this as a *positive association* between size and selling price. In addition, the points tend to cluster around a straight line from lower left to upper right. We describe this by saying that the relationship between the two variables is **linear**. Therefore, we can say that the scatterplot in Figure 4.1 exhibits a positive linear association between size and selling price.

In some cases, large values of one variable are associated with small values of another. An example is the weight of a car and its gas mileage. Large weights are associated with small gas mileages, and small weights are associated with large gas mileages. Therefore, we say that weight and gas mileage have a *negative association*.

Figure 4.2 presents examples of scatterplots that exhibit various kinds of association.

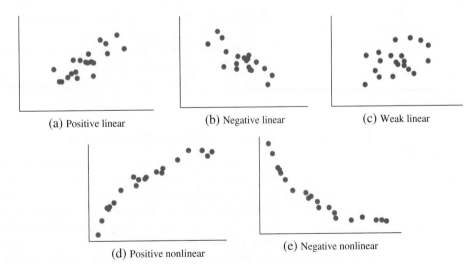

Figure 4.2 Scatterplots exhibiting various types of association

SUMMARY

- Two variables have a **positive association** if large values of one variable are associated with large values of the other.
- Two variables have a **negative association** if large values of one variable are associated with small values of the other.
- Two variables have a **linear relationship** if the data tend to cluster around a straight line when plotted on a scatterplot.

Check Your Understanding

1. Fill in the blank: If large values of one variable are associated with small values of another, then the two variables have a _____ association.

2. Fill in the blank: If two variables have a positive association, then large values of one variable are associated with _____ values of the other.

3. Fill in the blank: If the points on a scatterplot tend to cluster around a straight line, the variables plotted have a _____ association.

4. For each of the following scatterplots, state the type of association that is exhibited: *Choices: positive linear, negative linear, positive nonlinear, negative nonlinear, weak linear.*

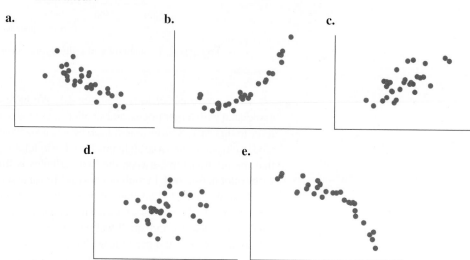

5. For each of the following pairs of variables, determine whether the association is positive or negative, and explain why.
   a. Would the association between outdoor temperature and consumption of heating oil be positive or negative? Explain.
   b. The number of years of education a person has and the person's income

Answers are on page 164.

Objective 2 Compute the correlation coefficient

The Correlation Coefficient

When two variables have a linear relationship, we want to measure how strong the relationship is. It isn't enough to look at the scatterplot, because the visual impression can be affected by the scales on the axes. Figure 4.3 presents two scatterplots of the house data presented in Table 4.1. The plots differ only in the scale used on the *y*-axis, yet the plot on the left appears to show a strong linear relationship, while the plot on the right appears to show a weak one.

We need a numerical measure of the strength of the linear relationship between two variables that is not affected by the scale of a plot. The appropriate quantity is the

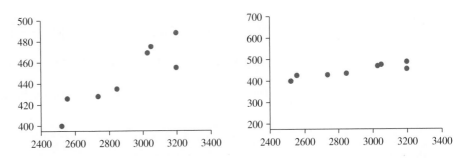

Figure 4.3 Both plots present the same data, yet the plot on the left appears to show a stronger linear relationship than the plot on the right. The reason is that the scales on the y-axis are different.

Explain It Again

Appearances can be misleading: The appearance of a scatterplot depends on the scales chosen for the axes. To get a reliable measure of the strength of a linear relationship, we must compute the correlation coefficient.

correlation coefficient. The formula for the correlation coefficient is a bit complicated, although calculating it does not involve much more than calculating sample means and standard deviations as was done in Chapter 3.

The Correlation Coefficient

Given ordered pairs (x, y), with sample means $\bar{x}$ and $\bar{y}$, sample standard deviations s_x and s_y, and sample size n, the correlation coefficient r is given by

$$r = \frac{1}{n-1} \sum \left(\frac{x - \bar{x}}{s_x}\right)\left(\frac{y - \bar{y}}{s_y}\right)$$

We often refer to r as the correlation between x and y.

Not surprisingly, most people nowadays use technology to compute the correlation coefficient. Procedures for the TI-84 Plus calculator, MINITAB, and Excel are presented in the Using Technology section on pages 159–160.

The correlation coefficient has several important properties, which we list.

Properties of the Correlation Coefficient

- The correlation coefficient is always between -1 and 1, inclusive. In other words, $-1 \leq r \leq 1$.
- The value of the correlation coefficient does not depend on the units of the variables. If we measure x and y in different units, the correlation will still be the same.
- It does not matter which variable is x and which is y.
- The correlation coefficient measures only the strength of the *linear* relationship between variables, and can be misleading when the relationship is nonlinear.
- The correlation coefficient is sensitive to outliers, and can be misleading when outliers are present.

EXAMPLE 4.2 Compute the correlation coefficient

Use the data in Table 4.1 to compute the correlation between size and selling price.

Solution

We will denote size by x and selling price by y. We compute the correlation coefficient using the following steps:

Step 1: Compute the sample means and standard deviations. We obtain $\bar{x} = 2891.25$, $\bar{y} = 447.0$, $s_x = 269.49357$, $s_y = 29.68405$.

Step 2: Compute the quantities $\dfrac{x - \bar{x}}{s_x}$ and $\dfrac{y - \bar{y}}{s_y}$.

Step 3: Compute the products $\left(\dfrac{x-\bar{x}}{s_x}\right)\left(\dfrac{y-\bar{y}}{s_y}\right)$.

Step 4: Add the products computed in Step 3, and divide the sum by $n-1$.

The calculations in Steps 2–4 are summarized in the following table.

x	y	$\dfrac{x-\bar{x}}{s_x}$	$\dfrac{y-\bar{y}}{s_y}$	$\left(\dfrac{x-\bar{x}}{s_x}\right)\left(\dfrac{y-\bar{y}}{s_y}\right)$
2521	400	−1.3738732	−1.5833419	2.1753110
2555	426	−1.2477106	−0.7074506	0.8826936
2735	428	−0.5797912	−0.6400744	0.3711095
2846	435	−0.1679075	−0.4042575	0.0678779
3028	469	0.5074333	0.7411387	0.3760785
3049	475	0.5853572	0.9432675	0.5521484
3198	488	1.1382461	1.3812131	1.5721604
3198	455	1.1382461	0.2695050	0.3067630

$$\frac{\sum\left(\dfrac{x-\bar{x}}{s_x}\right)\left(\dfrac{y-\bar{y}}{s_y}\right)}{n-1}=\frac{6.3041423}{7}$$
$$=0.9005918$$

CAUTION

Do not round the intermediate values used to calculate the correlation coefficient. Round only the final result.

We round our final answer to three decimal places. The correlation coefficient is $r=0.901$.

In general, we will round the correlation coefficient to three decimal places when it is the final result.

Objective 3 Interpret the correlation coefficient

Interpreting the correlation coefficient

The correlation coefficient measures the strength of the *linear* relationship between two variables. For this reason, it is meaningful only when the variables are linearly related. It can be misleading in other situations.

Interpreting the Correlation Coefficient

When two variables have a linear relationship, the correlation coefficient can be interpreted as follows:

- If r is positive, the two variables have a positive linear association.
- If r is negative, the two variables have a negative linear association.
- If r is close to 0, the linear association is weak.
- The closer r is to 1, the more strongly positive the linear association is.
- The closer r is to −1, the more strongly negative the linear association is.
- If $r=1$, then the points lie exactly on a straight line with positive slope; in other words, the variables have a perfect positive linear association.
- If $r=-1$, then the points lie exactly on a straight line with negative slope; in other words, the variables have a perfect negative linear association.

When two variables are not linearly related, the correlation coefficient does not provide a reliable description of the relationship between the variables.

CAUTION

Be sure that the relationship is linear before interpreting the correlation coefficient.

In Example 4.2, the two variables do have a linear relationship, as verified by the scatterplot in Figure 4.1. The value of the correlation coefficient is $r=0.901$, which indicates a strong positive linear association.

Check Your Understanding

6. The National Assessment for Educational Progress (NAEP) is a U.S. government organization that assesses the performance of students and schools at all levels across the United States. The following table presents the percentage of eighth-grade students who were found to be proficient in mathematics, and the percentage who were found to be proficient in reading in each of the ten most populous states.

State	Percentage Proficient in Reading	Percentage Proficient in Mathematics
California	60	59
Texas	73	78
New York	75	70
Florida	66	68
Illinois	75	70
Pennsylvania	79	77
Ohio	79	76
Michigan	73	66
Georgia	67	64
North Carolina	71	73

Source: National Assessment for Educational Progress

a. Construct a scatterplot with reading proficiency on the horizontal axis and math proficiency on the vertical axis. Is there a linear relationship?

b. Compute the correlation between reading proficiency and math proficiency. Is the linear association positive or negative? Weak or strong?

Answers are on page 164.

The correlation coefficient is not resistant

Recall: A statistic is resistant if its value is not affected much by extreme data values.

Figure 4.4 presents a scatterplot of the amount of farmland (including ranches) plotted against the total land area, for a selection of U.S. states. It is reasonable to suspect that states with larger land area would tend to have more farmland, and the scatterplot shows that, in general, this is true.

Recall: An outlier is a point that is detached from the main bulk of the data.

There is an outlier in the lower right corner of the plot, corresponding to the state of Alaska. Alaska is an outlier because it has a huge land area but very little farming. The correlation coefficient for the scatterplot in Figure 4.4 is $r = -0.119$. This suggests that there is actually a weak *negative* association between the total land area and the area of

Figure 4.4 Area of farmland versus total land area for a selection of U.S. states. Alaska is an outlier. Because of the outlier, the correlation coefficient for this plot is −0.119, which is misleading. If the outlier is removed, the correlation coefficient for the remaining points is $r = 0.710$.

farmland. If we ignore the Alaska point and compute the correlation coefficient for the remaining points, we get $r = 0.710$, a big difference.

With the Alaska outlier in the plot, the correlation is misleading. For the states other than Alaska, there is a strong positive association between total land area and farmland area. Because Alaska is such a big exception to this rule, it throws the correlation coefficient way off.

SUMMARY

The correlation coefficient is not resistant. It may be misleading when outliers are present.

Check Your Understanding

7. For which of the following scatterplots is the correlation coefficient an appropriate summary?

(a)

(b)

(c)

Answer is on page 164.

Objective 4 Understand that correlation is not the same as causation

Correlation Is Not Causation

A group of elementary school children took a vocabulary test. It turned out that children with larger shoe sizes tended to get higher scores on the test, and those with smaller shoe sizes tended to get lower scores. As a result, there was a large positive correlation between vocabulary and shoe size. Does this mean that learning new words causes one's feet to grow, or that growing feet cause one's vocabulary to increase? Obviously not. There is a third factor — age — that is related to both shoe size and vocabulary. Individuals with larger ages tend to have larger shoe sizes. Individuals with larger ages also tend to have larger vocabularies. It follows that individuals with larger shoe sizes will tend to have larger vocabularies.

Age is a confounder in this example. Age is related to both shoe size and vocabulary, which makes it appear as if shoe size and vocabulary are related to each other. The fact that shoe size and vocabulary are correlated does not mean that changing one variable will cause the other to change.

SUMMARY

Correlation is not the same as causation. In general, when two variables are correlated, we cannot conclude that changing the value of one variable will cause a change in the value of the other.

Check Your Understanding

8. An economist discovers that over the past several years, both the salaries of U.S. college professors and the amount of beer consumed in the United States have gone up. Thus, there is a positive correlation between the average salary of college professors and the amount of beer consumed. The economist concludes that the increase in beer consumption must be caused by professors spending their additional money on beer. Explain why this conclusion is not necessarily true.

Answer is on page 164.

USING TECHNOLOGY

We use the data in Table 4.1 to illustrate the technology steps.

TI-84 PLUS

Constructing a scatterplot

Step 1. Enter the *x*-values from Table 4.1 into **L1** and the *y*-values into **L2**.

Step 2. Press **2nd, Y=** to access the STAT PLOTS menu and select Plot1 by pressing **1**.

Step 3. Select **On** and the scatterplot icon (Figure A).

Step 4. Press **ZOOM** and then **9: ZoomStat** (Figure B).

Figure A

Figure B

Computing the correlation coefficient

The correlation coefficient is calculated as part of the procedure for computing the least-squares regression line and is presented at the end of Section 4.2 on page 171.

MINITAB

Constructing a scatterplot

Step 1. Enter the *x*-values from Table 4.1 into **Column C1** and the *y*-values into **Column C2**.

Step 2. Click on **Graph** and select **Scatterplot**. Choose the **Simple** option and press **OK**.

Step 3. Enter **C2** in the **Y variables** field and **C1** in the **X variables** field.

Step 4. Click **OK** (Figure C).

Figure C

Computing the correlation coefficient

Step 1. Enter the *x*-values from Table 4.1 into **Column C1** and the *y*-values into **Column C2**.

Step 2. Click on **Stat**, then **Basic Statistics**, then **Correlation**.

Step 3. Double-click **C1** and **C2** to enter these variables in the **Variables** field.

Step 4. Click **OK** (Figure D).

Correlations: C1, C2

Pearson correlation of C1 and C2 = 0.901

Figure D

EXCEL

Constructing a scatterplot

> **Step 1.** Enter the *x*-values from Table 4.1 into
> **Column A** and the *y*-values into **Column B**.
> **Step 2.** Highlight all values in **Columns A** and **B**.
> **Step 3.** Click on **Insert** and then **Scatter**.
> **Step 4.** Click **OK** (Figure E).

Figure E

Computing the correlation coefficient

> **Step 1.** Enter the data from Table 4.1 into the worksheet.
> **Step 2.** In an empty cell, select the **Insert Function** icon
> and highlight **Statistical** in the category field.
> **Step 3.** Click on the **CORREL** function and press **OK**.
> **Step 4.** Enter the range of cells that contain the *x*-values
> from Table 4.1 in the **Array1** field, and enter the
> range of cells that contain the *y*-values in the
> **Array2** field.
> **Step 5.** Click **OK** (Figure F).

	A	B	C
1	Size	Selling Price	
2	2521	400	
3	2555	426	
4	2735	428	
5	2846	435	
6	3028	469	
7	3049	475	
8	3198	488	
9	3198	455	
10			
11		=CORREL(A2:A9,B2:B9)	
12	Correlation	0.900591753	

Figure F

SECTION 4.1 **Exercises**

**Exercises 1–8 are the Check Your Understanding
exercises located within the section.**

Understanding the Concepts

**In Exercises 9–12, fill in each blank with the appropriate
word or phrase.**

9. Bivariate data consist of ordered _____ .

10. In a _____ , ordered pairs are plotted on a set of
    axes.

11. Two variables have a _____ relationship if the data
    tend to cluster around a straight line.

12. The correlation coefficient measures only the strength of
    the _____ relationship between variables.

**In Exercises 13–16, determine whether the statement is
true or false. If the statement is false, rewrite it as a true
statement.**

13. Two variables are negatively associated if large values
    of one variable are associated with large values of the
    other.

14. If the correlation coefficient *r* equals 1, then the
    points on a scatterplot lie exactly on a straight
    line.

15. The correlation coefficient is not resistant.

16. When two variables are correlated, changing the value of
    one variable will cause a change in value of the other
    variable.

Practicing the Skills

In Exercises 17–20, compute the correlation coefficient.

17.

x	1	2	3	4	5
y	2	1	4	3	7

18.

x	24	13	8	81	63	36	5
y	44	52	42	5	1	48	15

19.

x	5.5	4.2	4.7	5.6	6.0	3.9	6.3	5.7
y	4.9	4.8	4.8	4.7	5.5	5.1	5.8	6.5

20.

x	5	−8	−2	6	9	−10	13	7
y	−1	−3	−6	−7	−1	5	13	22

In Exercises 21–24, determine whether the correlation coefficient is an appropriate summary for the scatterplot and explain your reasoning.

21.

22.

23.

24.

In Exercises 25–30, determine whether the association between the two variables is positive or negative.

25. A person's age and the person's income

26. The age of a car and its resale value

27. The age of a car and the number of miles on its odometer

28. The number of times a pencil is sharpened and its length

29. The diameter of an apple and its weight

30. Weekly ice cream sales and weekly average temperature

Working with the Concepts

31. **Price of eggs and milk:** The following table presents the average price in dollars for a dozen eggs and a gallon of milk for each month from January through December 2012.

Dozen Eggs	Gallon of Milk
1.94	3.58
1.80	3.52
1.77	3.50
1.83	3.47
1.69	3.43
1.67	3.40
1.65	3.43
1.88	3.47
1.89	3.47
1.96	3.52
1.96	3.54
2.01	3.58

Source: Bureau of Labor Statistics

a. Construct a scatterplot of the price of milk (y) versus the price of eggs (x).

b. Compute the correlation coefficient between the price of eggs and the price of milk.

c. In a month where the price of eggs is above average, would you expect the price of milk to be above average or below average? Explain.

d. Which of the following is the best interpretation of the correlation coefficient?

  i. When the price of eggs rises, it causes the price of milk to rise.

  ii. When the price of milk rises, it causes the price of eggs to rise.

  iii. Changes in the price of eggs or milk do not cause changes in the price of the other; the correlation indicates that the prices of milk and eggs tend to go up and down together.

32. **Government funding:** The following table presents the budget (in millions of dollars) for selected organizations that received U.S. government funding for arts and culture in both 2006 and 2012.

Organization	2006	2012
Corporation for Public Broadcasting	460	444
Institute of Museum and Library Services	247	232
National Endowment for the Humanities	142	146
National Endowment for the Arts	124	146
National Gallery of Art	95	128
Kennedy Center for the Performing Arts	18	23
Commission of Fine Arts	2	2
Advisory Council on Historic Preservation	5	6

Source: National Endowment for the Arts

a. Construct a scatterplot of the funding in 2012 (y) versus the funding in 2006 (x).

b. Compute the correlation coefficient between the funding in 2006 and the funding in 2012.

c. For an organization whose funding in 2006 was above the average, would you expect their funding in 2012 to be above or below average? Explain.

d. Which of the following is the best interpretation of the correlation coefficient?

   i. If we increase the funding for an organization in 2006, this will cause the funding in 2012 to increase.

   ii. If we increase the funding for an organization in 2012, this will cause the funding in 2006 to increase.

   iii. Some organizations get more funding than others, and those that were more highly funded in 2006 were generally more highly funded in 2012 as well.

33. Pass the ball: The NFL Scouting Combine is an event at which football scouts evaluate the abilities of some top college prospects. Following are heights in inches and weights in pounds for some of the quarterbacks at the 2013 Combine.

Name	Height	Weight
Matt Barkley	75	227
Tyler Bray	78	232
Zac Dysert	75	231
Mike Glennon	79	225
Marqueis Gray	75	240
Landry Jones	76	225
Collin Klein	77	226
E. J. Manuel	77	237
Ryan Nassib	74	227
Sean Renfree	75	219
Geno Smith	74	218
Brad Sorensen	76	237
James Vandenberg	75	226
Tyler Wilson	74	215

Source: SB Nation

a. Construct a scatterplot of the weight (y) versus the height (x).

b. Compute the correlation coefficient between the height and weight of the quarterbacks.

c. If a quarterback is below average in height, would you expect him to be above average or below average in weight? Explain.

d. Which of the following is the best interpretation of the correlation coefficient?

   i. If a quarterback gains weight, he will grow taller.

   ii. Given two quarterbacks, the taller one is likely to be heavier than the shorter one.

   iii. Given two quarterbacks, the heavier one is likely to be shorter than the lighter one.

34. Carbon footprint: Carbon dioxide (CO_2) is produced by burning fossil fuels such as oil and natural gas, and has been connected to global warming. The following table presents the average amounts (in metric tons) of CO_2 emissions for the years 1990–2006 per person in the United States and per person in the rest of the world.

Year	Non-U.S.	U.S.
1990	3.6	19.2
1991	3.6	19.0
1992	3.5	18.8
1993	3.3	19.7
1994	3.3	19.8
1995	3.3	19.5
1996	3.4	19.7
1997	3.4	20.0
1998	3.3	19.5
1999	3.2	19.6
2000	3.3	20.0
2001	3.3	19.6
2002	3.3	19.6
2003	3.5	19.4
2004	3.7	19.6
2005	3.8	19.5
2006	3.7	19.0

Source: World Bank

a. Construct a scatterplot of U.S. carbon dioxide emissions (y) versus non-U.S. emissions (x).

b. Compute the correlation coefficient between U.S. carbon dioxide emissions and non-U.S. emissions.

c. In a year when U.S. emissions are above average, would you expect emissions in the rest of the world to be above average or below average? Explain.

d. As developing countries modernize, their use of fossil fuels increases. Countries that are already developed strive to reduce emissions. How does this fact explain the relationship between U.S. and non-U.S. emissions?

35. Foot temperatures: Foot ulcers are a common problem for people with diabetes. Higher skin temperatures on the foot indicate an increased risk of ulcers. In a study performed at the Colorado School of Mines, skin temperatures on both feet were measured, in degrees Fahrenheit, for 18 diabetic patients. The results are presented in the following table.

Left Foot	Right Foot	Left Foot	Right Foot
80	80	76	81
85	85	89	86
75	80	87	82
88	86	78	78
89	87	80	81
87	82	87	82
78	78	86	85
88	89	76	80
89	90	88	89

Source: Kimberly Anderson, M.S. thesis, Colorado School of Mines

a. Construct a scatterplot of the right foot temperature (y) versus the left foot temperature (x).

b. Compute the correlation coefficient between the temperatures of the left and right feet.

c. If a patient's left foot is cooler than the average, would you expect the patient's right foot to be warmer or cooler than average? Explain.

d. Which of the following is the best interpretation of the correlation coefficient?

i. Some patients have warmer feet than others. Those who have warmer left feet generally have warmer right feet as well.

ii. If we warm a patient's left foot, the patient's right foot will become warmer.

iii. If we cool a patient's left foot, the patient's right foot will become warmer.

36. Mortgage payments: The following table presents interest rates, in percent, for 30-year and 15-year fixed-rate mortgages, for January through December, 2012.

30-Year	15-Year	30-Year	15-Year
3.92	3.20	3.55	2.85
3.89	3.16	3.60	2.86
3.95	3.20	3.47	2.78
3.91	3.14	3.38	2.69
3.80	3.03	3.35	2.66
3.68	2.95	3.35	2.66

Source: Freddie Mac

a. Construct a scatterplot of the 15-year rate (y) versus the 30-year rate (x).

b. Compute the correlation coefficient between 30-year and 15-year rates.

c. When the 30-year rate is below average, would you expect the 15-year rate to be above or below average? Explain.

d. Which of the following is the best interpretation of the correlation coefficient?

i. When a bank increases the 30-year rate, that causes the 15-year rate to rise as well.

ii. Interest rates are determined by economic conditions. When economic conditions cause 30-year rates to increase, these same conditions cause 15-year rates to increase as well.

iii. When a bank increases the 15-year rate, that causes the 30-year rate to rise as well.

37. Blood pressure: A blood pressure measurement consists of two numbers: the systolic pressure, which is the maximum pressure taken when the heart is contracting, and the diastolic pressure, which is the minimum pressure taken at the beginning of the heartbeat. Blood pressures were measured, in millimeters, for a sample of 16 adults. The following table presents the results.

Systolic	Diastolic	Systolic	Diastolic
134	87	133	91
115	83	112	75
113	77	107	71
123	77	110	74
119	69	108	69
118	88	105	66
130	76	157	103
116	70	154	94

Based on results published in the *Journal of Human Hypertension*

a. Construct a scatterplot of the diastolic blood pressure (y) versus the systolic blood pressure (x).

b. Compute the correlation coefficient between systolic and diastolic blood pressure.

c. If someone's diastolic pressure is above average, would you expect that person's systolic pressure to be above or below average? Explain.

38. Butterfly wings: Do larger butterflies live longer? The wingspan (in millimeters) and the lifespan in the adult state (in days) were measured for 22 species of butterfly. Following are the results.

Wingspan	Lifespan	Wingspan	Lifespan
35.5	19.8	25.9	32.5
30.6	17.3	31.3	27.5
30.0	27.5	23.0	31.0
32.3	22.4	26.3	37.4
23.9	40.7	23.7	22.6
27.7	18.3	27.1	23.1
28.8	25.9	28.1	18.5
35.9	23.1	25.9	32.3
25.4	24.0	28.8	29.1
24.6	38.8	31.4	37.0
28.1	36.5	28.5	33.7

Source: *Oikos Journal of Ecology* 105:41–54

a. Construct a scatterplot of the lifespan (y) versus the wingspan (x).

b. Compute the correlation coefficient between wingspan and lifespan.

c. Do larger butterflies tend to live for a longer or shorter time than smaller butterflies? Explain.

39. Police and crime: In a survey of cities in the United States, it is discovered that there is a positive correlation between the number of police officers hired by the city and the number of crimes committed. Do you believe that increasing the number of police officers causes the crime rate to increase? Why or why not?

40. Age and education: A survey of U.S. adults showed that there is a negative correlation between age and education level. Does this mean that people become less educated as they become older? Why or why not?

41. What's the correlation? In a sample of adults, would the correlation between age and year graduated from high school be closest to −1, −0.5, 0, 0.5, or 1? Explain.

42. What's the correlation? In a sample of adults, would the correlation between year of birth and year graduated from high school be closest to −1, −0.5, 0, 0.5, or 1? Explain.

Extending the Concepts

43. Changing means and standard deviations: A small company has five employees. The following table presents the number of years each has been employed (x) and the hourly wage in dollars (y).

x (years)	0.5	1.0	1.75	2.5	3.0
y (dollars)	9.51	8.23	10.95	12.70	12.75

a. Compute $\bar{x}$, $\bar{y}$, s_x, and s_y.

b. Compute the correlation coefficient between years of service and hourly wage.

c. Each employee is given a raise of $1.00 per hour, so each y-value is increased by 1. Using these new y-values, compute the sample mean $\bar{y}$ and the sample standard deviation s_y.

d. In part (c), each y-value was increased by 1. What was the effect on $\bar{y}$? What was the effect on s_y?

e. Compute the correlation coefficient between years of service and the increased hourly wage. Explain why the correlation coefficient is unchanged even though the y-values have changed.

f. Convert x to months by multiplying each x-value by 12. So the new x-values are 6, 12, 21, 30, and 36.

Compute the sample mean $\bar{x}$ and the sample standard deviation s_x.

g. In part (f), each x-value was multiplied by 12. What was the effect on $\bar{x}$ and s_x?

h. Compute the correlation coefficient between months of service and hourly wage in dollars. Explain why the correlation coefficient is unchanged even though the x-values, $\bar{x}$, and s_x have changed.

i. Use the results of parts (a)–(h) to fill in the blank: If a constant is added to each x-value or to each y-value, the correlation coefficient is _____.

j. Use the results of parts (a)–(h) to fill in the blank: If each x-value or each y-value is multiplied by a positive constant, the correlation coefficient is _____.

Answers to Check Your Understanding Exercises for Section 4.1

1. negative

2. large

3. linear

4. **a.** Negative linear
   b. Positive nonlinear   **c.** Positive linear
   d. Weak linear
   e. Negative nonlinear

5. **a.** negative. When the temperature is higher, the amount of heating oil consumed tends to be lower.
   b. positive. People with more years of education tend to have larger incomes.

6. **a.**

The scatterplot shows a linear relationship.

b. $r = 0.809$. The linear relationship is positive and fairly strong.

7. Scatterplot (a) is the only one for which the correlation coefficient is an appropriate summary.

8. Correlation is not causation. It is possible that the college professors are drinking the same amount of beer as before, while people other than college professors are drinking more beer.

SECTION 4.2 The Least-Squares Regression Line

Objectives

1. Compute the least-squares regression line
2. Use the least-squares regression line to make predictions
3. Interpret predicted values, the slope, and the y-intercept of the least-squares regression line

Objective 1 Compute the least-squares regression line

The Least-Squares Regression Line

Table 4.2 presents the size in square feet and selling price in thousands of dollars for a sample of houses. These data were first presented in Section 4.1. A scatterplot of these data showed that they tend to cluster around a line, and we computed the correlation to be 0.901. We concluded that there is a strong positive linear association between size and price.

Table 4.2 Size and Selling Price
for a Sample of Houses

Size (square feet)	Selling Price ($1000s)
2521	400
2555	426
2735	428
2846	435
3028	469
3049	475
3198	488
3198	455

Source: Sue Bays Realty

We can use these data to predict the selling price of a house based on its size. The key is to summarize the data with a straight line. We want to find the line that fits the data best. We now explain what we mean by "best." Figure 4.5 presents scatterplots of the data in Table 4.2, each with a different line superimposed. For each line, we have drawn the vertical distances from the points to the line. It is clear that the line in Figure 4.5(a) fits better than the line in Figure 4.5(b). The reason is that the vertical distances are, on the whole, smaller for the line in Figure 4.5(a). We determine exactly how well a line fits the data by squaring the vertical distances and adding them up. The line that fits best is the line for which this sum of squared distances is as small as possible. This line is called the **least-squares regression line**.

(a)   (b)

Figure 4.5   The line in (a) fits better than the line in (b), because the vertical distances are generally smaller.

Fortunately, we don't have to worry about vertical distances when we calculate the least-squares regression line. We can use the following formula.

> ## Equation of the Least-Squares Regression Line
>
> Given ordered pairs (x, y), with sample means $\bar{x}$ and $\bar{y}$, sample standard deviations s_x and s_y, and correlation coefficient r, the equation of the least-squares regression line for predicting y from x is
>
> $$\hat{y} = b_0 + b_1 x$$
>
> where $b_1 = r\dfrac{s_y}{s_x}$ is the **slope** and $b_0 = \bar{y} - b_1\bar{x}$ is the **y-intercept**.

Explain It Again

Explanatory variable and outcome variable: The explanatory variable is used to explain or predict the value of the outcome variable.

In general, the variable we want to predict (in this case, selling price) is called the **outcome variable**, or **response variable**, and the variable we are given is called the **explanatory variable**, or **predictor variable**. In the equation of the least-squares regression line, x represents the explanatory variable and y represents the outcome variable.

Compute the least-squares regression line

Use the data in Table 4.2 to compute the least-squares regression line for predicting price from size.

CAUTION

When computing the least-squares regression line, be sure that x represents the variable you are given (the explanatory variable), and y represents the variable you want to predict (the outcome variable).

CAUTION

We don't round the value of r in this calculation because it is an intermediate value.

CAUTION

Do not confuse the slope b_1 of the least-squares regression line with the correlation coefficient r. In most cases, they are not equal.

Solution

We first find $\bar{x}$, $\bar{y}$, s_x, s_y, and r. We computed these quantities in Section 4.1:

$$\bar{x} = 2891.25, \quad \bar{y} = 447.0, \quad s_x = 269.49357, \quad s_y = 29.68405, \quad r = 0.9005918.$$

The slope of the least-squares regression line is

$$b_1 = r\frac{s_y}{s_x} = (0.9005918)\frac{29.68405}{269.49357} = 0.09919796$$

We use the value of b_1 just found to compute b_0, the y-intercept of the least-squares regression line:

$$b_0 = \bar{y} - b_1\bar{x} = 447.0 - (0.09919796)(2891.25) = 160.1939$$

The equation of the least-squares regression line is $\hat{y} = 160.1939 + 0.0992x$.

In general, we will round the slope and y-intercept values to four decimal places.

Figure 4.6 presents a scatterplot of the data in Table 4.2 with the least-squares regression line superimposed. It can be seen that the points tend to cluster around the least-squares regression line.

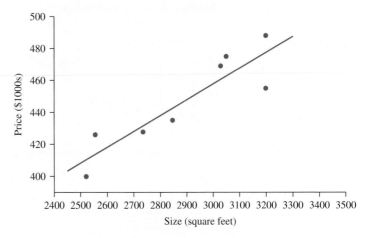

Figure 4.6   Scatterplot of the data in Table 4.2 with the least-squares regression line superimposed. The points tend to cluster around the least-squares regression line.

Computing the least-squares regression line with technology

Nowadays, least-squares regression lines are usually computed with technology rather than by hand. Figure 4.7 shows the least-squares regression line for predicting house price from house size, as presented by the TI-84 Plus calculator. There are four numbers in the output, labeled "a", "b", "r²", and "r". Of these, a is the y-intercept of the least-squares regression line and b is the slope. The calculator provides many more digits than are generally necessary for these quantities. If you round a and b, you will get a = 160.1939 and b = 0.0992, which agree with the results calculated by hand in Example 4.3. The last quantity, r, is the correlation coefficient. When rounded, this value is 0.901, which agrees with the value computed by hand in Example 4.2. Finally, the value r² is computed by squaring the correlation coefficient r. This is an important quantity that we will discuss in Section 4.3.

Step-by-step instructions for using the TI-84 Plus calculator to compute the least-squares line and correlation coefficient are presented in the Using Technology section on page 171.

Figure 4.7   The least-squares regression line as presented by the TI-84 Plus calculator. The *y*-intercept is denoted by a, and the slope of the line is denoted by b. The value r, in the last line, is the correlation coefficient.

Following is MINITAB output for the least-squares regression line for predicting house price from house size. The equation is given near the top of the output as Selling Price $= 160.19 + 0.09920$ Size. In the table below that, the values of the slope and intercept are presented again in the column labeled Coef.

Step-by-step instructions for using MINITAB to compute the least-squares line are presented in the Using Technology section on page 171.

```
The regression equation is
Selling Price = 160.19 + 0.09920 Size


Predictor           Coef     SE Coef       T       P
Constant          160.19       56.73    2.82   0.030
Size             0.09920     0.01955    5.08   0.002
```

Finally, we present the output from Excel for this least-squares regression line. The slope and intercept are found in the column labeled "Coefficients." The slope is labeled X Variable 1. Step-by-step instructions for using EXCEL to compare the least-squares line are presented in the Using Technology section on page 172.

	Coefficients	Standard Error	t Stat	P-value
Intercept	160.1939146	56.72636198	2.823976525	0.030195859
X Variable 1	0.099197954	0.019545859	5.075139241	0.00227642

Check Your Understanding

1. The following table presents the percentage of students who tested proficient in reading and the percentage who tested proficient in math for each of the ten most populous states in the United States. Compute the least-squares regression line for predicting math proficiency from reading proficiency.

State	Percent Proficient in Reading	Percent Proficient in Mathematics
California	60	59
Texas	73	78
New York	75	70
Florida	66	68
Illinois	75	70
Pennsylvania	79	77
Ohio	79	76
Michigan	73	66
Georgia	67	64
North Carolina	71	73

Source: National Assessment for Educational Progress

Answer is on page 176.

Objective 2 Use the least-squares regression line to make predictions

Using the Least-Squares Regression Line for Prediction

We can use the least-squares regression line to predict the value of the outcome variable if we are given the value of the explanatory variable. Simply substitute the value of the explanatory value for x in the equation of the least-squares regression line. The value of $\hat{y}$ that is computed is the **predicted value**.

EXAMPLE 4.4	Use the least-squares regression line for prediction

Use the least-squares regression line computed in Example 4.3 to predict the selling price of a house of size 2800 square feet.

Solution

The equation of the least-squares regression line is $\hat{y} = 160.1939 + 0.0992x$. Substituting 2800 for x yields

$$\hat{y} = 160.1939 + 0.0992(2800) = 438.0$$

We predict that the selling price of the house will be 438.0 thousand dollars, or $438,000.

We will round predicted values to one more decimal place than the outcome variable.

Figure 4.8 presents a scatterplot of the data with the point for the prediction added. The given value for the explanatory variable was $x = 2800$, and the predicted price is the y-value on the least-squares regression line corresponding to $x = 2800$.

Figure 4.8 The predicted price for a house with size $x = 2800$ is 438.0, which is the y-value on the least-squares regression line.

Explain It Again

Locating the predicted value on the least-squares regression line: For a given value of the explanatory variable x, the predicted value $\hat{y}$ is the y-value on the least-squares regression line corresponding to the given value of x.

The least-squares regression line goes through the point of averages

In the house data, the average size is $\bar{x} = 2891.25$ and the average selling price is $\bar{y} = 447.0$. What selling price do we predict for a house of average size? We substitute 2891.25 for x in the equation of the least-squares regression line to obtain

$$\hat{y} = 160.1939 + 0.0992(2891.25) = 447.0$$

For a house of average size, we predict that the selling price will be the average selling price. In general, when the explanatory variable x is equal to $\bar{x}$, the predicted value $\hat{y}$ is equal to $\bar{y}$.

SUMMARY

The least-squares regression line goes through the **point of averages** $(\bar{x}, \bar{y})$.

It makes sense that the least-squares regression line goes through $(\bar{x}, \bar{y})$. For example, if the correlation coefficient is positive, then above-average values of x are associated with above-average values of y, and below-average values of x are associated with below-average values of y. It follows that the value of x equal to its average would be associated with the average value of y.

Objective 3 Interpret predicted values, the slope, and the *y*-intercept of the least-squares regression line

Interpreting the Least-Squares Regression Line

The least-squares regression line has slope b_1 and y-intercept b_0. We use the least-squares regression line to compute a predicted value $\hat{y}$. We explain how to interpret these quantities.

Interpreting the predicted value $\hat{y}$

The predicted value $\hat{y}$ can be used to estimate the average outcome for a given value of the explanatory variable x. For any given value of x, the value $\hat{y}$ is an estimate of the average y-value for all points with that x-value.

EXAMPLE 4.5 ## Use the least-squares regression line to estimate the average outcome

Use the least-squares regression line computed in Example 4.3 to estimate the average price of all houses whose size is 3000 square feet.

Solution

The equation of the least-squares regression line is $\hat{y} = 160.1939 + 0.0992x$. Given $x = 3000$, we predict

$$\hat{y} = 160.1939 + 0.0992(3000) = 457.8$$

We estimate the average price for a house of 3000 square feet to be 457.8 thousand dollars, or \$457,800.

Interpreting the *y*-intercept b_0

The y-intercept b_0 is the point where the line crosses the y-axis. This has a practical interpretation only when the data contain both positive and negative values of x — in other words, only when the scatterplot contains points on both sides of the y-axis.

- If the data contain both positive and negative x-values, then the y-intercept is the estimated outcome when the value of the explanatory variable x is 0.
- If the x-values are all positive or all negative, then the y-intercept b_0 does not have a useful interpretation.

Interpreting the slope b_1

If the x-values of two points on a line differ by 1, their y-values will differ by an amount equal to the slope of the line. For example, if a line has a slope of 4, then two points whose x-values differ by 1 will have y-values that differ by 4. This fact enables us to interpret the slope b_1 of the least-squares regression line. If the values of the explanatory variable for two individuals differ by 1, their predicted values will differ by b_1. If the values of the explanatory variable differ by an amount d, then their predicted values will differ by $b_1 \cdot d$.

EXAMPLE 4.6 ## Compute the predicted difference in outcomes

Two houses differ in size by 150 square feet. By how much should we predict their prices to differ?

Solution

The slope of the least-squares regression line is $b_1 = 0.0992$. We predict the prices to differ by $(0.0992)(150) = 14.9$ thousand dollars, or \$14,900.

Check Your Understanding

2. At the final exam in a statistics class, the professor asks each student to indicate how many hours he or she studied for the exam. After grading the exam, the professor computes the least-squares regression line for predicting the final exam score from the number of hours studied. The equation of the line is $\hat{y} = 50 + 5x$.
   a. Antoine studied for 6 hours. What do you predict his exam score to be?
   b. Emma studied for 3 hours longer than Jeremy did. How much higher do you predict Emma's score to be?

3. For each of the following plots, interpret the y-intercept of the least-squares regression line if possible. If not possible, explain why not.

   a. The least-squares regression line is $\hat{y} = 1.98 + 0.039x$, where x is the temperature in a freezer in degrees Fahrenheit, and y is the time it takes to freeze a certain amount of water into ice.

   b. The least-squares regression line is $\hat{y} = -13.586 + 4.340x$, where x represents the age of an elementary school student and y represents the score on a standardized test.

   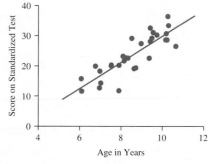

Answers are on page 176.

The least-squares regression line does not predict the effect of changing the explanatory variable

We have worded the interpretation of the slope of the least-squares regression line very carefully. The slope is the estimated difference in y-values for *two different* individuals whose x-values differ by 1. This does *not* mean that changing the value of x for a particular individual will cause that individual's y-value to change. The following example will help make this clear.

| EXAMPLE 4.7 | The least-squares regression line doesn't predict the result of changing the explanatory variable |

A study is done in which a sample of men were weighed, and then each man was tested to see how much weight he could lift. The explanatory variable (x) was the man's weight, and the outcome (y) was the amount he could lift. The least-squares regression line was computed to be $\hat{y} = 50 + 0.6x$.

Joe is a weightlifter. After looking at the equation of the least-squares regression line, he reasons as follows: "The slope of the least-squares regression line is 0.6. Therefore, if I gain 10 pounds, I'll be able to lift 6 pounds more, because $(0.6)(10) = 6$." Is he right?

Solution

No, he is not right. You can't improve your weightlifting ability simply by putting on weight. What the slope of 0.6 tells us is that if two men have weights that differ by

10 pounds, then, on the average, the heavier man will be able to lift 6 pounds more than the lighter man. This does not mean that an individual man can increase his weightlifting ability by increasing his weight.

Check Your Understanding

4. The Sanchez family wants to sell their house, which is 2800 square feet in size. Mr. Sanchez notices that the slope of the least-squares regression line for predicting price from size is 0.0992. He says that if they put a 250-square-foot addition on their house, the selling price will increase by $(0.0992)(250) = 24.8$ thousand dollars, or \$24,800. Is this necessarily a good prediction? Explain.

Answer is on page 176.

USING TECHNOLOGY

We use the data in Table 4.2 to illustrate the technology steps.

TI-84 PLUS

Computing the least-squares regression line

Note: Before computing the least-squares regression line, a one-time calculator setting should be modified to correctly configure the calculator to display the correlation coefficient. The following steps describe how to do this.

Step 1. Press **2nd, 0** to access the calculator catalog.

Step 2. Scroll down and select **DiagnosticOn**.

Step 3. Press **Enter** and then **Enter**.

The following steps describe how to compute the least-squares regression line.

Step 1. Enter the *x*-values from Table 4.2 into **L1** and the *y*-values into **L2**.

Step 2. Press **STAT** and then the right arrow key to access the **CALC** menu.

Step 3. Select **8: LinReg(a+bx)** (Figure A) and press **ENTER** (Figure B).

Using the TI-84 PLUS Stat Wizards (see Appendix B for more information)

Step 1. Enter the *x*-values from Table 4.2 into **L1** and the *y*-values into **L2**.

Step 2. Press **STAT** and then the right arrow key to access the **CALC** menu.

Step 3. Select **8:LinReg(a+bx)** (Figure A). Enter **L1** in the **Xlist** field and **L2** in the **Ylist** field. Select **Calculate** and press **ENTER** (Figure B).

Figure A

Figure B

MINITAB

Computing the least-squares regression line

Step 1. Enter the *x*-values from Table 4.2 into **Column C1** and the *y*-values into **Column C2**.

Step 2. Click **STAT**, then **Regression**, then **Regression**.

Step 3. Select the *y*-variable (**C2**) as the **Response variable** and the *x*-variable (**C1**) as the **Predictor** and click **OK** (Figure C).

Regression Analysis: C2 versus C1

The regression equation is
C2 = 160 + 0.0992 C1

Figure C

EXCEL

Computing the least-squares regression line

Step 1. Enter the x-values from Table 4.2 into **Column A** and the y-values into **Column B**.

Step 2. Select **Data**, then **Data Analysis**. Highlight **Regression** and press **OK**.

Step 3. Enter the range of cells that contain the x-values in the **Input X Range** field and the range of cells that contain the y-values in the **Input Y Range** field.

Step 4. Click **OK** (Figure D).

The least-squares regression line is given by
$$\hat{y} = (\text{X Variable 1})x + \text{Intercept}$$

	Coefficients
Intercept	160.1939146
X Variable 1	0.099197954

Figure D

SECTION 4.2 Exercises

Exercises 1–4 are the Check Your Understanding exercises located within the section.

Understanding the Concepts

In Exercises 5–7, fill in each blank with the appropriate word or phrase.

5. When we are given the value of the _____ variable, we can use the least-squares regression line to predict the value of the _____ variable.

6. If the correlation coefficient is equal to 0, the slope of the least-squares regression line will be equal to _____.

7. If the least-squares regression line has slope $b_1 = 5$, and two x-values differ by 3, the predicted difference in the y-values is _____.

In Exercises 8–12, determine whether the statement is true or false. If the statement is false, rewrite it as a true statement.

8. Substituting the value of the explanatory variable for x in the equation of the least-squares regression line results in a prediction for y.

9. The least-squares regression line passes through the point of averages $(\bar{x}, \bar{y})$.

10. In general, the slope of the least-squares regression line is equal to the correlation coefficient.

11. The least-squares regression line predicts the result of changing the value of the explanatory variable.

12. The y-intercept b_0 of a least-squares regression line has a useful interpretation only if the x-values are either all positive or all negative.

Practicing the Skills

In Exercises 13–16, compute the least-squares regression line for the given data set.

13.

x	1	2	3	4	5
y	5	6	9	8	7

14.

x	9	5	7	13	−8	−2	6	−10
y	3	3	31	36	0	3	−2	−14

15.

x	42	36	14	18	23	36	17
y	72	68	25	31	42	65	32

16.

x	5.7	4.1	6.2	4.4	6.5	5.8	4.9
y	1.9	4.8	0.8	3.9	1.2	1.7	3.0

17. Compute the least-squares regression line for predicting y from x given the following summary statistics:

$$\bar{x} = 5 \qquad s_x = 2 \qquad \bar{y} = 1350$$
$$s_y = 100 \qquad r = 0.70$$

18. Compute the least-squares regression line for predicting y from x given the following summary statistics:

$$\bar{x} = 8.1 \qquad s_x = 1.2 \qquad \bar{y} = 30.4$$
$$s_y = 1.9 \qquad r = -0.85$$

19. In a hypothetical study of the relationship between the income of parents (x) and the IQs of their children (y), the following summary statistics were obtained:

$$\bar{x} = 45{,}000 \qquad s_x = 20{,}000 \qquad \bar{y} = 100$$
$$s_y = 15 \qquad r = 0.40$$

Find the equation of the least-squares regression line for predicting IQ from income.

20. Assume that in a study of educational level in years (x) and income (y), the following summary statistics were obtained:

$$\bar{x} = 12.8 \qquad s_x = 2.3 \qquad \bar{y} = 41{,}000$$
$$s_y = 15{,}000 \qquad r = 0.60$$

Find the equation of the least-squares regression line for predicting income from educational level.

Working with the Concepts

21. Price of eggs and milk: The following table presents the average price in dollars for a dozen eggs and a gallon of milk for each month from January through December 2012.

Dozen Eggs	Gallon of Milk
1.94	3.58
1.80	3.52
1.77	3.50
1.83	3.47
1.69	3.43
1.67	3.40
1.65	3.43
1.88	3.47
1.89	3.47
1.96	3.52
1.96	3.54
2.01	3.58

Source: Bureau of Labor Statistics

a. Compute the least-squares regression line for predicting the price of milk from the price of eggs.

b. If the price of eggs differs by $0.25 from one month to the next, by how much would you expect the price of milk to differ?

c. Predict the price of milk in a month when the price of eggs is $1.95.

22. Government funding: The following table presents the budget (in millions) for selected organizations that received U.S. government funding for arts and culture in both 2006 and 2012.

Organization	2006	2012
Corporation for Public Broadcasting	460	444
Institute of Museum and Library Services	247	232
National Endowment for the Humanities	142	146
National Endowment for the Arts	124	146
National Gallery of Art	95	128
Kennedy Center for the Performing Arts	18	23
Commission of Fine Arts	2	2
Advisory Council on Historic Preservation	5	6

Source: National Endowment for the Arts

a. Compute the least-squares regression line for predicting the 2012 budget from the 2006 budget.

b. If two institutions have budgets that differ by 10 million dollars in 2006, by how much would you predict their budgets to differ in 2012?

c. Predict the 2012 budget for an organization whose 2006 budget was 100 million dollars.

23. Pass the ball: The NFL Scouting Combine is an event at which football scouts evaluate the abilities of some top college prospects. Following are heights in inches and weights in pounds for some of the quarterbacks at the 2013 Combine.

Name	Height	Weight
Matt Barkley	75	227
Tyler Bray	78	232
Zac Dysert	75	231
Mike Glennon	79	225
Marqueis Gray	75	240
Landry Jones	76	225
Collin Klein	77	226
E. J. Manuel	77	237
Ryan Nassib	74	227
Sean Renfree	75	219
Geno Smith	74	218
Brad Sorensen	76	237
James Vandenberg	75	226
Tyler Wilson	74	215

Source: SB Nation

a. Compute the least-squares regression line for predicting weight from height.

b. Is it possible to interpret the y-intercept? Explain.

c. If two quarterbacks differ in height by two inches, by how much would you predict their weights to differ?

d. Predict the weight of a quarterback who is 74.5 inches tall.

e. Geno Smith is 74 inches tall and weighs 218 pounds. Does he weigh more or less than the weight predicted by the least-squares regression line?

24. Carbon footprint: Carbon dioxide (CO_2) is produced by burning fossil fuels such as oil and natural gas, and has been connected to global warming. The following table presents the average amounts (in metric tons) of CO_2 emissions for the years 1990–2006 per person in the United States and per person in the rest of the world.

Year	Non-U.S.	U.S.
1990	3.6	19.2
1991	3.6	19.0
1992	3.5	18.8
1993	3.3	19.7
1994	3.3	19.8
1995	3.3	19.5
1996	3.4	19.7
1997	3.4	20.0
1998	3.3	19.5
1999	3.2	19.6
2000	3.3	20.0
2001	3.3	19.6
2002	3.3	19.6
2003	3.5	19.4
2004	3.7	19.6
2005	3.8	19.5
2006	3.7	19.0

Source: World Bank

a. Compute the least-squares regression line for predicting U.S. emissions from non-U.S. emissions.

b. If the non-U.S. emissions differ by 0.2 from one year to the next, by how much would you predict the U.S. emissions to differ?

c. Predict the U.S. emissions for a year when the non-U.S. emissions level is 3.4.

25. Foot temperatures: Foot ulcers are a common problem for people with diabetes. Higher skin temperatures on the foot indicate an increased risk of ulcers. In a study carried out at the Colorado School of Mines, skin temperatures on both feet were measured, in degrees Fahrenheit, for 18 diabetic patients. The results are presented in the following table.

Left Foot	Right Foot	Left Foot	Right Foot
80	80	76	81
85	85	89	86
75	80	87	82
88	86	78	78
89	87	80	81
87	82	87	82
78	78	86	85
88	89	76	80
89	90	88	89

Source: Kimberly Anderson, M.S. thesis, Colorado School of Mines

a. Compute the least-squares regression line for predicting the right foot temperature from the left foot temperature.
b. Construct a scatterplot of the right foot temperature (y) versus the left foot temperature (x). Graph the least-squares regression line on the same axes.
c. If the left foot temperatures of two patients differ by 2 degrees, by how much would you predict their right foot temperatures to differ?
d. Predict the right foot temperature for a patient whose left foot temperature is 81 degrees.

26. Mortgage payments: The following table presents interest rates, in percent, for 30-year and 15-year fixed-rate mortgages, for January through December, 2012.

30-Year	15-Year	30-Year	15-Year
3.92	3.20	3.55	2.85
3.89	3.16	3.60	2.86
3.95	3.20	3.47	2.78
3.91	3.14	3.38	2.69
3.80	3.03	3.35	2.66
3.68	2.95	3.35	2.66

Source: Freddie Mac

a. Compute the least-squares regression line for predicting the 15-year rate from the 30-year rate.
b. Construct a scatterplot of the 15-year rate (y) versus the 30-year rate (x). Graph the least-squares regression line on the same axes.

c. Is it possible to interpret the y-intercept? Explain.
d. If the 30-year rate differs by 0.3 percent from one month to the next, by how much would you predict the 15-year rate to differ?
e. Predict the 15-year rate for a month when the 30-year rate is 3.5 percent.

27. Blood pressure: A blood pressure measurement consists of two numbers: the systolic pressure, which is the maximum pressure taken when the heart is contracting, and the diastolic pressure, which is the minimum pressure taken at the beginning of the heartbeat. Blood pressures were measured, in millimeters, for a sample of 16 adults. The following table presents the results.

Systolic	Diastolic	Systolic	Diastolic
134	87	133	91
115	83	112	75
113	77	107	71
123	77	110	74
119	69	108	69
118	88	105	66
130	76	157	103
116	70	154	94

Based on results published in the *Journal of Human Hypertension*

a. Compute the least-squares regression line for predicting the diastolic pressure from the systolic pressure.
b. Is it possible to interpret the y-intercept? Explain.
c. If the systolic pressures of two patients differ by 10 millimeters, by how much would you predict their diastolic pressures to differ?
d. Predict the diastolic pressure for a patient whose systolic pressure is 125 millimeters.

28. Butterfly wings: Do larger butterflies live longer? The wingspan (in millimeters) and the lifespan in the adult state (in days) were measured for 22 species of butterfly. Following are the results.

Wingspan	Lifespan	Wingspan	Lifespan
35.5	19.8	25.9	32.5
30.6	17.3	31.3	27.5
30.0	27.5	23.0	31.0
32.3	22.4	26.3	37.4
23.9	40.7	23.7	22.6
27.7	18.3	27.1	23.1
28.8	25.9	28.1	18.5
35.9	23.1	25.9	32.3
25.4	24.0	28.8	29.1
24.6	38.8	31.4	37.0
28.1	36.5	28.5	33.7

Source: *Oikos Journal of Ecology* 105:41–54

a. Compute the least-squares regression line for predicting the lifespan from the wingspan.
b. Is it possible to interpret the y-intercept? Explain.
c. If the wingspans of two butterflies differ by 2 millimeters, by how much would you predict their lifespans to differ?
d. Predict the lifespan for a butterfly whose wingspan is 28.5 millimeters.

29. Interpreting technology: The following display from the TI-84 Plus calculator presents the least-squares regression line for predicting a student's score on a statistics exam (y) from the number of hours spent studying (x).

a. Write the equation of the least-squares regression line.
b. What is the correlation between the score and the time spent studying?
c. Predict the score for a student who studies for 10 hours.

30. **Interpreting technology:** The following display from the TI-84 Plus calculator presents the least-squares regression line for predicting the price of a certain stock (y) from the prime interest rate in percent (x).

a. Write the equation of the least-squares regression line.
b. What is the correlation between the interest rate and the yield of the stock?
c. Predict the price when the prime interest rate is 5%.

31. **Interpreting technology:** The following MINITAB output presents the least-squares regression line for predicting the concentration of ozone in the atmosphere from the concentration of oxides of nitrogen (NOx).

```
The regression equation is
Ozone = 33.8127 + 1.21015 NOx
```

Predictor	Coef	SE Coef	T	P
Constant	33.8127	1.06035	31.89	0.000
NOx	1.21015	0.09047	12.38	0.000

a. Write the equation of the least-squares regression line.
b. Predict the ozone concentration when the NOx concentration is 21.4.

32. **Interpreting technology:** The following MINITAB output presents the least-squares regression line for predicting the score on a final exam from the score on a midterm exam.

```
The regression equation is
Final = 23.7789 + 0.71384 Midterm
```

Predictor	Coef	SE Coef	T	P
Constant	23.7789	4.46723	5.323	0.000
Midterm	0.71384	0.13507	5.285	0.000

a. Write the equation of the least-squares regression line.
b. Predict the final exam score for a student who scored 75 on the midterm.

33. **Interpreting technology:** A business school professor computed a least-squares regression line for predicting the salary in $1000s for a graduate from the number of years of experience. The results are presented in the following Excel output.

	Coefficients
Intercept	55.91275257
Experience	2.58289361

a. Write the equation of the least-squares regression line.
b. Predict the salary for a graduate with 5 years of experience.

34. **Interpreting technology:** A biologist computed a least-squares regression line for predicting the brain weight in grams of a bird from its body weight in grams. The results are presented in the following Excel output.

	Coefficients
Intercept	3.79229595
Body weight	0.08063922

a. Write the equation of the least-squares regression line.
b. Predict the brain weight for a bird whose body weight is 300 grams.

Extending the Concepts

35. **Least-squares regression line for z-scores:** The following table presents math and verbal SAT scores for six freshmen.

Verbal (x)	428	386	653	316	438	323
Math (y)	373	571	686	319	607	440

a. Compute the correlation coefficient between math and verbal SAT score.
b. Compute the mean $\bar{x}$ and the standard deviation s_x for the verbal scores.
c. Compute the mean $\bar{y}$ and the standard deviation s_y for the math scores.
d. Compute the least-squares regression line for predicting math score from verbal score.
e. Compute the z-score for each x-value:
$$z_x = \frac{x - \bar{x}}{s_x}$$
f. Compute the z-score for each y-value:
$$z_y = \frac{y - \bar{y}}{s_y}$$
g. Compute the correlation coefficient r between z_x and z_y. Is it the same as the correlation between math and verbal SAT scores?
h. Compute the least-squares regression line for predicting z_y from z_x. Explain why the equation of the line is $\hat{z}_y = r z_x$.

Answers to Check Your Understanding Exercises for Section 4.2

1. $\hat{y} = 11.0358 + 0.8226x$

2. a. 80   **b.** 15 points higher

3. a. The length of time it takes to freeze water in a freezer set to 0°F

    b. No interpretation

4. No. The least-squares line does not predict the effect of changing the explanatory variable.

SECTION 4.3	Features and Limitations of the Least-Squares Regression Line

Objectives

1. Understand the importance of avoiding extrapolation
2. Compute residuals and state the least-squares property
3. Construct and interpret residual plots
4. Determine whether outliers are influential
5. Compute and interpret the coefficient of determination

Objective 1 Understand the importance of avoiding extrapolation

Don't Make Predictions Outside the Range of the Data

Table 4.2 in Section 4.2 presents the sizes and selling prices of houses in a certain neighborhood. The sizes of the houses range from 2521 to 3198 square feet. The least-squares regression line was computed to be $\hat{y} = 160.1939 + 0.0992x$. A new house of size 4000 square feet is built. Should we use the least-squares regression line to predict its selling price? The answer is no. We can compute the least-squares estimate, which is $\hat{y} = 160.1939 + 0.0992(4000) = 557.0$ thousand dollars, or \$557,000. But this prediction is unreliable because the value 4000 is outside the range of the data. While we know that the relationship between size and selling price is linear within the range 2521 to 3198 square feet, it may not be linear outside that range (see Figure 4.9).

Making predictions for values of the explanatory variable that are outside the range of the data is called **extrapolation**. In general, extrapolation can lead to unreliable predictions.

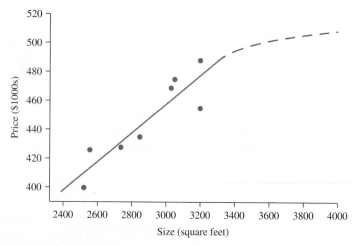

Figure 4.9 The size and selling price have a linear relationship within the range of observed values, but the relationship may become nonlinear outside that range. Therefore, predictions based on values outside the range of the data are not reliable.

SUMMARY

Do not use the least-squares regression line to make predictions for x-values that are outside the range of the data. The linear relationship may not hold there.

Check Your Understanding

1. A sample of adults was studied to determine the relationship between education level and annual income. The least-squares regression line for predicting income from education level was computed to be $\hat{y} = 2812 + 3375x$, where x is the number of years of education and $\hat{y}$ is the predicted annual income. The number of years of education among the people in the sample ranged from 8 to 18.
   a. If possible, use the least-squares regression line to predict the annual income of a person with 12 years of education. If not possible, explain why not.
   b. If possible, use the least-squares regression line to predict the annual income of a person with 6 years of education. If not possible, explain why not.

Answers are on page 189.

Objective 2 Compute residuals and state the least-squares property

Computing residuals

Figure 4.10 presents a scatterplot of selling price versus size for houses, with the least-squares regression line superimposed. For one point, (2555, 426), we have drawn a vertical line from the point to the least-squares regression line. The difference between the y-value of the point and the y-value on the line is called the *residual*. Note that residuals are positive for points above the line and negative for points below the line.

DEFINITION

Given a point (x, y) on a scatterplot, and the least-squares regression line $\hat{y} = b_0 + b_1x$, the **residual** for the point (x, y) is the difference between the observed value y and the predicted value $\hat{y}$.
$$\text{Residual} = y - \hat{y}$$

EXAMPLE 4.8   Compute a residual

Compute the residual for the point $(2555, 426)$ shown in Figure 4.10.

Figure 4.10   The residual is the difference between the observed value $y = 426$ and the predicted value $\hat{y} = 413.6$, which is $426 - 413.6 = 12.4$.

Solution

The y-value of the point is 426. The equation of the least-squares regression line was computed in Section 4.2 to be $\hat{y} = 160.1939 + 0.0992x$. The predicted value $\hat{y}$ is therefore $\hat{y} = 160.1939 + (0.0992)(2555) = 413.6$. The residual is

$$y - \hat{y} = 426 - 413.6 = 12.4$$

> ### Explain It Again
>
> **What the sum of squared residuals measures:** A residual for a point measures how close the predicted value is to the observed value for that point. The sum of squared residuals measures how close the predicted values are to the observed values overall. The smaller the sum of squared residuals, the better the predictions, on the whole.

Note that the magnitude of the residual is just the vertical distance from the point to the least-squares line. In Section 4.2, we described the least-squares line as the line for which the sum of the squared vertical distances is as small as possible. It follows that if we square each residual and add up the squares, the sum is less for the least-squares regression line than for any other line. This is the **least-squares property**.

The Least-Squares Property

The least-squares regression line satisfies the least-squares property. This means that the sum of the squared residuals is less for the least-squares regression line than for any other line.

Objective 3 Construct and interpret residual plots

Construct a Residual Plot to Determine Whether a Relationship Is Linear

In Section 4.1, we learned that the correlation coefficient should be used only when the relationship between x and y is linear. The same holds true for the least-squares regression line. When the scatterplot follows a curved pattern, it does not make sense to summarize it with a straight line. To illustrate this, Figure 4.11 presents a plot of a relationship that is nonlinear. The least-squares regression line does not fit the data well.

Figure 4.11   The relationship between the explanatory and outcome variables in this plot is not linear. The least-squares regression line does not fit the data well, and should not be used for prediction.

The residual plot

A **residual plot** is a plot in which the residuals are plotted against the values of the explanatory variable x. In other words, the points on the residual plot are $(x, y - \hat{y})$. When two variables have a linear relationship, the residual plot will not exhibit any noticeable pattern. If the residual plot does exhibit a pattern, such as a curved pattern, then the variables do not have a linear relationship, and the least-squares regression line should not be used.

Figure 4.12 presents a residual plot, produced by MINITAB, for the house data presented in Table 4.2. The residual plot does not exhibit any particular pattern, which suggests that the variables have a linear relationship. Note that a horizontal line is drawn through the value 0 on the y-axis. This can help make a pattern stand out, if there is one. Step-by-step instructions for constructing residual plots with the TI-84 Plus, with MINITAB, and with Excel are presented in the Using Technology section on page 184.

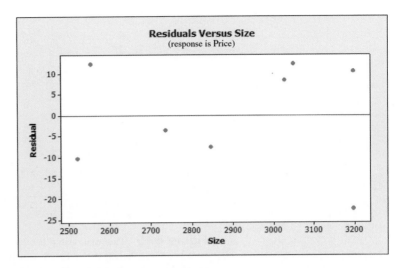

Figure 4.12   Residual plot for the data in Table 4.2. The plot exhibits no noticeable pattern, so use of the least-squares regression line is appropriate.

A nonlinear relationship can often be detected from a scatterplot, without constructing a residual plot. In some cases, however, curvature is easier to detect on a residual plot. As an example, Figure 4.13 presents a scatterplot with the least-squares regression line superimposed. The line appears to fit almost perfectly. Figure 4.14 presents the residual plot. The curvature is clearly evident.

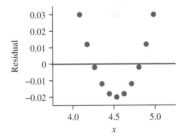

Figure 4.13   Scatterplot showing what appears to be a nearly perfect linear relationship.

Figure 4.14   The residual plot for the scatterplot in Figure 4.13. The nonlinearity of the relationship is now clear.

SUMMARY

- When a residual plot exhibits a noticeable pattern, the variables do not have a linear relationship, and the least-squares regression line should not be used.
- When a residual plot exhibits no noticeable pattern, the least-squares line may be used to describe the relationship between the variables.

The relationship can be nonlinear even when the correlation is large

If the correlation coefficient r is close to 1 or -1, it is tempting to assume that x and y have a linear relationship. However, this is not necessarily the case. Sometimes r can be close to 1 or -1 when the relationship is far from linear.

For example, consider the scatterplot in Figure 4.15 (page 180). The least-squares regression line is superimposed, and clearly does not describe the data well. However, the correlation coefficient is $r = 0.928$, which is fairly close to 1. The correlation coefficient is intended to be used only when variables have a linear relationship, and can be misleading otherwise. The only reliable way to determine whether two variables have a linear relationship is to look at a scatterplot or a residual plot.

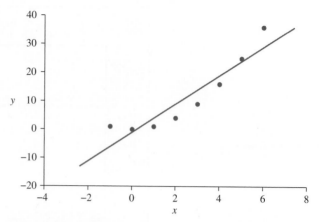

Figure 4.15 The relationship between x and y is clearly nonlinear. However, the correlation coefficient is $r = 0.928$, which is large. The correlation coefficient can be misleading when the relationship is not linear.

SUMMARY

Do not rely on the correlation coefficient to determine whether two variables have a linear relationship. Even when the correlation is close to 1 or to -1, the relationship may not be linear. To determine whether two variables have a linear relationship, construct a scatterplot or a residual plot.

Check Your Understanding

2. For each of the following residual plots, determine whether a linear model is appropriate.

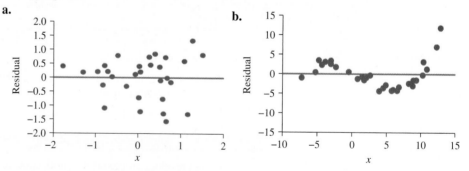

Answers are on page 189.

Objective 4 Determine whether outliers are influential

Outliers and Influential Points

Recall that an outlier is a point in a data set that is detached from the bulk of the data. Figure 4.16 presents a scatterplot of the total area of farmland (including ranches) plotted against the total land area, for each of the 50 states of the United States. Both Texas and Alaska are outliers. The blue line on the plot is the least-squares regression line computed for the 48 states not including Texas or Alaska. The red line is the least-squares regression line obtained when Texas is added to the 48 states. Including Texas moves the line somewhat. The green line is the least-squares regression line obtained when Alaska is added to the 48 states. Including Alaska causes a big shift in the position of the line.

An outlier that causes a big shift in the position of the line is said to be an *influential point*. In Figure 4.16, Alaska is an influential point, and Texas is somewhat influential as well. Influential points are troublesome, because the least-squares regression line is supposed to summarize all the data, rather than reflect the position of a single point.

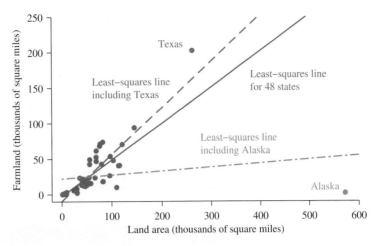

Figure 4.16 Scatterplot of farm area versus total land area for U.S. states. The blue solid line on the plot is the least-squares regression line computed for the 48 states not including Texas or Alaska. The red dashed line is the least-squares regression line for 49 states including Texas. Including Texas moves the line somewhat. The green dash-dot line is the least-squares regression line for 49 states including Alaska. Including Alaska causes a big shift in the position of the line.

DEFINITION

An **influential point** is a point that, when included in a scatterplot, strongly affects the position of the least-squares regression line.

When a scatterplot contains outliers, the least-squares regression line should be computed both with and without each outlier, to determine which outliers are influential. If there is an influential point, the least-squares regression line should be computed both with and without the point, and the equations of both lines should be reported.

SUMMARY

When a scatterplot contains outliers:

- Compute the least-squares regression line both with and without each outlier to determine which outliers are influential.
- Report the equations of the least-squares regression line both with and without each influential point.

Check Your Understanding

3. In each of the following plots, one point is an outlier. The blue solid line is the least-squares regression line computed without using the outlier, and the red dashed line is the least-squares regression line computed by including the outlier. State whether the outlier is influential.

a.

b.

Answers are on page 189.

Objective 5 Compute and interpret the coefficient of determination

The Coefficient of Determination

In Section 4.1, we saw that when two variables have a linear relationship, the correlation coefficient r tells us how strong the relationship is. Recall that when r is close to 1 or -1, the points are tightly clustered around the least-squares regression line. This means that the explanatory variable x tells us a lot about the outcome variable y, so the predictions of the least-squares regression line are likely to be close to the actual values. When r is close to 0, the points are more loosely clustered around the least-squares regression line. The explanatory variable x will not tell us much about the outcome variable y, and the least-squares predictions will not be as good. The quantity most often used to measure how well the least-squares regression line fits the data is r^2, the square of the correlation coefficient. The quantity r^2 is called the *coefficient of determination*. The closer r^2 is to 1, the closer the predictions made by the least-squares regression line are to the actual values, on average.

DEFINITION

The **coefficient of determination** is r^2, the square of the correlation coefficient r.

Explained variation and unexplained variation

Why is the coefficient of determination the best way to measure how close the predictions made by the least-squares regression line are to the actual values? As we will see, it turns out that some of the variation in the outcome variable is explained by the least-squares regression line and some is not. The coefficient of determination can be thought of as the proportion of the variation in the outcome variable that is explained by the least-squares regression line.

To illustrate this idea, Figure 4.17 presents a scatterplot on which two lines have been superimposed. One is the least-squares regression line and the other is the horizontal line $y = \bar{y}$, which goes through the sample mean $\bar{y}$. The sample mean $\bar{y}$ represents the center of the y-values. We have highlighted one particular point (x, y). Its y-value differs from the central value $\bar{y}$; the difference is $y - \bar{y}$. This difference can be divided into two parts. The first part, between the points $(x, \hat{y})$ and $(x, \bar{y})$ in Figure 4.17, is the difference between the central value $\bar{y}$ and the value $\hat{y}$ predicted for this particular x-value. This difference, $\hat{y} - \bar{y}$, is called the difference explained by the least-squares regression line, or the **explained difference**. The second part, between the points (x, y) and $(x, \hat{y})$ in Figure 4.17, is the difference between the observed data value y and the predicted value $\hat{y}$. This difference, which is the residual $y - \hat{y}$, is caused by factors unrelated to the least-squares regression line and is called the difference unexplained by the least-squares regression line, or the **unexplained difference**.

Now the better the least-squares predictions are, the smaller the unexplained differences will be, on the whole. We measure the size of the unexplained differences by squaring them

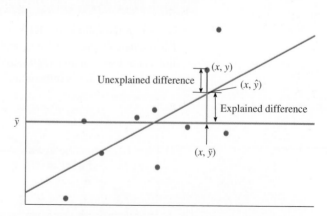

Figure 4.17   The red line between the points (x, y) and $(x, \hat{y})$ represents the difference $y - \hat{y}$, which is the unexplained difference. The green line between the points $(x, \hat{y})$ and $(x, \bar{y})$ represents the difference $\hat{y} - \bar{y}$, which is the explained difference.

and adding them together. The quantity thus obtained is called the **unexplained variation**. Then we do the same for the explained differences to obtain the **explained variation**. The sum of the explained variation and the unexplained variation is called the **total variation**.

The explained variation is

$$\text{Explained variation} = \sum (\hat{y} - \bar{y})^2$$

The unexplained variation is

$$\text{Unexplained variation} = \sum (y - \hat{y})^2$$

The total variation is

$$\text{Total variation} = \text{Unexplained variation} + \text{Explained variation}$$

It can be shown by advanced methods that the coefficient of determination is

$$r^2 = \frac{\text{Explained variation}}{\text{Total variation}}$$

Thus, r^2 measures the proportion of the total variation that is explained by the least-squares regression line.

Fortunately, we don't have to compute the explained and unexplained variation in order to compute r^2. All we need to do is to compute the correlation coefficient r and square it.

SUMMARY

- The coefficient of determination r^2 measures the proportion of the variation in the outcome variable that is explained by the least-squares regression line.
- The larger the value of r^2, the closer the predictions made by the least-squares regression line are to the actual values, on average.
- To compute the coefficient of determination, first compute the correlation coefficient r, then square it to obtain r^2.

EXAMPLE 4.9

Computing and interpreting the coefficient of determination

The correlation between size and selling price for the data in Table 4.1 was computed in Example 4.2 to be $r = 0.9005918$.

CAUTION

Don't round the correlation coefficient when computing the coefficient of determination.

a. What is the coefficient of determination?
b. How much of the variation in selling price is explained by the least-squares regression line?

Solution

a. To find the coefficient of determination, we square the correlation coefficient.

$$\text{Coefficient of determination} = r^2 = (0.9005918)^2 = 0.811$$

b. The coefficient of determination is 0.811. Therefore, 81.1% of the variation in selling price is explained by the least-squares regression line.

Check Your Understanding

4. For each of the following values of the correlation coefficient, determine how much of the variation in the outcome variable is explained by the least-squares regression line.

a. $r = 0.6$
b. $r = -0.9$
c. $r = 0$
d. $r = 1$

Answers are on page 189.

USING TECHNOLOGY

We use the data in Table 4.2 to illustrate the technology steps.

TI-84 PLUS

Constructing residual plots

Step 1. Enter the *x*-values from Table 4.2 into **L1** and the *y*-values into **L2**. Run the **LinReg(a+bx)** command, using the steps on page 171.

Step 2. Press **2nd, Y=** to access the STAT PLOTS menu and select Plot1 by pressing **1**.

Step 3. Select **On** and the scatterplot icon. Enter the residuals for the **Ylist** option by pressing **2nd, STAT** and then **7: RESID** (Figure A).

Step 4. Press **ZOOM** and then **9: ZoomStat** (Figure B).

Figure A

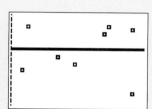

Figure B

MINITAB

Constructing residual plots

Step 1. Enter the *x*-values from Table 4.2 into **C1** and the *y*-values into **C2**.

Step 2. Click **Stat**, then **Regression**, then **Regression**.

Step 3. Select the *y*-variable (**C2**) as the **Response variable** and the *x*-variable (**C1**) as the **Predictor**.

Step 4. Click on **Graphs** and select *Regular* for **Residuals for Plots** and select *Individual Plots* for **Residual Plots**.

Step 5. In the **Residual versus the variables** field, enter **C1**.

Step 6. Click **OK** and then **OK**. Compute the least-squares regression line using the steps on page 171.

EXCEL

Constructing residual plots

Step 1. Enter the *x*-values from Table 4.2 into **Column A** and the *y*-values into **Column B**.

Step 2. Select **Data**, then **Data Analysis**. Highlight **Regression** and press **OK**.

Step 3. Enter the range of cells that contain the *x*-values in the **Input X Range** field and the range of cells that contain the *y*-values in the **Input Y Range** field.

Step 4. Select the **Residual Plots** option.

Step 5. Click **OK** (Figure C).

Figure C

SECTION 4.3   Exercises

Exercises 1–4 are the Check Your Understanding exercises located within the section.

Understanding the Concepts

In Exercises 5–10, fill in each blank with the appropriate word or phrase.

5. Making predictions for values of the explanatory variable that are outside of the range of the data is called _____.

6. A _____ is the difference between an observed value and a predicted value of the outcome variable.

7. The least-squares property says that the _____ is smaller for the least-squares regression line than for any other line.

8. An outlier that strongly affects the position of a least-squares regression line is said to be _____.

9. The coefficient of determination is the square of the _____.

10. The total variation is the sum of the _____ and the _____.

In Exercises 11–14, determine whether the statement is true or false. If the statement is false, rewrite it as a true statement.

11. When two values have a nonlinear relationship, the residual plot will exhibit a noticeable pattern.

12. When the correlation coefficient is close to 1 or −1, there is not necessarily a linear relationship between the variables.

13. The closer r^2 is to 0, the closer the predictions made by the least-squares regression line are to the actual values, on average.

14. The coefficient of determination may be interpreted as the proportion of variation in the outcome variable explained by the least-squares regression line.

Practicing the Skills

15. For the following data set:

x	1	2	3	4	5
y	5	6	9	8	7

    a. Compute the coefficient of determination.
    b. How much of the variation in the outcome variable is explained by the least-squares regression line?

16. For the following data set:

x	9	5	7	13	−8	−2	6	−10
y	3	3	31	36	0	3	−2	−14

    a. Compute the coefficient of determination.
    b. How much of the variation in the outcome variable is explained by the least-squares regression line?

17. For the following data set:

x	73	86	25	3	52	45	52
y	62	66	41	21	33	37	47

    a. Compute the least-squares regression line.
    b. Which point is an outlier?
    c. Remove the outlier and compute the least-squares regression line.
    d. Is the outlier influential? Explain.

18. For the following data set:

x	3.9	5.8	4.8	3.3	1.8	3.4	3.2
y	4.6	4.1	5.2	4.5	9.2	5.6	4.3

    a. Compute the least-squares regression line.
    b. Which point is an outlier?
    c. Remove the outlier and compute the least-squares regression line.
    d. Is the outlier influential? Explain.

19. **Least-squares OK?** Following is a residual plot produced by MINITAB. Was it appropriate to compute the least-squares regression line? Explain.

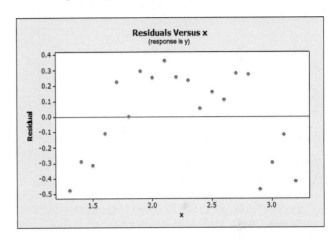

20. **Least-squares OK?** Following is a residual plot produced by MINITAB. Was it appropriate to compute the least-squares regression line? Explain.

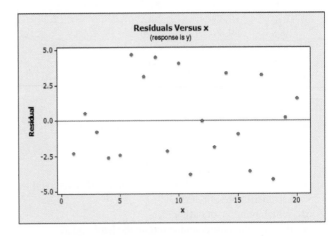

21. **Least-squares OK?** Following is a residual plot produced by MINITAB. Was it appropriate to compute the least-squares regression line? Explain.

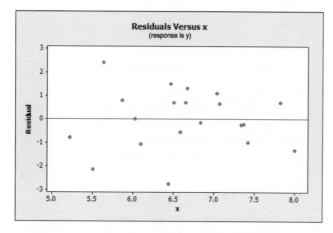

22. Least-squares OK? Following is a residual plot produced by MINITAB. Was it appropriate to compute the least-squares regression line? Explain.

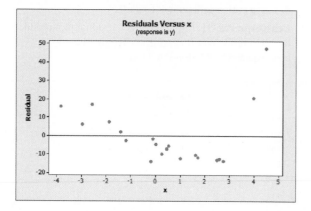

Working with the Concepts

23. Hot enough for you? The following table presents the temperature, in degrees Fahrenheit, and barometric pressure, in inches, on August 15 at 12 noon in Macon, Georgia, for the years 2004–2012.

Year	Pressure	Temperature
2004	30.16	80.1
2005	30.09	85.5
2006	29.99	87.0
2007	29.83	94.5
2008	29.86	84.0
2009	30.00	83.2
2010	29.98	80.1
2011	29.85	84.9
2012	30.01	84.0

Source: Weather Underground

a. Compute the least-squares regression line for predicting temperature from barometric pressure.
b. Compute the coefficient of determination.
c. Construct a scatterplot of the temperature (y) versus the barometric pressure (x).
d. Which point is an outlier?
e. Remove the outlier and compute the least-squares regression line for predicting temperature from barometric pressure.
f. Is the outlier influential? Explain.

g. Compute the coefficient of determination for the data set with the outlier removed. Is the proportion of variation explained by the least-squares regression line greater, less, or about the same without the outlier? Explain.

24. Presidents and first ladies: The following table presents the ages of the last 12 U.S. presidents and their wives on the first day of their presidencies.

Name	Her Age	His Age
Barack and Michelle Obama	45	47
George W. and Laura Bush	54	54
Bill and Hillary Clinton	45	46
George and Barbara Bush	55	64
Ronald and Nancy Reagan	59	69
Jimmy and Rosalynn Carter	49	52
Gerald and Betty Ford	56	61
Richard and Pat Nixon	56	56
Lyndon and Lady Bird Johnson	50	55
John and Jacqueline Kennedy	31	43
Dwight and Mamie Eisenhower	56	62
Harry and Bess Truman	60	60

a. Compute the least-squares regression line for predicting the president's age from the first lady's age.
b. Compute the coefficient of determination.
c. Construct a scatterplot of the presidents' ages (y) versus the first ladies' ages (x).
d. Which point is an outlier?
e. Remove the outlier and compute the least-squares regression line for predicting the president's age from the first lady's age.
f. Is the outlier influential? Explain.
g. Compute the coefficient of determination for the data set with the outlier removed. Is the proportion of variation explained by the least-squares regression line greater, less, or about the same without the outlier? Explain.

25. Mutant genes: In a study to determine whether the frequency of a certain mutant gene increases with age, the number of mutant genes per microgram of DNA was counted for each of 29 men. The results are presented in the following table.

Age	Number of Mutants	Age	Number of Mutants
44	100	53	49
46	52	80	41
43	93	57	66
46	193	59	109
45	77	65	264
53	35	68	69
58	105	47	29
54	68	61	101
68	71	61	148
55	82	70	78
55	267	70	195
51	43	72	83
52	181	82	449
57	29	83	202
52	58		

Source: *Proceedings of the National Academy of Sciences* 99:14952–14957

a. Compute the least-squares regression line for predicting number of mutants from age.

b. Construct a scatterplot of number of mutants (y) versus age (x).

c. Which point is an outlier?

d. Remove the outlier and compute the least-squares regression line for predicting number of mutants from age. Is the outlier influential? Explain.

e. Is the outlier influential? Explain.

26. Imports and exports: The following table presents the U.S. imports and exports (in billions of dollars) for each of 29 months.

Month	Imports	Exports	Month	Imports	Exports
Jan 2011	215.9	168.1	Apr 2012	230.9	184.3
Feb 2011	211.8	166.6	May 2012	230.5	184.2
Mar 2011	217.7	174.3	Jun 2012	227.6	185.2
Apr 2011	218.1	175.9	Jul 2012	226.8	183.4
May 2011	223.6	176.2	Aug 2012	226.1	182.1
Jun 2011	224.2	173.2	Sep 2012	228.4	186.8
Jul 2011	224.9	179.5	Oct 2012	225.3	182.7
Aug 2011	224.6	179.9	Nov 2012	231.6	185.2
Sep 2011	225.7	181.2	Dec 2012	227.0	188.7
Oct 2011	226.6	180.5	Jan 2013	229.4	186.7
Nov 2011	226.1	178.3	Feb 2013	231.0	187.1
Dec 2011	230.5	179.1	Mar 2013	222.3	185.2
Jan 2012	230.9	179.5	Apr 2013	227.7	187.6
Feb 2012	225.8	182.1	May 2013	232.1	187.1
Mar 2012	234.3	186.5			

Source: U.S. Census

a. Compute the least-squares regression line for predicting exports (y) from imports (x).

b. Compute the coefficient of determination.

c. The months with the two lowest exports are January and February 2011. Remove these points and compute the least-squares regression line. Is the result noticeably different?

d. Compute the coefficient of determination for the data set with January and February 2011 removed.

e. Two economists decide to study the relationship between imports and exports. One uses data from January 2011 through May 2013 and the other used data from March 2011 through May 2013. For which data set will the proportion of variance explained by the least-squares regression line be greater?

27. Energy consumption: The following table presents the average annual energy expenditures (in dollars) for housing units of various sizes (in square feet).

Size	Energy Expenditure
250	1087
750	1228
1250	1583
1750	1798
2250	1939
2750	2138
3250	2172
3750	2315

Source: Energy Information Administration

a. Compute the least-squares regression line for predicting energy consumption from house size.

b. Compute the coefficient of determination.

c. Construct a residual plot. Does the relationship appear to be linear?

d. True or false: If the coefficient of determination is large, the relationship must be linear.

28. Cost of health care: The following table presents the mean cost of a hospital stay, in $1000s, and the number of hospital stays, in millions, in the United States for each of 15 years.

Cost	Number	Cost	Number
9.8	34.3	15.0	37.2
10.3	34.1	17.3	37.8
10.3	34.2	19.7	38.2
10.6	34.2	20.5	38.7
11.3	34.7	22.3	39.2
11.8	34.9	24.0	39.5
12.5	35.5	26.1	39.5
13.7	36.4		

Source: Agency for Healthcare Research and Quality

a. Compute the least-squares regression line for predicting the number of stays from the cost.

b. Compute the coefficient of determination.

c. Construct a residual plot. Does the relationship appear to be linear?

d. True or false: If the coefficient of determination is large, the relationship must be linear.

29. Buy or rent? The following table presents the average sales price (in $1000s) and the average monthly rental price of homes in selected U.S. cities in July 2013.

City	Sales Price	Rental Price
Charlotte, NC	136.8	1156
Pensacola, FL	118.9	971
San Francisco, CA	590.6	2555
Washington, DC	331.6	2083
Springfield, MA	186.7	1404
Charleston, SC	160.7	1294
Cincinnati, OH	123.5	1110
Jacksonville, FL	127.3	1178
Trenton, NJ	202.0	1777
Warner Robins, GA	101.8	957
Santa Barbara, CA	438.5	2147
Denver, CO	237.9	1566
Tucson, AZ	146.8	1093
Concord, NH	184.1	1481
Providence, RI	211.2	1475

Source: Zillow

a. Compute the least-squares regression line for predicting average rental price (y) from average selling price (x).

b. Compute the coefficient of determination.

c. Construct a residual plot. Does the relationship appear to be linear?

d. True or false: If the coefficient of determination is large, the relationship must be linear.

30. Broadway: The following table presents the average ticket price (the average price paid per attendee) in dollars and the gross revenue (in millions of dollars) for Broadway productions for each season from 1993–1994 through 2012–2013.

Season	Ticket Price	Revenue	Season	Ticket Price	Revenue
'93–'94	43.90	356	'03–'04	66.41	771
'94–'95	44.91	406	'04–'05	66.70	769
'95–'96	46.09	436	'05–'06	71.83	862
'96–'97	47.21	499	'06–'07	76.28	939
'97–'98	48.61	558	'07–'08	76.45	938
'98–'99	50.39	588	'08–'09	77.61	943
'99–'00	52.99	603	'09–'10	85.79	1020
'00–'01	56.01	666	'10–'11	86.27	1081
'01–'02	58.72	643	'11–'12	92.38	1139
'02–'03	63.13	721	'12–'13	98.44	1139

Source: Broadway League

a. Compute the least-squares regression line for predicting gross revenue (y) from ticket price (x).

b. Compute the coefficient of determination.

c. Construct a residual plot. Does the relationship appear to be linear?

d. True or false: If the coefficient of determination is large, the relationship must be linear.

Extending the Concepts

31. Derive the coefficient of determination: For the house data in Table 4.1, the average selling price is $\bar{y} = 447.0$. Imagine that you do not know the size of any house, so you predict the selling price of each of them to be 447.0. This is equivalent to using the line $y = \bar{y}$ for prediction.

a. Compute the residual for each point using the line $y = \bar{y}$.

b. Compute the sum of squared residuals for the line $y = \bar{y}$.

c. Compute the residual for each point using the least-squares regression line.

d. Compute the sum of squared residuals for the least-squares regression line.

e. Determine how much better the least-squares predictions are by computing the difference

Decrease in sum of squared residuals

= (Sum of squared residuals for $y = \bar{y}$)
−(Sum of squared residuals for regression line)

f. Compute the ratio

$$\frac{\text{Decrease in sum of squared residuals}}{\text{Sum of squared residuals for } y = \bar{y}}$$

g. Show that the ratio obtained in part (f) is equal to the coefficient of determination.

32. Transforming a variable: The following table presents the speed (in mph) and the stopping distance (in feet) for a sample of cars.

Speed	Distance	Speed	Distance
4	12	16	44
7	14	17	48
8	19	18	49
9	20	19	57
10	18	20	59
11	22	22	75
12	25	23	78
13	32	24	87
14	38	25	90
15	38		

a. Compute the least-squares regression line for predicting stopping distance (y) from speed (x).

b. Construct a residual plot. Explain why the least-squares line is not an appropriate summary of the data.

c. For each data point, square the speed. This is x^2. Compute the least-squares regression line for predicting distance y from x^2.

d. Construct a residual plot. Is this line an appropriate summary?

e. Use the equation computed in part (c) to predict the stopping distance for a car whose speed is 15 mph.

33. Radon and cancer: Radon is a radioactive element that occurs naturally in soil as a result of the decay of uranium. Long-term exposure to radon, an odorless, invisible gas, is thought to increase the risk of cancer. Radon detectors are often placed in homes to determine whether radon levels are high enough to be dangerous. Two radon detectors were placed in different locations in the basement of a home in the eastern United States. Each provided an hourly measurement of the radon concentration, in units of picocuries/liter. The data are presented in the following table.

R_1	R_2	R_1	R_2	R_1	R_2	R_1	R_2
1.2	1.2	3.4	2.0	4.0	2.6	5.5	3.6
1.3	1.5	3.5	2.0	4.0	2.7	5.8	3.6
1.3	1.6	3.6	2.1	4.3	2.7	5.9	3.9
1.3	1.7	3.6	2.1	4.3	2.8	6.0	4.0
1.5	1.7	3.7	2.1	4.4	2.9	6.0	4.2
1.5	1.7	3.8	2.2	4.4	3.0	6.1	4.4
1.6	1.8	3.8	2.2	4.7	3.1	6.2	4.4
2.0	1.8	3.8	2.3	4.7	3.2	6.5	4.4
2.0	1.9	3.9	2.3	4.8	3.2	6.6	4.4
2.4	1.9	3.9	2.4	4.8	3.5	6.9	4.7
2.9	1.9	3.9	2.4	4.9	3.5	7.0	4.8
3.0	2.0	3.9	2.4	5.4	3.5		

a. Compute the least-squares line for predicting the radon concentration at location 2 from the concentration at location 1.

b. Construct a residual plot. Does the linear model seem appropriate?

c. Divide the data into two groups: points where $R_1 < 4$ in one group, and points where $R_1 \geq 4$ in the other. Compute the least-squares line and the residual plot for each group. Does the line describe either group well? Which one?

d. Explain why it might be a good idea to fit a linear model to part of these data, and a nonlinear model to the other.

34. How's your credit? The following table presents the average number of late payments and the average credit score for residents of selected U.S. cities in 2012.

Late Payments	Credit Score	Late Payments	Credit Score	Late Payments	Credit Score
0.50	729	0.24	777	0.33	762
0.50	727	0.34	776	0.60	733
0.45	758	0.25	771	0.33	769
0.52	731	0.53	724	0.39	760
0.65	755	0.51	740	0.51	737
0.26	778	0.73	702	0.33	772
0.30	772	0.65	729	0.37	767
0.58	729	0.68	714	0.33	762
0.67	739	0.40	739	0.39	761
0.41	761	0.53	746	0.33	776
0.58	717	0.57	711	0.21	784
0.46	752	0.45	766	0.41	767
0.42	767	0.25	787	0.43	752
0.50	742	0.52	731		

Source: Experian

a. Compute the correlation between the number of late payments and the credit score.
b. Compute the least-squares regression line for predicting average credit score (y) from average number of late payments (x).
c. Construct a residual plot. Are the residuals corresponding to $x \geq 0.5$ more spread out than those corresponding to $x < 0.5$?
d. Compute the least-squares regression line using only those points with $x < 0.5$.
e. Compute the least-squares regression line using only those points with $x \geq 0.5$.
f. Compute the coefficient of determination using only those points with $x < 0.5$.
g. Compute the coefficient of determination using only those points with $x \geq 0.5$.
h. In which situation is the average number of late payments more strongly related to average credit score, when the average number of late payments is less than 0.5 or when it is greater than or equal to 0.5?

Answers to Check Your Understanding Exercises for Section 4.3

1. a. $43,312
   b. It is not possible because 6 is outside the range of the data.

2. a. Appropriate   b. Not appropriate

3. a. Not influential   b. Influential
4. a. 36%   b. 81%   c. 0%   d. 100%

Chapter 4 Summary

Section 4.1: Bivariate data are data that consist of ordered pairs. A scatterplot provides a good graphical summary for bivariate data. When large values of one variable are associated with large values of the other, the variables are said to have a positive association. When large values of one variable are associated with small values of the other, the variables are said to have a negative association. When the points on a scatterplot tend to cluster around a straight line, the relationship is said to be linear.

The correlation coefficient r measures the strength of a linear relationship. The value of r is always between -1 and 1. Positive values of r indicate a positive linear association, while negative values of r indicate a negative linear association. Values near 1 or -1 indicate a strong linear association, while values near 0 indicate a weak linear association. The correlation coefficient should not be used when the relationship is not linear.

Correlation is not the same as causation. Even when two variables are highly correlated, it is not necessarily the case that changing the value of one of them will cause a change in the other.

Section 4.2: When two variables have a linear relationship, the points on a scatterplot tend to cluster around a straight line called the least-squares regression line. Given a value of the explanatory variable x, we can predict a value $\hat{y}$ for the outcome variable by substituting the value of x into the equation of the least-squares regression line. The slope of the least-squares regression line predicts the difference between the y-values for two points whose x-values differ by 1. The intercept of the least-squares regression line predicts the y-value of a point whose x-value is 0. The intercept can be interpreted only when the data set contains both positive and negative x-values.

Section 4.3: The least-squares regression line should be used for predictions only for values of x that lie within the range of the data used to compute the equation of the least-squares line. Making predictions outside the range of the data is called extrapolation, and such predictions are generally unreliable.

The difference between the observed value of an outcome variable and the value predicted by the least-squares regression line is called the residual. The sum of the squared residuals is smaller for the least-squares regression line than for any other line. This fact is summarized by saying that the least-squares line satisfies the least-squares property.

A plot of residuals versus values of the explanatory variable is called a residual plot. When a residual plot has no apparent pattern, a linear model is appropriate. The correlation coefficient can be misleading in this regard, because the correlation may be large even when the relationship is not linear.

The least-squares regression line explains some of the variation in the outcome variable. The proportion of variance explained is called the coefficient of determination, and is equal to the square of the correlation coefficient.

Vocabulary and Notation

bivariate data   152
coefficient of determination r^2 182
correlation coefficient r 155
explained difference   182
explained variation   183
explanatory variable (predictor variable)   165
extrapolation   176
influential point   181
least-squares property   178

least-squares regression line   165
linear   153
linear relationship   154
negative association   154
ordered pairs   152
outcome variable (response variable)   165
point of averages   168
positive association   154
predicted value   168

residual   177
residual plot   178
scatterplot   152
slope b_1 165
total variation   183
unexplained difference   182
unexplained variation   183
y-intercept b_0 165

Important Formulas

Correlation coefficient:

$$r = \frac{1}{n-1} \sum \left(\frac{x - \bar{x}}{s_x} \right) \left(\frac{y - \bar{y}}{s_y} \right)$$

Slope of least-squares regression line:

$$b_1 = r\frac{s_y}{s_x}$$

y-intercept of least-squares regression line:

$$b_0 = \bar{y} - b_1\bar{x}$$

Equation of least-squares regression line:

$$\hat{y} = b_0 + b_1 x$$

Chapter Quiz

1. Compute the correlation coefficient for the following data set.

x	2	5	6	7	11
y	15	9	6	4	1

2. The number of theaters showing the movie *Monsters University* x days after opening are presented in the following table.

x	Number of Theaters
5	4004
18	3739
22	3142
29	2186
36	1470
42	879

Source: http://www.the-numbers.com

Construct a scatterplot with number of days on the horizontal axis and number of theaters on the vertical axis.

3. Use the data in Exercise 2 to compute the correlation between the number of days after the opening of the movie and the number of theaters showing the movie. Is the association positive or negative? Weak or strong?

4. A scatterplot has a correlation of $r = -1$. Describe the pattern of the points.

5. In a survey of U.S. cities, it is discovered that there is a positive correlation between the number of paved streets in the city and the number of registered cars. Does this mean that paving more streets in the city will result in an increase in the number of registered cars? Explain.

6. The following table presents the average delay in minutes for departures and arrivals of domestic flights at O'Hare Airport in Chicago for selected years. (Data for 2013 are for January–July 2013.)

Year	Average Delay in Departures	Average Delay in Arrivals
2008	63.7	72.4
2009	57.9	64.4
2010	56.7	64.4
2011	59.4	68.5
2012	58.4	67.4
2013	60.7	70.8

Source: Bureau of Transportation Statistics.

Compute the least-squares regression line for predicting the delay in arrival time from the delay in departure time.

7. Use the least-squares regression line computed in Exercise 6 to predict the average delay in arrival time in a year when the average delay in departure time is 58.5 minutes.

8. Use the least-squares regression line computed in Exercise 6 to compute the residual for the year 2012.

9. Refer to Exercise 6. If the average delay in departure times differs by 2 minutes from one year to the next, by how much would you predict the average delay in arrival times to change?

10. A scatterplot has a least-squares regression line with a slope of 0. What is the correlation coefficient?

11. Compute the least-squares regression line for the following data set.

x	0	1	3	4	7	9
y	7	5	4	3	2	1

12. Two lines are drawn on a scatterplot. The sum of squared residuals for line A is 558.2, and the sum of squared residuals for line B is 723.1. Which of the following is true about the sum of squared residuals for the least-squares regression line?
    i. It will be greater than 723.1.
    ii. It will be between 558.2 and 723.1.
    iii. It will be less than 558.2.

13. A sample of students was studied to determine the relationship between sleeping habits and classroom performance. The least-squares regression line for predicting the score on a standardized exam from hours of sleep was computed to be $\hat{y} = 35.6 + 6.8x$, where x is the number of hours of sleep and $\hat{y}$ is the predicted exam score. The number of hours of sleep among the students in the sample ranged from 5 to 8. Should a prediction for the exam score be made for a student who slept for 10 hours? Why or why not?

14. In a scatterplot, the point $(-2, 7)$ is influential. If this point is removed from the scatterplot, which of the following describes the effect on the least-squares regression line?
    i. It will shift its position by a substantial amount.
    ii. It will shift its position slightly.
    iii. It will not shift its position at all.

15. The correlation coefficient for a data set is $r = -0.6$. How much of the variation in the outcome variable is explained by the least-squares regression line?

Review Exercises

1. **Predicting height:** The heights (y) and lengths of forearms (x) were measured in inches for a sample of men. The following summary statistics were obtained:

$$\bar{x} = 10.1 \quad s_x = 0.8 \quad \bar{y} = 70.1 \quad s_y = 2.5 \quad r = 0.81$$

   a. Compute the least-squares regression line for predicting height from forearm length.
   b. Joe's forearm is 1 inch longer than Sam's. How much taller than Sam do you predict Joe to be?
   c. Predict the height of a man whose forearm is 9.5 inches long.

2. **How much wood is in that tree?** For a sample of 12 trees, the volume of lumber (y) (in cubic meters) and the diameter (x) (in centimeters) at a fixed height above ground level was measured. The following summary statistics were obtained:

$$\bar{x} = 36.1 \quad s_x = 8.8 \quad \bar{y} = 0.86 \quad s_y = 0.49 \quad r = 0.94$$

   a. Compute the least-squares regression line for predicting volume from diameter.
   b. If the diameters of two trees differ by 8 centimeters, by how much do you predict their volumes to differ?
   c. Predict the volume for a tree whose diameter is 44 centimeters.

3. **How's your mileage?** Weight (in tons) and fuel economy (in mpg) were measured for a sample of seven diesel trucks. The results are presented in the following table.

Weight	8.00	24.50	27.00	14.50	28.50	12.75	21.25
Mileage	7.69	4.97	4.56	6.49	4.34	6.24	4.45

Source: Janet Yanowitz, Ph.D. thesis, Colorado School of Mines

   a. Compute the least-squares regression line for predicting mileage from weight.
   b. Construct a residual plot. Verify that a linear model is appropriate.
   c. If two trucks differ in weight by 5 tons, by how much would you predict their mileages to differ?
   d. Predict the mileage for trucks with a weight of 15 tons.

4. **How's your mileage?** Using the data in Exercise 3:
   a. Compute the correlation coefficient between weight and mileage.
   b. Compute the coefficient of determination.
   c. How much of the variation in mileage is explained by the least-squares regression line?

5. **Energy efficiency:** A sample of 10 households was monitored for one year. The household income (in $1000s) and the amount of energy consumed (in 10^{10} joules) were determined. The results follow.

Income	31	40	28	48	195	96	70	100	145	78
Energy	16	40	30	46	185	98	94	77	115	67

a. Compute the least-squares regression line for predicting energy consumption from income.
b. Construct a residual plot. Verify that a linear model is appropriate.
c. If two families differ in income by $12,000, by how much would you predict their energy consumptions to differ?
d. Predict the energy consumption for a family whose income is $50,000.

6. **Energy efficiency:** Using the data in Exercise 5:
a. Compute the correlation coefficient between income and energy consumption.
b. Compute the coefficient of determination.
c. How much of the variation in energy consumption is explained by the least-squares regression line?

7. **Pigskin:** In football, a turnover occurs when a team loses possession of the ball due to a fumble or an interception. Turnovers are bad when they happen to your team, but good when they happen to your opponent. The turnover margin for a team is the difference (Turnovers by opponent − Turnovers by team). The following table presents the turnover margin and the total number of wins for each team in the Southeastern Conference (SEC) in a recent season.

Team	Turnover Margin	Wins	Team	Turnover Margin	Wins
Florida	22	13	S. Carolina	−11	7
Alabama	6	12	Vanderbilt	9	7
Georgia	−3	10	Arkansas	−9	5
Ole Miss	−2	9	Auburn	−8	5
LSU	−1	8	Tennessee	2	5
Kentucky	5	7	Miss. State	−4	4

Source: www.secsports.com

a. Compute the least-squares regression line for predicting team wins from turnover margin.
b. Construct a residual plot. Verify that a linear model is appropriate.
c. Which teams won more games than would be predicted from their turnover margin?

8. **Pigskin:** Using the data in Exercise 7:
a. Compute the correlation coefficient between turnover margin and wins.
b. Compute the coefficient of determination.
c. How much of the variation in wins for this SEC college football season is explained by the least-squares regression line?

9. **SAT scores:** The following table presents the number of years of study in English and language arts and the average SAT writing score for students who took the 2008 SAT exam.

Years of Study	0.5	1.0	2.0	3.0	4.0
SAT Score	419	444	463	456	498

Source: The College Board

a. Compute the least-squares regression line for predicting mean SAT score from years of study.
b. Construct a residual plot. Verify that a linear model is appropriate.
c. Predict the mean SAT score for students with 2.5 years of study.

10. **SAT scores:** Using the data in Exercise 9:
a. Compute the correlation coefficient between years of study and SAT score.
b. Compute the coefficient of determination.
c. How much of the variation in SAT score is explained by the least-squares regression line?

11. **Baby weights:** The average gestational age (time from conception to birth) of a newborn infant is about 40 weeks. The following table presents the gestational age (in weeks) and corresponding mean birth weight (in pounds) for female infants born in Canada.

Gestational Age	36	37	38	39	40	41	42	43
Birth Weight	6.1	6.6	7.0	7.4	7.7	7.9	8.0	8.1

Source: *Pediatrics* Vol. 108 No. 2

a. Compute the least-squares regression line for predicting the birth weight from the gestational age.
b. Compute the coefficient of determination.

    c. Construct a residual plot. Does the relationship appear to be linear?

    d. True or false: If the coefficient of determination is large, the relationship must be linear.

12. Commute times: Every morning, Tania leaves for work a few minutes after 7:00 A.M. For eight days, she keeps track of the time she leaves (the number of minutes after 7:00) and the number of minutes it takes her to get to work. Following are the results.

Time Leaving	13	14	16	30	20	12	9	17	16	10	16
Length of Commute	27	20	23	45	20	21	20	28	27	23	30

    a. Construct a scatterplot of the length of commute (y) versus the time leaving (x).

    b. Compute the least-squares regression line for predicting the length of commute from the time leaving.

    c. Compute the coefficient of determination.

    d. Which point is an outlier?

    e. Remove the outlier and compute the least-squares regression line for predicting the length of commute from the time leaving.

    f. Is the outlier influential? Explain.

    g. Compute the coefficient of determination for the data set with the outlier removed. Is the relationship stronger, weaker, or about equally strong without the outlier?

13. Interpret technology: The following display from the TI-84 Plus calculator presents the results from computing a least-squares regression line.

```
LinReg
y=a+bx
a=.3946
b=3.643061791
r²=.0041179441
r=.0641712093
```

    a. Write the equation of the least-squares regression line.

    b. Predict the value of y when the x-value is 10.

    c. What is the correlation between x and y?

    d. Is the linear relationship between x and y strong or weak? Explain.

14. Interpret technology: The following display from the TI-84 Plus calculator presents the results from computing a least-squares regression line.

```
LinReg
y=a+bx
a=-1.04349
b=3.045141227
r²=.9871860312
r=.9935723583
```

    a. Write the equation of the least-squares regression line.

    b. Predict the value of y when the x-value is 50.

    c. What is the correlation between x and y?

    d. Is the linear relationship between x and y strong or weak? Explain.

15. Interpret technology: The following output from MINITAB presents the results from computing a least-squares regression line.

```
The regression equation is
Y = 4.99971 + 0.20462 X


Predictor           Coef      SE Coef       T        P
Constant         4.99971      0.02477   201.81    0.000
X                0.20462      0.01115    18.36    0.000
```

    a. Write the equation of the least-squares regression line.

    b. Predict the value of y when the x-value is 25.

Write About It

1. Describe an example in which two variables are strongly correlated, but changes in one do not cause changes in the other.
2. Two variables x and y have a positive association if large values of x are associated with large values of y. Write an equivalent definition that describes what small values of x are associated with. Then write a definition for negatively associated random variables that describes what small values of y are associated with.
3. Explain why the predicted value $\hat{y}$ is always equal to $\bar{y}$ when $r = 0$.
4. If the slope of the least-squares regression line is negative, can the correlation coefficient be positive? Explain why or why not.
5. Describe conditions under which the slope of the least-squares line will be equal to the correlation coefficient.
6. Describe circumstances under which the sum of the squared residuals will equal zero. What conclusions can be drawn about the least-squares regression line in this case?
7. Explain why extrapolation may lead to unreliable results.
8. Explain how it is possible for a point to be an outlier without being an influential point.
9. Consider the case where there are only two ordered pairs in the data set. What must be true about the residuals?

Case Study: How Are Inflation And Unemployment Related?

The following table, reproduced from the chapter introduction, presents the inflation rate and unemployment rate, both in percent, for the years 1985–2012.

Year	Inflation	Unemployment	Year	Inflation	Unemployment
1985	3.8	7.2	1999	2.7	4.2
1986	1.1	7.0	2000	3.4	4.0
1987	4.4	6.2	2001	1.6	4.7
1988	4.4	5.5	2002	2.4	5.8
1989	4.6	5.3	2003	1.9	6.0
1990	6.1	5.6	2004	3.3	5.5
1991	3.1	6.8	2005	3.4	5.1
1992	2.9	7.5	2006	2.5	4.6
1993	2.7	6.9	2007	4.1	4.6
1994	2.7	6.1	2008	0.1	5.8
1995	2.5	5.6	2009	2.7	9.3
1996	3.3	5.4	2010	1.5	9.6
1997	1.7	4.9	2011	3.0	8.9
1998	1.6	4.5	2012	1.7	8.1

Source: Bureau of Labor Statistics

We will investigate some methods for predicting unemployment. First, we will try to predict the unemployment rate from the inflation rate.

1. Construct a scatterplot of unemployment (y) versus inflation (x). Do you detect any strong nonlinearity?
2. Compute the least-squares line for predicting unemployment from inflation.
3. Predict the unemployment in a year when inflation is 3.0%.
4. Compute the correlation coefficient between inflation and unemployment.
5. What proportion of the variance in unemployment is explained by inflation?

The relationship between inflation and unemployment is not very strong. However, if we are interested in predicting unemployment, we would probably want to predict next year's unemployment from this year's inflation. We can construct an equation to do this by matching each year's inflation with the next year's unemployment, as shown in the following table.

Year	This Year's Inflation	Next Year's Unemployment	Year	This Year's Inflation	Next Year's Unemployment
1985	3.8	7.0	1999	2.7	4.0
1986	1.1	6.2	2000	3.4	4.7
1987	4.4	5.5	2001	1.6	5.8
1988	4.4	5.3	2002	2.4	6.0
1989	4.6	5.6	2003	1.9	5.5
1990	6.1	6.8	2004	3.3	5.1
1991	3.1	7.5	2005	3.4	4.6
1992	2.9	6.9	2006	2.5	4.6
1993	2.7	6.1	2007	4.1	5.8
1994	2.7	5.6	2008	0.1	9.3
1995	2.5	5.4	2009	2.7	9.6
1996	3.3	4.9	2010	1.5	8.9
1997	1.7	4.5	2011	3.0	8.1
1998	1.6	4.2			

Source: Bureau of Labor Statistics

6. Compute the least-squares line for predicting next year's unemployment from this year's inflation.

7. Predict next year's unemployment if this year's inflation is 3.0%.

8. Compute the correlation coefficient between this year's inflation and next year's unemployment.

9. What proportion of the variance in next year's unemployment is explained by this year's inflation?

   If we are going to use data from this year to predict unemployment next year, why not use this year's unemployment to predict next year's unemployment? A model like this, in which previous values of a variable are used to predict future values of the same variable, is called an *autoregressive* model. The following table presents the data needed to fit this model.

Year	This Year's Unemployment	Next Year's Unemployment	Year	This Year's Unemployment	Next Year's Unemployment
1985	7.2	7.0	1999	4.2	4.0
1986	7.0	6.2	2000	4.0	4.7
1987	6.2	5.5	2001	4.7	5.8
1988	5.5	5.3	2002	5.8	6.0
1989	5.3	5.6	2003	6.0	5.5
1990	5.6	6.8	2004	5.5	5.1
1991	6.8	7.5	2005	5.1	4.6
1992	7.5	6.9	2006	4.6	4.6
1993	6.9	6.1	2007	4.6	5.8
1994	6.1	5.6	2008	5.8	9.3
1995	5.6	5.4	2009	9.3	9.6
1996	5.4	4.9	2010	9.6	8.9
1997	4.9	4.5	2011	8.9	8.1
1998	4.5	4.2			

Source: Bureau of Labor Statistics

10. Compute the least-squares line for predicting next year's unemployment from this year's unemployment.

11. Predict next year's unemployment if this year's unemployment is 5.0%.

12. Compute the correlation coefficient between this year's unemployment and next year's unemployment.

13. What proportion of the variance in next year's unemployment is explained by this year's unemployment?

14. Which of the three models do you think provides the best prediction of unemployment, the one using inflation in the same year, the one using inflation in the previous year, or the one using unemployment in the previous year? Explain.

15. The U.S. economy was in recession during parts of 1990 and 1991, most of 2001, and from December 2007 through June 2009. Compute the residuals for the least-squares line found in Exercise 10. During times of recession, does the least-squares line predict the following year's unemployment well? Does it tend to overpredict or underpredict?

Probability

Introduction

How likely is it that you will live to be 100 years old? The following table, called a *life table*, can be used to answer this question.

United States Life Table, Total Population

Age Interval	Proportion Surviving	Age Interval	Proportion Surviving
0–10	0.99123	50–60	0.94010
10–20	0.99613	60–70	0.86958
20–30	0.99050	70–80	0.70938
30–40	0.98703	80–90	0.42164
40–50	0.97150	90–100	0.12248

Source: Centers for Disease Control and Prevention

The column labeled "Proportion Surviving" presents the proportion of people alive at the beginning of an age interval who will still be alive at the end of the age interval. For example, among those currently age 20, the proportion who will still be alive at age 30 is 0.99050, or 99.050%. With an understanding of some basic concepts of probability, one can use the life table to compute the probability that a person of a given age will still be alive a given number of years from now. Life insurance companies use this information to determine how much to charge for life insurance policies. In the case study at the end of the chapter, we will use the life table to study some further questions that can be addressed with the methods of probability.

This chapter presents an introduction to probability. Probability is perhaps the only branch of knowledge that owes its existence to gambling. In the seventeenth century, owners of gambling houses hired some of the leading mathematicians of the time to calculate the chances that players would win certain gambling games. Later, people realized that many real-world problems involve chance as well, and since then the methods of probability have been used in almost every area of knowledge.

SECTION 5.1	Basic Concepts of Probability

Objectives

1. Construct sample spaces
2. Compute and interpret probabilities
3. Approximate probabilities by using the Empirical Method
4. Approximate probabilities by using simulation

At the beginning of a football game, a coin is tossed to decide which team will get the ball first. There are two reasons for using a coin toss in this situation. First, it is impossible to predict which team will win the coin toss, because there is no way to tell ahead of time whether the coin will land heads or tails. The second reason is that in the long run, over the course of many football games, we know that the home team will win about half of the tosses, and the visiting team will win about half. In other words, although we don't know what the outcome of a single coin toss will be, we do know what the outcome of a long series of tosses will be — they will come out about half heads and half tails.

A coin toss is an example of a **probability experiment**. A probability experiment is one in which we do not know what any individual outcome will be, but we do know how a long series of repetitions will come out. Another familiar example of a probability experiment is the rolling of a die. A die has six faces; the faces have from one to six dots. We cannot predict which face will turn up on a single roll of a die, but, assuming the die is evenly balanced (not loaded), we know that in the long run, each face will turn up one-sixth of the time.

The *probability* of an event is the proportion of times that the event occurs in the long run. So, for a "fair" coin, that is, one that is equally likely to come up heads as tails, the probability of heads is 1/2 and the probability of tails is 1/2.

> ### DEFINITION
>
> The **probability** of an event is the proportion of times the event occurs in the long run, as a probability experiment is repeated over and over again.

The South African mathematician John Kerrich carried out a famous study that illustrates the idea of the long-run proportion. Kerrich was in Denmark when World War II broke out and spent the war interned in a prisoner-of-war camp. To pass the time, he carried out a series of probability experiments, including one in which he tossed a coin 10,000 times and recorded each toss as a head or a tail.

Figure 5.1 (page 199) summarizes a computer-generated re-creation of Kerrich's study, in which the proportion of heads is plotted against the number of tosses. For example, it turned out that after 5 tosses, 3 heads had appeared, so the proportion of heads was $3/5 = 0.6$. After 100 tosses, 49 heads had appeared, so the proportion of heads was $49/100 = 0.49$. After 10,000 tosses, the proportion of heads was 0.4994, which is very close to the true probability of 0.5. The figure shows that the proportion varies quite a bit within the first few tosses, but the proportion settles down very close to 0.5 as the number of tosses becomes larger.

The fact that the long-run proportion approaches the probability is called the *law of large numbers*.

Explain It Again

Law of large numbers: The law of large numbers is another way to state our definition of probability.

> ### Law of Large Numbers
>
> The **law of large numbers** says that as a probability experiment is repeated again and again, the proportion of times that a given event occurs will approach its probability.

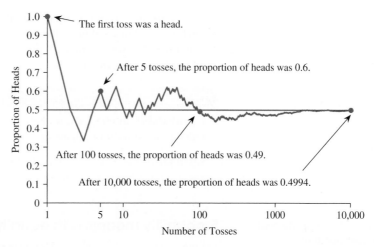

Figure 5.1 As the number of tosses increases, the proportion of heads fluctuates around the true probability of 0.5, and gets closer to 0.5. The horizontal axis is not drawn to scale.

Objective 1 Construct sample spaces

Probability Models

To study probability formally, we need some basic terminology. The collection of all the possible outcomes of a probability experiment is called a *sample space*.

DEFINITION

A **sample space** contains all the possible outcomes of a probability experiment.

EXAMPLE 5.1

Describe sample spaces

Describe a sample space for each of the following experiments.

    a. The toss of a coin
    b. The roll of a die
    c. Selecting a student at random from a list of 10,000 students at a large university
    d. Selecting a simple random sample of 100 students from a list of 10,000 students

Solution

    a. There are two possible outcomes for the toss of a coin: Heads and Tails. So a sample space is {Heads, Tails}.
    b. There are six possible outcomes for the roll of a die: the numbers from 1 to 6. So a sample space is {1, 2, 3, 4, 5, 6}.
    c. Each of the 10,000 students is a possible outcome for this experiment, so the sample space consists of the 10,000 students.
    d. This sample space consists of every group of 100 students that can be chosen from the population of 10,000 — in other words, every possible simple random sample of size 100. This is a huge number of outcomes; it can be written approximately as a 6 followed by 241 zeros. This is larger than the number of atoms in the universe.

We are often concerned with occurrences that consist of several outcomes. For example, when rolling a die, we might be interested in the possibility of rolling an odd number. Rolling an odd number corresponds to the collection of outcomes {1, 3, 5} from the sample space {1, 2, 3, 4, 5, 6}. In general, a collection of outcomes of a sample space is called an *event*.

DEFINITION

An **event** is an outcome or a collection of outcomes from a sample space.

Once we have a sample space for an experiment, we need to specify the probability of each event. This is done with a *probability model*. We use the letter "*P*" to denote probabilities. So, for example, if we toss a coin, we denote the probability that the coin lands heads by "*P*(Heads)."

DEFINITION

A **probability model** for a probability experiment consists of a sample space, along with a probability for each event.

Notation: If A denotes an event, the probability of the event A is denoted $P(A)$.

Objective 2 Compute and interpret probabilities

Probability models with equally likely outcomes

In many situations, the outcomes in a sample space are equally likely. For example, when we toss a coin, we usually assume that the two outcomes "Heads" and "Tails" are equally likely. We call such a coin a *fair* coin. Similarly, a fair die is one in which the numbers from 1 to 6 are equally likely to turn up. When the outcomes in a sample space are equally likely, we can use a simple formula to determine the probability of events.

Explain It Again

Fair and unfair: A fair coin or die is one for which all outcomes are equally likely. An unfair coin or die is one for which some outcomes are more likely than others.

Computing Probabilities with Equally Likely Outcomes

If a sample space has n **equally likely outcomes,** and an event A has k outcomes, then

$$P(A) = \frac{\text{Number of outcomes in } A}{\text{Number of outcomes in the sample space}} = \frac{k}{n}$$

EXAMPLE 5.2   Compute the probability of an event

A fair die is rolled. Find the probability that an odd number comes up.

Solution
The sample space has six equally likely outcomes: $\{1, 2, 3, 4, 5, 6\}$. The event of an odd number has three outcomes: $\{1, 3, 5\}$. The probability is

$$P(\text{odd number}) = \frac{3}{6} = \frac{1}{2}$$

EXAMPLE 5.3   Compute the probability of an event

In the Georgia Cash-4 Lottery game, a winning number between 0000 and 9999 is chosen at random, with all the possible numbers being equally likely. What is the probability that all four digits of the winning number are the same?

Solution
The outcomes in the sample space are the numbers from 0000 to 9999, so there are 10,000 equally likely outcomes in the sample space. There are 10 outcomes for which all the digits are the same: 0000, 1111, and so on up to 9999. The probability is

$$P(\text{all four digits the same}) = \frac{10}{10,000} = 0.001$$

The law of large numbers states that the probability of an event is the long-run proportion of times that the event occurs. An event that never occurs, even in the long run, has a probability of 0. This is the smallest probability an event can have. An event that occurs every time has a probability of 1. This is the largest probability an event can have.

SUMMARY

The probability of an event is always between 0 and 1. In other words, for any event A, $0 \leq P(A) \leq 1$.

If A cannot occur, then $P(A) = 0$.
If A is certain to occur, then $P(A) = 1$.

EXAMPLE 5.4   Computing probabilities

A family has three children. Denoting a boy by B and a girl by G, we can denote the genders of these children from oldest to youngest. For example, GBG means the oldest child is a girl, the middle child is a boy, and the youngest child is a girl. There are eight possible outcomes: BBB, BBG, BGB, BGG, GBB, GBG, GGB, and GGG. Assume these outcomes are equally likely.

a. What is the probability that there are two girls?
b. What is the probability that all three children are of the same gender?

Solution

a. Of the eight equally likely outcomes, the three outcomes BGG, GBG, and GGB correspond to having two girls. Therefore,

$$P(\text{Two girls}) = \frac{3}{8}$$

b. Of the eight equally likely outcomes, the two outcomes BBB and GGG correspond to having all children of the same gender. Therefore,

$$P(\text{All three have same gender}) = \frac{2}{8} = \frac{1}{4}$$

Check Your Understanding

1. In Example 5.4, what is the probability that the youngest child is a boy?

2. In Example 5.4, what is the probability that the oldest child and the youngest child are of the same gender?

Answers are on page 209.

EXAMPLE 5.5   Constructing a sample space

Cystic fibrosis is a disease of the mucous glands whose most common sign is progressive damage to the respiratory system and digestive system. This disease is inherited, as follows. A certain gene may be of type A or type a. Every person has two copies of the gene — one inherited from the person's mother, one from the person's father. If both copies are a, the person will have cystic fibrosis. Assume that a mother and father both have genotype Aa, that is, one gene of each type. Assume that each copy is equally likely to be transmitted to their child. What is the probability that the child will have cystic fibrosis?

Solution

Most of the work in solving this problem is in constructing the sample space. We'll do this in two ways. First, the tree diagram in Figure 5.2 on page 202 shows that there are four possible outcomes. In the tree diagram, the first two branches indicate the two possible outcomes, A and a, for the mother's gene. Then for each of these outcomes there are two branches indicating the possible outcomes for the father's gene. An alternate method is to construct a table like Table 5.1 on page 202.

Table 5.1

Mother's Gene	Father's Gene	Child's Genotype
A	A	AA
A	a	Aa
a	A	aA
a	a	aa

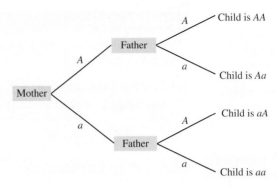

Figure 5.2 Tree diagram illustrating the four outcomes for the child's genotype

We can use either the table or the tree to list the outcomes. Listing the mother's gene first, the four outcomes are *AA*, *Aa*, *aA*, and *aa*. For one of the four outcomes, *aa*, the child will have cystic fibrosis. Therefore, the probability of cystic fibrosis is 1/4.

Check Your Understanding

3. A penny and a nickel are tossed. Each is a fair coin, which means that heads and tails are equally likely.
   a. Construct a sample space containing equally likely outcomes. Each outcome should specify the results for both coins.
   b. Find the probability that one coin comes up heads and the other comes up tails.

Answers are on page 209.

Sampling from a population is a probability experiment

In Section 1.1, we learned that statisticians collect data by drawing samples from populations. Sampling an individual from a population is a probability experiment. The population is the sample space, and the members of the population are equally likely outcomes. For this reason, the ideas of probability are fundamental to statistics.

EXAMPLE 5.6   Computing probabilities involving sampling

There are 10,000 families in a certain town. They are categorized by their type of housing as follows.

Own a house	4753
Own a condo	1478
Rent a house	912
Rent an apartment	2857

A pollster samples a single family at random from this population.

a. What is the probability that the sampled family owns a house?
b. What is the probability that the sampled family rents?

Solution

a. The sample space consists of the 10,000 households. Of these, 4753 own a house, so the probability that the sampled family owns a house is

$$P(\text{Owns a house}) = \frac{4753}{10,000} = 0.4753$$

b. The number of families who rent is $912 + 2857 = 3769$. Therefore, the probability that the sampled family rents is

$$P(\text{Rents}) = \frac{3769}{10,000} = 0.3769$$

In practice, of course, the pollster would sample many people, not just one. In fact, statisticians use the basic ideas of probability to draw conclusions about populations by studying samples drawn from them. In later chapters of this book, we will see how this is done.

Unusual events

As the name implies, an unusual event is one that is not likely to happen — in other words, an event whose probability is small. There are no hard-and-fast rules as to just how small a probability needs to be before an event is considered unusual, but 0.05 is commonly used.

> **SUMMARY**
>
> An **unusual event** is one whose probability is small.
>
> Sometimes people use the cutoff 0.05; that is, they consider any event whose probability is less than 0.05 to be unusual. But there are no hard-and-fast rules about this.

EXAMPLE 5.7

Determine whether an event is unusual

In a college of 5000 students, 150 are math majors. A student is selected at random, and turns out to be a math major. Is this an unusual event?

Solution

The sample space consists of 5000 students, each of whom is equally likely to be chosen. The event of choosing a math major consists of 150 students. Therefore,

$$P(\text{Math major is chosen}) = \frac{150}{5000} = 0.03$$

Since the probability is less than 0.05, then by the most commonly applied rule, this would be considered an unusual event.

Objective 3 Approximate probabilities by using the Empirical Method

Approximating Probabilities with the Empirical Method

The law of large numbers says that if we repeat a probability experiment a large number of times, then the proportion of times that a particular outcome occurs is likely to be close to the true probability of the outcome. The **Empirical Method** consists of repeating an experiment a large number of times, and using the proportion of times an outcome occurs to approximate the probability of the outcome.

EXAMPLE 5.8

Approximate the probability that a baby is a boy

The Centers for Disease Control and Prevention reports that in the year 2010, there were 2,046,935 boys and 1,952,451 girls born in the United States. Approximate the probability that a newborn baby is a boy.

Solution

We compute the number of times the experiment has been repeated:

$$
\begin{array}{ll}
2{,}046{,}935 & \text{boys} \\
+1{,}952{,}451 & \text{girls} \\
\hline
= 3{,}999{,}386 & \text{births}
\end{array}
$$

The proportion of births that are boys is

$$\frac{2,046,935}{3,999,386} = 0.5118$$

We approximate $P(\text{Boy}) \approx 0.5118$.

Example 5.8 is based on a very large number (3,999,386) of replications. The law of large numbers says that the proportion of outcomes approaches the true probability as the number of replications becomes large. For a number this large, we can be virtually certain that the proportion 0.5118 is very close to the true probability; in fact, we can be virtually certain that the true probability of having a boy is not 0.5 as many believe, but is actually greater than 0.5.

Check Your Understanding

4. There are 100,000 voters in a city. A pollster takes a simple random sample of 1000 of them and finds that 513 support a bond issue to support the public library, and 487 oppose it. Estimate the probability that a randomly chosen voter in this city supports the bond issue.

Answer is on page 209.

Objective 4 Approximate probabilities by using simulation

Simulation

In practice, it can be difficult or impossible to repeat an experiment many times in order to approximate a probability with the Empirical Method. In some cases, we can use technology to repeat an equivalent virtual experiment many times. Conducting a virtual experiment in this way is called **simulation**. Most statistical software packages, and many calculators, will perform simulations. Example 5.9 presents a step-by-step description of a simulation conducted using the software package MINITAB.

EXAMPLE 5.9

Approximating a probability with simulation

If three dice are rolled, the smallest possible total is 3, and the largest possible total is 18. We will perform a simulation to estimate the probability that the sum of three dice is 12 or less. If we had no technology, we could roll three dice a large number of times and compute the proportion of times that the sum was 12 or less. With technology, we can use a random number generator in place of dice. For this example, we will show how to use the software package MINITAB to simulate rolling three dice 1000 times.

Solution

Step 1: Click **Calc → Random Data → Integer.**
This brings up the window shown in Figure 5.3. Since we want 1000 replications of the rolling of three dice, we specify 1000 rows of data, columns C1–C3, a minimum value of 1, and a maximum value of 6. Click **OK.** This will generate 1000 rolls of three dice. Figure 5.4 shows the first 20 rolls of the three dice. The first roll is 6, 5, 5; the second is 4, 5, 6; and so on.

Step 2: Click **Calc → Row Statistics.**
This brings up the window shown in Figure 5.5. We fill out the dialog boxes as shown in the figure and click **OK.** This adds the values in columns C1–C3 and puts the sum in column C4. Column C4 then contains the 1000 sums of three dice, as shown in Figure 5.6.

Figure 5.3 Screenshot of window specifying how to generate the random data

Figure 5.5 Screenshot of window specifying how to sum the three dice

Figure 5.7 Screenshot of window specifying how to produce the cumulative counts

Figure 5.4 Screenshot of window showing the first 20 rolls of three dice

Figure 5.6 Screenshot of window showing the first 20 rolls of three dice. The sum of the dice appears in column C4.

Figure 5.8 Cumulative counts for the 1000 rolls of three dice

Step 3: Click **Stat → Tables → Tally**.

This brings up the window shown in Figure 5.7. We specify column C4 and cumulative counts as shown in the figure. Click **OK**. This produces the table shown in Figure 5.8. The table shows that 745 of the 1000 rolls of the dice produced a number of 12 or less.

We estimate the probability of obtaining a total of 12 or less in a roll of three dice to be

$$P(12 \text{ or less}) \approx \frac{745}{1000} = 0.745$$

USING TECHNOLOGY

TI-84 PLUS

Simulating 100 rolls of a die

Step 1. Press **MATH**, scroll to the **PRB** menu, and select **5: randInt(**

Step 2. Enter **1**, comma, **6**, comma, and then the number of rolls you wish to simulate (100). Close the parentheses (Figure A).

Step 3. To store the data in **L1**, press **STO**, then **2nd, 1**, then **ENTER** (Figure B).

Figure A

Figure B

MINITAB

Simulating 100 rolls of a die

Step 1. Click **Calc**, then **Random Data**, then **Integer**.

Step 2. Enter **100** as the number of rows of data.

Step 3. Enter **C1** in the **Store in column(s)** field.

Step 4. Enter **1** as the **Minimum value** and **6** as the **Maximum value**.

Step 5. Click **OK**.

EXCEL

Simulating 100 rolls of a die

Step 1. Click on a cell in the worksheet and type **=RANDBETWEEN(1, 6)**. Press **ENTER**.

Step 2. Copy and paste the formula into cells for the number of rolls you wish to simulate (100).

SECTION 5.1   Exercises

Exercises 1–4 are the Check Your Understanding exercises located within the section.

Understanding the Concepts

In Exercises 5–8, fill in each blank with the appropriate word or phrase.

5. If an event cannot occur, its probability is _____.

6. If an event is certain to occur, its probability is _____.

7. The collection of all possible outcomes of a probability experiment is called a _____.

8. An outcome or collection of outcomes from a sample space is called an _____.

In Exercises 9–12, determine whether the statement is true or false. If the statement is false, rewrite it as a true statement.

9. The law of large numbers states that as a probability experiment is repeated, the proportion of times that a given outcome occurs will approach its probability.

10. If A denotes an event, then the sample space is denoted by $P(A)$.

11. The Empirical Method can be used to calculate the exact probability of an event.

12. For any event A, $0 \le P(A) \le 1$.

Practicing the Skills

In Exercises 13–18, assume that a fair die is rolled. The sample space is $\{1, 2, 3, 4, 5, 6\}$, and all the outcomes are equally likely.

13. Find $P(2)$.

14. Find $P(\text{Even number})$.

15. Find $P(\text{Less than } 3)$.

16. Find $P(\text{Greater than } 2)$.

17. Find $P(7)$.

18. Find $P(\text{Less than } 10)$.

19. A fair coin has probability 0.5 of coming up heads.
   a. If you toss a fair coin twice, are you certain to get one head and one tail?
   b. If you toss a fair coin 100 times, are you certain to get 50 heads and 50 tails?
   c. As you toss the coin more and more times, will the proportion of heads approach 0.5?

20. Roulette wheels in Nevada have 38 pockets. They are numbered 0, 00, and 1 through 36. On each spin of the wheel, a ball lands in a pocket, and each pocket is equally likely.
   a. If you spin a roulette wheel 38 times, is it certain that each number will come up once?
   b. If you spin a roulette wheel 3800 times, is it certain that each number will come up 100 times?
   c. As the wheel is spun more and more times, will the proportion of times that each number comes up approach 1/38?

In Exercises 21–24, assume that a coin is tossed twice. The coin may not be fair. The sample space consists of the outcomes {HH, HT, TH, TT}.

21. Is the following a probability model for this experiment? Why or why not?

Outcome	HH	HT	TH	TT
Probability	0.55	0.42	0.31	0.25

22. Is the following a probability model for this experiment? Why or why not?

Outcome	HH	HT	TH	TT
Probability	0.36	0.24	0.24	0.16

23. Is the following a probability model for this experiment? Why or why not?

Outcome	HH	HT	TH	TT
Probability	0.09	0.21	0.21	0.49

24. Is the following a probability model for this experiment? Why or why not?

Outcome	HH	HT	TH	TT
Probability	0.33	0.46	−0.18	0.4

Working with the Concepts

25. How probable is it? Someone computes the probabilities of several events. The probabilities are listed on the left, and some verbal descriptions are listed on the right. Match each probability with the best verbal description. Some descriptions may be used more than once.

Probability	Verbal Description
(a) 0.50	i. This event is certain to happen.
(b) 0.00	ii. This event is as likely to happen as not.
(c) 0.90	
(d) 1.00	iii. This event may happen, but it isn't likely.
(e) 0.10	iv. This event is very likely to happen, but it isn't certain.
(f) −0.25	
(g) 0.01	v. It would be unusual for this event to happen.
(h) 2.00	vi. This event cannot happen.
	vii. Someone made a mistake.

26. Do you know SpongeBob? According to a survey by Nickelodeon TV, 88% of children under 13 in Germany recognized a picture of the cartoon character SpongeBob SquarePants. What is the probability that a randomly chosen German child recognizes SpongeBob?

27. Who will you vote for? In a survey of 500 likely voters in a certain city, 275 said that they planned to vote to reelect the incumbent mayor.
   a. What is the probability that a surveyed voter plans to vote to reelect the mayor?
   b. Interpret this probability by estimating the percentage of all voters in the city who plan to vote to reelect the mayor.

28. Job satisfaction: In a poll conducted by the General Social Survey, 497 out of 1769 people said that their main satisfaction in life comes from their work.
   a. What is the probability that a person who was polled finds his or her main satisfaction in life from work?
   b. Interpret this probability by estimating the percentage of all people whose main satisfaction in life comes from their work.

29. True–false exam: A section of an exam contains four true–false questions. A completed exam paper is selected at random, and the four answers are recorded.
   a. List all 16 outcomes in the sample space.
   b. Assuming the outcomes to be equally likely, find the probability that all the answers are the same.
   c. Assuming the outcomes to be equally likely, find the probability that exactly one of the four answers is "True."
   d. Assuming the outcomes to be equally likely, find the probability that two of the answers are "True" and two of the answers are "False."

30. A coin flip: A fair coin is tossed three times. The outcomes of the three tosses are recorded.
   a. List all eight outcomes in the sample space.
   b. Assuming the outcomes to be equally likely, find the probability that all three tosses are "Heads."
   c. Assuming the outcomes to be equally likely, find the probability that the tosses are all the same.
   d. Assuming the outcomes to be equally likely, find the probability that exactly one of the three tosses is "Heads."

31. Empirical Method: A coin is tossed 400 times and comes up heads 180 times. Use the Empirical Method to approximate the probability that the coin comes up heads.

32. Empirical Method: A die is rolled 600 times. On 85 of those rolls, the die comes up 6. Use the Empirical Method to approximate the probability that the die comes up 6.

33. Pitching: During part of a recent season, pitcher Ubaldo Jimenez threw 2825 pitches. Of these, 1912 were fastballs, 228 were curveballs, 457 were sliders, and 228 were changeups.
   a. What is the probability that Ubaldo Jimenez throws a fastball?
   b. What is the probability that Ubaldo Jimenez throws a breaking ball (curve or slider)?

34. More pitching: Pitcher Mark Buehrle threw 3103 pitches during part of a recent season. Of these, 1286 were thrown with no strikes on the batter, 946 were thrown with one strike, and 871 were thrown with two strikes.

   a. What is the probability that a Mark Buehrle pitch is thrown with no strikes?

   b. What is the probability that a Mark Buehrle pitch is thrown with fewer than two strikes?

35. Risky drivers: An automobile insurance company divides customers into three categories: good risks, medium risks, and poor risks. Assume that of a total of 11,217 customers, 7792 are good risks, 2478 are medium risks, and 947 are poor risks. As part of an audit, one customer is chosen at random.

   a. What is the probability that the customer is a good risk?

   b. What is the probability that the customer is not a poor risk?

36. Pay your bills: A company audit showed that of 875 bills that were sent out, 623 were paid on time, 155 were paid up to 30 days late, 78 were paid between 31 and 90 days late, and 19 were paid after 90 days. One bill is selected at random.

   a. What is the probability that the bill was paid on time?

   b. What is the probability that the bill was paid late?

37. Roulette: A Nevada roulette wheel has 38 pockets. Eighteen of them are red, eighteen are black, and two are green. Each time the wheel is spun, a ball lands in one of the pockets, and each pocket is equally likely.

   a. What is the probability that the ball lands in a red pocket?

   b. If you bet on red on every spin of the wheel, you will lose more than half the time in the long run. Explain why this is so.

38. More roulette: Refer to Exercise 37.

   a. What is the probability that the ball lands in a green pocket?

   b. If you bet on green on every spin of the wheel, you will lose more than 90% of the time in the long run. Explain why this is so.

39. Get an education: The General Social Survey asked 32,201 people how much confidence they had in educational institutions. The results were as follows.

Response	Number
A great deal	10,040
Some	17,890
Hardly any	4,271
Total	32,201

   a. What is the probability that a sampled person has either some or a great deal of confidence in educational institutions?

   b. Assume this is a simple random sample from a population. Use the Empirical Method to estimate the probability that a person has a great deal of confidence in educational institutions.

   c. If we use a cutoff of 0.05, is it unusual for someone to have hardly any confidence in educational institutions?

40. How many kids? The General Social Survey asked 46,349 women how many children they had. The results were as follows.

Number of Children	Number of Women
0	12,656
1	7,438
2	11,290
3	7,143
4	3,797
5	1,811
6	916
7	522
8 or more	776
Total	46,349

   a. What is the probability that a sampled woman has two children?

   b. What is the probability that a sampled woman has fewer than three children?

   c. Assume this is a simple random sample of U.S. women. Use the Empirical Method to estimate the probability that a U.S. woman has more than five children.

   d. Using a cutoff of 0.05, is it unusual for a woman to have no children?

41. Hospital visits: According to the Agency for Healthcare Research and Quality, there were 409,706 hospital visits for asthma-related illnesses in a recent year. The age distribution was as follows.

Age Range	Number
Less than 1 year	7,866
1–17	103,040
18–44	79,659
45–64	121,728
65–84	80,649
85 and up	16,764
Total	409,706

   a. What is the probability that an asthma patient is between 18 and 44 years old?

   b. What is the probability that an asthma patient is 65 or older?

   c. Using a cutoff of 0.05, is it unusual for an asthma patient to be less than 1 year old?

42. Don't smoke: The Centers for Disease Control and Prevention reported that there were 443,000 smoking-related deaths in the United States in a recent year. The numbers of deaths caused by various illnesses attributed to smoking are as follows:

Illness	Number
Lung cancer	128,900
Ischemic heart disease	126,000
Chronic obstructive pulmonary disease	92,900
Other	95,200
Total	443,000

   a. What is the probability that a smoking-related death was the result of lung cancer?

   b. What is the probability that a smoking-related death was the result of either ischemic heart disease or other?

43. Simulation: If five fair coins are tossed, the possible outcomes for the number of heads obtained are 0, 1, 2, 3, 4, and 5. Use a computer simulation to toss five fair coins 1000 times and estimate the probabilities of these outcomes. (Use 1 to represent a head and 0 to represent a tail.)

44. Simulation: In the game of Dungeons & Dragons, four-sided dice are rolled, with equally likely outcomes 1, 2, 3, and 4. If three four-sided dice are rolled, the possible totals are the numbers 3 through 12. Use a computer simulation to roll three four-sided dice 1000 times and estimate the probability of each outcome.

Extending the Concepts

Two dice are rolled. One is red and one is blue. Each will come up with a number between 1 and 6. There are 36 equally likely outcomes for this experiment. They are ordered pairs of the form (Red die, Blue die).

45. Find a sample space: Construct a sample space for this experiment that contains the 36 equally likely outcomes.

46. Find the probability: What is the probability that the sum of the dice is 5?

47. Find the probability: What is the probability that the sum of the dice is 7?

48. The red die has been rolled: Now assume that you have rolled the red die, and it has come up 3. How many of the original 36 outcomes are now possible?

49. Find a new sample space: Construct a sample space containing the outcomes that are still possible after the red die has come up 3.

50. New information changes the probability: Given that the red die came up 3, what is the probability that the sum of the dice is 5? Is the probability the same as it was before the red die was observed?

51. New information doesn't change the probability: Given that the red die came up 3, what is the probability that the sum of the dice is 7? Is the probability the same as it was before the red die was observed?

Answers to Check Your Understanding Exercises for Section 5.1

1. 0.5

2. 0.5

3. a.

Penny	Nickel
H	H
H	T
T	H
T	T

b. 0.5

4. 0.513

SECTION 5.2	The Addition Rule and the Rule of Complements

Objectives

1. Compute probabilities by using the General Addition Rule

2. Compute probabilities by using the Addition Rule for Mutually Exclusive Events

3. Compute probabilities by using the Rule of Complements

If you go out in the evening, you might go to dinner, or to a movie, or to both dinner and a movie. In probability terminology, "go to dinner and a movie" and "go to dinner or a movie" are referred to as *compound events*, because they are composed of combinations of other events — in this case the events "go to dinner" and "go to a movie."

> **DEFINITION**
>
> A **compound event** is an event that is formed by combining two or more events.

In this section, we will focus on compound events of the form "*A* or *B*." We will say that the event "*A* or *B*" occurs whenever *A* occurs, or *B* occurs, or both *A* and *B* occur. We will learn how to compute probabilities of the form $P(A \text{ or } B)$.

> **DEFINITION**
>
> $P(A \text{ or } B) = P(A \text{ occurs or } B \text{ occurs or both occur})$

Table 5.2 presents the results of a survey in which 1000 adults were asked whether they favored a law that would provide more government support for higher education. In addition, each person was asked whether he or she voted in the last election. Those who had voted were classified as "Likely to vote" and those who had not were classified as "Not likely to vote."

Table 5.2

	Favor	**Oppose**	**Undecided**
Likely to vote	372	262	87
Not likely to vote	151	103	25

Table 5.2 is called a **contingency table**. It categorizes people with regard to two variables: whether they are likely to vote, and their opinion on the law. There are six categories, and the numbers in the table present the frequencies for each category. For example, we can see that 372 people are in the row corresponding to "Likely to vote" and the column corresponding to "Favor." Thus, 372 people were likely to vote and favored the law. Similarly, 103 people were not likely to vote and opposed the law.

EXAMPLE 5.10   Compute probabilities by using equally likely outcomes

Use Table 5.2 to answer the following questions:

a. What is the probability that a randomly selected adult is likely to vote and favors the law?

b. What is the probability that a randomly selected adult is likely to vote?

c. What is the probability that a randomly selected adult favors the law?

Solution

We think of the adults in the survey as outcomes in a sample space. Each adult is equally likely to be the one chosen. We begin by counting the total number of outcomes in the sample space:

$$372 + 262 + 87 + 151 + 103 + 25 = 1000$$

To answer part (a), we observe that there are 372 people who are likely to vote and favor the law. There are 1000 people in the survey. Therefore,

$$P(\text{Likely to vote and Favor}) = \frac{372}{1000} = 0.372$$

To answer part (b), we count the total number of outcomes corresponding to adults who are likely to vote:

$$372 + 262 + 87 = 721$$

There are 1000 people in the survey, and 721 of them are likely to vote. Therefore,

$$P(\text{Likely to vote}) = \frac{721}{1000} = 0.721$$

To answer part (c), we count the total number of outcomes corresponding to adults who favor the law:

$$372 + 151 = 523$$

There are 1000 people in the survey, and 523 of them favor the law. Therefore,

$$P(\text{Favor}) = \frac{523}{1000} = 0.523$$

Objective 1 Compute probabilities by using the General Addition Rule

The General Addition Rule

EXAMPLE 5.11   Compute a probability of the form P(A or B)

Use the data in Table 5.2 to find the probability that a person is likely to vote or favors the law.

Solution

We will illustrate two approaches to this problem. In the first approach, we will use equally likely outcomes, and in the second, we will develop a method that is especially designed for probabilities of the form $P(A \text{ or } B)$.

Approach 1: To use equally likely outcomes, we reproduce Table 5.2 and circle the numbers that correspond to people who are either likely voters or who favor the law.

	Favor	Oppose	Undecided
Likely to vote	(372)	(262)	(87)
Not likely to vote	(151)	103	25

There are 1000 people altogether. The number of people who either are likely voters or favor the law is

$$372 + 262 + 87 + 151 = 872$$

Therefore,

$$P(\text{Likely to vote or Favor}) = \frac{372 + 262 + 87 + 151}{1000} = \frac{872}{1000} = 0.872$$

Approach 2: In this approach we will begin by computing the probabilities P(Likely to vote) and P(Favor) separately. We reproduce Table 5.2; this time we circle the numbers that correspond to likely voters and put rectangles around the numbers that correspond to favoring the law. Note that the number 372 has both a circle and a rectangle around it, because these 372 people are both likely to vote and favor the law.

	Favor	Oppose	Undecided
Likely to vote	[372]	(262)	(87)
Not likely to vote	[151]	103	25

There are $372 + 262 + 87 = 721$ likely voters and $372 + 151 = 523$ voters who favor the law. If we try to find the number of people who are likely to vote or who favor the law by adding these two numbers, we get $721 + 523 = 1244$, which is too large (there are only 1000 people in total). This happened because there are 372 people who are both likely voters and who favor the law, and these people are counted twice. We can still solve the problem by adding 721 and 523, but we must then subtract 372 to correct for the double counting.

We illustrate this reasoning, using probabilities.

$$P(\text{Likely to vote}) = \frac{721}{1000} = 0.721$$

$$P(\text{Favor}) = \frac{523}{1000} = 0.523$$

$$P(\text{Likely to vote AND Favor}) = \frac{372}{1000} = 0.372$$

$$P(\text{Likely to vote OR Favor}) = P(\text{Likely to vote}) + P(\text{Favor})$$
$$-P(\text{Likely to vote AND Favor})$$

$$= \frac{721}{1000} + \frac{523}{1000} - \frac{372}{1000}$$

$$= \frac{872}{1000} = 0.872$$

The method of subtracting in order to adjust for double counting is known as the General Addition Rule.

The General Addition Rule

For any two events A and B,

$$P(A \text{ or } B) = P(A) + P(B) - P(A \text{ and } B)$$

EXAMPLE 5.12 **Compute a probability by using the General Addition Rule**

Refer to Table 5.2. Use the General Addition Rule to find the probability that a randomly selected person is not likely to vote or is undecided.

Solution
Using the General Addition Rule, we compute

P(Not likely to vote or Undecided)

$= P$(Not likely to vote) $+ P$(Undecided) $- P$(Not likely to vote and Undecided)

There are $151 + 103 + 25 = 279$ people not likely to vote out of a total of 1000. Therefore,

$$P(\text{Not likely to vote}) = \frac{279}{1000} = 0.279$$

There are $87 + 25 = 112$ people who are undecided out of a total of 1000. Therefore,

$$P(\text{Undecided}) = \frac{112}{1000} = 0.112$$

Finally, there are 25 people who are both not likely to vote and undecided. Therefore,

$$P(\text{Not likely to vote and Undecided}) = \frac{25}{1000} = 0.025$$

Using the General Addition Rule,

$$P(\text{Not likely to vote or Undecided}) = 0.279 + 0.112 - 0.025 = 0.366$$

Check Your Understanding

1. The following table presents numbers of U.S. workers, in thousands, categorized by type of occupation and educational level.

Type of Occupation	Non-College Graduate	College Graduate
Managers and professionals	17,564	31,103
Service	15,967	2,385
Sales and office	22,352	7,352
Construction and maintenance	12,511	1,033
Production and transportation	14,597	1,308

Source: Bureau of Labor Statistics

a. What is the probability that a randomly selected worker is a college graduate?
b. What is the probability that the occupation of a randomly selected worker is categorized either as Sales and office or as Production and transportation?
c. What is the probability that a randomly selected worker is either a college graduate or has a service occupation?

Answers are on page 218.

Objective 2 Compute probabilities by using the Addition Rule for Mutually Exclusive Events

Mutually Exclusive Events

Sometimes it is impossible for two events both to occur. For example, when a coin is tossed, it is impossible to get both a head and a tail. Two events that cannot both occur are called mutually exclusive. The term *mutually exclusive* means that when one event occurs, it excludes the other.

> **DEFINITION**
>
> Two events are said to be **mutually exclusive** if it is impossible for both events to occur.

Explain It Again

Meaning of mutually exclusive events: Two events are mutually exclusive if the occurrence of one makes it impossible for the other to occur.

We can use **Venn diagrams** to illustrate mutually exclusive events. In a Venn diagram, the sample space is represented by a rectangle, and events are represented by circles drawn inside the rectangle. If two circles do not overlap, the two events cannot both occur. If two circles overlap, the overlap area represents the occurrence of both events. Figures 5.9 and 5.10 illustrate the idea.

   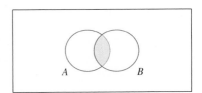

Figure 5.9 Venn diagram illustrating mutually exclusive events

Figure 5.10 Venn diagram illustrating events that are not mutually exclusive

EXAMPLE 5.13 Determine whether two events are mutually exclusive

In each of the following, determine whether events A and B are mutually exclusive:

    a. A die is rolled. Event A is that the die comes up 3, and event B is that the die comes up an even number.

    b. A fair coin is tossed twice. Event A is that one of the tosses is a head, and event B is that one of the tosses is a tail.

Solution

    a. These events are mutually exclusive. The die cannot both come up 3 and come up an even number.

    b. These events are not mutually exclusive. If the two tosses result in HT or TH, then both events occur.

Check Your Understanding

    2. A college student is chosen at random. Event A is that the student is older than 21 years, and event B is that the student is taking a statistics class. Are events A and B mutually exclusive?

    3. A college student is chosen at random. Event A is that the student is an only child, and event B is that the student has a brother. Are events A and B mutually exclusive?

Answers are on page 218.

If events A and B are mutually exclusive, then $P(A \text{ and } B) = 0$. This leads to a simplification of the General Addition Rule.

The Addition Rule for Mutually Exclusive Events

If A and B are mutually exclusive events, then

$$P(A \text{ or } B) = P(A) + P(B)$$

In general, three or more events are mutually exclusive if only one of them can happen. If A, B, C, ... are mutually exclusive, then

$$P(A \text{ or } B \text{ or } C \text{ or } ...) = P(A) + P(B) + P(C) + \cdots$$

EXAMPLE 5.14   Compute a probability by using the Addition Rule for Mutually Exclusive Events

In the 2012 Olympic Games, a total of 10,735 athletes participated. Of these, 530 represented the United States, 277 represented Canada, and 102 represented Mexico.

a. What is the probability that an Olympic athlete chosen at random represents the United States or Canada?

b. What is the probability that an Olympic athlete chosen at random represents the United States, Canada, or Mexico?

Solution

a. These events are mutually exclusive, because it is impossible to compete for both the United States and Canada. We compute $P(\text{U.S.})$ and $P(\text{Canada})$.

$$P(\text{U.S. or Canada}) = P(\text{U.S.}) + P(\text{Canada})$$

$$= \frac{530}{10{,}735} + \frac{277}{10{,}735}$$

$$= \frac{807}{10{,}735}$$

$$= 0.07517$$

b. These events are mutually exclusive, because it is impossible to compete for more than one country. Therefore

$$P(\text{U.S. or Canada or Mexico}) = P(\text{U.S.}) + P(\text{Canada}) + P(\text{Mexico})$$

$$= \frac{530}{10{,}735} + \frac{277}{10{,}735} + \frac{102}{10{,}735}$$

$$= \frac{909}{10{,}735}$$

$$= 0.08468$$

Check Your Understanding

4. In a statistics class of 45 students, 11 got a final grade of A, 22 got a final grade of B, and 8 got a final grade of C.
a. What is the probability that a randomly chosen student got an A or a B?
b. What is the probability that a randomly chosen student got an A, a B, or a C?

Answers are on page 218.

Objective 3 Compute probabilities by using the Rule of Complements

Complements

If there is a 60% chance of rain today, then there is a 40% chance that it will not rain. The events "Rain" and "No rain" are *complements*. The complement of an event A is the event that A does not occur.

DEFINITION

If A is any event, the **complement** of A is the event that A does not occur.
The complement of A is denoted A^c.

Some complements are straightforward. For example, the complement of "the plane was on time" is "the plane was not on time." In other cases, finding the complement requires some thought. Example 5.15 illustrates this.

EXAMPLE 5.15

Find the complement of an event

Two hundred students were enrolled in a statistics class. Find the complements of the following events.

    a. More than 50 of them are business majors.
    b. At least 50 of them are business majors.
    c. Fewer than 50 of them are business majors.
    d. Exactly 50 of them are business majors.

Solution

    a. If it is not true that more than 50 are business majors, then the number of business majors must be 50 or less than 50. The complement is that 50 or fewer of the students are business majors.
    b. If it is not true that at least 50 are business majors, then the number of business majors must be less than 50. The complement is that fewer than 50 of the students are business majors.
    c. If it is not true that fewer than 50 are business majors, then the number of business majors must be 50 or more than 50. Another way of saying this is that at least 50 of the students are business majors. The complement is that at least 50 of the students are business majors.
    d. If it is not true that exactly 50 are business majors, then the number of business majors must not equal 50. The complement is that the number of business majors is not equal to 50.

Explain It Again

The complement occurs when the event doesn't occur: If an event does not occur, then its complement occurs. If an event occurs, then its complement does not occur.

Two important facts about complements are:

1. Either A or A^c must occur. For example, it must either rain or not rain.
2. A and A^c are mutually exclusive; they cannot both occur. For example, it is impossible for it to both rain and not rain.

In probability notation, fact 1 says that $P(A \text{ or } A^c) = 1$, and fact 2 along with the Addition Rule for Mutually Exclusive Events says that $P(A \text{ or } A^c) = P(A) + P(A^c)$. Putting them together, we get

$$P(A) + P(A^c) = 1$$

Subtracting $P(A)$ from both sides yields

$$P(A^c) = 1 - P(A)$$

This is the Rule of Complements.

The Rule of Complements

$$P(A^c) = 1 - P(A)$$

EXAMPLE 5.16 Compute a probability by using the Rule of Complements

According to *The Wall Street Journal*, 40% of cars sold in July 2013 were small cars. What is the probability that a randomly chosen car sold in July 2013 is not a small car?

Solution

$$P(\text{Not a small car}) = 1 - P(\text{Small car}) = 1 - 0.40 = 0.60$$

SECTION 5.2 Exercises

Exercises 1–4 are the Check Your Understanding exercises located within the section.

Understanding the Concepts

In Exercises 5–8, fill in each blank with the appropriate word or phrase.

5. The General Addition Rule states that
   $P(A \text{ or } B) = P(A) + P(B) -$ _____ .

6. If events A and B are mutually exclusive, then
   $P(A \text{ and } B) =$ _____ .

7. Given an event A, the event that A does not occur is called the _____ of A.

8. The Rule of Complements states that
   $P(A^c) =$ _____ .

In Exercises 9–12, determine whether the statement is true or false. If the statement is false, rewrite it as a true statement.

9. The General Addition Rule is used for probabilities of the form $P(A \text{ or } B)$.

10. A compound event is formed by combining two or more events.

11. Two events are mutually exclusive if both events can occur.

12. If an event occurs, then its complement also occurs.

Practicing the Skills

13. If $P(A) = 0.75$, $P(B) = 0.4$, and $P(A \text{ and } B) = 0.25$, find $P(A \text{ or } B)$.

14. If $P(A) = 0.45$, $P(B) = 0.7$, and $P(A \text{ and } B) = 0.65$, find $P(A \text{ or } B)$.

15. If $P(A) = 0.2$, $P(B) = 0.5$, and A and B are mutually exclusive, find $P(A \text{ or } B)$.

16. If $P(A) = 0.7$, $P(B) = 0.1$, and A and B are mutually exclusive, find $P(A \text{ or } B)$.

17. If $P(A) = 0.3$, $P(B) = 0.4$, and $P(A \text{ or } B) = 0.7$, are A and B mutually exclusive?

18. If $P(A) = 0.5$, $P(B) = 0.4$, and $P(A \text{ or } B) = 0.8$, are A and B mutually exclusive?

19. If $P(A) = 0.35$, find $P(A^c)$.

20. If $P(B) = 0.6$, find $P(B^c)$.

21. If $P(A^c) = 0.27$, find $P(A)$.

22. If $P(B^c) = 0.64$, find $P(B)$.

23. If $P(A) = 0$, find $P(A^c)$.

24. If $P(A) = P(A^c)$, find $P(A)$.

In Exercises 25–30, determine whether events A and B are mutually exclusive.

25. A: Sophie is a member of the debate team; B: Sophie is the president of the theater club.

26. A: Jayden has a math class on Tuesdays at 2:00; B: Jayden has an English class on Tuesdays at 2:00.

27. A sample of 20 cars is selected from the inventory of a dealership. A: At least 3 of the cars in the sample are red; B: Fewer than 2 of the cars in the sample are red.

28. A sample of 75 books is selected from a library. A: At least 10 of the authors are female; B: At least 10 of the books are fiction.

29. A red die and a blue die are rolled. A: The red die comes up 2; B: The blue die comes up 3.

30. A red die and a blue die are rolled. A: The red die comes up 1; B: The total is 9.

In Exercises 31 and 32, find the complements of the events.

31. A sample of 225 Internet users was selected.
    a. More than 200 of them use Google as their primary search engine.
    b. At least 200 of them use Google as their primary search engine.
    c. Fewer than 200 of them use Google as their primary search engine.
    d. Exactly 200 of them use Google as their primary search engine.

32. A sample of 700 cell phone batteries was selected.
    a. Exactly 24 of the batteries were defective.
    b. At least 24 of the batteries were defective.
    c. More than 24 of the batteries were defective.
    d. Fewer than 24 of the batteries were defective.

Working with the Concepts

33. **Traffic lights:** A commuter passes through two traffic lights on the way to work. Each light is either red, yellow, or

green. An experiment consists of observing the colors of the two lights.

a. List the nine outcomes in the sample space.

b. Let A be the event that both colors are the same. List the outcomes in A.

c. Let B be the event that the two colors are different. List the outcomes in B.

d. Let C be the event that at least one of the lights is green. List the outcomes in C.

e. Are events A and B mutually exclusive? Explain.

f. Are events A and C mutually exclusive? Explain.

34. Dice: Two fair dice are rolled. The first die is red and the second is blue. An experiment consists of observing the numbers that come up on the dice.

a. There are 36 outcomes in the sample space. They are ordered pairs of the form (Red die, Blue die). List the 36 outcomes.

b. Let A be the event that the same number comes up on both dice. List the outcomes in A.

c. Let B be the event that the red die comes up 6. List the outcomes in B.

d. Let C be the event that one die comes up 6 and the other comes up 1. List the outcomes in C.

e. Are events A and B mutually exclusive? Explain.

f. Are events A and C mutually exclusive? Explain.

35. Car repairs: Let E be the event that a new car requires engine work under warranty and let T be the event that the car requires transmission work under warranty. Suppose that $P(E) = 0.10$, $P(T) = 0.02$, and $P(E \text{ and } T) = 0.01$.

a. Find the probability that the car needs work on either the engine, the transmission, or both.

b. Find the probability that the car needs no work on the engine.

36. Sick computers: Let V be the event that a computer contains a virus, and let W be the event that a computer contains a worm. Suppose $P(V) = 0.15$, $P(W) = 0.05$, and $P(V \text{ and } W) = 0.03$.

a. Find the probability that the computer contains either a virus or a worm or both.

b. Find the probability that the computer does not contain a virus.

37. Computer purchases: Out of 800 large purchases made at a computer retailer, 336 were personal computers, 398 were laptop computers, and 66 were printers. As part of an audit, one purchase record is sampled at random.

a. What is the probability that it is a personal computer?

b. What is the probability that it is not a printer?

38. Visit your local library: On a recent Saturday, a total of 1200 people visited a local library. Of these people, 248 were under age 10, 472 were aged 10–18, 175 were aged 19–30, and the rest were more than 30 years old. One person is sampled at random.

a. What is the probability that the person is less than 19 years old?

b. What is the probability that the person is more than 30 years old?

39. How are your grades? In a recent semester at a local university, 500 students enrolled in both Statistics I and Psychology I. Of these students, 82 got an A in statistics, 73 got an A in psychology, and 42 got an A in both statistics and psychology.

a. Find the probability that a randomly chosen student got an A in statistics or psychology or both.

b. Find the probability that a randomly chosen student did not get an A in psychology.

40. Statistics grades: In a statistics class of 30 students, there were 13 men and 17 women. Two of the men and three of the women received an A in the course. A student is chosen at random from the class.

a. Find the probability that the student is a woman.

b. Find the probability that the student received an A.

c. Find the probability that the student is a woman or received an A.

d. Find the probability that the student did not receive an A.

41. Sick children: There are 25 students in Mrs. Bush's sixth-grade class. On a cold winter day in February, many of the students had runny noses and sore throats. After examining each student, the school nurse constructed the following table.

	Sore Throat	No Sore Throat
Runny Nose	6	12
No Runny Nose	4	3

a. Find the probability that a randomly selected student has a runny nose.

b. Find the probability that a randomly selected student has a sore throat.

c. Find the probability that a randomly selected student has a runny nose or a sore throat.

d. Find the probability that a randomly selected student has neither a runny nose nor a sore throat.

42. Flawed parts: On a certain day, a foundry manufactured 500 cast aluminum parts. Some of these had major flaws, some had minor flaws, and some had both major and minor flaws. The following table presents the results.

	Minor Flaw	No Minor Flaw
Major Flaw	20	35
No Major Flaw	75	370

a. Find the probability that a randomly chosen part has a major flaw.

b. Find the probability that a randomly chosen part has a minor flaw.

c. Find the probability that a randomly chosen part has a flaw (major or minor).

d. Find the probability that a randomly chosen part has no major flaw.

e. Find the probability that a randomly chosen part has no flaw.

43. Senators: The following table displays the 100 senators of the 113th U.S. Congress on January 3, 2013, classified by political party affiliation and gender.

	Male	Female	Total
Democrat	36	16	52
Republican	42	4	46
Independent	2	0	2
Total	80	20	100

A senator is selected at random from this group. Compute the following probabilities.

a. The senator is a male Republican.

b. The senator is a Democrat or a female.

c. The senator is a Republican.

d. The senator is not a Republican.

e. The senator is a Democrat.

f. The senator is an Independent.

g. The senator is a Democrat or an Independent.

44. The following table presents the number of reports of graffiti in each of New York's five boroughs over a one-year period. These reports were classified as being open, closed, or pending.

Borough	Open Reports	Closed Reports	Pending Reports	Total
Bronx	1,121	1,622	80	2,823
Brooklyn	1,170	2,706	48	3,924
Manhattan	744	3,380	25	4,149
Queens	1,353	2,043	25	3,421
Staten Island	83	118	0	201
Total	4,471	9,869	178	14,518

Source: NYC OpenData

A graffiti report is selected at random. Compute the following probabilities.

a. The report is open and comes from Brooklyn.

b. The report is closed or comes from Queens.

c. The report comes from Manhattan.

d. The report does not come from Manhattan.

e. The report is pending.

f. The report is from the Bronx or Staten Island.

45. Add probabilities? In a certain community, 28% of the houses have fireplaces and 51% have garages. Is the probability that a house has either a fireplace or a garage equal to $0.51 + 0.28 = 0.79$? Explain why or why not.

46. Add probabilities? According to the National Health Statistics Reports, 16% of American women have one child, and 21% have two children. Is the probability that a woman has either one or two children equal to $0.16 + 0.21 = 0.37$? Explain why or why not.

Extending the Concepts

47. Mutual exclusivity is not transitive: Give an example of three events A, B, and C, such that A and B are mutually exclusive, B and C are mutually exclusive, but A and C are not mutually exclusive.

48. Complements: Let A and B be events. Express $(A \text{ and } B)^c$ in terms of A^c and B^c.

Answers to Check Your Understanding Exercises for Section 5.2

1. a. 0.342 **b.** 0.361 **c.** 0.469

2. No

3. Yes

4. a. 0.733 **b.** 0.911

SECTION 5.3	Conditional Probability and the Multiplication Rule

Objectives

1. Compute conditional probabilities
2. Compute probabilities by using the General Multiplication Rule
3. Compute probabilities by using the Multiplication Rule for Independent Events
4. Compute the probability that an event occurs at least once

Objective 1 Compute conditional probabilities

Conditional Probability

Approximately 15% of adult men in the United States are more than six feet tall. Therefore, if a man is selected at random, the probability that he is more than six feet tall is 0.15.

Now assume that you learn that the selected man is a professional basketball player. With this extra information, the probability that the man is more than six feet tall becomes much greater than 0.15. A probability that is computed with the knowledge of additional information is called a *conditional probability*; a probability computed without such knowledge is called an *unconditional probability*. As this example shows, the conditional probability of an event can be much different than the unconditional probability.

EXAMPLE 5.17

Compute an unconditional probability

Joe, Sam, Eliza, and Maria have been elected to the executive committee of their college's student government. They must choose a chairperson and a secretary. They decide to write each name on a piece of paper and draw two names at random. The first name drawn will be the chairperson and the second name drawn will be the secretary. What is the probability that Joe is the secretary?

Table 5.3 is a sample space for this experiment. The first name in each pair is the chairperson and the second name is the secretary.

Table 5.3 Twelve Equally Likely Outcomes

(Joe, Sam)	(Sam, Joe)	(Eliza, Joe)	(Maria, Joe)
(Joe, Eliza)	(Sam, Eliza)	(Eliza, Sam)	(Maria, Sam)
(Joe, Maria)	(Sam, Maria)	(Eliza, Maria)	(Maria, Eliza)

There are 12 equally likely outcomes. Three of them, (Sam, Joe), (Eliza, Joe), and (Maria, Joe), correspond to Joe's being secretary. Therefore, P(Joe is secretary) $= 3/12 = 1/4$.

EXAMPLE 5.18

Compute a conditional probability

Suppose that Eliza is the first name selected, so she is chairperson. Now what is the probability that Joe is secretary?

Solution

We'll answer this question with intuition first, then show the reasoning. Since Eliza was chosen to be chairperson, she won't be the secretary. That leaves Joe, Sam, and Maria. Each of these three is equally likely to be chosen. Therefore, the probability that Joe is chosen as secretary is 1/3. Note that this probability differs from the probability of 1/4 calculated in Example 5.17.

Now let's look at the reasoning behind this answer. The original sample space, shown in Table 5.3, had 12 outcomes. Once we know that Eliza is chairperson, we know that only three of those outcomes are now possible. Table 5.4 highlights these three outcomes from the original sample space.

Table 5.4

(Joe, Sam)	(Sam, Joe)	(Eliza, Joe)	(Maria, Joe)
(Joe, Eliza)	(Sam, Eliza)	(Eliza, Sam)	(Maria, Sam)
(Joe, Maria)	(Sam, Maria)	(Eliza, Maria)	(Maria, Eliza)

Of the three possible outcomes, only one, (Eliza, Joe), has Joe as secretary. Therefore, given that Eliza is chairperson, the probability that Joe is secretary is 1/3.

Example 5.18 asked us to compute the probability of an event (that Joe is secretary) after giving us information about another event (that Eliza is chairperson). A probability like this is called a *conditional probability*. The notation for this conditional probability is

$$P \text{(Joe is secretary} \mid \text{Eliza is chairperson)}$$

We read this as "the conditional probability that Joe is secretary, given that Eliza is chairperson." It denotes the probability that Joe is secretary, under the assumption that Eliza is chairperson.

DEFINITION

The **conditional probability** of an event B, given an event A, is denoted $P(B \mid A)$.

$P(B \mid A)$ is the probability that B occurs, under the assumption that A occurs.

We read $P(B \mid A)$ as "the probability of B, given A."

The General Method for computing conditional probabilities

In Example 5.18, we computed

$$P(\text{Joe is secretary} \mid \text{Eliza is chairperson}) = \frac{1}{3}$$

Let's take a closer look at the answer of $1/3$. The denominator is the number of outcomes that were left in the sample space after it was known that Eliza was chairperson. That is,

$$\text{Number of outcomes where Eliza is chairperson} = 3$$

The numerator is 1, and this corresponds to the one outcome in which Eliza is chairperson and Joe is secretary. That is,

$$\text{Number of outcomes where Eliza is chairperson and Joe is secretary} = 1$$

Therefore, we see that

$P(\text{Joe is secretary} \mid \text{Eliza is chairperson})$

$$= \frac{\text{Number of outcomes where Eliza is chairperson and Joe is secretary}}{\text{Number of outcomes where Eliza is chairperson}}$$

We can obtain another useful method by recalling that there were 12 outcomes in the original sample space. It follows that

$$P(\text{Eliza is chairperson}) = \frac{3}{12}$$

and

$$P(\text{Eliza is chairperson and Joe is secretary}) = \frac{1}{12}$$

We now see that

$$P(\text{Joe is secretary} \mid \text{Eliza is chairperson}) = \frac{P(\text{Eliza is chairperson and Joe is secretary})}{P(\text{Eliza is chairperson})}$$

This example illustrates the General Method for computing conditional probabilities, which we now state.

The General Method for Computing Conditional Probabilities

The probability of B given A is

$$P(B \mid A) = \frac{P(A \text{ and } B)}{P(A)}$$

Note that we cannot compute $P(B \mid A)$ if $P(A) = 0$.

When the outcomes in the sample space are equally likely, then

$$P(B \mid A) = \frac{\text{Number of outcomes corresponding to } (A \text{ and } B)}{\text{Number of outcomes corresponding to } A}$$

EXAMPLE 5.19   Use the General Method to compute a conditional probability

Table 5.5 presents the number of U.S. men and women (in millions) 25 years old and older who have attained various levels of education in a recent year.

Table 5.5 Number of Men and Women with Various Levels of Education (in millions)

	Not a high school graduate	High school graduate	Some college, no degree	Associate's degree	Bachelor's degree	Advanced degree
Men	14.0	29.6	15.6	7.2	17.5	10.1
Women	13.7	31.9	17.5	9.6	19.2	9.1

Source: U.S. Census Bureau

A person is selected at random.

a. What is the probability that the person is a man?

b. What is the probability that the person is a man with a bachelor's degree?

c. What is the probability that the person has a bachelor's degree, given that the person is a man?

Solution

a. Each person in the study is an outcome in the sample space. We first compute the total number of people in the study. We'll do this by computing the total number of men, then the total number of women.

$$\text{Total number of men} = 14.0 + 29.6 + 15.6 + 7.2 + 17.5 + 10.1 = 94.0$$

$$\text{Total number of women} = 13.7 + 31.9 + 17.5 + 9.6 + 19.2 + 9.1 = 101.0$$

There are 94.0 million men and 101.0 million women. The total number of people is $94.0 + 101.0 = 195.0$ million. We can now compute the probability that a randomly chosen person is a man.

$$P(\text{Man}) = \frac{94.0}{195.0} = 0.4821$$

b. The number of men with bachelor's degrees is found in Table 5.5 to be 17.5 million. The total number of people is 195.0 million. Therefore

$$P(\text{Man with a Bachelor's degree}) = \frac{17.5}{195.0} = 0.08974$$

c. We use the General Method for computing a conditional probability.

$$P(\text{Bachelor's degree}|\text{Man}) = \frac{P(\text{Man with a Bachelor's degree})}{P(\text{Man})} = \frac{17.5/195.0}{94.0/195.0} = 0.1862$$

Check Your Understanding

1. A person is selected at random from the population in Table 5.5.
   a. What is the probability that the person is a woman who is a high school graduate?
   b. What is the probability that the person is a high school graduate?
   c. What is the probability that the person is a woman, given that the person is a high school graduate?

Answers are on page 231.

Objective 2 Compute probabilities by using the General Multiplication Rule

The General Multiplication Rule

The General Method for computing conditional probabilities provides a way to compute probabilities for events of the form "*A* and *B*." If we multiply both sides of the equation by $P(A)$ we obtain the General Multiplication Rule.

The General Multiplication Rule

$$P(A \text{ and } B) = P(A)P(B \mid A)$$

or, equivalently,

$$P(A \text{ and } B) = P(B)P(A \mid B)$$

Explain It Again

The General Multiplication Rule:
Use the General Multiplication Rule to compute probabilities of the form $P(A \text{ and } B)$.

EXAMPLE 5.20 Use the General Multiplication Rule to compute a probability

Among those who apply for a particular job, the probability of being granted an interview is 0.1. Among those interviewed, the probability of being offered a job is 0.25. Find the probability that an applicant is offered a job.

Solution

Being offered a job involves two events. First, a person must be interviewed; then, given that the person has been interviewed, the person must be offered a job. Using the General Multiplication Rule, we obtain

$$P(\text{Offered a job}) = P(\text{Interviewed})P(\text{Offered a job} \mid \text{Interviewed})$$
$$= (0.1)(0.25)$$
$$= 0.025$$

Check Your Understanding

2. In a certain city, 70% of high school students graduate. Of those who graduate, 40% attend college. Find the probability that a randomly selected high school student will attend college.

Answer is on page 231.

Objective 3 Compute probabilities by using the Multiplication Rule for Independent Events

Independence

In some cases, the occurrence of one event has no effect on the probability that another event occurs. For example, if a coin is tossed twice, the occurrence of a head on the first toss does not make it any more or less likely that a head will come up on the second toss. Example 5.21 illustrates this fact.

EXAMPLE 5.21 Coin tossing probabilities

A fair coin is tossed twice.

CAUTION

Do not confuse independent events with mutually exclusive events. Two events are independent if the occurrence of one does not affect the probability of the occurrence of the other. Two events are mutually exclusive if the occurrence of one makes it impossible for the other to occur.

---

a. What is the probability that the second toss is a head?
b. What is the probability that the second toss is a head given that the first toss is a head?
c. Are the answers to (a) and (b) different? Does the probability that the second toss is a head change if the first toss is a head?

Solution

a. There are four equally likely outcomes for the two tosses. The sample space is {HH, HT, TH, TT}. Of these, there are two outcomes where the second toss is a head. Therefore, $P(\text{Second toss is H}) = 2/4 = 1/2$.

b. We use the General Method for computing conditional probabilities.

$$P(\text{Second toss is H} \mid \text{First toss is H})$$

$$= \frac{\text{Number of outcomes where first toss is H and second is H}}{\text{Number of outcomes where first toss is H}} = \frac{1}{2}$$

c. The two answers are the same. The probability that the second toss is a head does not change if the first toss is a head. In other words,

$$P(\text{Second toss is H} \mid \text{First toss is H}) = P(\text{Second toss is H})$$

In the case of two coin tosses, the outcome of the first toss does not affect the second toss. Events with this property are said to be *independent*.

> **DEFINITION**
>
> Two events are **independent** if the occurrence of one does not affect the probability that the other event occurs.
>
> If two events are not independent, we say they are **dependent**.

In many situations, we can determine whether events are independent just by understanding the circumstances surrounding the events. Example 5.22 illustrates this.

EXAMPLE 5.22

Determine whether events are independent

Determine whether the following pairs of events are independent:

    a. A college student is chosen at random. The events are "being a freshman" and "being less than 20 years old."

    b. A college student is chosen at random. The events are "born on a Sunday" and "taking a statistics class."

Solution

    a. These events are not independent. If the student is a freshman, the probability that the student is less than 20 years old is greater than for a student who is not a freshman.

    b. These events are independent. If a student was born on a Sunday, this has no effect on the probability that the student takes a statistics class.

When two events, A and B, are independent, then $P(B \mid A) = P(B)$, because knowing that A occurred does not affect the probability that B occurs. This leads to a simplified version of the Multiplication Rule.

Explain It Again

The Multiplication Rule for Independent Events: Use the Multiplication Rule for Independent Events to compute probabilities of the form $P(A \text{ and } B)$ when A and B are independent.

> **The Multiplication Rule for Independent Events**
>
> If A and B are independent events, then
>
> $$P(A \text{ and } B) = P(A)P(B)$$
>
> This rule can be extended to the case where there are more than two independent events. If $A, B, C, \ldots$ are independent events, then
>
> $$P(A \text{ and } B \text{ and } C \text{ and } \ldots) = P(A)P(B)P(C)\cdots$$

EXAMPLE 5.23   Using the Multiplication Rule for Independent Events

According to recent figures from the U.S. Census Bureau, the percentage of people under the age of 18 was 23.5% in New York City, 25.8% in Chicago, and 26.0% in Los Angeles. If one person is selected from each city, what is the probability that all of them are under 18? Is this an unusual event?

Solution

There are three events: person from New York is under 18, person from Chicago is under 18, and person from Los Angeles is under 18. These three events are independent, because the identity of the person chosen from one city does not affect who is chosen in the other cities. We therefore use the Multiplication Rule for Independent Events. Let N denote the event that the person from New York is under 18, and let C and L denote the corresponding events for Chicago and Los Angeles, respectively.

$$P(N \text{ and } C \text{ and } L) = P(N) \cdot P(C) \cdot P(L) = 0.235 \cdot 0.258 \cdot 0.260 = 0.0158$$

The probability is 0.0158. This is an unusual event, if we apply the most commonly used cutoff point of 0.05.

Distinguishing mutually exclusive from independent

Although the mutually exclusive property and the independence property are quite different, in practice it can be difficult to distinguish them. The following diagram can help you to determine whether two events are mutually exclusive, independent, or neither.

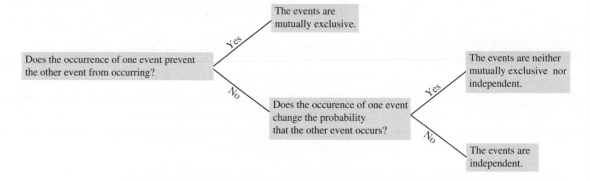

Check Your Understanding

3. Two dice are rolled. Each comes up with a number between 1 and 6. Let A be the event that the number on the first die is even, and let B be the event that the number on the second die is 6.
   a. Explain why events A and B are independent.
   b. Find $P(A)$, $P(B)$, and $P(A \text{ and } B)$.

Answers are on page 231.

Sampling with and without replacement

When we sample two items from a population, we can proceed in either of two ways. We can replace the first item drawn before sampling the second; this is known as **sampling with replacement**. When sampling with replacement, it is possible to draw the same item more than once. The other option is to leave the first item out when sampling the second one; this is known as **sampling without replacement**. When sampling without replacement, it is impossible to sample an item more than once.

When sampling with replacement, each draw is made from the entire population, so the probability of drawing a particular item on the second draw does not depend on the first draw. In other words, when sampling with replacement, the draws are independent. When sampling without replacement, the draws are not independent. Examples 5.24 and 5.25 illustrate this idea.

EXAMPLE 5.24

Sampling without replacement

A box contains two cards marked with a "0" and two cards marked with a "1" as shown in the following illustration. Two cards will be sampled without replacement from this population.

    a. What is the probability of drawing a $\boxed{1}$ on the second draw given that the first draw is a $\boxed{0}$?

    b. What is the probability of drawing a $\boxed{1}$ on the second draw given that the first draw is a $\boxed{1}$?

    c. Are the first and second draws independent?

Solution

    a. If the first draw is a $\boxed{0}$, then the second draw will be made from the population $\boxed{0}\ \boxed{1}\ \boxed{1}$. There are three equally likely outcomes, and two of them are $\boxed{1}$. The probability of drawing a $\boxed{1}$ is 2/3.

    b. If the first draw is a $\boxed{1}$, then the second draw will be made from the population $\boxed{0}\ \boxed{0}\ \boxed{1}$. There are three equally likely outcomes, and one of them is $\boxed{1}$. The probability of drawing a $\boxed{1}$ is 1/3.

    c. The first and second draws are not independent. The probability of drawing a $\boxed{1}$ on the second draw depends on the outcome of the first draw.

EXAMPLE 5.25

Sampling with replacement

Two items will be sampled with replacement from the population in Example 5.24. Does the probability of drawing a $\boxed{1}$ on the second draw depend on the outcome of the first draw? Are the first and second draws independent?

Solution

Since the sampling is with replacement, then no matter what the first draw is, the second draw will be made from the entire population $\boxed{0}\ \boxed{0}\ \boxed{1}\ \boxed{1}$. Therefore, the probability of drawing a $\boxed{1}$ on the second draw is 2/4 = 0.5 no matter what the first draw is. Since the probability on the second draw does not depend on the outcome of the first draw, the first and second draws are independent.

    The population in Examples 5.24 and 5.25 was very small — only four items. When the population is large, the draws will be nearly independent even when sampled without replacement, as illustrated in Example 5.26.

EXAMPLE 5.26

Sampling without replacement from a large population

A box contains 1000 cards marked with a "0" and 1000 cards marked with a "1," as shown in the following illustration. Two cards will be sampled without replacement from this population.

$$\boxed{1000\ \boxed{0}\text{'s}\qquad 1000\ \boxed{1}\text{'s}}$$

    a. What is the probability of drawing a $\boxed{1}$ on the second draw given that the first draw is a $\boxed{0}$?

b. What is the probability of drawing a $\boxed{1}$ on the second draw given that the first draw is a $\boxed{1}$?

c. Are the first and second draws independent? Are they approximately independent?

Solution

a. If the first draw is a $\boxed{0}$, then the second draw will be made from the population $\boxed{\text{999 }\boxed{0}\text{'s}\qquad\text{1000 }\boxed{1}\text{'s}}$. There are 1999 equally likely outcomes, and 1000 of them are $\boxed{1}$. The probability of drawing a $\boxed{1}$ is $1000/1999 = 0.50025$.

b. If the first draw is a $\boxed{1}$, then the second draw will be made from the population $\boxed{\text{1000 }\boxed{0}\text{'s}\qquad\text{999 }\boxed{1}\text{'s}}$. There are 1999 equally likely outcomes, and 999 of them are $\boxed{1}$. The probability of drawing a $\boxed{1}$ is $999/1999 = 0.49975$.

c. The probability of drawing a $\boxed{1}$ on the second draw depends slightly on the outcome of the first draw, so the draws are not independent. However, because the difference in the probabilities is so small (0.50025 versus 0.49975), the draws are approximately independent. In practice, it would be appropriate to treat the two draws as independent.

Example 5.26 shows that when the sample size is small compared to the population size, then items sampled without replacement may be treated as independent. A rule of thumb is that the items may be treated as independent so long as the sample comprises less than 5% of the population.

Explain It Again

Replacement doesn't matter when the population is large: When the sample size is less than 5% of the population, it doesn't matter whether the sampling is done with or without replacement. In either case, we will treat the sampled items as independent.

SUMMARY

- When sampling with replacement, the sampled items are independent.
- When sampling without replacement, if the sample size is less than 5% of the population, the sampled items may be treated as independent.
- When sampling without replacement, if the sample size is more than 5% of the population, the sampled items cannot be treated as independent.

Check Your Understanding

4. A pollster plans to sample 1500 voters from a city in which there are 1 million voters. Can the sampled voters be treated as independent? Explain.

5. Five hundred students attend a college basketball game. Fifty of them are chosen at random to receive a free T-shirt. Can the sampled students be treated as independent? Explain.

Answers are on page 231.

Objective 4 Compute the probability that an event occurs at least once

Solving "at least once" problems by using complements

Sometimes we need to find the probability that an event occurs **at least once** in several independent trials. We can calculate such probabilities by finding the probability of the complement and subtracting from 1. Examples 5.27 and 5.28 illustrate the method.

EXAMPLE 5.27   Find the probability that an event occurs at least once

A fair coin is tossed five times. What is the probability that it comes up heads at least once?

Solution

The tosses of a coin are independent, since the outcome of a toss is not affected by the outcomes of other tosses. The complement of coming up heads at least once is coming up tails all five times. We use the Rule of Complements to compute the probability.

P(Comes up heads at least once)

$$= 1 - P(\text{Does not come up heads at all})$$
$$= 1 - P(\text{Comes up tails all five times})$$
$$= 1 - P(\text{First toss is T and Second toss is T and . . . and Fifth toss is T})$$
$$= 1 - P(\text{First toss is T})P(\text{Second toss is T})\cdots P(\text{Fifth toss is T})$$
$$= 1 - \left(\frac{1}{2}\right)^5$$
$$= \frac{31}{32}$$

EXAMPLE 5.28

Find the probability that an event occurs at least once

Items are inspected for flaws by three inspectors. If a flaw is present, each inspector will detect it with probability 0.8. The inspectors work independently. If an item has a flaw, what is the probability that at least one inspector detects it?

Solution

The complement of the event that at least one of the inspectors detects the flaw is that none of the inspectors detects the flaw. We use the Rule of Complements to compute the probability.

We begin by computing the probability that an inspector fails to detect a flaw.

$$P(\text{Inspector fails to detect a flaw}) = 1 - P(\text{Inspector detects flaw}) = 1 - 0.8 = 0.2$$

P(At least one inspector detects the flaw)

$$= 1 - P(\text{None of the inspectors detects the flaw})$$
$$= 1 - P(\text{All three inspectors fail to detect the flaw})$$
$$= 1 - P(\text{First fails and second fails and third fails})$$
$$= 1 - P(\text{First fails})P(\text{Second fails})P(\text{Third fails})$$
$$= 1 - (0.2)^3$$
$$= 0.992$$

Check Your Understanding

6. An office has three smoke detectors. In case of fire, each detector has probability 0.9 of detecting it. If a fire occurs, what is the probability that at least one detector detects it?

Answer is on page 231.

Determining Which Method to Use

We have studied several methods for finding probabilities of events of the form $P(A \text{ and } B)$, $P(A \text{ or } B)$, and $P(\text{At least one})$. The following diagram can help you to determine the correct method to use for calculating these probabilities.

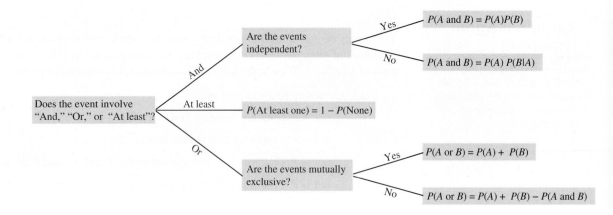

SECTION 5.3   Exercises

Exercises 1–6 are the Check Your Understanding exercises located within the section.

Understanding the Concepts

In Exercises 7–10, fill in each blank with the appropriate word or phrase.

7. A probability that is computed with the knowledge of additional information is called a _____ probability.

8. The General Multiplication Rule states that $P(A \text{ and } B) =$ _____.

9. When sampling without replacement, if the sample size is less than _____ % of the population, the sampled items may be treated as independent.

10. Two events are _____ if the occurrence of one does not affect the probability that the other event occurs.

In Exercises 11–14, determine whether the statement is true or false. If the statement is false, rewrite it as a true statement.

11. $P(B \mid A)$ represents the probability that A occurs under the assumption that B occurs.

12. If A and B are independent events, then $P(A \text{ and } B) = P(A)P(B)$.

13. When sampling without replacement, it is possible to draw the same item from the population more than once.

14. When sampling with replacement, the sampled items are independent.

Practicing the Skills

15. Let A and B be events with $P(A) = 0.4$, $P(B) = 0.7$, and $P(B \mid A) = 0.3$. Find $P(A \text{ and } B)$.

16. Let A and B be events with $P(A) = 0.6$, $P(B) = 0.4$, and $P(B \mid A) = 0.4$. Find $P(A \text{ and } B)$.

17. Let A and B be events with $P(A) = 0.2$ and $P(B) = 0.9$. Assume that A and B are independent. Find $P(A \text{ and } B)$.

18. Let A and B be events with $P(A) = 0.5$ and $P(B) = 0.7$. Assume that A and B are independent. Find $P(A \text{ and } B)$.

19. Let A and B be events with $P(A) = 0.8$, $P(B) = 0.1$, and $P(B \mid A) = 0.2$. Find $P(A \text{ and } B)$.

20. Let A and B be events with $P(A) = 0.3$, $P(B) = 0.5$, and $P(B \mid A) = 0.7$. Find $P(A \text{ and } B)$.

21. Let A, B, and C be independent events with $P(A) = 0.7$, $P(B) = 0.8$, and $P(C) = 0.5$. Find $P(A \text{ and } B \text{ and } C)$.

22. Let A, B, and C be independent events with $P(A) = 0.4$, $P(B) = 0.9$, and $P(C) = 0.7$. Find $P(A \text{ and } B \text{ and } C)$.

23. A fair coin is tossed four times. What is the probability that all four tosses are heads?

24. A fair coin is tossed four times. What is the probability that the sequence of tosses is HTHT?

25. A fair die is rolled three times. What is the probability that the sequence of rolls is 1, 2, 3?

26. A fair die is rolled three times. What is the probability that all three rolls are 6?

In Exercises 27–30, assume that a student is chosen at random from a class. Determine whether the events A and B are independent, mutually exclusive, or neither.

27. A: The student is a freshman.
    B: The student is a sophomore.

28. A: The student is on the basketball team.
    B: The student is more than six feet tall.

29. A: The student is a woman.
    B: The student belongs to a sorority.

30. A: The student is a woman.
    B: The student belongs to a fraternity.

31. Let A and B be events with $P(A) = 0.25$, $P(B) = 0.4$, and $P(A \text{ and } B) = 0.1$.
    a. Are A and B independent? Explain.
    b. Compute $P(A \text{ or } B)$.
    c. Are A and B mutually exclusive? Explain.

32. Let A and B be events with $P(A) = 0.6$, $P(B) = 0.9$, and $P(A \text{ and } B) = 0.5$.
  a. Are A and B independent? Explain.
  b. Compute $P(A \text{ or } B)$.
  c. Are A and B mutually exclusive? Explain.

33. Let A and B be events with $P(A) = 0.4$, $P(B) = 0.5$, and $P(A \text{ or } B) = 0.6$.
  a. Compute $P(A \text{ and } B)$.
  b. Are A and B mutually exclusive? Explain.
  c. Are A and B independent? Explain.

34. Let A and B be events with $P(A) = 0.5$, $P(B) = 0.3$, and $P(A \text{ or } B) = 0.8$.
  a. Compute $P(A \text{ and } B)$.
  b. Are A and B mutually exclusive? Explain.
  c. Are A and B independent? Explain.

35. A fair die is rolled three times. What is the probability that it comes up 6 at least once?

36. An unfair coin has probability 0.4 of landing heads. The coin is tossed four times. What is the probability that it lands heads at least once?

Working with the Concepts

37. Job interview: Seven people, named Anna, Bob, Chandra, Darlene, Ed, Frank, and Gina, will be interviewed for a job. The interviewer will choose two at random to interview on the first day. What is the probability that Anna is interviewed first and Darlene is interviewed second?

38. Shuffle: Charles has six songs on a playlist. Each song is by a different artist. The artists are Usher, Ke$ha, Lady Gaga, Eminem, the Black Eyed Peas, and Ludacris. He programs his player to play the songs in a random order, without repetition. What is the probability that the first song is by Lady Gaga and the second song is by Eminem?

39. Let's eat: A fast-food restaurant chain has 600 outlets in the United States. The following table categorizes them by city population size and location, and presents the number of restaurants in each category. A restaurant is to be chosen at random from the 600 to test market a new menu.

Population of city	Region			
	NE	**SE**	**SW**	**NW**
Under 50,000	30	35	15	5
50,000 – 500,000	60	90	70	30
Over 500,000	150	25	30	60

  a. Given that the restaurant is located in a city with a population over 500,000, what is the probability that it is in the Northeast?
  b. Given that the restaurant is located in the Southeast, what is the probability that it is in a city with a population under 50,000?
  c. Given that the restaurant is located in the Southwest, what is the probability that it is in a city with a population of 500,000 or less?
  d. Given that the restaurant is located in a city with a population of 500,000 or less, what is the probability that it is in the Southwest?
  e. Given that the restaurant is located in the South (either SE or SW), what is the probability that it is in a city with a population of 50,000 or more?

40. U.S. senators: The following table displays the 100 senators of the 113th U.S. Congress on January 3, 2013, viewed by political party affiliation and gender.

	Male	**Female**	**Total**
Democrat	36	16	52
Republican	42	4	46
Independent	2	0	2
Total	80	20	100

A senator is selected at random from this group.
  a. What is the probability that the senator is a woman?
  b. What is the probability that the senator is a Republican?
  c. What is the probability that the senator is a Republican and a woman?
  d. Given that the senator is a woman, what is the probability that she is a Republican?
  e. Given that the senator is a Republican, what is the probability that the senator is a woman?

41. Genetics: A geneticist is studying two genes. Each gene can be either dominant or recessive. A sample of 100 individuals is categorized as follows.

	Gene 2	
Gene 1	**Dominant**	**Recessive**
Dominant	56	24
Recessive	14	6

  a. What is the probability that in a randomly sampled individual, gene 1 is dominant?
  b. What is the probability that in a randomly sampled individual, gene 2 is dominant?
  c. Given that gene 1 is dominant, what is the probability that gene 2 is dominant?
  d. Two genes are said to be in linkage equilibrium if the event that gene 1 is dominant is independent of the event that gene 2 is dominant. Are these genes in linkage equilibrium?

42. Quality control: A population of 600 semiconductor wafers contains wafers from three lots. The wafers are categorized by lot and by whether they conform to a thickness specification, with the results shown in the following table. A wafer is chosen at random from the population.

Lot	Conforming	Nonconforming
A	88	12
B	165	35
C	260	40

  a. What is the probability that a wafer is from Lot A?
  b. What is the probability that a wafer is conforming?
  c. What is the probability that a wafer is from Lot A and is conforming?
  d. Given that the wafer is from Lot A, what is the probability that it is conforming?
  e. Given that the wafer is conforming, what is the probability that it is from Lot A?
  f. Let E_1 be the event that the wafer comes from Lot A, and let E_2 be the event that the wafer is conforming. Are E_1 and E_2 independent? Explain.

43. Stay in school: In a recent school year in the state of Washington, there were 326,000 high school students. Of these, 159,000 were girls and 167,000 were boys. Among the girls, 7800 dropped out of school, and among the boys, 10,300 dropped out. A student is chosen at random.

   a. What is the probability that the student is male?

   b. What is the probability that the student dropped out?

   c. What is the probability that the student is male and dropped out?

   d. Given that the student is male, what is the probability that he dropped out?

   e. Given that the student dropped out, what is the probability that the student is male?

44. Management: The Bureau of Labor Statistics reported that 64.5 million women and 74.6 million men were employed. Of the women, 25.8 million had management jobs, and of the men, 25.0 million had management jobs. An employed person is chosen at random.

   a. What is the probability that the person is a female?

   b. What is the probability that the person has a management job?

   c. What is the probability that the person is female and has a management job?

   d. Given that the person is female, what is the probability that she has a management job?

   e. Given that the person has a management job, what is the probability that the person is female?

45. GED: In a certain high school, the probability that a student drops out is 0.05, and the probability that a dropout gets a high-school equivalency diploma (GED) is 0.25. What is the probability that a randomly selected student gets a GED?

46. Working for a living: The Bureau of Labor Statistics reported that the probability that a randomly chosen employed adult worked in a service occupation was 0.17, and given that a person was in a service occupation, the probability that the person was a woman was 0.57. What is the probability that a randomly chosen employed person was a woman in a service occupation?

47. New car: At a certain car dealership, the probability that a customer purchases an SUV is 0.20. Given that a customer purchases an SUV, the probability that it is black is 0.25. What is the probability that a customer purchases a black SUV?

48. Do you know Squidward? According to a survey by Nickelodeon TV, 88% of children under 13 in Germany recognized a picture of the cartoon character SpongeBob SquarePants. Assume that among those children, 72% also recognized SpongeBob's cranky neighbor Squidward Tentacles. What is the probability that a German child recognized both SpongeBob and Squidward?

49. Target practice: Laura and Philip each fire one shot at a target. Laura has probability 0.5 of hitting the target, and Philip has probability 0.3. The shots are independent.

   a. Find the probability that both of them hit the target.

   b. Given that Laura hits the target, the probability is 0.1 that Philip's shot hits the target closer to the bull's-eye than Laura's. Find the probability that Laura hits the target and that Philip's shot is closer to the bull's-eye than Laura's shot is.

50. Bowling: Sarah and Thomas are going bowling. The probability that Sarah scores more than 175 is 0.4, and the probability that Thomas scores more than 175 is 0.2. Their scores are independent.

   a. Find the probability that both score more than 175.

   b. Given that Thomas scores more than 175, the probability that Sarah scores higher than Thomas is 0.3. Find the probability that Thomas scores more than 175 and Sarah scores higher than Thomas.

51. Defective components: A lot of 10 components contains 3 that are defective. Two components are drawn at random and tested. Let A be the event that the first component drawn is defective, and let B be the event that the second component drawn is defective.

   a. Find $P(A)$.

   b. Find $P(B\,|\,A)$.

   c. Find $P(A \text{ and } B)$.

   d. Are A and B independent? Explain.

52. More defective components: A lot of 1000 components contains 300 that are defective. Two components are drawn at random and tested. Let A be the event that the first component drawn is defective, and let B be the event that the second component drawn is defective.

   a. Find $P(A)$.

   b. Find $P(B\,|\,A)$.

   c. Find $P(A \text{ and } B)$.

   d. Are A and B independent? Is it reasonable to treat A and B as though they were independent? Explain.

53. Multiply probabilities? In a recent year, 21% of all vehicles in operation were pickup trucks. If someone owns two vehicles, is the probability that they are both pickup trucks equal to $0.21 \times 0.21 = 0.0441$? Explain why or why not.

54. Multiply probabilities? A traffic light at an intersection near Jamal's house is red 50% of the time, green 40% of the time, and yellow 10% of the time. Jamal encounters this light in the morning on his way to work and again in the evening on his way home. Is the probability that the light is green both times equal to $0.4 \times 0.4 = 0.16$? Explain why or why not.

55. Lottery: Every day, Jorge buys a lottery ticket. Each ticket has probability 0.2 of winning a prize. After seven days, what is the probability that Jorge has won at least one prize?

56. Car warranty: The probability that a certain make of car will need repairs in the first six months is 0.3. A dealer sells five such cars. What is the probability that at least one of them will require repairs in the first six months?

57. Tic-tac-toe: In the game of tic-tac-toe, if all moves are performed randomly the probability that the game will end in a draw is 0.127. Suppose ten random games of tic-tac-toe are played. What is the probability that at least one of them will end in a draw?

58. **Enter your PIN:** The technology consulting company DataGenetics suggests that 17.8% of all four-digit personal identification numbers, or PIN codes, have a repeating digits format such as 2525. Assuming this to be true, if the PIN codes of six people are selected at random, what is the probability that at least one of them will have repeating digits?

Extending the Concepts

Exercises 59–62 refer to the following situation:

A medical test is available to determine whether a patient has a certain disease. To determine the accuracy of the test, a total of 10,100 people are tested. Only 100 of these people have the disease, while the other 10,000 are disease free. Of the disease-free people, 9800 get a negative result, and 200 get a positive result. The 100 people with the disease all get positive results.

59. **Find the probability:** Find the probability that the test gives the correct result for a person who does not have the disease.

60. **Find the probability:** Find the probability that the test gives the correct result for a person who has the disease.

61. **Find the probability:** Given that a person gets a positive result, what is the probability that the person actually has the disease?

62. **Why are medical tests repeated?** For many medical tests, if the result comes back positive, the test is repeated. Why do you think this is done?

63. **Mutually exclusive and independent?** Let A and B be events. Assume that neither A nor B can occur; in other words, $P(A) = 0$ and $P(B) = 0$. Are A and B independent? Are A and B mutually exclusive? Explain.

64. **Still mutually exclusive and independent?** Let A and B be events. Now assume that $P(A) = 0$ but $P(B) > 0$. Are A and B always independent? Are A and B always mutually exclusive? Explain.

65. **Mutually exclusive and independent again?** Let A and B be events. Now assume that $P(A) > 0$ and $P(B) > 0$. Is it possible for A and B to be both independent and mutually exclusive? Explain.

Answers to Check Your Understanding Exercises for Section 5.3

1. **a.** 0.164 **b.** 0.315 **c.** 0.519
2. 0.28
3. **a.** The outcome on one die does not influence the outcome on the other die.
   b. $P(A) = 1/2$; $P(B) = 1/6$; $P(A \text{ and } B) = 1/12$

4. Yes, because the sample is less than 5% of the population.
5. No, because the sample is more than 5% of the population.
6. 0.999

SECTION 5.4	Counting

Objectives

1. Count the number of ways a sequence of operations can be performed
2. Count the number of permutations
3. Count the number of combinations

When computing probabilities, it is sometimes necessary to count the number of outcomes in a sample space without being able to list them all. In this section, we will describe several methods for doing this.

Objective 1 Count the number of ways a sequence of operations can be performed

The Fundamental Principle of Counting

The basic rule, which we will call the **Fundamental Principle of Counting**, is presented by means of the following example:

EXAMPLE 5.29	Using the Fundamental Principle of Counting

A certain make of automobile is available in any of three colors — red, blue, or green — and comes with either a large or small engine. In how many ways can a buyer choose a car?

Solution

There are 3 choices of color and 2 choices of engine. A complete list is shown in the tree diagram in Figure 5.11, and in the form of a table in Table 5.6. The total number of choices is $3 \cdot 2 = 6$.

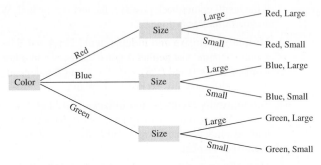

Figure 5.11   Tree diagram illustrating the six choices of color and engine size

Table 5.6 Six Outcomes for the Color and Engine Size

	Large	**Small**
Red	Red, Large	Red, Small
Blue	Blue, Large	Blue, Small
Green	Green, Large	Green, Small

To generalize Example 5.29, if there are m choices of color and n choices of engine, the total number of choices is mn. This leads to the Fundamental Principle of Counting.

The Fundamental Principle of Counting

If an operation can be performed in m ways, and a second operation can be performed in n ways, then the total number of ways to perform the sequence of two operations is mn.

If a sequence of several operations is to be performed, the number of ways to perform the sequence is found by multiplying together the numbers of ways to perform each of the operations.

EXAMPLE 5.30   Using the Fundamental Principle of Counting

License plates in a certain state contain three letters followed by three digits. How many different license plates can be made?

Solution

There are six operations in all: choosing three letters and choosing three digits. There are 26 ways to choose each letter and 10 ways to choose each digit. The total number of license plates is therefore

$$26 \cdot 26 \cdot 26 \cdot 10 \cdot 10 \cdot 10 = 17,576,000$$

Check Your Understanding

1. When ordering a certain type of computer, there are three choices of hard drive, four choices for the amount of memory, two choices of video card, and three choices of monitor. In how many ways can a computer be ordered?

2. A quiz consists of three true-false questions and two multiple-choice questions with five choices each. How many different sets of answers are there?

Answers are on page 239.

Objective 2 Count the number of permutations

Permutations

The word *permutation* is another word for *ordering*. When we count the number of permutations, we are counting the number of different ways that a group of items can be ordered.

EXAMPLE 5.31

Counting the number of permutations

Five runners run a race. One of them will finish first, another will finish second, and so on. In how many different orders can they finish?

Solution

We use the Fundamental Principle of Counting. There are five possible choices for the first-place finisher. Once the first-place finisher has been determined, there are four remaining choices for the second-place finisher. Then there are three possible choices for the third-place finisher, two choices for the fourth-place finisher, and only one choice for the fifth-place finisher. The total number of orders of five individuals is

$$\text{Number of orders} = 5 \cdot 4 \cdot 3 \cdot 2 \cdot 1 = 120$$

We say that there are 120 permutations of five individuals.

In Example 5.31, we computed a number of permutations by using the Fundamental Principle of Counting. We can generalize this method, but first we need some notation.

DEFINITION

For any positive integer n, the number $n!$ is pronounced "n factorial" and is equal to the product of all the integers from n down to 1.

$$n! = n(n-1)\ldots(2)(1)$$

By definition, $0! = 1$.

In Example 5.31, we found that the number of permutations of five objects is $5!$. This idea holds in general.

The number of permutations of n objects is $n!$.

Sometimes we want to count the number of permutations of a part of a group. Example 5.32 illustrates the idea.

EXAMPLE 5.32

Counting the number of permutations

Ten runners enter a race. The first-place finisher will win a gold medal, the second-place finisher will win a silver medal, and the third-place finisher will win a bronze medal. In how many different ways can the medals be awarded?

Solution

We use the Fundamental Principle of Counting. There are 10 possible choices for the gold-medal winner. Once the gold-medal winner is determined, there are nine remaining choices for the silver medal. Finally, there are eight choices for the bronze medal. The total number of ways the medals can be awarded is

$$10 \cdot 9 \cdot 8 = 720$$

In Example 5.32, three runners were chosen from a group of ten, then ordered as first, second, and third. This is referred to as a *permutation* of three items chosen from ten.

DEFINITION

A **permutation** of r items chosen from n items is an ordering of the r items. It is obtained by choosing r items from a group of n items, then choosing an order for the r items.

Notation: The number of permutations of r items chosen from n is denoted $_nP_r$.

In Example 5.32, we computed $_nP_r$ by using the Fundamental Principle of Counting. We can generalize this method by using factorial notation.

The number of permutations of r objects chosen from n is

$$_nP_r = n(n-1)\cdots(n-r+1) = \frac{n!}{(n-r)!}$$

EXAMPLE 5.33   Counting the number of permutations

Five lifeguards are available for duty one Saturday afternoon. There are three lifeguard stations. In how many ways can three lifeguards be chosen and ordered among the stations?

Solution

We are choosing three items from a group of five and ordering them. The number of ways to do this is

$$_5P_3 = \frac{5!}{(5-3)!} = \frac{5!}{2!} = \frac{5\cdot 4\cdot 3\cdot 2\cdot 1}{2\cdot 1} = 5\cdot 4\cdot 3 = 60$$

In some situations, computing the value of $_nP_r$ enables us to determine the number of outcomes in a sample space, and thereby compute a probability. Example 5.34 illustrates the idea.

EXAMPLE 5.34   Using counting to compute a probability

Refer to Example 5.33. The five lifeguards are named Abby, Bruce, Christopher, Donna, and Esmeralda. Of the three lifeguard stations, one is located at the north end of the beach, one in the middle of the beach, and one at the south end. The lifeguard assignments are made at random. What is the probability that Bruce is assigned to the north station, Donna is assigned to the middle station, and Abby is assigned to the south station?

Solution

The outcomes in the sample space consist of all the choices of three lifeguards chosen from five and ordered. From Example 5.33, we know that there are 60 such outcomes. Only one of the outcomes has Bruce, Donna, Abby, in that order. Thus, the probability is $\frac{1}{60}$.

EXAMPLE 5.35   Using counting to compute a probability

Refer to Example 5.34. What is the probability that Bruce is assigned to the north station, Abby is assigned to the south station, and either Donna or Esmeralda is assigned to the middle station?

Solution

As in Example 5.34, the sample space consists of the 60 permutations of three lifeguards chosen from five. Two of these permutations satisfy the stated conditions: Bruce, Donna, Abby; and Bruce, Esmeralda, Abby. So the probability is $\frac{2}{60} = \frac{1}{30}$.

Check Your Understanding

3. A committee of eight people must choose a president, a vice president, and a secretary. In how many ways can this be done?

4. Refer to Exercise 3. Two of the committee members are Ellen and Jose. Assume the assignments are made at random,
   a. What is the probability that Jose is president and Ellen is vice president?
   b. What is the probability that either Ellen or Jose is president and the other is vice president?

Answers are on page 239.

Objective 3 Count the number of combinations

Combinations

In some cases, when choosing a set of objects from a larger set, we don't care about the ordering of the chosen objects; we care only which objects are chosen. For example, we may not care which lifeguard occupies which station; we might care only which three lifeguards are chosen. Each distinct group of objects that can be selected, without regard to order, is called a **combination**. We will now show how to determine the number of combinations of r objects chosen from a set of n objects. We will illustrate the reasoning with the result of Example 5.33. In that example, we showed that there are 60 permutations of 3 objects chosen from 5. Denoting the objects A, B, C, D, E, Table 5.7 presents a list of all 60 permutations.

Table 5.7 The 60 Permutations of 3 Objects Chosen from 5

ABC	ABD	ABE	ACD	ACE	ADE	BCD	BCE	BDE	CDE
ACB	ADB	AEB	ADC	AEC	AED	BDC	BEC	BED	CED
BAC	BAD	BAE	CAD	CAE	DAE	CBD	CBE	DBE	DCE
BCA	BDA	BEA	CDA	CEA	DEA	CDB	CEB	DEB	DEC
CAB	DAB	EAB	DAC	EAC	EAD	DBC	EBC	EBD	ECD
CBA	DBA	EBA	DCA	ECA	EDA	DCB	ECB	EDB	EDC

Explain It Again

When to use combinations: Use combinations when the order of the chosen objects doesn't matter. Use permutations when the order does matter.

The 60 permutations in Table 5.7 are arranged in 10 columns of 6 permutations each. Within each column, the three objects are the same, and the column contains the 6 different permutations of those three objects. Therefore, each column represents a distinct combination of 3 objects chosen from 5, and there are 10 such combinations. Table 5.7 thus shows that the number of combinations of 3 objects chosen from 5 can be found by dividing the number of permutations of 3 objects chosen from 5, which is $\dfrac{5!}{(5-3)!}$, by the number of permutations of 3 objects, which is 3!. In summary:

The number of combinations of 3 objects chosen from 5 is $\dfrac{5!}{3!(5-3)!}$

The number of combinations of r objects chosen from n is often denoted by the symbol $_nC_r$. The reasoning above can be generalized to derive an expression for $_nC_r$.

The number of combinations of r objects chosen from a group of n objects is

$$_nC_r = \frac{n!}{r!(n-r)!}$$

EXAMPLE 5.36   Counting the number of combinations

Thirty people attend a certain event, and 5 will be chosen at random to receive prizes. The prizes are all the same, so the order in which the people are chosen does not matter. How many different groups of 5 people can be chosen?

Solution

Since the order of the 5 chosen people does not matter, we need to compute the number of combinations of 5 chosen from 30. This is

$$
\begin{aligned}
_{30}C_5 &= \frac{30!}{5!(30-5)!} \\
&= \frac{30 \cdot 29 \cdot 28 \cdot 27 \cdot 26}{5 \cdot 4 \cdot 3 \cdot 2 \cdot 1} \\
&= 142{,}506
\end{aligned}
$$

EXAMPLE 5.37 Using counting to compute a probability

Refer to Example 5.36. Of the 30 people in attendance, 12 are men and 18 are women.

    a. What is the probability that all the prize winners are men?

    b. What is the probability that at least one prize winner is a woman?

Solution

    a. The number of outcomes in the sample space is the number of combinations of 5 chosen from 30. We computed this in Example 5.36 to be $_{30}C_5 = 142{,}506$. The number of outcomes in which every prize winner is a man is the number of combinations of five men chosen from 12 men. This is

$$
_{12}C_5 = \frac{12!}{5!(12-5)!} = \frac{12 \cdot 11 \cdot 10 \cdot 9 \cdot 8}{5 \cdot 4 \cdot 3 \cdot 2 \cdot 1} = 792
$$

    The probability that all prize winners are men is

$$
P(\text{All men}) = \frac{792}{142{,}506} = 0.0056
$$

    b. This asks for the probability of at least one woman. We therefore find the probability of the complement; that is, we find the probability that none of the prize winners are women. The probability that none of the prize winners are women is the same as the probability that all of the prize winners are men. In part (a), we computed $P(\text{All men}) = 0.0056$. Therefore,

$$
P(\text{At least one woman}) = 1 - P(\text{All men}) = 1 - 0.0056 = 0.9944
$$

EXAMPLE 5.38 Using counting to compute a probability

A box of lightbulbs contains eight good lightbulbs and two burned-out bulbs. Four bulbs will be selected at random to put into a new lamp. What is the probability that all four bulbs are good?

Solution

The order in which the bulbs are chosen does not matter; all that matters is whether a burned-out bulb is chosen. Therefore, the outcomes in the sample space consist of all the combinations of four bulbs that can be chosen from 10. This number is

$$
_{10}C_4 = \frac{10!}{4!(10-4)!} = \frac{3{,}628{,}800}{24 \cdot 720} = 210
$$

To select four good bulbs, we must choose the four bulbs from the eight good bulbs. The number of outcomes that correspond to selecting four good bulbs is therefore the number of combinations of four bulbs that can be chosen from eight. This number is

$$
_8C_4 = \frac{8!}{4!(8-4)!} = \frac{40{,}320}{24 \cdot 24} = 70
$$

The probability that four good bulbs are selected is therefore

$$P(\text{Four good bulbs are selected}) = \frac{70}{210} = \frac{1}{3}$$

Check Your Understanding

5. Eight college students have applied for internships at a local firm. Three of them will be selected for interviews. In how many ways can this be done?

6. Refer to Exercise 5. Four of the eight students are from Middle Georgia State College. What is the probability that all three of the interviewed students are from Middle Georgia State College?

Answers are on page 239.

USING TECHNOLOGY

TI-84 PLUS

Evaluating a factorial

Step 1. To evaluate $n!$, enter n on the home screen.

Step 2. Press **MATH**, scroll to the **PRB** menu, and select **4: !**

Step 3. Press **ENTER**.

Permutations and combinations

Step 1. To evaluate $_nP_r$ or $_nC_r$, enter n on the home screen.

Step 2. Press **MATH** and scroll to the **PRB** menu.
- For permutations, select **2: nPr** and press **ENTER** (Figure A).
- For combinations, select **3: nCr** and press **ENTER**.

Step 3. Enter the value for r and press **ENTER**.

The results of $_{12}P_3$ and $_{12}C_3$ are shown in Figure B.

Figure A

Figure B

EXCEL

Evaluating a factorial

Step 1. To evaluate $n!$, click on a cell in the worksheet and type **=FACT(n)** and press **ENTER**. For example, to compute **12!**, type **=FACT(12)** and press **ENTER**.

Permutations

Step 1. To evaluate $_nP_r$, click on a cell in the worksheet and type **=PERMUT(n,r)**. Press **ENTER**.

Combinations

Step 1. To evaluate $_nC_r$, click on a cell in the worksheet and type **=COMBIN(n,r)**. Press **ENTER**.

SECTION 5.4   Exercises

Exercises 1–6 are the Check Your Understanding exercises located within the section.

Understanding the Concepts

In Exercises 7 and 8, fill in the blank with the appropriate word or phrase:

7. If an operation can be performed in m ways, and a second operation can be performed in n ways, then the total number of ways to perform the sequence of two operations is _____ .

8. The number of permutations of 6 objects is _____ .

In Exercises 9 and 10, determine whether the statement is true or false. If the statement is false, rewrite it as a true statement.

9. In a permutation, order is not important.

10. In a combination, order is not important.

Practicing the Skills

In Exercises 11–16, evaluate the factorial.

11. 9!   12. 5!   13. 0!
14. 12!   15. 1!   16. 3!

In Exercises 17–22, evaluate the permutation.

17. $_7P_3$ 18. $_8P_1$ 19. $_{35}P_2$
20. $_5P_4$ 21. $_{20}P_0$ 22. $_{45}P_5$

In Exercises 23–28, evaluate the combination.

23. $_9C_5$ 24. $_7C_1$ 25. $_{25}C_3$
26. $_{10}C_9$ 27. $_{12}C_0$ 28. $_{50}C_{50}$

Working with the Concepts

29. **Pizza time:** A local pizza parlor is offering a half-price deal on any pizza with one topping. There are eight toppings from which to choose. In addition, there are three different choices for the size of the pizza, and two choices for the type of crust. In how many ways can a pizza be ordered?

30. **Books:** Josephine has six chemistry books, three history books, and eight statistics books. She wants to choose one book of each type to study. In how many ways can she choose the three books?

31. **Playing the horses:** In horse racing, one can make a trifecta bet by specifying which horse will come in first, which will come in second, and which will come in third, in the correct order. One can make a box trifecta bet by specifying which three horses will come in first, second, and third, without specifying the order.
    a. In an eight-horse field, how many different ways can one make a trifecta bet?
    b. In an eight-horse field, how many different ways can one make a box trifecta bet?

32. **Ice cream:** A certain ice cream parlor offers 15 flavors of ice cream. You want an ice cream cone with three scoops of ice cream, all different flavors.
    a. In how many ways can you choose a cone if it matters which flavor is on the top, which is in the middle, and which is on the bottom?

b. In how many ways can you choose a cone if the order of the flavors doesn't matter?

33. **License plates:** In a certain state, license plates consist of four digits from 0 to 9 followed by three letters. Assume the numbers and letters are chosen at random. Replicates are allowed.
    a. How many different license plates can be formed?
    b. How many different license plates have the letters S-A-M in that order?
    c. If your name is Sam, what is the probability that your name is on your license plate?

34. **Committee:** The Student Council at a certain school has ten members. Four members will form an executive committee consisting of a president, a vice president, a secretary, and a treasurer.
    a. In how many ways can these four positions be filled?
    b. In how many ways can four people be chosen for the executive committee if it does not matter who gets which position?
    c. Four of the people on Student Council are Zachary, Yolanda, Xavier, and Walter. What is the probability that Zachary is president, Yolanda is vice president, Xavier is secretary, and Walter is treasurer?
    d. What is the probability that Zachary, Yolanda, Xavier, and Walter are the four committee members?

35. **Day and night shifts:** A company has hired 12 new employees, and must assign 8 to the day shift and 4 to the night shift.
    a. In how many ways can the assignment be made?
    b. Assume that the 12 employees consist of six men and six women and that the assignments to day and night shift are made at random. What is the probability that all four of the night-shift employees are men?
    c. What is the probability that at least one of the night-shift employees is a woman?

36. **Keep your password safe:** A computer password consists of eight characters. Replications are allowed.
    a. How many different passwords are possible if each character may be any lowercase letter or digit?
    b. How many different passwords are possible if each character may be any lowercase letter?
    c. How many different passwords are possible if each character may be any lowercase letter or digit, and at least one character must be a digit?
    d. A computer is generating passwords. The computer generates eight characters at random, and each is equally likely to be any of the 26 letters or 10 digits. Replications are allowed. What is the probability that the password will contain all letters?
    e. A computer system requires that passwords contain at least one digit. If eight characters are generated at random, what is the probability that they will form a valid password?

37. **It's in your genes:** Human genetic material (DNA) is made up of sequences of the molecules adenosine (A), guanine (G), cytosine (C), and thymine (T), which are called *bases*. A *codon* is a sequence of three bases. Replicates are allowed, so AAA, CGC, and so forth are codons. Codons are important because each codon causes a different protein to be created.
    a. How many different codons are there?
    b. How many different codons are there in which all three bases are different?
    c. The bases A and G are called *purines*, while C and T are called *pyrimidines*. How many different codons are there in which the first base is a purine and the second and third are pyrimidines?
    d. What is the probability that all three bases are different?
    e. What is the probability that the first base is a purine and the second and third are pyrimidines?

38. **Choosing officers:** A committee consists of ten women and eight men. Three committee members will be chosen as officers.
    a. How many different choices are possible?
    b. How many different choices are possible if all the officers are to be women?
    c. How many different choices are possible if all the officers are to be men?
    d. What is the probability that all the officers are women?
    e. What is the probability that at least one officer is a man?

39. **Texas hold 'em:** In the game of Texas hold 'em, a player is dealt two cards (called hole cards) from a standard deck of 52 playing cards. The order in which the cards are dealt does not matter.
    a. How many different combinations of hole cards are possible?
    b. The best hand consists of two aces. There are four aces in the deck. How many combinations are there in which both cards are aces?
    c. What is the probability that a hand consists of two aces?

40. **Blackjack:** In single-deck casino blackjack, the dealer is dealt two cards from a standard deck of 52. The first card is dealt face down and the second card is dealt face up.
    a. How many dealer hands are possible if it matters which card is face down and which is face up?
    b. How many dealer hands are possible if it doesn't matter which card is face down and which is face up?
    c. Of the 52 cards in the deck, four are aces and 16 others (kings, queens, jacks, and tens) are worth 10 points each. The dealer has a blackjack if one card is an ace and the other is worth 10 points; it doesn't matter which card is face up and which card is face down. How many different blackjack hands are there?
    d. What is the probability that a hand is a blackjack?

41. **Lottery:** In the Georgia Fantasy 5 Lottery, balls are numbered from 1 to 39. Five balls are drawn. To win the jackpot, you must mark five numbers from 1 to 39 on a ticket, and your numbers must match the numbers on the five balls. The order does not matter. What is the probability that you win?

42. **Lottery:** In the Colorado Lottery Lotto game, balls are numbered from 1 to 42. Six balls are drawn. To win the jackpot, you must mark six numbers from 1 to 42 on a ticket, and your numbers must match the numbers on the six balls. The order does not matter. What is the probability that you win?

Extending the Concepts

43. **Sentence completion:** Let A and B be events. Consider the following sentence:

    If A and B are ____(i)____ , then to find ____(ii)____ , ____(iii)____ $P(A)$ and $P(B)$.

    Each blank in the sentence can be filled in with either of two choices, as follows:
    (i) independent, mutually exclusive
    (ii) $P(A \text{ and } B)$, $P(A \text{ or } B)$
    (iii) multiply, add
    a. In how many ways can the sentence be completed?
    b. If choices are made at random for each of the blanks, what is the probability that the sentence is true?

Answers to Check Your Understanding Exercises for Section 5.4

1. 72	**4. a.** 1/56 **b.** 1/28
2. 200	**5.** 56
3. 336	**6.** 1/14

Chapter 5 Summary

Section 5.1: A probability experiment is an experiment that can result in any one of a number of outcomes. The collection of all possible outcomes is a sample space. Sampling from a population is a common type of probability experiment. The population is the sample space, and the individuals in the population are the outcomes. An event is a collection of outcomes from a sample space. The probability of an event is the proportion of times the event occurs in the long run, as the experiment is repeated over and over again. A probability model specifies a probability for every event.

An unusual event is one whose probability is small. There is no hard-and-fast rule about how small a probability has to be for an event to be unusual, but 0.05 is the most commonly used value. The Empirical Method allows us to approximate the probability of an event by repeating a probability experiment many times and computing the proportion of times the event occurs. This is generally done by simulation, in which technology is used to repeat virtual experiments many times.

Section 5.2: A compound event is an event that is formed by combining two or more events. An example of a compound event is one of the form "*A* or *B*." The General Addition Rule is used to compute probabilities of the form $P(A \text{ or } B)$. Two events are mutually exclusive if it is impossible for both events to occur. When two events are mutually exclusive, the Addition Rule for Mutually Exclusive Events can be used to find $P(A \text{ or } B)$. The complement of an event *A*, denoted A^c, is the event that *A* does not occur. The Rule of Complements states that $P(A^c)$ is found by subtracting $P(A)$ from 1.

Section 5.3: A conditional probability is a probability that is computed with the knowledge of additional information. Conditional probabilities can be computed with the General Method for computing conditional probabilities. Probabilities of the form $P(A \text{ and } B)$ can be computed with the General Multiplication Rule. If *A* and *B* are independent, then $P(A \text{ and } B)$ can be computed with the Multiplication Rule for Independent Events. Two events are independent if the occurrence of one does not affect the probability that the other occurs. When sampling from a population, sampled individuals are independent if the sampling is done with replacement, or if the sample size is less than 5% of the population.

Section 5.4: The Fundamental Principle of Counting states that the total number of ways to perform a sequence of operations is found by multiplying together the numbers of ways of performing each operation. We can compute the number of permutations and combinations of *r* items chosen from a group of *n* items. The number of ways that a group of *r* items can be chosen without regard to order is the number of combinations. The number of ways that a group of *r* items can be chosen and ordered is the number of permutations. Some sample spaces consist of the permutations or combinations of *r* items chosen from a group of *n* items. When working with these sample spaces, we can use the counting rules to compute probabilities.

Vocabulary and Notation

Important Formulas

General Addition Rule:

$P(A \text{ or } B) = P(A) + P(B) - P(A \text{ and } B)$

Multiplication Rule for Independent Events:

$P(A \text{ and } B) = P(A)P(B)$

Addition Rule for Mutually Exclusive Events:

$P(A \text{ or } B) = P(A) + P(B)$

Rule of Complements:

$P(A^c) = 1 - P(A)$

General Method for Computing Conditional Probability:

$P(B \mid A) = \dfrac{P(A \text{ and } B)}{P(A)}$

General Multiplication Rule:

$P(A \text{ and } B) = P(A)P(B \mid A) = P(B)P(A \mid B)$

Permutation of *r* items chosen from *n*:

$_nP_r = \dfrac{n!}{(n-r)!}$

Combination of *r* items chosen from *n*:

$_nC_r = \dfrac{n!}{r!(n-r)!}$

Chapter Quiz

1. Fill in the blank: The probability that a fair coin lands heads is 0.5. Therefore, we can be sure that if we toss a coin repeatedly, the proportion of times it lands heads will _____.
   i. approach 0.5
   ii. be equal to 0.5
   iii. be greater than 0.5
   iv. be less than 0.5

2. A pollster will draw a simple random sample of voters from a large city to ask whether they support the construction of a new light rail line. Assume that there are one million voters in the city, and that 560,000 of them support this proposition. One voter is sampled at random.

a. Identify the sample space.

b. What is the probability that the sampled voter supports the light rail line?

3. State each of the following rules:

  a. General Addition Rule

  b. Addition Rule for Mutually Exclusive Events

  c. Rule of Complements

  d. General Multiplication Rule

  e. Multiplication Rule for Independent Events

4. The following table presents the results of a survey in which 400 college students were asked whether they listen to music while studying.

	Listen	Do Not Listen
Male	121	78
Female	147	54

  a. Find the probability that a randomly selected student does not listen to music while studying.

  b. Find the probability that a randomly selected student listens to music or is male.

5. Which of the following pairs of events are mutually exclusive?

  i. *A*: A randomly chosen student is 18 years old. *B*: The same student is 20 years old.

  ii. *A*: A randomly chosen student owns a red car. *B*: The same student owns a blue car.

6. In a group of 100 teenagers, 61 received their driver's license on their first attempt on the driver's certification exam and 18 received their driver's license on their second attempt. What is the probability that a randomly selected teenager received their driver's license on their first or second attempt?

7. A certain neighborhood has 100 households. Forty-eight households have a dog as a pet. Of these, 32 also have a cat. Given that a household has a dog, what is the probability that it also has a cat?

8. The owner of a bookstore has determined that 80% of people who enter the store will buy a book. Of those who buy a book, 60% will pay with a credit card. Find the probability that a randomly selected person entering the store will buy a book and pay for it using a credit card.

9. A jar contains 4 red marbles, 3 blue marbles, and 5 green marbles. Two marbles are drawn from the jar one at a time without replacement. What is the probability that the second marble is red, given that the first was blue?

10. A student is chosen at random. Which of the following pairs of events are independent?

  i. *A*: The student was born on a Monday. *B*: The student's mother was born on a Monday.

  ii. *A*: The student is above average in height. *B*: The student's mother is above average in height.

11. Individual plays on a slot machine are independent. The probability of winning on any play is 0.38. What is the probability of winning 3 plays in a row?

12. Refer to Problem 11. Suppose that the slot machine is played 5 times in a row. What is the probability of winning at least once?

13. The Roman alphabet (the one used to write English) consists of five vowels (a, e, i, o, u), along with 21 consonants (we are considering y to be a consonant). Gregory needs to make up a computer password containing seven characters. He wants the first six characters to alternate — consonant, vowel, consonant, vowel, consonant, vowel — with repetitions allowed. Then he wants to use a digit for the seventh character.

  a. How many different passwords can he make up?

  b. If he makes up a password at random, what is the probability that his password is banana7?

14. A caterer offers 24 different types of dessert. In how many ways can 5 of them be chosen for a banquet if the order doesn't matter?

15. In a standard game of pool, there are 15 balls labeled 1 through 15.

  a. In how many ways can the 15 balls be ordered?

  b. In how many ways can 3 of the 15 balls be chosen and ordered?

Review Exercises

1. Colored dice: A six-sided die has one face painted red, two faces painted white, and three faces painted blue. Each face is equally likely to turn up when the die is rolled.

  a. Construct a sample space for the experiment of rolling this die.

  b. Find the probability that a blue face turns up.

2. How are your grades? There were 30 students in last semester's statistics class. Of these, 6 received a grade of A, and 12 received a grade of B. What is the probability that a randomly chosen student received a grade of A or B?

3. **Statistics, anyone?** Let S be the event that a randomly selected college student has taken a statistics course, and let C be the event that the same student has taken a chemistry course. Suppose $P(S) = 0.4$, $P(C) = 0.3$, and $P(S \text{ and } C) = 0.2$.
   a. Find the probability that a student has taken statistics or chemistry.
   b. Find the probability that a student has taken statistics given that the student has taken chemistry.

4. **Blood types:** Human blood may contain either or both of two antigens, A and B. Blood that contains only the A antigen is called type A, blood that contains only the B antigen is called type B, blood that contains both antigens is called type AB, and blood that contains neither antigen is called type O. A certain blood bank has blood from a total of 1200 donors. Of these, 570 have type O blood, 440 have type A, 125 have type B, and 65 have type AB.
   a. What is the probability that a randomly chosen blood donor is type O?
   b. A recipient with type A blood may safely receive blood from a donor whose blood does not contain the B antigen. What is the probability that a randomly chosen blood donor may donate to a recipient with type A blood?

5. **Start a business:** Suppose that start-up companies in the area of biotechnology have probability 0.2 of becoming profitable, and that those in the area of information technology have probability 0.15 of becoming profitable. A venture capitalist invests in one firm of each type. Assume the companies function independently.
   a. What is the probability that both companies become profitable?
   b. What is the probability that at least one of the two companies becomes profitable?

6. **Stop that car:** A drag racer has two parachutes, a main and a backup, that are designed to bring the vehicle to a stop at the end of a run. Suppose that the main chute deploys with probability 0.99, and that if the main fails to deploy, the backup deploys with probability 0.98.
   a. What is the probability that one of the two parachutes deploys?
   b. What is the probability that the backup parachute deploys?

7. **Defective parts:** A process manufactures microcircuits that are used in computers. Twelve percent of the circuits are defective. Assume that three circuits are installed in a computer. Denote a defective circuit by "D" and a good circuit by "G."
   a. List all eight items in the sample space.
   b. What is the probability that all three circuits are good?
   c. The computer will function so long as either two or three of the circuits are good. What is the probability that a computer will function?
   d. If we use a cutoff of 0.05, would it be unusual for all three circuits to be defective?

8. **Music to my ears:** Jeri is listening to the songs on a new CD in random order. She will listen to two different songs, and will buy the CD if she likes both of them. Assume there are 10 songs on the CD, and that she would like five of them.
   a. What is the probability that she likes the first song?
   b. What is the probability that she likes the second song, given that she liked the first song?
   c. What is the probability that she buys the CD?

9. **Female business majors:** At a certain university, the probability that a randomly chosen student is female is 0.55, the probability that the student is a business major is 0.20, and the probability that the student is female and a business major is 0.15.
   a. What is the probability that the student is female or a business major?
   b. What is the probability that the student is female given that the student is a business major?
   c. What is the probability that the student is a business major given that the student is female?
   d. Are the events "female" and "business major" independent? Explain.
   e. Are the events "female" and "business major" mutually exclusive? Explain.

10. **Heart attack:** The following table presents the number of hospitalizations for myocardial infarction (heart attack) for men and women in various age groups.

Age	Male	Female	Total
18–44	26,828	9,265	36,093
45–64	166,340	68,666	235,006
65–84	155,707	124,289	279,996
85 and up	35,524	57,785	93,309
Total	384,399	260,005	644,404

Source: Agency for Healthcare Research and Quality

   a. What is the probability that a randomly chosen patient is a woman?
   b. What is the probability that a randomly chosen patient is aged 45–64?
   c. What is the probability that a randomly chosen patient is a woman and aged 45–64?
   d. What is the probability that a randomly chosen patient is a woman or aged 45–64?

    e. What is the probability that a randomly chosen patient is a woman given that the patient is aged 45–64?

    f. What is the probability that a randomly chosen patient is aged 45–64 given that the patient is a woman?

11. Rainy weekend: Sally is planning to go away for the weekend this coming Saturday and Sunday. At the place she will be going, the probability of rain on any given day is 0.10. Sally says that the probability that it rains on both days is 0.01. She reasons as follows:

$$P(\text{Rain Saturday and Rain Sunday}) = P(\text{Rain Saturday})P(\text{Rain Sunday})$$
$$= (0.1)(0.1)$$
$$= 0.01$$

    a. What assumption is being made in this calculation?

    b. Explain why this assumption is probably not justified in the present case.

    c. Is the probability of 0.01 likely to be too high or too low? Explain.

12. Required courses: A college student must take courses in English, history, mathematics, biology, and physical education. She decides to choose three of these courses to take in her freshman year. In how many ways can this choice be made?

13. Required courses: Refer to Exercise 12. Assume the student chooses three courses at random. What is the probability that she chooses English, mathematics, and biology?

14. Bookshelf: Bart has six books: a novel, a biography, a dictionary, a self-help book, a statistics textbook, and a comic book.

    a. Bart's bookshelf has room for only three of the books. In how many ways can Bart choose and order three books?

    b. In how many ways may the books be chosen and ordered if he does not choose the comic book?

15. Bookshelf: Refer to Exercise 14. Bart chooses three books at random.

    a. What is the probability that the books on his shelf are statistics textbook, dictionary, and comic book, in that order?

    b. What is the probability that the statistics textbook, dictionary, and comic book are the three books chosen, in any order?

Write About It

1. Explain how you could use the law of large numbers to show that a coin is unfair by tossing it many times.

2. When it comes to betting, the chance of winning or losing may be expressed as odds. If there are n equally likely outcomes and m of them result in a win, then the odds of winning are $m:(n-m)$, read "m to $n-m$." For example, suppose that a player rolls a die and wins if the number of dots appearing is either 1 or 2. Since there are two winning outcomes out of six equally likely outcomes, the odds of winning are 2:4.

    Suppose that a pair of dice is rolled and the player wins if it comes up "doubles," that is, if the same number of dots appears on each die. What are the odds of winning?

3. If the odds of an event occurring are 5:8, what is the probability that the event will occur?

4. Explain why the General Addition Rule $P(A \text{ or } B) = P(A) + P(B) - P(A \text{ and } B)$ may be used even when A and B are mutually exclusive events.

5. Sometimes events are in the form "at least" a given number. For example, if a coin is tossed five times, an event could be getting at least two heads. What would be the complement of the event of getting at least two heads?

6. In practice, one must decide whether to treat two events as independent based on an understanding of the process that creates them. For example, in a manufacturing process that produces electronic circuit boards for calculators, assume that the probability that a board is defective is 0.01. You arrive at the manufacturing plant and sample the next two boards that come off the assembly line. Let A be the event that the first board is defective, and let B be the event that the second board is defective. Describe circumstances under which A and B would not be independent.

7. Describe circumstances under which you would use a permutation.

8. Describe circumstances under which you would use a combination.

Case Study: How Likely Are You To Live To Age 100?

The following table is a *life table*, reproduced from the chapter introduction. With an understanding of some basic concepts of probability, one can use the life table to compute the probability that a person of a given age will still be alive a given number of years from now. Life insurance companies use this information to determine how much to charge for life insurance policies.

United States Life Table, Total Population

Age Interval	Proportion Surviving	Age Interval	Proportion Surviving
0–10	0.99123	50–60	0.94010
10–20	0.99613	60–70	0.86958
20–30	0.99050	70–80	0.70938
30–40	0.98703	80–90	0.42164
40–50	0.97150	90–100	0.12248

Source: Centers for Disease Control and Prevention

The column labeled "Proportion Surviving" gives the proportion of people alive at the beginning of an age interval who will still be alive at the end of the age interval. For example, among those currently age 20, the proportion who will still be alive at age 30 is 0.99050, or 99.050%. We will begin by computing the probability that a person lives to any of the ages 10, 20, ... , 100.

The first number in the column is the probability that a person lives to age 10. So

$$P(\text{Alive at age } 10) = 0.99123$$

The key to using the life table is to realize that the rest of the numbers in the "Proportion Surviving" column are conditional probabilities. They are probabilities that a person is alive at the end of the age interval, given that they were alive at the beginning of the age interval. For example, the row labeled "20–30" contains the conditional probability that someone alive at age 20 will be alive at age 30:

$$P(\text{Alive at age } 30 \mid \text{Alive at age } 20) = 0.99050$$

In Exercises 1–5, compute the probability that a person lives to a given age.

1. From the table, find the conditional probability $P(\text{Alive at age } 20 \mid \text{Alive at age } 10)$.

2. Use the result from Exercise 1 along with the result $P(\text{Alive at age } 10) = 0.99123$ to compute $P(\text{Alive at age } 20)$.

3. Use the result from Exercise 2 along with the appropriate number from the table to compute $P(\text{Alive at age } 30)$.

4. Use the result from Exercise 3 along with the appropriate number from the table to compute $P(\text{Alive at age } 40)$.

5. Compute the probability that a person is alive at ages 50, 60, 70, 80, 90, and 100.

In Exercises 1–5, we computed the probability that a newborn lives to a given age. Now let's compute the probability that a person aged x lives to age y. We'll illustrate this with an example to compute the probability that a person aged 20 lives to age 100. This is the conditional probability that a person lives to age 100, given that the person has lived to age 20.

We want to compute the conditional probability

$$P(\text{Alive at age } 100 \mid \text{Alive at age } 20)$$

Using the definition of conditional probability, we have

$$P(\text{Alive at age } 100 \mid \text{Alive at age } 20) = \frac{P(\text{Alive at age } 100 \text{ and Alive at age } 20)}{P(\text{Alive at age } 20)}$$

You computed $P(\text{Alive at age } 20)$ in Exercise 2. Now we need to compute $P(\text{Alive at age } 100 \text{ and Alive at age } 20)$. The key is to realize that anyone who is alive at age 100 was also alive at age 20. Therefore,

$$P(\text{Alive at age } 100 \text{ and Alive at age } 20) = P(\text{Alive at age } 100)$$

Therefore,

$$P(\text{Alive at age } 100 \mid \text{Alive at age } 20) = \frac{P(\text{Alive at age } 100)}{P(\text{Alive at age } 20)}$$

In general, for $y > x$,

$$P(\text{Alive at age } y \mid \text{Alive at age } x) = \frac{P(\text{Alive at age } y)}{P(\text{Alive at age } x)}$$

6. Find the probability that a person aged 20 is still alive at age 100.

7. Find the probability that a person aged 50 is still alive at age 70.

8. Which is more probable, that a person aged 20 is still alive at age 50, or that a person aged 50 is still alive at age 60?

9. A life insurance company sells term insurance polices. These policies pay $100,000 if the policyholder dies before age 70, but pay nothing if a person is still alive at age 70. If a person buys a policy at age 40, what is the probability that the insurance company does not have to pay?

Discrete Probability Distributions

Introduction

How does the Internal Revenue Service detect a fraudulent tax return? One method involves the use of probability in a surprising way. A list of all the amounts claimed as deductions is made. Then a relative frequency distribution is constructed of the first digits of these amounts. This relative frequency distribution is compared to a theoretical distribution, called Benford's law, which gives the probability of each digit. If there is a large discrepancy, the return is suspected of being fraudulent.

The probabilities assigned by Benford's law are given in the following table:

Digit	1	2	3	4	5	6	7	8	9
Probability	0.301	0.176	0.125	0.097	0.079	0.067	0.058	0.051	0.046

Benford's law is an example of a probability distribution. It may be surprising that the first digits of amounts on tax returns are not all equally likely, but in fact the smaller digits occur much more frequently than the larger ones. It turns out that Benford's law describes many data sets that occur naturally. In the case study at the end of the chapter, we will learn more about Benford's law.

| SECTION 6.1 | Random Variables |

Objectives

1. Distinguish between discrete and continuous random variables
2. Determine a probability distribution for a discrete random variable
3. Describe the connection between probability distributions and populations
4. Construct a probability histogram for a discrete random variable
5. Compute the mean of a discrete random variable
6. Compute the variance and standard deviation of a discrete random variable

If we roll a fair die, the possible outcomes are the numbers 1, 2, 3, 4, 5, and 6, and each of these numbers has probability 1/6. Rolling a die is a probability experiment whose outcomes are numbers. The outcome of such an experiment is called a *random variable*. Thus, rolling a die produces a random variable whose possible values are the numbers 1 through 6, each having probability 1/6.

Mathematicians and statisticians like to use letters to represent numbers. Uppercase letters are often used to represent random variables. Thus, a statistician might say, "Let X be the number that comes up on the next roll of the die."

DEFINITION

A **random variable** is a numerical outcome of a probability experiment.
Notation: Random variables are usually denoted by uppercase letters.

In Section 1.2, we learned that numerical, or quantitative, variables can be discrete or continuous. The same is true for random variables.

Objective 1 Distinguish between discrete and continuous random variables

DEFINITION

- **Discrete random variables** are random variables whose possible values can be listed. The list may be infinite — for example, the list of all whole numbers.
- **Continuous random variables** are random variables that can take on any value in an interval. The possible values of a continuous variable are not restricted to any list.

| EXAMPLE 6.1 | Determining whether a random variable is discrete or continuous |

Which of the following random variables are discrete and which are continuous?

    a. The number that comes up on the roll of a die
    b. The height of a randomly chosen college student
    c. The number of siblings a randomly chosen person has
    d. Amount of electricity used to light a randomly chosen classroom

Solution

    a. The number that comes up on a die is discrete. The possible values are 1, 2, 3, 4, 5, and 6.
    b. Height is continuous. A person's height is not restricted to any list of values.
    c. The number of siblings is discrete. The possible values are 0, 1, 2, and so forth.
    d. The amount of electricity is continuous. It is not restricted to any list of values.

In this chapter, we will focus on discrete random variables. In Chapter 7, we will learn about an important continuous random variable.

Objective 2 Determine a probability distribution for a discrete random variable

DEFINITION

A **probability distribution** for a discrete random variable specifies the probability for each possible value of the random variable.

EXAMPLE 6.2

Determining a probability distribution

A fair coin is tossed twice. Let X be the number of heads that come up. Find the probability distribution of X.

Solution

There are four equally likely outcomes to this probability experiment, listed in Table 6.1. For each outcome, we count the number of heads, which is the value of the random variable X.

Table 6.1

First Toss	Second Toss	X = Number of Heads
H	H	2
H	T	1
T	H	1
T	T	0

There are three possible values for the number of heads: 0, 1, and 2. One of the four outcomes has the value "0," two of the outcomes have the value "1," and one outcome has the value "2." Therefore, the probabilities are

$$P(0) = \frac{1}{4} = 0.25 \qquad P(1) = \frac{2}{4} = 0.50 \qquad P(2) = \frac{1}{4} = 0.25$$

The probability distribution is presented in Table 6.2.

Table 6.2 Probability Distribution of X

x	0	1	2
$P(x)$	0.25	0.50	0.25

Discrete probability distributions satisfy two properties. First, since the values $P(x)$ are probabilities, they must all be between 0 and 1. Second, since the random variable always takes on one of the values in the list, the sum of the probabilities must equal 1.

Explain It Again

Properties of discrete probability distributions: In a probability distribution, each probability must be between 0 and 1, and the sum of all the probabilities must be equal to 1.

SUMMARY

Let $P(x)$ denote the probability that a random variable has the value x. Then

1. $0 \leq P(x) \leq 1$ for every possible value x.
2. $\sum P(x) = 1$

EXAMPLE 6.3

Identifying probability distributions

Which of the following tables represent probability distributions?

a.

x	$P(x)$
1	0.25
2	0.65
3	−0.30
4	0.11

b.

x	$P(x)$
−1	0.17
−0.5	0.25
0	0.31
0.5	0.22
1	0.05

c.

x	$P(x)$
1	1.02
10	0.31
100	0.90
1000	0.43

d.

x	$P(x)$
0	0.10
1	0.17
2	0.75
3	0.24

Solution

a. This is not a probability distribution. $P(3)$ is not between 0 and 1.

b. This is a probability distribution. All the probabilities are between 0 and 1, and they add up to 1.

c. This is not a probability distribution. $P(1)$ is not between 0 and 1.

d. This is not a probability distribution. Although all the probabilities are between 0 and 1, they do not add up to 1.

When we are given the probability distribution of a random variable, we can use the rules of probability to compute probabilities involving the random variable. Example 6.4 provides an illustration.

| EXAMPLE 6.4 | Computing probabilities |

Four patients have made appointments to have their blood pressure checked at a clinic. Let X be the number of them who have high blood pressure. Based on data from the National Health and Examination Survey, the probability distribution of X is

x	0	1	2	3	4
$P(x)$	0.23	0.41	0.27	0.08	0.01

a. Find $P(1)$

b. Find $P(2 \text{ or } 3)$.

c. Find $P(\text{More than } 1)$.

d. Find $P(\text{At least } 1)$.

Solution

Recall: The Addition Rule for Mutually Exclusive Events says that if A and B are mutually exclusive, then $P(A \text{ or } B) = P(A) + P(B)$.

a. From the probability distribution, we see that the event that exactly one of the patients had high blood pressure is $P(1) = 0.41$.

b. The events "2" and "3" are mutually exclusive, since they cannot both happen. We use the Addition Rule for Mutually Exclusive events:

$$P(2 \text{ or } 3) = P(2) + P(3) = 0.27 + 0.08 = 0.35$$

c. "More than 1" means "2 or 3 or 4." Again we use the Addition Rule for Mutually Exclusive events:

$$P(\text{More than } 1) = P(2 \text{ or } 3 \text{ or } 4) = 0.27 + 0.08 + 0.01 = 0.36$$

Recall: The Rule of Complements says that $P(A^c) = 1 - P(A)$.

d. We use the Rule of Complements:

$$P(\text{At least } 1) = 1 - P(0) = 1 - 0.23 = 0.77$$

Check Your Understanding

1. A family has three children. If the genders of these children are listed in the order they are born, there are eight possible outcomes: BBB, BBG, BGB, BGG, GBB, GBG, GGB, and GGG. Assume these outcomes are equally likely. Let X represent the number of children that are girls. Find the probability distribution of X.

2. Someone says that the following table shows the probability distribution for the number of boys in a family of four children. Is this possible? Explain why or why not.

| Explain It Again |

The probability of at least 1: The complement of "at least 1" is "none," or "0." Therefore $P(\text{at least } 1) = 1 - P(0)$.

x	0	1	2	3	4
$P(x)$	0.12	0.37	0.45	0.25	0.18

3. Which of the following tables represent probability distributions?

a.

x	$P(x)$
0	0.45
1	0.15
2	0.30
3	0.10

b.

x	$P(x)$
4	0.27
5	0.15
6	0.11
7	0.34
8	0.25

c.

x	$P(x)$
1	0.02
2	0.41
3	0.24
4	0.33

4. Following is the probability distribution of a random variable that represents the number of extracurricular activities a college freshman participates in.

x	0	1	2	3	4
$P(x)$	0.06	0.14	0.45	0.21	0.14

a. Find the probability that a student participates in exactly two activities.
b. Find the probability that a student participates in more than two activities.
c. Find the probability that a student participates in at least one activity.

Answers are on page 260.

Objective 3 Describe the connection between probability distributions and populations

Connection Between Probability Distributions and Populations

Statisticians are interested in studying samples drawn from populations. Random variables are important because when an item is drawn from a population, the value observed is the value of a random variable. The probability distribution of the random variable tells how frequently we can expect each of the possible values of the random variable to turn up in the sample. Example 6.5 presents the idea.

EXAMPLE 6.5 | Constructing a probability distribution that describes a population

An airport parking facility contains 1000 parking spaces. Of these, 142 are covered long-term spaces that cost \$2.00 per hour, 378 are covered short-term spaces that cost \$4.50 per hour, 423 are uncovered long-term spaces that cost \$1.50 per hour, and 57 are uncovered short-term spaces that cost \$4.00 per hour. A parking space is selected at random. Let X represent the hourly parking fee for the randomly sampled space. Find the probability distribution of X.

Explain It Again

A probability distribution describes a population: We can think of a probability distribution as describing a population. The probability of each value represents the proportion of population items that have that value.

Solution
To find the probability distribution, we must list the possible values of X and then find the probability of each of them. The possible values of X are 1.50, 2.00, 4.00, 4.50. We find their probabilities:

$$P(1.50) = \frac{\text{number of spaces costing \$1.50}}{\text{total number of spaces}} = \frac{423}{1000} = 0.423$$

$$P(2.00) = \frac{\text{number of spaces costing \$2.00}}{\text{total number of spaces}} = \frac{142}{1000} = 0.142$$

$$P(4.00) = \frac{\text{number of spaces costing \$4.00}}{\text{total number of spaces}} = \frac{57}{1000} = 0.057$$

$$P(4.50) = \frac{\text{number of spaces costing \$4.50}}{\text{total number of spaces}} = \frac{378}{1000} = 0.378$$

The probability distribution is

x	1.50	2.00	4.00	4.50
$P(x)$	0.423	0.142	0.057	0.378

5. There are 5000 undergraduates registered at a certain college. Of them, 478 are taking one course, 645 are taking two courses, 568 are taking three courses, 1864 are taking four courses, 1357 are taking five courses, and 88 are taking six courses. Let X be the number of courses taken by a student randomly sampled from this population. Find the probability distribution of X.

Answer is on page 260.

Objective 4 Construct a probability histogram for a discrete random variable

Probability histograms

In Section 2.2, we learned to summarize the data in a sample with a histogram. We can represent discrete probability distributions with histograms as well. A histogram that represents a discrete probability distribution is called a **probability histogram**. Constructing a probability histogram from a probability distribution is just like constructing a relative frequency histogram from a relative frequency distribution for discrete data. For each possible value of the random variable, we draw a rectangle whose height is equal to the probability of that value.

Table 6.3 presents the probability distribution for the number of boys in a family of five children, using the assumption that boys and girls are equally likely and that births are independent events. Figure 6.1 presents a probability histogram for this probability distribution.

Explain It Again

A probability histogram is like a histogram for a population: The height of each rectangle in a probability histogram tells us how frequently the value appears in the population.

Table 6.3

x	$P(x)$
0	0.03125
1	0.15625
2	0.31250
3	0.31250
4	0.15625
5	0.03125

Figure 6.1 Probability histogram for the distribution in Table 6.3

6. Following is a probability histogram for the number of children a woman has. The numbers on the tops of the rectangles are the heights.

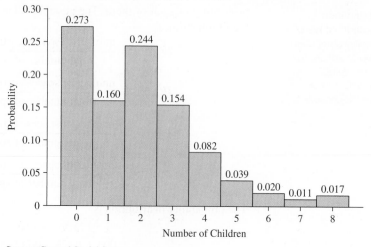

Source: General Social Survey

a. What is the probability that a randomly chosen woman has exactly two children?

b. What is the probability that a randomly chosen woman has fewer than two children?

c. What is the probability that a randomly chosen woman has either four or five children?

Answers are on page 260.

Objective 5 Compute the mean of a discrete random variable

The mean of a random variable

Recall that the mean is a measure of center. The mean of a random variable provides a measure of center for the probability distribution of a random variable.

DEFINITION

To find the **mean** of a discrete random variable, multiply each possible value by its probability, then add the products.

In symbols, $\mu_X = \sum [x \cdot P(x)]$.

Another name for the mean of a random variable is the **expected value**.

The notation for the expected value of X is $E(X)$.

EXAMPLE 6.6   Determining the mean of a discrete random variable

A computer monitor is composed of a very large number of points of light called pixels. It is not uncommon for a few of these pixels to be defective. Let X represent the number of defective pixels on a randomly chosen monitor. The probability distribution of X is as follows:

x	0	1	2	3
$P(x)$	0.2	0.5	0.2	0.1

Find the mean number of defective pixels.

Solution
The mean is

$$\mu_X = 0(0.2) + 1(0.5) + 2(0.2) + 3(0.1) = 1.2$$

The mean is 1.2.

The mean is a measure of the center of the probability distribution. Figure 6.2 presents a probability histogram for the distribution in Example 6.6 and shows the position of the mean. If we imagine each rectangle to be a weight, the mean is the point at which the histogram would balance.

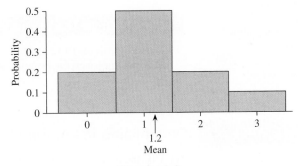

Figure 6.2

The symbol used to represent the mean of a random variable X is μ_X. It is not a coincidence that the mean of a population is also represented by μ. Recall that when we perform a probability experiment to obtain a value of a random variable, it is like sampling an item from a population. The probability distribution of the random variable tells how frequently each of the possible values of the random variable occurs in the population. The mean of the random variable is the same as the mean of the population.

Let's take a closer look at Example 6.6 to see how this is the case. Imagine that we had a population of ten computer monitors. Figure 6.3 presents a visualization of this population, with each monitor labeled with the number of defective pixels it has.

Figure 6.3 Population of ten computer monitors

The probability distribution in Example 6.6 represents this population. If we sample a monitor from this population, the probability that it will have 0 defective pixels is 0.2, the probability that it will have 1 defective pixel is 0.5, and so forth, just as in the probability distribution. Now we'll compute the mean of this population.

$$
\begin{aligned}
\mu &= \frac{0+0+1+1+1+1+1+2+2+3}{10} \\
&= \frac{0 \cdot 2 + 1 \cdot 5 + 2 \cdot 2 + 3 \cdot 1}{10} \\
&= 0 \cdot \frac{2}{10} + 1 \cdot \frac{5}{10} + 2 \cdot \frac{2}{10} + 3 \cdot \frac{1}{10} \\
&= 0(0.2) + 1(0.5) + 2(0.2) + 3(0.1) = 1.2
\end{aligned}
$$

The mean of the random variable is the same as the mean of the population.

Interpreting the mean of a random variable

In Example 6.6, imagine that we sampled a large number of computer monitors and counted the number of defective pixels on each. We would expect the mean number of defective pixels on our sampled monitors to be close to the population mean of 1.2. This idea provides an interpretation for the mean of a random variable.

Interpretation of the Mean of a Random Variable

If a probability experiment that produces a value of a random variable is repeated over and over again, the average of the values produced will approach the mean of the random variable.

When the probability experiment is sampling from a population, our interpretation of the mean says that as the sample size increases, the sample mean will approach the population mean. This is known as the **law of large numbers for means**.

Law of Large Numbers for Means

If we sample items from a population, then as the sample size grows larger, the sample mean will approach the population mean.

Explain It Again

Law of large numbers for means: The law of large numbers for means tells us that for a large sample, the sample mean will almost certainly be close to the population mean.

EXAMPLE 6.7

Determining the mean of a discrete random variable

Let X be the number of boys in a family of five children. The probability distribution of X is given in Table 6.4. Find the mean number of boys and interpret the result.

Table 6.4

x	P(x)
0	0.03125
1	0.15625
2	0.31250
3	0.31250
4	0.15625
5	0.03125

Solution

The calculations are presented in Table 6.5. We multiply each x by its corresponding $P(x)$, then add the products to obtain the mean. The mean is 2.5. We interpret this by saying that as we sample more and more families with five children, the mean number of boys in the sampled families will approach 2.5.

Table 6.5

x	P(x)	x · P(x)
0	0.03125	0.00000
1	0.15625	0.15625
2	0.31250	0.62500
3	0.31250	0.93750
4	0.15625	0.62500
5	0.03125	0.15625
		$\sum[x \cdot P(x)] = 2.5$

Check Your Understanding

7. A true–false quiz with 10 questions was given to a statistics class. Following is the probability distribution for the score of a randomly chosen student. Find the mean score and interpret the result.

x	5	6	7	8	9	10
P(x)	0.04	0.16	0.36	0.24	0.12	0.08

Answer is on page 260.

Objective 6 Compute the variance and standard deviation of a discrete random variable

The variance and standard deviation of a discrete random variable

The variance and standard deviation are measures of spread. The variance and standard deviation of a random variable measure the spread in the probability distribution of the random variable.

DEFINITION

The **variance** of a discrete random variable X is

$$\sigma_X^2 = \sum[(x - \mu_X)^2 \cdot P(x)]$$

An equivalent expression that is often easier when computing by hand is

$$\sigma_X^2 = \sum[x^2 \cdot P(x)] - \mu_X^2$$

The **standard deviation** of X is the square root of the variance:

$$\sigma_X = \sqrt{\sigma_X^2}$$

EXAMPLE 6.8   Computing the variance of a discrete random variable

In Example 6.6, we presented the following probability distribution for the number of defective pixels in a computer monitor. Compute the variance and standard deviation of the number of defective pixels.

x	0	1	2	3
$P(x)$	0.2	0.5	0.2	0.1

Solution

In Example 6.6, we computed $\mu_X = 1.2$. The calculations for the variance σ_X^2 are shown in Table 6.6. For each x, we subtract the mean μ_X to obtain $x - \mu_X$. Then we square these quantities and multiply by $P(x)$. We add the products to obtain the variance σ_X^2.

Table 6.6

x	$x - \mu_X$	$(x - \mu_X)^2$	$P(x)$	$(x - \mu_X)^2 \cdot P(x)$
0	−1.2	1.44	0.2	0.288
1	−0.2	0.04	0.5	0.020
2	0.8	0.64	0.2	0.128
3	1.8	3.24	0.1	0.324
				$\sigma_X^2 = \sum[(x - \mu_X)^2 \cdot P(x)] = 0.760$

The variance is $\sigma_X^2 = 0.760$. The standard deviation is

$$\sigma_X = \sqrt{0.760} = 0.872$$

EXAMPLE 6.9   Computing the mean and standard deviation by using technology

Use technology to compute the mean and standard deviation for the probability distribution in Example 6.8.

Figure 6.4

Solution

We use the TI-84 Plus calculator. Enter the possible values of the random variable into **L1** and their probabilities into **L2**. Then use the **1–Var Stats** command. Figure 6.4 presents the calculator display. The mean, μ_X, is denoted by x̄. The standard deviation is denoted σx. There is no value for the sample standard deviation Sx because we are not computing the standard deviation of a sample. Step-by-step instructions are presented in the Using Technology section on page 256.

Check Your Understanding

8. Following is the probability distribution for the age of a student at a certain public high school.

x	13	14	15	16	17	18
$P(x)$	0.08	0.24	0.23	0.28	0.14	0.03

  a. Find the variance of the ages.
  b. Find the standard deviation of the ages.

Answers are on page 260.

Applications of the mean

There are many occasions on which people want to predict how much they are likely to gain or lose if they make a certain decision or take a certain action. Often, this is done by computing the mean of a random variable. In such situations, the mean is sometimes called the "expected value." If the expected value is positive, it is an expected gain, and if it is negative, it is an expected loss. Examples 6.10–6.12 provide some illustrations.

Gambling

Probability was invented by mathematicians who were hired by gamblers to help them create games of chance. For this reason, probability is an extremely useful tool to analyze gambling games — this is what it was designed to do. Example 6.10 analyzes the game of roulette.

EXAMPLE 6.10

Find the expected loss at roulette

A Nevada roulette wheel contains 38 pockets, numbered 1 to 36 with a zero (0) and a double-zero (00). Eighteen of the numbers are colored red, eighteen are colored black, and two (the 0 and the 00) are colored green. Let's say you bet $1 on red. If a red number comes up, you get your dollar back, along with another dollar, so you win $1. If a black or green number comes up, you win −$1 (winning a negative amount is a mathematician's way of saying that you lose). Let X be the amount you win. Find the probability distribution of X and the expected value (mean) of X. Interpret the expected value.

Solution

The possible values of X are 1 and −1. There are 38 outcomes in the sample space, with 18 of them (the red ones) corresponding to 1 and the other 20 corresponding to −1. The probability distribution of X is therefore

x	−1	1
$P(x)$	20/38	18/38

The expected value is the mean of X:

$$\mu_X = (-1)\left(\frac{20}{38}\right) + 1\left(\frac{18}{38}\right) = -\frac{2}{38} = -0.0526$$

Since the expected value is negative, this is an expected loss. We can interpret the expected value by saying that, on the average, you can expect to lose 5.26¢ for every dollar you bet.

Business projections

When making business decisions, executives often use probability distributions to describe the amount of profit or loss that will result.

EXAMPLE 6.11

Computing the expected value of a business decision

A mineral economist estimated that a particular mining venture had probability 0.4 of a $30 million loss, probability 0.5 of a $20 million profit, and probability 0.1 of a $40 million profit. Let X represent the profit, in millions of dollars. Find the probability distribution of the profit and the expected value of the profit. Does this venture represent an expected gain or an expected loss?

Source: *Journal of the Australasian Institute of Mining and Metallurgy* 306:18–22

Solution

The probability distribution of the profit X is as follows:

x	−30	20	40
$P(x)$	0.4	0.5	0.1

The expected value is the mean of X:

$$\mu_X = (-30)(0.4) + 20(0.5) + 40(0.1) = 2.0$$

The expected value is positive, so this is an expected gain of $2 million.

Insurance premiums

Insurance companies must determine a price (called a *premium*) to charge for their policies. Computing an expected value is an important part of this process.

EXAMPLE 6.12	**Computing the expected value of an insurance premium**

An insurance company sells a one-year term life insurance policy to a 70-year-old man. The man pays a premium of $400. If he dies within one year, the company will pay $10,000 to his beneficiary. According to the U.S. Centers for Disease Control and Prevention, the probability that a 70-year-old man is still alive one year later is 0.9715. Let X be the profit made by the insurance company. Find the probability distribution and the expected value of the profit.

Solution

If the man lives, the insurance company keeps the $400 premium and doesn't have to pay anything. So its profit is $400. If the man dies, the insurance company still keeps the $400, but it must also pay $10,000. So its profit is $400 − $10,000 = −$9600, a loss of $9600. The probability that the man lives is 0.9715 and the probability that he dies is $1 − 0.9715 = 0.0285$. The probability distribution is

x	−9600	400
$P(x)$	0.0285	0.9715

The expected value is the mean of X:

$$\mu_X = (-9600)(0.0285) + 400(0.9715) = 115$$

The expected gain for the insurance company is $115. We can interpret this by saying that if the insurance company sells many policies like this, it can expect to earn $115 for each policy, on the average.

In this section, we introduced the concept of a random variable and its probability distribution. In the remaining sections of this chapter, we will discuss some specific probability distributions that are often used in practice, and describe some of their important applications.

USING TECHNOLOGY

We use Example 6.6 to illustrate the technology steps.

TI-84 PLUS

Computing the mean and standard deviation of a discrete random variable

Step 1. Enter the values of the random variable into **L1** in the data editor and the associated probabilities into **L2**. (Figure A).

Step 2. Press **STAT** and highlight the **CALC** menu.

Step 3. Select **1–Var Stats** and enter **L1**, comma, **L2**.

Step 4. Press **ENTER**. (Figure B).

Figure A   **Figure B**

Using the TI-84 PLUS Stat Wizards (see Appendix B for more information)

Step 1. Enter the values of the random variable into **L1** in the data editor and the associated probabilities into **L2**. (Figure A).

Step 2. Press **STAT** and highlight the **CALC** menu.

Step 3. Select **1–Var Stats**. Enter **L1** into the **List** field and **L2** into the **FreqList** field.

Step 4. Select **Calculate** and press **ENTER**. (Figure B).

SECTION 6.1   Exercises

Exercises 1–8 are the Check Your Understanding exercises located within the section.

Understanding the Concepts

In Exercises 9–12, fill in each blank with the appropriate word or phrase.

9. A numerical outcome of a probability experiment is called a _____.

10. The sum of all the probabilities in a discrete probability distribution must be equal to _____.

11. _____ random variables can take on any value in an interval.

12. As the sample size increases, the sample mean approaches the _____ mean.

In Exercises 13–16, determine whether the statement is true or false. If the statement is false, rewrite it as a true statement.

13. To find the mean of a discrete random variable, multiply each possible value of the random variable by its probability, then add the products.

14. The expected value is the mean amount gained or lost.

15. The possible values of a discrete random variable cannot be listed.

16. The standard deviation is found by squaring the variance.

Practicing the Skills

In Exercises 17–26, determine whether the random variable described is discrete or continuous.

17. The number of heads in 100 tosses of a coin
18. The number of people in line at the bank at a randomly chosen time
19. The weight of a randomly chosen student's backpack
20. The amount of rain during the next thunderstorm
21. The number of children absent from school on a randomly chosen day
22. The time it takes to drive to the airport
23. The final exam score of a randomly chosen student from last semester's statistics class
24. The amount of time you wait at a bus stop
25. The height of a randomly chosen college student
26. The number of songs on a randomly chosen iPod

In Exercises 27–32, determine whether the table represents a discrete probability distribution. If not, explain why not.

27.
x	P(x)
1	0.4
2	0.2
3	0.1
4	0.3

28.
x	P(x)
30	0.2
40	0.2
50	0.2
60	0.2
70	0.2

29.
x	P(x)
2.1	0.1
2.2	0.1
2.3	0.1
2.4	0.1

30.
x	P(x)
55	-0.3
65	0.6
75	0.4
85	0.2

31.
x	P(x)
100	0.2
200	0.3
300	0.5
400	0.4
500	0.1

32.
x	P(x)
-4	0.35
-1	0.25
0	0.15
2	0.25

In Exercises 33–38, compute the mean and standard deviation of the random variable with the given discrete probability distribution.

33.
x	P(x)
1	0.42
2	0.18
5	0.34
7	0.06

34.
x	P(x)
8	0.15
13	0.23
15	0.25
18	0.27
19	0.10

35.
x	P(x)
4.5	0.33
6	0.11
7	0.21
9.5	0.35

36.
x	P(x)
-3	0.10
0	0.17
1	0.56
3	0.17

37.
x	P(x)
15	0.15
17	0.23
19	0.25
22	0.27
26	0.10

38.
x	P(x)
120	0.30
150	0.30
170	0.15
180	0.25

39. Fill in the missing value so that the following table represents a probability distribution.

x	4	5	6	7
P(x)	0.3	?	0.3	0.2

40. Fill in the missing value so that the following table represents a probability distribution.

x	15	25	35	45	55
P(x)	0.25	0.15	?	0.05	0.15

Working with the Concepts

41. **Put some air in your tires:** Let X represent the number of tires with low air pressure on a randomly chosen car. The probability distribution of X is as follows.

x	0	1	2	3	4
P(x)	0.1	0.2	0.4	0.2	0.1

a. Find $P(1)$.
b. Find $P(\text{More than 2})$.
c. Find the probability that all four tires have low air pressure.
d. Find the probability that no tires have low air pressure.
e. Compute the mean μ_X.
f. Compute the standard deviation σ_X.

42. **Fifteen items or less:** The number of customers in line at a supermarket express checkout counter is a random variable with the following probability distribution.

x	0	1	2	3	4	5
P(x)	0.10	0.25	0.30	0.20	0.10	0.05

a. Find $P(2)$.

b. Find P(No more than 1).

c. Find the probability that no one is in line.

d. Find the probability that at least three people are in line.

e. Compute the mean μ_X.

f. Compute the standard deviation σ_X.

g. If each customer takes 3 minutes to check out, what is the probability that it will take more than 6 minutes for all the customers currently in line to check out?

43. Defective circuits: The following table presents the probability distribution of the number of defects X in a randomly chosen printed circuit board.

x	0	1	2	3
$P(x)$	0.5	0.3	0.1	0.1

a. Find $P(2)$.

b. Find P(1 or more).

c. Find the probability that at least two circuits are defective.

d. Find the probability that no more than two circuits are defective.

e. Compute the mean μ_X.

f. Compute the standard deviation σ_X.

g. A circuit will function if it has no defects or only one defect. What is the probability that a circuit will function?

44. Do you carpool? Let X represent the number of occupants in a randomly chosen car on a certain stretch of highway during morning commute hours. A survey of cars showed that the probability distribution of X is as follows.

x	1	2	3	4	5
$P(x)$	0.70	0.15	0.10	0.03	0.02

a. Find $P(2)$.

b. Find P(More than 3).

c. Find the probability that a car has only one occupant.

d. Find the probability that a car has fewer than four occupants.

e. Compute the mean μ_X.

f. Compute the standard deviation σ_X.

g. To save energy, a goal is set to have the mean number of occupants be at least two per car. Has this goal been met?

45. Dirty air: The federal government has enacted maximum allowable standards for air pollutants such as ozone. Let X be the number of days per year that the level of air pollution exceeds the standard in a certain city. The probability distribution of X is given by

x	0	1	2	3	4
$P(x)$	0.33	0.38	0.19	0.06	0.04

a. Find $P(1)$.

b. Find P(3 or fewer).

c. Find the probability that the standard is exceeded on at least one day.

d. Find the probability that the standard is exceeded on more than two days.

e. Compute the mean μ_X.

f. Compute the standard deviation σ_X.

46. Texting: Five teenagers are selected at random. Let X be the number of them who have sent text messages on their cell phones within the past 30 days. According to a study by the Nielsen Company, the probability distribution of X is as follows:

x	0	1	2	3	4	5
$P(x)$	0.015	0.097	0.258	0.343	0.227	0.060

a. Find $P(2)$.

b. Find P(More than 1).

c. Find the probability that three or more of the teenagers sent text messages.

d. Find the probability that fewer than two of the teenagers sent text messages.

e. Compute the mean μ_X.

f. Compute the standard deviation σ_X.

47. Relax! The General Social Survey asked 1676 people how many hours per day they were able to relax. The results are presented in the following table.

Number of Hours	Frequency
0	114
1	186
2	336
3	251
4	316
5	231
6	149
7	33
8	60
Total	1676

Consider these 1676 people to be a population. Let X be the number of hours of relaxation for a person sampled at random from this population.

a. Construct the probability distribution of X.

b. Find the probability that a person relaxes more than 4 hours per day.

c. Find the probability that a person doesn't relax at all.

d. Compute the mean μ_X.

e. Compute the standard deviation σ_X.

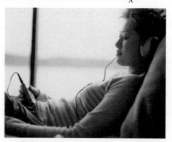

48. Pain: The General Social Survey asked 827 people how many days they would wait to seek medical treatment if they were suffering pain that interfered with their ability to work. The results are presented in the following table.

Number of Days	Frequency
0	27
1	436
2	263
3	72
4	19
5	10
Total	827

Consider these 827 people to be a population. Let X be the number of days for a person sampled at random from this population.

a. Construct the probability distribution of X.

b. Find the probability that a person would wait for 3 days.

c. Find the probability that a person would wait more than 2 days.

d. Compute the mean μ_X.

e. Compute the standard deviation σ_X.

49. **School days:** The following table presents the numbers of students enrolled in grades 1 through 8 in public schools in the United States.

Grade	Frequency (in 1000s)
1	3750
2	3640
3	3627
4	3585
5	3601
6	3660
7	3715
8	3765
Total	29,343

Source: *Statistical Abstract of the United States*

Consider these students to be a population. Let X be the grade of a student randomly chosen from this population.

a. Construct the probability distribution of X.

b. Find the probability that the student is in fourth grade.

c. Find the probability that the student is in seventh or eighth grade.

d. Compute the mean μ_X.

e. Compute the standard deviation σ_X.

50. **World Cup:** The World Cup soccer tournament has been held every four years since 1930, except for 1942 and 1946. The following table presents the number of goals scored by the winning team in each championship game.

Goals	Frequency
1	3
2	4
3	7
4	4
5	1
Total	19

Consider these 19 games to be a population. Let X be the number of goals scored in a game randomly chosen from this population.

a. Construct the probability distribution of X.

b. Find the probability that three goals were scored.

c. Find the probability that fewer than four goals were scored.

d. Compute the mean μ_X.

e. Compute the standard deviation σ_X.

51. **Lottery:** In the New York State Numbers Lottery, you pay $1 and pick a number from 000 to 999. If your number

comes up, you win $500, which is a profit of $499. If you lose, you lose $1. Your probability of winning is 0.001. What is the expected value of your profit? Is it an expected gain or an expected loss?

52. **Lottery:** In the New York State Numbers Lottery, you pay $1 and can bet that the sum of the numbers that come up is 13. The probability of winning is 0.075, and if you win, you win $6.50, which is a profit of $5.50. If you lose, you lose $1. What is the expected value of your profit? Is it an expected gain or an expected loss?

53. **Craps:** In the game of craps, two dice are rolled, and people bet on the outcome. For example, you can bet $1 that the dice will total 7. The probability that you win is 1/6, and if you win, your profit is $4. If you lose, you lose $1. What is the expected value of your profit? Is it an expected gain or an expected loss?

54. **More craps:** Another bet you can make in craps is that the sum of the dice will be 2 (also called "snake eyes"). The probability that you win is 1/36, and if you win, your profit is $30. If you lose, you lose $1. What is the expected value of your profit? Is it an expected gain or an expected loss?

55. **SAT:** You are trying to answer an SAT multiple choice question. There are five choices. If you get the question right, you gain one point, and if you get it wrong, you lose 1/4 point. Assume you have no idea what the right answer is, so you pick one of the choices at random.

a. What is the expected value of the number of points you get?

b. If you don't answer a question, you get 0 points. The test makers advise you not to answer a question if you have no idea which answer is correct. Do you think this is good advice? Explain.

56. **More SAT:** Refer to Exercise 55. Assume you can eliminate one of the five choices, and you choose one of the remaining four at random as your answer.

a. What is the expected value of the number of points you get?

b. If you don't answer a question, you get 0 points. The test makers advise you to guess if you can eliminate one or more answers. Do you think this is good advice? Explain.

57. **Business projection:** An investor is considering a $10,000 investment in a start-up company. She estimates that she has probability 0.25 of a $20,000 loss, probability 0.20 of a $10,000 profit, probability 0.15 of a $50,000 profit, and probability 0.40 of breaking even (a profit of $0). What is the expected value of the profit? Would you advise the investor to make the investment?

58. **Insurance:** An insurance company sells a one-year term life insurance policy to an 80-year-old woman. The woman pays a premium of $1000. If she dies within one year, the company will pay $20,000 to her beneficiary. According to the U.S. Centers for Disease Control and Prevention, the probability that an 80-year-old woman will be alive one year later is 0.9516. Let X be the profit made by the insurance company. Find the probability distribution and the expected value of the profit.

59. **Boys and girls:** A couple plans to have children until a girl is born, but they will have no more than three children.

Assume that each child is equally likely to be a boy or a girl. Let Y be the number of boys they have.
  a. Find the probability distribution of Y.
  b. Find the mean μ_Y.
  c. Find the standard deviation σ_Y.

60. Girls and boys: In Exercise 59, let X be the number of girls the couple has.
  a. Find the probability distribution of X.
  b. Find the mean μ_X.
  c. Find the standard deviation σ_X.

Extending the Concepts

61. Success and failure: Three components are randomly sampled, one at a time, from a large lot. As each component is selected, it is tested. If it passes the test, a success (S)

occurs; if it fails the test, a failure (F) occurs. Assume that 80% of the components in the lot will succeed in passing the test. Let X represent the number of successes among the three sampled components.
  a. What are the possible values for X?
  b. Find $P(3)$.
  c. The event that the first component fails and the next two succeed is denoted by FSS. Find $P(\text{FSS})$.
  d. Find $P(\text{SFS})$ and $P(\text{SSF})$.
  e. Use the results of parts (c) and (d) to find $P(2)$.
  f. Find $P(1)$.
  g. Find $P(0)$.
  h. Find μ_X.
  i. Find σ_X.

Answers to Check Your Understanding Exercises for Section 6.1

1.

x	0	1	2	3
$P(x)$	0.125	0.375	0.375	0.125

2. It is not possible. The probabilities do not add up to 1.

3. (a) and (c)

4. a. 0.45 **b.** 0.35 **c.** 0.94

5.

x	1	2	3	4	5	6
$P(x)$	0.0956	0.1290	0.1136	0.3728	0.2714	0.0176

6. a. 0.244 **b.** 0.433 **c.** 0.121

7. The mean is 7.48. Interpretation: If we were to give this quiz to more and more students, the mean score for these students would approach 7.48.

8. a. 1.6075 **b.** 1.2679

SECTION 6.2 | # The Binomial Distribution

Objectives

1. Determine whether a random variable is binomial
2. Determine the probability distribution of a binomial random variable
3. Compute binomial probabilities
4. Compute the mean and variance of a binomial random variable

Your favorite fast-food chain is giving away a coupon with every purchase of a meal. You scratch the coupon to reveal your prize. Twenty percent of the coupons entitle you to a free hamburger, and the rest of them say "better luck next time." You go to this restaurant in a group of ten people, and everyone orders lunch. What is the probability that three of you win a free hamburger? Let X be the number of people out of ten who win a free hamburger. What is the probability distribution of X? In this section, we will learn that X has a distribution called the **binomial distribution**, which is one of the most useful probability distributions.

Objective 1 Determine whether a random variable is binomial

Binomial Random Variables

In the situation just described, we are examining ten coupons. Each time we examine a coupon, we will call it a **trial**, so there are 10 trials. When a coupon is good for a free hamburger, we will call it a "success." The random variable X represents the number of successes in 10 trials.

Under certain conditions, a random variable that represents the number of successes in a series of trials has a probability distribution called the **binomial distribution**. The conditions are:

Conditions for the Binomial Distribution

1. A fixed number of trials are conducted.
2. There are two possible outcomes for each trial. One is labeled "success" and the other is labeled "failure."
3. The probability of success is the same on each trial.
4. The trials are independent. This means that the outcome of one trial does not affect the outcomes of the other trials.
5. The random variable X represents the number of successes that occur.

Notation: The following notation is commonly used:

• The number of trials is denoted by n.
• The probability of success is denoted by p, and the probability of failure is $1 - p$.

It is important to realize that the word *success* does not necessarily refer to a desirable outcome. For example, in medical studies that involve counting the number of people who suffer from a certain disease, the value of p is the probability that someone will come down with the disease. In these studies, disease is a "success," although it is certainly not a desirable outcome.

EXAMPLE 6.13

Determining whether a random variable is binomial

Determine which of the following are binomial random variables. For those that are binomial, state the two possible outcomes and specify which is a success. Also state the values of n and p.

    a. A fair coin is tossed ten times. Let X be the number of times the coin lands heads.
    b. Five basketball players each attempt a free throw. Let X be the number of free throws made.
    c. Ten cards are in a box. Five are red and five are green. Three of the cards are drawn at random without replacement. Let X be the number of red cards drawn.

Solution

    a. This is a binomial random variable. Each toss of the coin is a trial. There are two possible outcomes — heads and tails. Since X represents the number of heads, a head counts as a success. The trials are independent, because the outcome of one coin toss does not affect the other tosses. The number of trials is $n = 10$, and the success probability is $p = 0.5$.

    b. This is not a binomial random variable. The probability of success (making a shot) differs from player to player, because they will not all be equally skilled at making free throws.

    c. This is not a binomial random variable because the trials are not independent. If the first card is red, then four of the nine remaining cards will be red, and the probability of a red card on the second draw will be $4/9$. If the first card is not red, then five of the nine remaining cards will be red, and the probability of a red card on the second draw will be $5/9$. Since the probability of success on the second trial depends on the outcome of the first trial, the trials are not independent.

In part (c) of Example 6.13, the sampling was done without replacement, and the trials were not independent. If the sample is less than 5% of the population, however, then in most cases the lack of replacement will have only a negligible effect, and it is appropriate to consider the trials to be independent. (See the discussion in Section 5.3.) In particular, when a simple random sample is drawn from a population, we will consider the sampled individuals to be independent whenever the sample size is less than 5% of the population size.

> When a simple random sample comprises less than 5% of the population, we will consider the sampled individuals to be independent.

Check Your Understanding

1. Determine whether X is a binomial random variable.
   a. A fair die is rolled 20 times. Let X be the number of times the die comes up 6.
   b. A standard deck of 52 cards contains four aces. Four cards are dealt without replacement from this deck. Let X be the number that are aces.
   c. A simple random sample of 50 voters is drawn from the residents in a large city. Let X be the number who plan to vote for a proposition to increase spending on public schools.

Answers are on page 271.

Objective 2 Determine the probability distribution of a binomial random variable

The Binomial Probability Distribution

We will determine the probability distribution of a binomial random variable by considering a simple example. A biased coin has probability 0.6 of coming up heads. The coin is tossed three times. Let X be the number of heads that come up. Since X is the number of heads, coming up heads is a success. Then X is binomial, with $n = 3$ trials and success probability $p = 0.6$. We will compute $P(2)$, the probability that exactly two of the tosses are heads.

There are three arrangements of two heads in three tosses of a coin, HHT, HTH, and THH. We first compute the probability of HHT. The event HHT is a sequence of independent events: H on the first toss, H on the second toss, T on the third toss. We know the probabilities of each of these events separately:

$P(\text{H on the first toss}) = 0.6, \quad P(\text{H on the second toss}) = 0.6, \quad P(\text{T on the third toss}) = 0.4$

Because the events are independent, the Multiplication Rule for Independent Events tells us that the probability that they all occur is equal to the product of their probabilities. Therefore,

$$P(\text{HHT}) = (0.6)(0.6)(0.4) = (0.6)^2(0.4)^1$$

Similarly, $P(\text{HTH}) = (0.6)(0.4)(0.6) = (0.6)^2(0.4)^1$, and $P(\text{THH}) = (0.4)(0.6)(0.6) = (0.6)^2(0.4)^1$. We can see that all the different arrangements of two heads and one tail have the same probability. Now

$P(2) = P(\text{HHT or HTH or THH})$

$\quad = P(\text{HHT}) + P(\text{HTH}) + P(\text{THH})$ (Addition Rule for Mutually Exclusive Events)

$\quad = (0.6)^2(0.4)^1 + (0.6)^2(0.4)^1 + (0.6)^2(0.4)^1$

$\quad = 3(0.6)^2(0.4)^1$

Recall: $_nC_x = \dfrac{n!}{x!(n-x)!}$, where $n! = n(n-1)\cdots(2)(1)$

Examining this result, we see that the number 3 represents the number of arrangements of two successes (heads) and one failure (tails). In general, this number will be the number of arrangements of x successes in n trials, which is $_nC_x$. The number 0.6 is the success probability, which in general will be p. The exponent 2 is the number of successes, which in general will be x. The number 0.4 is the failure probability, which is $1 - p$, and the exponent 1 is the number of failures, which is $n - x$.

Formula for Binomial Probabilities

For a binomial random variable X that represents the number of successes in n trials with success probability p, the probability of obtaining x successes is

$$P(x) = {_nC_x}\,p^x(1-p)^{n-x}$$

The possible values of X are $0, 1, ..., n$.

Objective 3 Compute binomial probabilities

Computing Binomial Probabilities

The binomial probability distribution can require tedious calculations. While we can compute simple probabilities by hand, for more involved problems it is better to use a table or technology.

EXAMPLE 6.14	**Calculating probabilities by using the binomial probability distribution**

The Pew Research Center reported in June 2013 that approximately 30% of U.S. adults own a tablet computer such as an iPad, Samsung Galaxy Tab, or Kindle Fire. Suppose a simple random sample of 15 people is taken. Use the binomial probability distribution to find the following probabilities.

   a. Find the probability that exactly four of the sampled people own a tablet computer.
   b. Find the probability that fewer than three of the people own a tablet computer.
   c. Find the probability that more than one person owns a tablet computer.
   d. Find the probability that the number of people who own a tablet computer is between 1 and 4, inclusive.

Solution

a. We use the binomial probability distribution with $n = 15$, $p = 0.3$, and $x = 4$.

$$P(4) = {}_{15}C_4(0.3)^4(1 - 0.3)^{15-4}$$
$$= \frac{15!}{4!(15 - 4)!}(0.3)^4(0.7)^{11}$$
$$= 1365(0.3)^4(0.7)^{11}$$
$$= 0.219$$

b. The possible numbers of people that are fewer than three are 0, 1, and 2. So we need to find $P(0 \text{ or } 1 \text{ or } 2)$. We use the Addition Rule for Mutually Exclusive Events.

$$P(0 \text{ or } 1 \text{ or } 2) = P(0) + P(1) + P(2)$$
$$= {}_{15}C_0(0.3)^0(1 - 0.3)^{15-0} + {}_{15}C_1(0.3)^1(1 - 0.3)^{15-1}$$
$$+ {}_{15}C_2(0.3)^2(1 - 0.3)^{15-2}$$
$$= 0.0047 + 0.0305 + 0.0916$$
$$= 0.127$$

c. The possible numbers of people that are more than one are 2, 3, 4, and so forth up to 15. We could find $P(\text{More than } 1)$ by adding $P(2) + P(3) + \cdots + P(15)$, but fortunately there is an easier way. We will use the Rule of Complements. The complement of "more than 1" is "1 or fewer," or, equivalently, 0 or 1. We compute the probability of the complement, and subtract from 1.

$$P(0 \text{ or } 1) = P(0) + P(1)$$
$$= {}_{15}C_0(0.3)^0(1 - 0.3)^{15-0} + {}_{15}C_1(0.3)^1(1 - 0.3)^{15-1}$$
$$= 0.0047 + 0.0305$$
$$= 0.035$$

Now we use the Rule of Complements:

$$P(\text{More than } 1) = 1 - P(0 \text{ or } 1) = 1 - 0.035 = 0.965$$

d. Between 1 and 4 inclusive means 1, 2, 3, or 4.

$$P(1 \text{ or } 2 \text{ or } 3 \text{ or } 4) = P(1) + P(2) + P(3) + P(4)$$
$$= {}_{15}C_1(0.3)^1(1 - 0.3)^{15-1} + {}_{15}C_2(0.3)^2(1 - 0.3)^{15-2}$$
$$+ {}_{15}C_3(0.3)^3(1 - 0.3)^{15-3} + {}_{15}C_4(0.3)^4(1 - 0.3)^{15-4}$$
$$= 0.0305 + 0.0916 + 0.1700 + 0.2186$$
$$= 0.511$$

Explain It Again

When can we use a table?
Tables contain only a limited selection of values for n and p. Probabilities involving values not in the table must be computed by hand or with technology.

Using a table to compute binomial probabilities

Table A.1 contains probabilities for the binomial distribution. It can be used to compute binomial probabilities for values of n up to 20 and certain values of p.

EXAMPLE 6.15 Use a table to compute binomial probabilities

The Pew Research Center reported in June 2013 that approximately 30% of U.S. adults own a tablet computer such as an iPad, Samsung Galaxy Tab, or Kindle Fire. Suppose a simple random sample of 15 people is taken. Use the binomial probability distribution to find the following probabilities.

   a. Find the probability that exactly five of the sampled people own a tablet computer.
   b. Find the probability that fewer than four of the people own a tablet computer.
   c. Find the probability that the number of sampled people who own a tablet computer is between 6 and 8, inclusive.

Solution

   a. We have $n = 15$, so we go to the section of Table A.1 that corresponds to $n = 15$. This is shown in Figure 6.5. We look at the column corresponding to $p = 0.30$. Now for each value in the column labeled "x," the number in the table is the probability $P(x)$. We therefore look in the row corresponding to $x = 5$. The probability that exactly five people own a tablet computer is $P(5) = 0.206$.

| | | | | | | p | | | | | | | | |
n	x	0.05	0.10	0.20	0.25	0.30	0.40	0.50	0.60	0.70	0.75	0.80	0.90	0.95
15	0	0.463	0.206	0.035	0.013	0.005	0.000+	0.000+	0.000+	0.000+	0.000+	0.000+	0.000+	0.000+
	1	0.366	0.343	0.132	0.067	0.031	0.005	0.000+	0.000+	0.000+	0.000+	0.000+	0.000+	0.000+
	2	0.135	0.267	0.231	0.156	0.092	0.022	0.003	0.000+	0.000+	0.000+	0.000+	0.000+	0.000+
	3	0.031	0.129	0.250	0.225	0.170	0.063	0.014	0.002	0.000+	0.000+	0.000+	0.000+	0.000+
	4	0.005	0.043	0.188	0.225	0.219	0.127	0.042	0.007	0.001	0.000+	0.000+	0.000+	0.000+
	5	0.001	0.010	0.103	0.165	0.206	0.186	0.092	0.024	0.003	0.001	0.000+	0.000+	0.000+
	6	0.000+	0.002	0.043	0.092	0.147	0.207	0.153	0.061	0.012	0.003	0.001	0.000+	0.000+
	7	0.000+	0.000+	0.014	0.039	0.081	0.177	0.196	0.118	0.035	0.013	0.003	0.000+	0.000+
	8	0.000+	0.000+	0.003	0.013	0.035	0.118	0.196	0.177	0.081	0.039	0.014	0.000+	0.000+
	9	0.000+	0.000+	0.001	0.003	0.012	0.061	0.153	0.207	0.147	0.092	0.043	0.002	0.000+
	10	0.000+	0.000+	0.000+	0.001	0.003	0.024	0.092	0.186	0.206	0.165	0.103	0.010	0.001
	11	0.000+	0.000+	0.000+	0.000+	0.001	0.007	0.042	0.127	0.219	0.225	0.188	0.043	0.005
	12	0.000+	0.000+	0.000+	0.000+	0.000+	0.002	0.014	0.063	0.170	0.225	0.250	0.129	0.031
	13	0.000+	0.000+	0.000+	0.000+	0.000+	0.000+	0.003	0.022	0.092	0.156	0.231	0.267	0.135
	14	0.000+	0.000+	0.000+	0.000+	0.000+	0.000+	0.000+	0.005	0.031	0.067	0.132	0.343	0.366
	15	0.000+	0.000+	0.000+	0.000+	0.000+	0.000+	0.000+	0.000+	0.005	0.013	0.035	0.206	0.463

Figure 6.5

Explain It Again

Answers using technology may differ: If you find P(Fewer than 4) using technology, your answer will be 0.297 rather than 0.298. This difference isn't large enough to matter.

   b. P(Fewer than 4) $= P(0) + P(1) + P(2) + P(3)$. We find these probabilities in Table A.1 and add them. See Figure 6.6.

$$P(\text{Fewer than } 4) = 0.005 + 0.031 + 0.092 + 0.170$$
$$= 0.298$$

| | | | | | | p | | | | | | | | |
n	x	0.05	0.10	0.20	0.25	0.30	0.40	0.50	0.60	0.70	0.75	0.80	0.90	0.95
15	0	0.463	0.206	0.035	0.013	0.005	0.000+	0.000+	0.000+	0.000+	0.000+	0.000+	0.000+	0.000+
	1	0.366	0.343	0.132	0.067	0.031	0.005	0.000+	0.000+	0.000+	0.000+	0.000+	0.000+	0.000+
	2	0.135	0.267	0.231	0.156	0.092	0.022	0.003	0.000+	0.000+	0.000+	0.000+	0.000+	0.000+
	3	0.031	0.129	0.250	0.225	0.170	0.063	0.014	0.002	0.000+	0.000+	0.000+	0.000+	0.000+
	4	0.005	0.043	0.188	0.225	0.219	0.127	0.042	0.007	0.001	0.000+	0.000+	0.000+	0.000+
	5	0.001	0.010	0.103	0.165	0.206	0.186	0.092	0.024	0.003	0.001	0.000+	0.000+	0.000+
	6	0.000+	0.002	0.043	0.092	0.147	0.207	0.153	0.061	0.012	0.003	0.001	0.000+	0.000+
	7	0.000+	0.000+	0.014	0.039	0.081	0.177	0.196	0.118	0.035	0.013	0.003	0.000+	0.000+
	8	0.000+	0.000+	0.003	0.013	0.035	0.118	0.196	0.177	0.081	0.039	0.014	0.000+	0.000+
	9	0.000+	0.000+	0.001	0.003	0.012	0.061	0.153	0.207	0.147	0.092	0.043	0.002	0.000+
	10	0.000+	0.000+	0.000+	0.001	0.003	0.024	0.092	0.186	0.206	0.165	0.103	0.010	0.001
	11	0.000+	0.000+	0.000+	0.000+	0.001	0.007	0.042	0.127	0.219	0.225	0.188	0.043	0.005
	12	0.000+	0.000+	0.000+	0.000+	0.000+	0.002	0.014	0.063	0.170	0.225	0.250	0.129	0.031
	13	0.000+	0.000+	0.000+	0.000+	0.000+	0.000+	0.003	0.022	0.092	0.156	0.231	0.267	0.135
	14	0.000+	0.000+	0.000+	0.000+	0.000+	0.000+	0.000+	0.005	0.031	0.067	0.132	0.343	0.366
	15	0.000+	0.000+	0.000+	0.000+	0.000+	0.000+	0.000+	0.000+	0.005	0.013	0.035	0.206	0.463

Figure 6.6

c. $P(\text{Between 6 and 8 inclusive}) = P(6) + P(7) + P(8)$. We find these probabilities in Table A.1 and add them. See Figure 6.7.

$$P(\text{Between 6 and 8 inclusive}) = 0.147 + 0.081 + 0.035$$
$$= 0.263$$

| | | | | | | p | | | | | | | | |
n	x	0.05	0.10	0.20	0.25	0.30	0.40	0.50	0.60	0.70	0.75	0.80	0.90	0.95
15	0	0.463	0.206	0.035	0.013	0.005	0.000+	0.000+	0.000+	0.000+	0.000+	0.000+	0.000+	0.000+
	1	0.366	0.343	0.132	0.067	0.031	0.005	0.000+	0.000+	0.000+	0.000+	0.000+	0.000+	0.000+
	2	0.135	0.267	0.231	0.156	0.092	0.022	0.003	0.000+	0.000+	0.000+	0.000+	0.000+	0.000+
	3	0.031	0.129	0.250	0.225	0.170	0.063	0.014	0.002	0.000+	0.000+	0.000+	0.000+	0.000+
	4	0.005	0.043	0.188	0.225	0.219	0.127	0.042	0.007	0.001	0.000+	0.000+	0.000+	0.000+
	5	0.001	0.010	0.103	0.165	0.206	0.186	0.092	0.024	0.003	0.001	0.000+	0.000+	0.000+
	6	0.000+	0.002	0.043	0.092	0.147	0.207	0.153	0.061	0.012	0.003	0.001	0.000+	0.000+
	7	0.000+	0.000+	0.014	0.039	0.081	0.177	0.196	0.118	0.035	0.013	0.003	0.000+	0.000+
	8	0.000+	0.000+	0.003	0.013	0.035	0.118	0.196	0.177	0.081	0.039	0.014	0.000+	0.000+
	9	0.000+	0.000+	0.001	0.003	0.012	0.061	0.153	0.207	0.147	0.092	0.043	0.002	0.000+
	10	0.000+	0.000+	0.000+	0.001	0.003	0.024	0.092	0.186	0.206	0.165	0.103	0.010	0.001
	11	0.000+	0.000+	0.000+	0.000+	0.001	0.007	0.042	0.127	0.219	0.225	0.188	0.043	0.005
	12	0.000+	0.000+	0.000+	0.000+	0.000+	0.002	0.014	0.063	0.170	0.225	0.250	0.129	0.031
	13	0.000+	0.000+	0.000+	0.000+	0.000+	0.000+	0.003	0.022	0.092	0.156	0.231	0.267	0.135
	14	0.000+	0.000+	0.000+	0.000+	0.000+	0.000+	0.000+	0.005	0.031	0.067	0.132	0.343	0.366
	15	0.000+	0.000+	0.000+	0.000+	0.000+	0.000+	0.000+	0.000+	0.005	0.013	0.035	0.206	0.463

Figure 6.7

EXAMPLE 6.16 — Using technology to compute binomial probabilities

The Pew Research Center reported in June 2013 that approximately 30% of U.S. adults own a tablet computer such as an iPad, Samsung Galaxy Tab, or Kindle Fire. Suppose a simple random sample of 15 people is taken. Use the binomial probability distribution to find the following probabilities.

a. Find the probability that exactly four of the sampled people own a tablet computer.
b. Find the probability that five or fewer of the people own a tablet computer.
c. Find the probability that more than seven of the people own a tablet computer.

Solution

a. We will use the TI-84 Plus calculator. We use the **binompdf** command. We input 15 for n, .3 for p, and 4 for x. The following display shows the result. Step-by-step instructions are given in the Using Technology section on page 267.

```
binompdf(15,0.3,
4)
       .2186231313
```

b. We will use the TI-84 Plus calculator. Because we want to find the probability of "five or fewer," which is the same as "less than or equal to five," we will use the **binomcdf** command. We input 15 for n, .3 for p, and 5 for x. The following display shows the result. Step-by-step instructions are given in the Using Technology section on page 267.

```
binomcdf(15,0.3,
5)
       .7216214405
```

c. We will use MINITAB. MINITAB is not able to compute probabilities of the form "more than," or "greater than," but it can compute probabilities of the form "less than or equal to." We therefore note that the complement of "more than 7" is "less than or equal to 7." We will use MINITAB to compute P(Less than or equal to 7), and then subtract from 1. The following display shows the result.

Binomial with n = 15 and p = 0.3

x   P (X <= x)
7   0.949987

The display shows that P(Less than or equal to 7) = 0.949987. We conclude that P(More than 7) = 1 − 0.949987 = 0.050013. Step-by-step instructions for producing the MINITAB output are given in the Using Technology section on page 268.

Check Your Understanding

2. In June 2009, 20% of video gamers had heard of the game "Modern Warfare 2" (data from GamesBeat). Assume that 18 gamers are randomly sampled.
   a. Use the binomial probability distribution to compute the probability that exactly four of them have heard of "Modern Warfare 2."
   b. Use Table A.1 to find the probability that fewer than three of them have heard of "Modern Warfare 2."
   c. Use any valid method to find the probability that more than three of them have heard of "Modern Warfare 2."
   d. Use any valid method to find the probability that the number who have heard of "Modern Warfare 2" is between 2 and 6 inclusive.
3. The name of "Modern Warfare 2" was originally planned to be "Call to Duty: Modern Warfare 2." In June 2009, 40% of gamers had heard of "Call to Duty: Modern Warfare 2" (data from GamesBeat). Assume that 12 gamers are randomly sampled.
   a. Use the binomial probability distribution to compute the probability that exactly seven of them have heard of "Call to Duty: Modern Warfare 2."
   b. Use Table A.1 to find the probability that fewer than five of them have heard of "Call to Duty: Modern Warfare 2."
   c. Use any valid method to find the probability that more than six of them have heard of "Call to Duty: Modern Warfare 2."
   d. Use any valid method to find the probability that the number who have heard of "Call to Duty: Modern Warfare 2" is between 3 and 5 inclusive.

Answers are on page 271.

Objective 4 Compute the mean and variance of a binomial random variable

Mean and Variance of a Binomial Random Variable

A fair coin has probability 0.5 of landing heads. If we toss a fair coin ten times, we expect to get 5 heads, on the average. The reason is that 5 is half of 10, or, in symbols, $5 = 10 \cdot 0.5$. We can see that the mean number of successes was found by multiplying the number of trials by the success probability. This holds true in general.

The variance and standard deviation of a binomial random variable are straightforward to compute as well, although the reasoning behind them is not so obvious.

Mean, Variance, and Standard Deviation of a Binomial Random Variable

Let X be a binomial random variable with n trials and success probability p. Then the mean of X is
$$\mu_X = np$$
The variance of X is
$$\sigma_X^2 = np(1 - p)$$
The standard deviation of X is
$$\sigma_X = \sqrt{np(1 - p)}$$

EXAMPLE 6.17	Find the mean and standard deviation of a binomial random variable

The probability that a new car of a certain model will require repairs during the warranty period is 0.15. A particular dealership sells 25 such cars. Let X be the number that will require repairs during the warranty period. Find the mean and standard deviation of X.

Solution
There are $n = 25$ trials, with success probability $p = 0.15$. The mean is

$$\mu_X = np = 25 \cdot 0.15 = 3.75$$

The standard deviation is

$$\sigma_X = \sqrt{np(1 - p)} = \sqrt{25 \cdot 0.15 \cdot (1 - 0.15)} = 1.785$$

Check Your Understanding

4. Gregor Mendel discovered the basic laws of heredity by studying pea plants. In one experiment, he produced plants whose parent plants contained genes for both green and yellow pods. Mendel's theory states that the offspring of two such parents has probability 0.75 of having green pods. Assume that 80 such plants are produced.
   a. Find the mean number of plants that have green pods.
   b. Find the variance of the number of plants that have green pods.
   c. Find the standard deviation of the number of plants that have green pods.

Answers are on page 271.

USING TECHNOLOGY

We use Example 6.14 to illustrate the technology steps.

TI-84 PLUS

Computing binomial probabilities of the form $P(x)$ or P(Less than or equal to x)

Step 1. Press **2nd, VARS** to access the **DISTR** menu.

- To compute $P(x)$, select **binompdf** and enter the values for n, p, and x separated by commas and press **ENTER**.

- To compute P(Less than or equal to x), select **binomcdf** and enter the values for n, p, and x separated by commas, and press **ENTER**.

```
binompdf(15,0.3,
4)
          .2186231313
```
Figure A

```
binomcdf(15,0.3,
2)
          .1268277146
```
Figure B

Using the TI-84 PLUS Stat Wizards (see Appendix B for more information)

Step 1. Press **2nd, VARS** to access the **DISTR** menu.

- To compute $P(x)$, select **binompdf** and enter the value for n in the **trials** field, the value for p in the **p** field and the value for x in the **x value** field. Select **Paste** and press **ENTER** to paste the command to the home screen. Press **ENTER** again to run the command.

- To compute $P(x)$, select **binomcdf** and enter the value for n in the **trials** field, the value for p in the **p** field and the value for x in the **x value** field. Select **Paste** and press **ENTER** to paste the command to the home screen. Press **ENTER** again to run the command.

For Example 6.14, $n = 15$ and $p = 0.3$. Figure A displays the result of part (a), which asks for $P(4)$. Figure B displays the result of part (b), which asks for P(Less than or equal to 2).

MINITAB

Computing binomial probabilities of the form $P(x)$ or P(Less than or equal to x)

Step 1. Click **Calc**, then **Probability Distributions**, then **Binomial**.

Step 2. Enter the value for n in the **Number of trials** field and the value for p in the **Probability of success** field.

Step 3. To compute $P(x)$, select the **Probability** option and enter the value for x in the **Input constant** field. To compute P(Less than or equal to x), select the **Cumulative probability** option and enter the value for x in the **Input constant** field.

Step 4. Click **OK**.

Note: Binomial probabilities may be computed for a column of values by entering the column name in the **Input column** field.

EXCEL

Computing binomial probabilities of the form $P(x)$ or P(Less than or equal to x)

Step 1. In an empty cell, select the **Insert Function** icon and highlight **Statistical** in the category field.

Step 2. Click on the **BINOM.DIST** function and press **OK**.

Step 3. Enter the value for x in the **Number_s** field, the value for n in the **Trials** field, and the value for p in the **Probability_s** field.

Step 4. To compute $P(x)$, enter **FALSE** in the **Cumulative** field. To compute P(Less than or equal to x), enter **TRUE** in the **Cumulative field**.

Step 5. Click **OK**. Figure C illustrates computing $P(4)$ in part (a) of Example 6.14.

Figure C

SECTION 6.2 Exercises

Exercises 1–4 are the Check Your Understanding exercises located within the section.

Understanding the Concepts

In Exercises 5–7, fill in each blank with the appropriate word or phrase.

5. In a binomial distribution, there are _____ possible outcomes for each trial.

6. To compute a binomial probability, we must know both the success probability and the number n of _____.

7. If X is a binomial random variable with n trials and success probability p, the standard deviation of X is $\sigma_X = $ _____.

In Exercises 8–10, determine whether the statement is true or false. If the statement is false, rewrite it as a true statement.

8. The trials in a binomial distribution are independent.

9. A binomial random variable with n trials can sometimes have a value greater than n.

10. The mean of a binomial random variable is found by multiplying the number of trials by the success probability.

Practicing the Skills

In Exercises 11–16, determine whether the random variable X has a binomial distribution. If it does, state the number of trials n. If it does not, explain why not.

11. Ten students are chosen from a statistics class of 25 students. Let X be the number who got an A in the class.

12. Ten students are chosen from a statistics class of 300 students. Let X be the number who got an A in the class.

13. A coin is tossed seven times. Let X be the number of heads obtained.

14. A die is tossed three times. Let X be the sum of the three numbers obtained.

15. A coin is tossed until a head appears. Let X be the number of tosses.

16. A random sample of 250 voters is chosen from a list of 10,000 registered voters. Let X be the number who support the incumbent mayor for reelection.

In Exercises 17–26, determine the indicated probability for a binomial experiment with the given number of trials *n* and the given success probability *p*. Then find the mean, variance, and standard deviation.

17. $n = 5$, $p = 0.7$, $P(3)$

18. $n = 10$, $p = 0.2$, $P(1)$

19. $n = 20$, $p = 0.6$, $P(8)$

20. $n = 14$, $p = 0.3$, $P(8)$

21. $n = 3$, $p = 0.4$, $P(0)$

22. $n = 6$, $p = 0.8$, $P(6)$

23. $n = 8$, $p = 0.2$, $P(\text{Fewer than 3})$

24. $n = 15$, $p = 0.9$, $P(14 \text{ or more})$

25. $n = 50$, $p = 0.03$, $P(2 \text{ or fewer})$

26. $n = 30$, $p = 0.9$, $P(\text{More than 27})$

Working with the Concepts

27. Take a guess: A student takes a true–false test that has 10 questions and guesses randomly at each answer. Let *X* be the number of questions answered correctly.
 a. Find $P(4)$.
 b. Find $P(\text{Fewer than 3})$.
 c. To pass the test, the student must answer 7 or more questions correctly. Would it be unusual for the student to pass? Explain.

28. Take another guess: A student takes a multiple-choice test that has 10 questions. Each question has four choices. The student guesses randomly at each answer.
 a. Find $P(3)$.
 b. Find $P(\text{More than 2})$.
 c. To pass the test, the student must answer 7 or more questions correctly. Would it be unusual for the student to pass? Explain.

29. Your flight has been delayed: At Denver International Airport, 81% of recent flights have arrived on time. A sample of 12 flights is studied.
 a. Find the probability that all 12 of the flights were on time.
 b. Find the probability that exactly 10 of the flights were on time.

 c. Find the probability that 10 or more of the flights were on time.

d. Would it be unusual for 11 or more of the flights to be on time?
 Source: *The Denver Post*

30. Car inspection: Of all the registered automobiles in Colorado, 8% fail the state emissions test. Twelve automobiles are selected at random to undergo an emissions test.
 a. Find the probability that exactly three of them fail the test.
 b. Find the probability that fewer than three of them fail the test.
 c. Find the probability that more than two of them fail the test.
 d. Would it be unusual for none of them to fail the test?
 Source: *Air Care Colorado*

31. Google it: According to a report of the Nielsen Company, 67% of Internet searches in July 2013 used the Google search engine. Assume that a sample of 25 searches is studied.
 a. What is the probability that exactly 20 of them used Google?
 b. What is the probability that 15 or fewer used Google?
 c. What is the probability that more than 20 of them used Google?
 d. Would it be unusual if fewer than 12 used Google?

32. What should I buy? A study conducted by the Pew Research Center is 2013 reported that 58% of cell phone owners used their phones inside a store for guidance on purchasing decisions. A sample of 15 cell phone owners is studied.
 a. What is the probability that six or more of them used their phones for guidance on purchasing decisions?
 b. What is the probability that fewer than ten of them used their phones for guidance on purchasing decisions?
 c. What is the probability that exactly eight of them used their phones for guidance on purchasing decisions?
 d. Would it be unusual if more than 12 of them had used their phones for guidance on purchasing decisions?

33. Blood types: The blood type O negative is called the "universal donor" type, because it is the only blood type that may safely be transfused into any person. Therefore, when someone needs a transfusion in an emergency and their blood type cannot be determined, they are given type O negative blood. For this reason, donors with this blood type are crucial to blood banks. Unfortunately, this blood type is fairly rare; according to the Red Cross, only 7% of U.S. residents have type O negative blood. Assume that a blood bank has recruited 20 donors.
 a. What is the probability that two or more of them have type O negative blood?
 b. What is the probability that fewer than four of them will have type O negative blood?
 c. Would it be unusual if none of the donors had type O negative blood?
 d. What is the mean number of donors who have type O negative blood?

e. What is the standard deviation of the number of donors who have type O negative blood?

34. Coronary bypass surgery: The Agency for Healthcare Research and Quality reported that 53% of people who had coronary bypass surgery in a recent year were over the age of 65. Fifteen coronary bypass patients are sampled.

a. What is the probability that exactly 9 of them are over the age of 65?

b. What is the probability that more than 10 are over the age of 65?

c. What is the probability that fewer than 8 are over the age of 65?

d. Would it be unusual if all of them were over the age of 65?

e. What is the mean number of people over the age of 65 in a sample of 15 coronary bypass patients?

f. What is the standard deviation of the number of people over the age of 65 in a sample of 15 coronary bypass patients?

35. College bound: The *Statistical Abstract of the United States* reported that 66% of students who graduated from high school in 2012 enrolled in college. Thirty high school graduates are sampled.

a. What is the probability that exactly 18 of them enroll in college?

b. What is the probability that more than 15 enroll in college?

c. What is the probability that fewer than 12 enroll in college?

d. Would it be unusual if more than 25 of them enroll in college?

e. What is the mean number who enroll in college in a sample of 30 high school graduates?

f. What is the standard deviation of the number who enroll in college in a sample of 30 high school graduates?

36. Big babies: The Centers for Disease Control and Prevention reports that 25% of baby boys 6–8 months old in the United States weigh more than 20 pounds. A sample of 16 babies is studied.

a. What is the probability that exactly 5 of them weigh more than 20 pounds?

b. What is the probability that more than 6 weigh more than 20 pounds?

c. What is the probability that fewer than 3 weigh more than 20 pounds?

d. Would it be unusual if more than 8 of them weigh more than 20 pounds?

e. What is the mean number who weigh more than 20 pounds in a sample of 16 babies aged 6–8 months?

f. What is the standard deviation of the number who weigh more than 20 pounds in a sample of 16 babies aged 6–8 months?

37. High blood pressure: The National Health and Nutrition Survey reported that 30% of adults in the United States have hypertension (high blood pressure). A sample of 25 adults is studied.

a. What is the probability that exactly 6 of them have hypertension?

b. What is the probability that more than 8 have hypertension?

c. What is the probability that fewer than 4 have hypertension?

d. Would it be unusual if more than 10 of them have hypertension?

e. What is the mean number who have hypertension in a sample of 25 adults?

f. What is the standard deviation of the number who have hypertension in a sample of 25 adults?

38. Stress at work: In a poll conducted by the General Social Survey, 81% of respondents said that their jobs were sometimes or always stressful. Ten workers are chosen at random.

a. What is the probability that exactly 7 of them find their jobs stressful?

b. What is the probability that more than 6 find their jobs stressful?

c. What is the probability that fewer than 5 find their jobs stressful?

d. Would it be unusual if fewer than 4 of them find their jobs stressful?

e. What is the mean number who find their jobs stressful in a sample of 10 workers?

f. What is the standard deviation of the number who find their jobs stressful in a sample of 10 workers?

39. Testing a shipment: A certain large shipment comes with a guarantee that it contains no more than 15% defective items. If the proportion of items in the shipment is greater than 15%, the shipment may be returned. You draw a random sample of 10 items and test each one to determine whether it is defective.

a. If in fact 15% of the items in the shipment are defective (so that the shipment is good, but just barely), what is the probability that 7 or more of the 10 sampled items are defective?

b. Based on the answer to part (a), if 15% of the items in the shipment are defective, would 7 defectives in a sample of size 10 be an unusually large number?

c. If you found that 7 of the 10 sample items were defective, would this be convincing evidence that the shipment should be returned? Explain.

d. If in fact 15% of the items in the shipment are defective, what is the probability that 2 or more of the 10 sampled items are defective?

e. Based on the answer to part (d), if 15% of the items in the shipment are defective, would 2 defectives in a sample of size 10 be an unusually large number?

f. If you found that 2 of the 10 sample items were defective, would this be convincing evidence that the shipment should be returned? Explain.

40. Smoke detectors: An insurance company offers a discount to homeowners who install smoke detectors in their homes. A company representative claims that 80% or more of policy holders have smoke detectors. You draw a random sample of 8 policy holders.

a. If exactly 80% of the policy holders have smoke detectors (so the representative's claim is true, but just barely), what is the probability that at most 2 of the 8 sampled policy holders have smoke detectors?

b. Based on the answer to part (a), if 80% of the policy holders have smoke detectors, would 2 policy holders with smoke detectors in a sample of size 8 be an unusually small number?

c. If you found that 2 of the 8 sample policy holders had a smoke detector, would this be convincing evidence that the claim is false? Explain.

d. If exactly 80% of the policy holders have smoke detectors, what is the probability that at most 6 of the 8 sampled policy holders have smoke detectors?

e. Based on the answer to part (d), if 80% of the policy holders have smoke detectors, would 6 policy holders with smoke detectors in a sample of size 8 be an unusually small number?

f. If you found that 6 of the 8 sample policy holders had smoke detectors, would this be convincing evidence that the claim is false? Explain.

Extending the Concepts

41. Recursive computation of binomial probabilities: Binomial probabilities are often hard to compute by hand, because the computation involves factorials and numbers raised to large powers. It can be shown through algebraic manipulation that if X is a random variable whose distribution is binomial with n trials and success probability p, then

$$P(X = x + 1) = \left(\frac{p}{1 - p} \right) \left(\frac{n - x}{x + 1} \right) P(X = x)$$

If we know $P(X = x)$, we can use this equation to calculate $P(X = x + 1)$ without computing any factorials or powers.

a. Let X have the binomial distribution with $n = 25$ trials and success probability $p = 0.6$. It can be shown that $P(X = 14) = 0.14651$. Find $P(X = 15)$.

b. Let X have the binomial distribution with $n = 10$ trials and success probability $p = 0.35$. It can be shown that $P(X = 0) = 0.0134627$. Find $P(X = x)$ for $x = 1, 2, ..., 10$.

Answers to Check Your Understanding Exercises for Section 6.2

1. a. X is a binomial random variable.		**2. a.** 0.2153	**b.** 0.2713	**c.** 0.4990	**d.** 0.8496
b. X is not a binomial random variable.		**3. a.** 0.1009	**b.** 0.4382	**c.** 0.1582	**d.** 0.5818
c. X is a binomial random variable.		**4. a.** 60	**b.** 15	**c.** 3.8730	

SECTION 6.3 | **The Poisson Distribution**

Objectives

1. Compute probabilities for Poisson random variables

2. Compute the mean, variance, and standard deviation of a Poisson random variable

3. Use the Poisson distribution to approximate binomial probabilities

An advertising company has placed an ad on a website. The company's managers are interested to know how many hits the website gets, on the average, during a certain time interval. Assume that between 2:00 P.M. and 3:00 P.M., there are an average of three hits per minute.

If X is the number of hits that occur in a given time interval, then under certain conditions X will have a probability distribution known as the **Poisson distribution**. The Poisson distribution is used to describe certain events that occur in time or space.

Explain It Again

Distinguish the Poisson distribution from the binomial: We use the binomial distribution when there is a fixed number n of trials and we count the number of trials that result in success. We use the Poisson distribution when we count the number of events that occur in a given amount of time or space.

Conditions for the Poisson Distribution

Let X be a random variable that represents the number of events that occur in a time interval of length t. Then X will have the Poisson probability distribution if the following conditions are satisfied.

1. The average rate at which events occur is the same at all times.
2. The numbers of events that occur in nonoverlapping time intervals are independent.
3. For a very short interval of length t:
   a. It is essentially impossible for more than one event to occur within the time interval.
   b. The probability that one event occurs in the interval is approximately equal to λt, where λ is the average rate at which events occur.

EXAMPLE 6.18

Recall: $x! = x(x-1)\cdots(2)(1)$

Finding λ and t

The number of hits on a website during a given time interval has a Poisson distribution. If there are an average of three hits per minute, and X is the number of hits between 2:30 P.M. and 2:35 P.M., find λ and t.

Solution

The average number of hits per minute is 3, so $\lambda = 3$ hits per minute. The length of the time interval between 2:30 and 2:35 is five minutes, so $t = 5$ minutes.

Explain It Again

The number e: The number e is one of the most frequently occurring numbers in nature. Unfortunately, it is an infinitely long decimal, so it is almost impossible to use e in hand computations. Quantities that involve the number e, such as Poisson probabilities, must be computed with technology.

If the random variable X has a Poisson distribution with rate λ and time t, we can compute probabilities with the following probability distribution.

Formula for Poisson Probabilities

For a Poisson random variable X that represents the number of events that occur with average **rate** λ in a time **interval** of length t, the probability that x events occur is

$$P(x) = e^{-\lambda t}\frac{(\lambda t)^x}{x!}$$

The number e is a constant whose value is approximately 2.71828.

The possible values for X are 0, 1, 2, ...

EXAMPLE 6.19

Objective 1 Compute probabilities for Poisson random variables

Computing Poisson probabilities

The number of hits on a certain website follows a Poisson distribution with $\lambda = 3$ hits per minute. Let X be the number of hits in a 2-minute period.

a. Find $P(5)$.
b. Find $P(\text{Less than } 3)$.
c. Find $P(\text{More than } 2)$.

Solution

a. $P(5) = e^{-3\cdot 2}\dfrac{(3\cdot 2)^5}{5!} = 0.1606$

b. $P(\text{Less than } 3) = P(0) + P(1) + P(2)$

$$= e^{-3\cdot 2}\frac{(3\cdot 2)^0}{0!} + e^{-3\cdot 2}\frac{(3\cdot 2)^1}{1!} + e^{-3\cdot 2}\frac{(3\cdot 2)^2}{2!}$$

$$= 0.0025 + 0.0149 + 0.0446$$

$$= 0.0620$$

c. We note that the event "More than 2" is the complement of the event "Less than 3." We use the Rule of Complements:

$$P(\text{More than 2}) = 1 - P(\text{Less than 3}) = 1 - 0.0620 = 0.9380$$

Recall: The Rule of Complements says that $P(A^c) = 1 - P(A)$.

Check Your Understanding

1. The number of traffic accidents at a certain intersection follows a Poisson distribution with $\lambda = 2$ per year. Let X be the number of accidents that occur in a 4-year period.
   a. Find $P(7)$.
   b. Find $P(\text{Less than 4})$.
   c. Find $P(\text{More than 2})$.

2. The number of hits on a certain website follows a Poisson distribution with $\lambda = 12$ per minute. Let X be the number of hits that occur in a period of one-half minute.
   a. Find $P(5)$.
   b. Find $P(\text{Less than 2})$.
   c. Find $P(\text{More than 3})$.

Answers are on page 277.

In Example 6.19, the events were occurring in time. The Poisson distribution can also be used to compute probabilities involving events that occur in a spatial region. In this situation, the quantity t represents a spatial quantity such as length, area, or volume.

One of the earliest uses of the Poisson distribution involved events occurring in space, and was in fact an application to the brewing of beer. A crucial step in the brewing process is the addition of yeast culture to prepare mash for fermentation. The living yeast cells are kept suspended in a liquid medium. Because the cells are alive, their concentration in the medium changes over time. Therefore, just before the yeast is added, it is necessary to estimate the concentration of yeast cells in the medium, to be sure to add the right amount. Up until the early part of the 20th century, this posed a problem for brewers. The number of yeast cells in a given volume is a random variable, but no one knew how to find the probability distribution.

William Sealy Gosset, a young statistician in his mid-twenties who was employed by the Guinness Brewing Company of Dublin, Ireland, discovered in 1904 that the number of yeast cells in a sampled volume of liquid medium follows a Poisson distribution. Gosset's discovery not only enabled Guinness to produce a more consistent product, it also showed that the Poisson distribution could have important applications in many situations. Gosset wanted to publish his result, but his managers at Guinness considered his discovery to be proprietary information, and forbade publication. Gosset published it anyway, but to hide this fact from his employers, he used the pseudonym "Student."

In the next example, we will solve a problem similar to the one that made Student famous. Before we get to it, though, we will mention that soon after solving this problem, Student made another discovery that made him even more famous. This discovery solved one of the most important outstanding problems in statistics, and has profoundly influenced work in virtually all fields of science ever since. We will discuss this result in Section 8.2.

EXAMPLE 6.20   Computing probability of events in space

Yeast cells are suspended in a liquid medium at a concentration of 4 particles per milliliter. A volume of 2 milliliters is withdrawn. What is the probability that exactly 6 particles are contained in this volume?

Solution
Let X be the number of particles withdrawn. Then X has a Poisson distribution. The rate is $\lambda = 4$ particles per milliliter, and the volume is $t = 2$ milliliters. Therefore

$$P(6) = e^{-4 \cdot 2}\frac{(4 \cdot 2)^6}{6!} = 0.1221$$

Figure 6.8 presents the result in Example 6.20 on the TI-84 Plus calculator. Note that we enter 8, which is the product λt. Step-by-step instructions can be found in the Using Technology section on page 275.

```
poissonpdf(8,6)
     .1221382155
```

Figure 6.8

Check Your Understanding

3. An environmental scientist is counting the number of bacteria in a sample of wastewater. The number of bacteria has a Poisson distribution with $\lambda = 3$ per microliter. What is the probability that exactly 9 bacteria are found in a volume of $t = 4$ microliters?

Answer is on page 277.

Objective 2 Compute the mean, variance, and standard deviation of a Poisson random variable

Mean and Variance of the Poisson Distribution

Imagine that events occur at an average rate of two per minute, and you count the number of events that occur in a 5-minute period. How many events would you expect to see? It is reasonable to think that you would see $5 \cdot 2 = 10$ events, on the average, and this is in fact correct. In general, if X is a Poisson random variable with rate λ and interval length t, the mean of X is λt.

Mean, Variance, and Standard Deviation of a Poisson Random Variable

If X is a Poisson random variable with rate λ and interval length t, then the mean of X is

$$\mu_X = \lambda t$$

The variance of X is

$$\sigma_X^2 = \lambda t \quad \text{(the variance is equal to the mean)}$$

The standard deviation of X is

$$\sigma_X = \sqrt{\lambda t}$$

EXAMPLE 6.21 Computing the mean and standard deviation of a Poisson random variable

Yeast cells are suspended in a liquid medium at a concentration of 4 particles per milliliter. A volume of 2 milliliters is withdrawn. Let X be the number of particles contained in this volume. Find the mean and standard deviation of X.

Solution
We have $\lambda = 4$ and $t = 2$. Therefore,

$$\mu_X = 4 \cdot 2 = 8 \qquad \sigma_X = \sqrt{4 \cdot 2} = 2.8284$$

Objective 3 Use the Poisson distribution to approximate binomial probabilities

The Poisson Approximation to the Binomial Distribution

When the number of trials, n, is large, binomial probabilities can be hard to compute, even with technology. The reason is that when n is large, $n!$ is extremely large and will cause most forms of technology to overflow. To be specific, the largest factorial that most forms of technology can handle is 69!. If you try to compute 70!, you will get an error.

Fortunately, when n is large and p is small, the Poisson distribution can be used to approximate binomial probabilities. The Poisson approximation may be used for any binomial probability for which $n \geq 100$ and $p \leq 0.1$. To compute $P(x)$ by using the Poisson approximation, use the following steps:

Step 1: Check that $n \geq 100$ and $p \leq 0.1$.

Step 2: Compute $P(x)$ by using the Poisson distribution with p in place of λ and n in place of t.

EXAMPLE 6.22   Using the Poisson approximation to the binomial

A Nevada roulette wheel has 38 pockets, labeled 0, 00, and 1 through 36. If you bet on 00 (or any other number), the probability that you will win is $1/38 = 0.0263$. Assume you bet on 00 for 200 spins of the wheel. Find the probability that you win exactly six times.

Solution

Step 1: There are 200 trials, and each trial has success probability $1/38 = 0.0263$. The number of successes therefore has a binomial distribution with $n = 200$ and $p = 0.0263$. Since $n \geq 100$ and $p \leq 0.1$, the Poisson approximation is valid.

Step 2: We compute $P(6)$ by using the Poisson distribution with rate $\lambda = 0.0263$ and $t = 200$.

$$P(6) = e^{-0.0263 \cdot 200} \frac{(0.0263 \cdot 200)^6}{6!} = 0.153$$

Check Your Understanding

4. In the Georgia Cash 3 lottery game, your probability of winning is 0.001. The game is played twice daily, so 730 games are played in a year. If you play twice a day for a year, what is the probability that you win exactly twice?

Answer is on page 277.

USING TECHNOLOGY

We use Example 6.19 to illustrate the technology steps.

TI-84 PLUS

Computing Poisson probabilities of the form $P(x)$ or $P(\text{Less than or equal to } x)$

Step 1. Press **2nd, VARS** to access the **DISTR** menu.
  - To compute $P(x)$, select **poissonpdf** and enter the values for λt and x separated by commas and press **ENTER**.
  - To compute $P(\text{Less than or equal to } x)$, select **poissoncdf** and enter the values for λt and x separated by commas and press **ENTER**.

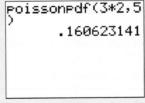

Figure A   Figure B

Using the TI-84 PLUS Stat Wizards (see Appendix B for more information)

Step 1. Press **2nd, VARS** to access the **DISTR** menu.
  - To compute $P(x)$, select **poissonpdf**, enter the value for λt in the λ field, and enter the value for x in the **x value** field. Select **Paste** and press **ENTER** to paste the command to the home screen. Press **ENTER** again to run the command.
  - To compute $P(\text{Less than or equal to } x)$, select **poissoncdf**, enter the value for λt in the λ field, and enter the value for x in the **x value** field. Select **Paste** and press **ENTER** to paste the command to the home screen. Press **ENTER** again to run the command.

For Example 6.19, $\lambda = 3$ and $t = 2$. Figure A displays the result of part (a), which asks for $P(5)$. Figure B displays the result of part (b), which asks for $P(\text{Less than or equal to } 2)$.

MINITAB

Computing Poisson probabilities of the form $P(x)$ or P(Less than or equal to x)

Step 1. Click **Calc**, then **Probability Distributions**, then **Poisson**.

Step 2. Enter the value for λt in the **Mean field**.

Step 3. To compute $P(x)$, select the **Probability option** and enter the value for x in the **Input constant** field. To compute P(Less than or equal to x), select the **Cumulative probability** option and enter the value for x in the **Input constant** field.

Step 4. Click **OK**.

Note: Poisson probabilities may be computed for a column of values by entering the column name in the **Input column** field.

EXCEL

Computing Poisson probabilities of the form $P(x)$ or P(Less than or equal to x)

Step 1. In an empty cell, select the **Insert Function** icon and highlight **Statistical** in the category field.

Step 2. Click on the **POISSON.DIST** function and click **OK**.

Step 3. Enter the value for x in the **X** field and the value for λt in the **Mean field**.

Step 4. To compute $P(x)$, enter **FALSE** in the **Cumulative field**. To compute P(Less than or equal to x), enter **TRUE** in the **Cumulative field**.

Step 5. Click **OK**.

SECTION 6.3   Exercises

Exercises 1–4 are the Check Your Understanding exercises located within the section.

Understanding the Concepts

In Exercises 5 and 6, fill in each blank with the appropriate word or phrase.

5. The Poisson distribution is used to describe events that occur in _____ or _____ .

6. The mean of the Poisson random variable X with rate λ and interval length t is _____ .

In Exercises 7 and 8, determine whether the statement is true or false. If the statement is false, rewrite it as a true statement.

7. For a Poisson random variable with rate λ and time interval t, the possible values of x are 0, 1, 2, ..., λ.

8. In a Poisson distribution, the mean and variance are equal.

Practicing the Skills

In Exercises 9–18, determine the indicated probability for a Poisson random variable with the given values of λ and t.

9. $\lambda = 2$, $t = 5$, $P(5)$

10. $\lambda = 0.5$, $t = 4$, $P(3)$

11. $\lambda = 0.1$, $t = 10$, $P(2)$

12. $\lambda = 1$, $t = 2$, $P(0)$

13. $\lambda = 0.2$, $t = 10$, P(At least one)

14. $\lambda = 0.3$, $t = 8$, P(At least one)

15. $\lambda = 1$, $t = 4$, P(No more than 5)

16. $\lambda = 0.2$, $t = 7$, P(More than 2)

17. $\lambda = 2$, $t = 4$, P(More than 9)

18. $\lambda = 0.3$, $t = 5$, P(Fewer than 3)

Working with the Concepts

19. **Potholes:** In a certain city, the number of potholes on a major street has a Poisson distribution with a rate of 3 per mile. Let X represent the number of potholes in a 2-mile stretch of road. Find
    a. $P(4)$
    b. P(More than 1)
    c. P(Between 5 and 7 inclusive)
    d. μ_X
    e. σ_X

20. **Flaws in aluminum foil:** The number of flaws in a given area of aluminum foil follows a Poisson distribution with a mean of 3 per square meter. Let X represent the number of flaws in a 1-square-meter sample of foil.
    a. $P(5)$
    b. $P(0)$
    c. P(Less than 2)
    d. P(Greater than 1)
    e. μ_X
    f. σ_X

21. **Flaws in lumber:** The number of flaws in a certain type of lumber has a Poisson distribution with a rate of 0.5 per meter of length.
    a. What is the probability that a board 3 meters in length has no flaws?

b. What is the probability that a board 4 meters in length has at least one flaw?

22. Flickering screen: The number of times that screen flicker appears in video recorded on a certain type of digital video recorder (DVR) has a Poisson distribution with a rate of 0.2 per minute.

   a. What is the probability that there are no screen flickers in 8 minutes of video?

   b. What is the probability that there are at least 2 screen flickers in 11 minutes of video?

23. Computer messages: The number of tweets received by a certain Twitter user is a Poisson random variable with a mean rate of 8 tweets per hour.

   a. What is the probability that 5 tweets are received in a given hour?

   b. What is the probability that 10 tweets are received in 1.5 hours?

   c. What is the probability that fewer than 3 tweets are received in one-half hour?

24. Grandma's cookies: Grandma bakes chocolate chip cookies in batches of 100. She puts 600 chips into each batch, so that the number of chips in a randomly chosen cookie has a Poisson distribution with $\lambda = 6$ chips per cookie.

   a. Grandma gives you one cookie. What is the probability that it contains exactly eight chips?

   b. Because you were extra good, Grandma will give you two cookies. What is the probability that the two cookies will contain a total of exactly 15 chips?

25. Trees in the forest: The number of trees of a certain species in a certain forest has a Poisson distribution with mean of 10 trees per acre.

   a. What is the probability that there will be exactly 18 trees in a 2-acre region?

   b. Find the mean number of trees in a 2-acre region.

   c. Find the standard deviation of the number of trees in a 2-acre region.

26. Bacteria: An environmental scientist obtains a sample of water from a pond that contains a certain type of bacteria at a concentration of 5 per milliliter.

   a. What is the probability that there will be exactly 12 bacteria in a 3-milliliter sample of water?

   b. Find the mean number of bacteria in a 3-milliliter sample.

   c. Find the standard deviation of the number of bacteria in a 3-milliliter sample.

27. Drive safely: In the year 2006, there were approximately 250,000,000 registered motor vehicles in the United States. Of these, approximately 4% were involved in an accident during the year. Assume that a simple random sample of $n = 200$ vehicles is drawn. Use the Poisson approximation to the binomial distribution to compute the following:

   a. The probability that exactly six of the 200 vehicles were involved in an accident

   b. The probability that fewer than three of the 200 vehicles were involved in an accident

   c. The probability that more than 10 of the 200 vehicles were involved in an accident

   d. The mean number of vehicles that were involved in an accident

   e. The standard deviation of the number of vehicles that were involved in an accident

28. Down Syndrome: A report from the Medical College of Georgia stated that the proportion of infants born with Down Syndrome is 0.0011. Assume a sample of 5000 births is studied. Use the Poisson approximation to the binomial distribution to compute the following:

   a. The probability that exactly 5 of the 5000 births involved Down Syndrome

   b. The probability that fewer than 4 of the 5000 births involved Down Syndrome

   c. The probability that more than 3 of the 5000 births involved Down Syndrome

   d. The mean number of births that involved Down Syndrome

   e. The standard deviation of the number of births that involved Down Syndrome

Extending the Concepts

The exponential distribution: *When the number of events occurring in a given time follows a Poisson distribution, the amount of time that elapses between two events is a random variable, and its distribution is called the **exponential distribution**.*

29. Assume that events occurring in time follow a Poisson distribution with rate λ. Let T be the amount of time, in seconds, that elapses between two events. This exercise will show how to compute probabilities involving T.

   a. Let X be the number of events that occur in a 1-second interval. Show that $P(X = 0) = e^{-\lambda}$.

   b. Explain why $X = 0$ is the same as $T > 1$.

   c. Show that $P(T > 1) = e^{-\lambda}$.

   d. Let X be the number of events that occur in a 2-second interval. Show that $P(X = 0) = e^{-2\lambda}$.

   e. Explain why $X = 0$ is the same as $T > 2$.

   f. Now let t be any amount of time, and let X be the number of events that occur in an interval of length t seconds. Show that $P(X = 0) = e^{-\lambda t}$.

   g. Explain why $X = 0$ is the same as $T > t$.

   h. Show that $P(T > t) = e^{-\lambda t}$.

Answers to Check Your Understanding Exercises for Section 6.3

1. a. 0.1396   **b.** 0.0424   **c.** 0.9862

2. a. 0.1606   **b.** 0.0174   **c.** 0.8488

3. 0.0874

4. 0.1284

Chapter 6 Summary

Section 6.1: A random variable is a numerical outcome of a probability experiment. Discrete random variables are random variables whose possible values can be listed, whereas continuous random variables can take on any value in some interval. A probability distribution for a discrete random variable specifies the probability for each possible value. A probability histogram is a histogram in which the heights of the rectangles are the probabilities for the possible values of the random variable. A probability histogram can also be thought of as a relative frequency histogram for a population. The law of large numbers for histograms states that as the sample size increases, the relative frequency histogram for the sample approaches the probability histogram.

 The mean of a random variable, also called the expected value, measures the center of the distribution. The standard deviation of a random variable measures the spread. The law of large numbers for means states that as the sample size increases, the sample mean approaches the population mean.

Section 6.2: The binomial distribution is an important discrete probability distribution. A random variable has a binomial distribution if it represents the number of successes in a fixed number n of independent trials, all of which have the same success probability p. Binomial probabilities can be found in a table or computed with technology. The mean of a binomial random variable is np, the number of trials multiplied by the success probability. The variance is $np(1 - p)$, and the standard deviation is $\sqrt{np(1 - p)}$.

Section 6.3: The Poisson distribution is used for certain random variables that represent the number of events that occur in time or space. Poisson probabilities are generally computed with technology. The mean and variance of a Poisson random variable are the same: Both are equal to the rate multiplied by the time. The Poisson distribution can also be used to approximate the binomial distribution when the number of trials, n, is large, and the success probability, p, is small.

Vocabulary and Notation

binomial distribution 260	mean 251	rate λ 272
continuous random variable 246	Poisson approximation to binomial 274	standard deviation 253
discrete random variable 246	Poisson distribution 271	trial 260
expected value 251	probability distribution 247	variance 253
interval length t 272	probability histogram 250	
law of large numbers for means 252	random variable 246	

Important Formulas

Mean of a discrete random variable:
$$\mu_X = \sum[x \cdot P(x)]$$

Variance of a discrete random variable:
$$\sigma_X^2 = \sum[(x - \mu_X)^2 \cdot P(x)] = \sum[x^2 \cdot P(x)] - \mu_X^2$$

Standard deviation of a discrete random variable:
$$\sigma_X = \sqrt{\sigma_X^2}$$

Mean of a binomial random variable:
$$\mu_X = np$$

Variance of a binomial random variable:
$$\sigma_X^2 = np(1 - p)$$

Standard deviation of a binomial random variable:
$$\sigma_X = \sqrt{np(1 - p)}$$

Mean of Poisson random variable:
$$\mu_X = \lambda t$$

Variance of Poisson random variable:
$$\sigma_X^2 = \lambda t$$

Standard deviation of Poisson random variable:
$$\sigma_X = \sqrt{\lambda t}$$

Chapter Quiz

1. Explain why the following is *not* a probability distribution.

x	6	7	8	9	10
$P(x)$	0.32	0.11	0.19	0.28	0.03

2. Find the mean of the random variable X with the following probability distribution.

x	−2	1	4	5
$P(x)$	0.3	0.2	0.1	0.4

3. Refer to Problem 2.
  a. Find the variance of the random variable X.
  b. Find the standard deviation of the random variable X.

4. Find the missing value that makes the following a valid probability distribution.

x	2	3	5	8	10
$P(x)$	0.23	0.12	0.09	?	0.37

5. The following table presents a probability distribution for the number of pets each family has in a certain neighborhood.

Number of pets	0	1	2	3	4
Probability	0.4	0.2	0.2	0.1	0.1

Construct a probability histogram.

6. Refer to Problem 5. Find the probability that a randomly selected family has:
  a. 1 or 2 pets
  b. More than 2 pets
  c. No more than 3 pets
  d. At least 1 pet

7. Refer to Problem 5. Find the mean number of pets.

8. Refer to Problem 5. Find the standard deviation of the number of pets.

9. At a cell phone battery plant, 5% of cell phone batteries produced are defective. A quality control engineer randomly collects a sample of 50 batteries from a large shipment from this plant and inspects them for defects. Find the probability that
  a. None of the batteries are defective.
  b. At least one of the batteries is defective.
  c. No more than 3 of the batteries are defective.

10. Refer to Problem 9. Find the mean and standard deviation for the number of defective batteries in the sample of size 50.

11. A meteorologist states that the probability of rain tomorrow is 0.4 and the probability of rain on the next day is 0.6. Assuming these probabilities are accurate, and that the rain events are independent, find the probability distribution for X, the number of days out of the next two that it rains.

12. The number of large packages delivered by a courier service follows a Poisson distribution with a rate of 5 per day. Find the probability that
  a. Exactly 4 large packages are delivered on a given day.
  b. Fewer than 6 large packages are delivered over a 2-day period.
  c. At least one large package is delivered over a half-day period.

13. The number of text messages received on a certain person's cell phone follows a Poisson distribution with the rate of 10 messages per hour. What is the mean number of messages received in an 8-hour period?

14. Refer to Problem 13. What are the variance and standard deviation of the number of messages received in an 8-hour period?

15. Huntington's disease is a genetically transmitted disease that causes degeneration of nerve cells in the brain. The probability that a person carries a gene for Huntington's disease is 0.00005 (i.e., 1/20,000). In a sample of 100,000 people, what is the probability that exactly four people carry the gene for Huntington's disease?

Review Exercises

1. Which are distributions? Which of the following tables represent probability distributions?

a.

x	$P(x)$
3	0.35
4	0.20
5	0.18
6	0.09
7	0.18

b.

x	$P(x)$
5	0.27
6	0.45
7	-0.06
8	0.44

c.

x	$P(x)$
0	0.02
1	0.34
2	1.02
3	0.01
4	0.43
5	0.14

d.

x	$P(x)$
2	0.10
3	0.07
4	0.75
5	0.08

2. Mean, variance, and standard deviation: A random variable X has the following probability distribution.

x	6	7	8	9	10	11
$P(x)$	0.21	0.12	0.29	0.11	0.01	0.26

a. Find the mean of X.
b. Find the variance of X.
c. Find the standard deviation of X.

3. **AP tests:** Advanced Placement (AP) tests are graded on a scale of 1 (low) through 5 (high). The College Board reported that the distribution of scores on the AP Statistics Exam in 2009 was as follows:

x	1	2	3	4	5
$P(x)$	0.34	0.25	0.18	0.16	0.07

A score of 3 or higher is generally required for college credit. What is the probability that a student scores 3 or higher?

4. **AP tests again:** During the 2008–2009 academic year, approximately 1.7 million students took one or more AP tests. Following is the frequency distribution of the number of AP tests taken by students who took one or more AP tests.

Number of Tests	Frequency (in 1000s)
1	953
2	423
3	194
4	80
5	29
6	9
7	3
8	1
Total	1692

Source: The College Board

Let X represent the number of exams taken by a student who took one or more.
a. Construct the probability distribution for X.
b. Find the probability that a student took exactly one exam.
c. Compute the mean μ_X.
d. Compute the standard deviation σ_X.

5. **Lottery tickets:** Several million lottery tickets are sold, and 60% of the tickets are held by women. Five winning tickets will be drawn at random.
a. What is the probability that three or fewer of the winners will be women?
b. What is the probability that three of the winners will be of one gender and two of the winners will be of the other gender?

6. **Genetic disease:** Sickle-cell anemia is a disease that results when a person has two copies of a certain recessive gene. People with one copy of the gene are called carriers. Carriers do not have the disease, but can pass the gene on to their children. A child born to parents who are both carriers has probability 0.25 of having sickle-cell anemia. A medical study samples 18 children in families where both parents are carriers.
a. What is the probability that four or more of the children have sickle-cell anemia?
b. What is the probability that fewer than three of the children have sickle-cell anemia?
c. Would it be unusual if none of the children had sickle-cell anemia?

7. **Craps:** In the game of craps, you may bet $1 that the next roll of the dice will be an 11. If the dice come up 11, your profit is $15. If the dice don't come up 11, you lose $1. The probability that the dice come up 11 is 1/18. What is the expected value of your profit? Is it an expected gain or an expected loss?

8. **Looking for a job:** According to the General Social Survey conducted at the University of Chicago, 59% of employed adults believe that if they lost their job, it would be easy to find another one with a similar salary. Suppose that 10 employed adults are randomly selected.
a. Find the probability that exactly three of them believe it would be easy to find another job.
b. Find the probability that more than two of them believe it would be easy to find another job.
c. Would it be unusual if all of them believed it would be easy to find another job?

9. **Reading tests:** According to the National Center for Education Statistics, 66% of fourth-graders could read at a basic level in 2007. Suppose that eight fourth-graders are randomly selected.
a. Find the probability that exactly five of them can read at a basic level.
b. Find the probability that more than six of them can read at a basic level.
c. Would it be unusual if all of them could read at a basic level?

10. **Rain, rain, go away:** Let X be the number of days during the next month that it rains. Does X have a binomial distribution? Why or why not?

11. **Survey sample:** In a college with 5000 students, 100 are randomly chosen to complete a survey in which they rate the quality of the cafeteria food. Let X be the number of freshmen who are chosen. Does X have a binomial distribution? Why or why not?

12. **Suspensions:** The concentration of particles in a suspension is 2 per milliliter. A volume of 3 milliliters is withdrawn. Find the following probabilities.
    a. P(Exactly five particles are withdrawn)
    b. P(Fewer than two particles are withdrawn)
    c. P(More than one particle is withdrawn)

13. **Mean:** Refer to Exercise 12. Find the mean number of particles that are withdrawn.

14. **Standard deviation:** Refer to Exercise 12. Find the standard deviation of the number of particles that are withdrawn.

15. **Congenital disease:** The connexin-26 mutation is a genetic mutation that results in deafness. The probability that a person carries a gene with this mutation is 0.0151. Find the probability that exactly 10 people in a sample of 500 carry this mutation.

Write About It

1. Provide an example of a discrete random variable and explain why it is discrete.

2. Provide an example of a continuous random variable and explain why it is continuous.

3. If a business decision has an expected gain, is it possible to lose money? Explain.

4. When a population mean is unknown, people will often approximate it with the mean of a large sample. Explain why this is justified.

5. Provide an example of a binomial random variable and explain how each condition for the binomial distribution is fulfilled.

6. Twenty percent of the men in a certain community are more than 6 feet tall. An anthropologist samples five men from a large family in the community and counts the number X who are more than 6 feet tall. Explain why the binomial distribution is not appropriate in this situation. Is $P(X = 0)$ likely to be greater than or less than the value predicted by the binomial distribution with $n = 10$ and $p = 0.4$? Explain your reasoning.

7. Provide an example of a Poisson random variable and explain how each condition for the Poisson distribution is fulfilled.

Case Study: Benford's Law: Do The Digits 1–9 Occur Equally Often?

One of the most surprising probability distributions found in practice is given by a rule known as Benford's law. This probability distribution concerns the first digits of numbers. The first digit of a number may be any of the digits 1, 2, 3, 4, 5, 6, 7, 8, or 9. It is reasonable to believe that, for most sets of numbers encountered in practice, these digits would occur equally often. In fact, it has been observed that for many naturally occurring data sets, smaller numbers occur more frequently as the first digit than larger numbers do. Benford's law is named for Frank Benford, an engineer at General Electric, who stated it in 1938.

Following are the populations of the 50 states in a recent census. The first digit of each population number is listed separately.

State	Population	First Digit	State	Population	First Digit	State	Population	First Digit
Alabama	4,557,808	4	Louisiana	4,523,628	4	Ohio	11,464,042	1
Alaska	663,661	6	Maine	1,321,505	1	Oklahoma	3,547,884	3
Arizona	5,939,292	5	Maryland	5,600,388	5	Oregon	3,641,056	3
Arkansas	2,779,154	2	Massachusetts	6,398,743	6	Pennsylvania	12,429,616	1
California	36,132,147	3	Michigan	10,120,860	1	Rhode Island	1,076,189	1
Colorado	4,665,177	4	Minnesota	5,132,799	5	South Carolina	4,255,083	4
Connecticut	3,510,297	3	Mississippi	2,921,088	2	South Dakota	775,933	7
Delaware	843,524	8	Missouri	5,800,310	5	Tennessee	5,962,959	5
Florida	17,789,864	1	Montana	935,670	9	Texas	22,859,968	2
Georgia	9,072,576	9	Nebraska	1,758,787	1	Utah	2,469,585	2
Hawaii	1,275,194	1	Nevada	2,414,807	2	Vermont	623,050	6
Idaho	1,429,096	1	New Hampshire	1,309,940	1	Virginia	7,567,465	7
Illinois	12,763,371	1	New Jersey	8,717,925	8	Washington	6,287,759	6
Indiana	6,271,973	6	New Mexico	1,928,384	1	West Virginia	1,816,856	1
Iowa	2,966,334	2	New York	19,254,630	1	Wisconsin	5,536,201	5
Kansas	2,744,687	2	North Carolina	8,683,242	8	Wyoming	509,294	5
Kentucky	4,173,405	4	North Dakota	636,677	6			

Here is a frequency distribution of the first digits of the state populations:

Digit	Frequency	Digit	Frequency
1	14	6	6
2	7	7	2
3	4	8	3
4	5	9	2
5	7		

For the state populations, the most frequent first digit is 1, with 7, 8, and 9 being the least frequent.

Now here is a table of the closing value of the Dow Jones Industrial Average for each of the years 1974–2008.

Year	Average	First Digit	Year	Average	First Digit
1974	616.24	6	1992	3301.11	3
1975	852.41	8	1993	3754.09	3
1976	1004.65	1	1994	3834.44	3
1977	831.17	8	1995	5117.12	5
1978	805.01	8	1996	6448.27	6
1979	838.74	8	1997	7908.25	7
1980	963.98	9	1998	9181.43	9
1981	875.00	8	1999	11497.12	1
1982	1046.55	1	2000	10786.85	1
1983	1258.64	1	2001	10021.50	1
1984	1211.57	1	2002	8341.63	8
1985	1546.67	1	2003	10453.92	1
1986	1895.95	1	2004	10783.01	1
1987	1938.83	1	2005	10717.50	1
1988	2168.57	2	2006	12463.15	1
1989	2753.20	2	2007	13264.82	1
1990	2633.66	2	2008	8776.39	8
1991	3168.83	3			

Here is a frequency distribution of the first digits of the stock market averages:

Digit	Frequency	Digit	Frequency
1	15	6	2
2	3	7	1
3	4	8	7
4	0	9	2
5	1		

For the stock market averages, the most frequent first digit by far is 1.

The stock market averages give a partial justification for Benford's law. Assume the stock market starts at 1000 and goes up 10% each year. It will take 8 years for the average to exceed 2000. Thus, the first eight averages will begin with the digit 1. Now imagine that the average starts at 5000. If it goes up 10% each year, it would take only 2 years to exceed 6000, so there would be only 2 years starting with the digit 5. In general, Benford's law applies well to data where increments occur as a result of multiplication rather than addition, and where there is a wide range of values. It does not apply to data sets where the range of values is small.

Here is the probability distribution of digits as predicted by Benford's law:

Digit	Frequency	Digit	Frequency
1	0.301	6	0.067
2	0.176	7	0.058
3	0.125	8	0.051
4	0.097	9	0.046
5	0.079		

The surprising nature of Benford's law makes it a useful tool to detect fraud. When people make up numbers, they tend to make the first digits approximately uniformly distributed; in other words, they have approximately equal numbers of 1s, 2s, and so on. Many tax agencies, including the Internal Revenue Service, use software to detect deviations from Benford's law in tax returns.

Following are results from three hypothetical corporate tax returns. Each purports to be a list of expenditures, in dollars, that the corporation is claiming as deductions. Two of the three are genuine, and one is a fraud. Which one is the fraud?

i.

79,386	17,988
203,374	80,535
11,967	3,037
100,229	132,056
46,428	59,727
7,012	38,354
957,559	137,648
551,284	4,163
97,439	1,279
780,216	91,404
22,443	323,547
1,023	194,288
738,527	24,346
634,814	695,236
850,840	160,546

ii.

1,393	165,648
47,689	601,981
75.854	262,971
5,395	65,407
53,079	6,892
7,791	748,151
93,401	45,054
129,906	83,821
568,823	228,976
4,693	913,337
21,902	252,378
337,122	82,581
162,182	538,342
7,942	99,613
31,121	78,175

iii.

64,888	374,242
1,643	12,338
832,618	14,204
126,811	31,484
13,545	1,818
2,332	104,625
29,288	34,178
81,074	3,684
401,437	11,665
3,040	15,376
244,676	541,894
49,273	65,928
112,111	250,601
56,776	650,316
262,359	90,852

The Normal Distribution

Introduction

Beverage cans are made from a very thin sheet of aluminum, only 1/80 inch thick. Yet they must withstand pressures of up to 90 pounds per square inch (approximately three times the pressure in an automobile tire). Beverage companies often purchase cans in large shipments. To ensure that can failures are rare, quality control inspectors sample several cans from each shipment and test their strength by placing them in testing machines that apply force until the can fails (is punctured or crushed). The testing process destroys the cans, so the number of cans that can be tested is limited.

Assume that a can is considered defective if it fails at a pressure of less than 90 pounds per square inch. The quality control inspectors want the proportion of defective cans to be no more than 0.001, or 1 in 1000. They test 10 cans, with the following results.

Can	1	2	3	4	5	6	7	8	9	10
Pressure at Failure	95	96	98	99	99	100	101	101	103	104

Although none of the 10 cans were defective, this is not enough by itself to determine whether the proportion of defective cans is less than 1 in 1000. To make this determination, we must know something about the probability distribution of the pressures at which cans fail. In this chapter, we will study the *normal distribution*, which is the most important distribution in statistics. In the case study at the end of this chapter, we will show that if the pressures follow a normal distribution, we can estimate the proportion of defective cans.

The Standard Normal Curve

Objectives

1. Use a probability density curve to describe a population
2. Use a normal curve to describe a normal population
3. Find areas under the standard normal curve
4. Find z-scores corresponding to areas under the normal curve

Objective 1 Use a probability density curve to describe a population

Figure 7.1 Relative frequency histogram for the emissions of a sample of 65 vehicles

Figure 7.1, first shown in Section 2.2, presents a relative frequency histogram for the emissions of a sample of 65 vehicles. The amount of emissions is a continuous variable, because its possible values are not limited to some discrete set. The class intervals are chosen so that each rectangle represents a reasonably large number of vehicles. If the sample were larger, we could make the rectangles narrower. In particular, if we had information on the entire population, containing millions of vehicles, we could make the rectangles extremely narrow. The histogram would then look quite smooth and could be approximated by a curve, which might look like Figure 7.2.

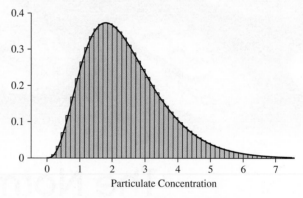

Figure 7.2 The histogram for a large population of vehicles could be drawn with extremely narrow rectangles, and could be represented by a curve.

If a vehicle were chosen at random from this population to have its emissions measured, the emissions level would be a continuous random variable. The curve used to describe the distribution of a continuous random variable is called the **probability density curve** of the random variable. The probability density curve tells us what proportion of the population falls within any given interval. For example, Figure 7.3 illustrates the proportion of the population of vehicles whose emissions levels are between 3 and 4. In general, the area under a probability density curve between any two values a and b has two interpretations: It represents the proportion of the population whose values are between a and b, and it also represents the probability that a randomly selected value from the population will be between a and b.

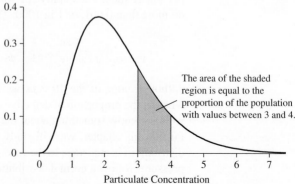

Figure 7.3 The area under a probability density curve between two values is equal to the proportion of the population that falls between the two values.

The region above a single point has zero width, and thus an area of 0. Therefore, when a population is represented with a probability density curve, the probability of obtaining a prespecified value exactly is equal to 0. For this reason, if X is a continuous random variable, then $P(X = a) = 0$ for any number a, and $P(a < X < b) = P(a \leq X \leq b)$ for any numbers a and b. For any probability density curve, the area under the entire curve is equal to 1, because this area represents the entire population.

SUMMARY

- A probability density curve represents the probability distribution of a continuous variable.
- The area under the entire curve is equal to 1.
- The area under the curve between two values a and b has two interpretations:
  1. It is the proportion of the population whose values are between a and b.
  2. It is the probability that a randomly selected individual will have a value between a and b.

EXAMPLE 7.1 Interpret the area under a probability density curve

Following is a probability density curve for a population.

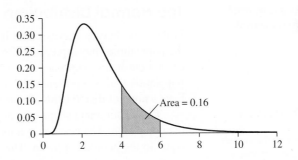

a. What proportion of the population is between 4 and 6?
b. If a value is chosen at random from this population, what is the probability that it will be between 4 and 6?
c. What proportion of the population is not between 4 and 6?
d. If a value is chosen at random from this population, what is the probability that it is not between 4 and 6?

Solution

a. The proportion of the population between 4 and 6 is equal to the area under the curve between 4 and 6, which is 0.16.
b. The probability that a randomly chosen value is between 4 and 6 is equal to the area under the curve between 4 and 6, which is 0.16.
c. The area under the entire curve is equal to 1. Therefore, the proportion that is not between 4 and 6 is equal to $1 - 0.16 = 0.84$.
d. The probability that a randomly chosen value is not between 4 and 6 is equal to the area under the curve that is not between 4 and 6, which is 0.84.

Recall: The Rule of Complements says that $P(\text{not } A) = 1 - P(A)$.

Another way to answer part (d) is to use the Rule of Complements:

$$P(\text{Not between 4 and 6}) = 1 - P(\text{Between 4 and 6}) = 1 - 0.16 = 0.84$$

Check Your Understanding

1. Following is a probability density curve with the area between 0 and 1 and the area between 1 and 2 indicated.

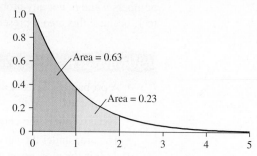

a. What proportion of the population is between 0 and 1?
b. What is the probability that a randomly selected value will be between 1 and 2?
c. What proportion of the population is between 0 and 2?
d. What is the probability that a randomly selected value will be greater than 2?

Answers are on page 299.

Objective 2 Use a normal curve to describe a normal population

The Normal Distribution

Probability density curves come in many varieties, depending on the characteristics of the populations they represent. Remarkably, many important statistical procedures can be carried out using only one type of probability density curve, called a **normal curve**. A population that is represented by a normal curve is said to be *normally distributed*, or to have a **normal distribution**. Figure 7.4 presents some examples of normal curves.

The location and shape of a normal curve reflect the mean and standard deviation of the population. The curve is symmetric around its peak, or mode. Therefore, the mode is equal to the population mean. The population standard deviation measures the spread of the population. Therefore, the normal curve is wide and flat when the population standard deviation is large, and tall and narrow when the population standard deviation is small.

Explain It Again

The mode of a curve: Recall that a peak in a histogram is called a *mode* of the histogram. Similarly, a peak in a probability density curve, such as a normal curve, is called a mode of the curve.

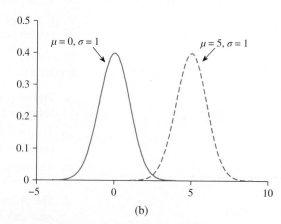

Figure 7.4 (a) Both populations have mean 0. The population with standard deviation 2 is more spread out than the population with standard deviation 1. (b) Both populations have the same spread, because they have the same standard deviation. The curves are centered over the population means.

Properties of Normal Distributions

1. Normal distributions have one mode.
2. Normal distributions are symmetric around the mode.
3. The mean and median of a normal distribution are both equal to the mode. In other words, the mean, median, and mode of a normal distribution are all the same.
4. The normal distribution follows the Empirical Rule (see Figure 7.5):
   - Approximately 68% of the population is within one standard deviation of the mean. In other words, approximately 68% of the population is in the interval $\mu - \sigma$ to $\mu + \sigma$.
   - Approximately 95% of the population is within two standard deviations of the mean. In other words, approximately 95% of the population is in the interval $\mu - 2\sigma$ to $\mu + 2\sigma$.
   - Approximately 99.7% of the population is within three standard deviations of the mean. In other words, approximately 99.7% of the population is in the interval $\mu - 3\sigma$ to $\mu + 3\sigma$.

Recall: The Empirical Rule holds for most unimodal symmetric distributions. See Section 3.2.

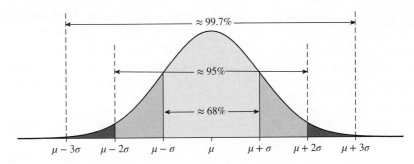

Figure 7.5 Normal curve with mean μ and standard deviation σ

Objective 3 Find areas under the standard normal curve

Finding Areas Under the Standard Normal Curve

A normal distribution can have any mean and any positive standard deviation, but it is only necessary to work with the normal distribution that has mean 0 and standard deviation 1, which is called the **standard normal distribution**. The probability density function for the standard normal distribution is called the **standard normal curve**.

For any interval, the area under a normal curve over the interval represents the proportion of the population that is contained within the interval. Finding an area under a normal curve is a crucial step in many statistical procedures. When finding an area under the standard normal curve, we use the letter z to indicate a value on the horizontal axis (see Figure 7.6). We refer to such a value as a *z-score*. Since the mean of the standard normal distribution, which is located at the mode, is 0, the z-score at the mode of the curve is 0. Points on the horizontal axis to the left of the mode have negative z-scores, and points to the right of the mode have positive z-scores.

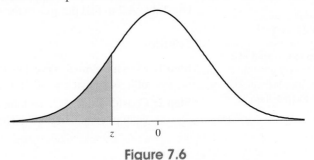

Figure 7.6

SUMMARY

- Points on the horizontal axis to the left of the mode have negative z-scores.
- Points on the horizontal axis to the right of the mode have positive z-scores.
- The mode itself has a z-score of 0.

Finding areas under the standard normal curve by using Table A.2

In general, areas under the standard normal curve can be found by using Table A.2 or by using technology. Table A.2 contains z-scores and areas. Each of the four-digit numbers in the body of the table is the area to the left of a z-score. Examples 7.2–7.4 will show how to use Table A.2.

EXAMPLE 7.2

Finding an area to the left of a z-score

Use Table A.2 to find the area to the left of $z = 1.26$.

Solution

Step 1: Sketch a normal curve, label the point $z = 1.26$, and shade in the area to the left of it. Note that $z = 1.26$ is located to the right of the mode, since it is positive.

Step 2: Consult Table A.2. To look up $z = 1.26$, find the row containing 1.2 and the column containing 0.06. The value in the intersection of the row and column is 0.8962. This is the area to the left of $z = 1.26$ (see Figure 7.7).

z	0.00	0.01	0.02	0.03	0.04	0.05	0.06	0.07	0.08	0.09
⋮	⋮	⋮	⋮	⋮	⋮	⋮	⋮	⋮	⋮	⋮
1.0	.8413	.8438	.8461	.8485	.8508	.8531	.8554	.8577	.8599	.8621
1.1	.8643	.8665	.8686	.8708	.8729	.8749	.8770	.8790	.8810	.8830
1.2	.8849	.8869	.8888	.8907	.8925	.8944	.8962	.8980	.8997	.9015
1.3	.9032	.9049	.9066	.9082	.9099	.9115	.9131	.9147	.9162	.9177
1.4	.9192	.9207	.9222	.9236	.9251	.9265	.9279	.9292	.9306	.9319
⋮	⋮	⋮	⋮	⋮	⋮	⋮	⋮	⋮	⋮	⋮

An area to the left of a z-score represents the proportion of a population that is less than a given value. Sometimes we need to know the proportion of a population that is greater than a given value. In these cases we need to find the area to the right of a z-score. Since the area under the entire curve is equal to 1, we can find the area to the right of a z-score by finding the area to the left and subtracting from 1.

EXAMPLE 7.3

Finding an area to the right of a z-score

Use Table A.2 to find the area to the right of $z = -0.58$.

Solution

Step 1: Sketch a normal curve, label the point $z = -0.58$, and shade in the area to the right of it. Note that $z = -0.58$ is located to the left of the mode, since it is negative.

Step 2: Consult Table A.2. To look up $z = -0.58$, find the row containing -0.5 and the column containing 0.08. The value in the intersection of the row and column is

---

Explain It Again

Looking up a z-score: In Table A.2, the units and tenths digits of the z-score correspond to a row, and the hundredths digit corresponds to a column. Thus, for the z-score 1.26, we find the row corresponding to 1.2 and the column corresponding to 0.06.

Area = 0.8962

1.26

Figure 7.7

Explain It Again

Finding an area to the right of a z-score: To find the area to the right of a z-score, use Table A.2 to find the area to the left, then subtract from 1.

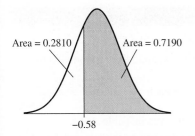

Area = 0.2810   Area = 0.7190

−0.58

Figure 7.8

0.2810. This is the area to the *left* of $z = -0.58$. To find the area to the right, we subtract from 1 (see Figure 7.8):

$$\text{Area to the right of } z = -0.58 = 1 - \text{Area to the left of } z = -0.58$$
$$= 1 - 0.2810$$
$$= 0.7190$$

z	**0.00**	**0.01**	**0.02**	**0.03**	**0.04**	**0.05**	**0.06**	**0.07**	**0.08**	**0.09**
⋮	⋮	⋮	⋮	⋮	⋮	⋮	⋮	⋮	⋮	⋮
−0.7	.2420	.2389	.2358	.2327	.2296	.2266	.2236	.2206	.2177	.2148
−0.6	.2743	.2709	.2676	.2643	.2611	.2578	.2546	.2514	.2483	.2451
−0.5	.3085	.3050	.3015	.2981	.2946	.2912	.2877	.2843	.2810	.2776
−0.4	.3446	.3409	.3372	.3336	.3300	.3264	.3228	.3192	.3156	.3121
−0.3	.3821	.3783	.3745	.3707	.3669	.3632	.3594	.3557	.3520	.3483
⋮	⋮	⋮	⋮	⋮	⋮	⋮	⋮	⋮	⋮	⋮

Sometimes we need to find the proportion of a population that falls between two values. In these cases, we need to find the area between two z-scores. We can do this using Table A.2 by finding the area to the left of each z-score. The area between the z-scores is found by subtracting the smaller area from the larger area.

EXAMPLE 7.4   Finding an area between two z-scores

Find the area between $z = -1.45$ and $z = 0.42$.

Solution

Step 1: Sketch a normal curve, label the points $z = -1.45$ and $z = 0.42$, and shade in the area between them. See Figure 7.9.

Step 2: Use Table A.2 to find the areas to the left of $z = -1.45$ and to the left of $z = 0.42$. The area to the left of $z = -1.45$ is 0.0735, and the area to the left of $z = 0.42$ is 0.6628.

Step 3: Subtract the smaller area from the larger area to find the area between the two z-scores:

−1.45   0.42

Figure 7.9

$$\text{Area between } z = -1.45 \text{ and } z = 0.42 = (\text{Area left of } 0.42) - (\text{Area left of } -1.45)$$
$$= 0.6628 - 0.0735$$
$$= 0.5893$$

The area between $z = -1.45$ and $z = 0.42$ is 0.5893. See Figure 7.10.

Area = 0.6628

−1.45   0.42

Area = 0.0735

−1.45

Area = 0.5893

−1.45   0.42

Figure 7.10   We start with the area to the left of $z = 0.42$ and subtract the area to the left of $z = -1.45$. This leaves the area between $z = 0.42$ and $z = -1.45$.

SUMMARY

To find an area under the standard normal curve by using Table A.2:

Step 1: Sketch a normal curve, label the z-score or scores, and shade in the area to be found.

Step 2: Look up the area in Table A.2 corresponding to each z-score.

Step 3:
- To find the area to the left of a z-score, use the area in Table A.2.
- To find the area to the right of a z-score, find the area to the left and subtract from 1.
- To find the area between two z-scores, find the area to the left of each, then subtract the smaller area from the larger area.

Check Your Understanding

2. Use Table A.2 to find the area to the left of $z = 0.25$.

3. Use Table A.2 to find the area to the right of $z = 2.31$.

4. Use Table A.2 to find the area between $z = -1.13$ and $z = 2.02$.

Answers are on page 299.

Finding areas under the standard normal curve by using technology

Areas under the standard normal curve can be found by using technology. Examples 7.5 and 7.6 illustrate the method.

EXAMPLE 7.5   Find the area between two z-scores by using technology

In Example 7.4, we found the area between $z = -1.45$ and $z = 0.42$. Find this area by using technology.

Solution

We present results from the TI-84 Plus calculator. The area is found by using the **normalcdf** command. We enter the left endpoint of the interval (-1.45), the right endpoint (0.42), the mean (0), and the standard deviation (1). Step-by-step instructions are given in the Using Technology section on page 295.

```
normalcdf(-1.45,
0.42,0,1)
          .5892279372
```

The TI-84 Plus gives the area as .5892279372. Rounding to four decimal places gives 0.5892.

EXAMPLE 7.6   Find the area to the right of a z-score by using technology

In Example 7.3, we found the area to the right of $z = -0.58$. Find this area by using technology.

Explain It Again

1E99: In Example 7.6, we want to find the area between −0.58 and infinity. Since we can't enter a value of infinity into the calculator, we enter the very large number **1E99** instead.

Solution

We present output from the TI-84 Plus calculator. We use the **normalcdf** command. We enter the left endpoint of the interval (-0.58). Since there is no right endpoint, we enter **1E99**, which represents the very large number that is written with a 1 followed by 99 zeros. Then we enter the mean (0) and the standard deviation (1). Step-by-step instructions are given in the Using Technology section on page 295.

```
normalcdf(-0.58,
1E99,0,1)
          .7190427366
```

The TI-84 Plus gives the area as .7190427366. Rounding to four decimal places gives 0.7190.

Objective 4 Find z-scores corresponding to areas under the normal curve

Finding a z-Score Corresponding to a Given Area

Often we are given an area and we need to find the z-score that corresponds to an area under the standard normal curve. Examples 7.7–7.11 show how this is done using either Table A.2 or technology. When using the table, it is useful to remember that the mode, $z = 0$, has an area of 0.5 both to its right and to its left.

| EXAMPLE 7.7 | ### Finding the z-score corresponding to an area to the left |

Use Table A.2 to find the z-score that has an area of 0.26 to its left.

Solution

Step 1: Sketch a normal curve and shade in the given area.

Step 2: In Table A.2, look through the body of the table to find the area closest to 0.26. This value is 0.2611. The values in the left-hand column and top row corresponding to 0.2611 are −0.6 and 0.04. Therefore, the z-score is $z = -0.64$. See Figure 7.11.

z	0.00	0.01	0.02	0.03	0.04	0.05	0.06	0.07	0.08	0.09
⋮	⋮	⋮	⋮	⋮	⋮	⋮	⋮	⋮	⋮	⋮
−0.8	.2119	.2090	.2061	.2033	.2005	.1977	.1949	.1922	.1894	.1867
−0.7	.2420	.2389	.2358	.2327	.2296	.2266	.2236	.2206	.2177	.2148
−0.6	.2743	.2709	.2676	.2643	.2611	.2578	.2546	.2514	.2483	.2451
−0.5	.3085	.3050	.3015	.2981	.2946	.2912	.2877	.2843	.2810	.2776
−0.4	.3446	.3409	.3372	.3336	.3300	.3264	.3228	.3192	.3156	.3121
⋮	⋮	⋮	⋮	⋮	⋮	⋮	⋮	⋮	⋮	⋮

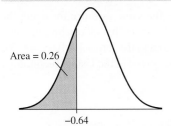

Area = 0.26

−0.64

Figure 7.11

| EXAMPLE 7.8 | ### Using technology to find the z-score corresponding to an area to the left |

In Example 7.7, we found the z-score that has an area of 0.26 to its left. Use technology to find this z-score.

Solution

We present results from the TI-84 Plus calculator. The z-score is found by using the **invNorm** command. We enter the area to the left (.26), the mean (0), and the standard deviation (1). Step-by-step instructions are given in the Using Technology section on page 296.

```
invNorm(.26,0,1)
         -.6433454021
```

The TI-84 Plus gives $z = -.6433454021$. Rounding to two decimal places gives $z = -0.64$.

Finding the z-score corresponding to an area to the right

Use Table A.2 to find the z-score that has an area of 0.68 to its right.

Solution

Step 1: Sketch a normal curve and shade in the given area.

Step 2: Determine the area to the left of the z-score. Since the area to the right is 0.68, the area to the left is $1 - 0.68 = 0.32$.

Step 3: In Table A.2, look through the body of the table to find the area closest to 0.32. This value is 0.3192. The z-score in Table A.2 corresponding to an area of 0.3192 is $z = -0.47$. See Figure 7.12.

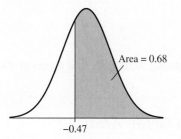

Area = 0.68

−0.47

Figure 7.12

z	0.00	0.01	0.02	0.03	0.04	0.05	0.06	0.07	0.08	0.09
⋮	⋮	⋮	⋮	⋮	⋮	⋮	⋮	⋮	⋮	⋮
−0.6	.2743	.2709	.2676	.2643	.2611	.2578	.2546	.2514	.2483	.2451
−0.5	.3085	.3050	.3015	.2981	.2946	.2912	.2877	.2843	.2810	.2776
−0.4	.3446	.3409	.3372	.3336	.3300	.3264	.3228	.3192	.3156	.3121
−0.3	.3821	.3783	.3745	.3707	.3669	.3632	.3594	.3557	.3520	.3483
−0.2	.4207	.4168	.4129	.4090	.4052	.4013	.3974	.3936	.3897	.3859
⋮	⋮	⋮	⋮	⋮	⋮	⋮	⋮	⋮	⋮	⋮

Using technology to find the z-score corresponding to an area to the right

In Example 7.9, we found the z-score that has an area of 0.68 to its right. Use technology to find this z-score.

Solution

We use the TI-84 Plus calculator. The first two steps are the same as those for using the table. We must first find the area to the *left* of the desired z-score, by subtracting the area to the right from 1. Since the area to the right is 0.68, the area to the left is $1 - 0.68 = 0.32$.

The results are as follows. Step-by-step instructions are given in the Using Technology section on page 296.

```
invNorm(.32,0,1)
         -.4676988012
```

The TI-84 Plus gives $z = -.4676988012$. Rounding to two decimal places gives $z = -0.47$.

Check Your Understanding

5. Use Table A.2 to find the z-score with an area of 0.45 to its left.

6. Use Table A.2 to find the z-score with an area of 0.37 to its right.

7. Use any method to find the z-score with an area of 0.74 to its left.

8. Use any method to find the z-score with an area of 0.09 to its right.

Answers are on page 299.

EXAMPLE 7.11	Finding the *z*-score corresponding to an area in the middle

Use Table A.2 to find the *z*-scores that bound the middle 95% of the area under the standard normal curve.

Solution

Step 1: Sketch a normal curve and shade in the given area. Label the *z*-score on the left z_1 and the *z*-score on the right z_2.

Step 2: Find the area to the left of z_1. Since the area in the middle is 0.95, the area in the two tails combined is 0.05. Half of that area, or 0.025, is to the left of z_1. We conclude that z_1 is the *z*-score that has an area of 0.025 to its left.

Step 3: In Table A.2, an area of 0.025 corresponds to a *z*-score of -1.96. Therefore $z_1 = -1.96$.

Step 4: To find z_2, note that the area to the left of z_2 is 0.9750. We can use Table A.2 to determine that $z_2 = 1.96$. Alternatively, note that z_1 and z_2 are equidistant from the mode. Since the normal curve is symmetric, z_1 must be the negative of z_2. Therefore $z_2 = 1.96$. See Figure 7.13.

Area = 0.025   Area = 0.025
Area = 0.95
-1.96 1.96

Figure 7.13

USING TECHNOLOGY

TI-84 PLUS

Finding areas under the standard normal curve

The **normalcdf** command is used to calculate area under a normal curve.

Step 1. Press **2nd**, then **VARS** to access the **DISTR** menu. Select **2:normalcdf** (Figure A).

Step 2. Enter the left endpoint, comma, the right endpoint, comma, the mean (**0** for the standard normal), comma, and the standard deviation (**1** for the standard normal).
- When finding the area to the right of a given value, use **1E99** as the right endpoint.
- When finding the area to the left of a given value, use **−1E99** as the left endpoint.

Step 3. Press **ENTER**.

Using the TI-84 PLUS Stat Wizards (see Appendix B for more information)

Step 1. Press **2nd**, then **VARS** to access the **DISTR** menu. Select **2:normalcdf** (Figure A).

Step 2. Enter the left endpoint in the **lower** field, the right endpoint in the **upper** field, the mean in the μ field (**0** for the standard normal), and the standard deviation in the σ field (**1** for the standard normal).
- When finding the area to the right of a given value, use **1E99** as the right endpoint.
- When finding the area to the left of a given value, use **−1E99** as the left endpoint.

Step 3. Select **Paste** and press **ENTER** to paste the command to the home screen. Press **ENTER** again to run the command.

Figure B illustrates finding the area to the left of $z = 1.26$ (Example 7.2).

Figure C illustrates finding the area to the right of $z = -0.58$ (Example 7.3).

Figure D illustrates finding the area between $z = -1.45$ and $z = 0.42$. (Example 7.4).

Note: The quantity 1E99 in the TI-84 PLUS calculator represents a very large number, specifically a 1 followed by 99 zeros.

```
DISTR DRAW
1:normalpdf(
2:normalcdf(
3:invNorm(
4:invT(
5:tpdf(
6:tcdf(
7↓X²pdf(
```
Figure A

```
normalcdf(-1E99,
1.26,0,1)
         .8961652533
```
Figure B

```
normalcdf(-0.58,
1E99,0,1)
         .7190427366
```
Figure C

```
normalcdf(-1.45,
0.42,0,1)
         .5892279372
```
Figure D

Finding a z-score corresponding to a given area

The **invNorm** command is used to calculate the *z*-score corresponding to an *area to the left*.

Step 1. Press **2nd**, then **VARS** to access the **DISTR** menu. Select **3:invNorm** (Figure E).

Step 2. Enter the area to the left of the desired *z*-score, comma, the mean (**0** for the standard normal), comma, and the standard deviation (**1** for the standard normal).

Step 3. Press **ENTER**.

Using the TI-84 PLUS Stat Wizards (see Appendix B for more information)

Step 1. Press **2nd**, then **VARS** to access the **DISTR** menu. Select **3:invNorm** (Figure E).

Step 2. Enter the area to the left of the desired *z*-score in the **area** field, the mean in the μ field (**0** for the standard normal), and the standard deviation in the σ field (**1** for the standard normal).

Step 3. Select **Paste** and press **ENTER** to paste the command to the home screen. Press **ENTER** again to run the command.

Figure F illustrates finding the *z*-score that has an area of 0.26 to its left (Example 7.8).

Figure E

Figure F

MINITAB

Finding areas under the standard normal curve

The following procedure computes the area to the *left* of a given value.

Step 1. Click **Calc**, then **Probability Distributions**, then **Normal**.

Step 2. Select the **Cumulative probability** option.

Step 3. Enter the value for the mean in the **Mean** field (**0** for the standard normal) and the value for the standard deviation in the **Standard deviation** field (**1** for the standard normal).

Step 4. To compute the area to the left of a given *x*, enter the value for *x* in the **Input constant** field.

Step 5. Click **OK**.

Figures G and H illustrate finding the area to the left of *z* = 1.26 (Example 7.2).

Figure G

Cumulative Distribution Function

```
Normal with mean = 0 and standard deviation = 1

  x    P( X <= x )
1.26    0.896165
```

Figure H

Finding a z-score corresponding to a given area

The following procedure is used to calculate the *z*-score corresponding to an *area to the left*.

Step 1. Click **Calc**, then **Probability Distributions**, then **Normal**.

Step 2. Select the **Inverse Cumulative Probability** option.

Step 3. Enter the value for the mean in the **Mean** field (**0** for the standard normal) and the value for the standard deviation in the **Standard deviation** field (**1** for the standard normal).

Step 4. Enter the area to the left of the desired *z*-score in the **Input constant** field and click **OK**.

Figure I illustrates finding the *z*-score that has an area of 0.26 to its left (Example 7.8).

Inverse Cumulative Distribution Function

```
Normal with mean = 0 and standard deviation = 1

P( X <= x )         x
     0.26    -0.643345
```

Figure I

EXCEL

Finding areas under the standard normal curve

The following procedure computes the area to the *left* of a given value.

Step 1. In an empty cell, select the **Insert Function** icon and highlight **Statistical** in the category field.

Step 2. Click on the **NORM.S.DIST** function and press **OK**.

Step 3. To compute the area to the left of a given *x*, enter the value of *x* in the **X** field.

Step 4. Enter **TRUE** in the **Cumulative** field and click **OK**.

Figure J illustrates finding the area to the left of $z = 1.26$ (Example 7.2).

=NORM.S.DIST(1.26,TRUE)

0.896165319

Figure J

Finding a z-score corresponding to a given area

The following procedure is used to calculate the z-score corresponding to an *area to the left*.

Step 1. In an empty cell, select the **Insert Function** icon and highlight **Statistical** in the category field.

Step 2. Click on the **NORM.S.INV** function and press **OK**.

Step 3. Enter the area to the left of the desired z-score in the **Probability** field.

Step 4. Click **OK**.

Figure K illustrates finding the z-score that has an area of 0.26 to its left (Example 7.8).

=NORM.S.INV(0.26)

-0.643345405

Figure K

SECTION 7.1   Exercises

Exercises 1–8 are the Check Your Understanding exercises located within the section.

Understanding the Concepts

In Exercises 9–14, fill in each blank with the appropriate word or phrase.

9. The proportion of a population that is contained within an interval corresponds to an area under the probability _____ curve.

10. If X is a continuous random variable, then $P(X = a) =$ _____ for any number a.

11. The area under the entire probability density curve is equal to _____ .

12. The mean, median, and mode of a normal distribution are _____ to each other.

13. A normal distribution with mean 0 and standard deviation 1 is called the _____ normal distribution.

14. Points on the horizontal axis to the left of the mode have _____ z-scores.

In Exercises 15–18, determine whether the statement is true or false. If the statement is false, rewrite it as a true statement.

15. The probability that a randomly selected value of a continuous random variable lies between a and b is given by the area under the probability density curve between a and b.

16. A normal curve is symmetric around its mode.

17. A normal curve is wide and flat when the standard deviation is small.

18. The area under the normal curve to the left of the mode is less than 0.5.

Practicing the Skills

19. The following figure is a probability density curve that represents the lifetime, in months, of a certain type of laptop battery.

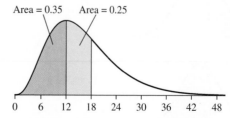

a. Find the proportion of batteries with lifetimes between 12 and 18 months.

b. Find the proportion of batteries with lifetimes less than 18 months.

c. What is the probability that a randomly chosen battery lasts more than 18 months?

20. The following figure is a probability density curve that represents the grade point averages (GPA) of the graduating seniors at a large university.

a. Find the proportion of seniors whose GPA is between 3.0 and 3.5.

b. What is the probability that a randomly chosen senior will have a GPA greater than 3.5?

21. Find each of the shaded areas under the standard normal curve.

a.

1.25

b.

0.36

c.

−0.71   1.62

d.

−0.28 0.94

22. Find each of the shaded areas under the standard normal curve.

a.

−0.38

b.

−0.48

c.

−2.48   −0.58

d.

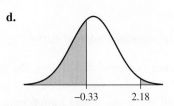

−0.33   2.18

23. Find the area under the standard normal curve to the left of
  a. $z = 0.74$
  b. $z = -2.16$
  c. $z = 1.02$
  d. $z = -0.15$

24. Find the area under the standard normal curve to the left of
  a. $z = 2.56$
  b. $z = 0.53$
  c. $z = -0.94$
  d. $z = -1.30$

25. Find the area under the standard normal curve to the right of
  a. $z = -1.55$
  b. $z = 0.32$
  c. $z = 3.20$
  d. $z = -2.39$

26. Find the area under the standard normal curve to the right of
  a. $z = 0.47$
  b. $z = -2.91$
  c. $z = 2.04$
  d. $z = 1.09$

27. Find the area under the standard normal curve that lies between
  a. $z = -0.75$ and $z = 1.70$
  b. $z = -2.30$ and $z = 1.08$
  c. $z = -3.27$ and $z = -1.44$
  d. $z = 1.26$ and $z = 2.32$

28. Find the area under the standard normal curve that lies between
  a. $z = -1.28$ and $z = 1.36$
  b. $z = -0.82$ and $z = -0.42$
  c. $z = 1.58$ and $z = 2.06$
  d. $z = -2.19$ and $z = 0.07$

29. Find the area under the standard normal curve that lies outside the interval between
  a. $z = -0.38$ and $z = 1.02$
  b. $z = -1.42$ and $z = 1.78$
  c. $z = 0.01$ and $z = 2.67$
  d. $z = -2.45$ and $z = -0.34$

30. Find the area under the standard normal curve that lies outside the interval between
  a. $z = -1.11$ and $z = 3.21$
  b. $z = -1.93$ and $z = 0.59$
  c. $z = 0.46$ and $z = 1.75$
  d. $z = -2.73$ and $z = -1.39$

31. Find the z-score for which the area to its left is 0.54.

32. Find the z-score for which the area to its left is 0.13.

33. Find the z-score for which the area to its left is 0.93.

34. Find the z-score for which the area to its left is 0.25.

35. Find the z-score for which the area to its right is 0.84.

36. Find the z-score for which the area to its right is 0.14.

37. Find the z-score for which the area to its right is 0.35.

38. Find the z-score for which the area to its right is 0.92.

39. Find the z-scores that bound the middle 50% of the area under the standard normal curve.

40. Find the z-scores that bound the middle 70% of the area under the standard normal curve.

41. Find the z-scores that bound the middle 80% of the area under the standard normal curve.

42. Find the z-scores that bound the middle 98% of the area under the standard normal curve.

Working with the Concepts

43. Symmetry: The area under the standard normal curve to the left of $z = -1.75$ is 0.0401. What is the area to the right of $z = 1.75$?

44. Symmetry: The area under the standard normal curve to the right of $z = -0.51$ is 0.6950. What is the area to the left of $z = 0.51$?

45. Symmetry: The area under the standard normal curve between $z = -1.93$ and $z = 0.59$ is 0.6956. What is the area between $z = -0.59$ and $z = 1.93$?

46. Symmetry: The area under the standard normal curve between $z = 1.32$ and $z = 1.82$ is 0.0590. What is the area between $z = -1.82$ and $z = -1.32$?

Extending the Concepts

47. No table, no technology: Let a be the number such that the area to the right of $z = a$ is 0.3. Without using a table or technology, find the area to the left of $z = -a$.

48. No table, no technology: Let a be the number such that the area to the right of $z = a$ is 0.21. Without using a table or technology, find the area between $z = -a$ and $z = a$.

Answers to Check Your Understanding Exercises for Section 7.1

1. a. 0.63	**b.** 0.23	**c.** 0.86	**d.** 0.14	**5.** −0.13
2. 0.5987				**6.** 0.33
3. 0.0104				**7.** 0.64
4. 0.8491				**8.** 1.34

SECTION 7.2 | Applications of the Normal Distribution

Objectives

1. Convert values from a normal distribution to z-scores
2. Find areas under a normal curve
3. Find the value from a normal distribution corresponding to a given proportion

In Section 7.1, we found areas under a standard normal curve, which has mean 0 and standard deviation 1. In this section, we will learn to find areas under normal curves with any mean and standard deviation.

Objective 1 Convert values from a normal distribution to z-scores

Converting Normal Values to z-Scores

Let x be a value from a normal distribution with mean μ and standard deviation σ. We can convert x to a z-score by using a method known as **standardization**. To standardize a value, subtract the mean and divide by the standard deviation. This produces the z-score.

Recall: We first described the method for finding the z-score in Section 3.3.

DEFINITION

Let x be a value from a normal distribution with mean μ and standard deviation σ. The z-**score** of x is

$$z = \frac{x - \mu}{\sigma}$$

The z-score satisfies the following properties.

Properties of the z-Score

1. The z-score follows a standard normal distribution.
2. Values below the mean have negative z-scores, and values above the mean have positive z-scores.
3. The z-score tells how many standard deviations the original value is above or below the mean.

Because the z-score follows a standard normal distribution, we can use the methods of Section 7.1 to find areas under any normal curve, by standardizing to convert the original values to z-scores.

Rounding Off z-Scores

The z-scores in Table A.2 are expressed to two decimal places. For this reason, when converting normal values to z-scores, we will round off the z-scores to two decimal places.

EXAMPLE 7.12   Finding and interpreting a z-score

Heights in a certain population of women follow a normal distribution with mean $\mu = 64$ inches and standard deviation $\sigma = 3$ inches.

   a. A randomly selected woman has a height of $x = 67$ inches. Find and interpret the z-score of this value.
   b. Another randomly selected woman has a height of $x = 63$ inches. Find and interpret the z-score of this value.

Solution

Explain It Again

Converting x-values to z-scores:
After we convert an x-value to a z-score, we use the standard normal curve. This allows us to find areas under the normal curve by using Table A.2.

   a. The z-score for $x = 67$ is
$$z = \frac{67 - \mu}{\sigma} = \frac{67 - 64}{3} = 1.00$$
   We interpret this by saying that a height of 67 inches is 1 standard deviation above the mean height of 64 inches.
   b. The z-score for $x = 63$ is
$$z = \frac{63 - \mu}{\sigma} = \frac{63 - 64}{3} = -0.33$$
   We interpret this by saying that a height of 63 inches is 0.33 standard deviation below the mean height of 64 inches.

Figure 7.14 illustrates the results of Example 7.12. Figure 7.14(a) is the normal curve that represents the heights of the population of women. It has a mean of 64. The heights of the two women are indicated at 63 and 67. Figure 7.14(b) is the standard normal curve. The mean is 0, and the heights are represented by their z-scores of -0.33 and 1.00.

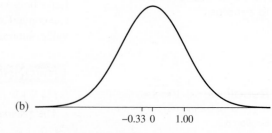

Figure 7.14   (a) This is the normal curve with mean 64 and standard deviation 3. It represents the population of heights of women. The heights of 63 and 67 are shown on the x-axis. (b) This is the standard normal curve. It also represents the population of heights of women, by using the z-scores instead of the actual heights. A height of 63 inches is represented by a z-score of -0.33, and a height of 67 inches is represented by a z-score of 1.00.

1. A normal distribution has mean $\mu = 15$ and standard deviation $\sigma = 4$. Find and interpret the z-score for $x = 11$.

2. A normal distribution has mean $\mu = 60$ and standard deviation $\sigma = 20$. Find and interpret the z-score for $x = 75$.

3. Compact fluorescent bulbs are more energy efficient than incandescent bulbs, but they take longer to reach full brightness. The time that it takes for a compact fluorescent bulb to reach full brightness is normally distributed with mean 29.8 seconds and standard deviation 4.5 seconds. A randomly selected bulb takes 28 seconds to reach full brightness. Find and interpret the z-score for $x = 28$.

Answers are on page 310.

Objective 2 Find areas under a normal curve

Finding Areas Under a Normal Curve

In Section 7.1, we used z-scores to compute areas under the standard normal curve. By standardizing, we can use z-scores to compute areas under a normal curve with any mean and standard deviation. An area under a normal curve over an interval can be interpreted in two ways: It represents the proportion of the population that is contained within the interval, and it also represents the probability that a randomly selected individual will have a value within the interval.

EXAMPLE 7.13

Finding an area under a normal curve

A study reported that the length of pregnancy from conception to birth is approximately normally distributed with mean $\mu = 272$ days and standard deviation $\sigma = 9$ days. What proportion of pregnancies last longer than 280 days?

Source: *Singapore Medical Journal* 35:1044–1048

Explain It Again

Probabilities and proportions: The probability that a randomly sampled value falls in a given interval is equal to the proportion of the population that is contained in the interval. So the area under a normal curve represents both probabilities and proportions.

Solution

The proportion of pregnancies lasting longer than 280 days is equal to the area under the normal curve corresponding to values of x greater than 280. We find this area as follows.

Step 1: Find the z-score for $x = 280$.

$$z = \frac{x - \mu}{\sigma} = \frac{280 - 272}{9} = 0.89$$

Step 2: Sketch a normal curve, label the mean, x-value, and z-score, and shade in the area to be found. See Figure 7.15.

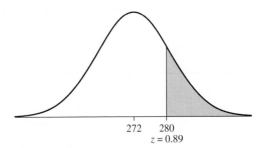

272   280
$z = 0.89$

Figure 7.15

Step 3: Find the area to the right of $z = 0.89$. Using Table A.2, we find the area to the *left* of $z = 0.89$ to be 0.8133. The area to the right is therefore $1 - 0.8133 = 0.1867$. We conclude that the proportion of pregnancies that last longer than 280 days is 0.1867.

| EXAMPLE 7.14 | Finding an area under a normal curve by using technology |

In Example 7.13, we used Table A.2 to compute the proportion of pregnancies that last longer than 280 days. Find this proportion by using technology.

Solution

We present output from the TI-84 Plus calculator. We use the **normalcdf** command. We enter the left endpoint of the interval (280). Since there is no right endpoint, we enter **1E99**, which represents the very large number that is written with a 1 followed by 99 zeros. Then we enter the mean (272) and the standard deviation (9). Step-by-step instructions are given in the Using Technology section on page 305.

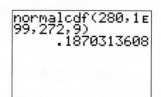

Explain It Again

Technology and tables can give slightly different answers: Answers obtained with technology sometimes differ slightly from those obtained by using tables, because the technology is more precise. The differences aren't large enough to matter.

In Example 7.13, we used Table A.2 and found the proportion of pregnancies that last longer than 280 days to be 0.1867. In Example 7.14, the TI-84 Plus calculator found the proportion to be 0.1870. Answers found with technology often differ somewhat from those obtained by using a table. The differences aren't large enough to matter. Whenever the answer obtained from technology differs from the answer obtained by using the table, we will present both answers.

| EXAMPLE 7.15 | Finding an area under a normal curve between two values |

The length of a pregnancy from conception to birth is approximately normally distributed with mean $\mu = 272$ days and standard deviation $\sigma = 9$ days. A pregnancy is considered full-term if it lasts between 252 days and 298 days. What proportion of pregnancies are full-term?

Solution

Step 1: Find the z-scores for $x = 252$ and $x = 298$.

For $x = 252$:
$$z = \frac{252 - 272}{9} = -2.22$$

For $x = 298$:
$$z = \frac{298 - 272}{9} = 2.89$$

Step 2: Sketch a normal curve, label the mean, the x-values, and the z-scores, and shade in the area to be found. See Figure 7.16.

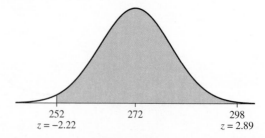

Figure 7.16

Step 3: Using Table A.2, we find that the area to the left of $z = 2.89$ is 0.9981 and the area to the left of $z = -2.22$ is 0.0132. The area between $z = -2.22$ and $z = 2.89$ is therefore $0.9981 - 0.0132 = 0.9849$.

We conclude that the proportion of pregnancies that are full-term is 0.9849.

Check Your Understanding

4. A normal population has mean $\mu = 3$ and standard deviation $\sigma = 1$. Find the proportion of the population that is less than 1.

5. A normal population has mean $\mu = 40$ and standard deviation $\sigma = 10$. Find the probability that a randomly sampled value is greater than 53.

6. A normal population has mean $\mu = 7$ and standard deviation $\sigma = 5$. Find the proportion of the population that is between -2 and 10.

Answers are on page 310.

Objective 3 Find the value from a normal distribution corresponding to a given proportion

Finding the Value from a Normal Distribution Corresponding to a Given Proportion

Sometimes we want to find the value from a normal distribution that has a given proportion of the population above or below it. The method for doing this is the reverse of the method for finding a proportion for a given value. In particular, we need to find the value from the distribution that has a given z-score.

Recall that the z-score tells how many standard deviations a value is above or below the mean. The value of x that corresponds to a given z-score is given by

$$x = \mu + z\sigma$$

EXAMPLE 7.16   Finding the value from a normal distribution with a given z-score

Heights in a group of men are normally distributed with mean $\mu = 69$ inches and standard deviation $\sigma = 3$ inches.

a. Find the height whose z-score is 1. Interpret the result.

b. Find the height whose z-score is -2.0. Interpret the result.

c. Find the height whose z-score is 0.6. Interpret the result.

Solution

Explain It Again

$x = \mu + z\sigma$: The z-score tells how many standard deviations x is above or below the mean. Therefore, the value of x that corresponds to a given z-score is equal to the mean (μ) plus z times the standard deviation (σ).

a. We want the height that is equal to the mean plus one standard deviation. Therefore, $x = \mu + z\sigma = 69 + (1)(3) = 72$. We interpret this by saying that a man 72 inches tall has a height one standard deviation above the mean.

b. We want the height that is equal to the mean minus two standard deviations. Therefore, $x = \mu + z\sigma = 69 + (-2)(3) = 63$. We interpret this by saying that a man 63 inches tall has a height two standard deviations below the mean.

c. We want the height that is equal to the mean plus 0.6 standard deviation. Therefore, $x = \mu + z\sigma = 69 + (0.6)(3) = 70.8$. We interpret this by saying that a man 70.8 inches tall has a height 0.6 standard deviation above the mean.

To find the value from a normal distribution that has a given proportion above or below it, we can use either Table A.2 or technology. Following are the steps to find the value that has a given proportion above or below it by using Table A.2.

Finding a Normal Value That Has a Given Proportion Above or Below It by Using Table A.2

Step 1: Sketch a normal curve, label the mean, label the value x to be found, and shade in and label the given area.

Step 2: If the given area is on the right, subtract it from 1 to get the area on the left.

Step 3: Look in the body of Table A.2 to find the area closest to the given area. Find the z-score corresponding to that area.

Step 4: Obtain the value from the normal distribution by computing $x = \mu + z\sigma$.

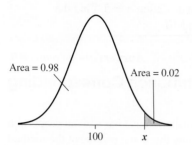

Figure 7.17

EXAMPLE 7.17   Finding a normal value corresponding to an area

Mensa is an organization whose membership is limited to people whose IQ is in the top 2% of the population. Assume that scores on an IQ test are normally distributed with mean $\mu = 100$ and standard deviation $\sigma = 15$. What is the minimum score needed to qualify for membership in Mensa?

Solution

Step 1: Figure 7.17 presents a sketch of the normal curve, showing the value x separating the upper 2% from the lower 98%.

Step 2: The area 0.02 is on the right, so we subtract from 1 and work with the area 0.98 on the left.

Step 3: The area closest to 0.98 in Table A.2 is 0.9798, which corresponds to a z-score of 2.05.

Step 4: The IQ score that separates the upper 2% from the lower 98% is

$$x = \mu + z\sigma = 100 + (2.05)(15) = 130.75$$

Since IQ scores are generally whole numbers, we will round this to $x = 131$.

Finding the normal value corresponding to a given area by using technology

To find the percentile of a normal distribution with technology, follow Steps 1 and 2 of the method for using the table. What is done after that depends on the technology being used. Example 7.18 illustrates the use of the TI-84 Plus calculator.

EXAMPLE 7.18   Finding the normal value corresponding to an area by using technology

IQ scores have a mean of 100 and a standard deviation of 15. Use technology to find the 90th percentile of IQ scores; in other words, find the IQ score that separates the upper 10% from the lower 90%.

Solution

Recall: The *p*th percentile of a population is the value that separates the lowest *p*% of the population from the highest (100 − *p*)%.

Step 1: Figure 7.18 presents a sketch of the normal curve, showing the value x separating the upper 10% from the lower 90%.

Step 2: We work with the area 0.90 on the left.

Step 3: For the TI-84 Plus calculator, use the **invNorm** command with area 0.90, mean 100, and standard deviation 15. Step-by-step instructions are given in the Using Technology section on page 305.

Figure 7.19 presents the results from the TI-84 Plus calculator. The IQ score corresponding to the top 10% is 119.

Figure 7.18

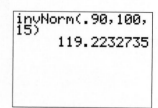

Figure 7.19

Check Your Understanding

7. A population has mean $\mu = 6.9$ and standard deviation $\sigma = 2.6$. Find the value that has 80% of the population below it (in other words, the 80th percentile).

8. A population has mean $\mu = 53$ and standard deviation $\sigma = 34$. Find the value that has 35% of the population above it.

Answers are on page 310.

USING TECHNOLOGY

TI-84 PLUS

Finding areas under a normal curve

The **normalcdf** command is used to calculate area under a normal curve.

Step 1. Press **2nd**, then **VARS** to access the **DISTR** menu. Select **2:normalcdf** (Figure A).

Step 2. Enter the left endpoint, comma, the right endpoint, comma, the mean, comma, and the standard deviation.
 • When finding the area to the right of a given value, use **1E99** as the right endpoint.
 • When finding the area to the left of a given value, use **−1E99** as the left endpoint.

Step 3. Press **ENTER**.

Figure A

Using the TI-84 PLUS Stat Wizards (see Appendix B for more information)

Step 1. Press **2nd**, then **VARS** to access the **DISTR** menu. Select **2:normalcdf** (Figure A).

Step 2. Enter the left endpoint in the **lower** field, the right endpoint in the **upper field**, the mean in the μ field, and the standard deviation in the σ field.
 • When finding the area to the right of a given value, use **1E99** as the right endpoint.
 • When finding the area to the left of a given value, use **−1E99** as the left endpoint.

Step 3. Select **Paste** and press **ENTER** to paste the command to the home screen. Press **ENTER** again to run the command.

Figure B illustrates finding the area to the right of $x = 280$ with $\mu = 272$ and $\sigma = 9$ (Example 7.13).

```
normalcdf(280,1E
99,272,9)
          .1870313608
```

Figure B

Finding a normal value corresponding to a given area

The **invNorm** command is used to calculate the z-score corresponding to an *area to the left*.

Step 1. Press **2nd**, then **VARS** to access the **DISTR** menu. Select **3:invNorm** (Figure C).

Step 2. Enter the area to the left of the desired normal value, comma, the mean, comma, and the standard deviation.

Step 3. Press **ENTER**.

Figure C

Using the TI-84 PLUS Stat Wizards (see Appendix B for more information)

Step 1. Press **2nd**, then **VARS** to access the **DISTR** menu. Select **3:invNorm** (Figure C).

Step 2. Enter the area to the left of the desired z-score in the **area** field, the mean in the μ field, and the standard deviation in the σ field.

Step 3. Select **Paste** and press **ENTER** to paste the command to the home screen. Press **ENTER** again to run the command.

Figure D illustrates finding the normal value that has an area of 0.98 to its left, where $\mu = 100$ and $\sigma = 15$ (Example 7.17).

```
invNorm(0.98,100
,15)
          130.8062337
```

Figure D

MINITAB

Finding areas under a normal curve

The following procedure computes the area to the *left* of a given value.

Step 1. Click **Calc**, then **Probability Distributions**, then **Normal**.

Step 2. Select the **Cumulative probability** option.

Step 3. Enter the value for the mean in the **Mean** field and the value for the standard deviation in the **Standard deviation** field.

Step 4. To compute the area to the left of a given x, enter the value for x in the **Input constant** field.

Step 5. Click **OK**.

Cumulative Distribution Function

```
Normal with mean = 272 and standard deviation = 9

   x   P( X <= x )
 280     0.812969
```

Figure E

Figure E illustrates finding the area to the *left* of $x = 280$ with $\mu = 272$ and $\sigma = 9$. To find the area to the right of $x = 280$, subtract this result from 1 (Example 7.13).

Finding a normal value corresponding to a given area

The following procedure is used to calculate a normal value corresponding to an *area to the left*.

Step 1. Click **Calc**, then **Probability Distributions**, then **Normal**.

Step 2. Select the **Inverse Cumulative Probability** option.

Step 3. Enter the value for the mean in the **Mean** field and the value for the standard deviation in the **Standard deviation** field.

Step 4. Enter the area to the left of the desired normal value and click **OK**.

Inverse Cumulative Distribution Function

```
Normal with mean = 100 and standard deviation = 15

P( X <= x )        x
      0.98   130.806
```

Figure F

Figure F illustrates finding the normal value that has an area of 0.98 to its left, where $\mu = 100$ and $\sigma = 15$ (Example 7.17).

EXCEL

Finding areas under a normal curve

The following procedure computes the area to the *left* of a given value.

Step 1. In an empty cell, select the **Insert Function** icon and highlight **Statistical** in the category field.

Step 2. Click on the **NORM.DIST** function and press **OK**.

Step 3. To compute the area to the left of a given x, enter the value of x in the **X** field.

Step 4. Enter the value for the mean in the **Mean** field and the value for the standard deviation in the **Standard deviation** field.

Step 5. Enter **TRUE** in the **Cumulative** field and click **OK**.

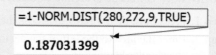

```
=1-NORM.DIST(280,272,9,TRUE)
```

0.187031399

Figure G

Figure G illustrates finding the area to the right of $x = 280$ with $\mu = 272$ and $\sigma = 9$ by subtracting the area on the left from 1 (Example 7.13).

Finding a normal value corresponding to a given area

The following procedure is used to calculate the normal value corresponding to an *area to the left*.

Step 1. In an empty cell, select the **Insert Function** icon and highlight **Statistical** in the category field.

Step 2. Click on the **NORM.INV** function and press **OK**.

Step 3. Enter the area to the left of the desired normal value in the **Probability** field.

Step 4. Enter the value for the mean in the **Mean** field and the value for the standard deviation in the **Standard deviation** field.

Step 5. Click **OK**.

Figure H

Figure H illustrates finding the normal value that has an area of 0.98 to its left, where $\mu = 100$ and $\sigma = 15$ (Example 7.17).

SECTION 7.2   Exercises

Exercises 1–8 are the Check Your Understanding exercises located within the section.

Understanding the Concepts

In Exercises 9–10, fill in each blank with the appropriate word or phrase.

9. The process of converting a value x from a normal distribution to a z-score is known as _____.

10. A value that is two standard deviations below the mean will have a z-score of _____.

In Exercises 11–16, determine whether the statement is true or false. If the statement is false, rewrite it as a true statement.

11. z-scores follow a standard normal distribution.

12. A z-score indicates how many standard deviations a value is above or below the mean.

13. If a normal population has a mean of μ and a standard deviation of σ, then the area to the left of μ is less than 0.5.

14. If a normal population has a mean of μ and a standard deviation of σ, then the area to the right of $\mu + \sigma$ is less than 0.5.

15. If a normal population has a mean of μ and a standard deviation of σ, then $P(X = \mu) = 1$.

16. If a normal population has a mean of μ and a standard deviation of σ, then $P(X = \mu) = 0$.

Practicing the Skills

17. A normal population has mean $\mu = 20$ and standard deviation $\sigma = 4$.
   a. What proportion of the population is less than 18?
   b. What is the probability that a randomly chosen value will be greater than 25?

18. A normal population has mean $\mu = 9$ and standard deviation $\sigma = 6$.
   a. What proportion of the population is less than 20?
   b. What is the probability that a randomly chosen value will be greater than 5?

19. A normal population has mean $\mu = 25$ and standard deviation $\sigma = 11$.
   a. What proportion of the population is greater than 34?
   b. What is the probability that a randomly chosen value will be less than 10?

20. A normal population has mean $\mu = 61$ and standard deviation $\sigma = 16$.
   a. What proportion of the population is greater than 100?
   b. What is the probability that a randomly chosen value will be less than 80?

21. A normal population has mean $\mu = 47$ and standard deviation $\sigma = 3$.
   a. What proportion of the population is between 40 and 50?
   b. What is the probability that a randomly chosen value will be between 50 and 55?

22. A normal population has mean $\mu = 35$ and standard deviation $\sigma = 8$.
   a. What proportion of the population is between 20 and 30?
   b. What is the probability that a randomly chosen value will be between 30 and 40?

23. A normal population has mean $\mu = 12$ and standard deviation $\sigma = 3$. What is the 40th percentile of the population?

24. A normal population has mean $\mu = 56$ and standard deviation $\sigma = 8$. What is the 85th percentile of the population?

25. A normal population has mean $\mu = 46$ and standard deviation $\sigma = 9$. What is the 19th percentile of the population?

26. A normal population has mean $\mu = 71$ and standard deviation $\sigma = 33$. What is the 91st percentile of the population?

Working with the Concepts

27. **Check your blood pressure:** In a recent study, the Centers for Disease Control and Prevention reported that diastolic blood pressures of adult women in the United States are approximately normally distributed with mean 80.5 and standard deviation 9.9.

a. What proportion of women have blood pressures lower than 70?

b. What proportion of women have blood pressures between 75 and 90?

c. A diastolic blood pressure greater than 90 is classified as hypertension (high blood pressure). What proportion of women have hypertension?

d. Is it unusual for a woman to have a blood pressure lower than 65?

28. Baby weights: According to a recent National Health Statistics Reports, the weight of male babies less than 2 months old in the United States is normally distributed with mean 11.5 pounds and standard deviation 2.7 pounds.

a. What proportion of babies weigh more than 13 pounds?

b. What proportion of babies weigh less than 15 pounds?

c. What proportion of babies weigh between 10 and 14 pounds?

d. Is it unusual for a baby to weigh more than 17 pounds?

29. Check your blood pressure: The Centers for Disease Control and Prevention reported that diastolic blood pressures of adult women in the United States are approximately normally distributed with mean 80.5 and standard deviation 9.9.

a. Find the 30th percentile of the blood pressures.

b. Find the 67th percentile of the blood pressures.

c. Find the third quartile of the blood pressures.

30. Baby weights: The weight of male babies less than 2 months old in the United States is normally distributed with mean 11.5 pounds and standard deviation 2.7 pounds.

a. Find the 81st percentile of the baby weights.

b. Find the 10th percentile of the baby weights.

c. Find the first quartile of the baby weights.

31. Fish story: According to a report by the U.S. Fish and Wildlife Service, the mean length of six-year-old rainbow trout in the Arolik River in Alaska is 481 millimeters with a standard deviation of 41 millimeters. Assume these lengths are normally distributed.

a. What proportion of six-year-old rainbow trout are less than 450 millimeters long?

b. What proportion of six-year-old rainbow trout are between 400 and 500 millimeters long?

c. Is it unusual for a six-year-old rainbow trout to be less than 400 millimeters long?

32. Big chickens: According to thepoultrysite.com, the weights of broilers (commercially raised chickens) are approximately normally distributed with mean 1387 grams and standard deviation 161 grams.

a. What proportion of broilers weigh between 1100 and 1200 grams?

b. What is the probability that a randomly selected broiler weighs more than 1500 grams?

c. Is it unusual for a broiler to weigh more than 1550 grams?

33. Fish story: The U.S. Fish and Wildlife Service reported that the mean length of six-year-old rainbow trout in the Arolik River in Alaska is 481 millimeters with a standard deviation of 41 millimeters. Assume these lengths are normally distributed.

a. Find the 58th percentile of the lengths.

b. Find the 76th percentile of the lengths.

c. Find the first quartile of the lengths.

d. A size limit is to be put on trout that are caught. What should the size limit be so that 15% of six-year-old trout have lengths shorter than the limit?

34. Big chickens: A report on thepoultrysite.com stated that the weights of broilers (commercially raised chickens) are approximately normally distributed with mean 1387 grams and standard deviation 161 grams.

a. Find the 22nd percentile of the weights.

b. Find the 93rd percentile of the weights.

c. Find the first quartile of the weights.

d. A chicken farmer wants to provide a money-back guarantee that his broilers will weigh at least a certain amount. What weight should he guarantee so that he will have to give his customers' money back only 1% of the time?

35. Radon: Radon is a naturally occurring radioactive substance that is found in the ground underneath many homes. Radon detectors are often placed in homes to determine whether radon levels are high enough to be dangerous. A radon level less than 4.0 picocuries is considered safe. Because levels fluctuate randomly, the levels measured by detectors are not exactly correct, but are instead normally distributed. It is known from physical theory that when the true level is 4.1 picocuries, the measurement made by a detector over a one-hour period will be normally distributed with mean 4.1 picocuries and standard deviation 0.2 picocurie.

a. If the true level is 4.1, what is the probability that a one-hour measurement will be less than 4.0?

b. If the true level is 4.1, would it be unusual for a one-hour measurement to indicate that the level is safe?

c. If a measurement is made for 24 hours, the mean will still be 4.1 picocuries, but the standard deviation will be only 0.04 picocurie. What is the probability that a 24-hour measurement will be below 4.0?

d. If the true level is 4.1, would it be unusual for a 24-hour measurement to indicate that the level is safe?

36. Electric bills: According to the U.S. Energy Information Administration, the mean monthly household electric bill in the United States in 2011 was $110.14. Assume the amounts are normally distributed with standard deviation $20.00.

a. What proportion of bills are greater than $130?

b. What proportion of bills are between $85 and $140?

c. What is the probability that a randomly selected household had a monthly bill less than $120?

37. Radon: Assume that radon measurements are normally distributed with mean 4.1 picocuries and standard deviation of 0.2.

a. Find the 35th percentile of the measurements.

b. Find the 92nd percentile of the measurements.

c. Find the median of the measurements.

38. Electric bills: The U.S. Energy Information Agency reported that the mean monthly household electric bill in the United States in 2011 was $110.14. Assume the amounts are normally distributed with standard deviation $20.00.

a. Find the 7th percentile of the bill amounts.

b. Find the 62nd percentile of the bill amounts.

c. Find the median of the bill amounts.

39. Tire lifetimes: The lifetime of a certain type of automobile tire (in thousands of miles) is normally distributed with mean $\mu = 40$ and standard deviation $\sigma = 5$.
  a. What is the probability that a randomly chosen tire has a lifetime greater than 48 thousand miles?
  b. What proportion of tires have lifetimes between 38 and 43 thousand miles?
  c. What proportion of tires have lifetimes less than 46 thousand miles?

40. Tree heights: Cherry trees in a certain orchard have heights that are normally distributed with mean $\mu = 112$ inches and standard deviation $\sigma = 14$ inches.
  a. What proportion of trees are more than 120 inches tall?
  b. What proportion of trees are less than 100 inches tall?
  c. What is the probability that a randomly chosen tree is between 90 and 100 inches tall?

41. Tire lifetimes: The lifetime of a certain type of automobile tire (in thousands of miles) is normally distributed with mean $\mu = 40$ and standard deviation $\sigma = 5$.
  a. Find the 15th percentile of the tire lifetimes.
  b. Find the 68th percentile of the tire lifetimes.
  c. Find the first quartile of the tire lifetimes.
  d. The tire company wants to guarantee that its tires will last at least a certain number of miles. What number of miles (in thousands) should the company guarantee so that only 2% of the tires violate the guarantee?

42. Tree heights: Cherry trees in a certain orchard have heights that are normally distributed with mean $\mu = 112$ inches and standard deviation $\sigma = 14$ inches.
  a. Find the 27th percentile of the tree heights.
  b. Find the 85th percentile of the tree heights.
  c. Find the third quartile of the tree heights.
  d. An agricultural scientist wants to study the tallest 1% of the trees to determine whether they have a certain gene that allows them to grow taller. To do this, she needs to study all the trees above a certain height. What height is this?

43. How much is in that can? The volume of beverage in a 12-ounce can is normally distributed with mean 12.05 ounces and standard deviation 0.02 ounce.
  a. What is the probability that a randomly selected can will contain more than 12.06 ounces?
  b. What is the probability that a randomly selected can will contain between 12 and 12.03 ounces?
  c. Is it unusual for a can to be underfilled (contain less than 12 ounces)?

44. How much do you study? A survey among freshmen at a certain university revealed that the number of hours spent studying the week before final exams was normally distributed with mean 25 and standard deviation 7.
  a. What proportion of students studied more than 40 hours?
  b. What is the probability that a randomly selected student spent between 15 and 30 hours studying?
  c. What proportion of students studied less than 30 hours?

45. How much is in that can? The volume of beverage in a 12-ounce can is normally distributed with mean 12.05 ounces and standard deviation 0.02 ounce.
  a. Find the 60th percentile of the volumes.
  b. Find the 4th percentile of the volumes.
  c. Between what two values are the middle 95% of the volumes?

46. How much do you study? A survey among freshmen at a certain university revealed that the number of hours spent studying the week before final exams was normally distributed with mean 25 and standard deviation 7.
  a. Find the 98th percentile of the number of hours studying.
  b. Find the 32nd percentile of the number of hours studying.
  c. Between what two values are the middle 80% of the hours spent studying?

47. Precision manufacturing: A process manufactures ball bearings with diameters that are normally distributed with mean 25.1 millimeters and standard deviation 0.08 millimeter.
  a. What proportion of the diameters are less than 25.0 millimeters?
  b. What proportion of the diameters are greater than 25.4 millimeters?
  c. To meet a certain specification, a ball bearing must have a diameter between 25.0 and 25.3 millimeters. What proportion of the ball bearings meet the specification?

48. Exam grades: Scores on a statistics final in a large class were normally distributed with a mean of 75 and a standard deviation of 8.
  a. What proportion of the scores were above 90?
  b. What proportion of the scores were below 65?
  c. What is the probability that a randomly chosen score is between 70 and 80?

49. Precision manufacturing: A process manufactures ball bearings with diameters that are normally distributed with mean 25.1 millimeters and standard deviation 0.08 millimeter.
  a. Find the 60th percentile of the diameters.
  b. Find the 32nd percentile of the diameters.
  c. A hole is to be designed so that 1% of the ball bearings will fit through it. The bearings that fit through the hole will be melted down and remade. What should the diameter of the hole be?
  d. Between what two values are the middle 50% of the diameters?

50. Exam grades: Scores on a statistics final in a large class were normally distributed with a mean of 75 and a standard deviation of 8.
  a. Find the 40th percentile of the scores.
  b. Find the 65th percentile of the scores.
  c. The instructor wants to give an A to the students whose scores were in the top 10% of the class. What is the minimum score needed to get an A?

d. Between what two values are the middle 60% of the scores?

Extending the Concepts

51. Tall men: Heights of men in a certain city are normally distributed with mean 70 inches. Sixteen percent of the men are more than 73 inches tall. What percentage of the men are between 67 and 70 inches tall?

52. Watch your speed: Speeds of automobiles on a certain stretch of freeway at 11:00 P.M. are normally distributed with mean 65 mph. Twenty percent of the cars are traveling at speeds between 55 and 65 mph. What percentage of the cars are going faster than 75 mph?

53. Contaminated wells: A study reported that the mean concentration of ammonium in water wells in the state of Iowa was 0.71 milligram per liter, and the standard deviation was 1.09 milligrams per liter. Is it possible to determine whether these concentrations are approximately normally distributed? If so, say whether they are normally

distributed, and explain how you know. If not, describe the additional information you would need to determine whether they are normally distributed.
Source: *Water Environment Research* 74:177–186

54. Heights: According to the National Health Statistics Reports, heights of adult women in the United States are normally distributed with mean 64 inches and standard deviation 4 inches. If three women are selected at random, what is the probability that at least one of them is more than 68 inches tall?

55. Exam scores: Scores on an exam were normally distributed. Ten percent of the scores were below 64 and 80% were below 81. Find the mean and standard deviation of the scores.

56. Commute to work: Megan drives to work each morning. Her commute time is normally distributed with mean 30 minutes and standard deviation 4 minutes. Her workday begins at 9:00 A.M. At what time should she leave for work so that the probability she is on time is 95%?

Answers to Check Your Understanding Exercises for Section 7.2

1. $z = -1$. Interpretation: A value of 11 is one standard deviation below the mean.

2. $z = 0.75$. Interpretation: A value of 75 is 0.75 standard deviation above the mean.

3. $z = -0.4$. Interpretation: The length of time for this bulb is 0.4 standard deviation below the mean.

4. 0.0228

5. 0.0968

6. 0.6898

7. 9.084 [Tech: 9.088]

8. 66.26 [Tech: 66.10]

SECTION 7.3   Sampling Distributions and the Central Limit Theorem

Objectives

1. Construct the sampling distribution of a sample mean
2. Use the Central Limit Theorem to compute probabilities involving sample means

In Section 7.2, we learned to compute probabilities for a randomly sampled individual from a normal population. In practice, statistical studies involve sampling several, perhaps many, individuals. As discussed in Chapter 3, we often compute numerical summaries of samples, and the most commonly used summary is the sample mean $\bar{x}$.

 If several samples are drawn from a population, they are likely to have different values for $\bar{x}$. Because the value of $\bar{x}$ varies each time a sample is drawn, $\bar{x}$ is a random variable, and it has a probability distribution. The probability distribution of $\bar{x}$ is called the *sampling distribution* of $\bar{x}$.

Objective 1 Construct the sampling distribution of a sample mean

An Example of a Sampling Distribution

Tetrahedral dice are four-sided dice, used in role-playing games such as Dungeons & Dragons. They are shaped like a pyramid, with four triangular faces. Each face corresponds to a number between 1 and 4, so that when you toss a tetrahedral die, it comes up with one of the numbers 1, 2, 3, or 4. Tossing a tetrahedral die is like sampling a value from the population

$$\boxed{1}\ \boxed{2}\ \boxed{3}\ \boxed{4}$$

The population mean, variance, and standard deviation are:

Population mean: $\mu = \dfrac{1 + 2 + 3 + 4}{4} = 2.5$

Population variance: $\sigma^2 = \dfrac{(1 - 2.5)^2 + (2 - 2.5)^2 + (3 - 2.5)^2 + (4 - 2.5)^2}{4} = 1.25$

Population standard deviation: $\sigma = \sqrt{\sigma^2} = \sqrt{1.25} = 1.118$

Now imagine tossing a tetrahedral die three times. The sequence of three numbers that is observed is a sample of size 3 drawn with replacement from the population just described. There are 64 possible samples, and they are all equally likely. Table 7.1 lists them and provides the value of the sample mean $\bar{x}$ for each.

Table 7.1 The 64 Possible Samples of Size 3 and Their Sample Means

Sample	$\bar{x}$	Sample	$\bar{x}$	Sample	$\bar{x}$	Sample	$\bar{x}$
1, 1, 1	1.00	2, 1, 1	1.33	3, 1, 1	1.67	4, 1, 1	2.00
1, 1, 2	1.33	2, 1, 2	1.67	3, 1, 2	2.00	4, 1, 2	2.33
1, 1, 3	1.67	2, 1, 3	2.00	3, 1, 3	2.33	4, 1, 3	2.67
1, 1, 4	2.00	2, 1, 4	2.33	3, 1, 4	2.67	4, 1, 4	3.00
1, 2, 1	1.33	2, 2, 1	1.67	3, 2, 1	2.00	4, 2, 1	2.33
1, 2, 2	1.67	2, 2, 2	2.00	3, 2, 2	2.33	4, 2, 2	2.67
1, 2, 3	2.00	2, 2, 3	2.33	3, 2, 3	2.67	4, 2, 3	3.00
1, 2, 4	2.33	2, 2, 4	2.67	3, 2, 4	3.00	4, 2, 4	3.33
1, 3, 1	1.67	2, 3, 1	2.00	3, 3, 1	2.33	4, 3, 1	2.67
1, 3, 2	2.00	2, 3, 2	2.33	3, 3, 2	2.67	4, 3, 2	3.00
1, 3, 3	2.33	2, 3, 3	2.67	3, 3, 3	3.00	4, 3, 3	3.33
1, 3, 4	2.67	2, 3, 4	3.00	3, 3, 4	3.33	4, 3, 4	3.67
1, 4, 1	2.00	2, 4, 1	2.33	3, 4, 1	2.67	4, 4, 1	3.00
1, 4, 2	2.33	2, 4, 2	2.67	3, 4, 2	3.00	4, 4, 2	3.33
1, 4, 3	2.67	2, 4, 3	3.00	3, 4, 3	3.33	4, 4, 3	3.67
1, 4, 4	3.00	2, 4, 4	3.33	3, 4, 4	3.67	4, 4, 4	4.00

The columns labeled "$\bar{x}$" contain the values of the sample mean for each of the 64 possible samples. Some of these values appear more than once, because several samples have the same mean. The mean of the sampling distribution is the average of these 64 values. The standard deviation of the sampling distribution is the population standard deviation of the 64 sample means, which can be computed by the method presented in Section 3.2. The mean and standard deviation are

$$\text{Mean: } \mu_{\bar{x}} = 2.5 \qquad \text{Standard deviation: } \sigma = 0.6455$$

Comparing the mean and standard deviation of the sampling distribution to the population mean and standard deviation, we see that the mean $\mu_{\bar{x}}$ of the sampling distribution is equal to the population mean μ. The standard deviation $\sigma_{\bar{x}}$ of the sampling distribution is 0.6455, which is less than the population standard deviation $\sigma = 1.118$. It is not immediately obvious how these two quantities are related. Note, however, that

$$\sigma_{\bar{x}} = 0.6455 = \frac{1.118}{\sqrt{3}} = \frac{\sigma}{\sqrt{3}}$$

The sample size is $n = 3$, so $\sigma_{\bar{x}} = \frac{\sigma}{\sqrt{n}}$.

These relationships hold in general. Note that the standard deviation $\sigma_{\bar{x}}$ is sometimes called the **standard error** of the mean.

SUMMARY

Let $\bar{x}$ be the mean of a simple random sample of size n, drawn from a population with mean μ and standard deviation σ.

The mean of the sampling distribution is $\mu_{\bar{x}} = \mu$.

The standard deviation of the sampling distribution is $\sigma_{\bar{x}} = \dfrac{\sigma}{\sqrt{n}}$.

The standard deviation $\sigma_{\bar{x}}$ is sometimes called the standard error of the mean.

EXAMPLE 7.19   Find the mean and standard deviation of a sampling distribution

Among students at a certain college, the mean number of hours of television watched per week is $\mu = 10.5$, and the standard deviation is $\sigma = 3.6$. A simple random sample of 16 students is chosen for a study of viewing habits. Let $\bar{x}$ be the mean number of hours of TV watched by the sampled students. Find the mean $\mu_{\bar{x}}$ and the standard deviation $\sigma_{\bar{x}}$ of $\bar{x}$.

Solution

The mean of $\bar{x}$ is

$$\mu_{\bar{x}} = \mu = 10.5$$

The sample size is $n = 16$. Therefore, the standard deviation of $\bar{x}$ is

$$\sigma_{\bar{x}} = \frac{\sigma}{\sqrt{n}} = \frac{3.6}{\sqrt{16}} = 0.9$$

It makes sense that the standard deviation of $\bar{x}$ is less than the population standard deviation σ. In a sample, it is unusual to get all large values or all small values. Samples usually contain both large and small values that cancel each other out when the sample mean is computed. For this reason, the distribution of $\bar{x}$ is less spread out than the population distribution. Therefore, the standard deviation of $\bar{x}$ is less than the population standard deviation.

Check Your Understanding

1. A population has mean $\mu = 6$ and standard deviation $\sigma = 4$. Find $\mu_{\bar{x}}$ and $\sigma_{\bar{x}}$ for samples of size $n = 25$.

2. A population has mean $\mu = 17$ and standard deviation $\sigma = 20$. Find $\mu_{\bar{x}}$ and $\sigma_{\bar{x}}$ for samples of size $n = 100$.

Answers are on page 318.

The probability histogram for the sampling distribution of $\bar{x}$

Consider again the example of the tetrahedral die. Let us compare the probability distribution for the population and the sampling distribution. The population consists of the numbers 1, 2, 3, and 4, each of which is equally likely. The sampling distribution for $\bar{x}$ can be determined from Table 7.1. The probability that the sample mean is 1.00 is 1/64, because out of the 64 possible samples, only one has a sample mean equal to 1.00. Similarly, the probability that $\bar{x} = 1.33$ is 3/64, because there are three samples out of 64 whose sample mean is 1.33. Figure 7.20 presents the probability histogram of the population and Figure 7.21 presents the sampling distribution for $\bar{x}$.

Figure 7.20   Probability histogram for the population

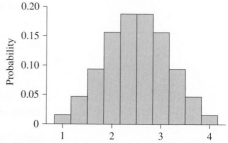

Figure 7.21   Probability histogram for the sampling distribution of $\bar{x}$ for samples of size 3

Note that the probability histogram for the sampling distribution looks a lot like the normal curve, whereas the probability histogram for the population does not. Remarkably, it is true that, for any population, if the sample size is large enough, the sample mean $\bar{x}$ will be approximately normally distributed. For a symmetric population like the one in Figure 7.20, the sample mean is approximately normally distributed even for a small sample size like $n = 3$.

In fact, when a population is normal, the sample mean will also be normal.

> For a normal population, the sample mean will be normal for any sample size.

For a skewed population, the sample size must be large for the sample mean to be approximately normal.

Computing the sampling distribution of $\bar{x}$ for a skewed population

For a certain make of car, the number of repairs needed while under warranty has the following probability distribution.

x	P(x)
0	0.60
1	0.25
2	0.10
3	0.03
4	0.02

Figure 7.22 presents the probability histogram for this distribution, along with probability histograms for the sampling distribution of $\bar{x}$ for samples of size 3, 10, and 30. The probability histograms for the sampling distributions were created by programming a computer to compute the probability for every possible value of $\bar{x}$.

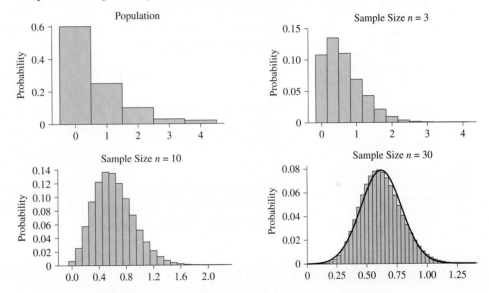

Figure 7.22 The probability histogram for the population distribution is highly skewed. As the sample size increases, the skewness decreases. For a sample size of 30, the probability histogram is reasonably well approximated by a normal curve.

The remarkable fact that the sampling distribution of $\bar{x}$ is approximately normal for a large sample from any distribution is called the **Central Limit Theorem**. The size of the sample needed to obtain approximate normality depends mostly on the skewness of the population. A sample of size $n > 30$ is large enough for most populations encountered in practice. Smaller sample sizes are adequate for distributions that are nearly symmetric.

The Central Limit Theorem

Let $\bar{x}$ be the mean of a large ($n > 30$) simple random sample from a population with mean μ and standard deviation σ.

Then $\bar{x}$ has an approximately normal distribution, with mean $\mu_{\bar{x}} = \mu$ and standard deviation $\sigma_{\bar{x}} = \dfrac{\sigma}{\sqrt{n}}$.

The Central Limit Theorem is the most important result in statistics, and forms the basis for much of the work that statisticians do.

Objective 2 Use the Central Limit Theorem to compute probabilities involving sample means

Computing Probabilities with the Central Limit Theorem

To compute probabilities involving a sample mean $\bar{x}$, use the following procedure:

> ### Procedure for Computing Probabilities with the Central Limit Theorem
>
> **Step 1:** Be sure the sample size is greater than 30. If so, it is appropriate to use the normal curve.
> **Step 2:** Find the mean $\mu_{\bar{x}}$ and standard deviation $\sigma_{\bar{x}}$.
> **Step 3:** Sketch a normal curve and shade in the area to be found.
> **Step 4:** Find the area using Table A.2 or technology.

EXAMPLE 7.20

Using the Central Limit Theorem to compute a probability

Based on data from the U.S. Census, the mean age of college students in 2011 was $\mu = 25$ years, with a standard deviation of $\sigma = 9.5$ years. A simple random sample of 125 students is drawn. What is the probability that the sample mean age of the students is greater than 26 years?

Solution

Step 1: The sample size is 125, which is greater than 30. We may use the normal curve.
Step 2: We compute $\mu_{\bar{x}}$ and $\sigma_{\bar{x}}$.

$$\mu_{\bar{x}} = \mu = 25 \qquad \sigma_{\bar{x}} = \frac{\sigma}{\sqrt{n}} = \frac{9.5}{\sqrt{125}} = 0.85$$

Step 3: Figure 7.23 presents the normal curve with the area of interest shaded.
Step 4: We will use Table A.2. We compute the z-score for 26.

$$z = \frac{x - \mu_{\bar{x}}}{\sigma_{\bar{x}}} = \frac{26 - 25}{0.85} = 1.18$$

The table gives the area to the *left* of $z = 1.18$ as 0.8810. The area to the right of $z = 1.18$ is $1 - 0.8810 = 0.1190$. The probability that the sample mean age of the students is greater than 26 years is 0.1190.

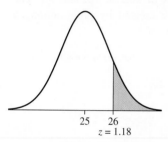

Figure 7.23

> **CAUTION**
>
> When computing the z-score for the distribution of $\bar{x}$, be sure to use the standard deviation $\sigma_{\bar{x}}$, rather than σ.

EXAMPLE 7.21

Using the Central Limit Theorem to determine whether a given value of $\bar{x}$ is unusual

Hereford cattle are one of the most popular breeds of beef cattle. Based on data from the Hereford Cattle Society, the mean weight of a one-year-old Hereford bull is 1135 pounds, with a standard deviation of 97 pounds. Would it be unusual for the mean weight of 100 head of cattle to be less than 1100 pounds?

Solution

We will compute the probability that the sample mean is less than 1100. We will say that this event is unusual if its probability is less than 0.05.

Step 1: The sample size is 100, which is greater than 30. We may use the normal curve.
Step 2: We compute $\mu_{\bar{x}}$ and $\sigma_{\bar{x}}$.

$$\mu_{\bar{x}} = \mu = 1135 \qquad \sigma_{\bar{x}} = \frac{\sigma}{\sqrt{n}} = \frac{97}{\sqrt{100}} = 9.7$$

Step 3: Figure 7.24 presents the normal curve. We are interested in the area to the left of 1100, which is too small to see.

Figure 7.24

Step 4: We will use Table A.2. We compute the z-score for 1100.

$$z = \frac{x - \mu_{\bar{x}}}{\sigma_{\bar{x}}} = \frac{1100 - 1135}{9.7} = -3.61$$

The area to the left of $z = -3.61$ is 0.0002. The probability that the sample mean weight is less than 1100 is 0.0002. This probability is less than 0.05, so it would be unusual for the sample mean to be less than 1100.

Check Your Understanding

3. A population has mean $\mu = 10$ and standard deviation $\sigma = 8$. A sample of size 50 is drawn.
   a. Find the probability that $\bar{x}$ is greater than 11.
   b. Would it be unusual for $\bar{x}$ to be less than 8? Explain.

4. A population has mean $\mu = 47.5$ and standard deviation $\sigma = 12.6$. A sample of size 112 is drawn.
   a. Find the probability that $\bar{x}$ is between 45 and 48.
   b. Would it be unusual for $\bar{x}$ to be greater than 48? Explain.

Answers are on page 318.

SECTION 7.3   Exercises

Exercises 1–4 are the Check Your Understanding exercises located within the section.

Understanding the Concepts

In Exercises 5 and 6, fill in each blank with the appropriate word or phrase.

5. The probability distribution of $\bar{x}$ is called a _____ distribution.

6. The _____ states that the sampling distribution of $\bar{x}$ is approximately normal when the sample is large.

In Exercises 7 and 8, determine whether the statement is true or false. If the statement is false, rewrite it as a true statement.

7. If $\bar{x}$ is the mean of a large ($n > 30$) simple random sample from a population with mean μ and standard deviation σ, then $\bar{x}$ is approximately normal with $\sigma_{\bar{x}} = \frac{\sigma}{\sqrt{n}}$.

8. As the sample size increases, the sampling distribution of $\bar{x}$ becomes more and more skewed.

Practicing the Skills

9. A sample of size 75 will be drawn from a population with mean 10 and standard deviation 12.
   a. Find the probability that $\bar{x}$ will be between 8 and 14.
   b. Find the 15th percentile of $\bar{x}$.

10. A sample of size 126 will be drawn from a population with mean 26 and standard deviation 3.
    a. Find the probability that $\bar{x}$ will be between 25 and 27.
    b. Find the 55th percentile of $\bar{x}$.

11. A sample of size 68 will be drawn from a population with mean 92 and standard deviation 24.
    a. Find the probability that $\bar{x}$ will be greater than 90.
    b. Find the 90th percentile of $\bar{x}$.

12. A sample of size 284 will be drawn from a population with mean 45 and standard deviation 7.
    a. Find the probability that $\bar{x}$ will be greater than 46.
    b. Find the 75th percentile of $\bar{x}$.

13. A sample of size 91 will be drawn from a population with mean 33 and standard deviation 17.
    a. Find the probability that $\bar{x}$ will be less than 30.
    b. Find the 25th percentile of $\bar{x}$.

14. A sample of size 82 will be drawn from a population with mean 24 and standard deviation 9.
    a. Find the probability that $\bar{x}$ will be less than 26.
    b. Find the 10th percentile of $\bar{x}$.

15. A sample of size 20 will be drawn from a population with mean 6 and standard deviation 3.
    a. Is it appropriate to use the normal distribution to find probabilities for $\bar{x}$?
    b. If appropriate find the probability that $\bar{x}$ will be greater than 4.
    c. If appropriate find the 30th percentile of $\bar{x}$.

16. A sample of size 42 will be drawn from a population with mean 52 and standard deviation 9.
    a. Is it appropriate to use the normal distribution to find probabilities for $\bar{x}$?
    b. If appropriate find the probability that $\bar{x}$ will be between 53 and 54.
    c. If appropriate find the 45th percentile of $\bar{x}$.

17. A sample of size 5 will be drawn from a normal population with mean 60 and standard deviation 12.
    a. Is it appropriate to use the normal distribution to find probabilities for $\bar{x}$?
    b. If appropriate find the probability that $\bar{x}$ will be between 50 and 70.

c. If appropriate find the 80th percentile of $\bar{x}$.

18. A sample of size 15 will be drawn from a population with mean 125 and standard deviation 28.

    a. Is it appropriate to use the normal distribution to find probabilities for $\bar{x}$?

    b. If appropriate find the probability that $\bar{x}$ will be less than 120.

    c. If appropriate find the 90th percentile of $\bar{x}$.

Working with the Concepts

19. Summer temperatures: Following are the temperatures, in degrees Fahrenheit, in Denver for five days in July 2009:

Date	Temperature
July 21	69
July 22	75
July 23	79
July 24	83
July 25	71

    a. Consider this to be a population. Find the population mean μ and the population standard deviation σ.

    b. List all samples of size 2 drawn with replacement. There are $5 \times 5 = 25$ different samples.

    c. Compute the sample mean $\bar{x}$ for each of the 25 samples of size 2. Compute the mean $\mu_{\bar{x}}$ and the standard deviation $\sigma_{\bar{x}}$ of the sample means.

    d. Verify that $\mu_{\bar{x}} = \mu$ and $\sigma_{\bar{x}} = \sigma/\sqrt{2}$.

20. Ages of winners: Following are the ages of the Grammy Award winners for Best New Artist for the years 2009–2013. (For the Zac Brown Band, the age given is that of lead singer Zac Brown. For Bon Iver, the age is that of lead singer Justin Vernon. For Fun., the age is that of guitarist Andrew Dost.)

Year: Winner	Age
2013: Fun.	29
2012: Bon Iver	30
2011: Esperanza Spalding	26
2010: Zac Brown Band	31
2009: Adele	20

    a. Consider this to be a population. Find the population mean μ and the population standard deviation σ.

    b. List all samples of size 2 drawn with replacement. There are $5 \times 5 = 25$ different samples.

    c. Compute the sample mean $\bar{x}$ for each of the 25 samples of size 2. Compute the mean $\mu_{\bar{x}}$ and the standard deviation $\sigma_{\bar{x}}$ of the sample means.

    d. Verify that $\mu_{\bar{x}} = \mu$ and $\sigma_{\bar{x}} = \sigma/\sqrt{2}$.

21. How's your mileage? The Environmental Protection Agency (EPA) rates the mean highway gas mileage of the 2013 Ford Edge to be 27 miles per gallon. Assume the standard deviation is 3 miles per gallon. A rental car company buys 60 of these cars.

    a. What is the probability that the average mileage of the fleet is greater than 26.5 miles per gallon?

    b. What is the probability that the average mileage of the fleet is between 26 and 26.8 miles per gallon?

    c. Would it be unusual if the average mileage of the fleet were less than 26 miles per gallon?

22. Watch your cholesterol: The National Health and Nutrition Examination Survey (NHANES) reported that in a recent year, the mean serum cholesterol level for U.S. adults was 202, with a standard deviation of 41 (the units are milligrams per deciliter). A simple random sample of 110 adults is chosen.

    a. What is the probability that the sample mean cholesterol level is greater than 210?

    b. What is the probability that the sample mean cholesterol level is between than 190 and 200?

    c. Would it be unusual for the sample mean to be less than 198?

23. TV sets: According to the Nielsen Company, the mean number of TV sets in a U.S. household in 2013 was 2.24. Assume the standard deviation is 1.2. A sample of 85 households is drawn.

    a. What is the probability that the sample mean number of TV sets is greater than 2?

    b. What is the probability that the sample mean number of TV sets is between 2.5 and 3?

    c. Find the 30th percentile of the sample mean.

    d. Would it be unusual for the sample mean to be less than 2?

    e. Can you tell whether it would be unusual for an individual household to have fewer than 2 TV sets? Explain.

24. SAT scores: The College Board reports that in 2013, the mean mathematics SAT score was 514, and the standard deviation was 118. A sample of 65 scores is chosen.

    a. What is the probability that the sample mean score is less than 500?

    b. What is the probability that the sample mean score is between 480 and 520?

    c. Find the 80th percentile of the sample mean.

    d. Would it be unusual if the sample mean were greater than 550?

    e. Can you tell whether it would be unusual for an individual to get a score greater than 550? Explain.

25. Taxes: The Internal Revenue Service reports that the mean federal income tax paid in the year 2010 was $8040. Assume that the standard deviation is $5000. The IRS plans to draw a sample of 1000 tax returns to study the effect of a new tax law.

    a. What is the probability that the sample mean tax is less than $8000?

    b. What is the probability that the sample mean tax is between $7600 and $7900?

    c. Find the 40th percentile of the sample mean.

    d. Would it be unusual if the sample mean were less than $7500?

    e. Can you tell whether it would be unusual for an individual to pay a tax of less than $7500? Explain.

26. High-rent district: The Real Estate Group NY reports that the mean monthly rent for a one-bedroom apartment without a doorman in Manhattan is $2631. Assume the standard deviation is $500. A real estate firm samples 100 apartments.

    a. What is the probability that the sample mean rent is greater than $2700?

    b. What is the probability that the sample mean rent is between $2500 and $2600?

c. Find the 60th percentile of the sample mean.

d. Can you tell whether it would be unusual if the sample mean were greater than $2800?

e. Do you think it would be unusual for an individual apartment to have a rent greater than $2800? Explain.

27. **Roller coaster ride:** A roller coaster is being designed that will accommodate 60 riders. The maximum weight the coaster can hold safely is 12,000 pounds. According to the National Health Statistics Reports, the weights of adult U.S. men have mean 194 pounds and standard deviation 68 pounds, and the weights of adult U.S. women have mean 164 pounds and standard deviation 77 pounds.

a. If 60 people are riding the coaster, and their total weight is 12,000 pounds, what is their average weight?

b. If a random sample of 60 adult men ride the coaster, what is the probability that the maximum safe weight will be exceeded?

c. If a random sample of 60 adult women ride the coaster, what is the probability that the maximum safe weight will be exceeded?

28. **Elevator ride:** Engineers are designing a large elevator that will accommodate 40 people. The maximum weight the elevator can hold safely is 8120 pounds. According to the National Health Statistics Reports, the weights of adult U.S. men have mean 194 pounds and standard deviation 68 pounds, and the weights of adult U.S. women have mean 164 pounds and standard deviation 77 pounds.

a. If 40 people are on the elevator, and their total weight is 8120 pounds, what is their average weight?

b. If a random sample of 40 adult men ride the elevator, what is the probability that the maximum safe weight will be exceeded?

c. If a random sample of 40 adult women ride the elevator, what is the probability that the maximum safe weight will be exceeded?

29. **Annual income:** The mean annual income for people in a certain city (in thousands of dollars) is 42, with a standard deviation of 30. A pollster draws a sample of 90 people to interview.

a. What is the probability that the sample mean income is less than 38?

b. What is the probability that the sample mean score is between 40 and 45?

c. Find the 60th percentile of the sample mean.

d. Would it be unusual for the sample mean to be less than 35?

e. Can you tell whether it would be unusual for an individual to have an income less than 35? Explain.

30. **Going to work:** An ABC News report stated that the mean distance that commuters in the United States travel each way to work is 16 miles. Assume the standard deviation is 8 miles. A sample of 75 commuters is chosen.

a. What is the probability that the sample mean commute distance is greater than 13 miles?

b. What is the probability that the sample mean commute distance is between 18 and 20 miles?

c. Find the 10th percentile of the sample mean.

d. Would it be unusual for the sample mean distance to be greater than 19 miles?

e. Can you tell whether it would be unusual for an individual to have a commute distance greater than 19 miles? Explain.

Extending the Concepts

31. **Eat your cereal:** A cereal manufacturer claims that the weight of a box of cereal labeled as weighing 12 ounces has a mean of 12.0 ounces and a standard deviation of 0.1 ounce. You sample 75 boxes and weigh them. Let $\bar{x}$ denote the mean weight of the 75 boxes.

a. If the claim is true, what is $P(\bar{x} \leq 11.99)$?

b. Based on the answer to part (a), if the claim is true, is 11.99 ounces an unusually small mean weight for a sample of 75 boxes?

c. If the mean weight of the boxes were 11.99 ounces, would you be convinced that the claim was false? Explain.

d. If the claim is true, what is $P(\bar{x} \leq 11.97)$?

e. Based on the answer to part (d), if the claim is true, is 11.97 ounces an unusually small mean weight for a sample of 75 boxes?

f. If the mean weight of the boxes were 11.97 ounces, would you be convinced that the claim was false? Explain.

32. **Battery life:** A battery manufacturer claims that the lifetime of a certain type of battery has a population mean of $\mu = 40$ hours and a standard deviation of $\sigma = 5$ hours. Let $\bar{x}$ represent the mean lifetime of the batteries in a simple random sample of size 100.

a. If the claim is true, what is $P(\bar{x} \leq 38.5)$?

b. Based on the answer to part (a), if the claim is true, is a sample mean lifetime of 38.5 hours unusually short?

c. If the sample mean lifetime of the 100 batteries were 38.5 hours, would you find the manufacturer's claim to be plausible? Explain.

d. If the claim is true, what is $P(\bar{x} \leq 39.8)$?

e. Based on the answer to part (d), if the claim is true, is a sample mean lifetime of 39.8 hours unusually short?

f. If the sample mean lifetime of the 100 batteries were 39.8 hours, would you find the manufacturer's claim to be plausible? Explain.

33. **Finite population correction:** The mean of a sample of size n has standard deviation $\sigma/\sqrt{n}$, where σ is the population standard deviation. When sampling without replacement, a more accurate expression can be obtained by multiplying by a correction factor. Specifically, if the sample size is more than 5% of the population size, it is better to compute the standard deviation of the sample mean as

$$\frac{\sigma}{\sqrt{n}} \sqrt{\frac{N-n}{N-1}}$$

where N is the population size and n is the sample size. The factor $\sqrt{\frac{N-n}{N-1}}$ is called the finite population correction factor.

a. One hundred students took an exam. The standard deviation of the 100 scores was 10. Twenty exams were chosen at random as part of a class assessment. Use the finite population correction to compute the standard deviation of the mean of the 20 exams.

b. In general, is the standard deviation computed with the correction smaller or larger than the standard deviation computed without it?

c. Use the finite population correction to show that if all 100 exams are sampled, the standard deviation of the sample mean is 0. Explain why this is so.

Answers to Check Your Understanding Exercises for Section 7.3

1. $\mu_{\bar{x}} = 6$, $\sigma_{\bar{x}} = 0.8$

2. $\mu_{\bar{x}} = 17$, $\sigma_{\bar{x}} = 2.0$

3. a. 0.1894 [Tech: 0.1884]

    b. The probability that $\bar{x}$ is less than 8 is 0.0384 [Tech: 0.0385]. If we define an event whose probability is less than 0.05 as unusual, then this is unusual.

4. a. 0.6449

    b. The probability that $\bar{x}$ is greater than 48 is 0.3372 [Tech: 0.3373]. This event is not unusual.

SECTION 7.4	The Central Limit Theorem for Proportions

Objectives

1. Construct the sampling distribution for a sample proportion

2. Use the Central Limit Theorem to compute probabilities for sample proportions

A computer retailer wants to estimate the proportion of people in her city who own laptop computers. She cannot survey everyone in the city, so she draws a sample of 100 people and surveys them. It turns out that 35 out of the 100 people in the sample own laptops. The proportion $35/100$ is called the *sample proportion* and is denoted $\hat{p}$. The proportion of people in the entire population who own laptops is called the *population proportion* and is denoted p.

DEFINITION

In a population, the proportion who have a certain characteristic is called the **population proportion**.

    In a simple random sample of n individuals, let x be the number in the sample who have the characteristic. The **sample proportion** is

$$\hat{p} = \frac{x}{n}$$

Notation:

- The population proportion is denoted by p.
- The sample proportion is denoted by $\hat{p}$.

If several samples are drawn from a population, they are likely to have different values for $\hat{p}$. Because the value of $\hat{p}$ varies each time a sample is drawn, $\hat{p}$ is a random variable, and it has a probability distribution. The probability distribution of $\hat{p}$ is called the *sampling distribution* of $\hat{p}$.

Objective 1 Construct the sampling distribution for a sample proportion

An Example of a Sampling Distribution

To present an example, consider tossing a fair coin five times. This produces a sample of size $n = 5$, where each item in the sample is either a head or a tail. The proportion of times the coin lands heads will be the sample proportion $\hat{p}$. Because the coin is fair, the probability that it lands heads each time is 0.5. Therefore, the population proportion of heads is $p = 0.5$. There are 32 possible samples. Table 7.2 lists them and presents the sample proportion $\hat{p}$ of heads for each.

Table 7.2 The 32 Possible Samples of Size 5 and Their Sample Proportions of Heads

Sample	$\hat{p}$	Sample	$\hat{p}$	Sample	$\hat{p}$	Sample	$\hat{p}$
TTTTT	0.0	THTTT	0.2	HTTTT	0.2	HHTTT	0.4
TTTTH	0.2	THTTH	0.4	HTTTH	0.4	HHTTH	0.6
TTTHT	0.2	THTHT	0.4	HTTHT	0.4	HHTHT	0.6
TTTHH	0.4	THTHH	0.6	HTTHH	0.6	HHTHH	0.8
TTHTT	0.2	THHTT	0.4	HTHTT	0.4	HHHTT	0.6
TTHTH	0.4	THHTH	0.6	HTHTH	0.6	HHHTH	0.8
TTHHT	0.4	THHHT	0.6	HTHHT	0.6	HHHHT	0.8
TTHHH	0.6	THHHH	0.8	HTHHH	0.8	HHHHH	1.0

The columns labeled "$\hat{p}$" contain the values of the sample proportion for each of the 32 possible samples. Some of these values appear more than once, because several samples have the same proportion. The mean of the sampling distribution is the average of these 32 values. The standard deviation of the sampling distribution is the population standard deviation of these 32 values, which can be computed by the method presented in Section 3.2. The mean and standard deviation are

$$\text{Mean: } \mu_{\hat{p}} = 0.5 \qquad \text{Standard deviation: } \sigma_{\hat{p}} = 0.2236$$

The values of $\mu_{\hat{p}}$ and $\sigma_{\hat{p}}$ are related to the values of the population proportion $p = 0.5$ and the sample size $n = 5$. Specifically,

$$\mu_{\hat{p}} = 0.5 = p$$

The mean of the sample proportion is equal to the population proportion. The relationship among $\sigma_{\hat{p}}$, p, and n is less obvious. However, note that

$$\sigma_{\hat{p}} = 0.2236 = \sqrt{\frac{0.5(1 - 0.5)}{5}} = \sqrt{\frac{p(1 - p)}{n}}$$

These relationships hold in general.

SUMMARY

Let $\hat{p}$ be the sample proportion of a simple random sample of size n, drawn from a population with population proportion p. The mean and standard deviation of the sampling distribution of $\hat{p}$ are

$$\mu_{\hat{p}} = p$$

$$\sigma_{\hat{p}} = \sqrt{\frac{p(1 - p)}{n}}$$

EXAMPLE 7.22   Find the mean and standard deviation of a sampling distribution

The soft-drink cups at a certain fast-food restaurant have tickets attached to them. Customers peel off the tickets to see whether they win a prize. The proportion of tickets that are winners is $p = 0.25$. A total of $n = 70$ people purchase soft drinks between noon and 1:00 P.M. on a certain day. Let $\hat{p}$ be the proportion that win a prize. Find the mean and standard deviation of $\hat{p}$.

Solution

The population proportion is $p = 0.25$ and the sample size is $n = 70$. Therefore,

$$\mu_{\hat{p}} = p = 0.25$$

$$\sigma_{\hat{p}} = \sqrt{\frac{0.25(1 - 0.25)}{70}} = 0.0518$$

Check Your Understanding

1. Find $\mu_{\hat{p}}$ and $\sigma_{\hat{p}}$ if $n = 20$ and $p = 0.82$.

2. Find $\mu_{\hat{p}}$ and $\sigma_{\hat{p}}$ if $n = 217$ and $p = 0.455$.

Answers are on page 324.

The probability histogram for the sampling distribution of $\hat{p}$

Figure 7.25 presents the probability histogram for the sampling distribution of $\hat{p}$ for the proportion of heads in five tosses of a fair coin, for which $n = 5$ and $p = 0.5$. The distribution is reasonably well approximated by a normal curve. Figure 7.26 presents the probability histogram for the sampling distribution of $\hat{p}$ for the proportion of heads in 50 tosses of a fair coin, for which $n = 50$ and $p = 0.5$. The distribution is very closely approximated by a normal curve.

Figure 7.25   The probability histogram for $\hat{p}$ when $n = 5$ and $p = 0.5$. The histogram is reasonably well approximated by a normal curve.

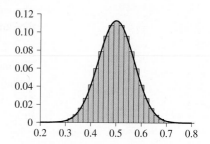

Figure 7.26   The probability histogram for $\hat{p}$ when $n = 50$ and $p = 0.5$. The histogram is very closely approximated by a normal curve.

When $p = 0.5$, the sampling distribution of $\hat{p}$ is somewhat close to normal even for a small sample size like $n = 5$. When p is close to 0 or close to 1, a larger sample size is needed before the distribution of $\hat{p}$ is close to normal. A common rule of thumb is that the distribution may be approximated with a normal curve whenever np and $n(1 - p)$ are both at least 10.

The Central Limit Theorem for Proportions

Let $\hat{p}$ be the sample proportion for a sample size of n and population proportion p. If

$$np \geq 10 \quad \text{and} \quad n(1 - p) \geq 10$$

then the distribution of $\hat{p}$ is approximately normal, with mean and standard deviation

$$\mu_{\hat{p}} = p \quad \text{and} \quad \sigma_{\hat{p}} = \sqrt{\frac{p(1 - p)}{n}}$$

Objective 2 Use the Central Limit Theorem to compute probabilities for sample proportions

Computing Probabilities with the Central Limit Theorem

To compute probabilities involving a sample proportion $\hat{p}$, use the following procedure:

Procedure for Computing Probabilities with the Central Limit Theorem

Step 1: Check to see that the conditions $np \geq 10$ and $n(1-p) \geq 10$ are both met. If so, it is appropriate to use the normal curve.

Step 2: Find the mean $\mu_{\hat{p}}$ and standard deviation $\sigma_{\hat{p}}$.

Step 3: Sketch a normal curve and shade in the area to be found.

Step 4: Find the area using Table A.2 or technology.

EXAMPLE 7.23

Using the Central Limit Theorem to compute a probability

According to a Harris poll taken in September 2013, chocolate is the favorite ice cream flavor for 27% of Americans. If a sample of 100 Americans is taken, what is the probablity that the sample proportion of those who prefer chocolate is greater than 0.30?

Explain It Again

Computing probabilities for sample proportions: Computing probabilities for sample proportions with the Central Limit Theorem is the same as computing probabilities for any normally distributed quantity. Use $\mu_{\hat{p}} = p$ for the mean and $\sigma_{\hat{p}} = \sqrt{p(1-p)/n}$ for the standard deviation.

Solution

Step 1: $np = (100)(0.27) = 27 \geq 10$, and $n(1-p) = (100)(1-0.27) = 73 \geq 10$. We may use the normal curve.

Step 2: $\mu_{\hat{p}} = p = 0.27$.

$$\sigma_{\hat{p}} = \sqrt{\frac{p(1-p)}{n}} = \sqrt{\frac{0.27(1-0.27)}{100}} = 0.044396$$

Step 3: Figure 7.27 presents the normal curve with the area shaded in.

Step 4: We will use Table A.2. We compute the z-score for 0.30.

$$z = \frac{\hat{p} - \mu_{\hat{p}}}{\sigma_{\hat{p}}} = \frac{0.30 - 0.27}{0.044396} = 0.68$$

The table gives the area to the *left* of $z = 0.68$ as 0.7517. The area to the right of $z = 0.68$ is $1 - 0.7517 = 0.2483$. The probability that the sample proportion of those who prefer chocolate is greater than 0.30 is 0.2483.

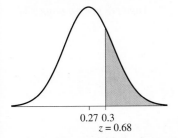

0.27 0.3
$z = 0.68$

Figure 7.27

EXAMPLE 7.24

Using the Central Limit Theorem to determine whether a given value of $\hat{p}$ is unusual

In the 2012 U.S. presidential election, 51% of voters voted for Barack Obama. If a sample of 75 voters were polled, would it be unusual if less than 40% of them had voted for Barack Obama?

Solution

We will compute the probability that the sample proportion is less than 0.40. If this probability is less than 0.05, we will say that the event is unusual.

Step 1: $np = (75)(0.51) = 38.25 \geq 10$, and $n(1-p) = (75)(1-0.51) = 36.75 \geq 10$. We may use the normal curve.

Step 2: $\mu_{\hat{p}} = p = 0.51$.

$$\sigma_{\hat{p}} = \sqrt{\frac{p(1-p)}{n}} = \sqrt{\frac{0.51(1-0.51)}{75}} = 0.057723$$

Step 3: Figure 7.28 presents the normal curve with the area shaded in.

Step 4: We will use Table A.2. We compute the z-score for 0.40.

$$z = \frac{\hat{p} - \mu_{\hat{p}}}{\sigma_{\hat{p}}} = \frac{0.40 - 0.51}{0.057723} = -1.91$$

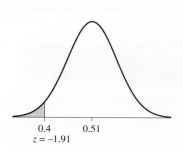

0.4   0.51
$z = -1.91$

Figure 7.28

The area to the left of $z = -1.91$ is 0.0281. It would be unusual for the sample proportion to be less than 0.40.

Check Your Understanding

3. The General Social Survey reported that 56% of American adults saw a doctor for an illness during the past year. A sample of 65 adults is drawn.
   a. What is the probability that more than 60% of them saw a doctor?
   b. Would it be unusual if more than 70% of them saw a doctor?

4. For a certain type of computer chip, the proportion of chips that are defective is 0.10. A computer manufacturer receives a shipment of 200 chips.
   a. What is the probability that the proportion of defective chips in the shipment is between 0.08 and 0.15?
   b. Would it be unusual for the proportion of defective chips to be less than 0.075?

Answers are on page 324.

SECTION 7.4   Exercises

Exercises 1–4 are the Check Your Understanding exercises located within the section.

Understanding the Concepts

In Exercises 5 and 6, fill in each blank with the appropriate word or phrase.

5. If n is the sample size and x is the number in the sample who have a certain characteristic, then x/n is called the sample _____.

6. The probability distribution of $\hat{p}$ is called a _____ distribution.

In Exercises 7 and 8, determine whether the statement is true or false. If the statement is false, rewrite it as a true statement.

7. The distribution of $\hat{p}$ is approximately normal if $np \geq 10$ and $n(1-p) \geq 10$.

8. If n is the sample size, p is the population proportion, and $\hat{p}$ is the sample proportion, then $\sigma_{\hat{p}} = np$.

Practicing the Skills

In Exercises 9–14, n is the sample size, p is the population proportion, and $\hat{p}$ is the sample proportion. If appropriate, use the Central Limit Theorem to find the indicated probability.

9. $n = 147, p = 0.13$; $P(\hat{p} < 0.11)$
10. $n = 65, p = 0.86$; $P(\hat{p} < 0.80)$
11. $n = 270, p = 0.57$; $P(\hat{p} > 0.61)$
12. $n = 103, p = 0.24$; $P(0.20 < \hat{p} < 0.23)$
13. $n = 145, p = 0.05$; $P(0.03 < \hat{p} < 0.08)$
14. $n = 234, p = 0.75$; $P(0.77 < \hat{p} < 0.81)$

Working with the Concepts

15. **Coffee:** The National Coffee Association reported that 63% of U.S. adults drink coffee daily. A random sample of 250 U.S. adults is selected.
    a. Find the mean $\mu_{\hat{p}}$.
    b. Find the standard deviation $\sigma_{\hat{p}}$.

c. Find the probability that more than 67% of the sampled adults drink coffee daily.

d. Find the probability that the proportion of the sampled adults who drink coffee daily is between 0.6 and 0.7.

e. Find the probability that less than 62% of the sampled adults drink coffee daily.

f. Would it be unusual if less than 57% of the sampled adults drink coffee daily?

16. Smartphones: A Pew Research report indicated that 37% of teenagers aged 12–17 own smartphones. A random sample of 150 teenagers is drawn.

a. Find the mean $\mu_{\hat{p}}$.

b. Find the standard deviation $\sigma_{\hat{p}}$.

c. Find the probability that more than 35% of the sampled teenagers own a smartphone.

d. Find the probability that the proportion of the sampled teenagers who own a smartphone is between 0.38 and 0.45.

e. Find the probability that less than 40% of the sampled teenagers own smartphones.

f. Would it be unusual if less than 31% of the sampled teenagers owned smartphones?

17. Student loans: The Institute for College Access and Success reported that 67% of college students in a recent year graduated with student loan debt. A random sample of 85 graduates is drawn.

a. Find the mean $\mu_{\hat{p}}$.

b. Find the standard deviation $\sigma_{\hat{p}}$.

c. Find the probability that less than 60% of the people in the sample were in debt.

d. Find the probability that between 65% and 80% of the people in the sample were in debt.

e. Find the probability that more than 75% of the people in the sample were in debt.

f. Would it be unusual if less than 65% of the people in the sample were in debt?

18. High school graduates: The National Center for Educational Statistics reported that 75% of freshmen entering public high schools in the United States in 2007 graduated with their class in 2011. A random sample of 135 freshmen is chosen.

a. Find the mean $\mu_{\hat{p}}$.

b. Find the standard deviation $\sigma_{\hat{p}}$.

c. Find the probability that less than 80% of freshmen in the sample graduated.

d. Find the probability that the sample proportion of students who graduated is between 0.65 and 0.80.

e. Find the probability that more than 65% of freshmen in the sample graduated.

f. Would it be unusual if the sample proportion of students who graduated was more than 0.85?

19. Government workers: The Bureau of Labor Statistics reported in 2013 that 16% of U.S. nonfarm workers are government employees. A random sample of 50 workers is drawn.

a. Is it appropriate to use the normal approximation to find the probability that less than 20% of the individuals in the sample are government employees? If so, find the probability. If not, explain why not.

b. A new sample of 90 workers is chosen. Find the probability that more than 20% of workers in this sample are government employees.

c. Find the probability that the proportion of workers in the sample of 90 who are government employees is between 0.15 and 0.18.

d. Find the probability that less than 25% of workers in the sample of 90 are government employees.

e. Would it be unusual if the proportion of government employees in the sample of 90 was greater than 0.25?

20. Working two jobs: The Bureau of Labor Statistics reported in a recent year that 5% of employed adults in the United States held multiple jobs. A random sample of 75 employed adults is chosen.

a. Is it appropriate to use the normal approximation to find the probability that less than 6.5% of the individuals in the sample hold multiple jobs? If so, find the probability. If not, explain why not.

b. A new sample of 350 employed adults is chosen. Find the probability that less than 6.5% of the individuals in this sample hold multiple jobs.

c. Find the probability that more than 6% of the individuals in the sample of 350 hold multiple jobs.

d. Find the probability that the proportion of individuals in the sample of 350 who hold multiple jobs is between 0.05 and 0.10.

e. Would it be unusual if less than 4% of the individuals in the sample of 350 held multiple jobs?

21. Future scientists: Education professionals refer to science, technology, engineering, and mathematics as the STEM disciplines. The Alliance for Science and Technology Research in America reported in 2013 that 28% of freshmen entering college planned to major in a STEM discipline. A random sample of 85 freshmen is selected.

a. Is it appropriate to use the normal approximation to find the probability that less than 30% of the freshmen in the sample are planning to major in a STEM discipline? If so, find the probability. If not, explain why not.

b. A new sample of 150 freshmen is selected. Find the probability that less than 30% of the freshmen in this sample are planning to major in a STEM discipline.

c. Find the probability that the proportion of freshmen in the sample of 150 who plan to major in a STEM discipline is between 0.30 and 0.35.

d. Find the probability that more than 32% of the freshmen in the sample of 150 are planning to major in a STEM discipline.

e. Would it be unusual if less than 25% of the freshmen in the sample of 150 were planning to major in a STEM discipline?

22. Blood pressure: High blood pressure has been identified as a risk factor for heart attacks and strokes. The National Health and Nutrition Examination Survey reported that the proportion of U.S. adults with high blood pressure is 0.3. A sample of 38 U.S. adults is chosen.

a. Is it appropriate to use the normal approximation to find the probability that more than 40% of the people in the sample have high blood pressure? If so, find the probability. If not, explain why not.

b. A new sample of 80 adults is drawn. Find the probability that more than 40% of the people in this sample have high blood pressure.

c. Find the probability that the proportion of individuals in the sample of 80 who have high blood pressure is between 0.20 and 0.35.

d. Find the probability that less than 25% of the people in the sample of 80 have high blood pressure.

e. Would it be unusual if more than 45% of the individuals in the sample of 80 had high blood pressure?

23. **Pay your taxes:** According to the Internal Revenue Service, the proportion of federal tax returns for which no tax was paid was $p = 0.326$. As part of a tax audit, tax officials draw a simple random sample of $n = 120$ tax returns.

a. What is the probability that the sample proportion of tax returns for which no tax was paid is less than 0.30?

b. What is the probability that the sample proportion of tax returns for which no tax was paid is between 0.35 and 0.40?

c. What is the probability that the sample proportion of tax returns for which no tax was paid is greater than 0.35?

d. Would it be unusual if the sample proportion of tax returns for which no tax was paid was less than 0.25?

24. **Weekly paycheck:** The Bureau of Labor Statistics reported that in 2009, the median weekly earnings for people employed full time in the United States was $755.

a. What proportion of full-time employees had weekly earnings of more than $755?

b. A sample of 150 full-time employees is chosen. What is the probability that more than 55% of them earned more than $755 per week?

c. What is the probability that less than 60% of the sample of 150 employees earned more than $755 per week?

d. What is the probability that between 45% and 55% of the sample of 150 employees earned more than $755 per week?

e. Would it be unusual if less than 45% of the sample of 150 employees earned more than $755 per week?

25. **Kidney transplants:** The Health Resources and Services Administration reported that 5% of people who received kidney transplants were under the age of 18. How large a sample of kidney transplant patients needs to be drawn so that the sample proportion $\hat{p}$ of those under the age of 18 is approximately normally distributed?

26. **How's your new car?** The General Social Survey reported that 91% of people who bought a car in the past five years were satisfied with their purchase. How large a sample of car buyers needs to be drawn so that the sample proportion $\hat{p}$ who are satisfied is approximately normally distributed?

Extending the Concepts

27. **Flawless tiles:** A new process has been designed to make ceramic tiles. The goal is to have no more than 5% of the tiles be nonconforming due to surface defects. A random sample of 1000 tiles is inspected. Let $\hat{p}$ be the proportion of nonconforming tiles in the sample.

a. If 5% of the tiles produced are nonconforming, what is $P(\hat{p} \geq 0.075)$?

b. Based on the answer to part (a), if 5% of the tiles are nonconforming, is a proportion of 0.075 nonconforming tiles in a sample of 1000 unusually large?

c. If the sample proportion of nonconforming tiles were 0.075, would it be plausible that the goal had been reached? Explain.

d. If 5% of the tiles produced are nonconforming, what is $P(\hat{p} \geq 0.053)$?

e. Based on the answer to part (d), if 5% of the tiles are nonconforming, is a proportion of 0.053 nonconforming tiles in a sample of 1000 unusually large?

f. If the sample proportion of nonconforming tiles were 0.053, would it be plausible that the goal had been reached? Explain.

Answers to Check Your Understanding Exercises for Section 7.4

1. $\mu_{\hat{p}} = 0.82$, $\sigma_{\hat{p}} = 0.08591$

2. $\mu_{\hat{p}} = 0.455$, $\sigma_{\hat{p}} = 0.03380$

3. a. 0.2578 [Tech: 0.2580]

b. The probability that $\hat{p}$ is greater than 0.70 is 0.0116 [Tech: 0.0115]. If we define an event whose probability is less than 0.05 as unusual, then this is unusual.

4. a. 0.8173 [Tech: 0.8179]

b. The probability that $\hat{p}$ is less than 0.075 is 0.1190 [Tech: 0.1193]. This event is not unusual.

| SECTION 7.5 | **The Normal Approximation to the Binomial Distribution** |

Objectives

1. Use the normal curve to approximate binomial probabilities

Objective 1 Use the normal curve to approximate binomial probabilities

We first introduced binomial random variables in Section 6.2. Recall that a binomial random variable represents the number of successes in a series of independent trials. The sample proportion is found by dividing the number of successes by the number of trials.

Since the sample proportion is approximately normally distributed whenever $np \geq 10$ and $n(1 - p) \geq 10$, the number of successes is also approximately normally distributed under these conditions. Therefore, the normal curve can also be used to compute approximate probabilities for the binomial distribution.

We begin by reviewing the conditions under which a random variable has a binomial distribution.

Conditions for the Binomial Distribution

1. A fixed number of trials are conducted.
2. There are two possible outcomes for each trial. One is labeled "success" and the other is labeled "failure."
3. The probability of success is the same on each trial.
4. The trials are independent. This means that the outcome of one trial does not affect the outcomes of the other trials.
5. The random variable X represents the number of successes that occur.

Notation: The following notation is commonly used:

- The number of trials is denoted by n.
- The probability of success is denoted by p, and the probability of failure is $1 - p$.

Mean, Variance, and Standard Deviation of a Binomial Random Variable

Let X be a binomial random variable with n trials and success probability p. Then the mean of X is

$$\mu_X = np$$

The variance of X is

$$\sigma_X^2 = np(1 - p)$$

The standard deviation of X is

$$\sigma_X = \sqrt{np(1 - p)}$$

Explain It Again

Calculating binomial probabilities: Binomial probabilities can be computed exactly by using the methods described in Section 6.2. Using these methods by hand is extremely difficult. The normal approximation provides an easier way to approximate these probabilities when computing by hand.

Binomial probabilities can be very difficult to compute exactly by hand, because many terms have to be calculated and added together. For example, imagine trying to compute the probability that the number of heads is between 75 and 125 when a coin is tossed 200 times. To do this, one would need to compute the following sum:

$$P(X = 75) + P(X = 76) + \cdots + P(X = 124) + P(X = 125)$$

This is nearly impossible to do without technology. Fortunately, probabilities like this can be approximated very closely by using the normal curve. In the days before cheap computing became available, use of the normal curve was the only feasible method for doing these calculations. The normal approximation is somewhat less important now, but is still useful for quick "back of the envelope" calculations.

Recall from Section 7.4 that a sample proportion $\hat{p}$ is approximately normally distributed whenever $np \geq 10$ and $n(1 - p) \geq 10$. Now if X is a binomial random variable representing the number of successes in n trials, the sample proportion is given by $\hat{p} = X/n$. Since $\hat{p}$ is obtained simply by dividing X by the number of trials, it is reasonable to expect that X will be approximately normal whenever $\hat{p}$ is approximately normal. This is in fact the case.

> ### The Normal Approximation to the Binomial
>
> Let X be a binomial random variable with n trials and success probability p. If $np \geq 10$ and $n(1-p) \geq 10$, then X is approximately normal with mean $\mu_X = np$ and standard deviation $\sigma_X = \sqrt{np(1-p)}$.

The continuity correction

The binomial distribution is discrete, whereas the normal distribution is continuous. The **continuity correction** is an adjustment, made when approximating a discrete distribution with a continuous one, that can improve the accuracy of the approximation. To see how it works, imagine that a fair coin is tossed 100 times. Let X represent the number of heads. Then X has the binomial distribution with $n = 100$ trials and success probability $p = 0.5$. Imagine that we want to compute the probability that X is between 45 and 55. This probability will differ depending on whether the endpoints, 45 and 55, are included or excluded. Figure 7.29 illustrates the case where the endpoints are included, that is, where we wish to compute $P(45 \leq X \leq 55)$. The exact probability is given by the total area of the rectangles of the binomial probability histogram corresponding to the integers 45 to 55, inclusive. The approximating normal curve is superimposed. To get the best approximation, we should compute the area under the normal curve between 44.5 and 55.5.

In contrast, Figure 7.30 illustrates the case where we wish to compute $P(45 < X < 55)$. Here the endpoints are excluded. The exact probability is given by the total area of the rectangles of the binomial probability histogram corresponding to the integers 46 to 54. The best normal approximation is found by computing the area under the normal curve between 45.5 and 54.5.

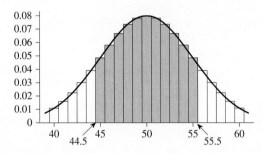

Figure 7.29 To compute $P(45 \leq X \leq 55)$, the areas of the rectangles corresponding to 45 and to 55 should be included. To approximate this probability with the normal curve, compute the area under the curve between 44.5 and 55.5.

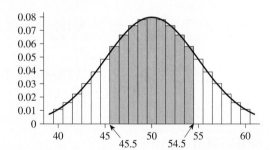

Figure 7.30 To compute $P(45 < X < 55)$, the areas of the rectangles corresponding to 45 and to 55 should be excluded. To approximate this probability with the normal curve, compute the area under the curve between 45.5 and 54.5.

In general, to apply the continuity correction, determine precisely which rectangles of the discrete probability histogram you wish to include, then compute the area under the normal curve corresponding to those rectangles.

EXAMPLE 7.25 Using the continuity correction to compute a probability

Let X be the number of heads that appear when a fair coin is tossed 100 times. Use the normal curve to find $P(45 \leq X \leq 55)$.

Solution

This situation is illustrated in Figure 7.29.

Step 1: Check the assumptions: The number of trials is $n = 100$. Since the coin is fair, the success probability is $p = 0.5$. Therefore, $np = (100)(0.5) = 50 \geq 10$ and $n(1 - p) = (100)(1 - 0.5) = 50 \geq 10$. We can use the normal approximation.

Step 2: We compute the mean and standard deviation of X:

$$\mu_X = np = (100)(0.5) = 50 \qquad \sigma_X = \sqrt{np(1 - p)} = \sqrt{(100)(0.5)(1 - 0.5)} = 5$$

Step 3: Because the probability is $P(45 \leq X \leq 55)$, we want to *include* both 45 and 55. Therefore, we set the left endpoint to 44.5 and the right endpoint to 55.5.

Step 4: We sketch a normal curve, label the mean of 50, and the endpoints 44.5 and 55.5.

Step 5: We use Table A.2 to find the area. The z-scores for 44.5 and 55.5 are

$$z = \frac{44.5 - 50}{5} = -1.1 \qquad z = \frac{55.5 - 50}{5} = 1.1$$

From Table A.2, we find that the probability is 0.7286. See Figure 7.31.

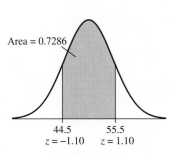

Area = 0.7286

44.5   55.5
$z = -1.10$ $z = 1.10$

Figure 7.31

In Example 7.25, we used the normal approximation to compute a probability of the form $P(a \leq X \leq b)$. We can also use the normal approximation to compute probabilities of the form $P(X \leq a)$, $P(X \geq a)$, and $P(X = a)$.

EXAMPLE 7.26

Illustrate areas to be found for the continuity correction

A fair coin is tossed 100 times. Let X be the number of heads that appear. Illustrate the area under the normal curve that represents each of the following probabilities.

    a. $P(X \leq 55)$
    b. $P(X \geq 55)$
    c. $P(X = 55)$

Solution

    a. We find the area to the left of 55.5, as illustrated in Figure 7.32.
    b. We find the area to the right of 54.5, as illustrated in Figure 7.33.
    c. We find the area between 54.5 and 55.5, as illustrated in Figure 7.34.

Figure 7.32   To approximate $P(X \leq 55)$, find the area to the left of 55.5.

Figure 7.33   To approximate $P(X \geq 55)$, find the area to the right of 54.5.

Figure 7.34   To approximate $P(X = 55)$, find the area between 54.5 and 55.5.

SUMMARY

Following are the areas under the normal curve to use when the continuity correction is applied.

$P(a \leq X \leq b)$

Find the area between $a - 0.5$ and $b + 0.5$.

$P(X \leq b)$

Find the area to the left of $b + 0.5$.

$P(X \geq a)$

Find the area to the right of $a - 0.5$.

$P(X = a)$

Find the area between $a - 0.5$ and $a + 0.5$.

Use the following steps to compute a binomial probability with the normal approximation.

Explain It Again

We will use Table A.2 when using the normal approximation to the binomial: The normal approximation is useful when computing by hand. If technology is to be used, the exact probability can be calculated, so the normal approximation is less useful.

Procedure for Computing Binomial Probabilities with the Normal Approximation

Step 1: Check to see that the conditions $np \geq 10$ and $n(1 - p) \geq 10$ are both met. If so, it is appropriate to use the normal approximation. If not, the probability must be calculated with the binomial distribution (see Section 6.2).

Step 2: Compute the mean μ_X and the standard deviation σ_X.

Step 3: For each endpoint, determine whether to add 0.5 or subtract 0.5.

Step 4: Sketch a normal curve, label the endpoints, and shade in the area to be found.

Step 5: Find the area using Table A.2 or technology. Note, however, that if you are using technology, you may be able to compute the probability exactly without using the normal approximation.

EXAMPLE 7.27   Using the continuity correction to compute a probability

The *Statistical Abstract of the United States* reported that 66% of students who graduated from high school in 2012 enrolled in college. One hundred high school graduates are sampled. Let X be the number who enrolled in college. Find $P(X \leq 75)$.

Solution

Step 1: Check the assumptions. The number of trials is $n = 100$ and the success probability is $p = 0.66$. Therefore $np = (100)(0.66) = 66 \geq 10$ and $n(1 - p) = (100)(1 - 0.66) = 34 \geq 10$. We can use the normal approximation.

Step 2: We compute the mean and standard deviation of X:

$$\mu_X = np = (100)(0.66) = 66 \quad \sigma_X = \sqrt{np(1 - p)} = \sqrt{(100)(0.66)(1 - 0.66)} = 4.73709$$

Step 3: Since the probability is $P(X \leq 75)$, we compute the area to the left of 75.5.

Step 4: We sketch a normal curve, and label the mean of 66 and the point 75.5.

Step 5: We use Table A.2 to find the area. The z-score for 75.5 is

$$z = \frac{75.5 - 66}{4.73709} = 2.01$$

From Table A.2 we find that the probability is 0.9778. See Figure 7.35.

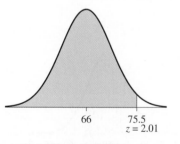

Figure 7.35

Sometimes we want to use the normal approximation to compute a probability of the form $P(X < a)$ or $P(X > a)$. One way to do this is by changing the inequality to a form involving $\leq$ or $\geq$. Example 7.28 provides an illustration.

EXAMPLE 7.28   Using the continuity correction to compute a probability

The *Statistical Abstract of the United States* reported that 66% of students who graduated from high school in 2012 enrolled in college. One hundred high school graduates are sampled. Approximate the probability that more than 60 enroll in college.

Solution

Let X be the number of students in the sample who enrolled in college. We need to find $P(X > 60)$. We change this to an inequality involving $\geq$ by noting that $P(X > 60) = P(X \geq 61)$. We therefore find $P(X \geq 61)$.

Step 1: Check the assumptions. The number of trials is $n = 100$ and the success probability is $p = 0.66$. Therefore $np = (100)(0.66) = 66 \geq 10$ and $n(1 - p) = (100)(1 - 0.66) = 34 \geq 10$. We can use the normal approximation.

Step 2: We compute the mean and standard deviation of X:

$$\mu_X = np = (100)(0.66) = 66$$
$$\sigma_X = \sqrt{np(1 - p)} = \sqrt{(100)(0.66)(1 - 0.66)} = 4.73709$$

Step 3: Since the probability is $P(X \geq 61)$, we compute the area to the right of 60.5.

Step 4: We sketch a normal curve, and label the mean of 66 and the point 60.5.

Step 5: We use Table A.2 to find the area. The z-score for 60.5 is

$$z = \frac{60.5 - 66}{4.73709} = -1.16$$

From Table A.2 we find that the probability is 0.8770. See Figure 7.36.

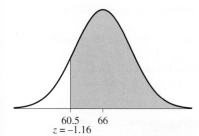

60.5   66
$z = -1.16$

Figure 7.36

Check Your Understanding

1. X is a binomial random variable with $n = 50$ and $p = 0.15$. Should the normal approximation be used to find $P(X > 10)$? Why or why not?

2. X is a binomial random variable with $n = 72$ and $p = 0.90$. Should the normal approximation be used to find $P(X \leq 60)$? Why or why not?

3. Let X have a binomial distribution with $n = 64$ and $p = 0.41$. If appropriate, use the normal approximation to find $P(X \leq 20)$. If not, explain why not.

4. Let X have a binomial distribution with $n = 379$ and $p = 0.09$. If appropriate, use the normal approximation to find $P(X > 40)$. If not, explain why not.

Answers are on page 331.

SECTION 7.5   Exercises

Exercises 1–4 are the Check Your Understanding exercises located within the section.

Understanding the Concepts

In Exercises 5 and 6, fill in each blank with the appropriate word or phrase.

5. If X is a binomial random variable and if $np \geq 10$ and $n(1 - p) \geq 10$, then X is approximately normal with $\mu_X = $ _____ and $\sigma_X = $ _____.

6. The adjustment made when approximating a discrete random distribution with a continuous one is called the _____ correction.

In Exercises 7 and 8, determine whether the statement is true or false. If the statement is false, rewrite it as a true statement.

7. If technology is to be used, exact binomial probabilities can be calculated and the normal approximation is not necessary.

8. If X is a binomial random variable with n trials and success probability p, then as n gets larger, the distribution of X becomes more skewed.

Practicing the Skills

In Exercises 9–14, n is the sample size, p is the population proportion of successes, and X is the number of successes in the sample. Use the normal approximation to find the indicated probability.

9. $n = 78$, $p = 0.43$; $P(X > 40)$

10. $n = 538$, $p = 0.86$; $P(X \leq 470)$

11. $n = 99$, $p = 0.57$; $P(X \geq 55)$

12. $n = 442$, $p = 0.54$; $P(X < 243)$

13. $n = 106$, $p = 0.14$; $P(14 < X < 18)$

14. $n = 61$, $p = 0.34$; $P(20 \leq X \leq 24)$

Working with the Concepts

15. **Google it:** According to Net Market Share, 83% of Internet searches in January 2013 used the Google search engine. A sample of 100 searches is studied.
    a. Approximate the probability that more than 85 of the searches used Google.
    b. Approximate the probability that 75 or fewer of the searches used Google.
    c. Approximate the probability that the number of searches that used Google is between 70 and 80, inclusive.

16. **Big babies:** The Centers for Disease Control and Prevention reports that 25% of baby boys 6–8 months old in the United States weigh more than 20 pounds. A sample of 150 babies is studied.
    a. Approximate the probability that more than 40 weigh more than 20 pounds.
    b. Approximate the probability that 35 or fewer weigh more than 20 pounds.
    c. Approximate the probability that the number who weigh more than 20 pounds is between 30 and 40, exclusive.

17. **High blood pressure:** The National Health and Nutrition Examination Survey reported that 30% of adults in the United States have hypertension (high blood pressure). A sample of 300 adults is studied.
    a. Approximate the probability that 85 or more have hypertension.
    b. Approximate the probability that fewer than 80 have hypertension.
    c. Approximate the probability that the number who have hypertension is between 75 and 85, exclusive.

18. **Stress at work:** In a poll conducted by the General Social Survey, 81% of respondents said that their jobs were sometimes or always stressful. Two hundred workers are chosen at random.
    a. Approximate the probability that 160 or fewer find their jobs stressful.
    b. Approximate the probability that more than 150 find their jobs stressful.
    c. Approximate the probability that the number who find their jobs stressful is between 155 and 162, inclusive.

19. **What's your opinion?** A pollster will interview a sample of 200 voters to ask whether they support a proposal to increase the sales tax to build a new light rail system. Assume that in fact 55% of the voters support the proposal.
    a. Approximate the probability that 100 or fewer of the sampled voters support the proposal.
    b. Approximate the probability that more than 105 voters support the proposal.
    c. Approximate the probability that the number of voters who support the proposal is between 100 and 110, inclusive.

20. **Gardening:** A gardener buys a package of seeds. Eighty percent of seeds of this type germinate. The gardener plants 90 seeds.
    a. Approximate the probability that fewer than 75 seeds germinate.
    b. Approximate the probability that 80 or more seeds germinate.
    c. Approximate the probability that the number of seeds that germinate is between 67 and 75, exclusive.

21. **The car is in the shop:** Among automobiles of a certain make, 23% require service during a one-year warranty period. A dealer sells 87 of these vehicles.
    a. Approximate the probability that 25 or fewer of these vehicles require repairs.
    b. Approximate the probability that more than 17 vehicles require repairs.
    c. Approximate the probability that the number of vehicles that require repairs is between 15 and 20, exclusive.

22. **Genetics:** Pea plants contain two genes for seed color, each of which may be Y (for yellow seeds) or G (for green seeds). Plants that contain one of each type of gene are called heterozygous. According to the Mendelian theory of genetics, if two heterozygous plants are crossed, each of their offspring will have probability 0.75 of having yellow seeds and probability 0.25 of having green seeds. One hundred such offspring are produced.
    a. Approximate the probability that more than 30 have green seeds.
    b. Approximate the probability that 80 or fewer have yellow seeds.
    c. Approximate the probability that the number with green seeds is between 30 and 35, inclusive.

23. **Getting bumped:** Airlines often sell more tickets for a flight than there are seats, because some ticket holders don't show up for the flight. Assume that an airplane has 100 seats for passengers and that the probability that a person holding a ticket appears for the flight is 0.90. If the airline sells

105 tickets, what is the probability that everyone who appears for the flight will get a seat?

24. **College admissions:** A small college has enough space to enroll 300 new students in its incoming freshman class. From past experience, the admissions office knows that 65% of students who are accepted actually enroll. If the admissions office accepts 450 students, what is the probability that there will be enough space for all the students who enroll?

Extending the Concepts

25. **Probability of a single number:** A fair coin is tossed 100 times. Use the normal approximation to approximate the probability that the coin comes up heads exactly 50 times.

Answers to Check Your Understanding Exercises for Section 7.5

1. No, $np = 7.5 < 10$.
2. No, $n(1 - p) = 7.2 < 10$.
3. 0.0721 [Tech: 0.0723]
4. 0.1251 [Tech: 0.1257]

SECTION 7.6 | Assessing Normality

Objectives

1. Use dotplots to assess normality
2. Use boxplots to assess normality
3. Use histograms to assess normality
4. Use stem-and-leaf plots to assess normality
5. Use normal quantile plots to assess normality

Many statistical procedures, some of which we will learn about in Chapters 8 and 9, require that we draw a sample from a population whose distribution is approximately normal. Often we don't know whether the population is approximately normal when we draw the sample. So the only way we have to assess whether the population is approximately normal is to examine the sample. In this section, we will describe some ways in which this can be done.

There are three important ideas to remember when assessing normality:

1. We are not trying to determine whether the population is *exactly* normal. No population encountered in practice is *exactly* normal. We are only trying to determine whether the population is *approximately* normal.

2. Assessing normality is more important for small samples than for large samples. When the sample size is large, say $n > 30$, the Central Limit Theorem ensures that $\bar{x}$ is approximately normal. Most statistical procedures designed for large samples rely on the Central Limit Theorem for their validity, so normality of the population is not so important in these cases.

3. Hard-and-fast rules do not work well. They are generally too lenient for very small samples (finding populations to be approximately normal when they are not) or too strict for larger samples (finding populations not to be approximately normal when they are). Informal judgment works as well as or better than hard-and-fast rules.

When a sample is very small, it is often impossible to be sure whether it came from an approximately normal population. The best we can do is to examine the sample for signs of nonnormality. If no such signs exist, we will treat the population as approximately normal. Because the normal curve is unimodal and symmetric, samples from normal populations rarely have more than one distinct mode, and rarely exhibit a large degree of skewness. In addition, samples from normal populations rarely contain outliers. We summarize the conditions under which we will reject the assumption that a population is approximately normal.

Recall: An outlier is a data value that is considerably larger or smaller than most of the rest of the data.

SUMMARY

We will reject the assumption that a population is approximately normal if a sample has *any* of the following features:

1. The sample contains an outlier.
2. The sample exhibits a large degree of skewness.
3. The sample is multimodal; in other words, it has more than one distinct mode.

If the sample has *none* of the preceding features, we will treat the population as being approximately normal.

Many methods have been developed for assessing normality; some of them are quite sophisticated. For our purposes, it will be sufficient to examine dotplots, boxplots, stem-and-leaf plots, and histograms of the sample. We will also describe normal quantile plots, which provide another useful method of assessment.

Objective 1 Use dotplots to assess normality

Dotplots

Dotplots are excellent for detecting outliers and multimodality. They can also be used to detect skewness, although they are not quite as effective as histograms for that purpose.

EXAMPLE 7.29

Recall: Dotplots were introduced in Section 2.3.

Using a dotplot to assess normality

The accuracy of an oven thermostat is being tested. The oven is set to 360°F, and the temperature when the thermostat turns off is recorded. A sample of size 7 yields the following results:

358 363 361 355 367 352 368

Is it reasonable to treat this as a sample from an approximately normal population? Explain.

Solution

Figure 7.37 presents a dotplot of the temperatures. The dotplot does not reveal any outliers. The plot does not exhibit a large degree of skewness, and there is no evidence that the population has more than one mode. Therefore, we can treat this as a sample from an approximately normal population.

Figure 7.37 The dotplot of the oven temperatures does not reveal any outliers. The plot does not exhibit a large degree of skewness, and there is no evidence that the population has more than one mode. Therefore, we can treat this as a sample from an approximately normal population.

EXAMPLE 7.30

Using a dotplot to assess normality

At a recent health fair, several hundred people had their pulse rates measured. A simple random sample of six records was drawn, and the pulse rates, in beats per minute, were

68 71 79 98 67 75

Is it reasonable to treat this as a sample from an approximately normal population? Explain.

Solution

Figure 7.38 presents a dotplot of the pulse rates. It is clear that the value 98 is an outlier. Therefore, we should not treat this as a sample from an approximately normal population.

Figure 7.38 The value 98 is an outlier. Therefore, we should not treat this as a sample from an approximately normal population.

Objective 2 Use boxplots to assess normality

Boxplots

Boxplots are very good for detecting outliers and skewness. They work best for data sets that are not too small. For very small samples, it is just as informative to plot all the points with a dotplot.

| EXAMPLE 7.31 | Using a boxplot to assess normality |

Recall: Boxplots were introduced in Section 3.3.

An insurance adjuster obtains a sample of 20 estimates, in hundreds of dollars, for repairs to cars damaged in collisions. Following are the data.

12.1	15.7	14.2	4.6	8.2	11.6	12.9	11.2	14.9	13.7
6.6	7.2	12.6	9.0	11.9	7.8	9.0	16.2	16.5	12.1

Is it reasonable to treat this as a sample from an approximately normal population? Explain.

Solution

Figure 7.39 presents a boxplot of the repair estimates. There are no outliers. Although the median is not exactly halfway between the quartiles, the skewness is not great. Therefore, we may treat this as a sample from an approximately normal population.

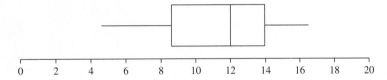

Figure 7.39   There are no outliers, and no evidence of strong skewness. Therefore, we may treat this as a sample from an approximately normal population.

| EXAMPLE 7.32 | Using a boxplot to assess normality |

A recycler determines the amount of recycled newspaper, in cubic feet, collected each week. Following are the results for a sample of 18 weeks.

2129	2853	2530	2054	2075	2011	2162	2285	2668
3194	4834	2469	2380	2567	4117	2337	3179	3157

Is it reasonable to treat this as a sample from an approximately normal population? Explain.

Solution

Figure 7.40 presents a boxplot of the amount of recycled newspaper. The value 4834 is an outlier. In addition, the upper whisker is much longer than the lower one, which indicates fairly strong skewness. Therefore, we should not treat this as a sample from an approximately normal population.

Figure 7.40   The value 4834 is an outlier, and there is evidence of fairly strong skewness as well. Therefore, we should not treat this as a sample from an approximately normal population.

Objective 3 Use histograms to assess normality

Histograms

Histograms are excellent for detecting strong skewness. They are more effective for data sets that are not too small (for very small data sets, a histogram is just like a dotplot with the dots replaced by rectangles).

EXAMPLE 7.33

Use a histogram to assess normality

Recall: Histograms were introduced in Section 2.2.

Diameters were measured, in millimeters, for a simple random sample of 20 grade A eggs from a certain farm. The results were

| 59 | 60 | 60 | 56 | 59 | 56 | 62 | 58 | 60 | 59 |
| 61 | 59 | 61 | 61 | 63 | 60 | 56 | 58 | 63 | 58 |

Construct a histogram for these data. Is it reasonable to treat this as a sample from an approximately normal population? Explain.

Solution

Figure 7.41 presents a relative frequency histogram of the diameters. The histogram does not reveal any outliers, nor does it exhibit a large degree of skewness. There is no evidence that the population has more than one mode. Therefore, we can treat this as a sample from an approximately normal population.

Figure 7.41 The histogram of the egg diameters does not reveal any outliers, nor a large degree of skewness, nor evidence of more than one mode. Therefore, we can treat this as a sample from an approximately normal population.

EXAMPLE 7.34

Use a histogram to assess normality

A shoe manufacturer is testing a new type of leather sole. A simple random sample of 22 people wore shoes with the new sole for a period of four months. The amount of wear on the right shoe was measured for each person. The results, in thousandths of an inch, were

| 24.1 | 2.2 | 11.8 | 2.7 | 4.1 | 13.9 | 33.6 | 2.4 | 36.2 | 16.8 | 5.4 |
| 4.6 | 4.5 | 4.1 | 6.1 | 6.3 | 22.6 | 29.1 | 12.2 | 4.6 | 15.8 | 7.7 |

Construct a histogram for these data. Is it reasonable to treat this as a sample from an approximately normal population? Explain.

Solution

Figure 7.42 presents a relative frequency histogram of the amounts of wear. The histogram reveals that the sample is strongly skewed to the right. We should not treat this as a sample from an approximately normal population.

Figure 7.42 The histogram is strongly skewed. Therefore, we should not treat this as a sample from an approximately normal population.

Objective 4 Use
stem-and-leaf plots to assess
normality

Recall: Stem-and-leaf plots
were introduced in Section 2.3.

Stem-and-Leaf Plots

Stem-and-leaf plots can be used in place of histograms when the number of stems is large enough to provide an idea of the shape of the sample. Like histograms, stem-and-leaf plots are excellent for detecting skewness. They are more useful for data sets that are not too small, so that some of the stems will contain more than one leaf. Stem-and-leaf plots are easier to construct by hand than histograms are, but histograms are sometimes easier to construct with technology. For example, the TI-84 Plus calculator will construct histograms, but cannot construct stem-and-leaf plots.

EXAMPLE 7.35

Use a stem-and-leaf plot to assess normality

A psychologist measures the time it takes for each of 20 rats to run a maze. The times, in seconds, are

54	48	49	54	63	54	66	32	45	52
41	37	56	56	52	53	41	45	48	43

Construct a stem-and-leaf plot for these data. Is it reasonable to treat this as a random sample from an approximately normal population?

Solution

Figure 7.43 presents a stem-and-leaf plot of the times. The stem-and-leaf plot reveals no outliers, strong skewness, or multimodality. We may treat this as a sample from an approximately normal population.

```
3 | 2
3 | 7
4 | 113
4 | 55889
5 | 223444
5 | 66
6 | 3
6 | 6
```

Figure 7.43 There are no outliers, strong skewness, or multimodality.

Check Your Understanding

1. For each of the following dotplots, determine whether it is reasonable to treat the sample as coming from an approximately normal population.

   a.

   b.

2. For each of the following histograms, determine whether it is reasonable to treat the sample as coming from an approximately normal population.

   a.   **b.**

3. The following stem-and-leaf plot represents a sample from a population. Is it reasonable to assume that this population is approximately normal?

```
1 | 34579
2 | 0278
3 | 25
4 | 37
5 | 38
6 | 4
7 |
8 | 1
9 | 6
```

4. The following boxplot represents a sample from a population. Is it reasonable to assume that this population is approximately normal?

Answers are on page 341.

Answers are on page 341.

Objective 5 Use normal quantile plots to assess normality

Normal Quantile Plots

Normal quantile plots are somewhat more complex than dotplots, histograms, and stem-and-leaf plots. We will present the idea behind normal quantile plots with an example. A simple random sample of size $n = 5$ is drawn, and we want to determine whether the population it came from is approximately normal. The five sample values, in increasing order, are

$$3.0 \quad 3.3 \quad 4.8 \quad 5.9 \quad 7.8$$

We proceed by using the following steps:

Step 1: Let n be the number of values in the data set. Spread the n values evenly over the interval from 0 to 1. This is done by assigning the value $1/(2n)$ to the first sample value, $3/(2n)$ to the second, and so forth. The last sample value will be assigned the value $(2n - 1)/(2n)$. These values, denoted a_i, represent areas under the normal curve. For $n = 5$, the values are 0.1, 0.3, 0.5, 0.7, and 0.9.

i	x_i	a_i
1	3.0	0.1
2	3.3	0.3
3	4.8	0.5
4	5.9	0.7
5	7.8	0.9

Step 2: The values assigned in Step 1 represent left-tail areas under the normal curve. We now find the z-scores corresponding to each of these areas. The results are shown in the following table.

i	x_i	a_i	z_i
1	3.0	0.1	−1.28
2	3.3	0.3	−0.52
3	4.8	0.5	0.00
4	5.9	0.7	0.52
5	7.8	0.9	1.28

Step 3: Plot the points (x_i, z_i). The plot is shown in Figure 7.44. A straight line has been added to the plot to help in interpreting the results. If the points approximately follow a straight line, then the population may be treated as being approximately normal. If the points deviate substantially from a straight line, the population should not be treated as normal. In this case, the points do approximately follow a straight line, so we may treat this population as approximately normal.

Why do the points on a normal quantile plot tend to follow a straight line when the population is normal? If the population is normal, then, on the average, the values z_i will be close to the actual z-scores of the x_i. Now for any sample, the actual z-scores will follow a straight line when plotted against the x_i. Therefore, if the population is approximately normal, it is likely that the points (x_i, z_i) will approximately follow a straight line.

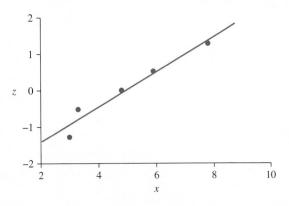

Figure 7.44   The points do not deviate substantially from a straight line. We can treat this population as approximately normal.

In practice, normal quantile plots are always constructed with technology. Figure 7.45 presents the normal quantile plot in Figure 7.44 as drawn by the TI-84 Plus calculator. The calculator does not add a line to the plot. Step-by-step instructions for constructing normal quantile plots with the TI-84 Plus calculator are presented in the Using Technology section on page 338.

Figure 7.45

EXAMPLE 7.36

Using a normal quantile plot to assess normality

A placement exam is given to each entering freshman at a large university. A simple random sample of 20 exam scores is drawn, with the following results.

| 61 | 60 | 60 | 68 | 63 | 63 | 94 | 66 | 65 | 98 |
| 61 | 71 | 74 | 63 | 66 | 61 | 61 | 65 | 72 | 85 |

Construct a normal probability plot. Is the distribution of exam scores approximately normal?

Figure 7.46

Solution

We use the TI-84 Plus calculator. The results are shown in Figure 7.46. The points do not closely follow a straight line. The distribution is not approximately normal.

Check Your Understanding

5. Is it reasonable to treat the sample in the following normal quantile plot as coming from an approximately normal population? Explain.

6. Is it reasonable to treat the sample in the following normal quantile plot as coming from an approximately normal population? Explain.

Answers are on page 341.

USING TECHNOLOGY

We use Example 7.36 to illustrate the technology steps.

TI-84 PLUS

Constructing normal quantile plots

Step 1. Enter the data from Example 7.36 into **L1** in the data editor.

Step 2. Press **2nd, Y=** to access the STAT PLOTS menu and select Plot1 by pressing **1**.

Step 3. Select **On** and the normal quantile plot icon.

Step 4. For **Data List**, select **L1**, and for **Data Axis**, choose the **X** option (Figure A).

Step 5. Press **ZOOM** and then **9: ZoomStat** (Figure B).

Figure A

Figure B

MINITAB

Constructing normal quantile plots

Step 1. Enter the data from Example 7.36 into **Column C1**.

Step 2. Click on **Graph** and select **Probability Plot**. Choose the **Single** option and press **OK**.

Step 3. Enter **C1** in the **Graph variables** field.

Step 4. Click **OK** (Figure C).

Figure C

Exercises 1–6 are the Check Your Understanding exercises located within the section.

Understanding the Concepts

In Exercise 7, fill in each blank with the appropriate word or phrase.

7. A population is rejected as being approximately normal if the sample contains an _____ , if the sample contains a large degree of _____ , or if the sample has more than one distinct _____ .

In Exercise 8, determine whether the statement is true or false. If the statement is false, rewrite it as a true statement.

8. If the points in a normal quantile plot deviate from a straight line, then the population can be treated as approximately normal.

Practicing the Skills

9. The following dotplot illustrates a sample. Is it reasonable to treat this as a sample from an approximately normal population? Explain.

10. The following dotplot illustrates a sample. Is it reasonable to treat this as a sample from an approximately normal population? Explain.

11. The following boxplot illustrates a sample. Is it reasonable to treat this as a sample from an approximately normal population? Explain.

12. The following boxplot illustrates a sample. Is it reasonable to treat this as a sample from an approximately normal population? Explain.

13. The following histogram illustrates a sample. Is it reasonable to treat this as a sample from an approximately normal population? Explain.

14. The following histogram illustrates a sample. Is it reasonable to treat this as a sample from an approximately normal population? Explain.

15. The following stem-and-leaf plot illustrates a sample. Is it reasonable to treat this as a sample from an approximately normal population? Explain.

0	35
1	46
2	0022358
3	18
4	3
5	4
6	36
7	19
8	34
9	4
10	
11	8

16. The following stem-and-leaf plot illustrates a sample. Is it reasonable to treat this as a sample from an approximately normal population? Explain.

7	35
8	47
9	024
10	379
11	37
12	34
13	0
14	1

17. The following normal quantile plot illustrates a sample. Is it reasonable to treat this as a sample from an approximately normal population? Explain.

18. The following normal quantile plot illustrates a sample. Is it reasonable to treat this as a sample from an approximately normal population? Explain.

Working with the Concepts

19. Drug concentrations: A sample of 10 people ingested a new formulation of a drug. Six hours later, the concentrations in their bloodstreams, in nanograms per milliliter, were as follows.

2.3 1.4 1.8 2.1 1.0 4.1 1.8 2.9 2.5 2.7

Construct a dotplot for this sample. Is it reasonable to treat the sample as coming from an approximately normal population? Explain.

20. Reading scores: A random sample of eight elementary school children were given a standardized reading test. Following are their scores.

72 77 65 85 68 83 73 79

Construct a dotplot for this sample. Is it reasonable to treat the sample as coming from an approximately normal population? Explain.

21. Timed task: The number of minutes needed to complete a certain spreadsheet task was measured for 20 clerical workers. The results were as follows.

4.5 5.8 3.7 4.9 4.3 4.7 5.8 3.2 3.0 5.1
3.6 4.3 3.6 5.4 4.7 3.0 3.4 4.3 4.4 3.5

Construct a boxplot for this sample. Is it reasonable to treat the sample as coming from an approximately normal population? Explain.

22. Impure cans: A manufacturer of aluminum cans measured the level of impurities in 24 cans. The amounts of impurities, in percent, were as follows.

2.1 1.5 1.9 1.3 2.7 4.9 2.1 1.8
1.3 1.0 3.2 4.4 8.0 4.5 2.5 1.6
2.8 8.2 9.5 3.8 1.9 1.5 2.8 1.6

Construct a boxplot for this sample. Is it reasonable to treat the sample as coming from an approximately normal population? Explain.

23. Defective items: The number of defective items produced on an assembly line during an hour is counted for a random sample of 20 hours. The results are as follows.

21 16 10 10 11   9 13 12 11 29
10 14 27 10 11 11 12 19 11   9

Construct a stem-and-leaf plot for this sample. Is it reasonable to treat the sample as coming from an approximately normal population? Explain.

24. Fish weights: A fish hatchery raises trout to stock streams and lakes. The weights, in ounces, of a sample of 18 trout at their time of release are as follows.

9.9 11.3 11.4 9.0 10.1 8.2 8.9 9.9 10.5
8.6   7.8 10.8 8.4   9.6 9.9 8.4 9.0   9.1

Construct a stem-and-leaf plot for this sample. Is it reasonable to treat the sample as coming from an approximately normal population? Explain.

25. Timed task: Construct a histogram for the data in Exercise 21. Explain how the histogram shows whether it is appropriate to treat this sample as coming from an approximately normal population.

26. Impure cans: Construct a histogram for the data in Exercise 22. Explain how the histogram shows whether it is appropriate to treat this sample as coming from an approximately normal population.

27. Defective items: Construct a normal quantile plot for the data in Exercise 23. Explain how the plot shows whether it is appropriate to treat this sample as coming from an approximately normal population.

28. Fish weights: Construct a normal quantile plot for the data in Exercise 24. Explain how the plot shows whether it is appropriate to treat this sample as coming from an approximately normal population.

Extending the Concepts

29. Transformation to normality: Consider the following data set:

2 37 67 108 148 40 1 9 3 237 12 80

a. Show that this data set does not come from an approximately normal population.

b. Take the square root of each value in the data set. This is called a *square-root transformation* of the data. Show that the square roots may be considered to be a sample from an approximately normal population.

30. Transformation to normality: Consider the following data set:

−0.5 0.8 1.7 −1.0 −10.0 1.7 0.5 0.3 −5.0

a. Show that this data set does not come from an approximately normal population.

b. Take the reciprocal of each value in the data set (the reciprocal of x is $1/x$). This is called a *reciprocal transformation* of the data. Show that the reciprocals may be considered to be a sample from an approximately normal population.

31. Transformation to normality: Consider the following data set:

4.1 2.7 1.2 10.3 0.9 2.4
1.5 1.9 2.1 16.1 1.4 1.0

a. Is it reasonable to treat it as a sample from an approximately normal population?

b. Perform a square-root transformation. Is it reasonable to treat the square-root-transformed data as a sample from an approximately normal population?

c. Perform a reciprocal transformation. Is it reasonable to treat the reciprocal-transformed data as a sample from an approximately normal population?

32. Transformation to normality: Consider the following data set:

28.0   6.7   8.6   2.3   25.0   12.5   4.4
37.3   12.0   48.0   0.7   11.6   0.1

a. Is it reasonable to treat it as a sample from an approximately normal population?

b. Perform a square-root transformation. Is it reasonable to treat the square-root-transformed data as a sample from an approximately normal population?

c. Perform a reciprocal transformation. Is it reasonable to treat the reciprocal-transformed data as a sample from an approximately normal population?

Answers to Check Your Understanding Exercises for Section 7.6

1. a. The plot contains an outlier. The population is not approximately normal.

b. We may treat this population as approximately normal.

2. a. We may treat this population as approximately normal.

b. The histogram has more than one mode. The population is not approximately normal.

3. The plot reveals that the sample is strongly skewed. The population is not approximately normal.

4. There are no outliers and no evidence of strong skewness. We may treat this population as approximately normal.

5. The points follow a straight line fairly closely. We may treat this population as approximately normal.

6. The points do not follow a straight line fairly closely. The population is not approximately normal.

Chapter 7 Summary

Section 7.1: Continuous random variables can be described with probability density curves. The area under a probability density curve over an interval can be interpreted in either of two ways. It represents the proportion of the population that is contained in the interval, and it also represents the probability that a randomly chosen value from the population falls within the interval. The normal curve is the most commonly used probability density curve. The standard normal curve represents a normal population with mean 0 and standard deviation 1. We can find areas under the standard normal curve by using Table A.2 or with technology.

Section 7.2: In practice, we need to work with normal distributions with different values for the mean and standard deviation. Technology can be used to compute probabilities for any normal distribution. We can also use Table A.2 to find probabilities for any normal distribution by standardization. Standardization involves computing the z-score by subtracting the mean and dividing by the standard deviation. The z-score has a standard normal distribution, so we can find probabilities by using Table A.2.

Section 7.3: The sampling distribution of a statistic such as a sample mean is the probability distribution of all possible values of the statistic. The Central Limit Theorem states that the sampling distribution of a sample mean is approximately normal so long as the sample size is large enough. Therefore, we can use the normal curve to compute approximate probabilities regarding the sample mean whenever the sample size is sufficiently large. For most populations, samples of size 30 or more are large enough.

Section 7.4: The Central Limit Theorem can also be used to compute approximate probabilities regarding sample proportions. The sampling distribution of a sample proportion is approximately normal so long as the sample size is large enough. The sample size is large enough if both np and $n(1 - p)$ are at least 10.

Section 7.5: A binomial random variable represents the number of successes in a series of independent trials. The number of successes is closely related to the sample proportion, because the sample proportion is found by dividing the number of successes by the number of trials. Since the sample proportion is approximately normally distributed whenever $np \geq 10$ and $n(1 - p) \geq 10$, the number of successes is also approximately normally distributed under these conditions. Therefore, the normal curve can also be used to compute approximate probabilities for the binomial distribution. Because the binomial distribution is discrete, the continuity correction can be used to provide more accurate approximations.

Section 7.6: Many statistical procedures require the assumption that a sample is drawn from a population that is approximately normal. Although it is very difficult to determine whether a small sample comes from such a population, we can examine the sample for outliers, multimodality, and large degrees of skewness. If a sample contains no outliers, is not strongly skewed, and has only one distinct mode, we will treat it as though it came from an approximately normal population. Dotplots, boxplots, histograms, stem-and-leaf plots, and normal quantile plots can be used to assess normality.

Vocabulary and Notation

Important Formulas

z-score:

$$z = \frac{x - \mu}{\sigma}$$

Convert z-score to raw score:

$$x = \mu + z\sigma$$

Standard deviation of the sample mean:

$$\sigma_{\bar{x}} = \frac{\sigma}{\sqrt{n}}$$

z-score for a sample mean:

$$z = \frac{\bar{x} - \mu}{\sigma_{\bar{x}}}$$

Standard deviation of the sample proportion:

$$\sigma_{\hat{p}} = \sqrt{\frac{p(1 - p)}{n}}$$

z-score for a sample proportion:

$$z = \frac{\hat{p} - p}{\sigma_{\hat{p}}}$$

Chapter Quiz

1. Following is a probability density curve for a population.

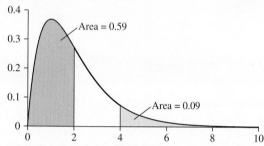

   a. What proportion of the population is between 2 and 4?
   b. If a value is chosen at random from this population, what is the probability that it will be greater than 2?

2. Find the area under the standard normal curve
   a. To the left of $z = 1.77$
   b. To the right of $z = 0.41$
   c. Between $z = -2.12$ and $z = 1.37$

3. Find the z-score that has
   a. An area of 0.33 to its left
   b. An area of 0.79 to its right

4. Find the z-scores that bound the middle 80% of the area under the normal curve.

5. Find $z_{0.15}$.

6. Suppose that salaries of recent graduates from a certain college are normally distributed with mean $\mu = \$42,650$ and standard deviation $\sigma = \$3800$. What two salaries bound the middle 50%?

7. A normal population has mean $\mu = 242$ and standard deviation $\sigma = 31$.
   a. What proportion of the population is greater than 233?
   b. What is the probability that a randomly chosen value will be less than 249?

8. Suppose that in a bowling league, the scores among all bowlers are normally distributed with mean $\mu = 182$ points and standard deviation $\sigma = 14$ points. A trophy is given to each player whose score is at or above the 97th percentile. What is the minimum score needed for a bowler to receive a trophy?

9. State the Central Limit Theorem.

10. A population has mean $\mu = 193$ and standard deviation $\sigma = 42$. Compute $\mu_{\bar{x}}$ and $\sigma_{\bar{x}}$ for samples of size $n = 64$.

11. The running time for videos submitted to YouTube in a given week is normally distributed with $\mu = 390$ seconds and standard deviation $\sigma = 148$ seconds.
    a. If a single video is randomly selected, what is the probability that the running time of the video exceeds 6 minutes (360 seconds)?
    b. Suppose that a sample of 40 videos is selected. What is the probability that the mean running time of the sample exceeds 6 minutes?

12. A sample of size $n = 55$ is drawn from a population with proportion $p = 0.34$. Let $\hat{p}$ be the sample proportion.
    a. Find $\mu_{\hat{p}}$ and $\sigma_{\hat{p}}$.
    b. Find $P(\hat{p} > 0.21)$.
    c. Find $P(\hat{p} < 0.40)$.

13. On a certain television channel, 18% of commercials are local advertisers. A sample of 120 commercials is selected.
    a. What is the probability that more than 20% of the commercials in the sample are local advertisers?
    b. Would it be unusual for more than 25% of the commercials to be local advertisers?

14. Let X have a binomial distribution with $n = 240$ and $p = 0.38$. Use the normal approximation to find:
    a. $P(X > 83)$
    b. $P(75 \leq X \leq 95)$
    c. $P(X < 96)$

15. Is it reasonable to treat the following sample as coming from an approximately normal population? Explain.

    5.5   8.7   9.3   10.1   15.2   3.5   11.9   7.6   13.7   8.7   14.3   5.8

Review Exercises

1. **Find the area:** Find the area under the standard normal curve
   a. To the left of $z = 0.35$
   b. To the right of $z = -1.56$
   c. Between $z = 0.35$ and $z = 2.47$

2. **Find the z-score:** Find the z-score for which the area to its right is 0.89.

3. **Your battery is dead:** The lifetimes of a certain type of automobile battery are normally distributed with mean 5.9 years and standard deviation 0.4 year. The batteries are guaranteed to last at least 5 years. What proportion of the batteries fail to meet the guarantee?

4. **Take your medicine:** Medication used to treat a certain condition is administered by syringe. The target dose in a particular application is 10 milligrams. Because of the variations in the syringe, in reading the scale, and in mixing the fluid suspension, the actual dose administered is normally distributed with mean $\mu = 10$ milligrams and standard deviation $\sigma = 1.6$ milligrams.
   a. What is the probability that the dose administered is between 9 and 11.5 milligrams?
   b. Find the 98th percentile of the administered dose.
   c. If a clinical overdose is defined as a dose larger than 15 milligrams, what is the probability that a patient will receive an overdose?

5. **Lightbulbs:** The lifetime of lightbulbs has a mean of 1500 hours and a standard deviation of 100 hours. A sample of 50 lightbulbs is tested.
   a. What is the probability that the sample mean lifetime is greater than 1520 hours?
   b. What is the probability that the sample mean lifetime is less than 1540 hours?
   c. What is the probability that the sample mean lifetime is between 1490 and 1550 hours?

6. **More lightbulbs:** Someone claims to have developed a new lightbulb whose mean lifetime is 1800 hours with a standard deviation of 100 hours. A sample of 100 of these bulbs is tested. The sample mean lifetime is 1770 hours.
   a. If the claim is true, what is the probability of obtaining a sample mean that is less than or equal to 1770 hours?
   b. If the claim is true, would it be unusual to obtain a sample mean that is less than or equal to 1770 hours?

7. **Pay your taxes:** Among all the state income tax forms filed in a particular state, the mean income tax paid was $\mu = \$2000$ and the standard deviation was $\sigma = \$500$. As part of a study of the impact of a new tax law, a sample of 80 income tax returns is examined. Would it be unusual for the sample mean of these 80 returns to be greater than $2150?

8. **Safe delivery:** A certain delivery truck can safely carry a load of 3400 pounds. The cartons that will be loaded onto the truck have a mean weight of 80 pounds with a standard deviation of 20 pounds. Forty cartons are loaded onto the truck.
   a. If the total weight of the 40 cartons is 3400 pounds, what is the sample mean weight?
   b. What is the probability that the truck can deliver the 40 cartons safely?

9. **Elementary school:** In a certain elementary school, 52% of the students are girls. A sample of 65 students is drawn.
   a. What is the probability that more than 60% of them are girls?
   b. Would it be unusual for more than 70% of them to be girls?

10. **Facebook:** Eighty percent of the students at a particular large university have logged on to Facebook at least once in the past week. A sample of 95 students is asked about their Internet habits.
    a. What is the probability that less than 75% of the sampled students have logged on to Facebook within the last week?
    b. What is the probability that more than 78% of the sampled students have logged on to Facebook within the last week?
    c. What is the probability that the proportion of the sampled students who have logged on to Facebook within the last week is between 0.82 and 0.85?

11. **It's all politics:** A politician in a close election race claims that 52% of the voters support him. A poll is taken in which 200 voters are sampled, and 44% of them support the politician.

a. If the claim is true, what is the probability of obtaining a sample proportion that is less than or equal to 0.44?

b. If the claim is true, would it be unusual to obtain a sample proportion less than or equal to 0.44?

c. If the claim is true, would it be unusual for less than half of the voters in the sample to support the politician?

12. **Side effects:** A new medical procedure produces side effects in 25% of the patients who receive it. In a clinical trial, 60 people undergo the procedure. What is the probability that 20 or fewer experience side effects?

13. **Defective rods:** A grinding machine used to manufacture steel rods produces rods, 5% of which are defective. When a customer orders 1000 rods, a package of 1060 rods is shipped, with a guarantee that at least 1000 of the rods are good. What is the probability that a package of 1060 rods contains 1000 or more that are good?

14. **Is it normal?** Is it reasonable to treat the following sample as though it comes from an approximately normal population? Explain.

$$2.6 \quad 4.2 \quad 1.5 \quad 2.0 \quad 0.6 \quad 0.7 \quad 6.6 \quad 2.2 \quad 9.7 \quad 1.8 \quad 4.2 \quad 4.4 \quad 0.6$$

15. **Is it normal?** Is it reasonable to treat the following sample as though it comes from an approximately normal population? Explain.

$$8.8 \quad 11.2 \quad 11.6 \quad 6.3 \quad 9.3 \quad 1.5 \quad 14.6 \quad 7.5 \quad 5.2 \quad 9.0 \quad 4.3 \quad 9.9 \quad 7.8 \quad 13.1$$

Write About It

1. Explain why $P(a < X < b)$ is equal to $P(a \le X \le b)$ when X is a continuous random variable.

2. Describe the information you must know to compute the area under the normal curve over a given interval.

3. Describe the information you must know to find the value corresponding to a given proportion of the area under a normal curve.

4. Suppose that in a large class, the instructor announces that the average grade on an exam is 75. Which is more likely to be closer to 75:
   i. The exam grade of a randomly selected student in the class?
   ii. The mean exam grade of a sample of 10 students?
   Explain.

5. Consider the formula for the standard deviation of the sampling distribution of $\hat{p}$ given by $\sigma_{\hat{p}} = \sqrt{\dfrac{p(1-p)}{n}}$. What happens to the standard deviation as n gets larger and larger? Explain what this means in terms of the spread of the sampling distribution.

6. Explain how to decide when it is appropriate to use the normal approximation to the binomial distribution.

7. Describe the effect, if any, that the size of a sample has in assessing the normality of a population.

Case Study: Testing The Strength Of Cans

In the chapter opener, we discussed a method used to determine whether shipments of aluminum cans are strong enough to withstand the pressure of containing a carbonated beverage. Several cans are sampled from a shipment and tested to determine the pressure they can withstand. Based on this small sample, quality inspectors must estimate the proportion of cans that will fail at or below a certain threshold, which we will take to be 90 pounds per square inch.

The quality control inspectors want the proportion of defective cans to be no more than 0.001, or 1 in 1000. They test 10 cans, with the following results.

Can	1	2	3	4	5	6	7	8	9	10
Pressure at failure	95	96	98	99	99	100	101	101	103	104

None of the cans in the sample were defective; in other words, none of them failed at a pressure of 90 or less. The quality control inspectors want to use these data to estimate the proportion of defective cans in the shipment. If the estimate is to be no more than 0.001, or 1 in 1000, they will accept the shipment; otherwise, they will return it for a refund.

The following exercises will lead you through the process used by the quality control inspectors. Assume the failure pressures are normally distributed.

1. Compute the sample mean $\bar{x}$ and the sample standard deviation s.

2. Estimate the population mean μ with $\bar{x}$ and the population standard deviation σ with s. In other words, assume that the data are a sample from a normal population with mean $\mu = \bar{x}$ and standard deviation $\sigma = s$. Under this assumption, what proportion of cans will fail at a pressure of 90 or less?

3. The shipment will be accepted if we estimate that the proportion of cans that fail at a pressure of 90 or less is less than 0.001. Will this shipment be accepted?

4. A second shipment of cans is received. Ten randomly sampled cans are tested with the following results.

Can	1	2	3	4	5	6	7	8	9	10
Pressure at failure	96	97	99	100	100	100	101	103	103	120

Explain why the second sample of cans is stronger than the first sample.

5. Compute the sample mean $\bar{x}$ and the sample standard deviation s for the second sample.

6. Using the same method as for the first sample, estimate the proportion of cans that will fail at a pressure of 90 or less.

7. The shipment will be accepted if we estimate that the proportion of cans that fail at a pressure of 90 or less is less than 0.001. Will this shipment be accepted?

8. Make a boxplot of the pressures for the second sample. Is the method appropriate for the second shipment?

Confidence Intervals

Introduction

Air pollution is a serious problem in many places. Particulate matter (PM), which consists of tiny particles in the air, is a form of air pollution that is suspected of causing respiratory illness. PM can come from many sources, such as ash from burning, and tiny particles of rubber from automobile and truck tires.

The town of Libby, Montana, has experienced high levels of PM, especially in the winter. Many houses in Libby are heated by wood stoves, which produce a lot of particulate pollution. In an attempt to reduce the winter levels of air pollution in Libby, a program was undertaken in which almost every wood stove in the town was replaced with a newer, cleaner-burning model. Most stoves were replaced in 2006 or early 2007. Measurements of several air pollutants, including PM, were taken both before and after the stove replacement. To determine PM levels after stoves were replaced, the level was measured on a sample of 20 days in the winter of 2007–2008. The units are micrograms per cubic meter. Following are the results:

| 21.7 | 27.8 | 24.7 | 15.3 | 18.4 | 14.4 | 19.0 | 23.7 | 22.4 | 25.6 |
| 15.0 | 17.0 | 23.2 | 17.7 | 11.1 | 29.8 | 20.0 | 21.6 | 14.8 | 21.0 |

Clearly, the amount varies from day to day. The sample mean is 20.21 and the sample standard deviation is 4.86. We would like to use this information to estimate the population mean, which is the mean over all days of the winter of 2007–2008. What is the best way to do this? If we had to pick a single number to estimate the population mean, the best choice would be the sample mean, which is 20.21. The estimate 20.21 is called a *point estimate*, because it is a single number. The problem with point estimates is that they are almost never exactly equal to the true values they are estimating. They are almost always off — sometimes by a little, sometimes by a lot. It is unlikely that the mean level of PM for all of 2007–2008 is exactly equal to 20.21; it is somewhat more or less than that. In order for a point estimate to be useful, it is necessary to describe just how close to the true value it is likely to be. To do this, statisticians construct *confidence intervals*. A confidence interval gives a range of values that is likely to contain the true value being estimated.

To construct a confidence interval, we put a plus-or-minus number on the point estimate. So, for example, we might estimate that the population mean is 20.21 ± 2.0, or equivalently, that the population mean is between 18.21 and 22.21. The interval 20.21 ± 2.0, or equivalently, (18.21, 22.21), is a confidence interval for the population mean.

One of the benefits of confidence intervals is that they come with a measure of the level of confidence we can have that they actually cover the true value being estimated. For example, we will show that we can be 95% confident that the population mean PM level during the winter of 2007–2008 is in the interval 20.21 ± 2.27, or equivalently, between 17.94 and 22.48. If we want more confidence, we can widen the interval. For example, we will learn how to show that we can be 99% confident that the population mean PM level is in the interval 20.21 ± 3.11, or equivalently, between 17.10 and 23.32. In the case study at the end of the chapter, we will use confidence intervals to further study the effects of the Libby stove replacement program.

There are many different situations in which confidence intervals can be constructed. The correct method to use varies from situation to situation. In this chapter, we will describe the methods that are appropriate in several of the most commonly encountered situations.

SECTION 8.1 | Confidence Intervals for a Population Mean, Standard Deviation Known

Objectives

1. Construct and interpret confidence intervals for a population mean when the population standard deviation is known
2. Find critical values for confidence intervals
3. Describe the relationship between the confidence level and the margin of error
4. Find the sample size necessary to obtain a confidence interval of a given width
5. Distinguish between confidence and probability

Estimating a Population Mean

Objective 1 Construct and interpret confidence intervals for a population mean when the population standard deviation is known

How can we measure the reading ability of elementary school students? The No Child Left Behind Act, signed into law in 2002, requires schools to regularly assess the proficiency of students in subjects such as reading and math. In a certain school district, administrators are trying out a new experimental approach to teach reading to fourth-graders. A simple random sample of 100 fourth-graders is selected to take part in the program. At the end of the program, the students are given a standardized reading test. On the basis of past results, it is known that scores on this test have a population standard deviation of $\sigma = 15$.

The sample mean score for the 100 students was $\bar{x} = 67.30$. The administrators want to estimate what the mean score μ would be if the entire population of fourth-graders in the district had enrolled in the program. The best estimate for the population mean is the sample mean, $\bar{x} = 67.30$. The sample mean is a *point estimate*, because it is a single number.

Recall: A parameter is a numerical summary of a population, such as a population mean μ or a population proportion p.

DEFINITION

A **point estimate** is a single number that is used to estimate the value of an unknown parameter.

It is very unlikely that the point estimate $\bar{x} = 67.30$ is exactly equal to the population mean μ. Therefore, in order for the estimate to be useful, we must describe how close it is likely to be. For example, if we think that $\bar{x} = 67.30$ is likely to be within 1 point of the population mean, we would estimate μ with the interval $66.30 < \mu < 68.30$. This could also be written

The symbol ±: The symbol ± means to form an interval by adding and subtracting. For example, 67.30 ± 1 means the interval from 67.30 − 1 to 67.30 + 1, or, in other words, from 66.30 to 68.30.

Recall: The quantity $\sigma/\sqrt{n}$ is called the standard error of the mean.

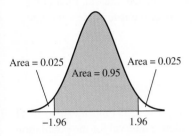

Figure 8.1 95% of the area under the standard normal curve is between $z = -1.96$ and $z = 1.96$.

as 67.30 ± 1. If we think that $\bar{x} = 67.30$ could be off by as much as 10 points from the population mean, we would estimate μ with the interval $57.30 < \mu < 77.30$, which could also be written as 67.30 ± 10. The plus-or-minus number is called the **margin of error**. We need to determine how large to make the margin of error so that the interval is likely to contain the population mean. To do this, we use the sampling distribution of $\bar{x}$.

Because the sample size is large ($n > 30$), the Central Limit Theorem tells us that the sampling distribution of $\bar{x}$ is approximately normal with mean μ and standard deviation (also called the **standard error**) given by

$$\text{Standard error} = \frac{\sigma}{\sqrt{n}} = \frac{15}{\sqrt{100}} = 1.5$$

We will now construct a *95% confidence interval* for μ. We begin with a normal curve, and using the method described in Section 7.1 (Example 7.11), we find the z-scores that bound the middle 95% of the area under the curve. These z-scores are 1.96 and −1.96 (see Figure 8.1). The value 1.96 is called the **critical value**. To obtain the margin of error, we multiply the critical value by the standard error.

$$\text{Margin of error} = \text{Critical value} \cdot \text{Standard error} = (1.96)(1.5) = 2.94$$

A 95% confidence interval for μ is therefore

$$\bar{x} - 2.94 < \mu < \bar{x} + 2.94$$
$$67.30 - 2.94 < \mu < 67.30 + 2.94$$
$$64.36 < \mu < 70.24$$

There are several ways to express this confidence interval. We can write $64.36 < \mu < 70.24$, 67.30 ± 2.94, or $(64.36, 70.24)$. In words, we would say, "We are 95% confident that the population mean is between 64.36 and 70.24."

Figures 8.2 and 8.3 help explain why this interval is called a 95% confidence interval. Figure 8.2 illustrates a sample whose mean $\bar{x}$ is in the middle 95% of its distribution. The 95% confidence interval constructed from this value of $\bar{x}$ covers the true population mean μ.

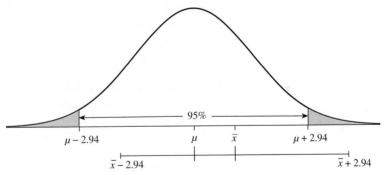

Figure 8.2 The sample mean $\bar{x}$ comes from the middle 95% of the distribution, so the 95% confidence interval $\bar{x} \pm 2.94$ succeeds in covering the population mean μ.

Figure 8.3 illustrates a sample whose mean $\bar{x}$ is in one of the tails of the distribution, outside the middle 95%. The 95% confidence interval constructed from this value of $\bar{x}$ does *not* cover the true population mean μ.

Figure 8.3 The sample mean $\bar{x}$ comes from the outer 5% of the distribution, so the 95% confidence interval $\bar{x} \pm 2.94$ fails to cover the population mean μ.

Now in the long run, 95% of the sample means we observe will be like the one in Figure 8.2. They will be in the middle 95% of the distribution and their confidence intervals will cover the population mean. Therefore, in the long run, 95% of the confidence intervals we construct will cover the population mean. Only 5% of the sample means we observe will be outside the middle 95% of the distribution like the one in Figure 8.3. So in the long run, only 5% of the confidence intervals we construct will fail to cover the population mean.

To summarize, the confidence interval we just constructed is a 95% confidence interval, because the method we used to construct it will cover the population mean μ for 95% of all the possible samples that might be drawn. We can also say that the interval has a *confidence level* of 95%.

DEFINITION

- A **confidence interval** is an interval that is used to estimate the value of a parameter.
- The **confidence level** is a percentage between 0% and 100% that measures the success rate of the method used to construct the confidence interval. If we were to draw many samples and use each one to construct a confidence interval, then in the long run, the percentage of confidence intervals that cover the true value would be equal to the confidence level.

EXAMPLE 8.1 Construct and interpret a 95% confidence interval

A large sample has mean $\bar{x} = 7.1$ and standard error $\sigma/\sqrt{n} = 2.3$. Find the margin of error for a 95% confidence interval. Construct a 95% confidence interval for the population mean μ and explain what it means to say that the confidence level is 95%.

Solution
As shown in Figure 8.1, the critical value for a 95% confidence interval is 1.96. Therefore, the margin of error is

$$\text{Margin of error} = \text{Critical value} \cdot \text{Standard error} = (1.96)(2.3) = 4.5$$

The point estimate of μ is $\bar{x} = 7.1$. To construct a confidence interval, we add and subtract the margin of error from the point estimate. So the 95% confidence interval is 7.1 ± 4.5. We can also write this as $7.1 - 4.5 < \mu < 7.1 + 4.5$, or $2.6 < \mu < 11.6$.

The level of this confidence interval is 95% because if we were to draw many samples and use this method to construct the corresponding confidence intervals, then in the long run, 95% of the intervals would cover the true value of the population mean μ. Unless we were unlucky in the sample we drew, the population mean μ will be between 2.6 and 11.6.

Objective 2 Find critical values for confidence intervals

Finding the critical value for a given confidence level

Although 95% is the most commonly used confidence level, sometimes we will want to construct a confidence interval with a different level. We can construct a confidence interval with any confidence level between 0% and 100% by finding the appropriate critical value for that level. We have seen that the critical value for a 95% confidence interval is $z = 1.96$, because 95% of the area under a normal curve is between $z = -1.96$ and $z = 1.96$. Similarly, the critical value for a 99% confidence interval is the z-score for which the area between z and $-z$ is 0.99, the critical value for a 98% confidence interval is the z-score for which the area between z and $-z$ is 0.98, and so on.

The row of Table A.3 labeled "z" presents critical values for several confidence levels. Following is part of that row, which presents critical values for four of the most commonly used confidence levels.

z	$\cdots$	1.645	1.96	2.326	2.576	$\cdots$
	$\cdots$	90%	95%	98%	99%	$\cdots$
			Confidence level			

| EXAMPLE 8.2 | Construct confidence intervals of various levels |

A large sample has mean $\bar{x} = 7.1$ and standard error $\sigma/\sqrt{n} = 2.3$. Construct confidence intervals for the population mean μ with the following levels:
a. 90%   **b.** 98%   **c.** 99%

Solution

The point estimate of μ is $\bar{x} = 7.1$.

a. From the bottom row of Table A.3, we see that the critical value for a 90% confidence interval is 1.645, so the margin of error is

Margin of error = Critical value · Standard error = (1.645)(2.3) = 3.8

The 90% confidence interval is 7.1 ± 3.8. We can also write this as

$$7.1 - 3.8 < \mu < 7.1 + 3.8, \text{ or } 3.3 < \mu < 10.9.$$

b. From the bottom row of Table A.3, we see that the critical value for a 98% confidence interval is 2.326, so the margin of error is

Margin of error = Critical value · Standard error = (2.326)(2.3) = 5.3

The 98% confidence interval is 7.1 ± 5.3. We can also write this as

$$7.1 - 5.3 < \mu < 7.1 + 5.3, \text{ or } 1.8 < \mu < 12.4.$$

c. From the bottom row of Table A.3, we see that the critical value for a 99% confidence interval is 2.576, so the margin of error is

Margin of error = Critical value · Standard error = (2.576)(2.3) = 5.9

The 99% confidence interval is 7.1 ± 5.9. We can also write this as

$$7.1 - 5.9 < \mu < 7.1 + 5.9, \text{ or } 1.2 < \mu < 13.0.$$

The notation z_α

Sometimes we may need to find a critical value for a confidence level not given in the last row of Table A.3. To do this, it is useful to learn a notation for a z-score with a given area to its right.

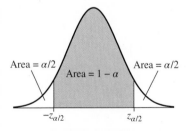

Figure 8.4

| DEFINITION |

Let α be any number between 0 and 1; in other words, $0 < \alpha < 1$.
- The notation z_α refers to the z-score with an area of α to its right.
- The notation $z_{\alpha/2}$ refers to the z-score with an area of $\alpha/2$ to its right.

To find the critical value for a confidence interval with a given level, let $1 - \alpha$ be the confidence level expressed as a decimal. The critical value is then $z_{\alpha/2}$, because the area under the standard normal curve between $-z_{\alpha/2}$ and $z_{\alpha/2}$ is $1 - \alpha$. See Figure 8.4.

| EXAMPLE 8.3 | Find a critical value |

Find the critical value $z_{\alpha/2}$ for a 92% confidence interval.

Solution

The level is 92%, so we have $1 - \alpha = 0.92$. It follows that $\alpha = 1 - 0.92 = 0.08$, so $\alpha/2 = 0.04$. The critical value is $z_{0.04}$. We now must find the value of $z_{0.04}$. To do this using Table A.2, we find the area to the left of $z_{0.04}$. Since the area to the right of $z_{0.04}$ is 0.04, the area to the left is $1 - 0.04 = 0.96$. See Figure 8.5.

Figure 8.5 The critical value $z_{0.04}$ contains an area of 0.92 under the standard normal curve between $-z_{0.04}$ and $z_{0.04}$.

We now look in the body of Table A.2 (a portion of which is shown in Figure 8.6) to find the closest value to 0.96. This value is 0.9599, and it corresponds to a z-score of 1.75. Therefore, the critical value is $z_{0.04} = 1.75$.

z	0.00	0.01	0.02	0.03	0.04	0.05	0.06	0.07	0.08	0.09
⋮	⋮	⋮	⋮	⋮	⋮	⋮	⋮	⋮	⋮	⋮
1.2	.8849	.8869	.8888	.8907	.8925	.8944	.8962	.8980	.8997	.9015
1.3	.9032	.9049	.9066	.9082	.9099	.9115	.9131	.9147	.9162	.9177
1.4	.9192	.9207	.9222	.9236	.9251	.9265	.9279	.9292	.9306	.9319
1.5	.9332	.9345	.9357	.9370	.9382	.9394	.9406	.9418	.9429	.9441
1.6	.9452	.9463	.9474	.9484	.9495	.9505	.9515	.9525	.9535	.9545
1.7	.9554	.9564	.9573	.9582	.9591	.9599	.9608	.9616	.9625	.9633
1.8	.9641	.9649	.9656	.9664	.9671	.9678	.9686	.9693	.9699	.9706
1.9	.9713	.9719	.9726	.9732	.9738	.9744	.9750	.9756	.9761	.9767
2.0	.9772	.9778	.9783	.9788	.9793	.9798	.9803	.9808	.9812	.9817
2.1	.9821	.9826	.9830	.9834	.9838	.9842	.9846	.9850	.9854	.9857
2.2	.9861	.9864	.9868	.9871	.9875	.9878	.9881	.9884	.9887	.9890
⋮	⋮	⋮	⋮	⋮	⋮	⋮	⋮	⋮	⋮	⋮

Figure 8.6

As an alternative to Table A.2, we can find $z_{0.04}$ with technology, for example, by using the TI-84 Plus calculator. Figure 8.5 shows that the area to the left of $z_{0.04}$ is 0.96. Therefore, we can find $z_{0.04}$ with the command **invNorm(.96,0,1)**. See Figure 8.7.

```
invNorm(.96,0,1)
            1.750686071
```

Figure 8.7

Check Your Understanding

1. Find the critical value $z_{\alpha/2}$ to construct a confidence interval with level
   a. 90%   **b.** 98%   **c.** 99.5%   **d.** 80%

2. Find the levels of the confidence intervals that have the following critical values.
   a. $z_{\alpha/2} = 1.96$ **b.** $z_{\alpha/2} = 2.17$ **c.** $z_{\alpha/2} = 1.28$
   d. $z_{\alpha/2} = 3.28$

3. Find the margin of error given the standard error and the confidence level.
   a. Standard error = 1.2, confidence level 95%
   b. Standard error = 0.4, confidence level 99%
   c. Standard error = 3.5, confidence level 90%
   d. Standard error = 2.75, confidence level 98%

Answers are on page 366.

Assumptions

The method we have described for constructing a confidence interval requires us to assume that the population standard deviation σ is known. In practice, it is more common that σ is not known. The advantage of first learning the method that assumes σ known is that it allows us to use the familiar normal distribution, so we can focus on the basic ideas of confidence intervals. We will learn how to construct confidence intervals when σ is unknown in Section 8.2.

Following are the assumptions for the method we describe.

Assumptions for Constructing a Confidence Interval for μ When σ Is Known

1. We have a simple random sample.
2. The sample size is large ($n > 30$), or the population is approximately normal.

When these assumptions are met, we can use the following steps to construct a confidence interval for μ when σ is known.

Explain It Again

$\bar{x}$ **must be approximately normal:** We need assumption 2 to be sure that the sampling distribution of $\bar{x}$ is approximately normal. This allows us to use $z_{\alpha/2}$ as the critical value.

> ## Procedure for Constructing a Confidence Interval for μ When σ Is Known
>
> Check to be sure the assumptions are satisfied. If they are, then proceed with the following steps.
>
> **Step 1:** Find the value of the point estimate $\bar{x}$, if it isn't given.
>
> **Step 2:** Find the critical value $z_{\alpha/2}$ corresponding to the desired confidence level from the last row of Table A.3, from Table A.2, or with technology.
>
> **Step 3:** Find the standard error $\sigma/\sqrt{n}$, and multiply it by the critical value to obtain the margin of error $z_{\alpha/2}\dfrac{\sigma}{\sqrt{n}}$.
>
> **Step 4:** Use the point estimate and the margin of error to construct the confidence interval:
>
> $$\text{Point estimate} \pm \text{Margin of error}$$
>
> $$\bar{x} \pm z_{\alpha/2}\frac{\sigma}{\sqrt{n}}$$
>
> $$\bar{x} - z_{\alpha/2}\frac{\sigma}{\sqrt{n}} < \mu < \bar{x} + z_{\alpha/2}\frac{\sigma}{\sqrt{n}}$$
>
> **Step 5:** Interpret the result.

Rounding off the final result

When constructing a confidence interval for a population mean, you may be given a value for $\bar{x}$, or you may be given the data and have to compute $\bar{x}$ yourself. If you are given the value of $\bar{x}$, round the final result to the same number of decimal places as $\bar{x}$. If you are given data, then round $\bar{x}$ and the final result to one more decimal place than is given in the data.

Although you should round off your final answer, do not round off the calculations you make along the way. Doing so may affect the accuracy of your final answer.

EXAMPLE 8.4   Construct a confidence interval

The mean test score for a simple random sample of $n = 100$ students was $\bar{x} = 67.30$. The population standard deviation of test scores is $\sigma = 15$. Construct a 98% confidence interval for the population mean test score μ.

Solution

First we check the assumptions. The sample is a simple random sample, and the sample size is large ($n > 30$). The assumptions are met, so we may proceed.

Step 1: Find the point estimate. The point estimate is the sample mean $\bar{x} = 67.30$.

Step 2: Find the critical value $z_{\alpha/2}$. The desired confidence level is 98%. We look on the last line of Table A.3 and find that the critical value is $z_{\alpha/2} = 2.326$.

Step 3: Find the standard error and the margin of error. The standard error is

$$\frac{\sigma}{\sqrt{n}} = \frac{15}{\sqrt{100}} = 1.5$$

We multiply the standard error by the critical value to obtain the margin of error:

$$\text{Margin of error} = z_{\alpha/2}\frac{\sigma}{\sqrt{n}} = 2.326(1.5) = 3.489$$

Step 4: Construct the confidence interval. The 98% confidence interval is

$$\bar{x} - z_{\alpha/2}\frac{\sigma}{\sqrt{n}} < \mu < \bar{x} + z_{\alpha/2}\frac{\sigma}{\sqrt{n}}$$

$$67.30 - 3.489 < \mu < 67.30 + 3.489$$

$$63.81 < \mu < 70.79 \quad \text{(rounded to two decimal places, like } \bar{x}\text{)}$$

Step 5: Interpret the result. We are 98% confident that the population mean score μ is between 63.81 and 70.79. Another way to say this is that we are 98% confident that μ is in the interval 67.30 ± 3.49. If we were to draw many different samples and use this method to construct the corresponding confidence intervals, then in the long run, 98% of the intervals would cover the true population mean μ. So unless we were quite unlucky in the sample we drew, the population mean μ will be between 63.81 and 70.79.

Check Your Understanding

4. An IQ test was given to a simple random sample of 75 students at a certain college. The sample mean score was 105.2. Scores on this test are known to have a standard deviation of $\sigma = 10$. It is desired to construct a 90% confidence interval for the mean IQ score of students at this college.
    a. What is the point estimate?
    b. Find the critical value.
    c. Find the standard error.
    d. Find the margin of error.
    e. Construct the 90% confidence interval.
    f. Is it likely that the population mean μ is greater than 100? Explain.

5. The lifetime of a certain type of battery is known to be normally distributed with standard deviation $\sigma = 20$ hours. A sample of 50 batteries had a mean lifetime of 120.1 hours. It is desired to construct a 95% confidence interval for the mean lifetime for this type of battery.
    a. What is the point estimate?
    b. Find the critical value.
    c. Find the standard error.
    d. Find the margin of error.
    e. Construct the 95% confidence interval.
    f. Is it likely that the population mean μ is greater than 130? Explain.

6. In a survey of a simple random sample of students at a certain college, the sample mean time per week spent watching television was 18.3 hours and the margin of error for a 95% confidence interval was 1.2 hours. True or false:
    a. A 95% confidence interval for the mean number of hours per week spent watching television by students at this college is $17.1 < \mu < 19.5$.
    b. Approximately 95% of the students at this university watch between 17.1 and 19.5 hours of television per week.

7. Use the data in Exercise 4 to construct a 95% confidence interval for the mean IQ score.

8. Use the data in Exercise 5 to construct a 98% confidence interval for the mean lifetime for this type of battery.

Answers are on page 366.

Constructing confidence intervals with technology

In Example 8.4, we found a 98% confidence interval for the mean test score, based on a sample size of $n = 100$, a sample mean of $\bar{x} = 67.30$, and a population standard deviation $\sigma = 15$. The following TI-84 Plus display presents the results.

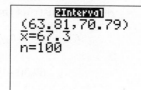

```
   ZInterval
(63.81,70.79)
x̄=67.3
n=100
```

The display presents the confidence interval $(63.81, 70.79)$, along with the values of $\bar{x}$ and n. Note that the confidence level (98%) is not given.

Following is MINITAB output for the same example.

```
The assumed sigma = 15.0000
Variable    N    Mean     StDev    SE Mean        98% CI
Score      100  67.3000  15.0000  1.50000   (63.8105, 70.7895 )
```

The output is fairly straightforward. Going from left to right, "N" represents the sample size, "Mean" is the sample mean $\bar{x}$, "StDev" is the population standard deviation σ, and "SE Mean" is the standard error $\sigma/\sqrt{n}$. The lower and upper confidence limits of the 98% confidence interval are given on the right. Note that neither the critical value nor the margin of error is given explicitly in the output.

Finally, we present EXCEL output for this example. The EXCEL function CONFIDENCE.NORM returns the margin of error. The inputs are the value of α, the population standard deviation σ, and the sample size n.

=CONFIDENCE.NORM(0.02, 15, 100)

Margin of Error   3.490

Step-by-step instructions for constructing confidence intervals with technology are given in the Using Technology section on page 361.

Check Your Understanding

9. To estimate the accuracy of a laboratory scale, a weight known to have a mass of 100 grams is weighed 32 times. The reading of the scale is recorded each time. The following MINITAB output presents a 95% confidence interval for the mean reading of the scale.

```
The assumed sigma = 2.5000
Variable         N    Mean      StDev   SE Mean        95% CI
Scale Reading   32  102.3527   2.5000  0.44194   (101.4865, 103.2189 )
```

A scientist claims that the mean reading μ is actually 100 grams. Is it likely that this claim is true?

10. Using the output in Exercise 9:
    a. Find the critical value $z_{\alpha/2}$ for a 99% confidence interval.
    b. Use the critical value along with the information in the output to construct a 99% confidence interval for the mean reading of the scale.

Answers are on page 366.

Objective 3 Describe the relationship between the confidence level and the margin of error

More Confidence Means a Bigger Margin of Error

Other things being equal, it is better to have more confidence than less. We would also rather have a smaller margin of error than a larger one. However, when it comes to confidence intervals, there is a trade-off. If we increase the level of confidence, we must increase the critical value, which in turn increases the margin of error. Examples 8.5 and 8.6 help explain this idea.

EXAMPLE 8.5   Construct a confidence interval

A machine that fills cereal boxes is supposed to put 20 ounces of cereal in each box. A simple random sample of 6 boxes is found to contain a sample mean of 20.25 ounces of cereal. It is known from past experience that the fill weights are normally distributed with a standard deviation of 0.2 ounce. Construct a 90% confidence interval for the mean fill weight.

Solution

We check the assumptions. The sample is a simple random sample, and the population is known to be normal. The assumptions are met, so we may proceed.

Step 1: Find the point estimate. The point estimate is the sample mean $\bar{x} = 20.25$.

Step 2: Find the critical value $z_{\alpha/2}$. The desired confidence level is 90%. We look on the last line of Table A.3 and find that the critical value is $z_{\alpha/2} = 1.645$.

Step 3: Find the standard error and the margin of error. The standard error is

$$\frac{\sigma}{\sqrt{n}} = \frac{0.2}{\sqrt{6}} = 0.08165$$

We multiply the standard error by the critical value to obtain the margin of error:

$$\text{Margin of error} = z_{\alpha/2}\frac{\sigma}{\sqrt{n}} = (1.645)(0.08165) = 0.1343$$

Step 4: Construct the confidence interval. The 90% confidence interval is

$$\bar{x} - z_{\alpha/2}\frac{\sigma}{\sqrt{n}} < \mu < \bar{x} + z_{\alpha/2}\frac{\sigma}{\sqrt{n}}$$

$$20.25 - 0.1343 < \mu < 20.25 + 0.1343$$

$$20.12 < \mu < 20.38 \quad \text{(rounded to two decimal places, like } \bar{x}\text{)}$$

Step 5: Interpret the result. We are 90% confident that the mean weight μ is between 20.12 and 20.38. Another way to say this is that we are 90% confident that the mean weight μ is in the interval 20.25 ± 0.13. If we were to draw many different samples and use this method to construct the corresponding confidence intervals, then in the long run, 90% of them would cover the true population mean μ. So unless we were somewhat unlucky in the sample we drew, the true mean weight is between 20.12 and 20.38 ounces.

A confidence level of 90% is the lowest level commonly used in practice. In Example 8.6, we will construct a 99% confidence interval.

| EXAMPLE 8.6 | Construct a confidence interval

Use the data in Example 8.5 to construct a 99% confidence interval for the mean fill weight. Compare the margin of error of this confidence interval to the 90% confidence interval constructed in Example 8.5.

Explain It Again

The relationship between confidence and the margin of error: If we want to increase our confidence that an interval contains the true value, we must increase the critical value. This increases the margin of error, which makes the confidence interval wider.

Solution

As in Example 8.5, the assumptions are met, so we may proceed.

Step 1: Find the point estimate. The point estimate is $\bar{x} = 20.25$.

Step 2: Find the critical value $z_{\alpha/2}$. The desired level is 99%. We look on the last line of Table A.3 and find that $z_{\alpha/2} = 2.576$.

Step 3: Find the standard error and the margin of error. The standard error is

$$\frac{\sigma}{\sqrt{n}} = \frac{0.2}{\sqrt{6}} = 0.08165$$

We multiply the standard error by the critical value to obtain the margin of error:

$$\text{Margin of error} = z_{\alpha/2}\frac{\sigma}{\sqrt{n}} = (2.576)(0.08165) = 0.2103$$

Step 4: Construct the confidence interval. The 99% confidence interval is

$$\bar{x} - z_{\alpha/2}\frac{\sigma}{\sqrt{n}} < \mu < \bar{x} + z_{\alpha/2}\frac{\sigma}{\sqrt{n}}$$

$$20.25 - 0.2103 < \mu < 20.25 + 0.2103$$

$$20.04 < \mu < 20.46 \quad \text{(rounded to two decimal places, like } \bar{x}\text{)}$$

Step 5: Interpret the result. We are 99% confident that the mean weight μ is between 20.04 and 20.46. Another way to say this is that we are 99% confident that the mean weight μ is in the interval 20.25 ± 0.21. If we were to draw many different samples and use this method to construct the corresponding confidence intervals, then in the long run, 99% of them would cover the true population mean μ. So unless we were very unlucky in the sample we drew, the true mean is between 20.04 and 20.46 ounces.

For this 99% confidence interval, the margin of error is 0.2103. For the 90% confidence interval in Example 8.5, the margin of error was only 0.1343. The reason is that for a 90% confidence interval, we used a critical value of 1.645, and for the 99% confidence interval, we must use a larger critical value of 2.576.

We can see that if we want to be more confident that our interval contains the true value, we must increase the critical value, which increases the margin of error. There is a trade-off. We would rather have a higher level of confidence than a lower level, but we would also rather have a smaller margin of error than a larger one. So we have to choose a level of confidence that strikes a good balance. The most common choice is 95%. In some cases where high confidence is very important, a larger confidence level such as 99% may be chosen. In general, intervals with confidence levels less than 90% are not considered to be reliable enough to be used in practical situations.

Figure 8.8 illustrates the trade-off between confidence level and margin of error. One hundred samples were drawn from a population with mean μ. The center diagram presents one hundred 95% confidence intervals, each based on one of these samples. The confidence intervals are all different, because each sample has a different mean $\bar{x}$. The diagram on the left presents 70% confidence intervals based on the same samples. These intervals are narrower because they have a smaller margin of error, but many of them fail to cover the population mean. These intervals are too unreliable to be of any practical value. The figure on the right presents 99.7% confidence intervals. These intervals are very reliable. In the long run, only 3 in 1000 of these intervals will fail to cover the population mean. However, they are wider due to the larger margin of error, so they do not convey as much information.

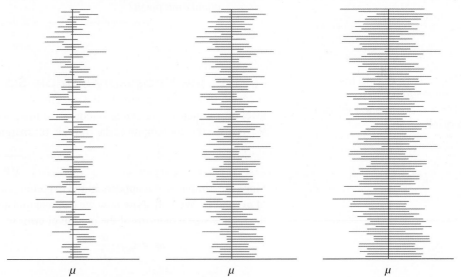

Figure 8.8 *Left:* One hundred 70% confidence intervals for a population mean, each constructed from a different sample. Although their margin of error is small, they cover the population mean only 70% of the time. This low success rate makes the 70% confidence interval unacceptable for practical purposes. *Center:* One hundred 95% confidence intervals constructed from these samples. This represents a good compromise between reliability and margin of error for many purposes. *Right:* One hundred 99.7% confidence intervals constructed from these samples. These intervals cover the population mean 997 times out of 1000. They almost always succeed in covering the population mean, but their margin of error is large.

The confidence level measures the success rate of the method used to construct the confidence interval

The center diagram in Figure 8.8 presents 100 different 95% confidence intervals. When we construct a confidence interval with level 95%, we are essentially getting a look at one of these confidence intervals. However, we don't get to see any of the other confidence intervals, nor do we get to see the vertical line that indicates where the true value μ is. Therefore, we cannot be sure whether we got one of the confidence intervals that covers μ, or whether we were unlucky enough to get one of the unsuccessful ones. What we do know is that our confidence interval was constructed by a method that succeeds 95% of the time. The confidence level describes the success rate of the *method* used to construct a confidence interval, not the success of any particular interval.

Check Your Understanding

11. To determine how well a new method of teaching vocabulary is working in a certain elementary school, education researchers plan to give a vocabulary test to a sample of 100 sixth-graders. It is known that scores on this test have a standard deviation of 8. The researchers plan to compute the sample mean $\bar{x}$, then construct a 95% confidence interval for the population mean test score.
    a. What is the critical value $z_{\alpha/2}$ for this confidence interval?
    b. Find the margin of error for this confidence interval.
    c. Let m represent the margin of error for this confidence interval. For what percentage of all samples will the confidence interval $\bar{x} \pm m$ cover the true population mean?

12. The researchers now plan to construct a 99% confidence interval for the test scores described in Exercise 11.
    a. What is the critical value $z_{\alpha/2}$ for this confidence interval?
    b. Find the margin of error for this confidence interval.
    c. Let m represent the margin of error for this confidence interval. For what percentage of all samples will the confidence interval $\bar{x} \pm m$ cover the true population mean?

Answers are on page 366.

Objective 4 Find the sample size necessary to obtain a confidence interval of a given width

Finding the Necessary Sample Size

We have seen that we can make the margin of error smaller if we are willing to reduce our level of confidence. We can also reduce the margin of error by increasing the sample size. We can see this by looking at the formula for margin of error:

$$m = z_{\alpha/2} \frac{\sigma}{\sqrt{n}}$$

Since the sample size n appears in the denominator, making it larger will make the value of m smaller. We will show how we can manipulate this formula using algebra to express the sample size n in terms of the margin of error m.

$$m = z_{\alpha/2} \frac{\sigma}{\sqrt{n}}$$

$$m\sqrt{n} = \frac{z_{\alpha/2} \cdot \sigma}{\sqrt{n}} \sqrt{n} \qquad \text{(Multiply both sides by } \sqrt{n}\text{)}$$

$$\frac{m\sqrt{n}}{m} = \frac{z_{\alpha/2} \cdot \sigma}{m} \qquad \text{(Divide both sides by } m\text{)}$$

$$n = \left(\frac{z_{\alpha/2} \cdot \sigma}{m} \right)^2 \qquad \text{(Square both sides)}$$

With this formula, if we know how small we want the margin of error to be, we can compute the sample size needed to achieve the desired margin of error.

CAUTION

Always round the sample size *up*. For example, if the value of *n* given by the formula is 84.01, round it *up* to 85.

SUMMARY

Let *m* be the desired margin of error. Let σ be the population standard deviation, and let $z_{\alpha/2}$ be the critical value for a confidence interval. The sample size *n* needed so that the confidence interval will have margin of error *m* is given by

$$n = \left(\frac{z_{\alpha/2} \cdot \sigma}{m} \right)^2$$

If the value of *n* given by the formula is not a whole number, round it *up* to the nearest whole number. By rounding up, we can be sure that the margin of error is no greater than the desired value *m*.

EXAMPLE 8.7   Finding the necessary sample size

Scientists want to estimate the mean weight of mice after they have been fed a special diet. From previous studies, it is known that the weight is normally distributed with standard deviation 3 grams. How many mice must be weighed so that a 95% confidence interval will have a margin of error of 0.5 gram?

Solution

Since we want a 95% confidence interval, we use $z_{\alpha/2} = 1.96$. We are also given $\sigma = 3$ and $m = 0.5$. We therefore use the formula as follows:

$$n = \left(\frac{z_{\alpha/2} \cdot \sigma}{m} \right)^2 = \left(\frac{1.96 \cdot 3}{0.5} \right)^2 = 138.30; \quad \text{round up to } 139$$

We must weigh 139 mice in order to obtain a 95% confidence interval with a margin of error of 0.5 gram.

Factors that limit sample size

Since larger sample sizes result in narrower confidence intervals, it is natural to wonder why we don't always collect a large sample when we want to construct a confidence interval. In practice, the size of the sample that is feasible to obtain is often limited. In some cases, an expensive experimental procedure must be repeated each time an observation is made. For example, studies of automobile safety that require the crashing of new cars are not likely to have large sample sizes. Sometimes ethical considerations restrict the sample size. For example, when a new drug is being tested, there is a risk of adverse health effects to the subjects who take the drug. It is important that the sample size not be larger than necessary, to limit the health risk to as few people as possible.

Check Your Understanding

13. A machine used to fill beverage cans is supposed to put exactly 12 ounces of beverage in each can, but the actual amount varies randomly from can to can. The population standard deviation is $\sigma = 0.05$ ounce. A simple random sample of filled cans will have their volumes measured, and a 95% confidence interval for the mean fill volume will be constructed. How many cans must be sampled for the margin of error to be equal to 0.01 ounce?

14. An IQ test is designed to have scores that have a standard deviation of $\sigma = 15$. A simple random sample of students at a large university will be given the test in order to construct a 98% confidence interval for the mean IQ of all students at the university. How many students must be tested so that the margin of error will be equal to 3 points?

Answers are on page 366.

Objective 5 Distinguish between confidence and probability

Distinguish Between Confidence and Probability

In Example 8.6, a 99% confidence interval for the population mean weight μ was computed to be $20.04 < \mu < 20.46$. It is tempting to say that the probability is 99% that μ is between 20.04 and 20.46. This, however, is not correct. The term *probability* refers to random events, which can come out differently when experiments are repeated. The numbers 20.04 and 20.46 are fixed, not random. The population mean is also fixed. The population mean weight is either between 20.04 and 20.46 or it is not. There is no randomness involved. Therefore, we say that we have 99% *confidence* (not probability) that the population mean is in this interval.

On the other hand, let's say that we are discussing a *method* used to construct a 99% confidence interval. The method will succeed in covering the population mean 99% of the time, and fail the other 1% of the time. In this case, whether the population mean is covered or not is a random event, because it can vary from experiment to experiment. Therefore it *is* correct to say that a *method* for constructing a 99% confidence interval has probability 99% of covering the population mean.

EXAMPLE 8.8

Interpreting a confidence level

A hospital administrator plans to draw a simple random sample of 100 records of patients who were admitted for cardiac bypass surgery. She will compute the sample mean number of days spent in the hospital, and construct a 95% confidence interval for the population mean, using an appropriate method. She claims that the probability is 0.95 that the confidence interval will cover the population mean. Is she right?

Solution

Yes, she is right. The probability that a 95% confidence interval constructed by an appropriate method will cover the true value is 0.95.

EXAMPLE 8.9

Interpreting a confidence interval

Refer to Example 8.8. After drawing the sample, the hospital administrator constructs the 95% confidence interval, and it turns out to be $7.1 < \mu < 7.5$. The administrator claims that the probability is 0.95 that the population mean is between 7.1 and 7.5. Is she right?

Solution

No, she is not right. Once a specific confidence interval has been constructed, there is no more probability. She should say that she is 95% confident that the population mean is between 7.1 and 7.5.

Check Your Understanding

15. A scientist plans to construct a 95% confidence interval for the mean length of steel rods that are manufactured by a certain process. She will draw a simple random sample of rods and compute the confidence interval using the methods described in this section. She says, "The probability is 95% that the population mean length will be covered by the confidence interval." Is she right? Explain.

16. The scientist in Exercise 15 constructs the 95% confidence interval for the mean length in centimeters, and it turns out to be $25.1 < \mu < 27.2$. She says, "The probability is 95% that the population mean length is between 25.1 and 27.2 centimeters." Is she right? Explain.

Answers are on page 366.

USING TECHNOLOGY

We use Example 8.4 to illustrate the technology steps.

TI-84 PLUS

Constructing a confidence interval for the mean when σ is known

Step 1. Press **STAT** and highlight the **TESTS** menu.

Step 2. Select **ZInterval** and press **ENTER** (Figure A). The **ZInterval** menu appears.

Step 3. For **Inpt**, select the **Stats** option and enter the values of σ, $\bar{x}$, and n. For Example 8.4, we use $\sigma = 15$, $\bar{x} = 67.30$, and $n = 100$.

Step 4. In the **C-Level** field, enter the confidence level. For Example 8.4, we use 0.98 (Figure B).

Step 5. Highlight **Calculate** and press **ENTER** (Figure C).

Note that if the raw data are given, the **ZInterval** command may be used by selecting **Data** as the **Inpt** option and entering the location of the data as the **List** option (Figure D).

Figure A

Figure B

Figure C

Figure D

MINITAB

Constructing a confidence interval for the mean when σ is known

Step 1. Click on **Stat**, then **Basic Statistics**, then **1-Sample Z**.

Step 2. Choose one of the following:
* If the summary statistics are given, click **Summarized Data** and enter the **Sample Size** (100), the **Sample Mean** (67.30), and the **Standard Deviation** (15) (Figure E).
* If the raw data are given, click **Samples in Columns** and select the column that contains the data. Enter the **Standard Deviation**.

Step 3. Click **Options** and enter the confidence level in the **Confidence Level** (98) field. Click OK.

Step 4. Click **OK** (Figure F).

Figure E

One-Sample Z

The assumed standard deviation = 15

```
  N     Mean   SE Mean        98% CI
100   67.3000   1.5000   (63.8105, 70.7895)
```

Figure F

EXCEL

Constructing a confidence interval for the mean when σ is known

The **CONFIDENCE.NORM** command returns the margin of error for a confidence interval when the population standard deviation σ is known.

Step 1. In an empty cell, select the **Insert Function** icon and highlight **Statistical** in the category field.

Step 2. Click on the **CONFIDENCE.NORM** function and press **OK**.

Step 3. Enter the value of α (0.02) in the **Alpha** field, the population standard deviation (15) in the **Standard_dev** field, and the sample size (100) in the **Size** field.

Step 4. Click **OK** (Figure G) to obtain the margin of error m.

The confidence interval is given by $\bar{x} - m < \mu < \bar{x} + m$.

=CONFIDENCE.NORM(0.02, 15, 100)
Margin of Error   3.490

Figure G

SECTION 8.1 Exercises

Exercises 1–16 are the Check Your Understanding exercises located within the section.

Understanding the Concepts

In Exercises 17–20, fill in each blank with the appropriate word or phrase.

17. A single number that estimates the value of an unknown parameter is called a _____ estimate.

18. The margin of error is the product of the standard error and the _____ .

19. In the confidence interval 24.3 ± 1.2, the quantity 1.2 is called the _____ .

20. If we increase the confidence level and keep the sample size the same, we _____ the margin of error.

In Exercises 21–24, determine whether the statement is true or false. If the statement is false, rewrite it as a true statement.

21. The confidence level is the proportion of all possible samples for which the confidence interval will cover the true value.

22. To construct a confidence interval for a population mean, we add and subtract the critical value from the point estimate.

23. Increasing the sample size while keeping the confidence level the same will result in a narrower confidence interval.

24. If a 95% confidence interval for a population mean is $1.7 < \mu < 2.3$, then the probability is 0.95 that the mean is between 1.7 and 2.3.

Practicing the Skills

In Exercises 25–28, find the critical value $z_{\alpha/2}$ needed to construct a confidence interval with the given level.

25. Level 95%

26. Level 85%

27. Level 96%

28. Level 99.7%

In Exercises 29–32, find the levels of the confidence intervals that have the given critical values.

29. 2.326   30. 2.576   31. 2.81   32. 1.04

33. A sample of size $n = 49$ is drawn from a population whose standard deviation is $\sigma = 4.8$.
   a. Find the margin of error for a 95% confidence interval for μ.
   b. If the sample size were $n = 60$, would the margin of error be larger or smaller?

34. A sample of size $n = 50$ is drawn from a population whose standard deviation is $\sigma = 26$.
   a. Find the margin of error for a 90% confidence interval for μ.
   b. If the sample size were $n = 40$, would the margin of error be larger or smaller?

35. A sample of size $n = 32$ is drawn from a population whose standard deviation is $\sigma = 12.1$.
   a. Find the margin of error for a 99% confidence interval for μ.
   b. If the confidence level were 90%, would the margin of error be larger or smaller?

36. A sample of size $n = 64$ is drawn from a population whose standard deviation is $\sigma = 24.18$.
   a. Find the margin of error for a 95% confidence interval for μ.
   b. If the confidence level were 98%, would the margin of error be larger or smaller?

37. A sample of size $n = 10$ is drawn from a normal population whose standard deviation is $\sigma = 2.5$. The sample mean is $\bar{x} = 7.92$.
   a. Construct a 95% confidence interval for μ.
   b. If the population were not approximately normal, would the confidence interval constructed in part (a) be valid? Explain.

38. A sample of size $n = 80$ is drawn from a normal population whose standard deviation is $\sigma = 6.8$. The sample mean is $\bar{x} = 40.41$.

a. Construct a 90% confidence interval for μ.

b. If the population were not approximately normal, would the confidence interval constructed in part (a) be valid? Explain.

39. A population has standard deviation $\sigma = 21.3$.

a. How large a sample must be drawn so that a 99% confidence interval for μ will have a margin of error equal to 2.5?

b. If the required confidence level were 95%, would the necessary sample size be larger or smaller?

40. A population has standard deviation $\sigma = 17.3$.

a. How large a sample must be drawn so that a 95% confidence interval for μ will have a margin of error equal to 1.4?

b. If the required confidence level were 98%, would the necessary sample size be larger or smaller?

41. A population has standard deviation $\sigma = 12.7$.

a. How large a sample must be drawn so that a 96% confidence interval for μ will have a margin of error equal to 2.5?

b. If the required margin of error were 1.5, would the necessary sample size be larger or smaller?

42. A population has standard deviation $\sigma = 9.2$.

a. How large a sample must be drawn so that a 92% confidence interval for μ will have a margin of error equal to 0.8?

b. If the required margin of error were 1.4, would the necessary sample size be larger or smaller?

Working with the Concepts

43. SAT scores: A college admissions officer takes a simple random sample of 100 entering freshmen and computes their mean mathematics SAT score to be 458. Assume the population standard deviation is $\sigma = 116$.

a. Construct a 99% confidence interval for the mean mathematics SAT score for the entering freshman class.

b. If the sample size were 75 rather than 100, would the margin of error be larger or smaller than the result in part (a)? Explain.

c. If the confidence level were 95% rather than 99%, would the margin of error be larger or smaller than the result in part (a)? Explain.

d. Based on the confidence interval constructed in part (a), is it likely that the mean mathematics SAT score for the entering freshman class is greater than 500?

44. How many computers? In a simple random sample of 150 households, the sample mean number of personal computers was 1.32. Assume the population standard deviation is $\sigma = 0.41$.

a. Construct a 95% confidence interval for the mean number of personal computers.

b. If the sample size were 100 rather than 150, would the margin of error be larger or smaller than the result in part (a)? Explain.

c. If the confidence level were 98% rather than 95%, would the margin of error be larger or smaller than the result in part (a)? Explain.

d. Based on the confidence interval constructed in part (a), is it likely that the mean number of personal computers is greater than 1.25?

45. Babies: According to the National Health Statistics Reports, a sample of 360 one-year-old baby boys in the United States had a mean weight of 25.5 pounds. Assume the population standard deviation is $\sigma = 5.3$ pounds.

a. Construct a 95% confidence interval for the mean weight of all one-year-old baby boys in the United States

b. Should this confidence interval be used to estimate the mean weight of all one-year-old babies in the United States? Explain.

c. Based on the confidence interval constructed in part (a), is it likely that the mean weight of all one-year-old boys is less than 28 pounds?

46. Watch your cholesterol: A sample of 314 patients between the ages of 38 and 82 were given a combination of the drugs ezetimibe and simvastatin. They achieved a mean reduction in total cholesterol of 0.94 millimole per liter. Assume the population standard deviation is $\sigma = 0.18$.

a. Construct a 98% confidence interval for the mean reduction in total cholesterol in patients who take this combination of drugs.

b. Should this confidence interval be used to estimate the mean reduction in total cholesterol for patients over the age of 85? Explain.

c. Based on the confidence interval constructed in part (a), is it likely that the mean reduction in cholesterol level is less than 0.90?

Source: *International Journal of Clinical Practice* 58:653–658

47. How smart is your phone? A random sample of 11 Samsung Galaxy smartphones being sold over the Internet in 2013 had the following prices, in dollars:

199	169	385	329	269	149
135	249	349	299	249	

Assume the population standard deviation is $\sigma = 85$.

a. Explain why it is necessary to check whether the population is approximately normal before constructing a confidence interval.

b. Following is a dotplot of these data. Is it reasonable to assume that the population is approximately normal?

c. If appropriate, construct a 95% confidence interval for the mean price for all phones of this type being sold on the Internet in 2013. If not appropriate, explain why not.

48. Stock prices: The Standard and Poor's (S&P) 500 is a group of 500 large companies traded on the New York Stock Exchange. Following are prices, in dollars, for a random sample of ten stocks on June 19, 2013.

84.86	8.11	74.23	35.25	13.19
53.55	84.25	201.94	24.68	53.47

Assume the population standard deviation is $\sigma = 50$.
a. Explain why it is necessary to check whether the population is approximately normal before constructing a confidence interval.
b. Following is a dotplot of these data. Is it reasonable to assume that the population is approximately normal?

c. If appropriate, construct a 95% confidence interval for the mean price for all S&P 500 stocks on this day. If not appropriate, explain why not.

49. High energy: A random sample of energy drinks had the following amounts of caffeine per fluid ounce.

14.2	8.3	80.8	6.7	3.6
13.7	12.9	11.5	24.7	9.5

Assume the population standard deviation is $\sigma = 24$.
a. Explain why it is necessary to check whether the population is approximately normal before constructing a confidence interval.
b. Following is a dotplot of these data. Is it reasonable to assume that the population is approximately normal?

c. If appropriate, construct a 95% confidence interval for the mean amount of caffeine in all energy drinks. If not appropriate, explain why not.

50. Let's shake on it: A random sample of 12-ounce milkshakes from 14 fast-food restaurants had the following number of calories.

504	399	580	476	450	591	510
700	608	472	642	613	473	375

Assume the population standard deviation is $\sigma = 90$.
a. Explain why it is necessary to check whether the population is approximately normal before constructing a confidence interval.
b. Following is a dotplot of these data. Is it reasonable to assume that the population is approximately normal?

c. If appropriate, construct a 95% confidence interval for the mean number of calories for all 12-ounce milkshakes sold at fast-food restaurants. If not appropriate, explain why not.

51. Lifetime of electronics: In a simple random sample of 100 electronic components produced by a certain method, the mean lifetime was 125 hours. Assume that component lifetimes are normally distributed with population standard deviation $\sigma = 20$ hours.
a. Construct a 98% confidence interval for the mean battery life.

b. Find the sample size needed so that a 99% confidence interval will have a margin of error of 3.

52. Efficient manufacturing: Efficiency experts study the processes used to manufacture items in order to make them as efficient as possible. One of the steps used to manufacture a metal clamp involves the drilling of three holes. In a sample of 75 clamps, the mean time to complete this step was 50.1 seconds. Assume that the population standard deviation is $\sigma = 10$ seconds.
a. Construct a 95% confidence interval for the mean time needed to complete this step.
b. Find the sample size needed so that a 98% confidence interval will have margin of error of 1.5.

53. Different levels: Joe and Sally are going to construct confidence intervals from the same simple random sample. Joe's confidence interval will have level 90% and Sally's will have level 95%.
a. Which confidence interval will have the larger margin of error? Or will they both be the same?
b. Which confidence interval is more likely to cover the population mean? Or are they both equally likely to do so?

54. Different levels: Bertha and Todd are going to construct confidence intervals from the same simple random sample. Bertha's confidence interval will have level 98% and Todd's will have level 95%.
a. Which confidence interval will have the larger margin of error? Or will they both be the same?
b. Which confidence interval is more likely to cover the population mean? Or are they both equally likely to do so?

55. Different standard deviations: Maria and Bob are going to construct confidence intervals from different simple random samples. Both confidence intervals will have level 95%. Maria's sample comes from a population with standard deviation $\sigma = 1$, and Bob's comes from a population with $\sigma = 2$. Both sample sizes are the same.
a. Which confidence interval will have the larger margin of error? Or will they both be the same?
b. Which confidence interval is more likely to cover the population mean? Or are they both equally likely to do so?

56. Different standard deviations: Martin and Bianca are going to construct confidence intervals from different simple random samples. Both confidence intervals will have level 99%. Martin's sample comes from a population with standard deviation $\sigma = 25$, and Bianca's comes from a population with $\sigma = 18$. Both sample sizes are the same.
a. Which confidence interval will have the larger margin of error? Or will they both be the same?
b. Which confidence interval is more likely to cover the population mean? Or are they both equally likely to do so?

57. Which interval is which? Sam constructed three confidence intervals, all from the same random sample. The confidence levels are 90%, 95%, and 99%. The confidence intervals are $5.6 < \mu < 14.4$, $7.2 < \mu < 12.8$, and $6.6 < \mu < 13.4$. Unfortunately, Sam has forgotten which confidence interval has which level. Match each confidence interval with its level.

58. Which interval is which? Matilda has constructed three confidence intervals, all from the same random sample. The confidence levels are 95%, 98%, and 99.9%. The confidence intervals are $6.4 < \mu < 12.3$, $5.1 < \mu < 13.6$, and $6.8 < \mu < 11.9$. Unfortunately, Matilda has forgotten which confidence interval has which level. Match each confidence interval with its level.

59. Don't construct a confidence interval: A psychology professor at a certain college gave a test to the students in her class. The test was designed to measure students' attitudes toward school, with higher scores indicating a more positive attitude. There were 30 students in the class, and their mean score was 78. Scores on this test are known to be normally distributed with a standard deviation of 10. Explain why these data should not be used to construct a confidence interval for the mean score for all the students in the college.

60. Don't construct a confidence interval: A college alumni organization sent a survey to all recent graduates to ask their annual income. Twenty percent of the alumni responded, and their mean annual income was $40,000. Assume the population standard deviation is $\sigma = \$10,000$. Explain why these data should not be used to construct a confidence interval for the mean annual income of all recent graduates.

61. Interpret a confidence interval: A dean at a certain college looked up the GPA for a random sample of 85 students. The sample mean GPA was 2.82, and a 95% confidence interval for the mean GPA of all students in the college was $2.76 < \mu < 2.88$. True or false, and explain:
a. We are 95% confident that the mean GPA of all students in the college is between 2.76 and 2.88.
b. We are 95% confident that the mean GPA of all students in the sample is between 2.76 and 2.88.
c. The probability is 0.95 that the mean GPA of all students in the college is between 2.76 and 2.88.
d. 95% of the students in the sample had a GPA between 2.76 and 2.88.

62. Interpret a confidence interval: A survey organization drew a simple random sample of 625 households from a city of 100,000 households. The sample mean number of people in the 625 households was 2.30, and a 95% confidence interval for the mean number of people in the 100,000 households was $2.16 < \mu < 2.44$. True or false, and explain:
a. We are 95% confident that the mean number of people in the 625 households is between 2.16 and 2.44.
b. We are 95% confident that the mean number of people in the 100,000 households is between 2.16 and 2.44.
c. The probability is 0.95 that the mean number of people in the 100,000 households is between 2.16 and 2.44.
d. 95% of the households in the sample contain between 2.16 and 2.44 people.

63. Interpret calculator display: The following display from a TI-84 Plus calculator presents a 95% confidence interval.

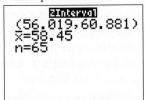

a. Fill in the blanks: We are _____ confident that the population mean is between _____ and _____ .
b. Assume the population is not normally distributed. Is the confidence interval still valid? Explain.

64. Interpret calculator display: The following display from a TI-84 Plus calculator presents a 99% confidence interval.

```
ZInterval
(17.012,20.048)
x̄=18.53
n=15
```

a. Fill in the blanks: We are _____ confident that the population mean is between _____ and _____ .
b. Assume the population is not normally distributed. Is the confidence interval still valid? Explain.

65. Interpret computer output: The following MINITAB output presents a 95% confidence interval.

The assumed sigma = 6.5000				
Variable	N	Mean	SE Mean	95% CI
X	23	12.352	1.3553	(9.6956, 15.0084)

a. Fill in the blanks: We are _____ confident that the population mean is between _____ and _____ .
b. Use the appropriate critical value along with the information in the computer output to construct a 99% confidence interval.
c. Find the sample size needed so that the 95% confidence interval will have a margin of error of 1.5.
d. Find the sample size needed so that the 99% confidence interval will have a margin of error of 1.5.

66. Interpret computer output: The following MINITAB output presents a 98% confidence interval.

The assumed sigma = 8.0000				
Variable	N	Mean	SE Mean	98% CI
X	58	2.657	1.0505	(0.1233, 5.1007)

a. Fill in the blanks: We are _____ confident that the population mean is between _____ and _____ .
b. Use the appropriate critical value along with the information in the computer output to construct a 95% confidence interval.
c. Find the sample size needed so that the 98% confidence interval will have a margin of error of 1.0.
d. Find the sample size needed so that the 95% confidence interval will have a margin of error of 1.0.

Extending the Concepts

One-sided confidence intervals: A confidence interval provides likely minimum and maximum values for a parameter. In some cases, we are interested only in a maximum or only in a minimum. In these cases, we construct a *one-sided confidence interval*. A one-sided confidence interval can be an *upper confidence bound*, which has the form $\bar{x} + z_\alpha \sigma/\sqrt{n}$, or a *lower confidence bound*, which has the form $\bar{x} - z_\alpha \sigma/\sqrt{n}$. Note that the critical value is z_α rather than $z_{\alpha/2}$.

67. **Computers in the classroom:** A simple random sample of 50 middle-school children participated in an experimental class designed to introduce them to computer programming. At the end of the class, the students took a final exam to

assess their learning. The sample mean score was 78 points, and the population standard deviation is 8 points. Compute a lower 99% confidence bound for the mean score.

68. **Charge it:** A random sample of 75 charges on a credit card had a mean of $56.85, and the population standard deviation is $21.08. Compute an upper 95% confidence bound for the mean amount charged.

Answers to Check Your Understanding Exercises for Section 8.1

1. **a.** 1.645 **b.** 2.326 **c.** 2.81 **d.** 1.28
2. **a.** 95% **b.** 97% **c.** 80% **d.** 99.9%
3. **a.** 2.352 **b.** 1.030 **c.** 5.758 [Tech: 5.757] **d.** 6.397
4. **a.** 105.2 **b.** 1.645 **c.** 1.1547 **d.** 1.90
   e. $103.3 < \mu < 107.1$
   f. Yes. We are 90% confident that μ is between 103.3 and 107.1, so it is likely that $\mu > 100$.
5. **a.** 120.1 **b.** 1.96 **c.** 2.8284 **d.** 5.54
   e. $114.6 < \mu < 125.6$
   f. No. We are 95% confident that μ is between 114.6 and 125.6, so it is not likely that $\mu > 130$.
6. **a.** True **b.** False
7. $102.9 < \mu < 107.5$

8. $113.5 < \mu < 126.7$
9. The confidence interval does not contain the value 100. Therefore, it is not likely that the claim that $\mu = 100$ is true.
10. **a.** 2.576 **b.** $101.2143 < \mu < 103.4911$
11. **a.** 1.96 **b.** 1.568 **c.** 95%
12. **a.** 2.576 **b.** 2.061 **c.** 99%
13. 97
14. 136
15. Yes, the probability that a 95% confidence interval constructed by an appropriate method will cover the true value is 0.95.
16. No. Once a specific confidence interval is constructed, there is no probability attached to it.

SECTION 8.2 Confidence Intervals for a Population Mean, Standard Deviation Unknown

Objectives

1. Describe the properties of the Student's *t* distribution
2. Construct confidence intervals for a population mean when the population standard deviation is unknown

Objective 1 Describe the properties of the Student's *t* distribution

The Student's *t* Distribution

In Section 8.1, we showed how to construct a confidence interval for the mean μ of a normal population when the population standard deviation σ is known. The confidence interval is

$$\bar{x} \pm z_{\alpha/2} \frac{\sigma}{\sqrt{n}}$$

The critical value is $z_{\alpha/2}$ because the quantity

$$\frac{\bar{x} - \mu}{\sigma/\sqrt{n}}$$

has a normal distribution.

In practice, it is more common that σ is unknown. When we don't know the value of σ, we replace it with the sample standard deviation s. However, we cannot then use $z_{\alpha/2}$ as the critical value, because the quantity

$$\frac{\bar{x} - \mu}{s/\sqrt{n}}$$

does not have a normal distribution. One reason is that s is, on the average, a bit smaller than σ, so replacing σ with s tends to increase the magnitude. Another reason is that s is random whereas σ is constant, so replacing σ with s increases the spread.

The distribution of this quantity is called the **Student's *t* distribution**. It was discovered in 1908 by William Sealy Gosset, a statistician who worked for the Guinness Brewing Company in Dublin, Ireland. The management at Guinness considered the discovery to be proprietary information, and forbade Gosset to publish it. He published it anyway, using the pseudonym "Student." Gosset had done this before; see Section 6.3.

In fact, there are many different Student's t distributions; they are distinguished by a quantity called the **degrees of freedom**. When using the Student's t distribution to construct a confidence interval for a population mean, the number of degrees of freedom is 1 less than the sample size.

Recall: Degrees of freedom were introduced in Section 3.2.

Degrees of Freedom for the Student's *t* Distribution

When constructing a confidence interval for a population mean, the number of degrees of freedom for the Student's t distribution is 1 less than the sample size n.

$$\text{number of degrees of freedom} = n - 1$$

Figure 8.9 presents t distributions for several different degrees of freedom, along with a standard normal distribution for comparison. The t distributions are symmetric and unimodal, just like the normal distribution. The t distribution is more spread out than the standard normal distribution, because the sample standard deviation s is, on the average, a bit less than σ. When the number of degrees of freedom is small, this tendency is more pronounced, so the t distributions are much more spread out than the normal. When the number of degrees of freedom is large, s tends to be very close to σ, so the t distribution is very close to the normal. Figure 8.9 shows that with 10 degrees of freedom, the difference between the t distribution and the normal is not great. If a t distribution with 30 degrees of freedom were plotted in Figure 8.9, it would be indistinguishable from the normal distribution.

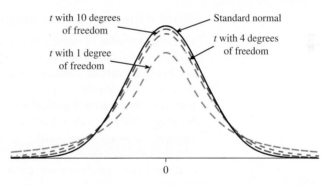

Figure 8.9   Plots of the Student's t distribution for 1, 4, and 10 degrees of freedom. The standard normal distribution is plotted for comparison. The t distributions are more spread out than the normal, but the amount of extra spread decreases as the number of degrees of freedom increases.

SUMMARY

The Student's t distribution has the following properties:

- It is symmetric and unimodal.
- It is more spread out than the standard normal distribution.
- If we increase the number of degrees of freedom, the Student's t curve becomes closer to the standard normal curve.

Finding the critical value

We use the Student's t distribution to construct confidence intervals for μ when σ is unknown. The idea behind the critical value is the same as for the normal distribution. To find the critical value for a confidence interval with a given level, let $1 - \alpha$ be the confidence level expressed as a decimal. The critical value is then $t_{\alpha/2}$, because the area under the Student's t curve between $-t_{\alpha/2}$ and $t_{\alpha/2}$ is $1 - \alpha$. See Figure 8.10. The critical value $t_{\alpha/2}$ can be found in Table A.3, in the row corresponding to the number of degrees of freedom and the column corresponding to the desired confidence level. Example 8.10 shows how to find a critical value.

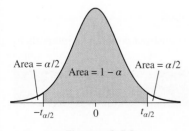

Figure 8.10

EXAMPLE 8.10

Finding a critical value

A simple random sample of size 10 is drawn from a normal population. Find the critical value $t_{\alpha/2}$ for a 95% confidence interval.

Solution

The sample size is $n = 10$, so the number of degrees of freedom is $n - 1 = 9$. We consult Table A.3, looking in the row corresponding to 9 degrees of freedom, and in the column with confidence level 95% (the confidence levels are listed along the bottom of the table). The critical value is $t_{\alpha/2} = 2.262$. See Figure 8.11.

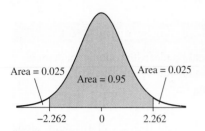

Figure 8.11 95% of the area under the Student's t curve with 9 degrees of freedom is between $t = -2.262$ and $t = 2.262$.

4	0.271	0.741	1.533	2.132	2.776	3.747	4.604	5.598	7.173	8.610
5	0.267	0.727	1.476	2.015	2.571	3.365	4.032	4.773	5.893	6.869
6	0.265	0.718	1.440	1.943	2.447	3.143	3.707	4.317	5.208	5.959
7	0.263	0.711	1.415	1.895	2.365	2.998	3.499	4.029	4.785	5.408
8	0.262	0.706	1.397	1.860	2.306	2.896	3.355	3.833	4.501	5.041
9	0.261	0.703	1.383	1.833	2.262	2.821	3.250	3.690	4.297	4.781
10	0.260	0.700	1.372	1.812	2.228	2.764	3.169	3.581	4.144	4.587
11	0.260	0.697	1.363	1.796	2.201	2.718	3.106	3.497	4.025	4.437
12	0.259	0.695	1.356	1.782	2.179	2.681	3.055	3.428	3.930	4.318
13	0.259	0.694	1.350	1.771	2.160	2.650	3.012	3.372	3.852	4.221
14	0.258	0.692	1.345	1.761	2.145	2.624	2.977	3.326	3.787	4.140
	20%	50%	80%	90%	95%	98%	99%	99.5%	99.8%	99.9%

Confidence Level

What if the number of degrees of freedom isn't in the table?

The largest number of degrees of freedom shown in Table A.3 is 200. If the desired number of degrees of freedom is less than 200 and is not shown in Table A.3, use the next smaller number of degrees of freedom in the table. If the desired number of degrees of freedom is more than 200, use the z-value found in the last row of Table A.3, or use Table A.2. This problem will not arise if a calculator or computer is used to construct a confidence interval, because technology can compute critical values for any number of degrees of freedom.

SUMMARY

If the desired number of degrees of freedom isn't listed in Table A.3, then

- If the desired number is less than 200, use the next smaller number that is in the table.
- If the desired number is greater than 200, use the z-value found in the last row of Table A.3, or use Table A.2.

EXAMPLE 8.11

Finding a critical value

Use Table A.3 to find the critical value for a 99% confidence interval for a sample of size 58.

Solution

Because the sample size is 58, there are 57 degrees of freedom. The number 57 doesn't appear in the degrees of freedom column in Table A.3, so we use the next smaller number that does appear, which is 50. The value of $t_{\alpha/2}$ corresponding to a confidence level of 99% is $t_{\alpha/2} = 2.678$.

Following are the assumptions that are necessary to construct confidence intervals by using the Student's t distribution.

Explain It Again

$\bar{x}$ must be approximately normal: We need assumption 2 to be sure that the sampling distribution of $\bar{x}$ is approximately normal.

Assumptions for Constructing a Confidence Interval for μ When σ Is Unknown

1. We have a simple random sample.
2. Either the sample size is large $(n > 30)$, or the population is approximately normal.

Checking the assumptions

When the sample size is small ($n \leq 30$), we must check to determine whether the sample comes from a population that is approximately normal. This can be done using the methods described in Section 7.6. A simple method is to draw a dotplot or boxplot of the sample. If there are no outliers, and if the sample is not strongly skewed, then it is reasonable to construct a confidence interval using the Student's t distribution.

Check Your Understanding

1. Use Table A.3 to find the critical value $t_{\alpha/2}$ needed to construct a confidence interval of the given level with the given sample size:
   a. Level 95%, sample size 15
   b. Level 99%, sample size 22
   c. Level 90%, sample size 63
   d. Level 95%, sample size 2

2. In each of the following situations, state whether the methods of this section should be used to construct a confidence interval for the population mean. Assume that σ is unknown.
   a. A simple random sample of size 8 is drawn from a distribution that is approximately normal.
   b. A simple random sample of size 15 is drawn from a distribution that is not close to normal.
   c. A simple random sample of size 150 is drawn from a distribution that is not close to normal.
   d. A nonrandom sample is drawn.

Answers are on page 378.

Objective 2 Construct confidence intervals for a population mean when the population standard deviation is unknown

Constructing a Confidence Interval for μ When σ Is Unknown

The ingredients for a confidence interval for a population mean μ when σ is unknown are the point estimate $\bar{x}$, the critical value $t_{\alpha/2}$, and the standard error $s/\sqrt{n}$. The margin of error is $t_{\alpha/2}s/\sqrt{n}$. When the assumptions for the Student's t distribution are met, we can use the following step-by-step procedure for constructing a confidence interval for a population mean.

Procedure for Constructing a Confidence Interval for μ When σ Is Unknown

Check to be sure the assumptions are satisfied. If they are, then proceed with the following steps.

Step 1: Compute the sample mean $\bar{x}$ and sample standard deviation s, if they are not given.

Step 2: Find the number of degrees of freedom $n - 1$ and the critical value $t_{\alpha/2}$.

Step 3: Compute the standard error $s/\sqrt{n}$ and multiply it by the critical value to obtain the margin of error $t_{\alpha/2} \dfrac{s}{\sqrt{n}}$.

Step 4: Use the point estimate and the margin of error to construct the confidence interval:

$$\text{Point estimate} \pm \text{Margin of error}$$

$$\bar{x} \pm t_{\alpha/2} \frac{s}{\sqrt{n}}$$

$$\bar{x} - t_{\alpha/2} \frac{s}{\sqrt{n}} < \mu < \bar{x} + t_{\alpha/2} \frac{s}{\sqrt{n}}$$

Step 5: Interpret the result.

EXAMPLE 8.12   Constructing a confidence interval

A food chemist analyzed the calorie content for a popular type of chocolate cookie. Following are the numbers of calories in a sample of eight cookies.

$$113 \quad 114 \quad 111 \quad 116 \quad 115 \quad 120 \quad 118 \quad 116$$

Find a 98% confidence interval for the mean number of calories in this type of cookie.

Solution

We check the assumptions. We have a simple random sample. Because the sample size is small, the population must be approximately normal. We check this with a dotplot of the data.

There is no evidence of strong skewness, and no outliers. Therefore, we may proceed.

Step 1:   Find the sample mean and sample standard deviation. We compute the mean and standard deviation of the sample values. We obtain

$$\bar{x} = 115.375 \quad s = 2.8253$$

Step 2:   Find the number of degrees of freedom and the critical value $t_{\alpha/2}$. The number of degrees of freedom is $n - 1 = 8 - 1 = 7$. Using Table A.3, we find that the critical value corresponding to a level of 98% is $t_{\alpha/2} = 2.998$.

Step 3:   Compute the margin of error. The margin of error is

$$t_{\alpha/2} \frac{s}{\sqrt{n}}$$

We substitute $t_{\alpha/2} = 2.998$, $s = 2.8253$, and $n = 8$ to obtain

$$t_{\alpha/2} \frac{s}{\sqrt{n}} = 2.998 \frac{2.8253}{\sqrt{8}} = 2.9947$$

Step 4:   Construct the confidence interval. The 98% confidence interval is given by

$$\bar{x} - t_{\alpha/2} \frac{s}{\sqrt{n}} < \mu < \bar{x} + t_{\alpha/2} \frac{s}{\sqrt{n}}$$

$$115.375 - 2.9947 < \mu < 115.375 + 2.9947$$

$$112.4 < \mu < 118.4$$

Note that we round the final result to one decimal place, because the sample values were whole numbers.

Step 5:   Interpret the result. We are 98% confident that the mean number of calories per cookie is between 112.4 and 118.4.

EXAMPLE 8.13   Constructing a confidence interval

The General Social Survey is a survey of opinions and lifestyles of U.S. adults, conducted by the National Opinion Research Center at the University of Chicago. A sample of 123 people aged 18–22 reported the number of hours they spent on the Internet in an average week. The sample mean was 8.20 hours, with a sample standard deviation of 9.84 hours. Assume this is a simple random sample from the population of people aged 18–22 in the United States. Construct a 95% confidence interval for μ, the population mean number of hours per week spent on the Internet by people aged 18–22 in the United States.

Solution

We check the assumptions. We have a simple random sample. Now either the sample size must be greater than 30, or the population must be approximately normal. Since the sample size is $n = 123$, the assumptions are met.

Step 1: Find the sample mean and sample standard deviation. These are given as $\bar{x} = 8.20$ and $s = 9.84$.

Step 2: Find the number of degrees of freedom and the critical value $t_{\alpha/2}$. The number of degrees of freedom is $n - 1 = 123 - 1 = 122$. Since this number of degrees of freedom does not appear in Table A.3, we use the next smaller value in the table, which is 100. The critical value corresponding to a level of 95% is $t_{\alpha/2} = 1.984$.

Step 3: Compute the margin of error. The margin of error is

$$t_{\alpha/2} \frac{s}{\sqrt{n}}$$

We substitute $t_{\alpha/2} = 1.984$, $s = 9.84$, and $n = 123$ to obtain

$$t_{\alpha/2} \frac{s}{\sqrt{n}} = 1.984 \frac{9.84}{\sqrt{123}} = 1.7603$$

Step 4: Construct the confidence interval. The 95% confidence interval is given by

$$\bar{x} - t_{\alpha/2} \frac{s}{\sqrt{n}} < \mu < \bar{x} + t_{\alpha/2} \frac{s}{\sqrt{n}}$$

$$8.20 - 1.7603 < \mu < 8.20 + 1.7603$$

$$6.44 < \mu < 9.96$$

Note that we round the final result to two decimal places, because the value of $\bar{x}$ was given to two decimal places.

Step 5: Interpret the result. We are 95% confident that the mean number of hours per week spent on the Internet by people 18–22 years old is between 6.44 and 9.96.

Figure 8.12

Note that in Example 8.13, the sample standard deviation of 9.84 is larger than the sample mean of 8.20. Since the minimum possible time to spend on the Internet is 0, the smallest sample value is less than one standard deviation below the mean. This indicates that the sample is fairly skewed. Figure 8.12 confirms this. Even though the sample is skewed, the t statistic is still appropriate, because the sample size of 123 is large.

Constructing confidence intervals with technology

The following TI-84 Plus display presents the results of Example 8.13.

```
      TInterval
(6.4436,9.9564)
x̄=8.2
Sx=9.84
n=123
```

> **CAUTION**
>
> Confidence intervals constructed using technology may differ from those constructed by hand due to rounding. The differences are never large enough to matter.

The display is fairly straightforward. The quantity Sx is the sample standard deviation s. The TI-84 Plus uses the exact number of degrees of freedom, 122, rather than 100 as we did in the solution to Example 8.13. This does not make a difference when the answer is rounded to two decimal places. Note that the confidence level (95%) is not given in the display.

The following MINITAB output presents the results of the same example.

Variable	N	Mean	StDev	SE Mean	95% CI
Hours	123	8.20000	9.84000	0.88724	(6.44361, 9.95639)

The output is fairly straightforward. Going from left to right, "N" represents the sample size, "Mean" is the sample mean $\bar{x}$, and "StDev" is the sample standard deviation s. The quantity labeled "SE Mean" is the standard error $s/\sqrt{n}$. Note that neither the critical value nor the margin of error is given explicitly in the output. Finally, the lower and upper limits of the 95% confidence interval are given on the right. Like the TI-84 Plus, MINITAB uses 122 degrees of freedom rather than 100 as we did in the solution to the example. This does not make a difference when the answer is rounded to two decimal places.

Finally, we present EXCEL output for this example. The EXCEL function CONFIDENCE.T returns the margin of error. The inputs are the value of α, the population standard deviation σ, and the sample size n.

=CONFIDENCE.T(0.05, 9.84, 123)	
Margin of Error	1.756

Step-by-step instructions for constructing confidence intervals with technology are given in the Using Technology section on page 373.

Check Your Understanding

3. A potato chip company wants to evaluate the accuracy of its potato chip bag-filling machine. Bags are labeled as containing 8 ounces of potato chips. A simple random sample of 12 bags had mean weight 8.12 ounces with a sample standard deviation of 0.1 ounce. Assume the weights are approximately normally distributed. Construct a 99% confidence interval for the population mean weight of bags of potato chips.

4. A company has developed a new type of lightbulb, and wants to estimate its mean lifetime. A simple random sample of 100 bulbs had a sample mean lifetime of 750.2 hours with a sample standard deviation of 30 hours. Construct a 95% confidence interval for the population mean lifetime of all bulbs manufactured by this new process.

5. The following TI-84 Plus display presents a 95% confidence interval for a population mean.

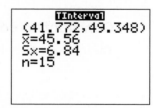

```
    TInterval
(41.772,49.348)
x=45.56
Sx=6.84
n=15
```

    a. How many degrees of freedom are there?
    b. Find the critical value $t_{\alpha/2}$ for a 99% confidence interval.
    c. Use the critical value along with the information in the display to construct a 99% confidence interval for the population mean.

6. The following MINITAB output presents a confidence interval for a population mean.

Variable	N	Mean	StDev	SE Mean	95% CI
X	10	8.5963	0.11213	0.03546	(8.5161, 8.6765)

    a. How many degrees of freedom are there?
    b. Find the critical value $t_{\alpha/2}$ for a 99% confidence interval.
    c. Use the critical value along with the information in the computer output to construct a 99% confidence interval for the population mean.

Answers are on page 378.

USING TECHNOLOGY

We use Example 8.13 to illustrate the technology steps.

TI-84 PLUS

Constructing a confidence interval for the mean when σ is unknown

Step 1. Press **STAT** and highlight the **TESTS** menu.

Step 2. Select **TInterval** and press **ENTER** (Figure A). The **TInterval** menu appears.

Step 3. For **Inpt**, select the **Stats** option and enter the values of $\bar{x}$, s, and n. For Example 8.13, we use $\bar{x} = 8.20$, $s = 9.84$, and $n = 123$.

Step 4. In the **C-Level** field, enter the confidence level. For Example 8.13 we use 0.95 (Figure B).

Step 5. Highlight **Calculate** and press **ENTER** (Figure C).

Note that if the raw data are given, the **TInterval** command may be used by selecting **Data** as the **Inpt** option and entering the location of the data as the **List** option (Figure D).

Figure A

Figure B

Figure C

Figure D

MINITAB

Constructing a confidence interval for the mean when σ is unknown

Step 1. Click on **Stat**, then **Basic Statistics**, then **1-Sample t**.

Step 2. Choose one of the following:
- If the summary statistics are given, click **Summarized Data** and enter the **Sample Size** (123), the **Mean** (8.20), and the **Standard Deviation** (9.84) (Figure E).
- If the raw data are given, click **Samples in Columns** and select the column that contains the data.

Step 3. Click **Options** and enter the confidence level in the **Confidence Level** (95) field. For Example 8.13, enter **95%**. Click **OK**.

Step 4. Click **OK** (Figure F).

One-Sample T

N	Mean	StDev	SE Mean	95% CI
123	8.20000	9.84000	0.88724	(6.44361, 9.95639)

Figure F

Figure E

EXCEL

Constructing a confidence interval for the mean when σ is unknown

The **CONFIDENCE.T** command returns the margin of error for a confidence interval when the population standard deviation σ is unknown.

Step 1. In an empty cell, select the **Insert Function** icon and highlight **Statistical** in the category field.

Step 2. Click on the **CONFIDENCE.T** function and press **OK**.

Step 3. Enter the value of α (0.05) in the **Alpha** field, the sample standard deviation (9.84) in the **Standard_dev** field, and the sample size (123) in the **Size** field.

Step 4. Click **OK** (Figure G) to obtain the margin of error m.

The confidence interval is given by $\bar{x} - m < \mu < \bar{x} + m$.

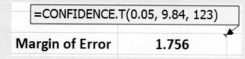

=CONFIDENCE.T(0.05, 9.84, 123)

| **Margin of Error** | 1.756 |

Figure G

SECTION 8.2   Exercises

Exercises 1–6 are the Check Your Understanding exercises located within the section.

Understanding the Concepts

In Exercises 7 and 8, fill in each blank with the appropriate word or phrase.

7. When constructing a confidence interval for a population mean μ from a sample of size 12, the number of degrees of freedom for the critical value $t_{\alpha/2}$ is _____ .

8. When the number of degrees of freedom is large, the Student's t distribution is close to the _____ distribution.

In Exercises 9 and 10, determine whether the statement is true or false. If the statement is false, rewrite it as a true statement.

9. The Student's t curve is less spread out than the standard normal curve.

10. The Student's t distribution should not be used to find a confidence interval for μ if outliers are present in a small sample.

Practicing the Skills

11. Find the critical value $t_{\alpha/2}$ needed to construct a confidence interval of the given level with the given sample size.
  a. Level 95%, sample size 23
  b. Level 90%, sample size 3
  c. Level 98%, sample size 18
  d. Level 99%, sample size 29

12. Find the critical value $t_{\alpha/2}$ needed to construct a confidence interval of the given level with the given sample size.
  a. Level 90%, sample size 6
  b. Level 98%, sample size 12
  c. Level 95%, sample size 32
  d. Level 99%, sample size 10

13. A sample of size $n = 18$ is drawn from a normal population.

a. Find the critical value $t_{\alpha/2}$ needed to construct a 95% confidence interval.
  b. If the sample size were $n = 25$, would the critical value be smaller or larger?

14. A sample of size $n = 22$ is drawn from a normal population.
  a. Find the critical value $t_{\alpha/2}$ needed to construct a 90% confidence interval.
  b. If the sample size were $n = 15$, would the critical value be smaller or larger?

15. A sample of size $n = 12$ is drawn from a normal population.
  a. Find the critical value $t_{\alpha/2}$ needed to construct a 98% confidence interval.
  b. If the sample size were $n = 50$, would it be necessary for the population to be approximately normal?

16. A sample of size $n = 61$ is drawn.
  a. Find the critical value $t_{\alpha/2}$ needed to construct a 95% confidence interval.
  b. If the sample size were $n = 15$, what additional assumption would need to be made for the confidence interval to be valid?

17. A sample of size $n = 15$ has sample mean $\bar{x} = 2.1$ and sample standard deviation $s = 1.7$.
  a. Construct a 95% confidence interval for the population mean μ.
  b. If the sample size were $n = 25$, would the confidence interval be narrower or wider?

18. A sample of size $n = 44$ has sample mean $\bar{x} = 56.9$ and sample standard deviation $s = 9.1$.
  a. Construct a 98% confidence interval for the population mean μ.
  b. If the sample size were $n = 30$, would the confidence interval be narrower or wider?

19. A sample of size $n = 89$ has sample mean $\bar{x} = 87.2$ and sample standard deviation $s = 5.3$.
  a. Construct a 95% confidence interval for the population mean μ.

b. If the confidence level were 99%, would the confidence interval be narrower or wider?

20. A sample of size $n = 35$ has sample mean $\bar{x} = 34.85$ and sample standard deviation $s = 17.9$.

a. Construct a 98% confidence interval for the population mean μ.

b. If the confidence level were 95%, would the confidence interval be narrower or wider?

Working with the Concepts

21. Online courses: A sample of 263 students who were taking online courses were asked to describe their overall impression of online learning on a scale of 1–7, with 7 representing the most favorable impression. The average score was 5.53, and the standard deviation was 0.92.

a. Construct a 95% confidence interval for the mean score.

b. Assume that the mean score for students taking traditional courses is 5.55. A college that offers online courses claims that the mean scores for online courses and traditional courses are the same. Does the confidence interval contradict this claim? Explain.

Source: *Innovations in Education and Teaching International* 45:115–126

22. Get an education: In 2012 the General Social Survey asked 1972 adults how many years of education they had. The sample mean was 13.37 years with a standard deviation of 3.13 years.

a. Construct a 98% confidence interval for the mean length of time workers have been employed in their current job.

b. Data collected in an earlier study suggests that the mean in 2000 was 13.26 years. A sociologist believes that the mean 2012 is the same. Does the confidence interval contradict this belief? Explain.

23. Fake Twitter followers: Many celebrities and public figures have Twitter accounts with large numbers of followers. However, some of these followers are fake, resulting from accounts generated by spamming computers. In a sample of 46 twitter audits conducted in July 2013, the mean percentage of fake followers was 14.1 with a standard deviation of 9.6.

a. Construct a 90% confidence interval for the mean percentage of fake Twitter followers.

b. Based on the confidence interval, is it reasonable that the mean percentage of fake Twitter followers is less than 10?

Source: *www.twitteraudit.com*

24. Let's go to the movies: A random sample of 35 Hollywood movies made since the year 2010 had a mean length of 121.6 minutes, with a standard deviation of 20.4 minutes.

a. Construct a 95% confidence interval for the true mean length of all Hollywood movies made since 2000.

b. As of June 2013, three Ironman movies have been released, and their mean length is 127 minutes. Someone claims that the mean length of Ironman movies is actually less than the mean length of all Hollywood movies. Does the confidence interval contradict this claim? Explain.

Based on data from Box Office Mojo.

25. Hip surgery: In a sample of 123 hip surgeries of a certain type, the average surgery time was 136.9 minutes with a standard deviation of 22.6 minutes.

a. Construct a 95% confidence interval for the mean surgery time for this procedure.

b. If a 99% confidence interval were constructed with these data, would it be wider or narrower than the interval constructed in part (a)? Explain.

Source: *Journal of Engineering in Medicine* 221:699–712

26. Sound it out: Phonics is an instructional method in which children are taught to connect sounds with letters or groups of letters. A sample of 134 first-graders who were learning English were asked to identify as many letter sounds as possible in a period of one minute. The average number of letter sounds identified was 34.06 with a standard deviation of 23.83.

a. Construct a 98% confidence interval for the mean number of letter sounds identified in one minute.

b. If a 95% confidence interval were constructed with these data, would it be wider or narrower than the interval constructed in part (a)? Explain.

Source: *School Psychology Review* 37:5–17

27. Software instruction: A hybrid course is one that contains both online and classroom instruction. In a study performed at Middle Georgia State College, a software package was used as the main source of instruction in a hybrid college algebra course. The software tracked the number of hours it took for each student to meet the objectives of the course. In a sample of 45 students, the mean number of hours was 80.5, with a standard deviation of 51.2.

a. Construct a 95% confidence interval for the mean number of hours it takes for a student to meet the course objectives.

b. If a sample of 90 students had been studied, would you expect the confidence interval to be wider or narrower than the interval constructed in part (a)? Explain.

28. Baby talk: In a sample of 77 children, the mean age at which they first began to combine words was 16.51 months, with a standard deviation of 9.59 months.

a. Construct a 95% confidence interval for the mean age at which children first begin to combine words.

b. If a sample of 50 children had been studied, would you expect the confidence interval to be wider or narrower than the interval constructed in part (a)? Explain.

Source: *Proceedings of the 4th International Symposium on Bilingualism*, pp. 58–77

29. Baby weights: Following are weights, in pounds, of 12 two-month-old baby girls. It is reasonable to assume that the population is approximately normal.

| 12.23 | 12.32 | 11.87 | 12.34 | 11.48 | 12.66 |
| 8.51 | 14.13 | 12.95 | 10.30 | 9.34 | 8.63 |

a. Construct a 98% confidence interval for the mean weight of two-month-old baby girls.

b. According to the National Health Statistics Reports, the mean weight of two-month-old baby boys is 11.5 pounds. Based on the confidence interval, is it reasonable to believe that the mean weight of two-month-old baby girls may be the same as that of two-month-old baby boys? Explain.

30. **Eat your cereal:** Boxes of cereal are labeled as containing 14 ounces. Following are the weights, in ounces, of a sample of 12 boxes: It is reasonable to assume that the population is approximately normal.

| 14.02 | 13.97 | 14.11 | 14.12 | 14.10 | 14.02 |
| 14.15 | 13.97 | 14.05 | 14.04 | 14.11 | 14.12 |

a. Construct a 98% confidence interval for the mean weight.

b. The quality control manager is concerned that the mean weight is actually less than 14 ounces. Based on the confidence interval, is there a reason to be concerned? Explain.

31. **Eat your spinach:** Six measurements were made of the mineral content (in percent) of spinach, with the following results. It is reasonable to assume that the population is approximately normal.

            19.1   20.8   20.8   21.4   20.5   19.7

a. Construct a 95% confidence interval for the mean mineral content.

b. Based on the confidence interval, is it reasonable to believe that the mean mineral content of spinach may be greater than 21%? Explain.

Source: *Journal of Nutrition* 66:55–66

32. **Mortgage rates:** Following are interest rates (annual percentage rates) for a 30-year fixed rate mortgage from a sample of lenders in Macon, Georgia on June 20, 2013. It is reasonable to assume that the population is approximately normal.

| 4.750 | 4.375 | 4.176 | 4.679 | 4.426 | 4.227 |
| 4.125 | 4.250 | 3.950 | 4.191 | 4.299 | 4.415 |

Source: www.bankrate.com

a. Construct a 99% confidence interval for the mean rate.

b. One week earlier, the mean rate was 4.050%. A mortgage broker claims that the mean rate is now higher. Based on the confidence interval, is this a reasonable claim? Explain.

33. **Hi-def:** Following are prices of a random sample of 18 smart TVs sold on shopper.cnet.com in 2013 with screen sizes between 46" and 50", along with a dotplot of the data.

548	598	697	699	749	799
829	849	928	1050	1098	1169
1198	1269	1299	1399	1455	1599

a. Is it reasonable to assume that the conditions for constructing a confidence interval for the mean price are satisfied? Explain.

b. If appropriate, construct a 95% confidence interval for the mean price of all smart TVs in this size range.

34. **Big salary for the boss:** Following is the total 2012 compensation, in millions of dollars, for Chief Executive Officers at 20 large U.S. corporations, along with a boxplot of the data.

2.75	9.15	2.32	23.84	13.12	9.91	6.39
2.19	4.28	0.70	3.18	8.20	1.68	4.64
6.05	6.59	11.31	1.35	25.84	2.85	

a. Is it reasonable to assume that the conditions for constructing a confidence interval for the mean compensation are satisfied? Explain.

b. If appropriate, construct a 95% confidence interval for the mean compensation of a Chief Executive Officer.

35. **Pain relief:** One of the ways in which doctors try to determine how long a single dose of pain-reliever will provide relief is to measure the drug's half-life, which is the length of time it takes for one-half of the dose to be eliminated from the body. Following are half-lives (in hours) of the pain reliever oxycodone for a sample of 18 individuals, along with a boxplot of the data, based on a report of the National Institutes of Health.

| 3.3 | 1.7 | 2.0 | 5.0 | 1.2 | 2.8 | 3.7 | 3.5 | 4.8 |
| 4.7 | 4.9 | 2.5 | 5.1 | 6.0 | 3.9 | 4.3 | 2.1 | 3.0 |

a. Is it reasonable to assume that the conditions for constructing a confidence interval for the mean half-life are satisfied? Explain.

b. If appropriate, construct a 95% confidence interval for the mean half-life.

c. The National Institutes of Health report states that the mean half-life is 3.51 hours. If appropriate, explain whether this confidence interval contradicts this claim.

36. **How's your mileage?** Gas mileage, in miles per gallon, was measured for five 2012 Ford Explorer SUVs. Following are the results, along with a dotplot of the data.

            20.4   19.6   17.1   17.9   18.7

a. Is it reasonable to assume that the conditions for constructing a confidence interval for the mean gas mileage are satisfied? Explain.

b. If appropriate, construct a 95% confidence interval for the mean gas mileage.

c. The EPA gas mileage rating for the 2012 Ford Explorer is 19 miles per gallon. If appropriate, explain whether the rated mileage may be equal to the mean mileage, based on the confidence interval.

Source: TrueDelta.com

37. Eat your kale: Kale is a type of cabbage commonly found in salad and used in cooking in many parts of the world. Six measurements were made of the mineral content (in percent) of kale, with the following results.

26.1   17.5   15.4   16.4   15.1   12.8

It turns out that the value 26.1 came from a specimen that the investigator forgot to wash before measuring.

a. The data contain an outlier that is clearly a mistake. Eliminate the outlier, then construct a 95% confidence interval for the mean mineral content from the remaining values.

b. Leave the outlier in and construct the 95% confidence interval. Are the results noticeably different? Explain why it is important to check data for outliers.

Source: *Journal of Nutrition* 66:55–66

38. Sleeping outlier: A simple random sample of eight college freshmen were asked how many hours of sleep they typically got per night. The results were

7.5   8.0   6.5   24   8.5   6.5   7.0   7.5

Notice that one joker said that he sleeps 24 hours a day.

a. The data contain an outlier that is clearly a mistake. Eliminate the outlier, then construct a 95% confidence interval for the mean amount of sleep from the remaining values.

b. Leave the outlier in and construct the 95% confidence interval. Are the results noticeably different? Explain why it is important to check data for outliers.

39. How much confidence? In a sample of 100 U.S. adults aged 18–24 who celebrate Halloween, the mean amount spent on a costume was $37.51 with a standard deviation of $16.44. A retail specialist claims that the mean amount spent on Halloween costumes for all U.S. adults aged 18–24 is between $35.07 and $39.95. With what level of confidence can this claim be made?

40. How much confidence? In a survey of 200 adult women in a certain city, the mean number of children they had was 2.3 with a standard deviation of 1.2. A sociologist states that the mean number of children per woman in this city is between 2.13 and 2.47. With what level of confidence can this claim be made?

41. Don't construct a confidence interval: There have been 43 presidents of the United States. Their mean height is 70.8 inches, with a standard deviation of 2.7 inches. Explain why these data should not be used to construct a confidence interval for the mean height of the presidents.

42. Don't construct a confidence interval: As of July 1, 2012, the mean population of the 50 states of the United States was 6.3 million, with a standard deviation of 7.0 million. Explain why these data should not be used to construct a confidence interval for the mean population of the states.

43. Interpret calculator display: The following display from a TI-84 Plus calculator presents a 98% confidence interval.

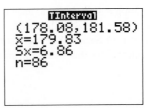

a. Fill in the blanks: We are _____ confident that the population mean is between _____ and _____ .

b. Assume the population is not normally distributed. Is the confidence interval still valid? Explain.

44. Interpret calculator display: The following display from a TI-84 Plus calculator presents a 95% confidence interval.

```
    TInterval
(.85638,2.2836)
x̄=1.57
Sx=.68
n=6
```

a. Fill in the blanks: We are _____ confident that the population mean is between _____ and _____ .

b. Assume the population is not normally distributed. Is the confidence interval still valid? Explain.

45. Interpret computer output: The following MINITAB output presents a confidence interval for a population mean.

Variable	N	Mean	StDev	SE Mean	95% CI
X	15	5.9373	2.0387	0.5264	(4.8083, 7.0663)

a. How many degrees of freedom are there?

b. If the population were not approximately normal, would this confidence interval be valid? Explain.

c. Find the critical value $t_{\alpha/2}$ for a 98% confidence interval.

d. Use the critical value and the information in the output to construct a 98% confidence interval.

46. Interpret computer output: The following MINITAB output presents a confidence interval for a population mean.

Variable	N	Mean	StDev	SE Mean	99% CI
X	71	23.8760	3.9385	0.4674	(22.638, 25.114)

a. How many degrees of freedom are there?

b. If the population were not approximately normal, would this confidence interval be valid? Explain.

c. Find the critical value $t_{\alpha/2}$ for a 95% confidence interval.

d. Use the critical value and the information in the output to construct a 95% confidence interval.

Extending the Concepts

47. Sample of size 1: The concentration of carbon monoxide in parts per million is believed to be normally distributed with a standard deviation $\sigma = 8$. A single measurement of the concentration is made, and its value is 85.

a. Use the methods of Section 8.1 to construct a 95% confidence interval for the mean concentration.

b. Would it be possible to construct a confidence interval using the methods of this section if the population standard deviation were unknown? Explain.

Answers to Check Your Understanding Exercises for Section 8.2

1. a. 2.145   **b.** 2.831   **c.** 1.671   **d.** 12.706

2. a. Yes   **b.** No   **c.** Yes   **d.** No

3. $8.03 < \mu < 8.21$

4. $744.2 < \mu < 756.2$

5. a. 14   **b.** 2.977   **c.** $40.30 < \mu < 50.82$

6. a. 9   **b.** 3.250   **c.** $8.4811 < \mu < 8.7115$

SECTION 8.3	Confidence Intervals for a Population Proportion

Objectives

1. Construct a confidence interval for a population proportion
2. Find the sample size necessary to obtain a confidence interval of a given width
3. Describe a method for constructing confidence intervals with small samples

Objective 1 Construct a confidence interval for a population proportion

Construct a Confidence Interval for a Population Proportion

Are you a Guitar Hero? The music organization Little Kids Rock surveyed 517 music teachers, and 403 of them said that video games like Guitar Hero and Rock Band, in which players try to play music in time with a video image, have a positive effect on music education. Assuming these teachers to be a random sample of U.S. music teachers, we would like to construct a confidence interval for the proportion of music teachers who believe that music video games have a positive effect on music classrooms.

This is an example of a population whose items fall into two categories. In this example, the categories are those teachers who believe that video games have a positive effect, and those who do not. We are interested in the population proportion of those who believe there is a positive effect. We will use the following notation.

NOTATION

- p is the population proportion of individuals who are in a specified category.
- x is the number of individuals in the sample who are in the specified category.
- n is the sample size.
- $\hat{p}$ is the sample proportion of individuals who are in the specified category.
  $\hat{p} = x/n$

To construct a confidence interval, we need a point estimate and a margin of error. The point estimate we use for the population proportion p is the sample proportion

$$\hat{p} = \frac{x}{n}$$

Explain It Again

The population proportion and the sample proportion: The population proportion p is unknown. The sample proportion $\hat{p}$ is known, and we use the value of $\hat{p}$ to estimate the unknown value p.

To compute the margin of error, we multiply the standard error of the point estimate by the critical value. The standard error and the critical value are determined by the sampling distribution of $\hat{p}$. In Section 7.4 we found that when the sample size n is large enough, the sample proportion $\hat{p}$ is approximately normal with standard deviation

$$\sqrt{\frac{p(1-p)}{n}}$$

In practice, we don't know the value of p, so we substitute $\hat{p}$ instead to obtain the standard error we use for the confidence interval:

$$\text{Standard error} = \sqrt{\frac{\hat{p}(1-\hat{p})}{n}}$$

Since the point estimate $\hat{p}$ is approximately normal with standard error $\sqrt{\hat{p}(1-\hat{p})/n}$, the appropriate margin of error is

$$\text{Margin of error} = z_{\alpha/2}\sqrt{\frac{\hat{p}(1-\hat{p})}{n}}$$

The confidence interval is

$$\text{Point estimate} \pm \text{Margin of error}$$

Explain It Again

Reasons for the assumptions:
The population must be much larger than the sample (at least 20 times as large), so that the sampled items are independent. The assumption that there are at least 10 items in each category is an approximate check on the assumption that both np and $n(1-p)$ are at least 10, which ensures that the sampling distribution of $\hat{p}$ is approximately normal.

$$\hat{p} \pm z_{\alpha/2}\sqrt{\frac{\hat{p}(1-\hat{p})}{n}}$$

The method we have just described requires certain assumptions, which we now state.

Assumptions for Constructing a Confidence Interval for p

1. We have a simple random sample.
2. The population is at least 20 times as large as the sample.
3. The items in the population are divided into two categories.
4. The sample must contain at least 10 individuals in each category.

Following is a step-by-step description of the procedure for constructing a confidence interval for a population proportion p.

Procedure for Constructing a Confidence Interval for p

Check to be sure the assumptions are satisfied. If they are, then proceed with the following steps.

Step 1: Compute the value of the point estimate $\hat{p}$.

Step 2: Find the critical value $z_{\alpha/2}$ corresponding to the desired confidence level, either from the last line of Table A.3, from Table A.2, or with technology.

Step 3: Compute the standard error $\sqrt{\hat{p}(1-\hat{p})/n}$ and multiply it by the critical value to obtain the margin of error $z_{\alpha/2}\sqrt{\hat{p}(1-\hat{p})/n}$.

Step 4: Use the point estimate and the margin of error to construct the confidence interval:

$$\text{Point estimate} \pm \text{Margin of error}$$

$$\hat{p} \pm z_{\alpha/2}\sqrt{\frac{\hat{p}(1-\hat{p})}{n}}$$

$$\hat{p} - z_{\alpha/2}\sqrt{\frac{\hat{p}(1-\hat{p})}{n}} < p < \hat{p} + z_{\alpha/2}\sqrt{\frac{\hat{p}(1-\hat{p})}{n}}$$

Step 5: Interpret the result.

Explain It Again

Round-off rule: When constructing a confidence interval for a proportion, round the final result to three decimal places.

Rounding off the final result

When constructing confidence intervals for a proportion, we will round the final result to three decimal places. Note that you should round only the final result, and not the calculations you have made along the way.

EXAMPLE 8.14   ## Construct a confidence interval for a proportion

In a survey of 517 music teachers, 403 said that the video games Guitar Hero and Rock Band have a positive effect on music education.

a. Construct a 95% confidence interval for the proportion of music teachers who believe that these video games have a positive effect.

b. A video game manufacturer claims that 80% of music teachers believe that Guitar Hero and Rock Band have a positive effect. Does the confidence interval contradict this claim?

Solution

a. We begin by checking the assumptions. We have a simple random sample. It is reasonable to believe that the population of music teachers in the United States is at least 20 times as large as the sample. The items in the population can be divided into two categories: those who believe that the games have a positive effect, and those who do not. There are 403 teachers who believe that the games have a positive effect, and $517 - 403 = 114$ who do not, so there are 10 or more items in each category. The assumptions are met.

Step 1: Compute the point estimate $\hat{p}$. The sample size is $n = 517$ and the number who believe that video games have a positive effect is $x = 403$. The point estimate is

$$\hat{p} = \frac{403}{517} = 0.779497$$

Step 2: Find the critical value. The critical value for a 95% confidence interval is $z_{\alpha/2} = 1.96$.

Step 3: Compute the margin of error. The margin of error is

$$z_{\alpha/2}\sqrt{\frac{\hat{p}(1-\hat{p})}{n}}$$

We substitute $z_{\alpha/2} = 1.96$, $\hat{p} = 0.779497$, and $n = 517$ to obtain

$$z_{\alpha/2}\sqrt{\frac{\hat{p}(1-\hat{p})}{n}} = 1.96\sqrt{\frac{0.779497(1 - 0.779497)}{517}} = 0.035738$$

Step 4: Construct the 95% confidence interval. The point estimate is 0.779497 and the margin of error is 0.035738. The 95% confidence interval is

$$0.779497 - 0.035738 < p < 0.779497 + 0.035738$$
$$0.744 < p < 0.815 \quad \text{(rounded to three decimal places)}$$

Sept 5: Interpret the result. We are 95% confident that the proportion of music teachers who believe that video games have a positive effect is between 0.744 and 0.815.

b. Because the value 0.80 is within the confidence interval, the confidence interval does not contradict the claim.

Constructing confidence intervals with technology

Example 8.14 presented a 95% confidence interval for the proportion of music teachers who believe that video games have a positive effect on music education. The confidence interval obtained was $0.744 < p < 0.815$. Following are the results from a TI-84 Plus calculator.

```
1-PropZInt
(.74376,.81523)
p̂=.7794970986
n=517
```

The display presents the confidence interval, followed by the point estimate $\hat{p}$ and the sample size n. Note that the level (95%) is not displayed.

Following is MINITAB output for the same example.

Sample	X	N	Sample p	95% CI
1	403	517	0.779497	(0.743760, 0.815234)

Step-by-step instructions for constructing confidence intervals with technology are presented in the Using Technology section on page 386.

Check Your Understanding

1. A simple random sample of 200 third-graders in a large school district was chosen to participate in an after-school program to improve reading skills. After completing the program, the children were tested, and 142 of them showed improvement in their reading skills.
   a. Find a point estimate for the proportion of third-graders in the school district whose reading scores would improve after completing the program.
   b. Construct a 95% confidence interval for the proportion of third-graders in the school district whose reading scores would improve after completing the program.
   c. Is it reasonable to conclude that more than 60% of the students would improve their reading scores after completing the program? Explain.
   d. The district superintendent wants to construct a confidence interval for the proportion of all elementary schoolchildren in the district whose scores would improve. Should the sample of 200 third-graders be used for this purpose? Explain.

Answers are on page 390.

Objective 2 Find the sample size necessary to obtain a confidence interval of a given width

Finding the Necessary Sample Size

If we wish to make the margin of error of a confidence interval smaller while keeping the confidence level the same, we can do this by making the sample size larger. Sometimes we have a specific value m that we would like the margin of error to attain, and we wish to compute a sample size n that is likely to give us a margin of error of that size.

The method for computing the sample size is based on the formula for the margin of error:

$$m = z_{\alpha/2}\sqrt{\frac{\hat{p}(1-\hat{p})}{n}}$$

By manipulating this formula algebraically, we can solve for n:

$$n = \hat{p}(1-\hat{p})\left(\frac{z_{\alpha/2}}{m}\right)^2$$

This formula shows that the sample size n depends not only on the margin of error m but also on the sample proportion $\hat{p}$. Therefore, in order to compute the sample size n, we need a value for $\hat{p}$ as well as a value for m. Now of course we don't know ahead of time what $\hat{p}$ is going to be. The approach, therefore, is to determine a value for $\hat{p}$.

There are two ways to determine a value for $\hat{p}$. One is to use a value that is available from a previously drawn sample. The other is to assume that $\hat{p} = 0.5$. The value $\hat{p} = 0.5$ makes the margin of error as large as possible for any sample size. Therefore, if we assume that $\hat{p} = 0.5$, we will always get a margin of error that is less than or equal to the desired value.

SUMMARY

Let m be the desired margin of error, and let $z_{\alpha/2}$ be the critical value. The sample size n needed so that a confidence interval for a proportion will have margin of error approximately equal to m is

$$n = \hat{p}(1 - \hat{p}) \left(\frac{z_{\alpha/2}}{m} \right)^2 \quad \text{if a value for } \hat{p} \text{ is available}$$

$$n = 0.25 \left(\frac{z_{\alpha/2}}{m} \right)^2 \quad \text{if no value for } \hat{p} \text{ is available}$$

(This is equivalent to assuming that $\hat{p} = 0.5$.)

If the value of n given by the formula is not a whole number, round it *up* to the nearest whole number. By rounding up, we can be sure that the margin of error is no greater than the desired value m.

EXAMPLE 8.15

Find the necessary sample size

Example 8.14 described a sample of 517 music teachers, 403 of whom believe that video games have a positive effect on music education. Estimate the sample size needed so that a 95% confidence interval will have a margin of error of 0.03.

Solution

The desired level is 95%. The critical value is therefore $z_{\alpha/2} = 1.96$. We now compute $\hat{p}$:

$$\hat{p} = \frac{403}{517} = 0.779497$$

The desired margin of error is $m = 0.03$. Since we have a value for $\hat{p}$, we substitute $\hat{p} = 0.779497$, $z_{\alpha/2} = 1.96$, and $m = 0.03$ into the formula

$$n = \hat{p}(1 - \hat{p}) \left(\frac{z_{\alpha/2}}{m} \right)^2$$

and obtain

$$n = (0.779497)(1 - 0.779497) \left(\frac{1.96}{0.03} \right)^2 = 733.67 \quad \text{(round up to 734)}$$

We estimate that we need to sample 734 teachers to obtain a 95% confidence interval with a margin of error of 0.03.

EXAMPLE 8.16

Find the necessary sample size

We plan to sample music teachers in order to construct a 95% confidence interval for the proportion who believe that listening to hip-hop music has a positive effect on music education. We have no value of $\hat{p}$ available. Estimate the sample size needed so that a 95% confidence interval will have a margin of error of 0.03. Explain why the estimated sample size in this example is larger than the one in Example 8.15.

Solution

The desired level is 95%. The critical value is therefore $z_{\alpha/2} = 1.96$. The desired margin of error is $m = 0.03$. Since we have no value of $\hat{p}$, we substitute the values $z_{\alpha/2} = 1.96$ and $m = 0.03$ into the formula

$$n = 0.25 \left(\frac{z_{\alpha/2}}{m} \right)^2$$

and obtain

$$n = 0.25 \left(\frac{1.96}{0.03} \right)^2 = 1067.1 \quad \text{(round up to 1068)}$$

We estimate that we need to sample 1068 teachers to obtain a 95% confidence interval with a margin of error of 0.03. This estimate is larger than the one in Example 8.15 because we used a value of 0.5 for $\hat{p}$, which provides a sample size large enough to guarantee that the margin of error will be no greater than 0.03 no matter what the true value of p is.

Check Your Understanding

2. In a preliminary study, a simple random sample of 100 computer chips was tested, and 17 of them were found to be defective. Now another sample will be drawn in order to construct a 95% confidence interval for the proportion of chips that are defective. Use the results of the preliminary study to estimate the sample size needed so that the confidence interval will have a margin of error of 0.06.

3. A pollster is going to sample a number of voters in a large city and construct a 98% confidence interval for the proportion who support the incumbent candidate for mayor. Find a sample size so that the margin of error will be no larger than 0.05.

Answers are on page 390.

The margin of error does not depend on the population size

In 2013, there were about 18 million registered voters in the state of California, and about 0.27 million registered voters in the state of Wyoming. A simple random sample of 1000 Wyoming voters is selected to estimate the proportion of voters who favor the Democratic candidate for president. Another simple random sample of 1000 California voters is selected to determine the proportion of Democratic voters in that state. Which estimate has the smaller standard error? Because California has a much larger population of registered voters than does Wyoming, it might seem that a larger sample would be needed in California to produce the same standard error. Surprisingly enough, this is not the case. In fact, the standard errors for the two estimates will be about the same. This is clear from the formula for the standard error: The population size does not enter into the calculation. Since the standard errors are about the same, the margins of error will be about the same if confidence intervals of the same level are constructed for both population proportions.

Intuitively, we can see that population size doesn't matter by considering an analogy with testing the water in a swimming pool. To determine whether the chemical balance is correct, one withdraws a few drops of water to test. As long as the contents of the pool are well mixed, so that the water removed constitutes a simple random sample of molecules from the pool, it doesn't matter how large the pool is. One doesn't need to sample more water from a bigger pool.

EXAMPLE 8.17

The margin of error does not depend on the population size

A pollster has conducted a poll using a sample of 500 drawn from a town with population 25,000. He now wants to conduct the poll in a larger town with population 250,000, and to obtain approximately the same margin of error as in the smaller town. How large a sample must he draw?

Solution
He should draw a sample of 500, just as in the small town. The population size does not affect the margin of error.

Check Your Understanding

4. A pollster is planning to draw a simple random sample of 500 people in Colorado (population 5.2 million). He then will conduct a similar poll in Texas (population 26.1 million). He wants to have approximately the same standard error in both polls. True or false:

    a. The pollster needs a sample in Texas that is about 5 times as large as the one in Colorado.

    b. The pollster needs a sample in Texas that is about the same size as the one in Colorado.

5. A marketing firm in New York City (population 8.3 million) plans to draw a simple random sample of 1000 people to estimate the proportion who have heard about a new product. The firm then plans to take a simple random sample of 500 in Denver (population 634,000) for the same purpose. True or false:

    a. The margin of error for a 95% confidence interval will be larger in New York.

    b. The margin of error for a 95% confidence interval will be larger in Denver.

    c. The margin of error for a 95% confidence interval will be about the same in both cities.

Answers are on page 390.

Objective 3 Describe a method for constructing confidence intervals with small samples

A Method for Constructing Confidence Intervals with Small Samples

The method that we have presented for constructing a confidence interval for a proportion requires that we have at least 10 individuals in each category. When this condition is not met, we can still construct a confidence interval by adjusting the sample proportion a bit. We increase the number of individuals in each category by 2, so that the sample size increases by 4. Thus, instead of using the sample proportion $\hat{p} = x/n$, we use the *adjusted sample proportion*

$$\tilde{p} = \frac{x + 2}{n + 4}$$

The standard error and critical value are calculated in the same way as in the traditional method, except that we use the adjusted sample proportion $\tilde{p}$ in place of $\hat{p}$, and $n + 4$ in place of n.

Constructing Confidence Intervals for a Proportion with Small Samples

If x is the number of individuals in a sample of size n who have a certain characteristic, and p is the population proportion, then:

The adjusted sample proportion is

$$\tilde{p} = \frac{x + 2}{n + 4}$$

A confidence interval for p is

$$\tilde{p} - z_{\alpha/2}\sqrt{\frac{\tilde{p}(1 - \tilde{p})}{n + 4}} < p < \tilde{p} + z_{\alpha/2}\sqrt{\frac{\tilde{p}(1 - \tilde{p})}{n + 4}}$$

Another way to write this is

$$\tilde{p} \pm z_{\alpha/2}\sqrt{\frac{\tilde{p}(1 - \tilde{p})}{n + 4}}$$

| EXAMPLE 8.18 | Construct a confidence interval with a small sample |

In a random sample of 10 businesses in a certain city, 6 of them had more than 15 employees. Use the small-sample method to construct a 95% confidence interval for the proportion of businesses in this city that have more than 15 employees.

Solution
The adjusted sample proportion is

$$\tilde{p} = \frac{x+2}{n+4} = \frac{6+2}{10+4} = 0.5714$$

The critical value is $z_{\alpha/2} = 1.96$. The 95% confidence interval is therefore

$$\tilde{p} - z_{\alpha/2}\sqrt{\frac{\tilde{p}(1-\tilde{p})}{n+4}} < p < \tilde{p} + z_{\alpha/2}\sqrt{\frac{\tilde{p}(1-\tilde{p})}{n+4}}$$

$$0.5714 - 1.96\sqrt{\frac{0.5714(1-0.5714)}{10+4}} < p < 0.5714 + 1.96\sqrt{\frac{0.5714(1-0.5714)}{10+4}}$$

$$0.312 < p < 0.831$$

Check Your Understanding

6. In a simple random sample of 15 seniors from a certain college, 8 of them had found jobs. Use the small-sample method to construct a 95% confidence interval for the proportion of seniors at that college who have found jobs.

Answer is on page 390.

Using technology to implement the small-sample method

Because the only difference between the small-sample method and the traditional method is the use of $\tilde{p}$ rather than $\hat{p}$, a software package or calculator such as the TI-84 Plus that uses the traditional method can be made to produce a confidence interval using the small-sample method. Simply input $x + 2$ for the number of individuals in the category of interest, and $n + 4$ for the sample size.

The small-sample method is better overall

The small-sample method can be used for any sample size, and recent research has shown that it has two advantages over the traditional method. First, the margin of error is smaller, because we divide by $n + 4$ rather than n. Second, the actual probability that the small-sample confidence interval covers the true population proportion is almost always at least as great as, or greater than, that of the traditional method. This holds for confidence levels of 90% or more, which are the levels commonly used in practice. For more information on this method, see the article "Approximate is Better Than 'Exact' for Interval Estimation of Binomial Proportions" (A. Agresti and B. Coull, *The American Statistician*, 52:119–126).

USING TECHNOLOGY

We use Example 8.14 to illustrate the technology steps.

TI-84 PLUS

Constructing a confidence interval for a proportion

Step 1. Press **STAT** and highlight the **TESTS** menu.

Step 2. Select **1–PropZInt** and press **ENTER** (Figure A). The **1–PropZInt** menu appears.

Step 3. Enter the values of x and n. For Example 8.14, we use $x = 403$ and $n = 517$.

Step 4. In the **C-Level** field, enter the confidence level. For Example 8.14, we use 0.95 (Figure B).

Step 5. Highlight **Calculate** and press **ENTER** (Figure C).

Note: The preceding steps produce the traditional confidence interval. To produce the small-sample interval, enter the value of $x + 2$ for x and the value of $n + 4$ for n.

Figure A

Figure B

Figure C

MINITAB

Constructing a confidence interval for a proportion

Step 1. Click on **Stat**, then **Basic Statistics**, then **1 Proportion**.

Step 2. Click **Summarized Data** and enter the value of n in the **Number of trials** field and the value of x in the **Number of events** field. For Example 8.14, we use $x = 403$ and $n = 517$ (Figure D).

Step 3. Click **Options** and enter the confidence level in the **Confidence Level** (95) field. Click **OK**.

Step 4. Click **OK** (Figure E).

Note: The preceding steps produce the traditional confidence interval. To produce the small-sample interval, enter the value of $x + 2$ for x and the value of $n + 4$ for n.

Figure D

```
Sample    X    N   Sample p          95% CI
1        403  517  0.779497  (0.741246, 0.814517)
```

Figure E

EXCEL

Constructing a confidence interval for a proportion

This procedure requires the **MegaStat** EXCEL add-in to be loaded. The **MegaStat** add-in may be downloaded from www.mhhe.com/megastat.

Step 1. Click on the **MegaStat** menu and select **Confidence Intervals/Sample Size...**

Step 2. Click on the **Confidence interval – p** option and enter the value of x in the **p** field (note that p automatically changes to x) and the value of n in the **n** field. For Example 8.14, we use $x = 403$ and $n = 517$.

Step 3. In the **Confidence Level** field, enter the confidence level. For Example 8.14, enter **95%** (Figure F).

Step 4. Click **Preview** (Figure F).

Note: The preceding steps produce the traditional confidence interval. To produce the small-sample interval, enter the value of $x + 2$ for x and the value of $n + 4$ for n.

Figure F

SECTION 8.3   Exercises

Exercises 1–6 are the Check Your Understanding exercises located within the section.

Understanding the Concepts

In Exercises 7 and 8, fill in each blank with the appropriate word or phrase.

7. If $\hat{p}$ is the sample proportion and n is the sample size, then
$$\sqrt{\frac{\hat{p}(1 - \hat{p})}{n}}$$ is the _____ .

8. To estimate the necessary sample size when no value of $\hat{p}$ is available, we use $\hat{p} = $ _____ .

In Exercises 9 and 10, determine whether the statement is true or false. If the statement is false, rewrite it as a true statement.

9. If we estimate the necessary sample size and no value for $\hat{p}$ is available, the estimated sample size will be larger than if a value for $\hat{p}$ were available.

10. The margin of error does not depend on the sample size.

Practicing the Skills

In Exercises 11–14, find the point estimate, the standard error, and the margin of error for the given confidence levels and values of x and n.

11. $x = 146$, $n = 762$, confidence level 95%

12. $x = 46$, $n = 97$, confidence level 99%

13. $x = 236$, $n = 474$, confidence level 90%

14. $x = 29$, $n = 80$, confidence level 92%

In Exercises 15–18, use the given data to construct a confidence interval of the requested level.

15. $x = 28$, $n = 64$, confidence level 93%

16. $x = 52$, $n = 71$, confidence level 97%

17. $x = 125$, $n = 317$, confidence level 95%

18. $x = 178$, $n = 531$, confidence level 90%

Working with the Concepts

19. Smart phone: Among 238 cell phone owners aged 18–24 surveyed by the Pew Research Center, 102 said their phone was an Android phone.

   a. Find a point estimate for the proportion of cell phone owners aged 18–24 who have an Android phone.

   b. Construct a 95% confidence interval for the proportion of cell phone owners aged 18–24 who have an Android phone.

   c. Assume that an advertisement claimed that 45% of cell phone owners aged 18–24 have an Android phone. Does the confidence interval contradict this claim?

20. **Working at home:** According to the U.S. Census Bureau, 43% of men who worked at home were college graduates. In a sample of 500 women who worked at home, 162 were college graduates.
    a. Find a point estimate for the proportion of college graduates among women who work at home.
    b. Construct a 98% confidence interval for the proportion of women who work at home who are college graduates.
    c. Based on the confidence interval, is it reasonable to believe that the proportion of college graduates among women who work at home is the same as the proportion of college graduates among men who work at home? Explain.

21. **Sleep apnea:** Sleep apnea is a disorder in which there are pauses in breathing during sleep. People with this condition must wake up frequently to breathe. In a sample of 427 people aged 65 and over, 104 of them had sleep apnea.
    a. Find a point estimate for the population proportion of those aged 65 and over who have sleep apnea.
    b. Construct a 99% confidence interval for the proportion of those aged 65 and over who have sleep apnea.
    c. In another study, medical researchers concluded that more than 9% of elderly people have sleep apnea. Based on the confidence interval, does it appear that more than 9% of people aged 65 and over have sleep apnea? Explain.
    Sources: *Sleep* 14:486–495; *Mayo Clinic Proceedings* 76:897–905

22. **Internet service:** An Internet service provider sampled 540 customers and found that 75 of them experienced an interruption in high-speed service during the previous month.
    a. Find a point estimate for the population proportion of all customers who experienced an interruption.
    b. Construct a 90% confidence interval for the proportion of all customers who experienced an interruption.
    c. The company's quality control manager claims that no more than 10% of its customers experienced an interruption during the previous month. Does the confidence interval contradict this claim? Explain.

23. **Volunteering:** In 2012 the General Social Survey asked 1294 people whether they performed any volunteer work during the past year. A total of 517 people said they did.
    a. Find a point estimate for the proportion of people who performed volunteer work during the past year.
    b. Construct a 95% confidence interval for the proportion of people who performed volunteer work during the past year.
    c. A sociologist states that 50% of Americans perform volunteer work in a given year. Does the confidence interval contradict this statement? Explain.

24. **SAT scores:** A college admissions officer sampled 120 entering freshmen and found that 42 of them scored more than 550 on the math SAT.
    a. Find a point estimate for the proportion of all entering freshmen at this college who scored more than 550 on the math SAT.
    b. Construct a 98% confidence interval for the proportion of all entering freshmen at this college who scored more than 550 on the math SAT.

    c. According to the College Board, 39% of all students who took the math SAT in 2009 scored more than 550. The admissions officer believes that the proportion at her university is also 39%. Does the confidence interval contradict this belief? Explain.

25. **WOW:** In the computer game *World of Warcraft*, some of the strikes are critical strikes, which do more damage. Assume that the probability of a critical strike is the same for every attack, and that attacks are independent. During a particular fight, a character has 242 critical strikes out of 595 attacks.
    a. Construct a 95% confidence interval for the proportion of strikes that are critical strikes.
    b. Construct a 98% confidence interval for the proportion of strikes that are critical strikes.
    c. What is the effect of increasing the level of confidence on the width of the interval?

26. **Contaminated water:** In a sample of 42 water specimens taken from a construction site, 26 contained detectable levels of lead.
    a. Construct a 90% confidence interval for the proportion of water specimens that contain detectable levels of lead.
    b. Construct a 95% confidence interval for the proportion of water specimens that contain detectable levels of lead.
    c. What is the effect of increasing the level of confidence on the width of the interval?
    Source: *Journal of Environmental Engineering* 128:237–245

27. **Call me:** A sociologist wants to construct a 95% confidence interval for the proportion of children aged 8–10 living in New York who own a cell phone.
    a. A survey by the National Consumers League taken in 2012 estimated the nationwide proportion to be 0.32. Using this estimate, what sample size is needed so that the confidence interval will have a margin of error of 0.02?
    b. Estimate the sample size needed if no estimate of p is available.
    c. If the sociologist wanted to estimate the proportion in the entire United States rather than in New York, would the necessary sample size be larger, smaller, or about the same? Explain.

28. **Reading proficiency:** An educator wants to construct a 98% confidence interval for the proportion of elementary schoolchildren in Colorado who are proficient in reading.
    a. The results of a recent statewide test suggested that the proportion is 0.70. Using this estimate, what sample size is needed so that the confidence interval will have a margin of error of 0.05?
    b. Estimate the sample size needed if no estimate of p is available.
    c. If the educator wanted to estimate the proportion in the entire United States rather than in Colorado, would the necessary sample size be larger, smaller, or about the same? Explain.

29. **Surgical complications:** A medical researcher wants to construct a 99% confidence interval for the proportion of knee replacement surgeries that result in complications.
    a. An article in the *Journal of Bone and Joint Surgery* suggested that approximately 8% of such operations

result in complications. Using this estimate, what sample size is needed so that the confidence interval will have a margin of error of 0.04?

b. Estimate the sample size needed if no estimate of *p* is available.

Source: *Journal of Bone and Joint Surgery* 87:1719–1724

30. How's the economy? A pollster wants to construct a 95% confidence interval for the proportion of adults who believe that economic conditions are getting better.

a. A Gallup poll taken in February 2013 estimates this proportion to be 0.34. Using this estimate, what sample size is needed so that the confidence interval will have a margin of error of 0.03?

b. Estimate the sample size needed if no estimate of *p* is available.

31. Changing jobs: A sociologist sampled 200 people who work in computer-related jobs, and found that 42 of them have changed jobs in the past 6 months.

a. Construct a 95% confidence interval for the proportion of those who work in computer-related jobs who have changed jobs in the past six months.

b. Among the 200 people, 120 of them are under the age of 35. These constitute a simple random sample of workers under the age of 35. If this sample were used to construct a 95% confidence interval for the proportion of workers under the age of 35 who have changed jobs in the past six months, is it likely that the margin of error would be larger, smaller, or about the same as the one in part (a)?

32. Political polling: A simple random sample of 300 voters was polled several months before a presidential election. One of the questions asked was: "Are you satisfied with the choice of candidates for president?" A total of 123 of them said that they were not satisfied.

a. Construct a 99% confidence interval for the proportion of voters who are not satisfied with the choice of candidates.

b. Among the 300 voters were 158 women. These constitute a simple random sample of women voters. If this sample were used to construct a 99% confidence interval for the proportion of women voters who are satisfied with the choice of candidates for president, is it likely that the margin of error would be larger, smaller, or about the same as the one in part (a)?

33. Small sample: Eighteen concrete blocks were sampled and tested for crushing strength in order to estimate the proportion that were sufficiently strong for a certain application. Sixteen of the 18 blocks were sufficiently strong. Use the small-sample method to construct a 95% confidence interval for the proportion of blocks that are sufficiently strong.

34. Small sample: During an economic downturn, 20 companies were sampled and asked whether they were planning to increase their workforce. Only 3 of the 20 companies were planning to increase their workforce. Use the small-sample method to construct a 98% confidence interval for the proportion of companies that are planning to increase their workforce.

35. Calculator display: The following TI-84 Plus display presents a 99% confidence interval for a proportion.

a. Fill in the blanks. We are _____ confident that the population proportion is between _____ and _____ .

b. Use the information in the display to construct a 95% confidence interval for *p*.

36. Calculator display: The following TI-84 Plus display presents a 95% confidence interval for a proportion.

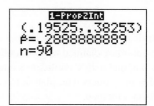

a. Fill in the blanks. We are _____ confident that the population proportion is between _____ and _____ .

b. Use the information in the display to construct a 98% confidence interval for *p*.

37. Computer output: The following MINITAB output presents a 98% confidence interval for a proportion.

Sample	X	N	Sample p	98% CI
1	145	181	0.801105	(0.732082, 0.870128)

a. Fill in the blanks. We are _____ confident that the population proportion is between _____ and _____ .

b. Use the information in the display to construct a 90% confidence interval for *p*.

38. Computer output: The following MINITAB output presents a 95% confidence interval for a proportion.

Sample	X	N	Sample p	95% CI
1	31	58	0.534483	(0.406111, 0.662854)

a. Fill in the blanks. We are _____ confident that the population proportion is between _____ and _____ .

b. Use the information in the display to construct a 99% confidence interval for *p*.

39. Don't construct a confidence interval: The United States Senate consists of 100 senators. In January 2013, 20 of them were women. Explain why these data should not be used to construct a 95% confidence interval for the proportion of senators who are women.

40. Don't construct a confidence interval: At the end of a television documentary on the nature of government, viewers are invited to tweet an answer to the question, "Do you believe that women are more effective at governing than men are?" A total of 2348 viewers answer the question, and 1247 of them answer "Yes." Explain why these data should not be used to construct a confidence interval for the proportion of people who believe that women are more effective at governing than men are.

Extending the Concepts

Wilson's interval: *The small-sample method for constructing a confidence interval is a simple approximation of a more complicated interval known as Wilson's interval. Let $\hat{p} = x/n$. Wilson's confidence interval for p is given by*

$$\frac{\hat{p} + \dfrac{z_{\alpha/2}^2}{2n} \pm z_{\alpha/2} \sqrt{\dfrac{\hat{p}(1-\hat{p})}{n} + \dfrac{z_{\alpha/2}^2}{4n^2}}}{1 + \dfrac{z_{\alpha/2}^2}{n}}$$

41. **College-bound:** In a certain high school, 9 out of 15 tenth-graders said they planned to go to college after graduating. Construct a 95% confidence interval for the proportion of tenth-graders who plan to attend college:
    a. Using Wilson's method
    b. Using the small-sample method
    c. Using the traditional method

42. **Comparing the methods:** Refer to Exercise 41.
    a. Which of the three confidence intervals is the narrowest?
    b. Does the small-sample method provide a good approximation to Wilson's interval in this case?
    c. Explain why the traditional interval is the widest of the three.

43. **Approximation depends on the level:** The small-sample method is a good approximation to Wilson's method for all confidence levels commonly used in practice, but is best when $z_{\alpha/2}$ is close to 2. Refer to Exercise 41.
    a. Use Wilson's method to construct a 90% confidence interval, a 95% confidence interval, and a 99% confidence interval for the proportion of tenth-graders who plan to attend college.
    b. Use the small-sample method to construct a 90% confidence interval, a 95% confidence interval, and a 99% confidence interval for the proportion of tenth-graders who plan to attend college.
    c. For which level is the small-sample method the closest to Wilson's method? Explain why this is the case.

Answers to Check Your Understanding Exercises for Section 8.3

1. **a.** 0.710   **b.** 0.647 < p < 0.773

   c. Yes. We are 95% confident that the proportion who would improve their scores is between 0.647 and 0.773. Therefore, it is reasonable to conclude that the proportion is greater than 0.60.

   d. No. Because the sample contains only third-graders, it should not be used to construct a confidence interval for all elementary schoolchildren.

2. 151

3. 542

4. **a.** False   **b.** True

5. **a.** False   **b.** True   **c.** False

6. 0.302 < p < 0.751

| SECTION 8.4 | Confidence Intervals for a Standard Deviation |

Objectives

1. Find critical values of the chi-square distribution
2. Construct confidence intervals for the variance and standard deviation of a normal distribution

Objective 1 Find critical values of the chi-square distribution

The Chi-Square Distribution

Most confidence intervals constructed in practice are for means and proportions. However, when the population is normal, it is possible to construct confidence intervals for the standard deviation or variance. These confidence intervals are based on a distribution known as the **chi-square distribution**, denoted χ^2. The symbol χ is the Greek letter chi (pronounced "kigh"; rhymes with sky). We will begin by describing this distribution.

There are actually many different chi-square distributions, each with a different number of degrees of freedom. Figure 8.13 on page 391 presents chi-square distributions for several different degrees of freedom. There are two important points to notice:

- The chi-square distributions are not symmetric. They are skewed to the right.
- Values of the χ^2 statistic are always greater than or equal to 0. They are never negative.

CAUTION

The methods of this section apply only for samples drawn from a normal distribution. If the distribution differs even slightly from normal, these methods should not be used.

Figure 8.13

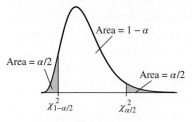

Figure 8.14

Confidence intervals for variances and standard deviations follow a somewhat different pattern than those for means and proportions. Whereas confidence intervals for means and proportions consist of a point estimate, a critical value, and a standard error, confidence intervals for variances and standard deviations consist of a point estimate and two critical values. The point estimate for the population variance σ^2 is s^2, the sample variance. The critical values come from the chi-square distribution.

The critical values for a level $100(1 - \alpha)\%$ confidence interval are the values that contain the middle $100(1 - \alpha)\%$ of the area under the curve between them. The notation for the critical values tells how much area is to the *right* of the critical value. Thus, for a level $1 - \alpha$ confidence interval, the critical values are denoted $\chi^2_{1-\alpha/2}$ and $\chi^2_{\alpha/2}$. See Figure 8.14.

Example 8.19 shows how to use Table A.4 to find critical values.

EXAMPLE 8.19

Figure 8.15

Find critical values

Find the critical values for a 95% confidence interval using the chi-square distribution with 10 degrees of freedom.

Solution

Figure 8.15 presents the chi-square distribution with 10 degrees of freedom and shows the locations of the critical values. The confidence level is 95%, so the critical values are the values that contain the middle 95% of the area under the curve between them. The lower critical value, denoted $\chi^2_{0.975}$, has an area of 0.975 to its right, and the upper critical value, denoted $\chi^2_{0.025}$, has an area of 0.025 to its right. These critical values are found in Table A.4, at the intersection of the row corresponding to 10 degrees of freedom and the columns corresponding to 0.975 and 0.025. The critical values are $\chi^2_{0.975} = 3.247$ and $\chi^2_{0.025} = 20.483$.

Degrees of Freedom	\multicolumn{10}{c}{Area in Right Tail}
	0.995	0.99	0.975	0.95	0.90	0.10	0.05	0.025	0.01	0.005
⋮	⋮	⋮	⋮	⋮	⋮	⋮	⋮	⋮	⋮	⋮
8	1.344	1.646	2.180	2.733	3.490	13.362	15.507	17.535	20.090	21.955
9	1.735	2.088	2.700	3.325	4.168	14.684	16.919	19.023	21.666	23.589
10	2.156	2.558	3.247	3.940	4.865	15.987	18.307	20.483	23.209	25.188
11	2.603	3.053	3.816	4.575	5.578	17.275	19.675	21.920	24.725	26.757
12	3.074	3.571	4.404	5.226	6.304	18.549	21.026	23.337	26.217	28.300
⋮	⋮	⋮	⋮	⋮	⋮	⋮	⋮	⋮	⋮	⋮

Check Your Understanding

1. Find the critical values for a 95% confidence interval using the chi-square distribution with 18 degrees of freedom.

2. Find the critical values for a 99% confidence interval using the chi-square distribution with 25 degrees of freedom.

Answers are on page 395.

Objective 2 Construct confidence intervals for the variance and standard deviation of a normal distribution

Confidence Intervals for the Variance and Standard Deviation

We will present the method for constructing confidence intervals for the variance and the standard deviation of a normal distribution. We will follow with an example, then present a justification for the method.

Confidence Intervals for the Variance and Standard Deviation

Let s^2 be the sample variance from a simple random sample of size n from a normal population. A level $100(1 - \alpha)\%$ confidence interval for the population variance σ^2 is

$$\frac{(n-1)s^2}{\chi^2_{\alpha/2}} < \sigma^2 < \frac{(n-1)s^2}{\chi^2_{1-\alpha/2}}$$

A level $100(1 - \alpha)\%$ confidence interval for the population standard deviation σ is

$$\sqrt{\frac{(n-1)s^2}{\chi^2_{\alpha/2}}} < \sigma < \sqrt{\frac{(n-1)s^2}{\chi^2_{1-\alpha/2}}}$$

The critical values are taken from a chi-square distribution with $n - 1$ degrees of freedom.

Explain It Again

Degrees of freedom: When constructing a confidence interval for a variance or standard deviation, the number of degrees of freedom is always 1 less than the sample size.

Step-by-step procedure for constructing confidence intervals

We summarize the procedure for constructing confidence intervals for the variance and standard deviation of a normal population.

Procedure for Constructing Confidence Intervals for the Variance and Standard Deviation of a Normal Distribution

Step 1: Compute the sample variance s^2, if it isn't given.

Step 2: Find the critical values $\chi^2_{1-\alpha/2}$ and $\chi^2_{\alpha/2}$, using the chi-square distribution with $n - 1$ degrees of freedom.

Step 3: Compute the lower and upper confidence bounds:

$$\text{Lower bound} = \frac{(n-1)s^2}{\chi^2_{\alpha/2}} \qquad \text{Upper bound} = \frac{(n-1)s^2}{\chi^2_{1-\alpha/2}}$$

Step 4: The level $100(1 - \alpha)\%$ confidence interval for σ^2 is

$$\frac{(n-1)s^2}{\chi^2_{\alpha/2}} < \sigma^2 < \frac{(n-1)s^2}{\chi^2_{1-\alpha/2}}$$

To find a level $100(1 - \alpha)\%$ confidence interval for the population standard deviation σ, take the square roots of the confidence bounds for the variance:

$$\sqrt{\frac{(n-1)s^2}{\chi^2_{\alpha/2}}} < \sigma < \sqrt{\frac{(n-1)s^2}{\chi^2_{1-\alpha/2}}}$$

Step 5: Interpret the result.

EXAMPLE 8.20   ## Constructing a confidence interval

The compressive strengths of seven concrete blocks, in pounds per square inch, are measured, with the following results.

$$1989.9 \quad 1993.8 \quad 2074.5 \quad 2070.5 \quad 2070.9 \quad 2033.6 \quad 1939.6$$

Assume these values are a simple random sample from a normal population. Construct a 95% confidence interval for the population standard deviation σ.

Solution

Step 1: Find s^2.

$$s^2 = \frac{\sum(x - \bar{x})^2}{7 - 1} = 2699.8648$$

Step 2: Find the critical values. We have $7 - 1 = 6$ degrees of freedom. Since the confidence level is 95%, the critical values are $\chi^2_{0.975}$ and $\chi^2_{0.025}$. From Table A.4, we find

$$\chi^2_{0.975} = 1.237 \qquad \chi^2_{0.025} = 14.449$$

Step 3: Compute the lower and upper confidence bounds.

$$\text{Lower bound} = \frac{(n - 1)s^2}{\chi^2_{\alpha/2}} = \frac{(7 - 1)(2699.8648)}{14.449} = 1121.129$$

$$\text{Upper bound} = \frac{(n - 1)s^2}{\chi^2_{1-\alpha/2}} = \frac{(7 - 1)(2699.8648)}{1.237} = 13,095.545$$

Step 4: The 95% confidence interval for σ^2 is

$$1121.129 < \sigma^2 < 13,095.545$$

To find the confidence interval for σ, we take square roots. We find that $\sqrt{1121.129} = 33.48$, and $\sqrt{13,095.545} = 114.44$. The 95% confidence interval for σ is

$$33.48 < \sigma < 114.44$$

Step 5: Interpret the result. We are 95% confident that the population standard deviation of the strengths of the concrete blocks is between 33.48 and 114.44.

Check Your Understanding

3. Construct a 95% confidence interval for the population standard deviation σ if a sample of size 10 has standard deviation $s = 6$.

4. Construct a 99% confidence interval for the population standard deviation σ if a sample of size 23 has standard deviation $s = 12$.

Answers are on page 395.

Justification for the method

Confidence intervals for the variance of a normal distribution are based on the fact that when a sample of size n is drawn from a normal distribution, the quantity $(n - 1)s^2/\sigma^2$ follows a chi-square distribution with $n - 1$ degrees of freedom. Therefore, for a proportion $1 - \alpha$ of all possible samples,

$$\chi^2_{\alpha/2} < \frac{(n - 1)s^2}{\sigma^2} < \chi^2_{1-\alpha/2}$$

Through algebraic manipulation, we can solve for σ^2, to obtain a level $100(1 - \alpha)\%$ confidence interval for σ^2:

$$\frac{(n-1)s^2}{\chi^2_{\alpha/2}} < \sigma^2 < \frac{(n-1)s^2}{\chi^2_{1-\alpha/2}}$$

USING TECHNOLOGY

We use Example 8.20 to illustrate the technology steps.

MINITAB

Constructing a confidence interval for σ

Step 1. Enter the data in **Column C1**. For Example 8.20, we use **1989.9, 1993.8, 2074.5, 2070.5, 2070.9, 2033.6, 1939.6**.

Step 2. Click on **Stat**, then **Basic Statistics**, then **Graphical Summary**.

Step 3. Enter **C1** in the **Variables** field and enter the confidence level in the **Confidence Level** field. For Example 8.20, we use **95**.

Step 4. Click **OK** (Figure A).

```
95% Confidence Interval for Mean
    1976.6              2072.7
95% Confidence Interval for Median
    1976.4              2071.9
95% Confidence Interval for StDev
    33.5               114.4
```
Figure A

SECTION 8.4   Exercises

Exercises 1–4 are the Check Your Understanding exercises located within the section.

Understanding the Concepts

In Exercises 5 and 6, fill in each blank with the appropriate word or phrase.

5. To find a confidence interval for a standard deviation from a sample of size 15, we use a chi-square distribution with _____ degrees of freedom.

6. The method described for finding confidence intervals should be used only when the distribution of the population is almost exactly _____ .

In Exercises 7 and 8, determine whether the statement is true or false. If the statement is false, rewrite it as a true statement.

7. When constructing a confidence interval for a standard deviation, we must find two critical values.

8. If we have a confidence interval for a variance, we can obtain a confidence interval for the standard deviation by squaring the confidence bounds.

Practicing the Skills

9. Find the critical values for a 95% confidence interval using the chi-square distribution with 15 degrees of freedom.

10. Find the critical values for a 99% confidence interval using the chi-square distribution with 5 degrees of freedom.

11. Construct a 95% confidence interval for the population standard deviation σ if a sample of size 25 has standard deviation $s = 15$.

12. Construct a 99% confidence interval for the population standard deviation σ if a sample of size 8 has standard deviation $s = 7.5$.

Working with the Concepts

13. **SAT scores:** Scores on the math SAT are normally distributed. A sample of 20 SAT scores had standard deviation $s = 87$.
   a. Construct a 98% confidence interval for the population standard deviation σ.
   b. Someone says that the scoring system for the SAT is designed so that the population standard deviation will be $\sigma = 100$. Does this confidence interval contradict this claim? Explain.

14. **IQ scores:** Scores on an IQ test are normally distributed. A sample of 25 IQ scores had standard deviation $s = 8$.
   a. Construct a 95% confidence interval for the population standard deviation σ.
   b. The developer of the test claims that the population standard deviation is $\sigma = 15$. Does this confidence interval contradict this claim? Explain.

15. **Baby weights:** Following are weights of 12 two-month-old baby girls. Assume that the population is normally distributed.

12.23	12.32	11.87	12.34	11.48	12.66
8.51	14.13	12.95	10.30	9.34	8.63

   a. Find the sample standard deviation s.
   b. Construct a 95% confidence interval for the population standard deviation σ.
   c. According to the National Health Statistics Reports, the standard deviation of the weight of two-month-old baby boys is 2.7 pounds. Based on the confidence interval, is it reasonable to believe that the standard deviation of the

weights of two-month-old baby girls is the same as that of two-month-old baby boys? Explain.

16. Eat your cereal: Boxes of cereal are labeled as containing 14 ounces. Following are the weights of a sample of 12 boxes. Assume that the population is normally distributed.

14.02	13.97	14.11	14.12	14.10	14.02
14.15	13.97	14.05	14.04	14.11	14.12

a. Find the sample standard deviation s.
b. Construct a 98% confidence interval for the population standard deviation σ.
c. The goal of the quality control manager is for the population standard deviation of the weights to be less than 0.03. Based on the confidence interval, is it reasonable to believe that the goal has been met?

17. Eat your spinach: Six measurements were made of the mineral content (in percent) of spinach, with the following results. Assume that the population is normally distributed.

$$19.1 \quad 20.8 \quad 20.8 \quad 21.4 \quad 20.5 \quad 19.7$$

a. Find the sample standard deviation s.
b. Construct a 99% confidence interval for the population standard deviation σ.
c. Based on the confidence interval, is it reasonable to believe that the population standard deviation might be less than 1.5? Explain.
Source: *Journal of Nutrition* 66:55–66

18. Mortgage rates: Following are interest rates (annual percentage rates) for a 30-year fixed rate mortgage from a sample of lenders in Macon, Georgia on June 20, 2013. It is reasonable to assume that the population is approximately normal.

4.750	4.375	4.176	4.679	4.426	4.227
4.125	4.250	3.950	4.191	4.299	4.415

Source: www.bankrate.com

a. Find the sample standard deviation s.
b. Construct a 95% confidence interval for population standard deviation σ.

c. Based on the confidence interval, is it reasonable to believe that the population standard deviation is greater than 0.5? Explain.

Extending the Concepts

The chi-square distribution is skewed, but as the number of degrees of freedom becomes large, the skewness diminishes. If the number of degrees of freedom, k, is large enough, the chi-square distribution is reasonably well approximated by a normal distribution with mean k and variance 2k.

19. Exact confidence interval: A sample of size 101 from a normal population has sample standard deviation $s = 40$. Use Table A.4 to find the exact critical values $\chi^2_{0.025}$ and $\chi^2_{0.975}$ for a 95% confidence interval, and construct a 95% confidence interval for σ.

20. Using the normal approximation: Refer to Exercise 19. Use the normal approximation to estimate the critical values $\chi^2_{0.025}$ and $\chi^2_{0.975}$ for a 95% confidence interval, and construct a 95% confidence interval for σ.

21. Comparing results: How close is the confidence interval based on the normal approximation constructed in Exercise 20 to the exact confidence interval constructed in Exercise 19?

A more accurate normal approximation to χ^2_α is given by

$$\chi^2_\alpha \approx 0.5 \left(z_\alpha + \sqrt{2k-1} \right)^2, \text{ where } z_\alpha \text{ is the } z\text{-score that has area } \alpha \text{ to its right.}$$

22. More accuracy: Refer to Exercise 19. Use the more accurate normal approximation to estimate the critical values $\chi^2_{0.025}$ and $\chi^2_{0.975}$ for a 95% confidence interval, and construct a 95% confidence interval for σ.

23. Comparing results: How close is the confidence interval based on the more accurate normal approximation to the exact one?

Answers to Check Your Understanding Exercises for Section 8.4

1. $\chi^2_{0.975} = 8.231$, $\chi^2_{0.025} = 31.526$

2. $\chi^2_{0.995} = 10.520$, $\chi^2_{0.005} = 46.928$

3. $4.13 < \sigma < 10.95$

4. $8.60 < \sigma < 19.15$

SECTION 8.5 | **Determining Which Method to Use**

Objectives

1. Determine which method to use when constructing a confidence interval

Objective 1 Determine which method to use when constructing a confidence interval

One of the challenges in constructing a confidence interval is to determine which method to use. The first step is to determine which type of parameter we are estimating. There are three types of parameters for which we have learned to construct confidence intervals:

- Population mean μ
- Population proportion p
- Population standard deviation σ or variance σ^2

Once you have determined which type of parameter you are estimating, proceed as follows:

- **Population mean:** There are two methods for constructing a confidence interval for a population mean, the z method (Section 8.1) and the t method (Section 8.2). To determine which method to use, we must determine whether the population standard deviation is known, whether the population is approximately normal, and whether the sample size is large ($n > 30$). The following diagram can help you make the correct choice.

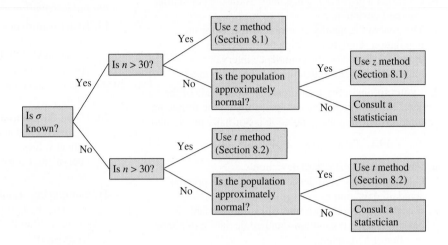

- **Population proportion:** To construct a confidence interval for a population proportion, use the method described in Section 8.3.
- **Population standard deviation or variance:** To construct a confidence interval for a population standard deviation or variance, use the method described in Section 8.4.

EXAMPLE 8.21   Determining which method to use

A random sample of 41 pumpkins harvested from a pumpkin patch has a mean weight of $\bar{x} = 8.53$ pounds with a sample standard deviation of $s = 1.32$ pounds. Construct a 95% confidence interval for the mean weight of pumpkins from this patch. Determine the type of parameter that is to be estimated and construct the confidence interval.

Solution
We are asked to find a confidence interval for the mean weight; this is a population mean. We consult the diagram to determine the correct method. We must first determine whether σ is known. There is no information given about σ, so σ is unknown. We follow the "No" path. Next we must determine whether $n > 30$. There are 41 pumpkins in the sample, so $n > 30$. We follow the "Yes" path, and find that we should use the t method described in Section 8.2.

To construct the confidence interval, there are $41 - 1 = 40$ degrees of freedom. Because the confidence level is 95%, the critical value is $t_{0.025} = 2.021$. The 95% confidence interval is

$$8.53 \pm 2.021 \frac{1.32}{\sqrt{41}}$$

$$8.11 < \mu < 8.95$$

Check Your Understanding

In Exercises 1–4, state which type of parameter is to be estimated, then construct the confidence interval.

1. A simple random sample of size 15 has mean $\bar{x} = 10.34$ and standard deviation $s = 3.48$. The population is normally distributed. Construct a 95% confidence interval for the population standard deviation.

2. A simple random sample of size 80 has mean $\bar{x} = 7.31$. The population standard deviation is $\sigma = 6.26$. Construct a 99% confidence interval for the population mean.

3. In a simple random sample of 100 children, 22 had reading skills above their grade level. Construct a 99% confidence interval for the proportion of children who have reading skills above their grade level.

4. A simple random sample of size 25 has mean $\bar{x} = 17.4$ and standard deviation $s = 5.3$. The population is approximately normally distributed. Construct a 95% confidence interval for the population mean.

Answers are on page 398.

SECTION 8.5   Exercises

Exercises 1–4 are the Check Your Understanding exercises located within the section.

Practicing the Skills

In Exercises 5–12, state which type of parameter is to be estimated, then construct the confidence interval.

5. A simple random sample of size 18 has mean $\bar{x} = 71.32$ and standard deviation $s = 15.78$. The population is approximately normally distributed. Construct a 95% confidence interval for the population mean.

6. In a simple random sample of 400 voters, 220 said that they were planning to vote for the incumbent mayor in the next election. Construct a 99% confidence interval for the proportion of voters who plan to vote for the incumbent mayor in the next election.

7. A simple random sample of size 8 has mean $\bar{x} = 3.21$ and standard deviation $s = 1.69$. The population is normally distributed. Construct a 99% confidence interval for the population standard deviation.

8. A simple random sample of size 12 has mean $\bar{x} = 3.37$. The population standard deviation is $\sigma = 1.62$. The population is approximately normally distributed. Construct a 95% confidence interval for the population mean.

9. In a survey of 250 employed adults, 185 said that they had missed one or more days of work in the past six months. Construct a 95% confidence interval for the proportion of employed adults who missed one or more days of work in the past six months.

10. A simple random sample of size 17 has mean $\bar{x} = 8.44$ and standard deviation $s = 5.38$. The population is normally distributed. Construct a 95% confidence interval for the population standard deviation.

11. A simple random sample of size 120 has mean $\bar{x} = 8.45$. The population standard deviation is $\sigma = 4.81$. Construct a 99% confidence interval for the population mean.

12. A simple random sample of size 23 has mean $\bar{x} = 1.48$ and standard deviation $s = 1.32$. The population is approximately normally distributed. Construct a 99% confidence interval for the population mean.

Working with the Concepts

13. **Football players:** The weights of 52 randomly selected NFL football players in 2013 are presented below. The sample mean is $\bar{x} = 248.38$ and the sample standard deviation is $s = 46.68$.

305	265	287	285	290	235	300	230	195
236	244	194	190	307	218	315	265	210
194	216	255	300	315	190	185	183	313
246	212	201	308	270	241	242	306	237
315	215	200	295	187	204	257	185	255
318	230	316	200	324	245	185		

Source: *Chicago Tribune*

Construct a 95% confidence interval for the mean weight of NFL football players in 2010.

14. **Ages of students:** A simple random sample of 100 U.S. college students had a mean age of 22.68 years. Assume the population standard deviation is $\sigma = 4.74$ years. Construct a 99% confidence interval for the mean age of U.S. college students.

15. **Calories in bread:** Following are the numbers of calories in a random sample of 10 slices of bread. Assume the population is normally distributed.

55	51	49	48	68	52	62	70	67	70

Construct a 95% confidence interval for the standard deviation of the number of calories.

16. **Credit card debt:** In a survey of 1118 U.S. adults conducted in 2012 by the Financial Industry Regulatory Authority, 626 said they always pay their credit cards in full each month. Construct a 95% confidence interval for the proportion of U.S. adults who pay their credit cards in full each month.

17. **Windy place:** Mt. Washington, New Hampshire, is one of the windiest places in the United States. Wind speed measurements on a simple random sample of 50 days had a sample mean of 45.01 mph. Assume the population standard deviation is $\sigma = 25.6$ mph. Construct a 95% confidence interval for the mean wind speed on Mt. Washington.

18. **An apple a day:** Following are the numbers of grams of sugar per 100 grams of apple in a random sample of 6 Red Delicious apples. Assume the population is normally distributed.

12.0	12.6	13.1	13.5	12.1	10.5

    Construct a 95% confidence interval for the standard deviation of the number of grams of sugar.

19. **Pneumonia:** In a simple random sample of 1500 patients admitted to the hospital with pneumonia, 145 were under the age of 18. Construct a 99% confidence interval for the proportion of pneumonia patients who are under the age of 18.

20. **College tuition:** A simple random sample of 35 colleges and universities in the United States had a mean tuition of $18,702 with a standard deviation of $10,653. Construct a 95% confidence interval for the mean tuition for all colleges and universities in the United States.

Answers to Check Your Understanding Exercises for Section 8.5

1. The parameter is the population standard deviation. The 95% confidence interval is $2.55 < \sigma < 5.49$.

2. The parameter is the population mean. The 99% confidence interval is $5.51 < \mu < 9.11$.

3. The parameter is the population proportion. The 99% confidence interval is $0.113 < p < 0.327$.

4. The parameter is the population mean. The 95% confidence interval is $15.2 < \mu < 19.6$.

Chapter 8 Summary

Section 8.1: In this section, we presented the basic ideas behind confidence intervals. We learned that a point estimate is a single number that is used to estimate the value of an unknown parameter. For example, the sample mean $\bar{x}$ is a point estimate of the population mean μ. The standard error of a point estimate tells us roughly how far from the true value the point estimate is likely to be. We multiply the standard error by a critical value to obtain the margin of error. By adding and subtracting the margin of error from the point estimate, we obtain a confidence interval. The level of a confidence interval is the proportion of samples for which the confidence interval will contain the true value. If the sample size is large ($n > 30$) or if the population is approximately normal, then the confidence interval for the population mean is $\bar{x} \pm z_{\alpha/2}\sigma/\sqrt{n}$ if the population mean σ is known.

Section 8.2: When the population is approximately normal, or the sample size is large ($n > 30$), we can use the Student's t distribution to construct a confidence interval for a population mean μ when the population standard deviation is unknown. Let s be the sample standard deviation. The confidence interval is $\bar{x} \pm t_{\alpha/2}s/\sqrt{n}$ if σ is unknown.

Section 8.3: In this section, we learned to construct confidence intervals for population proportions. The assumptions are that the individuals in the population can be divided into two categories, that the sample contains at least 10 individuals in each category, and that the population is at least 20 times as large as the sample. We denote the sample size by n and the number of individuals in the sample who fall into the specified category by x. When the assumptions are met, the confidence interval for the population proportion p is $\hat{p} \pm z_{\alpha/2}\sqrt{\hat{p}(1-\hat{p})/n}$, where $\hat{p} = x/n$. A small-sample method can also be used. In the small-sample method, we define $\tilde{p} = \dfrac{x+2}{n+4}$. The confidence interval is $\tilde{p} \pm z_{\alpha/2}\sqrt{\dfrac{\tilde{p}(1-\tilde{p})}{n+4}}$. The small-sample confidence interval is actually valid for any sample size.

Section 8.4: When the population is almost exactly normal, it is possible to find a confidence interval for the variance or standard deviation of the population using the chi-square distribution. This method is very sensitive to the assumption of normality, and should not be used unless it is certain that the population is almost exactly normal. The lower bound of the confidence interval is $(n-1)s^2/\chi^2_{\alpha/2}$ and the upper bound is $(n-1)s^2/\chi^2_{1-\alpha/2}$.

Section 8.5: We have learned to construct confidence intervals for a population mean, a population proportion, and a population standard deviation or variance. There are two methods for constructing a confidence interval for a population mean, the z method and the t method. The method to use depends on whether the population standard deviation σ is known.

Vocabulary and Notation

chi-square distribution   390
confidence interval   350
confidence level   350

critical value   349
degrees of freedom   367
margin of error   349

point estimate   348
standard error   349
Student's t distribution   366

Important Formulas

Confidence interval for a mean, standard deviation known:

$$\bar{x} - z_{\alpha/2}\frac{\sigma}{\sqrt{n}} < \mu < \bar{x} + z_{\alpha/2}\frac{\sigma}{\sqrt{n}}$$

Sample size to construct an interval for μ with margin of error m:

$$n = \left(\frac{z_{\alpha/2} \cdot \sigma}{m}\right)^2$$

Confidence interval for a mean, standard deviation unknown:

$$\bar{x} - t_{\alpha/2}\frac{s}{\sqrt{n}} < \mu < \bar{x} + t_{\alpha/2}\frac{s}{\sqrt{n}}$$

Confidence interval for a proportion:

$$\hat{p} - z_{\alpha/2}\sqrt{\frac{\hat{p}(1-\hat{p})}{n}} < p < \hat{p} + z_{\alpha/2}\sqrt{\frac{\hat{p}(1-\hat{p})}{n}}$$

Sample size to construct an interval for p with margin of error m:

$$n = \hat{p}(1-\hat{p})\left(\frac{z_{\alpha/2}}{m}\right)^2 \quad \text{if a value for } \hat{p} \text{ is available}$$

$$n = 0.25\left(\frac{z_{\alpha/2}}{m}\right)^2 \quad \text{if no value for } \hat{p} \text{ is available}$$

Confidence interval for the variance of a normal distribution:

$$\frac{(n-1)s^2}{\chi^2_{\alpha/2}} < \sigma^2 < \frac{(n-1)s^2}{\chi^2_{1-\alpha/2}}$$

Confidence interval for the standard deviation of a normal distribution:

$$\sqrt{\frac{(n-1)s^2}{\chi^2_{\alpha/2}}} < \sigma < \sqrt{\frac{(n-1)s^2}{\chi^2_{1-\alpha/2}}}$$

Chapter Quiz

1. Define the following terms:
   a. Point estimate
   b. Confidence interval
   c. Confidence level

2. Find the critical value $t_{\alpha/2}$ needed to construct a 90% confidence interval for a population mean with sample size 27.

3. An owner of a fleet of taxis wants to estimate the mean gas mileage, in miles per gallon, of the cars in the fleet. A random sample of 40 cars is followed for one month, and the sample mean gas mileage is 23.2 with a standard deviation of 5.8. Construct a 90% confidence interval for the mean gas mileage in the fleet.

4. Construct a 95% confidence interval for the population standard deviation σ if a sample of size 20 has standard deviation $s = 10$.

5. A cookie manufacturer wants to estimate the length of time that her boxes of cookies spend in the store before they are bought. She visits a sample of 15 supermarkets and determines the number of days since manufacture of the oldest box of cookies in the store. The mean is 54.8 days with a standard deviation of 11.3 days. A dotplot of the data indicates that the assumptions for constructing a confidence interval for the mean are satisfied. Construct a 99% confidence interval for the mean number of days.

6. A person selects a random sample of 15 credit cards and determines the annual interest rate, in percent, of each. The sample mean is 12.42 with a sample standard deviation of 1.3. Construct a 95% confidence interval for the mean credit card annual interest rate, assuming that the rates are approximately normally distributed.

7. Construct a 90% confidence interval for the population standard deviation σ if a sample of size 6 has standard deviation $s = 22$.

8. Find the critical value $z_{\alpha/2}$ needed to construct a confidence interval for a population proportion with confidence level 92%.

9. Find the critical values for a 98% confidence interval using the chi-square distribution with 18 degrees of freedom.

10. The amount of time that a certain cell phone will keep a charge is known to be normally distributed with standard deviation $\sigma = 16$ hours. A sample of 40 cell phones had a mean time of 141 hours. Let μ represent the population mean time that a cell phone will keep a charge.
    a. What is the point estimate of μ?
    b. What is the standard error of the point estimate?

11. Refer to Exercise 10. Suppose that a 95% confidence interval is to be constructed for the mean time.
    a. What is the critical value?
    b. What is the margin of error?
    c. Construct the 95% confidence interval.

12. Refer to Exercise 10. What sample size is necessary so that a 95% confidence interval will have a margin of error of 1 hour?

13. In a survey of 802 U.S. adult drivers, 265 state that traffic is getting worse in their community. Construct a 99% confidence interval for the proportion of adult drivers who think that traffic is getting worse.

14. Refer to Exercise 13. How large a sample is needed so that a 99% confidence interval will have margin of error of 0.08, using the sample proportion for $\hat{p}$?

15. Refer to Exercise 13. How large a sample is needed so that a 99% confidence interval will have margin of error of 0.08, assuming no estimate of $\hat{p}$ is available?

Review Exercises

1. Build more parking? A survey is to be conducted in which a random sample of residents in a certain city will be asked whether they favor or oppose the building of a new parking structure downtown. How many residents should be polled to be sure that a 90% confidence interval for the proportion who favor the construction will have a margin of error no greater than 0.05?

2. Drill lifetime: A sample of 50 drills had a mean lifetime of 12.68 holes drilled when drilling a low-carbon steel. Assume the population standard deviation is 6.83.
   a. Construct a 95% confidence interval for the mean lifetime of this type of drill.
   b. The manufacturer of the drills claims that the mean lifetime is greater than 13. Does this confidence interval contradict this claim? Explain.
   c. How large would the sample need to be so that a 95% confidence interval would have a margin of error of 1.0?
   Source: *Journal of Engineering Manufacture* 216:301–305

3. Cost of environmental restoration: In a survey of 189 Scottish voters, 61 said they would be willing to pay additional taxes in order to restore the Affric forest.
   a. Assuming that the 189 voters who responded constitute a random sample, construct a 99% confidence interval for the proportion of voters who would be willing to pay to restore the Affric forest.
   b. Use the results from the sample of size 189 to estimate the sample size needed so that the 99% confidence interval will have a margin of error of 0.03.
   c. Another survey is planned, in which voters will be asked whether they would be willing to pay in order to restore the Strathsprey forest. At this point, no estimate of this proportion is available. Find an estimate of the sample size needed so that the margin of error of a 99% confidence interval will be 0.03.
   Source: *Environmental and Resource Economics* 18:391–410

4. More repairs: A sample of six records for repairs of a component showed the following costs:

93   97   27   79   81   87

   a. Construct a 90% confidence interval for the mean cost of a repair for this type of component.
   b. Is there any evidence to suggest that this confidence interval may not be reliable? Explain.

5. More repairs: Refer to Exercise 4. Would it be appropriate to use these data to construct a confidence interval for σ using the methods of Section 8.4? Explain.

6. Contaminated water: Polychlorinated biphenyls (PCBs) are a group of synthetic oil-like chemicals that were at one time widely used as insulation in electrical equipment and were discharged into rivers. They were discovered to be a health hazard, and were banned in the 1970s. Assume that water samples are being drawn from a river in order to estimate the PCB concentration. Suppose that a random sample of size 60 has a sample mean of 1.96 parts per billion (ppb). Assume the population standard deviation is $\sigma = 0.35$ ppb.
   a. Construct a 98% confidence interval for the PCB concentration.
   b. EPA standards require that the PCB concentration in drinking water be no more than 0.5 ppb. Based on the confidence interval, is it reasonable to believe that this water meets the EPA standard for drinking water? Explain.
   c. Estimate the sample size needed so that a 98% confidence interval will have a margin of error of 0.03.

7. Defective electronics: A simple random sample of 200 electronic components was tested, and 17 of them were found to be defective.
   a. Construct a 99% confidence interval for the proportion of components that are defective.
   b. Use the results from the sample of 200 to estimate the sample size needed so that the 99% confidence interval will have a margin of error equal to 0.04.
   c. A simple random sample of a different type of component will be tested. At this point, there is no estimate of the proportion defective. Find a sample size so that the 99% confidence interval will have a margin of error no greater than 0.04.

8. Cost of repairs: A sample of eight repair records for a certain fiber-optic component was drawn, and the cost of each repair, in dollars, was recorded, with the following results:

30   35   19   23   27   22   26   16

   a. Construct a dotplot for these data. Are the assumptions for constructing a confidence interval for the mean satisfied? Explain.
   b. If appropriate, construct a 98% confidence interval for the mean cost of a repair.

9. **Cost of repairs:** Refer to Exercise 8. Assume the population is normal. Construct a 95% confidence interval for the population standard deviation σ.

10. **Super Bowl:** A simple random sample of 140 residents in a certain town was polled the week after the Super Bowl, and 75 of them said they had watched the game on television.
    a. Construct a 95% confidence interval for the proportion of people in the town who watched the Super Bowl on television.
    b. The 2013 Super Bowl between the Baltimore Ravens and the San Francisco 49ers was the third most watched television program in history, with 48.1% of television sets in the U.S. tuned to the game. Someone claims that the percentage of people who watched the game in this town was less than 48.1%. Does the confidence interval contradict this claim? Explain.
    c. Use the results from the sample of 140 to estimate the sample size necessary for a 95% confidence interval to have a margin of error of 0.025.

11. **Testing math skills:** In order to test the effectiveness of a program to improve mathematical skills, a simple random sample of 45 fifth-graders was chosen to participate in the program. The students were given an exam at the beginning of the program and again at the end. The sample mean increase in the exam score was 12.2 points, with a sample standard deviation of 4.7 points.
    a. Construct a 99% confidence interval for the mean increase in score.
    b. The developers of the program claim that the program will produce a mean increase of more than 15 points. Does the confidence interval contradict this claim? Explain.

12. **Sleep time:** In a sample of 87 young adults, the average time per day spent in bed asleep was 7.06 hours. Assume the population sample standard deviation is 1.11 hours.
    a. Construct a 99% confidence interval for the mean time spent in bed asleep.
    b. Some health experts recommend that people get 8 hours or more of sleep per night. Based on the confidence interval, is it reasonable to believe that the mean number of hours of sleep for young adults is 8 or more? Explain.
    c. How large would the sample have to be so that a 99% confidence interval would have a margin of error of 0.1?
    Source: *Behavioral Medicine* 27:71–76

13. **Leaking tanks:** Leakage from underground fuel tanks has been a source of water pollution. In a random sample of 107 gasoline stations, 18 were found to have at least one leaking underground tank.
    a. Find a point estimate for the proportion of gasoline stations with at least one leaking underground tank.
    b. Construct a 95% confidence interval for the proportion of gasoline stations with at least one leaking underground tank.
    c. Use the point estimate computed in part (a) to determine the number of stations that must be sampled so that a 95% confidence interval will have a margin of error of 0.03.

14. **Waist size:** According to the National Health Statistics Reports, a sample of 783 men aged 20–29 years had a mean waist size of 36.9 inches with a standard deviation of 8.8 inches.
    a. Construct a 95% confidence interval for the mean waist size.
    b. The results of another study suggest that the mean waist size for men aged 30–39 is 38.7 inches. Based on the confidence interval, is it reasonable to believe that the mean waist size for men aged 20–29 may be 38.7 inches? Explain.

15. **Don't construct a confidence interval:** A meteorology student examines precipitation records for a certain city and discovers that of the last 365 days, it rained on 46 of them. Explain why these data cannot be used to construct a confidence interval for the proportion of days in this city that are rainy.

Write About It

1. When constructing a confidence interval for μ when σ is known, we assume that we have a simple random sample, that σ is known, and that either the sample size is large or the population is approximately normal. Why is it necessary for these assumptions to be met?

2. What factors can you think of that may affect the width of a confidence interval? In what way does each factor affect the width?

3. Explain the difference between confidence and probability.

In Exercises 4 and 5, express the following survey results in terms of confidence intervals for p:

4. According to a survey of 1000 American adults, 55% of Americans do not have a will specifying the handling of their estate. The survey's margin of error was plus or minus 3%.
   Source: FindLaw.com

5. In a survey of 5050 U.S. adults, 29% would consider traveling abroad for medical care because of medical costs. The survey's margin of error was plus or minus 2%.
   Source: The Gallup Poll

6. When constructing a confidence interval for μ, how do you decide whether to use the t distribution or the normal distribution? Are there any circumstances when it is acceptable to use either distribution?

7. It is stated in the text that there are many different t distributions. Explain how this is so.

Case Study: Do Newer Wood Stoves Produce Less Pollution?

The town of Libby, Montana, has experienced high levels of air pollution in the winter because many of the houses in Libby are heated by wood stoves that produce a lot of pollution. In an attempt to reduce the level of air pollution in Libby, a program was undertaken in which almost every wood stove in the town was replaced with a newer, cleaner-burning model. Measurements of several air pollutants were taken both before and after the stove replacement. They included particulate matter (PM), total carbon (TC), organic carbon (OC), which is carbon bound in organic molecules, and levoglucosan (LE), which is a compound found in charcoal and is thus an indicator of the amount of wood smoke in the atmosphere.

In order to determine how much the pollution levels were reduced, scientists measured the levels of these pollutants for three winters prior to the replacement. The mean levels over this period of time are referred to as the *baseline* levels. Following are the baseline levels for these pollutants. The units are micrograms per cubic meter.

PM: 27.08   OC: 17.41   TC: 18.87   LE: 2.57

The following table presents values measured on samples of winter days during the two years following replacement.

Year 1				Year 2			
PM	OC	TC	LE	PM	OC	TC	LE
21.7	15.6	17.73	1.78	27.0	15.79	19.46	2.06
27.8	15.6	17.87	2.25	24.7	13.61	15.98	3.10
24.7	17.2	18.75	1.98	21.8	12.94	15.79	2.68
15.3	8.3	9.21	0.67	23.2	12.97	16.32	2.80
18.4	11.3	12.46	0.86	23.3	11.19	13.49	2.07
14.4	8.4	9.66	1.93	16.2	9.61	12.44	2.14
19.0	13.2	14.73	1.51	13.4	6.97	8.40	2.32
23.7	11.4	13.23	1.98	13.0	7.96	10.02	2.18
22.4	13.8	17.08	1.69	16.9	8.43	11.08	2.06
25.6	13.2	15.86	2.30	26.3	14.92	21.46	1.94
15.0	15.7	17.27	1.24	31.4	17.15	20.57	1.85
17.0	9.3	10.21	1.44	40.1	15.13	19.64	2.11
23.2	10.5	11.47	1.43	28.0	8.66	10.75	2.50
17.7	14.2	15.64	1.07	4.2	15.95	20.36	2.27
11.1	11.6	13.48	0.59	15.9	11.73	14.59	2.17
29.8	7.0	7.795	2.10	20.5	14.34	17.64	2.74
20.0	19.9	21.20	1.73	23.8	8.99	11.75	2.45
21.6	14.8	15.65	1.56	14.6	10.63	13.12	
14.8	12.6	13.51	1.10	17.8			
21.0	9.1	9.94					

1. For each of the four pollutants, construct a boxplot for the values for Year 1 to verify that the assumptions for constructing a confidence interval are satisfied.

2. Construct a 95% confidence interval for the mean level of each pollutant for the Year 1.

3. Is it reasonable to conclude that the mean levels in Year 1 were lower than the baseline levels for some or all of the pollutants? Which ones, if any?

The investigators were concerned that the reduction in pollution levels might be only temporary. Specifically, they were concerned that people might use their new stoves carefully at first, thus obtaining the full advantage of their cleaner burning, but then become more casual in their operation, leading to an increase in pollution levels. We will investigate this issue by constructing confidence intervals for the mean levels in Year 2.

4. Repeat Exercises 1 and 2 for the Year 2 data.

5. Is it reasonable to conclude that the mean levels in Year 2 were lower than the baseline levels for some or all of the pollutants? Which ones, if any?

Hypothesis Testing

Introduction

Is global warming real? There is much evidence to suggest that temperatures have been increasing since the early part of the 20th century. Whether the increase is caused by human activity or whether it is part of a natural cycle is a topic of debate. Studies of global warming involve analyzing temperature records. The following table presents the record high and low temperatures in degrees Fahrenheit in Washington, D.C., along with the year in which the record was set, for a selection of days in April through September. The column labeled "More Recent" tells which record, high or low, occurred more recently.

Date	High	Year	Low	Year	More Recent	Date	High	Year	Low	Year	More Recent	Date	High	Year	Low	Year	More Recent
Apr 2	89	1963	23	1907	High	Jun 4	99	1925	46	1929*	Low	Aug 6	106	1918	53	1912	High
9	90	1959	28	1972*	Low	11	101	1911	45	1913	Low	13	101	1881	55	1930*	Low
16	92	2002	29	1928	High	18	97	1944	51	1965*	Low	20	101	1983	50	1896	High
23	95	1960	33	1933*	High	25	100	1997	53	1902	High	27	100	1987	51	1885	High
30	92	1942*	34	1874	High	Jul 2	101	1898	55	1940	Low	Sep 3	98	1953	48	1909*	High
May 7	95	1930	38	1970	Low	9	104	1936	55	1891	High	10	98	1983	44	1883	High
14	93	1956	41	1928	High	16	104	1988	56	1930*	High	17	96	1991	44	1923	High
21	95	1934	41	1907	High	23	102	2011	56	1890	High	24	99	2010	39	1963	High
28	97	1941	42	1961	Low	30	99	1953	56	1914	High						

*Indicates that the record occurred more than once; only the most recent year is given.
Source: National Weather Service

If there were no temperature trend, we would expect that the proportion of days on which the high temperature record was more recent to be about one-half. It would not be surprising if this proportion were somewhat different from one-half, because we would expect to see some difference just by chance. However, we would not expect the proportion to be much different from one-half. If the proportion were much different from one-half, we would conclude that there was a temperature trend.

There are 26 days in the table. Half of 26 is 13. If the record high had been more recent on 13 of the days, there would be no reason to believe there was a warming trend. If the record high had been more recent on 14 or 15 days, this would be slightly more than one-half, but it would seem reasonable to believe that a difference this small was just due to chance.

In fact, the record high temperature was more recent for 18 of the 26 days. The question we need to address is whether this difference from one-half is too large for us to believe that it is simply due to chance. This is the sort of question that hypothesis tests are designed to answer. In this chapter, we will learn to perform hypothesis tests in a variety of commonly occurring situations. In the case study at the end of the chapter, we will study a data set that includes the data in the preceding table, and investigate the possibility of a warming trend in Washington, D.C.

SECTION 9.1 | Basic Principles of Hypothesis Testing

Objectives

1. Define the null and alternate hypotheses
2. State conclusions to hypothesis tests
3. Distinguish between Type I and Type II errors

Objective 1 Define the null and alternate hypotheses

The Null Hypothesis and the Alternate Hypothesis

Air pollution has become a serious health problem in many cities. One of the forms of air pollution that health officials are most concerned about is particulate matter (PM), which refers to fine particles that can be trapped in the lungs, increasing the risk of respiratory disease. Some of the PM in the atmosphere comes from car exhaust, so one important way to reduce PM pollution is to design automobile engines that produce less PM. The following example will show how hypothesis testing can play a part in this effort.

A study published in the *Journal of the Air and Waste Management Association* reported that the mean amount of PM produced by cars and light trucks in an urban setting is 35 milligrams of PM per mile of travel. Suppose that a new engine design is proposed that is intended to reduce this level. Now there are two possibilities: either the new design will reduce the level, or it will not. These possibilities are called *hypotheses*. To be specific,

1. The *null hypothesis* says that the new design will not reduce the level, so the mean for the new engines will be $\mu = 35$.
2. The *alternate hypothesis* says that the new design will reduce the level, so $\mu < 35$.

In general, the null hypothesis says that a parameter is equal to a certain value, while the alternate hypothesis says that the parameter differs from this value. Often the null hypothesis is a statement of no change or no difference, while the alternate hypothesis states that a change or difference has occurred.

DEFINITION

- The **null hypothesis** about a parameter states that the parameter is equal to a specific value, for example, $H_0: \mu = 35$. The null hypothesis is denoted H_0.
- The **alternate hypothesis** about a parameter states that the value of the parameter differs from the value specified by the null hypothesis. The alternate hypothesis is denoted H_1.

There are three types of alternate hypothesis, which we now define.

DEFINITION

- **A left-tailed** alternate hypothesis states that the parameter is less than the value specified by the null hypothesis, for example, $H_1: \mu < 35$.
- **A right-tailed** alternate hypothesis states that the parameter is greater than the value specified by the null hypothesis, for example, $H_1: \mu > 35$.
- **A two-tailed** alternate hypothesis states that the parameter is not equal to the value specified by the null hypothesis, for example, $H_1: \mu \neq 35$.

Left-tailed and right-tailed hypotheses are called **one-tailed** hypotheses.

EXAMPLE 9.1

State the null and alternate hypotheses

Boxes of a certain kind of cereal are labeled as containing 20 ounces. An inspector thinks that the mean weight may be less than this. State the appropriate null and alternate hypotheses.

Solution

The null hypothesis says that there is no difference, so the null hypothesis is $H_0: \mu = 20$. The inspector thinks that the mean weight may be less than 20, so the alternate hypothesis is $H_1: \mu < 20$.

EXAMPLE 9.2

State the null and alternate hypotheses

Last year, the mean monthly rent for an apartment in a certain city was $800. A real estate agent believes that the mean rent is higher this year. State the appropriate null and alternate hypotheses.

Solution

The null hypothesis says that there is no change, so the null hypothesis is $H_0: \mu = 800$. The real estate agent wants to know whether the mean is higher, so the alternate hypothesis is $H_1: \mu > 800$.

EXAMPLE 9.3

State the null and alternate hypotheses

Scores on a standardized test have a mean of 70. Some modifications are made to the test, and an educator believes that the mean may have changed. State the appropriate null and alternate hypotheses.

Solution

The null hypothesis says that there is no change, so the null hypothesis is $H_0: \mu = 70$. The educator wants to know whether the mean has changed, without specifying whether it has increased or decreased. Therefore, the alternate hypothesis is $H_1: \mu \neq 70$.

Check Your Understanding

1. Last year, the mean amount spent by customers at a certain restaurant was $35. The restaurant owner believes that the mean may be higher this year. State the appropriate null and alternate hypotheses.

2. In a recent year, the mean weight of newborn boys in a certain country was 6.6 pounds. A doctor wants to know whether the mean weight of newborn girls differs from this. State the appropriate null and alternate hypotheses.

3. A certain model of car can be ordered with either a large or small engine. The mean number of miles per gallon for cars with a small engine is 25.5. An automotive

engineer thinks that the mean for cars with the larger engine will be less than this. State the appropriate null and alternate hypotheses.

Answers are on page 409.

A hypothesis test is like a trial

The purpose of a **hypothesis test** is to determine how likely it is that the null hypothesis is true. The idea behind hypothesis testing is the same as the idea behind a criminal trial. At the start of a trial, the defendant is assumed to be innocent. Then the evidence is presented. If the evidence strongly indicates that the defendant is guilty, we abandon the assumption of innocence and find the defendant guilty. In a hypothesis test, the null hypothesis plays the role of the defendant. At the start of a hypothesis test, we assume that the null hypothesis is true. Then we look at the evidence, which comes from data that have been collected. If the data strongly indicate that the null hypothesis is false, we abandon our assumption that it is true and believe the alternate hypothesis instead. This is referred to as **rejecting** the null hypothesis.

SUMMARY

- We begin a hypothesis test by assuming the null hypothesis to be true.
- If the data provide strong evidence against the null hypothesis, we reject it, and believe the alternate hypothesis.

Objective 2 State conclusions to hypothesis tests

Stating Conclusions

If the null hypothesis is rejected, we conclude that H_1 is true. We can state this conclusion by expressing H_1 in words. We should not simply say "we reject the null hypothesis."

EXAMPLE 9.4 — State a conclusion when the null hypothesis is rejected

Boxes of a certain kind of cereal are labeled as containing 20 ounces. An inspector thinks that the mean weight may be less than this, so he performs a test of H_0: $\mu = 20$ versus H_1: $\mu < 20$. He rejects the null hypothesis. State an appropriate conclusion.

Solution

Because the null hypothesis is rejected, we conclude that the alternate hypothesis is true. We express the alternate hypothesis in words: "We conclude that the mean weight of cereal boxes is less than 20 ounces."

Explain It Again

The conclusion of a hypothesis test is like the verdict of a jury: Not rejecting H_0 is like a jury verdict of not guilty. A not guilty verdict doesn't mean that the defendant is innocent; it just means that the evidence wasn't strong enough to be sure of guilt. Not rejecting H_0 does not mean that H_0 is true; it just means that the evidence wasn't strong enough to reject it.

If the null hypothesis is rejected, the conclusion is straightforward: We conclude that the null hypothesis is false and the alternate hypothesis is true. However, if the null hypothesis is not rejected, we do *not* conclude that the null hypothesis is true. In our formulation, the null hypothesis says that a parameter, such as μ, is equal to a certain value. Now we can never be sure that a parameter is *exactly* equal to a particular value. Therefore, we can never be sure that the null hypothesis is true. When we do not reject the null hypothesis, this just means that the evidence wasn't strong enough to reject it. An appropriate way to state a conclusion when the null hypothesis is not rejected is to state that there is not sufficient evidence to conclude that H_1 is true.

SUMMARY

- If there is sufficient evidence to reject the null hypothesis, we conclude that the alternate hypothesis is true.
- If there is not sufficient evidence to reject the null hypothesis, we conclude that the null hypothesis *might* be true, but we never conclude that the null hypothesis *is* true.

EXAMPLE 9.5 State a conclusion when the null hypothesis is not rejected

Boxes of a certain kind of cereal are labeled as containing 20 ounces. An inspector thinks that the mean weight may be less than this, so he performs a test of $H_0: \mu = 20$ versus $H_1: \mu < 20$. He does not reject the null hypothesis. State an appropriate conclusion.

Solution

The null hypothesis is not rejected, so we do not have sufficient evidence to conclude that the alternate hypothesis is true. We can express this as follows: "There is not enough evidence to conclude that the mean weight of cereal boxes is less than 20 ounces." Another way to state this is: "The mean weight of cereal boxes may be equal to 20 ounces."

Objective 3 Distinguish between Type I and Type II errors

Explain It Again

Type I and Type II errors in a trial: In a trial, the null hypothesis is that the defendant is innocent. A Type I error occurs if an innocent defendant is found guilty. A Type II error occurs if a guilty defendant is found not guilty.

Type I and Type II Errors

Whenever a decision is made, there is a possibility that it is the wrong decision. There are two ways to make a wrong decision with a hypothesis test. First, if H_0 is true, we might mistakenly reject it. Second, if H_0 is false, we might mistakenly decide not to reject it. These two types of errors have names. Rejecting H_0 when it is true is called a **Type I error**. Failing to reject H_0 when it is false is called a **Type II error**. We summarize the possibilities in the following table.

Decision	Reality	
	H_0 **True**	H_0 **False**
Reject H_0	Type I error	Correct decision
Don't reject H_0	Correct decision	Type II error

EXAMPLE 9.6 Determining which type of error has been made

The dean of a business school wants to determine whether the mean starting salary of graduates of her school is greater than $50,000. She will perform a hypothesis test with the following null and alternate hypotheses:

$$H_0: \mu = \$50,000 \qquad H_1: \mu > \$50,000$$

a. Suppose that the true mean is $\mu = \$50,000$, and the dean rejects H_0. Is this a Type I error, a Type II error, or a correct decision?

b. Suppose that the true mean is $\mu = \$55,000$, and the dean rejects H_0. Is this a Type I error, a Type II error, or a correct decision?

c. Suppose that the true mean is $\mu = \$55,000$, and the dean does not reject H_0. Is this a Type I error, a Type II error, or a correct decision?

Solution

a. The true mean is $\mu = \$50,000$, so H_0 is true. Because the dean rejects H_0, this is a Type I error.

b. The true mean is $\mu = \$55,000$, so H_0 is false. Because the dean rejects H_0, this is a correct decision.

c. The true mean is $\mu = \$55,000$, so H_0 is false. Because the dean does not reject H_0, this is a Type II error.

Check Your Understanding

4. A test is made of $H_0: \mu = 100$ versus $H_1: \mu \neq 100$. The true value of μ is 150, and H_0 is rejected. Is this a Type I error, a Type II error, or a correct decision?

5. A test is made of $H_0: \mu = 18$ versus $H_1: \mu > 18$. The true value of μ is 20, and H_0 is not rejected. Is this a Type I error, a Type II error, or a correct decision?

6. A test is made of H_0: $\mu = 3$ versus H_1: $\mu < 3$. The true value of μ is 3, and H_0 is rejected. Is this a Type I error, a Type II error, or a correct decision?

Answers are on page 409.

SECTION 9.1 Exercises

Exercises 1–6 are the Check Your Understanding exercises located within the section.

Understanding the Concepts

In Exercises 7 and 8, fill in each blank with the appropriate word or phrase.

7. The _____ hypothesis states that a parameter is equal to a certain value while the _____ hypothesis states that the parameter differs from this value.

8. Rejecting H_0 when it is true is called a _____ error, and failing to reject H_0 when it is false is called a _____ error.

In Exercises 9–12, determine whether the statement is true or false. If the statement is false, rewrite it as a true statement.

9. H_1: $\mu > 50$ is an example of a left-tailed alternate hypothesis.

10. If we reject H_0, we conclude that H_0 is false.

11. If we do not reject H_0, then we conclude that H_1 is false.

12. If we do not reject H_0, we conclude that H_0 is true.

Practicing the Skills

In Exercises 13–16, determine whether the alternate hypothesis is left-tailed, right-tailed, or two-tailed.

13. H_0: $\mu = 5$ H_1: $\mu < 5$
14. H_0: $\mu = 10$ H_1: $\mu > 10$
15. H_0: $\mu = 1$ H_1: $\mu \neq 1$
16. H_0: $\mu = 26$ H_1: $\mu \neq 26$

In Exercises 17–20, determine whether the outcome is a Type I error, a Type II error, or a correct decision.

17. A test is made of H_0: $\mu = 20$ versus H_1: $\mu \neq 20$. The true value of μ is 25, and H_0 is rejected.

18. A test is made of H_0: $\mu = 5$ versus H_1: $\mu < 5$. The true value of μ is 5, and H_0 is rejected.

19. A test is made of H_0: $\mu = 63$ versus H_1: $\mu > 63$. The true value of μ is 75, and H_0 is not rejected.

20. A test is made of H_0: $\mu = 45$ versus H_1: $\mu < 45$. The true value of μ is 40, and H_0 is rejected.

Working with the Concepts

21. Fertilizer: A new type of fertilizer is being tested on a plot of land in an orange grove, to see whether it increases the amount of fruit produced. The mean number of pounds of fruit on this plot of land with the old fertilizer was 400 pounds. Agriculture scientists believe that the new fertilizer may increase the yield. State the appropriate null and alternate hypotheses.

22. Big fish: A sample of 100 flounder of a certain species have sample mean weight 21.5 grams. Scientists want to perform a hypothesis test to determine how strong the evidence is that the mean weight differs from 20 grams. State the appropriate null and alternate hypotheses.

23. Check, please: A restaurant owner claims that the mean amount spent by diners at his restaurant is more than $30. A test is made of H_0: $\mu = 30$ versus H_1: $\mu > 30$. The null hypothesis is rejected. State an appropriate conclusion.

24. Coffee: The mean caffeine content per cup of regular coffee served at a certain coffee shop is supposed to be 100 milligrams. A test is made of H_0: $\mu = 100$ versus H_1: $\mu \neq 100$. The null hypothesis is rejected. State an appropriate conclusion.

25. Big dogs: A veterinarian claims that the mean weight of adult German shepherd dogs is 75 pounds. A test is made of H_0: $\mu = 75$ versus H_1: $\mu \neq 75$. The null hypothesis is not rejected. State an appropriate conclusion.

26. Business trips: A sales manager believes that the mean number of days per year her company's sales representatives spend traveling is less than 50. A test is made of H_0: $\mu = 50$ versus H_1: $\mu < 50$. The null hypothesis is not rejected. State an appropriate conclusion.

27. Type I error: A company that manufactures steel wires guarantees that the mean breaking strength (in kilonewtons) of the wires is greater than 50. They measure the strengths for a sample of wires and test H_0: $\mu = 50$ versus H_1: $\mu > 50$.
a. If a Type I error is made, what conclusion will be drawn regarding the mean breaking strength?
b. If a Type II error is made, what conclusion will be drawn regarding the mean breaking strength?
c. This test uses a one-tailed alternate hypothesis. Explain why a one-tailed hypothesis is more appropriate than a two-tailed hypothesis in this situation.

28. Type I error: Washers used in a certain application are supposed to have a thickness of 2 millimeters. A quality control engineer measures the thicknesses for a sample of washers and tests H_0: $\mu = 2$ versus H_1: $\mu \neq 2$.
a. If a Type I error is made, what conclusion will be drawn regarding the mean washer thickness?
b. If a Type II error is made, what conclusion will be drawn regarding the mean washer thickness?
c. This test uses a two-tailed hypothesis. Explain why a two-tailed hypothesis is more appropriate than a one-tailed hypothesis in this situation.

29. Scales: It is desired to check the calibration of a scale by weighing a standard 10-gram weight 100 times. Let μ be the population mean reading on the scale, so that the scale is in calibration if $\mu = 10$ and out of calibration if $\mu \neq 10$. A test is made of the hypotheses H_0: $\mu = 10$ versus H_1: $\mu \neq 10$. Consider three possible conclusions: (i) The scale is in calibration. (ii) The scale is not in calibration. (iii) The scale might be in calibration.

a. Which of the three conclusions is best if H_0 is rejected?

b. Which of the three conclusions is best if H_0 is not rejected?

c. Assume that the scale is in calibration, but the conclusion is reached that the scale is not in calibration. Which type of error is this?

d. Assume that the scale is not in calibration. Is it possible to make a Type I error? Explain.

e. Assume that the scale is not in calibration. Is it possible to make a Type II error? Explain.

30. IQ: Scores on a certain IQ test are known to have a mean of 100. A random sample of 60 students attend a series of coaching classes before taking the test. Let μ be the

population mean IQ score that would occur if every student took the coaching classes. The classes are successful if $\mu > 100$. A test is made of the hypotheses H_0: $\mu = 100$ versus H_1: $\mu > 100$. Consider three possible conclusions: (i) The classes are successful. (ii) The classes are not successful. (iii) The classes might not be successful.

a. Which of the three conclusions is best if H_0 is rejected?

b. Which of the three conclusions is best if H_0 is not rejected?

c. Assume that the classes are successful but the conclusion is reached that the classes might not be successful. Which type of error is this?

d. Assume that the classes are not successful. Is it possible to make a Type I error? Explain.

e. Assume that the classes are not successful. Is it possible to make a Type II error? Explain.

Answers to Check Your Understanding Exercises for Section 9.1

1. H_0: $\mu = 35$, H_1: $\mu > 35$

2. H_0: $\mu = 6.6$, H_1: $\mu \neq 6.6$

3. H_0: $\mu = 25.5$, H_1: $\mu < 25.5$

4. Correct decision

5. Type II error

6. Type I error

SECTION 9.2

Hypothesis Tests for a Population Mean, Standard Deviation Known

Objectives

1. Perform hypothesis tests with the critical value method

2. Perform hypothesis tests with the *P*-value method

3. Describe the relationship between hypothesis tests and confidence intervals

4. Describe the relationship between α and the probability of error

5. Report the *P*-value or the test statistic value

6. Distinguish between statistical significance and practical significance

Does coaching improve SAT scores? The College Board reported that the mean math SAT score in 2009 was 515, with a standard deviation of 116. Results of an earlier study (*Preparing for the SAT — An Update*, College Board Report 98–5) suggest that coached students should have a mean SAT score of approximately 530. A teacher who runs an online coaching program thinks that students coached by his method have a higher mean score than this. We will see how to perform a hypothesis test to determine whether the teacher is right.

There are two ways to perform hypothesis tests; both methods produce the same results. The first one we will discuss is called the *critical value method*. Then we will discuss the second method, known as the *P-value method*.

Objective 1 Perform hypothesis tests with the critical value method

The Critical Value Method

In the SAT example, the teacher believes that the mean score for his students is greater than 530. Therefore, the null hypothesis says that the mean μ is equal to 530, and the alternate hypothesis says that μ is greater than 530. In symbols,

$$H_0: \mu = 530 \qquad H_1: \mu > 530$$

Now assume that the teacher draws a random sample of 100 students who are planning to take the SAT, and enrolls them in the online coaching program. After completing the program, their sample mean SAT score is $\bar{x} = 562$. This is higher than 530. Can he reject H_0 and conclude that the mean SAT math score for his students is greater than 530?

Recall: The z-score tells us how many standard deviations $\bar{x}$ is from μ.

Recall: When the sample size is large ($n > 30$), the sample mean $\bar{x}$ is approximately normally distributed with mean μ and standard deviation $\sigma/\sqrt{n}$.

The sample mean, $\bar{x} = 562$, differs somewhat from the null hypothesis value for the population mean, $\mu = 530$. The key idea behind a hypothesis test is to measure how large this difference is. If the sample differs from H_0 only slightly, then H_0 may well be true, because slight differences can easily be due to chance. However, if the difference is larger, it is less likely to be due to chance, and H_0 is less likely to be true.

We must now determine how strong the disagreement is between the sample mean $\bar{x} = 562$ and the null hypothesis $\mu = 530$. We do this by calculating the value of a **test statistic**. In this example, the test statistic is the z-score of the sample mean $\bar{x}$. We now show how to compute the z-score.

Recall that in a hypothesis test, we begin by assuming that H_0 is true. We therefore assume that the mean of $\bar{x}$ is $\mu = 530$. Because the sample size is large ($n = 100$), we know that $\bar{x}$ is approximately normally distributed. Suppose the population standard deviation is known to be $\sigma = 116$. The standard deviation of $\bar{x}$ is

$$\frac{\sigma}{\sqrt{n}} = \frac{116}{\sqrt{100}} = 11.6$$

The z-score for $\bar{x}$ is

$$z = \frac{\bar{x} - \mu}{\sigma/\sqrt{n}} = \frac{562 - 530}{116/\sqrt{100}} = 2.76$$

We have found that the value of the test statistic is $z = 2.76$. Does this present strong evidence against H_0? Figure 9.1 presents the distribution of the sample mean under the assumption that H_0 is true. The value $\bar{x} = 562$ that we observed has a z-score of 2.76, which means that our observed mean is 2.76 standard deviations away from the assumed mean of 530. Visually, we can see from Figure 9.1 that our observed value $\bar{x} = 562$ is pretty far out in the tail of the distribution — far from the null hypothesis value $\mu = 530$. Intuitively, therefore, it appears that the evidence against H_0 is fairly strong.

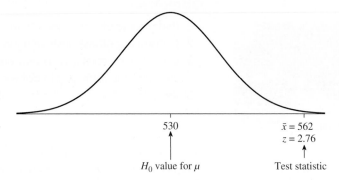

530
H_0 value for μ

$\bar{x} = 562$
$z = 2.76$
Test statistic

Figure 9.1   If H_0 is true, then the value of $\bar{x}$ is in the tail of the distribution, and far from the null hypothesis mean $\mu = 530$. Visually, it appears that the evidence against H_0 is fairly strong.

The critical value method is based on the idea that we should reject H_0 if the value of the test statistic is unusual when we assume H_0 to be true. In this method, we choose a **critical value**, which forms a boundary between values that are considered unusual and values that are not. The region that contains the unusual values is called the **critical region**. If the value of the test statistic is in the critical region, we reject H_0.

The critical value we choose depends on how small we believe a probability should be for an event to be considered unusual. Let's say that an event with a probability of 0.05 or less is unusual. Figure 9.2 illustrates a critical value of 1.645 and a critical region consisting of z-scores greater than or equal to 1.645. The probability that a z-score is in the critical region is 0.05, so the critical region contains the z-scores that are considered unusual. We have observed a z-score of 2.76, which is in the critical region. Therefore, we reject H_0. We conclude that the mean SAT math score for students completing the online coaching program is greater than 530.

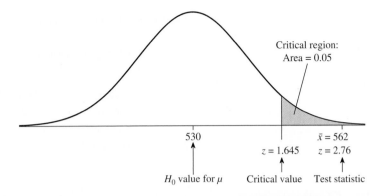

Figure 9.2 The critical value is 1.645. The critical region contains all z-scores greater than or equal to 1.645. The value of our test statistic is $z = 2.76$. This value is in the critical region, so we reject H_0.

The probability that we use to determine whether an event is unusual is called the **significance level** of the test, and is denoted with the letter α. In Figure 9.2, we used $\alpha = 0.05$. This is the most commonly used value for α, but other values are sometimes used as well. Next to $\alpha = 0.05$, the most commonly used value is $\alpha = 0.01$.

The choice of α is determined by how strong we require the evidence against H_0 to be in order to reject it. The smaller the value of α, the stronger we require the evidence to be. For example, if we choose $\alpha = 0.05$, we will reject H_0 if the test statistic is in the most extreme 5% of its distribution. However, if we choose $\alpha = 0.01$, we will not reject H_0 unless the test statistic is in the most extreme 1% of its distribution.

DEFINITION

If we reject H_0 after choosing a significance level α, we say that the result is **statistically significant** at the α level.

We also say that H_0 *is rejected at the α level.*

In our SAT example, we rejected H_0 at the $\alpha = 0.05$ level, and the result was statistically significant at the $\alpha = 0.05$ level.

Our alternate hypothesis of $\mu > 530$ was a right-tailed alternative. For this reason, the critical region was in the right tail of the distribution. The location of the critical region depends on whether the alternate hypothesis is left-tailed, right-tailed, or two-tailed.

Critical Values for Hypothesis Tests

Let α denote the chosen significance level. The critical value depends on whether the alternate hypothesis is left-tailed, right-tailed, or two-tailed.

For left-tailed H_1: The critical value is $-z_\alpha$, which has area α to its left. Reject H_0 if $z \le -z_\alpha$.

For right-tailed H_1: The critical value is z_α, which has area α to its right. Reject H_0 if $z \ge z_\alpha$.

For two-tailed H_1: The critical values are $z_{\alpha/2}$, which has area $\alpha/2$ to its right, and $-z_{\alpha/2}$, which has area $\alpha/2$ to its left. Reject H_0 if $z \ge z_{\alpha/2}$ or $z \le -z_{\alpha/2}$.

Table 9.1 presents critical values for some commonly used significance levels α.

Table 9.1 Table of Critical Values

H_1	Significance Level α			
	0.10	**0.05**	**0.02**	**0.01**
Left-tailed	−1.282	−1.645	−2.054	−2.326
Right-tailed	1.282	1.645	2.054	2.326
Two-tailed	±1.645	±1.96	±2.326	±2.576

EXAMPLE 9.7   Find the critical region for a right-tailed alternate hypothesis

A test is made of H_0: $\mu = 1$ versus H_1: $\mu > 1$. The value of the test statistic is $z = 1.85$.

    a. Is H_0 rejected at the $\alpha = 0.05$ level?

    b. Is H_0 rejected at the $\alpha = 0.01$ level?

Solution

The alternate hypothesis is H_1: $\mu > 1$, so this is a right-tailed test.

    a. From Table 9.1, we see that the critical value for $\alpha = 0.05$ is $z_\alpha = 1.645$. For a right-tailed test, we reject H_0 if $z \geq z_\alpha$. Because $1.85 > 1.645$, we reject H_0 at the $\alpha = 0.05$ level.

    b. The critical value for $\alpha = 0.01$ is 2.326. Because $1.85 < 2.326$, we do not reject H_0 at the $\alpha = 0.01$ level.

Check Your Understanding

  1. A test is made of H_0: $\mu = 25$ versus H_1: $\mu < 25$. The value of the test statistic is $z = -1.84$.

    a. Find the critical value and the critical region for a significance level of $\alpha = 0.05$.

    b. Do you reject H_0 at the $\alpha = 0.05$ level?

    c. Find the critical value and the critical region for a significance level of $\alpha = 0.01$.

    d. Do you reject H_0 at the $\alpha = 0.01$ level?

  2. A test is made of H_0: $\mu = 7.5$ versus H_1: $\mu > 7.5$. The value of the test statistic is $z = 2.71$.

    a. Find the critical value and the critical region for a significance level of $\alpha = 0.05$.

    b. Do you reject H_0 at the $\alpha = 0.05$ level?

    c. Find the critical value and the critical region for a significance level of $\alpha = 0.01$.

    d. Do you reject H_0 at the $\alpha = 0.01$ level?

  3. A test is made of H_0: $\mu = 12$ versus H_1: $\mu \neq 12$. The value of the test statistic is $z = 1.78$.

    a. Find the critical value and the critical region for a significance level of $\alpha = 0.05$.

    b. Do you reject H_0 at the $\alpha = 0.05$ level?

    c. Find the critical value and the critical region for a significance level of $\alpha = 0.01$.

    d. Do you reject H_0 at the $\alpha = 0.01$ level?

Answers are on page 432.

The method we have described requires certain assumptions, which we now state.

Assumptions for Performing a Hypothesis Test About μ When σ Is Known

1. We have a simple random sample.
2. The sample size is large ($n > 30$), or the population is approximately normal.

When these assumptions are met, a hypothesis test can be performed using the following steps.

Performing a Hypothesis Test for a Population Mean with σ Known Using the Critical Value Method

Check to be sure the assumptions are satisfied. If they are, then proceed with the following steps.

Step 1: State the null and alternate hypotheses. The null hypothesis specifies a value for the population mean μ. We will call this value μ_0. So the null hypothesis is of the form $H_0: \mu = \mu_0$. The alternate hypothesis can be stated in one of three ways:

Left-tailed: $H_1: \mu < \mu_0$
Right-tailed: $H_1: \mu > \mu_0$
Two-tailed: $H_1: \mu \neq \mu_0$

Step 2: Choose a significance level α and find the critical value or values.

Step 3: Compute the test statistic $z = \dfrac{\bar{x} - \mu_0}{\sigma/\sqrt{n}}$.

Step 4: Determine whether to reject H_0, as follows:

Left-tailed: $H_1: \mu < \mu_0$ Reject if $z \leq -z_\alpha$.
Right-tailed: $H_1: \mu > \mu_0$ Reject if $z \geq z_\alpha$.
Two-tailed: $H_1: \mu \neq \mu_0$ Reject if $z \geq z_{\alpha/2}$ or $z \leq -z_{\alpha/2}$.

Step 5: State a conclusion.

EXAMPLE 9.8   Performing a hypothesis test with the critical value method

The American Automobile Association reported that the mean price of a gallon of regular grade gasoline in the city of Los Angeles in July 2013 was \$4.04. A recently taken simple random sample of 50 gas stations in Los Angeles had an average price of \$3.99 for a gallon of regular grade gasoline. Assume that the standard deviation of prices is \$0.15. An economist is interested in determining whether the mean price is less than \$3.99. Use the critical value method to perform a hypothesis test at the $\alpha = 0.05$ level of significance.

Solution

We first check the assumptions. We have a simple random sample, the sample size is large ($n > 30$), and the population standard deviation σ is known. The assumptions are satisfied.

Step 1: State H_0 and H_1. The null hypothesis says that the mean price is \$4.04. Therefore, we have

$$H_0: \mu = 4.04$$

We are interested in knowing whether the mean price is less than \$4.04. Therefore, the alternate hypothesis is

$$H_1: \mu < 4.04$$

At this point, we assume H_0 to be true.

Step 2: Choose a significance level and find the critical value. The significance level is $\alpha = 0.05$. Since the alternate hypothesis is $\mu < 4.04$, this is a left-tailed test. The critical value corresponding to $\alpha = 0.05$ is -1.645.

Step 3: Compute the test statistic. The test statistic is the z-score of the sample mean $\bar{x}$. The population standard deviation is $\sigma = 0.15$. Since we assume H_0 to be true, the population mean is $\mu_0 = 4.04$. The sample size is $n = 50$. Therefore, the test statistic is

$$z = \frac{\bar{x} - \mu_0}{\sigma/\sqrt{n}} = \frac{3.99 - 4.04}{0.15/\sqrt{50}} = -2.36$$

Step 4: Determine whether to reject H_0. This is a left-tailed test, so we reject H_0 if $z < -1.645$. Since $-2.36 < -1.645$, we reject H_0 at the $\alpha = 0.05$ level. See Figure 9.3.

Critical region:
Area = 0.05

$z = -2.36$ -1.645
Critical value

Figure 9.3   The value of the test statistic, $z = -2.36$, is in the level $\alpha = 0.05$ critical region. Therefore, we reject H_0 at the $\alpha = 0.05$ level.

Step 5: State a conclusion. We conclude that the mean price of a gallon of regular gasoline in Los Angeles is less than \$4.04.

Check Your Understanding

4. A test is made of H_0: $\mu = 15$ versus H_1: $\mu > 15$. The sample mean is $\bar{x} = 16.5$, the sample size is $n = 50$, and the population standard deviation is $\sigma = 5$.
   a. Find the value of the test statistic z.
   b. Find the critical region for a level $\alpha = 0.05$ test.
   c. Do you reject H_0 at the $\alpha = 0.05$ level?

5. A test is made of H_0: $\mu = 125$ versus H_1: $\mu < 125$. The sample mean is $\bar{x} = 123$, the sample size is $n = 100$, and the population standard deviation is $\sigma = 20$.
   a. Find the value of the test statistic z.
   b. Find the critical region for a level $\alpha = 0.02$ test.
   c. Do you reject H_0 at the $\alpha = 0.02$ level?

6. A test is made of H_0: $\mu = 100$ versus H_1: $\mu \neq 100$. The sample mean is $\bar{x} = 97$, the sample size is $n = 75$, and the population standard deviation is $\sigma = 8$.
   a. Find the value of the test statistic z.
   b. Find the critical region for a level $\alpha = 0.01$ test.
   c. Do you reject H_0 at the $\alpha = 0.01$ level?

Answers are on page 432.

With the critical value method, the value of the test statistic is considered to be unusual if it is in the critical region, and not unusual if it is not in the critical region. We will now describe the **P-value method**, which provides more information than the critical value method. Whereas the critical value method tells us only whether the test statistic was unusual or not, the P-value method tells us exactly how unusual the test statistic is. For this reason, the P-value method is the one more often used in practice. In particular, almost all forms of technology use the P-value method.

Objective 2 Perform
hypothesis tests with the
P-value method

The *P*-Value Method

We will introduce the *P*-value method with our SAT example. An online coaching program is supposed to increase the mean SAT math score to a value greater than 530. The null and alternate hypotheses are

$$H_0: \mu = 530 \qquad H_1: \mu > 530$$

Now assume that 100 students are randomly chosen to participate in the program, and their sample mean score is $\bar{x} = 562$. Suppose that the population standard deviation for SAT math scores is known to be $\sigma = 116$. Does this provide strong evidence against the null hypothesis $\mu = 530$?

To measure just how strong the evidence against H_0 is, we compute a quantity called the **P-value**. The *P*-value is the probability that a number drawn from the distribution of the sample mean would be as extreme as or more extreme than our observed value of 562. The more extreme the value, the stronger is the evidence against H_0. Because our alternate hypothesis is $H_1: \mu > 530$, this is a right-tailed test, so values of $\bar{x}$ greater than 562 are more extreme than our observed value is. We find the *P*-value by computing the z-score of our observed sample mean $\bar{x} = 562$. We now explain how to do this.

Recall that we begin by assuming that H_0 is true. We therefore assume that the mean of $\bar{x}$ is $\mu = 530$. The sample size is large ($n = 100$), so we know that $\bar{x}$ is approximately normally distributed. The standard deviation of $\bar{x}$ is

$$\frac{\sigma}{\sqrt{n}} = \frac{116}{\sqrt{100}} = 11.6$$

Recall: When the sample size is large ($n > 30$), the sample mean $\bar{x}$ is approximately normally distributed with mean μ and standard deviation $\sigma/\sqrt{n}$.

Therefore, the *P*-value is the probability that $\bar{x}$ is greater than 562 when μ is assumed to be 530 and the standard deviation is 11.6.

The z-score for $\bar{x}$ is

$$z = \frac{\bar{x} - \mu}{\sigma/\sqrt{n}} = \frac{562 - 530}{116/\sqrt{100}} = 2.76$$

The *P*-value is therefore the area under the normal curve to the right of $z = 2.76$. Using Table A.2, we see that the area to the *left* of $z = 2.76$ is 0.9971. Therefore, the area to the right of $z = 2.76$ is $1 - 0.9971 = 0.0029$ (see Figure 9.4). Therefore, the *P*-value for this test is 0.0029.

Explain It Again

Using technology: The *P*-value is the area to the right of $\bar{x} = 562$ when the mean is 530 and the standard deviation is 11.6. This area can be found with technology. The following display illustrates the **normalcdf** command on the TI-84 Plus calculator.

```
normalcdf(562,1E
99,530,11.6)
        .0029023496
```

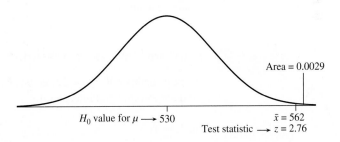

Area = 0.0029

H_0 value for $\mu \longrightarrow$ 530

$\bar{x} = 562$
Test statistic $\longrightarrow z = 2.76$

Figure 9.4 If H_0 is true, the probability that $\bar{x}$ takes on a value as extreme as or more extreme than the observed value of 562 is 0.0029. This is the *P*-value.

This *P*-value tells us that if H_0 were true, the probability of observing a value of $\bar{x}$ as large as 562 is only 0.0029. Therefore, there are only two possibilities:

- H_0 is false.
- H_0 is true, and we got an unusual sample, whose mean lies in the most extreme 0.0029 of its distribution.

In practice, events in the most extreme 0.0029 of their distributions are very unusual. This means that a *P*-value as small as 0.0029 is very unlikely to occur if H_0 is true. A *P*-value of 0.0029 is very strong evidence against H_0.

EXAMPLE 9.9 Find and interpret a *P*-value

A test is made of H_0: $\mu = 10$ versus H_1: $\mu > 10$. The value of the test statistic is $z = 2.25$. Find the *P*-value and interpret it.

Solution

The alternate hypothesis is H_1: $\mu > 10$, so this is a right-tailed test. Therefore, values of z greater than our observed value of 2.25 are more extreme than our value is. The *P*-value is the area under the normal curve to the right of the test statistic $z = 2.25$. Using Table A.2, we see that the area to the *left* of $z = 2.25$ is 0.9878. Therefore, the area to the right of $z = 2.25$ is $1 - 0.9878 = 0.0122$. See Figure 9.5.

The *P*-value of 0.0122 tells us that if H_0 is true, then the probability of observing a test statistic of 2.25 or more is only 0.0122. This result is fairly unusual if we assume H_0 to be true. Therefore, this is fairly strong evidence against H_0.

Area = 0.0122

$z = 2.25$

Figure 9.5

EXAMPLE 9.10 Find and interpret a *P*-value

A test is made of H_0: $\mu = 5$ versus H_1: $\mu < 5$. The value of the test statistic is $z = -0.63$. Find the *P*-value and interpret it.

Solution

The alternate hypothesis is H_1: $\mu < 5$, so this is a left-tailed test. Therefore, values of z less than our value of -0.63 are more extreme than our value is. The *P*-value is the area under the normal curve to the left of the test statistic $z = -0.63$. Using Table A.2, we see that the area to the left of $z = -0.63$ is 0.2643. See Figure 9.6.

The *P*-value of 0.2643 tells us that if H_0 is true, then the probability of observing a test statistic of -0.63 or less is 0.2643. This is not particularly unusual, so this is not strong evidence against H_0.

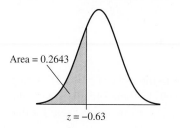

Area = 0.2643

$z = -0.63$

Figure 9.6

EXAMPLE 9.11 Find and interpret a *P*-value

A test is made of H_0: $\mu = 20$ versus H_1: $\mu \neq 20$. The value of the test statistic is $z = -2.70$. Find the *P*-value and interpret it.

Solution

The alternate hypothesis is H_1: $\mu \neq 20$, so this is a two-tailed test. Therefore, values of z less than our value of -2.70 and values greater than 2.70 are both more extreme than our value is. The *P*-value is the sum of the areas under the normal curve to the right of $z = 2.70$ and to the left of $z = -2.70$. Using Table A.2, we see that the area to the left of $z = -2.70$ is 0.0035. The area to the right of $z = 2.70$ is also 0.0035. The sum of the areas is therefore $0.0035 + 0.0035 = 0.0070$. See Figure 9.7.

The *P*-value of 0.0070 tells us that if H_0 is true, then the probability of observing a test statistic greater than 2.70 or less than -2.70 is only 0.0070. This result is quite unusual if we assume H_0 to be true. Therefore, this is very strong evidence against H_0.

Area = 0.0035   Area = 0.0035

$z = -2.70$ $z = 2.70$

Figure 9.7

Check Your Understanding

7. Which provides stronger evidence against H_0: a P-value of 0.05 or a P-value of 0.50?

8. A test is made of H_0: $\mu = 30$ versus H_1: $\mu < 30$. The test statistic is $z = -1.28$. Find and interpret the P-value.

9. A test is made of H_0: $\mu = 6$ versus H_1: $\mu \neq 6$.
   a. The test statistic is $z = 0.75$. Find and interpret the P-value.
   b. The test statistic is $z = -2.20$. Find and interpret the P-value.
   c. Which provides stronger evidence against H_0: $z = 0.75$ or $z = -2.20$?

Answers are on page 432.

The *P*-value is not the probability that H_0 is true

Because the P-value is a probability and small P-values indicate that H_0 should be rejected, it is tempting to think that the P-value represents the probability that H_0 is true. This is not the case. The P-value is the probability that a test statistic such as z would take on an extreme value. Probability is used for events that can be different for different samples. Therefore, it makes sense to talk about the probability that the value of z will be more extreme than an observed value, because the value of z can come out differently for different samples. The null hypothesis, however, is either true or not true. The truth of H_0 does not change from sample to sample. For this reason, it does not make sense to talk about the probability that H_0 is true.

SUMMARY

The P-value is the probability, under the assumption that H_0 is true, that the test statistic takes on a value as extreme as or more extreme than the value actually observed.

The P-value is not the probability that the null hypothesis is true.

Check Your Understanding

10. If $P = 0.02$, which is the best conclusion?
   i. The probability that H_0 is true is 0.02.
   ii. If H_0 is true, the probability of obtaining a test statistic more extreme than the one actually observed is 0.02.
   iii. The probability that H_1 is true is 0.02.
   iv. If H_1 is true, the probability of obtaining a test statistic more extreme than the one actually observed is 0.02.

Answer is on page 433.

Choosing a significance level

We have seen that the smaller the P-value, the stronger the evidence against H_0. In practice, people often do not choose a significance level. They simply report the P-value and let the reader decide whether the evidence is strong enough to reject H_0. Sometimes, however, we need to make a firm decision whether to reject H_0. We then choose a significance level α between 0 and 1 before performing the test, and reject H_0 if the P-value is less than or equal to α. The most commonly used value is $\alpha = 0.05$, but other values are sometimes used as well. Next to $\alpha = 0.05$, the most commonly used value is $\alpha = 0.01$.

SUMMARY

To make a decision whether to reject H_0 when using the P-value method:

- Choose a significance level α between 0 and 1.
- Compute the P-value.
- If $P \leq \alpha$, reject H_0. If $P > \alpha$, do not reject H_0.

If $P \leq \alpha$, we say that H_0 is rejected at the α level, or that the result is statistically significant at the α level.

EXAMPLE 9.12 Find the *P*-value

In Example 9.9, the P-value was $P = 0.0122$.

    a. Do you reject H_0 at the $\alpha = 0.05$ level?
    b. Do you reject H_0 at the $\alpha = 0.01$ level?
    c. Is the result statistically significant at the $\alpha = 0.05$ level?
    d. Is the result statistically significant at the $\alpha = 0.01$ level?

Solution

    a. Because $P \leq 0.05$, we reject H_0 at the $\alpha = 0.05$ level.
    b. Because $P > 0.01$, we do not reject H_0 at the $\alpha = 0.01$ level.
    c. We reject H_0 at the $\alpha = 0.05$ level, so the result is statistically significant at the $\alpha = 0.05$ level.
    d. We do not reject H_0 at the $\alpha = 0.01$ level, so the result is not statistically significant at the $\alpha = 0.01$ level.

Check Your Understanding

11. A hypothesis test is performed with a significance level of $\alpha = 0.05$.
    a. If the P-value is 0.08, is H_0 rejected?
    b. If the P-value is 0.08, are the results statistically significant at the 0.05 level?
    c. If the P-value is 0.03, is H_0 rejected?
    d. If the P-value is 0.03, are the results statistically significant at the 0.05 level?

12. For each of the following P-values, state whether H_0 will be rejected at the 0.10 level.
    a. $P = 0.12$
    b. $P = 0.07$
    c. $P = 0.05$
    d. $P = 0.20$

13. For each of the following P-values, state whether the result is statistically significant at the 0.10 level.
    a. $P = 0.08$
    b. $P = 0.15$
    c. $P = 0.01$
    d. $P = 0.50$

Answers are on page 433.

The assumptions for using the P-value method are the same as for the critical value method. We repeat these assumptions here.

Assumptions for Performing a Hypothesis Test About μ When σ Is Known

1. We have a simple random sample.
2. The sample size is large ($n > 30$), or the population is approximately normal.

We now summarize the steps in testing a hypothesis with the P-value method.

Performing a Hypothesis Test for a Population Mean with σ Known Using the P-Value Method

Check to be sure the assumptions are satisfied. If they are, then proceed with the following steps.

Step 1: State the null and alternate hypotheses. The null hypothesis specifies a value for the population mean μ. We will call this value μ_0. So the null hypothesis is of the form $H_0: \mu = \mu_0$. The alternate hypothesis can be stated in one of three ways:

Left-tailed: $H_1: \mu < \mu_0$
Right-tailed: $H_1: \mu > \mu_0$
Two-tailed: $H_1: \mu \neq \mu_0$

Step 2: If making a decision, choose a significance level α.

Step 3: Compute the test statistic $z = \dfrac{\bar{x} - \mu_0}{\sigma/\sqrt{n}}$.

Step 4: Compute the P-value of the test statistic. The P-value is the probability, assuming that H_0 is true, of observing a value for the test statistic that is as extreme or more extreme than the value actually observed. The P-value is an area under the standard normal curve; it depends on the type of alternate hypothesis. Note that the inequality in the alternate hypothesis points in the direction of the tail that contains the area for the P-value.

The P-value is the area to the left of z.

Left-tailed: $H_1: \mu < \mu_0$

The P-value is the area to the right of z.

Right-tailed: $H_1: \mu > \mu_0$

The P-value is the sum of the areas in the two tails.

Two-tailed: $H_1: \mu \neq \mu_0$

Step 5: Interpret the P-value. If making a decision, reject H_0 if the P-value is less than or equal to the significance level α.

Step 6: State a conclusion.

EXAMPLE 9.13   Perform a hypothesis test

The National Health and Nutrition Examination Surveys (NHANES) are designed to assess the health and nutritional status of adults and children in the United States. According to a recent NHANES survey, the mean height of adult men in the United States is 69.7 inches, with a standard deviation of 3 inches. A sociologist believes that taller men may be more likely to be promoted to positions of leadership, so the mean height μ of male business executives may be greater than the mean height of the entire male population. A simple random sample of 100 male business executives has a mean height of 69.9 in. Assume that the standard deviation of male executive heights is $\sigma = 3$ inches. Can we conclude that male business executives are taller, on the average, than the general male population at the $\alpha = 0.05$ level?

Solution

We first check the assumptions. We have a simple random sample, the sample size is large ($n > 30$), and the population standard deviation is known. The assumptions are satisfied.

Step 1: State H_0 and H_1. The null hypothesis, H_0, says that there is no difference between the mean heights of executives and others. Therefore, we have

$$H_0: \mu = 69.7$$

We are interested in determining whether the mean height of executives is greater than 69.7. Therefore, we have

$$H_1: \mu > 69.7$$

At this point, we assume that H_0 is true.

Step 2: Choose a level of significance. The level of significance is $\alpha = 0.05$.

Step 3: Compute the test statistic. Because the sample size is large ($n = 100$), the sample mean $\bar{x}$ is approximately normally distributed. The test statistic is the z-score for $\bar{x}$. To find the z-score, we first need to find the mean and standard deviation of $\bar{x}$. Because we are assuming that H_0 is true, we assume that the mean of $\bar{x}$ is $\mu = 69.7$. We know that the population standard deviation is $\sigma = 3$. The standard deviation of $\bar{x}$ is therefore

$$\frac{\sigma}{\sqrt{n}} = \frac{3}{\sqrt{100}} = 0.3$$

It follows that $\bar{x}$ is normally distributed with mean 69.7 and standard deviation 0.3. We observed a value of $\bar{x} = 69.9$. The z-score is

$$z = \frac{\bar{x} - \mu_0}{\sigma/\sqrt{n}} = \frac{69.9 - 69.7}{0.3} = 0.67$$

Step 4: Compute the P-value. Since the alternate hypothesis is $\mu > 69.7$, this is a right-tailed test. The P-value is the area under the curve to the right of $z = 0.67$. Using Table A.2, we see that the area to the *left* of $z = 0.67$ is 0.7486. Therefore, the area to the right of $z = 0.67$ is $1 - 0.7486 = 0.2514$. The P-value is 0.2514. See Figure 9.8.

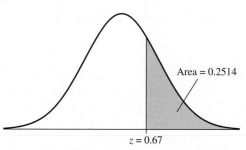

$z = 0.67$

Figure 9.8

Step 5: Interpret the P-value. The P-value of 0.2514 says that if H_0 is true, the probability of observing a test statistic as large or larger than 0.67 is 0.2514. This is not unusual; it will happen for about one out of every four samples. Therefore, this sample does not strongly disagree with H_0. In particular, $P > 0.05$, so we do not reject the null hypothesis at the $\alpha = 0.05$ level.

Step 6: State a conclusion. There is not enough evidence to conclude that male executives have a greater mean height than adult males in general. The mean height of male executives may be the same as the mean height of adult males in general.

EXAMPLE 9.14   Perform a two-tailed hypothesis test

At a large company, the attitudes of workers are regularly measured with a standardized test. The scores on the test range from 0 to 100, with higher scores indicating greater satisfaction with their jobs. The mean score over all of the company's employees was 74, with a standard deviation of $\sigma = 8$. Some time ago, the company adopted a policy of telecommuting. Under this policy, workers could spend one day per week working from home. After the policy had been in place for some time, a random sample of 80 workers was given the test to determine whether their mean level of satisfaction had changed since the policy was put into effect. The sample mean was 76. Assume the standard deviation is still $\sigma = 8$. Can we conclude that the mean level of satisfaction is different since the policy change at the $\alpha = 0.05$ level?

Solution

We first check the assumptions. We have a simple random sample, the sample size is large ($n > 30$), and the population standard deviation is known. The assumptions are satisfied.

Step 1: State H_0 and H_1. The null hypothesis, H_0, says that there is no difference between the mean level of satisfaction before and after telecommuting. Therefore, we have

$$H_0: \mu = 74$$

We are interested in knowing whether the mean level has changed. We are not specifically interested in whether it went up or down. Therefore, the alternate hypothesis is

$$H_1: \mu \neq 74$$

At this point, we assume that H_0 is true.

Step 2: Choose a level of significance. The level of significance is $\alpha = 0.05$.

Step 3: Compute the test statistic. Since the sample size, $n = 80$, is large, $\bar{x}$ is approximately normally distributed. The test statistic is the z-score for the sample mean $\bar{x}$. The population standard deviation is $\sigma = 8$. Because we assume H_0 to be true, the population mean is $\mu = 74$. Therefore, $\bar{x}$ is normally distributed with mean 74 and standard error

$$\frac{\sigma}{\sqrt{n}} = \frac{8}{\sqrt{80}} = 0.8944$$

We observed a value of $\bar{x} = 76$. The z-score is

$$z = \frac{76 - 74}{0.8944} = 2.24$$

Step 4: Compute the P-value. The alternate hypothesis is $\mu \neq 74$, so this is a two-tailed test. The P-value is thus the sum of two areas: the area to the right of $z = 2.24$ and an equal area to the left of $z = -2.24$. Using Table A.2, we see that the area to the left of $z = -2.24$ is 0.0125. Therefore, the area to the right of $z = 2.24$ is also 0.0125. The P-value is therefore $0.0125 + 0.0125 = 0.0250$. See Figure 9.9.

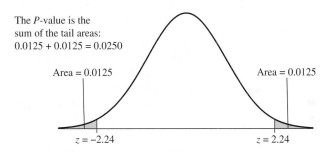

The P-value is the sum of the tail areas: $0.0125 + 0.0125 = 0.0250$

Area = 0.0125   Area = 0.0125

$z = -2.24$ $z = 2.24$

Figure 9.9

Step 5: Interpret the P-value. The P-value says that if H_0 is true, then the probability of observing a test statistic as extreme as the one we actually observed is only 0.0250. In practice, this would generally be considered fairly strong evidence against H_0. In particular, $P < 0.05$, so we reject H_0 at the $\alpha = 0.05$ level.

Step 6: State a conclusion. We conclude that the mean score among employees has changed since the adoption of telecommuting.

Check Your Understanding

14. A social scientist suspects that the mean number of years of education μ for adults in a certain large city is greater than 12 years. She will test the null hypothesis $H_0: \mu = 12$ against the alternate hypothesis $H_1: \mu > 12$. She surveys a random sample of 100 adults and finds that the sample mean number of years is $\bar{x} = 12.98$. Assume that the standard deviation for the number of years of education is $\sigma = 3$ years.
   a. Compute the value of the test statistic.
   b. Compute the P-value.

c. Interpret the *P*-value.
d. Is H_0 rejected at the $\alpha = 0.05$ level?
e. Is H_0 rejected at the $\alpha = 0.01$ level?

Answers are on page 433.

Performing hypothesis tests with technology

Following are the results of Example 9.14, as presented by the TI-84 Plus calculator.

Z-Test
μ≠74
z=2.236067977
p=.0253472347
x̄=76
n=80

The first line presents the alternate hypothesis, $\mu \neq 74$. Following that are the test statistic (z), the *P*-value (p), the sample mean (x̄), and the sample size (n). Note that the *P*-value differs slightly from the value obtained in Example 9.14 by using Table A.2. This is common. Results given by technology are more precise, and therefore often differ slightly from results obtained from tables. The differences are never large enough to matter.

Following are the results of Example 9.14 as presented by MINITAB.

```
Test of mu = 74.0 vs not = 74.0
The assumed standard deviation = 8.0
```

N	Mean	SE Mean	95% CI	Z	P
80	76.00	0.8944	(74.247, 77.753)	2.236	0.025

The second line of the output presents both the null and alternate hypotheses. The quantity labeled "SE Mean" is the standard error of the mean, $\sigma/\sqrt{n}$, which is the standard deviation of x̄. Notice that MINITAB provides a 95% confidence interval for μ along with the hypothesis test.

Step-by-step instructions for performing hypothesis tests with technology are presented in the Using Technology section on page 427.

Check Your Understanding

15. The following display from a TI-84 Plus calculator presents the results of a hypothesis test for a population mean.

a. What are the null and alternate hypotheses?
b. What is the value of the test statistic?
c. What is the *P*-value?
d. Do you reject H_0 at the $\alpha = 0.05$ level?

16. The following output from MINITAB presents the results of a hypothesis test for a population mean.

```
Test of mu = 53.5 vs not = 53.5
The assumed standard deviation = 2.3634

   N    Mean   SE Mean       95% CI          Z      P
 145   53.246   0.1962   (52.861, 53.631)  -1.29   0.196
```

   a. What are the null and alternate hypotheses?
   b. What is the value of the test statistic?
   c. What is the P-value?
   d. Do you reject H_0 at the $\alpha = 0.05$ level?

Answers are on page 433.

Objective 3 Describe the relationship between hypothesis tests and confidence intervals

The Relationship Between Hypothesis Tests and Confidence Intervals

In Example 9.14, we tested the hypotheses H_0: $\mu = 74$ versus H_1: $\mu \neq 74$ and obtained a P-value of 0.025. Because $P < 0.05$, H_0 is rejected at the 0.05 level. Informally, this says that the value 74 is not plausible for μ.

Another way to express information about μ is through a confidence interval. A 95% confidence interval for μ is $74.247 < \mu < 77.753$. (This confidence interval is displayed in the MINITAB output following Example 9.14.) Note that the 95% confidence interval does not contain the null hypothesis value of 74. In this way, the 95% confidence interval agrees with the results of the hypothesis test. Informally, a confidence interval for μ contains all the values that are plausible for μ. Because 74 is not in the confidence interval, 74 is not a plausible value for μ.

This relationship holds for any confidence interval for a population mean, and any two-tailed hypothesis test.

> If we test H_0: $\mu = \mu_0$ versus H_1: $\mu \neq \mu_0$, then
>
> - If the 95% confidence interval contains μ_0, then H_0 will not be rejected at the 0.05 level.
> - If the 95% confidence interval does not contain μ_0, then H_0 will be rejected at the 0.05 level.

This relationship between hypothesis tests and confidence intervals holds exactly for population means, but only approximately for other parameters such as population proportions. The reason is that the standard error that is used in a hypothesis test for a proportion differs somewhat from the standard error that is used in a confidence interval for a proportion.

Hypothesis tests and confidence intervals address different questions

Although hypothesis tests are closely related to confidence intervals, the two address different questions. A confidence interval provides all of the values that are plausible at a specified level. A hypothesis test tells us about only one value, but it tells us much more precisely how plausible that one value is.

For example, consider the 95% confidence interval $74.247 < \mu < 77.753$ previously mentioned for the mean satisfaction level in Example 9.14. The value $\mu = 74$ is not in the confidence interval, so we can conclude that the hypothesis H_0: $\mu = 74$ will be rejected at the $\alpha = 0.05$ level with a two-tailed test. However, this tells us only that $P < 0.05$. It does not tell us exactly how much less than 0.05 the P-value is. By performing the hypothesis test, we find that $P = 0.025$. This tells us much more precisely just how plausible or implausible the value of 74 is for μ.

SUMMARY

- A confidence interval contains all the values that are plausible at a particular level.
- A hypothesis test tells us precisely how plausible a particular value is.

Check Your Understanding

17. A 95% confidence interval for μ is computed to be $(1.75, 3.25)$. For each of the following hypotheses, state whether H_0 will be rejected at the 0.05 level.
  a. H_0: $\mu = 3$ versus H_1: $\mu \neq 3$
  b. H_0: $\mu = 4$ versus H_1: $\mu \neq 4$
  c. H_0: $\mu = 1.7$ versus H_1: $\mu \neq 1.7$
  d. H_0: $\mu = 3.5$ versus H_1: $\mu \neq 3.5$

18. You want to test H_0: $\mu = 4$ versus H_1: $\mu \neq 4$, so you compute a 95% confidence interval for μ. The 95% confidence interval is $5.1 < \mu < 7.2$.
  a. Do you reject H_0 at the $\alpha = 0.05$ level?
  b. Your friend thinks that $\alpha = 0.01$ is a more appropriate significance level. Can you tell from the confidence interval whether to reject at this level?

Answers are on page 433.

Objective 4 Describe the relationship between α and the probability of error

The Relationship Between α and the Probability of an Error

Recall that a Type I error occurs if we reject H_0 when it is true, and a Type II error occurs if we do not reject H_0 when it is false (see Table 9.2). When designing a hypothesis test, we would like to make the probabilities of these two errors small. In order to do this, we need to know how to calculate the probabilities of these errors. It is straightforward to find the probability of a Type I error: It is equal to the significance level. So, for example, if we perform a test at a significance level of $\alpha = 0.05$, the probability of a Type I error is 0.05.

Table 9.2

	H_0 true	H_0 false
Reject H_0	Type I error	Correct
Don't reject H_0	Correct	Type II error

SUMMARY

When a test is performed with a significance level α, the probability of a Type I error is α.

The probability of a Type II error is denoted by the letter β. Computing the probability of a Type II error is more difficult than finding the probability of a Type I error. A Type II error occurs when H_0 is false, and a decision is made not to reject. The probability of a Type II error depends on the true value of the parameter being tested. We will learn how to compute these probabilities in Section 9.7.

Because α is the probability of a Type I error, why don't we always choose a very small value for α? The reason is that the smaller a value we choose for α, the larger the value of β, the probability of making a Type II error, becomes (unless we increase the sample size).

SUMMARY

The smaller a value we choose for the significance level α:

- The smaller the probability of a Type I error becomes.
- The larger the probability of a Type II error becomes.

In general, making a Type I error is more serious than making a Type II error. When a Type I error is much more serious, a smaller value of α is appropriate. When a Type I error is only slightly more serious, a larger value of α can be justified.

Check Your Understanding

19. A hypothesis test is performed at a significance level $\alpha = 0.05$. What is the probability of a Type I error?

20. Charlie will perform a hypothesis test at the $\alpha = 0.05$ level. Felice will perform the same test at the $\alpha = 0.01$ level.
   a. If H_0 is true, who has a greater probability of making a Type I error?
   b. If H_0 is false, who has a greater probability of making a Type II error?

Answers are on page 433.

Objective 5 Report the *P*-value or the test statistic value

Report the *P*-Value or the Test Statistic Value

Sometimes people report only that a test result was statistically significant at a certain level, without giving the *P*-value. It is common, for example, to read that a result was "statistically significant at the 0.05 level" or "statistically significant ($P \leq 0.05$)." It is much better to report the *P*-value along with the decision whether to reject. There are two reasons for this.

The first reason is that there is a big difference between a *P*-value that is just barely small enough to reject, say $P = 0.049$, and a *P*-value that is extremely small, say $P = 0.0001$. If $P = 0.049$, the evidence is just barely strong enough to reject H_0 at the $\alpha = 0.05$ level, whereas if $P = 0.0001$, the evidence against H_0 is overwhelming. Thus, reporting the *P*-value describes exactly how strong the evidence against H_0 is.

The second reason is that not everyone may agree with your choice of α. For example, let's say you have chosen a significance level of $\alpha = 0.05$. You obtain a *P*-value of $P = 0.03$. Since $P < 0.05$, you reject H_0. Let's say that you report only that H_0 is rejected at the $\alpha = 0.05$ level, without stating the *P*-value. Now imagine that the person reading your report believes that a Type I error would be very serious, so that a significance level of $\alpha = 0.01$ would be more appropriate. This reader cannot tell whether to reject H_0 at the $\alpha = 0.01$ level, because you have not reported the *P*-value. It is much more helpful to report that $P = 0.03$, so that people can decide for themselves whether or not to reject H_0.

When using the critical value method, you should report the value of the test statistic, rather than simply stating whether the test statistic was in the critical region. In this way, the reader can tell whether the value of the test statistic was just barely inside the critical region, or well inside. In addition, reporting the value of the test statistic gives the reader the opportunity to choose a different critical value and determine whether H_0 can be rejected at a different level.

SUMMARY

When presenting the results of a hypothesis test, state the *P*-value or the value of the test statistic. Don't just state whether or not H_0 was rejected.

Check Your Understanding

21. A test was made of the hypotheses H_0: $\mu = 15$ versus H_1: $\mu > 15$. Four statisticians wrote summaries of the results. For each summary, state whether it contains enough information. If there is not enough information, indicate what needs to be added.
   a. The *P*-value was 0.02, so we reject H_0 at the $\alpha = 0.05$ level.
   b. The critical value was 1.645. Because $z > 1.645$, we reject H_0 at the $\alpha = 0.05$ level.
   c. The critical value was 1.645. Because $z = 2.05$, we reject H_0 at the $\alpha = 0.05$ level.
   d. Because $P < 0.05$, we reject H_0 at the $\alpha = 0.05$ level.

Answers are on page 433.

Objective 6 Distinguish between statistical significance and practical significance

Statistical Significance Is Not the Same as Practical Significance

When a result has a small *P*-value, we say that it is "statistically significant." In common usage, the word *significant* means "important." It is therefore tempting to think that statistically significant results must always be important. This is not the case. Sometimes statistically significant results do not have any practical importance. Example 9.15 illustrates the idea.

EXAMPLE 9.15

Determining practical significance

At a large company, employee satisfaction is measured with a standardized test for which scores range from 0 to 100. The mean score on this test was 74. The company then implemented a new policy that allowed telecommuting, so that employees could work from home. After the policy change, the mean score for a sample of employees was 76. In order to determine whether the mean score for all employees, μ, had changed after the new policy was implemented, a hypothesis test was performed of

$$H_0: \mu = 74 \qquad H_1: \mu \neq 74$$

We performed this test in Example 9.14. The standard error of $\bar{x}$ was 0.8944 and the *P*-value was 0.0250, so we rejected H_0 at the $\alpha = 0.05$ level. We concluded that the mean satisfaction level changed after the new policy was implemented. The human resources manager now writes a report stating that the new policy resulted in a large change in employee satisfaction. Explain why the human resources manager is not interpreting the result correctly.

Solution

The increase in mean score was from 74 to 76. Although this is statistically significant, it is only two points out of 100. It is unlikely that this difference is large enough to matter. The lesson here is that a result can be statistically significant without being large enough to be of practical importance. How can this happen? A difference is statistically significant when it is large compared to its standard error. In the example, a difference of two points was statistically significant because the standard error of $\bar{x}$ was small — only 0.8944. When the standard error is small, even a small difference can be statistically significant.

SUMMARY

When a result is statistically significant, we can only conclude that the true value of the parameter is different from the value specified by H_0. We cannot conclude that the difference is large enough to be important.

Check Your Understanding

22. A certain type of calculator battery has a mean lifetime of 100 hours and a standard deviation of $\sigma = 10$ hours. A company has developed a new battery and claims it has a longer mean life. A random sample of 1000 batteries is tested, and their sample mean lifetime is $\bar{x} = 101$ hours. A test was made of the hypotheses

    $$H_0: \mu = 100 \qquad H_1: \mu > 100$$

    a. Show that H_0 is rejected at the $\alpha = 0.01$ level.
    b. The battery manufacturer says that because the evidence is strong that $\mu > 100$, you should be willing to pay a much higher price for its battery than for the old type of battery. Do you agree? Why or why not?

Answers are on page 433.

USING TECHNOLOGY

We use Example 9.14 to illustrate the technology steps.

TI-84 PLUS

Testing a hypothesis about the population mean when σ is known

Step 1. Press **STAT** and highlight the **TESTS** menu.

Step 2. Select **Z–Test** and press **ENTER** (Figure A). The **Z–Test** menu appears.

Step 3. For **Inpt**, select the **Stats** option and enter the values of μ_0, σ, $\bar{x}$, and n. For Example 9.14, we use $\mu_0 = 74$, $\sigma = 8$, $\bar{x} = 76$, and $n = 80$.

Step 4. Select the form of the alternate hypothesis. For Example 9.14, the alternate hypothesis has the form $\mu \neq \mu_0$ (Figure B).

Step 5. Highlight **Calculate** and press **ENTER** (Figure C).

Note that if the raw data are given, the **Z–Test** command can be used by selecting **Data** as the **Inpt** option and entering the location of the data as the **List** option (Figure D).

Figure A

Figure B

Figure C

Figure D

MINITAB

Testing a hypothesis about the population mean when σ is known

Step 1. Click on **Stat**, then **Basic Statistics**, then **1-Sample Z**.

Step 2. Choose one of the following:
- If the summary statistics are given, click **Summarized Data** and enter the **Sample Size** (80), the **Sample Mean** (76), the **Standard Deviation** (8), and the **Test Mean** (74) (Figure E).
- If the raw data are given, click **Samples in Columns** and select the column that contains the data. Enter the **Standard Deviation**.

Step 3. Click **Options** and select the form of the alternate hypothesis. For Example 9.14, we select **Not Equal**. Given significance level α, enter $100(1 - \alpha)$ as the **Confidence Level**. For Example 9.14, $\alpha = 0.05$ and the confidence level is $100(1 - 0.05) = 95$. Click **OK**.

Step 4. Click **OK** (Figure F).

Figure E

```
Test of mu = 74 vs not = 74
The assumed standard deviation = 8


 N    Mean   SE Mean      95% CI          Z     P
80  76.0000   0.8944  (74.2470, 77.7530)  2.24  0.025
```

Figure F

EXCEL

Testing a hypothesis about the population mean when σ is known

This procedure requires the **MegaStat** EXCEL add-in to be loaded. The **MegaStat** add-in can be downloaded from www.mhhe.com/megastat.

Step 1. Load the **MegaStat** EXCEL add-in.

Step 2. Click on the **MegaStat** menu and select **Hypothesis Tests**, then **Mean vs. Hypothesized Value...**

Step 3. Choose one of the following:
- If the summary statistics are given, choose **summary input** and enter the range of the cells that contains, in the following order, the **variable name**, $\bar{x}$, σ, and n. Figure G illustrates the range of cells for Example 9.14 using *Satisfaction* as the variable name.
- If the raw data are given, choose **data input** and select the range of cells that contains the data in the **Input Range** field.

Step 4. Enter the **Hypothesized mean** (74) and select the form of the alternate hypothesis (not equal).

Step 5. Choose the **z-test** option (Figure H).

Step 6. Click **OK** (Figure I).

	A	B
1	Variable	Satisfaction
2	Sample Mean	76
3	Standard Deviation	8
4	Sample Size	80

Figure G

Figure H

Hypothesis Test: Mean vs. Hypothesized Value

```
74.00  hypothesized value
76.00  mean Satisfaction
 8.00  std. dev.
 0.89  std. error
 80    n

 2.24  z
.0253  p-value (two-tailed)
```

Figure I

SECTION 9.2   Exercises

Exercises 1–22 are the Check Your Understanding exercises located within the section.

Understanding the Concepts

In Exercises 23–28, fill in each blank with the appropriate word or phrase.

23. The _____ is the probability, assuming H_0 is true, of observing a value for the test statistic that is as extreme as or more extreme than the value actually observed.

24. The smaller the *P*-value is, the stronger the evidence against the _____ hypothesis becomes.

25. When using the critical value method, the region that contains the unusual values is called the _____ region.

26. If we decrease the value of the significance level α, we _____ the probability of a Type I error.

27. If we decrease the value of the significance level α, we _____ the probability of a Type II error.

28. When results are statistically significant, they do not necessarily have _____ significance.

In Exercises 29–34, determine whether the statement is true or false. If the statement is false, rewrite it as a true statement.

29. The smaller the *P*-value, the stronger the evidence against H_0.

30. If the *P*-value is less than the significance level, we reject H_0.

31. The probability of a Type II error is α, the significance level.

32. If the *P*-value is very small, we can be sure that the results have practical significance.

33. The P-value represents the probability that H_0 is true.

34. When presenting the results of a hypothesis test, one should report the P-value or the value of the test statistic.

Practicing the Skills

35. A test is made of H_0: $\mu = 50$ versus H_1: $\mu > 50$. A sample of size $n = 75$ is drawn, and $\bar{x} = 56$. The population standard deviation is $\sigma = 20$.
   a. Compute the value of the test statistic z.
   b. Is H_0 rejected at the $\alpha = 0.05$ level?
   c. Is H_0 rejected at the $\alpha = 0.01$ level?

36. A test is made of H_0: $\mu = 14$ versus H_1: $\mu \neq 14$. A sample of size $n = 48$ is drawn, and $\bar{x} = 12$. The population standard deviation is $\sigma = 6$.
   a. Compute the value of the test statistic z.
   b. Is H_0 rejected at the $\alpha = 0.05$ level?
   c. Is H_0 rejected at the $\alpha = 0.01$ level?

37. A test is made of H_0: $\mu = 130$ versus H_1: $\mu \neq 130$. A sample of size $n = 63$ is drawn, and $\bar{x} = 135$. The population standard deviation is $\sigma = 40$.
   a. Compute the value of the test statistic z.
   b. Is H_0 rejected at the $\alpha = 0.05$ level?
   c. Is H_0 rejected at the $\alpha = 0.01$ level?

38. A test is made of H_0: $\mu = 5$ versus H_1: $\mu < 5$. A sample of size $n = 87$ is drawn, and $\bar{x} = 4.5$. The population standard deviation is $\sigma = 25$.
   a. Compute the value of the test statistic z.
   b. Is H_0 rejected at the $\alpha = 0.05$ level?
   c. Is H_0 rejected at the $\alpha = 0.01$ level?

39. A test of the hypothesis H_0: $\mu = 65$ versus H_1: $\mu \neq 65$ was performed. The P-value was 0.035. Fill in the blank: If $\mu = 65$, then the probability of observing a test statistic as extreme as or more extreme than the one actually observed is _____.

40. A test of the hypothesis H_0: $\mu = 150$ versus H_1: $\mu < 150$ was performed. The P-value was 0.28. Fill in the blank: If $\mu = 150$, then the probability of observing a test statistic as extreme as or more extreme than the one actually observed is _____.

41. True or false: If $P = 0.02$, then
   a. The result is statistically significant at the $\alpha = 0.05$ level.
   b. The result is statistically significant at the $\alpha = 0.01$ level.
   c. The null hypothesis is rejected at the $\alpha = 0.05$ level.
   d. The null hypothesis is rejected at the $\alpha = 0.01$ level.

42. True or false: If $P = 0.08$, then
   a. The result is statistically significant at the $\alpha = 0.05$ level.
   b. The result is statistically significant at the $\alpha = 0.10$ level.
   c. The null hypothesis is rejected at the $\alpha = 0.05$ level.
   d. The null hypothesis is rejected at the $\alpha = 0.10$ level.

43. A test of H_0: $\mu = 17$ versus H_1: $\mu < 17$ is performed using a significance level of $\alpha = 0.01$. The value of the test statistic is $z = -2.68$.

   a. Is H_0 rejected?
   b. If the true value of μ is 17, is the result a Type I error, a Type II error, or a correct decision?
   c. If the true value of μ is 10, is the result a Type I error, a Type II error, or a correct decision?

44. A test of H_0: $\mu = 50$ versus H_1: $\mu \neq 50$ is performed using a significance level of $\alpha = 0.01$. The value of the test statistic is $z = 1.23$.
   a. Is H_0 rejected?
   b. If the true value of μ is 50, is the result a Type I error, a Type II error, or a correct decision?
   c. If the true value of μ is 65, is the result a Type I error, a Type II error, or a correct decision?

45. A test of H_0: $\mu = 0$ versus H_1: $\mu \neq 0$ is performed using a significance level of $\alpha = 0.05$. The P-value is 0.15.
   a. Is H_0 rejected?
   b. If the true value of μ is 1, is the result a Type I error, a Type II error, or a correct decision?
   c. If the true value of μ is 0, is the result a Type I error, a Type II error, or a correct decision?

46. A test of H_0: $\mu = 6$ versus H_1: $\mu > 6$ is performed using a significance level of $\alpha = 0.01$. The P-value is 0.002.
   a. Is H_0 rejected?
   b. If the true value of μ is 8, is the result a Type I error, a Type II error, or a correct decision?
   c. If the true value of μ is 6, is the result a Type I error, a Type II error, or a correct decision?

47. If H_0 is rejected at the $\alpha = 0.05$ level, which of the following is the best conclusion?
   i. H_0 is also rejected at the $\alpha = 0.01$ level.
   ii. H_0 is not rejected at the $\alpha = 0.01$ level.
   iii. We cannot determine whether H_0 is rejected at the $\alpha = 0.01$ level.

48. If H_0 is rejected at the $\alpha = 0.01$ level, which of the following is the best conclusion?
   i. H_0 is also rejected at the $\alpha = 0.05$ level.
   ii. H_0 is not rejected at the $\alpha = 0.05$ level.
   iii. We cannot determine whether H_0 is rejected at the $\alpha = 0.05$ level.

49. If $P = 0.03$, which of the following is the best conclusion?
   i. If H_0 is true, the probability of obtaining a test statistic as extreme as or more extreme than the one actually observed is 0.03.
   ii. The probability that H_0 is true is 0.03.
   iii. The probability that H_0 is false is 0.03.
   iv. If H_0 is false, the probability of obtaining a test statistic as extreme as or more extreme than the one actually observed is 0.03.

50. If $P = 0.25$, which of the following is the best conclusion?
   i. The probability that H_0 is true is 0.25.
   ii. If H_0 is false, the probability of obtaining a test statistic as extreme as or more extreme than the one actually observed is 0.25.
   iii. If H_0 is true, the probability of obtaining a test statistic as extreme as or more extreme than the one actually observed is 0.25.
   iv. The probability that H_0 is false is 0.25.

Working with the Concepts

51. Facebook: A study by the Web metrics firm Experian showed that in August 2011, the mean time spent per visit to Facebook was 20.8 minutes. Assume the standard deviation is $\sigma = 8$ minutes. Suppose that a simple random sample of 100 visits in August 2013 has a sample mean of $\bar{x} = 23$ minutes. A social scientist is interested to know whether the mean time of Facebook visits has increased.
   a. State the appropriate null and alternate hypotheses.
   b. Compute the value of the test statistic.
   c. State a conclusion. Use the $\alpha = 0.05$ level of significance.

52. Are you smarter than a second-grader? A random sample of 60 second-graders in a certain school district are given a standardized mathematics skills test. The sample mean score is $\bar{x} = 52$. Assume the standard deviation of test scores is $\sigma = 15$. The nationwide average score on this test is 50. The school superintendent wants to know whether the second-graders in her school district have greater math skills than the nationwide average.
   a. State the appropriate null and alternate hypotheses.
   b. Compute the value of the test statistic.
   c. State a conclusion. Use the $\alpha = 0.01$ level of significance.

53. Height and age: Are older men shorter than younger men? According to the National Health Statistics Reports, the mean height for U.S. men is 69.4 inches. In a sample of 300 men between the ages of 60 and 69, the mean height was $\bar{x} = 69.0$ inches. Public health officials want to determine whether the mean height μ for older men is less than the mean height of all adult men.
   a. State the appropriate null and alternate hypotheses.
   b. Assume the population standard deviation to be $\sigma = 2.84$ inches. Compute the value of the test statistic.
   c. State a conclusion. Use the $\alpha = 0.01$ level of significance.

54. Calibrating a scale: Making sure that the scales used by businesses in the United States are accurate is the responsibility of the National Institute for Standards and Technology (NIST) in Washington, D.C. Suppose that NIST technicians are testing a scale by using a weight known to weigh exactly 1000 grams. They weigh this weight on the scale 50 times and read the result each time. The 50 scale readings have a sample mean of $\bar{x} = 1000.6$ grams. The scale is out of calibration if the mean scale reading differs from 1000 grams. The technicians want to perform a hypothesis test to determine whether the scale is out of calibration.
   a. State the appropriate null and alternate hypotheses.
   b. The standard deviation of scale reading is known to be $\sigma = 2$. Compute the value of the test statistic.
   c. State a conclusion. Use the $\alpha = 0.05$ level of significance.

55. Measuring lung function: One of the measurements used to determine the health of a person's lungs is the amount of air a person can exhale under force in one second. This is called the forced expiratory volume in one second, and is abbreviated FEV_1. Assume the mean FEV_1 for 10-year-old boys is 2.1 liters and that the population standard deviation is $\sigma = 0.3$. A random sample of 100 10-year-old boys who live in a community with high levels of ozone pollution are found to have a sample mean FEV_1 of 1.95 liters. Can you conclude that the mean FEV_1 in the high-pollution community is less than 2.1 liters? Use the $\alpha = 0.05$ level of significance.

56. Heavy children: Are children heavier now than they were in the past? The National Health Examination and Nutrition Survey (NHANES) published in 2004 reported that the mean weight of six-year-old girls in the United States was 49.3 pounds. Another NHANES survey, published in 2012, reported that a sample of 177 six-year-old girls had an average weight of 51.9 pounds. Assume the population standard deviation is $\sigma = 17$ pounds. Can you conclude that the mean weight of six-year-old girls is higher in 2012 than in 2004? Use the $\alpha = 0.01$ level of significance.

57. House prices: Data from the National Association of Realtors indicates that the mean price of a home in Denver, Colorado, during April through June of 2012 was 260.7 thousand dollars. A random sample of 50 homes sold in 2013 had a mean price of 290.5 thousand dollars.
   a. Assume the population standard deviation is $\sigma = 150$. Can you conclude that the mean price in 2013 differs from the mean price in April through June of 2008? Use the $\alpha = 0.05$ level of significance.
   b. Following is a boxplot of the data. Explain why it is not reasonable to assume that the population is approximately normally distributed.

   c. Explain why the assumptions for the hypothesis test are satisfied even though the population is not normal.

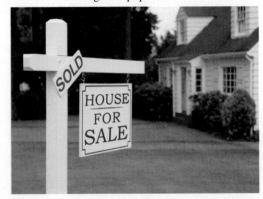

58. SAT scores: The College Board reports that in 2012 the mean score on the math SAT was 516 and the population standard deviation was $\sigma = 117$. A random sample of 20 students who took the test in 2013 had a mean score of 521. Following is a dotplot of the 20 scores.

   a. Are the assumptions for a hypothesis test satisfied? Explain.

b. If appropriate, perform a hypothesis test to investigate whether the mean score in 2013 differs from the mean score in 2012. Assume the population standard deviation is $\sigma = 117$. What can you conclude? Use the $\alpha = 0.05$ level of significance.

59. What are you drinking? Environmental Protection Agency standards require that the amount of lead in drinking water be less than 15 micrograms per liter. Twelve samples of water from a particular source have the following concentrations, in units of micrograms per liter:

11.4	13.9	11.2	14.5	15.2	8.1
12.4	8.6	10.5	17.1	9.8	15.9

a. Explain why it is necessary to check that the population is approximately normal before performing a hypothesis test.

b. Following is a dotplot of the data. Is it reasonable to assume that the population is approximately normal?

c. Assume that the population standard deviation is $\sigma = 3$. If appropriate, perform a hypothesis test at the $\alpha = 0.01$ level to determine whether you can conclude that the mean concentration of lead meets the EPA standard. What do you conclude?

60. GPA: The mean GPA at a certain university is 2.80. Following are GPAs for a random sample of 16 business students from this university.

2.27	3.05	2.57	3.36	3.10	3.03	3.19	3.08
2.60	2.92	2.77	3.55	2.63	2.79	2.70	2.92

a. Following is a boxplot of the data. Is it reasonable to assume that the population is approximately normal?

b. Assume that the population standard deviation is $\sigma = 0.3$. If appropriate, perform a hypothesis test at the $\alpha = 0.05$ level to determine whether the mean GPA for business students differs from the mean GPA at the whole university. What do you conclude?

61. Interpreting calculator display: The following TI-84 Plus display presents the results of a hypothesis test.

```
        Z-Test
μ≠45
z=3.094063348
p=.0019744896
x̄=48.78
n=67
```

a. What are the null and alternate hypotheses?
b. What is the value of the test statistic?
c. What is the P-value?
d. Do you reject H_0 at the $\alpha = 0.05$ level?
e. Do you reject H_0 at the $\alpha = 0.01$ level?

62. Interpreting calculator display: The following TI-84 Plus display presents the results of a hypothesis test.

```
        Z-Test
μ≠125
z=1.73116377
p=.0834224813
x̄=131.6
n=43
```

a. What are the null and alternate hypotheses?
b. What is the value of the test statistic?
c. What is the P-value?
d. Do you reject H_0 at the $\alpha = 0.05$ level?
e. Do you reject H_0 at the $\alpha = 0.01$ level?

63. Interpreting computer output: The following output from MINITAB presents the results of a hypothesis test.

```
Test of mu = 225.0 vs not = 225.0
The assumed standard deviation = 35.0
```

N	Mean	SE Mean	95% CI	Z	P
50	235.32	4.9497	(225.619, 245.021)	2.085	0.037

a. What are the null and alternate hypotheses?
b. What is the value of the test statistic?
c. What is the P-value?
d. Do you reject H_0 at the $\alpha = 0.05$ level?
e. Do you reject H_0 at the $\alpha = 0.01$ level?
f. Use the results of the output to compute the value of the test statistic z for a test of H_0: $\mu = 230$ versus H_1: $\mu > 230$.
g. Find the P-value.
h. Do you reject the null hypothesis in part (f) at the $\alpha = 0.05$ level?

64. Interpreting computer output: The following output from MINITAB presents the results of a hypothesis test.

```
Test of mu = 20.0 vs > 20.0
The assumed standard deviation = 6.5
```

N	Mean	SE Mean	95% CI	Z	P
45	21.324	0.9690	(19.425, 23.223)	1.366	0.086

a. What are the null and alternate hypotheses?
b. What is the value of the test statistic?
c. What is the P-value?
d. Do you reject H_0 at the $\alpha = 0.05$ level?
e. Do you reject H_0 at the $\alpha = 0.01$ level?
f. Use the results of the output to compute the value of the test statistic z for a test of H_0: $\mu = 24$ versus H_0: $\mu \neq 24$.
g. Find the P-value.
h. Do you reject the null hypothesis in part (f) at the $\alpha = 0.05$ level?

65. Statistical or practical significance: A new method of teaching arithmetic to elementary school students was evaluated. The students who were taught by the new method were given a standardized test with a maximum score of 100 points. They scored an average of one point higher than students taught by the old method. A hypothesis test was performed in which the null hypothesis stated that there was no difference between the two groups, and the alternate hypothesis stated that the mean score for the new method was higher. The P-value was 0.001. True or false:

a. Because the P-value is very small, we can conclude that the mean score for students taught by the new method is higher than for students taught by the old method.

b. Because the P-value is very small, we can conclude that the new method represents an important improvement over the old method.

66. **Statistical or practical significance:** A new method of postoperative treatment was evaluated for patients undergoing a certain surgical procedure. Under the old method, the mean length of hospital stay was 6.3 days. The sample mean for the new method was 6.1 days. A hypothesis test was performed in which the null hypothesis stated that the mean length of stay was the same for both methods, and the alternate hypothesis stated that the mean stay was lower for the new method. The P-value was 0.002. True or false:

a. Because the P-value is very small, we can conclude that the new method provides an important reduction in the mean length of hospital stay.

b. Because the P-value is very small, we can conclude that the mean length of hospital stay is less for patients treated by the new method than for patients treated by the old method.

67. **Test scores:** A math teacher has developed a new program to help high school students prepare for the math SAT. A sample of 100 students enroll in the program. They take a math SAT exam before the program starts and again at the end to measure their improvement. The mean number of points improved was $\bar{x} = 2.5$. Assume the standard deviation is $\sigma = 10$. Let μ be the population mean number of points improved. To determine whether the program is effective, a test is made of the hypotheses H_0: $\mu = 0$ versus H_1: $\mu > 0$.

a. Compute the value of the test statistic.
b. Compute the P-value.
c. Do you reject H_0 at the $\alpha = 0.05$ level?
d. Is the result of practical significance? Explain.

68. **Weight loss:** A doctor has developed a new diet to help people lose weight. A random sample of 500 people went on

the diet for six weeks. The mean number of pounds lost was $\bar{x} = 0.5$. Assume the standard deviation is $\sigma = 5$. Let μ be the population mean number of pounds lost. To determine whether the diet is effective, a test is made of the hypotheses H_0: $\mu = 0$ versus H_1: $\mu > 0$.

a. Compute the value of the test statistic.
b. Compute the P-value.
c. Do you reject H_0 at the $\alpha = 0.05$ level?
d. Is the result of practical significance? Explain.

69. **Enough information?** A test was made of the hypotheses H_0: $\mu = 70$ versus H_1: $\mu \neq 70$. A report of the results stated: "$P < 0.05$, so we reject H_0 at the $\alpha = 0.05$ level." Is there any additional information that should have been included in the report? If so, what is it?

70. **Enough information?** A test was made of the hypotheses H_0: $\mu = 10$ versus H_1: $\mu > 10$. A report of the results stated: "The critical value was 1.645. Since $z > 1.645$, we reject H_0 at the $\alpha = 0.05$ level." Is there any additional information that should have been included in the report? If so, what is it?

Extending the Concepts

71. **Somebody's wrong:** Cindy computes a 95% confidence interval for μ and obtains (94.6, 98.3). Luis performs a test of the hypotheses H_0: $\mu = 100$ versus H_1: $\mu \neq 100$ and obtains a P-value of 0.12. Explain why they can't both be right.

72. **Large samples and practical significance:** A sample of size $n = 100$ is used to test H_0: $\mu = 20$ versus H_1: $\mu > 20$. The value of μ will not have practical significance unless $\mu > 25$. The population standard deviation is $\sigma = 10$. The value of $\bar{x}$ is 21.

a. Assume the sample size is $n = 100$. Compute the P-value. Show that you do not reject H_0 at the $\alpha = 0.05$ level.

b. Assume the sample size is $n = 1000$. Compute the P-value. Show that you reject H_0 at the $\alpha = 0.05$ level.

c. Do you think the difference is likely to be of practical significance? Explain.

d. Explain why a larger sample can be more likely to produce a statistically significant result that is not practically significant.

Answers to Check Your Understanding Exercises for Section 9.2

1. **a.** Critical value is -1.645, critical region is $z \leq -1.645$.
   b. Yes
   c. Critical value is -2.326, critical region is $z \leq -2.326$.
   d. No

2. **a.** Critical value is 1.645, critical region is $z \geq 1.645$.
   b. Yes
   c. Critical value is 2.326, critical region is $z \geq 2.326$.
   d. Yes

3. **a.** Critical values are 1.96 and -1.96, critical region is $z \leq -1.96$ or $z \geq 1.96$.
   b. No

c. Critical values are 2.576 and -2.576, critical region is $z \leq -2.576$ or $z \geq 2.576$.
   d. No

4. **a.** $z = 2.12$ **b.** $z \geq 1.645$ **c.** Yes

5. **a.** $z = -1.00$ **b.** $z \leq -2.054$ **c.** No

6. **a.** $z = -3.25$ **b.** $z \leq -2.576$ or $z \geq 2.576$ **c.** Yes

7. $P = 0.05$

8. $P = 0.1003$. If H_0 is true, then the probability of observing a test statistic less than or equal to the value we actually observed is 0.1003. This result is not very unusual, so the evidence against H_0 is not strong.

9. a. $P = 0.4532$. If H_0 is true, then the probability of observing a test statistic as extreme as or more extreme than the value we actually observed is 0.4532. This result is not unusual, so the evidence against H_0 is not strong.

  b. $P = 0.0278$. If H_0 is true, then the probability of observing a test statistic as extreme as or more extreme than the value we actually observed is 0.0278. This result is fairly unusual, so the evidence against H_0 is fairly strong.

  c. $z = -2.20$

10. ii

11. a. No   **b.** No   **c.** Yes   **d.** Yes

12. a. No   **b.** Yes   **c.** Yes   **d.** No

13. a. Yes   **b.** No   **c.** Yes   **d.** No

14. a. $z = 3.27$ **b.** $P = 0.0005$

  c. If H_0 is true, then the probability of observing a test statistic greater than or equal to the value we actually observed is 0.0005. This result is very unusual, so the evidence against H_0 is very strong.

  d. Yes   **e.** Yes

15. a. H_0: $\mu = 15$, H_1: $\mu \neq 15$ **b.** $z = 2.072750901$

  c. $P = 0.0381953363$ **d.** Yes

16. a. H_0: $\mu = 53.5$, H_1: $\mu \neq 53.5$ **b.** $z = -1.29$

  c. $P = 0.196$ **d.** No

17. a. No   **b.** Yes   **c.** Yes   **d.** Yes

18. a. Yes   **b.** No

19. 0.05

20. a. Charlie   **b.** Felice

21. a. Contains enough information

  b. The value of the test statistic z needs to be added.

  c. Contains enough information

  d. The P-value needs to be added.

22. a. $z = 3.16$, $P = 0.0008$, so H_0 is rejected at the $\alpha = 0.01$ level.

  b. No. The difference between $\bar{x} = 101$ and 100 is not large enough to be of practical significance.

SECTION 9.3   Hypothesis Tests for a Population Mean, Standard Deviation Unknown

Objectives

1. Test a hypothesis about a mean using the P-value method
2. Test a hypothesis about a mean using the critical value method

Objective 1 Test a hypothesis about a mean using the P-value method

Do low-fat diets work? The following study was reported in the *Journal of the American Medical Association* (297:969–977). A total of 76 subjects were placed on a low-fat diet. After 12 months, their sample mean weight loss was $\bar{x} = 2.2$ kilograms, with a sample standard deviation of $s = 6.1$ kilograms. How strong is the evidence that people who adhere to this diet will lose weight, on the average?

To answer this question, we need to perform a hypothesis test on a population mean. Assume that the subjects in the study constitute a simple random sample from a population of interest. We are interested in their population mean weight loss μ. We know the sample mean $\bar{x} = 2.2$. We do not know the population standard deviation σ, but we know that the sample standard deviation is $s = 6.1$.

Because we do not know σ, we cannot use the z-score

$$z = \frac{\bar{x} - \mu}{\sigma / \sqrt{n}}$$

as our test statistic. Instead, we replace σ with the sample standard deviation s and use the t statistic

$$t = \frac{\bar{x} - \mu}{s / \sqrt{n}}$$

Recall: When $\bar{x}$ is the mean of a sample from a normal population, the quantity $\frac{\bar{x} - \mu}{s / \sqrt{n}}$ has a Student's t distribution with $n - 1$ degrees of freedom.

When the null hypothesis is true, the t statistic has a Student's t distribution with $n - 1$ degrees of freedom. We described the Student's t distribution in Section 8.2. When we perform a test using the t statistic, we call the test a **t-test**. We can perform a t-test for a population mean whenever the following assumptions are satisfied.

Assumptions for a Test of a Population Mean μ When σ Is Unknown

1. We have a simple random sample.
2. The sample size is large ($n > 30$), or the population is approximately normal.

When these assumptions are met, a hypothesis test can be performed. Either the critical value method or the *P*-value method may be used. Following are the steps for the *P*-value method.

Performing a Hypothesis Test on a Population Mean with σ Unknown Using the *P*-Value Method

Check to determine whether the assumptions are satisfied. If they are, then proceed with the following steps.

Step 1: State the null and alternate hypotheses. The null hypothesis specifies a value for the population mean μ. We will call this value μ_0. So the null hypothesis is of the form H_0: $\mu = \mu_0$. The alternate hypothesis can be stated in one of three ways:

Left-tailed: $\qquad H_1$: $\mu < \mu_0$
Right-tailed: $\qquad H_1$: $\mu > \mu_0$
Two-tailed: $\qquad H_1$: $\mu \neq \mu_0$

Step 2: If making a decision, choose a significance level α.

Step 3: Compute the test statistic $t = \dfrac{\bar{x} - \mu_0}{s/\sqrt{n}}$.

Step 4: Compute the *P*-value of the test statistic. The *P*-value is the probability, assuming that H_0 is true, of observing a value for the test statistic that disagrees as strongly as or more strongly with H_0 than the value actually observed. The *P*-value is an area under the Student's *t* curve with $n - 1$ degrees of freedom. The area is in the left tail, the right tail, or in both tails, depending on the type of alternate hypothesis. Note that the inequality points in the direction of the tail that contains the area for the *P*-value.

The *P*-value is the area to the left of *t*.

Left-tailed: H_1: $\mu < \mu_0$

The *P*-value is the area to the right of *t*.

Right-tailed: H_1: $\mu > \mu_0$

The *P*-value is the sum of the areas in the two tails.

Two-tailed: H_1: $\mu \neq \mu_0$

Step 5: Interpret the *P*-value. If making a decision, reject H_0 if the *P*-value is less than or equal to the significance level α.

Step 6: State a conclusion.

EXAMPLE 9.16   Perform a hypothesis test

In a recent medical study, 76 subjects were placed on a low-fat diet. After 12 months, their sample mean weight loss was $\bar{x} = 2.2$ kilograms, with a sample standard deviation of $s = 6.1$ kilograms. Can we conclude that the mean weight loss is greater than 0? Use the $\alpha = 0.05$ level of significance.

Source: *Journal of the American Medical Association* 297:969–977

Solution

We first check the assumptions. We have a simple random sample. The sample size is 76, so $n > 30$. The assumptions are satisfied.

Step 1: State H_0 and H_1. The issue is whether the mean weight loss μ is greater than 0. So the null and alternate hypotheses are

$$H_0: \mu = 0 \qquad H_1: \mu > 0$$

Note that we have a right-tailed test, because we are particularly interested in whether the diet results in a weight loss.

Step 2: Choose a level of significance. The level of significance is $\alpha = 0.05$.

Step 3: Compute the test statistic. The test statistic is

$$t = \frac{\bar{x} - \mu_0}{s/\sqrt{n}}$$

To compute its value, we note that $\bar{x} = 2.2$, $s = 6.1$, and $n = 76$. We set $\mu_0 = 0$, the value for μ specified by H_0. The value of the test statistic is

$$t = \frac{2.2 - 0}{6.1/\sqrt{76}} = 3.144$$

Step 4: Compute the *P*-value. When H_0 is true, the test statistic t has the Student's t distribution with $n - 1$ degrees of freedom. In this case, the sample size is $n = 76$, so there are $n - 1 = 75$ degrees of freedom. To obtain the *P*-value, note that the alternate hypothesis is H_1: $\mu > 0$. Therefore, values of the t statistic in the right tail of the Student's t distribution provide evidence against H_0. The *P*-value is the probability that a value as extreme as or more extreme than the observed value of 3.144 is observed from a t distribution with 75 degrees of freedom. To find the *P*-value exactly, it is necessary to use technology. The *P*-value is 0.0012. Figure 9.10 illustrates the *P*-value as an area under the Student's t curve, and presents the results from the TI-84 Plus calculator. Step-by-step instructions for performing hypothesis tests with technology are given in the Using Technology section on page 442.

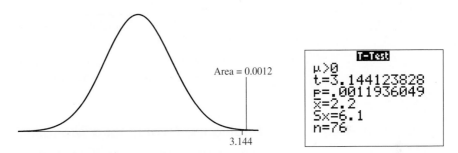

Figure 9.10 The *P*-value is the area to the right of the observed value of the test statistic, 3.144. The TI-84 Plus display shows that $P = 0.0012$, rounded to four decimal places.

Step 5: Interpret the *P*-value. The *P*-value is 0.0012. Because $P < 0.05$, we reject H_0.

Step 6: State a conclusion. We conclude that the mean weight loss of people who adhered to this diet for 12 months is greater than 0.

Estimating the *P*-Value from a Table

If no technology is available to compute the *P*-value, the t table (Table A.3) can be used to provide an approximation. When using a t table, we cannot find the *P*-value exactly. Instead, we can only specify that P is between two values. We now show how to use Table A.3 to bracket P between two values.

In Example 9.16, there are 75 degrees of freedom. We consult Table A.3 and find that the number 75 does not appear in the degrees of freedom column. We therefore use the next smallest number, which is 60. Now look across the row for two numbers that bracket the observed value 3.144. These are 2.915 and 3.232. The upper-tail probabilities are 0.0025 for 2.915 and 0.001 for 3.232. The *P*-value must therefore be between 0.001

and 0.0025 (see Figure 9.11). We can conclude that the *P*-value is small enough to reject H_0.

Degrees of freedom	\multicolumn{10}{c}{Area in the Right Tail}
	0.40	0.25	0.10	0.05	0.025	0.01	0.005	0.0025	0.001	0.0005
1	0.325	1.000	3.078	6.314	12.706	31.821	63.657	127.321	318.309	636.619
2	0.289	0.816	1.886	2.920	4.303	6.965	9.925	14.089	22.327	31.599
3	0.277	0.765	1.638	2.353	3.182	4.541	5.841	7.453	10.215	12.924
⋮	⋮	⋮	⋮	⋮	⋮	⋮	⋮	⋮	⋮	⋮
38	0.255	0.681	1.304	1.686	2.024	2.429	2.712	2.980	3.319	3.566
39	0.255	0.681	1.304	1.685	2.023	2.426	2.708	2.976	3.313	3.558
40	0.255	0.681	1.303	1.684	2.021	2.423	2.704	2.971	3.307	3.551
50	0.255	0.679	1.299	1.676	2.009	2.403	2.678	2.937	3.261	3.496
60	0.254	0.679	1.296	1.671	2.000	2.390	2.660	2.915	3.232	3.460
80	0.254	0.678	1.292	1.664	1.990	2.374	2.639	2.887	3.195	3.416
100	0.254	0.677	1.290	1.660	1.984	2.364	2.626	2.871	3.174	3.390
200	0.254	0.676	1.289	1.653	1.972	2.345	2.601	2.839	3.131	3.340

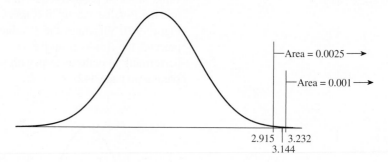

Figure 9.11 The *P*-value is the area to the right of the observed value of the test statistic, 3.144. The *P*-value is between 0.001 and 0.0025.

Finding the *P*-value for a two-tailed test from a table

In the previous example, what if the alternate hypothesis were H_1: $\mu \neq 0$? The *P*-value would be the sum of the areas in two tails. We know that the area in the right tail is 0.0012 (see Figure 9.10). Since the *t* distribution is symmetric, the sum of the areas in two tails is twice as much: $0.0012 + 0.0012 = 0.0024$. This is shown in Figure 9.12.

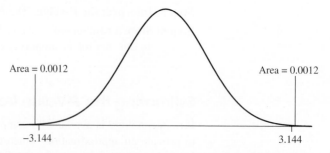

Figure 9.12 The *P*-value for a two-tailed test is the sum of the areas in the two tails. Each tail has area 0.0012. The *P*-value is $0.0012 + 0.0012 = 0.0024$.

If we are using Table A.3, we can only specify that *P* is between two values. We know that the area in one tail is between 0.001 and 0.0025. Therefore, the area in both tails is between $2(0.001) = 0.002$ and $2(0.0025) = 0.005$. This is shown in Figure 9.13.

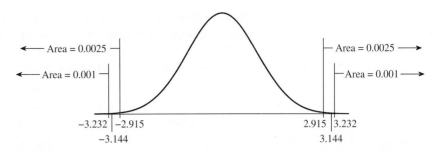

Figure 9.13   The *P*-value for a two-tailed test is the sum of the areas in the two tails. The area in each tail is between 0.001 and 0.0025. The sum of the areas in both tails is therefore between $2(0.001) = 0.002$ and $2(0.0025) = 0.005$.

Check Your Understanding

1. Find the *P*-value for the following values of the test statistic *t*, sample size *n*, and alternate hypothesis H_1. If you use Table A.3, you may specify that *P* is between two values.
   a. $t = 2.584, n = 12, H_1: \mu > \mu_0$
   b. $t = -1.741, n = 21, H_1: \mu < \mu_0$
   c. $t = 3.031, n = 14, H_1: \mu \neq \mu_0$
   d. $t = -2.584, n = 31, H_1: \mu \neq \mu_0$

2. In Example 9.16, the sample size was $n = 76$, and we observed $\bar{x} = 2.2$ and $s = 6.1$. We tested $H_0: \mu = 0$ versus $H_1: \mu > 0$, and the *P*-value was 0.0012. Assume that the sample size was 41 instead of 76, but that the values of $\bar{x}$ and s were the same.
   a. Find the value of the test statistic *t*.
   b. How many degrees of freedom are there?
   c. Find the *P*-value.
   d. Is the evidence against H_0 stronger or weaker than the evidence from the sample of 76? Explain.

Answers are on page 447.

EXAMPLE 9.17   ## Perform a hypothesis test

Generic drugs are lower-cost substitutes for brand-name drugs. Before a generic drug can be sold in the United States, it must be tested and found to perform equivalently to the brand-name product. The U.S. Food and Drug Administration is now supervising the testing of a new generic antifungal ointment. The brand-name ointment is known to deliver a mean of 3.5 micrograms of active ingredient to each square centimeter of skin.

As part of the testing, seven subjects apply the ointment. Six hours later, the amount of drug that has been absorbed into the skin is measured. The amounts, in micrograms, are

2.6   3.2   2.1   3.0   3.1   2.9   3.7

How strong is the evidence that the mean amount absorbed differs from 3.5 micrograms? Use the $\alpha = 0.01$ level of significance.

Solution

We first check the assumptions. Because the sample is small, the population must be approximately normal. We check this with a dotplot of the data.

There is no evidence of strong skewness, and no outliers. Therefore, we can proceed.

Step 1: State the null and alternate hypotheses. The issue is whether the mean μ differs from 3.5. Therefore, the null and alternate hypotheses are

$$H_0: \mu = 3.5 \qquad H_1: \mu \neq 3.5$$

Step 2: Choose a significance level α. The significance level is $\alpha = 0.01$.

Step 3: Compute the value of the test statistic t. To compute t, we need to know the sample mean $\bar{x}$, the sample standard deviation s, the null hypothesis mean μ_0, and the sample size n. We compute $\bar{x}$ and s from the sample. The values are

$$\bar{x} = 2.9429 \qquad s = 0.4995$$

The null hypothesis mean is $\mu_0 = 3.5$. The sample size is $n = 7$. The value of the t statistic is

$$t = \frac{\bar{x} - \mu_0}{s/\sqrt{n}}$$
$$= \frac{2.9429 - 3.5}{0.4995/\sqrt{7}}$$
$$= -2.951$$

Step 4: Compute the P-value. The number of degrees of freedom is $n - 1 = 7 - 1 = 6$. The alternate hypothesis is two-tailed, so the P-value is the sum of the area to the left of the observed t statistic -2.951 and the area to the right of 2.951, in a t distribution with 6 degrees of freedom. We can use technology to find that $P = 0.0256$. The following TI-84 Plus display presents the results. Step-by-step instructions for performing hypothesis tests with technology are given in the Using Technology section on page 442.

The P-value is given on the third line of the display. Rounding off to four decimal places, we see that $P = 0.0256$.

Alternatively, we can use Table A.3 to specify that the P-value is between two numbers. In the row corresponding to 6 degrees of freedom, the two values closest to 2.951 are 2.447 and 3.143. The area to the right of 2.447 is 0.025 and the area to the right of 3.143 is 0.01. Therefore the area in the right tail is between 0.01 and 0.025. The P-value is twice the area in the right tail, so we conclude that P is between $2(0.01) = 0.02$ and $2(0.025) = 0.05$. See Figure 9.14.

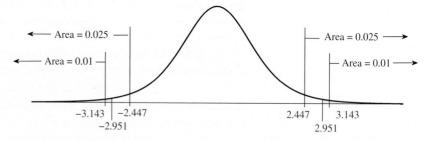

Figure 9.14 The P-value for a two-tailed test is the sum of the areas in the two tails. The area in each tail is between 0.01 and 0.025. The sum of the areas in both tails is therefore between $2(0.01) = 0.02$ and $2(0.025) = 0.05$.

Step 5: Interpret the P-value. The P-value of 0.0256 tells us that if H_0 is true, the probability of observing a value of the test statistic as extreme as or more extreme than the value of -2.951 that we observed is 0.0256. The P-value is small enough to give us doubt about the truth of H_0. However, because $P > 0.01$, we do not reject H_0 at the 0.01 level.

Step 6: State a conclusion. There is not enough evidence to conclude that the mean amount of drug absorbed differs from 3.5 micrograms. The mean may be equal to 3.5 micrograms.

Performing hypothesis tests with technology

The following output (from MINITAB) presents the results of Example 9.17.

```
Test of mu = 3.5 vs not = 3.5

N    Mean    StDev    SE Mean       99% CI         T        P
7   2.9429   0.4995   0.1888   (2.2430, 3.6248)   -2.95   0.026
```

Most of the output is straightforward. The first line specifies the null and alternate hypotheses. The sample size, sample mean, and sample standard deviation are given as "N," "Mean," and "StDev," respectively. The quantity labeled "SE Mean" is the standard error of the mean, which is the quantity $s/\sqrt{n}$ that appears in the denominator of the t statistic. Next, MINITAB provides a 99% confidence interval for μ. Finally, the value of the t statistic and the P-value are given.

The following TI-84 Plus display presents the results of Example 9.17. This display was also shown in the solution to Example 9.17.

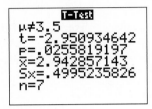

The first line states the alternate hypothesis. The quantity labeled "Sx" is the sample standard deviation s. Step-by-step instructions for performing hypothesis tests with technology are given in the Using Technology section on page 442.

Check Your Understanding

3. In Example 9.17, the alternate hypothesis was H_1: $\mu \neq 3.5$ and the P-value for the two-tailed test was $P = 0.0256$. What would the P-value be for the alternate hypothesis H_1: $\mu < 3.5$?

4. The following TI-84 Plus display presents the results of a t-test.

    a. What are the null and alternate hypotheses?
    b. What is the sample size?
    c. How many degrees of freedom are there?
    d. What is the value of $\bar{x}$?
    e. What is the value of s?
    f. What is the value of the test statistic?
    g. What is the P-value?
    h. Do you reject H_0 at the $\alpha = 0.01$ level?

5. Refer to the display in Exercise 4. If the sample mean were 6.80 instead of 6.68, what would the P-value be? If you use Table A.3, you may specify that P is between two values.

Answers are on page 447.

Objective 2 Test a hypothesis about a mean using the critical value method

Testing a Hypothesis About a Population Mean Using the Critical Value Method

The critical value method when σ is unknown is the same as that when σ is known, except that we use the Student's t distribution rather than the normal distribution. The critical value can be found in Table A.3 or with technology. The procedure depends on whether the alternate hypothesis is left-tailed, right-tailed, or two-tailed.

Critical Values for the *t* Statistic

Let α denote the chosen significance level and let n denote the sample size. The critical value depends on whether the alternate hypothesis is left-tailed, right-tailed, or two-tailed. We use the Student's t distribution with $n-1$ degrees of freedom.

For left-tailed H_1: The critical value is $-t_\alpha$, which has area α to its left. Reject H_0 if $t \leq -t_\alpha$.

For right-tailed H_1: The critical value is t_α, which has area α to its right. Reject H_0 if $t \geq t_\alpha$.

For two-tailed H_1: The critical values are $t_{\alpha/2}$, which has area $\alpha/2$ to its right, and $-t_{\alpha/2}$, which has area $\alpha/2$ to its left. Reject H_0 if $t \geq t_{\alpha/2}$ or $t \leq -t_{\alpha/2}$.

Check Your Understanding

6. Find the critical value or values for the following values of the significance level α, sample size n, and alternate hypothesis H_1.
   a. $\alpha = 0.05$, $n = 3$, $H_1: \mu > \mu_0$
   b. $\alpha = 0.01$, $n = 26$, $H_1: \mu \neq \mu_0$
   c. $\alpha = 0.10$, $n = 81$, $H_1: \mu < \mu_0$
   d. $\alpha = 0.05$, $n = 14$, $H_1: \mu \neq \mu_0$

Answers are on page 447.

The assumptions for the critical value method are the same as those for the P-value method. We repeat them here.

Assumptions for a Test of a Population Mean μ When σ Is Unknown

1. We have a simple random sample.
2. The sample size is large ($n > 30$), or the population is approximately normal.

When these assumptions are satisfied, a hypothesis test can be performed using the following steps.

Performing a Hypothesis Test on a Population Mean with σ Unknown Using the Critical Value Method

Check to be sure that the assumptions are satisfied. If they are, then proceed with the following steps:

Step 1: State the null and alternate hypotheses. The null hypothesis specifies a value for the population mean μ. We will call this value μ_0, so the null hypothesis is of the form $H_0: \mu = \mu_0$. The alternate hypothesis can be stated in one of three ways:
   Left-tailed: $H_1: \mu < \mu_0$
   Right-tailed: $H_1: \mu > \mu_0$
   Two-tailed: $H_1: \mu \neq \mu_0$

Step 2: Choose a significance level α and find the critical value or values. Use $n - 1$ degrees of freedom, where n is the sample size.

Step 3: Compute the test statistic $t = \dfrac{\bar{x} - \mu_0}{s/\sqrt{n}}$.

Step 4: Determine whether to reject H_0, as follows:

Left-tailed: $H_1: \mu < \mu_0$ Reject if $t \leq -t_\alpha$.

Right-tailed: $H_1: \mu > \mu_0$ Reject if $t \geq t_\alpha$.

Two-tailed: $H_1: \mu \neq \mu_0$ Reject if $t \geq t_{\alpha/2}$ or $t \leq -t_{\alpha/2}$.

Step 5: State a conclusion.

EXAMPLE 9.18   Test a hypothesis using the critical value method

A computer software vendor claims that a new version of its operating system will crash fewer than six times per year on average. A system administrator installs the operating system on a random sample of 41 computers. At the end of a year, the sample mean number of crashes is 7.1, with a standard deviation of 3.6. Can you conclude that the vendor's claim is false? Use the $\alpha = 0.05$ significance level.

Solution

We first check the assumptions. We have a large ($n > 30$) random sample, so the assumptions are satisfied.

Step 1:   State the null and alternate hypotheses. To conclude that the vendor's claim is false, we must conclude that $\mu > 6$. This is H_1. The hypotheses are $H_0: \mu = 6$ versus $H_1: \mu > 6$.

Step 2:   Choose a significance level α and find the critical value. We will use a significance level of $\alpha = 0.05$. We use Table A.3. The number of degrees of freedom is $41 - 1 = 40$. This is a right-tailed test, so the critical value is the t-value with area 0.05 above it in the right tail. Thus, the critical value is $t_\alpha = 1.684$.

Step 3:   Compute the test statistic. We have $\bar{x} = 7.1$, $\mu_0 = 6$, $s = 3.6$, and $n = 41$. The test statistic is

$$t = \frac{7.1 - 6}{3.6/\sqrt{41}} = 1.957$$

Step 4:   Determine whether to reject H_0. Because this is a right-tailed test, we reject H_0 if $t \geq t_\alpha$. Because $t = 1.957$ and $t_\alpha = 1.684$, we reject H_0. Figure 9.15 illustrates the critical region and the test statistic.

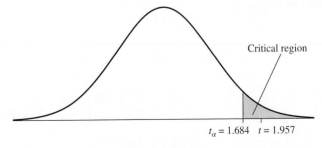

$t_\alpha = 1.684$ $t = 1.957$

Figure 9.15

Step 5:   State a conclusion. We conclude that the mean number of crashes is greater than six per year.

USING TECHNOLOGY

We use Example 9.17 to illustrate the technology steps.

TI-84 PLUS

Testing a hypothesis about a population mean when σ is unknown

Step 1. Press **STAT** and highlight the **TESTS** menu.

Step 2. Select **T–Test** and press **ENTER** (Figure A). The **T–Test** menu appears.

Step 3. Choose one of the following:
- If the summary statistics are given, select **Stats** as the **Inpt** option and enter μ_0, $\bar{x}$, s, and n.
- If the raw data are given, select **Data** as the **Inpt** option and enter the location of the data as the **List** option. For Example 9.17, the sample has been entered in list **L1**.

Step 4. Select the form of the alternate hypothesis. For Example 9.17, the alternate hypothesis has the form $\neq \mu_0$ (Figure B).

Step 5. Highlight **Calculate** and press **ENTER** (Figure C).

Figure A

Figure B

Figure C

MINITAB

Testing a hypothesis about a population mean when σ is unknown

Step 1. Click on **Stat**, then **Basic Statistics**, then **1-Sample t**.

Step 2. Choose one of the following:
- If the summary statistics are given, click **Summarized Data** and enter the **Sample Size**, the **Mean**, and the **Standard Deviation** for the sample.
- If the raw data are given, click **Samples in Columns** and select the column that contains the data. For Example 9.17, the sample has been entered in column **C1**.

Step 3. Enter μ_0 in the **Test Mean** field.

Step 4. Click **Options** and select the form of the alternate hypothesis. For Example 9.17, we select **Not Equal**. Given significance level α, enter $100(1 - \alpha)$ as the **Confidence Level**. For Example 9.17, since $\alpha = 0.01$, the confidence level is $100(1 - 0.01) = 99$. Click **OK**.

Step 5. Click **OK** (Figure D).

```
Test of mu = 3.5 vs not = 3.5


Variable   N     Mean    StDev   SE Mean        99% CI           T      P
C1         7   2.94286  0.49952  0.18880  (2.24289, 3.64283)   -2.95  0.026
```

Figure D

EXCEL

Testing a hypothesis about a population mean when σ is unknown

This procedure requires the **MegaStat** EXCEL add-in to be loaded. The **MegaStat** add-in can be downloaded from www.mhhe.com/megastat.

Step 1. Load the **MegaStat** EXCEL add-in.

Step 2. Click on the **MegaStat** menu and select **Hypothesis Tests**, then **Mean vs. Hypothesized Value...**

Step 3. Choose one of the following:
- If the summary statistics are given, choose **summary input** and enter the range of the cells that contains, in the following order, the **variable name**, $\bar{x}$, s, and n.
- If the raw data are given, choose **data input** and select the range of cells that contains the data in the **Input Range** field. For Example 9.17, the sample has been entered in column **A** (Figure E).

Step 4. Enter the **Hypothesized mean** (3.5) and select the form of the alternate hypothesis (not equal).

Step 5. Choose the **t-test** option (Figure F).

Step 6. Click **OK** (Figure G)

	A
1	**Drug Absorbed**
2	2.6
3	3.2
4	2.1
5	3.0
6	3.1
7	2.9
8	3.7

Figure E

Figure F

Hypothesis Test: Mean vs. Hypothesized Value

3.5000 hypothesized value
2.9429 mean Drug Absorbed
0.4995 std. dev.
0.1888 std. error
   7 n
   6 df

-2.95 t
.0256 p-value (two-tailed)

Figure G

SECTION 9.3   Exercises

Exercises 1–6 are the Check Your Understanding exercises located within the section.

Understanding the Concepts

In Exercises 7 and 8, fill in each blank with the appropriate word or phrase.

7. To perform a *t*-test when the sample size is small, the sample must show no evidence of strong _____ and must contain no _____.

8. The number of degrees of freedom for the Student's *t*-test of a population mean is always 1 less than the _____.

In Exercises 9 and 10, determine whether the statement is true or false. If the statement is false, rewrite it as a true statement.

9. A *t*-test is used when the population standard deviation is unknown.

10. A *t*-test is used when the number of degrees of freedom is unknown.

Practicing the Skills

11. Find the *P*-value for the following values of the test statistic *t*, sample size *n*, and alternate hypothesis H_1. If you use Table A.3, you may specify that *P* is between two values.
- **a.** $t = 2.336$, $n = 5$, H_1: $\mu > \mu_0$
- **b.** $t = 1.307$, $n = 18$, H_1: $\mu \neq \mu_0$
- **c.** $t = -2.864$, $n = 51$, H_1: $\mu < \mu_0$
- **d.** $t = -2.031$, $n = 3$, H_1: $\mu \neq \mu_0$

12. Find the *P*-value for the following values of the test statistic *t*, sample size *n*, and alternate hypothesis H_1. If you use Table A.3, you may specify that *P* is between two values.
- **a.** $t = -1.584$, $n = 19$, H_1: $\mu \neq \mu_0$
- **b.** $t = -2.473$, $n = 41$, H_1: $\mu < \mu_0$
- **c.** $t = 1.491$, $n = 30$, H_1: $\mu \neq \mu_0$
- **d.** $t = 3.635$, $n = 4$, H_1: $\mu > \mu_0$

13. Find the critical value or values for the following values of the significance level α, sample size *n*, and alternate hypothesis H_1.
- **a.** $\alpha = 0.05$, $n = 27$, H_1: $\mu \neq \mu_0$
- **b.** $\alpha = 0.01$, $n = 61$, H_1: $\mu > \mu_0$

c. $\alpha = 0.10$, $n = 16$, H_1: $\mu \neq \mu_0$
d. $\alpha = 0.05$, $n = 11$, H_1: $\mu < \mu_0$

14. Find the critical value or values for the following values of the significance level α, sample size n, and alternate hypothesis H_1.
   a. $\alpha = 0.05$, $n = 39$, H_1: $\mu > \mu_0$
   b. $\alpha = 0.01$, $n = 34$, H_1: $\mu < \mu_0$
   c. $\alpha = 0.10$, $n = 6$, H_1: $\mu \neq \mu_0$
   d. $\alpha = 0.05$, $n = 25$, H_1: $\mu \neq \mu_0$

Working with the Concepts

15. Is there a doctor in the house? The market research firm Salary.com reported that in June of 2013, the mean annual earnings of all family practitioners in the United States was $178,258. A random sample of 55 family practitioners in Los Angeles that month had mean earnings of $\bar{x} = \$192,340$ with a standard deviation of $42,387. Do the data provide sufficient evidence to conclude that the mean salary for family practitioners in Los Angeles is greater than the national average?
   a. State the null and alternate hypotheses.
   b. Compute the value of the t statistic. How many degrees of freedom are there?
   c. State your conclusion. Use the $\alpha = 0.05$ level of significance.

16. College tuition: The mean annual tuition and fees in the 2013–2014 academic year for a sample of 14 private colleges in California was $37,900 with a standard deviation of $7,200. A dotplot shows that it is reasonable to assume that the population is approximately normal. Can you conclude that the mean tuition and fees for private institutions in California differs from $35,000?
   a. State the null and alternate hypotheses.
   b. Compute the value of the t statistic. How many degrees of freedom are there?
   c. State your conclusion. Use the $\alpha = 0.01$ level of significance.
   Based on data from collegeprowler.com

17. Big babies: The National Health Statistics Reports described a study in which a sample of 360 one-year-old baby boys were weighed. Their mean weight was 25.5 pounds with standard deviation 5.3 pounds. A pediatrician claims that the mean weight of one-year-old boys is greater than 25 pounds. Do the data provide convincing evidence that the pediatrician's claim is true? Use the $\alpha = 0.01$ level of significance.

18. Good credit: The Fair Isaac Corporation (FICO) credit score is used by banks and other lenders to determine whether someone is a good credit risk. Scores range from 300 to 850, with a score of 720 or more indicating that a person is a very good credit risk. An economist wants to determine whether the mean FICO score is lower than the cutoff of 720. She finds that a random sample of 100 people had a mean FICO score of 703 with a standard deviation of 92. Can the economist conclude that the mean FICO score is less than 720? Use the $\alpha = 0.05$ level of significance.

19. Commuting to work: The 2011 American Community Survey sampled 1923 people in Colorado and asked them how long it took them to commute to work each day. The sample mean one-way commute time was 24.5 minutes with a standard deviation of 13.0 minutes. A transportation engineer claims that the mean commute time is less than 25 minutes. Do the data provide convincing evidence that the engineer's claim is true? Use the $\alpha = 0.05$ level of significance.

20. Watching TV: In 2012, the General Social Survey asked a sample of 1298 people how much time they spent watching TV each day. The mean number of hours was 3.09 with a standard deviation of 2.87. A sociologist claims that people watch a mean of 3 hours of TV per day. Do the data provide sufficient evidence to disprove the claim? Use the $\alpha = 0.01$ level of significance.

21. Weight loss: In a study to determine whether counseling could help people lose weight, a sample of people experienced a group-based behavioral intervention, which involved weekly meetings with a trained interventionist for a period of six months. The following data are the numbers of pounds lost for 14 people, based on means and standard deviations given in the article.

| 18.2 | 24.8 | 3.9 | 20.0 | 17.1 | 8.8 | 13.4 |
| 17.3 | 33.8 | 29.7 | 8.5 | 31.2 | 19.3 | 15.1 |

Source: *Journal of the American Medical Association* 299:1139–1148

   a. Following is a boxplot for these data. Is it reasonable to assume that the conditions for performing a hypothesis test are satisfied? Explain.

   b. If appropriate, perform a hypothesis test to determine whether the mean weight loss is greater than 10 pounds. Use the $\alpha = 0.05$ level of significance. What do you conclude?

22. How much is in that can? A machine that fills beverage cans is supposed to put 12 ounces of beverage in each can. Following are the amounts measured in a simple random sample of eight cans.

11.96   12.10   12.04   12.13   11.98   12.05   11.91   12.03

   a. Following is a dotplot for these data. Is it reasonable to assume that the conditions for performing a hypothesis test are satisfied? Explain.

11.90   11.95   12.00   12.05   12.10   12.15

b. If appropriate, perform a hypothesis test to determine whether the mean volume differs from 12 ounces. Use the $\alpha = 0.05$ level of significance. What do you conclude?

23. Credit card debt: Following are outstanding credit card balances for a sample of 16 college seniors at a large university.

870	419	1021	723	152	387	335	334
2618	529	593	769	502	485	1213	347

a. Following is a dotplot for these data. Is it reasonable to assume that the conditions for performing a hypothesis test are satisfied? Explain.

0   500   1000   1500   2000   2500   3000

b. According to the report *How America Pays for College*, by Sallie Mae, the mean outstanding balance for college seniors in 2012 was $515. If appropriate, perform a hypotheis test to determine whether the mean debt for seniors at this university differs from $515.

24. Rats: A psychologist is designing an experiment in which rats will navigate a maze. Ten rats run the maze, and the time it takes for each to complete the maze is recorded. The results are as follows.

66.3	68.1	52.5	68.3	62.6
55.6	42.1	60.9	69.2	69.3

a. Following is a boxplot for these data. Is it reasonable to assume that the conditions for performing a hypothesis test are satisfied? Explain.

40   45   50   55   60   65   70

b. The psychologist hopes that the mean time for a rat to run the maze will be greater than 60 seconds. If appropriate, perform a hypotheis test to determine whether the mean time is greater than 60 seconds.

25. Keep cool: Following are prices, in dollars, of a random sample of ten 7.5-cubic-foot refrigerators.

314	377	330	285	319
274	332	350	299	306

a. Following is a dotplot for these data. Is it reasonable to assume that the conditions for performing a hypothesis test are satisfied? Explain.

260   280   300   320   340   360   380

b. A consumer organization reports that the mean price of 7.5-cubic-foot refrigerators is greater than $300. Do the data provide convincing evidence of this claim? Use the $\alpha = 0.01$ level of significance.

26. Free dessert: In an attempt to increase business on Monday nights, a restaurant offers a free dessert with every dinner order. Before the offer, the mean number of dinner customers on Monday was 150. Following are the numbers

of diners on a random sample of 12 days while the offer was in effect.

206	169	191	152	212	139
142	151	174	220	192	153

a. Following is a boxplot for these data. Is it reasonable to assume that the conditions for performing a hypothesis test are satisfied? Explain.

120   140   160   180   200   220   240

b. Can you conclude that the mean number of diners increased while the free dessert offer was in effect? Use the $\alpha = 0.01$ level of significance.

27. Effective drugs: When testing a new drug, scientists measure the amount of the active ingredient that is absorbed by the body. In a study done at the Colorado School of Mines, a new antifungal medication that was designed to be applied to the skin was tested. The medication was applied to the skin of eight adult subjects. One hour later, the amount of active ingredient that had been absorbed into the skin was measured for each subject. The results, in micrograms, were

1.28   1.81   2.71   3.13   1.55   2.55   3.36   3.86

a. Construct a boxplot for these data. Is it appropriate to perform a hypothesis test?

b. If appropriate, perform a hypothesis test to determine whether the mean amount absorbed is greater than 2 micrograms. Use the $\alpha = 0.05$ level of significance. What do you conclude?

28. More effective drugs: An antifungal medication was applied to the skin of eight adult subjects. One hour later, the amount of active ingredient that had been absorbed into the skin was measured for each subject. The results, in micrograms, were

2.13   1.88   2.07   1.19   2.51   5.61   2.81   3.05

a. Construct a boxplot for these data. Is it appropriate to perform a hypothesis test?

b. If appropriate, perform a hypothesis test to determine whether the mean amount absorbed is less than 3 micrograms. Use the $\alpha = 0.05$ level of significance. What do you conclude?

29. Interpret calculator display: The following display from a TI-84 Plus calculator presents the results of a hypothesis test for a population mean μ.

a. State the null and alternate hypotheses.

b. What is the value of $\bar{x}$?

c. What is the value of s?

d. How many degrees of freedom are there?

e. Do you reject H_0 at the 0.05 level? Explain.

f. Someone wants to test the hypothesis $H_0: \mu = 1.8$ versus $H_1: \mu > 1.8$. Use the information in the display to compute the t statistic for this test.

g. Compute the P-value for this test.

h. Can the null hypothesis in part (f) be rejected at the 0.05 level? Explain.

30. Interpret calculator display: The following display from a TI-84 Plus calculator presents the results of a hypothesis test for a population mean μ.

```
          T-Test
μ≠20
t=1.206976206
p=.2347112421
x̄=20.75
Sx=3.93
n=40
```

a. State the null and alternate hypotheses.
b. What is the value of $\bar{x}$?
c. What is the value of s?
d. How many degrees of freedom are there?
e. Do you reject H_0 at the 0.05 level? Explain.
f. Someone wants to test the hypothesis H_0: $\mu = 22.5$ versus H_1: $\mu \neq 22.5$. Use the information in the display to compute the t statistic for this test.
g. Compute the P-value for this test.
h. Can the null hypothesis in part (f) be rejected at the 0.05 level? Explain.

31. Interpret computer output: The following computer output (from MINITAB) presents the results of a hypothesis test for a population mean μ.

Test of mu = 5.5 vs > 5.5

N	Mean	StDev	SE Mean	95% Lower Bound	T	P
5	5.92563	0.15755	0.07046	5.77542	6.04	0.002

a. State the null and alternate hypotheses.
b. What is the value of $\bar{x}$?
c. What is the value of s?
d. How many degrees of freedom are there?
e. Do you reject H_0 at the 0.05 level? Explain.
f. Someone wants to test the hypothesis H_0: $\mu = 6.5$ versus H_1: $\mu < 6.5$. Use the information in the output to compute the t statistic for this test.
g. Compute the P-value for this test.
h. Can the null hypothesis in part (f) be rejected at the 0.05 level? Explain.

32. Interpret computer output: The following computer output (from MINITAB) presents the results of a hypothesis test for a population mean μ.

Test of mu = 16 vs not = 16

N	Mean	StDev	SE Mean	95% CI	T	P
11	13.2874	6.0989	1.8389	(9.1901, 17.3847)	−1.48	0.171

a. State the null and alternate hypotheses.
b. What is the value of $\bar{x}$?
c. What is the value of s?
d. How many degrees of freedom are there?
e. Do you reject H_0 at the 0.05 level? Explain.
f. Someone wants to test the hypothesis H_0: $\mu = 9$ versus H_1: $\mu > 9$. Use the information in the output to compute the t statistic for this test.

g. Compute the P-value for this test.
h. Can the null hypothesis in part (f) be rejected at the 0.05 level? Explain.

33. Does this diet work? In a study of the effectiveness of a certain diet, 100 subjects went on the diet for a period of six months. The sample mean weight loss was 0.5 pound, with a sample standard deviation of 4 pounds.
a. Find the t statistic for testing H_0: $\mu = 0$ versus H_1: $\mu > 0$.
b. Find the P-value for testing H_0: $\mu = 0$ versus H_1: $\mu > 0$.
c. Can you conclude that the diet produces a mean weight loss that is greater than 0? Use the $\alpha = 0.05$ level of significance.

34. Effect of larger sample size: The study described in Exercise 33 is repeated with a larger sample of 1000 subjects. Assume that the sample mean is once again 0.5 pound and the sample standard deviation is once again 4 pounds.
a. Find the t statistic for testing H_0: $\mu = 0$ versus H_1: $\mu > 0$. Is the value of the t statistic greater than or less than the value obtained with a smaller sample of 100?
b. Find the P-value for testing H_0: $\mu = 0$ versus H_1: $\mu > 0$.
c. Can you conclude that the diet produces a mean weight loss that is greater than 0? Use the $\alpha = 0.05$ level of significance.
d. Explain why the mean weight loss is not of practical significance, even though the results are statistically significant at the 0.05 level.

35. Perform a hypothesis test? A sociologist wants to test the null hypothesis that the mean number of people per household in a given city is equal to 3. He surveys 50 households on a certain block in the city and finds that the sample mean number of people is 3.4 with a standard deviation of 1.2. Should these data be used to perform a hypothesis test? Explain why or why not.

36. Perform a hypothesis test? A health professional wants to test the null hypothesis that the mean length of hospital stay for a certain surgical procedure is 4 days. She obtains records for all the patients who have undergone the procedure at a certain hospital during a given year, and finds that the mean length of stay is 4.7 days with a standard deviation of 1.1 days. Should these data be used to perform a hypothesis test? Explain why or why not.

37. Larger or smaller P-value? In a study of sleeping habits, a researcher wants to test the null hypothesis that adults in a certain community get a mean of 8 hours of sleep versus the alternative that the mean is not equal to 8. In a sample of 250 adults, the mean number of hours of sleep was 8.2. A second researcher repeated the study with a different sample of 250, and obtained a sample mean of 7.5. Both researchers obtained the same standard deviation. Will the P-value of the second researcher be greater than or less than that of the first researcher? Explain.

38. Larger or smaller P-value? Juan and Mary want to test the null hypothesis that the mean length of text messages sent by students at their school is 10 characters versus the alternative that it is less. Juan samples 100 text messages and finds the mean length to be 8.4 characters. Mary samples 100 messages and finds the mean length to be 7.3 characters. Both Juan and Mary obtained the same standard deviation. Will Juan's *P*-value be greater than, less than, or the same as Mary's *P*-value? Explain.

39. **Interpret a *P*-value:** A real estate agent believes that the mean size of houses in a certain city is greater than 1500 square feet. He samples 100 houses, and performs a test of H_0: $\mu = 1500$ versus H_1: $\mu > 1500$. He obtains a *P*-value of 0.0002.
    a. The real estate agent concludes that because the *P*-value is very small, the mean house size must be much greater than 1500. Is this conclusion justified?
    b. Another real estate agent says that because the *P*-value is very small, we can be fairly certain that the mean size is greater than 1500, but we cannot conclude that it is a lot greater. Is this conclusion justified?

40. **Interpret a *P*-value:** The manufacturer of a medication designed to lower blood pressure claims that the mean systolic blood pressure for people taking their medication is less than 135. To test this claim, blood pressure is measured for a sample of 500 people who are taking the medication. The *P*-value for testing H_0: $\mu = 135$ versus H_1: $\mu < 135$ is $P = 0.001$.
    a. The manufacturer concludes that because the *P*-value is very small, we can be fairly certain that the mean pressure

is less than 135, but we cannot conclude that it is a lot smaller. Is this conclusion justified?
    b. Someone else says that because the *P*-value is very small, we can conclude that the mean pressure is a lot less than 135. Is this conclusion justified?

Extending the Concepts

41. **Using *z* instead of *t*:** When the sample size is large, some people treat the sample standard deviation *s* as if it were the population standard deviation σ, and use the standard normal distribution rather than the Student's *t* distribution, to find a critical value. Assume that a right-tailed test will be made with a sample of size 100 from a normal population, using the $\alpha = 0.05$ significance level.
    a. Find the critical value under the assumption that σ is known.
    b. In fact, σ is unknown. How many degrees of freedom should be used for the Student's *t* distribution?
    c. What is the probability of rejecting H_0 when it is true if the critical value in part (a) is used? You will need technology to find the answer.

Answers to Check Your Understanding Exercises for Section 9.3

1. **a.** *P*-value is between 0.01 and 0.025 [Tech: 0.0127]
   b. *P*-value is between 0.025 and 0.05 [Tech: 0.0485]
   c. *P*-value is between 0.005 and 0.01 [Tech: 0.0096]
   d. *P*-value is between 0.01 and 0.02 [Tech: 0.0148]

2. **a.** $t = 2.309$ **b.** 40
   c. Between 0.01 and 0.025 [Tech: 0.0131]
   d. Weaker; the *P*-value is larger.

3. 0.0128

4. **a.** H_0: $\mu = 7$, H_1: $\mu < 7$ **b.** 6 **c.** 5 **d.** 6.68
   e. 0.205 **f.** -3.823593745 **g.** 0.00616394 **h.** Yes

5. 0.0312

6. **a.** 2.920 **b.** -2.787, 2.787
   c. 1.292 **d.** -2.160, 2.160

SECTION 9.4 | Hypothesis Tests for Proportions

Objectives

1. Test a hypothesis about a proportion using the *P*-value method
2. Test a hypothesis about a proportion using the critical value method

Objective 1 Test a hypothesis about a proportion using the *P*-value method

How cool is Facebook? In a recent GenX2Z American College Student Survey, 90% of female college students rated the social network site Facebook as "cool." The other 10% rated it as "lame." Assume that the survey was based on a sample of 500 students. A marketing executive at Facebook wants to advertise the site with the slogan "More than 85% of female college students think Facebook is cool." Before launching the ad campaign, he wants to be confident that the slogan is true. Can he conclude that the proportion of female college students who think Facebook is cool is greater than 0.85?

This is an example of a problem that calls for a hypothesis test about a population proportion. There are two categories, "cool" and "lame." The quantity 0.85 represents the proportion in the "cool" category. To perform the test, we will need some notation, which we summarize as follows.

NOTATION

- p is the population proportion of individuals who are in a specified category.
- p_0 is the population proportion specified by H_0.
- x is the number of individuals in the sample who are in the specified category.
- n is the sample size.
- $\hat{p}$ is the sample proportion of individuals who are in the specified category. $\hat{p} = x/n$.

Explain It Again

Reasons for the assumptions:
The population must be much larger than the sample (at least 20 times as large), so that the sampled items are independent. The assumption that both np_0 and $n(1 - p_0)$ are at least 10 ensures that the sampling distribution of $\hat{p}$ is approximately normal when we assume that H_0 is true.

We can perform a test whenever the sample proportion $\hat{p}$ is approximately normally distributed. This will occur when the following assumptions are met.

Assumptions for Performing a Hypothesis Test for a Population Proportion

1. We have a simple random sample.
2. The population is at least 20 times as large as the sample.
3. The individuals in the population are divided into two categories.
4. The values np_0 and $n(1 - p_0)$ are both at least 10.

Either the critical value method or the P-value method may be used to perform a hypothesis test for a population proportion. We will present the steps for the P-value method first.

Performing a Hypothesis Test for a Population Proportion Using the *P*-Value Method

Check to be sure the assumptions are satisfied. If they are, then proceed with the following steps:

Step 1: State the null and alternate hypotheses. The null hypothesis will have the form $H_0: p = p_0$. The alternate hypothesis will be $p < p_0$, $p > p_0$, or $p \neq p_0$.

Step 2: If making a decision, choose a significance level α.

Step 3: Compute the test statistic $z = \dfrac{\hat{p} - p_0}{\sqrt{\dfrac{p_0(1 - p_0)}{n}}}$.

Step 4: Compute the P-value. The P-value is an area under the standard normal curve; it depends on the alternate hypothesis as follows:

The P-value is the area to the left of z.

Left-tailed: $H_1: p < p_0$

The P-value is the area to the right of z.

Right-tailed: $H_1: p > p_0$

The P-value is the sum of the areas in the two tails.

Two-tailed: $H_1: p \neq p_0$

Step 5: Interpret the P-value. If making a decision, reject H_0 if the P-value is less than or equal to the significance level α.

Step 6: State a conclusion.

EXAMPLE 9.19   Perform a hypothesis test

In a recent GenX2Z American College Student Survey, 90% of female college students rated the social network site Facebook as "cool." Assume that the survey was based on a random sample of 500 students. A marketing executive at Facebook wants to advertise the site with the slogan "More than 85% of female college students think Facebook is cool." Can you conclude that the proportion of female college students who think Facebook is cool is greater than 0.85? Use the $\alpha = 0.05$ level of significance.

Solution

We first check the assumptions. We have a simple random sample of students. The members of the population fall into two categories: those who think that Facebook is cool and those who don't. The size of the population of female college students is more than 20 times the sample size of $n = 500$. The proportion specified by the null hypothesis is $p_0 = 0.85$. Now $np_0 = (500)(0.85) = 425 > 10$ and $n(1 - p_0) = (500)(1 - 0.85) = 75 > 10$. The assumptions are satisfied.

Step 1: State H_0 and H_1. We are asked whether we can conclude that the population proportion p is greater than 0.85. The null and alternate hypotheses are therefore

$$H_0: p = 0.85 \qquad H_1: p > 0.85$$

Step 2: Choose a significance level. The significance level is $\alpha = 0.05$.

Step 3: Compute the test statistic. The sample proportion $\hat{p}$ is 0.90. The value of p specified by the null hypothesis is $p_0 = 0.85$. The test statistic is the z-score for $\hat{p}$:

$$z = \frac{\hat{p} - p_0}{\sqrt{\dfrac{p_0(1 - p_0)}{n}}} = \frac{0.90 - 0.85}{\sqrt{\dfrac{0.85(1 - 0.85)}{500}}} = 3.13$$

Step 4: Compute the P-value. The alternate hypothesis is $H_1: \mu > 0.85$, which is right-tailed. The P-value is therefore the area to the right of $z = 3.13$. Using Table A.2, we see that the area to the left of $z = 3.13$ is 0.9991. The area to the right of $z = 3.13$ is therefore $1 - 0.9991 = 0.0009$. The P-value is $P = 0.0009$. See Figure 9.16.

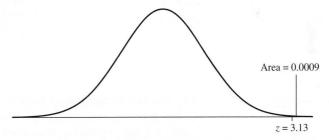

Area = 0.0009

$z = 3.13$

Figure 9.16

Step 5: Interpret the P-value. A P-value of $P = 0.0009$ is very small. This is very strong evidence against H_0. In particular, because $P < 0.05$, we reject H_0 at the $\alpha = 0.05$ level.

Step 6: State a conclusion. We conclude that more than 85% of female college students think Facebook is cool.

Check Your Understanding

1. The Pew Research Center reported that only 15% of 18- to 24-year-olds read a daily newspaper. The publisher of a local newspaper wants to know whether the percentage of newspaper readers among students at a nearby large university differs from the percentage among 18- to 24-year-olds in general. She surveys a simple random sample of 200 students at the university and finds that 40 of them, or 20%, read a newspaper each day. Can she conclude that the proportion of students who read a daily newspaper differs from 0.15? Use the $\alpha = 0.05$ level of significance.

a. State the null and alternate hypotheses.
b. Compute the test statistic.
c. Compute the P-value.
d. State a conclusion.

Answers are on page 458.

Performing a hypothesis test with technology

The following computer output (from MINITAB) presents the results of Example 9.19.

```
Test of p = 0.85 vs p > 0.85
                                      95%
                                     Lower
Sample     X     N    Sample p      Bound    Z-Value   P-Value
1        450   500    0.900000    0.877932     3.13     0.001
```

Most of the output is straightforward. The first line specifies the null and alternate hypotheses. The quantity labeled "X" is the number of people in the sample who think Facebook is cool, and N is the sample size. The quantity labeled "Sample p" is the sample proportion $\hat{p}$. The quantity labeled "95% Lower Bound" is a 95% lower confidence bound for the population proportion p. The interpretation of this quantity is that we are 95% confident that the population proportion p is greater than or equal to 0.877932. Next is the value of the test statistic z, labeled "Z-value," and finally at the end of the row is the P-value.

The following display from a TI-84 Plus calculator presents the results of Example 9.19.

```
    1-PropZTest
  prop>.85
  z=3.131121455
  p=8.7076769E-4
  p̂=.9
  n=500
```

The first line in the display presents the alternate hypothesis. The word "prop" refers to the population proportion p. Note that the letter "p" in the third line is the P-value, not the population proportion. This number is written in scientific notation as 8.7076769E-4. This indicates that we should move the decimal point four places to the left, so $P = 0.00087076769$.

Step-by-step instructions for performing hypothesis tests with technology are presented in the Using Technology section on page 453.

Check Your Understanding

2. The following output from MINITAB presents the results of a hypothesis test.

```
Test of p = 0.60 vs p < 0.60
                                      95%
                                     Lower
Sample     X     N    Sample p      Bound    Z-Value   P-Value
1         72   150    0.480000    0.547097    -3.00     0.001
```

a. What are the null and alternate hypotheses?
b. What is the sample size?
c. What is the value of $\hat{p}$?
d. What is the value of the test statistic?
e. Do you reject H_0 at the 0.05 level?
f. Do you reject H_0 at the 0.01 level?

3. The following display from a TI-84 Plus calculator presents the results of a hypothesis test.

```
┌─────────────────────────┐
│       1-PropZTest       │
│  prop>.75               │
│  z=.4783759373          │
│  p=.3161913337          │
│  p̂=.7559183673          │
│  n=1225                 │
│                         │
└─────────────────────────┘
```

a. What are the null and alternate hypotheses?
b. What is the sample size?
c. What is the value of $\hat{p}$?
d. What is the value of the test statistic?
e. Do you reject H_0 at the 0.05 level?
f. Do you reject H_0 at the 0.01 level?

Answers are on page 458.

Objective 2 Test a hypothesis about a proportion using the critical value method

Testing Hypotheses for a Proportion Using the Critical Value Method

To use the critical value method, compute the test statistic as before. Because the test statistic is a z-score, critical values can be found in Table A.2, in the last line of Table A.3, or with technology. The assumptions for the critical value method are the same as for the P-value method.

Assumptions for Performing a Hypothesis Test for a Population Proportion

1. We have a simple random sample.
2. The population is at least 20 times as large as the sample.
3. The items in the population are divided into two categories.
4. The values np_0 and $n(1 - p_0)$ are both at least 10.

Following are the steps for the critical value method.

Performing a Hypothesis Test for a Proportion Using the Critical Value Method

Check to be sure the assumptions are satisfied. If they are, then proceed with the following steps:

Step 1: State the null and alternate hypotheses. The null hypothesis will have the form $H_0: p = p_0$. The alternate hypothesis will be $p < p_0$, $p > p_0$, or $p \neq p_0$.

Step 2: Choose a significance level α and find the critical value or values.

Step 3: Compute the test statistic $z = \dfrac{\hat{p} - p_0}{\sqrt{\dfrac{p_0(1 - p_0)}{n}}}$.

Step 4: Determine whether to reject H_0, as follows:

Left-tailed: $H_1: p < p_0$
Reject if $z \leq -z_\alpha$.

Right-tailed: $H_1: p > p_0$
Reject if $z \geq z_\alpha$.

Two-tailed: $H_1: p \neq p_0$
Reject if $z \geq z_{\alpha/2}$ or $z \leq -z_{\alpha/2}$.

Step 5: State a conclusion.

EXAMPLE 9.20

Test a hypothesis about a population proportion using the critical value method

A nationwide survey of working adults indicates that only 50% of them are satisfied with their jobs. The president of a large company believes that more than 50% of employees at his company are satisfied with their jobs. To test his belief, he surveys a random sample of 100 employees, and 54 of them report that they are satisfied with their jobs. Can he conclude that more than 50% of employees at the company are satisfied with their jobs? Use the $\alpha = 0.05$ level of significance.

Solution

We first check the assumptions. We have a simple random sample from the population of employees. Each employee is categorized as being satisfied or not satisfied. The sample size is $n = 100$ and the proportion p_0 specified by H_0 is 0.5. Therefore, we calculate that $np_0 = 100(0.5) = 50 > 10$, and $n(1 - p_0) = 100(1 - 0.5) = 50 > 10$. If the total number of employees in the company is more than 2000, as we shall assume, then the population is more than 20 times as large as the sample. All the assumptions are therefore satisfied.

Step 1: State the null and alternate hypotheses. The issue is whether the proportion of employees that are satisfied with their jobs is more than 0.5. Therefore, the null and alternate hypotheses are

$$H_0: p = 0.5 \qquad H_1: p > 0.5$$

Step 2: Choose a significance level and find the critical value. The significance level is $\alpha = 0.05$. The alternate hypothesis is $p > 0.5$, so this is a right-tailed test. The critical value corresponding to $\alpha = 0.05$ is $z_\alpha = 1.645$.

Step 3: Compute the test statistic. The test statistic is

$$z = \frac{\hat{p} - p_0}{\sqrt{\dfrac{p_0(1 - p_0)}{n}}}$$

The value of the sample proportion $\hat{p}$ is

$$\hat{p} = \frac{\text{Number of satisfied employees}}{\text{Sample size}} = \frac{54}{100} = 0.54$$

The quantity p_0 is the value of p specified by H_0, so $p_0 = 0.5$. The sample size is $n = 100$. Therefore, the value of the test statistic is

$$z = \frac{0.54 - 0.5}{\sqrt{\dfrac{0.5(1 - 0.5)}{100}}} = 0.80$$

Step 4: Determine whether to reject H_0. Because this is a right-tailed test, we reject H_0 if $z \geq 1.645$. Because $0.80 < 1.645$, we do not reject H_0. See Figure 9.17.

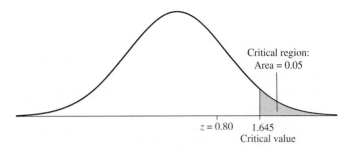

Figure 9.17

Step 5: State a conclusion. There is not enough evidence to conclude that the company president is correct in his belief that the proportion of employees who are satisfied with their jobs is greater than 0.5. The proportion may be equal to 0.5.

Check Your Understanding

4. A Gallup poll taken in December 2009 sampled 1000 adults in the United States. Of these people, 770 said they enjoyed situations in which they competed with other people. Can you conclude that less than 80% of U.S. adults like to compete? Use the critical value method with significance level $\alpha = 0.05$.
   a. State the null and alternate hypotheses.
   b. Compute the test statistic.
   c. Find the critical value.
   d. State a conclusion.

Answers are on page 458.

USING TECHNOLOGY

We use Example 9.20 to illustrate the technology steps.

TI-84 PLUS

Testing a hypothesis about a proportion

Step 1. Press **STAT** and highlight the **TESTS** menu.
Step 2. Select **1–PropZTest** and press **ENTER** (Figure A). The **1–PropZTest** menu appears.
Step 3. Enter the values of p_0, x, and n. For Example 9.20, we use $p_0 = 0.5$, $x = 54$, and $n = 100$.
Step 4. Select the form of the alternate hypothesis. For Example 9.20, the alternate hypothesis has the form $> p_0$ (Figure B).
Step 5. Highlight **Calculate** and press **ENTER** (Figure C).

Figure A

Figure B

Figure C

MINITAB

Testing a hypothesis about a proportion

Step 1. Click on **Stat**, then **Basic Statistics**, then **1 Proportion**.

Step 2. Click **Summarized Data**, and enter the value of n in the **Number of Trials** field and the value of x in the **Number of Events** field. For Example 9.20, we use $x = 54$ and $n = 100$ (Figure D).

Step 3. Click **Options**, and enter the value of p_0 in the **Test proportion** field and the form of the alternate hypothesis in the **Alternative** field. For Example 9.20, we enter 0.5 for p_0 and enter **greater than** as the alternate hypothesis.

Step 4. Given significance level α, enter $100(1 - \alpha)$ as the **Confidence Level**. For Example 9.20, $\alpha = 0.05$, so the confidence level is $100(1 - 0.05) = 95$ (Figure E).

Step 5. Check the **Use test and interval based on normal distribution** option and click **OK**.

Step 6. Click **OK** (Figure F).

Figure D

Figure E

Test and CI for One Proportion

```
Test of p = 0.5 vs p > 0.5


                                95% Lower
Sample    X     N   Sample p       Bound   Z-Value  P-Value
1        54   100   0.540000     0.458021     0.80    0.212

Using the normal approximation.
```

Figure F

EXCEL

Testing a hypothesis about a proportion

This procedure requires the **MegaStat** EXCEL add-in to be loaded. The **MegaStat** add-in can be downloaded from www.mhhe.com/megastat.

Step 1. Load the **MegaStat** EXCEL add-in.

Step 2. Click on the **MegaStat** menu and select **Hypothesis Tests**, then **Proportion vs. Hypothesized Value...**

Step 3. Under the *Observed* column, enter the value of x in the **p** field (note that p automatically changes to x) and the sample size n in the **n** field. For Example 9.20, we use $x = 54$ and $n = 100$.

Step 4. Under the *Hypothesized* column, enter the value of p_0. We use $p_0 = 0.5$.

Step 5. Select the form of the alternate hypothesis. For Example 9.20, we use **greater than** (Figure G).

Step 6. Click **OK** (Figure H).

Figure G

Hypothesis test for proportion vs hypothesized value

Observed	Hypothesized	
0.54	0.5	p (as decimal)
54/100	50/100	p (as fraction)
54.	50.	X
100	100	n
	0.05	std. error
	0.80	z
	.2119	p-value (one-tailed, upper)

Figure H

SECTION 9.4   Exercises

Exercises 1–4 are the Check Your Understanding exercises located within the section.

Understanding the Concepts

In Exercises 5 and 6, fill in each blank with the appropriate word or phrase.

5. To test $H_0: p = p_0$ with the methods in this section, the values np_0 and $n(1 - p_0)$ must both be at least _____.

6. To test $H_0: p = p_0$ with the methods in this section, the population size must be at least _____ times as large as the sample size.

In Exercises 7 and 8, determine whether the statement is true or false. If the statement is false, rewrite it as a true statement.

7. When testing a hypothesis for a proportion, we assume that the items in the population are divided into two categories.

8. When testing a hypothesis for a proportion, the alternate hypothesis is always two-tailed.

Practicing the Skills

9. In a simple random sample of size 80, there were 54 individuals in the category of interest.
   a. Compute the sample proportion $\hat{p}$.
   b. Are the assumptions for a hypothesis test satisfied? Explain.
   c. It is desired to test $H_0: p = 0.8$ versus $H_1: p < 0.8$. Compute the test statistic z.
   d. Do you reject H_0 at the 0.05 level?

10. In a simple random sample of size 60, there were 38 individuals in the category of interest.
   a. Compute the sample proportion $\hat{p}$.
   b. Are the assumptions for a hypothesis test satisfied? Explain.
   c. It is desired to test $H_0: p = 0.7$ versus $H_1: p \neq 0.7$. Compute the test statistic z.

   d. Do you reject H_0 at the 0.05 level?

11. In a simple random sample of size 75, there were 42 individuals in the category of interest.
   a. Compute the sample proportion $\hat{p}$.
   b. Are the assumptions for a hypothesis test satisfied? Explain.
   c. It is desired to test $H_0: p = 0.6$ versus $H_1: p \neq 0.6$. Compute the test statistic z.
   d. Do you reject H_0 at the 0.05 level?

12. In a simple random sample of size 150, there were 90 individuals in the category of interest.
   a. Compute the sample proportion $\hat{p}$.
   b. Are the assumptions for a hypothesis test satisfied? Explain.
   c. It is desired to test $H_0: p = 0.5$ versus $H_1: p > 0.5$. Compute the test statistic z.
   d. Do you reject H_0 at the 0.05 level?

Working with the Concepts

13. Spam: According to SecureList, 71.8% of all email sent in July 2012 was spam. A system manager at a large corporation believes that the percentage at his company may be 80%. He examines a random sample of 500 emails received at an email server, and finds that 382 of the messages are spam.
   a. State the appropriate null and alternate hypotheses.
   b. Compute the test statistic z.
   c. Using $\alpha = 0.05$, can you conclude that the percentage of emails that are spam differs from 80%?
   d. Using $\alpha = 0.01$, can you conclude that the percentage of emails that are spam differs from 80%?

14. Confidence in banks: A poll conducted in 2012 by the General Social Survey asked a random sample of 1325 adults in the United States how much confidence they had in banks and other financial institutions. A total of 149 adults

said that they had a great deal of confidence. An economist claims that less than 15% of U.S. adults have a great deal of confidence in banks.
a. State the appropriate null and alternate hypotheses.
b. Compute the test statistic z.
c. Using $\alpha = 0.05$, can you conclude that the executive's claim is true?
d. Using $\alpha = 0.01$, can you conclude that the executive's claim is true?

15. **Kids with cell phones:** A marketing manager for a cell phone company claims that more than 55% of children aged 8–12 have cell phones. In a 2012 survey of 802 children aged 8–12 by the National Consumers League, 449 of them had cell phones. Can you conclude that the manager's claim is true? Use the $\alpha = 0.01$ level of significance.

16. **Internet tax:** In 2013, the Gallup Poll asked 1015 U.S. adults whether they believed that people should pay sales tax on items purchased over the Internet. Of these, 437 said they supported such a tax. Does the survey provide convincing evidence that less than 45% of U.S. adults favor an Internet sales tax? Use the $\alpha = 0.05$ level of significance.

17. **Quit smoking:** In a survey of 444 HIV-positive smokers, 170 reported that they had used a nicotine patch to try to quit smoking. Can you conclude that less than half of HIV-positive smokers have used a nicotine patch? Use the $\alpha = 0.05$ level of significance.
Source: *American Journal of Health Behavior* 32:3–15

18. **Game consoles:** A poll taken by the Software Usability Research Laboratory in 2012 surveyed 341 video gamers, and 110 of them said that they prefer playing games on a console, rather than a computer or hand-held device. An executive at a game console manufacturing company claims that more than 25% of gamers prefer consoles. Does the poll provide convincing evidence that the claim is true? Use the $\alpha = 0.01$ level of significance.

19. **Tattoo:** A Harris poll taken in 2012 surveyed 2016 adults and found that 423 of them had one or more tattoos. Can you conclude that the percentage of adults who have a tattoo is less than 25%? Use the $\alpha = 0.01$ level of significance.

20. **Curing diabetes:** Vertical banded gastroplasty is a surgical procedure that reduces the volume of the stomach in order to produce weight loss. In a recent study, 82 patients with Type 2 diabetes underwent this procedure, and 59 of them experienced a recovery from diabetes. Does this study

provide convincing evidence that more than 60% of those with diabetes who undergo this surgery will recover from diabetes? Use the $\alpha = 0.05$ level of significance.
Source: *New England Journal of Medicine* 357:753–761

21. **Tweet tweet:** An article in *Forbes* magazine in December 2012 reported that 73% of Fortune 500 companies have Twitter accounts. A economist thinks the percentage is higher at technology companies. She samples 70 technology companies and finds that 55 of them have Twitter accounts. Can she conclude that more than 73% of technology companies have Twitter accounts? Use the $\alpha = 0.05$ level of significance.

22. **Online photos:** A Pew poll taken in December 2012 surveyed 1802 Internet users and found that 829 of them had posted a photo or video online. Can you conclude that less than half of Internet users have posted photos or videos online? Use the $\alpha = 0.05$ level of significance.

23. **Choosing a doctor:** Which do patients value more when choosing a doctor: interpersonal skills or technical ability? In a recent study, 304 people were asked to choose a physician based on two hypothetical descriptions. One physician was described as having high technical skills and average interpersonal skills, and the other was described as having average technical skills and high interpersonal skills. The physician with high interpersonal skills was chosen by 116 of the people. Can you conclude that less than half of patients prefer a physician with high interpersonal skills? Use the $\alpha = 0.01$ level of significance.
Source: *Health Services Research* 40:957–977

24. **Cable TV choices:** A telecommunications company provided its cable TV subscribers with free access to a new sports channel for a period of one month. It then chose a sample of 400 television viewers and asked them whether they would be willing to pay an extra $10 per month to continue to access the channel. A total of 25 of the 400 replied that they would be willing to pay. The marketing director of the company claims that more than 5% of all its subscribers would pay for the channel. Can you conclude that the director's claim is true? Use the $\alpha = 0.01$ level of significance.

25. **Interpret calculator display:** The following display from a TI-84 Plus calculator presents the results of a hypothesis test for a population proportion p.

a. What are the null and alternate hypotheses?
b. What is the value of the sample proportion $\hat{p}$?
c. Can H_0 be rejected at the 0.05 level? Explain.
d. Someone wants to use these data to test H_0: $p = 0.25$ versus H_1: $p \neq 0.25$. Find the test statistic z and use the method of this section to find the P-value. Do you reject H_0 at the $\alpha = 0.05$ level?

26. **Interpret calculator display:** The following display from a TI-84 Plus calculator presents the results of a hypothesis test for a population proportion p.

a. What are the null and alternate hypotheses?
b. What is the value of the sample proportion $\hat{p}$?
c. Can H_0 be rejected at the 0.05 level? Explain.
d. Someone wants to use these data to test H_0: $p = 0.5$ versus H_1: $p < 0.5$. Find the test statistic z and use the method of this section to find the P-value. Do you reject H_0 at the $\alpha = 0.05$ level?

27. Interpret computer output: The following MINITAB output presents the results of a hypothesis test for a population proportion p.

Test of p = 0.6 vs p > 0.6

X	N	Sample p	95% Lower Bound	Z-Value	P-Value
539	871	0.618829	0.591760	1.13	0.129

a. What are the null and alternate hypotheses?
b. What is the value of the sample proportion $\hat{p}$?
c. Can H_0 be rejected at the 0.05 level? Explain.
d. Someone wants to use these data to test H_0: $p = 0.65$ versus H_1: $p < 0.65$. Find the test statistic z and use the method of this section to find the P-value. Do you reject H_0 at the $\alpha = 0.05$ level?

28. Interpret computer output: The following MINITAB output presents the results of a hypothesis test for a population proportion p.

Test of p = 0.7 vs p not equal 0.7

X	N	Sample p	95% CI	Z-Value	P-Value
27	52	0.519231	(0.383432, 0.655029)	−2.84	0.004

a. What are the null and alternate hypotheses?
b. What is the value of the sample proportion $\hat{p}$?
c. Can H_0 be rejected at the 0.05 level? Explain.
d. Someone wants to use these data to test H_0: $p = 0.6$ versus H_1: $p \neq 0.6$. Find the test statistic z and use the method of this section to find the P-value. Do you reject H_0 at the $\alpha = 0.05$ level?

29. Satisfied with college? A simple random sample of 500 students at a certain college were surveyed and asked whether they were satisfied with college life. Two hundred eighty of them replied that they were satisfied. The Dean of Students claims that more than half of the students at the college are satisfied. To test this claim, a test of the hypotheses H_0: $p = 0.5$ versus H_1: $p > 0.5$ is performed.

a. Show that the P-value is 0.004.
b. The P-value is very small, so H_0 is rejected. Someone claims that because P is very small, the population proportion p must be much greater than 0.5. Is this a correct interpretation of the P-value?
c. Someone else claims that because the P-value is very small, we can be fairly certain that the population proportion p is greater than 0.5, but we cannot be certain that it is a lot greater. Is this a correct interpretation of the P-value?

30. Who will you vote for? A simple random sample of 1500 voters were surveyed and asked whether they were planning to vote for the incumbent mayor for re-election. Seven hundred ninety-eight of them replied that they were planning to vote for the mayor. The mayor claims that more than half of all voters are planning to vote for her. To test this claim, a test of the hypotheses H_0: $p = 0.5$ versus H_1: $p > 0.5$ is performed.

a. Show that the P-value is 0.007.
b. The P-value is very small, so H_0 is rejected. A pollster claims that because the P-value is very small, we can be fairly certain that the population proportion p is greater than 0.5, but we cannot be certain that it is a lot greater. Is this a correct interpretation of the P-value?
c. The mayor's campaign manager claims that because the P-value is very small, the population proportion of voters who plan to vote for the mayor must be much greater than 0.5. Is this a correct interpretation of the P-value?

31. Don't perform a test: A few weeks before election day, a TV station broadcast a debate between the two leading candidates for governor. Viewers were invited to send a tweet to indicate which candidate they plan to vote for. A total of 3125 people tweeted, and 1800 of them said that they planned to vote for candidate A. Explain why these data should not be used to test the claim that more than half of the voters plan to vote for candidate A.

32. Don't perform a test: Over the past 100 days, the price of a certain stock went up on 60 days and went down on 40 days. Explain why these data should not be used to test the claim that this stock price goes down on less than half of the days.

Extending the Concepts

33. Exact test: When $np_0 < 10$ or $n(1 - p_0) < 10$, we cannot use the normal approximation, but we can use the binomial distribution to perform what is known as an *exact test*. Let p be the probability that a given coin lands heads. The coin is tossed 10 times and comes up heads 9 times. Test H_0: $p = 0.5$ versus H_1: $p > 0.5$, as follows.

a. Let n be the number of tosses and let X denote the number of heads. Find the values of n and X in this example.
b. The distribution of X is binomial. Assuming H_0 is true, find n and p.
c. Because the alternate hypothesis is $p > 0.5$, large values of X support H_1. Find the probability of observing a value of X as extreme as or more extreme than the value actually observed, assuming H_0 to be true. This is the P-value.
d. Do you reject H_0 at the $\alpha = 0.05$ level?

Answers to Check Your Understanding Exercises for Section 9.4

1. a. $H_0: p = 0.15$, $H_1: p \neq 0.15$ **b.** $z = 1.98$
c. $P = 0.0478$ [Tech: 0.0477]
d. We conclude that the proportion of students who read a newspaper differs from 0.15.
2. a. $H_0: p = 0.6$, $H_1: p < 0.6$ **b.** 150 **c.** 0.48
d. -3.00 **e.** Yes **f.** Yes

3. a. $H_0: p = 0.75$, $H_1: p > 0.75$ **b.** 1225
c. 0.7559183673 **d.** 0.4783759373 **e.** No **f.** No
4. a. $H_0: p = 0.80$, $H_1: p < 0.80$ **b.** $z = -2.37$
c. -1.645
d. We conclude that less than 80% of U.S. adults enjoy competing with others.

SECTION 9.5 | ## Hypothesis Tests for a Standard Deviation

Objectives

1. Find critical values of the chi-square distribution
2. Test hypotheses about the standard deviation of a normal distribution

Objective 1 Find critical values of the chi-square distribution

The chi-square distribution was introduced in Section 8.4. The critical value χ^2_α represents the value that has area α to its right. We review the method for finding critical values for the chi-square distribution from Table A.4.

EXAMPLE 9.21 | ### Find a critical value

Find the critical value $\chi^2_{0.05}$ for a chi-square distribution with 10 degrees of freedom.

Solution

We consult Table A.4. The critical value is located at the intersection of the row corresponding to 10 degrees of freedom and the column corresponding to $\alpha = 0.05$. The critical value is $\chi^2_{0.05} = 18.307$.

CAUTION

The methods of this section apply only for samples drawn from a normal distribution. If the distribution differs even slightly from normal, these methods should not be used.

Degrees of Freedom	0.995	0.99	0.975	0.95	0.90	0.10	0.05	0.025	0.01	0.005
$\vdots$	$\vdots$	$\vdots$	$\vdots$	$\vdots$	$\vdots$	$\vdots$	$\vdots$	$\vdots$	$\vdots$	$\vdots$
8	1.344	1.646	2.180	2.733	3.490	13.362	15.507	17.535	20.090	21.955
9	1.735	2.088	2.700	3.325	4.168	14.684	16.919	19.023	21.666	23.589
10	2.156	2.558	3.247	3.940	4.865	15.987	18.307	20.483	23.209	25.188
11	2.603	3.053	3.816	4.575	5.578	17.275	19.675	21.920	24.725	26.757
12	3.074	3.571	4.404	5.226	6.304	18.549	21.026	23.337	26.217	28.300
$\vdots$	$\vdots$	$\vdots$	$\vdots$	$\vdots$	$\vdots$	$\vdots$	$\vdots$	$\vdots$	$\vdots$	$\vdots$

(Header spanning columns 0.995 through 0.005: **Area in Right Tail**)

Objective 2 Test hypotheses about the standard deviation of a normal distribution

Hypothesis Tests for the Standard Deviation

The null hypothesis for a standard deviation σ is of the form $H_0: \sigma = \sigma_0$. The test is based on the fact that if H_0 is true, then the test statistic

$$\chi^2 = \frac{(n-1) \cdot s^2}{\sigma_0^2}$$

has a chi-square distribution with $n - 1$ degrees of freedom. We will focus on how to perform a test using the critical value method and Table A.4.

Performing a Hypothesis Test for a Standard Deviation

Check to be sure that the assumptions are satisfied. We must have a simple random sample from a normal population.

Step 1: State the null and alternate hypotheses. The null hypothesis will be of the form $H_0: \sigma = \sigma_0$. The alternate hypothesis can be stated in one of three ways:
Left-tailed: $H_1: \sigma < \sigma_0$
Right-tailed: $H_1: \sigma > \sigma_0$
Two-tailed: $H_1: \sigma \neq \sigma_0$

Step 2: Choose a significance level α, and find the critical value, using Table A.4 with $n - 1$ degrees of freedom.
Left-tailed: $H_1: \sigma < \sigma_0$:   The critical value is $\chi^2_{1-\alpha}$.
Right-tailed: $H_1: \sigma > \sigma_0$:   The critical value is χ^2_{α}.
Two-tailed: $H_1: \sigma \neq \sigma_0$:   The critical values are $\chi^2_{\alpha/2}$ and $\chi^2_{1-\alpha/2}$.

Step 3: Compute the test statistic $\chi^2 = \dfrac{(n-1) \cdot s^2}{\sigma_0^2}$.

Step 4: Determine whether to reject H_0, as follows:
Left-tailed: $H_1: \sigma < \sigma_0$:   Reject if $\chi^2 \leq \chi^2_{1-\alpha}$.
Right-tailed: $H_1: \sigma > \sigma_0$:   Reject if $\chi^2 \geq \chi^2_{\alpha}$.
Two-tailed: $H_1: \sigma \neq \sigma_0$:   Reject if $\chi^2 \geq \chi^2_{\alpha/2}$ or $\chi^2 \leq \chi^2_{1-\alpha/2}$.

Step 5: State a conclusion.

Explain It Again

Degrees of freedom: When a hypothesis test for a standard deviation is performed, the degrees of freedom is always 1 less than the sample size.

EXAMPLE 9.22

Performing a hypothesis test

To check the reliability of a scale in a butcher shop, a test weight known to weigh 400 grams was weighed 16 times. For the scale to be considered reliable, the standard deviation of repeated measurements must be less than 1 gram. The standard deviation of the 16 measured weights was $s = 0.8$ gram. Assume that the measured weights are independent and follow a normal distribution. Can we conclude that the population standard deviation of the measurements is less than 1 gram? Use the $\alpha = 0.05$ level of significance.

Solution

We first check the assumptions. We have a random sample from a normal population, so the assumptions are satisfied.

Step 1: State the null and alternate hypotheses. Because we are interested in determining whether $\sigma < 1$, the hypotheses are
$$H_0: \sigma = 1 \qquad H_1: \sigma < 1$$

Step 2: Choose a significance level and find the critical value. We will use a significance level of $\alpha = 0.05$. We have $16 - 1 = 15$ degrees of freedom. This is a left-tailed test, so the critical value is $\chi^2_{1-\alpha} = \chi^2_{0.95} = 7.261$.

Step 3: Compute the test statistic. The sample size is $n = 16$. The sample variance is $s^2 = (0.8)^2 = 0.64$, and the value specified by H_0 is $\sigma_0 = 1$. The value of the test statistic is
$$\chi^2 = \frac{(n-1)s^2}{\sigma_0^2} = \frac{(16-1)(0.64)}{1^2} = 9.6$$

Step 4: Determine whether to reject H_0. The value of the test statistic is $\chi^2 = 9.6$. The critical value is $\chi^2_{0.95} = 7.261$. This is a left-tailed test, so we reject H_0 if $\chi^2 < \chi^2_{0.95}$. Because $9.6 > 7.261$, we do not reject H_0 at the $\alpha = 0.05$ level.

Step 5: State a conclusion. There is not enough evidence to conclude that the population standard deviation σ is less than 1 gram. We cannot consider the scale to be reliable.

Check Your Understanding

1. A random sample of size 20 from a normal distribution has standard deviation $s = 50$. Test H_0: $\sigma = 45$ versus H_1: $\sigma > 45$ at the $\alpha = 0.05$ level. Do you reject H_0?

2. A random sample of size 12 from a normal distribution has standard deviation $s = 7$. Test H_0: $\sigma = 15$ versus H_1: $\sigma < 15$ at the $\alpha = 0.01$ level. Do you reject H_0?

3. A random sample of size 28 from a normal distribution has standard deviation $s = 5$. Test H_0: $\sigma = 3$ versus H_1: $\sigma \neq 3$ at the $\alpha = 0.01$ level. Do you reject H_0?

4. A random sample of size 5 from a normal distribution has standard deviation $s = 28$. Test H_0: $\sigma = 40$ versus H_1: $\sigma \neq 40$ at the $\alpha = 0.05$ level. Do you reject H_0?

Answers are on page 461.

SECTION 9.5 Exercises

Exercises 1–4 are the Check Your Understanding exercises located within the section.

Understanding the Concepts

In Exercises 5 and 6, fill in each blank with the appropriate word or phrase.

5. To test a hypothesis about a standard deviation using a sample of size 15, we use a chi-square distribution with _____ degrees of freedom.

6. The method described for testing hypotheses about standard deviations should be used only when the distribution of the population is almost exactly _____.

In Exercises 7 and 8, determine whether the statement is true or false. If the statement is false, rewrite it as a true statement.

7. When a test for a standard deviation is performed, it does not matter whether the population is normal, so long as the sample is large.

8. Hypothesis tests for a standard deviation may be either one- or two-tailed.

Practicing the Skills

9. A random sample of size 11 from a normal distribution has standard deviation $s = 98$. Test H_0: $\sigma = 70$ versus H_1: $\sigma > 70$. Use the $\alpha = 0.05$ level of significance.

10. A random sample of size 29 from a normal distribution has standard deviation $s = 49$. Test H_0: $\sigma = 55$ versus H_1: $\sigma < 55$. Use the $\alpha = 0.01$ level of significance.

11. A random sample of size 24 from a normal distribution has standard deviation $s = 29$. Test H_0: $\sigma = 35$ versus H_1: $\sigma < 35$. Use the $\alpha = 0.01$ level of significance.

12. A random sample of size 13 from a normal distribution has standard deviation $s = 83$. Test H_0: $\sigma = 60$ versus H_1: $\sigma > 60$. Use the $\alpha = 0.05$ level of significance.

13. A random sample of size 25 from a normal distribution has standard deviation $s = 51$. Test H_0: $\sigma = 30$ versus H_1: $\sigma \neq 30$. Use the $\alpha = 0.05$ level of significance.

14. A random sample of size 8 from a normal distribution has standard deviation $s = 75$. Test H_0: $\sigma = 50$ versus H_1: $\sigma \neq 50$. Use the $\alpha = 0.01$ level of significance.

Working with the Concepts

15. **Babies:** A sample of 25 one-year-old girls had a mean weight of 24.1 pounds with a standard deviation of 4.3 pounds. Assume that the population of weights is normally distributed. A pediatrician claims that the standard deviation of the weights of one-year-old girls is less than 5 pounds. Do the data provide convincing evidence that the pediatrician's claim is true? Use the $\alpha = 0.05$ level of significance.

Based on data from the National Health Statistics Reports

16. **Watching TV:** The 2012 General Social Survey asked a large number of people how much time they spent watching TV each day. The mean number of hours was 3.09 with a standard deviation of 2.87. Assume that in a sample of 40 teenagers, the sample standard deviation of daily TV time is 2.0 hours, and that the population of TV watching times is normally distributed. Can you conclude that the population standard deviation of TV watching times for teenagers is less than 2.87? Use the $\alpha = 0.01$ level of significance.

17. **IQ scores:** Scores on an IQ test are normally distributed. A sample of 25 IQ scores had standard deviation $s = 8$. The developer of the test claims that the population standard deviation is $\sigma = 15$. Do these data provide sufficient evidence to contradict this claim? Use the $\alpha = 0.05$ level of significance.

18. **SAT scores:** Scores on the math SAT are normally distributed. A sample of 20 SAT scores had standard deviation $s = 87$. Someone says that the scoring system for the SAT is designed so that the population standard deviation will be $\sigma = 100$. Do these data provide sufficient evidence to contradict this claim? Use the $\alpha = 0.05$ level of significance.

19. **How much is in that can?** A machine that fills beverage cans is supposed to put 12 ounces of beverage in each can. The standard deviation of the amount in each can is

0.1 ounce. The machine is moved to a new location. To determine whether the standard deviation has changed, ten cans are filled. Following are the amounts in the ten cans. Assume them to be a random sample from a normal population.

12.18	11.77	12.09	12.03	11.87
11.96	12.03	12.36	12.28	11.85

Perform a hypothesis test to determine whether the standard deviation differs from 0.1 ounce. Use the $\alpha = 0.05$ level of significance. What do you conclude?

20. **Long-lasting drugs:** One of the ways in which doctors try to determine how long a single dose of pain reliever will provide relief is to measure the drug's half-life, which is the length of time it takes for one-half of the dose to be eliminated from the body. A report of the National Institutes of Health states that the standard deviation of the half-life of the pain reliever oxycodone is $\sigma = 1.43$ hours. Assume that a sample of 25 patients is given the drug, and the sample standard deviation of the half-lives was $s = 1.5$ hours. Assume the population is normally distributed. Can you conclude that the true standard deviation is greater than the value reported by the National Institutes of Health? Use the $\alpha = 0.01$ level of significance.

Extending the Concepts

The chi-square distribution is skewed, but as the number of degrees of freedom becomes large, the skewness diminishes. If the number of degrees of freedom, k, is large enough, the chi-square distribution is reasonably well approximated by a normal distribution with mean k and variance 2k.

21. **Exact test:** A sample of size 101 from a normal population has sample standard deviation $s = 40$. Use Table A.4 to find the exact critical values $\chi^2_{0.025}$ and $\chi^2_{0.975}$ to test $H_0: \sigma = 30$ versus $H_1: \sigma \neq 30$. Can you reject H_0 at the $\alpha = 0.05$ level?

22. **Using the normal approximation:** Refer to Exercise 21. Use the normal approximation to estimate the critical values $\chi^2_{0.025}$ and $\chi^2_{0.975}$. Using these critical values, can you reject H_0 at the $\alpha = 0.05$ level?

A more accurate normal approximation to χ^2_α is given by
$$\chi^2_\alpha \approx 0.5 \left(z_\alpha + \sqrt{2k-1} \right)^2, \text{ where } z_\alpha \text{ is the z-score that has area } \alpha \text{ to its right.}$$

23. **More accuracy:** Refer to Exercise 21. Use the more accurate normal approximation to estimate the critical values $\chi^2_{0.025}$ and $\chi^2_{0.975}$. Using these critical values, can you reject H_0 at the $\alpha = 0.05$ level?

Answers to Check Your Understanding Exercises for Section 9.5

1. $\chi^2 = 23.457$, critical value is 30.144. Do not reject H_0.
2. $\chi^2 = 2.396$, critical value is 3.053. Reject H_0.
3. $\chi^2 = 75.000$, critical values are 11.808 and 49.645. Reject H_0.
4. $\chi^2 = 1.960$, critical values are 0.484 and 11.143. Do not reject H_0.

SECTION 9.6 | Determining Which Method to Use

Objectives

1. Determine which method to use when performing a hypothesis test

Objective 1 Determine which method to use when performing a hypothesis test

One of the challenges in performing a hypothesis test is to determine which method to use. The first step is to determine which type of parameter we are testing. There are three types of parameters about which we have learned to perform hypothesis tests:

- Population mean μ
- Population proportion p
- Population standard deviation σ or variance σ^2

Once you have determined which type of parameter you are testing, proceed as follows:

- **Population mean:** There are two methods for performing a hypothesis test for a population mean, the z-test (Section 9.2) and the t-test (Section 9.3). To determine which method to use, we must determine whether the population is approximately normal, and whether the sample size is large ($n > 30$). The following diagram can help you make the correct choice.

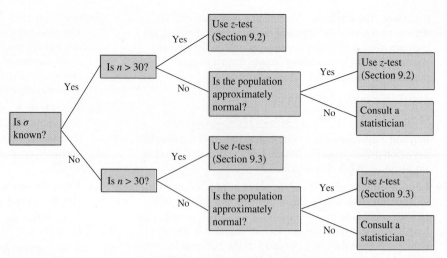

- **Population proportion:** To perform a hypothesis test for a population proportion, use the method described in Section 9.4.
- **Population standard deviation or variance:** To perform a hypothesis test for a population standard deviation or variance, use the method described in Section 9.5.

EXAMPLE 9.23	Determining which method to use

Starting salaries for a random sample of 51 physicians had a mean of $103,000. Assume that the population standard deviation is $10,500. Can you conclude that the mean starting salary for physicians is greater than $100,000? Determine the type of parameter that is to be tested and perform the hypothesis test. Use the $\alpha = 0.05$ level of significance.

Solution

We are asked to perform a hypothesis test for the mean salary; this is a population mean. We consult the diagram to determine the correct method. We must first determine whether σ is known. We are told that the population standard deviation is $10,500. Therefore, $\sigma = 10,500$. We follow the "Yes" path. Next we must determine whether $n > 30$. The sample size is 51, so $n > 30$. We follow the "Yes" path, and find that we should use the z-test described in Section 9.2. To perform the test, we compute the value of the test statistic:

$$z = \frac{103,000 - 100,000}{10,500/\sqrt{51}} = 2.04$$

Because the alternate hypothesis is H_1: $\mu > 100,000$, this is a right-tailed test. The P-value is the area under the normal curve to the right of $z = 2.04$. Using Table A.2, we find that the area to the left of $z = 2.04$ is 0.9793. The area to the right is therefore $1 - 0.9793 = 0.0207$. The P-value is 0.0207. Because $P < 0.05$, we conclude that the mean starting salary is greater than $100,000.

Check Your Understanding

In Exercises 1–4, state which type of parameter is to be tested; then perform the hypothesis test.

1. In a simple random sample of 150 cars undergoing emissions testing, 23 failed the test. Can you conclude that the proportion of cars that fail the test is less than 20%? Use the $\alpha = 0.05$ level of significance.

2. A simple random sample of size 15 has mean $\bar{x} = 27.72$ and standard deviation $s = 8.21$. The population is approximately normally distributed. Can you conclude that the population mean differs from 35? Use the $\alpha = 0.01$ level of significance.

3. A simple random sample of size 7 has mean $\bar{x} = 32.5$ and standard deviation $s = 8.27$. The population is normally distributed. Can you conclude that the population standard deviation is greater than 6? Use the $\alpha = 0.05$ level of significance.

4. A simple random sample of size 65 has mean $\bar{x} = 38.16$. The population standard deviation is $\sigma = 5.95$. Can you conclude that the mean is less than 40? Use the $\alpha = 0.01$ level of significance.

Answers are on page 464.

SECTION 9.6   Exercises

Exercises 1–4 are the Check Your Understanding exercises located within the section.

Practicing the Skills

In Exercises 5–12, state which type of parameter is to be tested; then perform the hypothesis test.

5. A simple random sample of size 13 has mean $\bar{x} = 7.26$ and standard deviation $s = 2.45$. The population is approximately normally distributed. Can you conclude that the population mean differs from 9? Use the $\alpha = 0.01$ level of significance.

6. A simple random sample of size 65 has mean $\bar{x} = 57.3$. The population standard deviation is $\sigma = 12.6$. Can you conclude that the population mean is less than 60? Use the $\alpha = 0.05$ level of significance.

7. A simple random sample of size 16 has mean $\bar{x} = 4.1$ and standard deviation $s = 6.2$. The population is normally distributed. Can you conclude that the population standard deviation is greater than 5? Use the $\alpha = 0.05$ level of significance.

8. In a simple random sample of 120 law students, 54 were women. Can you conclude that less than half of law students are women? Use the $\alpha = 0.01$ level of significance.

9. A simple random sample of size 23 has mean $\bar{x} = 41.8$. The population standard deviation is $\sigma = 3.72$. Can you conclude that the population mean differs from 40? Use the $\alpha = 0.05$ level of significance.

10. A simple random sample of size 6 has mean $\bar{x} = 5.49$ and standard deviation $s = 2.37$. The population is approximately normally distributed. Can you conclude that the population mean is greater than 4? Use the $\alpha = 0.05$ level of significance.

11. In a simple random sample of 95 families, 70 had one or more pets at home. Can you conclude that the proportion of families with one or more pets differs from 0.6? Use the $\alpha = 0.01$ level of significance.

12. A simple random sample of size 22 has mean $\bar{x} = 91.7$ and standard deviation $s = 19.3$. The population is normally distributed. Can you conclude that the population standard deviation is greater than 15? Use the $\alpha = 0.05$ level of significance.

Working with the Concepts

13. **Saving for college:** In a survey of 909 U.S. adults with children conducted in 2012 by the Financial Industry Regulatory Authority, 309 said that they had saved money for their children's college education. Can you conclude that more than 30% of U.S. adults with children have saved money for college? Use the $\alpha = 0.05$ level of significance.

14. **Big houses:** The U.S. Census Bureau reported that the mean area of U.S. homes built in 2012 was 2505 square feet. Assume that a simple random sample of 20 homes built in 2013 had a mean area of 2581 square feet, with a standard deviation of 225 square feet. Can you conclude that the mean area of homes built in 2013 is greater than that of homes built in 2012? Use the $\alpha = 0.01$ level of significance.

15. **Cookies:** Following are the weights of 8 boxes of cookies, each of which is labeled as containing 16 ounces. Assume that the population of weights is normally distributed.

   15.91   16.06   16.08   15.97   16.02   15.88   15.89   16.01

   Can you conclude that the population standard deviation is less than 0.1? Use the $\alpha = 0.01$ level of significance.

16. **Careers in science:** The 2012 General Social Survey asked 514 people whether they had ever considered a career in science, and 178 said that they had. Can you conclude that more than 30% of people have considered a career in science? Use the $\alpha = 0.05$ level of significance.

17. **Teacher salaries:** A random sample of 50 public school teachers in Georgia had a mean annual salary of $48,300. Assume the population standard deviation is $\sigma = \$8,000$. Can you conclude that the mean salary of public school teachers in Georgia differs from $50,000? Use the $\alpha = 0.01$ level of significance.

18. **Highway speeds:** Speeds for a sample of nine cars were measured by radar along a stretch of highway. The results, in miles per hour, were as follows. Assume that the population of speeds is normally distributed.

   56   60   53   55   54   51   54   51   56

   Can you conclude that the population standard deviation is greater than 2? Use the $\alpha = 0.05$ level of significance.

19. **Mercury pollution:** Mercury is a toxic metal that is used in many industrial applications. Seven measurements, in milligrams per cubic meter, were taken of the mercury concentration in a lake, with the following results. Assume

that the population of measurements is approximately normally distributed.

| 1.02 | 1.23 | 0.91 | 1.29 | 1.01 | 1.35 | 1.43 |

Can you conclude that the mean concentration is greater than 1 milligram per cubic meter? Use the $\alpha = 0.05$ level of significance.

20. Ladies' shoes: A random sample of 100 pairs of ladies' shoes had a mean size of 8.3. Assume the population standard deviation is $\sigma = 1.5$. Can you conclude that the mean size of ladies' shoes differs from 8? Use the $\alpha = 0.01$ level of significance.

Answers to Check Your Understanding Exercises for Section 9.6

1. The parameter is the population proportion. The test statistic is $z = -1.43$. The P-value is 0.0764 [Tech: 0.0765]. Do not reject H_0. There is not enough evidence to conclude that the proportion of cars that fail the test is less than 0.20.

2. The parameter is the population mean. The test statistic is $t = -3.43$. The P-value is $0.002 < P < 0.005$ [Tech: 0.004]. Reject H_0. We conclude that the mean differs from 35.

3. The parameter is the population standard deviation. The test statistic is $\chi^2 = 11.399$. The critical value is $\chi^2_{0.975} = 12.59$. Do not reject H_0. There is not enough evidence to conclude that the population standard deviation is greater than 6.

4. The parameter is the population mean. The test statistic is $z = -2.49$. The P-value is 0.0064 [Tech: 0.0063]. Reject H_0. We conclude that the population mean is less than 40.

SECTION 9.7 | Power

Objectives

1. Compute the power of a test

Objective 1 Compute the power of a test

Table 9.3

	H_0 true	H_0 false
Reject H_0	Type I error	Correct
Don't reject H_0	Correct	Type II error

Recall that there are two types of errors that can be made with a hypothesis test. A Type I error occurs when H_0 is true and we reject it. A Type II error occurs when H_0 is false and we fail to reject it. Table 9.3 summarizes the possibilities. The probability of a Type I error is α, the significance level of the test. The probability of a Type II error is denoted β.

The *power* of a test is the probability that we do not make a Type II error. In other words, the power of a test is the probability that we reject H_0 when it is false.

> **DEFINITION**
>
> The **power** of a test is the probability of rejecting H_0 when it is false.
>
> $$\text{Power} = 1 - \beta$$
>
> where $\beta = P(\text{Type II error})$.

We would like to have a small probability of rejecting H_0 when it is true, and a large probability of rejecting H_0 when it is false. In other words, we would like the probability of a Type I error to be small, and the power to be large. We can make the probability of a Type I error small by choosing a small value for the significance level α. As we have seen, the values $\alpha = 0.05$ and $\alpha = 0.01$ are considered to be appropriate in many situations. Once the value of α has been chosen, we can increase the power only by increasing the sample size n. What is done in practice is to choose values for α and n and then compute the power. Values of 0.8 or 0.9 are considered to be acceptably large for the power. If the power is not large enough, then if sufficient time and resources are available, the sample size can be increased.

The power of a test about a population mean depends on the true value of the population mean. To compute the power, we specify a value μ_1 for the population mean that satisfies the alternate hypothesis. The power is the probability that the test statistic falls in the critical region when μ_1 is the true value of the population mean. The following steps can be used to compute the power of a hypothesis test for the population mean when the population standard deviation σ is known.

Computing the Power of a Test

Let μ_0 be the value of μ specified by H_0, let α be the significance level, let n be the sample size, and let σ be the population standard deviation.

Step 1: Find the critical value: z_α for a one-tailed test or $z_{\alpha/2}$ for a two-tailed test.

Step 2:
- For a one-tailed test, find the value of $\bar{x}$ whose z-score is equal to the critical value. We call this value $\bar{x}^*$. Find the value of $\bar{x}^*$ as follows:

  Left-tailed: $H_1: \mu < \mu_0$ $\bar{x}^* = \mu_0 - z_\alpha \cdot \dfrac{\sigma}{\sqrt{n}}$

  Right-tailed: $H_1: \mu > \mu_0$ $\bar{x}^* = \mu_0 + z_\alpha \cdot \dfrac{\sigma}{\sqrt{n}}$

- For a two-tailed test, there are two values of $\bar{x}^*$. We call them $\bar{x}^*_{\text{left}}$ and $\bar{x}^*_{\text{right}}$. They are computed as follows:

$$\bar{x}^*_{\text{left}} = \mu_0 - z_{\alpha/2} \cdot \frac{\sigma}{\sqrt{n}} \qquad \bar{x}^*_{\text{right}} = \mu_0 + z_{\alpha/2} \cdot \frac{\sigma}{\sqrt{n}}$$

Step 3: Let μ_1 be a specific value that satisfies the alternate hypothesis. Sketch a normal curve with mean μ_1.

Step 4: The power is an area under the normal curve sketched in Step 3. The area depends on the form of the alternate hypothesis, as follows:

Left-tailed: $H_1: \mu < \mu_0$ Area to the left of $\bar{x}^*$

Right-tailed: $H_1: \mu > \mu_0$ Area to the right of $\bar{x}^*$

Two-tailed: $H_1: \mu \neq \mu_0$ Sum of the area to the left of $\bar{x}^*_{\text{left}}$ and the area to the right of $\bar{x}^*_{\text{right}}$

EXAMPLE 9.24   Compute the power of a test

The 2008 General Social Survey indicates that Americans watch an average of 2.98 hours of television per day, with a standard deviation of $\sigma = 2.66$ hours. A sociologist believes that the mean viewing time for college students is less, because students spend more time on the Internet and playing video games. The sociologist will sample 75 college students and test the hypotheses

$$H_0: \mu = 2.98 \qquad H_1: \mu < 2.98$$

at the $\alpha = 0.05$ level. Assume the population standard deviation for college students is also $\sigma = 2.66$. Find the power of the test against the alternative $\mu_1 = 2$.

Solution

Step 1: This is a one-tailed test, with significance level $\alpha = 0.05$. Therefore, we use the critical value $z_\alpha = 1.645$.

Step 2: We have $\mu_0 = 2.98$, $z_\alpha = 1.645$, $\sigma = 2.66$, and $n = 75$. This is a left-tailed test. Therefore,

$$\bar{x}^* = 2.98 - (1.645)\frac{2.66}{\sqrt{75}} = 2.475$$

Step 3: The following figure presents a normal curve with mean 2. The value of $\bar{x}^* = 2.475$ is indicated as well.

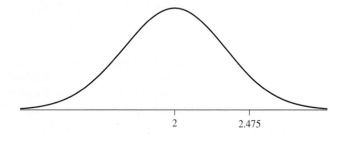

Step 4: This is a left-tailed test, so the power is the area to the left of $\bar{x}^* = 2.475$. To find this area, we find the z-score for 2.475, using the value $\mu_1 = 2$.

$$z = \frac{2.475 - 2}{2.66/\sqrt{75}} = 1.55$$

The power is the area under the normal curve to the left of $z = 1.55$. This area is 0.9394, as the following figure shows.

Figure 9.18 presents two distributions for $\bar{x}$. The curve on the right is the distribution under the assumption that H_0 is true. This distribution has mean $\mu_0 = 2.98$. The curve on the left is the distribution under the assumption that the mean is equal to the alternate value $\mu_1 = 2$. The critical region is the region to the left of $\bar{x}^* = 2.475$. The area of the critical region under the null hypothesis distribution is the significance level $\alpha = 0.05$. The area under the alternate distribution is the power, 0.9394.

Figure 9.18 The critical region is shaded. The area of the critical region is $\alpha = 0.05$ when the curve for H_0 is used, and is equal to the power, 0.9394, when the curve with the alternate mean of 2 is used.

Check Your Understanding

1. For testing the hypotheses H_0: $\mu = 10$ versus H_1: $\mu < 10$ at the $\alpha = 0.05$ level, assume the population standard deviation is $\sigma = 3$ and the sample size is $n = 50$. Find the power against the alternative $\mu_1 = 9$.

2. For testing the hypotheses H_0: $\mu = 35$ versus H_1: $\mu > 35$ at the $\alpha = 0.01$ level, assume the population standard deviation is $\sigma = 10$ and the sample size is $n = 100$. Find the power against the alternative $\mu_1 = 40$.

Answers are on page 468.

SECTION 9.7   Exercises

Exercises 1 and 2 are the Check Your Understanding exercises located within the section.

Understanding the Concepts

In Exercises 3–5, fill in each blank with the appropriate word or phrase.

3. The greater the power, the less likely we are to make a _____ error.

4. The power is the probability of rejecting H_0 when it is _____.

5. If the sample size is increased, the power will _____.

In Exercises 6–8, determine whether the statement is true or false. If the statement is false, rewrite it as a true statement.

6. Once the significance level α has been chosen, the only way to increase the power of a test is to decrease the sample size.

7. The power is the probability of making a Type I error.

8. For a test about the population mean, the power of a test depends on a specified value that satisfies the alternate hypothesis.

Practicing the Skills

9. A test has power 0.90 when $\mu_1 = 15$. True or false:
   a. The probability of rejecting H_0 when $\mu_1 = 15$ is 0.90.
   b. The probability of making a correct decision when $\mu_1 = 15$ is 0.90.
   c. The probability of making a correct decision when $\mu_1 = 15$ is 0.10.
   d. The probability that H_0 is true when $\mu_1 = 15$ is 0.10.

10. A test has power 0.80 when $\mu_1 = 3.5$. True or false:
   a. The probability of rejecting H_0 when $\mu_1 = 3.5$ is 0.80.
   b. The probability of making a Type I error when $\mu_1 = 3.5$ is 0.80.
   c. The probability of making a Type I error when $\mu_1 = 3.5$ is 0.20.
   d. The probability of making a Type II error when $\mu_1 = 3.5$ is 0.80.
   e. The probability of making a Type II error when $\mu_1 = 3.5$ is 0.20.
   f. The probability that H_0 is false when $\mu_1 = 3.5$ is 0.80.

Working with the Concepts

11. **Tire lifetimes:** A tire company claims that the lifetimes of its tires average 50,000 miles. The standard deviation of tire lifetimes is known to be $\sigma = 5000$ miles. You sample 100 tires and will test the hypotheses H_0: $\mu = 50{,}000$ versus H_1: $\mu < 50{,}000$ at the $\alpha = 0.05$ level of significance.
   a. Find the power of the test against the alternative $\mu_1 = 49{,}500$.
   b. Find the power of the test against the alternative $\mu_1 = 49{,}000$.
   c. Find the power of the test against the alternative $\mu_1 = 49{,}500$ if the test is made at level $\alpha = 0.01$.
   d. Find the power of the test against the alternative $\mu_1 = 49{,}000$ if the test is made at level $\alpha = 0.01$.

12. **Coffee beans:** Shipments of coffee beans are checked for moisture content. A high moisture content indicates water contamination and will result in the shipment being rejected. Let μ represent the mean water content (in percent by weight) in a shipment. Fifty moisture measurements will be made on beans chosen at random from the shipment. A test of the hypotheses H_0: $\mu = 10$ versus H_1: $\mu > 10$ will be made at the $\alpha = 0.05$ level of significance. Assume the standard deviation of moisture content is $\sigma = 5.0$.
   a. Find the power of the test against the alternative $\mu_1 = 11$.
   b. Find the power of the test against the alternative $\mu_1 = 12$.
   c. Find the power of the test against the alternative $\mu_1 = 11$ if the test is made at level $\alpha = 0.01$.
   d. Find the power of the test against the alternative $\mu_1 = 12$ if the test is made at level $\alpha = 0.01$.

13. **SAT scores:** A college admissions officer will draw a simple random sample of 100 mathematics SAT scores from the entering freshman class. The admissions officer will perform a test of the hypotheses H_0: $\mu = 500$ versus H_1: $\mu > 500$ at the $\alpha = 0.05$ level of significance. Assume the population standard deviation is $\sigma = 116$.
   a. Find the power of the test against the alternative $\mu_1 = 520$.
   b. Find the power of the test against the alternative $\mu_1 = 550$.
   c. Find the power of the test against the alternative $\mu_1 = 520$ if the test is made at level $\alpha = 0.01$.
   d. Find the power of the test against the alternative $\mu_1 = 550$ if the test is made at level $\alpha = 0.01$.

14. **Watch your cholesterol:** An article in the *International Journal of Clinical Practice* described a study in which a sample of 314 patients took a combination of the drugs ezetimbe and simvastatin in order to reduce their total blood cholesterol levels. A test of the hypotheses H_0: $\mu = 1$ versus H_1: $\mu > 1$ will be made at the $\alpha = 0.05$ level of significance. Assume the population standard deviation is $\sigma = 0.2$.
   a. Find the power of the test against the alternative $\mu_1 = 1.02$.

b. Find the power of the test against the alternative $\mu_1 = 1.05$.

c. Find the power of the test against the alternative $\mu_1 = 1.02$ if the test is made at level $\alpha = 0.01$.

d. Find the power of the test against the alternative $\mu_1 = 1.05$ if the test is made at level $\alpha = 0.01$.

15. Tire lifetimes: Refer to Exercise 11. A test of the hypotheses $H_0: \mu = 50,000$ versus $H_1: \mu \neq 50,000$ will be made at the $\alpha = 0.05$ level of significance. Assume that the standard deviation of tire lifetimes is $\sigma = 5000$ and that the sample size is $n = 100$. Find the power of the test against the alternative $\mu_1 = 49,000$.

16. Coffee beans: Refer to Exercise 12. A test of the hypotheses $H_0: \mu = 10$ versus $H_1: \mu \neq 10$ will be made at the $\alpha = 0.05$ level of significance. Assume that the standard deviation of moisture content is $\sigma = 5.0$ and that the sample size is $n = 50$. Find the power of the test against the alternative $\mu_1 = 12$.

17. SAT scores: Refer to Exercise 13. The admissions officer will perform a test of the hypotheses $H_0: \mu = 500$ versus $H_1: \mu \neq 500$ at the $\alpha = 0.01$ level of significance. Assume that the population standard deviation is $\sigma = 116$ and that the sample size is $n = 100$. Find the power of the test against the alternative $\mu_1 = 550$.

18. Watch your cholesterol: Refer to Exercise 14. A test of the hypotheses $H_0: \mu = 1$ versus $H_1: \mu \neq 1$ will be made at the $\alpha = 0.01$ level of significance. Assume that the population standard deviation is $\sigma = 0.2$ and that the sample size is $n = 314$. Find the power of the test against the alternative $\mu_1 = 0.98$.

Extending the Concepts

The power for a test of a population proportion can be found by a method similar to that for a population mean.

19. Power for test of a proportion: A marketing firm samples 150 residents of a certain town to determine the sample proportion $\hat{p}$ that have seen a new advertisement. A test of the hypotheses $H_0: p = 0.4$ versus $H_1: p < 0.4$ will be performed, using a significance level of $\alpha = 0.05$.

a. Find the critical value z_α.

b. Find the value of $\hat{p}$ whose z-score is $-z_\alpha$. Call this value p^*. Note that the population standard deviation is $\sigma = \sqrt{p_0(1 - p_0)}$.

c. We want to find the power against the alternative $p_1 = 0.3$. The power is the area to the left of p^* under the assumption that the true proportion is $p_1 = 0.3$. Find the power. [Note that the standard deviation is now $\sigma = \sqrt{p_1(1 - p_1)}$.]

Answers to Check Your Understanding Exercises for Section 9.7

1. 0.7611 [Tech: 0.7618]
2. 0.9962

Chapter 9 Summary

Section 9.1: A hypothesis test involves a null hypothesis, H_0, which makes a statement about one or more population parameters, and an alternate hypothesis, which contradicts H_0. We begin by assuming that H_0 is true. If the data provide strong evidence against H_0, we then reject H_0 and believe H_1. A Type I error occurs when a true null hypothesis is rejected. A Type II error occurs when a false H_0 is not rejected.

Section 9.2: We follow one of two methods in performing a hypothesis test. In the critical value method, we choose a significance level α, then find a critical region. We reject H_0 if the test statistic falls inside the critical region. The probability of a Type I error is α, the significance level of the test. In the P-value method, we compute a P-value, which is the probability of observing a value for the test statistic that is as extreme as or more extreme than the value actually observed, under the assumption that H_0 is true. The smaller the P-value, the stronger the evidence against H_0. If we want to make a firm decision about the truth of H_0, we choose a significance level α and reject H_0 if $P \leq \alpha$.

When testing a hypothesis about a population mean with the population standard deviation σ known, the test statistic, z, has a standard normal distribution. If the sample size is not large, the population must be approximately normal. We can check normality with a boxplot or dotplot.

Statistical significance is not the same as practical significance. When a result is statistically significant, we can conclude only that the true value of the parameter is different from the value specified by H_0. We cannot conclude that the difference is large enough to be important.

When presenting the results of a hypothesis test, it is important to state the P-value or the value of the test statistic, so that others can decide for themselves whether to reject H_0. It isn't enough simply to state whether or not H_0 was rejected.

Section 9.3: When testing a hypothesis about a population mean with the population standard deviation σ unknown, the test statistic has a Student's t distribution. The number of degrees of freedom is 1 less than the sample size. The population must be approximately normal, or the sample size must be large ($n > 30$). We can check normality with a boxplot or dotplot.

Section 9.4: When testing a hypothesis about a population proportion, the test statistic is z. The sample proportion must be approximately normal. We check this by requiring that both np_0 and $n(1 - p_0)$ are at least 10.

Section 9.5: When a population is almost exactly normal, we can test a hypothesis about a population standard deviation. We use the chi-square distribution, with degrees of freedom 1 less than the sample size.

Section 9.6: We have learned to perform hypothesis tests for a population mean, a population proportion, and a population standard deviation or variance. There are two tests for a population mean, the z-test and the t-test. The test to use depends on whether the population standard deviation σ is known.

Section 9.7: The power of a test is the probability that a false H_0 is rejected. It is desirable for a test to have a high degree of power. If the sample size remains the same, however, increasing the power also increases the probability of a Type I error. In order to increase the power without increasing the probability of a Type I error, it is necessary to increase the sample size.

Vocabulary and Notation

alternate hypothesis   404
critical region   410
critical value   410
critical value method   409
hypothesis test   406
left-tailed hypothesis   405
null hypothesis   404

one-tailed hypothesis   405
P-value   415
P-value method   414
power   464
rejecting H_0 406
right-tailed hypothesis   405
significance level   411

statistically significant   411
t-test   433
test statistic   410
two-tailed hypothesis   405
Type I error   407
Type II error   407

Important Formulas

Test statistic for a mean, standard deviation known:

$$z = \frac{\bar{x} - \mu_0}{\sigma/\sqrt{n}}$$

Test statistic for a mean, standard deviation unknown:

$$t = \frac{\bar{x} - \mu_0}{s/\sqrt{n}}$$

Test statistic for a proportion:

$$z = \frac{\hat{p} - p_0}{\sqrt{\dfrac{p_0(1 - p_0)}{n}}}$$

Test statistic for a standard deviation:

$$\chi^2 = \frac{(n - 1) \cdot s^2}{\sigma_0^2}$$

Chapter Quiz

1. Fill in the blank: A test of the hypotheses H_0: $\mu = 65$ versus H_1: $\mu \neq 65$ was performed. The P-value was 0.035. Fill in the blank: If $\mu = 65$, then the probability of observing a test statistic as extreme as or more extreme than the one actually observed is _____.

2. A hypothesis test results in a P-value of 0.008. Which is the best conclusion?
   i. H_0 is definitely false.
   ii. H_0 is definitely true.
   iii. H_0 is plausible.
   iv. H_0 might be true, but it's very unlikely.
   v. H_0 might be false, but it's very unlikely.

3. True or false: If $P = 0.03$, then
   a. The result is statistically significant at the $\alpha = 0.05$ level.
   b. The result is statistically significant at the $\alpha = 0.01$ level.
   c. The null hypothesis is rejected at the $\alpha = 0.05$ level.
   d. The null hypothesis is rejected at the $\alpha = 0.01$ level.

4. A null hypothesis is rejected at the $\alpha = 0.05$ level. True or false:
   a. The P-value is greater than 0.05.
   b. The P-value is less than or equal to 0.05.
   c. The result is statistically significant at the $\alpha = 0.05$ level.
   d. The result is statistically significant at the $\alpha = 0.10$ level.

5. A sample of size 8 is drawn from a normal population with mean μ, and the population standard deviation is unknown.
   a. Is it appropriate to perform a z-test? Explain.
   b. Is it appropriate to perform a t-test? Explain.

6. A test will be made of H_0: $\mu = 4$ versus H_1: $\mu > 4$, using a sample of size 25. The population standard deviation is unknown. Find the critical value of the test statistic if the significance level is $\alpha = 0.05$.

7. True or false: We never conclude that H_0 is true.

8. In a random sample of 500 people who took their driver's test, 445 passed. Let p be the population proportion who pass. A test will be made of H_0: $p = 0.85$ versus H_1: $p > 0.85$.
   a. Compute the value of the test statistic.
   b. Do you reject H_0 at the $\alpha = 0.05$ level?
   c. State a conclusion.

9. For testing H_0: $\mu = 3$ versus H_1: $\mu < 3$, a P-value of 0.024 is obtained.
   a. If the significance level is $\alpha = 0.05$, would you conclude that $\mu < 3$? Explain.
   b. If the significance level is $\alpha = 0.01$, would you conclude that $\mu < 3$? Explain.

10. True or false: When we reject H_0, we are certain that H_1 is true.

11. The result of a hypothesis test is reported as follows: "We reject H_0 at the $\alpha = 0.05$ level." What additional information should be included?

12. In a test of H_0: $\mu = 5$ versus H_1: $\mu > 5$, the value of the test statistic is $t = 2.96$. There are 17 degrees of freedom. Do you reject H_0 at the $\alpha = 0.05$ level?

13. True or false: We can perform a test for a standard deviation only when the population is almost exactly normal.

14. A random sample of size 20 from a normal population has sample standard deviation $s = 10$. Test H_0: $\sigma = 15$ versus H_1: $\sigma < 15$. Use the $\alpha = 0.05$ level.

15. A test of H_0: $\mu = 50$ versus H_1: $\mu > 50$ will be made at a significance level of $\alpha = 0.05$. The population standard deviation is $\sigma = 10$ and the sample size is $n = 60$. Find the power of the test against the alternative $\mu_1 = 55$.

Review Exercises

1. **What's the conclusion?** A hypothesis test is performed, and $P = 0.02$. Which of the following is the best conclusion?
   i. H_0 is rejected at the 0.05 level.
   ii. H_0 is rejected at the 0.01 level.
   iii. H_1 is rejected at the 0.05 level.
   iv. H_1 is rejected at the 0.01 level.

2. **Scoring runs:** In 2012, the mean number of runs scored by both teams in a Major League Baseball game was 8.62. Following are the numbers of runs scored in a sample of 24 games in 2013.

2	10	3	9	15	10	7	4	3	7	5	9
5	9	15	15	4	5	13	6	14	11	6	12

   a. Construct a boxplot of the data. Is it appropriate to perform a hypothesis test?
   b. If appropriate, perform a hypothesis test to determine whether the mean number of runs in 2013 is less than it was in 2012. Use the $\alpha = 0.05$ level.

3. **Facebook:** A popular blog reports that 60% of college students log in to Facebook on a daily basis. The Dean of Students at a certain university thinks that the proportion may be different at her university. She polls a simple random sample of 200 students, and 134 of them report that they log in to Facebook daily. Can you conclude that the proportion of students who log in to Facebook daily differs from 0.60?
   a. State the null and alternate hypotheses.
   b. Compute the value of the test statistic.
   c. Do you reject H_0? Use the $\alpha = 0.05$ level.
   d. State a conclusion.

4. **Playing the market:** The Russell 2000 is a group of 2000 small-company stocks. On June 21, 2013, a random sample of 35 of these stocks had a mean price of $26.89, with a standard deviation of $23.41. A stock market analyst predicted that the mean price of all 2000 stocks would be $25.00. Can you conclude that the mean price differs from $25.00?
   a. State the null and alternate hypotheses.
   b. Should we perform a z-test or a t-test? Explain.
   c. Compute the value of the test statistic.
   d. Do you reject H_0? Use the $\alpha = 0.05$ level.
   e. State a conclusion.

5. **Power:** A test of H_0: $\mu = 100$ versus H_1: $\mu > 100$ will be made at a significance level of $\alpha = 0.01$. The population standard deviation is $\sigma = 50$ and the sample size is $n = 75$. Find the power of the test against the alternative $\mu_1 = 110$.

6. **More power:** Refer to Exercise 5. If the test is made at the $\alpha = 0.05$ level with the same sample size, would the power be greater than or less than in Exercise 5? Explain.

7. **Household size:** For the past several years, the mean number of people in a household has been declining. A social scientist believes that in a certain large city, the mean number of people per household is less than 2.5. To investigate this, she takes a simple

random sample of 150 households in the city, and finds that the sample mean number of people is 2.3 with a sample standard deviation of 1.5. Can you conclude that the mean number of people per household is less than 2.5?

   a. State the null and alternate hypotheses.
   b. Should we perform a z-test or a t-test? Explain.
   c. Compute the value of the test statistic.
   d. Do you reject H_0? Use the $\alpha = 0.01$ level.
   e. State a conclusion.

8. **Job satisfaction:** In 2012, the General Social Survey sampled 762 employed people and asked them how satisfied they were with their jobs. Of the 762 people sampled, 386 said that they were completely satisfied or very satisfied with their jobs. Can you conclude that more than 45% of employed people in the United States are completely or very satisfied with their jobs?

   a. State the null and alternate hypotheses.
   b. Compute the value of the test statistic.
   c. Do you reject H_0? Use the $\alpha = 0.01$ level.
   d. State a conclusion.

9. **Sugar content:** The sugar content in milligrams of a syrup used to pack canned fruit is measured for 8 batches of syrup. The measurements are normally distributed, and the sample standard deviation is $s = 3$. Can you conclude that the population standard deviation is greater than 2? Use the $\alpha = 0.01$ level of significance.

10. **Interpret computer output:** The following output from MINITAB presents the results of a hypothesis test.

```
Test of mu = 4.7 vs not = 4.7
The assumed standard deviation = 2.0

 N    Mean    SE Mean      95% CI        Z       P
35   5.401    0.3381   (4.738, 6.064)  2.074   0.038
```

   a. What are the null and alternate hypotheses?
   b. What is the value of the test statistic?
   c. What is the P-value?
   d. Do you reject H_0 at the $\alpha = 0.05$ level?
   e. Do you reject H_0 at the $\alpha = 0.01$ level?

11. **Interpret calculator display:** The following TI-84 Plus display presents the results of a hypothesis test.

   a. Is this a test for a mean, a proportion, or a standard deviation?
   b. What are the null and alternate hypotheses?
   c. What is the value of the test statistic?
   d. What is the P-value?
   e. Do you reject H_0 at the $\alpha = 0.05$ level?
   f. Do you reject H_0 at the $\alpha = 0.01$ level?

12. **How many TV sets?** A survey organization sampled 60 households in a community and found that the sample mean number of TV sets per household was 3.1. The population standard deviation is $\sigma = 1.5$. Can you conclude that the mean number of TV sets per household is greater than 3?

   a. State the null and alternate hypotheses.
   b. Should we perform a z-test or a t-test? Explain.
   c. Compute the value of the test statistic.
   d. Do you reject H_0? Use the $\alpha = 0.01$ level.
   e. State a conclusion.

13. **Crackers:** Boxes of crackers are filled by a machine. The weights of a random sample of 25 boxes of crackers have standard deviation $s = 5$ grams. Assume the weights are normally distributed. Can you conclude that the population standard deviation is less than 10 grams? Use the $\alpha = 0.05$ level of significance.

14. **How much rent?** A housing official in a certain city claims that the mean monthly rent for apartments in the city is more than $1000. To verify this claim, a simple random sample of 40 renters in the city was taken, and the sample mean rent paid was $1100 with a sample standard deviation of $300. Can you conclude that the mean monthly rent in the city is greater than $1000?

472   Chapter 9   Hypothesis Testing

a. State the null and alternate hypotheses.
b. Should we perform a z-test or a t-test? Explain.
c. Compute the value of the test statistic.
d. Do you reject H_0? Use the $\alpha = 0.05$ level.
e. State a conclusion.

15. **What's the news?** The Pew Research Center reported in 2012 that 23% of 18- to 29-year-olds watch a cable news channel regularly. The director of media relations at a large university wants to know whether the population proportion of cable news viewers among students at her university is greater than the proportion among 18- to 29-year-olds in general. She surveys a simple random sample of 200 students at the university and finds that 66 of them watch cable news regularly. Can she conclude that the proportion of students at her university who watch cable news regularly is greater than 0.23?
a. State the null and alternate hypotheses.
b. Compute the value of the test statistic.
c. Do you reject H_0? Use the $\alpha = 0.01$ level.
d. State a conclusion.

Write About It

1. A result is significant at the 0.01 level. Explain why it must also be significant at the 0.05 level.
2. What does the P-value represent?
3. Why is it important to report the P-value or the test statistic when presenting the results of a hypothesis test?
4. Why don't we need to know the population standard deviation when performing a test about a population proportion?
5. In what ways are hypothesis tests for a population mean different from hypothesis tests for a proportion? In what ways are they similar?
6. A test will be made of H_0: $\mu = 12$ versus H_1: $\mu > 12$. Explain why the power will be greater when the true value of μ is 20 than when the true value of μ is 15.

Case Study: Is It Getting Warmer In Washington, D.C.?

There is substantial evidence to indicate that temperatures on the surface of the Earth have been increasing for the past 100 years or so. We will investigate the possibility of warming trends in one location: Washington, D.C. Table 9.4 presents the record high and low temperatures, along with the year they occurred, for every seventh day at the site of Reagan National Airport in Washington, D.C. The data span the years 1871–2013.

Table 9.4 Dates of Record Temperatures in Washington, D.C.

Date	High	Year	Low	Year	More Recent	Date	High	Year	Low	Year	More Recent	Date	High	Year	Low	Year	More Recent
Jan 1	69	2005	−14	1881	High	May 7	95	1930	38	1970	Low	Sep 3	98	1953	48	1909*	High
8	73	2008	0	1878	High	14	93	1956	41	1928	High	10	98	1983	44	1883	High
15	77	1932	4	1886	High	21	95	1934	41	1907	High	17	96	1991	44	1923	High
22	76	1927	1	1893	High	28	97	1941	42	1961	Low	24	99	2010	39	1963	High
29	76	1975	2	1873	High	Jun 4	99	1925	46	1929*	Low	Oct 1	93	1941*	36	1899	High
Feb 5	70	1991*	−2	1918*	High	11	101	1911	45	1913	Low	8	91	2007	36	1964*	High
12	74	1999	4	1899	High	18	97	1944	51	1965*	Low	15	87	1975	32	1874	High
19	74	1939	4	1903	High	25	100	1997	53	1902	High	22	84	1979*	29	1895	High
26	74	1932	12	1970	Low	Jul 2	101	1898	55	1940	Low	29	82	1918	30	1976*	Low
Mar 5	83	1976	6	1872	High	9	104	1936	55	1891	High	Nov 5	81	2003*	20	1879	High
12	89	1990	11	1900	High	16	104	1988	56	1930*	High	12	77	1912*	24	1926	Low
19	87	1945	12	1876	High	23	102	2011	56	1890	High	19	77	1928	18	1891	High
26	87	1921	23	1955	Low	30	99	1953	56	1914	High	26	74	1979	17	1950	High
Apr 2	89	1963	23	1907	High	Aug 6	106	1918	53	1912	High	Dec 3	71	2012	15	1976	High
9	90	1959	28	1972*	Low	13	101	1881	55	1930*	Low	10	67	1966*	4	1876	High
16	92	2002	29	1928	High	20	101	1983	50	1896	High	17	64	1984*	10	1876	High
23	95	1960	33	1933*	High	27	100	1987	51	1885	High	24	69	1933	5	1983	Low
30	92	1942*	34	1874	High												

*Indicates that the record occurred more than once; only the most recent year is given.
Source: National Weather Service

1. If there have been no temperature trends over the years, then it will be equally likely for the record high or the record low to be more recent. If there has been a warming trend, it might be more likely for the record high to be more recent. Let p be the probability that the record high occurred more recently than the record low. Use the sample proportion of dates where the high occurred more recently to test $H_0: p = 0.5$ versus $H_1: p > 0.5$. What do you conclude?

2. The following table presents the records for every day in June. The data show that it is common for records to be set on two or more consecutive days in the same year. This is due to hot spells and cold spells in the weather. For example, five consecutive record highs, from June 2 through June 6, occurred in 1925. Explain why using data for every day may violate the assumption, used in Exercise 1, that the data are a simple random sample.

Date	High	Year	Low	Year	Date	High	Year	Low	Year	Date	High	Year	Low	Year
Jun 1	98	2011	45	1938*	Jun 11	101	1911	45	1913	Jun 21	99	2012	51	1940
2	97	1925*	43	1897	12	95	2002*	50	1907*	22	101	1988	51	1992*
3	99	1925	45	1910	13	96	1954	51	1887	23	98	1988*	51	1918
4	99	1925	46	1929*	14	98	1994	49	1933	24	100	2010	46	1902
5	100	1925	48	1926	15	101	1994	47	1933	25	100	1997	53	1902
6	97	1925	46	1945*	16	99	1994	50	1917	26	101	1952	56	1979
7	98	2008*	47	1894	17	95	1991*	50	1926	27	99	2010	57	1927*
8	99	2011	49	1977*	18	97	1944	51	1965*	28	100	1969	54	1927
9	102	2011*	45	1913*	19	99	1994	51	1909	29	104	2012	54	1888
10	100	1964	46	1913	20	99	1931	54	1926*	30	100	1959	50	1919

*Indicates that the record occurred more than once; only the most recent year is given.

3. We will perform another test to determine whether record highs are more likely to have occurred recently. If a record high is equally likely to occur in any year of observation, the mean year in which a record is observed would occur at the midpoint of the observation period, which is $(1871 + 2013)/2 = 1942$. Use the data in Table 9.4 to test the hypothesis that the mean year in which a record high occurred is 1942 against the alternative that it is greater. What do you conclude?

4. For some records, marked with a *, the record temperature occurred more than once. In these cases, only the most recent year is listed. Explain how this might cause the mean to be greater than the midpoint of 1942, even if records are equally likely to occur in any year.

5. Using the data in Table 9.4, drop the dates in which the record high occurred more than once, and test the hypothesis in Exercise 3 again. Does your conclusion change?

6. Perform a hypothesis test on the record lows, after dropping dates on which the record low occurred more than once, in which the alternate hypothesis is that the mean year is less than 1942. What do you conclude?

7. Using the analyses you have performed, write a summary of your findings. Describe how strong you believe the evidence to be that record highs have tended to occur more recently than record lows.

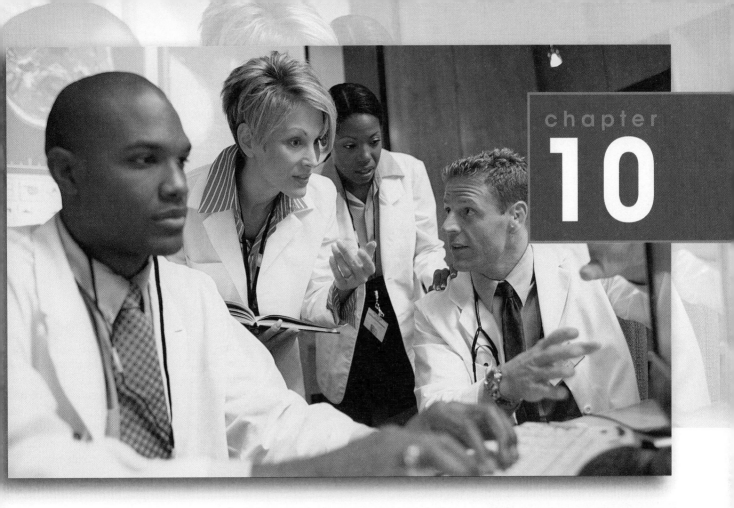

Two-Sample
Confidence Intervals

Introduction

When a new medical treatment is proposed, a clinical trial is conducted to determine whether the treatment is safe and effective. In a clinical trial, patients are assigned to receive either the new treatment or an existing treatment. If the patients receiving the new treatment tend to have better outcomes, this is evidence that the new treatment represents an improvement over the old one. When assigning patients to treatments, it is important that the two groups be approximately equal with regard to prior health status. If one group is much healthier than the other, this can bias the results of the trial.

An article in the *New England Journal of Medicine* (361:1329–1338) reported the results of a clinical trial to compare the effectiveness of a new type of heart pacemaker in preventing cardiac failure in patients with heart disease. A total of 1820 patients participated, with 1089 receiving the new treatment and 731 receiving the standard treatment. The assignment to treatments was not made by simple random sampling, but instead by an algorithm that was designed to balance the two groups. The following tables present some of the important characteristics of the two groups.

Characteristic	Standard Treatment		New Treatment	
	Mean	Standard Deviation	Mean	Standard Deviation
Age	64	11	65	11
Systolic blood pressure	121	18	124	17
Diastolic blood pressure	71	10	72	10

Characteristic	Standard Treatment	New Treatment
	Percent with the Characteristic	Percent with the Characteristic
Treatment for hypertension	63.2	63.7
Atrial fibrillation	12.6	11.1
Diabetes	30.6	30.2
Cigarette smoking	12.8	11.4
Coronary bypass surgery	28.5	29.1

There are differences between the two groups in all of these characteristics. This is not surprising, because we would expect to see some differences just by chance. The question is whether the differences are large enough to suggest that they may be due to the assignment procedure, and if so, whether the differences may be large enough to be of concern when evaluating the results of the trial.

In this chapter, we will learn to construct confidence intervals that will address questions like this. In the case study at the end of the chapter, we will investigate the differences in the table, to determine whether any differences that result from the assignment procedure might be large enough to be of concern.

SECTION 10.1   Confidence Intervals for the Difference Between Two Means: Independent Samples

Objectives

1. Distinguish between independent and paired samples
2. Construct confidence intervals for the difference between two population means
3. Describe the pooled standard deviation and the known standard deviation methods

Objective 1 Distinguish between independent and paired samples

How can we tell whether a new drug reduces blood pressure better than an old one? A drug company has developed a new drug that is designed to reduce high blood pressure. The researchers wish to design a study to compare the effectiveness of the new drug to that of the old drug. Here are two ways in which the study can be designed.

Design 1: Two samples of individuals are chosen. One sample is given the old drug and the other sample is given the new drug. After several months, blood pressures of the members of both samples are measured. We compare the blood pressures in the first sample to the blood pressures in the second sample to determine which drug is more effective.

In design 1, we have **independent samples**. This means that the observations in one sample do not influence the observations in the other.

Design 2: A single group of individuals is chosen. They are given the old drug for a month, then their blood pressures are measured. They then switch to the new drug for a month, after which their blood pressures are measured again. This produces two samples of measurements, the first one from the old drug and the second one from the new drug. We compare the blood pressures in the first sample to the blood pressures in the second sample to determine which drug is more effective.

In design 2, we have **paired samples**. Each observation in one sample can be paired with an observation in the second.

> **SUMMARY**
>
> - Two samples are independent if the observations in one sample do not influence the observations in the other.
> - Two samples are paired if each observation in one sample can be paired with an observation in the other. Typically the samples consist of pairs of measurements on the same individual, or on pairs of individuals who are related, such as husbands and wives or brothers and sisters.

In this section, we will learn how to construct confidence intervals from independent samples. In Section 10.3, we will learn how to construct confidence intervals from paired samples.

Check Your Understanding

1. A sample of students is enrolled in a speed-reading class. Each takes a reading test before and again after the class. The two samples of scores are compared to determine how large an improvement in reading speed occurred. Are these samples independent or paired?

2. A sample of students is enrolled in an online statistics class, and another sample is enrolled in a traditional statistics class. At the end of the semester, the students are given a test. The scores from each sample are compared to determine which class was more effective. Are these samples independent or paired?

Answers are on page 487.

Objective 2 Construct confidence intervals for the difference between two population means

Construct Confidence Intervals for the Difference Between Two Population Means

Imagine that we will compare the effectiveness of a new drug designed to reduce blood pressure to the effectiveness of a standard drug. We will draw two independent samples. The first sample will get the new drug, and the second sample will get the standard drug. We can imagine these samples as coming from two populations — a population of patients who take the new drug, and a population of patients who take the standard drug. For each population, there is a mean reduction in blood pressure. We are interested in estimating the difference between the population means.

Let μ_1 be the population mean reduction for the new drug, and let μ_2 be the population mean reduction for the standard drug. We wish to construct a confidence interval for the difference $\mu_1 - \mu_2$. We need a point estimate, a standard error, and a critical value.

The point estimate for the population mean μ_1 is the sample mean $\bar{x}_1$, and the point estimate for the population mean μ_2 is the sample mean $\bar{x}_2$. It follows that the point estimate for the difference $\mu_1 - \mu_2$ between the population means is the difference between the sample means:

$$\text{Point estimate for } \mu_1 - \mu_2 \text{ is } \bar{x}_1 - \bar{x}_2$$

The sample means $\bar{x}_1$ and $\bar{x}_2$ have variances σ_1^2/n_1 and σ_2^2/n_2 respectively, where σ_1^2 and σ_2^2 are the population variances. It is a fact that when samples are independent, the variance of the difference $\bar{x}_1 - \bar{x}_2$ is the *sum* of the variances, so

$$\text{Variance of } \bar{x}_1 - \bar{x}_2 = \frac{\sigma_1^2}{n_1} + \frac{\sigma_2^2}{n_2}$$

The standard error of $\bar{x}_1 - \bar{x}_2$ is the square root of the variance. We don't know the values of σ_1^2 and σ_2^2, so we approximate them with the sample variances s_1^2 and s_2^2. The standard error is

$$\text{Standard error of } \bar{x}_1 - \bar{x}_2 = \sqrt{\frac{s_1^2}{n_1} + \frac{s_2^2}{n_2}}$$

Now we need a critical value. Let $1 - \alpha$ be the confidence level, expressed as a decimal. It can be shown by advanced methods that the appropriate critical value for a confidence interval is $t_{\alpha/2}$, based on the Student's t distribution. We need to determine the number of degrees of freedom for $t_{\alpha/2}$. There are two ways to do this: a simple method that is easier when computing by hand, and a more complicated method that is used by software packages. The simple method is:

$$\text{Degrees of freedom} = \text{Smaller of } n_1 - 1 \text{ and } n_2 - 1$$

The margin of error is obtained by multiplying the critical value by the standard error:

$$\text{Margin of error} = t_{\alpha/2}\sqrt{\frac{s_1^2}{n_1} + \frac{s_2^2}{n_2}}$$

Finally, we construct the confidence interval by adding and subtracting the margin of error from the point estimate:

$$\bar{x}_1 - \bar{x}_2 - t_{\alpha/2}\sqrt{\frac{s_1^2}{n_1} + \frac{s_2^2}{n_2}} < \mu_1 - \mu_2 < \bar{x}_1 - \bar{x}_2 + t_{\alpha/2}\sqrt{\frac{s_1^2}{n_1} + \frac{s_2^2}{n_2}}$$

This method is often referred to as Welch's t method, or more simply, as **Welch's method**. The method we have just described requires some assumptions, which we now state.

Explain It Again

Reason for assumption 3:
Assumption 3 is necessary to ensure that the sampling distributions of $\bar{x}_1$ and $\bar{x}_2$ are approximately normal. This justifies the use of the Student's t distribution.

Assumptions for Constructing a Confidence Interval for $\mu_1 - \mu_2$ with Independent Samples (Welch's method)

1. We have simple random samples from two populations.
2. The samples are independent of one another.
3. Each sample size is large ($n > 30$), *or* its population is approximately normal.

When these assumptions are satisfied, we can construct a confidence interval by using the following steps.

Procedure for Constructing a Confidence Interval for $\mu_1 - \mu_2$ with Independent Samples (Welch's method)

Check to be sure that the assumptions are satisfied. If they are, then proceed with the following steps.

Step 1: Compute the sample means $\bar{x}_1$ and $\bar{x}_2$ if they are not given; then compute the point estimate $\bar{x}_1 - \bar{x}_2$.

Step 2: Find the number of degrees of freedom, which is the smaller of $n_1 - 1$ and $n_2 - 1$, and the critical value $t_{\alpha/2}$.

Step 3: Compute the sample standard deviations s_1 and s_2 if they are not given, and compute the standard error $\sqrt{\frac{s_1^2}{n_1} + \frac{s_2^2}{n_2}}$. Multiply the standard error by the critical value to obtain the margin of error: $t_{\alpha/2}\sqrt{\frac{s_1^2}{n_1} + \frac{s_2^2}{n_2}}$

Step 4: Use the point estimate and the margin of error to construct the confidence interval:

$$\text{Point estimate} \pm \text{Margin of error}$$

$$\bar{x}_1 - \bar{x}_2 \pm t_{\alpha/2}\sqrt{\frac{s_1^2}{n_1} + \frac{s_2^2}{n_2}}$$

$$\bar{x}_1 - \bar{x}_2 - t_{\alpha/2}\sqrt{\frac{s_1^2}{n_1} + \frac{s_2^2}{n_2}} < \mu_1 - \mu_2 < \bar{x}_1 - \bar{x}_2 + t_{\alpha/2}\sqrt{\frac{s_1^2}{n_1} + \frac{s_2^2}{n_2}}$$

Step 5: Interpret the result.

EXAMPLE 10.1 Constructing a confidence interval

A drug company has developed a new drug that is designed to reduce high blood pressure. To test the drug, a sample of 15 patients is recruited to take the drug. Their systolic blood pressures are reduced by an average of 28.3 millimeters, with a standard deviation of 12.0 millimeters. In addition, another sample of 20 patients takes a standard drug. The blood pressures in this group are reduced by an average of 17.1 millimeters with a standard deviation of 9.0 millimeters. Assume that blood pressure reductions are approximately normally distributed. Find a 95% confidence interval for the difference between the population mean reduction for the new drug and that of the standard drug.

Solution

To help us keep track of the relevant information, we present it in the following table:

	New Drug	Standard Drug
Sample mean	$\bar{x}_1 = 28.3$	$\bar{x}_2 = 17.1$
Sample standard deviation	$s_1 = 12$	$s_2 = 9$
Sample size	$n_1 = 15$	$n_2 = 20$
Population mean	μ_1 (unknown)	μ_2 (unknown)

We check the assumptions. We have two independent random samples, and the populations are approximately normally distributed. The assumptions are satisfied.

Step 1: Compute the point estimate.

$$\bar{x}_1 - \bar{x}_2 = 28.3 - 17.1 = 11.2$$

Step 2: Find the critical value. In this example, $n_1 = 15$ and $n_2 = 20$, so the degrees of freedom is the smaller of $n_1 - 1 = 14$ and $n_2 - 1 = 19$, which is 14. We look up the critical value $t_{\alpha/2}$ in Table A.3. The value corresponding to 14 degrees of freedom with a confidence level of 95% is $t_{\alpha/2} = 2.145$.

Step 3: Compute the standard error and the margin of error. The standard deviations are $s_1 = 12.0$ and $s_2 = 9.0$. The sample sizes are $n_1 = 15$ and $n_2 = 20$. The standard error is

$$\sqrt{\frac{s_1^2}{n_1} + \frac{s_2^2}{n_2}} = \sqrt{\frac{12.0^2}{15} + \frac{9.0^2}{20}} = 3.6946$$

The margin of error is obtained by multiplying the standard error by the critical value.

$$\text{Margin of error} = t_{\alpha/2}\sqrt{\frac{s_1^2}{n_1} + \frac{s_2^2}{n_2}}$$

In this example, the margin of error is

$$t_{\alpha/2}\sqrt{\frac{s_1^2}{n_1} + \frac{s_2^2}{n_2}} = 2.145\sqrt{\frac{12.0^2}{15} + \frac{9.0^2}{20}} = 7.925$$

Step 4: Construct the confidence interval. The 95% confidence interval is

$$11.2 - 7.925 < \mu_1 - \mu_2 < 11.2 + 7.925$$
$$3.3 < \mu_1 - \mu_2 < 19.1$$

Note that we have rounded the final result to one decimal place, because each of the sample means (28.3 and 17.1) was given to one decimal place.

Step 5: Interpret the result. We are 95% confident that the new drug provides a greater reduction in systolic blood pressure, and that the improvement due to the new drug is between 3.3 and 19.1 millimeters.

Check Your Understanding

3. **Big fish:** A sample of 87 one-year-old spotted flounder had a mean length of 126.31 millimeters with a sample standard deviation of 18.10 millimeters, and a sample of 132 two-year-old spotted flounder had a mean length of 162.41 millimeters with a sample standard deviation of 28.49 millimeters. Construct a 95% confidence interval for the mean length difference between two-year-old flounder and one-year-old flounder.

   Source: *Turkish Journal of Veterinary and Animal Science* 29:1013–1018

4. **Traffic speed:** The mean speed for a sample of 39 cars at a certain intersection was 26.50 kilometers per hour with a standard deviation of 2.37 kilometers per hour, and the mean speed for a sample of 142 motorcycles was 37.14 kilometers per hour with a standard deviation of 3.66 kilometers per hour. Construct a 99% confidence interval for the difference between the mean speeds of motorcycles and cars at this intersection.

   Source: *Journal of Transportation Engineering* 121:317–323

Answers are on page 487.

Technology calculates the degrees of freedom differently

If you construct a confidence interval for the difference between two means with technology, you will usually get a somewhat different answer than you will get using the method we have presented here. The reason is that computers and calculators compute the number of degrees of freedom differently, using a more accurate but rather complicated formula. We present this formula, but you don't need to use it when computing by hand. When computing by hand, it is acceptable just to use the smaller of $n_1 - 1$ and $n_2 - 1$ for the degrees of freedom.

More Accurate Formula for the Degrees of Freedom

Most computer packages compute the degrees of freedom as follows:

$$\text{Degrees of freedom} = \frac{\left[\dfrac{s_1^2}{n_1} + \dfrac{s_2^2}{n_2}\right]^2}{\dfrac{(s_1^2/n_1)^2}{n_1 - 1} + \dfrac{(s_2^2/n_2)^2}{n_2 - 1}}$$

When computing by hand, it is acceptable, and simpler, just to use the smaller of $n_1 - 1$ and $n_2 - 1$ for the degrees of freedom.

Constructing confidence intervals with technology

The following TI-84 Plus display presents results for Example 10.1.

```
    2-SampTInt
(3.5912,18.809)
df=25.02267409
x̄1=28.3
x̄2=17.1
Sx1=12
↓Sx2=9
```

The results differ from those we obtained, because the degrees of freedom (labeled "df") has been calculated by the more accurate formula. Note that the degrees of freedom is not a whole number.

The following MINITAB output presents results for Example 10.1.

```
        N     Mean    StDev    SE Mean
A      15     28.3    12.0       3.1
B      20     17.1     9.0       2.0

Difference = mu (A) − mu (B)
Estimate for difference: 11.2
95% CI for difference: (3.5908, 18.8092)   DF = 25
```

Most of the output is straightforward. The quantities labeled "SE Mean" are the values of $s_1/\sqrt{n_1}$ and $s_2/\sqrt{n_2}$. MINITAB uses the more accurate formula for the degrees of freedom but rounds the degrees of freedom down to the nearest whole number, which in this case is 25. For this reason, the MINITAB confidence interval differs slightly from the one calculated by the TI-84 Plus.

Step-by-step instructions for constructing confidence intervals with technology are presented in the Using Technology section on page 482.

In most situations in practice, Welch's method is the method of choice for constructing confidence intervals for the difference between two means with independent samples. There are two other methods that have sometimes been used. We describe them here, because they are often offered as options with statistical software. They are generally not the best to use in practice, however, so we will always use Welch's method.

Objective 3 Describe the pooled standard deviation and the known standard deviation methods

Constructing confidence intervals by using the pooled standard deviation

When the two population variances, σ_1 and σ_2, are known to be equal, there is an alternate method for computing a confidence interval. This alternate method was widely used in the past, and is still an option in many forms of technology. We recommend against using it, for reasons that we will discuss.

A Method for Constructing a Confidence Interval When $\sigma_1 = \sigma_2$ (Not Recommended)

Step 1: Compute the **pooled standard deviation**, s_p, as follows:

$$s_p = \sqrt{\frac{(n_1 - 1)s_1^2 + (n_2 - 1)s_2^2}{n_1 + n_2 - 2}}$$

Step 2: Compute the degrees of freedom:

$$\text{Degrees of freedom} = n_1 + n_2 - 2$$

A level $100(1 - \alpha)\%$ confidence interval is

$$\bar{x}_1 - \bar{x}_2 - t_{\alpha/2}s_p\sqrt{\frac{1}{n_1} + \frac{1}{n_2}} < \mu_1 - \mu_2 < \bar{x}_1 - \bar{x}_2 + t_{\alpha/2}s_p\sqrt{\frac{1}{n_1} + \frac{1}{n_2}}$$

The major problem with this method is that the assumption that the population variances are equal is very strict. The method can be quite unreliable if it is used when the population variances are not equal. Now in practice, it is rarely possible to be sure that the variances are equal. There is a test for equality of variances (the F-test, Section 11.4), for which the null hypothesis is that the variances are equal. One could perform this test and assume that the variances are equal if the null hypothesis is not rejected. We recommend against this for two reasons. First, the F-test is unreliable unless the populations are almost exactly normal. Second, failure to reject the null hypothesis does not allow one to assume that the null hypothesis is true. Finally, even when the variances are equal, this method is usually only slightly better than Welch's method. In summary, the best practice is not to use the method that assumes the population variances are equal unless you are very sure that they are.

Constructing confidence intervals when the population standard deviations are known

When the two population variances, σ_1 and σ_2, are known, we can use $z_{\alpha/2}$, rather than $t_{\alpha/2}$, as the critical value. In practice, σ_1 and σ_2 are rarely known, so this method is not often applicable. We present it here because it is often offered as an option in statistical calculators and software. The assumptions for this method are the same as for Welch's method, with the additional assumption that the population standard deviations are known.

A Method for Constructing a Confidence Interval When σ_1 and σ_2 Are Known

A level $100(1 - \alpha)\%$ confidence interval when σ_1 and σ_2 are known is given by

$$\bar{x}_1 - \bar{x}_2 \pm z_{\alpha/2}\sqrt{\frac{\sigma_1^2}{n_1} + \frac{\sigma_2^2}{n_2}}$$

$$\bar{x}_1 - \bar{x}_2 - z_{\alpha/2}\sqrt{\frac{\sigma_1^2}{n_1} + \frac{\sigma_2^2}{n_2}} < \mu_1 - \mu_2 < \bar{x}_1 - \bar{x}_2 + z_{\alpha/2}\sqrt{\frac{\sigma_1^2}{n_1} + \frac{\sigma_2^2}{n_2}}$$

Note that this method is the same as Welch's except that the sample standard deviations s_1 and s_2 are replaced with the population standard deviations σ_1 and σ_2, and $t_{\alpha/2}$ is replaced with $z_{\alpha/2}$.

USING TECHNOLOGY

We use Example 10.1 to illustrate the technology steps.

TI-84 PLUS

Constructing a confidence interval for the difference between two means

Step 1. Press **STAT** and highlight the **TESTS** menu.
Step 2. Select **2–SampTInt** and press **ENTER** (Figure A). The **2–SampTInt** menu appears.
Step 3. Choose one of the following:
- If the summary statistics are given, select **Stats** as the **Inpt** option and enter $\bar{x}_1$, s_1, n_1, $\bar{x}_2$, s_2, and n_2. For Example 10.1, we use $\bar{x}_1 = 28.3$, $s_1 = 12$, $n_1 = 15$, $\bar{x}_2 = 17.1$, $s_2 = 9$, $n_2 = 20$ (Figure B).
- If the raw data are given, select **Data** as the **Inpt** option and enter the location of the data as the **List1** and **List2** options.

Step 4. In the **C-Level** field, enter the confidence level. For Example 10.1, we use 0.95.
Step 5. Select **No** for the **Pooled** option (Figure C).
Step 6. Highlight **Calculate** and press **ENTER** (Figure D).

```
EDIT CALC TESTS
4↑2-SampTTest…
5:1-PropZTest…
6:2-PropZTest…
7:ZInterval…
8:TInterval…
9:2-SampZInt…
0:2-SampTInt…
```
Figure A

```
2-SampTInt
Inpt:Data Stats
x̄1:28.3
Sx1:12
n1:15
x̄2:17.1
Sx2:9
↓n2:20
```
Figure B

```
2-SampTInt
↑Sx1:12
n1:15
x̄2:17.1
Sx2:9
n2:20
C-Level:.95
↓Pooled:No Yes
```
Figure C

```
2-SampTInt
(3.5912,18.809)
df=25.02267409
x̄1=28.3
x̄2=17.1
Sx1=12
↓Sx2=9
```
Figure D

MINITAB

Constructing a confidence interval for the difference between two means

Step 1. Click on **Stat**, then **Basic Statistics**, then **2-Sample t**.

Step 2. Choose one of the following:
- If the summary statistics are given, click **Summarized Data** and enter the **Sample Size**, the **Mean**, and the **Standard Deviation** for each sample. For Example 10.1, we use $\bar{x}_1 = 28.3$, $s_1 = 12$, $n_1 = 15$, $\bar{x}_2 = 17.1$, $s_2 = 9$, $n_2 = 20$.
- If the raw data are given, click **Samples in different columns** and select the columns that contain the data.

Step 3. Click **Options**, and enter the confidence level in the **Confidence Level** field (95) and choose **not equal** in the **Alternative field**. Click **OK**.

Step 4. Click **OK** (Figure E).

```
Difference = mu (1) - mu (2)
Estimate for difference:  11.2000
95% CI for difference:  (3.5908, 18.8092)
T-Test of difference = 0 (vs not =): T-Value = 3.03  P-Value = 0.006  DF = 25
```

Figure E

EXCEL

Constructing a confidence interval for the difference between two means

This procedure requires the **MegaStat** EXCEL add-in to be loaded. The **MegaStat** add-in can be downloaded from www.mhhe.com/megastat.

Step 1. Click on the **MegaStat** menu, select **Hypothesis Tests**, then **Compare Two Independent Groups**...

Step 2. Choose one of the following:
- If the summary statistics are given, click **summary input** and select the ranges of cells that contain the data label, the mean, the standard deviation, and the sample size (Figure F).
- If the raw data are given, click **data input** and select the ranges of cells for each sample.

Step 3. Enter **0** in the **Hypothesized difference** field and select **not equal** in the **Alternative field**.

Step 4. Choose the **t-test (unequal variance)** option and select the **Display confidence interval** option with the desired confidence level (Figure G).

Step 5. Click **OK** (Figure H).

	A	B	C
1		**New Drug**	**Standard Drug**
2	**Sample Mean**	28.3	17.1
3	**Sample St Dev**	12	9
4	**Sample Size**	15	20

Figure F

3.5908	confidence interval 95.% lower
18.8092	confidence interval 95.% upper
7.6092	margin of error

Figure H

Figure G

Exercises 1–4 are the Check Your Understanding exercises located within the section.

Understanding the Concepts

In Exercises 5 and 6, fill in each blank with the appropriate word or phrase.

5. If observations in one sample do not influence the observations in another sample, the samples are said to be _____.

6. When determining the number of degrees of freedom by hand with sample sizes n_1 and n_2, we choose the smaller of _____ and _____.

In Exercises 7 and 8, determine whether the statement is true or false. If the statement is false, rewrite it as a true statement.

7. The point estimate for $\mu_1 - \mu_2$ is $\bar{x}_1 + \bar{x}_2$.

8. The number of degrees of freedom calculated with technology is generally different from the number calculated by hand.

Practicing the Skills

In Exercises 9–14, construct the confidence interval for the difference $\mu_1 - \mu_2$ for the given level and values of $\bar{x}_1, \bar{x}_2, s_1, s_2, n_1$, and n_2.

9. Level 90%: $\bar{x}_1 = 104.6$, $\bar{x}_2 = 92.9$, $s_1 = 4.8$, $s_2 = 6.9$, $n_1 = 26$, $n_2 = 19$

10. Level 95%: $\bar{x}_1 = 478.81$, $\bar{x}_2 = 322.49$, $s_1 = 42.84$, $s_2 = 25.17$, $n_1 = 14$, $n_2 = 16$

11. Level 99%: $\bar{x}_1 = 603.55$, $\bar{x}_2 = 516.63$, $s_1 = 54.7$, $s_2 = 45.2$, $n_1 = 15$, $n_2 = 24$

12. Level 98%: $\bar{x}_1 = 77.3$, $\bar{x}_2 = 72.6$, $s_1 = 9.1$, $s_2 = 8.8$, $n_1 = 12$, $n_2 = 16$

13. Level 95%: $\bar{x}_1 = 47.7$, $\bar{x}_2 = 42.6$, $s_1 = 33.9$, $s_2 = 17.6$, $n_1 = 13$, $n_2 = 19$

14. Level 99.5%: $\bar{x}_1 = 82.9$, $\bar{x}_2 = 64.1$, $s_1 = 9.8$, $s_2 = 6.2$, $n_1 = 19$, $n_2 = 10$

Working with the Concepts

15. Does this diet help? A group of 78 people enrolled in a weight-loss program that involved adhering to a special diet and to a daily exercise program. After six months, their mean weight loss was 25 pounds, with a sample standard deviation of 9 pounds. A second group of 43 people went on the diet but didn't exercise. After six months, their mean weight loss was 14 pounds, with a sample standard deviation of 7 pounds. Construct a 95% confidence interval for the mean difference in weight losses.

16. Contaminated water: The concentration of benzene was measured in units of milligrams per liter for a simple random sample of five specimens of untreated wastewater produced at a gas field. The sample mean was 7.8 with a

sample standard deviation of 1.4. Seven specimens of treated wastewater had an average benzene concentration of 3.2 with a standard deviation of 1.7. It is reasonable to assume that both samples come from populations that are approximately normal. Construct a 99% confidence interval for the reduction in benzene concentration after treatment.

17. Fertilizer: In an agricultural experiment, the effects of two fertilizers on the production of oranges were measured. Sixteen randomly selected plots of land were treated with fertilizer A, and 12 randomly selected plots were treated with fertilizer B. The number of pounds of harvested fruit was measured from each plot. Following are the results.

Fertilizer A							
445	523	464	483	441	491	403	466
448	457	437	516	417	420	400	506

Fertilizer B					
362	414	408	398	382	368
393	437	387	373	424	384

a. Explain why it is necessary to check whether the populations are approximately normal before constructing a confidence interval.

b. Following are boxplots of these data. Is it reasonable to assume that the populations are approximately normal?

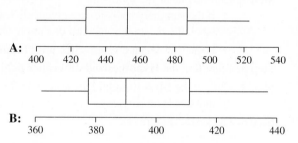

c. If appropriate, construct a 98% confidence interval for the difference between the mean yields for the two types of fertilizer. If not appropriate, explain why not.

18. Computer crashes: A computer system administrator notices that computers running a particular operating system seem to crash more often as the installation of the operating system ages. She measures the time (in minutes) before crash for seven computers one month after installation, and for nine computers seven months after installation. The results are as follows:

One month after installation						
209	230	217	230	221	243	247

Seven months after installation								
85	59	129	201	176	240	149	154	105

a. Explain why it is necessary to check whether the populations are approximately normal before constructing a confidence interval.

b. Following are dotplots of these data. Is it reasonable to assume that the populations are approximately normal?

One month after installation

Seven months after installation

c. If appropriate, construct a 95% confidence interval for the mean difference in time to crash between the first month after installation and the seventh. If not appropriate, explain why not.

19. Are you smarter than your older brother? In a study of birth order and intelligence, IQ tests were given to 18- and 19-year-old men to estimate the size of the difference, if any, between the mean IQs of firstborn sons and of secondborn sons. The following data for 10 firstborn sons and 10 secondborn sons are consistent with the means and standard deviations reported in the article. It is reasonable to assume that the samples come from populations that are approximately normal.

Firstborn				
104	82	102	96	129
89	114	107	89	103

Secondborn				
103	103	91	113	102
103	92	90	114	113

a. Construct a 95% confidence interval for the difference in mean IQ between firstborn and secondborn sons.

b. Based on the confidence interval, is it reasonable to believe that firstborn sons and secondborn sons may have the same mean IQ?

Source: Based on data in *Science* 316:1717

20. Effectiveness of distance learning: A study was done to compare the effectiveness of distance learning with traditional classroom instruction. Twelve students took a business administration course online, while 14 students took it in a classroom. The final exam scores were as follows. It is reasonable to assume that the samples come from populations that are approximately normal.

Online					
66	75	85	64	88	77
74	91	72	69	77	83

Classroom						
80	83	64	81	75	80	86
81	91	64	99	85	74	77

a. Construct a 95% confidence interval for the difference between the mean scores for the two types of instruction.

b. An educator claims that both methods of instruction are equally effective. Does the confidence interval contradict this claim?

21. Boys and girls: The National Health Statistics Reports published in 2012 reported that a sample of 318 one-year-old boys had a mean weight of 25.0 pounds with a standard deviation of 3.6 pounds. In addition, a sample of 297 one-year-old girls had a mean weight of 24.1 pounds with a standard deviation of 3.8 pounds.

a. Construct a 95% confidence interval for the difference between the mean weights.

b. A magazine article states that the mean weight of one-year-old boys is the same as that of one-year-old girls. Does the confidence interval contradict this statement?

22. Body mass index: In a survey of adults with diabetes, the average body mass index (BMI) in a sample of 1924 women was 31.1 with a standard deviation of 0.2. The BMI in a sample of 1559 men was 30.4, with a standard deviation of 0.6.

a. Construct a 99% confidence interval for the difference in the mean BMI between women and men with diabetes.

b. Does the confidence interval contradict the claim that the mean BMI is the same for both men and women with diabetes?

Source: *Journal of Women's Health* 16:1421–1428

23. Energy drinks: A survey of college students reported that in a sample of 413 male college students, the average number of energy drinks consumed per month was 2.49 with a standard deviation of 4.87, and in a sample of 382 female college students, the average was 1.22 with a standard deviation of 3.24.

a. Construct a 99% confidence interval for the difference between men and women in the mean number of energy drinks consumed.

b. Based on the confidence interval, is it reasonable to believe that the mean number of energy drinks consumed may be the same for both male and female college students?

Source: *Journal of American College Health* 56:481–489

24. Low-fat or low-carb? Are low-fat diets or low-carb diets more effective for weight loss? A sample of 77 subjects went on a low-carbohydrate diet for six months. At the end of that time, the sample mean weight loss was 4.7 kilograms with a sample standard deviation of 7.2 kilograms. A second sample of 79 subjects went on a low-fat diet. Their sample mean weight loss was 2.6 kilograms with a standard deviation of 5.9 kilograms.

a. Construct a 98% confidence interval for the difference in mean weight loss between the low-fat and low-carb diets.

b. A dietitian claims that both diets are equally effective. Does the confidence interval contradict this claim?

Source: *Journal of the American Medical Association* 297:969–977

25. Online testing: Do you prefer taking tests on paper or online? A college instructor gave identical tests to two randomly sampled groups of 35 students. One group took the test on paper and the other took it online. Following are the test scores.

Paper						
79	75	49	78	73	81	70
63	79	65	60	74	64	64
74	69	57	65	61	58	92
71	69	66	71	68	67	81
78	80	76	67	49	56	45

Online						
79	75	71	81	56	72	49
75	63	81	74	72	71	73
83	59	78	65	53	47	63
82	81	76	65	82	78	76
65	72	85	84	81	50	79

a. Construct a 95% confidence interval for the difference in mean scores between paper and online tests.

b. The instructor claims that the mean scores are the same for both the paper and the online versions of the test. Does the confidence interval contradict this claim?

26. **Drive safely:** How often does the average driver have an accident? The Allstate Insurance Company determined the average number of years between accidents for drivers in a large number of U.S. cities. Following are the results for 32 cities east of the Mississippi River and 32 cities west of the Mississippi River.

East							
10.2	9.5	11.2	11.4	9.8	10.0	11.6	10.7
9.5	9.0	7.2	11.5	7.7	9.0	8.9	8.4
6.8	5.3	8.0	8.8	9.0	10.0	9.7	11.6
7.9	9.9	10.5	11.9	10.9	8.6	6.7	9.6

West							
9.5	7.9	13.4	12.6	9.7	8.2	14.0	7.1
10.2	9.4	10.8	9.5	7.5	9.1	8.4	7.6
7.4	7.7	7.1	11.9	10.2	11.9	7.7	7.0
10.0	8.1	11.5	8.4	9.6	9.5	9.9	11.5

a. Construct a 95% confidence interval for the difference in mean scores between western and eastern cities.

b. An insurance company executive claims that the mean number of years between accidents for western cities is 1.5 years greater than the mean for eastern cities. Does the confidence interval contradict this claim?

27. **Interpret calculator display:** The following TI-84 Plus calculator display presents a 95% confidence interval for the difference between two means. The sample sizes are $n_1 = 7$ and $n_2 = 10$.

```
2-SampTInt
(20.904,74.134)
df=12.28537157
x̄1=150.375
x̄2=102.856
Sx1=25.724
↓Sx2=23.548
```

a. Compute the point estimate of $\mu_1 - \mu_2$.

b. How many degrees of freedom did the calculator use?

c. Fill in the blanks: We are 95% confident that the difference between the means is between _____ and _____.

28. **Interpret calculator display:** The following TI-84 Plus calculator display presents a 99% confidence interval for the difference between two means. The sample sizes are $n_1 = 50$ and $n_2 = 42$.

```
2-SampTInt
(3.0101,4.6114)
df=86.51750655
x̄1=6.83562
x̄2=3.02487
Sx1=1.72541
↓Sx2=1.17482
```

a. Compute the point estimate of $\mu_1 - \mu_2$.

b. How many degrees of freedom did the calculator use?

c. Fill in the blanks: We are 99% confident that the difference between the means is between _____ and _____.

29. **Interpret computer output:** The following MINITAB output display presents a 98% confidence interval for the difference between two means.

	N	Mean	StDev	SE Mean
A	17	72.9172	10.7134	2.5984
B	25	52.1743	9.1237	1.8247

Difference = mu (A) − mu (B)
Estimate for difference: 20.7429
98% CI for difference: (12.9408, 28.5450)   DF = 30

a. What is the point estimate of $\mu_1 - \mu_2$?

b. How many degrees of freedom did MINITAB use?

c. Fill in the blanks: We are _____ confident that the difference between the means is between _____ and _____.

30. **Interpret computer output:** The following MINITAB output display presents a 95% confidence interval for the difference between two means.

	N	Mean	StDev	SE Mean
A	48	33.827	8.423	1.2157
B	57	10.372	9.314	1.2337

Difference = mu (A) − mu (B)
Estimate for difference: 23.455
95% CI for difference: (20.019, 29.891)   DF = 102

a. What is the point estimate of $\mu_1 - \mu_2$?

b. How many degrees of freedom did MINITAB use?

c. Fill in the blanks: We are _____ confident that the difference between the means is between _____ and _____.

Extending the Concepts

31. **Calculator display:** The following TI-84 Plus display presents a 95% confidence interval, using the more accurate formula for the degrees of freedom. The sample sizes are $n_1 = 12$ and $n_2 = 10$.

b. Find the critical value $t_{\alpha/2}$ for a 95% confidence interval using this value for the degrees of freedom.

c. Use this critical value along with the results on the TI-84 Plus display to construct the 95% confidence interval using this value for the degrees of freedom.

a. Use the simpler method to compute the degrees of freedom as the smaller of $n_1 - 1$ and $n_2 - 1$.

Answers to Check Your Understanding Exercises for Section 10.1

1. Paired

2. Independent

3. $29.83 < \mu_1 - \mu_2 < 42.37$ [Tech: $29.89 < \mu_1 - \mu_2 < 42.31$]

4. $9.32 < \mu_1 - \mu_2 < 11.96$ [Tech: $9.36 < \mu_1 - \mu_2 < 11.92$]

SECTION 10.2	Confidence Intervals for the Difference Between Two Proportions

Objectives

1. Construct confidence intervals for the difference between two proportions

Objective 1 Construct confidence intervals for the difference between two proportions

Construct Confidence Intervals for the Difference Between Two Proportions

Does breathing polluted air reduce lung capacity? In a hypothetical study of the effect of air pollution on lung function, a sample of 50 children living in a community with a high level of ozone pollution had their lung functions tested, and 14 of them had lung capacities that were below normal for their size. A second sample of 80 children was drawn from a community with a low level of ozone pollution, and 12 of them had lung capacities that were below normal for their size.

These samples come from two different populations — the populations of children living in the two communities. Each population is one whose individuals fall into two categories: those with reduced lung function, and those with normal lung function. The samples are independent, because the choice of the individuals to be in one sample did not affect the choice of individuals to be in the other.

If we compute the sample proportions, we see that the sample proportion of children in the high-pollution community with reduced lung function is $14/50 = 0.28$, while the proportion in the low-pollution community is only $12/80 = 0.15$. We would like to estimate the difference between the population proportions. Specifically, we would like to construct a 95% confidence interval for the difference between the population proportions.

We will need some notation for the population proportions, the numbers of individuals in each category, and the sample sizes:

NOTATION

- p_1 and p_2 are the population proportions of the category of interest in the two populations.

- x_1 and x_2 are the numbers of individuals in the category of interest in the two samples.

- n_1 and n_2 are the two sample sizes.

We wish to construct a 95% confidence interval for the difference $p_1 - p_2$. As usual, we need three ingredients: a point estimate, a critical value, and a standard error.

We begin with the point estimate. The natural point estimates for the population proportions are the sample proportions

$$\hat{p}_1 = \frac{x_1}{n_1} \qquad \hat{p}_2 = \frac{x_2}{n_2}$$

The point estimate of $p_1 - p_2$ is $\hat{p}_1 - \hat{p}_2$.

Now we compute the standard error. The variances of $\hat{p}_1$ and $\hat{p}_2$ are

$$\text{Variance of } \hat{p}_1 = \frac{p_1(1 - p_1)}{n_1} \qquad \text{Variance of } \hat{p}_2 = \frac{p_2(1 - p_2)}{n_2}$$

The samples are independent, so the variance of $\hat{p}_1 - \hat{p}_2$ is the sum of the variances of $\hat{p}_1$ and $\hat{p}_2$.

$$\text{Variance of } \hat{p}_1 - \hat{p}_2 = \frac{p_1(1 - p_1)}{n_1} + \frac{p_2(1 - p_2)}{n_2}$$

The standard error is the square root of the variance. In practice, we don't know the values of p_1 and p_2, so we approximate them with $\hat{p}_1$ and $\hat{p}_2$. The standard error is

$$\text{Standard error of } \hat{p}_1 - \hat{p}_2 = \sqrt{\frac{\hat{p}_1(1 - \hat{p}_1)}{n_1} + \frac{\hat{p}_2(1 - \hat{p}_2)}{n_2}}$$

Now when the sample sizes are large enough, $\hat{p}_1$ and $\hat{p}_2$ are approximately normally distributed, so the critical value is $z_{\alpha/2}$, found in Table A.2. We multiply the standard error by the critical value to obtain the margin of error

$$\text{Margin of error} = z_{\alpha/2}\sqrt{\frac{\hat{p}_1(1 - \hat{p}_1)}{n_1} + \frac{\hat{p}_2(1 - \hat{p}_2)}{n_2}}$$

The level $100(1 - \alpha)\%$ confidence interval is

$$\text{Point estimate} \pm \text{Margin of error}$$

$$\hat{p}_1 - \hat{p}_2 \pm z_{\alpha/2}\sqrt{\frac{\hat{p}_1(1 - \hat{p}_1)}{n_1} + \frac{\hat{p}_2(1 - \hat{p}_2)}{n_2}}$$

The method we have just described requires some assumptions, which we now list.

Assumptions for Constructing a Confidence Interval for the Difference Between Proportions

1. We have two independent simple random samples.

2. Each population is at least 20 times as large as the sample drawn from it.

3. The individuals in each sample are divided into two categories.

4. Both samples contain at least 10 individuals in each category.

When these assumptions are met, we can construct a confidence interval for the difference between two proportions by using the following steps.

Procedure for Constructing a Confidence Interval for $p_1 - p_2$

Check to be sure that the assumptions are satisfied. If they are, then proceed with the following steps:

Step 1: Compute the value of the point estimate $\hat{p}_1 - \hat{p}_2$.

Step 2: Find the critical value $z_{\alpha/2}$ corresponding to the desired confidence level from the last line of Table A.3, from Table A.2, or with technology.

Step 3: Compute the standard error

$$\sqrt{\frac{\hat{p}_1(1-\hat{p}_1)}{n_1} + \frac{\hat{p}_2(1-\hat{p}_2)}{n_2}}$$

and multiply it by the critical value to obtain the margin of error

$$z_{\alpha/2}\sqrt{\frac{\hat{p}_1(1-\hat{p}_1)}{n_1} + \frac{\hat{p}_2(1-\hat{p}_2)}{n_2}}$$

Step 4: Use the point estimate and the margin of error to construct the confidence interval:

$$\text{Point estimate} \pm \text{Margin of error}$$

$$\hat{p}_1 - \hat{p}_2 \pm z_{\alpha/2}\sqrt{\frac{\hat{p}_1(1-\hat{p}_1)}{n_1} + \frac{\hat{p}_2(1-\hat{p}_2)}{n_2}}$$

$$\hat{p}_1 - \hat{p}_2 - z_{\alpha/2}\sqrt{\frac{\hat{p}_1(1-\hat{p}_1)}{n_1} + \frac{\hat{p}_2(1-\hat{p}_2)}{n_2}} < p_1 - p_2 < \hat{p}_1 - \hat{p}_2 + z_{\alpha/2}\sqrt{\frac{\hat{p}_1(1-\hat{p}_1)}{n_1} + \frac{\hat{p}_2(1-\hat{p}_2)}{n_2}}$$

Step 5: Interpret the results.

EXAMPLE 10.2 Constructing a confidence interval

In a study of the effect of air pollution on lung function, a sample of 50 children living in a community with a high level of ozone pollution had their lung capacities measured, and 14 of them had capacities that were below normal for their size. A second sample of 80 children was drawn from a community with a low level of ozone pollution, and 12 of them had lung capacities that were below normal for their size. Construct a 95% confidence interval for the difference between the proportions of children with lung capacities below normal in the two communities.

Solution

We begin by summarizing the available information in a table:

	High Pollution	Low Pollution
Sample size	$n_1 = 50$	$n_2 = 80$
Number with below-normal lung capacity	$x_1 = 14$	$x_2 = 12$
Population proportion	p_1 (unknown)	p_2 (unknown)

We check the assumptions: We have two independent random samples. The populations of children are more than 20 times as large as the samples. The individuals are divided into two categories. In the first sample, there are 14 children with lung capacity below normal, and $50 - 14 = 36$ whose lung capacity is not below normal. In the second sample, the corresponding numbers are 12 and 68. Therefore, each sample contains at least 10 individuals in each category.

Step 1: Compute the value of the point estimate. The sample proportions are

$$\hat{p}_1 = \frac{14}{50} = 0.280 \qquad \hat{p}_2 = \frac{12}{80} = 0.150$$

The point estimate is

$$\hat{p}_1 - \hat{p}_2 = 0.280 - 0.150 = 0.130$$

Step 2: Find the critical value. The desired confidence level is 95%, so the critical value is $z_{\alpha/2} = 1.96$.

Step 3: Compute the standard error and the margin of error. The standard error is

$$\sqrt{\frac{\hat{p}_1(1 - \hat{p}_1)}{n_1} + \frac{\hat{p}_2(1 - \hat{p}_2)}{n_2}} = \sqrt{\frac{0.280(1 - 0.280)}{50} + \frac{0.150(1 - 0.150)}{80}}$$

$$= 0.075005$$

The margin of error is

$$z_{\alpha/2}\sqrt{\frac{\hat{p}_1(1 - \hat{p}_1)}{n_1} + \frac{\hat{p}_2(1 - \hat{p}_2)}{n_2}} =$$

$$1.96\sqrt{\frac{0.280(1 - 0.280)}{50} + \frac{0.150(1 - 0.150)}{80}} = 0.14701$$

Step 4: Construct the confidence interval. The 95% confidence interval is

$$\text{Point estimate} \pm \text{Margin of error}$$

$$0.130 \pm 0.14701$$

$$0.130 - 0.14701 < p_1 - p_2 < 0.130 + 0.14701$$

$$-0.017 < p_1 - p_2 < 0.277$$

Step 5: Interpret the results. We are 95% confident that the difference between the proportions is between -0.017 and 0.277. This confidence interval contains 0. Therefore, we cannot be sure that the proportions of children with diminished lung capacity differ between the two communities.

In Example 10.2, we rounded the final result to three decimal places. We will follow this rule in general.

Recall: Some commonly used critical values are shown in the following table:

Level	Critical Value
95%	1.96
98%	2.326
99%	2.576

Explain It Again

Round-off rule: When computing a confidence interval for the difference between proportions, round the final result to three decimal places.

Constructing confidence intervals with technology

Example 10.2 presented a 95% confidence interval for the difference between the proportions of children with diminished lung capacity in two communities. We now present the results from a TI-84 Plus calculator and the software package MINITAB.

Following is the TI-84 Plus display.

```
   2-PropZInt
(-.017,.27701)
p̂₁=.28
p̂₂=.15
n₁=50
n₂=80
```

In addition to the confidence interval, the display presents the two sample proportions, $\hat{p}_1$ and $\hat{p}_2$, along with the sample sizes n_1 and n_2.

Following is the MINITAB output.

```
Sample    X    N    Sample p
1        14   50    0.280000
2        12   80    0.150000

Difference = p (1) − p (2)
Estimate for difference: 0.130000
95% CI for difference: (−0.01701, 0.27701)
```

The MINITAB output is mostly straightforward. The quantities listed under the heading "Sample p" are $\hat{p}_1$ and $\hat{p}_2$. The quantity "Estimate for difference" is $\hat{p}_1 - \hat{p}_2$.

Step-by-step instructions for constructing confidence intervals with technology are given in the Using Technology section on page 492.

Check Your Understanding

1. **Teaching methods:** A class of 30 computer science students were taught introductory computer programming class with an innovative teaching method that used a graphical interface and drag-and-drop methods of creating computer programs. At the end of the class, 23 of these students said that they felt confident in their ability to write computer programs. Another class of 40 students were taught the same material using a standard method. At the end of class, 25 of these students said they felt confident. Assume that each class contained a simple random sample of students. Construct a 95% confidence interval for the difference between the proportions of students who felt confident.

2. **Damp electrical connections:** In a test of the effect of dampness on electrical connections, 80 electrical connections were tested under damp conditions and 130 were tested under dry conditions. Twenty of the damp connections failed and only 8 of the dry ones failed. If possible, construct a 90% confidence interval for the difference between the proportions of connections that fail when damp as opposed to dry. If not possible, explain why.

Answers are on page 495.

USING TECHNOLOGY

We use Example 10.2 to illustrate the technology steps.

TI-84 PLUS

Constructing a confidence interval for the difference between two proportions

Step 1. Press **STAT** and highlight the **TESTS** menu.

Step 2. Select **2–PropZInt** and press **ENTER** (Figure A). The **2–PropZInt** menu appears.

Step 3. Enter x_1, n_1, x_2, and n_2. For Example 10.2, we use $x_1 = 14$, $n_1 = 50$, $x_2 = 12$, and $n_2 = 80$.

Step 4. In the **C-Level** field, enter the confidence level. For Example 10.2, we use 0.95 (Figure B).

Step 5. Highlight **Calculate** and press **ENTER** (Figure C).

Figure A

Figure B

```
   2-PropZInt
(-.017,.27701)
p̂₁=.28
p̂₂=.15
n₁=50
n₂=80
```

Figure C

MINITAB

Constructing a confidence interval for the difference between two proportions

Step 1. Click on **Stat**, then **Basic Statistics**, then **2-Proportions**.

Step 2. Choose one of the following:

- If the summary statistics are given, click **Summarized Data** and enter the values of n_1 and n_2 in the **Trials** column and the values of x_1 and x_2 in the **Events** column. For Example 10.2, we use $x_1 = 14$, $n_1 = 50$, $x_2 = 12$, and $n_2 = 80$.
- If the raw data are given, click **Samples in different columns** and select the columns that contain the data.

Step 3. Click **Options** and enter the confidence level in the **Confidence Level** field (95) and choose **not equal** in the **Alternative field**. Click **OK**.

Step 4. Click **OK** (Figure D).

```
Sample   X   N   Sample p
1       14  50   0.280000
2       12  80   0.150000


Difference = p (1) - p (2)
Estimate for difference:  0.13
95% CI for difference:  (-0.0170071, 0.277007)
Test for difference = 0 (vs not = 0):   Z = 1.73   P-Value = 0.083
```

Figure D

EXCEL

Constructing a confidence interval for the difference between two proportions

This procedure requires the **MegaStat** EXCEL add-in to be loaded. The **MegaStat** add-in can be downloaded from www.mhhe.com/megastat.

Step 1. Click on the **MegaStat** menu, select **Hypothesis Tests**, then **Compare Two Independent Proportions...**

Step 2. In the **p** field for each group, enter the values of x_1 and x_2 (note that p changes to x). In the **n** field for each group, enter the values of n_1 and n_2. For Example 10.2, we use $x_1 = 14$, $n_1 = 50$, $x_2 = 12$, and $n_2 = 80$.

Step 3. Enter **0** in the **Hypothesized difference** field and select **not equal** in the **Alternative** field.

Step 4. Select the **Display confidence interval** option with the desired confidence level (95) (Figure E).

Step 5. Click **OK** (Figure F).

Figure E

```
-0.017 confidence interval 95.% lower
 0.277 confidence interval 95.% upper
 0.147   margin of error
```

Figure F

SECTION 10.2 Exercises

Exercises 1 and 2 are the Check Your Understanding exercises located within the section.

Understanding the Concepts

In Exercises 3 and 4, fill in each blank with the appropriate word or phrase.

3. To construct a confidence interval for $p_1 - p_2$, we must have two _____ samples.

4. When constructing a confidence interval for $p_1 - p_2$, we assume that items in each sample are divided into _____ categories and that there are at least _____ items in each category.

In Exercises 5 and 6, determine whether the statement is true or false. If the statement is false, rewrite it as a true statement.

5. The point estimate for $p_1 - p_2$ is $\hat{p}_1 - \hat{p}_2$, where $\hat{p}_1 = x_1/n_1$ and $\hat{p}_2 = x_2/n_2$.

6. The margin of error for $\hat{p}_1 - \hat{p}_2$ is

$$\sqrt{\frac{\hat{p}_1(1 - \hat{p}_1)}{n_1} + \frac{\hat{p}_2(1 - \hat{p}_2)}{n_2}}.$$

Practicing the Skills

In Exercises 7–12, construct the confidence interval for the difference $p_1 - p_2$ for the given level and values of $x_1, n_1, x_2,$ and n_2.

7. Level 95%: $x_1 = 42$, $n_1 = 80$, $x_2 = 18$, $n_2 = 60$

8. Level 90%: $x_1 = 14$, $n_1 = 25$, $x_2 = 12$, $n_2 = 40$

9. Level 99%: $x_1 = 57$, $n_1 = 147$, $x_2 = 86$, $n_2 = 118$

10. Level 95%: $x_1 = 63$, $n_1 = 106$, $x_2 = 70$, $n_2 = 126$

11. Level 98%: $x_1 = 49$, $n_1 = 74$, $x_2 = 62$, $n_2 = 153$

12. Level 99%: $x_1 = 24$, $n_1 = 53$, $x_2 = 17$, $n_2 = 41$

Working with the Concepts

13. Traffic accidents: Traffic engineers compared rates of traffic accidents at intersections with raised medians with rates at intersections with two-way left-turn lanes. They found that out of 4644 accidents at intersections with raised medians, 2280 were rear-end accidents, and out of 4584 accidents at two-way left-turn lanes, 1982 were rear-end accidents.

  a. Assuming these to be random samples of accidents from the two types of intersection, construct a 95% confidence interval for the difference between the proportions of accidents that are of the rear-end type at the two types of intersection.

  b. Does the confidence interval contradict the claim that the proportion of rear-end accidents is the same at both types of intersection?

  Source: *Journal of Transportation Engineering* 121:317–323

14. Computers in the classroom: In a new experimental teaching method, when students have questions in class, they send them to the instructor via a laptop computer. From time to time, the instructor pauses to read the questions and to provide the answers, without revealing the identities of the students who are asking the questions. This new method is supposed to eliminate the reluctance of students to ask questions for fear of revealing their lack of understanding in public. In a study of this new method, 67 male students and 72 female students participated. Treat these as if they were simple random samples. Of the male students, 38 said that they liked the new method better than the traditional one in which students raise their hands to ask questions. Of the female students, only 22 said they liked the new method better.

a. Construct a 95% confidence interval for the difference between the proportions of male and female students who like the new method better.

b. An educator claims that the proportion of students who like the new method better is the same for males and females. Does the confidence interval contradict this claim?

15. **Pain after surgery:** In a random sample of 50 patients undergoing a standard surgical procedure, 15 required medication for postoperative pain. In a random sample of 90 patients undergoing a new procedure, only 16 required pain medication.

a. Construct a 95% confidence interval for the difference in the proportions of patients needing pain medication between the old and new procedures.

b. A physician claims that the proportion of patients who need pain medication is the same for both procedures. Does the confidence interval contradict this claim?

16. **Pretzels:** In order to judge the effectiveness of an advertising campaign for a certain brand of pretzel, a company obtained a simple random sample of 90 convenience store receipts the week before the ad campaign began, and found that 21 of them showed a purchase of the pretzels. Another simple random sample of 70 receipts was taken the week after the ad campaign, and 39 of them showed a pretzel purchase.

a. Construct a 98% confidence interval for the difference between the proportions of customers purchasing pretzels before and after the ad campaign.

b. A marketing manager claims that the proportion of customers who purchased pretzels did not change after the ad campaign. Does the confidence interval contradict this claim?

17. **Defective electronics:** A team of designers was given the task of reducing the defect rate in the manufacture of a certain printed circuit board. The team decided to reconfigure the cooling system. A total of 973 boards were produced the week before the reconfiguration was

implemented, and 254 of these were defective. A total of 847 boards were produced the week after reconfiguration, and 95 of these were defective.

a. Construct a 90% confidence interval for the decrease in the defective rate after the reconfiguration.

b. A quality control engineer claims that the reconfiguration has decreased the proportion of defective parts by more than 0.15. Does the confidence interval contradict this claim?

Source: *The American Statistician* 56:312–315

18. **Satisfied?** A poll taken by the General Social Survey in 2010 asked people in the United States whether they were satisfied with their financial situation. A total of 478 out of 2038 people said they were satisfied. The same question was asked in 2012, and 537 out of 1967 people said they were satisfied.

a. Construct a 95% confidence interval for the difference between the proportion of adults who said they were satisfied in 2012 and proportion in 2010.

b. A sociologist claims that the proportion of people who are satisfied increased from 2010 to 2012 by more than 0.05. Does the confidence interval contradict this claim?

19. **Cancer prevention:** Colonoscopy is a medical procedure that is designed to find and remove precancerous lesions in the colon before they become cancerous. In a sample of 51,460 people without colorectal cancer, 5043 had previously had a colonoscopy, and in a sample of 10,292 people diagnosed with colorectal cancer, 720 had previously had a colonoscopy.

a. Construct a 95% confidence interval for the difference in the proportions of people who had colonoscopies between those who were diagnosed with colorectal cancer and those who were not.

b. Does the confidence interval contradict the claim that the proportion of people who have had colonoscopies is the same among those with colorectal cancer and those without?

Source: *Annals of Internal Medicine* 150:1–8

20. **Smartphones:** The Smartphone Ownership survey study found that of 1914 cell phone owners in 2011, 632 owned a smartphone. The study was repeated in 2013, and 1142 out of 2076 cell phone owners owned smartphones.

a. Construct a 95% confidence interval for the increase in the proportion of cell phone owners with smartphones between 2011 and 2013.

b. A smartphone marketing executive claims that the proportion of cell phone owners with smartphones increased by less than 0.10 between 2005 and 2009. Does the confidence interval contradict this claim?

21. **Interpret calculator display:** The following TI-84 Plus calculator display presents a 95% confidence interval for the difference between two proportions.

a. Compute the point estimate of $p_1 - p_2$.
b. Fill in the blanks: We are 95% confident that the difference between the proportions is between _____ and _____.

22. Interpret calculator display: The following TI-84 Plus calculator display presents a 99% confidence interval for the difference between two proportions.

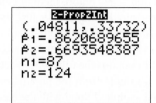

```
    2-PropZInt
 (.04811,.33732)
 p̂₁=.8620689655
 p̂₂=.6693548387
 n₁=87
 n₂=124
```

a. Compute the point estimate of $p_1 - p_2$.
b. Fill in the blanks: We are 99% confident that the difference between the proportions is between _____ and _____.

23. Interpret computer output: The following MINITAB output presents a confidence interval for the difference between two proportions.

```
Sample   X    N    Sample p
1        32   59   0.542373
2        23   63   0.365079


Difference = p(1) − p(2)
Estimate for difference: 0.177294
99% CI for difference: (−0.051451, 0.406038 )
```

a. What is the point estimate of $p_1 - p_2$?
b. Fill in the blanks: We are _____ confident that the difference between the proportions is between _____ and _____.

24. Interpret computer output: The following MINITAB output presents a confidence interval for the difference between two proportions.

```
Sample   X    N    Sample p
1        16   546  0.029304
2        18   935  0.019251


Difference = p(1) − p(2)
Estimate for difference: 0.010053
95% CI for difference: (−0.006612, 0.026717)
```

a. What is the point estimate of $p_1 - p_2$?
b. Fill in the blanks: We are _____ confident that the difference between the proportions is between _____ and _____.

Extending the Concepts

25. Finding the sample size: Polls are to be conducted in two cities to determine the difference in the proportions of residents who believe that the economy will improve over the next year. A 95% confidence interval will be constructed for the difference between the proportions. If the sample sizes are to be equal in the two cities, how large should each sample be so that the margin of error will be no more than 0.08?

Answers to Check Your Understanding Exercises for Section 10.2

1. $-0.071 < p_1 - p_2 < 0.355$

2. It is not possible to construct a confidence interval, because the sample of dry connections contains fewer than 10 that failed.

SECTION 10.3   Confidence Intervals for the Difference Between Two Means: Paired Samples

Objectives

1. Construct confidence intervals with paired samples

Objective 1 Construct confidence intervals with paired samples

Construct Confidence Intervals with Paired Samples

Does drinking a small amount of alcohol reduce reaction time noticeably? Sixteen volunteers were given a test in which they had to push a button in response to the appearance of an image on a screen. Their reaction times were measured. Then the subjects consumed enough alcohol to raise their blood alcohol level to 0.05%. (In most states, a person is not considered to be "under the influence" until the blood alcohol level reaches 0.08%.) They then took the reaction time test again. Their reaction times, in milliseconds, are presented in Table 10.1 on page 496. The row labeled "Difference" is the increase in reaction time after consuming alcohol. A negative difference occurs when the reaction time after consuming alcohol is less.

Table 10.1 Reaction Times Before and After Consuming Alcohol

	1	2	3	4	5	6	7	8	9	10	11	12	13	14	15	16	Sample Mean
Blood alcohol 0.05%	102	100	77	61	85	50	95	115	64	98	107	44	47	92	70	94	81.3
Blood alcohol 0	103	99	69	50	96	26	71	109	53	89	103	27	50	100	66	86	74.8
Difference	−1	1	8	11	−11	24	24	6	11	9	4	17	−3	−8	4	8	6.5

We have two samples, a sample of reaction times before alcohol consumption and a sample after alcohol consumption. These are paired samples, because each value in one sample can be paired with the value from the same person in the other sample. For example, the first pair is (102, 103), which are the two values from subject 1. These pairs are called **matched pairs**. The bottom row of Table 10.1 contains the differences between the values in each matched pair. These differences are a sample from the population of differences. We can compute the means of the two original samples, along with the mean of the sample of differences. Denote the means of the original samples by $\bar{x}_1$ and $\bar{x}_2$. Denote the mean of the sample of differences by $\bar{d}$. The sample means are presented in the rightmost column of Table 10.1. They are

$$\bar{x}_1 = 81.3 \qquad \bar{x}_2 = 74.8 \qquad \bar{d} = 6.5$$

Simple arithmetic shows that the mean of the differences is the same as the difference between the sample means. In other words, $\bar{d} = \bar{x}_1 - \bar{x}_2$. The same relationship holds for the population means. If we denote the population means by μ_1 and μ_2, and denote the population mean of the differences by μ_d, then $\mu_d = \mu_1 - \mu_2$. This is a very useful fact. It means that a confidence interval for the mean μ_d is also a confidence interval for the difference $\mu_1 - \mu_2$. The matched pairs reduce the two-sample problem to a one-sample problem.

The data show that the sample mean reaction time increased after the subjects consumed alcohol. We would like to construct a 95% confidence interval for the population mean increase μ_d. The method for computing a confidence interval for μ_d is the usual method for computing a confidence interval for a population mean. This method was presented in Section 8.2. We now list the assumptions for this method, when applied to matched pairs.

Assumptions for Constructing a Confidence Interval Using Matched Pairs

1. We have two paired random samples.
2. Either the sample size is large ($n > 30$), *or* the differences between the matched pairs come from a population that is approximately normal.

Notation:

- $\bar{d}$ is the sample mean of the differences between the values in the matched pairs.
- s_d is the sample standard deviation of the differences between the values in the matched pairs.
- μ_d is the population mean difference for the matched pairs.

Constructing a Confidence Interval Using Matched Pairs

Let $\bar{d}$ be the sample mean of the differences between matched pairs, and let s_d be the sample standard deviation. Let μ_d be the population mean difference between matched pairs.

A level $100(1 - \alpha)\%$ confidence interval for μ_d is

$$\bar{d} - t_{\alpha/2}\frac{s_d}{\sqrt{n}} < \mu_d < \bar{d} + t_{\alpha/2}\frac{s_d}{\sqrt{n}}$$

Another way to write this is

$$\bar{d} \pm t_{\alpha/2}\frac{s_d}{\sqrt{n}}$$

EXAMPLE 10.3   Construct a confidence interval

Use the data in Table 10.1 to construct a 95% confidence for μ_d, the mean difference in reaction times.

We check the assumptions. Because the sample size is small ($n = 16$), we construct a boxplot for the differences to check for outliers or strong skewness.

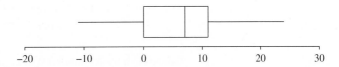

There are no outliers and no evidence of strong skewness, so we can proceed.

Step 1: Compute the sample mean difference $\bar{d}$, and the sample standard deviation of the differences s_d. The sample mean and standard deviation are

$$\bar{d} = 6.500 \qquad s_d = 9.93311$$

Step 2: Compute the critical value. We use the t statistic. The sample size is $n = 16$, so the number of degrees of freedom is $16 - 1 = 15$. The confidence level is 95%. From Table A.3, we find the critical value to be

$$t_{\alpha/2} = 2.131$$

Step 3: Compute the standard error and the margin of error. The standard error is

$$\frac{s_d}{\sqrt{n}} = \frac{9.93311}{\sqrt{16}} = 2.48328$$

The margin of error is

$$t_{\alpha/2}\frac{s_d}{\sqrt{n}} = 2.131(2.48328) = 5.292$$

Step 4: Construct the confidence interval. The 95% confidence interval is

$$\text{Point estimate} \pm \text{Margin of error}$$

$$6.5 - 5.292 < \mu_d < 6.5 + 5.292$$

$$1.2 < \mu_d < 11.8$$

Note that we have rounded the final result to one decimal place, because the original data (the differences) were given as whole numbers (no places after the decimal point).

Step 5: Interpret the result. We are 95% confident that the mean difference is between 1.2 and 11.8. In particular, the confidence interval does not contain 0, and all the values in the confidence interval are positive. We can be fairly certain that the mean reaction time is greater when the blood alcohol level is 0.05%.

Check Your Understanding

1. **High blood pressure:** A group of five individuals with high blood pressure were given a new drug that was designed to lower blood pressure. Systolic blood pressure was measured before and after treatment for each individual. The results follow. Construct a 95% confidence interval for the mean reduction in systolic blood pressure.

	Individual				
	1	**2**	**3**	**4**	**5**
Before	170	164	168	158	183
After	145	132	129	135	145

2. **Extra help:** The statistics department at a large university instituted a program in which students could get extra help with statistics in the evening. The following table presents scores for tests taken before and after the program for a random sample of six students.

	Student					
	1	**2**	**3**	**4**	**5**	**6**
Before	67	58	78	61	75	80
After	73	66	85	69	80	82

Construct a 99% confidence interval for the mean increase in test score.

Answers are on page 502.

Constructing confidence intervals with technology

The following TI-84 Plus display presents results for Example 10.3.

The following output from MINITAB presents the results for Example 10.3.

	N	Mean	StDev	SE Mean
Difference	16	6.50000	9.93311	2.48328

95% CI for mean difference: (1.20702, 11.79298)

Most of the output is straightforward. The quantity labeled "StDev" is the standard deviation of the differences, s_d, and the quantity labeled "SE Mean" is $s_d/\sqrt{n}$, which is the standard error.

Step-by-step instructions for constructing confidence intervals with technology are given in the Using Technology section on page 499.

Matched pairs usually have a smaller margin of error than independent samples

When designing a study to estimate the difference between two means, there is an advantage to using matched pairs when possible. The reason is that the margin of error with matched pairs is much less in most cases than the margin of error for two independent samples. To see this, we will compute the sample standard deviations for the reaction time data in Table 10.1. Let s_1 denote the sample standard deviation for the blood level 0.05% sample and let s_2 denote the sample standard deviation for the blood level 0 sample. The values of s_1 and s_2 are

$$s_1 = 22.71 \qquad s_2 = 27.39$$

If the samples had been independent, the standard error would have been

$$\sqrt{\frac{s_1^2}{16} + \frac{s_2^2}{16}} = \sqrt{\frac{22.71^2}{16} + \frac{27.39^2}{16}} = 8.90$$

Because we were able to use the sample of differences, the standard error was only 2.48.

The number of degrees of freedom for matched pairs is usually somewhat less than for independent samples. This makes the critical value a bit larger for matched pairs. However, the slight increase in the critical value is in most cases more than made up for by a larger reduction in the standard error.

USING TECHNOLOGY

We use Example 10.3 to illustrate the technology steps.

TI-84 PLUS

Constructing a confidence interval for the difference using matched pairs

Step 1. Enter the data into **L1** and **L2** in the data editor. On the home screen, enter (**L1−L2**) STO **L3** to assign the differences in list **L3** (Figure A).

Step 2. Press **STAT** and highlight the **TESTS** menu.

Step 3. Select **TInterval** and press **ENTER** (Figure B). The **TInterval** menu appears.

Step 4. For **Inpt**, select the **Data** option and enter **L3** as the **List** option.

Step 5. In the **C-Level** field, enter the confidence level. For Example 10.3, we use 0.95 (Figure C).

Step 6. Highlight **Calculate** and press **ENTER** (Figure D).

Figure A

Figure B

Figure C

Figure D

MINITAB

Constructing a confidence interval for the difference using matched pairs

Step 1. Enter the data from Example 10.3 into **Columns C1** and **C2**.

Step 2. Click on **Stat**, then **Basic Statistics**, then **Paired t**.

Step 3. Select **Samples in Columns** and enter **C1** in the **First Sample** field and **C2** in the **Second Sample** field.

Step 4. Click **Options** and enter the confidence level in the **Confidence Level** (95) field. Enter **0** in the **Test Mean** field and choose **not equal** in the **Alternative** field. Click **OK** (Figure E).

Step 5. Click **OK** (Figure F).

Figure E

```
              N     Mean     StDev   SE Mean
Difference   16   6.50000   9.93311  2.48328


95% CI for mean difference: (1.20702, 11.79298)
```

Figure F

EXCEL

Constructing a confidence interval for the difference using matched pairs

This procedure requires the **MegaStat** EXCEL add-in to be loaded. The **MegaStat** add-in can be downloaded from www.mhhe.com/megastat.

Step 1. Enter the data from Example 10.3 into **Columns A** and **B** in the worksheet.

Step 2. Click on the **MegaStat** menu, select **Hypothesis Tests**, then **Paired Observations...**

Step 3. Select the **data input** option and enter the range of cells for the first sample in the **Group 1** field and the range of cells for the second sample in the **Group 2** field.

Step 4. Enter **0** in the **Hypothesized difference** field and select **not equal** in the **Alternative** field.

Step 5. Choose the **t-test (unequal variances)** option and select the **Display confidence interval** option with the desired confidence level (Figure G).

Step 6. Click **OK** (Figure H).

Figure G

1.207	confidence interval 95.% lower
11.793	confidence interval 95.% upper
5.293	margin of error

Figure H

SECTION 10.3   Exercises

Exercises 1 and 2 are the Check Your Understanding exercises located within the section.

Understanding the Concepts

In Exercises 3 and 4, fill in each blank with the appropriate word or phrase.

3. With _____ samples, each value in one sample can be matched with a corresponding value from another sample.

4. The margin of error for the mean difference between matched pairs is usually _____ than the margin of error for independent samples.

In Exercises 5 and 6, determine whether the statement is true or false. If the statement is false, rewrite it as a true statement.

5. Paired data reduce a two-sample problem to a one-sample problem.

6. To construct a confidence interval using matched pairs, we must compute the standard deviation of each sample.

Working with the Concepts

7. Fast computer: Two microprocessors are compared on a sample of 6 benchmark codes to determine whether there is a difference in speed. The times (in seconds) used by each processor on each code are as follows:

	Code					
	1	**2**	**3**	**4**	**5**	**6**
Processor A	27.2	18.1	27.2	19.7	24.5	22.1
Processor B	24.1	19.3	26.8	20.1	27.6	29.8

a. Find a 95% confidence interval for the difference between the mean speeds.

b. A computer scientist claims that the mean speed is the same for both processors. Does the confidence interval contradict this claim?

8. Brake wear: For a sample of 9 automobiles, the mileage (in 1000s of miles) at which the original front brake pads were worn to 10% of their original thickness was measured, as was the mileage at which the original rear brake pads were worn to 10% of their original thickness. The results were as follows:

Car	Rear	Front
1	41.2	32.5
2	35.8	26.5
3	46.6	35.6
4	46.9	36.2
5	39.2	29.8
6	51.5	40.9
7	51.0	40.7
8	46.0	34.5
9	47.3	36.5

a. Construct a 99% confidence interval for the difference in mean lifetime between the front and rear brake pads.

b. An automotive engineer claims that the mean lifetime for rear brake pads is more than 10,000 miles more than the mean lifetime for front brake pads. Does the confidence interval contradict this claim?

9. Strength of concrete: The compressive strength, in kilopascals, was measured for concrete blocks from five different batches of concrete, both three and six days after pouring. The data are as follows:

	Batch				
	1	2	3	4	5
After 6 days	1376	1373	1366	1384	1358
After 3 days	1341	1316	1352	1355	1327

a. Construct a 95% confidence interval for the difference between the mean strengths of blocks cured three days and blocks cured six days.

b. A civil engineer claims that the mean strength of blocks cured three days is the same as that of blocks cured six days. Does the confidence interval contradict this claim?

10. Truck pollution: In an experiment to determine the effect of ambient temperature on the emissions of oxides of nitrogen (NO_x) of diesel trucks, ten trucks were run at temperatures of 40°F and 80°F. The emissions, in parts per billion, are presented in the following table.

Truck	40°F	80°F
1	834.7	815.2
2	753.2	765.2
3	855.7	842.6
4	901.2	797.1
5	785.4	764.3
6	862.9	819.5
7	882.7	783.6
8	740.3	694.5
9	748.0	772.9
10	848.6	794.7

a. Construct a 99% confidence interval for the difference in mean emissions between trucks running at 40°F and trucks running at 80°F.

b. Based on the confidence interval, is it reasonable to believe that the mean emissions at 40°F may be the same as the mean at 80°F?

11. High cholesterol: A group of eight individuals with high cholesterol levels were given a new drug that was designed to lower cholesterol levels. Cholesterol levels, in milligrams per deciliter, were measured before and after treatment for each individual, with the following results:

Individual	Before	After
1	283	215
2	299	206
3	274	187
4	284	212
5	248	178
6	275	212
7	293	192
8	277	196

a. Construct a 90% confidence interval for the mean reduction in cholesterol level.

b. A physician claims that the mean reduction in cholesterol level is more than 80 milligrams per deciliter. Does the confidence interval contradict this claim?

12. Tires and fuel economy: A tire manufacturer is interested in testing the fuel economy for two different tread patterns. Tires of each tread type were driven for 1000 miles on each of 9 different cars. The mileages, in miles per gallon, were as follows:

Car	Tread A	Tread B
1	24.7	20.3
2	22.5	19.0
3	24.0	22.5
4	26.9	23.1
5	22.5	20.9
6	23.5	23.6
7	22.7	21.4
8	19.7	18.2
9	27.5	25.9

a. Construct a 95% confidence interval for the mean difference in fuel economy.

b. Based on the confidence interval, is it reasonable to believe that the mean mileage may be the same for both types of tread?

13. Growth spurt: It is generally known that boys grow at an unusually fast rate between the ages of about 12 and 14. Following are heights, in inches of 40 boys, measured at age 12 and again at age 14.

Age 12	Age 14	Age 12	Age 14
57.9	62.9	55.4	61.8
61.1	65.9	58.7	64.2
62.7	67.5	64.3	69.8
67.5	73.7	58.1	63.3
59.2	64.9	63.3	69.6
61.4	67.0	61.2	66.4
60.7	65.5	64.5	69.2
55.9	62.1	55.9	62.0
59.7	65.4	60.4	65.7
56.3	61.5	57.8	62.8
63.0	68.5	68.3	73.6
58.6	63.9	63.0	67.7
61.1	65.8	64.4	69.2
59.5	64.5	58.2	64.6
61.6	66.3	59.7	66.1
59.3	64.2	60.2	65.9
62.1	67.1	63.7	68.3
62.8	68.1	60.2	65.2
65.3	69.9	62.7	67.9
60.4	66.7	55.6	61.7

a. Construct a 95% confidence interval for the mean increase in height for boys between the ages of 12 and 14.

b. A pediatrician claims that the mean increase in height is 5.5 inches. Does the confidence interval contradict this claim?

14. SAT coaching: A sample of 32 students took a class designed to improve their SAT math scores. Following are their scores before and after the class.

Before	After	Before	After
383	420	394	430
334	368	513	525
378	396	483	482
467	488	447	482
470	489	440	479
473	473	439	451
443	448	435	431
459	473	451	454
426	428	453	463
493	525	491	511
382	382	526	529
473	474	473	493
408	407	440	466
433	434	481	482
478	490	459	455
502	508	399	404

a. Construct a 95% confidence interval for the mean increase in scores after the class.
b. The class instructor claims that the mean increase is greater than 20 points. Does the confidence interval contradict this claim?

15. **Interpret calculator display:** The following TI-84 Plus calculator display presents a 95% confidence interval for the mean difference between matched pairs.

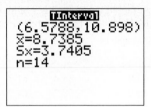

```
    TInterval
(6.5788,10.898)
x̄=8.7385
Sx=3.7405
n=14
```

a. What is the point estimate of μ_d?
b. How many degrees of freedom are there?
c. Fill in the blanks: We are 95% confident that the mean difference is between _____ and _____.

16. **Interpret calculator display:** The following TI-84 Plus calculator display presents a 99% confidence interval for the mean difference between matched pairs.

```
    TInterval
(-4.798,9.9348)
x̄=2.56842
Sx=5.2569
n=7
```

a. What is the point estimate of μ_d?
b. How many degrees of freedom are there?
c. Fill in the blanks: We are 99% confident that the mean difference is between _____ and _____.

17. **Interpret computer output:** The following output from MINITAB presents a confidence interval for the mean difference between matched pairs.

	N	Mean	StDev	SE Mean
Difference	6	2.5324	3.6108	1.47410

99% CI for mean difference: (−3.411394, 8.476194)

a. What is the point estimate of μ_d?
b. How many degrees of freedom are there?
c. Fill in the blanks: We are _____ confident that the mean difference is between _____ and _____.

18. **Interpret computer output:** The following output from MINITAB presents a confidence interval for the mean difference between matched pairs.

	N	Mean	StDev	SE Mean
Difference	12	16.412	3.626	1.04674

95% CI for mean difference: (14.10815, 18.71585)

a. What is the point estimate of μ_d?
b. How many degrees of freedom are there?
c. Fill in the blanks: We are _____ confident that the mean difference is between _____ and _____.

Extending the Concepts

19. **Advantage of matched pairs:** Refer to Exercise 12.
    a. Assume that the measurements for tread A and for tread B come from independent samples, and construct a 95% confidence interval for the difference in mean tread wear using the methods of Section 10.1.
    b. Which confidence interval is narrower, the one using matched pairs or the one constructed under the assumption that the samples are independent? Explain.

20. **Paired or independent?** To construct a confidence interval for each of the following quantities, say whether it would be better to use paired samples or independent samples, and explain why.
    a. The mean difference in height between identical twins.
    b. The mean difference in test scores between students taught by different methods.
    c. The mean difference in height between men and women.
    d. The mean difference in apartment rents in a certain town between this year and last year.

Answers to Check Your Understanding Exercises for Section 10.3

1. $22.3 < \mu_d < 40.5$

2. $2.2 < \mu_d < 9.8$

Chapter 10 Summary

Section 10.1: When we want to estimate the difference between two population means, we may use either independent samples or paired samples. Two samples are independent if the observations in one sample do not influence the observations in the other. Two samples are paired if each observation in one sample can be paired with an observation in the other. When the samples are independent, Welch's method can be used to construct a confidence interval for the difference between the population means. If both sample sizes are large or both populations are approximately normal, a level $100(1 - \alpha)\%$ confidence interval for the difference between two population means is $\bar{x}_1 - \bar{x}_2 \pm t_{\alpha/2}\sqrt{s_1^2/n_1 + s_2^2/n_2}$.

Section 10.2: Confidence intervals can also be constructed for the difference between two proportions. The assumptions necessary to construct a confidence interval for a single population proportion must hold for both populations. A level $100(1 - \alpha)\%$ confidence interval for the difference between two proportions is given by $\hat{p}_1 - \hat{p}_2 \pm z_{\alpha/2}\sqrt{\hat{p}_1(1 - \hat{p}_1)/n_1 + \hat{p}_2(1 - \hat{p}_2)/n_2}$.

Section 10.3: When the samples are paired, we can construct a confidence interval for the mean difference by computing the difference for each pair and constructing a confidence interval for the mean. The assumptions for the population of differences are the same as the assumptions for constructing a confidence interval for a population mean. A level $100(1 - \alpha)\%$ confidence interval is $\bar{d} \pm t_{\alpha/2}s_d/\sqrt{n}$, where $\bar{d}$ is the average difference and s_d is the standard deviation of the sample of differences.

Vocabulary and Notation

independent samples   476

matched pairs   496

paired samples   476

pooled standard deviation   481

Welch's method   478

Important Formulas

Confidence interval for the difference between two means, independent samples:

$$\bar{x}_1 - \bar{x}_2 - t_{\alpha/2}\sqrt{\frac{s_1^2}{n_1} + \frac{s_2^2}{n_2}} < \mu_1 - \mu_2 < \bar{x}_1 - \bar{x}_2 + t_{\alpha/2}\sqrt{\frac{s_1^2}{n_1} + \frac{s_2^2}{n_2}}$$

Confidence interval for the difference between two proportions:

$$\hat{p}_1 - \hat{p}_2 - z_{\alpha/2}\sqrt{\frac{\hat{p}_1(1 - \hat{p}_1)}{n_1} + \frac{\hat{p}_2(1 - \hat{p}_2)}{n_2}} < p_1 - p_2 < \hat{p}_1 - \hat{p}_2 + z_{\alpha/2}\sqrt{\frac{\hat{p}_1(1 - \hat{p}_1)}{n_1} + \frac{\hat{p}_2(1 - \hat{p}_2)}{n_2}}$$

Confidence interval for the difference between two means, matched pairs:

$$\bar{d} - t_{\alpha/2}\frac{s_d}{\sqrt{n}} < \mu_d < \bar{d} + t_{\alpha/2}\frac{s_d}{\sqrt{n}}$$

Chapter Quiz

In Exercises 1 and 2, determine whether the samples described are paired or independent.

1. A sample of 15 weight lifters is tested to see how much weight they can bench press. They then follow a special training program for three weeks, after which they are tested again. The samples are the amounts of weights that were lifted before and after the training program.

2. A sample of 20 weight lifters is tested to see how much weight they can bench press. Ten of them are chosen at random as the treatment group. They participate in a special training program for three weeks. The remaining 10 are the control group. They follow their usual program. At the end of three weeks, all 20 weight lifters are tested again and the increases in the amounts they can lift are recorded. The two samples are the increases in the amounts lifted by the treatment group and the control group.

3. In a survey of 300 randomly selected female and 240 male holiday shoppers, 87 of the females and 98 of the males stated that they will wait until the last week before Christmas to finish buying gifts. Let p_1 be the population proportion of males who will wait until the last week and let p_2 be the population proportion of females. Compute a point estimate for the difference $p_1 - p_2$.

4. Refer to Exercise 3. Find the critical value for a 95% confidence interval for the difference $p_1 - p_2$.

5. Refer to Exercise 3. Compute the margin of error for a 95% confidence interval for the difference $p_1 - p_2$.

6. Refer to Exercise 3. Construct a 95% confidence interval for the difference $p_1 - p_2$.

7. Eight students in a particular college course are given a pre-test at the beginning of the semester and are then given the same exam at the end to test what they have learned. The exam scores at the beginning and at the end are given in the following table.

	Student							
	1	2	3	4	5	6	7	8
End	83	71	79	95	84	72	69	78
Beginning	72	58	76	81	69	63	71	77

Let μ_d be the mean difference End − Beginning. Compute a point estimate for μ_d.

8. Refer to Exercise 7. Find the critical value for a 99% confidence interval for μ_d.

9. Refer to Exercise 7. Compute the margin of error for a 99% confidence interval for the difference μ_d.

10. Refer to Exercise 7. Construct a 99% confidence interval for the difference μ_d.

11. A random sample of 76 residents in a small town had a mean annual income of $34,214, with a sample standard deviation of $2171. In a neighboring town, a random sample of 88 residents had a mean annual income of $31,671 with a sample standard deviation of $3279. Let μ_1 be the population mean annual income in the first town, and let μ_2 be the population mean annual income in the neighboring town. Compute a point estimate for the difference $\mu_1 - \mu_2$.

12. Refer to Exercise 11. Find the critical value for a 90% confidence interval for the difference $\mu_1 - \mu_2$.

13. Refer to Exercise 11. Compute the margin of error for a 90% confidence interval for the difference $\mu_1 - \mu_2$.

14. Refer to Exercise 11. Construct a 90% confidence interval for the difference $\mu_1 - \mu_2$.

15. In a poll of 100 voters, 57 said they were planning to vote for the incumbent governor, and 48 said they were planning to vote for the incumbent mayor. Explain why these data should not be used to construct a confidence interval for the difference between the proportions of voters who plan to vote for the governor and those who plan to vote for the mayor.

Review Exercises

1. **Hatching shrimp eggs:** A study of the effect of water pollution on the habitat of shrimp reported that out of 1985 eggs produced by shrimp at the Diesel Creek site in Charleston, South Carolina, 1919 hatched, and at the Shipyard Creek site, also in Charleston, 4561 out of 4988 eggs hatched. Construct a 95% confidence interval for the difference between the proportions of eggs that hatch at the two sites.

2. **Don't construct a confidence interval:** An investor is trying to decide in which of two stocks to invest. He examines records for the past 365 days, and finds that the price of stock A increased on 197 of them and the price of stock B increased on 158 of them. Explain why these data should not be used to construct a confidence interval for the difference between the proportions of days that the stock prices increase.

3. **Exercise and heart rate:** A simple random sample of 7 people embarked on a program of regular aerobic exercise. Their heart rates, in beats per minute, were measured before and after, with the following results:

	Person						
	1	2	3	4	5	6	7
Before	81	84	79	85	79	84	87
After	73	77	73	78	71	75	80

Construct a 95% confidence interval for the mean reduction in heart rate.

4. **College tuition:** Independent random samples of private college and university tuition and fees for the 2012–2013 academic year from California and New York yielded the following data, in dollars.

California:	29,940	35,650	27,231	21,000	32,610	37,850	36,992	34,584	22,850	25,788
New York:	29,283	40,870	28,460	33,835	24,410	30,868	26,400	37,882	22,656	38,980

Based on data from www.collegprowler.com.

Construct a 95% confidence interval for the difference between the mean amounts of tuition in the two states.

5. **Recovery time from surgery:** A new postsurgical treatment is being compared with a standard treatment. Seven subjects receive the new treatment, while seven others (the controls) receive the standard treatment. The recovery times, in days, are given below.

Control: 18 23 24 30 32 35 39   Treatment: 12 13 15 19 20 21 24

Construct a 98% confidence interval for the reduction in the mean recovery times associated with treatment.

6. **Polling results:** A simple random sample of 400 voters in the town of East Overshoe was polled, and 242 said they planned to vote in favor of a bond issue to raise money for elementary schools. A simple random sample of 300 voters in West Overshoe was polled, and 161 said they were in favor. Construct a 98% confidence interval for the difference between the proportions of voters in the two towns who favor the bond issue.

7. **Treating bean plants:** In a study to measure the effect of an herbicide on the phosphate content of bean plants, a sample of 75 plants treated with the herbicide had a mean phosphate concentration (in percent) of 3.52 with a standard deviation of 0.41, and 100 untreated plants had a mean phosphate concentration of 5.82 with a standard deviation of 0.52. Construct a 95% confidence interval for the difference in mean phosphate concentration between treated and untreated plants.

8. **Dueling scales:** In an experiment to determine whether there is a systematic difference between the weights obtained with two different scales, ten rock specimens were weighed, in grams, on each scale. The following data were obtained:

					Specimen					
	1	2	3	4	5	6	7	8	9	10
Weight on scale 1	11.24	14.36	8.31	10.50	23.44	9.12	13.46	6.44	12.41	19.35
Weight on scale 2	11.27	14.40	8.35	10.53	23.41	9.16	13.50	6.43	12.44	19.34

Construct a 98% confidence interval for mean difference in weight between the two scales.

9. **Automobile pollution:** In a random sample of 340 cars driven at low altitudes, 46 exceeded a threshold for the amount of particulate pollution. In a random sample of 85 cars driven at high altitudes, 21 exceeded a threshold for the amount of particulate pollution. Construct a 99% confidence interval for the difference between the proportions for high-altitude and low-altitude vehicles.

10. **Interpret calculator display:** The following TI-84 Plus calculator display presents a 95% confidence interval for the difference between two means. The sample sizes are $n_1 = 85$ and $n_2 = 71$.

a. Find the point estimate of $\mu_1 - \mu_2$.
b. How many degrees of freedom did the calculator use?
c. Fill in the blanks: We are 95% confident that the difference between the means is between _____ and _____.

11. **Interpret computer output:** The following MINITAB output display presents a 95% confidence interval for the difference between two means.

```
       N    Mean    StDev   SE Mean
A     12   9.5713   1.025    0.2959
B      8   7.2198   5.173    1.8289


Difference = mu (A) − mu (B)
Estimate for difference: 2.3515
95% CI for difference: (−2.02947, 6.73247)   DF = 7
```

a. Find the point estimate of $\mu_1 - \mu_2$.
b. How many degrees of freedom did MINITAB use?
c. Fill in the blanks: We are _____ confident that the difference between the means is between _____ and _____.

12. **Interpret calculator display:** The following TI-84 Plus calculator display presents a 95% confidence interval for the difference between two proportions.

a. Find the point estimate of $p_1 - p_2$.

b. Fill in the blanks: We are 95% confident that the difference between the proportions is between _____ and _____.

13. **Interpret computer output:** The following MINITAB output presents a confidence interval for the difference between two proportions.

Sample	X	N	Sample p
1	85	103	0.825243
2	72	125	0.576000

Difference = p(1) − p(2)
Estimate for difference: 0.249243
95% CI for difference: (0.135735, 0.362751)

a. Find the point estimate of $p_1 - p_2$.

b. Fill in the blanks: We are _____ confident that the difference between the proportions is between _____ and _____.

14. **Interpret calculator display:** The following TI-84 Plus calculator display presents a 95% confidence interval for the mean difference between matched pairs.

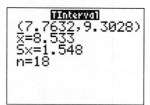

a. Find the point estimate of μ_d.

b. How many degrees of freedom are there?

c. Fill in the blanks: We are 95% confident that the mean difference is between _____ and _____.

15. **Interpret computer output:** The following output from MINITAB presents a confidence interval for the mean difference between matched pairs.

	N	Mean	StDev	SE Mean
Difference	15	9.8612	3.7149	0.95918

95% CI for mean difference: (7.803957, 11.918443)

a. Find the point estimate of μ_d.

b. How many degrees of freedom are there?

c. Fill in the blanks: We are _____ confident that the mean difference is between _____ and _____.

Write About It

1. Why is it necessary for all values in the confidence interval to be positive to conclude that $\mu_1 > \mu_2$? What would have to be true to conclude that $\mu_1 < \mu_2$?

2. In what ways is the procedure for constructing a confidence interval for the difference between two proportions similar to constructing a confidence interval for one proportion? In what ways is it different?

3. Provide an example of two samples that are independent. Explain why these samples are independent.

4. Provide an example of two samples that are paired. Explain why these samples are paired.

Case Study: Evaluating The Assignment Of Subjects In A Clinical Trial

In the chapter opener, we described a study in which patients were assigned to receive either a new treatment or a standard treatment for the prevention of heart failure. A total of 1820 patients participated, with 1089 receiving the new treatment and 731 receiving the standard treatment. The assignment was not made by simple random sampling; instead, an algorithm was used that was designed to produce balance between the groups. The following table presents summary statistics describing several health characteristics of the people in each group.

Characteristic	Standard Treatment		New Treatment	
	Mean	Standard Deviation	Mean	Standard Deviation
Age	64	11	65	11
Systolic blood pressure	121	18	124	17
Diastolic blood pressure	71	10	72	10

Characteristic	Standard Treatment	New Treatment
	Percentage with the Characteristic	Percentage with the Characteristic
Treatment for hypertension	63.2	63.7
Atrial fibrillation	12.6	11.1
Diabetes	30.6	30.2
Cigarette smoking	12.8	11.4
Coronary bypass surgery	28.5	29.1

1. For each health characteristic, construct a 95% confidence interval for the difference between the group means or group proportions.
2. Based on the confidence intervals in Exercise 1, for which health characteristics does it appear that the assignment to groups is not balanced?
3. Based on the confidence intervals in Exercise 1, does it appear that the imbalance is large enough to be of concern? Explain.

Two-Sample Hypothesis Tests

Introduction

Does exposure to electrical fields cause cancer? This question was addressed in a study carried out at the University of Southern California. Three hundred children who suffered from brain cancer (referred to as "cases") were sampled, along with an equal number of children who did not have cancer (referred to as "controls"). The investigators counted the number of cases and controls who lived in neighborhoods in which the electrical wiring was underground, rather than overhead. In addition, they collected information about ethnicity and income. The following table presents some of the information collected.

	Cases	Controls
Proportion with high income	0.09	0.13
Proportion owning a home	0.72	0.76
Proportion with underground wiring	0.14	0.07
Proportion with father white non-Hispanic	0.43	0.36
Proportion with father Hispanic	0.38	0.47

Cases and controls differ in their proportions for all of these categories. This is not surprising. After all, if we were to sample two groups of people at random, we would expect to see some differences, just by chance. The question we need to address is whether the differences between the cases and controls are just due to chance, or whether they reflect a real association. The basic principle is simple: The smaller the difference, the more likely it is just due to chance. Larger differences are more likely to reflect a real association. Now the question becomes: How do we determine when a difference is large enough to justify concluding that it reflects a real association? That is the question that hypothesis tests are designed to answer. In this chapter, we will learn to perform hypothesis tests in a variety of commonly occurring situations. In the case study at the end of the chapter, we will investigate the differences in the preceding table, and determine which factors seem to be associated with cancer.

SECTION 11.1 | Hypothesis Tests for the Difference Between Two Means: Independent Samples

Objectives

1. Perform a hypothesis test for the difference between two means using the P-value method
2. Perform a hypothesis test for the difference between two means using the critical value method

Objective 1 Perform a hypothesis test for the difference between two means using the P-value method

Perform a Hypothesis Test for the Difference Between Two Means Using the P-Value Method

Do computers help high-school students to learn math? One way to address this question is to give a test to two samples of students — one from the population of students who used computers in their math classes, and another from the population of students who did not. If the difference between the sample mean scores is large enough, we can conclude that there is a real difference between the two populations.

The National Assessment of Educational Progress (NAEP) has been testing students for the past 30 years. Scores on the NAEP mathematics test range from 0 to 500. In a recent year, the sample mean score for students using a computer was 309, with a sample standard deviation of 29. For students not using a computer, the sample mean was 303, with a sample standard deviation of 32. Assume there were 60 students in the computer sample, and 40 students in the sample that didn't use a computer. We can see that the sample mean scores differ by 6 points: $309 - 303 = 6$. Now, we are interested in the difference between the population means, which will not be exactly the same as the difference between the sample means. Is it plausible that the difference between the population means could be 0? How strong is the evidence that the population mean scores are different?

This is an example of a situation in which the data consist of two **independent samples**. We will describe a method for performing a hypothesis test to determine whether the population means are equal. We will need some notation for the population means, the sample means, the sample standard deviations, and the sample sizes:

Recall: Two samples are independent if the observations in one sample do not influence the observations in the other.

NOTATION

- μ_1 and μ_2 are the population means.
- $\bar{x}_1$ and $\bar{x}_2$ are the sample means.
- s_1 and s_2 are the sample standard deviations.
- n_1 and n_2 are the sample sizes.

We will now describe how to perform the hypothesis test.

The null and alternate hypotheses

The issue is whether the population means μ_1 and μ_2 are equal. The null hypothesis says that the population means are equal:

$$H_0: \mu_1 = \mu_2$$

There are three possibilities for the alternate hypothesis.

$$H_1: \mu_1 < \mu_2 \qquad H_1: \mu_1 > \mu_2 \qquad H_1: \mu_1 \neq \mu_2$$

The test statistic

The test statistic is based on the difference between the two sample means $\bar{x}_1 - \bar{x}_2$. The mean of $\bar{x}_1 - \bar{x}_2$ is $\mu_1 - \mu_2$. We approximate the standard deviation of $\bar{x}_1 - \bar{x}_2$ with the standard error, which we derived in Section 10.1.

$$\text{Standard error of } \bar{x}_1 - \bar{x}_2 = \sqrt{\frac{s_1^2}{n_1} + \frac{s_2^2}{n_2}}$$

The test statistic is

$$t = \frac{(\bar{x}_1 - \bar{x}_2) - (\mu_1 - \mu_2)}{\sqrt{\frac{s_1^2}{n_1} + \frac{s_2^2}{n_2}}}$$

Under the assumption that H_0 is true, the test statistic has approximately a Student's t distribution. The number of degrees of freedom can be calculated with a rather complicated formula. (The formula is presented in Section 10.1, although you don't need to know it.) A simpler method, which is acceptable, is

$$\text{Degrees of freedom} = \text{Smaller of } n_1 - 1 \text{ and } n_2 - 1$$

The method just described requires certain assumptions, which we now state:

> ## Assumptions for Performing a Hypothesis Test for the Difference Between Two Means with Independent Samples
>
> 1. We have simple random samples from two populations.
> 2. The samples are independent of one another.
> 3. Each sample size is large ($n > 30$), *or* its population is approximately normal.

We now summarize the steps in testing a hypothesis about the difference between two means with independent samples, using the P-value method. Later we will describe the critical value method.

> ## Performing a Hypothesis Test for the Difference Between Two Means Using the P-Value Method
>
> Check to be sure the assumptions are satisfied. If they are, then proceed with the following steps.
>
> **Step 1:** State the null and alternate hypotheses. The null hypothesis specifies that the population means are equal: H_0: $\mu_1 = \mu_2$. The alternate hypothesis will be $\mu_1 < \mu_2$, $\mu_1 > \mu_2$, or $\mu_1 \neq \mu_2$.
>
> **Step 2:** If making a decision, choose a significance level α.
>
> **Step 3:** Compute the test statistic $\quad t = \dfrac{(\bar{x}_1 - \bar{x}_2) - (\mu_1 - \mu_2)}{\sqrt{\dfrac{s_1^2}{n_1} + \dfrac{s_2^2}{n_2}}}.$
>
> **Step 4:** Compute the P-value. The P-value is an area under the Student's t curve. If using Table A.3, approximate the number of degrees of freedom with the smaller of $n_1 - 1$ and $n_2 - 1$. The P-value depends on the alternate hypothesis as follows:
>
>
> The P-value is the area to the left of t.
> Left-tailed: H_1: $\mu_1 < \mu_2$
>
>
> The P-value is the area to the right of t.
> Right-tailed: H_1: $\mu_1 > \mu_2$
>
>
> The P-value is the sum of the areas in the two tails.
> Two-tailed: H_1: $\mu_1 \neq \mu_2$
>
> **Step 5:** Interpret the P-value. If making a decision, reject H_0 if the P-value is less than or equal to the significance level α.
>
> **Step 6:** State a conclusion.

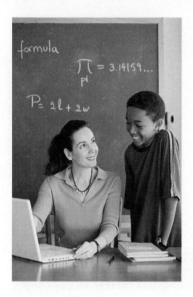

EXAMPLE 11.1   Perform a hypothesis test

The National Assessment of Educational Progress (NAEP) tested a sample of students who had used a computer in their mathematics classes, and another sample of students who had not used a computer. The sample mean score for students using a computer was 309, with a sample standard deviation of 29. For students not using a computer, the sample mean was 303, with a sample standard deviation of 32. Assume there were 60 students in the computer sample, and 40 students in the sample that hadn't used a computer. Can you conclude that the population mean scores differ? Use the $\alpha = 0.05$ level.

Solution

We first check the assumptions. We have two independent random samples. Both sample sizes are larger than 30. The assumptions are satisfied.

Step 1: State the null and alternate hypotheses. We are asked whether we can conclude that the two means differ. Therefore, this is a two-tailed test. The null and alternate hypotheses are

$$H_0: \mu_1 = \mu_2 \qquad H_1: \mu_1 \neq \mu_2$$

Step 2: Choose a significance level. The significance level is $\alpha = 0.05$.

Step 3: Compute the test statistic. The test statistic is

$$t = \frac{(\bar{x}_1 - \bar{x}_2) - (\mu_1 - \mu_2)}{\sqrt{\dfrac{s_1^2}{n_1} + \dfrac{s_2^2}{n_2}}}$$

To help keep track of things, we'll begin by organizing the relevant information in the following table:

	With Computer	**Without Computer**
Sample mean	$\bar{x}_1 = 309$	$\bar{x}_2 = 303$
Sample standard deviation	$s_1 = 29$	$s_2 = 32$
Sample size	$n_1 = 60$	$n_2 = 40$
Population mean	μ_1 (unknown)	μ_2 (unknown)

Under the assumption that H_0 is true, $\mu_1 - \mu_2 = 0$. The value of the test statistic is

$$t = \frac{(309 - 303) - (0)}{\sqrt{\dfrac{29^2}{60} + \dfrac{32^2}{40}}} = 0.953$$

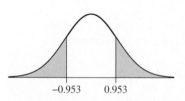

Figure 11.1 The P-value for a two-tailed test is the sum of the areas in the two tails. These areas can be approximated by using Table A.3, or found more precisely with technology.

Step 4: Compute the P-value. We will approximate the P-value by using Table A.3. We begin by finding the number of degrees of freedom. The sample sizes are $n_1 = 60$ and $n_2 = 40$, so the number of degrees of freedom is $40 - 1 = 39$. The value of the test statistic is $t = 0.953$. This is a two-tailed test, so the P-value is the sum of the areas to the right of 0.953 and to the left of -0.953. Figure 11.1 illustrates the P-value.

We consult Table A.3 and look at the row corresponding to 39 degrees of freedom. We see that the value of the test statistic, 0.953, is between 0.681 and 1.304. These are the values that correspond to tail areas of 0.25 and 0.10. Therefore, the area in each tail is between 0.10 and 0.25. The P-value is the sum of the areas in both tails, so we double these numbers:

$$0.20 < P\text{-value} < 0.50$$

Step 5: Interpret the P-value. If H_0 were true, we would expect to observe a value of the test statistic as extreme as or more extreme than our value of 0.953 between 20% and 50% of the time. This is not unusual, so there is no strong evidence against H_0. In particular, $P > 0.05$, so we do not reject H_0 at the $\alpha = 0.05$ level.

Step 6: State a conclusion. There is not enough evidence to conclude that the mean scores differ between those students who use a computer and those who do not. The mean scores may be the same.

Performing a hypothesis test with technology

The following computer output (from MINITAB) presents the results of Example 11.1.

```
Two-sample T for Computer vs No Computer

                N     Mean    StDev    SE Mean
Computer       60     309.0    29.0    3.74388
No computer    40     303.0    32.0    5.05964

Difference = mu (Treatment1) - mu (Treatment2)
Estimate for difference: 6.000
95% CI for difference: (-6.5333, 18.5333 )
T-Test of difference = 0 (vs not =):   T-Value = 0.95
                                       P-Value = 0.343    DF = 77
```

The output presents the sample sizes (N), the sample means (Mean), and sample standard deviations (StDev). The column labeled "SE Mean" presents the standard errors of $\bar{x}_1$ and $\bar{x}_2$, which are $s_1/\sqrt{n_1}$ and $s_2/\sqrt{n_2}$, respectively. The row labeled "Estimate for difference" presents the difference between the sample means: $\bar{x}_1 - \bar{x}_2$. The next row contains a 95% confidence interval for $\mu_1 - \mu_2$. The row after that specifies that the alternate hypothesis is two-tailed, then presents the value of the test statistic (T-Value), the P-value, and the number of degrees of freedom. The number of degrees of freedom (DF) is 77, which differs from the value of 39 that we used in the solution to Example 11.1. The reason is that MINITAB uses a more complicated formula to compute the number of degrees of freedom, then rounds the value down to the nearest whole number.

The following TI-84 Plus display presents results for Example 11.1.

As with the MINITAB output, the results differ somewhat from those we obtained, because the degrees of freedom (labeled "df") have been calculated by the more complicated formula. Note that the degrees of freedom is not a whole number. Unlike MINITAB, the TI-84 Plus does not round the degrees of freedom.

Step-by-step instructions for performing hypothesis tests with technology are presented in the Using Technology section on page 517.

Check Your Understanding

1. A test was made of H_0: $\mu_1 = \mu_2$ versus H_1: $\mu_1 > \mu_2$. Independent random samples were drawn from approximately normal populations. The sample means were $\bar{x}_1 = 6.8$ and $\bar{x}_2 = 4.9$. The sample standard deviations were $s_1 = 1.6$ and $s_2 = 1.3$. The sample sizes were $n_1 = 12$ and $n_2 = 10$.
   a. Find the value of the test statistic t.
   b. Find the number of degrees of freedom for t.
   c. Find the P-value.
   d. Interpret the P-value. Do you reject H_0 at the $\alpha = 0.01$ level?

2. A test was made of H_0: $\mu_1 = \mu_2$ versus H_1: $\mu_1 \neq \mu_2$. Independent random samples were drawn from approximately normal populations. The sample means were $\bar{x}_1 = 73.9$ and $\bar{x}_2 = 71.8$. The sample standard deviations were $s_1 = 4.2$ and $s_2 = 3.8$. The sample sizes were $n_1 = 23$ and $n_2 = 17$.

a. Find the value of the test statistic t.

b. Find the number of degrees of freedom for t.

c. Find the P-value.

d. Interpret the P-value. Do you reject H_0 at the $\alpha = 0.05$ level?

Answers are on page 521.

Objective 2 Perform a hypothesis test for the difference between two means using the critical value method

Performing Hypothesis Tests Using the Critical Value Method

To use the critical value method, compute the test statistic as before. The procedure for finding the critical value is the same as for hypothesis tests for a mean with σ unknown. The critical value can be found in Table A.3 or with technology. The procedure depends on whether the alternate hypothesis is left-tailed, right-tailed, or two-tailed. The assumptions for the critical value method are the same as for the P-value method.

Following are the steps for the critical value method.

Performing a Hypothesis Test for the Difference Between Two Means Using the Critical Value Method

Check to be sure the assumptions are satisfied. If they are, then proceed with the following steps.

Step 1: State the null and alternate hypotheses. The null hypothesis will have the form H_0: $\mu_1 = \mu_2$. The alternate hypothesis will be $\mu_1 < \mu_2$, $\mu_1 > \mu_2$, or $\mu_1 \neq \mu_2$.

Step 2: Choose a significance level α, and find the critical value or values.

Step 3: Compute the test statistic

$$t = \frac{(\bar{x}_1 - \bar{x}_2) - (\mu_1 - \mu_2)}{\sqrt{\dfrac{s_1^2}{n_1} + \dfrac{s_2^2}{n_2}}}.$$

Step 4: Determine whether to reject H_0, as follows:

Left-tailed: H_1: $\mu_1 < \mu_2$
Reject H_0 if $t \leq -t_\alpha$.

Right-tailed: H_1: $\mu_1 > \mu_2$
Reject H_0 if $t \geq t_\alpha$.

Two-tailed: H_1: $\mu_1 \neq \mu_2$
Reject H_0 if $t \geq t_{\alpha/2}$ or $t \leq -t_{\alpha/2}$.

Step 5: State a conclusion.

EXAMPLE 11.2   **Test a hypothesis using the critical value method**

Treatment of wastewater is important to reduce the concentration of undesirable pollutants. One such substance is benzene, which is used as an industrial solvent. Two methods of water treatment are being compared. Treatment 1 is applied to five specimens of wastewater, and treatment 2 is applied to seven specimens. The benzene concentrations, in units of milligrams per liter, for each specimen are as follows:

> Treatment 1:   7.8   7.6   5.6   6.8   6.4
> Treatment 2:   4.1   6.5   3.7   7.7   7.3   4.7   5.9

How strong is the evidence that the mean concentration is less for treatment 2 than for treatment 1? We will test at the $\alpha = 0.05$ significance level.

Solution

We first check the assumptions. Because the samples are small, we must check for strong skewness and outliers. We construct dotplots for each sample.

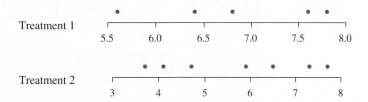

There are no outliers, and no evidence of strong skewness, in either sample. Therefore, we can proceed.

Step 1: State the null and alternate hypotheses. The issue is whether the mean for treatment 2 is less than the mean for treatment 1. Let μ_1 denote the mean for treatment 1 and μ_2 denote the mean for treatment 2. The hypotheses are

$$H_0: \mu_1 = \mu_2 \qquad H_1: \mu_1 > \mu_2$$

Step 2: Choose a significance level and find the critical value. The significance level is $\alpha = 0.05$. We will find the critical value in Table A.3. The sample sizes are $n_1 = 5$ and $n_2 = 7$. For the number of degrees of freedom, we use the smaller of $5 - 1 = 4$ and $7 - 1 = 6$, which is 4. Because the alternate hypothesis, $\mu_1 - \mu_2 > 0$, is right-tailed, the critical value is the value with area 0.05 to its right. We consult Table A.3 with 4 degrees of freedom and find that $t_\alpha = 2.132$.

Step 3: Compute the test statistic. To compute the test statistic, we first compute the sample means and standard deviations. These are

$$\bar{x}_1 = 6.84 \qquad \bar{x}_2 = 5.70 \qquad s_1 = 0.8989 \qquad s_2 = 1.5706$$

The sample sizes are $n_1 = 5$ and $n_2 = 7$. The test statistic is

$$t = \frac{(\bar{x}_1 - \bar{x}_2) - (\mu_1 - \mu_2)}{\sqrt{\dfrac{s_1^2}{n_1} + \dfrac{s_2^2}{n_2}}}$$

Under the assumption that H_0 is true, $\mu_1 - \mu_2 = 0$. The value of the test statistic is

$$t = \frac{(6.84 - 5.70) - 0}{\sqrt{\dfrac{0.8989^2}{5} + \dfrac{1.5706^2}{7}}} = 1.590$$

Step 4: Determine whether to reject H_0. This is a right-tailed test, so we reject H_0 if $t \geq t_\alpha$. Because $t = 1.590$ and $t_\alpha = 2.132$, we do not reject H_0.

Step 5: State a conclusion. There is not enough evidence to conclude that the mean benzene concentration with treatment 1 is greater than that with treatment 2. The concentrations may be the same.

An alternate method when the population variances are equal

When the two population standard deviations, σ_1 and σ_2, are known to be equal, there is an alternate method for testing hypotheses about $\mu_1 - \mu_2$. This alternate method was widely used in the past, and is still an option in many computer packages. We will describe the method here, because it is still sometimes used. However, the method is rarely appropriate, for the same reasons that the pooled method for constructing confidence intervals is rarely appropriate. These reasons were discussed in Section 10.1.

A Method for Testing a Hypothesis About $\mu_1 - \mu_2$ When $\sigma_1 = \sigma_2$ (Not Recommended)

Step 1: Compute the **pooled standard deviation**, s_p, as follows:

$$s_p = \sqrt{\frac{(n_1 - 1)s_1^2 + (n_2 - 1)s_2^2}{n_1 + n_2 - 2}}$$

Step 2: Compute the test statistic:

$$t = \frac{(\bar{x}_1 - \bar{x}_2) - (\mu_1 - \mu_2)}{s_p\sqrt{\dfrac{1}{n_1} + \dfrac{1}{n_2}}}$$

Step 3: Compute the degrees of freedom:

$$\text{Degrees of freedom} = n_1 + n_2 - 2$$

Step 4: Compute the P-value using a Student's t distribution with $n_1 + n_2 - 2$ degrees of freedom.

When using technology, you may be offered a choice of whether to assume that the variances are equal. The best practice is not to assume that the variances are equal.

Performing hypothesis tests when the population standard deviations are known

When the two population standard deviations, σ_1 and σ_2, are known, we can modify the test statistic presented here by replacing s_1 and s_2 with σ_1 and σ_2, and using the standard normal distribution rather than the Student's t distribution to find the P-value or critical values. In practice, σ_1 and σ_2 are rarely known, so this method is not often applicable. We present it here because it is sometimes offered as an option in statistical calculators and software. The assumptions for this method are the same as for the method using the Student's t distribution, with the additional assumption that the population standard deviations are known.

SUMMARY

When the population standard deviations σ_1 and σ_2 are known, tests for the difference between two means can be carried out by using the following test statistic:

$$z = \frac{(\bar{x}_1 - \bar{x}_2) - (\mu_1 - \mu_2)}{\sqrt{\dfrac{\sigma_1^2}{n_1} + \dfrac{\sigma_2^2}{n_2}}}$$

The test statistic has a standard normal distribution when H_0 is true.

USING TECHNOLOGY

We use Example 11.2 to illustrate the technology steps.

TI-84 PLUS

Testing a hypothesis about the difference between two means

Step 1. Press **STAT** and highlight the **TESTS** menu.
Step 2. Select **2–SampTTest** and press **ENTER** (Figure A). The **2–SampTTest** menu appears.
Step 3. Choose one of the following:
  • If the summary statistics are given, select **Stats** as the **Inpt** option and enter $\bar{x}_1, s_1, n_1, \bar{x}_2, s_2, n_2$.
  • If the raw data are given, select **Data** as the **Inpt** option and enter the location of the data as the **List1** and **List2** options. For Example 11.2, the sample has been entered in lists **L1** and **L2**.
Step 4. Select the form of the alternate hypothesis. For Example 11.2, the alternate hypothesis has the form $>\mu_2$.
Step 5. Select **No** for the **Pooled** option (Figure B).
Step 6. Highlight **Calculate** and press **ENTER** (Figure C).

Figure A

Figure B

Figure C

MINITAB

Testing a hypothesis about the difference between two means

Step 1. Click on **Stat**, then **Basic Statistics**, then **2-Sample t**.
Step 2. Choose one of the following:
  • If the summary statistics are given, click **Summarized Data** and enter the **Sample Size**, the **Mean**, and the **Standard Deviation** for each sample.
  • If the raw data are given, click **Samples in different columns** and select the columns that contain the data. For Example 11.2, the two samples have been entered in columns **C1** and **C2**.
Step 3. Click **Options**, and enter the difference between the means in the **Test difference** field and select the form of the alternate hypothesis. Given significance level α, enter $100(1 - \alpha)$ as the **Confidence Level**. For Example 11.2, we use **95** as the **Confidence Level**, **0** as the **Test difference**, and **greater than** as the **Alternative**. Click **OK**.
Step 4. Click **OK** (Figure D).

```
Two-sample T for C1 vs C2

      N   Mean   StDev  SE Mean
C1    5   6.840  0.899    0.40
C2    7   5.70   1.57     0.59


Difference = mu (C1) - mu (C2)
Estimate for difference:  1.14000
95% lower bound for difference:  -0.17420
T-Test of difference = 0 (vs >): T-Value = 1.59  P-Value = 0.073  DF = 9
```

Figure D

EXCEL

Testing a hypothesis about the difference between two means

This procedure requires the **MegaStat** EXCEL add-in to be loaded. The **MegaStat** add-in can be downloaded from www.mhhe.com/megastat.

Step 1. Load the **MegaStat** EXCEL add-in.

Step 2. Click on the **MegaStat** menu and select **Hypothesis Tests**, then **Compare Two Independent Groups...**

Step 3. Choose one of the following:
- If the summary statistics are given, choose **summary input** and enter the range of the cells that contains, in the following order, the **variable name**, $\bar{x}$, s, and n.
- If the raw data are given, choose **data input** and select the range of cells that contains the data in the **Input Range** field. For Example 11.2, the samples have been entered in columns **A** and **B** (Figure E).

Step 4. Enter the **Hypothesized difference (0)** and select the form of the alternate hypothesis **(greater than)**.

Step 5. Choose the **t-test (unequal variance)** option (Figure F).

Step 6. Click **OK** (Figure G)

	A	B
1	**Treatment 1**	**Treatment 2**
2	7.8	4.1
3	7.6	6.5
4	5.6	3.7
5	6.8	7.7
6	6.4	7.3
7		4.7
8		5.9

Figure E

Figure F

Hypothesis Test: Independent Groups (t-test, unequal variance)

Treatment 1	Treatment 2	
6.840	5.700	mean
0.899	1.571	std. dev.
5	7	n

9 df
1.1400 difference (Treatment 1 - Treatment 2)
0.7169 standard error of difference
0 hypothesized difference

1.59 t
.0731 p-value (one-tailed, upper)

Figure G

SECTION 11.1   Exercises

Exercises 1 and 2 are the Check Your Understanding exercises located within the section.

Understanding the Concepts

In Exercises 3 and 4, fill in each blank with the appropriate word or phrase.

3. To use the methods of this section to test a hypothesis about the difference between two means when the samples are small, the samples must show no evidence of strong _____ and must contain no _____.

4. If the sample sizes are n_1 and n_2, the simplest way to compute the degrees of freedom is to use the smaller of _____ and _____.

In Exercises 5 and 6, determine whether the statement is true or false. If the statement is false, rewrite it as a true statement.

5. To use the methods of this section to test a hypothesis about the difference between two means, the population standard deviations must be known.

6. In general, it is not recommended to use a pooled standard deviation when testing a hypothesis for the difference between means.

Practicing the Skills

7. A test was made of $H_0: \mu_1 = \mu_2$ versus $H_1: \mu_1 < \mu_2$. The sample means were $\bar{x}_1 = 6$ and $\bar{x}_2 = 11$, the sample standard deviations were $s_1 = 3$ and $s_2 = 5$, and the sample sizes were $n_1 = 10$ and $n_2 = 20$.
   a. How many degrees of freedom are there for the test statistic, using the simple method?
   b. Compute the value of the test statistic.
   c. Is H_0 rejected at the 0.05 level? Explain.

8. A test was made of $H_0: \mu_1 = \mu_2$ versus $H_1: \mu_1 \neq \mu_2$. The sample means were $\bar{x}_1 = 10$ and $\bar{x}_2 = 8$, the sample standard deviations were $s_1 = 4$ and $s_2 = 7$, and the sample sizes were $n_1 = 15$ and $n_2 = 27$.
   a. How many degrees of freedom are there for the test statistic, using the simple method?
   b. Compute the value of the test statistic.
   c. Is H_0 rejected at the 0.05 level? Explain.

Working with the Concepts

9. **More time on the Internet:** The General Social Survey polled a sample of 1048 adults in the year 2010, asking them how many hours per week they spent on the Internet. The sample mean was 9.79 with a standard deviation of 13.41. A second sample of 1018 adults was taken in the year 2012. For this sample, the mean was 10.47 with a standard deviation of 14.52. Assume these are simple random samples from populations of adults. Can you conclude that the mean number of hours per week spent on the Internet increased between 2010 and 2012? Use the $\alpha = 0.05$ level.
   a. State the appropriate null and alternate hypotheses.
   b. Compute the test statistic.
   c. How many degrees of freedom are there, using the simple method?
   d. Do you reject H_0? State a conclusion.

10. **Low-fat or low-carb?** Are low-fat diets or low-carb diets more effective for weight loss? A sample of 77 subjects went on a low-carbohydrate diet for six months. At the end of that time, the sample mean weight loss was 4.7 kilograms with a sample standard deviation of 7.16 kilograms. A second sample of 79 subjects went on a low-fat diet. Their sample mean weight loss was 2.6 kilograms with a standard deviation of 5.90 kilograms. Can you conclude that the mean weight loss differs between the two diets? Use the $\alpha = 0.01$ level.
   a. State the appropriate null and alternate hypotheses.
   b. Compute the test statistic.
   c. How many degrees of freedom are there, using the simple method?
   d. Do you reject H_0? State a conclusion.
      Source: *Journal of the American Medical Association* 297:969–977

11. **Are you smarter than your older brother?** In a study of birth order and intelligence, IQ tests were given to 18- and 19-year-old men to estimate the size of the difference, if any, between the mean IQs of firstborn sons and secondborn sons. The following data for 10 firstborn sons and 10 secondborn sons are consistent with the means and standard deviations reported in the article. It is reasonable to assume that the samples come from populations that are approximately normal.

Firstborn				
104	82	102	96	129
89	114	107	89	103

Secondborn				
103	103	91	113	102
103	92	90	114	113

Can you conclude that there is a difference in mean IQ between firstborn and secondborn sons? Use the $\alpha = 0.01$ level.
Based on data in *Science* 316:1717

12. **Recovering from surgery:** A new postsurgical treatment was compared with a standard treatment. Seven subjects received the new treatment, while seven others (the controls) received the standard treatment. The recovery times, in days, are given below.

Treatment:	12	13	15	19	20	21	24
Control:	18	23	24	30	32	35	39

Can you conclude that the mean recovery time for those receiving the new treatment is less than the mean for those receiving the standard treatment? Use the $\alpha = 0.05$ level.

13. **Contaminated water:** The concentration of benzene was measured in units of milligrams per liter for a simple random sample of five specimens of untreated wastewater produced at a gas field. The sample mean was 7.8 with a sample standard deviation of 1.4. Seven specimens of treated wastewater had an average benzene concentration of 3.2 with a standard deviation of 1.7. It is reasonable to assume that both samples come from populations that are approximately normal. Can you conclude that the mean benzene concentration is less in treated water than in untreated water? Use the $\alpha = 0.05$ level of significance.

14. **Exercise:** Medical researchers conducted a study to determine whether treadmill exercise could improve the walking ability of patients suffering from claudication, which is pain caused by insufficient blood flow to the muscles of the legs. A sample of 48 patients walked on a treadmill for six minutes every day. After six months, the mean distance walked in six minutes was 348 meters, with a standard deviation of 80 meters. For a control group of 46 patients who did not walk on a treadmill, the mean distance was 309 meters with a standard deviation of 89 meters. Can you conclude that the mean distance walked for patients using a treadmill is greater than the mean for the controls? Use the $\alpha = 0.05$ level of significance.
Source: *Journal of the American Medical Association* 301:165–174

15. **Mummy's curse:** King Tut was an ancient Egyptian ruler whose tomb was discovered and opened in 1923. Legend has it that the archaeologists who opened the tomb were subject

to a "mummy's curse," which would shorten their life spans. A team of scientists conducted an investigation of the mummy's curse. They reported that the 25 people exposed to the curse had a mean life span of 70.0 years with a standard deviation of 12.4 years, while a sample of 11 Westerners in Egypt at the time who were not exposed to the curse had a mean life span of 75.0 years with a standard deviation of 13.6 years. Assume that the populations are approximately normal. Can you conclude that the mean life span of those exposed to the mummy's curse is less than the mean of those not exposed? Use the $\alpha = 0.05$ level.
Source: *British Medical Journal* 325:1482

16. Baby weights: Following are weights in pounds for random samples of 25 newborn baby boys and baby girls born in Denver in 2011. Boxplots indicate that the samples come from populations that are approximately normal.

Boys								
6.6	5.9	6.4	7.6	6.4	8.1	7.9	8.3	7.3
6.4	8.4	8.5	6.9	6.3	7.4	7.8	7.5	6.9
7.8	8.6	7.7	7.4	7.7	8.1	6.4		

Girls								
7.7	7.0	8.2	7.4	6.0	6.7	8.2	7.5	5.7
6.6	6.4	8.5	7.2	6.9	8.2	6.5	6.7	7.2
6.3	5.9	8.1	8.2	6.7	6.2	7.7		

Can you conclude that the mean weights differ between boys and girls? Use the $\alpha = 0.05$ level of significance.

17. Empathy: The Interpersonal Reactivity Index is a survey designed to assess four different types of empathy. One type of empathy, called Empathetic Concern, measures the tendency to feel sympathy and compassion for people who are less fortunate. The index ranges from 0 (less empathetic) to 28 (more empathetic). The following data, representing random samples of 16 males and 16 females, are consistent with results reported in psychological studies. Boxplots show that it is reasonable to assume that the populations are approximately normal.

Males							
13	20	12	16	13	26	21	23
8	15	18	25	15	23	17	22

Females							
22	20	26	25	28	24	16	19
20	23	21	23	15	26	19	25

Can you conclude that there is a difference in mean empathy score between men and women? Use the $\alpha = 0.05$ level.

18. Text me: The Pew Internet and American Life Project surveyed 2227 adults in 2011, asking them about their texting habits. Among men, the mean number of text messages sent per day was 40.9, and among women it was 42.0. Assume that 1085 men and 1142 women were sampled, and that the standard deviations were 56.5 for the men and 51.4 for the women. Can you conclude that the mean number of text messages differs between men and women? Use the $\alpha = 0.05$ level of significance.

19. Online testing: Do you prefer taking tests on paper or online? A college instructor gave identical tests to two randomly sampled groups of 35 students. One group took the test on paper and the other took it online. Following are the test scores.

Paper						
79	75	49	78	73	81	70
63	79	65	60	74	64	64
74	69	57	65	61	58	92
71	69	66	71	68	67	81
78	80	76	67	49	56	45

Online						
79	75	71	81	56	72	49
75	63	81	74	72	71	73
83	59	78	65	53	47	63
82	81	76	65	82	78	76
65	72	85	84	81	50	79

Can you conclude that there is a difference in the mean score between paper and online testing? Use the $\alpha = 0.05$ level of significance.

20. Drive safely: How often does the average driver have an accident? The Allstate Insurance Company determined the average number of years between accidents for drivers in a large number of U.S. cities. Following are the results for 32 cities east of the Mississippi River and 32 cities west of the Mississippi River.

East							
10.2	9.5	11.2	11.4	9.8	10.0	11.6	10.7
9.5	9.0	7.2	11.5	7.7	9.0	8.9	8.4
6.8	5.3	8.0	8.8	9.0	10.0	9.7	11.6
7.9	9.9	10.5	11.9	10.9	8.6	6.7	9.6

West							
9.5	7.9	13.4	12.6	9.7	8.2	14.0	7.1
10.2	9.4	10.8	9.5	7.5	9.1	8.4	7.6
7.4	7.7	7.1	11.9	10.2	11.9	7.7	7.0
10.0	8.1	11.5	8.4	9.6	9.5	9.9	11.5

Can you conclude that the mean time between accidents is greater for western cities than for eastern cities? Use the $\alpha = 0.05$ level of significance.

21. Interpret calculator display: The following TI-84 Plus calculator display presents the results of a hypothesis test for the difference between two means. The sample sizes are $n_1 = 12$ and $n_2 = 15$.

a. Is this a left-tailed test, a right-tailed test, or a two-tailed test?

b. How many degrees of freedom did the calculator use?

c. What is the P-value?

d. Can you reject H_0 at the $\alpha = 0.05$ level?

22. Interpret calculator display: The following TI-84 Plus calculator display presents the results of a hypothesis test for the difference between two means. The sample sizes are $n_1 = 25$ and $n_2 = 28$.

```
2-SampTTest
μ1≠μ2
t=2.39806437
P=.0209030051
df=42.91558207
x̄1=26.80805
↓x̄2=24.66451
```

a. Is this a left-tailed test, a right-tailed test, or a two-tailed test?

b. How many degrees of freedom did the calculator use?

c. What is the *P*-value?

d. Can you reject H_0 at the $\alpha = 0.05$ level?

23. Interpret computer output: The following computer output (from MINITAB) presents the results of a hypothesis test for the difference $\mu_1 - \mu_2$ between two population means:

```
Two-sample T for X1 vs X2

        N     Mean    StDev   SE Mean
X1     58   35.848   10.233    1.3437
X2     36   26.851   15.329    2.5548


Difference = mu (X1) - mu (X2)
Estimate for difference: 8.997
95% CI for difference: (4.16605, 13.827951)
T-Test of difference = 0 (vs not =):   T-Value = 3.12
                                        P-Value = 0.003    DF = 54
```

a. What is the alternate hypothesis?

b. Can H_0 be rejected at the $\alpha = 0.05$ level? Explain.

c. How many degrees of freedom are there for the test statistic?

d. Use the simpler method to compute the degrees of freedom as the smaller of $n_1 - 1$ and $n_2 - 1$.

e. Find the *P*-value using this value for the degrees of freedom.

24. Interpret computer output: The following computer output (from MINITAB) presents the results of a hypothesis test for the difference $\mu_1 - \mu_2$ between two population means:

```
Two-sample T for X1 vs X2

        N    Mean    StDev   SE Mean
X1     20    3.44     2.65      0.23
X2     25    4.43     2.38      0.18


Difference = mu (X1) - mu (X2)
Estimate for difference: -0.99
95% upper bound for difference: 0.291437
T-Test of difference = 0 (vs <):   T-Value = -1.30
                                    P-Value = 0.100    DF = 38
```

a. What is the alternate hypothesis?

b. Can H_0 be rejected at the $\alpha = 0.05$ level? Explain.

c. How many degrees of freedom are there for the test statistic?

d. Use the simpler method to compute the degrees of freedom as the smaller of $n_1 - 1$ and $n_2 - 1$.

e. Find the *P*-value using this value for the degrees of freedom.

Extending the Concepts

25. More accurate degrees of freedom: A test will be made of $H_0: \mu_1 = \mu_2$ versus $H_1: \mu_1 > \mu_2$. The sample sizes were $n_1 = 10$ and $n_2 = 20$. The sample standard deviations were $s_1 = 3$ and $s_2 = 8$.

a. Compute the critical value for a level $\alpha = 0.05$ test, using 1 less than the smaller of the two sample sizes for the degrees of freedom.

b. Use the expression in Section 10.1 to compute a more accurate number of degrees of freedom.

c. Use the more accurate number of degrees of freedom to compute the probability of rejecting H_0 when it is true when the critical value found in part (a) is used. You will need technology to find the answer.

Answers to Check Your Understanding Exercises for Section 11.1

1. a. 3.073   **b.** 9

c. The *P*-value is between 0.005 and 0.01. [Tech: 0.0066] [TI-84 Plus: 0.0030] [MINITAB: 0.0031]

d. If H_0 is true, the probability of observing a value for the test statistic as extreme as or more extreme than the value actually observed is 0.0066. This is unusual, so the evidence against H_0 is strong. Because $P < 0.01$, we reject H_0 at the $\alpha = 0.01$ level.

2. a. 1.652   **b.** 16

c. The *P*-value is between 0.10 and 0.20. [Tech: 0.1180] [TI-84 Plus: 0.1072] [MINITAB: 0.1073]

d. If H_0 is true, the probability of observing a value for the test statistic as extreme as or more extreme than the value actually observed is 0.1073. This is not very unusual, so the evidence against H_0 is not strong. Because $P > 0.05$, we do not reject H_0 at the $\alpha = 0.05$ level.

SECTION 11.2 # Hypothesis Tests for the Difference Between Two Proportions

Objectives

1. Perform a hypothesis test for the difference between two proportions using the *P*-value method

2. Perform a hypothesis test for the difference between two proportions using the critical value method

Objective 1 Perform a
hypothesis test for the
difference between two
proportions using the *P*-value
method

Perform a Hypothesis Test for the Difference Between Two Proportions Using the *P*-Value Method

Are older, more experienced workers more likely to use a computer at work than younger workers? The General Social Survey took a poll to address this question. They asked 350 employed people aged 25–40 whether they used a computer at work, and 259 said they did. They also asked the same question of 500 employed people aged 41–65, and 384 of them said that they used a computer at work.

We can compute the sample proportions of people who used a computer at work in each of these age groups. Among those 25–40, the sample proportion was $259/350 = 0.740$, and among those aged 41–65 the sample proportion was $384/500 = 0.768$. So the sample proportion is larger among older workers. The question of interest, however, involves the population proportions. There are two populations involved; the population of all employed people aged 25–40, and the population of all employed people aged 41–65. The question is whether the population proportion of people aged 41–65 who use a computer at work is greater than the population proportion among those aged 25–40.

This is an example of a situation in which we have two independent samples, with the sample proportion computed for each one. We will describe a method for performing a hypothesis test to determine whether the two population proportions are equal. We will need some notation for the population proportions, the sample proportions, the numbers of individuals in each category, and the sample sizes.

NOTATION

- p_1 and p_2 are the proportions of the category of interest in the two populations.
- $\hat{p}_1$ and $\hat{p}_2$ are the proportions of the category of interest in the two samples.
- x_1 and x_2 are the numbers of individuals in the category of interest in the two samples.
- n_1 and n_2 are the two sample sizes.

We will now describe how to perform the hypothesis test.

The null and alternate hypotheses

The issue is whether the population proportions p_1 and p_2 are equal. The null hypothesis says that they are equal:

$$H_0: p_1 = p_2$$

There are three possibilities for the alternate hypothesis:

$$H_1: p_1 < p_2 \qquad H_1: p_1 > p_2 \qquad H_1: p_1 \neq p_2$$

The test statistic

The test statistic is based on the difference between the sample proportions, $\hat{p}_1 - \hat{p}_2$. When the sample size is large, this difference is approximately normally distributed.

The mean and standard deviation of this distribution are

Recall: Because the samples are independent, the variance of $\hat{p}_1 - \hat{p}_2$ is the sum of the variances of $\hat{p}_1$ and $\hat{p}_2$. The standard deviation is the square root of the variance.

$$\text{Mean} = p_1 - p_2 \qquad \text{Standard deviation} = \sqrt{\frac{p_1(1-p_1)}{n_1} + \frac{p_2(1-p_2)}{n_2}}$$

To compute the test statistic, we must find values for the mean and standard deviation. The mean is straightforward: Under the assumption that H_0 is true, $p_1 - p_2 = 0$. The standard deviation is a bit more involved. The standard deviation depends on the population proportions p_1 and p_2, which are unknown. We need to estimate p_1 and p_2. Under H_0, we assume that $p_1 = p_2$. Therefore, we need to estimate p_1 and p_2 with the same value. The value to use is the **pooled proportion**, which we will denote by $\hat{p}$. The pooled proportion is found by treating the two samples as though they were one big sample. We divide the total number of individuals in the category of interest in the two samples by the sum of the two sample sizes:

$$\hat{p} = \frac{x_1 + x_2}{n_1 + n_2}$$

The standard deviation is estimated with the standard error:

$$\text{Standard error} = \sqrt{\frac{\hat{p}(1 - \hat{p})}{n_1} + \frac{\hat{p}(1 - \hat{p})}{n_2}} = \sqrt{\hat{p}(1 - \hat{p})\left(\frac{1}{n_1} + \frac{1}{n_2}\right)}$$

The test statistic is the z-score for $\hat{p}_1 - \hat{p}_2$:

$$z = \frac{(\hat{p}_1 - \hat{p}_2) - (p_1 - p_2)}{\sqrt{\hat{p}(1 - \hat{p})\left(\frac{1}{n_1} + \frac{1}{n_2}\right)}} = \frac{(\hat{p}_1 - \hat{p}_2) - 0}{\sqrt{\hat{p}(1 - \hat{p})\left(\frac{1}{n_1} + \frac{1}{n_2}\right)}} = \frac{\hat{p}_1 - \hat{p}_2}{\sqrt{\hat{p}(1 - \hat{p})\left(\frac{1}{n_1} + \frac{1}{n_2}\right)}}$$

The method just described requires certain assumptions, which we now state:

Explain It Again

Reason for assumption 4: We require that both samples contain at least 10 individuals in each category in order to be sure that $\hat{p}_1$ and $\hat{p}_2$ are both approximately normally distributed.

Assumptions for Performing a Hypothesis Test for the Difference Between Two Proportions

1. There are two simple random samples that are independent of one another.
2. Each population is at least 20 times as large as the sample drawn from it.
3. The individuals in each sample are divided into two categories.
4. Both samples contain at least 10 individuals in each category.

We summarize the steps for testing a hypothesis about the difference between two proportions using the P-value method. Later we will describe the critical value method.

Performing a Hypothesis Test for the Difference Between Two Proportions Using the *P*-Value Method

Check to be sure the assumptions are satisfied. If they are, then proceed with the following steps.

Step 1: State the null and alternate hypotheses. The null hypothesis will have the form $H_0: p_1 = p_2$. The alternate hypothesis will be $p_1 < p_2, p_1 > p_2$, or $p_1 \neq p_2$.

Step 2: If making a decision, choose a significance level α.

Step 3: Compute the test statistic $z = \dfrac{\hat{p}_1 - \hat{p}_2}{\sqrt{\hat{p}(1 - \hat{p})\left(\dfrac{1}{n_1} + \dfrac{1}{n_2}\right)}}$

where $\hat{p}$ is the pooled proportion: $\hat{p} = \dfrac{x_1 + x_2}{n_1 + n_2}$

Step 4: Compute the P-value. The P-value is an area under the normal curve. The P-value depends on the alternate hypothesis as follows:

The P-value is the area to the left of z.

Left-tailed: $H_1: p_1 < p_2$

The P-value is the area to the right of z.

Right-tailed: $H_1: p_1 > p_2$

The P-value is the sum of the areas in the two tails.

Two-tailed: $H_1: p_1 \neq p_2$

Step 5: Interpret the P-value. If making a decision, reject H_0 if the P-value is less than or equal to the significance level α.

Step 6: State a conclusion.

EXAMPLE 11.3 ## Perform a hypothesis test

Are older, more experienced workers more likely to use a computer at work than younger workers? The General Social Survey took a poll to address this question. They asked 350 employed people aged 25–40 whether they used a computer at work, and 259 said they did. They also asked the same question of 500 employed people aged 41–65, and 384 of them said that they used a computer at work. Can you conclude that the proportion of people who use a computer at work is greater among those aged 41–65 than among those aged 25–40? Use the $\alpha = 0.05$ level.

Solution

We first check the assumptions. We have two independent random samples, and the populations are more than 20 times as large as the samples. The individuals in each sample are divided into two categories: those who use a computer at work and those who do not. Finally, each sample contains more than 10 individuals in each category. The assumptions are satisfied.

Step 1: **State the null and alternate hypotheses:** We'll let p_1 be the population proportion of people aged 25–40 who used a computer at work, and p_2 be the proportion among those aged 41–65. The issue is whether the proportions are the same, or whether the proportion of those aged 41–65, p_2, is greater than the proportion of those aged 25–40, p_1. Therefore the null and alternate hypotheses are

$$H_0: p_1 = p_2 \qquad H_1: p_1 < p_2$$

Step 2: **Choose a significance level:** The significance level is $\alpha = 0.05$.

Step 3: **Compute the test statistic:** We'll begin by summarizing the necessary information in a table.

	Ages 25–40	Ages 41–65
Sample size	$n_1 = 350$	$n_2 = 500$
Number of individuals	$x_1 = 259$	$x_2 = 384$
Sample proportion	$\hat{p}_1 = 259/350 = 0.740$	$\hat{p}_2 = 384/500 = 0.768$
Population proportion	p_1 (unknown)	p_2 (unknown)

Next, we compute the pooled proportion $\hat{p}$.

$$\hat{p} = \frac{x_1 + x_2}{n_1 + n_2} = \frac{259 + 384}{350 + 500} = 0.756471$$

The value of the test statistic is

$$z = \frac{\hat{p}_1 - \hat{p}_2}{\sqrt{\hat{p}(1 - \hat{p})\left(\dfrac{1}{n_1} + \dfrac{1}{n_2}\right)}} = \frac{0.740 - 0.768}{\sqrt{0.756471(1 - 0.756471)\left(\dfrac{1}{350} + \dfrac{1}{500}\right)}}$$

$$= -0.94$$

Step 4: **Compute the P-value:** The alternate hypothesis, $p_1 < p_2$, is left-tailed. Therefore the P-value is the area to the left of $z = -0.94$. Using Table A.2, we find this area to be 0.1736. Figure 11.2 illustrates the P-value.

Step 5: **Interpret the P-value:** The P-value of 0.1736 is greater than the significance level $\alpha = 0.05$. Therefore we do not reject H_0.

Step 6: **State a conclusion:** We cannot conclude that the proportion of workers aged 41–65 who use a computer at work is greater than the proportion among those aged 25–40.

P–value = 0.1736

−0.94

Figure 11.2

Performing a hypothesis test with technology

The following computer output (from MINITAB) presents the results of Example 11.3.

```
Test and CI for Two Proportions: 25-40, 41-65

Sample        X        N      Sample p
25-40        259      350     0.740000
41-65        384      500     0.768000

Difference = p (25-40) − p (41-65)
Estimate for difference: −0.028000
95% upper bound for difference: 0.021516
Test of difference = 0 (vs < 0): Z = −0.94 P-Value = 0.175
```

The numbers of individuals in the category of interest (X), the sample sizes (N), and the sample proportions (Sample p) are given. The quantity "Estimate for difference" is the difference between the sample proportions, $\hat{p}_1 - \hat{p}_2$. Next is the 95% upper bound for the difference. We can be 95% confident that the difference $p_1 - p_2$ is less than or equal to this value of 0.021516. The last line in the output presents the alternate hypothesis, the value of the test statistic (Z), and the *P*-value.

Following are the results as presented by the TI-84 Plus calculator.

The letter "p" in the fourth line is the *P*-value.

Step-by-step instructions for performing hypothesis tests with technology are presented on page 528.

Check Your Understanding

1. In a clinical trial to compare the effectiveness of two pain relievers, a sample of 100 patients was given drug 1 and an independent sample of 200 patients was given drug 2. Of the patients on drug 1, 76 experienced substantial relief, while of the patients on drug 2, 128 experienced substantial relief. Investigators want to know whether the proportion of patients experiencing substantial relief is greater for drug 1. They will use the $\alpha = 0.05$ level of significance.
   a. Let p_1 be the population proportion of patients experiencing substantial relief from drug 1, and let p_2 be the population proportion of patients experiencing substantial relief from drug 2. State the appropriate null and alternate hypotheses about p_1 and p_2.
   b. Compute the sample proportions $\hat{p}_1$ and $\hat{p}_2$.
   c. Compute the value of the test statistic.
   d. Compute the *P*-value.
   e. Interpret the *P*-value.
   f. State a conclusion.

2. A sample of 200 voters over the age of 60 were asked whether they thought Social Security benefits should be increased for people over the age of 65. A total of 95 of them answered yes. A sample of 150 voters aged 18–25 were asked the same question and 63 of them answered yes. A pollster wants to know whether the proportion of voters who support an increase in Social Security benefits is greater among older voters. He will use the $\alpha = 0.05$ level of significance.

a. Let p_1 be the population proportion of older voters expressing support, and let p_2 be the population proportion of younger voters expressing support. State the appropriate null and alternate hypotheses about p_1 and p_2.

b. Compute the sample proportions $\hat{p}_1$ and $\hat{p}_2$.

c. Compute the value of the test statistic.

d. Compute the P-value.

e. Interpret the P-value.

f. State a conclusion.

Answers are on page 531.

Objective 2 Perform a hypothesis test for the difference between two proportions using the critical value method

Using the Critical Value Method

To use the critical value method, compute the test statistic as before. Because the test statistic is a z-score, critical values can be found in Table A.2, in Table 9.1 in Section 9.2, or with technology. The assumptions for the critical value method are the same as for the P-value method.

Following are the steps for the critical value method.

Performing a Hypothesis Test for the Difference Between Two Proportions Using the Critical Value Method

Check to be sure that the assumptions are satisfied. If they are, then proceed with the following steps.

Step 1: State the null and alternate hypotheses. The null hypothesis will have the form $H_0: p_1 = p_2$. The alternate hypothesis will be $p_1 < p_2$, $p_1 > p_2$, or $p_1 \neq p_2$.

Step 2: Choose a significance level α and find the critical value or values.

Step 3: Compute the test statistic $z = \dfrac{\hat{p}_1 - \hat{p}_2}{\sqrt{\hat{p}(1 - \hat{p})\left(\dfrac{1}{n_1} + \dfrac{1}{n_2}\right)}}$

where $\hat{p}$ is the pooled proportion: $\hat{p} = \dfrac{x_1 + x_2}{n_1 + n_2}$

Step 4: Determine whether to reject H_0, as follows:

Left-tailed: $H_1: p_1 < p_2$
Reject if $z \leq -z_\alpha$.

Right-tailed: $H_1: p_1 > p_2$
Reject if $z \geq z_\alpha$.

Two-tailed: $H_1: p_1 \neq p_2$
Reject if $z \geq z_{\alpha/2}$ or $z \leq -z_{\alpha/2}$.

Step 5: State a conclusion.

EXAMPLE 11.4   Testing a hypothesis using the critical value method

Are younger drivers more likely to have accidents in their driveways? Traffic engineers tabulated types of car accidents by drivers of various ages. Out of a total of 82,486 accidents involving drivers aged 15–24 years, 4243 of them, or 5.1%, occurred in a driveway. Out of a total of 219,170 accidents involving drivers aged 25–64 years, 10,701 of them, or 4.9%, occurred in a driveway. Can you conclude that accidents involving drivers aged 15–24 are more likely to occur in driveways than accidents involving drivers aged 25–64? Use the $\alpha = 0.05$ significance level.

Source: *Journal of Transportation Engineering* 125:502–507

Solution

We first check the assumptions. We have two independent samples, and the individuals in each sample fall into two categories. Each sample contains at least 10 individuals in each category. The assumptions are satisfied.

Step 1: State H_0 and H_1. We are interested in whether the proportion of accidents in driveways is greater among younger drivers. Let p_1 be the population proportion for younger drivers, and let p_2 be the population proportion for older drivers. Then the null and alternate hypotheses are:

$$H_0: p_1 = p_2 \qquad H_1: p_1 > p_2$$

Step 2: Choose a significance level α and find the critical value. We will use $\alpha = 0.05$. Because this is a right-tailed test, the critical value is the value for which the area to the right is 0.05. This value is $z_\alpha = 1.645$.

Step 3: Compute the test statistic. We begin by organizing the available information in a table:

	Ages 15–24	**Ages 25–64**
Sample size	$n_1 = 82{,}486$	$n_2 = 219{,}170$
Number of individuals	$x_1 = 4{,}243$	$x_2 = 10{,}701$
Sample proportion	$\hat{p}_1 = 4{,}243/82{,}486$ $= 0.051439$	$\hat{p}_2 = 10{,}701/219{,}170$ $= 0.048825$
Population proportion	p_1 (unknown)	p_2 (unknown)

Next, we compute the pooled proportion, $\hat{p}$.

$$\hat{p} = \frac{x_1 + x_2}{n_1 + n_2} = \frac{4{,}243 + 10{,}701}{82{,}486 + 219{,}170} = 0.049540$$

The test statistic is

$$z = \frac{\hat{p}_1 - \hat{p}_2}{\sqrt{\hat{p}(1 - \hat{p})\left(\dfrac{1}{n_1} + \dfrac{1}{n_2}\right)}} = 2.95$$

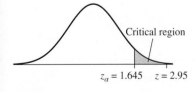

Critical region

$z_\alpha = 1.645 \quad z = 2.95$

Figure 11.3 The value of the test statistic, $z = 2.95$, is inside the critical region. Therefore, we reject H_0.

Step 4: Determine whether to reject H_0. This is a right-tailed test, so we reject H_0 if $z > z_\alpha$. Because $z = 2.95$ and $z_\alpha = 1.645$, $z > z_\alpha$. We reject H_0 at the $\alpha = 0.05$ level. Figure 11.3 illustrates the critical region and the test statistic.

Step 5: State a conclusion. We conclude that accidents involving drivers aged 15–24 are more likely to occur in a driveway than accidents involving drivers aged 25–64.

USING TECHNOLOGY

We use Example 11.4 to illustrate the technology steps.

TI-84 PLUS

Testing a hypothesis about the difference between two proportions

Step 1. Press **STAT** and highlight the **TESTS** menu.

Step 2. Select **2–PropZTest** and press **ENTER** (Figure A). The **2–PropZTest** menu appears.

Step 3. Enter the values of x_1, n_1, x_2, and n_2. For Example 11.4, we use $x_1 = 4243$, $n_1 = 82486$, $x_2 = 10701$, and $n_2 = 219170$.

Step 4. Select the form of the alternate hypothesis. For Example 11.4, the alternate hypothesis has the form $> p_2$ (Figure B).

Step 5. Highlight **Calculate** and press **ENTER** (Figure C).

Figure A

Figure B

Figure C

MINITAB

Testing a hypothesis about the difference between two proportions

Step 1. Click on **Stat**, then **Basic Statistics**, then **2 Proportions**.

Step 2. Click **Summarized Data**, and enter the value of n_1 and n_2 in the **Trials** fields in the **First** and **Second** rows. Enter the values of x_1 and x_2 in the **Events** field in the **First** and **Second** rows. For Example 11.4, we use $x_1 = 4243$, $n_1 = 82486$, $x_2 = 10701$, and $n_2 = 219170$ (Figure D).

Step 3. Click **Options**, and enter **0** in the **Test difference** field and select the form of the alternate hypothesis. Given significance level α, enter $100(1 - \alpha)$ as the **Confidence Level**. For Example 11.4, we use **95** as the **Confidence Level**, **0** as the **Test difference**, and **greater than** as the **Alternative**. Check the box for **Use pooled proportion**. Click **OK**.

Step 4. Click **OK** (Figure E).

Figure D

```
Sample      X         N    Sample p
1        4243     82486    0.051439
2       10701    219170    0.048825


Difference = p (1) - p (2)
Estimate for difference:  0.00261392
95% lower bound for difference:  0.00113957
Test for difference = 0 (vs > 0):  Z = 2.95  P-Value = 0.002
```

Figure E

EXCEL

Testing a hypothesis about the difference between two proportions

This procedure requires the **MegaStat** EXCEL add-in to be loaded. The **MegaStat** add-in can be downloaded from www.mhhe.com/megastat.

Step 1. Load the **MegaStat** EXCEL add-in.

Step 2. Click on the **MegaStat** menu and select **Hypothesis Tests**, then **Compare Two Independent Proportions...**

Step 3. Under the **Group1** column, enter the value of x_1 in the **p** field (note that p automatically changes to x) and the sample size n_1 in the **n** field. Under the **Group2** column, repeat for x_2 and n_2. For Example 11.4, we use $x_1 = 4243$, $n_1 = 82486$, $x_2 = 10701$, and $n_2 = 219170$.

Step 4. Enter the **Hypothesized difference (0)** and select the form of the alternate hypothesis (**greater than**) (Figure F).

Step 5. Click **OK** (Figure G).

Figure F

Hypothesis test for two independent proportions

	p1	p2	p_c	
	0.051439032	0.048825113	0.04954	p (as decimal)
	0.051439032	0.048825113	0.04954	p (as fraction)
	4243	10701	14944	X
	82486	219170	301656	n
			0.002613919	difference
			0	hypothesized difference
			0.00088638	std. error
			2.948984202	z
			0.001594	p-value (one-tailed, upper)

Figure G

SECTION 11.2 Exercises

Exercises 1 and 2 are the Check Your Understanding exercises located within the section.

Understanding the Concepts

In Exercises 3 and 4, fill in each blank with the appropriate word or phrase.

3. To use the method of this section to test a hypothesis about the difference between two proportions, each population must be at least _____ times as large as the sample drawn from it.

4. To use the method of this section to test a hypothesis about the difference between two proportions, each sample must contain at least _____ individuals in each category.

In Exercises 5 and 6, determine whether the statement is true or false. If the statement is false, rewrite it as a true statement.

5. The individuals in each sample are divided into three or more categories.

6. To compute the test statistic, it is necessary to compute the pooled proportion.

Practicing the Skills

7. In a test for the difference between two proportions, the sample sizes were $n_1 = 120$ and $n_2 = 85$, and the numbers of events were $x_1 = 55$ and $x_2 = 45$. A test is made of the hypotheses H_0: $p_1 = p_2$ versus H_1: $p_1 \neq p_2$.

a. Compute the value of the test statistic.

b. Can you reject H_0 at the $\alpha = 0.05$ level of significance?

c. Can you reject H_0 at the $\alpha = 0.01$ level of significance?

8. In a test for the difference between two proportions, the sample sizes were $n_1 = 68$ and $n_2 = 76$, and the numbers of events were $x_1 = 41$ and $x_2 = 25$. A test is made of the hypotheses H_0: $p_1 = p_2$ versus H_1: $p_1 > p_2$.

a. Compute the value of the test statistic.

b. Can you reject H_0 at the $\alpha = 0.05$ level of significance?

c. Can you reject H_0 at the $\alpha = 0.01$ level of significance?

Working with the Concepts

9. Childhood obesity: The National Health and Nutrition Examination Survey (NHANES) weighed a sample of 546 boys aged 6–11 and found that 87 of them were overweight. They weighed a sample of 508 girls aged 6–11 and found that 74 of them were overweight. Can you conclude that the proportion of boys who are overweight differs from the proportion of girls who are overweight?

a. State the appropriate null and alternate hypotheses.

b. Compute the value of the test statistic.

c. State a conclusion at the $\alpha = 0.05$ level of significance.

10. **Pollution and altitude:** In a random sample of 340 cars driven at low altitudes, 46 of them exceeded a standard of 10 grams of particulate pollution per gallon of fuel consumed. In an independent random sample of 85 cars driven at high altitudes, 21 of them exceeded the standard. Can you conclude that the proportion of high-altitude vehicles exceeding the standard is greater than the proportion of low-altitude vehicles exceeding the standard?

   a. State the appropriate null and alternate hypotheses.

   b. Compute the value of the test statistic.

   c. State a conclusion at the $\alpha = 0.01$ level of significance.

11. **Preventing heart attacks:** Medical researchers performed a comparison of two drugs, clopidogrel and ticagrelor, which are designed to reduce the risk of heart attack or stroke in coronary patients. A total of 6676 patients were given clopidogrel, and 6732 were given ticagrelor. Of the clopidogrel patients, 668 suffered a heart attack or stroke within one year, and of the ticagrelor patients, 569 suffered a heart attack or stroke. Can you conclude that the proportion of patients suffering a heart attack or stroke is less for ticagrelor? Use the $\alpha = 0.01$ level.

   Source: *Lancet* 375:283–293

12. **Cholesterol:** An article in the *Archives of Internal Medicine* reported that in a sample of 244 men, 73 had elevated total cholesterol levels (more than 200 milligrams per deciliter). In a sample of 232 women, 44 had elevated cholesterol levels. Can you conclude that the proportion of people with elevated cholesterol levels differs between men and women? Use the $\alpha = 0.05$ level.

13. **Treating circulatory disease:** Angioplasty is a medical procedure in which an obstructed blood vessel is widened. In some cases, a wire mesh tube, called a stent, is placed in the vessel to help it remain open. A study was conducted to compare the effectiveness of a bare metal stent with one that has been coated with a drug designed to prevent reblocking of the vessel. A total of 5320 patients received bare metal stents, and of these, 841 needed treatment for reblocking within a year. A total of 1120 received drug-coated stents, and 134 of them required treatment within a year. Can you conclude that the proportion of patients who needed retreatment is less for those who received drug-coated stents? Use the $\alpha = 0.05$ level.

   Source: *Canadian Medical Association Journal* 180:167–174

14. **Hurricane damage:** In August and September 2005, Hurricanes Katrina and Rita caused extraordinary flooding in New Orleans, Louisiana. Many homes were severely damaged or destroyed, and of those that survived, many required extensive cleaning. It was thought that cleaning flood-damaged homes might present a health hazard due to the large amounts of mold present in many of the homes. In a sample of 365 residents of Orleans Parish who had participated in the cleaning of one or more homes, 77 had experienced symptoms of wheezing, and in a sample of 179 residents who had not participated in cleaning, 23 reported wheezing symptoms (numbers read from a graph). Can you conclude that the proportion of residents with wheezing symptoms is greater among those who participated in the cleaning of flood-damaged homes? Use the $\alpha = 0.05$ level.

Source: *American Journal of Public Health* 98:869–875

15. **Don't perform a hypothesis test:** In a certain year, there was measurable snowfall on 80 out of 365 days in Denver, and 63 out of 365 days in Chicago. A meteorologist proposes to perform a test of the hypothesis that the proportions of days with snow are equal in the two cities. Explain why this cannot be done using the method presented in this section.

16. **Don't perform a hypothesis test:** A new reading program is being tested. Parents are asked whether they would like to enroll their children, and 50 children are enrolled in the program. There are 45 children whose parents do not choose to enroll their children. At the end of the school year, the children are tested. Of the 50 children who participated in the program, 38 are found to be reading at grade level. Of the 45 children who did not participate, 24 were reading at grade level. Explain why these data should not be used to test the hypothesis that the proportion of children reading at grade level is higher for those who participate in the program.

17. **Interpret calculator display:** The following TI-84 Plus calculator display presents the results of a hypothesis test for the difference between two proportions. The sample sizes are $n_1 = 165$ and $n_2 = 152$.

   a. Is this a left-tailed test, a right-tailed test, or a two-tailed test?

   b. What is the *P*-value?

   c. Can you reject H_0 at the $\alpha = 0.05$ level?

18. Interpret calculator display: The following TI-84 Plus calculator display presents the results of a hypothesis test for the difference between two proportions. The sample sizes are $n_1 = 71$ and $n_2 = 62$.

```
      2-PropZTest
 P1≠P2
 z=.8141749055
 p=.4155446361
 p̂1=.6338028169
 p̂2=.564516129
↓p̂=.6015037594
```

a. Is this a left-tailed test, a right-tailed test, or a two-tailed test?
b. What is the P-value?
c. Can you reject H_0 at the $\alpha = 0.05$ level?

19. Interpret computer output: The following computer output (from MINITAB) presents the results of a hypothesis test on the difference between two proportions.

```
Test and CI for Two Proportions: P1, P2

Variable     X     N    Sample p
P1          22    43    0.511628
P2          55    93    0.591398

Difference = p (P1) – p (P2)
Estimate for difference: –0.079770
95% upper bound for difference: 0.071079
Test for difference = 0 (vs < 0):   Z = –0.87   P-Value = 0.192
```

a. Is this a left-tailed test, a right-tailed test, or a two-tailed test?
b. What is the P-value?
c. Can H_0 be rejected at the 0.05 level? Explain.

20. Interpret computer output: The following computer output (from MINITAB) presents the results of a hypothesis test on the difference between two proportions.

```
Test and CI for Two Proportions: P1, P2

Variable     X     N    Sample p
P1         405   577    0.701906
P2         363   578    0.628028

Difference = p (P1) – p (P2)
Estimate for difference: 0.073879
95% CI for difference: (0.0194191, 0.127827)
Test of difference = 0 (vs not = 0):   Z = –2.66   P-Value = 0.008
```

a. Is this a left-tailed test, a right-tailed test, or a two-tailed test?
b. What is the P-value?
c. Can H_0 be rejected at the 0.05 level? Explain.

Extending the Concepts

Null difference other than 0: *Occasionally someone may wish to test a hypothesis of the form $H_0\colon p_1 - p_2 = p_d$, where $p_d \neq 0$. In this situation, the null hypothesis says that the population proportions are unequal, so we do not compute the pooled proportion, which assumes the population proportions are equal. One approach to testing this hypothesis is to use the test statistic*

$$z = \frac{(p_1 - p_2) - p_d}{\sqrt{\dfrac{\hat{p}_1(1 - \hat{p}_1)}{n_1} + \dfrac{\hat{p}_2(1 - \hat{p}_2)}{n_2}}}$$

When the assumptions of this section are met, this statistic has approximately a standard normal distribution when H_0 is true.

21. Computer chips: A computer manufacturer has a choice of two machines, a less expensive one and a more expensive one, to manufacture a particular computer chip. Out of 500 chips manufactured on the less expensive machine, 70 were defective. Out of 400 chips manufactured on the more expensive machine, only 20 were defective. The manufacturer will buy the more expensive machine if he is convinced that the proportion of defectives is more than 5% less than on the less expensive machine. Let p_1 represent the proportion of defectives produced by the less expensive machine, and let p_2 represent the proportion of defectives produced by the more expensive machine.
a. State appropriate null and alternate hypotheses.
b. Compute the value of the test statistic.
c. Can you reject H_0 at the $\alpha = 0.05$ level?
d. Which machine should the manufacturer buy?

Answers to Check Your Understanding Exercises for Section 11.2

1. a. $H_0\colon p_1 = p_2$, $H_1\colon p_1 > p_2$
  b. $\hat{p}_1 = 0.76$, $\hat{p}_2 = 0.64$ **c.** $z = 2.10$ **d.** $P = 0.0179$
  e. If H_0 is true, the probability of observing a value for the test statistic as extreme as or more extreme than the value actually observed is 0.0179. This is unusual, so the evidence against H_0 is strong. Because $P < 0.05$, we reject H_0 at the $\alpha = 0.05$ level.
  f. We conclude that the proportion of patients experiencing substantial relief is greater for drug 1.

2. a. $H_0\colon p_1 = p_2$, $H_1\colon p_1 > p_2$

b. $\hat{p}_1 = 0.475$, $\hat{p}_2 = 0.420$ **c.** $z = 1.02$
d. $P = 0.1539$ [Tech: 0.1531]
e. If H_0 is true, the probability of observing a value for the test statistic as extreme as or more extreme than the value actually observed is 0.1539. This is not unusual, so the evidence against H_0 is not strong. Because $P > 0.05$, we do not reject H_0 at the $\alpha = 0.05$ level.
f. We conclude that the proportions of younger and older voters that support an increase in Social Security benefits may be the same.

Hypothesis Tests for the Difference Between Two Means: Paired Samples

Objectives

1. Perform a hypothesis test with matched pairs using the *P*-value method
2. Perform a hypothesis test with matched pairs using the critical value method

Objective 1 Perform a hypothesis test with matched pairs using the *P*-value method

Hypothesis Tests with Matched Pairs Using the *P*-Value Method

Does tuning a car engine improve the gas mileage? A sample of eight automobiles were run to determine their mileage, in miles per gallon. Then each car was given a tune-up, and run again to measure the mileage a second time. The results are presented in Table 11.1.

Table 11.1 Gas Mileage Before and After Tune-up for Eight Automobiles

	Automobile							
	1	**2**	**3**	**4**	**5**	**6**	**7**	**8**
After Tune-up	35.44	35.17	31.07	31.57	26.48	23.11	25.18	32.39
Before Tune-up	33.76	34.30	29.55	30.90	24.92	21.78	24.30	31.25
Difference	1.68	0.87	1.52	0.67	1.56	1.33	0.88	1.14

The sample mean mileage was higher after tune-up. We would like to determine how strong the evidence is that the population mean mileage is higher after tune-up.

These are paired samples, because each value before tune-up is paired with the value from the same car after tune-up. By computing the difference between the values in each matched pair, we construct a sample of differences, shown in the last row of Table 11.1.

If we denote the population mean mileage before tune-up by μ_1, and the population mean mileage after tune-up by μ_2, then we are interested in the difference $\mu_1 - \mu_2$. Because these are paired samples, the population mean of the differences, μ_d, is the same as $\mu_1 - \mu_2$. Therefore, performing a hypothesis test on μ_d is the same as performing a hypothesis test on the difference of the population means $\mu_1 - \mu_2$.

We summarize the notation we will use in this section.

Recall: When we have paired samples, the pairs are called matched pairs.

NOTATION

- $\bar{d}$ is the sample mean of the differences between the values in the matched pairs.
- s_d is the sample standard deviation of the differences between the values in the matched pairs.
- μ_d is the population mean difference for the matched pairs.

The method for testing a hypothesis about μ_d is the usual method for testing a hypothesis about a population mean, as presented in Section 9.3. We now list the assumptions for this method, when applied to paired samples.

Assumptions for Performing a Hypothesis Test for the Difference Between Two Means with Matched Pairs

1. We have a simple random sample of matched pairs.
2. Either the sample size is large ($n > 30$), *or* the differences between items in the matched pairs show no evidence of strong skewness and no outliers. This is required to be sure that $\bar{d}$ will be approximately normally distributed.

When these assumptions are satisfied, a hypothesis test can be performed in the same way as for a test for a population mean, using the following steps.

Performing a Hypothesis Test with Matched-Pair Data Using the *P*-Value Method

Check to be sure the assumptions are satisfied. If they are, then proceed with the following steps.

Step 1: State the null and alternate hypotheses. The null hypothesis will have the form $H_0: \mu_d = \mu_0$. The alternate hypothesis will be of the form $\mu_d < \mu_0$, $\mu_d > \mu_0$, or $\mu_d \neq \mu_0$.

Step 2: If making a decision, choose a significance level α.

Step 3: Compute the test statistic $t = \dfrac{\bar{d} - \mu_0}{s_d / \sqrt{n}}$.

Step 4: Compute the *P*-value. The *P*-value is an area under the *t* curve with $n - 1$ degrees of freedom. The *P*-value depends on the alternate hypothesis as follows:

The *P*-value is the area to the left of *t*.

The *P*-value is the area to the right of *t*.

The *P*-value is the sum of the areas in the two tails.

Left-tailed: $H_1: \mu_d < \mu_0$ Right-tailed: $H_1: \mu_d > \mu_0$ Two-tailed: $H_1: \mu_d \neq \mu_0$

Step 5: Interpret the *P*-value. If making a decision, reject H_0 if the *P*-value is less than or equal to the significance level α.

Step 6: State a conclusion.

EXAMPLE 11.5   Test a hypothesis with matched-pair data

Test $H_0: \mu_d = 0$ versus $H_1: \mu_d > 0$, using the data in Table 11.1. Use the $\alpha = 0.01$ significance level.

Solution

We first check the assumptions. We have a simple random sample of differences. Because the sample size is small ($n = 8$), we must check for signs of strong skewness or outliers. Following is a dotplot of the differences.

```
   •        ••          •          •        • •      •
 ├────────┼──────────┼──────────┼────────┼────────┼────────┤
0.6      0.8        1.0        1.2      1.4      1.6      1.8
```

The dotplot does not reveal any outliers or strong skewness. Therefore, we can proceed.

Step 1: State H_0 and H_1. The issue is whether the mileage is greater after tune-up, so the null and alternate hypotheses are

$$H_0: \mu_d = 0 \qquad H_1: \mu_d > 0$$

Step 2: Choose a significance level. We will use $\alpha = 0.01$.

Step 3: Compute the test statistic. First we compute the sample mean and sample standard deviation of the differences. These are

$$\bar{d} = 1.20625 \qquad s_d = 0.37317$$

The test statistic is

$$t = \frac{\bar{d} - \mu_0}{s_d / \sqrt{n}}$$

Under the assumption that H_0 is true, $\mu_d = \mu_0 = 0$. The value of the test statistic is therefore

$$t = \frac{\bar{d} - \mu_0}{s_d / \sqrt{n}} = \frac{1.20625 - 0}{0.37317 / \sqrt{8}} = 9.1427$$

Step 4: Compute the *P*-value. Under the assumption that H_0 is true, the test statistic has a *t* distribution. The number of degrees of freedom is $n - 1 = 8 - 1 = 7$. The alternate hypothesis is $\mu_d > 0$, so the *P*-value is the area to the right of the observed value of 9.1427. Technology gives $P = 0.0000193$.

Step 5: Interpret the *P*-value. The *P*-value is nearly 0. If H_0 were true, there would be virtually no chance of observing a test statistic as extreme as the value of 9.1427 that we observed. Because $P < 0.01$, we reject H_0 at the $\alpha = 0.01$ level.

Step 6: State a conclusion. We conclude that the gas mileage increased after a tune-up.

Performing hypothesis tests with technology

The following computer output (from MINITAB) presents the results of the hypothesis test performed in Example 11.5.

```
Paired T for After - Before

              N       Mean      StDev     SE Mean
After         8    30.05125    4.60928    1.62963
Before        8    28.84500    4.63519    1.63879
Difference    8     1.20625    0.37317    0.13194

99% lower bound for mean difference: 0.81071
T-Test of mean difference = 0 (vs > 0):   T-Value = 9.14  P-Value = 0.000
```

The column labeled "SE Mean" presents the standard errors for the means of the two samples, and for the mean of the differences. The standard error is the standard deviation divided by the square root of the sample size. Note that the sample standard deviation of the differences is much smaller than the sample standard deviations of the original samples. This is the case for most matched-pair data, and is the reason that tests based on matched pairs have more power than tests based on independent samples.

Following are the results of Example 11.5 as displayed on a TI-84 Plus calculator.

The second line states the alternate hypothesis. This is followed by the value of the test statistic *t*, and the *P*-value. The quantity x̄ is the sample mean of the differences $\bar{d}$, and the quantity Sx is the sample standard deviation of the differences, s_d.

Step-by-step instructions for performing hypothesis tests with technology are presented in the Using Technology section on page 537.

Check Your Understanding

1. A sample of five third-graders took a reading test. They then participated in a reading improvement program, and took the test again to determine whether their reading ability had improved. Following are the test scores for each of the students both before and after the program. Can you conclude that the mean reading score increased after the program?

	1	2	3	4	5
After	67	68	78	75	84
Before	59	63	81	74	78

    a. Let μ_d denote the population mean difference After − Before. State the appropriate null and alternate hypotheses about μ_d.

    b. Compute the differences After − Before.

    c. Compute the value of the test statistic.

    d. Compute the *P*-value.

    e. Do you reject H_0 at the $\alpha = 0.05$ level?

    f. State a conclusion.

2. Following are the annual amounts of rainfall, in inches, in six randomly chosen cities for two consecutive years. Can you conclude that the mean rainfall was greater in year 2 than in year 1?

	1	2	3	4	5	6
Year 2	34.6	18.7	42.6	41.3	60.6	29.9
Year 1	25.1	15.3	46.4	31.2	51.7	24.2

    a. Let μ_d denote the population mean difference Year 2 − Year 1. State the appropriate null and alternate hypotheses about μ_d.

    b. Compute the differences Year 2 − Year 1.

    c. Compute the value of the test statistic.

    d. Compute the *P*-value.

    e. Do you reject H_0 at the $\alpha = 0.05$ level?

    f. State a conclusion.

Answers are on page 540.

Objective 2 Perform a hypothesis test with matched pairs using the critical value method

Testing a Hypothesis with Matched-Pair Data Using the Critical Value Method

The critical value method for matched-pair data is essentially the same as that for a population mean with σ unknown. We can find the critical value in Table A.3 or with technology. The assumptions for the critical value method are the same as for the *P*-value method. When the assumptions are satisfied, a hypothesis test can be performed using the following steps.

Performing a Hypothesis Test with Matched-Pair Data Using the Critical Value Method

Check to be sure the assumptions are satisfied. If they are, then proceed with the following steps.

Step 1: State the null and alternate hypotheses. The null hypothesis will have the form $H_0: \mu_d = \mu_0$. The alternate hypothesis will be of the form $\mu_d < \mu_0$, $\mu_d > \mu_0$, or $\mu_d \neq \mu_0$.

Step 2: Choose a significance level α and find the critical value or values.

Step 3: Compute the test statistic $t = \dfrac{\bar{d} - \mu_0}{s_d/\sqrt{n}}$.

Step 4: Determine whether to reject H_0, as follows:

Left-tailed: $H_1: \mu_d < \mu_0$

Right-tailed: $H_1: \mu_d > \mu_0$

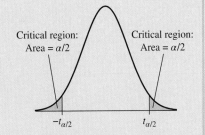

Two-tailed: $H_1: \mu_d \neq \mu_0$

Step 5: State a conclusion.

EXAMPLE 11.6

Testing hypotheses with the critical value method

For a sample of nine automobiles, the mileage (in 1000s of miles) at which the original front brake pads were worn to 10% of their original thickness was measured, as was the mileage at which the original rear brake pads were worn to 10% of their original thickness. The results are given below.

	Automobile								
	1	**2**	**3**	**4**	**5**	**6**	**7**	**8**	**9**
Rear	42.7	36.7	46.1	46.0	39.9	51.7	51.6	46.1	47.3
Front	32.8	26.6	35.6	36.4	29.2	40.9	40.9	34.8	36.6
Difference	9.9	10.1	10.5	9.6	10.7	10.8	10.7	11.3	10.7

The differences in the last line of the table are Rear − Front. Can you conclude that the mean time for the rear brake pads to wear out is longer than the mean time for the front pads? Use the $\alpha = 0.05$ significance level.

Solution

We first check the assumptions. Because the sample size is small, we will construct a dotplot.

The dotplot shows no evidence of outliers or extreme skewness, so we can proceed.

Step 1: State the null and alternate hypotheses. We are interested in determining whether the mean time for the rear pads is longer than for the front. Therefore, the hypotheses are

$$H_0: \mu_d = 0 \qquad H_1: \mu_d > 0$$

Step 2: Choose a significance level α and find the critical value. We will use $\alpha = 0.05$. Because this is a right-tailed test, the critical value is the value for which the area to the right is 0.05. The sample size is $n = 9$, so there are $9 - 1 = 8$ degrees of freedom. The critical value is $t_\alpha = 1.860$.

Step 3: Compute the test statistic. The sample size is $n = 9$. We compute the sample mean and standard deviation of the differences:

$$\bar{d} = 10.478 \qquad s_d = 0.5215$$

The test statistic is

$$t = \frac{\bar{d} - 0}{s_d/\sqrt{n}} = \frac{10.478 - 0}{0.5215/\sqrt{9}} = 60.28$$

Step 4: Determine whether to reject H_0. This is a right-tailed test, so we reject H_0 if $t \geq t_\alpha$. Because $t = 60.28$ and $t_\alpha = 1.860$, we reject H_0 at the $\alpha = 0.05$ level.

Step 5: State a conclusion. We conclude that the mean time for rear brake pads to wear out is longer than the mean time for front brake pads.

USING TECHNOLOGY

We use Example 11.5 to illustrate the technology steps.

TI-84 PLUS

Testing a hypothesis about a difference using matched pairs

Step 1. Enter the data for into **L1** and **L2** in the data editor. On the home screen, enter **(L1 – L2) STO L3** to assign the differences in list **L3** (Figure A).

Step 2. Press **STAT** and highlight the **TESTS** menu.

Step 3. Select **T–Test** and press **ENTER** (Figure B). The **T–Test** menu appears.

Step 4. For **Inpt**, select the **Data** option and enter **L3** as the **List** option.

Step 5. Enter the null hypothesis mean for μ_0 and select the form of the alternate hypothesis. For Example 11.5, we have $\mu_0 = 0$ and the alternate hypothesis has the form $>\mu_0$ (Figure C).

Step 6. Highlight **Calculate** and press **ENTER** (Figure D).

Figure A

Figure B

Figure C

Figure D

MINITAB

Testing a hypothesis about a difference using matched pairs

Step 1. Enter the data from Example 11.5 into **Columns C1** and **C2**.

Step 2. Click on **Stat**, then **Basic Statistics**, then **Paired t**.

Step 3. Select Samples in **Columns**, and enter **C1** in the **First Sample** field and **C2** in the **Second Sample** field.

Step 4. Click **Options**, and enter the null hypothesis difference between the means in the **Test difference** field and select the form of the alternate hypothesis. Given significance level α, enter $100(1 - \alpha)$ as the **Confidence Level**. For Example 11.5, we use **99** as the **Confidence Level**, **0** as the **Test difference**, and **greater than** as the **Alternative**. Click **OK**.

Step 5. Click **OK** (Figure E).

```
Paired T for C1 - C2

             N     Mean    StDev   SE Mean
C1           8   30.0513   4.6093   1.6296
C2           8   28.8450   4.6352   1.6388
Difference   8   1.20625  0.37317  0.13194


99% lower bound for mean difference: 0.81071
T-Test of mean difference = 0 (vs > 0): T-Value = 9.14   P-Value = 0.000
```

Figure E

EXCEL

Testing a hypothesis about a difference using matched pairs

This procedure requires the **MegaStat** EXCEL add-in to be loaded. The **MegaStat** add-in can be downloaded from www.mhhe.com/megastat.

Step 1. Enter the data from Example 11.5 into **Columns A** and **B** in the worksheet.

Step 2. Click on the **MegaStat** menu, select **Hypothesis Tests**, then **Paired Observations...**

Step 3. Select the **data input** option, and enter the range of cells for the first sample in the **Group 1** field and the range of cells for the second sample in the **Group 2** field.

Step 4. Enter **0** in the **Hypothesized difference** field and select **greater than** in the **Alternative field**.

Step 5. Choose the **t-test** option (Figure F).

Step 6. Click **OK** (Figure G).

Figure F

Hypothesis Test: Paired Observations

0.00000	hypothesized value
30.05125	mean After Tune-up
28.84500	mean Before Tune-up
1.20625	mean difference (After Tune-up - Before Tune-up)
0.37317	std. dev.
0.13194	std. error
8	n
7	df
9.14	t
1.93E-05	p-value (one-tailed, upper)

Figure G

SECTION 11.3   Exercises

Exercises 1 and 2 are the Check Your Understanding exercises located within the section.

Understanding the Concepts

In Exercises 3 and 4, fill in each blank with the appropriate word or phrase.

3. If the sample size is small, the differences between the items in the matched pairs must show no evidence of strong _____ and must contain no _____.

4. With matched pairs, the test for the difference between population means is the same as the test for a single population _____.

In Exercises 5 and 6, determine whether the statement is true or false. If the statement is false, rewrite it as a true statement.

5. Paired data are data for which each value in one sample can be matched with a corresponding value in another sample.

6. To compute the test statistic for a test with matched pairs, we must compute the standard deviations of the samples.

Practicing the Skills

7. Following is a sample of five matched pairs.

Sample 1	19	15	16	23	24
Sample 2	18	19	10	14	17

Let μ_1 and μ_2 represent the population means and let $\mu_d = \mu_1 - \mu_2$. A test will be made of the hypotheses H_0: $\mu_d = 0$ versus H_1: $\mu_d > 0$.
a. Compute the differences.
b. Compute the test statistic.
c. Can you reject H_0 at the $\alpha = 0.05$ level of significance?
d. Can you reject H_0 at the $\alpha = 0.01$ level of significance?

8. Following is a sample of ten matched pairs.

Sample 1	28	29	22	25	26	29	27	24	27	28
Sample 2	34	30	31	26	31	30	31	32	29	37

Let μ_1 and μ_2 represent the population means and let $\mu_d = \mu_1 - \mu_2$. A test will be made of the hypotheses H_0: $\mu_d = 0$ versus H_1: $\mu_d \neq 0$.
a. Compute the differences.
b. Compute the test statistic.
c. Can you reject H_0 at the $\alpha = 0.05$ level of significance?
d. Can you reject H_0 at the $\alpha = 0.01$ level of significance?

Working with the Concepts

9. **Crossover trial:** A crossover trial is a type of experiment used to compare two drugs. Subjects take one drug for a

period of time, then switch to the other. The responses of the subjects are then compared using matched-pair methods. In an experiment to compare two pain relievers, seven subjects took one pain reliever for two weeks, then switched to the other. They rated their pain level from 1 to 10, with larger numbers representing higher levels of pain. The results were:

			Subject				
	1	2	3	4	5	6	7
Drug A	6	3	4	5	7	1	4
Drug B	5	1	5	5	5	2	2

Can you conclude that the mean pain level is less with drug B?
a. State the null and alternate hypotheses.
b. Compute the test statistic.
c. State a conclusion. Use the $\alpha = 0.05$ level of significance.

10. **Comparing scales:** In an experiment to determine whether there is a systematic difference between the weights obtained with two different scales, 10 rock specimens were weighed, in grams, on each scale. The following data were obtained:

Specimen	Weight on Scale 1	Weight on Scale 2
1	11.23	11.27
2	14.36	14.41
3	8.33	8.35
4	10.50	10.52
5	23.42	23.41
6	9.15	9.17
7	13.47	13.52
8	6.47	6.46
9	12.40	12.45
10	19.38	19.35

Can you conclude that the mean weight differs between the scales?
a. State the null and alternate hypotheses.
b. Compute the test statistic.
c. State a conclusion. Use the $\alpha = 0.01$ level of significance.

11. **Strength of concrete:** The compressive strength, in kilopascals, was measured for concrete blocks from five different batches of concrete, both three and six days after pouring. The data are as follows:

			Block		
	1	2	3	4	5
After 3 days	1341	1316	1352	1355	1327
After 6 days	1376	1373	1366	1384	1358

Can you conclude that the mean strength after three days differs from the mean strength after six days?
a. State the null and alternate hypotheses.
b. Compute the test statistic.
c. State a conclusion. Use the $\alpha = 0.05$ level of significance.

12. **Truck pollution:** In an experiment to determine the effect of ambient temperature on the emissions of oxides of

nitrogen (NO_x) of diesel trucks, ten trucks were run at temperatures of 40°F and 80°F. The emissions, in parts per billion, are presented in the following table.

Truck	40°F	80°F
1	834.7	815.2
2	753.2	765.2
3	855.7	842.6
4	901.2	797.1
5	785.4	764.3
6	862.9	819.5
7	882.7	783.6
8	740.3	694.5
9	748.0	772.9
10	848.6	794.7

Can you conclude that the mean emissions are higher at 40°?
a. State the null and alternate hypotheses.
b. Compute the test statistic.
c. State a conclusion. Use the $\alpha = 0.05$ level of significance.

13. **Growth spurt:** It is generally known that boys grow at an unusually fast rate between the ages of about 12 and 14. Following are heights, in inches of 40 boys, measured at age 12 and again at age 14.

Age 12	Age 14	Age 12	Age 14
57.9	62.9	55.4	61.8
61.1	65.9	58.7	64.2
62.7	67.5	64.3	69.8
67.5	73.7	58.1	63.3
59.2	64.9	63.3	69.6
61.4	67.0	61.2	66.4
60.7	65.5	64.5	69.2
55.9	62.1	55.9	62.0
59.7	65.4	60.4	65.7
56.3	61.5	57.8	62.8
63.0	68.5	68.3	73.6
58.6	63.9	63.0	67.7
61.1	65.8	64.4	69.2
59.5	64.5	58.2	64.6
61.6	66.3	59.7	66.1
59.3	64.2	60.2	65.9
62.1	67.1	63.7	68.3
62.8	68.1	60.2	65.2
65.3	69.9	62.7	67.9
60.4	66.7	55.6	61.7

Can you conclude that the mean increase in height is greater than 5 inches?
a. State the null and alternate hypotheses.
b. Compute the test statistic.
c. State a conclusion. Use the $\alpha = 0.05$ level of significance.

14. **SAT coaching:** A sample of 32 students took a class designed to improve their SAT math scores. Following are their scores before and after the class.

Before	After	Before	After
383	420	394	430
334	368	513	525
378	396	483	482
467	488	447	482
470	489	440	479
473	473	439	451
443	448	435	431
459	473	451	454
426	428	453	463
493	525	491	511
382	382	526	529
473	474	473	493
408	407	440	466
433	434	481	482
478	490	459	455
502	508	399	404

Can you conclude that the mean increase in score is less than 15 points?
a. State the null and alternate hypotheses.
b. Compute the test statistic.
c. State a conclusion. Use the $\alpha = 0.05$ level of significance.

15. **Interpret calculator display:** The following TI-84 Plus calculator display presents the results of a hypothesis test for the mean difference between matched pairs.

```
        T-Test
μ≠0
t=2.40285205
p=.0296591111
x̄=3.37
Sx=5.61
n=16
```

a. Is this a left-tailed test, a right-tailed test, or a two-tailed test?
b. How many degrees of freedom are there?
c. What is the P-value?
d. Can H_0 be rejected at the 0.05 level? Explain.

16. **Interpret calculator display:** The following TI-84 Plus calculator display presents the results of a hypothesis test for the mean difference between matched pairs.

```
        T-Test
μ>0
t=.9627814908
p=.1745766435
x̄=1.564
Sx=6.892
n=18
```

a. Is this a left-tailed test, a right-tailed test, or a two-tailed test?
b. How many degrees of freedom are there?
c. What is the P-value?
d. Can H_0 be rejected at the 0.05 level? Explain.

17. **Interpret computer output:** The following MINITAB output presents the results of a hypothesis test for a mean difference.

```
Paired T for X - Y

               N      Mean     StDev    SE Mean
X             12   134.233    68.376     19.739
Y             12   100.601    94.583     27.304
Difference    12   33.6316    59.511     17.179


95% lower bound for mean difference: 2.7793
T-Test of mean difference = 0 (vs > 0 ):   T-Value = 1.96
                                           P-Value = 0.038
```

a. Is this a left-tailed test, a right-tailed test, or a two-tailed test?
b. Can H_0 be rejected at the 0.05 level? Explain.
c. Can H_0 be rejected at the 0.01 level? Explain.

18. **Interpret computer output:** The following MINITAB output presents the results of a hypothesis test for a mean difference.

```
Paired T for X - Y

               N      Mean     StDev    SE Mean
X              7   21.4236    10.145     3.8344
Y              7   19.2587    8.4049     3.1767
Difference     7   2.16485    5.3707     2.0299


95% CI for mean difference: (−2.802239, 7.131933)
T-Test of mean difference = 0 (vs not = 0):   T-Value = 1.07
                                              P-Value = 0.327
```

a. Is this a left-tailed test, a right-tailed test, or a two-tailed test?
b. Can H_0 be rejected at the 0.05 level? Explain.
c. Can H_0 be rejected at the 0.01 level? Explain.

Extending the Concepts

19. Refer to Exercise 12. Assume you did not know that the two samples were paired, so you used the methods of Section 11.1 to perform the test.
a. What is the P-value?
b. Explain why the P-value is greater when the methods of Section 11.1 are used.

Answers to Check Your Understanding Exercises for Section 11.3

1. a. $H_0: \mu_d = 0$, $H_1: \mu_d > 0$ b. $8, 5, -3, 1, 6$
   c. $t = 1.731$
   d. P-value is between 0.05 and 0.10 [Tech: 0.0793].
   e. If H_0 is true, the probability of observing a value for the test statistic as extreme as or more extreme than the value actually observed is 0.0793. This is somewhat unusual, so there is some evidence against H_0. However, because $P > 0.05$, we do not reject H_0 at the $\alpha = 0.05$ level.
   f. We conclude that the mean reading score may have remained the same after the reading program.

2. a. $H_0: \mu_d = 0$, $H_1: \mu_d > 0$
   b. $9.5, 3.4, -3.8, 10.1, 8.9, 5.7$ c. $t = 2.612$
   d. P-value is between 0.01 and 0.025 [Tech: 0.0238].
   e. If H_0 is true, the probability of observing a value for the test statistic as extreme as or more extreme than the value actually observed is 0.0238. This is fairly unusual, so there is fairly strong evidence against H_0. In particular, because $P < 0.05$, we reject H_0 at the $\alpha = 0.05$ level.
   f. We conclude that the mean rainfall was greater in year 2 than in year 1.

| SECTION 11.4 | Hypothesis Tests for Two Population Standard Deviations |

Objectives

1. Find critical values of the F distribution
2. Perform a hypothesis test for two population standard deviations

Objective 1 Find critical values of the F distribution

Find Critical Values of the F Distribution

The tests we have studied so far have involved means or proportions. Occasionally it may be desirable to test a null hypothesis that two populations have equal standard deviations. In general, there is no good way to do this. In the special case where both populations are normal, however, a method is available.

CAUTION

The test for two standard deviations is sensitive to the assumption of normality. If the populations differ even slightly from normal, the test should not be used.

Assumptions for Performing a Hypothesis Test for Two Population Standard Deviations

1. We have independent random samples from two populations.
2. Both populations are normal.

In what follows, σ_1 and σ_2 will denote the standard deviations of the two populations, s_1 and s_2 will denote the sample standard deviations, and n_1 and n_2 will denote the sample sizes.

Any of three hypotheses may be tested. They are

$$H_0: \sigma_1 = \sigma_2 \qquad H_1: \sigma_1 > \sigma_2$$
$$H_0: \sigma_1 = \sigma_2 \qquad H_1: \sigma_1 < \sigma_2$$
$$H_0: \sigma_1 = \sigma_2 \qquad H_1: \sigma_1 \neq \sigma_2$$

The test statistic used in this test is one that we have not seen before; it is denoted by the letter F. We compute F using the sample variances s_1^2 and s_2^2. We divide the larger sample variance by the smaller sample variance:

$$F = \frac{\text{Larger of } s_1^2 \text{ and } s_2^2}{\text{Smaller of } s_1^2 \text{ and } s_2^2}$$

When H_0 is true, $\sigma_1 = \sigma_2$. Therefore, the sample standard deviations s_1 and s_2 are, on average, approximately the same size, so the test statistic F is likely to be near 1. If the test statistic is much greater than 1, we will reject H_0. In order to use F as a test statistic, we must know its distribution when H_0 is true. This distribution is called an **F distribution**, which we now describe.

Figure 11.4 The F distribution is skewed to the right.

The F distribution

Statistics that have an F distribution are ratios of quantities, such as the ratio of two sample variances. The F distribution, therefore, has two values for the degrees of freedom: one associated with the numerator, and one associated with the denominator. The degrees of freedom are indicated with subscripts under the letter F. For example, the symbol $F_{3,16}$ denotes the F distribution with 3 degrees of freedom for the numerator and 16 degrees of freedom for the denominator. Note that the degrees of freedom for the numerator are always listed first. Unlike the normal and Student's t distributions, which are symmetric, the F distribution is skewed to the right. Figure 11.4 presents an F curve.

The level α critical value of the F distribution is the value for which the area in the tail to the right of the critical value is α. We will denote the level α critical value by f_α. We will not need to find critical values for left tails.

| EXAMPLE 11.7 | **Find a critical value of an *F* distribution** |

Use Table A.5 to find $f_{0.05}$, the $\alpha = 0.05$ critical value, for the $F_{3,16}$ distribution.

Solution
In Table A.5, we find the column corresponding to 3 degrees of freedom for the numerator, and the rows corresponding to 16 degrees of freedom for the denominator. We choose the row for which the tail area is 0.05. The critical value is $f_{0.05} = 3.24$.

Denominator Degrees of Freedom	Area	\multicolumn{9}{c}{**Numerator Degrees of Freedom**}
		1	**2**	**3**	**4**	**5**	**6**	**7**	**8**	**9**
15	0.100	3.07	2.70	2.49	2.36	2.27	2.21	2.16	2.12	2.09
15	0.050	4.54	3.68	3.29	3.06	2.90	2.79	2.71	2.64	2.59
15	0.025	6.20	4.77	4.15	3.80	3.58	3.41	3.29	3.20	3.12
15	0.010	8.68	6.36	5.42	4.89	4.56	4.32	4.14	4.00	3.89
15	0.001	16.59	11.34	9.34	8.25	7.57	7.09	6.74	6.47	6.26
16	0.100	3.05	2.67	2.46	2.33	2.24	2.18	2.13	2.09	2.06
16	0.050	4.49	3.63	3.24	3.01	2.85	2.74	2.66	2.59	2.54
16	0.025	6.12	4.69	4.08	3.73	3.50	3.34	3.22	3.12	3.05
16	0.010	8.53	6.23	5.29	4.77	4.44	4.20	4.03	3.89	3.78
16	0.001	16.12	10.97	9.01	7.94	7.27	6.80	6.46	6.19	5.98
17	0.100	3.03	2.64	2.44	2.31	2.22	2.15	2.10	2.06	2.03
17	0.050	4.45	3.59	3.20	2.96	2.81	2.70	2.61	2.55	2.49
17	0.025	6.04	4.62	4.01	3.66	3.44	3.28	3.16	3.06	2.98
17	0.010	8.40	6.11	5.18	4.67	4.34	4.10	3.93	3.79	3.68
17	0.001	15.72	10.66	8.73	7.68	7.02	6.56	6.22	5.96	5.75

Figure 11.5 presents the probability density curve of the $F_{3,16}$ distribution with the $\alpha = 0.05$ critical region shaded. The $\alpha = 0.05$ critical value is $f_{0.05} = 3.24$.

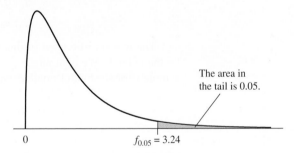

The area in the tail is 0.05.

0 $f_{0.05} = 3.24$

Figure 11.5 Probability density curve of the $F_{3,16}$ distribution. The $\alpha = 0.05$ critical value is $f_{0.05} = 3.24$.

Check Your Understanding

1. Find $f_{0.05}$ for $F_{8,18}$.

2. Find $f_{0.01}$ for $F_{12,8}$.

Answers are on page 547.

Objective 2 Perform a hypothesis test for two population standard deviations

Perform a Hypothesis Test for Two Population Standard Deviations

We will first describe how to perform a hypothesis test using the critical value method and Table A.5. Then we will describe how to use the *P*-value method with technology.

Because we compute the test statistic by dividing the larger sample variance by the smaller sample variance, the value of the test statistic is never less than 1. Therefore, the test statistic cannot fall into the left tail of the distribution. The critical region for a one-tailed test (either right- or left-tailed) is $F \geq f_\alpha$. The critical region for a two-tailed test differs from that of other tests. For most other two-tailed tests, the critical region consists of an area of $\alpha/2$ in both left and right tails. For the F-test, however, the test statistic never falls into the left tail, so the critical region for a two-tailed test consists of only the right tail. We reject H_0 when $F > f_{\alpha/2}$.

We can perform a hypothesis test for two population standard deviations using the following steps.

Performing a Hypothesis Test for Two Standard Deviations

Check to be sure the assumptions are satisfied. If they are, then proceed with the following steps.

Step 1: State the null and alternate hypotheses. The null hypothesis is H_0: $\sigma_1 = \sigma_2$. The alternate hypothesis can be stated in one of three ways:

Left-tailed: H_1: $\sigma_1 < \sigma_2$
Right-tailed: H_1: $\sigma_1 > \sigma_2$
Two-tailed: H_1: $\sigma_1 \neq \sigma_2$

Step 2: Choose a significance level α and find the critical value, as follows:

One-tailed: H_1: $\sigma_1 < \sigma_2$ or H_1: $\sigma_1 > \sigma_2$ The critical value is f_α.
Two-tailed: H_1: $\sigma_1 \neq \sigma_2$ The critical value is $f_{\alpha/2}$.

Step 3: Compute the test statistic $F = \dfrac{\text{Larger of } s_1^2 \text{ and } s_2^2}{\text{Smaller of } s_1^2 \text{ and } s_2^2}$.

Step 4: Determine whether to reject H_0, as follows:

One-tailed: H_1: $\sigma_1 < \sigma_2$ or H_1: $\sigma_1 > \sigma_2$ Reject if $F \geq f_\alpha$.
Two-tailed: H_1: $\sigma_1 \neq \sigma_2$ Reject if $F \geq f_{\alpha/2}$.

Step 5: State a conclusion.

EXAMPLE 11.8   Perform a hypothesis test

In a series of experiments to determine the absorption rate of certain pesticides into skin, two pesticides were applied to several skin specimens. After a time, the amounts absorbed (in micrograms) were measured. For pesticide 1, the variance of the amounts absorbed in 6 specimens was 2.3, while for pesticide 2, the variance of the amounts absorbed in 10 specimens was 0.6. Assume that for each pesticide, the amounts absorbed are a simple random sample from a normal population. Can we conclude that the standard deviation of the amount absorbed is greater for pesticide 1 than for pesticide 2? Use the $\alpha = 0.05$ significance level.

Solution

We first check the assumptions. We have two independent samples from normal populations, so the assumptions are satisfied.

Step 1: State the null and alternate hypotheses. Because we are interested in determining whether $\sigma_1 > \sigma_2$, the hypotheses are

$$H_0\text{: } \sigma_1 = \sigma_2 \qquad H_1\text{: } \sigma_1 > \sigma_2$$

Step 2: Choose a significance level and find the critical value. We will use a significance level of $\alpha = 0.05$. To find the critical value, we first find the degrees of freedom. The value 2.3 in the numerator of the test statistic came from a sample of size 6. Therefore, there are $6 - 1 = 5$ degrees of freedom for the numerator. The value 0.6 in the denominator of the test statistic came from a sample of size 10. Therefore, there are $10 - 1 = 9$ degrees of freedom for the denominator. This is a one-tailed test, so the critical value is $f_{0.05}$. We consult Table A.5 with 5 and 9 degrees of freedom and a tail area of 0.05. The critical value is $f_{0.05} = 3.48$.

Denominator Degrees of Freedom	Area	Numerator Degrees of Freedom								
		1	2	3	4	5	6	7	8	9
8	0.100	3.46	3.11	2.92	2.81	2.73	2.67	2.62	2.59	2.56
8	0.050	5.32	4.46	4.07	3.84	3.69	3.58	3.50	3.44	3.39
8	0.025	7.57	6.06	5.42	5.05	4.82	4.65	4.53	4.43	4.36
8	0.010	11.26	8.65	7.59	7.01	6.63	6.37	6.18	6.03	5.91
8	0.001	25.41	18.49	15.83	14.39	13.48	12.86	12.40	12.05	11.77
9	0.100	3.36	3.01	2.81	2.69	2.61	2.55	2.51	2.47	2.44
9	0.050	5.12	4.26	3.86	3.63	3.48	3.37	3.29	3.23	3.18
9	0.025	7.21	5.71	5.08	4.72	4.48	4.32	4.20	4.10	4.03
9	0.010	10.56	8.02	6.99	6.42	6.06	5.80	5.61	5.47	5.35
9	0.001	22.86	16.39	13.90	12.56	11.71	11.13	10.70	10.37	10.11
10	0.100	3.29	2.92	2.73	2.61	2.52	2.46	2.41	2.38	2.35
10	0.050	4.96	4.10	3.71	3.48	3.33	3.22	3.14	3.07	3.02
10	0.025	6.94	5.46	4.83	4.47	4.24	4.07	3.95	3.85	3.78
10	0.010	10.04	7.56	6.55	5.99	5.64	5.39	5.20	5.06	4.94
10	0.001	21.04	14.91	12.55	11.28	10.48	9.93	9.52	9.20	8.96

Step 3: Compute the test statistic. The larger sample variance is 2.3 and the smaller one is 0.6. The test statistic is

$$F = \frac{2.3}{0.6} = 3.83$$

Step 4: Determine whether to reject H_0. The value of the test statistic is $F = 3.83$. The critical value is $f_{0.05} = 3.48$. Because $F > f_{0.05}$, we reject H_0 at the $\alpha = 0.05$ level.

Step 5: State a conclusion. We conclude that the standard deviation of the amount absorbed is greater for pesticide 1 than for pesticide 2.

Check Your Understanding

3. For testing H_0: $\sigma_1 = \sigma_2$ versus H_1: $\sigma_1 < \sigma_2$, we observe $s_1 = 0.9$, $s_2 = 1.5$, $n_1 = 16$, and $n_2 = 11$. Test at the significance level $\alpha = 0.01$.
  a. Find the critical value.
  b. Do you reject H_0?

4. For testing H_0: $\sigma_1 = \sigma_2$ versus H_1: $\sigma_1 \neq \sigma_2$, we observe $s_1 = 9.3$, $s_2 = 2.1$, $n_1 = 6$, and $n_2 = 13$. Test at the significance level $\alpha = 0.05$.
  a. Find the critical value.
  b. Do you reject H_0?

Answers are on page 547.

Performing hypothesis tests using the *P*-value method

We show how to perform tests for two population standard deviations using the *P*-value method. We calculate the *P*-value with technology.

The following TI-84 Plus display presents the results for the test described in Example 11.8.

The first line specifies the alternate hypothesis. Then follows the test statistic F, the P-value, and the sample standard deviations ($1.5166 = \sqrt{2.3}$ and $0.7746 = \sqrt{0.6}$). The P-value is 0.039, rounded to three decimal places. We reject H_0 at the $\alpha = 0.05$ level.

The *F*-test is sensitive to the normality assumption

The F-test, like the t-test, requires that the samples come from normal populations. Unlike the t-test, the F-test for comparing variances is very sensitive to this assumption. If the shapes of the populations differ much from the normal curve, the F-test may give misleading results. For this reason, the F-test for comparing variances must be used with caution.

USING TECHNOLOGY

We use Example 11.8 to illustrate the technology steps. The TI-84 Plus calculator and EXCEL require as input the standard deviations. In Example 11.8, the variances $s_1^2 = 2.3$ and $s_2^2 = 0.6$ are given. We find the standard deviations by taking the square root: $s_1 = \sqrt{2.3} = 1.5166$; $s_2 = \sqrt{0.6} = 0.7746$.

TI-84 PLUS

Testing a hypothesis about two population standard deviations

Step 1. Press **STAT** and highlight the **TESTS** menu.

Step 2. Select **2–SampFTest** and press **ENTER** (Figure A). The **2–SampFTest** menu appears.

Step 3. Choose one of the following:
- If the summary statistics are given, select **Stats** as the **Inpt** option and enter s_1, n_1, s_2, n_2. For Example 11.8, we use $s_1 = 1.5166$, $n_1 = 6$, $s_2 = 0.7746$, $n_2 = 10$.
- If the raw data are given, select **Data** as the **Inpt** option and enter the location of the data as the **List1** and **List2** options.

Step 4. Select the form of the alternate hypothesis. For Example 11.8, the alternate hypothesis has the form $>\sigma 2$ (Figure B).

Step 5. Highlight **Calculate** and press **ENTER** (Figure C).

Figure A

Figure B

Figure C

MINITAB

Testing a hypothesis about two population standard deviations

Step 1. Click on **Stat**, then **Basic Statistics**, then **2 Variances**.

Step 2. Choose one of the following:
- If the summary statistics are given, click **Summarized Data**, and enter the **Sample Size** and the **Variance** for each sample. For Example 11.8, we use $s_1^2 = 2.3$, $n_1 = 6$, $s_2^2 = 0.6$, $n_2 = 10$.
- If the raw data are given, click **Samples in different columns** and select the columns that contain the data.

Step 3. Click **OK** (Figure D).

Note: MINITAB presents the P-value for a two-tailed test by default. For a one-tailed test, such as in Example 11.8, divide this value by 2. For Example 11.8, the P-value is $0.078/2 = 0.039$.

```
F-Test (normal distribution)
Test statistic = 3.83, p-value = 0.078
```

Figure D

EXCEL

Testing a hypothesis about two population standard deviations

This procedure requires the **MegaStat** EXCEL add-in to be loaded. The **MegaStat** add-in can be downloaded from www.mhhe.com/megastat.

Step 1. Load the **MegaStat** EXCEL add-in.

Step 2. Click on the **MegaStat** menu and select **Hypothesis Tests**, then **Compare Two Independent Groups...**

Step 3. Choose one of the following:
- If the summary statistics are given, choose **summary input** and enter, for each sample, the range of the cells that contains, in the following order, the **variable name**, $\bar{x}$, s, and n. If $\bar{x}$ is not known, an arbitrary value can be used. Figure E illustrates the values of the summary statistics for Example 11.8 entered in columns **B** and **C**. Note that **0** is used for the mean.
- If the raw data are given, choose **data input** and select the range of cells, for each sample, that contains the data in the **Input Range** field.

Step 4. Choose the **Test for equality of variances** option (Figure F).

Step 5. Click **OK** (Figure G).

	A	B	C
1		**Pesticide 1**	**Pesticide 2**
2	**Mean**	0	0
3	**Standard Deviation**	1.5166	0.7746
4	**Sample Size**	6	10

Figure E

Figure F

F-test for equality of variance
2.3000756 variance: Pesticide 1
0.6000052 variance: Pesticide 2
3.83 F
.0775 p-value

Figure G

SECTION 11.4 Exercises

Exercises 1–4 are the Check Your Understanding exercises located within the section.

Understanding the Concepts

In Exercises 5 and 6, fill in each blank with the appropriate word or phrase.

5. To use an F-test, both populations must have a _____ distribution.

6. The F statistic is the quotient of the sample _____.

In Exercises 7 and 8, determine whether the statement is true or false. If the statement is false, rewrite it as a true statement.

7. The test statistic F is never greater than 1.

8. The F-test for two standard deviations is very sensitive to the normality assumption.

Practicing the Skills

9. Find the critical value $f_{0.05}$ for $F_{7,20}$.

10. Find the critical value $f_{0.01}$ for $F_{2,5}$.

11. An F-test with 12 degrees of freedom in the numerator and 6 degrees of freedom in the denominator produced a test statistic whose value was 3.42. The null and alternate hypotheses were H_0: $\sigma_1 = \sigma_2$ versus H_1: $\sigma_1 < \sigma_2$.
  a. Do you reject H_0 at the $\alpha = 0.05$ level?
  b. Do you reject H_0 at the $\alpha = 0.01$ level?

12. An F-test with 5 degrees of freedom in the numerator and 7 degrees of freedom in the denominator produced a test statistic whose value was 5.31. The null and alternate hypotheses were H_0: $\sigma_1 = \sigma_2$ versus H_1: $\sigma_1 > \sigma_2$.
  a. Do you reject H_0 at the $\alpha = 0.05$ level?
  b. Do you reject H_0 at the $\alpha = 0.01$ level?

Working with the Concepts

13. Sugar content: A broth used to manufacture a pharmaceutical product has its sugar content, in milligrams per milliliter, measured several times on each of three successive days.

Day 1:   5.0 4.8 5.1 5.1 4.8 5.1 4.8 4.8 5.0
         5.2 4.9 4.9 5.0
Day 2:   5.8 4.7 4.7 4.9 5.1 4.9 5.4 5.3 5.3
         4.8 5.7 5.1 5.7
Day 3:   6.3 4.7 5.1 5.9 5.1 5.9 4.7 6.0 5.3
         4.9 5.7 5.3 5.6

a. Can you conclude that the variability of the process is greater on the second day than on the first day? Use the $\alpha = 0.05$ level of significance.
b. Can you conclude that the variability of the process is greater on the third day than on the second day? Use the $\alpha = 0.01$ level of significance.

14. Hockey sticks: The breaking strength of hockey stick shafts made of two different graphite-kevlar composites yields the following results (in newtons).

Composite 1:   487.3 444.5 467.7 456.3 449.7
               459.2 478.9 461.5 477.2
Composite 2:   488.5 501.2 475.3 467.2 462.5
               499.7 470.0 469.5 481.5 485.2
               509.3 479.3 498.3

Can you conclude that the standard deviation of the breaking strength differs between the two composites? Use the $\alpha = 0.05$ level of significance.

15. Frozen computer: A computer system administrator notices that computers running a particular operating system seem to freeze up more often as the installation of the operating system ages. She measures the time (in minutes) before freeze-up for 7 computers one month after installation, and for 9 computers seven months after installation. The results are as follows.

One month: 207.4 233.1 215.9 235.1 225.6
                244.4 245.3
Seven months: 84.3 53.2 127.3 201.3 174.2
                246.2 149.4 156.4 103.3

Can you conclude that the time to freeze-up is more variable in the seventh month than in the first month after installation? Use the $\alpha = 0.01$ level of significance.

16. Are you smarter than your older brother? In a study of birth order and intelligence, IQ tests were given to 18- and 19-year-old men to estimate the size of the difference, if any, between the mean IQs of firstborn sons and secondborn sons. The following data for 10 firstborn sons and 10 secondborn sons are consistent with the means and standard deviations reported in the article. Assume that the samples come from normal populations.

Firstborn				
104	82	102	96	129
89	114	107	89	103

Secondborn				
103	103	91	113	102
103	92	90	114	113

Can you conclude that the standard deviation of IQ differs between firstborn and secondborn sons? Use the $\alpha = 0.05$ level.

Source: *Science* 316:1717

Extending the Concepts

17. Left tail: It is desired to find the value $f_{0.99}$ for an $F_{10,5}$ distribution. Use Table A.5 and the $F_{5,10}$ distribution to find this value.

Answers to Check Your Understanding Exercises for Section 11.4

1. 2.51
2. 5.67
3. a. 3.80   **b.** No
4. a. 3.89   **b.** Yes

SECTION 11.5   The Multiple Testing Problem

Objectives

1. Describe the multiple testing problem
2. Use the Bonferroni method to adjust the *P*-value

Objective 1 Describe the multiple testing problem

The Multiple Testing Problem

Rates of diseases such as cancer are monitored by the Centers for Disease Control and Prevention (CDC). According to a recent CDC study, the nationwide rate of new cancer cases in a recent year was 487 per 100,000 people. Many scientists believe that some of these cancers are caused by exposure to substances in the environment. How can we determine which environmental substances increase the risk of cancer? One way is to measure cancer rates in various local regions. Some regions turn out to have cancer rates that are significantly larger than the nationwide rate. These regions are sometimes referred to as cancer "hot spots." When a hot spot is discovered, scientists can investigate the environmental conditions there to try to find possible cancer-causing agents.

A natural way to discover hot spots is through hypothesis testing. A region whose cancer rate is statistically significant and higher than the national average is a candidate for a hot spot. It would seem, then, that a good way to proceed would be to perform hypothesis tests on a large number of regions, then look for cancer-causing agents in those whose rates are significantly higher than the nationwide rate. Unfortunately, this approach doesn't work very well, because of a phenomenon known as the **multiple testing problem**.

The multiple testing problem is this: *As more hypothesis tests are performed, small P-values become less meaningful.* We illustrate this point with the cancer hot spot example. Suppose that 20 cities are chosen, and the cancer rates are measured in each. Because we are interested in determining the cities in which the rate per 100,000 people is above the national average of 487, we perform, in each city, a test of the hypotheses

$$H_0: \mu = 487 \qquad H_1: \mu > 487$$

We will use a significance level of 0.05, so we will reject H_0 whenever $P \leq 0.05$.

Now suppose that for 19 of the 20 cities, the P-value is greater than 0.05, so H_0 is not rejected. For one city, $P \leq 0.05$, so H_0 is rejected. It might seem reasonable to conclude that this city really does have a cancer rate that is above the national average. However, this conclusion is not justified. Examples 11.9 and 11.10 show why.

EXAMPLE 11.9   Find the probability of a false rejection

If only one city were tested, and the true mean for this city was equal to the national average of 487, what is the probability that H_0 would be rejected, leading to a wrong conclusion?

Solution

If the true mean is 487, H_0 is true. Rejecting H_0 is then a Type I error. The question is therefore asking for the probability of a Type I error. This probability is equal to the significance level of the test, which in this case is 0.05. The probability is 0.05 that H_0 will be rejected.

EXAMPLE 11.10   Find the probability of at least one false rejection with multiple tests

If 20 cities are tested, and H_0 is rejected for one of the 20 cities, is it plausible that the mean rate for this city is actually equal to 487?

Solution

Yes. It is quite plausible that all of the cities, including the one for which H_0 was rejected, have mean rates equal to 487. There were 20 hypothesis tests made. For each test, there was a probability of 0.05 (that is, 1 chance in 20) of a Type I error. Therefore, we expect, on the average, that out of every 20 true null hypotheses, one will be rejected. So rejecting H_0 in one out of the 20 tests is exactly what one would expect in the case that all of the cities had a mean equal to 487.

Examples 11.9 and 11.10 illustrate the multiple testing problem. Put simply, the multiple testing problem is this: When H_0 is rejected, we have strong evidence that it is false. But strong evidence is not certainty. Occasionally a true null hypothesis will be rejected. When many tests are performed, it is more likely that some true null hypotheses will be rejected, resulting in Type I errors. Thus, when many tests are performed, it is difficult to tell which of the rejected null hypotheses are really false and which correspond to Type I errors.

Objective 2 Use the Bonferroni method to adjust the *P*-value

The Bonferroni Method

The **Bonferroni method** provides a way to adjust *P*-values upward when several hypothesis tests are performed. If a *P*-value remains small after the adjustment, the null hypothesis may be rejected. To make the Bonferroni adjustment, simply multiply the *P*-value by the number of tests performed. Here are two examples.

<table>
<tr><td>EXAMPLE 11.11</td><td></td></tr>
</table>

Use the Bonferroni method to adjust the P-value

Four different cities are tested to determine whether their cancer rates are above 487 per 100,000 people. The null hypothesis $H_0: \mu = 487$ versus $H_1: \mu > 487$ is tested for each city, and the results are

$$
\begin{array}{ll}
\text{City A:} & P = 0.37 \\
\text{City B:} & P = 0.41 \\
\text{City C:} & P = 0.005 \\
\text{City D:} & P = 0.21
\end{array}
$$

Scientists suspect that city C may be a hot spot, but they know that the P-value of 0.005 is unreliable, because several tests have been performed. Use the Bonferroni adjustment to produce a reliable P-value.

Solution

Four tests were performed, so the Bonferroni adjustment yields $P = (4)(0.005) = 0.02$ for city C. So the evidence is reasonably strong that city C is in fact a hot spot.

<table>
<tr><td>EXAMPLE 11.12</td><td></td></tr>
</table>

Interpret an adjusted P-value

In Example 11.11, assume the P-value for city C had been 0.03 instead of 0.005. What conclusion would you reach in that case?

Solution

The Bonferroni adjustment would yield $P = (4)(0.03) = 0.12$. This is probably not strong enough evidence to conclude that city C is in fact a hot spot. Because the original P-value was small, however, it is likely that one would still be suspicious about city C.

The Bonferroni adjustment is conservative; in other words, the P-value it produces is never smaller than the true P-value. So when the Bonferroni-adjusted P-value is small, the null hypothesis can be rejected conclusively. Unfortunately, as Example 11.12 shows, there are many occasions in which the original P-value is small enough to arouse a strong suspicion that a null hypothesis may be false, but the Bonferroni adjustment does not allow the hypothesis to be rejected.

When the Bonferroni-adjusted P-value is too large to reject a null hypothesis, yet the original P-value is small enough to lead one to suspect that the hypothesis is in fact false, often the best thing to do is to retest the hypothesis that appears to be false, using data from a new experiment. For a suspected cancer hot spot, an appropriate procedure would be to collect new data from that city at a future time, and test the hypothesis again. If the P-value is again small, this time without multiple tests, this provides real evidence against the null hypothesis.

Check Your Understanding

1. Five tests are performed, and the P-values are 0.02, 0.11, 0.23, 0.38, and 0.45. Use the Bonferroni adjustment to adjust the P-value of 0.02.

2. Six tests are performed, and the smallest P-value is 0.03. Which is the best conclusion?
   i. Reject H_0 because $P < 0.05$.
   ii. Do not reject H_0 because the Bonferroni adjustment yields $P = 6(0.03) = 0.18$, which is greater than 0.05.
   iii. Do not reject H_0, but collect new data and repeat the test.

Answers are on page 550.

SECTION 11.5 Exercises

Exercises 1 and 2 are the Check Your Understanding exercises located within the section.

Understanding the Concepts

In Exercises 3 and 4, fill in each blank with the appropriate word or phrase.

3. The multiple testing problem states that as more hypothesis tests are performed, small P-values become _____ meaningful.

4. The Bonferroni adjustment is made by multiplying the P-value by the number of _____.

In Exercises 5 and 6, determine whether the statement is true or false. If the statement is false, rewrite it as a true statement.

5. The Bonferroni-adjusted P-value is always greater than the uncorrected P-value.

6. When testing several hypotheses, it is reasonable to reject every hypothesis for which $P < 0.05$.

Practicing the Skills

7. Five null hypotheses were tested, and the P-values were:

Hypothesis	1	2	3	4	5
P-value	0.35	0.019	0.012	0.008	0.042

   a. Which hypotheses, if any, can be rejected at the $\alpha = 0.05$ level?
   b. Which hypotheses, if any, can be rejected at the $\alpha = 0.01$ level?

8. Six null hypotheses were tested, and the P-values were:

Hypothesis	1	2	3	4	5	6
P-value	0.003	0.012	0.002	0.001	0.024	0.032

   a. Which hypotheses, if any, can be rejected at the $\alpha = 0.05$ level?
   b. Which hypotheses, if any, can be rejected at the $\alpha = 0.01$ level?

Working with the Concepts

9. **Gender differences:** A sociologist surveys a sample of college students to determine whether there are differences in the attitudes and behaviors of male and female students. The survey contains 20 questions. For one question, which asks how much time students spend studying each week, the difference between males and females is statistically significant with a P-value of 0.01. On all the other questions, the differences are not statistically significant. The sociologist concludes at the $\alpha = 0.05$ level that the time spent studying differs between males and females. Explain why this conclusion is not justified.

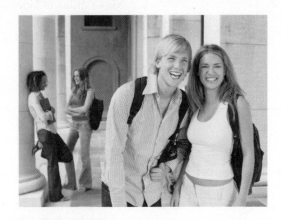

10. **Defective parts:** Six different settings are tried on a machine to determine whether any of them will reduce the proportion of defective parts. For each setting, an appropriate null hypothesis is tested to determine whether the proportion of defective parts has been reduced. The six P-values are 0.34, 0.27, 0.002, 0.45, 0.03, and 0.19. A quality engineer concludes at the $\alpha = 0.05$ level that the method whose P-value is 0.002 reduces the proportion of defective parts. Explain why this conclusion is justified.

11. **More gender differences:** Refer to Exercise 9. For the result that was statistically significant, the P-value was 0.01. What P-value would be needed to conclude at the $\alpha = 0.05$ level that the time spent studying differs between males and females after applying the Bonferroni correction?

12. **More defective parts:** Refer to Exercise 10. The quality engineer suspects that the setting with a P-value of 0.03 may actually reduce the proportion of defective parts. What P-value would be needed to make this conclusion at the $\alpha = 0.05$ level after applying the Bonferroni correction?

Extending the Concepts

13. **Many multiple tests:** Five hundred null hypotheses are tested at the $\alpha = 0.05$ level, and 40 of them are rejected. Assume the tests are independent,
   a. Assuming that all 500 null hypotheses are true, what is the probability that a given hypothesis is rejected?
   b. Let X be the number of hypotheses that are rejected, so that the observed value of X is 40. Under the assumption that all 500 null hypotheses are true, what is the distribution of X?
   c. Under the assumption that all 500 null hypotheses are true, what is the probability of observing a value of X as extreme as or more extreme than the value actually observed?
   d. Can you conclude that some of the null hypotheses are false? Explain.

Answers to Check Your Understanding Exercises for Section 11.5

1. 0.10   **2.** iii

Chapter 11 Summary

Section 11.1: We can test hypotheses about the difference between two population means. The test statistic has a Student's t distribution. The number of degrees of freedom can be taken to be 1 less than the smaller sample size. A more complicated formula, used by technology, provides a greater number of degrees of freedom. The assumptions that are necessary for one-sample tests must hold for both populations.

Section 11.2: We can test hypotheses about the difference between two proportions. The test statistic has a standard normal distribution. The assumptions that are necessary for tests involving a single proportion must hold for both populations.

Section 11.3: When the data consist of matched pairs, we can test hypotheses about the difference between the population means by computing the difference between the values in each pair, then following the procedure for hypotheses about a single population mean. The assumptions required for a test of a population mean must hold for the population of differences.

Section 11.4: We can use the F distribution to test whether two population standard deviations are equal. The validity of this test is sensitive to the assumption that both populations are normal. Therefore, this test should not be used unless it is certain that both populations are very close to normally distributed.

Section 11.5: When several tests are made, we have several chances to make an error. The Bonferroni adjustment can be used to correct P-values for the fact that several tests have been made. To make the adjustment, we multiply each P-value by the number of tests that are made.

Vocabulary and Notation

Bonferroni method   548
F distribution   541

independent samples   510
multiple testing problem   548

pooled proportion   522
pooled standard deviation   516

Important Formulas

Test statistic for the difference between two means, independent samples:

$$t = \frac{(\bar{x}_1 - \bar{x}_2) - (\mu_1 - \mu_2)}{\sqrt{\dfrac{s_1^2}{n_1} + \dfrac{s_2^2}{n_2}}}$$

Test statistic for the difference between two proportions:

$$z = \frac{\hat{p}_1 - \hat{p}_2}{\sqrt{\hat{p}(1 - \hat{p})\left(\dfrac{1}{n_1} + \dfrac{1}{n_2}\right)}}$$

where $\hat{p}$ is the pooled proportion $\hat{p} = \dfrac{x_1 + x_2}{n_1 + n_2}$

Test statistic for the difference between two means, matched pairs:

$$t = \frac{\bar{d} - \mu_0}{s_d / \sqrt{n}}$$

Test statistic for two standard deviations:

$$F = \frac{\text{Larger of } s_1^2 \text{ and } s_2^2}{\text{Smaller of } s_1^2 \text{ and } s_2^2}$$

Chapter Quiz

1. A simple random sample of 75 people are given a new drug that is designed to relieve pain. A second sample of 50 people are given a standard drug. The question of interest is whether the proportion of people experiencing relief is greater among those taking the new drug. To address this question, which of the following is the most appropriate type of hypothesis test?
   i. A test for the difference between two population means using independent samples
   ii. A test for the difference between two population proportions
   iii. A test for the difference between two population means using matched pairs
   iv. A test for the difference between two population standard deviations

2. Two machines are used to fill cans. The machines are supposed to fill each can with 12 ounces of liquid. The amounts actually filled follow a normal distribution. A simple random sample of 10 cans is filled by each machine. The question of interest is whether the variability in the fill volume differs between the two machines. To address this question, which of the following is the most appropriate type of hypothesis test?
   i. A test for the difference between two population means using independent samples
   ii. A test for the difference between two population proportions
   iii. A test for the difference between two population means using matched pairs
   iv. A test for the difference between two population standard deviations

3. A fleet of 100 taxis is divided into two groups of 50 cars each to determine whether premium gasoline reduces maintenance costs. Premium unleaded fuel is used in group A, while regular unleaded fuel is used in group B. The total maintenance cost for each vehicle during a one-year period is recorded. The question of interest is whether the mean maintenance cost is less for vehicles using premium fuel. To address this question, which of the following is the most appropriate type of hypothesis test?
   i. A test for the difference between two population means using independent samples
   ii. A test for the difference between two population proportions
   iii. A test for the difference between two population means using matched pairs
   iv. A test for the difference between two population standard deviations

4. A simple random sample of 75 people are given a new drug that is designed to relieve pain. After taking this drug for a month, they switch to a standard drug. The question of interest is whether the proportion of people who experienced relief is greater when taking the new drug. To address this question, which of the following is the most appropriate type of hypothesis test?
   i. A test for the difference between two population means using independent samples
   ii. A test for the difference between two population proportions
   iii. A test for the difference between two population means using matched pairs
   iv. A test for the difference between two population standard deviations

5. In a test of H_0: $p_1 = p_2$ versus H_1: $p_1 \neq p_2$, the value of the test statistic is $z = -1.21$. What do you conclude about the difference $p_1 - p_2$ at the $\alpha = 0.05$ level of significance?

6. A sample of size 15 is drawn from a normal population. The sample standard deviation is $s_1 = 5.2$. A sample of size 10 is drawn from another normal population. The sample standard deviation is $s_2 = 9.3$. Can you conclude that the variances of these two populations are different? Use the $\alpha = 0.05$ level of significance.

7. For a test of H_0: $\mu_1 = \mu_2$ versus H_1: $\mu_1 \neq \mu_2$, the sample sizes were $n_1 = 15$ and $n_2 = 25$. How many degrees of freedom are there for the test statistic? Use the simple method.

8. In a set of 12 matched pairs, the mean difference was $\bar{d} = 18$ and the standard deviation of the differences was $s_d = 4$. Find the value of the test statistic for testing H_0: $\mu_d = 15$ versus H_1: $\mu_d > 15$. Can you reject H_0 at the $\alpha = 0.05$ level?

9. Two suppliers of machine parts delivered large shipments. A simple random sample of 150 parts was chosen from each shipment. For supplier A, 12 of the 150 parts were defective. For supplier B, 28 of the 150 parts were defective. The question of interest is whether the proportion of defective parts is greater for supplier B than for supplier A. Let p_1 be the population proportion of defective parts for supplier A, and let p_2 be the population proportion of defective parts for supplier B. State appropriate null and alternate hypotheses.

10. Refer to Exercise 9. Compute the value of the test statistic.

11. Refer to Exercise 9. Can you reject H_0 at the $\alpha = 0.01$ level? State a conclusion.

12. A simple random sample of 17 business majors from a certain university had a mean GPA of 2.81 with a standard deviation of 0.27. A simple random sample of 23 psychology majors was selected from the same university, and their mean GPA was 2.97 with a standard deviation of 0.23. Boxplots show that it is reasonable to assume that the populations are approximately normal. The question of interest is whether the mean GPAs differ between business majors and psychology majors. Let μ_1 be the population mean GPA for business majors, and let μ_2 be the population mean GPA for psychology majors. State the null and alternate hypotheses.

13. Refer to Exercise 12. Compute the value of the test statistic.

14. Refer to Exercise 12. Can you reject H_0 at the $\alpha = 0.05$ level? State a conclusion.

15. Five null hypotheses were tested, and the P-values were 0.24, 0.17, 0.03, 0.002, and 0.02. How many of the hypotheses are rejected at the $\alpha = 0.05$ level if the Bonferroni adjustment is made?

Review Exercises

1. **Sick days:** A large company is considering a policy of flextime, in which employees can choose their own work schedules within broad limits. The company is interested to determine whether this policy would reduce the number of sick days taken. They chose two simple random samples of 100 employees each. The employees in one sample were allowed to choose their own schedules. The other sample was a control group. Employees in that sample were required to come to work according to a schedule set by management. At the end of one year, the employees in the flextime group had a sample mean of 4.7 days missed, with a sample standard deviation of 3.1 days. The employees in the control group had a sample mean of 5.9 days missed, with a sample standard deviation of 3.9 days. Perform a hypothesis test to measure the strength of the evidence that the mean number of days missed is less in the flextime group. State the null and alternate hypotheses, find the P-value, and state your conclusion. Use the $\alpha = 0.01$ level of significance.

2. **Political polling:** In a certain state, a referendum is being held to determine whether the transportation authority should issue additional highway bonds. A sample of 500 voters is taken in county A, and 285 say that they favor the bond proposal. A sample of 600 voters is taken in county B, and 305 say that they favor the bond issue. Perform a hypothesis test to measure the strength of the evidence that the proportion of voters who favor the proposal is greater in county A than in county B. State the null and alternate hypotheses, find the P-value, and state your conclusion. Use the $\alpha = 0.05$ level of significance.

3. **Contaminated water:** The concentration of benzene was measured in units of milligrams per liter for a simple random sample of five specimens of untreated wastewater produced at a gas field. The sample mean was 7.8 with a sample standard deviation of 1.4. Seven specimens of treated wastewater had an average benzene concentration of 3.2 with a standard deviation of 1.7. Assume that both samples come from normal populations. Can you conclude that the standard deviation of benzene concentration differs between treated water and untreated water? Use the $\alpha = 0.05$ level of significance.

4. **Sales commissions:** A company studied two programs for compensating its sales staff. Nine salespeople participated in the study. In program A, salespeople were paid a higher salary, plus a small commission for each item they sold. In program B, they were paid a lower salary with a larger commission. Following are the amounts sold, in thousands of dollars, for each salesperson on each program.

Program	Salesperson								
	1	2	3	4	5	6	7	8	9
A	55	22	34	22	25	61	55	36	68
B	53	24	36	28	31	61	58	38	72

Can you conclude that the mean sales differ between the two programs? Use the $\alpha = 0.05$ level of significance.

5. **Exercise:** Medical researchers conducted a study to determine whether treadmill exercise could improve the walking ability of patients suffering from claudication, which is pain caused by insufficient blood flow to the muscles of the legs. A sample of 8 patients walked on a treadmill for six minutes every day. After six months, the mean distance walked in six minutes was 348 meters, with a standard deviation of 80 meters. For a control group of 7 patients who did not walk on a treadmill, the mean distance was 305 meters with a standard deviation of 93 meters. Can you conclude that the standard deviations of the distances walked differ between treatment and control? Use the $\alpha = 0.05$ level of significance.
Based on data in the *Journal of the American Medical Association* 301:165–174

6. **Watching television:** The General Social Survey reported in 2012 that in a sample of 68 men aged 18–25, the mean number of hours of television watched per day was 2.76 with a standard deviation of 2.21. In a sample of 72 women aged 18–25, the mean number of hours of television watched per day was 2.88 with a standard deviation of 2.43. Can you conclude that the mean number of hours of television watched differs between men and women? Use the $\alpha = 0.05$ level of significance.

7. **Spreadsheets:** An accounting firm tested two spreadsheet applications to determine whether there is a difference between them in the mean speed with which a standard accounting problem can be solved. The times needed for each of twelve accountants to solve the problem are as follows:

Application	Accountant											
	1	2	3	4	5	6	7	8	9	10	11	12
A	42	34	45	35	40	44	48	49	41	33	31	38
B	50	43	48	31	39	56	45	60	46	47	33	50

Can you conclude that the mean time differs between the two applications? Use the $\alpha = 0.05$ level of significance.

8. **Meeting specifications:** Two extrusion machines that manufacture steel rods are being compared. In a sample of 1000 rods taken from machine A, 960 met specifications regarding length and diameter. In a sample of 600 rods taken from machine B, 582 met the specifications. Machine B is more expensive to run, so it is decided that machine A will be used unless it can be convincingly shown that machine B produces a larger proportion of rods meeting specifications.
   a. State the appropriate null and alternate hypotheses for making the decision as to which machine to use.
   b. Perform a hypothesis test at the $\alpha = 0.05$ level to determine which machine to use. State your conclusion.

9. **Interpret calculator display:** The following TI-84 Plus calculator display presents the results of a hypothesis test for the difference between two means. The sample sizes are $n_1 = 18$ and $n_2 = 16$.

   a. Is this a left-tailed test, a right-tailed test, or a two-tailed test?
   b. How many degrees of freedom did the calculator use?
   c. What is the *P*-value?
   d. Can you reject H_0 at the $\alpha = 0.05$ level?

10. **Interpret computer output:** The following MINITAB output presents a 95% confidence interval for the difference between two means.

```
Two-sample T for Population1 vs Population2

              N      Mean     StDev  SE Mean
Population1   55  16.48435  10.23430  1.52564
Population2   47  18.32197   8.38450  1.22301


Difference = mu (Treatment1) − mu (Treatment2)
Estimate for difference: −1.83762
95% upper bound for difference: 2.22314
T-Test of difference = 0 (vs < 0): T-Value = −0.997  P-Value = 0.161   DF = 99
```

    a. Is this a left-tailed test, a right-tailed test, or a two-tailed test?
    b. How many degrees of freedom did MINITAB use?
    c. What is the P-value?
    d. Can you reject H_0 at the $\alpha = 0.05$ level?

11. **Interpret calculator display:** The following TI-84 Plus calculator display presents the results of a hypothesis test for the difference between two proportions. The sample sizes are $n_1 = 125$ and $n_2 = 150$.

    a. Is this a left-tailed test, a right-tailed test, or a two-tailed test?
    b. What is the P-value?
    c. Can you reject H_0 at the $\alpha = 0.05$ level?

12. **Interpret computer output:** The following MINITAB output presents the results of a hypothesis test for the difference between two proportions.

```
Test and CI for Two Proportions: A, B

Variable      X      N     Sample p
A            31     45     0.688889
B            23     58     0.396552


Difference = p (A) − p (B)
Estimate for difference: 0.292337
95% Lower Bound for difference: 0.128108
Test for difference = 0 (vs > 0): Z = 2.95  P-Value = 0.003
```

    a. Is this a left-tailed test, a right-tailed test, or a two-tailed test?
    b. What is the P-value?
    c. Can you reject H_0 at the $\alpha = 0.05$ level?

13. **Interpret calculator display:** The following TI-84 Plus calculator display presents the results of a hypothesis test for the mean difference between matched pairs.

    a. Is this a left-tailed test, a right-tailed test, or a two-tailed test?
    b. How many degrees of freedom are there?
    c. What is the P-value?
    d. Can you reject H_0 at the $\alpha = 0.05$ level?

14. Interpret computer output: The following output from MINITAB presents a confidence interval for the mean difference between matched pairs.

```
Paired T-Test and CI: Before, After

Paired T for Before - After

                N      Mean     StDev    SE Mean
Before         12   20.50337   9.69467   2.79861
After          12   17.82398   6.36154   1.83642
Difference     12    2.67399   3.09085   0.89225
95% confidence interval for mean difference: ( 0.71016, 4.63782 )
T-Test of mean difference = 0 (vs ne 0):   T-Value = 3.00  P-Value = 0.012
```

    a. Is this a left-tailed test, a right-tailed test, or a two-tailed test?
    b. How many degrees of freedom are there?
    c. What is the P-value?
    d. Can you reject H_0 at the $\alpha = 0.05$ level?

15. Strong bolts: Five different variations of a bolt-making process are run to determine whether any of them can increase the mean breaking strength of the bolts over that of the current process. The P-values are 0.13, 0.34, 0.03, 0.28, and 0.38. Of the following choices, which is the best thing to do next?
    i. Implement the process whose P-value was 0.03, because it performed the best.
    ii. Because none of the processes had Bonferroni-adjusted P-values less than 0.05, we should continue with the current process.
    iii. Rerun the process whose P-value was 0.03 to determine whether it remains small in the absence of multiple testing.
    iv. Rerun all five variations again, to determine whether any of them produce a small P-value the second time around.

Write About It

1. Provide an example, real or imagined, of a hypothesis test for the difference between two means.

2. Describe under what circumstances a hypothesis test for the difference between two proportions would be performed. Provide an example.

3. Describe the differences between performing a hypothesis test for $\mu_1 - \mu_2$ with paired samples and performing the same test for independent samples.

4. How do hypothesis tests for two population standard deviations differ from other tests presented in this chapter? In what ways are they similar?

5. Several null hypotheses are tested, and one of them is rejected after the Bonferroni adjustment. Explain why this null hypothesis would also have been rejected if it had been the only one tested.

Case Study: Does Exposure To Electrical Fields Cause Brain Cancer?

In the chapter opener, we presented the results of a study that was designed to determine whether exposure to electrical fields could cause cancer. Three hundred children with brain cancer (cases) and three hundred healthy children (controls) were studied. The proportions with underground wiring in each group were computed, along with proportions in certain income and ethnic groups. The results for five factors are presented in the following table.

	Cases	Controls
Proportion with high income	0.09	0.13
Proportion owning a home	0.72	0.76
Proportion with underground wiring	0.14	0.07
Proportion with father white non-Hispanic	0.43	0.36
Proportion with father Hispanic	0.38	0.47

1. For each of the five factors, test the null hypothesis that the population proportions are equal against the alternate hypothesis that they are not equal. Find the P-value for each test.

2. Which factors are significant at the $\alpha = 0.05$ level?

3. For the P-values that are less than 0.05, use the Bonferroni adjustment to adjust them for the fact that several tests have been performed.

4. For each factor, describe how strong the evidence is that the proportions differ for that factor.

Tests with Qualitative Data

Introduction

Do graduate schools discriminate against women? This issue was addressed in a famous study carried out at the University of California at Berkeley. The following table presents the numbers of male and female applicants to six of the most popular departments at the University of California at Berkeley. Out of 2691 male applicants, 1198, or 44.5%, were accepted. Out of 1835 female applicants, only 557, or 30.4%, were accepted. Is this difference due to discrimination against women? In the case study at the end of the chapter, we will use the methods presented in this chapter to determine the real reason for this difference.

Gender	Accept	Reject	Total
Male	1198	1493	2691
Female	557	1278	1835
Total	1755	2771	4526

SECTION 12.1	Testing Goodness of Fit

Objectives

1. Find critical values of the chi-square distribution
2. Perform goodness-of-fit tests

Objective 1 Find critical values of the chi-square distribution

The Chi-Square Distribution

In this chapter, we will introduce hypothesis tests for qualitative data, also called categorical data. These tests are based on the **chi-square distribution**, which we first introduced in Section 8.4. The test statistic used for these tests is called the **chi-square statistic**, denoted χ^2. The symbol χ is the Greek letter chi (pronounced "kigh"; rhymes with sky). We find critical values for this statistic by using the chi-square distribution. We will begin by reviewing the features of this distribution.

There are actually many different chi-square distributions, each with a different number of degrees of freedom. Figure 12.1 presents chi-square distributions for several different degrees of freedom. There are two important points to notice.

Recall: Qualitative data classify individuals into categories.

- The chi-square distribution is not symmetric. It is skewed to the right.
- Values of the χ^2 statistic are always greater than or equal to 0. They are never negative.

Figure 12.1 Chi-square distributions with various degrees of freedom

Finding critical values for the chi-square distribution

We find right-tail critical values for the chi-square distribution. These values can be found using Table A.4.

EXAMPLE 12.1 Find a critical value

Find the $\alpha = 0.05$ critical value for the chi-square distribution with 12 degrees of freedom.

Solution

The critical value is found at the intersection of the row corresponding to 12 degrees of freedom and the column corresponding to $\alpha = 0.05$. The critical value is 21.026.

Degrees of Freedom	Area in Right Tail									
	0.995	0.99	0.975	0.95	0.90	0.10	0.05	0.025	0.01	0.005
⋮	⋮	⋮	⋮	⋮	⋮	⋮	⋮	⋮	⋮	⋮	⋮
10	2.156	2.558	3.247	3.940	4.865	15.987	18.307	20.483	23.209	25.188
11	2.603	3.053	3.816	4.575	5.578	17.275	19.675	21.920	24.725	26.757
12	3.074	3.571	4.404	5.226	6.304	18.549	21.026	23.337	26.217	28.300
13	3.565	4.107	5.009	5.892	7.042	19.812	22.362	24.736	27.688	29.819
14	4.075	4.660	5.629	6.571	7.790	21.064	23.685	26.119	29.141	31.319
15	4.601	5.229	6.262	7.261	8.547	22.307	24.996	27.488	30.578	32.801
⋮	⋮	⋮	⋮	⋮	⋮	⋮	⋮	⋮	⋮	⋮

Figure 12.2 presents the chi-square distribution with 12 degrees of freedom, with the $\alpha = 0.05$ critical value labeled.

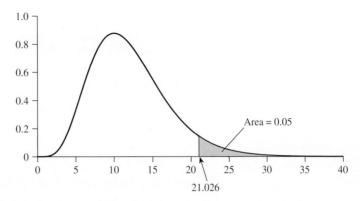

Figure 12.2 The area to the right of 21.026 is 0.05, so 21.026 is the $\alpha = 0.05$ critical value.

Check Your Understanding

1. Find the $\alpha = 0.05$ critical value for the chi-square distribution with 18 degrees of freedom.

2. Find the $\alpha = 0.10$ critical value for the chi-square distribution with 4 degrees of freedom.

3. Find the area to the right of 29.141 under the chi-square distribution with 14 degrees of freedom.

4. Find the area to the right of 46.979 under the chi-square distribution with 30 degrees of freedom.

Answers are on page 567.

Objective 2 Perform goodness-of-fit tests

Goodness-of-Fit Tests

Imagine that you want to determine whether a coin is fair. You could toss the coin a number of times, and compute the sample proportion $\hat{p}$ of heads. You could then use a test for a population proportion (Section 9.4) to test the hypotheses H_0: $p = 0.5$ versus H_1: $p \neq 0.5$. The test for a population proportion is designed for an experiment with two possible outcomes, such as the toss of a coin. Sometimes we work with experiments that have more than two possible outcomes. For example, imagine that a gambler wants to test a die to determine whether it is fair. The roll of a die has six possible outcomes: 1, 2, 3, 4, 5, and 6; and the die is fair if each of these outcomes is equally likely. The gambler rolls the die 60 times, and counts the number of times each number comes up. These counts, which are called the **observed frequencies**, are presented in Table 12.1.

Table 12.1 Observed Frequencies for 60 Rolls of a Die

Outcome	1	2	3	4	5	6
Observed	12	7	14	15	4	8

Explain It Again

Hypotheses for goodness-of-fit tests: The null hypothesis always specifies a probability for each category. The alternate hypothesis says that some or all of these probabilities differ from the true probabilities of the categories.

The gambler wants to perform a hypothesis test to determine whether the die is fair. The null hypothesis for this test says that the die is fair; in other words, it says that each of the six outcomes has probability 1/6 of occurring. Let p_1 be the probability of rolling a 1, p_2 be the probability of rolling a 2, and so on. Then the null hypothesis is

$$H_0: p_1 = p_2 = p_3 = p_4 = p_5 = p_6 = 1/6$$

The alternate hypothesis says that the roll of a die does not follow the distribution specified by H_0; in other words, it states that not all of the p_i are equal to 1/6.

Computing expected frequencies

To test H_0, we begin by computing **expected frequencies**. The expected frequencies are the mean counts that would occur if H_0 were true.

> **DEFINITION**
>
> If the probabilities specified by H_0 are $p_1, p_2, ...,$ and the total number of trials is n, the expected frequencies are
>
> $$E_1 = np_1, \quad E_2 = np_2, \quad \text{and so on}$$

EXAMPLE 12.2   Computing expected frequencies

Compute the expected frequencies for the die example.

Solution

The probabilities specified by H_0 are $p_1 = p_2 = \cdots = p_6 = 1/6$. The total number of trials is $n = 60$. Therefore, the expected frequencies are

$$E_1 = E_2 = E_3 = E_4 = E_5 = E_6 = (60)(1/6) = 10$$

Table 12.2 presents both observed and expected frequencies for the die example.

Table 12.2 Observed and Expected Frequencies

Outcome	1	2	3	4	5	6
Observed	12	7	14	15	4	8
Expected	10	10	10	10	10	10

Check Your Understanding

5. A researcher wants to determine whether children are more likely to be born on certain days of the week. She will sample 350 births and record the day of the week for each. The null hypothesis is that a birth is equally likely to occur on any day of the week. Compute the expected frequencies.

6. A researcher wants to test the hypothesis that births are more likely to occur on weekdays. The null hypothesis is that 17% of births occur on each of the days Monday through Friday, 10% occur on Saturday, and 5% occur on Sunday. If 350 births are sampled, find the expected frequencies.

Answers are on page 567.

If H_0 is true, the observed and expected frequencies should be fairly close. The larger the differences are between the observed and expected frequencies, the stronger the evidence is against H_0. We compute a test statistic that measures how large these differences are. As mentioned previously, the statistic is called the chi-square statistic, denoted χ^2.

> **DEFINITION**
>
> Let k be the number of categories, let $O_1, ..., O_k$ be the observed frequencies, and let $E_1, ..., E_k$ be the expected frequencies. The chi-square statistic is
>
> $$\chi^2 = \sum \frac{(O - E)^2}{E}$$

When H_0 is true, the chi-square statistic has approximately a chi-square distribution, provided that all the expected frequencies are 5 or more.

Explain It Again

Degrees of freedom and the number of categories: For a goodness-of-fit test, the number of degrees of freedom for the chi-square statistic is always 1 less than the number of categories.

CAUTION

The assumption that the expected frequencies are at least 5 does not apply to the observed frequencies. There are no assumptions on the observed frequencies.

When H_0 is true, the statistic

$$\chi^2 = \sum \frac{(O - E)^2}{E}$$

has a chi-square distribution with $k - 1$ degrees of freedom, where k is the number of categories, provided that all the expected frequencies are greater than or equal to 5.

We will first describe how to perform a hypothesis test using the critical value method and Table A.4. Then we will describe how to use the P-value method with technology.

Performing a Goodness-of-Fit Test

Step 1: State the null and alternate hypotheses. The null hypothesis specifies a probability for each category. The alternate hypothesis says that some or all of the actual probabilities differ from those specified by H_0.

Step 2: Compute the expected frequencies and check to be sure that all of them are 5 or more. If they are, then proceed.

Step 3: Choose a significance level α.

Step 4: Compute the test statistic $\chi^2 = \sum \dfrac{(O - E)^2}{E}$.

Step 5: Find the critical value from Table A.4, using $k - 1$ degrees of freedom, where k is the number of categories. If χ^2 is greater than or equal to the critical value, reject H_0. Otherwise, do not reject H_0.

Step 6: State a conclusion.

EXAMPLE 12.3   Perform a goodness-of-fit test

Table 12.3 presents the observed frequencies for the die example. Can you conclude at the $\alpha = 0.05$ level that the die is not fair?

Table 12.3

Category	Observed
1	12
2	7
3	14
4	15
5	4
6	8

Table 12.4

Category	Observed	Expected
1	12	10
2	7	10
3	14	10
4	15	10
5	4	10
6	8	10

Solution

Step 1: State the null and alternate hypotheses. The null hypothesis says that the die is fair, and the alternate hypothesis says that the die is not fair, so we have

$$H_0: p_1 = p_2 = p_3 = p_4 = p_5 = p_6 = 1/6 \qquad H_1: \text{Some or all of the } p_i \text{ differ from } 1/6$$

Step 2: Compute the expected frequencies. The expected frequencies were computed in Example 12.2. We present them in Table 12.4. They are all greater than 5, so we proceed.

Step 3: Choose a level of significance. We will use $\alpha = 0.05$.

Step 4: Compute the value of the test statistic. The following table presents the calculations.

Category	O	E	$O - E$	$(O - E)^2$	$\dfrac{(O - E)^2}{E}$
1	12	10	2	4	0.4
2	7	10	−3	9	0.9
3	14	10	4	16	1.6
4	15	10	5	25	2.5
5	4	10	−6	36	3.6
6	8	10	−2	4	0.4

$$\chi^2 = \sum \frac{(O - E)^2}{E} = 9.4$$

The value of the test statistic is $\chi^2 = 9.4$.

Step 5: Find the critical value. There are six categories, so there are $6 - 1 = 5$ degrees of freedom. From Table A.4, we find that the $\alpha = 0.05$ critical value for 5 degrees of freedom is 11.070. The value of the test statistic is $\chi^2 = 9.4$. Because $9.4 < 11.070$, we do not reject H_0.

Step 6: State a conclusion. There is not enough evidence to conclude that the die is unfair.

<table>
<tr><td>EXAMPLE 12.4</td><td>

Perform a goodness-of-fit test using technology

</td></tr>
</table>

The following TI-84 Plus display presents the results of Example 12.3.

Most of the output is straightforward. The first, second, and third lines present the value of the chi-square statistic, the P-value, and the degrees of freedom, respectively. The last line, labeled "CNTRB," presents the quantities $\dfrac{(O - E)^2}{E}$ for the first two categories.

The P-value is 0.0941. Because $P > 0.05$, we do not reject H_0 at the $\alpha = 0.05$ level. This conclusion is the same as the one we reached when we used the critical value method in Example 12.3.

In EXCEL, the function **CHISQ.TEST** returns the P-value for a goodness-of-fit test. The inputs are the range of cells that contain the observed frequencies and the range of cells that contain the expected frequencies. Following is the EXCEL output. The P-value is shown to be 0.0941.

	A	B
1	**Observed**	**Expected**
2	12	10
3	7	10
4	14	10
5	15	10
6	4	10
7	8	10

0.0941

Step-by-step instructions for using technology are presented in the Using Technology section on page 564.

<table>
<tr><td>EXAMPLE 12.5</td><td>

Perform a goodness-of-fit test

</td></tr>
</table>

A poll conducted by the General Social Survey asked 1155 people whether they thought that people with high incomes should pay a greater or smaller percentage of their income in tax than low-income people. The results are presented in the following table.

Category	Observed
Pay much more	218
Pay somewhat more	497
Pay the same	425
Pay less	15

Five years earlier, it was determined that 18.5% believed that the rich should pay much more, 39.2% believed they should pay somewhat more, 41.2% believed they should pay the same, and 1.1% believed they should pay less. Can we conclude that the current percentages differ from these? Use the $\alpha = 0.05$ level of significance.

Solution

Step 1: State the null and alternate hypotheses. The null hypothesis is H_0: $p_1 = 0.185$, $p_2 = 0.392, p_3 = 0.412, p_4 = 0.011$. The alternate hypothesis states that some of the probabilities are not equal to the values specified by H_0.

Step 2: Compute the expected frequencies. The number of trials is $n = 1155$. The expected frequencies are

$$E_1 = np_1 = (1155)(0.185) = 213.675 \qquad E_2 = np_2 = (1155)(0.392) = 452.76$$

$$E_3 = np_3 = (1155)(0.412) = 475.86 \qquad E_4 = np_4 = (1155)(0.011) = 12.705$$

All the expected frequencies are 5 or more, so we proceed.

Step 3: Choose a significance level. We will use $\alpha = 0.05$.

Step 4: Compute the test statistic. Using the observed and expected frequencies, we compute the value of the test statistic to be

$$\chi^2 = \frac{(218 - 213.675)^2}{213.675} + \frac{(497 - 452.76)^2}{452.76} + \frac{(425 - 475.86)^2}{475.86} + \frac{(15 - 12.705)^2}{12.705} = 10.261$$

Step 5: Find the critical value. There are four categories, so there are $4 - 1 = 3$ degrees of freedom. From Table A.4, we find that the $\alpha = 0.05$ critical value for 3 degrees of freedom is 7.815. The value of the test statistic is $\chi^2 = 10.261$. Because $10.261 > 7.815$, we reject H_0.

Step 6: State a conclusion. We conclude that the distribution of opinions on this issue changed during the 5 years prior to the survey.

Check Your Understanding

7. Following are observed frequencies for five categories:

Category	1	2	3	4	5
Observed	25	14	23	6	2

    a. Compute the expected frequencies for testing
       H_0: $p_1 = 0.3$, $p_2 = 0.25, p_3 = 0.2$, $p_4 = 0.15$, $p_5 = 0.1$.
    b. One of the observed frequencies is less than 5. Is the chi-square test appropriate?
    c. Compute the value of χ^2.
    d. How many degrees of freedom are there?
    e. Find the level $\alpha = 0.05$ critical value.
    f. Do you reject H_0 at the 0.05 level?
    g. Find the level $\alpha = 0.01$ critical value.
    h. Do you reject H_0 at the 0.01 level?

8. For the data in Exercise 7:
    a. Compute the expected frequencies for testing
       H_0: $p_1 = 0.4, p_2 = 0.3, p_3 = 0.1, p_4 = 0.15, p_5 = 0.05$.
    b. Is it appropriate to perform a chi-square test for the hypothesis in part (a)? Explain.

Answers are on page 567.

USING TECHNOLOGY

We use Example 12.3 to illustrate the technology steps.

TI-84 PLUS

Testing goodness-of-fit

Step 1. Enter observed frequencies into **L1** in the data editor and expected frequencies into **L2**. Figure A illustrates this for Example 12.3.

Step 2. Press **STAT** and highlight the **TESTS** menu. Select χ^2**GOF–Test** and press **ENTER** (Figure B). The χ^2**GOF–Test** menu appears.

Step 3. In the **Observed** field, enter **L1**, and in the **Expected** field, enter **L2**.

Step 4. Enter the degrees of freedom in the **df** field. For Example 12.3, there are 5 degrees of freedom (Figure C).

Step 5. Highlight **Calculate** and press **ENTER** (Figure D).

Figure A

Figure B

Figure C

Figure D

EXCEL

Testing goodness-of-fit

The **CHISQ.TEST** command returns the *P*-value for a goodness-of-fit test given the observed and expected frequencies.

Step 1. Enter the observed and expected frequencies into EXCEL. For Example 12.3, the observed frequencies are contained in A2:A7 and the expected frequencies are contained in B2:B7 (Figure E).

Step 2. In an empty cell, select the **Insert Function** icon and highlight **Statistical** in the category field.

Step 3. Click on the **CHISQ.TEST** function and press **OK**.

Step 4. In the **Actual_range** field, enter or select the range of cells containing the observed frequencies. In the **Expected_range** field, enter or select the range of cells containing the expected frequencies.

Step 5. Click **OK** to obtain the *P*-value (Figure F).

	A	B
1	**Observed**	**Expected**
2	12	10
3	7	10
4	14	10
5	15	10
6	4	10
7	8	10

Figure E

=CHISQ.TEST(A2:A7, B2:B7)

0.0941

Figure F

SECTION 12.1 Exercises

Exercises 1–8 are the Check Your Understanding exercises located within the section.

Understanding the Concepts

In Exercises 9 and 10, fill in each blank with the appropriate word or phrase.

9. For the goodness-of-fit test to be valid, each of the _____ frequencies must be at least 5.

10. In a goodness-of-fit test, we reject H_0 if the _____ frequencies are much different from the expected frequencies.

In Exercises 11 and 12, determine whether the statement is true or false. If the statement is false, rewrite it as a true statement.

11. The chi-square distribution is symmetric.

12. The alternate hypothesis for a goodness-of-fit test says that some of the probabilities differ from those specified by the null hypothesis.

Practicing the Skills

13. Find the $\alpha = 0.05$ critical value for the chi-square statistic with 14 degrees of freedom.

14. Find the $\alpha = 0.01$ critical value for the chi-square statistic with 5 degrees of freedom.

15. Find the area to the right of 24.725 under the chi-square distribution with 11 degrees of freedom.

16. Find the area to the right of 40.256 under the chi-square distribution with 30 degrees of freedom.

17. For the following observed and expected frequencies:

Observed	9	22	53	9	7
Expected	15	20	40	15	10

   a. Compute the value of χ^2.
   b. How many degrees of freedom are there?
   c. Test the hypothesis that the distribution of the observed frequencies is as given by the expected frequencies. Use the $\alpha = 0.05$ level of significance.

18. For the following observed and expected frequencies:

Observed	43	42	31	19	34	32
Expected	44	44	33	15	36	29

   a. Compute the value of χ^2.
   b. How many degrees of freedom are there?
   c. Test the hypothesis that the distribution of the observed frequencies is as given by the expected frequencies. Use the $\alpha = 0.01$ level of significance.

19. Following are observed frequencies. The null hypothesis is H_0: $p_1 = 0.5$, $p_2 = 0.3$, $p_3 = 0.15$, $p_4 = 0.05$.

Category	1	2	3	4
Observed	106	64	24	6

   a. Compute the expected frequencies.
   b. Compute the value of χ^2.

   c. How many degrees of freedom are there?
   d. Test the hypothesis that the distribution of the observed frequencies is as given by the null hypothesis. Use the $\alpha = 0.01$ level of significance.

20. Following are observed frequencies. The null hypothesis is H_0: $p_1 = 0.4$, $p_2 = 0.25$, $p_3 = 0.05$, $p_4 = 0.1$, $p_5 = 0.2$.

Category	1	2	3	4	5
Observed	50	51	14	12	23

   a. Compute the expected frequencies.
   b. Compute the value of χ^2.
   c. How many degrees of freedom are there?
   d. Test the hypothesis that the distribution of the observed frequencies is as given by the null hypothesis. Use the $\alpha = 0.05$ level of significance.

Working with the Concepts

21. **Is the lottery fair?** Powerball is a multistate lottery in which players try to guess the numbers that will turn up in a drawing of numbered balls. One of the balls drawn is the "Powerball." Matching the number drawn on the Powerball increases one's winnings. Between January 18, 2012, and June 26, 2013, the Powerball was drawn from a collection of 35 balls numbered 1 through 35. A total of 151 drawings were made. For the purposes of this exercise, we grouped the numbers in to five categories: 1–7, 8–14, and so on. If the lottery is fair, then the winning number is equally likely to occur in any category. Following are the observed frequencies.

Category	1–7	8–14	15–21	22–28	29–35
Observed	33	26	28	27	37

Source: www.usamega.com

   a. Compute the expected frequencies.
   b. Compute the value of χ^2.
   c. How many degrees of freedom are there?
   d. Test the hypothesis that each of the categories is equally likely. Use the $\alpha = 0.05$ level of significance.

22. **Grade distribution:** A statistics teacher claims that, on the average, 20% of her students get a grade of A, 35% get a B, 25% get a C, 10% get a D, and 10% get an F. The grades of a random sample of 100 students were recorded. The following table presents the results.

Grade	A	B	C	D	F
Observed	29	42	20	5	4

   a. How many of the students in the sample got an A? How many got an F?
   b. Compute the expected frequencies.
   c. Which grades were given more often than expected? Which grades were given less often than expected?
   d. What is the value of χ^2?
   e. How many degrees of freedom are there?

f. Test the hypothesis that the grades follow the distribution claimed by the teacher. Use the 0.05 level of significance.

23. False alarm: The numbers of false fire alarms were counted each month at a number of sites. The results are given in the following table.

Month	Number of Alarms
January	32
February	15
March	37
April	38
May	45
June	48
July	46
August	42
September	34
October	36
November	28
December	26

Source: *Journal of Architectural Engineering* 5:62–65

Test the hypothesis that false alarms are equally likely to occur in any month. Use the $\alpha = 0.01$ level of significance.

24. Crime rates: The FBI computed the proportion of violent crimes in the United States in 2012 falling into each of four categories. A simple random sample of 500 violent crimes committed in California during 2012 were categorized in the same way. The following table presents the results.

Category	U.S. Proportion	California Frequency
Murder	0.013	5
Forcible Rape	0.051	23
Robbery	0.360	206
Aggravated Assault	0.576	266

Source: FBI 2012 Annual Uniform Crime Report

Can you conclude that the proportions of crimes in the various categories in California differ from the United States as a whole? Use the 0.05 level of significance.

25. Where do you live? The U.S. Census Bureau computed the proportion of U.S. residents who lived in each of four geographic regions in 2000. Then a simple random sample was drawn of 1000 people living in the United States in 2011. The following table presents the results.

Region	2000 Proportion	2011 Frequency
Northeast	0.190	150
Midwest	0.229	234
South	0.356	390
West	0.225	226

Can you conclude that the proportions of people living in the various regions changed between 2000 and 2011? Use the 0.05 level of significance.

26. Abortion policy: A Gallup poll taken in May 2013 asked 1535 adult Americans to state their opinion on the availability of abortions. The following table presents the results, along with the proportions of people who held these views in 2010.

Opinion	2010 Proportion	2013 Frequency
Generally available	0.24	399
Available with limits	0.54	798
Should not be permitted	0.19	307
Don't know/No answer	0.03	31

Can you conclude that the proportions of people giving the various responses changed between 2010 and 2013? Use the 0.01 level of significance.

27. Economic future: A CNN/ORC poll taken in July 2012 obtained responses from 1517 adult Americans to the question "How would you rate the economic conditions in the country today — as very good, somewhat good, somewhat poor, or very poor?" The following table presents the results, along with the proportions of people who gave these responses in 2011.

View	2011 Proportion	2012 Frequency
Very good	0.01	30
Somewhat good	0.24	379
Somewhat poor	0.40	592
Very poor	0.35	516

Can you conclude that the proportions of people giving the various responses changed between 2011 and 2012? Use the 0.01 level of significance.

28. Guess the answer: A statistics instructor gave a four-question true–false quiz to his class of 150 students. The results were as follows.

Number correct	0	1	2	3	4
Observed	2	13	29	71	35

The instructor thinks that the students may have answered the questions by guessing, so that the probability that any given answer is correct is 0.5. Under this null hypothesis, the number of correct answers has a binomial distribution with 4 trials and success probability 0.5. Perform a chi-square test of this hypothesis. Can you reject H_0 at the $\alpha = 0.05$ level?

Extending the Concepts

29. Fair die? A gambler rolls a die 600 times to determine whether or not it is fair. Following are the results.

Outcome	1	2	3	4	5	6
Observed	113	101	106	81	108	91

a. Let p_1 be the probability that the die comes up 1, let p_2 be the probability that the die comes up 2, and so on. Use the chi-square distribution to test the null hypothesis, at the $\alpha = 0.05$ level of significance, that the die is fair:

b. The gambler decides to use the test for proportions (discussed in Section 9.4) to test $H_0: p_i = 1/6$ for each p_i. Find the P-values for each of these tests.

c. Show that the hypothesis $H_0: p_4 = 1/6$ is rejected at level 0.05.

d. The gambler now reasons as follows: "I reject $H_0: p_4 = 1/6$, and I conclude that $p_4 \neq 1/6$. Therefore,

I can reject the null hypothesis that the die is fair." Explain why this reasoning is incorrect. (*Hint:* See Section 11.5 on the multiple testing problem.)

e. Use the Bonferroni correction (Section 11.5) to adjust the *P*-value for the test of $H_0: p_4 = 1/6$. Can you now conclude that the die is not fair? Explain.

Answers to Check Your Understanding Exercises for Section 12.1

1. 28.869

2. 7.779

3. 0.01

4. 0.025

5.

Sunday	Monday	Tuesday	Wednesday	Thursday	Friday	Saturday
50	50	50	50	50	50	50

6.

Sunday	Monday	Tuesday	Wednesday	Thursday	Friday	Saturday
17.5	59.5	59.5	59.5	59.5	59.5	35.0

7. a.

1	2	3	4	5
21.0	17.5	14.0	10.5	7.0

b. Yes, because all the expected frequencies are 5 or more.
c. 12.748   **d.** 4   **e.** 9.488   **f.** Yes
g. 13.277   **h.** No

8. a.

1	2	3	4	5
28.0	21.0	7.0	10.5	3.5

b. No, one of the expected frequencies is less than 5.

SECTION 12.2 — Tests for Independence and Homogeneity

Objectives

1. Interpret contingency tables
2. Perform tests of independence
3. Perform tests of homogeneity

Objective 1 Interpret contingency tables

Contingency Tables

Do some college majors require more studying than others? The 2009 National Survey of Student Engagement asked a number of college freshmen what their major was and how many hours per week they spent studying, on average. A sample of 1000 of these students was chosen, and the numbers of students in each category are presented in Table 12.5.

Table 12.5 Observed Frequencies

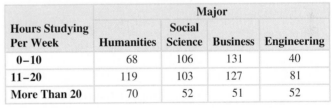

Hours Studying Per Week	Major			
	Humanities	Social Science	Business	Engineering
0–10	68	106	131	40
11–20	119	103	127	81
More Than 20	70	52	51	52

Table 12.5 is called a **contingency table**. A contingency table relates two qualitative variables. One of the variables, called the **row variable**, has one category for each row of the table. The other variable, called the **column variable**, has one category for each column of the table. In Table 12.5, hours studying is the row variable and major is the column variable. In general, it does not matter which variable is the row variable and which is the column variable. We could just as well have made major the row variable and hours studying the column variable. The intersection of a row and a column is called a **cell**. For example, the number 68 appears in the upper left cell, which tells us that 68 students were humanities majors who study 0–10 hours per week.

Objective 2 Perform tests of independence

Performing a Test of Independence

We are interested in determining whether the distribution of one variable differs, depending on the value of the other variable. If so, the variables are *dependent*. If the distribution of one variable is the same for all the values of the other variable, the variables are *independent*. For Table 12.5, the null and alternate hypotheses are

H_0: Hours studying and major are independent.

H_1: Hours studying and major are not independent.

We will use the chi-square statistic to test the null hypothesis that major and hours studying are independent. If we reject H_0, we will conclude that the variables are dependent. The values in Table 12.5 are the observed frequencies. To compute the value of χ^2, we must compute the expected frequencies.

Computing the expected frequencies

The first step in computing the expected frequencies is to compute the row and column totals. For example, the total in the first row is

Total number of students studying 0–10 hours = 68 + 106 + 131 + 40 = 345

The total in the first column is

Total number of humanities majors = 68 + 119 + 70 = 257

Table 12.6 presents Table 12.5 with the row and column totals included. The total number of individuals in the table, 1000, is included as well. This total is often called the **grand total**.

Table 12.6 Observed Frequencies with Row and Column Totals

Hours Studying Per Week	Major				
	Humanities	Social Science	Business	Engineering	Row Total
0–10	68	106	131	40	345
11–20	119	103	127	81	430
More Than 20	70	52	51	52	225
Column Total	257	261	309	173	1000

As with any hypothesis test, we begin by assuming the null hypothesis to be true. The null hypothesis says that hours studying and major are independent. We can now use the Multiplication Rule for Independent Events to compute the expected frequencies. For example,

P(Study 0–10 hours and Humanities major) = P(Study 0–10 hours)P(Humanities major)

Now out of a total of 1000 students, 345 studied 0–10 hours per week. Therefore,

$$P(\text{Study 0–10 hours}) = \frac{345}{1000}$$

Out of a total of 1000 students, 257 were humanities majors. Therefore,

$$P(\text{Humanities major}) = \frac{257}{1000}$$

The Multiplication Rule for Independent Events tells us that if H_0 is true, then

$$P(\text{Study 0–10 hours and Humanities major}) = \left(\frac{345}{1000}\right)\left(\frac{257}{1000}\right)$$

We obtain the expected frequency for those who study 0–10 hours and are humanities majors by multiplying this probability by the grand total, which is 1000.

$$\text{Expected frequency} = 1000\left(\frac{345}{1000}\right)\left(\frac{257}{1000}\right)$$

Before calculating this quantity, simplify it as follows:

$$\text{Expected frequency} = 1000\left(\frac{345}{1000}\right)\left(\frac{257}{1000}\right) = \frac{345 \cdot 257}{1000} = 88.665$$

We see that the expected frequency can be computed as

$$\text{Expected frequency} = \frac{\text{Row total} \cdot \text{Column total}}{\text{Grand total}}$$

Explain It Again

Independence: If hours studied and major are independent, then the distribution of hours studied will be the same for all majors, and the distribution of majors will be the same for all categories of hours studied.

Recall: The Multiplication Rule for Independent Events says that if events E and F are independent, then $P(E \text{ and } F) = P(E) \cdot P(F)$.

SUMMARY

To find the expected frequency for a cell, multiply the row total by the column total, then divide by the grand total.

$$E = \frac{\text{Row total} \cdot \text{Column total}}{\text{Grand total}}$$

The expected frequency for a cell represents the number of individuals we would expect to find in that cell under the assumption that the two variables are independent. If the differences between the observed and expected frequencies tend to be large, we will reject the null hypothesis of independence.

Check Your Understanding

1. The following contingency table presents observed frequencies. Compute the expected frequencies.

	1	2	3
A	13	8	27
B	18	21	35
C	19	13	15
D	20	17	27

Answer is on page 577.

Once the expected frequencies are computed, we check to determine whether all of them are at least 5. If so, we use the chi-square statistic as a test statistic. The number of degrees of freedom is $(r-1)(c-1)$, where r is the number of rows and c is the number of columns. We will first describe how to perform a hypothesis test using the critical value method and Table A.4. Then we will describe how to use the P-value method with technology.

Recall: We can use the chi-square statistic whenever the expected frequencies are all at least 5.

Performing a Test of Independence

Step 1: State the null and alternate hypotheses. The null hypothesis says that the row and column variables are independent. The alternate hypothesis says that they are not independent.

Step 2: Compute the row and column totals.

Step 3: Compute the expected frequencies:

$$E = \frac{\text{Row total} \cdot \text{Column total}}{\text{Grand total}}$$

Check to be sure that all the expected frequencies are at least 5.

Step 4: Choose a level of significance α, and compute the test statistic:

$$\chi^2 = \sum \frac{(O - E)^2}{E}$$

Step 5: Find the critical value from Table A.4, using $(r-1)(c-1)$ degrees of freedom, where r is the number of rows and c is the number of columns. If χ^2 is greater than or equal to the critical value, reject H_0. Otherwise, do not reject H_0.

Step 6: State a conclusion.

EXAMPLE 12.6   Perform a test of independence

Perform a test of the null hypothesis that major and hours studying are independent, using the data in Table 12.6. Use $\alpha = 0.01$.

Solution

Step 1: State the null and alternate hypotheses. The hypotheses are

H_0: Major and hours studying are independent.

H_1: Major and hours studying are not independent.

Step 2: Compute the row and column totals. These are shown in Table 12.6.

Step 3: Compute the expected frequencies. As an example, we compute the expected frequency for the cell corresponding to business major, studying $11-20$ hours. The row total is 430, the column total is 309, and the grand total is 1000. The expected frequency is

$$E = \frac{430 \cdot 309}{1000} = 132.87$$

Table 12.7 presents the expected frequencies. All the expected frequencies are at least 5, so we can proceed.

Table 12.7 Expected Frequencies

Hours Studying Per Week	Major			
	Humanities	Social Science	Business	Engineering
0–10	88.665	90.045	106.605	59.685
11–20	110.510	112.230	132.870	74.390
More Than 20	57.825	58.725	69.525	38.925

Step 4: Choose a level of significance and compute the test statistic. We will use the $\alpha = 0.01$ level of significance. To compute the test statistic, we use the observed frequencies in Table 12.6 and the expected frequencies in Table 12.7:

$$\chi^2 = \frac{(68 - 88.665)^2}{88.665} + \cdots + \frac{(52 - 38.925)^2}{38.925} = 34.638$$

Step 5: Find the critical value. There are $r = 3$ rows and $c = 4$ columns, so the number of degrees of freedom is $(3 - 1)(4 - 1) = 6$. From Table A.4, we find that the critical value corresponding to 6 degrees of freedom and $\alpha = 0.01$ is 16.812. The value of the test statistic is 34.638. Because $34.638 > 16.812$, we reject H_0.

Step 6: State a conclusion. We conclude that the choice of major and the number of hours spent studying are not independent. The numbers of hours that students study varies among majors.

Check Your Understanding

2. The following contingency table presents observed frequencies.

	Observed		
	1	**2**	**3**
A	8	13	14
B	18	1	15
C	16	17	15
D	19	19	8

a. Compute the expected frequencies.

b. One of the observed frequencies is less than 5. Is it appropriate to perform a test of independence? Explain.

c. Compute the value of the test statistic.

d. How many degrees of freedom are there?

e. Do you reject H_0 at the $\alpha = 0.05$ level?

Answers are on page 577.

| EXAMPLE 12.7 | **Perform a test of independence with technology** |

Use technology to test the hypothesis of independence of major and hours studied.

Solution

We present the results from MINITAB.

In the MINITAB output, each cell (intersection of row and column) contains three numbers. The top number is the observed frequency, the middle number is the expected frequency, and the bottom number is the contribution $\dfrac{(O-E)^2}{E}$ to the chi-square statistic from that cell. The P-value is given as 0.000. This means that the P-value is less than 0.0005, so when rounded to three decimal places, the value is 0.000.

Because $P < 0.01$, we reject H_0 at the $\alpha = 0.01$ level, and conclude that the choice of major and the number of hours studying are not independent. This conclusion is the same as the one we reached in Example 12.6 using the critical value method.

Step-by-step instructions for using technology are presented in the Using Technology section on page 573.

```
Chi-Square Test: Humanities, Science, Business, Engineering

Expected counts are printed below observed counts
Chi-Square contributions are printed below expected counts

         Humanities    Science    Business   Engineering    Total
    1            68        106         131            40       345
             88.67      90.05      106.61         59.69
             4.816      2.827       5.582         6.492

    2           119        103         127            81       430
            110.51     112.23      132.87         74.39
             0.652      0.759       0.259         0.587

    3            70         52          51            52       225
             57.83      58.73       69.53         38.93
             2.563      0.770       4.936         4.392

  Total         257        261         309           173      1000

Chi-Sq= 34.638, DF = 6, P-Value = 0.000
```

Tests of Homogeneity

| Objective 3 Perform tests of homogeneity |

In the contingency tables we have seen so far, the individuals in the table were sampled from a single population. For each individual, the values of both the row and column variables were random. In some cases, values of one of the variables (say, the row variable) are assigned by the investigator, and are not random. In these cases, we consider the rows as representing separate populations, and we are interested in testing the hypothesis that the distribution of the column variable is the same for each row. This is known as a **test of homogeneity**. Following is an example.

The drugs telmisartan and ramipril are designed to reduce high blood pressure. In a clinical trial to compare the effectiveness of these drugs in preventing heart attacks, 25,620 patients were divided into three groups. One group took one telmisartan tablet each day, another took one ramipril tablet each day, and the third group took one tablet of each drug each day. The patients were followed for 56 months, and the numbers who suffered fatal and nonfatal heart attacks were counted. Table 12.8 on page 572 presents the results.

In this table, the patients were assigned a row category, so only the column variable is random. We are interested in performing a test of homogeneity, to test the hypothesis that the distribution of outcomes is the same for each row. We have already seen the method for performing a test of homogeneity. It is the same as the method for performing a test of independence.

Table 12.8

	Fatal Heart Attack	Nonfatal Heart Attack	No Heart Attack
Telmisartan only	598	431	7513
Ramipril only	603	400	7573
Both drugs	620	424	7458

Source: www.clinicaltrials.gov

> The method for performing a test of homogeneity is identical to the method for performing a test of independence.

EXAMPLE 12.8	**Perform a test of homogeneity**

Explain It Again

Interpretation of a test of homogeneity: If we reject the null hypothesis, we conclude that the distributions are not all the same, but we cannot tell which ones are different.

Refer to Table 12.8. Test the hypothesis that the distribution of outcomes is the same for all three treatment groups. Use the $\alpha = 0.05$ level.

Solution

We follow the same steps as for a test of independence.

Step 1: State the null and alternate hypotheses. The null hypothesis says that the distribution of outcomes (fatal heart attack, nonfatal heart attack, no heart attack) is the same for all the drug treatments. The alternate hypothesis says that the distribution of outcomes is not the same for all the treatments.

Step 2: Compute the row and column totals. These are shown in the following table.

	Fatal Heart Attack	Nonfatal Heart Attack	No Heart Attack	Row Total
Telmisartan only	598	431	7513	8542
Ramipril only	603	400	7573	8576
Both drugs	620	424	7458	8502
Column Total	1821	1255	22,544	25,620

Step 3: Compute the expected frequencies. As an example, we compute the expected frequency for the cell corresponding to Telmisartan, Fatal heart attack. The row total is 8542, the column total is 1821, and the grand total is 25,620. The expected frequency is

$$\frac{8542 \cdot 1821}{25,620} = 607.14$$

The following table presents the expected frequencies. All the expected frequencies are at least 5, so we can proceed.

	Fatal Heart Attack	Nonfatal Heart Attack	No Heart Attack
Telmisartan only	607.14	418.43	7516.43
Ramipril only	609.56	420.10	7546.34
Both drugs	604.30	416.47	7481.23

Step 4: Choose a level of significance and compute the test statistic. We will use the $\alpha = 0.05$ level of significance. The test statistic is

$$\chi^2 = \frac{(598 - 607.14)^2}{607.14} + \cdots + \frac{(8502 - 7481.23)^2}{7481.23} = 2.259$$

Step 5: Find the critical value. There are $r = 3$ rows and $c = 3$ columns, so the number of degrees of freedom is $(3 - 1)(3 - 1) = 4$. From Table A.4, we find that the critical value corresponding to 4 degrees of freedom and $\alpha = 0.05$ is 9.488. The value of the test statistic is 2.259. Since $2.259 < 9.488$, we do not reject H_0.

Step 6: State a conclusion. There is not enough evidence to conclude that the distribution of outcomes is different for different drug treatments.

USING TECHNOLOGY

We use Table 12.5 to illustrate the technology steps.

TI-84 PLUS

Testing for independence

Step 1. Press **2nd**, then **MATRIX** to access the Matrix menu. Highlight **EDIT** and press **ENTER**. Select **1:[A]**.

Step 2. To input the data from Table 12.5, enter the size of the matrix as **3 × 4**. Enter each of the data values (Figure A; note that the last column does not show).

Step 3. Press **STAT** and highlight the **TESTS** menu. Select χ^2–**Test** and press **ENTER** (Figure B). The χ^2–**Test** menu appears.

Step 4. Enter **[A]** in the **Observed** field. The default value for the **Expected** field is **[B]** (Figure C).

Step 5. Highlight **Calculate** and press **ENTER** (Figure D).

Note: The expected frequencies can be viewed by accessing matrix **[B]** from the Matrix menu.

Figure A

Figure B

Figure C

Figure D

MINITAB

Testing for independence

Step 1. Enter the data from Table 12.5 as shown in Figure E.

Step 2. Click on **Stat**, then **Tables**, then **Chi-Square Test**.

Step 3. Select all columns (**Humanities–Engineering**) in the **Columns containing the table** field (Figure F).

Step 4. Click **OK** (Figure G).

↓	C1	C2	C3	C4
	Humanities	Social Science	Business	Engineering
1	68	106	131	40
2	119	103	127	81
3	70	52	51	52

Figure E

Chi-Square Test (Table in Worksheet)

C1	Humanities	**Columns containing the table:**
C2	Social Scie	Humanities-Engineering
C3	Business	
C4	Engineering	

Select

Help   OK   Cancel

Figure F

Chi-Square Test: Humanities, Social Science, Business, Engineering

Expected counts are printed below observed counts
Chi-Square contributions are printed below expected counts

```
                      Social
          Humanities  Science  Business  Engineering  Total
    1          68       106      131          40       345
             88.67     90.05   106.61       59.69
             4.816     2.827    5.582        6.492

    2         119       103      127          81       430
            110.51    112.23   132.87       74.39
             0.652     0.759    0.259        0.587

    3          70        52       51          52       225
             57.83     58.73    69.53       38.93
             2.563     0.770    4.936        4.392

Total        257       261      309         173      1000

Chi-Sq = 34.638, DF = 6, P-Value = 0.000
```

Figure G

EXCEL

Testing independence

The **CHISQ.TEST** command returns the *P*-value for a test of independence given the observed and expected frequencies.

Step 1. Enter the observed and expected frequencies into EXCEL. For Table 12.5, the observed frequencies are contained in B3:E5 and the expected frequencies are in B9:E11 (Figure E).

Step 2. In an empty cell, select the **Insert Function** icon and highlight **Statistical** in the category field.

Step 3. Click on the **CHISQ.TEST** function and press **OK**.

Step 4. In the **Actual_range** field, enter or select the range of cells containing the observed frequencies. In the **Expected_range** field, enter or select the range of cells containing the expected frequencies.

Step 5. Click **OK** to obtain the *P*-value (Figure F).

	A	B	C	D	E
1			**OBSERVED**		
2	**Hours**	**Humanities**	**Social Science**	**Business**	**Engineering**
3	**0 - 10**	68	106	131	40
4	**11 - 20**	119	103	127	81
5	**More than 20**	70	52	51	52
6					
7			**EXPECTED**		
8	**Hours**	**Humanities**	**Social Science**	**Business**	**Engineering**
9	**0 - 10**	88.665	90.045	106.605	59.685
10	**11 - 20**	110.51	112.23	132.87	74.39
11	**More than 20**	57.825	58.725	69.525	38.925

```
=CHISQ.TEST(B3:E5,B9:E11)

        5.06487E-06
```

Figure E   **Figure F**

SECTION 12.2   Exercises

Exercises 1 and 2 are the Check Your Understanding exercises located within the section.

Understanding the Concepts

In Exercises 3–5, fill in each blank with the appropriate word or phrase.

3. To calculate the expected frequencies, we must know the row totals, the column totals, and the _____ total.

4. We reject H_0 if the value of the test statistic is _____ the critical value.

5. In the test for _____, the null hypothesis is that the distribution of the column variable is the same in each row.

In Exercises 6–8, determine whether the statement is true or false. If the statement is false, rewrite it as a true statement.

6. A contingency table containing observed values has three rows and four columns. The number of degrees of freedom for the chi-square statistic is 7.

7. In a test of homogeneity, the alternate hypothesis says that the distributions in the rows are the same.

8. The procedure for testing homogeneity is the same as the procedure for testing independence.

Practicing the Skills

9. For the given table of observed frequencies:

	1	2	3
A	15	10	12
B	3	11	11
C	9	14	12

a. Compute the row totals, the column totals, and the grand total.

b. Construct the corresponding table of expected frequencies.

c. Compute the value of the chi-square statistic.

d. How many degrees of freedom are there?

e. If appropriate, perform a test of independence, using the $\alpha = 0.05$ level of significance. If not appropriate, explain why.

10. For the given table of observed frequencies:

	1	2	3
A	25	4	11
B	3	3	4
C	42	3	5

a. Compute the row totals, the column totals, and the grand total.

b. Construct the corresponding table of expected frequencies.

c. Compute the value of the chi-square statistic.
d. How many degrees of freedom are there?
e. If appropriate, perform a test of independence, using the $\alpha = 0.05$ level of significance. If not appropriate, explain why.

Working with the Concepts

11. Carbon monoxide: A recent study examined the effects of carbon monoxide exposure on a group of construction workers. The following table presents the numbers of workers who reported various symptoms, along with the shift (morning, evening, or night) that they worked.

	Shift		
	Morning	**Evening**	**Night**
Influenza	16	13	18
Headache	24	33	6
Weakness	11	16	5
Shortness of Breath	7	9	9

Source: *Journal of Environmental Science and Health* A39:1129–1139

a. Compute the expected frequencies under the null hypothesis.
b. Compute the value of the chi-square statistic.
c. How many degrees of freedom are there?
d. Test the hypothesis of independence. Use the $\alpha = 0.05$ level of significance. What do you conclude?

12. Beryllium disease: Beryllium is an extremely lightweight metal that is used in many industries, such as aerospace and electronics. Long-term exposure to beryllium can cause people to become sensitized. Once an individual is sensitized, continued exposure can result in chronic beryllium disease, which involves scarring of the lungs. In a study of the effects of exposure to beryllium, workers were categorized by their duration of exposure (in years) and by their disease status (diseased, sensitized, or normal). The results were as follows:

	Duration of Exposure		
	< 1	**1 to < 5**	**≥ 5**
Diseased	10	8	23
Sensitized	9	19	11
Normal	70	136	206

Source: *Environmental Health Perspectives* 113:1366–1372

a. Compute the expected frequencies under the null hypothesis.
b. Compute the value of the chi-square statistic.
c. How many degrees of freedom are there?
d. Test the hypothesis of independence. Use the $\alpha = 0.01$ level of significance. What do you conclude?

13. No smoking: The General Social Survey conducted a poll of 668 adults in which the subjects were asked whether they agree that the government should prohibit smoking in public places. In addition, each was asked how many people lived in his or her household. The results are summarized in the following contingency table.

	Household Size				
	1	**2**	**3**	**4**	**5**
Agree	73	109	48	37	37
No Opinion	31	52	29	20	18
Disagree	42	71	38	40	23

a. Compute the expected frequencies.
b. Compute the value of the test statistic.
c. How many degrees of freedom are there?
d. Test the hypothesis of independence. Use the $\alpha = 0.01$ level of significance. What do you conclude?

14. How big is your family? The General Social Survey asked a sample of 2780 adults how many children they had, and also how many siblings (brothers and sisters) they had. The results are summarized in the following contingency table.

	Number of Children					
Siblings	**0**	**1**	**2**	**3**	**4**	**More Than 4**
0	60	19	34	15	10	6
1	189	97	163	63	29	12
2	179	98	137	81	41	20
3	124	85	132	83	31	16
4	85	53	88	61	32	22
More Than 4	128	121	175	156	77	58

a. Compute the expected frequencies.
b. Compute the value of the test statistic.
c. How many degrees of freedom are there?
d. Test the hypothesis of independence. Use the $\alpha = 0.05$ level of significance. What do you conclude?

15. Age discrimination: The following table presents the numbers of employees, by age group, who were promoted, or not promoted, in a sample drawn from a certain industry during the past year.

	Age			
	Under 30	**30–39**	**40–49**	**50 and Over**
Promoted	16	22	26	15
Not Promoted	38	39	45	38

Can you conclude that the people in some age groups are more likely to be promoted than those in other age groups? Use the $\alpha = 0.05$ level of significance.

16. Schools and museums: Do people who are interested in environmental issues visit museums more often than people who are not? In 2012, the General Social Survey asked 990 people how interested they were in environmental issues, and how often they had visited a science museum in the past year. The following table presents the results.

	Number of Visits		
	None	**One**	**More Than One**
Very interested	331	96	40
Moderately interested	347	57	25
Not interested	70	15	9

Can you conclude that the number of visits to a science museum is related to interest in environmental issues? Use the $\alpha = 0.01$ level of significance.

17. Genes: At a certain genetic locus on a chromosome, each individual has one of three different DNA sequences (alleles). The three alleles are denoted A, B, C. At another genetic locus on the same chromosome, each organism has one of three alleles, denoted 1, 2, 3. Each individual therefore has one of nine possible allele pairs: A1, A2, A3, B1, B2, B3, C1, C2, or C3. These allele pairs are called *haplotypes*. The loci are said to be in *linkage equilibrium* if the two alleles in an individual's haplotype are independent. Haplotypes were determined for 316 individuals. The following MINITAB output presents the results of a chi-square test for independence.

```
Chi-Square Test: A, B, C

Expected counts are printed below observed counts
Chi-Square contributions are printed below expected counts

              A        B        C    Total
   1         66       44       34      144
          61.06    47.39    35.54
          0.399    0.243    0.067

   2         36       38       20       94
          39.86    30.94    23.20
          0.374    1.613    0.442

   3         32       22       24       78
          33.08    25.67    19.25
          0.035    0.525    1.170

Total      134      104       78      316

Chi-Sq= 4.868, DF = 4,  P-Value = 0.301
```

a. How many individuals were observed to have the haplotype B3?

b. What is the expected number of individuals with the haplotype A2?

c. Which of the nine haplotypes was least frequently observed?

d. Which of the nine haplotypes has the smallest expected count?

e. What is the *P*-value?

f. Can you conclude at the $\alpha = 0.05$ level that the loci are not in linkage equilibrium (i.e., not independent)? Explain.

g. Can you conclude that the loci are in linkage equilibrium (i.e., independent)? Explain.

18. Product rating: A firm that is planning to market a new cleaning product surveyed 1268 users of the leading competitor's product. Each person rated the product as fair, good, or excellent. In addition, each person stated whether they use the product rarely (1), occasionally (2), or frequently (3). The firm is interested in determining whether the rating given to the product is independent of the frequency with which the product is used. The following MINITAB output presents the results of a chi-square test for independence.

```
Chi-Square Test: Fair, Good, Excellent

Expected counts are printed below observed counts
Chi-Square contributions are printed below expected counts

            Fair     Good  Excellent    Total
   1          97      136         92      325
           79.97   141.23     103.81
           3.627    0.193      1.343

   2         128      234        155      517
          127.21   224.66     165.13
           0.005    0.388      0.621

   3          87      181        158      426
          104.82   185.12     136.06
           3.030    0.091      3.536

Total       312      551        405     1268

Chi-Sq= 12.835, DF = 4,  P-Value = 0.012
```

a. How many individuals who rarely use the product rated it good?

b. What is the expected number of individuals who used the product occasionally and rated it excellent?

c. Which of the nine combinations was least frequently observed?

d. Which of the nine combinations has the smallest expected count?

e. What is the *P*-value?

f. Can you conclude at the $\alpha = 0.05$ level that rating and frequency of use are not independent? Explain.

Extending the Concepts

19. Degrees of freedom: In the following contingency table, the row and column totals are presented, but the data for one row and one column have been omitted. Thus, there are data for only $r - 1$ rows and $c - 1$ columns. Show that you can calculate the missing values.

	1	2	3	4	Row Total
A	10	52	29	—	98
B	25	38	10	—	92
C	—	—	—	—	147
Column Total	65	117	83	72	

Conclude that if $(r - 1)(c - 1)$ entries in the table are known, the remaining ones are automatically determined, so there are no degrees of freedom left.

20. Are you an optimist? The General Social Survey asked 1373 men and 993 women in the United States whether they agreed that they were generally optimistic about the future. The results are presented in the following table.

	Male	Female
Optimistic	1148	815
Pessimistic	225	178

a. Compute the value of the χ^2 statistic for testing the hypothesis that the opinion on optimism is independent of gender.

b. Compute the proportion of men who were optimistic.

c. Compute the proportion of women who were optimistic.

d. Compute the test statistic z for testing the null hypothesis that the two proportions are equal versus the alternative that they are not equal.

e. Show that $\chi^2 = z^2$.

f. Use technology to find the P-value for each test. Show that the P-values are equal.

g. Conclude that when a contingency table has two rows and two columns, the chi-square test is equivalent to the test for the difference between proportions.

Answers to Check Your Understanding Exercises for Section 12.2

1.

	1	2	3
A	14.421	12.155	21.425
B	22.232	18.738	33.030
C	14.120	11.901	20.979
D	19.227	16.206	28.567

2. a.

	1	2	3
A	13.098	10.736	11.166
B	12.724	10.429	10.847
C	17.963	14.724	15.313
D	17.215	14.110	14.675

b. Yes, because all the expected frequencies are 5 or more.

c. 20.973   **d.** 6   **e.** Yes

Chapter 12 Summary

Section 12.1: In this section, we introduced the chi-square test for goodness of fit. This test is based on the chi-square statistic. The data consist of observed frequencies in several categories. The null hypothesis specifies a distribution, which consists of a probability for each category. The chi-square statistic is used to test this null hypothesis. The degrees of freedom is 1 less than the number of categories. We reject H_0 if the value of the test statistic is greater than or equal to the critical value.

Section 12.2: In this section, we studied contingency tables. A contingency table relates the values of two qualitative variables. When both row and column totals are random, we test the hypothesis that the two variables are independent. When one of the variables is assigned, and the other is random, we test the hypothesis of homogeneity. The procedure is the same for both tests. The test statistic is the chi-square statistic with $(r-1)(c-1)$ degrees of freedom, where r is the number of rows and c is the number of columns. We reject H_0 if the value of the test statistic is greater than or equal to the critical value.

Vocabulary and Notation

chi-square distribution   558
chi-square statistic   558
cell   567
column variable   567

contingency table   567
expected frequency   560
goodness-of-fit test   559
grand total   568

observed frequency   559
row variable   567
test of homogeneity   571
test of independence   567

Important Formulas

Chi-square statistic:

$$\chi^2 = \Sigma \frac{(O-E)^2}{E}$$

Expected frequency for goodness-of-fit:

$E = np$

Expected frequency for independence or homogeneity:

$$E = \frac{\text{Row total} \cdot \text{Column total}}{\text{Grand total}}$$

Chapter Quiz

1. A contingency table containing observed values has four rows and five columns. The value of the chi-square statistic for testing independence is 22.87. Is H_0 rejected at the $\alpha = 0.05$ level?

2. A goodness-of-fit test is performed to test the null hypothesis that each of the six faces on a die has probability $1/6$ of coming up. The null hypothesis is rejected. True or false: We can conclude that the probability that a 6 comes up is not equal to $1/6$.

3. A test of homogeneity is performed and the null hypothesis is rejected. True or false: We can conclude that the distributions are not the same in every row.

Exercises 4–9 refer to the following data:

Electric motors are assembled on four different production lines. Random samples of motors are taken from each line and inspected. The numbers that pass and that fail the inspection are counted for each line, with the following results:

	Line			
	1	**2**	**3**	**4**
Pass	482	467	458	404
Fail	57	59	37	47

4. State the appropriate null and alternate hypotheses for determining whether to conclude that the failure rates differ among the four lines.

5. Compute the expected frequencies.

6. Compute the value of the chi-square statistic.

7. How many degrees of freedom are there?

8. Find the critical value for the $\alpha = 0.05$ level of significance.

9. State a conclusion.

Exercises 10–15 refer to the following data:

Anthropologists can estimate the birthrate of an ancient society by studying the age distribution of skeletons found in ancient cemeteries. An article in the journal *Current Anthropology* presented the following numbers of skeletons of various ages found at two such sites.

	Ages of Skeletons		
Site	**0–4 Years**	**5–19 Years**	**20 Years or More**
Casa da Moura	27	61	126
Wandersleben	38	60	118

Source: *Current Anthropology* 43:637–650

10. State the appropriate null and alternate hypotheses for determining whether to conclude that the age distributions differ between the two sites.

11. Compute the expected frequencies.

12. Compute the value of the chi-square statistic.

13. How many degrees of freedom are there?

14. Find the critical value for the $\alpha = 0.01$ level of significance.

15. State a conclusion.

Review Exercises

Exercises 1–3 refer to the following data:

A hypothetical sample of 200 families, each with four children, was selected, and the number of boys was recorded for each. The results are presented in the following table.

Number of boys	0	1	2	3	4
Observed	10	57	80	38	15

1. Null hypothesis: Let p_0 be the probability that a family with four children has no boys, p_1 be the probability that a family has one boy, and so on. The null hypothesis is that the number of boys in a four-child family follows a binomial distribution with $n = 4$ and $p = 0.5$. State the null hypothesis in terms of p_0, p_1, p_2, p_3, and p_4.

2. Expected frequencies: Compute the expected frequencies.

3. State a conclusion: Can you conclude that the distribution of boys differs from the binomial with $p = 0.5$? Use the $\alpha = 0.05$ level of significance.

Exercises 4–6 refer to the following data:

The General Social Survey polled 1280 men and 1531 women to determine their level of education. The results are presented in the following table.

	Educational Level				
	No High School Diploma	High School Diploma	Associate's Degree	Bachelor's Degree	Graduate Degree
Men	178	608	96	248	150
Women	186	827	128	259	131

4. **Expected frequencies:** Compute the expected frequencies under the null hypothesis of independence.

5. **Test statistic:** Compute the value of the chi-square statistic.

6. **State a conclusion:** Can you conclude that education level is independent of gender? Use the $\alpha = 0.01$ level of significance.

Exercises 7–9 refer to the following data:

At an assembly plant for light trucks, routine monitoring of the quality of welds yielded the following data.

	Number of Welds		
	High Quality	Moderate Quality	Low Quality
Day Shift	467	191	42
Evening Shift	445	171	34
Night Shift	254	129	17

7. **Expected frequencies:** Compute the expected frequencies under the null hypothesis of homogeneity.

8. **Test statistic:** Compute the value of the chi-square statistic.

9. **State a conclusion:** Can you conclude that the quality varies among shifts? Use the $\alpha = 0.01$ level of significance.

Exercises 10–15 refer to the following data:

In a hypothetical study, four hospitals were compared with regard to the outcome of a particular type of surgery. For each patient at each hospital, the outcome was classified as Substantial improvement, Some improvement, or No improvement. The results are presented in the following contingency table.

	Outcome		
Hospital	Substantial Improvement	Some Improvement	No Improvement
A	114	64	22
B	132	52	16
C	100	64	36
D	126	56	18

10. **Expected frequencies:** Compute the expected frequencies under the null hypothesis of homogeneity.

11. **Test statistic:** Compute the value of the chi-square statistic.

12. **State a conclusion:** Can you conclude that the distribution of outcomes differs among the hospitals? Use the $\alpha = 0.05$ level of significance.

13. **Can't read the numbers:** Because of printer failure, none of the observed frequencies in the following table were printed, but some of the row and column totals were. Is it possible to construct the corresponding table of expected frequencies from the information given? If so, construct it. If not, describe the additional information you would need.

	Observed			
	1	2	3	Total
A	—	—	—	25
B	—	—	—	—
C	—	—	—	40
D	—	—	—	75
Total	50	20	—	150

14. **Lottery:** Powerball is a multistate lottery in which players try to guess the numbers that will turn up in a drawing of numbered balls. Between January 18, 2012, and June 26, 2013, a total of 755 balls numbered 1 through 59 were drawn. Following are the observed frequencies.

Number	Frequency	Number	Frequency	Number	Frequency	Number	Frequency
1	15	16	14	31	11	46	16
2	7	17	13	32	12	47	9
3	16	18	10	33	11	48	13
4	11	19	13	34	11	49	13
5	16	20	14	35	11	50	11
6	14	21	9	36	17	51	9
7	17	22	14	37	6	52	12
8	17	23	21	38	8	53	12
9	10	24	8	39	15	54	12
10	14	25	6	40	10	55	17
11	16	26	19	41	16	56	18
12	9	27	9	42	9	57	13
13	16	28	14	43	13	58	13
14	17	29	15	44	15	59	16
15	10	30	11	45	11		

Source: www.usamega.com

Perform a test of the null hypothesis that each of the numbers is equally likely to come up. Use the $\alpha = 0.05$ level.

15. **Absent from school:** Following are the total numbers of absences on each day of the week for the academic year 2008–2009 at a Montana elementary school.

Monday	Tuesday	Wednesday	Thursday	Friday
844	909	781	795	837

Test the null hypothesis that absences are equally likely to occur on any day of the week. Use the $\alpha = 0.05$ level of significance.

Write About It

1. Why do large values of χ^2 provide evidence against H_0? Why don't small values of χ^2 provide evidence against H_0?

2. Explain what the expected frequencies represent in a goodness-of-fit test.

3. If the row variable and column variable are interchanged, how are the expected frequencies affected? Is the value of the chi-square statistic affected? Is the number of degrees of freedom affected?

4. Explain what the expected frequencies represent in a test of independence.

Case Study: Gender Bias In Graduate Admissions

The chapter opener described a famous study carried out at the University of California at Berkeley in the 1970s, regarding what appeared to be a case of gender discrimination in admissions to graduate school. The information in this case study was taken from the article "Sex Bias in Graduate Admissions Data from Berkeley," which appeared in *Science* magazine.

Table 12.9 presents the numbers of male and female applicants to six of the most popular departments at the University of California at Berkeley. Out of 2691 male applicants, 1198, or 44.5%, were accepted. Out of 1835 female applicants, only 557, or 30.4%, were accepted.

Table 12.9

Gender	Accept	Reject	Total
Male	1198	1493	2691
Female	557	1278	1835
Total	1755	2771	4526

In graduate school, each department conducts its own admissions process. Table 12.10 presents the numbers of applicants accepted and rejected by each of the six departments.

Table 12.10

Department	A	B	C	D	E	F	Total
Accept	601	370	322	269	147	46	1755
Reject	332	215	596	523	437	668	2771
Total	933	585	918	792	584	714	4526

University policy does not allow these departments to be identified by name.

Table 12.11 presents the numbers of men and women who applied to each of the six departments.

Table 12.11

Department	A	B	C	D	E	F	Total
Male	825	560	325	417	191	373	2691
Female	108	25	593	375	393	341	1835
Total	933	585	918	792	584	714	4526

University policy does not allow these departments to be identified by name.

1. Use the data in Table 12.9 to test the null hypothesis that acceptance to graduate school is independent of gender. Show that this hypothesis is rejected at the $\alpha = 0.01$ level.

2. Use the data in Table 12.10 to test the null hypothesis that acceptance to graduate school is independent of department. Show that this hypothesis is rejected at the $\alpha = 0.01$ level.

3. Use the data in Table 12.11 to test the null hypothesis that gender is independent of department. Show that this hypothesis is rejected at the $\alpha = 0.01$ level.

   We conclude that department is associated with both gender and admissions. Therefore, department is a confounder for the relationship between gender and admissions. This means that it is possible that the differences in admissions rates between men and women may be due to differences in the departments they apply to, rather than to gender. To determine whether this is the case, we must look at the admissions rates for men and women in each department separately. The following six tables present these data.

Department A			
Gender	Accept	Reject	Total
Male	512	313	825
Female	89	19	108
Total	601	332	933

Department B			
Gender	Accept	Reject	Total
Male	353	207	560
Female	17	8	25
Total	370	215	585

Department C			
Gender	Accept	Reject	Total
Male	120	205	325
Female	202	391	593
Total	322	596	918

Department D			
Gender	Accept	Reject	Total
Male	138	279	417
Female	131	244	375
Total	269	523	792

Department E			
Gender	Accept	Reject	Total
Male	53	138	191
Female	94	299	393
Total	147	437	584

Department F			
Gender	Accept	Reject	Total
Male	22	351	373
Female	24	317	341
Total	46	668	714

4. For each department, test the hypothesis that gender and admissions are independent. Use the $\alpha = 0.01$ level of significance. Show that the only department for which this hypothesis is rejected is department A. Show that, in fact, the admissions rate for women in department A is significantly higher than the rate for men.

5. For each department, compute the proportion of all applicants who are accepted.

6. For each department, compute the proportion of all applicants who are men.

7. In general, do the departments with the higher acceptance proportions tend to have the higher proportions of male applicants?

8. Which of the following is the best explanation for the fact that the graduate admissions rate is lower for women than for men?
  i. The admissions process discriminates against women.
  ii. Women tend to apply to departments that are harder to get into.

    The data set in this case study provides an example of a result known as *Simpson's Paradox*. Simpson's Paradox occurs when two variables in a contingency table are dependent, but the dependence vanishes or reverses when separate tables are made for each level of a confounder. The lesson to be learned is that lack of independence is not the same as causation. When we reject the null hypothesis of independence in an observational study, it is possible that the result may be due to confounding. To determine whether a certain variable is a confounder, it is necessary to make a separate table for each value of the confounder.

Inference in Linear Models

Introduction

Renewable energy sources have the potential to produce clean energy indefinitely. One form of renewable energy that has been extensively studied is wind. Windmills are used to produce electric power from the energy in the wind. Of course, stronger winds produce more power. The following data set, based on a model relating wind speed and generated power, represents measurements collected at a particular site on 42 different days. Wind speed is measured in miles per hour and power is measured in watts.

Wind Speed	Power	Wind Speed	Power	Wind Speed	Power	Wind Speed	Power
4.2	10.3	4.2	5.3	2.6	9.1	4.6	5.0
1.4	1.4	3.7	1.1	7.7	46.1	8.0	65.1
6.6	39.4	5.9	29.2	6.1	26.9	7.7	54.6
4.7	12.3	6.0	24.4	5.5	19.7	1.6	0.6
2.6	9.0	5.3	13.9	4.7	5.4	5.1	17.9
5.8	20.6	5.1	9.5	4.0	12.4	6.6	32.5
1.8	3.8	4.9	2.9	2.3	5.0	2.3	9.4
5.8	21.1	8.3	63.7	8.6	67.4	5.8	10.8
7.3	41.3	7.1	37.1	5.6	15.2	6.9	29.3
7.1	45.2	9.2	87.5	4.2	11.0	6.2	21.4
6.4	28.4	4.4	0.5				

It is important to be able to predict the amount of power that will be obtained from a given wind speed. In the case study at the end of the chapter, we will use the methods of the chapter to determine how precisely we can predict power from wind speed.

SECTION 13.1   Inference on the Slope of the Regression Line

Objectives

1. State the assumptions of the linear model
2. Check the assumptions of the linear model
3. Construct confidence intervals for the slope
4. Test hypotheses about the slope

Objective 1 State the assumptions of the linear model

The Linear Model

In Chapter 4, we discussed bivariate data — data that are in the form of ordered pairs. We learned to construct scatterplots, and we found that, in many cases, the points on a scatterplot tend to cluster around a straight line. We learned to summarize a scatterplot by computing the least-squares regression line.

Before continuing, we will review these ideas with a data set based on a nutritional analysis of more than 700 candy products conducted by the United States Department of Agriculture. Table 13.1 presents the number of calories and the number of grams of fat per 100 grams of product for a sample of 18 candy products.

Table 13.1 Grams of Fat and Number of Calories per 100 Grams of Product

Product	Fat (x)	Calories (y)	Product	Fat (x)	Calories (y)
3 Musketeers	12.75	436	Mr. Goodbar	33.21	538
Kit Kat	25.99	518	100 Grand	19.33	468
M&M Plain	21.13	492	Baby Ruth	21.60	459
M&M Peanut	26.13	515	Bit O' Honey	7.50	375
Milky Way	17.23	456	Butterfinger	18.90	459
Skittles	4.37	405	Oh Henry!	23.00	462
Snickers	23.85	491	Reese's Pieces	24.77	497
Starburst	8.36	408	Tootsie Roll	3.31	387
Twix	24.85	502	Twizzlers	2.32	350

Source: USDA National Nutrient Database

The data consist of 18 points (x, y). To summarize the data, we compute the least-squares regression line, which was introduced in Section 4.2. We repeat the definition here.

The Least-Squares Regression line

Let x be an explanatory variable and y an outcome variable, with means $\bar{x}$ and $\bar{y}$, standard deviations s_x and s_y, and correlation coefficient r. The equation of the least-squares regression line for predicting y from x is

$$\hat{y} = b_0 + b_1 x$$

Formulas for b_1 and b_0 are

$$b_1 = r\frac{s_y}{s_x}, \text{ where } r \text{ is the correlation coefficient}$$
$$b_0 = \bar{y} - b_1\bar{x}$$

An alternate formula for b_1 is

$$b_1 = \frac{\sum(x - \bar{x})(y - \bar{y})}{\sum(x - \bar{x})^2}$$

EXAMPLE 13.1   Compute the least-squares regression line

Use technology to find the least-squares regression line for the data in Table 13.1.

Solution

The following TI-84 Plus display shows that the equation of the least-squares regression line is $\hat{y} = 356.7392193 + 5.639341034x$.

Note that the display labels the y-intercept as "a" and the slope as "b," whereas in our notation, we refer to the y-intercept as b_0 and the slope as b_1. In our notation, the intercept of the line is $b_0 = 356.7392193$ and the slope is $b_1 = 5.639341034$.

Figure 13.1 presents a scatterplot of the data, from MINITAB, with the least-squares regression line superimposed.

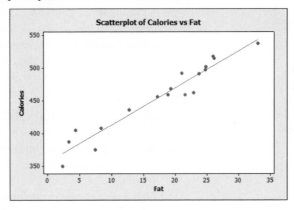

Figure 13.1   A scatterplot of the data in Table 13.1. The least-squares regression line is superimposed.

When the points on a scatterplot are a random sample from a population, we can imagine plotting every point in the population on a scatterplot. Then, if certain assumptions are met, we say that the population follows a **linear model**. The intercept b_0 and the slope b_1 of the least-squares regression line are then estimates of a population intercept β_0 and a population slope β_1. We cannot determine the exact values of β_0 and β_1, because we cannot observe the entire population. However, we can use the sample points to construct confidence intervals and test hypotheses about β_0 and β_1. We will focus on β_1.

The assumptions for a linear model are illustrated in Figure 13.2 on page 586, which presents a hypothetical scatterplot of an entire population. The scatterplot is divided up into narrow vertical strips, so that the points within each strip have approximately the same x-value. The following three conditions hold:

Assumptions for the Linear Model

1. The mean of the y-values within a strip is denoted $\mu_{y|x}$. As x varies, the values of $\mu_{y|x}$ follow a straight line: $\mu_{y|x} = \beta_0 + \beta_1 x$.

2. The amount of vertical spread is approximately the same in each strip, except perhaps near the ends.

3. The y-values within a strip are approximately normally distributed. (This is not obvious from the scatterplot.) This assumption is not necessary if the sample size is large ($n > 30$).

Figure 13.2 A scatterplot of a population that follows the linear model

When the assumptions of the linear model hold, the points (x, y) satisfy the following linear model equation:

$$y = \beta_0 + \beta_1 x + \varepsilon$$

In this equation, β_0 is the y-intercept, β_1 is the slope of the line that summarizes the entire population, and ε is a random error. The y-intercept b_0 and the slope b_1 of the least-squares line are estimates of β_0 and β_1.

In the remainder of this section, we will describe how to check the assumptions of the linear model and how to construct confidence intervals and perform hypothesis tests when these assumptions are met. The computations are rather lengthy, and technology is almost always used in practice. We will present the hand calculations first, to illustrate the ideas behind the methods, and then we will present examples using technology.

Objective 2 Check the assumptions of the linear model

Checking Assumptions

Before we can compute confidence intervals and perform hypothesis tests, we must check that the assumptions of the linear model are satisfied. In practice, we do not see the entire population, so we must use the sample to do the checking.

We check the assumptions with a residual plot. Residual plots were first discussed in Section 4.3. We review the definition here.

DEFINITION

Given a point (x, y) and the least-squares regression line $\hat{y} = b_0 + b_1 x$, the **residual** for the point (x, y) is the difference between the observed value y and the predicted value $\hat{y}$:

$$\text{Residual} = y - \hat{y}$$

A **residual plot** is a plot in which the residuals are plotted against the values of the explanatory variable x. In other words, the points on the residual plot are $(x, \; y - \hat{y})$.

Figure 13.3 presents the residual plot, as drawn by MINITAB, for the data in Table 13.1.

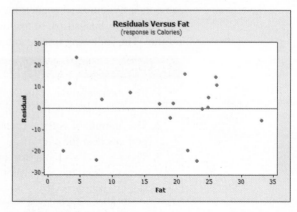

Figure 13.3 Residual plot for the data in Table 13.1

Explain It Again

Reasons for the conditions:
Condition 1 checks that $\mu_{y|x}$ follows a straight line. Condition 2 checks that the variance of the y-values is the same for every x-value. Condition 3 checks that the y-values corresponding to a given x are approximately normally distributed.

When the linear model assumptions are satisfied, the residual plot will not show any obvious pattern. In particular, the residual plot must satisfy the following conditions.

Conditions for the Residual Plot

1. The residual plot must not exhibit an obvious pattern.
2. The vertical spread of the points in the residual plot must be roughly the same across the plot.
3. There must be no outliers.

Figure 13.3 satisfies all three conditions, so we can assume that the linear model is valid, and we can construct confidence intervals and test hypotheses.

EXAMPLE 13.2

Checking the assumptions with a residual plot

For each of the following residual plots, determine whether the assumptions of the linear model are satisfied. If they are not, specify which assumptions are violated.

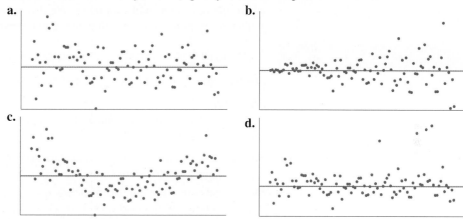

Solution

a. The plot shows no obvious pattern. The assumptions of the linear model are satisfied.
b. The vertical spread varies. The linear model is not valid.
c. The plot shows an obvious curved pattern. The linear model is not valid.
d. The plot contains outliers. The linear model is not valid.

Check Your Understanding

1. For each of the following residual plots, determine whether the assumptions of the linear model are satisfied. If they are not, specify which assumptions are violated.

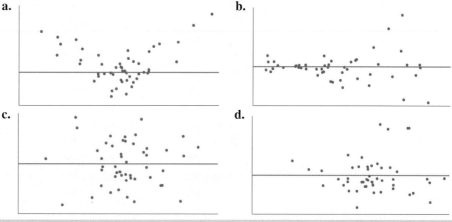

Answers are on page 599.

Objective 3 Construct confidence intervals for the slope

Constructing Confidence Intervals for the Slope

The slope b_1 of the least-squares regression line is a point estimate of the population slope β_1. When the assumptions of the linear model are satisfied, we can construct a confidence interval for β_1. To form a confidence interval, we need a point estimate, a standard error, and a critical value. The point estimate for β_1 is b_1.

The standard error of b_1

To compute the standard error of b_1, we first compute a quantity called the **residual standard deviation**. This quantity, denoted s_e, measures the spread of the points on the scatterplot around the least-squares regression line. The formula for s_e is

$$s_e = \sqrt{\frac{\sum(y - \hat{y})^2}{n - 2}}$$

To understand this formula, note that the predicted value $\hat{y}$ is an estimate of the mean of y. The residual $y - \hat{y}$ estimates the deviation between the observed data value y and its mean. The formula for s_e is similar to the formula for the sample standard deviation, in that we square the deviations and sum them. We divide by $n - 2$ rather than $n - 1$ because the predicted values involve the estimation of two parameters, β_0 and β_1.

EXAMPLE 13.3

Compute the residual standard deviation

Compute the residual standard deviation for the data in Table 13.1.

Solution

In Example 13.1, we computed the equation of the least-squares line to be $\hat{y} = 356.7392193 + 5.639341034x$.

Step 1: Compute the residuals $y - \hat{y}$, and the sum of squared residuals $\sum(y - \hat{y})^2$. Table 13.2 illustrates the calculations. We obtain $\sum(y - \hat{y})^2 = 3398.6787$.

Step 2: Substitute the number of points $n = 18$, and the value $\sum(y - \hat{y})^2 = 3398.6787$ into the formula for s_e:

$$s_e = \sqrt{\frac{\sum(y - \hat{y})^2}{n - 2}} = \sqrt{\frac{3398.6787}{18 - 2}} = 14.574547$$

Table 13.2

x	y	Predicted Value $\hat{y} = 356.73921925 + 5.63934103x$	Residual $y - \hat{y}$	Residual^2 $(y - \hat{y})^2$
12.75	436	428.640817	7.359183	54.15756880
25.99	518	503.305693	14.694307	215.92266939
21.13	492	475.898495	16.101505	259.25845638
26.13	515	504.095200	10.904800	118.91465510
17.23	456	453.905065	2.094935	4.38875183
4.37	405	381.383140	23.616860	557.75609746
23.85	491	491.237503	−0.237503	0.05640759
8.36	408	403.884110	4.115890	16.94054835
24.85	502	496.876844	5.123156	26.24672898
33.21	538	544.021735	−6.021735	36.26129068
19.33	468	465.747681	2.252319	5.07293926
21.60	459	478.548985	−19.548985	382.16283400
7.50	375	399.034277	−24.034277	577.64646971
18.90	459	463.322765	−4.322765	18.68629480
23.00	462	486.444063	−24.444063	597.51221301
24.77	497	496.425697	0.574303	0.32982444
3.31	387	375.405438	11.594562	134.43386660
2.32	350	369.822490	−19.822490	392.93112723

$$\sum(y - \hat{y})^2 = 3398.6787$$

Once we have computed s_e, we divide it by the quantity $\sqrt{\sum(x-\bar{x})^2}$ to obtain the standard error of b_1. The quantity $\sum(x-\bar{x})^2$ is called the **sum of squares for x**. The standard error of b_1 is

$$s_b = \frac{s_e}{\sqrt{\sum(x-\bar{x})^2}}$$

EXAMPLE 13.4

Compute the standard error of b_1

Compute the standard error s_b for the data in Table 13.1.

Solution

Step 1: Compute s_e, the residual standard deviation. We did this in Example 13.3, and obtained $s_e = 14.574547$.

Step 2: Compute $\sum(x-\bar{x})^2$. The calculations are presented in Table 13.3. We obtain $\sum(x-\bar{x})^2 = 1415.7452$.

Table 13.3

x	$x-\bar{x}$	$(x-\bar{x})^2$
12.75	−4.95	24.5025
25.99	8.29	68.7241
21.13	3.43	11.7649
26.13	8.43	71.0649
17.23	−0.47	0.2209
4.37	−13.33	177.6889
23.85	6.15	37.8225
8.36	−9.34	87.2356
24.85	7.15	51.1225
33.21	15.51	240.5601
19.33	1.63	2.6569
21.60	3.90	15.2100
7.50	−10.20	104.0400
18.90	1.20	1.4400
23.00	5.30	28.0900
24.77	7.07	49.9849
3.31	−14.39	207.0721
2.32	−15.38	236.5444

$$\bar{x} = \frac{\sum x}{n} = 17.7 \qquad \sum(x-\bar{x})^2 = 1415.7452$$

We now compute the standard error s_b:

$$s_b = \frac{s_e}{\sqrt{\sum(x-\bar{x})^2}} = \frac{14.574547}{\sqrt{1415.7452}} = 0.387349$$

The critical value and the margin of error

Under the assumptions of the linear model, the quantity

$$\frac{b_1 - \beta_1}{s_b}$$

has a Student's t distribution with $n-2$ degrees of freedom. Therefore, the critical value for a level $100(1-\alpha)\%$ confidence interval is the value $t_{\alpha/2}$ for which the area under the t curve with $n-2$ degrees of freedom between $-t_{\alpha/2}$ and $t_{\alpha/2}$ is $1-\alpha$.

The margin of error for a level $100(1 - \alpha)\%$ confidence interval is

$$\text{Margin of error} = t_{\alpha/2} \cdot s_b$$

A level $100(1 - \alpha)\%$ confidence interval for β_1 is

$$b_1 \pm t_{\alpha/2} \cdot s_b$$

$$b_1 - t_{\alpha/2} \cdot s_b < \beta_1 < b_1 + t_{\alpha/2} \cdot s_b$$

The steps for constructing a confidence interval for β_1 are as follows:

Constructing a Confidence Interval for β_1

Step 1: Compute the least-squares regression line $\hat{y} = b_0 + b_1 x$.

Step 2: Compute the residuals and construct a residual plot to be sure the assumptions of the linear model are satisfied.

Step 3: Compute the residual standard deviation: $s_e = \sqrt{\dfrac{\sum(y - \hat{y})^2}{n - 2}}$.

Step 4: Compute the standard error of b_1: $s_b = \dfrac{s_e}{\sqrt{\sum(x - \bar{x})^2}}$.

Step 5: Find the critical value $t_{\alpha/2}$ from the Student's t curve with $n - 2$ degrees of freedom and multiply it by the standard error to obtain the margin of error $t_{\alpha/2} \cdot s_b$.

Step 6: Use the point estimate b_1 and the margin of error $t_{\alpha/2} \cdot s_b$ to compute the confidence interval.

$$\text{Point estimate} \pm \text{Margin of error}$$

$$b_1 \pm t_{\alpha/2} \cdot s_b$$

$$b_1 - t_{\alpha/2} \cdot s_b < \beta_1 < b_1 + t_{\alpha/2} \cdot s_b$$

Step 7: Interpret the result.

EXAMPLE 13.5   Construct a confidence interval

Construct a 95% confidence interval for the slope β_1 for the candy data.

Solution

Step 1: Compute the least-squares regression line. We computed the equation of the least-squares regression line in Example 13.1 and obtained $\hat{y} = 356.7392193 + 5.639341034x$.

Step 2: Compute the residuals and construct a residual plot to be sure the assumptions of the linear model are satisfied. The residual plot was presented in Figure 13.3. The assumptions of the linear model are satisfied.

Step 3: Compute the residual standard deviation. We computed the residual standard deviation in Example 13.3, and obtained $s_e = 14.574547$.

Step 4: Compute the standard error of b_1. We computed the standard error of b_1 in Example 13.4, and obtained $s_b = 0.387349$.

Step 5: Find the critical value $t_{\alpha/2}$ from the Student's t curve with $n - 2$ degrees of freedom and multiply it by the standard error to obtain the margin of error $t_{\alpha/2} \cdot s_b$. There are $n - 2 = 18 - 2 = 16$ degrees of freedom for the Student's t distribution. From Table A.3, we find that the critical value for a 95% confidence interval is $t_{\alpha/2} = 2.120$. The margin of error is therefore

$$\text{Margin of error} = t_{\alpha/2} \cdot s_b = 2.120 \cdot 0.387349 = 0.82118$$

Step 6: Use the point estimate b_1 and the margin of error to construct the confidence interval. The point estimate is $b_1 = 5.63934103$. The margin of error is 0.82118. The 95% confidence interval is

$$\text{Point estimate} \pm \text{Margin of error}$$

$$5.63934103 \pm 0.82118$$

$$4.8182 < \beta_1 < 6.4605$$

Step 7: Interpret the result. We are 95% confident that the mean difference in calories for items that differ by 1 gram in fat content is between 4.8182 and 6.4605.

EXAMPLE 13.6   Interpreting a confidence interval

A nutritionist believes that if two candies differ in their fat content by 1 gram, that their calorie count will differ, on average, by 5 calories. Is the confidence interval constructed in Example 13.5 consistent with this belief?

Solution

The parameter β_1 represents the mean difference in calories corresponding to a difference of one gram of fat. The confidence interval constructed in Example 13.5 is $4.8182 < \beta_1 < 6.4605$. This interval contains 5, so it is consistent with the belief.

Construct confidence intervals with technology

EXAMPLE 13.7   Construct a confidence interval with technology

Use the TI-84 Plus calculator to construct the confidence interval in Example 13.5.

Solution

The TI-84 Plus display is as follows:

The confidence interval is shown on the second line of the display. Step-by-step instructions for constructing confidence intervals with technology are presented in the Using Technology section on page 595.

Check Your Understanding

2. A certain data set contains 27 points. The least-squares regression line is computed, with the following results: $b_1 = 5.78$, $s_e = 1.35$, and $\sum (x - \bar{x})^2 = 3.4$. Construct a 95% confidence interval for β_1.

3. Following is a TI-84 Plus display showing a 95% confidence interval for β_1.

   a. What is the slope of the least-squares regression line?
   b. How many degrees of freedom are there?
   c. How many points are in the data set?
   d. What is the 95% confidence interval for β_1?

Answers are on page 599.

Objective 4 Test hypotheses about the slope

Explain It Again

$\beta_1 = 0$ **means no linear relationship:** If $\beta_1 = 0$, then the mean of the outcome variable is $\mu_{y|x} = \beta_0$ for every value of x. Therefore, no matter what the value of x, the mean of y stays the same.

Testing Hypotheses About the Slope

We can use the values of b_1 and s_b to test hypotheses about the population slope β_1. If $\beta_1 = 0$, then there is no linear relationship between the explanatory variable x and the outcome variable y. For this reason, the null hypothesis most often tested is $H_0: \beta_1 = 0$. If this null hypothesis is rejected, we conclude that there is a linear relationship between x and y, and that the explanatory variable x is useful in predicting the outcome variable y.

Recall that the quantity

$$\frac{b_1 - \beta_1}{s_b}$$

has a Student's t distribution with $n - 2$ degrees of freedom. We construct the test statistic for testing $H_0: \beta_1 = 0$ by setting $\beta_1 = 0$. The test statistic is

$$t = \frac{b_1}{s_b}$$

When H_0 is true, the test statistic has a Student's t distribution with $n - 2$ degrees of freedom. If the assumptions of the linear model are satisfied, a test of the hypothesis $\beta_1 = 0$ can be performed. The steps are as follows:

Testing $H_0: \beta_1 = 0$

Step 1: Compute the least-squares regression line. Verify that the assumptions of the linear model are satisfied.

Step 2: State the null and alternate hypotheses. The null hypothesis is $H_0: \beta_1 = 0$. The alternate hypothesis may be stated in any of three ways:

Left-tailed: $H_1: \beta_1 < 0$
Right-tailed: $H_1: \beta_1 > 0$
Two-tailed: $H_1: \beta_1 \neq 0$

Step 3: If making a decision, choose a significance level α.

Step 4: Compute the standard error of the slope s_b.

Step 5: Compute the value of the test statistic $t = \dfrac{b_1}{s_b}$ and the number of degrees of freedom $n - 2$.

The *P*-Value Method

Step 6: Compute the *P*-value of the test statistic.

The *P*-value is the area to the left of t. / The *P*-value is the area to the right of t. / The *P*-value is the sum of the areas in the two tails.

Left-tailed   Right-tailed   Two-tailed

Step 7: Interpret the *P*-value. If making a decision, reject H_0 if the *P*-value is less than or equal to the significance level α.

Step 8: State a conclusion.

The Critical Value Method

Step 6: Find the critical value.

Step 7: Determine whether to reject H_0, as follows:

Left-tailed: $H_1: \beta_1 < 0$ Reject if $t \leq -t_\alpha$.
Right-tailed: $H_1: \beta_1 > 0$ Reject if $t \geq t_\alpha$.
Two-tailed: $H_1: \beta_1 \neq 0$ Reject if $t \geq t_{\alpha/2}$ or $t \leq -t_{\alpha/2}$.

Step 8: State a conclusion.

EXAMPLE 13.8   Test a hypothesis about the slope

For the data in Table 13.1, perform a test of $H_0: \beta_1 = 0$ versus $H_1: \beta_1 > 0$. Use the $\alpha = 0.05$ level of significance.

Solution

Step 1: Compute the least-squares regression line and verify that the assumptions of the linear model are satisfied. The least-squares regression line was computed in Example 13.1. We obtained $b_1 = 10.0079$. The assumptions were verified in Figure 13.3.

Step 2: State the null and alternate hypotheses. The hypotheses are

$$H_0: \beta_1 = 0 \qquad H_1: \beta_1 > 0$$

Step 3: Choose a significance level. We will choose $\alpha = 0.05$.

Step 4: Compute s_b. In Example 13.4, we computed $s_b = 0.387349$.

Step 5: Compute the value of the test statistic. The value of s_b is 0.387349. The point estimate is $b_1 = 5.639341034$. We compute

$$t = \frac{b_1}{s_b} = \frac{5.639341034}{0.387349} = 14.56$$

Because the sample size is $n = 18$, the number of degrees of freedom is $n - 2 = 16$.

P-Value Method

Step 6: Compute the P-value of the test statistic. We use technology to compute the P-value. The following TI-84 Plus calculator display presents the P-value in scientific notation. The notation **E-11** indicates that the decimal point should be moved eight places to the left. Thus, the P-value is $P = 0.0000000000597$.

```
LinRegTTest
y=a+bx
β>0 and ρ>0
t=14.55880877
p=5.972739E-11
df=16
↓a=356.7392193
```

Step 7: Interpret the P-value. Because $P < 0.05$, we reject H_0. We conclude that $\beta_1 > 0$.

Step 8: State a conclusion. There is a linear relationship between the amount of fat and the number of calories in candy products. Because we conclude that $\beta_1 > 0$, we conclude that products with more fat tend to have more calories.

Critical Value Method

Step 6: Find the critical value. This is a right-tailed test, so the critical value is the value t_α for which the area to the right is $\alpha = 0.05$. We use Table A.3 with 16 degrees of freedom. The critical value is $t_\alpha = 1.746$.

Step 7: Determine whether to reject H_0. The value of the test statistic is $t = 14.56$. Because $t > t_\alpha$, we reject H_0. We conclude that $\beta_1 > 0$.

Step 8: State a conclusion. There is a linear relationship between the amount of fat and the number of calories in candy products. Because we conclude that $\beta_1 > 0$, we conclude that products with more fat tend to have more calories.

Testing the correlation

In Section 4.1, we introduced the sample correlation coefficient r, which is computed from a sample. If we knew the entire population and computed the correlation from it, we would obtain the **population correlation**, which is denoted with the Greek letter ρ (rho). The correlation measures the strength of the linear relationship between two variables. The population correlation ρ and the population slope β_1 always have the same sign. In particular, whenever one of them is equal to 0, the other is equal to 0 as well. For this reason, a test of the hypothesis $\beta_1 = 0$ is also a test of the hypothesis $\rho = 0$. A specialized test of $H_0: \rho = 0$ is available, but it always produces the same result as the test of $H_0: \beta_1 = 0$ that we have presented.

EXAMPLE 13.9   Test a hypothesis with technology

For the data in Table 13.1, perform a test of $H_0: \beta_1 = 0$ versus $H_1: \beta_1 > 0$. Use the $\alpha = 0.05$ level of significance.

Solution
The following TI-84 Plus display was presented in Example 13.8. We explain it in more detail.

```
LinRegTTest
y=a+bx
β>0 and ρ>0
t=14.55880877
p=5.972739E-11
df=16
↓a=356.7392193
```

The second line specifies the alternate hypothesis. The third line presents the value of the test statistic, and the fourth line shows the *P*-value. As explained in Example 13.8, the symbol **E-11** means that the decimal point should be moved over eight places to the left.

Now we will present the results from MINITAB.

```
Regression Analysis: Calories versus Fat

The regression equation is
Calories = 356.7 + 5.6 Fat

Predictor         Coef     SE Coef        T        P
Constant        356.74       7.669    46.52    0.000
Fat             5.6393       0.387    14.56    0.000
```

The *P*-value is given in the column headed "P," in the row labeled "Fat."

Check Your Understanding

4. For a given data set containing 18 points, the assumptions of the linear model are satisfied. The following values are computed: $b_1 = 5.58$ and $s_b = 4.42$. Perform a test of the hypothesis $H_0: \beta_1 = 0$ versus $H_1: \beta_1 \neq 0$. Use the $\alpha = 0.05$ level of significance. Can you conclude that the explanatory variable is useful in predicting the outcome variable? Explain.

5. For a given data set containing 26 points, the assumptions of the linear model are satisfied. The following values are computed: $b_1 = 46.8$ and $s_b = 15.2$. Perform a test of the hypothesis $H_0: \beta_1 = 0$ versus $H_1: \beta_1 > 0$. Use the $\alpha = 0.01$ level of significance. Can you conclude that the explanatory variable is useful in predicting the outcome variable? Explain.

6. The following TI-84 Plus display presents the results of a test of the null hypothesis $H_0: \beta_1 = 0$.

```
LinRegTTest
y=a+bx
β≠0 and ρ≠0
t=2.906470757
p=.0156590864
df=10
↓a=1.49955157
```

a. What is the alternate hypothesis?
b. What is the value of the test statistic?
c. How many degrees of freedom are there?
d. What is the *P*-value?
e. Can you conclude that the explanatory variable is useful in predicting the outcome variable? Answer this question using the $\alpha = 0.05$ significance level.

Answers are on page 599.

USING TECHNOLOGY

We use the data in Table 13.1 and Examples 13.5 and 13.8 to illustrate the technology steps.

TI-84 PLUS

Constructing a confidence interval for the slope of the least-squares regression line

Step 1. Enter the *x*-values from Table 13.1 into **L1** and the *y*-values into **L2** (Figure A).

Step 2. Press **STAT** and highlight the **TESTS** menu. Select **LinRegTInt** and press **ENTER** (Figure B). The **LinRegTInt** menu appears.

Step 3. Enter **L1** in the **Xlist** field and **L2** in the **Ylist** field.

Step 4. Enter the confidence level in the **C-Level** field. For Example 13.5, we enter **.95** (Figure C).

Step 5. Highlight **Calculate** and press **ENTER** (Figure D).

Figure A   Figure B

Figure C   Figure D

TI-84 PLUS

Testing a hypothesis about the slope of the least-squares regression line

Step 1. Enter the *x*-values from Table 13.1 into **L1** and the *y*-values into **L2** (Figure E).

Step 2. Press **STAT** and highlight the **TESTS** menu. Select **LinRegTTest** and press **ENTER** (Figure F). The **LinRegTTest** menu appears.

Step 3. Enter **L1** in the **Xlist** field and **L2** in the **Ylist** field.

Step 4. Select the form of the alternate hypothesis. For Example 13.8, the alternate hypothesis has the form **>0** (Figure G).

Step 5. Highlight **Calculate** and press **ENTER** (Figure H).

Figure E   Figure F

Figure G   Figure H

MINITAB

Testing a hypothesis about the slope of the least-squares regression line

Step 1. Label **Column C1** as **Fat** and **Column C2** as **Calories**. Enter the *x*-values from Table 13.1 into the **Fat** column and the *y*-values into the **Calories** column.

Step 2. Click on **Stat**, then **Regression**, then **Regression**.

Step 3. Select the *y*-variable (**Calories**) as the **Response variable** and the *x*-variable as the **Predictor (Fat)**.

Step 4. Click **OK** (Figure I).

```
The regression equation is
Calories = 356.7 + 5.6 Fat

Predictor      Coef      SE Coef       T        P
Constant      356.74       7.669    46.52    0.000
Fat           5.6393       0.387    14.56    0.000
```

Figure I

Note: MINITAB presents the *P*-value for a two-tailed test by default. For a one-tailed test, divide this value by 2.

EXCEL

Constructing a confidence interval and testing a hypothesis for the slope of the least-squares regression line

This procedure requires the **MegaStat** EXCEL add-in to be loaded. The **MegaStat** add-in may be downloaded from www.mhhe.com/megastat.

Step 1. Enter the x-values from Table 13.1 into **Column A** and the y-values into **Column B** (Figure J).

Step 2. Click on the **MegaStat** menu, select **Correlation/Regression**, then **Regression Analysis...**

Step 3. Enter the range of cells containing the x-values in the **X, Independent variable(s)** field and the range of cells containing the y-values in the **Y, Dependent variable** field. To find the confidence interval, enter the confidence level in the **Confidence Level** field. For Example 13.5, we use **95%** (Figure K).

Step 4. Click **OK** (Figure L).

	A	B
1	**Fat**	**Calories**
2	12.75	436
3	25.99	518
4	21.13	492
5	26.13	515
6	⋮	⋮

Figure J

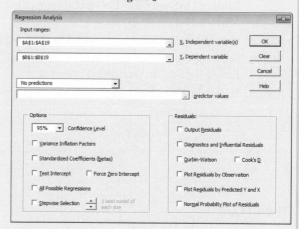

Figure K

Regression output

variables	coefficients	std. error	t (df=16)	p-value	95% lower	95% upper
				confidence interval		
Intercept	356.7392					
Fat	5.6393	0.3873	14.559	**1.19E-10**	4.8182	6.4605

Figure L

Note: EXCEL presents the *P*-value for a two-tailed test by default. For a one-tailed test, divide this value by 2.

SECTION 13.1   Exercises

Exercises 1–6 are the Check Your Understanding exercises located within the section.

Understanding the Concepts

In Exercises 7 and 8, fill in each blank with the appropriate word or phrase.

7. If there are 20 pairs (x, y) in a data set, then the number of degrees of freedom for the critical value is _____ .

8. Under the assumptions of the linear model, the values of $\mu_{y|x}$ follow a _____ .

In Exercises 9 and 10, determine whether the statement is true or false. If the statement is false, rewrite it as a true statement.

9. Under the assumptions of the linear model, the residual plot will exhibit a linear pattern.

10. Under the assumptions of the linear model, the vertical spread in a residual plot will be about the same across the plot.

Practicing the Skills

11. The summary statistics for a certain set of points are: $n = 30$, $s_e = 3.975$, $\sum(x - \bar{x})^2 = 15.425$, and $b_1 = 1.212$. Assume the conditions of the linear model hold. A 95% confidence interval for β_1 will be constructed.
  a. How many degrees of freedom are there for the critical value?

b. What is the critical value?
c. What is the margin of error?
d. Construct the 95% confidence interval.

12. The summary statistics for a certain set of points are: $n = 20$, $s_e = 4.65$, $\sum(x - \bar{x})^2 = 118.26$, and $b_1 = 1.62$. Assume the conditions of the linear model hold. A 99% confidence interval for β_1 will be constructed.
  a. How many degrees of freedom are there for the critical value?
  b. What is the critical value?
  c. What is the margin of error?
  d. Construct the 99% confidence interval.

13. Use the summary statistics in Exercise 11 to test the null hypothesis H_0: $\beta_1 = 0$ versus H_1: $\beta_1 \neq 0$. Use the $\alpha = 0.01$ level of significance.

14. Use the summary statistics in Exercise 12 to test the null hypothesis H_0: $\beta_1 = 0$ versus H_1: $\beta_1 > 0$. Use the $\alpha = 0.05$ level of significance.

In Exercises 15–18, use the given set of points to
  a. Compute b_1.
  b. Compute the residual standard deviation s_e.
  c. Compute the sum of squares for x, $\sum(x - \bar{x})^2$.
  d. Compute the standard error of b_1, s_b.
  e. Find the critical value for a 95% confidence interval for β_1.

f. Compute the margin of error for a 95% confidence interval for β_1.

g. Construct a 95% confidence interval for β_1.

h. Test the null hypothesis $H_0: \beta_1 = 0$ versus $H_1: \beta_1 \neq 0$. Use the $\alpha = 0.05$ level of significance.

15.

x	12	21	27	27	10	15
y	52	90	113	111	45	65

16.

x	12	17	3	17	16	11	14	9
y	13	14	16	13	14	14	13	14

17.

x	18	20	17	12	10
y	71	77	68	52	46

18.

x	12	13	15	13	12	14	13
y	18	19	18	16	16	15	20

Working with the Concepts

19. Calories and protein: The following table presents the number of grams of protein and the number of calories per 100 grams for each of 18 fast-food products.

Protein	Calories	Protein	Calories
17.25	289	5.62	367
13.73	376	3.58	188
12.96	315	11.97	275
21.24	268	5.84	408
22.54	221	12.22	460
16.23	309	8.05	189
10.64	226	5.15	366
6.16	344	15.93	334
3.69	163	16.52	241

Source: United States Department of Agriculture

a. Compute the least-squares regression line for predicting calories (y) from protein (x).

b. Construct a 95% confidence interval for the slope.

c. Test $H_0: \beta_1 = 0$ versus $H_1: \beta_1 \neq 0$. Can you conclude that the amount of protein is useful in predicting the number of calories? Use the $\alpha = 0.05$ level of significance.

20. Like father, like son: In 1906, the statistician Karl Pearson measured the heights of 1078 pairs of fathers and sons. The following table presents a sample of 16 pairs, with height measured in inches, simulated from the distribution specified by Pearson.

Father's height	Son's height	Father's height	Son's height
70.8	69.8	72.4	69.1
65.4	66.0	65.7	65.3
65.7	70.9	69.1	71.8
69.0	69.1	70.7	71.0
73.6	74.9	72.3	71.9
66.7	68.8	73.6	76.5
70.1	73.3	69.3	71.4
68.3	68.3	64.5	68.5

a. Compute the least-squares regression line for predicting son's height (y) from father's height (x).

b. Construct a 95% confidence interval for the slope.

c. Test $H_0: \beta_1 = 0$ versus $H_1: \beta_1 \neq 0$. Can you conclude that father's height is useful in predicting son's height? Use the $\alpha = 0.05$ level of significance.

21. Butterfly wings: Do larger butterflies live longer? The wingspan (in millimeters) and the lifespan in the adult state (in days) were measured for 22 species of butterfly. Following are the results.

Wingspan	Lifespan	Wingspan	Lifespan
35.5	19.8	25.9	32.5
30.6	17.3	31.3	27.5
30.0	27.5	23.0	31.0
32.3	22.4	26.3	37.4
23.9	40.7	23.7	22.6
27.7	18.3	27.1	23.1
28.8	25.9	28.1	18.5
35.9	23.1	25.9	32.3
25.4	24.0	28.8	29.1
24.6	38.8	31.4	37.0
28.1	36.5	28.5	33.7

Source: *Oikos Journal of Ecology* 105:41–54

a. Compute the least-squares regression line for predicting lifespan (y) from wingspan (x).

b. Construct a 99% confidence interval for the slope.

c. Test $H_0: \beta_1 = 0$ versus $H_1: \beta_1 < 0$. Can you conclude that wingspan is useful in predicting lifespan? Use the $\alpha = 0.05$ level of significance.

d. Do larger butterflies tend to live for a longer or shorter time than smaller butterflies? Explain.

22. Blood pressure: A blood pressure measurement consists of two numbers: the systolic pressure, which is the maximum pressure taken when the heart is contracting, and the diastolic pressure, which is the minimum pressure taken at the beginning of the heartbeat. Blood pressures were measured, in millimeters, for a sample of 16 adults. The following table presents the results.

Systolic	Diastolic	Systolic	Diastolic
134	87	133	91
115	83	112	75
113	77	107	71
123	77	110	74
119	69	108	69
118	88	105	66
130	76	157	103
116	70	154	94

Based on results published in the *Journal of Hypertension* 26:199–209

a. Compute the least-squares regression line for predicting diastolic pressure (y) from systolic pressure (x).

b. Construct a 99% confidence interval for the slope.

c. Test $H_0: \beta_1 = 0$ versus $H_1: \beta_1 > 0$. Can you conclude that systolic blood pressure is useful in predicting diastolic blood pressure? Use the $\alpha = 0.01$ level of significance.

d. Do people with higher diastolic pressure tend to have higher or lower systolic pressures? Explain.

23. Noisy streets: How much noisier are streets where cars travel faster? The following table presents noise levels in decibels and average speed in kilometers per hour for a sample of roads.

Speed	Noise
28.26	78.1
36.22	79.6
38.73	81.0
29.07	78.7
30.28	78.6
30.25	78.5
29.03	78.4
33.17	79.6

Source: *Journal of Transportation Engineering* 125:152–159

a. Compute the least-squares regression line for predicting noise level (y) from speed (x).
b. Construct a 95% confidence interval for the slope.
c. Test $H_0: \beta_1 = 0$ versus $H_1: \beta_1 \neq 0$. Can you conclude that speed is useful in predicting noise level? Use the $\alpha = 0.01$ level of significance.

24. Fast reactions: In a study of reaction times, the time to respond to a visual stimulus (x) and the time to respond to an auditory stimulus (y) were recorded for each of 10 subjects. Times were measured in thousandths of a second. The results are presented in the following table.

Visual	Auditory
161	159
203	206
235	241
176	163
201	197
188	193
228	209
211	189
191	169
178	201

a. Compute the least-squares regression line for predicting auditory response time (y) from visual response time (x).
b. Construct a 95% confidence interval for the slope.
c. Test $H_0: \beta_1 = 0$ versus $H_1: \beta_1 \neq 0$. Can you conclude that visual response is useful in predicting auditory response? Use the $\alpha = 0.01$ level of significance.

25. Getting bigger: Concrete expands both horizontally and vertically over time. Measurements of horizontal and vertical expansion (in units of parts per hundred thousand) were made at several locations on a bridge in Quebec City in Canada. The results are presented in the following table.

Horizontal	Vertical
20	58
15	58
43	55
5	80
18	58
24	68
32	57
10	69
21	63

Source: *Canadian Journal of Civil Engineering* 32:463–479

a. Compute the least-squares line for predicting vertical expansion (y) from horizontal expansion (x).
b. Construct a 95% confidence interval for the slope β_1.
c. Test $H_0: \beta_1 = 0$ versus $H_1: \beta_1 \neq 0$. Can you conclude that horizontal expansion is useful in predicting vertical expansion? Use the $\alpha = 0.05$ level of significance.

26. Dry up: In a study to determine the relationship between ambient outdoor temperature and the rate of evaporation of water from soil, measurements of average daytime temperature in °C and evaporation in millimeters per day were taken for 10 days. The results are shown in the following table.

Temperature	Evaporation
11.8	2.4
21.5	4.4
16.5	5.0
23.6	4.1
19.1	6.0
21.6	5.9
31.0	4.8
18.9	3.0
24.2	7.1
19.1	1.6

a. Compute the least-squares line for predicting evaporation (y) from temperature (x).
b. Construct a 99% confidence interval for β_1.
c. Test $H_0: \beta_1 = 0$ versus $H_1: \beta_1 \neq 0$. Can you conclude that temperature is useful in predicting evaporation? Use the $\alpha = 0.05$ level of significance.

27. Calculator display: The following TI-84 Plus display presents the results of a test of the null hypothesis $H_0: \beta_1 = 0$.

a. What is the alternate hypothesis?
b. What is the value of the test statistic?
c. How many degrees of freedom are there?
d. What is the P-value?
e. Can you conclude that the explanatory variable is useful in predicting the outcome variable? Answer this question using the $\alpha = 0.05$ level of significance.

28. Calculator display: The following TI-84 Plus display presents the results of a test of the null hypothesis $H_0: \beta_1 = 0$.

a. What is the alternate hypothesis?
b. What is the value of the test statistic?
c. How many degrees of freedom are there?
d. What is the P-value?
e. Can you conclude that the explanatory variable is useful in predicting the outcome variable? Answer this question using the $\alpha = 0.05$ level of significance.

29. **Air pollution:** Ozone is a major component of air pollution in many cities. Atmospheric ozone levels are influenced by many factors, including weather. In one study, the mean percent relative humidity (x) and the ozone levels (y) were measured for 120 days in a western city. Ozone levels were measured in parts per billion. The following MINITAB output describes the fit of a linear model to these data. Assume that the assumptions of the linear model are satisfied.

```
The regression equation is
Ozone = 88.761 - 0.7524 Humidity

Predictor     Coef     SE Coef      T         P
Constant     88.761      7.288    12.18     0.000
Humidity    -0.7524     0.13024   -5.78     0.000
```

a. What are the slope and intercept of the least-squares regression line?
b. Can you conclude that relative humidity is useful in predicting ozone levels? Answer this question using the $\alpha = 0.05$ level of significance.

30. **Cholesterol:** Serum cholesterol levels (y) and age in years (x) were recorded for several men in a medical center.

Cholesterol levels were measured in milligrams per deciliter. The following MINITAB output describes the fit of a linear model to these data. Assume that the assumptions of the linear model are satisfied.

```
The regression equation is
Cholesterol = 162.15 + 1.2499 Age

Predictor     Coef     SE Coef      T         P
Constant     162.15     16.439     9.863     0.000
Age          1.2499     0.38708    3.772     0.007
```

a. What are the slope and intercept of the least-squares regression line?
b. Can you conclude that age is useful in predicting cholesterol levels? Answer this question using the $\alpha = 0.05$ level of significance.

Extending the Concepts

31. **Confidence interval for the conditional mean:** In Example 13.5, we constructed a 95% confidence interval for the slope β_1 in the model to predict the number of calories from the number of grams of fat. The 95% confidence interval is $4.8182 < \beta_1 < 6.4605$. Let $\mu_{y|15}$ be the mean number of calories for food products containing 15 grams of fat, and let $\mu_{y|20}$ be the mean number of calories for food products containing 20 grams of fat. Construct a 95% confidence interval for the difference $\mu_{y|20} - \mu_{y|15}$.

Answers to Check Your Understanding Exercises for Section 13.1

1. a. Not satisfied; plot exhibits a pattern.
   b. Not satisfied; vertical spread varies.
   c. Satisfied
   d. Not satisfied; plot contains outliers.
2. $4.27 < \beta_1 < 7.29$
3. a. 0.1971681416   b. 10   c. 12
   d. $0.17354 < \beta_1 < 0.2208$

4. No. $P > 0.05$, so we do not reject H_0. It is possible that $\beta_1 = 0$ so that the explanatory variable is not helpful in predicting the outcome variable.
5. Yes. $P < 0.01$, so we reject H_0. We conclude that $\beta_1 > 0$, so the explanatory variable is helpful in predicting the outcome variable.
6. a. $\beta_1 \neq 0$ b. $t = 2.906470757$ c. 10
   d. 0.0156590864   e. Yes

| **SECTION 13.2** | Inference About the Response |

Objectives

1. Construct confidence intervals for the mean response
2. Construct prediction intervals for an individual response

Objective 1 Construct confidence intervals for the mean response

Confidence Intervals for the Mean Response

In Section 13.1, we learned how to construct confidence intervals for the slope in a linear model that was used to predict the number of calories in a candy product from the number of grams of fat. The least-squares regression line was $\hat{y} = 356.7392193 + 5.639341034x$.

In this section, we will consider two further problems.

1. Given that the number of grams of fat is x, estimate the mean number of calories for all candy products whose fat content is x.
2. Given that the number of grams of fat is x, predict the number of calories for a particular candy product whose fat content is x.

Figure 13.4 presents an intuitive picture of these two problems. Imagine that this figure represents the entire population of candy products. Each point represents a particular product. The *x*-value of the point represents the fat content, and the *y*-value of the point represents the number of calories. The vertical strip contains all the points for which the fat content is *x*. We can visualize the two problems by looking at this figure.

1. The mean number of calories for all the products whose fat content is *x* is the value $\mu_{y|x}$. This is the *y*-value of the line $\mu_{y|x} = \beta_0 + \beta_1 x$ through the middle of the vertical strip.

2. The number of calories in a particular candy product whose fat content is *x* is the *y*-value of a randomly chosen point in the vertical strip.

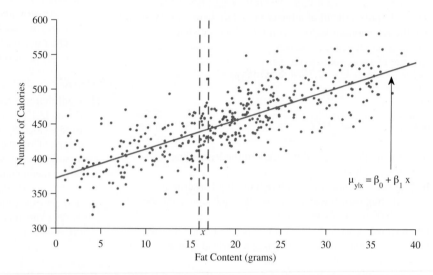

Figure 13.4   Imagine that this scatterplot represents the entire population of candy products. The vertical strip contains all the products for which the fat content is *x*.

The point estimate is the same for both of these problems. The point estimate is the *y*-value on the least-squares regression line: $\hat{y} = 356.7392193 + 5.639341034x$. However, if we want to construct intervals around these point estimates, the interval for the predicted number of calories in a particular product will have a larger margin of error than the interval for the estimated mean number of calories of all the products. The reason is that there is less variation in the mean of all the values in a vertical strip than in the distribution of the individual points.

Constructing a confidence interval for the mean response

The mean *y*-value corresponding to a given *x*-value is called the **mean response**. We show how to construct a confidence interval for the mean response.

CAUTION

Be sure that the assumptions of the linear model are satisfied before constructing a confidence interval for the mean response.

Constructing a Confidence Interval for the Mean Response

Let x^* be a value of the explanatory variable *x*, let $\hat{y} = b_0 + b_1 x^*$ be the predicted value corresponding to x^*, and let *n* be the sample size. A level $100(1 - \alpha)\%$ confidence interval for the mean response is

$$\hat{y} \pm t_{\alpha/2} \cdot s_e \sqrt{\frac{1}{n} + \frac{(x^* - \bar{x})^2}{\sum(x - \bar{x})^2}}$$

The critical value $t_{\alpha/2}$ is based on $n - 2$ degrees of freedom.

| **EXAMPLE 13.10** | Constructing a confidence interval for the mean response |

Construct a 95% confidence interval for the mean number of calories for candy products containing 18 grams of fat.

Solution

The least-squares regression line is $\hat{y} = 356.7392193 + 5.639341034x$. To obtain the point estimate $\hat{y}$, we replace x with $x^* = 18$ to obtain

$$\hat{y} = 356.7392193 + 5.639341034(18) = 458.24736$$

The sample size is $n = 18$, so there are 16 degrees of freedom. The critical value is $t_{\alpha/2} = 2.120$.

To obtain the margin of error, recall that we computed $\bar{x} = 17.7$, $s_e = 14.574547$, and $\sum(x - \bar{x})^2 = 1415.7452$ in Example 13.4 in Section 13.1. The margin of error is therefore

Recall: The quantity s_e is the residual standard deviation:

$$s_e = \sqrt{\frac{\sum(y - \hat{y})^2}{n - 2}}$$

$$t_{\alpha/2} \cdot s_e \sqrt{\frac{1}{n} + \frac{(x^* - \bar{x})^2}{\sum(x - \bar{x})^2}} = 2.120 \cdot 14.574547 \sqrt{\frac{1}{18} + \frac{(18 - 17.7)^2}{1415.7452}} = 7.286578$$

The 95% confidence interval is

$$458.24736 \pm 7.286578$$

$$450.960 < \text{Mean response} < 465.534$$

We are 95% confident that the mean number of calories for candy products containing 18 grams of fat is between 450.960 and 465.534.

| **EXAMPLE 13.11** | Interpreting a confidence interval |

In Example 13.10, we found that we are 95% confident that the mean number of calories for candy products containing 18 grams of fat is between 450.960 and 465.534. You are planning to purchase a particular product that contains 18 grams of fat. Can you be 95% confident that the number of calories in your particular product will be between 450.960 and 465.534? Explain why or why not.

Solution

No, you cannot be 95% confident that the number of calories in your particular product will be between 450.960 and 465.534. The confidence interval for the mean response provides information about the mean number of calories for all fast-food products with 18 grams of fat. To estimate the number of calories in a particular product, we need a prediction interval for an **individual response**. The prediction interval will have a larger margin of error than the confidence interval for the mean response. We will learn to construct prediction intervals later in this section.

Check Your Understanding

1. For a sample of size $n = 20$, the following values were obtained: $b_0 = 1.05$, $b_1 = 4.50$, $s_e = 0.54$, $\sum(x - \bar{x})^2 = 10.9$, $\bar{x} = 8.52$. Construct a 95% confidence interval for the mean response when $x = 10$.

Answer is on page 605.

Objective 2 Construct prediction intervals for an individual response

Prediction Intervals for an Individual Response

Following is the method for constructing a **prediction interval**.

Constructing a Prediction Interval

Let x^* be a value of the explanatory variable x, let $\hat{y} = b_0 + b_1 x^*$ be the predicted value corresponding to x^*, and let n be the sample size. A level $100(1 - \alpha)\%$ prediction interval for an individual response is

$$\hat{y} \pm t_{\alpha/2} \cdot s_e \sqrt{1 + \frac{1}{n} + \frac{(x^* - \bar{x})^2}{\sum(x - \bar{x})^2}}$$

The critical value $t_{\alpha/2}$ is based on $n - 2$ degrees of freedom.

CAUTION

Be sure that the assumptions of the linear model are satisfied before constructing a prediction interval.

Note that the standard error for the prediction interval is similar to the one for the confidence interval for a mean response. The difference is that the standard error for prediction has a "1" added to the quantity under the square root. This reflects the extra variability in the prediction interval.

EXAMPLE 13.12   Constructing a prediction interval

A particular candy product has a fat content of 18 grams. Construct a 95% prediction interval for the number of calories in this product.

Solution
The least-squares regression line is $\hat{y} = 356.7392193 + 5.639341034x$. To obtain the point estimate $\hat{y}$, we replace x with $x^* = 18$ to obtain

$$\hat{y} = 356.7392193 + 5.639341034(18) = 458.24736$$

The sample size is $n = 18$, so there are 16 degrees of freedom. The critical value is $t_{\alpha/2} = 2.120$.

To obtain the margin of error, recall that we computed $\bar{x} = 17.7$, $s_e = 14.574547$, and $\sum(x - \bar{x})^2 = 1415.7452$ in Example 13.4 in Section 13.1. The margin of error is therefore

$$t_{\alpha/2} \cdot s_e \sqrt{1 + \frac{1}{n} + \frac{(x^* - \bar{x})^2}{\sum(x - \bar{x})^2}} = 2.120 \cdot 14.574547 \sqrt{1 + \frac{1}{18} + \frac{(18 - 17.7)^2}{1415.7452}}$$

$$= 31.744256$$

The 95% prediction interval is

$$458.24736 \pm 31.744256$$

$$426.503 < \text{Number of calories} < 489.992$$

We are 95% confident that a particular candy product with a fat content of 18 grams will have between 426.503 and 489.992 calories.

EXAMPLE 13.13   Interpret a prediction interval

Refer to Example 13.12. You are planning to eat a candy bar, and you want to consume less than 500 calories. If you choose an item that contains 18 grams of fat, can you be reasonably sure that it will contain less than 500 calories?

Solution
Yes. We are 95% confident that a particular candy bar with a fat content of 18 grams will have between 426.503 and 489.992 calories. Therefore, we can be reasonably sure that it will contain less than 500 calories.

Check Your Understanding

2. For a sample of size $n = 15$, the following values were obtained: $b_0 = 3.71$, $b_1 = 8.38$, $s_e = 1.13$, $\sum(x - \bar{x})^2 = 7.71$, $\bar{x} = 13.16$. Construct a 95% prediction interval for an individual response when $x = 8$.

Answer is on page 605.

EXAMPLE 13.14   Constructing intervals with technology

Use technology to construct a 95% confidence interval for the mean number of calories for candy products containing 18 grams of fat, and to construct a 95% prediction interval for the number of calories in a particular candy product that contains 18 grams of fat.

Solution

We will use MINITAB. The output is as follows.

Fit	StDev Fit	95.0% CI	95.0% PI
458.247	3.437	(450.960, 465.534)	(426.503, 489.992)

The 95% confidence interval and the 95% prediction interval are shown. The prediction interval constructed with technology differs slightly from the one computed by hand in Example 13.12 because MINITAB uses a more precise critical value. Step-by-step instructions for constructing confidence intervals for the mean response and prediction intervals are given in the Using Technology section that follows.

USING TECHNOLOGY

We use the data in Table 13.1 and Example 13.12 to illustrate the technology steps.

MINITAB

Constructing confidence and prediction intervals

Step 1. Follow Steps 1–3 in the MINITAB procedure described on page 595 in Section 13.1.

Step 2. Click **Options...** and enter the response value in the **Prediction intervals for new observations** field. For Example 13.12, we use **18**.

Step 3. Enter the confidence level in the **Confidence level** field. We use **95**. Click **OK**.

Step 4. Click **OK** (Figure A).

Fit	StDev Fit	95.0% CI	95.0% PI
458.247	3.437	(450.960, 465.534)	(426.503, 489.992)

Figure A

EXCEL

Constructing confidence and prediction intervals

Step 1. Follow Steps 1–3 in the EXCEL procedure described on page 596 in Section 13.1.

Step 2. Select **Type in predictor values** from the pull-down menu and enter the response value in the **predictor values** field. For Example 13.12, we use **18**.

Step 3. Click **OK** (Figure B).

X1	Predicted	95% Confidence Interval		95% Prediction Interval	
		lower	upper	lower	upper
18	458.247	450.960	465.534	426.503	489.992

Figure B

SECTION 13.2 Exercises

Exercises 1 and 2 are the Check Your Understanding exercises located within the section.

Understanding the Concepts

In Exercises 3 and 4, fill in each blank with the appropriate word or phrase.

3. A _____ interval estimates the mean y-value for all individuals with a given x-value.

4. A _____ interval estimates the y-value for a particular individual with a given x-value.

In Exercises 5 and 6, determine whether the statement is true or false. If the statement is false, rewrite it as a true statement.

5. For a given x-value, the 95% confidence interval for the mean response will always be wider than the 95% prediction interval.

6. For a given x-value, the point estimate for a 95% confidence interval for the mean response is the same as the one for the 95% prediction interval.

Practicing the Skills

7. For a sample of size 25, the following values were obtained: $b_0 = 3.25$, $b_1 = 2.32$, $s_e = 3.53$, $\sum(x - \bar{x})^2 = 224.05$, and $\bar{x} = 0.98$.
   a. Construct a 95% confidence interval for the mean response when $x = 2$.
   b. Construct a 95% prediction interval for an individual response when $x = 2$.

8. For a sample of size 18, the following values were obtained: $b_0 = 2.27$, $b_1 = -1.46$, $s_e = 5.72$, $\sum(x - \bar{x})^2 = 360.26$, and $\bar{x} = 1.95$.
   a. Construct a 99% confidence interval for the mean response when $x = 2$.
   b. Construct a 99% prediction interval for an individual response when $x = 2$.

In Exercises 9 and 10, use the given set of points to
   a. Compute b_0 and b_1.
   b. Compute the predicted value $\hat{y}$ for the given value of x.
   c. Compute the residual standard deviation s_e.
   d. Compute the sum of squares for x, $\sum(x - \bar{x})^2$.
   e. Find the critical value for a 95% confidence or prediction interval.
   f. Construct a 95% confidence interval for the mean response for the given value of x.
   g. Construct a 95% prediction interval for an individual response for the given value of x.

9.

x	15	11	17	15	11	16
y	30	15	33	27	22	37

$x = 12$

10.

x	23	16	17	19	30	19	18	27
y	51	22	56	34	67	59	55	25

$x = 25$

Working with the Concepts

11. **Calories and protein:** Use the data in Exercise 19 in Section 13.1 for the following:
   a. Compute a point estimate for the mean number of calories in fast-food products that contain 15 grams of protein.
   b. Construct a 95% confidence interval for the mean number of calories in fast-food products that contain 15 grams of protein.
   c. Predict the number of calories in a particular product that contains 15 grams of protein.
   d. Construct a 95% prediction interval for the number of calories in a particular product that contains 15 grams of protein.

12. **Like father, like son:** Use the data in Exercise 20 in Section 13.1 for the following.
   a. Compute a point estimate of the mean height of sons whose fathers are 70 inches tall.
   b. Construct a 95% confidence interval for the mean height of sons whose fathers are 70 inches tall.
   c. Predict the height of a particular son whose father is 70 inches tall.
   d. Construct a 95% prediction interval for the height of a particular son whose father is 70 inches tall.

13. **Butterfly wings:** Use the data in Exercise 21 in Section 13.1 for the following.
   a. Compute a point estimate of the mean lifespan of butterflies with a wingspan of 30 millimeters.
   b. Construct a 95% confidence interval for the mean lifespan of butterflies with a wingspan of 30 millimeters.
   c. Predict the lifespan of a particular butterfly whose wingspan is 30 millimeters.
   d. Construct a 95% prediction interval for the lifespan of a particular butterfly whose wingspan is 30 millimeters.

14. **Blood pressure:** Use the data in Exercise 22 in Section 13.1 for the following.
   a. Compute a point estimate of the mean diastolic pressure for people whose systolic pressure is 120.
   b. Construct a 95% confidence interval for the mean diastolic pressure for people whose systolic pressure is 120.
   c. Predict the diastolic pressure of a particular person whose systolic pressure is 120.
   d. Construct a 95% prediction interval for the diastolic pressure of a particular person whose systolic pressure is 120.

15. **Noisy streets:** Use the data in Exercise 23 in Section 13.1 for the following.
   a. Compute a point estimate for the mean noise level for streets with a mean speed of 35 kilometers per hour.
   b. Construct a 99% confidence interval for the mean noise level for streets with a mean speed of 35 kilometers per hour.
   c. Predict the noise level for a particular street with a mean speed of 35 kilometers per hour.
   d. Construct a 99% prediction interval for the noise level of a particular street with a mean speed of 35 kilometers per hour.

16. Fast reactions: Use the data in Exercise 24 in Section 13.1 for the following.

a. Compute a point estimate for the mean auditory response time for subjects with a visual response time of 200.

b. Construct a 99% confidence interval for the mean auditory response time for subjects with a visual response time of 200.

c. Predict the auditory response time for a particular subject whose visual response time is 200.

d. Construct a 99% prediction interval for the auditory response time for a particular subject whose visual response time is 200.

17. Getting bigger: Use the data in Exercise 25 in Section 13.1 for the following.

a. Compute a point estimate for the mean vertical expansion at locations where the horizontal expansion is 25.

b. Construct a 99% confidence interval for the mean vertical expansion at locations where the horizontal expansion is 25.

c. Predict the vertical expansion at a particular location where the horizontal expansion is 25.

d. Construct a 99% prediction interval for the vertical expansion at a particular location where the horizontal expansion is 25.

18. Dry up: Use the data in Exercise 26 in Section 13.1 for the following.

a. Compute a point estimate for the mean evaporation rate when the temperature is 20°C.

b. Construct a 99% confidence interval for the mean evaporation rate for all days with a temperature of 20°C.

c. Predict the evaporation rate when the temperature is 20°C.

d. Construct a 99% prediction interval for the evaporation rate on a given day with a temperature of 20°C.

19. Air pollution: The following MINITAB output presents a 95% confidence interval for the mean ozone level on days when the relative humidity is 60%, and a 95% prediction interval for the ozone level on a particular day when the relative humidity is 60%. The units of ozone are parts per billion.

```
Predicted Values for New Observations

New Obs     Fit   SE Fit     95.0% CI         95.0% PI
1         43.62     1.20   (41.23, 46.00)   (20.86, 66.37)

Values of Predictors for New Observations
New Obs     Humidity
1               60.0
```

a. What is the point estimate for the mean ozone level for days when the relative humidity is 60%?

b. What is the 95% confidence interval for the mean ozone level for days when the relative humidity is 60%?

c. Predict the ozone level for a day when the relative humidity is 60%.

d. Upon learning that the relative humidity on a certain day is 60%, someone predicts that the ozone level that day will be 80 parts per billion. Is this a reasonable prediction? If so, explain why. If not, give a reasonable range of predicted values.

20. Cholesterol: The following MINITAB output presents a 95% confidence interval for the mean cholesterol levels for men aged 50 years, and a 95% prediction interval for an individual man aged 50. The units of cholesterol are milligrams per deciliter.

```
Predicted Values for New Observations

New Obs    Fit   SE Fit      95.0% CI            95.0% PI
1       224.64     6.08   (211.40, 237.89)   (182.66, 266.42)

Values of Predictors for New Observations
New Obs        Age
1             50.0
```

a. What is the point estimate for the mean cholesterol level for men aged 50?

b. What is the 95% confidence interval for the mean cholesterol level for men aged 50?

c. Predict the cholesterol level for a man aged 50.

d. Upon learning that a man is 50 years old, someone predicts that his cholesterol level is 160. Is this a reasonable prediction? If so, explain why. If not, give a reasonable range of predicted values.

Extending the Concepts

21. Margin of error: Several 95% confidence intervals for the mean response will be constructed, based on a data set for which the sample mean value for the explanatory variable is $\bar{x} = 10$. The values of x^* for which the confidence intervals will be constructed are $x^* = 9$, $x^* = 12$, and $x^* = 14$.

a. For which of these values of x^* will the margin of error be the smallest?

b. For which of these values of x^* will the margin of error be the largest?

c. If one wanted to construct a 95% confidence interval with the smallest possible margin of error, which value of x^* would one use?

Answers to Check Your Understanding Exercises for Section 13.2

1. 45.482 < Mean response < 46.618

2. 65.561 < Individual response < 75.939

Multiple Regression

Objectives

1. Use technology to compute the multiple regression equation
2. Interpret the multiple regression coefficients
3. Test the multiple regression coefficients
4. Construct confidence intervals and prediction intervals
5. Test goodness-of-Fit
6. Eliminate unnecessary variables from a multiple regression model

Objective 1 Use technology to compute the multiple regression equation

The Multiple Regression Equation

In Sections 13.1 and 13.2 we studied the least-squares regression line, which is used to predict a value of an outcome variable y from the value of a single explanatory variable x. Sometimes we need several explanatory variables to predict the value of the response variable. In these situations we use **multiple regression**. We refer to the explanatory variables as x_1, x_2, and so on. If there are p explanatory variables, the multiple regression equation is

$$\hat{y} = b_0 + b_1 x_1 + \cdots + b_p x_p$$

Table 13.4 presents the results of a hypothetical study in which the lung capacities of 16 children were measured. Lung capacity is measured by having a person take as deep a breath as possible, then blowing into a tube connected to a machine called a spirometer, which measures the volume of air exhaled. Variables that might be useful to predict a child's lung capacity are height, weight, and age, along with the temperature and barometric pressure at the time the lung capacity was measured. These variables were measured as well.

Table 13.4 Lung Capacity, Height, Weight, Age, Barometric Pressure, and Temperature for Measurements on Sixteen Children

Lung Capacity (y)	Height (x_1)	Weight (x_2)	Age (x_3)	Barometric Pressure (x_4)	Temperature (x_5)
2.68	61	105	13	30.0	70.4
2.22	49	50	9	30.3	71.6
2.29	61	128	15	29.9	73.6
2.16	54	81	10	29.6	70.8
3.19	68	120	13	29.6	69.1
1.97	56	92	10	30.0	70.1
2.98	68	140	16	30.2	67.7
1.96	57	86	10	30.7	56.3
3.05	62	120	16	30.5	63.8
2.48	60	114	13	30.3	69.9
2.09	57	94	12	30.1	73.0
3.27	67	145	16	30.6	70.3
2.26	52	66	10	28.8	75.5
3.20	72	159	16	30.0	72.1
2.10	53	65	10	30.5	67.5
1.79	51	56	9	29.7	65.6

In Table 13.4, lung capacity (y) is measured in liters, height (x_1) in inches, weight (x_2) in pounds, age (x_3) in years, barometric pressure (x_4) in inches, and temperature (x_5) in degrees Fahrenheit. To predict the lung capacity for a person of a given height, weight, age, and given values of the pressure and temperature, we must compute the multiple regression equation

$$\hat{y} = b_0 + b_1 x_1 + b_2 x_2 + b_3 x_3 + b_4 x_4 + b_5 x_5$$

The calculations involved in computing the multiple regression equation are quite complex, and almost impossible to do by hand. Technology is always used in practice. Figure 13.5 presents MINITAB output for the computation of the multiple regression equation. Step-by-step instructions are given in the Using Technology section on page 615.

```
The regression equation is
Lung Capacity = -7.82 + 0.119 Height - 0.0246 Weight + 0.164 Age
                 + 0.076 Pressure + 0.0202 Temperature


Predictor          Coef      SE Coef       T       P
Constant         -7.817        6.288   -1.24   0.242
Height          0.11927      0.03292    3.62   0.005
Weight        -0.024627     0.008965   -2.75   0.021
Age             0.16411      0.06252    2.62   0.025
Pressure         0.0764       0.1645    0.46   0.652
Temperature     0.02022      0.01716    1.18   0.266


S = 0.214478     R-Sq = 88.0%     R-Sq(adj) = 81.9%

Analysis of Variance

Source           DF         SS         MS        F       P
Regression        5    3.35908    0.67182    14.60   0.000
Residual Error   10    0.46001    0.04600
Total            15    3.81909
```

Figure 13.5 Multiple regression equation and related output for data in Table 13.4, from MINITAB.

The multiple regression equation is presented at the top of the output. The column labeled "Coef" presents the coefficients of the multiple regression equation. The rest of the information in the output describes how well the equation fits the data; we will explain this as we go along.

The residual plot

For each individual, the residual is the difference $y - \hat{y}$ between their outcome variable y and the predicted value $\hat{y}$. As with the least-squares regression line, we determine whether the multiple regression equation is appropriate by constructing a residual plot. This is a plot of residuals against the fitted values $\hat{y}$. When the multiple regression is appropriate, the residual plot will not exhibit any noticeable pattern. If the residual plot does exhibit a pattern, such as a curved pattern, then the multiple regression equation should not be used.

Figure 13.6 presents a residual plot, produced by MINITAB, for the data in Table 13.4. The residual plot does not exhibit any particular pattern, which suggests that the multiple regression equation is appropriate. Step-by-step instructions for constructing residual plots with MINITAB are presented in the Using Technology section on page 615.

Figure 13.6   Residual plot for the data in Table 13.4. The plot exhibits no noticeable pattern, so use of the multiple regression equation is appropriate.

EXAMPLE 13.15	Predict the value of an outcome variable

Predict the lung capacity for a 12-year-old who is 61 inches tall and weighs 105 pounds, at a pressure of 30.1 inches and a temperature of 70 degrees.

Solution

We use the coefficients of the multiple regression equation, presented in Figure 13.5, to compute

$$\hat{y} = -7.817 + 0.11927(61) - 0.024627(105) + 0.16411(12)$$
$$+ 0.0764(30.1) + 0.02022(70)$$
$$= 2.557$$

The predicted value is 2.557 liters.

Objective 2 Interpret the multiple regression coefficients

Interpret the Multiple Regression Coefficients

The interpretation of the coefficients in multiple regression is similar to the interpretation of the slope of the least-squares regression line. If the values of an explanatory variable x_i for two individuals differ by 1, and we assume that the values of the other explanatory values are the same, then the predicted values will differ by b_i. If the values of the explanatory variable differ by an amount d, then their predicted values will differ by $b_i \cdot d$.

EXAMPLE 13.16	Compute the predicted difference in outcomes

Two people differ in height by 2 inches. Their weights and ages are the same, and their lung capacities are measured at the same pressure and temperature. By how much should we predict their lung capacities to differ?

Solution

The coefficient of height, shown in Figure 13.5, is 0.11927. We predict their lung capacities to differ by $(0.11927)(2) = 0.23854$ liters.

Check Your Understanding

1. Use the coefficients of the multiple regression equation in Figure 13.5 to predict the lung capacity for a 10-year-old who is 57 inches tall and weighs 90 pounds, at a pressure of 30.2 inches and a temperature of 65 degrees.

2. Two people differ in age by 1.5 years. Their heights and weights are the same, and their lung capacities are measured at the same pressure and temperature. By how much should we predict their lung capacities to differ?

Answers are on page 622.

Objective 3 Test the multiple regression coefficients

Test the Multiple Regression Coefficients

The 16 individuals described in Table 13.4 are a sample from a large population. When we compute the multiple regression equation for this sample, it is an estimate of what the multiple regression equation would be for the whole population. The coefficients of the multiple regression equation for the sample are denoted b_0, b_1, and so forth. The coefficients of the multiple regression equation for the population are denoted β_0, β_1, and so forth. Thus, b_0 is an estimate of β_0, b_1 is an estimate of β_1, and so forth.

When a coefficient β_i is equal to 0, the variable x_i has no linear relation to the outcome variable, and can be omitted from the multiple regression equation. The most important hypothesis test to perform on β_i is therefore H_0: $\beta_i = 0$. This test is automatically performed for each β_i by most computer software packages. For each coefficient, the column labeled "T" in the MINITAB output presents test statistic, and the column labeled "P" presents the P-value for the test of H_0: $\beta_i = 0$ versus H_1: $\beta_i \neq 0$.

EXAMPLE 13.17

Test hypotheses about the multiple regression coefficients

Following is part of the MINITAB output shown in Figure 13.5, which presents the coefficients, test statistics and P-values for the multiple regression equation for the data in Table 13.4.

Predictor	Coef	SE Coef	T	P
Constant	−7.817	6.288	−1.24	0.242
Height	0.11927	0.03292	3.62	0.005
Weight	−0.024627	0.008965	−2.75	0.021
Age	0.16411	0.06252	2.62	0.025
Pressure	0.0764	0.1645	0.46	0.652
Temperature	0.02022	0.01716	1.18	0.266

a. For the data in Table 13.4, β_1 is the coefficient of height. Test the hypothesis H_0: $\beta_1 = 0$ versus H_1: $\beta_1 \neq 0$ at the $\alpha = 0.05$ level. What do you conclude?

b. For the data in Table 13.4, β_5 is the coefficient of temperature. Test the hypothesis H_0: $\beta_5 = 0$ versus H_1: $\beta_1 \neq 0$ at the $\alpha = 0.05$ level. What do you conclude?

Solution

a. In Figure 13.5, the P-value for height is 0.005. This value is less than 0.05, so we reject H_0 and conclude that $\beta_1 \neq 0$. The explanatory variable height has a linear relationship with the outcome variable, lung capacity.

b. In Figure 13.5, the P-value for temperature is 0.266. This value is greater than 0.05, so we do not reject H_0. There is not enough evidence to conclude that $\beta_1 \neq 0$. We cannot conclude that the explanatory variable temperature may not a linear relationship with the outcome variable, lung capacity.

Objective 4 Construct confidence intervals and prediction intervals

Construct Confidence Intervals and Prediction Intervals

In Section 13.2 we learned to construct confidence intervals for a mean response and prediction intervals for an individual response based on the least-squares regression line. With technology we can do the same with the multiple regression equation.

Recall the difference between a confidence interval and a prediction interval. A confidence interval presents plausible values for the mean value of the response for all individuals with a specified set of values for their explanatory variables. A prediction interval presents a plausible range of values for a particular individual with a specified set of values for the explanatory variables.

| EXAMPLE 13.18 | Find a confidence interval and a prediction interval |

a. Use the data in Table 13.4 to construct a 95% confidence interval for the mean lung capacity for a 12-year-old who is 65 inches tall and weighs 110 pounds, if the measurement is taken at a pressure of 30 inches and a temperature of 70 degrees.

b. Use the data in Table 13.4 to construct a 95% prediction interval for the lung capacity of a particular 12-year-old who is 65 inches tall and weighs 110 pounds, if the measurement is taken at a pressure of 30 inches and a temperature of 70 degrees.

Solution

The MINITAB output follows:

```
Predicted Values for New Observations

New
Obs      Fit     SE Fit       95% CI              95% PI
  1   2.9032    0.1510    (2.5667, 3.2397)    (2.3187, 3.4876)


Values of Predictors for New Observations

New
Obs  Height  Weight  Age  Pressure  Temperature
  1    65.0     110  12.0    30.0        70.0
```

The values of the explanatory variables are listed under the heading "Values of Predictors for New Observations."

a. The 95% confidence interval is given under the heading "95% CI." The confidence interval is (2.5667, 3.2397). We are 95% confident that the mean lung function over all individuals with the given values for the explanatory variables is between 2.5667 and 3.2397 liters.

b. The 95% prediction interval is given under the heading "95% PI." The prediction interval is (2.3187, 3.4876). We are 95% confident that the lung function for a particular individual with the given values for the explanatory variables will be between 2.3187 and 3.4876 liters.

Check Your Understanding

3. The following MINITAB output presents a multiple regression equation $\hat{y} = b_0 + b_1 x_1 + b_2 x_2 + b_3 x_3$. Test $H_0: \beta_i = 0$ versus $H_1: \beta_i \neq 0$ for $i = 1, 2, 3$. Use the $\alpha = 0.05$ level.

```
The regression equation is
Y = 3.4105 + 0.9543 X1 + 0.2970 X2 + 2.6126 X3

Predictor          Coef    SE Coef       T       P
Constant         3.4105     0.5832  5.8475   0.000
X1               0.9543     0.2975  3.2074   0.008
X2               0.2970     0.5796  0.5124   0.619
X3               2.6136     0.4252  6.1461   0.000

S = 1.8421      R-Sq = 81.2%     R-Sq (adj) = 76.0%

Analysis of Variance

Source             DF       SS       MS      F       P
Regression          3   160.71   53.570  15.787   0.000
Residual Error     11    37.327   3.3934
Total              14   198.04
```

4. The following MINITAB output presents a confidence interval for a mean response and a prediction interval for an individual response.

```
New Obs      Fit     SE Fit       95.0% CI          95.0% PI
1         10.523     1.3996   (7.443, 13.604)   (5.4315, 15.615)

Values of Predictors for New Observations

New Obs       X1        X2        X3
1           1.32      1.58      2.06
```

a. Predict the value of y when $x_1 = 1.32$, $x_2 = 1.58$, and $x_3 = 2.06$.
b. We are 95% confident that an individual whose values are $x_1 = 1.32$, $x_2 = 1.58$, and $x_3 = 2.06$ will have a response between _____ and _____ .
c. We are 95% confident that the mean response when $x_1 = 1.32$, $x_2 = 1.58$, and $x_3 = 2.06$ is between _____ and _____ .

Answers are on page 622.

Objective 5 Test goodness-of-fit

Testing Goodness-of-Fit

The coefficient of determination

For each individual, we can compute the difference $y - \bar{y}$ between that individual's observed value y and the average observation $\bar{y}$. In multiple regression, as with the least-squares regression line, we can divide the difference into two parts. The difference $\hat{y} - \bar{y}$ between $\hat{y}$, the predicted value for the individual, and $\bar{y}$ is called the explained difference and the difference $y - \hat{y}$ is called the unexplained difference. The quantity $\sum(\hat{y} - \bar{y})^2$ is called the **regression sum of squares**, and the quantity $\sum(y - \hat{y})^2$ is called the **error sum of squares**. We add the regression sum of squares and the error sum of squares to obtain the **total sum of squares**.

The **coefficient of determination**, denoted R^2, is defined by

$$R^2 = \frac{\text{regression sum of squares}}{\text{total sum of squares}}$$

The coefficient of determination represents the proportion of variation in the response variable explained by the multiple regression equation.

Explain It Again

Proportion of variation explained: The regression sum of squares is the same as the explained variation in the least-squares regression line, and the total sum of squares is the total variation. For this reason, R^2 is the proportion of variation explained by the multiple regression equation.

The coefficient of determination has a disadvantage as a measure of goodness-of-fit. If more explanatory variables are added to the model, the value of R^2 will not decrease, and will almost certainly increase, even if the explanatory variables are unrelated to the outcome variable. To compensate for this, a quantity known as **adjusted R^2** is used. This quantity adjusts the value of R^2 by using the number of observations n and the number of explanatory variables p. The value of adjusted R^2 is

$$\text{adjusted } R^2 = R^2 - \frac{p}{n-p-1}(1 - R^2)$$

Because the value of adjusted R^2 is found by subtracting a positive quantity from R^2, its value is always somewhat less than the value of R^2. The values of R^2 and adjusted R^2 are routinely provided in the output from computer software packages.

EXAMPLE 13.19

Find R^2 and adjusted R^2

Find the values of R^2 and adjusted R^2 for the data in Table 13.4.

Solution
Figure 13.5 presents the MINITAB output. The value of R^2 is labeled "R-sq." The value of R^2 is 88.0%. The value of adjusted R^2 is labeled "R-sq(adj)." The value of adjusted R^2 is 81.9%.

The *F*-test for goodness-of-fit

To determine whether a multiple regression equation is useful for prediction, we test the null hypothesis

$$H_0: \beta_1 = \beta_2 = \cdots = \beta_p = 0 \quad \text{versus} \quad H_1: \text{at least one of the } \beta_i \neq 0$$

If H_0 is false, then one or more of the β_i is not zero, and the model is useful for prediction. If H_0 is true, the model is not useful for prediction. The test statistic is based on the regression and error sums of squares, as follows.

Each sum of squares has a number of degrees of freedom associated with it. The degrees of freedom for the regression sum of squares is p, the number of explanatory variables in the model. The degrees of freedom for the error sum of squares is $n - p - 1$. When a sum of squares is divided by its degrees of freedom, the result is known as a mean square. The regression mean square is denoted MSR and the error mean square is denoted MSE. The statistic for testing goodness-of-fit is

$$F = \frac{MSR}{MSE}$$

When H_0 is true, the test statistic has an F distribution with p and $n - p - 1$ degrees of freedom. We could use the F-table to determine whether to reject H_0, but computer packages routinely provide the P-value, so in practice this is not necessary.

EXAMPLE 13.20

Test the goodness-of-fit of a multiple regression equation

For the data in Table 13.4, perform an F-test. Is the model useful for prediction? Use the $\alpha = 0.01$ level.

Solution
Figure 13.5 presents the MINITAB output. Information regarding the F-test is found under the heading "Analysis of Variance." The value of the test statistic, labeled "F," is 14.60.

The P-value, labeled "P," is given as 0.000. This means that the P-value is 0 when rounded to three decimal places. Since $P < 0.01$, we reject H_0. We conclude that at least one of the coefficients is not equal to 0, and the model is useful for prediction.

Check Your Understanding

5. The following MINITAB output presents a multiple regression equation.

```
The regression equation is
Y = 5.3296 + 3.6204 X1 + 1.5721 X2 + 2.3328 X3

Predictor        Coef        SE Coef            T           P
Constant       5.3296         1.5051       3.5410       0.000
X1             3.6204         1.2845       2.8186       0.010
X2             1.5721         1.3634       1.1531       0.261
X3             2.3328         1.4494       1.6095       0.122

S = 7.4857          R-Sq = 32.6%        R-Sq (adj) = 23.3%


Analysis of Variance

Source          DF          SS           MS           F           P
Regression       3      595.14       198.38      3.5402       0.031
Residual Error  22      1232.8       56.036
Total           25      1827.9
```

    a. What percentage of the variation in the response is explained by the multiple regression equation?

    b. What percentage of the variation in the response is explained by the multiple regression equation, if we account for the number of explanatory variables in the model?

6. Refer to Exercise 5.

    a. Is the multiple regression equation useful for prediction? Explain. Use the $\alpha = 0.05$ level.

    b. Is the multiple regression equation useful for prediction? Explain. Use the $\alpha = 0.01$ level.

Answers are on page 622.

Objective 6 Eliminate unnecessary variables from a multiple regression model

Eliminating Unnecessary Variables from a Multiple Regression Model

In practice, there are often many explanatory variables that could potentially be included in a multiple regression equation. When using multiple regression, it is important to build a model that includes only as many explanatory variables as necessary, to avoid needless complexity. There are many methods available for choosing variables for multiple regression models; entire books have been written on the subject. Here, we will show how to use one basic principle to eliminate unnecessary variables.

> ## Principle for Determining Whether a Variable Should be Removed
>
> If removing a variable increases the adjusted R^2, then the variable should be removed.
>
> *Note:* You cannot remove the variable labeled "Constant." This is the intercept b_0, which should always be part of the equation.

We show how to use this principle to eliminate variables from the equation shown in Figure 13.5. The variable pressure has the largest *P*-value, which means that it explains less of the variation in the response variable than any of the other variables. We will remove pressure to see what happens to the adjusted R^2. Figure 13.7 presents the results of fitting the multiple regression equation using only the variables height, weight, age, and temperature.

```
The regression equation is
Lung Capacity = -5.01 + 0.114 Height - 0.0237 Weight + 0.169 Age
                 + 0.0149 Temperature


Predictor          Coef     SE Coef      T       P
Constant         -5.009       1.662   -3.01   0.012
Height          0.11423     0.02994    3.81   0.003
Weight        -0.023688    0.008418   -2.81   0.017
Age             0.16915     0.05934    2.85   0.016
Temperature     0.01485     0.01223    1.21   0.250


S = 0.206692    R-Sq = 87.7%     R-Sq (adj) = 83.2%


Analysis of Variance

Source            DF        SS        MS       F       P
Regression         4   3.34916   0.83729   19.60   0.000
Residual Error    11   0.46994   0.04272
Total             15   3.81909
```

Figure 13.7 Multiple regression equation and related output using explanatory variables height, weight, age, and temperature, for data in Table 13.4, from MINITAB

Comparing the adjusted R^2 value in Figure 13.7 to the value in Figure 13.5, we see that eliminating pressure has increased the value of adjusted R^2 from 81.9% to 83.2%. This tells us that including pressure in the model does not significantly improve the fit, so we should leave it out. Note that removing pressure from the model caused the values of the coefficients for the remaining variables to change somewhat. This is typical.

We now see that temperature has the largest *P*-value of the remaining variables. We therefore fit the model using only the variables height, weight, and age. Figure 13.8 presents the results.

Comparing the adjusted R^2 value in Figure 13.8 to the value in Figure 13.7, we see that eliminating temperature has decreased the value of adjusted R^2 from 83.2% to 82.6%. This tells us that we should leave temperature in the model. We would therefore use as our final model the equation that contains the explanatory variables height, weight, age, and temperature, shown in Figure 13.7.

```
The regression equation is
Lung Capacity = -3.65 + 0.105 Height - 0.0217 Weight + 0.169 Age
```

Predictor	Coef	SE Coef	T	P
Constant	−3.647	1.252	−2.91	0.013
Height	0.10519	0.02957	3.56	0.004
Weight	−0.021737	0.008425	−2.58	0.024
Age	0.16950	0.06050	2.80	0.016

```
S = 0.210748    R-Sq = 86.0%    R-Sq (adj) = 82.6%
```

Analysis of Variance

Source	DF	SS	MS	F	P
Regression	3	3.2861	1.0954	24.66	0.000
Residual Error	12	0.5330	0.0444		
Total	15	3.8191			

Figure 13.8 Multiple regression equation and related output using explanatory variables height, weight, and age, for data in Table 13.4, from MINITAB

USING TECHNOLOGY

We use Table 13.4 to illustrate the technology steps.

MINITAB

Constructing the multiple regression equation and residual plot

Step 1. Enter the values for Lung Capacity (*y*) from Table 13.4 into column **C1** and the values for Height, Weight, Age, Barometric Pressure, and Temperature into columns **C2** through **C6**.

Step 2. Select **Stat**, then **Regression**, and then **Regression**.

Step 3. Select the *y*-variable (**C1**) as the Response and the x-variables (**C2, C3, C4, C5, C6**) as the Predictors. To generate the residual plot, select **Graphs...** and select **Residuals versus fits** and click **OK**.

Step 4. Click **OK** (Figure A).

Regression Analysis: Capacity versus Height, Weight, ...

```
The regression equation is
Capacity = - 7.82 + 0.119 Height - 0.0246 Weight + 0.164 Age + 0.076 Pressure
          + 0.0202 Temperature
```

Predictor	Coef	SE Coef	T	P
Constant	−7.817	6.288	−1.24	0.242
Height	0.11927	0.03292	3.62	0.005
Weight	−0.024627	0.008965	−2.75	0.021
Age	0.16411	0.06252	2.62	0.025
Pressure	0.0764	0.1645	0.46	0.652
Temperature	0.02022	0.01716	1.18	0.266

```
S = 0.214478    R-Sq = 88.0%    R-Sq(adj) = 81.9%
```

Analysis of Variance

Source	DF	SS	MS	F	P
Regression	5	3.35908	0.67182	14.60	0.000
Residual Error	10	0.46001	0.04600		
Total	15	3.81909			

Figure A

EXCEL

Constructing the multiple regression equation and residual plot

Step 1. Enter the values for Lung Capacity (y) from Table 13.4 into **Column A** and the values for Height, Weight, Age, Barometric Pressure, and Temperature, into **Columns B** through **F** (Figure B)

Step 2. Select **Data**, then **Data Analysis**. Highlight **Regression** and click **OK**.

Step 3. Enter the range of cells that contain the x-values in the **Input X Range** field and the range of cells that contain the y-values in the **Input Y Range** field. The residuals may be generated by selecting the **Residuals** option.

Step 4. Click **OK** (Figure C).

A residual plot may be generated using the procedure presented in Section 4.3 on page 184.

	A	B	C	D	E	F
1	Lung Cap	Height	Weight	Age	Pressure	Temp
2	2.68	61	105	13	30	70.4
3	2.22	49	50	9	30.3	71.6
4	2.29	61	128	15	29.9	73.6
5	2.16	54	81	10	29.6	70.8
6	3.19	68	120	13	29.6	69.1
7	1.97	56	92	10	30	70.1
8	2.98	68	140	16	30.2	67.7
9	1.96	57	86	10	30.7	56.3
10	3.05	62	120	16	30.5	63.8
11	2.48	60	114	13	30.3	69.9
12	2.09	57	94	12	30.1	73
13	3.27	67	145	16	30.6	70.3
14	2.26	52	66	10	28.8	75.5
15	3.2	72	159	16	30	72.1
16	2.1	53	65	10	30.5	67.5
17	1.79	51	56	9	29.7	65.6

Figure B

Regression Statistics

Multiple R	0.93784
R Square	0.87955
Adjusted R Square	0.81933
Standard Error	0.21448
Observations	16

ANOVA

	df	SS	MS	F	Significance F
Regression	5	3.35908	0.67182	14.60440	0.00025
Residual	10	0.46001	0.04600		
Total	15	3.81909			

	Coefficients	Standard Error	t Stat	P-value
Intercept	-7.81721	6.28758	-1.24328	0.24212
Height	0.11927	0.03292	3.62364	0.00466
Weight	-0.02463	0.00897	-2.74683	0.02059
Age	0.16411	0.06252	2.62480	0.02539
Pressure	0.07640	0.16447	0.46451	0.65223
Temp	0.02022	0.01716	1.17827	0.26597

Figure C

SECTION 13.3   Exercises

Exercises 1–6 are the Check Your Understanding exercises located within the section.

Understanding the Concepts

In Exercises 7 and 8, fill in each blank with the appropriate word or phrase.

7. A _____ plot can be used to determine whether a multiple regression equation is appropriate.

8. We should leave a variable out of a multiple regression equation when removing it _____ the value of adjusted R^2.

In Exercises 9 and 10, determine whether the statement is true or false. If the statement is false, rewrite it as a true statement.

9. The coefficient of determination R^2 measures the percentage of variation in the outcome that is explained by the model.

10. If the value of the F-statistic is large, the multiple regression equation is not useful for making predictions.

Practicing the Skills

11. For the following data set:

y	x_1	x_2	x_3
13.9	4.1	0.6	0.1
47.2	8.4	4.4	9.3
21.9	6.9	2.8	1.1
27.9	6.0	0.0	7.2
48.1	6.2	2.9	9.1
44.1	0.9	8.0	7.7
49.7	6.9	7.6	9.0
32.7	7.3	2.2	4.7
30.6	9.7	4.3	3.7
41.1	6.9	1.5	7.4

a. Construct the multiple regression equation $\hat{y} = b_0 + b_1x_1 + b_2x_2 + b_3x_3$.

b. Predict the value of y when $x_1 = 1$, $x_2 = 4.5$, $x_3 = 6.2$.

c. What percentage of the variation in y is explained by the model?

d. Is the model useful for prediction? Why or why not? Use the $\alpha = 0.05$ level.

e. Test $H_0: \beta_1 = 0$ versus $H_1: \beta_1 \neq 0$ at the $\alpha = 0.05$ level. Can you reject H_0? Repeat for β_2 and β_3.

12. For the following data set:

y	x_1	x_2	x_3
49.4	35.1	11.7	36.0
33.2	16.3	6.2	32.8
31.9	38.7	2.3	32.3
23.3	0.6	5.7	16.5
48.0	23.7	8.9	37.7
49.1	35.7	32.9	27.7
31.7	34.8	17.7	20.7
28.3	11.1	25.1	36.1
34.5	21.9	4.4	25.9
17.1	12.9	21.8	3.9
36.6	37.5	29.4	7.4
22.9	27.1	0.4	12.1

a. Construct the multiple regression equation
$\hat{y} = b_0 + b_1 x_1 + b_2 x_2 + b_3 x_3$.

b. Predict the value of y when $x_1 = 10.1$, $x_2 = 8.5$, $x_3 = 26.2$.

c. What percentage of the variation in y is explained by the model?

d. Is the model useful for prediction? Why or why not? Use the $\alpha = 0.05$ level.

e. Test $H_0: \beta_1 = 0$ versus $H_1: \beta_1 \neq 0$ at the $\alpha = 0.05$ level. Can you reject H_0? Repeat for β_2 and β_3.

13. For the following data set:

y	x_1	x_2	x_3	x_4
27.4	12.3	4.7	3.4	4.2
55.2	9.7	12.9	10.7	2.1
35.3	12.4	2.8	9.1	9.7
36.3	5.2	13.6	5.0	0.7
42.7	1.0	9.3	2.6	13.5
58.5	14.8	9.1	3.2	10.1
58.9	5.9	11.1	3.7	4.2
54.0	11.7	11.8	8.6	0.8
64.2	3.7	14.7	1.8	14.6
61.1	13.3	12.1	0.4	2.3
53.3	14.5	9.4	6.3	10.0
59.5	5.0	9.6	13.4	14.7
43.3	4.1	7.0	10.4	9.4
40.3	2.4	9.0	8.3	6.2
43.3	3.8	6.9	7.8	11.3

a. Construct the multiple regression equation
$\hat{y} = b_0 + b_1 x_1 + b_2 x_2 + b_3 x_3 + b_4 x_4$.

b. Predict the value of y when $x_1 = 5.2$, $x_2 = 9.1$, $x_3 = 8.7$, $x_4 = 2.8$.

c. What percentage of the variation in y is explained by the model?

d. Is the model useful for prediction? Why or why not? Use the $\alpha = 0.01$ level.

e. Test $H_0: \beta_1 = 0$ versus $H_1: \beta_1 \neq 0$ at the $\alpha = 0.05$ level. Can you reject H_0? Repeat for β_2 and β_3.

14. For the following data set:

y	x_1	x_2	x_3	x_4
163.4	2.9	24.2	4.3	4.4
120.9	22.2	12.9	10.6	13.7
116.7	5.5	6.8	17.3	13.9
63.9	8.8	12.0	0.9	1.2
86.9	5.1	18.3	2.5	13.4
74.2	14.7	8.0	8.5	5.6
72.8	7.1	17.2	0.2	24.7
84.1	1.6	5.6	0.5	4.3
136.4	21.6	2.0	3.9	15.7
122.3	14.5	13.3	14.7	13.1
78.9	17.7	12.7	3.2	8.0
108.0	16.4	15.0	1.6	23.3
124.5	24.3	20.6	9.1	11.3
90.9	21.7	0.6	5.7	5.8
143.5	18.6	11.2	13.9	14.7

a. Construct the multiple regression equation
$\hat{y} = b_0 + b_1 x_1 + b_2 x_2 + b_3 x_3 + b_4 x_4$.

b. Predict the value of y when $x_1 = 15.3$, $x_2 = 4.7$, $x_3 = 0.6$, $x_4 = 8.2$.

c. What percentage of the variation in y is explained by the model?

d. Is the model useful for prediction? Why or why not? Use the $\alpha = 0.05$ level.

e. Test $H_0: \beta_1 = 0$ versus $H_1: \beta_1 \neq 0$ at the $\alpha = 0.05$ level. Can you reject H_0? Repeat for β_2, β_3 and β_4.

Working with the Concepts

15. Engine emissions: In a laboratory test of a new automobile engine design carried out at the Colorado School of Mines, the emission rate (in milligrams per second) of oxides of nitrogen was measured for 32 engines at various engine speeds (in rpm) and engine torque (in foot-pounds). The following MINITAB output presents the multiple regression equation Emission rate $= b_0 + b_1$ Speed $+ b_2$ Torque.

```
The regression equation is
Emission rate = -321 + 0.378 Speed - 0.160 Torque


Predictor        Coef        StDev          T           P
Constant        -320.59       98.14       -3.27        0.003
Speed            0.37820     0.06861       5.51        0.000
Torque          -0.16047     0.06082      -2.64        0.013


S = 67.13        R-Sq = 51.6%        R-Sq (adj) = 48.3%


Analysis of Variance
Source              DF          SS          MS          F          P
Regression           2       139419.6     69079.8     15.469      0.000
Residual Error      29       130685.2      4506.4
Total               31       270104.8
```

a. Predict the emissions rate for an engine with speed 1500 and torque 500.

b. Two engines are run at the same torque, but their speeds differ by 300 rpm. By how much should we predict their emissions rates to differ?

c. Is the model useful for prediction? Explain. Use the $\alpha = 0.05$ level.

d. How much of the variation in the emission rate is explained by the model?

e. Let β_1 be the coefficient of speed. Test H_0: $\beta_1 = 0$ versus H_1: $\beta_1 \neq 0$ at the $\alpha = 0.05$ level. What do you conclude?

f. Let β_2 be the coefficient of torque. Test H_0: $\beta_2 = 0$ versus H_1: $\beta_2 \neq 0$ at the $\alpha = 0.01$ level. What do you conclude?

16. Fracking: Natural gas is found in rock formations underground. In order to extract the gas, a procedure known as hydraulic fracturing, or "fracking," is often used. In this procedure, fluid mixed with sand is pumped into the gas well and forced through cracks in the rock. The sand holds open the cracks, allowing the gas to flow through. A study was performed at the Colorado School of Mines to determine whether increasing the amounts of fluid or sand would increase the production of a well. Varying amounts of fluid and sand were pumped into 255 gas wells, and the amount of gas recovered was measured. The amount of gas produced is measured in units of 100 cubic feet per each foot of well depth, fluid is measured in thousands of gallons per foot, and sand is measured in thousands of pounds per foot. The following MINITAB output presents the multiple regression equation Production = $b_0 + b_1$ Fluid + b_2 Sand.

```
The regression equation is
Production = 1.0729 + 1.1752 Fluid - 0.024886 Sand


Predictor        Coef         StDev          T           P
Constant        1.0729       0.23348      4.5954       0.000
Fluid           1.1752       0.24609      4.7753       0.000
Sand           -0.024886    0.096792     -0.25711      0.797


S = 2.0197        R-Sq = 29.1%        R-Sq (adj) = 28.5%


Analysis of Variance
Source              DF          SS          MS          F          P
Regression           2        421.70      210.85      51.689      0.000
Residual Error     252        1028.0      4.0792
Total              254        1449.7
```

a. Predict the gas production for a well treated with 3 thousand gallons of fluid and 7 thousand pounds of sand.

b. Two wells have the same amount of sand pumped in, and the amount of fluid differs by 2 thousand pounds. By how much should we predict their productions to differ?

c. Is the model useful for prediction? Explain. Use the $\alpha = 0.01$ level.

d. How much of the variation in the emission rate is explained by the model?

e. Let β_1 be the coefficient of fluid. Test $H_0: \beta_1 = 0$ versus $H_1: \beta_1 \neq 0$ at the $\alpha = 0.01$ level. What do you conclude?

f. Let β_2 be the coefficient of sand. Test $H_0: \beta_2 = 0$ versus $H_1: \beta_2 \neq 0$ at the $\alpha = 0.05$ level. What do you conclude?

17. Drop that variable: The following MINITAB output presents a multiple regression equation $\hat{y} = b_0 + b_1x_1 + b_2x_2 + b_3x_3 + b_4x_4$.

```
The regression equation is
Y = 0.742 + 3.58 X1 - 2.18 X2 - 0.818 X3 - 1.58 X4

Predictor        Coef        StDev           T          P
Constant        0.7424      0.7557        0.9824      0.333
X1              3.5829      1.0602        3.3794      0.002
X2             -2.1772      0.6896       -3.1574      0.003
X3             -0.8177      0.8606       -0.9501      0.349
X4             -1.5840      0.8851       -1.7897      0.082

S = 4.6846       R-sq = 46.0%       R-sq (adj) = 39.8%

Analysis of Variance
Source           DF          SS           MS          F          P
Regression        4        654.38       163.59      7.4547     0.000
Residual Error   35        768.08        21.945
Total            39       1422.5
```

It is desired to drop one of the explanatory variables. Which of the following is the most appropriate action?

i. Drop x_3, then see whether R^2 increases.

ii. Drop x_3, then see whether adjusted R^2 increases.

iii. Drop x_1, then see whether R^2 increases.

iv. Drop x_1, then see whether adjusted R^2 increases.

18. Drop that variable: The following MINITAB output presents a multiple regression equation $\hat{y} = b_0 + b_1x_1 + b_2x_2 + b_3x_3 + b_4x_4$.

```
The regression equation is
Y = 0.742 + 3.58 X1 - 2.18 X2 - 0.818 X3 - 1.58 X4

Predictor        Coef        StDev           T          P
Constant        1.3774      0.7317        1.8823      0.066
X1              1.1307      0.6932        1.6311      0.110
X2              4.3897      0.6755        6.4987      0.000
X3             -1.4479      0.7978       -1.8149      0.076
X4              0.7414      0.6447        1.1500      0.256
X5              1.7095      0.6266        2.7283      0.009

S = 4.5160       R-sq = 63.7%       R-sq (adj) = 59.6%

Analysis of Variance
Source           DF          SS           MS          F          P
Regression        5       1573.38       314.79      15.435     0.000
Residual Error   44        897.35        20.394
Total            49       2471.3
```

It is desired to drop one of the explanatory variables. Which of the following is the most appropriate action?

i. Drop x_2, then see whether R^2 increases.

ii. Drop x_4, then see whether R^2 increases.

iii. Drop x_4, then see whether adjusted R^2 increases.

iv. Drop x_2, then see whether adjusted R^2 increases.

19. Predicting GPA: Twenty college students were sampled after their freshman year. Following are their freshman GPAs, their high school GPAs, their SAT reading scores, and their SAT math scores.

Freshman GPA	High School GPA	SAT Reading	SAT Math
2.98	2.90	506	497
3.59	3.58	572	521
2.26	2.18	402	498
2.43	2.44	490	474
2.14	2.69	440	398
3.77	3.61	645	625
3.26	3.61	541	481
3.18	3.40	569	490
2.80	3.05	447	626
3.20	3.37	477	594
3.02	2.41	486	508
2.84	2.95	555	466
2.66	3.30	486	425
3.46	3.47	535	523
2.52	2.47	397	453
2.01	1.89	482	388
2.79	2.41	459	294
3.01	2.62	421	319
3.31	3.54	525	549
3.17	3.10	514	542

a. Let y represent freshman GPA, x_1 represent high school GPA, x_2 represent SAT reading score, and x_3 represent SAT math score, Construct the multiple regression equation $\hat{y} = b_0 + b_1 x_1 + b_2 x_2 + b_3 x_3$.

b. An applicant for next year's freshman class has a high school GPA of 3.05, an SAT reading score of 510, and an SAT math score of 515. Predict the freshman GPA for this student.

c. Refer to part (b). Construct a 95% confidence interval for the freshman GPA.

d. Refer to part (b). Construct a 95% prediction interval for the freshman GPA.

e. What percentage of the variation in freshman GPA is explained by the model?

f. Is the model useful for prediction? Why or why not? Use the $\alpha = 0.01$ level.

g. Test $H_0: \beta_1 = 0$ versus $H_1: \beta_1 \neq 0$ at the $\alpha = 0.05$ level. Can you reject H_0? Repeat for β_2 and β_3.

20. Paint lifetime: A paint company collected data on the lifetime (in years) of its paint in eleven United States cities. The data are in the following table.

City	Paint Lifetime	Average January Temperature	Average July Temperature	Average Annual Precipitation (inches)
Atlanta, GA	11.5	41.9	78.6	48.6
Boston, MA	11.7	29.6	73.5	43.8
Kansas City, KS	12.3	28.4	80.9	29.3
Minneapolis, MN	10.5	11.2	73.1	26.4
Dallas, TX	11.2	45.0	86.3	34.2
Denver, CO	15.2	29.5	73.3	15.3
Miami, FL	8.7	67.1	82.4	57.5
Phoenix, AZ	11.1	52.3	92.3	7.1
San Francisco, CA	16.7	48.5	62.2	19.7
Seattle, WA	14.2	40.6	65.3	38.9
Washington, DC	12.6	35.2	78.9	39.0

a. Let y represent paint lifetime, x_1 represent January temperature, x_2 represent July temperature, and x_3 represent annual precipitation. Construct the multiple regression equation $\hat{y} = b_0 + b_1 x_1 + b_2 x_2 + b_3 x_3$.

b. In Cheyenne, Wyoming, the average January temperature is 26.1, the average July temperature is 68.9, and the average annual precipitation is 13.3. Predict the lifetime of this paint in Cheyenne.

c. Refer to part (b). Construct a 95% confidence interval for the paint lifetime in Cheyenne.

d. Refer to part (b). Construct a 95% prediction interval for the paint lifetime in Cheyenne.

e. What percentage of the variation in the paint lifetime in Cheyenne is explained by the model?

f. Is the model useful for prediction? Why or why not? Use the $\alpha = 0.05$ level.

g. Test $H_0: \beta_1 = 0$ versus $H_1: \beta_1 \neq 0$ at the $\alpha = 0.05$ level. Can you reject H_0? Repeat for β_2 and β_3.

21. Chemical reaction: A chemical reaction was run 48 times. In each run, different values were chosen for the temperature in degrees Celsius (x_1), the concentration of the primary reactant (x_2), and the number of hours the reaction was allowed to run (x_3). The outcome variable (y) is the amount of product.

y	x_1	x_2	x_3	y	x_1	x_2	x_3	y	x_1	x_2	x_3	y	x_1	x_2	x_3
31.3	50	19	4.0	37.4	60	30	5.0	24.0	60	24	4.0	51.0	80	34	7.5
56.9	90	38	8.0	43.3	60	26	7.0	36.2	50	21	7.0	34.3	60	22	2.5
43.1	70	28	6.5	36.3	70	25	7.5	26.5	50	17	2.0	31.5	60	24	5.0
41.5	70	25	5.5	38.4	70	31	5.5	47.1	80	34	8.5	33.2	60	23	4.0
39.0	60	26	6.5	41.5	60	27	7.5	38.1	70	27	5.5	39.2	60	29	6.5
40.9	70	29	5.0	36.1	60	23	6.0	33.5	60	24	2.5	46.7	70	27	7.5
35.9	60	23	5.5	38.5	60	23	6.0	43.6	70	27	10.0	30.4	70	32	4.0
43.5	70	28	5.5	42.4	60	24	9.0	41.0	80	32	6.5	43.2	60	25	5.5
47.9	80	34	6.5	46.5	70	31	5.5	50.2	50	26	9.0	30.6	60	26	3.5
33.8	70	26	4.5	43.1	80	32	6.0	34.4	50	22	4.0	43.3	70	28	7.5
41.1	70	26	8.0	50.8	60	26	10.0	47.9	80	34	6.5	32.4	60	22	5.5
38.7	70	26	8.0	44.2	70	28	4.5	46.2	50	21	10.0	35.5	60	22	4.5

  a. Construct the multiple regression equation $\hat{y} = b_0 + b_1x_1 + b_2x_2 + b_3x_3$.
  b. A reaction is run at a temperature of 50 degrees Celsius, with the concentration of the primary reactant set to 30, and the reaction is allowed to run for 6 hours. Predict the amount of product produced.
  c. Refer to part (b). Construct a 95% confidence interval for the amount produced.
  d. Refer to part (b). Construct a 95% prediction interval for the amount produced.
  e. What percentage of the variation in the amount of product is explained by the model?
  f. Is the model useful for prediction? Why or why not? Use the $\alpha = 0.05$ level.
  g. Test H_0: $\beta_1 = 0$ versus H_1: $\beta_1 \neq 0$ at the $\alpha = 0.05$ level. Can you reject H_0? Repeat for β_2 and β_3.

22. Old Faithful: The following table lists values measured for 60 consecutive eruptions of the geyser Old Faithful in Yellowstone National Park. They are the duration of the eruption (x_1), the duration of the dormant period immediately before the eruption (x_2), and the duration of the dormant period immediately after the eruption (y). All the times are in minutes.

x_1	x_2	y	x_1	x_2	y	x_1	x_2	y	x_1	x_2	y
3.5	80	84	1.8	42	91	4.7	88	51	4.1	70	79
4.1	84	50	4.1	91	51	1.8	51	80	3.7	79	60
2.3	50	93	1.8	51	79	4.6	80	49	3.8	60	86
4.7	93	55	3.2	79	53	1.9	49	82	3.4	86	71
1.7	55	76	1.9	53	82	3.5	82	75	4.0	71	67
4.9	76	58	4.6	82	51	4.0	75	73	2.3	67	81
1.7	58	74	2.0	51	76	3.7	73	67	4.4	81	76
4.6	74	75	4.5	76	82	3.7	67	68	4.1	76	83
3.4	75	80	3.9	82	84	4.3	68	86	4.3	83	76
4.3	80	56	4.3	84	53	3.6	86	72	3.3	76	55
1.7	56	80	2.3	53	86	3.8	72	75	2.0	55	73
3.9	80	69	3.8	86	51	3.8	75	75	4.3	73	56
3.7	69	57	1.9	51	85	3.8	75	66	2.9	56	83
3.1	57	90	4.6	85	45	2.5	66	84	4.6	83	57
4.0	90	42	1.8	45	88	4.5	84	70	1.9	57	71

  a. Construct the multiple regression equation $\hat{y} = b_0 + b_1x_1 + b_2x_2$.
  b. Predict the duration of dormant period immediately after an eruption, if the duration of the eruption is 3.0 minutes and the duration of the dormant period immediately before the eruption is 70 minutes.
  c. Refer to part (b). Construct a 95% confidence interval for the duration of the dormant period immediately following the eruption.
  d. Refer to part (b). Construct a 95% prediction interval for the duration of the dormant period immediately following the eruption.
  e. What percentage of the variation in the duration is explained by the model?
  f. Is the model useful for prediction? Why or why not? Use the $\alpha = 0.01$ level.
  g. Test H_0: $\beta_1 = 0$ versus H_1: $\beta_1 \neq 0$ at the $\alpha = 0.05$ level. Can you reject H_0? Repeat for β_2.

23. How's your credit? Credit data were collected on a random sample of 25 U.S. cities in a recent year. Following are the average credit scores, the average debt, the average number of late payments, the average percentage of credit available, and the average number of open credit cards.

Credit Score	Debt	Late Payments	% of Credit Available	Credit Cards	Credit Score	Debt	Late Payments	% of Credit Available	Credit Cards
748	24,713	0.40	68.16	1.85	759	23,602	0.36	73.03	2.10
755	23,393	0.36	71.63	1.73	779	21,113	0.30	76.53	1.91
713	25,054	0.66	65.88	1.65	727	25,733	0.50	66.41	1.86
764	24,166	0.35	74.21	1.85	767	26,555	0.41	69.79	2.14
751	25,076	0.49	69.68	1.91	747	23,762	0.48	69.98	1.77
710	26,074	0.61	66.55	1.60	756	24,376	0.45	72.51	1.97
766	24,854	0.35	71.44	2.13	770	24,884	0.33	74.85	1.83
739	23,104	0.40	68.74	2.05	724	26,242	0.53	69.35	2.01
761	24,982	0.41	71.08	2.14	767	24,955	0.37	70.69	2.00
755	24,705	0.39	68.26	2.02	729	23,602	0.58	71.98	1.55
729	26,055	0.65	66.36	1.72	781	23,622	0.33	75.58	1.86
742	23,767	0.48	71.61	1.73	768	24,842	0.33	72.06	1.92
702	26,414	0.73	65.76	1.36					

a. Let y represent average credit score, x_1 represent average debt, x_2 represent average number of late payments, x_3 represent average percentage of credit available, and x_4 represent average number of open credit cards. Construct the multiple regression equation $\hat{y} = b_0 + b_1 x_1 + b_2 x_2 + b_3 x_3 + b_4 x_4$.

b. In Denver, Colorado, the average debt was $26,775, the average number of late payments was 0.42, the average percentage of available credit was 70.24, and the average number of open credit cards was 2.18. Predict the average credit score in Denver.

c. Refer to part (b). Construct a 95% confidence interval for the average credit score.

d. Refer to part (b). Construct a 95% prediction interval for the average credit score.

e. What percentage of the variation in the average credit score is explained by the model?

f. Is the model useful for prediction? Why or why not? Use the $\alpha = 0.01$ level.

g. Test $H_0: \beta_1 = 0$ versus $H_1: \beta_1 \neq 0$ at the $\alpha = 0.05$ level. Can you reject H_0? Repeat for β_2, β_3, and β_4.

Answers to Check Your Understanding Exercises for Section 13.3

1. The predicted value is 2.0276 liters.

2. By 0.2462 liters.

3. The P-value for β_1 is 0.008; reject H_0. The P-value for β_2 is 0.619; do not reject H_0. The P-value for β_3 is 0.000; reject H_0.

4. (a) The predicted value is 10.523.   (b) 5.4315, 15.615   (c) 7.443, 13.604

5. (a) 32.6%   (b) 23.3%

6. (a) The P-value is 0.031. We reject $H_0: \beta_1 = \beta_2 = \beta_3$ at the 0.05 level. We conclude that the model is useful for prediction. (b) We do not reject $H_0: \beta_1 = \beta_2 = \beta_3$ at the 0.01 level. There is not enough evidence to conclude at the $\alpha = 0.01$ level that the model is useful for prediction.

Chapter 13 Summary

Section 13.1: When the assumptions of the linear model are satisfied, the intercept b_0 and slope b_1 of the least-squares regression line are estimates of a true intercept β_0 and a true slope β_1. The linear model assumptions can be checked by constructing a residual plot. When the residual plot exhibits no obvious pattern, the vertical spread is approximately the same across the plot, and there are no outliers, we may conclude that the assumptions of the linear model are satisfied. We may then compute confidence intervals and test hypotheses about β_1. If we reject $H_0: \beta_1 = 0$, we may conclude that the explanatory variable is useful to help predict the value of the outcome variable.

Section 13.2: When the assumptions of the linear model are satisfied, we may construct confidence intervals for the mean response and prediction intervals for an individual response. A confidence interval for the mean response is an interval that is likely to contain the mean value of the response variable y for a given value of the explanatory variable x. A prediction interval for an individual response is an interval that is likely to contain the value of the response variable y for a randomly chosen individual whose value of the explanatory variable is x.

Section 13.3: Sometimes we need several explanatory variables to predict the value of the response variable. In these situations we use multiple regression. As with the least-squares regression line, we can construct confidence intervals for the mean response and prediction intervals for an individual response. The coefficient of determination, R^2, measures the proportion of variation in the response variable that is explained by the multiple regression equation. The adjusted R^2 value can be used to determine which explanatory variables can be dropped from the model. If removing a variable increases the adjusted R^2, then the variable should be removed.

Vocabulary and Notation

adjusted R^2 612
coefficient of determination R^2 611
error sum of squares $\sum(y - \hat{y})^2$ 611
F-test for goodness-of-fit   612
individual response   601
linear model   585

mean response   600
multiple regression   606
population correlation   593
prediction interval   601
regression sum of squares $\sum(y - \bar{y})^2$ 611
residual   586

residual plot   586
residual standard deviation   588
sum of squares for x: $\sum(x - \bar{x})^2$ 589
total sum of squares   611

Important Formulas

Residual standard deviation:

$$s_e = \sqrt{\frac{\sum(y - \hat{y})^2}{n - 2}}$$

Standard error for b_1:

$$s_b = \frac{s_e}{\sqrt{\sum(x - \bar{x})^2}}$$

Confidence interval for slope:

$$b_1 - t_{\alpha/2} \cdot s_b < \beta_1 < b_1 + t_{\alpha/2} \cdot s_b$$

Confidence interval for the mean response:

$$\hat{y} \pm t_{\alpha/2} \cdot s_e \sqrt{\frac{1}{n} + \frac{(x^* - \bar{x})^2}{\sum(x - \bar{x})^2}}$$

Test statistic for slope b_1:

$$t = \frac{b_1}{s_b}$$

Prediction interval for an individual response:

$$\hat{y} \pm t_{\alpha/2} \cdot s_e \sqrt{1 + \frac{1}{n} + \frac{(x^* - \bar{x})^2}{\sum(x - \bar{x})^2}}$$

Chapter Quiz

1. A confidence interval for β_1 is to be constructed from a sample of 20 points. How many degrees of freedom are there for the critical value?

2. A confidence interval for a mean response and a prediction interval for an individual response are to be constructed from the same data. True or false: The number of degrees of freedom for the critical value is the same for both intervals.

3. True or false: If we fail to reject the null hypothesis H_0: $\beta_1 = 0$, we can conclude that there is no linear relationship between the explanatory variable and the outcome variable.

4. True or false: When the sample size is large, confidence intervals and hypothesis tests for β_1 are valid even when the assumptions of the linear model are not met.

5. A statistics student has constructed a confidence interval for the mean height of daughters whose mothers are 66 inches tall, and a prediction interval for the height of a particular daughter whose mother is 66 inches tall. One of the intervals is (65.3, 68.2) and the other is (63.8, 69.7). Unfortunately, the student has forgotten which interval is which. Can you tell which is the confidence interval and which is the prediction interval? Explain.

Exercises 6–10 refer to the following data set:

x	25	13	16	19	29	19	16	30
y	40	20	33	30	50	37	34	37

6. Compute the point estimates b_0 and b_1.

7. Construct a 95% confidence interval for β_1.

8. Test the hypotheses H_0: $\beta_1 = 0$ versus H_1: $\beta_1 \neq 0$. Use the $\alpha = 0.01$ level of significance.

9. Construct a 95% confidence interval for the mean response when $x = 20$.

10. Construct a 95% prediction interval for an individual response when $x = 20$.

Exercises 11–15 refer to the following data set:

y	x_1	x_2	x_3
69.8	7.9	37.3	62.4
32.3	9.3	20.2	40.7
66.9	13.3	30.5	48.7
87.5	27.4	38.8	35.8
93.5	9.3	40.1	28.9
65.1	23.3	29.2	46.9
26.3	24.1	11.5	50.4
48.7	21.5	19.0	58.2
52.8	25.5	32.2	34.7
58.0	8.3	37.9	23.7
65.6	30.1	34.6	56.2
42.3	21.1	23.9	42.6

11. Construct the multiple regression equation $\hat{y} = b_0 + b_1 x_1 + b_2 x_2 + b_3 x_3$.

12. Predict the value of y when $x_1 = 20$, $x_2 = 20$, and $x_3 = 30$.

13. What percentage of the variation in y is explained by the model?

14. Is this model useful for prediction? Why or why not? Use the $\alpha = 0.05$ level.

15. Test H_0: $\beta_1 = 0$ versus H_1: $\beta_1 \neq 0$ at the $\alpha = 0.05$ level. Repeat for β_2 and β_3.

Review Exercises

1. How's your mileage? Weight (in tons) and fuel economy (in mpg) were measured for a sample of seven diesel trucks. The results are presented in the following table.

Weight	Mileage
8.00	7.69
24.50	4.97
27.00	4.56
14.50	6.49
28.50	4.34
12.75	6.24
21.25	4.45

Source: J. Yanowitz, Ph.D. thesis, Colorado School of Mines

a. Compute the least-squares regression line for predicting mileage (y) from weight (x).

b. Construct a 95% confidence interval for the slope.

c. Test H_0: $\beta_1 = 0$ versus H_1: $\beta_1 < 0$. Can you conclude that weight is useful in predicting mileage? Use the $\alpha = 0.05$ level of significance.

2. How's your mileage? Use the data in Exercise 1 for the following.

a. Compute a point estimate for the mean mileage for trucks that weigh 20 tons.

b. Construct a 95% confidence interval for the mean mileage of trucks that weigh 20 tons.

c. Predict the mileage of a particular truck that weighs 20 tons.

d. Construct a 95% prediction interval for the mileage of a particular truck that weighs 20 tons.

3. How much wood is in that tree? For a sample of 12 trees, the volume of lumber (in cubic meters) and the diameter (in centimeters) at a fixed height above ground level were measured. The results were as follows.

Diameter	Volume	Diameter	Volume
35.1	0.81	33.8	0.80
48.4	1.39	45.3	1.69
47.9	1.31	25.2	0.30
35.3	0.67	28.5	0.19
47.3	1.46	30.1	0.63
26.4	0.47	30.0	0.64

a. Compute the least-squares regression line for predicting volume (y) from diameter (x).

b. Construct a 95% confidence interval for the slope.

c. Test H_0: $\beta_1 = 0$ versus H_1: $\beta_1 > 0$. Can you conclude that diameter is useful in predicting volume? Use the $\alpha = 0.01$ level of significance.

4. **How much wood is in that tree?** Use the data in Exercise 3 for the following.
   a. Compute a point estimate for the mean volume for trees with a diameter of 44 centimeters.
   b. Construct a 95% confidence interval for the mean volume of trees with a diameter of 44 centimeters.
   c. Predict the volume for a particular tree whose diameter is 44 centimeters.
   d. Construct a 95% prediction interval for the volume of a particular tree whose diameter is 44 centimeters.

5. **Watching paint dry:** In tests designed to measure the effect of the concentration (in percent) of a certain additive on the drying time (in hours) of paint, the following data were obtained.

Concentration of Additive	Drying Time
4.0	8.7
4.2	8.8
4.4	8.3
4.6	8.7
4.8	8.1
5.0	8.0
5.2	8.1
5.4	7.7
5.6	7.5
5.8	7.2

   a. Compute the least-squares regression line for predicting drying time (y) from concentration (x).
   b. Construct a 95% confidence interval for the slope.
   c. Test $H_0: \beta_1 = 0$ versus $H_1: \beta_1 < 0$. Can you conclude that concentration is useful in predicting drying time? Use the $\alpha = 0.05$ level of significance.

6. **Watching paint dry:** Use the data in Exercise 5 for the following.
   a. Compute a point estimate for the mean drying time for paint with a concentration of 5.1.
   b. Construct a 95% confidence interval for the mean drying time for paint whose concentration is 5.1.
   c. Predict the mean drying time for a particular can of paint with a concentration of 5.1.
   d. Construct a 95% prediction interval for the drying time of a particular can of paint whose concentration is 5.1.

7. **Energy use:** A sample of 10 households was monitored for one year. The household income (in $1000s) and the amount of energy consumed (in 10^{10} joules) were determined. The results follow.

Income	Energy	Income	Energy
31	16.0	96	98.3
40	40.2	70	93.8
28	29.8	100	77.1
48	45.6	145	114.8
195	184.6	78	67.0

   a. Compute the least-squares line for predicting energy consumption (y) from income (x).
   b. Compute a 95% confidence interval for the slope.
   c. Test $H_0: \beta_1 = 0$ versus $H_1: \beta_1 > 0$. Can you conclude that income is useful in predicting energy consumption? Use the $\alpha = 0.01$ level of significance.

8. **Energy use:** Use the data in Exercise 7 for the following.
   a. Compute a point estimate for the mean energy use for families with an income of $50,000.
   b. Construct a 95% confidence interval for the mean energy use for families with an income of $50,000.
   c. Predict the energy use for a particular family with an income of $50,000.
   d. Construct a 95% prediction interval for the energy use for a particular family with an income of $50,000.

9. **Interpret calculator display:** The following TI-84 Plus display presents the results of a test of the null hypothesis $H_0: \beta_1 = 0$.

```
LinRegTTest
y=a+bx
β≠0 and ρ≠0
t=4.953149177
P=.0011164587
df=8
↓a=23.22340023
```

   a. What is the alternate hypothesis?
   b. What is the value of the test statistic?
   c. How many degrees of freedom are there?

d. What is the P-value?

e. Can you conclude that the explanatory variable is useful in predicting the outcome variable? Answer this question using the $\alpha = 0.01$ level of significance.

10. **Interpret computer output:** The following MINITAB output presents the results of a test of the null hypothesis $H_0: \beta_1 = 0$.

```
Regression Analysis : Y versus X

The regression equation is
Y = 1.92 + 5.56 X

Predictor       Coef      SE Coef        T          P
Constant      1.9167       0.1721     11.14      0.000
X             5.5582       2.1158      2.63      0.006
```

a. What are the slope and intercept of the least-squares regression line?

b. Can you conclude that x is useful for predicting y? Use the $\alpha = 0.05$ level of significance.

11. **Air pollution:** Following are measurements of particulate matter (PM) concentration (in micrograms per cubic meter), temperature, in degrees Fahrenheit, wind speed in miles per hour, and humidity in percent for 38 days in Denver, Colorado.

PM	Temperature	Wind Speed	Humidity	PM	Temperature	Wind Speed	Humidity
24.6	35.8	4.8	37.8	12.6	36.0	3.0	48.6
71.7	41.5	13.3	32.6	7.3	47.9	2.5	19.7
13.7	38.1	9.5	39.2	17.4	33.5	3.5	50.9
19.5	35.6	6.0	37.2	7.7	25.1	4.5	55.1
79.6	35.1	12.7	37.5	14.0	28.5	3.1	47.7
6.4	30.8	1.3	52.0	20.8	32.8	4.4	38.7
13.9	39.7	2.5	43.5	19.8	36.9	2.7	36.1
25.1	44.6	2.6	46.4	13.1	43.7	3.8	25.2
21.3	53.8	1.4	33.8	10.5	48.3	4.5	19.4
19.0	52.5	3.3	27.9	22.7	54.9	3.6	20.0
33.5	48.2	11.9	35.1	24.0	57.8	2.7	20.5
37.8	29.6	3.2	53.0	39.1	49.2	4.0	39.2
8.0	31.7	3.1	37.8	28.2	51.6	2.5	42.5
15.1	41.5	2.0	29.5	41.9	54.2	2.6	38.1
27.9	55.1	5.2	22.6	25.7	53.8	2.8	41.6
35.4	59.9	8.0	19.4	30.1	60.6	5.7	26.8
22.0	62.6	9.3	15.3	49.1	44.2	8.7	43.9
45.4	51.3	10.7	17.9	11.5	40.9	3.9	39.5
23.7	40.5	6.5	29.1	27.1	39.6	7.3	43.8

a. Let y represent PM, x_1 represent temperature, x_2 represent wind speed, and x_3 represent humidity. Construct the multiple regression equation $\hat{y} = b_0 + b_1 x_1 + b_2 x_2 + b_3 x_3$.

b. Predict the particulate concentration on a day where the temperature is 40 degrees, the wind speed is 5 miles per hour, and the humidity is 30 percent.

c. Refer to part (b). Construct a 95% confidence interval for the particulate concentration.

d. Refer to part (b). Construct a 95% prediction interval for the particulate concentration.

e. Are all the values in the prediction interval reasonable? Explain.

f. What percentage of the variation in particulate concentration is explained by the model?

g. Is the model useful for prediction? Why or why not? Use the $\alpha = 0.05$ level.

h. Test $H_0: \beta_1 = 0$ versus $H_1: \beta_1 \neq 0$ at the $\alpha = 0.05$ level. Can you reject H_0? Repeat for β_2 and β_3.

12. **Icy lakes:** Following are data on maximum ice thickness in millimeters (y), average number of days per year of ice cover (x_1), average number of days the bottom temperature is lower than $8°C$ (x_2), and the average snow depth in millimeters (x_3) for 13 lakes in Minnesota.

y	x_1	x_2	x_3
730	152	198	91
760	173	201	81
850	166	202	69
840	161	202	72
720	152	198	91
730	153	205	91
840	166	204	70
730	157	204	90
650	136	172	47
850	142	218	59
740	151	207	88
720	145	209	60
710	147	190	63

a. Construct the multiple regression equation $\hat{y} = b_0 + b_1x_1 + b_2x_2 + b_3x_3$.

b. Predict the ice thickness for a lake which is covered by ice an average of 140 days per year, the bottom temperature is less than 8°C an average of 190 days per year, and the average snow depth is 60 millimeters.

c. Refer to part (b). Construct a 95% confidence interval for the ice thickness.

d. Refer to part (b). Construct a 95% prediction interval for the ice thickness.

e. What percentage of the variation in ice thickness is explained by the model?

f. Is the model useful for prediction? Why or why not? Use the $\alpha = 0.05$ level.

g. Test $H_0: \beta_1 = 0$ versus $H_1: \beta_1 \neq 0$ at the $\alpha = 0.05$ level. Can you reject H_0? Repeat for β_2 and β_3.

13. **Percentage of variation:** The following MINITAB output presents a multiple regression equation.

```
The regression equation is
Y = 5.3296 + 3.6204 X1 + 1.5721 X2 + 2.3328 X3

Predictor          Coef          SE Coef              T              P
Constant         -52.042          63.533        -0.8191          0.422
X1                1.4686          1.2076         1.2161          0.238
X2               -1.0333          1.1427        -0.9043          0.377
X3                2.0944          0.8783         2.3847          0.027


S = 44.953           R-sq = 40.0%         R-sq (adj) = 31.0%


Analysis of Variance

Source              DF            SS             MS             F              P
Regression           3         269.17         89.723          4.44          0.015
Residual Error      20         404.16         20.208
Total               23         673.33         29.275
```

a. What percentage of the variation in the response is explained by the multiple regression equation?

b. What percentage of the variation in the response is explained by the multiple regression equation, if we account for the number of explanatory variables in the model?

c. Is the multiple regression equation useful for prediction? Explain. Use the $\alpha = 0.05$ level.

14. **Confidence and prediction intervals:** The following MINITAB output presents a confidence interval for a mean response and a prediction interval for an individual response.

```
New Obs          Fit          SE Fit         95.0% CI              95.0% PI
1              10.933         1.9789      (6.805, 15.061)      (0.6874, 21.178)


Values of Predictors for New Observations

New Obs          X1             X2             X3
1                2.5            4.1            3.2
```

a. Predict the value of y when $x_1 = 2.5$, $x_2 = 4.1$, and $x_3 = 3.2$.

b. We are 95% confident that an individual whose values are $x_1 = 2.5$, $x_2 = 4.1$, and $x_3 = 3.2$ will have a response between _____ and _____ .

c. We are 95% confident that the mean response when $x_1 = 2.5$, $x_2 = 4.1$, and $x_3 = 3.2$ is between _____ and _____ .

15. **Drop that variable:** The following MINITAB output presents a multiple regression equation $\hat{y} = b_0 + b_1x_1 + b_2x_2 + b_3x_3 + b_4x_4$.

```
The regression equation is
Y = 0.742 + 3.58 X1 - 2.18 X2 - 0.818 X3 - 1.58 X4


Predictor           Coef       StDev          T          P
Constant          23.310      20.447      1.1400      0.263
X1                 1.2736      0.3893      3.2713      0.003
X2                 0.1954      0.3708      0.5270      0.602
X3                -0.6788      0.2830     -2.3981      0.023
X4                -4.5620      3.3893     -1.3460      0.188


S = 17.718        R-Sq = 35.0%        R-Sq (adj) = 26.6%


Analysis of Variance
Source              DF          SS          MS          F          P
Regression           4       523.37      130.84      4.1678     0.008
Residual Error      31       973.21      31.394
Total               35       1496.6      42.760
```

It is desired to drop one of the explanatory variables. Which of the following is the most appropriate action?
   i. Drop x_1, then see whether R^2 increases.
  ii. Drop x_2, then see whether adjusted R^2 increases.
 iii. Drop x_4, then see whether R^2 increases.
  iv. Drop x_1, then see whether adjusted R^2 increases.

Write About It

1. The quantity $\sum(x - \bar{x})^2$ measures the spread in the x-values. Explain why using x-values that are more spread out will result in a narrower confidence interval for β_1.

2. Suppose you are planning to buy a car that weighs 3500 pounds, and you have a linear model that predicts gas mileage (y) given the weight of the car (x). Which do you think would be more appropriate, a confidence interval for the mean gas mileage of cars that weigh 3500 pounds, or a prediction interval for the mileage of a particular car that weighs 3500 pounds? Explain.

Case Study: How Much Electric Power Is Generated By Wind Turbines?

In the chapter opener, we presented the following table of wind speeds and amounts of electric power generated by a windmill on 42 different days. It is important to be able to predict the amount of power that will be obtained from a given wind speed. To do this, we will compute an appropriate least-squares regression line to predict power from wind speed, then use the methods of this chapter to construct confidence intervals and prediction intervals for the amount of power produced.

Wind Speed	Power	Wind Speed	Power	Wind Speed	Power	Wind Speed	Power
4.2	10.3	4.2	5.3	2.6	9.1	4.6	5.0
1.4	1.4	3.7	1.1	7.7	46.1	8.0	65.1
6.6	39.4	5.9	29.2	6.1	26.9	7.7	54.6
4.7	12.3	6.0	24.4	5.5	19.7	1.6	0.6
2.6	9.0	5.3	13.9	4.7	5.4	5.1	17.9
5.8	20.6	5.1	9.5	4.0	12.4	6.6	32.5
1.8	3.8	4.9	2.9	2.3	5.0	2.3	9.4
5.8	21.1	8.3	63.7	8.6	67.4	5.8	10.8
7.3	41.3	7.1	37.1	5.6	15.2	6.9	29.3
7.1	45.2	9.2	87.5	4.2	11.0	6.2	21.4
6.4	28.4	4.4	0.5				

1. Compute the least-squares regression line for predicting power (y) from wind speed (x).

2. Construct a residual plot and explain how it shows that the assumptions of the linear model are violated.

3. We cannot use the least-squares regression line to predict y from x. In these situations, statisticians often try to manipulate one or both of the variables to try to find a linear fit. This is called **transforming the variables**. For each data point, compute the cube of the wind speed to obtain x^3, then compute the least-squares regression line for predicting power y from x^3.

4. Construct a residual plot to verify that the assumptions of the linear model are now satisfied.

5. Use the least-squares regression line computed in Exercise 3 to predict the power when the wind speed is 5.0 mph. (Note that the value of the explanatory variable is $5^3 = 125$.)

6. Use the least-squares regression line computed in Exercise 3 to compute a 95% confidence interval for the mean power on days when the wind speed is 5.0 mph.

7. Use the least-squares regression line computed in Exercise 3 to compute a 95% prediction interval for the power on a given day when the wind speed is 5.0 mph.

Analysis of Variance

Introduction

Almost all electronic devices rely on integrated circuits, which are complex electrical circuits built onto wafers constructed of semiconductor material, typically silicon. The silicon wafers are coated with a very thin layer of a metal called tungsten. To determine whether the correct amount of tungsten has been applied, the wafers are weighed on an extremely accurate scale called a microbalance. In a study published in the collection *Statistical Case Studies for Industrial Process Improvement*, the performance of a microbalance was examined.

The policy in the laboratory where this study was conducted was that the balance was to be turned off when not in use, and powered up only when it was about to be used. Each of three operators made two weighings of several silicon wafers. Results are presented in the following table for three of the wafers. All the wafers had weights very close to 54 grams, so the weights are reported in units of micrograms above 54 grams.

	Operator A		Operator B		Operator C	
Wafer 1	11	15	10	6	14	10
Wafer 2	210	208	205	201	208	207
Wafer 3	111	113	102	105	108	111

A new policy was then instituted to leave the balance powered up continuously. The three operators then made two weighings of three different wafers. The results are presented in the following table.

	Operator A		Operator B		Operator C	
Wafer 1	152	156	156	155	152	157
Wafer 2	443	440	442	439	435	439
Wafer 3	229	227	229	232	225	228

Which policy leads to more accurate results? In the case study at the end of this chapter, we will use the methods described in the chapter to find the answer.

SECTION 14.1   One-Way Analysis of Variance

Objectives

1. Describe the data and hypotheses for one-way ANOVA
2. Check the assumptions for one-way ANOVA
3. Perform a one-way ANOVA hypothesis test
4. Perform the Tukey–Kramer test of pairwise comparisons
5. Perform a one-way ANOVA test with technology

Objective 1 Describe the data and hypotheses for one-way ANOVA

Introduction to One-Way Analysis of Variance

In Section 11.1, we studied tests for a null hypothesis that two populations had equal means. In some situations, we have three or more populations to compare. In such situations, when appropriate assumptions are satisfied, we may test the hypothesis that all the population means are equal. This is done with a method called **analysis of variance** (ANOVA for short).

We illustrate the idea with an example. Welds, used to join metal parts, are made by heating a powdered material called flux. In a study conducted by G. Fredrickson at the Colorado School of Mines, four welding fluxes, with differing chemical compositions, were prepared. The purpose of the study was to determine whether the welds made from the fluxes would have different degrees of hardness. Five welds using each flux were made, and the hardness of each was measured in Brinell units; the higher the number, the harder the weld. The data are presented in Table 14.1.

Table 14.1 Hardness of Welds Using Four Different Fluxes

Flux	Sample Values					Sample Mean	Sample Standard Deviation
A	250	264	256	260	239	253.8	9.7570
B	263	254	267	265	267	263.2	5.4037
C	257	279	269	273	277	271.0	8.7178
D	253	258	262	264	273	262.0	7.4498

Source: G. Fredrickson, M.Sc. thesis, Colorado School of Mines

Each of the four samples in Table 14.1 can be thought of as coming from a different population. Figure 14.1 presents dotplots for the sample hardnesses using the four fluxes. Each sample mean is marked with an "X." It is clear that the sample means differ. In particular, the welds made using flux C have the largest sample mean and those using flux A have the smallest. Of course, the sample means are random, and if the experiment were repeated, different fluxes might have the largest and the smallest means. The question we want to answer is this: Can we conclude that there are differences in the population means among the four flux types?

This experiment involves an outcome variable, hardness, and an explanatory variable, flux type. The explanatory variable is qualitative, and is called a **factor**. The outcome variable is called the **response variable**. The different values for the factor (in this example, the different fluxes) are called **treatments**. Because there is only one factor (flux type), this is called a **one-factor experiment**, Table 14.1 is called a **one-way table**, and the method of determining whether the population means differ is called **one-way ANOVA**.

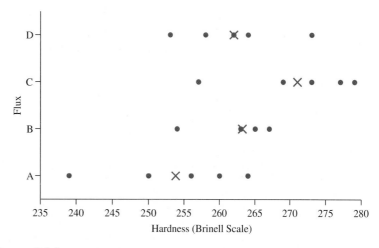

Figure 14.1   Dotplots for each sample in Table 14.1. Each sample mean is marked with an "X." The sample means differ somewhat, but the sample values overlap considerably.

Objective 2 Check the assumptions for one-way ANOVA

Assumptions for One-Way ANOVA

Data for a one-way analysis of variance consist of samples from several populations. The methods of one-way ANOVA require that the populations satisfy certain assumptions.

> ### Assumptions for One-Way ANOVA
>
> **1.** We have independent simple random samples from three or more populations.
> **2.** Each of the populations must be approximately normal.
> **3.** The populations must all have the same variance, which we will denote by σ^2.

Balanced versus unbalanced designs

When the sample sizes are all the same, the data are said to have a **balanced design**. Although one-way analysis of variance can be used with both balanced and unbalanced designs, balanced designs offer a big advantage. A balanced design is much less sensitive to violations of the assumption of equality of variance than an unbalanced one. Because moderate departures from this assumption can be difficult to detect, it is best to use a balanced design whenever possible, so that undetected violations of the assumption will not seriously compromise the validity of the results. When a balanced design is impossible to achieve, a slightly unbalanced design is preferable to a severely unbalanced one.

> ### Explain It Again
>
> **Balanced designs are better:** Whenever possible, use a balanced design. If balance is not possible, an unbalanced design is acceptable in one-way ANOVA.

> ### SUMMARY
>
> • With a balanced design, the effect of unequal variances is generally not great.
> • With an unbalanced design, the effect of unequal variances can be substantial.
> • The more unbalanced the design, the greater the effect of unequal variances.

Checking the assumptions

When sample sizes are small, which is often the case in ANOVA, there is really no very good way to check the assumptions of normality and equality of variance. Fortunately, the methods of ANOVA work well unless the assumptions are severely violated. We will check the assumption of normality with dotplots.

EXAMPLE 14.1   Check the assumption of normality

Check the assumption of normality for the data in Table 14.1.

Solution

Figure 14.2 presents a dotplot for each sample in Table 14.1. There are no indications of severe violations of the assumption of approximate normality.

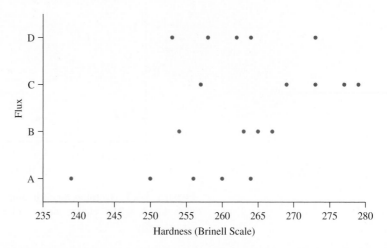

Figure 14.2 Dotplots for the samples in Table 14.1. There are no indications of severe violations of the assumption of approximate normality.

When the design is balanced, the method of ANOVA is not very sensitive to the assumption of equal population variances. When the design is unbalanced, we can check the assumption of equal variances by comparing the largest sample standard deviation to the smallest. If the largest sample standard deviation is no more than twice as large as the smallest, we conclude that the assumption of equal variances is satisfied.

> When the design is unbalanced, we can check the assumption of equal population variances by computing the following quotient:
>
> $$\frac{\text{Largest sample standard deviation}}{\text{Smallest sample standard deviation}}$$
>
> If the result is less than 2, we conclude that the assumption of equal population variances is satisfied.

Objective 3 Perform a one-way ANOVA hypothesis test

The One-Way ANOVA Hypothesis Test

In practice, ANOVA tests are almost always performed with technology. To make clear the ideas behind the procedure, we will explain the hand calculations first. Then we will present examples in which technology is used.

Notation

We begin by describing our notation. We have I samples, each from a different population. The population means are denoted

$$\mu_1, \mu_2, \ldots, \mu_I$$

The sample means are denoted

$$\bar{x}_1, \bar{x}_2, \ldots, \bar{x}_I$$

The sample standard deviations are denoted

$$s_1, s_2, \ldots, s_I$$

The sample sizes are denoted

$$n_1, n_2, \ldots, n_I$$

The total number in all the samples combined is denoted by N:

$$N = n_1 + n_2 + \cdots + n_I$$

Finally, the average of all the items in all the samples taken together is called the **grand mean**, and is denoted $\bar{\bar{x}}$.

Table 14.2 summarizes the notation.

Table 14.2 Notation for One-Way ANOVA

Sample	Population Mean	Sample Mean	Sample Standard Deviation	Sample Size
1	μ_1	$\bar{x}_1$	s_1	n_1
2	μ_2	$\bar{x}_2$	s_2	n_2
$\vdots$	$\vdots$	$\vdots$	$\vdots$	$\vdots$
I	μ_I	$\bar{x}_I$	s_I	n_I

N is the total number of items in all samples combined.
$\bar{\bar{x}}$ is the grand mean, the average of all the items in all samples combined.

EXAMPLE 14.2

Compute means and variances

For the data in Table 14.1, find $I, n_1, n_2, ..., n_I, \bar{x}_3, s_2, N, \bar{\bar{x}}$.

Solution

There are 4 samples, so $I = 4$. Each sample contains 5 observations, so $n_1 = n_2 = n_3 = n_4 = 5$. The quantity $\bar{x}_3$ is the sample mean of the third sample; this value is presented in Table 14.1, and $\bar{x}_3 = 271.0$. The total number of observations is $N = 20$. Finally, the grand mean $\bar{\bar{x}}$ is found by averaging all 20 observations; the grand mean is $\bar{\bar{x}} = 262.5$.

The hypotheses for one-way ANOVA

In one-way ANOVA, we test the null hypothesis that all the population means are equal, versus the alternate hypothesis that two or more of the population means differ.

SUMMARY

The hypotheses in one-way ANOVA are

$$H_0: \mu_1 = \cdots = \mu_I \qquad H_1: \text{Two or more of the } \mu_i \text{ are different.}$$

Figure 14.3 presents the idea behind one-way ANOVA. The figure illustrates several hypothetical samples from different treatments, along with their sample means and the sample grand mean. The sample means are spread out around the sample grand mean. One-way ANOVA provides a way to measure this spread. If the sample means are highly spread out, then it is likely that the treatment means are different, and we will reject H_0.

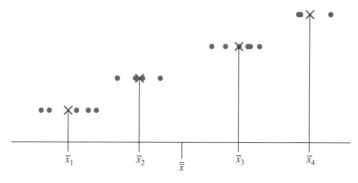

Figure 14.3 The spread within each sample is influenced only by the size of the population standard deviation σ. The spread of the sample means is influenced by both σ and the differences among the population means.

The spread of the sample means around the sample grand mean is measured by a quantity called the **treatment sum of squares** (*SSTr* for short), which is given by

$$SSTr = n_1(\bar{x}_1 - \bar{\bar{x}})^2 + n_2(\bar{x}_2 - \bar{\bar{x}})^2 + \cdots + n_I(\bar{x}_I - \bar{\bar{x}})^2$$

Each term in *SSTr* involves the distance from the sample mean to the sample grand mean. Note that each squared distance is multiplied by the sample size corresponding to its sample mean, so that the means for the larger samples count more. *SSTr* provides an indication of how different the treatment means are from each other. If *SSTr* is large, then the sample means are spread out widely, and it is reasonable to conclude that the treatment means differ and to reject H_0. If, on the other hand, *SSTr* is small, then the sample means are all close to each other, so it is plausible that the treatment means are equal.

To determine whether *SSTr* is large enough to reject H_0, we compare it to another sum of squares, called the **error sum of squares** (*SSE* for short). *SSE* measures the spread in the individual sample points around their respective sample means. This spread is measured by combining the sample variances. *SSE* is given by

$$SSE = (n_1 - 1)s_1^2 + (n_2 - 1)s_2^2 + \cdots + (n_I - 1)s_I^2$$

The size of *SSE* depends only on the standard deviations of the samples, and is not affected by the location of treatment means relative to one another. *SSE* is influenced only by the population standard deviation σ.

EXAMPLE 14.3

Compute *SSTr* and *SSE*

For the data in Table 14.1, compute *SSTr* and *SSE*.

Solution

The sample means are presented in Table 14.1. They are:

$$\bar{x}_1 = 253.8 \qquad \bar{x}_2 = 263.2 \qquad \bar{x}_3 = 271.0 \qquad \bar{x}_4 = 262.0$$

The sample grand mean was computed in Example 14.2 to be $\bar{\bar{x}} = 262.5$. The sample sizes are $n_1 = n_2 = n_3 = n_4 = 5$. We now compute *SSTr*:

$$SSTr = 5(253.8 - 262.5)^2 + 5(263.2 - 262.5)^2 + 5(271.0 - 262.5)^2 + 5(262.0 - 262.5)^2$$
$$= 743.4$$

Now we compute *SSE*:

$$SSE = (5 - 1)(9.7570)^2 + (5 - 1)(5.4037)^2 + (5 - 1)(8.7178)^2 + (5 - 1)(7.4498)^2$$
$$= 1023.6$$

The sums of squares *SSTr* and *SSE* have different degrees of freedom. *SSTr* involves the deviations of the *I* sample means around the grand mean, so *SSTr* has $I - 1$ degrees of freedom. *SSE* involves the *I* sample variances. The degrees of freedom for the sample variances are $n_1 - 1, n_2 - 1, ..., n_I - 1$. The degrees of freedom for *SSE* is the sum of these, which is $N - I$:

$$(n_1 - 1) + (n_2 - 1) + \cdots + (n_I - 1) = (n_1 + \cdots + n_I) - I = N - I$$

When a sum of squares is divided by its degrees of freedom, the quantity obtained is called a **mean square**. The **treatment mean square** is denoted *MSTr*, and the **error mean square** is denoted *MSE*. They are defined by

$$MSTr = \frac{SSTr}{I - 1} \qquad MSE = \frac{SSE}{N - I}$$

Now *MSE* measures how spread out the samples are around their own means. In particular, *MSE* is an estimate of σ^2, which is the variance of each of the populations. On the other hand, *MSTr* measures how spread out the sample means are around the grand mean. This spread is affected by σ^2, and also by how spread out the population means are. Now if

H_0 is true, and all populations have the same mean, then $MSTr$ is an estimate of σ^2, just as MSE is. When the population means differ, $MSTr$ tends to be larger than σ^2, and thus larger than MSE.

We can use $MSTr$ and MSE to construct a test statistic. The test statistic for testing $H_0: \mu_1 = \cdots = \mu_I$ is

$$F = \frac{MSTr}{MSE}$$

When H_0 is true, the numerator and denominator of F are, on average, the same size, so F tends to be near 1. In fact, when H_0 is true, this test statistic has an F distribution with $I - 1$ and $N - I$ degrees of freedom, denoted $F_{I-1, N-I}$. When H_0 is false, $MSTr$ tends to be larger, but MSE does not, so F tends to be greater than 1.

We will now describe how to perform a hypothesis test using the critical value method and Table A.5. Later we will describe how to use the P-value method with technology.

The *F*-test for One-Way ANOVA

To test $H_0: \mu_1 = \mu_2 = \cdots = \mu_I$ versus H_1: Two or more of the μ_i are different:

Step 1: Check that the assumptions are satisfied.

Step 2: Choose a level of significance α.

Step 3: Compute $SSTr$ and SSE.

Step 4: Compute $MSTr$ and MSE.

Step 5: Compute the test statistic: $F = \dfrac{MSTr}{MSE}$.

Step 6: Find the critical value by using Table A.5 with $I - 1$ and $N - I$ degrees of freedom. If the value of the test statistic is greater than or equal to the critical value, reject H_0. Otherwise, do not reject H_0.

Step 7: State a conclusion.

We now apply the method of analysis of variance to the example with which we introduced this section.

EXAMPLE 14.4   Perform a one-way ANOVA test

For the data in Table 14.1, test the null hypothesis that all the means are equal. Use the $\alpha = 0.05$ level of significance. What do you conclude?

Solution

Step 1: Check the assumptions. We checked the assumption of normality in Example 14.1. The design is balanced, so the procedure is not very sensitive to violations of the assumption of equal variances. In any event, the largest sample standard deviation is less than twice as large as the smallest. The assumptions are satisfied.

Step 2: Choose a level of significance. We will use $\alpha = 0.05$.

Step 3: Compute *SSTr* and *SSE*. We computed these quantities in Example 14.3: $SSTr = 743.4$ and $SSE = 1023.6$.

Step 4: Compute *MSTr* and *MSE*. We have $I = 4$ samples and $N = 20$ observations in all the samples taken together. Therefore,

$$MSTr = \frac{743.4}{4 - 1} = 247.8 \qquad MSE = \frac{1023.6}{20 - 4} = 63.975$$

Step 5: Compute the test statistic.

$$F = \frac{MSTr}{MSE} = \frac{247.8}{63.975} = 3.8734$$

Step 6: Find the critical value. The degrees of freedom are $I - 1 = 4 - 1 = 3$ and $N - I = 20 - 4 = 16$. We find the column corresponding to 3 degrees of freedom for the numerator, and the rows corresponding to 16 degrees of freedom for the denominator. We choose the row for which the tail area is 0.05. The critical value is 3.24. The value of the test statistic is $F = 3.8734$. Because $3.8734 > 3.24$, we reject H_0.

Denominator Degrees of Freedom	Area	Numerator Degrees of Freedom								
		1	**2**	**3**	**4**	**5**	**6**	**7**	**8**	**9**
15	0.100	3.07	2.70	2.49	2.36	2.27	2.21	2.16	2.12	2.09
15	0.050	4.54	3.68	3.29	3.06	2.90	2.79	2.71	2.64	2.59
15	0.025	6.20	4.77	4.15	3.80	3.58	3.41	3.29	3.20	3.12
15	0.010	8.68	6.36	5.42	4.89	4.56	4.32	4.14	4.00	3.89
15	0.001	16.59	11.34	9.34	8.25	7.57	7.09	6.74	6.47	6.26
16	0.100	3.05	2.67	2.46	2.33	2.24	2.18	2.13	2.09	2.06
16	0.050	4.49	3.63	3.24	3.01	2.85	2.74	2.66	2.59	2.54
16	0.025	6.12	4.69	4.08	3.73	3.50	3.34	3.22	3.12	3.05
16	0.010	8.53	6.23	5.29	4.77	4.44	4.20	4.03	3.89	3.78
16	0.001	16.12	10.97	9.01	7.94	7.27	6.80	6.46	6.19	5.98
17	0.100	3.03	2.64	2.44	2.31	2.22	2.15	2.10	2.06	2.03
17	0.050	4.45	3.59	3.20	2.96	2.81	2.70	2.61	2.55	2.49
17	0.025	6.04	4.62	4.01	3.66	3.44	3.28	3.16	3.06	2.98
17	0.010	8.40	6.11	5.18	4.67	4.34	4.10	3.93	3.79	3.68
17	0.001	15.72	10.66	8.73	7.68	7.02	6.56	6.22	5.96	5.75

Explain It Again

Interpreting the hypothesis test: If we reject the null hypothesis, we conclude that not all the population means are the same. However, we cannot tell which means are different.

Step 7: State a conclusion. We conclude that the mean hardness is not the same for all fluxes. There are at least two fluxes with different means.

Check Your Understanding

1. A certain experiment consisted of $I = 5$ treatments, with sample sizes $n_1 = n_2 = n_3 = n_4 = n_5 = 4$. The sums of squares were $SSTr = 43.7$ and $SSE = 79.8$.
   a. How many degrees of freedom are there for $SSTr$?
   b. How many degrees of freedom are there for SSE?
   c. Find the treatment mean square $MSTr$.
   d. Find the error mean square MSE.
   e. Find the value of the test statistic for testing the hypothesis $H_0: \mu_1 = \mu_2 = \mu_3 = \mu_4 = \mu_5$.
   f. Find the level $\alpha = 0.05$ critical value.
   g. State your conclusion regarding H_0. Use the $\alpha = 0.05$ level of significance.

2. A certain experiment consisted of $I = 3$ treatments, with sample sizes $n_1 = n_2 = n_3 = 5$. The sums of squares were $SSTr = 7.3$ and $SSE = 3.9$.
   a. How many degrees of freedom are there for $SSTr$?
   b. How many degrees of freedom are there for SSE?
   c. Find the treatment mean square $MSTr$.
   d. Find the error mean square MSE.
   e. Find the value of the test statistic for testing the hypothesis $H_0: \mu_1 = \mu_2 = \mu_3$.
   f. Find the level $\alpha = 0.01$ critical value.
   g. State your conclusion regarding H_0. Use the $\alpha = 0.01$ level of significance.

Answers are on page 647.

Objective 4 Perform the Tukey–Kramer test of pairwise comparisons

The Tukey–Kramer Test of Pairwise Comparisons

When we reject H_0 with the F-test, we conclude that the population means are not all equal. However, the F-test does not tell us which means are different from the rest. We can perform the **Tukey–Kramer test** on each pair of means to determine whether the two means in the pair are significantly different. Because this test is performed on pairs of means, it is called a test of **pairwise comparisons**.

The Tukey–Kramer Test Statistic

Let μ_i and μ_j represent two population means. The test statistic for the Tukey–Kramer test of H_0: $\mu_i = \mu_j$ versus H_1: $\mu_i \neq \mu_j$ is

$$q = \frac{|\bar{x}_i - \bar{x}_j|}{\sqrt{\dfrac{MSE}{2}\left(\dfrac{1}{n_i} + \dfrac{1}{n_j}\right)}}$$

We can perform the Tukey–Kramer test on every pair of population means. We conclude that the pairs of means for which H_0 is rejected are not equal.

When H_0 is true, the test statistic q has a distribution called the **Studentized range distribution**. Critical values for this distribution are given in Table A.6. The Studentized range distribution has two values for the degrees of freedom, I and $N - I$, where I is the number of samples and N is the total number of observations in all samples combined. Example 14.5 illustrates the method for finding a critical value for the Tukey–Kramer statistic q using the Studentized range table, Table A.6.

EXAMPLE 14.5   Finding a critical value for the Tukey–Kramer statistic

Find the $\alpha = 0.05$ critical value for the Tukey–Kramer statistic for the data in Table 14.1.

Solution

We first find the degrees of freedom. The number of samples is $I = 4$, and the total number of observations in all samples combined is $N = 20$. Therefore, the degrees of freedom are $I = 4$ and $N - I = 20 - 4 = 16$. We find the column corresponding to $I = 4$ and the rows corresponding to $N - I = 16$. We choose the row for which $\alpha = 0.05$. The critical value is 4.05.

$N-I$	Area	2	3	4	5	6	7	8	9	10	11	12	13	14	15
15	0.10	2.48	3.14	3.54	3.83	4.05	4.23	4.39	4.52	4.64	4.75	4.84	4.93	5.01	5.08
15	0.05	3.01	3.67	4.08	4.37	4.59	4.78	4.94	5.08	5.20	5.31	5.40	5.49	5.57	5.65
15	0.01	4.17	4.84	5.25	5.56	5.80	5.99	6.16	6.31	6.44	6.55	6.66	6.76	6.84	6.93
16	0.10	2.47	3.12	3.52	3.80	4.03	4.21	4.36	4.49	4.61	4.71	4.81	4.89	4.97	5.04
16	0.05	3.00	3.65	4.05	4.33	4.56	4.74	4.90	5.03	5.15	5.26	5.35	5.44	5.52	5.59
16	0.01	4.13	4.79	5.19	5.49	5.72	5.92	6.08	6.22	6.35	6.46	6.56	6.66	6.74	6.82
17	0.10	2.46	3.11	3.50	3.78	4.00	4.18	4.33	4.46	4.58	4.68	4.77	4.86	4.93	5.01
17	0.05	2.98	3.63	4.02	4.30	4.52	4.70	4.86	4.99	5.11	5.21	5.31	5.39	5.47	5.54
17	0.01	4.10	4.74	5.14	5.43	5.66	5.85	6.01	6.15	6.27	6.38	6.48	6.57	6.66	6.73

(The column headers 2–15 are grouped under the heading *I*.)

If the value of I or $N - I$ is not in the table, use the closest smaller value.

Following are the steps for performing the Tukey–Kramer test.

Performing the Tukey–Kramer Test

To test $H_0: \mu_i = \mu_j$ versus $H_1: \mu_i \neq \mu_j$ for every pair of population means:

Step 1: Check that the assumptions are satisfied.

Step 2: Choose a level of significance α.

Step 3: Compute the sample means $\bar{x}_1, ..., \bar{x}_I$.

Step 4: Compute *MSE*.
Now perform Steps 5–7 for every pair of sample means $\bar{x}_i$ and $\bar{x}_j$.

Step 5: Compute the test statistic: $q = \dfrac{|\bar{x}_i - \bar{x}_j|}{\sqrt{\dfrac{MSE}{2}\left(\dfrac{1}{n_i} + \dfrac{1}{n_j}\right)}}$

Step 6: Find the critical value by using Table A.6 with I and $N - I$ degrees of freedom. If the value of I or $N - I$ is not in the table, use the closest smaller value. If the value of the test statistic is greater than or equal to the critical value, reject H_0. Otherwise, do not reject H_0.

Step 7: State a conclusion.

EXAMPLE 14.6

Perform the Tukey–Kramer test

Perform the Tukey–Kramer test on the data in Table 14.1 to determine which pairs of population means, if any, can be concluded to differ. Use the $\alpha = 0.05$ level of significance.

Solution

We will illustrate the method in detail for $H_0: \mu_1 = \mu_2$, where μ_1 is the mean for flux A and μ_2 is the mean for flux B.

Step 1: Check the assumptions. We checked the assumption of normality in Example 14.1. Since the design is balanced, the procedure is not very sensitive to violations of the assumption of equal variances. In any event, the largest sample standard deviation is less than twice as large as the smallest. The assumptions are satisfied.

Step 2: Choose a level of significance. We will use $\alpha = 0.05$.

Step 3: Compute the sample means. The sample means appear in Table 14.1: $\bar{x}_1 = 253.8$, $\bar{x}_2 = 263.2$, $\bar{x}_3 = 271.0$, $\bar{x}_4 = 262.0$.

Step 4: Compute MSE. We computed *MSE* in Example 14.4: $MSE = 63.975$.

Step 5: Compute the test statistic. We will compare means μ_1 and μ_2. The sample sizes are $n_1 = 5$ and $n_2 = 5$. We compute the statistic

$$q = \frac{|\bar{x}_1 - \bar{x}_2|}{\sqrt{\dfrac{MSE}{2}\left(\dfrac{1}{n_1} + \dfrac{1}{n_2}\right)}} = \frac{|253.8 - 263.2|}{\sqrt{\dfrac{63.975}{2}\left(\dfrac{1}{5} + \dfrac{1}{5}\right)}} = 2.63$$

Step 6: Find the critical value. In Example 14.5, we found the critical value to be 4.05. The value of the test statistic is 2.63, which is not greater than 4.05. Therefore, we do not reject H_0.

We repeat Steps 5 and 6 for the other pairs of means. The following table presents the results for all pairs.

Means	Test Statistic	Critical Value	Decision
A and B	2.63	4.05	Do not reject H_0
A and C	4.81	4.05	Reject H_0
A and D	2.29	4.05	Do not reject H_0
B and C	2.18	4.05	Do not reject H_0
B and D	0.34	4.05	Do not reject H_0
C and D	2.52	4.05	Do not reject H_0

Step 7: State a conclusion. We conclude that the means for flux A and flux C differ. There is not enough evidence to conclude that any two other means differ.

EXAMPLE 14.7

Use technology to perform the Tukey–Kramer test

Use technology to perform the Tukey–Kramer test on the data in Table 14.1 to determine which pairs of population means, if any, can be concluded to differ. Use the $\alpha = 0.05$ level of significance.

Solution

The following MINITAB output presents the results. MINITAB produces confidence intervals. For each pair of means, we reject H_0 if the confidence interval for that pair of means does not contain 0.

To interpret the output, the values labeled "Center" are the differences between pairs of treatment means. The quantities labeled "Lower" and "Upper" are the lower and upper bounds, respectively, of the confidence interval. The first group of confidence intervals is labeled "A subtracted from:". These intervals estimate the difference between the population mean for flux A and the population means for fluxes B, C, and D. The confidence interval "A subtracted from C" does not contain 0. Therefore, we conclude that the population means for fluxes A and C are different. The confidence intervals "A subtracted from B" and "A subtracted from D" do contain 0. Therefore, we cannot conclude that the population mean for flux A differs from the mean for B or D. In the other two groups of confidence intervals, all the intervals contain 0. Therefore, the only two fluxes for which we can conclude that the means are different are A and C.

```
A subtracted from:

     Lower   Center   Upper      ------+---------+---------+---------+---
B   -5.087    9.400   23.887               (---------*---------)
C    2.713   17.200   31.687                  (---------*---------)
D   -6.287    8.200   22.687               (---------*---------)
                                 ------+---------+---------+---------+---
                                     -15        0        15        30

B subtracted from:

     Lower   Center   Upper      ------+---------+---------+---------+---
C   -6.687    7.800   22.287                  (--------*---------)
D  -15.687   -1.200   13.287            (---------*---------)
                                 ------+---------+---------+---------+---
                                     -15        0        15        30

C subtracted from:

     Lower   Center   Upper      ------+---------+---------+---------+---
D  -23.487   -9.000    5.487     (---------*---------)
                                 ------+---------+---------+---------+---
                                     -15        0        15        30
```

Objective 5 Perform a one-way ANOVA test with technology

One-Way ANOVA with Technology: The ANOVA Table

In practice, ANOVA calculations are almost always performed with technology. The results can be summarized in an **analysis of variance (ANOVA) table**. The ANOVA table presents the degrees of freedom, the sums of squares $SSTr$ and SSE, the mean squares $MSTr$ and MSE, the value of the F statistic, and the P-value for testing the hypothesis that all the population means are equal.

EXAMPLE 14.8   Construct an ANOVA table with technology

Use technology to construct an ANOVA table for the data in Table 14.1 and to test the null hypothesis that all the population means are equal.

Solution

We will present results from MINITAB and from the TI-84 Plus calculator. Following is the ANOVA table constructed by MINITAB:

```
One-way ANOVA: Hardness

Source   DF        SS        MS      F      P
Flux      3     743.40   247.800   3.87   0.029
Error    16    1023.60    63.975
Total    19    1767.00
```

In the row labeled "Flux," we find the degrees of freedom for the treatments, the treatment sum of squares $SSTr$, the treatment mean square $MSTr$, the value of the F statistic, and the P-value for the null hypothesis that all the population means are equal. The next row, labeled "Error," contains the degrees of freedom for error, the error sum of squares SSE, and the error mean square MSE. Finally, the row labeled "Total" contains the total number of degrees of freedom and the sum $SSTr + SSE$. This is called the **total sum of squares**. The P-value in the MINITAB output is 0.029, which indicates that we reject the hypothesis that all the population means are equal at the $\alpha = 0.05$ level. This agrees with the result calculated by hand in Example 14.4.

Following is the display from the TI-84 Plus calculator.

The TI-84 Plus presents the same information as the ANOVA table constructed by MINITAB. In addition, at the bottom of the right-hand screen, the quantity Sxp is an estimate of the population standard deviation.

Step-by-step instructions for performing one-way ANOVA tests with technology are presented in the Using Technology section on pages 643–644.

Check Your Understanding

3. The following MINITAB output presents the results of a one-way ANOVA.

```
One-way ANOVA: A, B, C

Source   DF        SS       MS       F       P
Factor    2     8.0150   4.0075   7.1057   0.007
Error    15     8.4598   0.5640
Total    17    16.475
```

   a. State the null hypothesis.
   b. How many levels were there for the factor?
   c. Assume the design was balanced. What was the sample size for each factor?
   d. What are the values of $SSTr$, SSE, $MSTr$, and MSE?
   e. What is the value of the test statistic?

f. What is the *P*-value?

g. State your conclusion regarding H_0. Use the $\alpha = 0.01$ significance level.

4. The following TI-84 Plus display presents the results of a one-way ANOVA.

a. State the null hypothesis.

b. How many levels were there for the factor?

c. Assume the design was balanced. What was the sample size for each factor?

d. What are the values of *SSTr*, *SSE*, *MSTr*, and *MSE*?

e. What is the value of the test statistic?

f. What is the *P*-value?

g. State your conclusion regarding H_0. Use the $\alpha = 0.05$ significance level.

5. In a one-way ANOVA with three samples, the sample means were $\bar{x}_1 = 24.03$, $\bar{x}_2 = 14.88$, and $\bar{x}_3 = 12.76$. The sample sizes were $n_1 = n_2 = n_3 = 6$, and the error mean square was $MSE = 10.53$. Perform the Tukey–Kramer test on each pair of means. Which pairs can you conclude to be different? Use the $\alpha = 0.05$ level.

6. In a one-way ANOVA with four samples, the sample means were $\bar{x}_1 = 86.8$, $\bar{x}_2 = 82.4$, $\bar{x}_3 = 85.8$, and $\bar{x}_4 = 89.1$. The sample sizes were $n_1 = n_2 = n_3 = n_4 = 4$, and the error mean square was $MSE = 1.3$. Perform the Tukey–Kramer test on each pair of means. Which pairs can you conclude to be different? Use the $\alpha = 0.01$ level.

Answers are on page 647.

USING TECHNOLOGY

We use Table 14.1 and Example 14.4 to illustrate the technology steps.

TI-84 PLUS

Performing a one-way ANOVA test

Step 1. Enter the values from flux A in Table 14.1 into **L1**, the values from flux B into **L2**, the values from flux C into **L3**, and the values from flux D into **L4**.

Step 2. Press **STAT** and highlight the **TESTS** menu. Select **ANOVA** and press **ENTER**.

Step 3. Enter **L1, comma, L2, comma, L3, comma, L4** (Figure A), and then press **Enter** (Figures B and C).

Figure A

Figure B

Figure C

MINITAB

Performing a one-way ANOVA test

Step 1. Enter the values from flux A in Table 14.1 into **Column C1**, the values from flux B into **Column C2**, the values from flux C into **Column C3**, and the values from flux D into **Column C4**.

Step 2. Click on **Stat**, then **ANOVA**, and then **One-Way (Unstacked)**.

Step 3. In the **Responses** field, enter **C1, C2, C3, C4**.

Step 4. Click **OK** (Figure D).

```
Source   DF      SS      MS     F      P
Factor    3   743.4   247.8  3.87  0.029
Error    16  1023.6    64.0
Total    19  1767.0

S = 7.998    R-Sq = 42.07%    R-Sq(adj) = 31.21%
```

Figure D

EXCEL

Performing a one-way ANOVA test

Step 1. Enter the values from flux A in Table 14.1 into **Column A**, the values from flux B into **Column B**, the values from flux C into **Column C**, and the values from flux D into **Column D**.

Step 2. Select **Data**, then **Data Analysis**. Highlight **Anova: Single Factor** and press **OK**.

Step 3. Enter the range of cells containing the data in the **Input range** field and select the **Grouped by Columns** option (Figure E).

Step 4. Click **OK** (Figure F).

Figure E

ANOVA

Source of Variation	SS	df	MS	F	P-value	F crit
Between Groups	743.4	3	247.8	3.873388	0.029437	3.238872
Within Groups	1023.6	16	63.975			
Total	1767	19				

Figure F

SECTION 14.1   Exercises

Exercises 1–6 are the **Check Your Understanding** exercises located within the section.

Understanding the Concepts

In Exercises 7 and 8, fill in each blank with the appropriate word or phrase.

7. In one-way ANOVA, the null hypothesis states that all the population means are _____ .

8. In one-way ANOVA, the effect of unequal variances can be substantial when the design is _____ .

In Exercises 9–12, determine whether the statement is true or false. If the statement is false, rewrite it as a true statement.

9. If the sample means are widely spread, the value of $SSTr$ will tend to be large.

10. In one-way ANOVA, all the samples must be the same size.

11. In one-way ANOVA, we have samples from several populations.

12. If the sample means are widely spread, the value of SSE will tend to be large.

Practicing the Skills

13. In a one-way ANOVA, the following data were collected: $SSTr = 0.25$, $SSE = 2.11$, $N = 34$, $I = 4$.
   a. How many samples are there?
   b. How many degrees of freedom are there for $SSTr$ and SSE?
   c. Compute the mean squares $MSTr$ and MSE.
   d. Compute the value of the test statistic F.

e. Can you conclude that two or more of the population means are different? Use the $\alpha = 0.05$ level of significance.

14. In a one-way ANOVA, the following data were collected: $SSTr = 145.34$, $SSE = 38.45$, $N = 9$, $I = 3$.
   a. How many samples are there?
   b. How many degrees of freedom are there for $SSTr$ and SSE?
   c. Compute the mean squares $MSTr$ and MSE.
   d. Compute the value of the test statistic F.
   e. Can you conclude that two or more of the population means are different? Use the $\alpha = 0.01$ level of significance.

15. Samples were drawn from three populations. The sample sizes were $n_1 = 8$, $n_2 = 6$, $n_3 = 9$. The sample means were $\bar{x}_1 = 1.04$, $\bar{x}_2 = 1.25$, $\bar{x}_3 = 1.87$. The sample standard deviations were $s_1 = 0.25$, $s_2 = 0.43$, $s_3 = 0.34$. The grand mean was $\bar{\bar{x}} = 1.42$.
   a. Compute the sums of squares $SSTr$ and SSE.
   b. How many degrees of freedom are there for $SSTr$ and SSE?
   c. Compute the mean squares $MSTr$ and MSE.
   d. Compute the value of the test statistic F.
   e. Can you conclude that two or more of the population means are different? Use the $\alpha = 0.01$ level of significance.

16. Samples were drawn from four populations. The sample sizes were $n_1 = 9$, $n_2 = 7$, $n_3 = 10$, $n_4 = 8$. The sample means were $\bar{x}_1 = 73.5$, $\bar{x}_2 = 74.8$, $\bar{x}_3 = 75.1$, $\bar{x}_4 = 78.2$. The sample standard deviations were $s_1 = 2.5$, $s_2 = 4.5$, $s_3 = 3.8$, $s_4 = 3.2$. The grand mean was $\bar{\bar{x}} = 75.34$.
   a. Compute the sums of squares $SSTr$ and SSE.
   b. How many degrees of freedom are there for $SSTr$ and SSE?
   c. Compute the mean squares $MSTr$ and MSE.
   d. Compute the value of the test statistic F.
   e. Can you conclude that two or more of the population means are different? Use the $\alpha = 0.05$ level of significance.

Working with the Concepts

17. **Pesticide danger:** One of the factors that determines the degree of risk a pesticide poses to human health is the rate at which the pesticide is absorbed into skin after contact. An important question is whether the amount in the skin continues to increase with the length of the contact, or whether it increases for only a short time before leveling off. To investigate this, measured amounts of a certain pesticide were applied to 20 samples of rat skin. Four skins were analyzed at each of the time intervals 1, 2, 4, 10, and 24 hours. The amounts of the chemical (in micrograms) that were in the skin are given in the following table.

Duration	Amounts Absorbed			
1	1.7	1.5	1.2	1.5
2	1.7	1.6	1.9	1.9
4	1.9	1.7	2.1	2.0
10	2.2	1.9	1.7	1.6
24	2.1	2.2	2.5	2.3

   a. Construct an ANOVA table.

b. Can you conclude that the amount in the skin varies with time? Use the $\alpha = 0.05$ level of significance.

18. **Life-saving drug:** Penicillin is produced by the *Penicillium* fungus, which is grown in a broth whose sugar content must be carefully controlled. Several samples of broth were taken on three successive days, and the amount of dissolved sugars, in milligrams per milliliter, was measured on each sample. The results were as follows.

Day 1:	4.8 5.1 5.1 4.8 5.2 4.9 5.0
	4.9 5.0 4.8 4.8 5.1 5.0
Day 2:	5.4 5.0 5.0 5.1 5.2 5.1 5.3
	5.2 5.2 5.1 5.4 5.2 5.4
Day 3:	5.7 5.1 5.3 5.5 5.3 5.5 5.1
	5.6 5.3 5.2 5.5 5.3 5.4

   a. Construct an ANOVA table.
   b. Can you conclude that the mean sugar concentration differs among the three days? Use the $\alpha = 0.05$ level of significance.

19. **Pesticide danger:** Using the data in Exercise 17, perform the Tukey–Kramer test to determine which pairs of means, if any, differ. Use the $\alpha = 0.05$ level of significance.

20. **Life-saving drug:** Using the data in Exercise 18, perform the Tukey–Kramer test to determine which pairs of means, if any, differ. Use the $\alpha = 0.05$ level of significance.

21. **Artificial hips:** Artificial hip joints consist of a ball and socket. As the joint wears, the ball (head) becomes rough. Investigators performed wear tests on metal artificial hip joints. Joints with several different diameters were tested. The following table presents measurements of head roughness (in nanometers).

Diameter	Head Roughness				
16	0.83	2.25	0.40	2.78	3.23
28	2.72	2.48	3.80		
36	6.49	5.32	4.59		

Source: Based on data in *Proceedings of the Institution of Mechanical Engineers* 215:483–493

   a. Because the design is unbalanced, check that the assumption of equal variances is satisfied by showing that the largest sample standard deviation is less than twice as large as the smallest one.
   b. Construct an ANOVA table.
   c. Can you conclude that mean roughness varies with diameter? Use the $\alpha = 0.01$ level of significance.

22. **Floods:** Rapid drainage of floodwater is crucial to prevent damage during heavy rains. Several designs for a drainage canal were considered for a certain city. Each design was tested five times, to determine how long it took to drain the water in a reservoir. The following table presents the drainage times, in minutes.

Channel Type	Drainage Time				
1	41.4	43.4	50.0	41.2	
2	38.7	48.3	51.1	38.3	
3	32.6	33.7	34.8	22.5	
4	26.3	30.9	33.3	23.8	
5	44.9	47.2	48.5	37.1	

   a. Construct an ANOVA table.

b. Can you conclude that there is a difference in the mean drainage times for the different channel designs? Use the $\alpha = 0.01$ level of significance.

23. Artificial hips: Using the data in Exercise 21, perform the Tukey–Kramer test to determine which pairs of means, if any, differ. Use the $\alpha = 0.01$ level of significance.

24. Floods: Using the data in Exercise 22, perform the Tukey–Kramer test to determine which pairs of means, if any, differ. Use the $\alpha = 0.01$ level of significance.

25. Polluting power plants: Power plants can emit levels of pollution that can lower air quality. For this reason, it is important to monitor the levels of pollution they produce. Investigators measured dust emissions, in milligrams per cubic meter, at four power plants. Thirty measurements were taken for each plant. The sample means and standard deviations are presented in the following table.

Plant	Mean	Standard Deviation	Sample Size
A	211.50	24.85	30
B	214.00	35.26	30
C	211.75	33.53	30
D	236.08	23.09	30

Source: *Journal of Air and Waste Management* 55:1042–1049

a. What are the sample sizes n_1, n_2, n_3, n_4?
b. What are the sample standard deviations s_1, s_2, s_3, s_4?
c. Compute *SSTr*.
d. Compute *SSE*.
e. How many degrees of freedom are there for *SSTr*? For *SSE*?
f. Compute *MSTr* and *MSE*.
g. Compute the value of the test statistic for testing the null hypothesis that the population means are all equal.
h. Can you conclude that there are differences among the mean emission levels? Use the $\alpha = 0.05$ level of significance.

26. Breaking boards: The strength of wood products used in construction is measured to be sure that they are suitable for the purpose for which they are used. Measurements, in megapascals, of the strength needed to break green mixed oak boards were made. The sample means, standard deviations, and sample sizes for four different grades of lumber are presented in the following table.

Grade	Mean	Standard Deviation	Sample Size
Select	45.1	8.52	32
No. 1	42.0	5.50	11
No. 2	33.2	6.71	15
Below grade	38.1	8.04	42

Source: *Journal of Materials in Civil Engineering* 11:91–97

a. What are the sample sizes n_1, n_2, n_3, n_4?
b. What are the sample standard deviations s_1, s_2, s_3, s_4?
c. Compute *SSTr*.
d. Compute *SSE*.
e. How many degrees of freedom are there for *SSTr*? For *SSE*?
f. Compute *MSTr* and *MSE*.
g. Compute the value of the test statistic for testing the null hypothesis that the population means are all equal.
h. Can you conclude that the strength differs for different grades of lumber? Use the $\alpha = 0.05$ level of significance.

27. Small, medium, or large? Several large-sized sodas were ordered at each of several fast-food restaurants, and the volume of beverage in each was measured. The following TI-84 Plus display presents the results of a one-way ANOVA to determine whether the mean volume differs among the restaurants.

a. State the null hypothesis.
b. How many restaurants were involved in the study?
c. Assume the design was balanced. How many sodas were measured in each restaurant?
d. What are the values of *SSTr*, *SSE*, *MSTr*, and *MSE*?
e. What is the value of the test statistic?
f. What is the *P*-value?
g. Can you conclude that the mean volume differs among restaurants? Use the $\alpha = 0.05$ level of significance.

28. Spreadsheets: Several spreadsheet programs were tested by performing a certain task several times on each. The following TI-84 Plus display presents the results of a one-way ANOVA to determine whether the mean time required to perform a certain task differs among them.

a. State the null hypothesis.
b. How many spreadsheets were involved in the study?
c. Assume the design was balanced. How many times was the task performed for each spreadsheet?
d. What are the values of *SSTr*, *SSE*, *MSTr*, and *MSE*?
e. What is the value of the test statistic?
f. What is the *P*-value?
g. Can you conclude that the mean time differs among spreadsheets? Use the $\alpha = 0.01$ level of significance.

29. Watch your head: Automotive engineers compared three types of A-pillars in automobiles to determine which provided the greatest protection to occupants of automobiles during a collision. Following is a one-way ANOVA table. The response variable is the head injury criterion (HIC), which is a number that measures the amount of impact energy absorbed by the pillar.

```
One-way ANOVA:  Pillar

Source    DF      SS        MS        F       P
Pillar    2     50946.6   25473.3   5.071   0.015
Error     24    120550.9   5023.0
Total     26    171497.4
```

a. What are the values of $SSTr$ and SSE?
b. What are the values of $MSTr$ and MSE?
c. What is the value of the test statistic?
d. Can you conclude that the mean energy absorbed varies with the type of pillar? Use the $\alpha = 0.01$ level of significance.
Source: *Proceedings of the Institution of Mechanical Engineers* 215:1161–1169

30. Soil conservation: High levels of phosphorus in soil can lead to severe reductions in water quality and animal populations. Investigators treated soil specimens with six different treatments, and the acid phosphatase activity, which is related to the production of phosphorus, was recorded. Results are presented in the following one-way ANOVA table. (The units are micromoles of phosphatase per gram per hour.)

```
One-way ANOVA:  Treatments A, B, C, D, E, F

Source     DF      SS        MS        F       P
Treatment  5     1.18547   0.23709   46.64   0.000
Error      6     0.03050   0.00508
Total      11    1.21597
```

a. What are the values of $SSTr$ and SSE?
b. What are the values of $MSTr$ and MSE?

c. What is the value of the test statistic?
d. Can you conclude that the mean level of acid phosphatase activity varies among the treatments? Use the $\alpha = 0.05$ level of significance.
Source: *Journal of Agronomy and Crop Science* 184:137–142

Extending the Concepts

In Section 11.5, we studied the Bonferroni correction for multiple tests. Because the Bonferroni correction is valid for any collection of tests, it can be used as an alternative to the Tukey–Kramer test for pairwise comparisons. The Bonferroni test statistic is

$$\frac{|\bar{x}_i - \bar{x}_j|}{\sqrt{MSE\left(\frac{1}{n_i} + \frac{1}{n_j}\right)}}$$

The critical value for the Bonferroni test statistic comes from the Student's t distribution with $N - I$ degrees of freedom. If the desired level of significance is α, we reject H_0 if the P-value is less than $\alpha/(2K)$, where K is the number of differences $\mu_i - \mu_j$ to be tested.

31. More pairwise comparisons: Refer to Table 14.1, which presents data on the hardnesses of welds produced from four different fluxes.
a. There are six possible hypotheses to test regarding pairs of fluxes. They are: $H_0: \mu_1 - \mu_2 = 0$, $H_0: \mu_1 - \mu_3 = 0$, $H_0: \mu_1 - \mu_4 = 0$, $H_0: \mu_2 - \mu_3 = 0$, $H_0: \mu_2 - \mu_4 = 0$, and $H_0: \mu_3 - \mu_4 = 0$. Compute the value of the Bonferroni test statistic for each of these pairs.
b. Find the P-value for each of the tests, using the Student's t distribution with $N - I$ degrees of freedom. The alternatives are two-tailed.
c. Use the Bonferroni correction to determine which pairs of means, if any, differ at the $\alpha = 0.05$ level of significance.
d. Compare the results of the Bonferroni correction with the results of the Tukey–Kramer test.

Answers to Check Your Understanding Exercises for Section 14.1

1. a. 4   **b.** 15   **c.** 10.925   **d.** 5.32   **e.** 2.05
f. 3.06
g. We do not reject H_0. It is possible that the means are all the same.

2. a. 2   **b.** 12   **c.** 3.65   **d.** 0.325   **e.** 11.23   **f.** 6.93
g. We reject H_0 and conclude that two or more of the means differ.

3. a. $H_0: \mu_1 = \mu_2 = \mu_3$ **b.** 3   **c.** 6
d. $SSTr = 8.0150$, $SSE = 8.4598$, $MSTr = 4.0075$, $MSE = 0.5640$
e. 7.1057   **f.** 0.007

g. We reject H_0 and conclude that two or more of the means differ.

4. a. $H_0: \mu_1 = \mu_2 = \mu_3 = \mu_4$ **b.** 4   **c.** 8
d. $SSTr = 1070.5$, $SSE = 15{,}711.5$, $MSTr = 356.833333$, $MSE = 561.125$
e. 0.6359248533   **f.** 0.5981346619
g. We do not reject H_0. There is not enough evidence to conclude that two or more of the means differ.

5. We can conclude that $\mu_1 \neq \mu_2$ and $\mu_1 \neq \mu_3$.

6. We can conclude that $\mu_1 \neq \mu_2$, $\mu_2 \neq \mu_3$, $\mu_2 \neq \mu_4$, and $\mu_3 \neq \mu_4$.

SECTION 14.2   Two-Way Analysis of Variance

Objectives

1. Interpret a two-way table
2. Describe main effects and interactions
3. Test hypotheses about interactions and main effects
4. Construct interaction plots

Objective 1 Interpret a two-way table

Two-Way Tables

Analysis of variance is concerned with data sets that consist of samples from several populations. In Section 14.1, we discussed one-factor experiments, in which each population corresponds to a different level of a single factor. In this section, we will discuss **two-factor experiments**, in which there are two factors that vary. Each combination of levels of the two factors corresponds to a population, and we have a sample from each population.

We will illustrate these ideas with an example. In a hypothetical study, a college instructor conducted an experiment to determine whether the instructional method and the class size affect the amount that students learn. Three instructional methods were used in the study: a traditional lecture method, an interactive method in which students worked with a computer program, and a third method that combined lectures with interactivity. Two classes were taught with each method. One class was large in size; the other was small. Four students were sampled from each class. Their final exam scores are shown in Table 14.3.

Table 14.3 Final Exam Scores

	Teaching Method											
	Lecture				**Interactive**				**Combination**			
Large Class	62	74	68	69	87	75	69	84	66	66	81	75
Small Class	75	77	82	74	75	87	89	94	76	84	90	93

There are two factors, teaching method and class size, so this is a **two-way table**. The table contains two rows, corresponding to the two class sizes, and three columns, corresponding to the three teaching methods. Therefore, this is a 2×3 table. The two rows and three columns produce six **cells** in the table; each cell corresponds to a particular combination of class size and teaching method. The numbers in each cell are the exam scores of four students taught with that combination of teaching method and class size. For example, the exam scores for the students taught with a traditional lecture in a large class are 62, 74, 68, and 69. The exam scores for the students taught with the interactive method in a small class are 75, 87, 89, and 94. Because exam score is the quantity we are measuring, exam score is the response variable.

The purpose of the experiment that produced Table 14.3 would be to determine whether the factors, teaching method and class size, have an effect on the response variable, exam score. It turns out that there are two kinds of effects that a factor can have on the response: main effects and interaction effects (often referred to as interactions). We now describe these two kinds of effects.

Objective 2 Describe main effects and interactions

Main Effects and Interactions

We illustrate main effects and interactions with an example.

A chemist is running a chemical reaction in order to produce a chemical product that can be sold. She has two choices for the temperature at which to run the reaction: a high temperature or a low one. The reaction must be stirred, and the stirring rate can be set to either fast or slow. The chemist wants to determine whether the factors temperature and stirring rate have an effect on the yield, or amount of product, that is produced. She runs three reactions at each combination of temperature and stirring rate, and measures the yield each time. The results are shown in Table 14.4. The units of yield are grams.

Table 14.4 Yield of a Chemical Reaction

	Stirring Rate					
	Fast			Slow		
High Temperature	85	91	73	19	36	26
Low Temperature	34	25	20	52	45	42

To determine the effect of stirring rate on yield, we will compute the **main effect** of stirring rate. The main effect of stirring rate is the difference between the average yield at a fast rate and the average yield at the slow rate. There are six yields at the fast rate: 85, 91, 73, 34, 25, and 20. There are also six yields at the slow rate: 19, 36, 26, 52, 45, and 42. The main effect of stirring rate is therefore

Main effect of stirring rate
$$= \frac{85 + 91 + 73 + 34 + 25 + 20}{6} - \frac{19 + 36 + 26 + 52 + 45 + 42}{6} = 18$$

The main effect of stirring rate is 18. This says that the mean yield with the fast rate of stirring is about 18 grams higher than the mean yield with the slow rate.

It turns out that this main effect is misleading. To see why, we will compute the effect of stirring rate when the temperature is high. When the temperature is high, there are three yields with the fast stirring rate: 85, 91, and 73; and three yields at the slow rate: 19, 36, and 26. The effect of stirring rate at high temperature is

$$\text{Effect of stirring rate at high temperature} = \frac{85 + 91 + 73}{3} - \frac{19 + 36 + 26}{3} = 56$$

This says that when the high temperature is used, the mean yield with the fast stirring rate is 56 grams higher than the mean yield with the slow stirring rate. This is quite a bit different from the main effect of 18. Now let's compute the effect of stirring rate when the low temperature is used.

When the low temperature is used, there are three yields at the fast stirring rate: 34, 25, and 20; and three yields at the slow stirring rate: 52, 45, and 42. The effect of stirring rate when the temperature is low is

$$\text{Effect of stirring rate at low temperature} = \frac{34 + 25 + 20}{3} - \frac{52 + 45 + 42}{3} = -20$$

The effect of stirring rate at the low temperature is negative. The effect of -20 says that when the low temperature is used, the mean yield with the fast stirring rate is 20 grams *less* than the mean yield with the slow stirring rate. This is also quite a bit different from the main effect, which says that the mean yield with fast stirring is 18 grams *higher*.

> **Explain It Again**
>
> **Main effects and interactions:** If a large interaction is present, the main effects may be misleading.

We see that the actual effect of stirring rate differs considerably, depending on whether we use high temperature or low temperature. Because of this, the main effect of stirring rate is misleading. The main effect says that the mean yield at the high stirring rate will be 18 grams higher than the mean yield at the low stirring rate. But if we use a high temperature, the mean yield will actually be about 56 grams higher, and if we use the low temperature, the mean strength will actually be 20 grams lower. So the main effect does not describe the actual effect of stirring rate for either choice of temperature.

When the effect of one factor depends on the level of another factor, we say that there is an **interaction** between the two factors. When a large interaction exists, we should not try to interpret the main effects, because they may be misleading.

Objective 3 Test hypotheses about interactions and main effects

Testing Hypotheses in Two-Way Analysis of Variance

There are three hypotheses to test in two-way analysis of variance. First, we perform a test about the interaction.

H_0: There is no interaction between the factors

H_1: There is an interaction between the factors

If we reject H_0, we conclude that there is an interaction, and we do not perform any tests on the main effects. We must choose a level of significance at which to test the hypothesis of no interaction. It is reasonable to choose a fairly large level of significance, for the following reason: If H_0 is false, and we do not reject H_0, then we will perform a test of the main effects whose results may be misleading. Therefore, we want to use a somewhat higher significance level than usual, so that this type of error will be less likely. We will use the significance level $\alpha = 0.10$ for the test of interaction.

If we do not reject the hypothesis of no interaction, we test the following hypotheses about the main effects.

H_0: There is no main effect of factor A on the response variable

H_1: There is a main effect of factor A on the response variable

and

H_0: There is no main effect of factor B on the response variable

H_1: There is a main effect of factor B on the response variable

Tests about main effects are usually conducted with a typical level of significance such as $\alpha = 0.05$ or $\alpha = 0.01$.

The tests for two-way ANOVA require certain assumptions.

Assumptions for Two-Way ANOVA

1. The populations must be approximately normal.
2. The sample sizes must all be the same.
3. The populations must all have the same variance (because the sample sizes must be the same, this assumption needs to be satisfied only very approximately).

The assumption of approximate normality can be checked with dotplots. The second assumption says that the design must be balanced. Although it is possible to analyze unbalanced designs, the analysis is much more difficult than for balanced designs. We present only the methods for balanced designs. The third assumption says that the populations have the same variance. As in one-way ANOVA, the methods are not very sensitive to violations of this assumption when the design is balanced.

Procedure for Performing Tests in Two-Way ANOVA

Step 1: Test the null hypothesis that no interaction is present. Use a fairly large level of significance such as $\alpha = 0.10$.

Step 2: If the hypothesis of no interaction is rejected, stop. Do not perform tests about the main effects, as the results could be misleading.

Step 3: If the hypothesis of no interaction is not rejected, then perform tests about the main effects, and interpret the results.

The calculations for two-way ANOVA tests are involved, and are invariably performed with technology. We illustrate with some examples. Step-by-step instructions for performing two-way ANOVA tests with technology are presented in the Using Technology section on page 656.

EXAMPLE 14.9 Test hypotheses in two-way ANOVA

The following table presents final exam scores for students taught by one of three teaching methods and one of two class sizes. These data were first presented in Table 14.3.

	Teaching Method		
	Lecture	**Interactive**	**Combination**
Large Class	62 74 68 69	87 75 69 84	66 66 81 75
Small Class	75 77 82 74	75 87 89 94	76 84 90 93

a. Test the null hypothesis of no interaction between teaching method and class size.

b. If the null hypothesis of no interaction is not rejected, test the hypothesis that there is no main effect of class size. Use the $\alpha = 0.05$ level of significance. Interpret the result.

c. If the null hypothesis of no interaction is not rejected, test the hypothesis that there is no main effect of teaching method. Use the $\alpha = 0.05$ level of significance. Interpret the result.

Solution

The MINITAB output is as follows.

```
Analysis of Variance for Exam Score

Source            DF       SS         MS        F        P
Class Size         1    600.00    600.000    12.84    0.002
Teaching Method    2    399.25    199.625     4.27    0.030
Interaction        2     43.75     21.875     0.47    0.634
Error             18    841.00     46.722
Total             23   1884.00
```

Explain It Again

Interpreting the hypothesis test: If we reject the null hypothesis of no main effects, we conclude that not all the levels of the factor have the same effect. However, we cannot tell which levels are different.

We use the output to perform the tests.

a. The P-value for interaction is 0.634, which is greater than 0.10. We therefore do not reject the null hypothesis of no interaction.

b. We continue by checking the P-values for the main effects. The P-value for the main effect of class size is 0.002. This value is less than 0.05, so we conclude that there are differences in mean exam scores between large and small classes.

c. The P-value for teaching method is $P = 0.030$. This value is less than 0.05, so we conclude that there are differences in mean scores among the three teaching methods.

EXAMPLE 14.10   **Test hypotheses in two-way ANOVA**

A construction engineer is experimenting with three different concrete mixtures, A, B, and C. He forms nine blocks with each mixture. Three of the blocks are allowed to cure for two days, three others cure for four days, and the last three cure for six days. The engineer measures the strength of each block by measuring the pressure (in megapascals) needed to crush the block. The results are shown in the following table.

	Curing Time		
	2 Days	**4 Days**	**6 Days**
Mixture A	11 15 13	15 18 16	13 15 12
Mixture B	9 11 12	12 13 14	14 16 16
Mixture C	11 12 14	15 14 16	15 17 18

a. Test the null hypothesis of no interaction between curing time and mixture.

b. If the null hypothesis of no interaction is not rejected, test the hypothesis that there is no main effect of mixture. Use the $\alpha = 0.05$ level of significance. Interpret the result.

c. If the null hypothesis of no interaction is not rejected, test the hypothesis that there is no main effect of curing time. Use the $\alpha = 0.05$ level of significance. Interpret the result.

Solution

The MINITAB output is as follows.

Analysis of Variance for Strength

Source	DF	SS	MS	F	P
Mixture	2	13.407	6.7037	3.175	0.066
Curing Time	2	52.519	26.259	12.44	0.000
Interaction	4	29.037	7.2593	3.349	0.030
Error	18	38.000	2.1111		
Total	26	132.96			

We use the output to perform the tests.

a. The *P*-value for interaction is 0.030, which is less than 0.10. We therefore reject the null hypothesis of no interaction.

b. and **c.** Because the factors interact, a hypothesis test about the main effects could be misleading. Therefore, we do not perform hypothesis tests on the main effects.

Check Your Understanding

1. Dental researchers conducted an investigation of the manufacture of dental resins in which the response variable was the hardness of the resin. The factors were the distance between the curing tip and the surface of the resin during manufacture, and the depth of the resin at which the hardness was measured. The following MINITAB output presents the results.

Analysis of Variance for Hardness

Source	DF	SS	MS	F	P
Depth	5	3855.4	771.08	33.874	0.000
Distance	2	4111.8	2055.9	90.314	0.000
Interaction	10	47.486	4.7846	0.2102	0.994
Error	54	1229.2	22.764		
Total	71	9244.2			

a. Test the null hypothesis of no interaction between depth and distance.

b. If the null hypothesis of no interaction is not rejected, test the hypothesis that there is no main effect of depth. Interpret the result.

c. If the null hypothesis of no interaction is not rejected, test the hypothesis that there is no main effect of distance. Interpret the result.

Source: *Brazilian Dental Journal* 11:11–17

2. Environmental scientists performed a two-factor experiment in which the response variable was the amount of an herbicide absorbed by a certain type of plant. The factors were the concentration of the herbicide and the ratio of the delivery rates of iron and water. The following MINITAB output presents the results.

Analysis of Variance for Sorption

Source	DF	SS	MS	F	P
Concentration	2	0.3793	0.1897	3.873	0.040
Delivery Ratio	2	7.3400	3.6700	74.95	0.000
Interaction	4	3.4447	0.8612	17.59	0.000
Error	18	0.8814	0.0490		
Total	26	12.045			

a. Test the null hypothesis of no interaction between concentration and delivery ratio.

b. If the null hypothesis of no interaction is not rejected, test the hypothesis that there is no main effect of concentration. Interpret the result.

c. If the null hypothesis of no interaction is not rejected, test the hypothesis that there is no main effect of delivery ratio. Interpret the result.

Source: *Journal of Environmental Science and Health* 36:261–274

Answers are on page 660.

Objective 4 Construct interaction plots

Construct Interaction Plots

An **interaction plot** is a graph that allows us to visualize the magnitude of the interactions in a two-way ANOVA. Example 14.11 explains the steps in constructing an interaction plot.

EXAMPLE 14.11

Construct an interaction plot

Construct an interaction plot for the data in Example 14.10.

Solution

We perform the following steps:

Step 1: For each cell, compute the mean of the values in that cell. This is called the **cell mean**. List the cell means in a two-way table. The following table presents the cell means for the data in Example 14.10.

	Curing Time		
	2 Days	**4 Days**	**6 Days**
Mixture A	13.00	16.33	13.33
Mixture B	10.67	13.00	15.33
Mixture C	12.33	15.00	16.67

Step 2: Construct a horizontal and a vertical axis. On the horizontal axis, place a tick mark for each level of the column factor (curing time in our example). The tick marks should be evenly spaced. Label the vertical axis with a scale that will allow the plotting of the cell means.

Figure 14.4 presents the axes for the data in Example 14.10. We have placed evenly spaced tick marks to represent the levels of curing time and have labeled the vertical axis with a range of values that will allow us to plot the cell means.

Figure 14.4

Step 3: For each mean in the first row of cell means, plot the value of the mean on the vertical axis above the tick mark for its column. Connect the points with straight lines. Figure 14.5 on page 654 presents the results for plotting the three means in the first row (corresponding to mixture A).

Figure 14.5

Step 4: Repeat for each row, adding the lines to the existing plot. The completed plot is shown in Figure 14.6.

Figure 14.6

Step 5: Interpret the plot. The interaction plot contains a broken line for each row of the two-way table. The magnitude of the interactions is reflected in the directions of these lines. When the interactions are small, the lines are nearly parallel. When interactions are large, some of the lines will differ substantially in their direction from others.

In Figure 14.6, we can see that the line for mixture A has a very different direction from the other two lines. This indicates the presence of large interactions. In particular, it indicates that mixture A responds differently to the levels of curing time than the other mixtures. This is consistent with the results of the test for interaction, which showed that the interactions were statistically significant.

EXAMPLE 14.12   Use technology to construct an interaction plot

Use technology to construct an interaction plot for the data in Example 14.9. Interpret the plot.

Solution

Following is the interaction plot from MINITAB. The lines are not very different in their directions, which suggests that the interactions are not large. This is consistent with the results of Example 14.9.

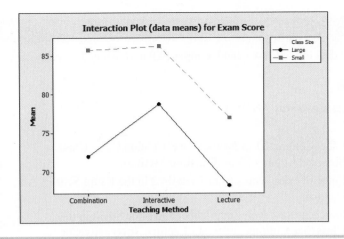

Step-by-step instructions for constructing interaction plots using MINITAB are presented on page 656.

Check Your Understanding

3. A psychiatrist studied the effects of three antidepressants on subjects in three different age groups. Each subject was rated on a scale of 0 to 100, with higher numbers indicating greater relief from depression. The following table presents the results.

	Age Group		
	18–30	**31–50**	**51–80**
Drug A	43 43 41 42	53 51 53 52	57 55 58 56
Drug B	52 51 51 50	62 61 60 61	65 66 66 67
Drug C	71 72 70 71	81 83 82 82	86 84 87 86

  a. Compute the nine cell means.
  b. Construct an interaction plot. Do there seem to be strong interactions?
  c. Use technology to perform a test for interactions. Do you reject the hypothesis of no interactions? Are the results of the test consistent with the interaction plot?

4. Following are interaction plots for three data sets. Which data set has the largest interactions? Which has the smallest?

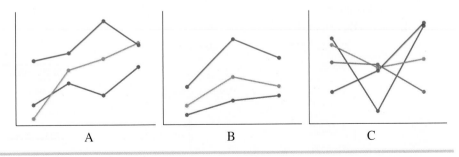

Answers are on page 660.

USING TECHNOLOGY

We use the data in Table 14.3 and Example 14.9 to illustrate the technology steps.

MINITAB

Performing a two-way ANOVA test

Step 1. Label Column C1 as **Exam Score**, Column C2 as **Class Size**, and Column C3 as **Teaching Method**.

Step 2. Enter all exam scores from Table 14.3 in the **Exam Score** Column.

Step 3. For each exam score, enter either **Large** or **Small** in the **Class Size Column** and either **Lecture**, **Interactive**, or **Combination** in the **Teaching Method Column**. The first ten exam scores are shown in Figure A.

Step 4. Click on **Stat**, then **ANOVA**, and then **Two-Way**.

Step 5. In the **Response field**, double-click on **Exam Score**. In the **Row factor field**, double-click on **Class Size**. In the **Column factor** field, double-click on **Teaching Method**.

Step 6. Click **OK** (Figure B).

↓	C1	C2-T	C3-T
	Exam Score	Class Size	Teaching Method
1	62	Large	Lecture
2	74	Large	Lecture
3	68	Large	Lecture
4	69	Large	Lecture
5	87	Large	Interactive
6	75	Large	Interactive
7	69	Large	Interactive
8	84	Large	Interactive
9	66	Large	Combination
10	66	Large	Combination

Figure A

Note: To generate the interaction plot, click on **Stat**, then **ANOVA**, and then **Interaction Plot**. In the **Response field**, double-click on **Exam Score**. In the **Factors field**, double-click on **Class Size** and **Teaching Method**.

```
Two-way ANOVA: Exam Score versus Class Size, Teaching Method

Source            DF      SS       MS      F      P
Class Size         1  600.00  600.000  12.84  0.002
Teaching Method    2  399.25  199.625   4.27  0.030
Interaction        2   43.75   21.875   0.47  0.634
Error             18  841.00   46.722
Total             23 1884.00

S = 6.835   R-Sq = 55.36%   R-Sq(adj) = 42.96%
```

Figure B

EXCEL

Performing a two-way ANOVA test

Step 1. Enter the exam scores from Table 14.3 as shown in Figure C.

Step 2. Select **Data**, then **Data Analysis**. Highlight **Anova: Two–Factor with Replication** and press **OK**.

Step 3. Enter the range of cells containing the data (including labels) in the **Input range** field and enter the number of data elements per sample in the **Rows per sample** field. For Example 14.9, we use **4** (Figure D).

Step 4. Click **OK** (Figure E).

	A	B	C	D
1		Lecture	Interactive	Combination
2	Large Class	62	87	66
3		74	75	66
4		68	69	81
5		69	84	75
6	Small Class	75	75	76
7		77	87	84
8		82	89	90
9		74	94	93

Figure C

Figure D

ANOVA					
Source of Variation	SS	df	MS	F	P-value
Sample	600	1	600	12.84185	0.002123
Columns	399.25	2	199.625	4.272592	0.03031
Interaction	43.75	2	21.875	0.468193	0.633547
Within	841	18	46.72222		
Total	1884	23			

Figure E

SECTION 14.2   Exercises

Exercises 1–4 are the Check Your Understanding exercises located within the section.

Understanding the Concepts

In Exercises 5 and 6, fill in each blank with the appropriate word or phrase.

5. When the effect of one factor depends on the level of another factor, an _____ is present.

6. In two-way ANOVA, the factors can have two kinds of effects on the response: _____ and _____ .

In Exercises 7 and 8, determine whether the statement is true or false. If the statement is false, rewrite it as a true statement.

7. In two-way ANOVA, we should not interpret the main effects when there are no interactions.

8. In two-way ANOVA, the results can be difficult to interpret if the design is unbalanced.

Practicing the Skills

9. In a two-way ANOVA, the P-value for interactions is $P = 0.45$, and the P-values for the main effects are $P = 0.023$ for the row factor and $P = 0.34$ for the column factor.
   a. Can you reject the hypothesis of no interactions? Explain.
   b. Is it appropriate to interpret the main effect of the row factor? If so, state whether there are differences in the mean response among different levels of the row factor, using the $\alpha = 0.05$ level of significance. If not, explain why not.
   c. Is it appropriate to interpret the main effect of the column factor? If so, state whether there are differences in the mean response among different levels of the column factor, using the $\alpha = 0.05$ level of significance. If not, explain why not.

10. In a two-way ANOVA, the P-value for interactions is $P = 0.07$, and the P-values for the main effects are $P = 0.003$ for the row factor and $P = 0.014$ for the column factor.
    a. Can you reject the hypothesis of no interactions? Explain.
    b. Is it appropriate to interpret the main effect of the row factor? If so, state whether there are differences in the mean response among different levels of the row factor, using the $\alpha = 0.05$ level of significance. If not, explain why not.
    c. Is it appropriate to interpret the main effect of the column factor? If so, state whether there are differences in the mean response among different levels of the column factor, using the $\alpha = 0.05$ level of significance. If not, explain why not.

In Exercises 11–14, interpret the interaction plot by explaining whether there appear to be large interactions between the factors.

11.

12.

13.

14.

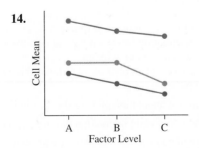

Working with the Concepts

15. **What's your favorite subject?** An education professor wants to find out which subject students find most difficult. She obtains final exam scores for three randomly selected freshmen and three randomly selected sophomores from three subjects, statistics, psychology, and history. The results are as follows.

	Statistics	Psychology	History
Freshmen	76 88 72	69 84 75	77 87 76
Sophomores	82 99 80	85 97 87	85 96 90

a. Can you reject the hypothesis of no interactions? Explain.

b. Can the main effect of class (freshmen or sophomores) on final exam score be interpreted? If so, interpret the main effect, using the $\alpha = 0.05$ level of significance. If not, explain why not.

c. Can the main effect of subject on final exam score be interpreted? If so, interpret the main effect, using the $\alpha = 0.05$ level of significance. If not, explain why not.

16. **The sap is running:** Vermont maple sugar producers sponsored a testing program to determine the benefit of a new fertilizer. A random sample of 27 maple trees in Vermont were chosen and treated with one of three levels of fertilizer. In this experimental setup, nine trees (three in each of three climatic zones) were treated with each fertilizer level and the amount of sap produced (in milliliters) by the trees was measured. The results are as follows.

	Southern Zone	Central Zone	Northern Zone
Low Fertilizer	76.2 80.4 74.2	81.4 85.9 86.9	86.5 85.2 80.1
Medium Fertilizer	88.0 94.1 93.0	98.2 94.7 96.2	88.4 90.4 92.2
High Fertilizer	84.2 87.5 83.1	90.3 89.9 93.2	81.4 84.7 82.2

a. Can you reject the hypothesis of no interactions? Explain.

b. Can the main effect of fertilizer on the amount of sap be interpreted? If so, interpret the main effect, using the $\alpha = 0.05$ level of significance. If not, explain why not.

c. Can the main effect of climatic zone on the amount of sap be interpreted? If so, interpret the main effect, using the $\alpha = 0.05$ level of significance. If not, explain why not.

17. **What's your favorite subject?** Construct and interpret an interaction plot for the data in Exercise 15.

18. **The sap is running:** Construct and interpret an interaction plot for the data in Exercise 16.

19. **Lather up:** A study was done to determine the effects of two factors on the latherability of soap. The two factors were type of water (tap or distilled) and glycerol (present or absent). The outcome measured was the amount of foam produced (in milliliters). The experiment was repeated three times for each combination of the factors. The results are as follows.

	No Glycerol	Glycerol
Distilled Water	165 181 168	170 197 190
Tap Water	155 142 139	139 160 160

a. Can you reject the hypothesis of no interactions? Explain.

b. Can the main effect of water type on the amount of foam be interpreted? If so, interpret the main effect, using the $\alpha = 0.01$ level of significance. If not, explain why not.

c. Can the main effect of glycerol on the amount of foam be interpreted? If so, interpret the main effect, using the $\alpha = 0.01$ level of significance. If not, explain why not.

20. **Cleaning solution:** An experiment was performed to measure the effects of two factors on the ability of cleaning solutions to remove oil from a piece of cloth. The factors were the concentration of soap (in percent by weight) and the percentage of lauric acid in the solution. The experiment was repeated twice for each combination of factors. The outcome measured was the percentage of oil removed. The results are as follows.

	10% acid	25% acid
15% soap	52.3 54.5	57.4 53.7
25% soap	56.4 58.4	42.7 45.1

a. Can you reject the hypothesis of no interactions? Explain.

b. Can the main effect of soap concentration on the amount of oil removed be interpreted? If so, interpret the main effect, using the $\alpha = 0.05$ level of significance. If not, explain why not.

c. Can the main effect of lauric acid on the amount of oil removed be interpreted? If so, interpret the main effect, using the $\alpha = 0.05$ level of significance. If not, explain why not.

21. **Lather up:** Construct and interpret an interaction plot for the data in Exercise 19.

22. **Cleaning solution:** Construct and interpret an interaction plot for the data in Exercise 20.

23. **Let's stick together:** Two parts will be fastened together with an adhesive. The adhesive will be applied to each part, and then the parts will be pressed together. Two types of adhesive are available. The amount of pressure applied can be either high or low. An experiment is run in which three sets of parts are glued together with each combination of adhesive and pressure. The strengths of the bond (in megapascals) are measured, with the following results.

	High Pressure	Low Pressure
Adhesive A	8 9 7	1 3 2
Adhesive B	3 3 2	5 4 4

a. Can you reject the hypothesis of no interactions? Explain.

b. Can the main effect of adhesive on the strength be interpreted? If so, interpret the main effect, using the $\alpha = 0.05$ level of significance. If not, explain why not.

c. Can the main effect of pressure on the strength be interpreted? If so, interpret the main effect, using the $\alpha = 0.05$ level of significance. If not, explain why not.

24. **Strong beams:** The following table presents measurements of the tensile strength (in kilopascals) of asphalt-rubber concrete beams for three levels of binder content and three levels of rubber content.

	Low Rubber Content	Medium Rubber Content	High Rubber Content
Low Binder Content	212.5 138.5 246.0	200.9 144.9 236.1	280.9 97.8 152.2
Medium Binder Content	201.5 170.0 249.5	179.5 229.8 187.7	167.3 236.2 163.4
High Binder Content	184.1 132.6 211.3	114.0 180.3 209.7	77.5 161.9 183.5

Source: Based on data in *Journal of Materials in Civil Engineering* 11:287–294

a. Can you reject the hypothesis of no interactions? Explain.

b. Can the main effect of the binder content be interpreted? If so, interpret the main effect. Use the $\alpha = 0.05$ level of significance.

c. Can the main effect of the rubber content be interpreted? If so, interpret the main effect. Use the $\alpha = 0.05$ level of significance.

25. **Let's stick together:** Construct and interpret an interaction plot for the data in Exercise 23.

26. **Strong beams:** Construct and interpret an interaction plot for the data in Exercise 24.

27. **Fruit yields:** An agricultural scientist performed a two-way ANOVA to determine the effects of three different pesticides on the yield of lemons, in pounds, from four different varieties of lemon tree. The following MINITAB output presents the results.

```
Analysis of Variance for Yield

Source        DF        SS         MS        F         P
Pesticide      2      1114.19     557.10    11.406    0.000
Variety        3      108.527     695.09     2.2219   0.103
Interaction    6      356.280      59.380    1.2157   0.321
Error         36      1758.37      48.844
Total         47      5314.11
```

a. Can you reject the hypothesis of no interactions? Explain.

b. Can the main effect of pesticide on yield be interpreted? If so, interpret the main effect, using the $\alpha = 0.05$ level of significance. If not, explain why not.

c. Can the main effect of variety on yield be interpreted? If so, interpret the main effect, using the $\alpha = 0.05$ level of significance. If not, explain why not.

28. **Fast shipping:** An online retailer sends many packages to customers. To investigate the quality of shipping, the retailer employed three firms to ship packages. Each firm was given four packages to ship at each of three times of day, morning, afternoon, and evening. The outcome variable was the number of hours elapsed during shipment. The following MINITAB output presents the results.

```
Analysis of Variance for Hours Elapsed

Source         DF        SS         MS        F         P
Firm            2      47.494     23.747    4.6546    0.018
Time of day     2      62.779     31.390    6.1526    0.006
Interaction     4      65.170     16.293    4.1735    0.009
Error          27      137.75      5.1019
Total          35      333.19
```

a. Can you reject the hypothesis of no interactions? Explain.

b. Can the main effect of firm on number of hours elapsed be interpreted? If so, interpret the main effect, using the $\alpha = 0.05$ level of significance. If not, explain why not.

c. Can the main effect of time of day on number of hours elapsed be interpreted? If so, interpret the main effect, using the $\alpha = 0.05$ level of significance. If not, explain why not.

Extending the Concepts

29. Watch those interactions: A component can be manufactured according to either of two designs, and with either of two types of material. Two components are manufactured with each combination of design and material, and the lifetimes of each are measured (in hours). The results are as follows.

	Material 1	Material 2
Design A	51 49	43 41
Design B	43 41	51 49

a. Explain why, if one wishes to maximize the lifetime, one should choose either design A and material 1, or design B and material 2.

b. Show that the main effect of design is equal to 0.

c. Show that the main effect of material is equal to 0.

d. Is it reasonable to conclude that neither design nor material has an effect on lifetime? Explain.

Answers to Check Your Understanding Exercises for Section 14.2

1. a. $P = 0.994$. Do not reject the hypothesis of no interaction.

b. $P = 0.000$. Reject the hypothesis of no main effect. We conclude that the mean hardness varies with depth.

c. $P = 0.000$. Reject the hypothesis of no main effect. We conclude that the mean hardness varies with distance.

2. a. $P = 0.000$. Reject the hypothesis of no interaction.

b. and **c.** Since the main effects interact, we do not perform hypothesis tests on the main effects.

3. a.

	18–30	31–50	51–80
Drug A	42.25	52.25	56.5
Drug B	51	61	66
Drug C	71	82	85.75

b.

There do not appear to be strong interactions.

c. The MINITAB output is as follows.

Source	DF	SS	MS	F	P
Drug	2	5386.50	2693.25	2879.91	0.000
Age Group	2	1362.67	618.333	728.554	0.000
Interaction	4	2.33333	0.58333	0.62376	0.650
Error	27	25.2500	0.93519		
Total	35	6776.75			

The P-value for interactions is 0.650; we do not reject the hypothesis of no interaction. The results of the test are consistent with the interaction plot.

4. Plot C indicates the largest interactions. Plot B indicates the smallest.

Chapter 14 Summary

Section 14.1: Analysis of variance (ANOVA) is used to test the hypothesis that several population means are all equal. In one-way ANOVA, the populations correspond to different levels of a single factor. The levels of the factor are called treatments. We have a simple random sample from each population. Under the assumption that the populations are all approximately normal with the same variance, we construct a test statistic by comparing the spread of the sample means ($SSTr$) with the spread of the individual samples (SSE). When the quotient $\dfrac{MSTr}{MSE}$ is large, we reject H_0. Rejecting H_0 allows us to conclude that two or more of the population means are different, but it does not tell us which ones are different. We can perform the Tukey–Kramer test to determine which pairs of means differ from each other.

Section 14.2: In two-way ANOVA, there are two factors, and each population corresponds to a combination of levels of the two factors. The main effect of a factor is the difference in the mean response between two levels of that factor. When the effect of a factor depends on the level of the other factor, the factors are said to interact. When large interactions are present, the main effects can be misleading. We therefore begin by testing the hypothesis of no interaction. If we do not reject this hypothesis, we then test the hypothesis of no main effects. Interaction plots allow us to visualize the strength of the interactions.

Vocabulary and Notation

analysis of variance (ANOVA)   632
ANOVA table   641
balanced design   633
cell   648
cell mean   653
error mean square   636
error sum of squares   636
factor   632
grand mean   635

interaction   649
interaction plot   653
main effect   649
mean square   636
one-factor experiment   632
one-way ANOVA   632
one-way table   632
pairwise comparison   639
response variable   632

Studentized range distribution   639
total sum of squares   642
treatment   632
treatment mean square   636
treatment sum of squares   636
Tukey–Kramer test   639
two-factor experiment   648
two-way table   648

Important Formulas

Treatment sum of squares:

$$SSTr = n_1(\bar{x}_1 - \bar{\bar{x}})^2 + n_2(\bar{x}_2 - \bar{\bar{x}})^2 + \cdots + n_I(\bar{x}_I - \bar{\bar{x}})^2$$

Treatment mean square:

$$MSTr = \frac{SSTr}{I - 1}$$

Error sum of squares:

$$SSE = (n_1 - 1)s_1^2 + (n_2 - 1)s_2^2 + \cdots + (n_I - 1)s_I^2$$

Error mean square:

$$MSE = \frac{SSE}{N - I}$$

F statistic for one-way ANOVA:

$$F = \frac{MSTr}{MSE}$$

Test statistic for Tukey–Kramer test:

$$q = \frac{|\bar{x}_i - \bar{x}_j|}{\sqrt{\dfrac{MSE}{2}\left(\dfrac{1}{n_i} + \dfrac{1}{n_j}\right)}}$$

Chapter Quiz

Exercises 1–4 refer to the following data:

At a certain college, random samples of students with various majors were taken, and their grade point averages (GPA) were computed. The following table presents the sample means, standard deviations, and sample sizes.

Major	Mean	Standard Deviation	Sample Size
English	3.15	0.45	50
Statistics	3.05	0.56	48
Psychology	3.19	0.35	45
Business	3.01	0.49	55

1. How many degrees of freedom are there for $SSTr$ and for SSE?

2. Compute the sums of squares $SSTr$ and SSE and the mean squares $MSTr$ and MSE.

3. Compute the value of the test statistic F.

4. Can you conclude that there are differences in mean GPA among the majors? Use the $\alpha = 0.05$ level of significance.

Exercises 5–7 refer to the following data:

A machine shop has four machines used in precision grinding of bearings. Three machinists are employed to grind bearings on the machines. In an experiment to determine whether there are differences in output among the operators or the machines, each operator worked on each machine for three days. The response variable was the number of parts produced. The results are presented in the following table.

	Machine 1	Machine 2	Machine 3	Machine 4
Operator A	38 33 20	40 43 47	43 31 40	40 38 39
Operator B	33 34 34	39 37 47	35 43 31	36 46 46
Operator C	28 34 40	49 36 50	39 37 35	34 39 38

5. Can you reject the null hypothesis of no interactions? Explain.

6. Can the main effect of operator on the number of parts produced be interpreted? If so, interpret the main effect, using the $\alpha = 0.05$ level of significance. If not, explain why not.

7. Can the main effect of machine on the number of parts produced be interpreted? If so, interpret the main effect, using the $\alpha = 0.05$ level of significance. If not, explain why not.

Exercises 8–10 refer to the following data:

As part of a study on weight loss, random samples of men and women were assigned to follow one of three diets. After several weeks, the amount of weight lost, in pounds, was recorded for each person. A two-way analysis was performed, with gender and diet as the factors, and weight loss as the response. The following MINITAB output presents the ANOVA table.

```
Analysis of Variance for Weight Loss


Source        DF        SS        MS        F         P
Gender        1     496.13    496.13    19.649     0.000
Diet          2      40.2      20.1     0.79604    0.463
Interaction   2     534.47    267.23    10.583     0.001
Error        24       606      25.25
Total        29     1676.8
```

8. Can you reject the null hypothesis of no interactions? Explain.

9. Can the main effect of gender on weight loss be interpreted? If so, interpret the main effect, using the $\alpha = 0.01$ level of significance. If not, explain why not.

10. Can the main effect of diet on weight loss be interpreted? If so, interpret the main effect, using the $\alpha = 0.01$ level of significance. If not, explain why not.

Review Exercises

1. Soil quality: Soil scientists conducted an experiment in which the mineral gypsum was added in four different amounts to soil specimens to try to control the acid level of the soil. Three soil specimens received each amount added. The response variable is pH, which is a measurement of the acid level. The results are as follows. Construct an ANOVA table.

Gypsum Amount	pH
None	7.88 7.72 7.68
Low	7.81 7.64 7.85
Medium	7.84 7.63 7.87
High	7.80 7.73 8.00

Source: *Journal of The Soil Science Society of America* 66:92–98

2. Soil quality: Using the data in Exercise 1, can you conclude that the pH differs with the amount of gypsum added? Use the $\alpha = 0.05$ level of significance.

3. Electrical conductor: The scientists referred to in Exercise 1 also considered the effect of gypsum on the electrical conductivity (in decisiemens per meter) of soil. Two types of soil were treated with three different amounts of gypsum. The results are as follows.

	Soil Type A	Soil Type B
No Gypsum	1.52 1.05	1.01 0.92
Low Gypsum	1.49 0.91	1.12 0.92
High Gypsum	0.99 0.92	0.88 0.92

Source: *Journal of The Soil Science Society of America* 66:92–98

a. Can you reject the hypothesis of no interactions? Explain.

b. Can the main effect of the amount of gypsum be interpreted? If so, interpret the main effect. Use the $\alpha = 0.05$ level of significance.

c. Can the main effect of soil type be interpreted? If so, interpret the main effect. Use the $\alpha = 0.05$ level of significance.

4. Chemicals in the water: Earth scientists measured several chemical properties in water samples taken from wells at three locations. Following are sample means, standard deviations, and sample sizes for pH measurements.

Location	Mean	Standard Deviation	Sample Size
Upstream	6.8	0.6	16
Midstream	6.2	0.4	31
Downstream	6.4	0.6	30

Source: *Geosciences Journal* 6:35–46

   a. What are the sample sizes n_1, n_2, n_3?
   b. What are the sample standard deviations s_1, s_2, s_3?
   c. Compute *SSTr*.
   d. Compute *SSE*.

5. Chemicals in the water: Refer to Exercise 4.
   a. How many degrees of freedom are there for *SSTr*? For *SSE*?
   b. Compute *MSTr* and *MSE*.
   c. Compute the value of the test statistic for testing the null hypothesis that the population means are all equal.
   d. Can you conclude that there are differences among the mean pH levels? Use the $\alpha = 0.05$ level of significance.

6. Chemicals in the water: Using the data in Exercise 4, perform the Tukey–Kramer test to determine which pairs of means, if any, differ. Use the $\alpha = 0.05$ level of significance.

7. Electrical conductor: The scientists referred to in Exercise 3 measured electrical conductivity (in microsiemens per centimeter) in several water samples. Following are sample means, standard deviations, and sample sizes.

Location	Mean	Standard Deviation	Sample Size
Upstream	463	208	49
Midstream	363	98	31
Downstream	647	878	30

Source: *Geosciences Journal* 6:35–46

Can a one-way ANOVA be used to determine whether conductivity varies with location? Or is one of the necessary assumptions violated? Explain.

8. Orange trees: An agricultural scientist wants to determine how the type of fertilizer and the type of soil affect the yield of oranges in an orange grove. He has two types of fertilizer and three types of soil. For each of the six combinations of fertilizer and soil, the scientist plants four stands of trees, and measures the yield of oranges (in tons per acre) from each stand. The data are shown in the following table.

	Soil Type 1	Soil Type 2	Soil Type 3
Fertilizer A	20 24 22 9	22 22 27 25	29 25 23 28
Fertilizer B	23 23 30 24	23 29 31 32	25 29 30 31

   a. Can you reject the hypothesis of no interactions? Explain.
   b. Can the main effect of fertilizer on the yield of oranges be interpreted? If so, interpret the main effect, using the $\alpha = 0.05$ level of significance. If not, explain why not.
   c. Can the main effect of soil on the yield of oranges be interpreted? If so, interpret the main effect, using the $\alpha = 0.05$ level of significance. If not, explain why not.

9. Orange trees: Construct and interpret an interaction plot for the data in Exercise 8.

10. Power plant pollution: Sulfur dioxide (SO_2) is a chemical produced by many industrial processes. It is one of the ingredients in acid rain, so it is important to monitor the amount of sulfur dioxide that is released into the atmosphere. In an article published in the journal *Atmospheric Environment*, several measurements of the maximum hourly concentrations (in milligrams per cubic meter) of SO_2 were reported for each of four power plants. The results were as follows.

Plant	Concentration					
A	438	619	732	638		
B	857	1014	1153	883	1053	
C	925	786	1179	786		
D	893	891	917	695	675	595

   a. Because the design is unbalanced, check that the assumption of equal variances is satisfied by showing that the largest sample standard deviation is less than twice as large as the smallest one.
   b. Construct an ANOVA table.

11. Power plant pollution: Using the data in Exercise 10, perform the Tukey–Kramer test to determine which pairs of means, if any, differ. Use the $\alpha = 0.05$ level of significance.

12. Car tires: Three brands of tires were tested on each of four makes of car. Three cars of each make were used. The number of miles traveled, in thousands, is given as follows for each car.

	Tire 1	Tire 2	Tire 3
Car Type A	56 58 59	59 61 62	54 56 56
Car Type B	60 61 62	61 64 65	58 59 59
Car Type C	57 59 59	60 62 62	56 57 58
Car Type D	58 61 59	58 59 59	55 57 56

   a. Can you reject the hypothesis of no interactions? Explain.
   b. Can the main effect of car type on number of miles traveled be interpreted? If so, interpret the main effect, using the $\alpha = 0.05$ level of significance. If not, explain why not.
   c. Can the main effect of tire on number of miles traveled be interpreted? If so, interpret the main effect, using the $\alpha = 0.05$ level of significance. If not, explain why not.

13. **Car tires:** Construct and interpret an interaction plot for the data in Exercise 12.

14. **Quality control:** Four machines are used with three different materials to fabricate a certain part. Each combination of machine and material is run for three days, and the number of satisfactory parts made each day is counted. A two-way ANOVA is performed to determine whether the mean number of satisfactory parts differs among machines or materials. The following MINITAB output presents the ANOVA table.

```
Analysis of Variance for Parts


Source        DF        SS        MS         F         P
Material       2     39.056    19.528     5.618     0.010
Machine        3     120.11    40.037     11.52     0.000
Interaction    6     20.722    3.4537     0.994     0.452
Error         24     83.423    3.4760
Total         35     263.31
```

   a. Can you reject the null hypothesis of no interactions? Explain.
   b. Can the main effect of material on number of satisfactory parts be interpreted? If so, interpret the main effect, using the $\alpha = 0.05$ level of significance. If not, explain why not.
   c. Can the main effect of machine on number of satisfactory parts be interpreted? If so, interpret the main effect, using the $\alpha = 0.05$ level of significance. If not, explain why not.

15. **Interactions:** The following table presents the cell means for a two-way ANOVA. Construct and interpret an interaction plot.

		Column Factor		
		Low	**Medium**	**High**
	Low	6.4	26.7	62.6
Row Factor	**Medium**	17.1	42.3	71.0
	High	34.1	54.4	87.1

Write About It

1. Why does a large value of the F statistic provide evidence against the null hypothesis that all the population means are equal? Why doesn't a small value of F statistic provide evidence against H_0?

2. One-way analysis of variance requires the assumption that the populations are approximately normal. However, if the sample sizes are large ($n > 30$), this assumption is not so important. Why is this so?

3. Why are the main effects misleading when interactions are present?

4. When testing the null hypothesis of no interaction, describe the consequences of a Type I error and of a Type II error.

5. Explain why it is reasonable to use a larger level of significance for a test for interaction than for most other hypothesis tests.

Case Study: Optimal Procedure For Using A Microbalance

In the chapter opener, we presented data collected in an experiment designed to evaluate the performance of a microbalance. The purpose of the experiment was to determine whether it was better to turn off the microbalance when not in use, or to leave it powered up. The following table presents measurements of weights of three silicon wafers. The weight of each wafer was measured twice by each of three operators. These measurements were made under the policy that the balance was to be turned off when not in use, and powered up only when it was about to be used.

	Operator A		Operator B		Operator C	
Wafer 1	11	15	10	6	14	10
Wafer 2	210	208	205	201	208	207
Wafer 3	111	113	102	105	108	111

1. Construct an ANOVA table.

2. Can it be determined from the ANOVA table whether there are differences in the measured weights among the operators? If so, provide the value of the test statistic and the *P*-value. If not, explain why not.

   It turned out that the measurements made by operator B were taken shortly after the balance had been powered up, while the others were taken after the balance had been powered up for some time. A new policy was then instituted to leave the balance powered up continuously. The three operators then made two weighings of three different wafers. The results are presented in the following table.

	Operator A		Operator B		Operator C	
Wafer 1	152	156	156	155	152	157
Wafer 2	443	440	442	439	435	439
Wafer 3	229	227	229	232	225	228

3. Construct an ANOVA table.

4. Compare the ANOVA table in Exercise 3 with the ANOVA table in Exercise 1. Would you recommend leaving the balance powered up continuously? Explain your reasoning.

Nonparametric Statistics

Introduction

Modern water conservation efforts depend in large part on the treatment of sewage, or wastewater. Wastewater treatment is used to remove contaminants so that the water can be used for purposes such as irrigation. Microorganisms play an important role in wastewater treatment. In fact, wastewater treatment technology depends on beneficial microorganisms to remove organic material, transforming it into carbon dioxide and water.

An important feature in any wastewater treatment system is the average amount of time that solid organic matter resides in the treatment tank, which is called the solids retention time. A study was performed at the Colorado School of Mines to determine the relationship between the solids retention time and the diversity of the microorganism population. Because microorganisms feed on solid organic matter, it was thought that longer retention times could provide more opportunity for genetic mutations to occur, resulting in a greater diversity of microorganisms.

Following are counts of the numbers of species of microorganisms for solid retention times of 3 days, 12 days, and 30 days.

3-Day Cycle								
282	273	238	240	241	253	230	266	253

12-Day Cycle											
266	299	287	289	272	284	292	279	274	306	291	272

30-Day Cycle								
293	271	293	296	271	292	265	262	278
272	246	278	276	283	273	286	303	289

To determine whether there are differences in the numbers of species, we could perform *t*-tests, as described in Section 11.1. Of course, *t*-tests require the assumption that the populations are approximately normal. Tests that make assumptions about the distribution of a population are called **parametric** tests. As an alternative, we can use **nonparametric** tests. Nonparametric tests are tests that do not assume the data to have any specific distribution. In this chapter, we will discuss three nonparametric tests. In Section 15.1, we will discuss the **sign test**, which can be used to test hypotheses about the median of a population. The sign test is a nonparametric version of the *t*-test. Unlike the *t*-test, it does not require the assumption that the population is approximately normal. In Section 15.2 and Section 15.3 we discuss the **Wilcoxon rank-sum test** and **Wilcoxon signed-rank test**, respectively. These tests are nonparametric versions of the two-sample *t* tests for independent and for paired samples.

Differences Between Nonparametric Tests and Parametric Tests

- Nonparametric tests do not require the population to have any specific distribution.
- Parametric tests generally do require the population to have a specific distribution, such as the normal distribution.

SECTION 15.1 | The Sign Test

Objectives

1. Test hypotheses using the sign test

Objective 1 Test hypotheses using the sign test

The Sign Test

The sign test can be used to test a hypothesis about the median of a population. The corresponding parametric test is the *t*-test for a population mean discussed in Section 9.3. When samples are small, the *t*-test requires that the population be approximately normal. The primary advantage of the sign test, therefore, is that the only assumption required is that the population be continuous. The primary disadvantage is that when the population is approximately normal, the sign test is less likely to reject a false null hypothesis.

Primary Advantage and Disadvantage of the Sign Test

Advantage: The sign test is valid for any continuous population. It does not require the assumption of normality.

Disadvantage: When the population is approximately normal, the sign test is less likely to reject a false null hypothesis than the *t*-test.

Performing the Sign Test

Many of the hypothesis tests we have studied in previous chapters have been about the population mean. The sign test is used to test hypotheses about a population median m. The null hypothesis is of the form H_0: $m = m_0$. Sample values less than m_0 are assigned a minus sign, values above m_0 are assigned a plus sign, and values equal to m_0 are assigned a 0.

We then count the number of plus signs and the number of minus signs. If H_0 is true, and m_0 is the true population median, we would expect to see approximately equal numbers of plus and minus signs. If H_0 is false, we would expect to see more of one type of sign than the other.

To construct the test statistic, let x be the number of times the *less* frequent sign appears. For example, if there are 12 plus signs and 5 minus signs, then $x = 5$. The value of the test statistic is based on x, as follows:

Test Statistic for the Sign Test

Let x be the number of times the less frequent sign occurs. Let n be the total number of plus and minus signs.

- If $n \leq 25$, the test statistic is x.
- If $n > 25$, the test statistic is

$$z = \frac{x + 0.5 - n/2}{\sqrt{n}/2}$$

For $n \leq 25$, critical values of the test statistic are found in Table A.7.
For $n > 25$, critical values are found in Table A.2.

For sample sizes $n \leq 25$, the table is not detailed enough to allow the P-value to be computed. We therefore focus on the critical value method for the sign test. Following are the steps for performing the test:

Performing the Sign Test

Step 1: State the null and alternate hypotheses. The null hypothesis specifies a value for the population median m. We will call this value m_0. So the null hypothesis is of the form $H_0: m = m_0$.
The alternate hypothesis may be stated in one of three ways:

Left-tailed:	$H_1: m < m_0$
Right-tailed:	$H_1: m > m_0$
Two-tailed:	$H_1: m \neq m_0$

Step 2: Choose a significance level α.

Step 3: For each sample value greater than m_0, assign a plus sign. For each sample value less than m_0, assign a minus sign. For each sample value equal to m_0, assign a 0. Count the number of plus and minus signs.

Step 4: Let x be the number of times the less frequent sign appears. Let n be the total number of plus and minus signs (not counting any 0s). Compute the test statistic:
- If $n \leq 25$, the test statistic is x.
- If $n > 25$, the test statistic is $z = \dfrac{x + 0.5 - n/2}{\sqrt{n}/2}$

Step 5: Determine whether to reject H_0, as follows:
- If $n \leq 25$, find the critical value from Table A.7. If x is less than or equal to the critical value, reject H_0. Otherwise do not reject H_0.
- If $n > 25$, find the critical value from Table A.2. If z is less than or equal to the critical value, reject H_0. Otherwise do not reject H_0.

Step 6: State a conclusion.

Explain It Again

When to reject H_0: Because we use the frequency of the sign that occurs less often, we always reject when the test statistic is less than the critical value.

EXAMPLE 15.1

Performing the sign test

Annual income, in thousands of dollars, was recorded for a random sample of fifteen families in a certain town. The results were as follows.

75 10 92 49 58 25 96 40 90 87 10 49 20 200 34

An economist is interested in determining whether the median income in the town is greater than 40 thousand dollars. Perform a hypothesis test at the $\alpha = 0.05$ level of significance.

Solution

Step 1: State H_0 and H_1: The issue is whether the median is greater than 40. Therefore the null and alternate hypotheses are

$$H_0: m = 40 \qquad H_1: m > 40$$

Step 2: Choose a significance level: We choose a significance level of $\alpha = 0.05$.

Step 3: Assign plus and minus signs: The following table shows a plus sign for each value greater than 40 and a minus sign for each value less than 40.

75	10	92	49	58	25	96	40	90	87	10	49	20	200	34
+	−	+	+	+	−	+	0	+	+	−	+	−	+	−

Step 4: Compute the test statistic: There are 9 plus signs and 5 minus signs. The sample size is $n = 14$. Since $n \leq 25$, the test statistic is the number of times the less frequent sign appears. The value of the test statistic is $x = 5$.

Step 5: Determine whether to reject H_0: Because $n \leq 25$, we use Table A.7 to find the critical value. We have a one-tailed test, $n = 14$, and significance level $\alpha = 0.05$. The critical value is 3. Because $x > 3$, we do not reject H_0.

	One-tailed			
	$\alpha = 0.005$	$\alpha = 0.01$	$\alpha = 0.025$	$\alpha = 0.05$
	Two-tailed			
n	$\alpha = 0.01$	$\alpha = 0.02$	$\alpha = 0.05$	$\alpha = 0.10$
⋮	⋮	⋮	⋮	⋮
12	1	1	2	2
13	1	1	2	3
14	1	2	3	3
15	2	2	3	3
16	2	2	3	4
⋮	⋮	⋮	⋮	⋮

Step 6: State a conclusion: We cannot conclude that the median income is greater than 40 thousand dollars.

EXAMPLE 15.2

Performing the sign test

The length of time, in minutes, needed to perform a certain type of surgery was recorded for 32 operations. The results were as follows.

149	144	218	153	134	152	148	144
178	107	199	135	171	110	160	119
86	127	106	153	169	153	153	173
156	145	205	132	169	174	130	175

A hospital administrator is interested in determining whether the median time for surgery is less than 170 minutes. Perform a hypothesis test at the $\alpha = 0.01$ level of significance.

Solution

Step 1: **State H_0 and H_1:** The issue is whether the median is less than 170. Therefore the null and alternate hypotheses are

$$H_0: m = 170 \qquad H_1: m < 170$$

Step 2: **Choose a significance level:** We choose a significance level of $\alpha = 0.01$.

Step 3: **Assign plus and minus signs:** The following table shows a plus sign for each value greater than 170 and a minus sign for each value less than 170.

149	144	218	153	134	152	148	144
−	−	+	−	−	−	−	−
178	107	199	135	171	110	160	119
+	−	+	−	+	−	−	−
86	127	106	153	169	153	153	173
−	−	−	−	−	−	−	+
156	145	205	132	169	174	130	175
−	−	+	−	−	+	−	+

Step 4: **Compute the test statistic:** There are 8 plus signs and 24 minus signs. The number of times the less frequent sign appears is $x = 8$. The sample size is $n = 32$. Since $n > 25$, the test statistic is

$$z = \frac{x + 0.5 - n/2}{\sqrt{n}/2}$$

We substitute $x = 8$ and $n = 32$ to obtain

$$z = \frac{x + 0.5 - n/2}{\sqrt{n}/2} = \frac{8 + 0.5 - 32/2}{\sqrt{32}/2} = -2.65$$

Step 5: **Determine whether to reject H_0:** Because $n > 25$, we use Table A.2 to find the critical value. We have a one-tailed test with significance level $\alpha = 0.01$. The critical value is -2.326. Because $-2.65 < -2.326$, we reject H_0.

Step 6: **State a conclusion:** We conclude that the median time in surgery is less than 170 minutes.

Check Your Understanding

1. Prices for a gallon of regular gas were recorded for a sample of twelve gas stations, as shown below. Can you conclude that the median price is less than \$4.00? Use the $\alpha = 0.05$ level of significance.

3.92	3.95	3.76	3.89	3.57	4.06
3.61	3.66	3.98	3.79	3.96	3.74

   a. State the null and alternate hypotheses.
   b. Compute the test statistic.
   c. Find the critical value.
   d. State a conclusion.

2. The registrar at a college sampled 30 students and recorded the GPA for each, as shown below. Can you conclude that the median GPA is less than 2.90? Use the $\alpha = 0.05$ level of significance.

2.55	2.07	3.35	2.86	2.90	3.22	3.12	2.63	3.55	3.39
2.25	2.26	2.18	2.02	2.85	3.31	3.45	3.06	2.22	3.26
2.25	2.27	2.20	2.28	2.34	3.39	2.63	2.63	2.44	2.50

   a. State the null and alternate hypotheses.
   b. Compute the test statistic.
   c. Find the critical value.
   d. State a conclusion.

Answers are on page 673.

USING TECHNOLOGY

We use Example 15.1 to illustrate the technology steps.

MINITAB

Performing the sign test

Note: MINITAB returns a *P*-value for the test.

Step 1. Enter the data in column **C1**.

Step 2. Select **Stat**, then **Nonparametrics**, and then **1-Sample Sign**.

Step 3. Enter **C1** in the **Variables** field. Select the **Test median** option and enter the null hypothesis value for the median. For Example 15.1, we enter 40.

Step 4. In the **Alternative field**, select the form of the alternate hypothesis. We use **greater than**.

Step 5. Click **OK** (Figure A).

Sign Test for Median: C1

Sign test of median = 40.00 versus > 40.00

	N	Below	Equal	Above	P	Median
C1	15	5	1	9	0.2120	49.00

Figure A

SECTION 15.1   Exercises

Exercises 1–2 are the Check Your Understanding exercises located within the section.

Understanding the Concepts

In Exercises 3 and 4, fill in each blank with the appropriate word or phrase.

3. For the sign test, we reject H_0 when the value of the test statistic is _____ than the critical value.

4. A disadvantage of the sign test is that when the population is approximately normal the sign test is _____ likely to reject a false null hypothesis than the *t*-test.

In Exercises 5 and 6, determine whether the statement is true or false. If the statement is false, rewrite it as a true statement.

5. The sign test is valid for any continuous population.

6. The test statistic for the sign test is the frequency of the most frequently appearing sign.

Practicing the Skills

7. Compute the value of the test statistic for the following sample, for testing H_0: $m = 10$ versus H_1: $m \neq 10$.

14	11	9	3	4	2	14	11	9	5

8. Compute the value of the test statistic for the following sample, for testing H_0: $m = 50$ versus H_1: $m < 50$.

38	47	55	38	40	43	36	49
43	55	43	47	38	50	38	43

9. The sign test is performed to test H_0: $m = 35$ versus H_1: $m \neq 35$. There are 16 plus signs, 4 minus signs, and 3 zeros.

   a. What is the value of the test statistic?

   b. Is H_0 rejected at the $\alpha = 0.05$ level?

   c. Is H_0 rejected at the $\alpha = 0.01$ level?

10. The sign test is performed to test H_0: $m = 500$ versus H_1: $m < 500$. There are 9 plus signs, 26 minus signs, and 2 zeros.

   a. What is the value of the test statistic?

   b. Is H_0 rejected at the $\alpha = 0.05$ level?

   c. Is H_0 rejected at the $\alpha = 0.01$ level?

Working with the Concepts

11. Weight loss: A weight loss company claims that the median weight loss for people who follow their program is at least 15 pounds. Weight losses for a random sample of 20 people in the program were recorded, and the results were as follows. Can you conclude that the claim is false? Use the $\alpha = 0.05$ level of significance.

14	8	12	17	15	7	7	6	8	9
6	11	2	0	11	9	6	7	6	17

   a. State the appropriate null and alternate hypotheses.

   b. Compute the value of the test statistic.

   c. Find the critical value.

   d. State a conclusion.

12. At the movies: A sample of 12 movies released in a given year had the following running times, in minutes.

118	125	113	156	168	121
115	136	194	149	105	161

Can you conclude that the median running time for movies released that year was greater than 120 minutes? Use the $\alpha = 0.01$ level of significance.

   a. State the appropriate null and alternate hypotheses.

   b. Compute the value of the test statistic.

   c. Find the critical value.

   d. State a conclusion.

13. Tall trees: Heights, in feet, of a sample of 18 mature oak trees in a forest were measured. The results were as follows.

65	71	75	68	67	60
64	71	77	72	51	70
63	63	54	74	60	57

Can you conclude that the median height of oak trees in this forest is greater than 60 feet? Use the $\alpha = 0.05$ level of significance.

14. Wheat yields: The highest non-irrigated wheat yields in the United States are in the Palouse region of Eastern Washington and Northern Idaho. Following are hypothetical yields, in bushels, for 15 randomly selected acres of farmland.

74	75	82	82	76	76	81	86
98	89	76	93	65	62	63

Can you conclude that the median yield differs from 75 bushels per acre? Use the $\alpha = 0.01$ level of significance.

15. Paying the rent: Monthly rents were recorded for a sample of 36 apartments in a certain city. The results were as follows.

660	980	1210	440	530	1110
610	350	810	670	700	680
1030	1100	910	920	1000	620
820	980	620	1300	360	1290
1090	1250	380	820	1050	1020
500	390	1150	1290	720	710

Can you conclude that the median rent is less than $1000 per month? Use the $\alpha = 0.05$ level of significance.

16. Morning commute: A random sample of 30 commuters were asked how long their typical morning commute is. The results, in minutes, were as follows.

8	10	5	33	20	43
69	63	63	26	14	14
35	33	60	46	31	36
2	39	19	66	2	4
4	35	42	43	61	20

Can you conclude that the median commute time is less than 60 minutes? Use the $\alpha = 0.05$ level of significance.

17. Scam alert: The rates of fraud complaints (per 100,000 people) for the 20 largest metropolitan areas in the United States in 2012 are as follows.

404.4	525.6	409.2	385.9
461.7	486.4	525.3	390.7
451.6	426.9	370.7	485.8
532.0	432.5	506.3	398.3
453.5	393.7	607.8	402.4

Can you conclude that the median fraud complaint rate is less than 480? Use the $\alpha = 0.05$ level of significance.

18. Stamp of approval: The Gallup poll approval ratings for the United States Congress from January 2011 to December 2013 are as follows.

12	17	18	17	11	17
9	16	18	15	13	24
11	15	21	17	13	17
19	13	13	12	15	18
14	15	10	10	13	23
15	14	16	13	18	20

Can you conclude that the median approval rating differs from 17? Use the $\alpha = 0.05$ level of significance.

Extending the Concepts

Exercise 19 demonstrates that the sign test will generally not work when the population is discrete.

19. Discrete population: In a certain city, 45% of the households have no dog, 30% have one dog, 10% have two dogs, 10% have three dogs, and 5% have four dogs. The population median number of dogs is therefore 1. In a sample of 100 households, 45 have 0 dogs, 30 have 1 dog, 10 have two dogs, 10 have three dogs, and 5 have four dogs. Note that the sample reflects the population perfectly. Consider the hypotheses H_0: $m = 1$ versus H_1: $m < 1$. Compute the value of the test statistic for the sign test and show that H_0 is rejected at the $\alpha = 0.05$ level.

Answers to Check Your Understanding Exercises for Section 15.1

1. a. H_0: $m = 4$, H_1: $m < 4$
b. $x = 1$ **c.** The critical value is 2.
d. Reject H_0. We conclude that the median price is less than $4.00.

2. a. H_0: $m = 2.90$, H_1: $m < 2.90$
b. $z = -1.49$ **c.** $z = -1.645$
d. Do not reject H_0. We cannot conclude that the median GPA is less than 2.90.

SECTION 15.2   The Rank-Sum Test

Objectives

1. Assign ranks to a sample
2. Test hypotheses using the rank-sum test

Objective 1 Assign ranks to a sample

Ranks

The nonparametric tests we will discuss in this section are based on **ranks**. To rank the values in a sample, we order them from smallest to largest. The smallest value is assigned a rank of 1, the next smallest is assigned a rank of 2, and so forth. For values in the data set whose ranks are tied, each value is assigned the average of the corresponding ranks. Examples 15.3 and 15.4 illustrate the method.

EXAMPLE 15.3 Assign ranks to a sample

The weights of six first-graders, in pounds, are 35, 43, 38, 51, 45, 49. Assign ranks to these values.

Solution

We arrange the values in increasing order. The smallest value, 35, is assigned a rank of 1. The next smallest, 38, is assigned a rank of 2, and so forth. The results are as follows.

Weight	35	38	43	45	49	51
Rank	1	2	3	4	5	6

In some data sets, there are ties, that is, two or more values are the same. When two or more values are tied, each receives the average of the ranks. Example 15.4 illustrates the method.

EXAMPLE 15.4 Assign ranks to a sample when ties are present

The heights of ten fourth-graders, in inches, are 48, 46, 45, 54, 46, 52, 50, 46, 52, 49. Assign ranks to these values.

Solution

We arrange the values in increasing order. We notice there are three values tied at 46, and two tied at 52. We assign the ranks 1 through 10 as preliminary ranks. The three values of 46 have preliminary ranks 2, 3, 4. We compute the average of these preliminary ranks: $(2 + 3 + 4)/3 = 9/3 = 3$. The rank assigned to each of these values is therefore 3. The two values of 52 have preliminary ranks 8 and 9. We assign each the rank $(8 + 9)/2 = 17/2 = 8.5$. The results are shown in the following table.

Height	45	46	46	46	48	49	50	52	52	54
Preliminary Ranks	1	2	3	4	5	6	7	8	9	10
Ranks	1	3	3	3	5	6	7	8.5	8.5	10

Objective 2 Test hypotheses using the rank-sum test

The Rank-Sum Test

Wilcoxon's rank-sum test, also known as the Mann-Whitney test, is a nonparametric test to determine whether two population medians are equal. It is a nonparametric alternative to the two-sample t-test, which is discussed in Section 11.1. The test requires that the populations have approximately the same shape. When two populations have the same shape and their medians are equal, they have the same distribution. For this reason the rank-sum test is sometimes described as a test to determine whether two populations differ.

The main advantage of the rank-sum test is that it does not require that the populations be approximately normal. The main disadvantage is that the null hypothesis can be falsely rejected when the population medians are equal, if the shapes of the two populations differ substantially.

> ## Main Advantage and Disadvantage of the Rank-Sum Test
> **Advantage:** The rank-sum test does not require the assumption of normality.
> **Disadvantage:** The null hypothesis can be rejected when the two populations differ in shape, even if their medians are equal.

Performing the Rank-Sum Test

The rank-sum test is used to test the null hypothesis that two population medians are equal. Independent samples are drawn from each population. The combined samples are then ranked. The smallest value in the two samples combined is assigned a rank of 1, the second smallest is assigned a rank of 2, and so forth. If the two populations are the same, the ranks for the values assigned to one sample should not be substantially larger or smaller than the ranks assigned to the other. Thus the rank-sum test rejects the null hypothesis when the sum of the ranks for one of the samples is exceptionally large or small.

To construct the test statistic, let n_1 be the smaller of the two sample sizes and let n_2 be the larger. Let S be the sum of the ranks for the smaller sample. If both samples are the same size, then n_1 and n_2 are both equal to that size, and S may be the sum of the ranks of either sample. When H_0 is true, it can be shown that S has mean μ_S and standard deviation σ_S given by

$$\mu_S = \frac{n_1(n_1 + n_2 + 1)}{2} \qquad \sigma_S = \sqrt{\frac{n_1 n_2 (n_1 + n_2 + 1)}{12}}$$

We construct the test statistic as

$$z = \frac{S - \mu_S}{\sigma_S}$$

When both samples sizes are 10 or more, the test statistic has an approximately standard normal distribution, so P-values and critical values can be found using the z-table (Table A.2). When one or both of the sample sizes is less than 10, special tables can be used. Since most samples in practice consist of at least 10 values, we do not consider this situation here; we will only consider the situation in which both sample sizes are at least 10.

Performing the Rank-Sum Test

Let m_1 be the population median corresponding to the sample of size n_1, and let m_2 be the population median corresponding to the sample of size n_2.

Step 1: State the null and alternate hypotheses.
The null hypothesis is H_0: $m_1 = m_2$
The alternate hypothesis can be stated in one of three ways:

Left-tailed:	H_1: $m_1 < m_2$
Right-tailed:	H_1: $m_1 > m_2$
Two-tailed:	H_1: $m_1 \neq m_2$

Step 2: Choose a significance level α.

Step 3: Combine the two samples, arrange the combined sample in increasing order, and assign ranks to the values.

Step 4: Let n_1 be the smaller sample size and let n_2 be the larger. Compute S, the sum of the ranks of the smaller sample. If both samples are the same size, then n_1 and n_2 are both equal to that size, and S may be the sum of the ranks of either sample. Then compute μ_S and σ_S, as follows.

$$\mu_S = \frac{n_1(n_1 + n_2 + 1)}{2} \qquad \sigma_S = \sqrt{\frac{n_1 n_2 (n_1 + n_2 + 1)}{12}}$$

Step 5: Compute the test statistic $z = \dfrac{S - \mu_S}{\sigma_S}$.

Step 6: Compute the P-value, as follows. H_1: $m_1 < m_2$: The P-value is the area to the left of $-z$. H_1: $m_1 > m_2$: The P-value is the area to the right of z. H_1: $m_1 \neq m_2$: The P-value is the sum of the areas to the right of $|z|$ and to the left of $-|z|$.

Step 7: Interpret the P-value. If making a decision, reject H_0 if the P-value is less than or equal to the significance level α.

Step 8: State a conclusion.

EXAMPLE 15.5 **Performing the rank-sum test**

A sample of 14 students took a statistics class online, and another sample of 12 students took an equivalent class in a traditional classroom. Both classes were given the same final exam at the end of the course. The scores were as follows.

Online	78	82	83	87	75	63	78	60	94	62	98	90	97	81
Traditional	73	72	92	100	74	90	64	84	77	89	70	64		

Can you conclude that there is a difference between the median scores from the two classes? Use the $\alpha = 0.05$ level of significance.

Solution

Step 1: State H_0 and H_1: The null hypothesis is that there is no difference in the scores between the two teaching methods. The alternate hypothesis is that there is a difference.

Step 2: Choose a significance level: We will use a significance level of $\alpha = 0.05$.

Step 3: Arrange the combined samples in increasing order and assign ranks: Following are the results. It is important to label each value with the sample it came from.

Score	60	62	63	64	64	70	72	73	74	75	77	78	78
Sample	O	O	O	T	T	T	T	T	T	O	T	O	O
Rank	1	2	3	4.5	4.5	6	7	8	9	10	11	12.5	12.5

Score	81	82	83	84	87	89	90	90	92	94	97	98	100
Sample	O	O	O	T	O	T	T	O	T	O	O	O	T
Rank	14	15	16	17	18	19	20.5	20.5	22	23	24	25	26

Step 4: Compute S, μ_S, and σ_S: The sample sizes are $n_1 = 12$ and $n_2 = 14$. The smaller sample belongs to the traditional class, so S is the sum of the ranks corresponding to that sample. We compute $S = 154.5$, $\mu_S = 162$, and $\sigma_S = 19.442$.

Step 5: Compute the test statistic: The value of the test statistic is

$$z = \frac{S - \mu_S}{\sigma_S} = \frac{154.5 - 162}{19.442} = -0.39$$

Step 6: Compute the P-value: From Table A.2, we find that the P-value is 0.6965. (Technology gives 0.6997.)

Step 7: Interpret the P-value: Because $P > 0.05$, we do not reject H_0.

Step 8: State a conclusion: We cannot conclude that there are differences in the test scores between the two methods of instruction.

Check Your Understanding

1. Heart rates, in beats per minute, were measured for samples of 12 track athletes and 15 swimmers. The results are shown below. Can you conclude that the median heart rate is greater for swimmers than for track athletes? Use the $\alpha = 0.05$ level of significance.

Track	68	62	65	72	70	68	64	77	77	66	72	76			
Swim	82	81	71	69	79	65	66	70	80	78	75	82	75	63	79

a. State the null and alternate hypotheses.
b. Compute the value of the test statistic.
c. Compute the P-value.
d. State a conclusion.

2. Battery lifetimes, in hours, were measured for two types of batteries commonly used in laptop computers. Twelve batteries of one type and 13 of another type were tested. The results are shown below. Can you conclude that the lifetimes differ between the two types of batteries? Use the $\alpha = 0.01$ level of significance.

Type A
4.1 6.8 5.8 5.2 3.2 2.5
3.9 2.1 4.9 5.2 2.1 4.0

Type B
6.7 4.1 5.6 7.0 3.1 2.3 4.6
3.8 4.4 4.9 7.1 4.7 4.0

a. State the null and alternate hypotheses.
b. Compute the value of the test statistic.
c. Compute the P-value.
d. State a conclusion.

Answers are on page 679.

USING TECHNOLOGY

We use Example 15.5 to illustrate the technology steps.

MINITAB

Performing the rank-sum test

Note: MINITAB returns a P-value for the test.

Step 1. Enter the data for Sample 1 in column **C1**. and the data for Sample 2 in column **C2**. For Example 15.5, we enter the online class data into **C1** and the traditional class data into **C2**.

Step 2. Click **Stat**, then **Nonparametrics**, then **Mann-Whitney**.

Step 3. Enter **C1** in the **First Sample** field and **C2** in the **Second Sample** field. In the **Alternative** field, select the form of the alternate hypothesis. We use **not equal**.

Step 4. Click **OK** (Figure A).

```
Mann-Whitney Test and CI: C1, C2

         N   Median
C1  14    81.50
C2  12    75.50

Test of ETA1 = ETA2 vs ETA1 not = ETA2 is significant at 0.7188
The test is significant at 0.7187 (adjusted for ties)
```

Figure A

SECTION 15.2   Exercises

Exercises 1–2 are the **Check Your Understanding** exercises located within the section.

Understanding the Concepts

In Exercises 3 and 4, fill in each blank with the appropriate word or phrase.

3. If two or more values are the same when assigning ranks, we assign them the _____ of the preliminary ranks.

4. To use the normal approximation to find the P-value for a rank-sum test, both sample sizes must be at least

_____ .

In Exercises 5 and 6, determine whether the statement is true or false. If the statement is false, rewrite it as a true statement.

5. To compute the test statistic for the rank-sum test, we sum the ranks of the larger sample.

6. The null hypothesis for the rank-sum test is that the two population medians are equal.

Practicing the Skills

In Exercises 7–10, compute μ_S, σ_S, and the value of the test statistic z. Then find the P-value for the specified alternate hypothesis and values of n_1, n_2, and S.

7. $n_1 = 15$, $n_2 = 18$, $S = 244$, H_1: $\mu_1 < \mu_2$.

8. $n_1 = 20$, $n_2 = 25$, $S = 607$, H_1: $\mu_1 \neq \mu_2$.

9. $n_1 = 17$, $n_2 = 23$, $S = 272$, H_1: $\mu_1 \neq \mu_2$.

10. $n_1 = 14$, $n_2 = 27$, $S = 308$, H_1: $\mu_1 > \mu_2$.

Working with the Concepts

11. **How's your gas mileage?** Gas mileages, in miles per gallon, were measured for samples of compact cars and midsize cars. The results were as follows.

Compact				
34.5	33.7	26.1	28.5	27.4
30.6	31.1	28.0	33.0	33.0
32.8	28.5	25.5	32.1	34.9

Midsize				
21.6	21.1	29.1	24.8	28.5
28.1	21.9	22.5	20.5	26.1

Can you conclude that the median mileage is less for compact cars than for midsize cars? Use the $\alpha = 0.05$ level of significance.
  a. State the null and alternate hypotheses.

  b. Compute the value of the test statistic.
  c. Compute the P-value.
  d. State a conclusion.

12. **Coffee prices:** Prices of a 12-ounce latte (including tax) were determined at samples of coffee shops in two cities. The results were as follows.

City 1						
2.93	2.75	2.95	3.15	3.14	2.89	2.87
2.79	2.63	3.05	2.68	2.81	2.84	3.13

City 2						
2.41	2.91	2.57	2.55	2.69	2.84	2.52
2.53	2.99	2.76	2.58	2.98	2.94	2.51

Can you conclude that the median price differs between the two cities? Use the $\alpha = 0.01$ level of significance.
  a. State the null and alternate hypotheses.

  b. Compute the value of the test statistic.
  c. Compute the P-value.
  d. State a conclusion.

13. **Compressive strength:** In an experiment to determine the effect of curing time on the compressive strength of concrete blocks, two samples of 15 blocks each were prepared identically except for curing time. The blocks in one sample were cured for 2 days, while the blocks in the other were cured for 6 days. The compressive strengths of the blocks, in tons per square inch, are presented below.

Cured 2 Days				
90	84	87	75	79
95	88	82	85	96
70	86	70	80	77

Cured 6 Days				
94	73	91	97	98
92	82	84	76	88
79	97	90	81	91

Can you conclude that the median strength differs between the two curing times? Use the $\alpha = 0.05$ level of significance.

14. **Recovery times:** A new postsurgical treatment was compared to a standard treatment. Ten subjects received the new treatment, while 12 others (the controls) received the

standard treatment. The recovery times, in days, were as follows.

Treatment				
12	20	15	22	20
21	27	13	19	21

Control					
30	23	33	30	32	35
40	18	32	24	29	28

Can you conclude that the median recovery time is less for treatment than for control? Use the $\alpha = 0.01$ level of significance.

15. **Let's watch a Blu-ray:** Times, in seconds, for a Blu-ray to load were measured for samples of standard Blu-ray players and 3-D Blu-ray players. The results were as follows.

3-D Blu-ray Players			
27	21	22	26
21	24	28	24
28	19	19	22
25	32	22	24

Standard Blu-ray Players			
25	20	20	23
24	23	24	19
22	24	21	18

Can you conclude that the median load time for 3-D Blu-ray players is greater than for standard Blu-ray players? Use the $\alpha = 0.05$ level of significance.

16. **Laundry time:** The costs, in dollars, per load of laundry were measured for samples of high-efficiency detergents like those used in many front-loading washing machines and standard detergents. The results were as follows.

Standard Laundry Detergents			
0.23	0.16	0.18	0.18
0.13	0.18	0.12	0.14
0.28	0.14	0.20	0.13
0.30	0.13	0.56	0.20
0.25	0.29		

High-Efficiency Laundry Detergents			
0.18	0.18	0.13	0.30
0.19	0.15	0.15	0.18
0.15	0.09	0.09	0.08
0.06	0.05	0.08	0.12
0.18	0.13		

Can you conclude that the median cost per load for high-efficiency detergents differs from the median cost per load for standard detergents? Use the $\alpha = 0.05$ level of significance.

Extending the Concepts

The rank-sum test is valid under the assumption that the populations have similar shapes. Exercise 17 shows that when the shapes differ considerably, the null hypothesis may be rejected even when the sample medians are equal.

17. Populations of differing shapes: Consider the following two samples.

Sample 1				
11	12	13	14	15
16	17	30	50	60
70	80	90	100	110

Sample 2				
0	1	2	3	4
5	6	30	31	32
33	34	35	36	37

a. Compute the median of each sample to show that they are equal.

b. Compute μ_S, σ_S, and the value of the test statistic z.

c. Compute the P-value for testing H_0: $m_1 = m_2$ versus H_1: $m_1 \neq m_2$. Do you reject H_0? Use the $\alpha = 0.05$ level of significance.

Answers to Check Your Understanding Exercises for Section 15.2

1. a. H_0: $\mu_1 = \mu_2$, H_1: $\mu_1 < \mu_2$ **b.** $S = 129.5$, $\mu_S = 168$, $\sigma_S = 20.49$, $z = -1.88$ **c.** $P = 0.0301$
   d. Reject H_0. We conclude that the median heart rate for track athletes is less than the median for swimmers.

2. a. H_0: $\mu_1 = \mu_2$, H_1: $\mu_1 \neq \mu_2$ **b.** $S = 139.5$, $\mu_S = 156$, $\sigma_S = 18.38$, $z = -0.90$. **c.** $P = 0.3682$ [Tech: 0.3695]
   d. Do not reject H_0. There is not enough evidence to conclude that the median lifetime differs between the two types of batteries.

SECTION 15.3 | The Signed-Rank Test

Objectives

1. Test hypotheses using the signed-rank test

Objective 1 Test hypotheses using the signed-rank test

The Signed-Rank Test

Wilcoxon's signed-rank test is a nonparametric test designed to determine whether population means differ when the data is in the form of paired samples. It is therefore a nonparametric alternative to the t-test for matched pairs discussed in Section 11.3. As with other nonparametric tests, its main advantage is that it does not require that the population of differences have a specified distribution such as the normal distribution. However, it does require that the distribution be approximately symmetric.

The primary advantage and disadvantage of the signed-rank test are similar to those of the sign test. The advantage is that it does not require that the population of differences be approximately normal. The disadvantage is that when the population of differences is close to normal, the signed-rank test is less likely to reject a false null hypothesis than the t-test.

Primary Advantage and Disadvantage of the Signed-Rank Test

Advantage: The signed-rank test does not require the assumption of normality.
Disadvantage: The assumption of symmetry, while less restrictive than normality, is still required. When the population of differences is close to normal, the signed-rank test is less likely to reject a false null hypothesis than the t-test.

Performing the Signed-Rank Test

To perform the signed-rank test, we compute the differences between the values in each matched pair. It is likely that some of the differences will be positive and some will be negative. For each negative difference, we take the absolute value, so, for example, we would replace -3 with 3, -5 with 5, and so forth. If the true median of the differences is 0, the sum of the absolute values of the negative ranks should be close to the sum of the positive ranks.

Example 15.6 illustrates the method used to perform the signed rank test.

EXAMPLE 15.6 Performing the signed-rank test

A group of eight individuals with high cholesterol levels participated in an exercise program that was designed to lower cholesterol levels. Cholesterol levels, in milligrams per deciliter, were measured before and after exercise for each individual, with the following results:

	Subject							
	1	**2**	**3**	**4**	**5**	**6**	**7**	**8**
Before	283	299	274	284	248	275	293	277
After	290	281	262	287	253	287	267	271

Can you conclude that the mean cholesterol level after treatment is different from the mean before treatment? Use the $\alpha = 0.05$ level of significance.

Solution

Step 1: State H_0 and H_1: The hypotheses are H_0: $m_d = 0$ and H_1: $m_d \neq 0$, where m_d is the population median of the differences.

Step 2: Choose a significance level: We choose a level of $\alpha = 0.05$.

Step 3: Compute the differences for matched pairs: The following table presents these differences.

	Subject							
	1	**2**	**3**	**4**	**5**	**6**	**7**	**8**
Before	283	299	274	284	248	275	293	277
After	290	281	262	287	253	287	267	271
Difference	−7	18	12	−3	−5	−12	26	6

Step 4: Compute the absolute values of the differences: We do this simply by making the negative differences positive. The following table presents the results.

	Subject							
	1	**2**	**3**	**4**	**5**	**6**	**7**	**8**
Before	283	299	274	284	248	275	293	277
After	290	281	262	287	253	287	267	271
Difference	−7	18	12	−3	−5	−12	26	6
Absolute Value	7	18	12	3	5	12	26	6

Step 5: Rank the absolute values, ignoring any differences of 0: The following table presents the results. There are no differences of 0 to ignore.

	Subject							
	1	**2**	**3**	**4**	**5**	**6**	**7**	**8**
Before	283	299	274	284	248	275	293	277
After	290	281	262	287	253	287	267	271
Difference	−7	18	12	−3	−5	−12	26	6
Absolute Value	7	18	12	3	5	12	26	6
Rank	4	7	5.5	1	2	5.5	8	3

Step 6: Give each rank a plus or minus sign, according to the sign of the difference it corresponds to: These are called the *signed ranks*. The following table presents the results.

	Subject							
	1	**2**	**3**	**4**	**5**	**6**	**7**	**8**
Before	283	299	274	284	248	275	293	277
After	290	281	262	287	253	287	267	271
Difference	−7	18	12	−3	−5	−12	26	6
Absolute Value	7	18	12	3	5	12	26	6
Rank	4	7	5.5	1	2	5.5	8	3
Signed Rank	−4	7	5.5	−1	−2	−5.5	8	3

Step 7: **Compute the sum of the positive ranks and the sum of the negative ranks:** The positive sum is $7 + 5.5 + 8 + 3 = 23.5$. The negative sum is $-4 - 1 - 2 - 5.5 = -12.5$.

Step 8: **Compute the test statistic:** The test statistic S is the sum that is smaller in absolute value. Because 12.5 is smaller than 23.5, the test statistic is $S = 12.5$.

Step 9: **Find the critical value and determine whether to reject H_0:** Critical values are found in Table A.8. For a sample size $n = 8$ and a significance level of $\alpha = 0.05$, the critical value is 4. We reject H_0 if the value of the test statistic is less than or equal to the critical value. Because $12.5 > 4$, we do not reject H_0.

n	$\alpha = 0.10$	$\alpha = 0.05$	$\alpha = 0.02$	$\alpha = 0.01$
⋮	⋮	⋮	⋮	⋮
6	2	1	*	*
7	4	2	0	*
8	6	4	2	0
9	8	6	3	2
10	11	8	5	3
⋮	⋮	⋮	⋮	⋮

Step 10: **State a conclusion:** We cannot conclude that cholesterol levels before and after exercise are different.

In summary, the signed-rank test is performed using the following steps.

Performing the Signed-Rank Test

Step 1: State the null and alternate hypotheses. These are
$H_0: m_d = 0$ and $H_1: m_d \neq 0$.

Step 2: Choose a significance level α.

Step 3: Compute the differences for the matched pairs.

Step 4: Compute the absolute values of the differences. To do this, simply make the negative differences positive.

Step 5: Rank the absolute values, ignoring any differences of 0.

Step 6: Compute the signed ranks. To do this, give each rank a plus or minus sign, according to the sign of the difference it corresponds to.

Step 7: Compute the sum of the positive ranks and the sum of the negative ranks.

Step 8: Compute the test statistic S. The test statistic S is the sum that is smaller in absolute value.

Step 9: Find the critical value and determine whether to reject H_0. Critical values are found in Table A.8. Reject H_0 if the value of the test statistic is less than or equal to the critical value.

Step 10: State a conclusion.

Check Your Understanding

1. Following are the current prices and last year's prices of a gallon of regular gas at a sample of 14 gas stations. Can you conclude that the median price is different now from what it was a year ago? Use the $\alpha = 0.01$ level of significance.

	1	2	3	4	5	6	7	8	9	10	11	12	13	14
Current	3.85	3.88	3.81	3.58	3.48	3.55	3.59	3.80	4.11	3.51	3.86	3.93	3.64	3.54
Last year	3.90	3.79	3.73	3.55	3.42	3.54	3.62	3.78	3.99	3.52	3.85	3.93	3.68	3.50

    a. Compute the value of the test statistic.
    b. Find the critical value.
    c. State a conclusion.

2. Corn was grown on twelve plots of land. Half of each plot was treated with fertilizer A, and the other half was treated with fertilizer B. Yields in bushels per acre were as follows. Can you conclude that the median yield differs between the fertilizers? Use the $\alpha = 0.05$ level of significance.

	1	2	3	4	5	6	7	8	9	10	11	12
Fertilizer A	170	145	139	146	138	132	174	164	151	158	145	153
Fertilizer B	155	128	125	142	133	135	152	154	147	137	146	129

    a. Compute the value of the test statistic.
    b. Find the critical value.
    c. State a conclusion.

Answers are on page 684.

USING TECHNOLOGY

We use Example 15.6 to illustrate the technology steps.

MINITAB

Performing the signed-rank test

Note: MINITAB returns a *P*-value for the test.

Step 1. Enter the data for Sample 1 in column **C1** and the data for Sample 2 in column **C2**. For Example 15.6, we enter the before data into **C1** and the after data into **C2**.

Step 2. Calculate the differences by selecting **Calc**, then **Calculator**. Type **Difference** in the **Store result in variable** field and enter **C2 – C1** in the **Expression** field.

Step 3. Click on **Stat**, then **Nonparametrics**, then **1-Sample Wilcoxon**.

Step 4. Enter **Difference** in the **Variables** field and select the **Test median** option.

Step 5. Enter **C1** in the **First Sample** field and **C2** in the **Second Sample** field. In the **Alternative** field, select the form of the alternate hypothesis. We use **not equal**.

Step 4. Click **OK** (Figure A).

```
Wilcoxon Signed Rank Test: Difference

Test of median = 0.000000 versus median not = 0.000000

                    N
                  for   Wilcoxon              Estimated
            N   Test  Statistic      P      Median
Difference  8     8       12.5   0.484      -4.000
```

Figure A

Exercises 1–2 are the Check Your Understanding exercises located within the section.

Understanding the Concepts

In Exercises 3 and 4, fill in each blank with the appropriate word or phrase.

3. For the signed-rank test, we reject H_0 when the value of the test statistic is _____ than the critical value.

4. The signed-rank test does not require that the population be normal, but is does require that it be _____ .

In Exercises 5 and 6, determine whether the statement is true or false. If the statement is false, rewrite it as a true statement.

5. When performing the signed-rank test, we rank the absolute values of the differences.

6. When performing the signed-rank test, we reject H_0 if the test statistic is greater than or equal to the critical value.

Practicing the Skills

In Exercises 7–10, compute the test statistic and the critical value, and determine whether to reject H_0 at the $\alpha = 0.05$ level.

7.
Sample 1	73	48	38	45	42	32	75	66	56	58	46	54
Sample 2	59	32	22	15	30	32	55	59	49	34	50	33

8.
Sample 1	342	264	263	253	260	183	249	257	190
Sample 2	324	219	248	223	214	144	211	218	175

9.
Sample 1	67	51	54	52	53	51	47	56	52	64	58
Sample 2	75	60	56	44	62	53	55	54	54	55	66

10.
Sample 1	614	567	691	588	620	644	636	543
Sample 2	438	449	751	556	559	573	668	539

Working with the Concepts

11. **Test scores:** A sample of 12 students participated in an online tutoring program in mathematics. Each student took a pretest before the program started, and a posttest after completing the program. Following are the scores.

						Student						
	1	**2**	**3**	**4**	**5**	**6**	**7**	**8**	**9**	**10**	**11**	**12**
Pretest	73	76	45	77	65	44	51	53	78	70	46	79
Posttest	83	89	57	79	75	39	62	62	76	79	39	94

Can you conclude that the median scores differ between pretest and posttest? Use the $\alpha = 0.05$ level of significance.

12. **The rent is due:** To determine whether rents are increasing in a certain town, an economist compared the current monthly rents for eight studio apartments with the rents for those same apartments one year ago. The results were as follows.

				Apartment				
	1	**2**	**3**	**4**	**5**	**6**	**7**	**8**
Now	975	710	810	640	1040	800	820	615
One Year Ago	950	690	800	630	1060	790	820	630

Can you conclude that the median rent now differs from that of a year ago? Use the $\alpha = 0.01$ level of significance.

13. **Traffic:** To determine whether traffic levels differ between the morning and evening rush hours, a traffic engineer counted the number of cars passing through a certain intersection during five-minute periods in both the morning and evening for ten days. The results were as follows.

					Day					
	1	**2**	**3**	**4**	**5**	**6**	**7**	**8**	**9**	**10**
Morning	65	68	44	64	63	38	36	55	78	47
Evening	70	51	59	49	63	50	59	83	81	25

Can you conclude that the median traffic level differs between morning and evening? Use the $\alpha = 0.01$ level of significance.

14. **Exercise and weight:** A sample of nine men participated in a regular exercise program at a local gym. They were weighed both before and after the program. The results were as follows.

					Subject				
	1	**2**	**3**	**4**	**5**	**6**	**7**	**8**	**9**
Before	165	181	234	175	244	174	243	202	170
After	155	176	233	169	231	177	245	185	161

Can you conclude that the median weight differs before and after the exercise program? Use the $\alpha = 0.05$ level of significance.

15. **Balance your checkbook:** A sample of 14 people attended a financial seminar. Each participant was given a 20-point financial literacy quiz both before and after the seminar with the following results.

							Subject							
	1	**2**	**3**	**4**	**5**	**6**	**7**	**8**	**9**	**10**	**11**	**12**	**13**	**14**
Before	3	15	2	17	20	16	4	9	13	4	2	15	3	16
After	6	15	6	18	18	15	11	7	19	9	1	18	5	10

Can you conclude that the median score on the quiz before the seminar differs from the median score after? Use the $\alpha = 0.05$ level of significance.

16. **Hurricane Katrina:** Hurricane Katrina made landfall in August 2005 evacuating a numerous people from the area. The total number of births was observed each month for a year following the hurricane in the counties directly affected by it. These were compared to the number of births in the same months for the previous year.

	Sep.	Oct.	Nov.	Dec.	Jan.	Feb.
Before Katrina	3113	3111	3000	3056	2926	2615
After Katrina	2288	2177	2134	2204	2424	2112

	Mar.	Apr.	May	Jun.	Jul.	Aug.
Before Katrina	2928	2602	2751	2848	2806	2446
After Katrina	2375	2105	2239	2334	2608	2629

Can you conclude that the median number of number of births differed after the hurricane from the median number of births before? Use the $\alpha = 0.05$ level of significance.

Extending the Concepts

Exercise 17 demonstrates that the results of the signed-rank test may be misleading when the assumption of symmetry is violated.

17. Asymmetric differences: Consider the following paired samples and their differences.

Sample 1												
68	59	66	67	64	50	60	55	74	59	66	55	68
58	70	71	54	68	65	82	34	89	97	73	78	

Sample 2												
80	70	76	76	72	57	66	60	78	62	68	56	68
45	56	56	38	51	47	63	14	68	75	50	54	

Differences												
−12	−11	−10	−9	−8	−7	−6	−5	−4	−3	−2	−1	0
13	14		15	16	17	18	19	20	21	22	23	24

a. Verify that the median of the sample differences is equal to 0.

b. Compute the test statistic for the signed-rank test.

c. Do you reject the null hypothesis that the median is equal to 0? Use the $\alpha = 0.05$ level of significance.

Answers to Check Your Understanding Exercises for Section 15.3

1. **a.** $S = 24$ **b.** The critical value is 13.

   c. Do not reject H_0. We cannot conclude that the median price is different today from what it was a year ago.

2. **a.** $S = 3$ **b.** The critical value is 14.

   c. Reject H_0. We conclude that the median yields differ between the two fertilizers.

Chapter 15 Summary

Section 15.1: In this section we presented the sign test, which is a nonparametric alternative to the t-test. The sign test is valid for any continuous population, and does not require the assumption of normality. The major disadvantage of the sign test is that when the population is approximately normal, it is less likely to reject a false null hypothesis than the t-test.

Section 15.2: Wilcoxon's rank-sum test is a nonparametric test used to determine whether two population medians differ. The null hypothesis is that the medians are equal. The rank-sum test is a nonparametric alternative to the two-sample t-test. Its major advantage is that it does not require the assumption of normality. Its major disadvantage is that the null hypothesis can be rejected when the two populations differ in shape, even if their medians are equal.

Section 15.3: In this section we presented the signed-rank test, which is a nonparametric alternative to the t-test for paired differences. Although the signed-rank test does not require that the population of differences be normally distributed, it does require that the distribution be approximately symmetric. Like the sign test, the signed-rank test has the disadvantage that when the population is approximately normal, it is less likely to reject a false null hypothesis than the t-test.

Vocabulary and Notation

nonparametric   668

parametric   668

sign test   668

ranks   674

Wilcoxon rank-sum test   668

Wilcoxon signed-rank test   668

Important Formulas

Test statistic for the sign test:

$$z = \frac{x + 0.5 - n/2}{\sqrt{n}/2} \quad \text{if } n > 25$$

If $n \le 25$, the test statistic is x, the number of times the less frequent sign occurs.

Mean of S, the sum of the ranks for the rank-sum test:

$$\mu_S = \frac{n_1(n_1 + n_2 + 1)}{2}$$

Standard deviation of S, the sum of the ranks for the rank-sum test:

$$\sigma_S = \sqrt{\frac{n_1 n_2 (n_1 + n_2 + 1)}{12}}$$

Test statistic for the rank-sum test:

$$z = \frac{S - \mu_S}{\sigma_S}$$

Chapter Quiz

1. Following is a sample from a population with median m. Use the sign test to test H_0: $m = 30$ versus H_1: $m \neq 30$.

| 50 | 38 | 23 | 29 | 57 | 26 | 53 | 42 | 60 | 23 |

    a. What is the value of the test statistic?
    b. Is H_0 rejected at the $\alpha = 0.05$ level?
    c. Is H_0 rejected at the $\alpha = 0.01$ level?

2. Following is a sample from a population with median m. Use the sign test to test H_0: $m = 20$ versus H_1: $m > 20$.

| 38 | 47 | 55 | 38 | 40 | 43 | 36 | 49 | 43 | 55 | 43 | 47 | 38 | 50 | 38 | 43 |

    a. What is the value of the test statistic?
    b. Is H_0 rejected at the $\alpha = 0.05$ level?
    c. Is H_0 rejected at the $\alpha = 0.01$ level?

3. The sign test is performed to test H_0: $m = 42$ versus H_1: $m \neq 42$. There are 19 plus signs, 5 minus signs, and 4 zeros.
    a. What is the value of the test statistic?
    b. Is H_0 rejected at the $\alpha = 0.05$ level?
    c. Is H_0 rejected at the $\alpha = 0.01$ level?

4. The sign test is performed to test H_0: $m = 250$ versus H_1: $m < 250$. There are 15 plus signs, 36 minus signs, and 6 zeros.
    a. What is the value of the test statistic?
    b. Is H_0 rejected at the $\alpha = 0.05$ level?
    c. Is H_0 rejected at the $\alpha = 0.01$ level?

Exercises 5–7 present sample sizes and the sum of ranks for the rank-sum test. Compute μ_S, σ_S, and the value of the test statistic z. Then find the P-value.

5. $n_1 = 20$, $n_2 = 30$, $S = 400$, H_1: $m_1 < m_2$.

6. $n_1 = 25$, $n_2 = 32$, $S = 850$, H_1: $m_1 \neq m_2$.

7. $n_1 = 15$, $n_2 = 18$, $S = 280$, H_1: $m_1 > m_2$.

In Exercises 8–10, use the signed-rank test to test the null hypothesis that the population medians are equal. Compute the test statistic, the critical value, and determine whether to reject H_0 at the $\alpha = 0.05$ level.

8.

| Sample 1 | 61 | 37 | 36 | 28 | 59 | 30 | 16 | 52 | 31 | 23 | 32 | 15 | 17 | 61 | 62 | 41 |
| Sample 2 | 56 | 35 | 48 | 18 | 73 | 19 | 27 | 57 | 47 | 37 | 53 | 36 | 15 | 74 | 55 | 27 |

9.

| Sample 1 | 159 | 193 | 213 | 295 | 160 | 156 | 181 | 273 | 272 | 276 | 239 | 250 | 209 | 279 |
| Sample 2 | 187 | 217 | 237 | 317 | 180 | 184 | 206 | 287 | 301 | 287 | 254 | 270 | 227 | 304 |

10.

| Sample 1 | 73 | 74 | 64 | 75 | 61 | 56 | 81 | 80 | 64 | 70 | 69 | 49 |
| Sample 2 | 72 | 66 | 55 | 72 | 61 | 54 | 78 | 75 | 62 | 65 | 63 | 48 |

Review Exercises

1. Freshman weights: Do college students tend to gain or lose weight during their freshman year? Following are a sample of weights of 15 college freshmen at the beginning and at the end of their freshman year.

	Student														
	1	**2**	**3**	**4**	**5**	**6**	**7**	**8**	**9**	**10**	**11**	**12**	**13**	**14**	**15**
Beginning	174	150	148	190	161	162	186	181	158	118	124	189	103	149	117
End	193	161	153	194	153	172	177	173	164	111	139	204	115	143	127

Can you conclude that the median weight at the end of the year differs from the median at the beginning of the year? Use the $\alpha = 0.05$ level of significance.

2. **Keep cool:** Following are the prices, in dollars, of a sample of 20 air conditioners.

377	559	430	490	352	343	500	203	223	239
326	503	491	194	330	381	337	433	421	287

Can you conclude that the median price differs from $300? Use the $\alpha = 0.01$ level of significance.

3. **Slow Mondays:** The owner of a restaurant believes that she serves the same number of customers on Mondays as on Tuesdays. For twelve weeks she counts the number of customers who come in on Monday and on Tuesday, with the following results.

	Week											
	1	**2**	**3**	**4**	**5**	**6**	**7**	**8**	**9**	**10**	**11**	**12**
Monday	163	122	218	137	131	157	140	169	154	215	212	125
Tuesday	173	113	215	139	149	154	159	161	162	222	214	133

Can you conclude that the median number of customers differs between Mondays and Tuesdays? Use the $\alpha = 0.01$ level of significance.

4. **Exercise and blood pressure:** Systolic blood pressure, in millimeters of mercury, was measured for a sample of 15 subjects who reported exercising regularly and for a sample of 12 patients who reported having a sedentary lifestyle. The results were as follows.

Exercise														
127	107	105	140	107	101	122	135	127	108	115	118	139	106	134

Sedentary											
154	142	122	134	110	150	149	146	118	135	106	126

Can you conclude that the median systolic blood pressure is less for those who exercise than for those who are sedentary? Use the $\alpha = 0.05$ level of significance.

5. **Scoring runs:** Following are the number of runs scored by both teams in a sample of 20 major league baseball games played in 2012.

8	6	11	3	5	1	8	13	16	6
4	10	17	16	22	15	10	8	10	2

Can you conclude that the median is different from 8? Use the $\alpha = 0.05$ level of significance.

6. **Recovery times:** The number of days spent in the hospital was determined for 12 patients who underwent coronary bypass surgery.

7	10	4	10	6	8	5	8	8	3	4	6

Can you conclude that the median is less than 5 days? Use the $\alpha = 0.01$ level of significance.

7. **Boys and girls:** Following are weights, in pounds, for samples of 16 newborn boys and 14 newborn girls.

Boys															
8.3	7.8	8.2	8.0	8.8	8.4	8.5	7.9	7.5	7.0	6.6	6.4	5.9	6.7	7.4	7.6

Girls													
7.7	8.8	8.9	6.2	6.0	7.9	5.8	7.3	7.4	8.5	7.2	6.9	6.2	8.1

Can you conclude that the median weight differs between boys and girls? Use the $\alpha = 0.01$ level of significance.

8. **Morning commute:** A woman who has moved into a new house is trying to determine which of two routes to work has the shorter driving time. Times in minutes for 12 trips on route A and 10 trips on route B are presented below.

Route A											
15.9	16.3	17.0	15.7	16.3	16.7	16.2	16.9	16.0	16.5	16.7	16.4

Route B									
17.0	16.6	16.9	19.8	17.4	20.0	17.2	16.1	17.6	16.7

Can you conclude that the median time differs between the two routes? Use the $\alpha = 0.05$ level of significance.

9. **Tire tread:** Two gauges that measure tire tread depth were compared. Twelve different locations on a tire were measured once by each gauge. The results, in millimeters, were as follows.

	Location											
	1	**2**	**3**	**4**	**5**	**6**	**7**	**8**	**9**	**10**	**11**	**12**
Gauge 1	3.95	3.23	3.60	3.48	3.89	3.76	3.56	3.01	3.82	3.44	3.58	3.78
Gauge 2	3.80	3.30	3.59	3.61	3.88	3.73	3.45	3.02	3.77	3.49	3.42	3.66

Can you conclude that the median measurement differs between the gauges? Use the $\alpha = 0.05$ level of significance.

10. How much is in that can? A machine that fills beverage cans is supposed to put 12 ounces of beverage in each can. Following are the amounts measured in a sample of 14 cans.

12.00	11.97	11.93	12.02	11.95	11.91	12.05	11.95	11.99	12.04	11.97	12.05	11.98	12.04

Can you conclude that the median amount differs from 12 ounces? Use the $\alpha = 0.01$ level of significance.

Write About It

1. Would it be possible to use the number of times the more frequent sign appears as the test statistic for the sign test? Explain why you would reject the null hypothesis when the value of this test statistic is greater than a critical value.

2. Describe the assumptions necessary for the rank-sum test. How do they differ from the assumptions necessary for the two-sample t-test?

3. Describe a situation in which the signed-rank test would be preferable to the paired t-test. Describe a situation in which the paired t-test would be preferable to the signed-rank test.

Case Study: Diversity of Microorganism Populations

Microorganisms play a crucial role in wastewater treatment by transforming solid organic matter into carbon dioxide and water. In a wastewater treatment system, the solids retention time is average amount of time that solid organic matter resides in the treatment tank. A study was performed at the Colorado School of Mines to determine the relationship between the solids retention time and the diversity of the microorganism population. It was thought, because microorganisms feed on the solid matter, that longer retention times would provide more opportunity for genetic mutations to occur, resulting in a greater diversity of microorganisms.

Following are counts of the numbers of species of microorganism for retention times of 3 days, 12 days, and 30 days.

3 Days								
282	273	238	240	241	253	230	266	253

12 Days											
266	299	287	289	272	284	292	279	274	306	291	272

30 Days																		
293	271	293	296	271	292	265	262	278	272	246	278	276	283	273	286	303	289

1. Can you conclude at the $\alpha = 0.05$ level that the median number of species is greater for a retention time of 12 days than for a retention time of 3 days? Compute the appropriate test statistic, find the P-value, and state a conclusion.
2. Can you conclude at the $\alpha = 0.05$ level that the median number of species is greater for a retention time of 30 days than for a retention time of 3 days? Compute the appropriate test statistic, find the P-value, and state a conclusion.
3. Can you conclude at the $\alpha = 0.05$ level that the median number of species for a retention time of 30 days differs from the median for a retention time of 12 days? Compute the appropriate test statistic, find the P-value, and state a conclusion.
4. Based on the results of the tests performed in Exercises 1–3, which of the following is the best conclusion?
    i. We can conclude that the median number of species increases as the solids retention time increases from 3 to 12 and from 12 to 30 days.
    ii. We cannot conclude that the median number of species increases as the solids retention time increases.
    iii. We can conclude that the median number of species increases as the solids retention time increases from 3 to 12 days, but it may not continue to increase as the retention time increases to 30 days.
    iv. We can conclude that the median number of species increases as the solid retention time increases from 12 to 30 days, but it may not increase as the retention time increases from 3 to 12 days.

Tables

Table A.1 Binomial Probabilities

$$P(x) = \frac{n!}{x!(n-x)!} p^x (1-p)^{(n-x)}$$

								p							
n	x	0.05	0.10	0.20	0.25	0.30	0.40	0.50	0.60	0.70	0.75	0.80	0.90	0.95
2	0	0.903	0.810	0.640	0.563	0.490	0.360	0.250	0.160	0.090	0.063	0.040	0.010	0.003
	1	0.095	0.180	0.320	0.375	0.420	0.480	0.500	0.480	0.420	0.375	0.320	0.180	0.095
	2	0.003	0.010	0.040	0.063	0.090	0.160	0.250	0.360	0.490	0.563	0.640	0.810	0.903
3	0	0.857	0.729	0.512	0.422	0.343	0.216	0.125	0.064	0.027	0.016	0.008	0.001	0.000+
	1	0.135	0.243	0.384	0.422	0.441	0.432	0.375	0.288	0.189	0.141	0.096	0.027	0.007
	2	0.007	0.027	0.096	0.141	0.189	0.288	0.375	0.432	0.441	0.422	0.384	0.243	0.135
	3	0.000+	0.001	0.008	0.016	0.027	0.064	0.125	0.216	0.343	0.422	0.512	0.729	0.857
4	0	0.815	0.656	0.410	0.316	0.240	0.130	0.063	0.026	0.008	0.004	0.002	0.000+	0.000+
	1	0.171	0.292	0.410	0.422	0.412	0.346	0.250	0.154	0.076	0.047	0.026	0.004	0.000+
	2	0.014	0.049	0.154	0.211	0.265	0.346	0.375	0.346	0.265	0.211	0.154	0.049	0.014
	3	0.000+	0.004	0.026	0.047	0.076	0.154	0.250	0.346	0.412	0.422	0.410	0.292	0.171
	4	0.000+	0.000+	0.002	0.004	0.008	0.026	0.063	0.130	0.240	0.316	0.410	0.656	0.815
5	0	0.774	0.590	0.328	0.237	0.168	0.078	0.031	0.010	0.002	0.001	0.000+	0.000+	0.000+
	1	0.204	0.328	0.410	0.396	0.360	0.259	0.156	0.077	0.028	0.015	0.006	0.000+	0.000+
	2	0.021	0.073	0.205	0.264	0.309	0.346	0.313	0.230	0.132	0.088	0.051	0.008	0.001
	3	0.001	0.008	0.051	0.088	0.132	0.230	0.313	0.346	0.309	0.264	0.205	0.073	0.021
	4	0.000+	0.000+	0.006	0.015	0.028	0.077	0.156	0.259	0.360	0.396	0.410	0.328	0.204
	5	0.000+	0.000+	0.000+	0.001	0.002	0.010	0.031	0.078	0.168	0.237	0.328	0.590	0.774
6	0	0.735	0.531	0.262	0.178	0.118	0.047	0.016	0.004	0.001	0.000+	0.000+	0.000+	0.000+
	1	0.232	0.354	0.393	0.356	0.303	0.187	0.094	0.037	0.010	0.004	0.002	0.000+	0.000+
	2	0.031	0.098	0.246	0.297	0.324	0.311	0.234	0.138	0.060	0.033	0.015	0.001	0.000+
	3	0.002	0.015	0.082	0.132	0.185	0.276	0.313	0.276	0.185	0.132	0.082	0.015	0.002
	4	0.000+	0.001	0.015	0.033	0.060	0.138	0.234	0.311	0.324	0.297	0.246	0.098	0.031
	5	0.000+	0.000+	0.002	0.004	0.010	0.037	0.094	0.187	0.303	0.356	0.393	0.354	0.232
	6	0.000+	0.000+	0.000+	0.000+	0.001	0.004	0.016	0.047	0.118	0.178	0.262	0.531	0.735
7	0	0.698	0.478	0.210	0.133	0.082	0.028	0.008	0.002	0.000+	0.000+	0.000+	0.000+	0.000+
	1	0.257	0.372	0.367	0.311	0.247	0.131	0.055	0.017	0.004	0.001	0.000+	0.000+	0.000+
	2	0.041	0.124	0.275	0.311	0.318	0.261	0.164	0.077	0.025	0.012	0.004	0.000+	0.000+
	3	0.004	0.023	0.115	0.173	0.227	0.290	0.273	0.194	0.097	0.058	0.029	0.003	0.000+
	4	0.000+	0.003	0.029	0.058	0.097	0.194	0.273	0.290	0.227	0.173	0.115	0.023	0.004
	5	0.000+	0.000+	0.004	0.012	0.025	0.077	0.164	0.261	0.318	0.311	0.275	0.124	0.041
	6	0.000+	0.000+	0.000+	0.001	0.004	0.017	0.055	0.131	0.247	0.311	0.367	0.372	0.257
	7	0.000+	0.000+	0.000+	0.000+	0.000+	0.002	0.008	0.028	0.082	0.133	0.210	0.478	0.698
8	0	0.663	0.430	0.168	0.100	0.058	0.017	0.004	0.001	0.000+	0.000+	0.000+	0.000+	0.000+
	1	0.279	0.383	0.336	0.267	0.198	0.090	0.031	0.008	0.001	0.000+	0.000+	0.000+	0.000+
	2	0.051	0.149	0.294	0.311	0.296	0.209	0.109	0.041	0.010	0.004	0.001	0.000+	0.000+
	3	0.005	0.033	0.147	0.208	0.254	0.279	0.219	0.124	0.047	0.023	0.009	0.000+	0.000+
	4	0.000+	0.005	0.046	0.087	0.136	0.232	0.273	0.232	0.136	0.087	0.046	0.005	0.000+
	5	0.000+	0.000+	0.009	0.023	0.047	0.124	0.219	0.279	0.254	0.208	0.147	0.033	0.005
	6	0.000+	0.000+	0.001	0.004	0.010	0.041	0.109	0.209	0.296	0.311	0.294	0.149	0.051
	7	0.000+	0.000+	0.000+	0.000+	0.001	0.008	0.031	0.090	0.198	0.267	0.336	0.383	0.279
	8	0.000+	0.000+	0.000+	0.000+	0.000+	0.001	0.004	0.017	0.058	0.100	0.168	0.430	0.663
9	0	0.630	0.387	0.134	0.075	0.040	0.010	0.002	0.000+	0.000+	0.000+	0.000+	0.000+	0.000+
	1	0.299	0.387	0.302	0.225	0.156	0.060	0.018	0.004	0.000+	0.000+	0.000+	0.000+	0.000+
	2	0.063	0.172	0.302	0.300	0.267	0.161	0.070	0.021	0.004	0.001	0.000+	0.000+	0.000+
	3	0.008	0.045	0.176	0.234	0.267	0.251	0.164	0.074	0.021	0.009	0.003	0.000+	0.000+
	4	0.001	0.007	0.066	0.117	0.172	0.251	0.246	0.167	0.074	0.039	0.017	0.001	0.000+
	5	0.000+	0.001	0.017	0.039	0.074	0.167	0.246	0.251	0.172	0.117	0.066	0.007	0.001
	6	0.000+	0.000+	0.003	0.009	0.021	0.074	0.164	0.251	0.267	0.234	0.176	0.045	0.008
	7	0.000+	0.000+	0.000+	0.001	0.004	0.021	0.070	0.161	0.267	0.300	0.302	0.172	0.063
	8	0.000+	0.000+	0.000+	0.000+	0.000+	0.004	0.018	0.060	0.156	0.225	0.302	0.387	0.299
	9	0.000+	0.000+	0.000+	0.000+	0.000+	0.000+	0.002	0.010	0.040	0.075	0.134	0.387	0.630

A value of 0.000+ indicates that the probability is 0.000 when rounded to three decimal places. The actual probability is slightly greater than 0.

Table A.1 Binomial Probabilities (continued)

n	x	0.05	0.10	0.20	0.25	0.30	0.40	0.50	0.60	0.70	0.75	0.80	0.90	0.95
10	0	0.599	0.349	0.107	0.056	0.028	0.006	0.001	0.000+	0.000+	0.000+	0.000+	0.000+	0.000+
	1	0.315	0.387	0.268	0.188	0.121	0.040	0.010	0.002	0.000+	0.000+	0.000+	0.000+	0.000+
	2	0.075	0.194	0.302	0.282	0.233	0.121	0.044	0.011	0.001	0.000+	0.000+	0.000+	0.000+
	3	0.010	0.057	0.201	0.250	0.267	0.215	0.117	0.042	0.009	0.003	0.001	0.000+	0.000+
	4	0.001	0.011	0.088	0.146	0.200	0.251	0.205	0.111	0.037	0.016	0.006	0.000+	0.000+
	5	0.000+	0.001	0.026	0.058	0.103	0.201	0.246	0.201	0.103	0.058	0.026	0.001	0.000+
	6	0.000+	0.000+	0.006	0.016	0.037	0.111	0.205	0.251	0.200	0.146	0.088	0.011	0.001
	7	0.000+	0.000+	0.001	0.003	0.009	0.042	0.117	0.215	0.267	0.250	0.201	0.057	0.010
	8	0.000+	0.000+	0.000+	0.000+	0.001	0.011	0.044	0.121	0.233	0.282	0.302	0.194	0.075
	9	0.000+	0.000+	0.000+	0.000+	0.000+	0.002	0.010	0.040	0.121	0.188	0.268	0.387	0.315
	10	0.000+	0.000+	0.000+	0.000+	0.000+	0.000+	0.001	0.006	0.028	0.056	0.107	0.349	0.599
11	0	0.569	0.314	0.086	0.042	0.020	0.004	0.000+	0.000+	0.000+	0.000+	0.000+	0.000+	0.000+
	1	0.329	0.384	0.236	0.155	0.093	0.027	0.005	0.001	0.000+	0.000+	0.000+	0.000+	0.000+
	2	0.087	0.213	0.295	0.258	0.200	0.089	0.027	0.005	0.001	0.000+	0.000+	0.000+	0.000+
	3	0.014	0.071	0.221	0.258	0.257	0.177	0.081	0.023	0.004	0.001	0.000+	0.000+	0.000+
	4	0.001	0.016	0.111	0.172	0.220	0.236	0.161	0.070	0.017	0.006	0.002	0.000+	0.000+
	5	0.000+	0.002	0.039	0.080	0.132	0.221	0.226	0.147	0.057	0.027	0.010	0.000+	0.000+
	6	0.000+	0.000+	0.010	0.027	0.057	0.147	0.226	0.221	0.132	0.080	0.039	0.002	0.000+
	7	0.000+	0.000+	0.002	0.006	0.017	0.070	0.161	0.236	0.220	0.172	0.111	0.016	0.001
	8	0.000+	0.000+	0.000+	0.001	0.004	0.023	0.081	0.177	0.257	0.258	0.221	0.071	0.014
	9	0.000+	0.000+	0.000+	0.000+	0.001	0.005	0.027	0.089	0.200	0.258	0.295	0.213	0.087
	10	0.000+	0.000+	0.000+	0.000+	0.000+	0.001	0.005	0.027	0.093	0.155	0.236	0.384	0.329
	11	0.000+	0.000+	0.000+	0.000+	0.000+	0.000+	0.000+	0.004	0.020	0.042	0.086	0.314	0.569
12	0	0.540	0.282	0.069	0.032	0.014	0.002	0.000+	0.000+	0.000+	0.000+	0.000+	0.000+	0.000+
	1	0.341	0.377	0.206	0.127	0.071	0.017	0.003	0.000+	0.000+	0.000+	0.000+	0.000+	0.000+
	2	0.099	0.230	0.283	0.232	0.168	0.064	0.016	0.002	0.000+	0.000+	0.000+	0.000+	0.000+
	3	0.017	0.085	0.236	0.258	0.240	0.142	0.054	0.012	0.001	0.000+	0.000+	0.000+	0.000+
	4	0.002	0.021	0.133	0.194	0.231	0.213	0.121	0.042	0.008	0.002	0.001	0.000+	0.000+
	5	0.000+	0.004	0.053	0.103	0.158	0.227	0.193	0.101	0.029	0.011	0.003	0.000+	0.000+
	6	0.000+	0.000+	0.016	0.040	0.079	0.177	0.226	0.177	0.079	0.040	0.016	0.000+	0.000+
	7	0.000+	0.000+	0.003	0.011	0.029	0.101	0.193	0.227	0.158	0.103	0.053	0.004	0.000+
	8	0.000+	0.000+	0.001	0.002	0.008	0.042	0.121	0.213	0.231	0.194	0.133	0.021	0.002
	9	0.000+	0.000+	0.000+	0.000+	0.001	0.012	0.054	0.142	0.240	0.258	0.236	0.085	0.017
	10	0.000+	0.000+	0.000+	0.000+	0.000+	0.002	0.016	0.064	0.168	0.232	0.283	0.230	0.099
	11	0.000+	0.000+	0.000+	0.000+	0.000+	0.000+	0.003	0.017	0.071	0.127	0.206	0.377	0.341
	12	0.000+	0.000+	0.000+	0.000+	0.000+	0.000+	0.000+	0.002	0.014	0.032	0.069	0.282	0.540
13	0	0.513	0.254	0.055	0.024	0.010	0.001	0.000+	0.000+	0.000+	0.000+	0.000+	0.000+	0.000+
	1	0.351	0.367	0.179	0.103	0.054	0.011	0.002	0.000+	0.000+	0.000+	0.000+	0.000+	0.000+
	2	0.111	0.245	0.268	0.206	0.139	0.045	0.010	0.001	0.000+	0.000+	0.000+	0.000+	0.000+
	3	0.021	0.100	0.246	0.252	0.218	0.111	0.035	0.006	0.001	0.000+	0.000+	0.000+	0.000+
	4	0.003	0.028	0.154	0.210	0.234	0.184	0.087	0.024	0.003	0.001	0.000+	0.000+	0.000+
	5	0.000+	0.006	0.069	0.126	0.180	0.221	0.157	0.066	0.014	0.005	0.001	0.000+	0.000+
	6	0.000+	0.001	0.023	0.056	0.103	0.197	0.209	0.131	0.044	0.019	0.006	0.000+	0.000+
	7	0.000+	0.000+	0.006	0.019	0.044	0.131	0.209	0.197	0.103	0.056	0.023	0.001	0.000+
	8	0.000+	0.000+	0.001	0.005	0.014	0.066	0.157	0.221	0.180	0.126	0.069	0.006	0.000+
	9	0.000+	0.000+	0.000+	0.001	0.003	0.024	0.087	0.184	0.234	0.210	0.154	0.028	0.003
	10	0.000+	0.000+	0.000+	0.000+	0.001	0.006	0.035	0.111	0.218	0.252	0.246	0.100	0.021
	11	0.000+	0.000+	0.000+	0.000+	0.000+	0.001	0.010	0.045	0.139	0.206	0.268	0.245	0.111
	12	0.000+	0.000+	0.000+	0.000+	0.000+	0.000+	0.002	0.011	0.054	0.103	0.179	0.367	0.351
	13	0.000+	0.000+	0.000+	0.000+	0.000+	0.000+	0.000+	0.001	0.010	0.024	0.055	0.254	0.513
14	0	0.488	0.229	0.044	0.018	0.007	0.001	0.000+	0.000+	0.000+	0.000+	0.000+	0.000+	0.000+
	1	0.359	0.356	0.154	0.083	0.041	0.007	0.001	0.000+	0.000+	0.000+	0.000+	0.000+	0.000+
	2	0.123	0.257	0.250	0.180	0.113	0.032	0.006	0.001	0.000+	0.000+	0.000+	0.000+	0.000+
	3	0.026	0.114	0.250	0.240	0.194	0.085	0.022	0.003	0.000+	0.000+	0.000+	0.000+	0.000+
	4	0.004	0.035	0.172	0.220	0.229	0.155	0.061	0.014	0.001	0.000+	0.000+	0.000+	0.000+
	5	0.000+	0.008	0.086	0.147	0.196	0.207	0.122	0.041	0.007	0.002	0.000+	0.000+	0.000+
	6	0.000+	0.001	0.032	0.073	0.126	0.207	0.183	0.092	0.023	0.008	0.002	0.000+	0.000+
	7	0.000+	0.000+	0.009	0.028	0.062	0.157	0.209	0.157	0.062	0.028	0.009	0.000+	0.000+
	8	0.000+	0.000+	0.002	0.008	0.023	0.092	0.183	0.207	0.126	0.073	0.032	0.001	0.000+
	9	0.000+	0.000+	0.000+	0.002	0.007	0.041	0.122	0.207	0.196	0.147	0.086	0.008	0.000+
	10	0.000+	0.000+	0.000+	0.000+	0.001	0.014	0.061	0.155	0.229	0.220	0.172	0.035	0.004
	11	0.000+	0.000+	0.000+	0.000+	0.000+	0.003	0.022	0.085	0.194	0.240	0.250	0.114	0.026
	12	0.000+	0.000+	0.000+	0.000+	0.000+	0.001	0.006	0.032	0.113	0.180	0.250	0.257	0.123
	13	0.000+	0.000+	0.000+	0.000+	0.000+	0.000+	0.001	0.007	0.041	0.083	0.154	0.356	0.359
	14	0.000+	0.000+	0.000+	0.000+	0.000+	0.000+	0.000+	0.001	0.007	0.018	0.044	0.229	0.488

A value of 0.000+ indicates that the probability is 0.000 when rounded to three decimal places. The actual probability is slightly greater than 0.

Table A.1 Binomial Probabilities (continued)

n	x	0.05	0.10	0.20	0.25	0.30	0.40	0.50	0.60	0.70	0.75	0.80	0.90	0.95
15	0	0.463	0.206	0.035	0.013	0.005	0.000+	0.000+	0.000+	0.000+	0.000+	0.000+	0.000+	0.000+
	1	0.366	0.343	0.132	0.067	0.031	0.005	0.000+	0.000+	0.000+	0.000+	0.000+	0.000+	0.000+
	2	0.135	0.267	0.231	0.156	0.092	0.022	0.003	0.000+	0.000+	0.000+	0.000+	0.000+	0.000+
	3	0.031	0.129	0.250	0.225	0.170	0.063	0.014	0.002	0.000+	0.000+	0.000+	0.000+	0.000+
	4	0.005	0.043	0.188	0.225	0.219	0.127	0.042	0.007	0.001	0.000+	0.000+	0.000+	0.000+
	5	0.001	0.010	0.103	0.165	0.206	0.186	0.092	0.024	0.003	0.001	0.000+	0.000+	0.000+
	6	0.000+	0.002	0.043	0.092	0.147	0.207	0.153	0.061	0.012	0.003	0.001	0.000+	0.000+
	7	0.000+	0.000+	0.014	0.039	0.081	0.177	0.196	0.118	0.035	0.013	0.003	0.000+	0.000+
	8	0.000+	0.000+	0.003	0.013	0.035	0.118	0.196	0.177	0.081	0.039	0.014	0.000+	0.000+
	9	0.000+	0.000+	0.001	0.003	0.012	0.061	0.153	0.207	0.147	0.092	0.043	0.002	0.000+
	10	0.000+	0.000+	0.000+	0.001	0.003	0.024	0.092	0.186	0.206	0.165	0.103	0.010	0.001
	11	0.000+	0.000+	0.000+	0.000+	0.001	0.007	0.042	0.127	0.219	0.225	0.188	0.043	0.005
	12	0.000+	0.000+	0.000+	0.000+	0.000+	0.002	0.014	0.063	0.170	0.225	0.250	0.129	0.031
	13	0.000+	0.000+	0.000+	0.000+	0.000+	0.000+	0.003	0.022	0.092	0.156	0.231	0.267	0.135
	14	0.000+	0.000+	0.000+	0.000+	0.000+	0.000+	0.000+	0.005	0.031	0.067	0.132	0.343	0.366
	15	0.000+	0.000+	0.000+	0.000+	0.000+	0.000+	0.000+	0.000+	0.005	0.013	0.035	0.206	0.463
16	0	0.440	0.185	0.028	0.010	0.003	0.000+	0.000+	0.000+	0.000+	0.000+	0.000+	0.000+	0.000+
	1	0.371	0.329	0.113	0.053	0.023	0.003	0.000+	0.000+	0.000+	0.000+	0.000+	0.000+	0.000+
	2	0.146	0.275	0.211	0.134	0.073	0.015	0.002	0.000+	0.000+	0.000+	0.000+	0.000+	0.000+
	3	0.036	0.142	0.246	0.208	0.146	0.047	0.009	0.001	0.000+	0.000+	0.000+	0.000+	0.000+
	4	0.006	0.051	0.200	0.225	0.204	0.101	0.028	0.004	0.000+	0.000+	0.000+	0.000+	0.000+
	5	0.001	0.014	0.120	0.180	0.210	0.162	0.067	0.014	0.001	0.000+	0.000+	0.000+	0.000+
	6	0.000+	0.003	0.055	0.110	0.165	0.198	0.122	0.039	0.006	0.001	0.000+	0.000+	0.000+
	7	0.000+	0.000+	0.020	0.052	0.101	0.189	0.175	0.084	0.019	0.006	0.001	0.000+	0.000+
	8	0.000+	0.000+	0.006	0.020	0.049	0.142	0.196	0.142	0.049	0.020	0.006	0.000+	0.000+
	9	0.000+	0.000+	0.001	0.006	0.019	0.084	0.175	0.189	0.101	0.052	0.020	0.000+	0.000+
	10	0.000+	0.000+	0.000+	0.001	0.006	0.039	0.122	0.198	0.165	0.110	0.055	0.003	0.000+
	11	0.000+	0.000+	0.000+	0.000+	0.001	0.014	0.067	0.162	0.210	0.180	0.120	0.014	0.001
	12	0.000+	0.000+	0.000+	0.000+	0.000+	0.004	0.028	0.101	0.204	0.225	0.200	0.051	0.006
	13	0.000+	0.000+	0.000+	0.000+	0.000+	0.001	0.009	0.047	0.146	0.208	0.246	0.142	0.036
	14	0.000+	0.000+	0.000+	0.000+	0.000+	0.000+	0.002	0.015	0.073	0.134	0.211	0.275	0.146
	15	0.000+	0.000+	0.000+	0.000+	0.000+	0.000+	0.000+	0.003	0.023	0.053	0.113	0.329	0.371
	16	0.000+	0.000+	0.000+	0.000+	0.000+	0.000+	0.000+	0.000+	0.003	0.010	0.028	0.185	0.440
17	0	0.418	0.167	0.023	0.008	0.002	0.000+	0.000+	0.000+	0.000+	0.000+	0.000+	0.000+	0.000+
	1	0.374	0.315	0.096	0.043	0.017	0.002	0.000+	0.000+	0.000+	0.000+	0.000+	0.000+	0.000+
	2	0.158	0.280	0.191	0.114	0.058	0.010	0.001	0.000+	0.000+	0.000+	0.000+	0.000+	0.000+
	3	0.041	0.156	0.239	0.189	0.125	0.034	0.005	0.000+	0.000+	0.000+	0.000+	0.000+	0.000+
	4	0.008	0.060	0.209	0.221	0.187	0.080	0.018	0.002	0.000+	0.000+	0.000+	0.000+	0.000+
	5	0.001	0.017	0.136	0.191	0.208	0.138	0.047	0.008	0.001	0.000+	0.000+	0.000+	0.000+
	6	0.000+	0.004	0.068	0.128	0.178	0.184	0.094	0.024	0.003	0.001	0.000+	0.000+	0.000+
	7	0.000+	0.001	0.027	0.067	0.120	0.193	0.148	0.057	0.009	0.002	0.000+	0.000+	0.000+
	8	0.000+	0.000+	0.008	0.028	0.064	0.161	0.185	0.107	0.028	0.009	0.002	0.000+	0.000+
	9	0.000+	0.000+	0.002	0.009	0.028	0.107	0.185	0.161	0.064	0.028	0.008	0.000+	0.000+
	10	0.000+	0.000+	0.000+	0.002	0.009	0.057	0.148	0.193	0.120	0.067	0.027	0.001	0.000+
	11	0.000+	0.000+	0.000+	0.001	0.003	0.024	0.094	0.184	0.178	0.128	0.068	0.004	0.000+
	12	0.000+	0.000+	0.000+	0.000+	0.001	0.008	0.047	0.138	0.208	0.191	0.136	0.017	0.001
	13	0.000+	0.000+	0.000+	0.000+	0.000+	0.002	0.018	0.080	0.187	0.221	0.209	0.060	0.008
	14	0.000+	0.000+	0.000+	0.000+	0.000+	0.000+	0.005	0.034	0.125	0.189	0.239	0.156	0.041
	15	0.000+	0.000+	0.000+	0.000+	0.000+	0.000+	0.001	0.010	0.058	0.114	0.191	0.280	0.158
	16	0.000+	0.000+	0.000+	0.000+	0.000+	0.000+	0.000+	0.002	0.017	0.043	0.096	0.315	0.374
	17	0.000+	0.000+	0.000+	0.000+	0.000+	0.000+	0.000+	0.000+	0.002	0.008	0.023	0.167	0.418

A value of 0.000+ indicates that the probability is 0.000 when rounded to three decimal places. The actual probability is slightly greater than 0.

Table A.1 Binomial Probabilities (continued)

							p							
n	x	0.05	0.10	0.20	0.25	0.30	0.40	0.50	0.60	0.70	0.75	0.80	0.90	0.95
18	0	0.397	0.150	0.018	0.006	0.002	0.000+	0.000+	0.000+	0.000+	0.000+	0.000+	0.000+	0.000+
	1	0.376	0.300	0.081	0.034	0.013	0.001	0.000+	0.000+	0.000+	0.000+	0.000+	0.000+	0.000+
	2	0.168	0.284	0.172	0.096	0.046	0.007	0.001	0.000+	0.000+	0.000+	0.000+	0.000+	0.000+
	3	0.047	0.168	0.230	0.170	0.105	0.025	0.003	0.000+	0.000+	0.000+	0.000+	0.000+	0.000+
	4	0.009	0.070	0.215	0.213	0.168	0.061	0.012	0.001	0.000+	0.000+	0.000+	0.000+	0.000+
	5	0.001	0.022	0.151	0.199	0.202	0.115	0.033	0.004	0.000+	0.000+	0.000+	0.000+	0.000+
	6	0.000+	0.005	0.082	0.144	0.187	0.166	0.071	0.015	0.001	0.000+	0.000+	0.000+	0.000+
	7	0.000+	0.001	0.035	0.082	0.138	0.189	0.121	0.037	0.005	0.001	0.000+	0.000+	0.000+
	8	0.000+	0.000+	0.012	0.038	0.081	0.173	0.167	0.077	0.015	0.004	0.001	0.000+	0.000+
	9	0.000+	0.000+	0.003	0.014	0.039	0.128	0.185	0.128	0.039	0.014	0.003	0.000+	0.000+
	10	0.000+	0.000+	0.001	0.004	0.015	0.077	0.167	0.173	0.081	0.038	0.012	0.000+	0.000+
	11	0.000+	0.000+	0.000+	0.001	0.005	0.037	0.121	0.189	0.138	0.082	0.035	0.001	0.000+
	12	0.000+	0.000+	0.000+	0.000+	0.001	0.015	0.071	0.166	0.187	0.144	0.082	0.005	0.000+
	13	0.000+	0.000+	0.000+	0.000+	0.000+	0.004	0.033	0.115	0.202	0.199	0.151	0.022	0.001
	14	0.000+	0.000+	0.000+	0.000+	0.000+	0.001	0.012	0.061	0.168	0.213	0.215	0.070	0.009
	15	0.000+	0.000+	0.000+	0.000+	0.000+	0.000+	0.003	0.025	0.105	0.170	0.230	0.168	0.047
	16	0.000+	0.000+	0.000+	0.000+	0.000+	0.000+	0.001	0.007	0.046	0.096	0.172	0.284	0.168
	17	0.000+	0.000+	0.000+	0.000+	0.000+	0.000+	0.000+	0.001	0.013	0.034	0.081	0.300	0.376
	18	0.000+	0.000+	0.000+	0.000+	0.000+	0.000+	0.000+	0.000+	0.002	0.006	0.018	0.150	0.397
19	0	0.377	0.135	0.014	0.004	0.001	0.000+	0.000+	0.000+	0.000+	0.000+	0.000+	0.000+	0.000+
	1	0.377	0.285	0.068	0.027	0.009	0.001	0.000+	0.000+	0.000+	0.000+	0.000+	0.000+	0.000+
	2	0.179	0.285	0.154	0.080	0.036	0.005	0.000+	0.000+	0.000+	0.000+	0.000+	0.000+	0.000+
	3	0.053	0.180	0.218	0.152	0.087	0.017	0.002	0.000+	0.000+	0.000+	0.000+	0.000+	0.000+
	4	0.011	0.080	0.218	0.202	0.149	0.047	0.007	0.001	0.000+	0.000+	0.000+	0.000+	0.000+
	5	0.002	0.027	0.164	0.202	0.192	0.093	0.022	0.002	0.000+	0.000+	0.000+	0.000+	0.000+
	6	0.000+	0.007	0.095	0.157	0.192	0.145	0.052	0.008	0.001	0.000+	0.000+	0.000+	0.000+
	7	0.000+	0.001	0.044	0.097	0.153	0.180	0.096	0.024	0.002	0.000+	0.000+	0.000+	0.000+
	8	0.000+	0.000+	0.017	0.049	0.098	0.180	0.144	0.053	0.008	0.002	0.000+	0.000+	0.000+
	9	0.000+	0.000+	0.005	0.020	0.051	0.146	0.176	0.098	0.022	0.007	0.001	0.000+	0.000+
	10	0.000+	0.000+	0.001	0.007	0.022	0.098	0.176	0.146	0.051	0.020	0.005	0.000+	0.000+
	11	0.000+	0.000+	0.000+	0.002	0.008	0.053	0.144	0.180	0.098	0.049	0.017	0.000+	0.000+
	12	0.000+	0.000+	0.000+	0.000+	0.002	0.024	0.096	0.180	0.153	0.097	0.044	0.001	0.000+
	13	0.000+	0.000+	0.000+	0.000+	0.001	0.008	0.052	0.145	0.192	0.157	0.095	0.007	0.000+
	14	0.000+	0.000+	0.000+	0.000+	0.000+	0.002	0.022	0.093	0.192	0.202	0.164	0.027	0.002
	15	0.000+	0.000+	0.000+	0.000+	0.000+	0.001	0.007	0.047	0.149	0.202	0.218	0.080	0.011
	16	0.000+	0.000+	0.000+	0.000+	0.000+	0.000+	0.002	0.017	0.087	0.152	0.218	0.180	0.053
	17	0.000+	0.000+	0.000+	0.000+	0.000+	0.000+	0.000+	0.005	0.036	0.080	0.154	0.285	0.179
	18	0.000+	0.000+	0.000+	0.000+	0.000+	0.000+	0.000+	0.001	0.009	0.027	0.068	0.285	0.377
	19	0.000+	0.000+	0.000+	0.000+	0.000+	0.000+	0.000+	0.000+	0.001	0.004	0.014	0.135	0.377
20	0	0.358	0.122	0.012	0.003	0.001	0.000++	0.000++	0.000++	0.000++	0.000++	0.000++	0.000++	0.000++
	1	0.377	0.270	0.058	0.021	0.007	0.000+	0.000+	0.000+	0.000+	0.000+	0.000+	0.000+	0.000+
	2	0.189	0.285	0.137	0.067	0.028	0.003	0.000+	0.000+	0.000+	0.000+	0.000+	0.000+	0.000+
	3	0.060	0.190	0.205	0.134	0.072	0.012	0.001	0.000+	0.000+	0.000+	0.000+	0.000+	0.000+
	4	0.013	0.090	0.218	0.190	0.130	0.035	0.005	0.000+	0.000+	0.000+	0.000+	0.000+	0.000+
	5	0.002	0.032	0.175	0.202	0.179	0.075	0.015	0.001	0.000+	0.000+	0.000+	0.000+	0.000+
	6	0.000+	0.009	0.109	0.169	0.192	0.124	0.037	0.005	0.000+	0.000+	0.000+	0.000+	0.000+
	7	0.000+	0.002	0.055	0.112	0.164	0.166	0.074	0.015	0.001	0.000+	0.000+	0.000+	0.000+
	8	0.000+	0.000+	0.022	0.061	0.114	0.180	0.120	0.035	0.004	0.001	0.000+	0.000+	0.000+
	9	0.000+	0.000+	0.007	0.027	0.065	0.160	0.160	0.071	0.012	0.003	0.000+	0.000+	0.000+
	10	0.000+	0.000+	0.002	0.010	0.031	0.117	0.176	0.117	0.031	0.010	0.002	0.000+	0.000+
	11	0.000+	0.000+	0.000+	0.003	0.012	0.071	0.160	0.160	0.065	0.027	0.007	0.000+	0.000+
	12	0.000+	0.000+	0.000+	0.001	0.004	0.035	0.120	0.180	0.114	0.061	0.022	0.000+	0.000+
	13	0.000+	0.000+	0.000+	0.000+	0.001	0.015	0.074	0.166	0.164	0.112	0.055	0.002	0.000+
	14	0.000+	0.000+	0.000+	0.000+	0.000+	0.005	0.037	0.124	0.192	0.169	0.109	0.009	0.000+
	15	0.000+	0.000+	0.000+	0.000+	0.000+	0.001	0.015	0.075	0.179	0.202	0.175	0.032	0.002
	16	0.000+	0.000+	0.000+	0.000+	0.000+	0.000+	0.005	0.035	0.130	0.190	0.218	0.090	0.013
	17	0.000+	0.000+	0.000+	0.000+	0.000+	0.000+	0.001	0.012	0.072	0.134	0.205	0.190	0.060
	18	0.000+	0.000+	0.000+	0.000+	0.000+	0.000+	0.000+	0.003	0.028	0.067	0.137	0.285	0.189
	19	0.000+	0.000+	0.000+	0.000+	0.000+	0.000+	0.000+	0.000+	0.007	0.021	0.058	0.270	0.377
	20	0.000+	0.000+	0.000+	0.000+	0.000+	0.000+	0.000+	0.000+	0.001	0.003	0.012	0.122	0.358

A value of 0.000+ indicates that the probability is 0.000 when rounded to three decimal places. The actual probability is slightly greater than 0.

Table A.2 Cumulative Normal Distribution

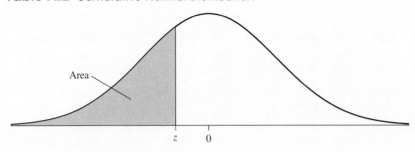

Area

z	0.00	0.01	0.02	0.03	0.04	0.05	0.06	0.07	0.08	0.09
−3.7 or less	.0001									
−3.6	.0002	.0002	.0001	.0001	.0001	.0001	.0001	.0001	.0001	.0001
−3.5	.0002	.0002	.0002	.0002	.0002	.0002	.0002	.0002	.0002	.0002
−3.4	.0003	.0003	.0003	.0003	.0003	.0003	.0003	.0003	.0003	.0002
−3.3	.0005	.0005	.0005	.0004	.0004	.0004	.0004	.0004	.0004	.0003
−3.2	.0007	.0007	.0006	.0006	.0006	.0006	.0006	.0005	.0005	.0005
−3.1	.0010	.0009	.0009	.0009	.0008	.0008	.0008	.0008	.0007	.0007
−3.0	.0013	.0013	.0013	.0012	.0012	.0011	.0011	.0011	.0010	.0010
−2.9	.0019	.0018	.0018	.0017	.0016	.0016	.0015	.0015	.0014	.0014
−2.8	.0026	.0025	.0024	.0023	.0023	.0022	.0021	.0021	.0020	.0019
−2.7	.0035	.0034	.0033	.0032	.0031	.0030	.0029	.0028	.0027	.0026
−2.6	.0047	.0045	.0044	.0043	.0041	.0040	.0039	.0038	.0037	.0036
−2.5	.0062	.0060	.0059	.0057	.0055	.0054	.0052	.0051	.0049	.0048
−2.4	.0082	.0080	.0078	.0075	.0073	.0071	.0069	.0068	.0066	.0064
−2.3	.0107	.0104	.0102	.0099	.0096	.0094	.0091	.0089	.0087	.0084
−2.2	.0139	.0136	.0132	.0129	.0125	.0122	.0119	.0116	.0113	.0110
−2.1	.0179	.0174	.0170	.0166	.0162	.0158	.0154	.0150	.0146	.0143
−2.0	.0228	.0222	.0217	.0212	.0207	.0202	.0197	.0192	.0188	.0183
−1.9	.0287	.0281	.0274	.0268	.0262	.0256	.0250	.0244	.0239	.0233
−1.8	.0359	.0351	.0344	.0336	.0329	.0322	.0314	.0307	.0301	.0294
−1.7	.0446	.0436	.0427	.0418	.0409	.0401	.0392	.0384	.0375	.0367
−1.6	.0548	.0537	.0526	.0516	.0505	.0495	.0485	.0475	.0465	.0455
−1.5	.0668	.0655	.0643	.0630	.0618	.0606	.0594	.0582	.0571	.0559
−1.4	.0808	.0793	.0778	.0764	.0749	.0735	.0721	.0708	.0694	.0681
−1.3	.0968	.0951	.0934	.0918	.0901	.0885	.0869	.0853	.0838	.0823
−1.2	.1151	.1131	.1112	.1093	.1075	.1056	.1038	.1020	.1003	.0985
−1.1	.1357	.1335	.1314	.1292	.1271	.1251	.1230	.1210	.1190	.1170
−1.0	.1587	.1562	.1539	.1515	.1492	.1469	.1446	.1423	.1401	.1379
−0.9	.1841	.1814	.1788	.1762	.1736	.1711	.1685	.1660	.1635	.1611
−0.8	.2119	.2090	.2061	.2033	.2005	.1977	.1949	.1922	.1894	.1867
−0.7	.2420	.2389	.2358	.2327	.2296	.2266	.2236	.2206	.2177	.2148
−0.6	.2743	.2709	.2676	.2643	.2611	.2578	.2546	.2514	.2483	.2451
−0.5	.3085	.3050	.3015	.2981	.2946	.2912	.2877	.2843	.2810	.2776
−0.4	.3446	.3409	.3372	.3336	.3300	.3264	.3228	.3192	.3156	.3121
−0.3	.3821	.3783	.3745	.3707	.3669	.3632	.3594	.3557	.3520	.3483
−0.2	.4207	.4168	.4129	.4090	.4052	.4013	.3974	.3936	.3897	.3859
−0.1	.4602	.4562	.4522	.4483	.4443	.4404	.4364	.4325	.4286	.4247
−0.0	.5000	.4960	.4920	.4880	.4840	.4801	.4761	.4721	.4681	.4641

Table A.2 Cumulative Normal Distribution (continued)

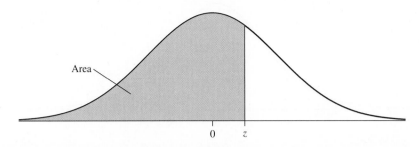

z	0.00	0.01	0.02	0.03	0.04	0.05	0.06	0.07	0.08	0.09
0.0	.5000	.5040	.5080	.5120	.5160	.5199	.5239	.5279	.5319	.5359
0.1	.5398	.5438	.5478	.5517	.5557	.5596	.5636	.5675	.5714	.5753
0.2	.5793	.5832	.5871	.5910	.5948	.5987	.6026	.6064	.6103	.6141
0.3	.6179	.6217	.6255	.6293	.6331	.6368	.6406	.6443	.6480	.6517
0.4	.6554	.6591	.6628	.6664	.6700	.6736	.6772	.6808	.6844	.6879
0.5	.6915	.6950	.6985	.7019	.7054	.7088	.7123	.7157	.7190	.7224
0.6	.7257	.7291	.7324	.7357	.7389	.7422	.7454	.7486	.7517	.7549
0.7	.7580	.7611	.7642	.7673	.7704	.7734	.7764	.7794	.7823	.7852
0.8	.7881	.7910	.7939	.7967	.7995	.8023	.8051	.8078	.8106	.8133
0.9	.8159	.8186	.8212	.8238	.8264	.8289	.8315	.8340	.8365	.8389
1.0	.8413	.8438	.8461	.8485	.8508	.8531	.8554	.8577	.8599	.8621
1.1	.8643	.8665	.8686	.8708	.8729	.8749	.8770	.8790	.8810	.8830
1.2	.8849	.8869	.8888	.8907	.8925	.8944	.8962	.8980	.8997	.9015
1.3	.9032	.9049	.9066	.9082	.9099	.9115	.9131	.9147	.9162	.9177
1.4	.9192	.9207	.9222	.9236	.9251	.9265	.9279	.9292	.9306	.9319
1.5	.9332	.9345	.9357	.9370	.9382	.9394	.9406	.9418	.9429	.9441
1.6	.9452	.9463	.9474	.9484	.9495	.9505	.9515	.9525	.9535	.9545
1.7	.9554	.9564	.9573	.9582	.9591	.9599	.9608	.9616	.9625	.9633
1.8	.9641	.9649	.9656	.9664	.9671	.9678	.9686	.9693	.9699	.9706
1.9	.9713	.9719	.9726	.9732	.9738	.9744	.9750	.9756	.9761	.9767
2.0	.9772	.9778	.9783	.9788	.9793	.9798	.9803	.9808	.9812	.9817
2.1	.9821	.9826	.9830	.9834	.9838	.9842	.9846	.9850	.9854	.9857
2.2	.9861	.9864	.9868	.9871	.9875	.9878	.9881	.9884	.9887	.9890
2.3	.9893	.9896	.9898	.9901	.9904	.9906	.9909	.9911	.9913	.9916
2.4	.9918	.9920	.9922	.9925	.9927	.9929	.9931	.9932	.9934	.9936
2.5	.9938	.9940	.9941	.9943	.9945	.9946	.9948	.9949	.9951	.9952
2.6	.9953	.9955	.9956	.9957	.9959	.9960	.9961	.9962	.9963	.9964
2.7	.9965	.9966	.9967	.9968	.9969	.9970	.9971	.9972	.9973	.9974
2.8	.9974	.9975	.9976	.9977	.9977	.9978	.9979	.9979	.9980	.9981
2.9	.9981	.9982	.9982	.9983	.9984	.9984	.9985	.9985	.9986	.9986
3.0	.9987	.9987	.9987	.9988	.9988	.9989	.9989	.9989	.9990	.9990
3.1	.9990	.9991	.9991	.9991	.9992	.9992	.9992	.9992	.9993	.9993
3.2	.9993	.9993	.9994	.9994	.9994	.9994	.9994	.9995	.9995	.9995
3.3	.9995	.9995	.9995	.9996	.9996	.9996	.9996	.9996	.9996	.9997
3.4	.9997	.9997	.9997	.9997	.9997	.9997	.9997	.9997	.9997	.9998
3.5	.9998	.9998	.9998	.9998	.9998	.9998	.9998	.9998	.9998	.9998
3.6	.9998	.9998	.9999	.9999	.9999	.9999	.9999	.9999	.9999	.9999
3.7 or more	.9999									

Table A.3 Critical Values for the Student's *t* Distribution

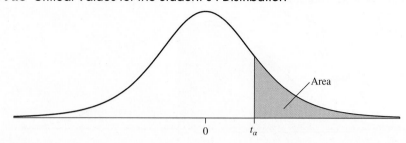

Degrees of Freedom	Area in Right Tail									
	0.40	0.25	0.10	0.05	0.025	0.01	0.005	0.0025	0.001	0.0005
1	0.325	1.000	3.078	6.314	12.706	31.821	63.657	127.321	318.309	636.619
2	0.289	0.816	1.886	2.920	4.303	6.965	9.925	14.089	22.327	31.599
3	0.277	0.765	1.638	2.353	3.182	4.541	5.841	7.453	10.215	12.924
4	0.271	0.741	1.533	2.132	2.776	3.747	4.604	5.598	7.173	8.610
5	0.267	0.727	1.476	2.015	2.571	3.365	4.032	4.773	5.893	6.869
6	0.265	0.718	1.440	1.943	2.447	3.143	3.707	4.317	5.208	5.959
7	0.263	0.711	1.415	1.895	2.365	2.998	3.499	4.029	4.785	5.408
8	0.262	0.706	1.397	1.860	2.306	2.896	3.355	3.833	4.501	5.041
9	0.261	0.703	1.383	1.833	2.262	2.821	3.250	3.690	4.297	4.781
10	0.260	0.700	1.372	1.812	2.228	2.764	3.169	3.581	4.144	4.587
11	0.260	0.697	1.363	1.796	2.201	2.718	3.106	3.497	4.025	4.437
12	0.259	0.695	1.356	1.782	2.179	2.681	3.055	3.428	3.930	4.318
13	0.259	0.694	1.350	1.771	2.160	2.650	3.012	3.372	3.852	4.221
14	0.258	0.692	1.345	1.761	2.145	2.624	2.977	3.326	3.787	4.140
15	0.258	0.691	1.341	1.753	2.131	2.602	2.947	3.286	3.733	4.073
16	0.258	0.690	1.337	1.746	2.120	2.583	2.921	3.252	3.686	4.015
17	0.257	0.689	1.333	1.740	2.110	2.567	2.898	3.222	3.646	3.965
18	0.257	0.688	1.330	1.734	2.101	2.552	2.878	3.197	3.610	3.922
19	0.257	0.688	1.328	1.729	2.093	2.539	2.861	3.174	3.579	3.883
20	0.257	0.687	1.325	1.725	2.086	2.528	2.845	3.153	3.552	3.850
21	0.257	0.686	1.323	1.721	2.080	2.518	2.831	3.135	3.527	3.819
22	0.256	0.686	1.321	1.717	2.074	2.508	2.819	3.119	3.505	3.792
23	0.256	0.685	1.319	1.714	2.069	2.500	2.807	3.104	3.485	3.768
24	0.256	0.685	1.318	1.711	2.064	2.492	2.797	3.091	3.467	3.745
25	0.256	0.684	1.316	1.708	2.060	2.485	2.787	3.078	3.450	3.725
26	0.256	0.684	1.315	1.706	2.056	2.479	2.779	3.067	3.435	3.707
27	0.256	0.684	1.314	1.703	2.052	2.473	2.771	3.057	3.421	3.690
28	0.256	0.683	1.313	1.701	2.048	2.467	2.763	3.047	3.408	3.674
29	0.256	0.683	1.311	1.699	2.045	2.462	2.756	3.038	3.396	3.659
30	0.256	0.683	1.310	1.697	2.042	2.457	2.750	3.030	3.385	3.646
31	0.256	0.682	1.309	1.696	2.040	2.453	2.744	3.022	3.375	3.633
32	0.255	0.682	1.309	1.694	2.037	2.449	2.738	3.015	3.365	3.622
33	0.255	0.682	1.308	1.692	2.035	2.445	2.733	3.008	3.356	3.611
34	0.255	0.682	1.307	1.691	2.032	2.441	2.728	3.002	3.348	3.601
35	0.255	0.682	1.306	1.690	2.030	2.438	2.724	2.996	3.340	3.591
36	0.255	0.681	1.306	1.688	2.028	2.434	2.719	2.990	3.333	3.582
37	0.255	0.681	1.305	1.687	2.026	2.431	2.715	2.985	3.326	3.574
38	0.255	0.681	1.304	1.686	2.024	2.429	2.712	2.980	3.319	3.566
39	0.255	0.681	1.304	1.685	2.023	2.426	2.708	2.976	3.313	3.558
40	0.255	0.681	1.303	1.684	2.021	2.423	2.704	2.971	3.307	3.551
50	0.255	0.679	1.299	1.676	2.009	2.403	2.678	2.937	3.261	3.496
60	0.254	0.679	1.296	1.671	2.000	2.390	2.660	2.915	3.232	3.460
80	0.254	0.678	1.292	1.664	1.990	2.374	2.639	2.887	3.195	3.416
100	0.254	0.677	1.290	1.660	1.984	2.364	2.626	2.871	3.174	3.390
200	0.254	0.676	1.286	1.653	1.972	2.345	2.601	2.839	3.131	3.340
z	0.253	0.674	1.282	1.645	1.960	2.326	2.576	2.807	3.090	3.291
	20%	50%	80%	90%	95%	98%	99%	99.5%	99.8%	99.9%
	Confidence Level									

Table A.4 Critical Values for the χ^2 Distribution

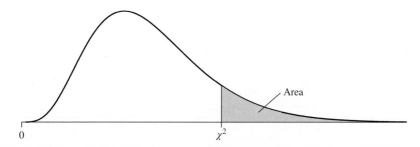

Degrees of	Area in Right Tail									
Freedom	0.995	0.99	0.975	0.95	0.90	0.10	0.05	0.025	0.01	0.005
1	0.000	0.000	0.001	0.004	0.016	2.706	3.841	5.024	6.635	7.879
2	0.010	0.020	0.051	0.103	0.211	4.605	5.991	7.378	9.210	10.597
3	0.072	0.115	0.216	0.352	0.584	6.251	7.815	9.348	11.345	12.838
4	0.207	0.297	0.484	0.711	1.064	7.779	9.488	11.143	13.277	14.860
5	0.412	0.554	0.831	1.145	1.610	9.236	11.070	12.833	15.086	16.750
6	0.676	0.872	1.237	1.635	2.204	10.645	12.592	14.449	16.812	18.548
7	0.989	1.239	1.690	2.167	2.833	12.017	14.067	16.013	18.475	20.278
8	1.344	1.646	2.180	2.733	3.490	13.362	15.507	17.535	20.090	21.955
9	1.735	2.088	2.700	3.325	4.168	14.684	16.919	19.023	21.666	23.589
10	2.156	2.558	3.247	3.940	4.865	15.987	18.307	20.483	23.209	25.188
11	2.603	3.053	3.816	4.575	5.578	17.275	19.675	21.920	24.725	26.757
12	3.074	3.571	4.404	5.226	6.304	18.549	21.026	23.337	26.217	28.300
13	3.565	4.107	5.009	5.892	7.042	19.812	22.362	24.736	27.688	29.819
14	4.075	4.660	5.629	6.571	7.790	21.064	23.685	26.119	29.141	31.319
15	4.601	5.229	6.262	7.261	8.547	22.307	24.996	27.488	30.578	32.801
16	5.142	5.812	6.908	7.962	9.312	23.542	26.296	28.845	32.000	34.267
17	5.697	6.408	7.564	8.672	10.085	24.769	27.587	30.191	33.409	35.718
18	6.265	7.015	8.231	9.390	10.865	25.989	28.869	31.526	34.805	37.156
19	6.844	7.633	8.907	10.117	11.651	27.204	30.144	32.852	36.191	38.582
20	7.434	8.260	9.591	10.851	12.443	28.412	31.410	34.170	37.566	39.997
21	8.034	8.897	10.283	11.591	13.240	29.615	32.671	35.479	38.932	41.401
22	8.643	9.542	10.982	12.338	14.041	30.813	33.924	36.781	40.289	42.796
23	9.260	10.196	11.689	13.091	14.848	32.007	35.172	38.076	41.638	44.181
24	9.886	10.856	12.401	13.848	15.659	33.196	36.415	39.364	42.980	45.559
25	10.520	11.524	13.120	14.611	16.473	34.382	37.652	40.646	44.314	46.928
26	11.160	12.198	13.844	15.379	17.292	35.563	38.885	41.923	45.642	48.290
27	11.808	12.879	14.573	16.151	18.114	36.741	40.113	43.195	46.963	49.645
28	12.461	13.565	15.308	16.928	18.939	37.916	41.337	44.461	48.278	50.993
29	13.121	14.256	16.047	17.708	19.768	39.087	42.557	45.722	49.588	52.336
30	13.787	14.953	16.791	18.493	20.599	40.256	43.773	46.979	50.892	53.672
31	14.458	15.655	17.539	19.281	21.434	41.422	44.985	48.232	52.191	55.003
32	15.134	16.362	18.291	20.072	22.271	42.585	46.194	49.480	53.486	56.328
33	15.815	17.074	19.047	20.867	23.110	43.745	47.400	50.725	54.776	57.648
34	16.501	17.789	19.806	21.664	23.952	44.903	48.602	51.966	56.061	58.964
35	17.192	18.509	20.569	22.465	24.797	46.059	49.802	53.203	57.342	60.275
36	17.887	19.233	21.336	23.269	25.643	47.212	50.998	54.437	58.619	61.581
37	18.586	19.96	22.106	24.075	26.492	48.363	52.192	55.668	59.893	62.883
38	19.289	20.691	22.878	24.884	27.343	49.513	53.384	56.896	61.162	64.181
39	19.996	21.426	23.654	25.695	28.196	50.660	54.572	58.120	62.428	65.476
40	20.707	22.164	24.433	26.509	29.051	51.805	55.758	59.342	63.691	66.766
41	21.421	22.906	25.215	27.326	29.907	52.949	56.942	60.561	64.950	68.053
42	22.138	23.650	25.999	28.144	30.765	54.090	58.124	61.777	66.206	69.336
43	22.859	24.398	26.785	28.965	31.625	55.230	59.304	62.990	67.459	70.616
44	23.584	25.148	27.575	29.787	32.487	56.369	60.481	64.201	68.710	71.893
45	24.311	25.901	28.366	30.612	33.350	57.505	61.656	65.410	69.957	73.166
50	27.991	29.707	32.357	34.764	37.689	63.167	67.505	71.420	76.154	79.490
60	35.534	37.485	40.482	43.188	46.459	74.397	79.082	83.298	88.379	91.952
70	43.275	45.442	48.758	51.739	55.329	85.527	90.531	95.023	100.425	104.215
80	51.172	53.540	57.153	60.391	64.278	96.578	101.879	106.629	112.329	116.321
90	59.196	61.754	65.647	69.126	73.291	107.565	113.145	118.136	124.116	128.299
100	67.328	70.065	74.222	77.929	82.358	118.498	124.342	129.561	135.807	140.169

Table A.5 Critical Values for the *F* Distribution

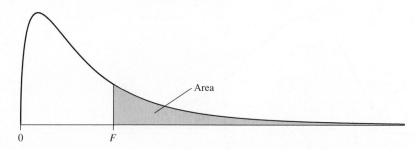

Denominator Degrees of Freedom	Area	Numerator Degrees of Freedom								
		1	**2**	**3**	**4**	**5**	**6**	**7**	**8**	**9**
1	0.100	39.86	49.50	53.59	55.83	57.24	58.20	58.91	59.44	59.86
1	0.050	161.45	199.50	215.71	224.58	230.16	233.99	236.77	238.88	240.54
1	0.025	647.79	799.50	864.16	899.58	921.85	937.11	948.22	956.66	963.28
1	0.010	4052.18	4999.50	5403.35	5624.58	5763.65	5858.99	5928.36	5981.07	6022.47
2	0.100	8.53	9.00	9.16	9.24	9.29	9.33	9.35	9.37	9.38
2	0.050	18.51	19.00	19.16	19.25	19.30	19.33	19.35	19.37	19.38
2	0.025	38.51	39.00	39.17	39.25	39.30	39.33	39.36	39.37	39.39
2	0.010	98.50	99.00	99.17	99.25	99.30	99.33	99.36	99.37	99.39
2	0.001	998.50	999.00	999.17	999.25	999.30	999.33	999.36	999.37	999.39
3	0.100	5.54	5.46	5.39	5.34	5.31	5.28	5.27	5.25	5.24
3	0.050	10.13	9.55	9.28	9.12	9.01	8.94	8.89	8.85	8.81
3	0.025	17.44	16.04	15.44	15.10	14.88	14.73	14.62	14.54	14.47
3	0.010	34.12	30.82	29.46	28.71	28.24	27.91	27.67	27.49	27.35
3	0.001	167.03	148.50	141.11	137.10	134.58	132.85	131.58	130.62	129.86
4	0.100	4.54	4.32	4.19	4.11	4.05	4.01	3.98	3.95	3.94
4	0.050	7.71	6.94	6.59	6.39	6.26	6.16	6.09	6.04	6.00
4	0.025	12.22	10.65	9.98	9.60	9.36	9.20	9.07	8.98	8.90
4	0.010	21.20	18.00	16.69	15.98	15.52	15.21	14.98	14.80	14.66
4	0.001	74.14	61.25	56.18	53.44	51.71	50.53	49.66	49.00	48.47
5	0.100	4.06	3.78	3.62	3.52	3.45	3.40	3.37	3.34	3.32
5	0.050	6.61	5.79	5.41	5.19	5.05	4.95	4.88	4.82	4.77
5	0.025	10.01	8.43	7.76	7.39	7.15	6.98	6.85	6.76	6.68
5	0.010	16.26	13.27	12.06	11.39	10.97	10.67	10.46	10.29	10.16
5	0.001	47.18	37.12	33.20	31.09	29.75	28.83	28.16	27.65	27.24
6	0.100	3.78	3.46	3.29	3.18	3.11	3.05	3.01	2.98	2.96
6	0.050	5.99	5.14	4.76	4.53	4.39	4.28	4.21	4.15	4.10
6	0.025	8.81	7.26	6.60	6.23	5.99	5.82	5.70	5.60	5.52
6	0.010	13.75	10.92	9.78	9.15	8.75	8.47	8.26	8.10	7.98
6	0.001	35.51	27.00	23.70	21.92	20.80	20.03	19.46	19.03	18.69
7	0.100	3.59	3.26	3.07	2.96	2.88	2.83	2.78	2.75	2.72
7	0.050	5.59	4.74	4.35	4.12	3.97	3.87	3.79	3.73	3.68
7	0.025	8.07	6.54	5.89	5.52	5.29	5.12	4.99	4.90	4.82
7	0.010	12.25	9.55	8.45	7.85	7.46	7.19	6.99	6.84	6.72
7	0.001	29.25	21.69	18.77	17.20	16.21	15.52	15.02	14.63	14.33
8	0.100	3.46	3.11	2.92	2.81	2.73	2.67	2.62	2.59	2.56
8	0.050	5.32	4.46	4.07	3.84	3.69	3.58	3.50	3.44	3.39
8	0.025	7.57	6.06	5.42	5.05	4.82	4.65	4.53	4.43	4.36
8	0.010	11.26	8.65	7.59	7.01	6.63	6.37	6.18	6.03	5.91
8	0.001	25.41	18.49	15.83	14.39	13.48	12.86	12.40	12.05	11.77
9	0.100	3.36	3.01	2.81	2.69	2.61	2.55	2.51	2.47	2.44
9	0.050	5.12	4.26	3.86	3.63	3.48	3.37	3.29	3.23	3.18
9	0.025	7.21	5.71	5.08	4.72	4.48	4.32	4.20	4.10	4.03
9	0.010	10.56	8.02	6.99	6.42	6.06	5.80	5.61	5.47	5.35
9	0.001	22.86	16.39	13.90	12.56	11.71	11.13	10.70	10.37	10.11

Table A.5 Critical Values for the *F* Distribution (continued)

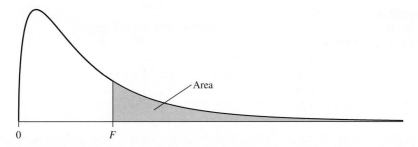

Denominator Degrees of Freedom	Area	Numerator Degrees of Freedom								
		10	12	15	20	25	30	40	50	60
1	0.100	60.19	60.71	61.22	61.74	62.05	62.26	62.53	62.69	62.79
1	0.050	241.88	243.91	245.95	248.01	249.26	250.10	251.14	251.77	252.20
1	0.025	968.63	976.71	984.87	993.10	998.08	1001.41	1005.60	1008.12	1009.80
1	0.010	6055.85	6106.32	6157.28	6208.73	6239.83	6260.65	6286.78	6302.52	6313.03
2	0.100	9.39	9.41	9.42	9.44	9.45	9.46	9.47	9.47	9.47
2	0.050	19.40	19.41	19.43	19.45	19.46	19.46	19.47	19.48	19.48
2	0.025	39.40	39.41	39.43	39.45	39.46	39.46	39.47	39.48	39.48
2	0.010	99.40	99.42	99.43	99.45	99.46	99.47	99.47	99.48	99.48
2	0.001	999.40	999.42	999.43	999.45	999.46	999.47	999.47	999.48	999.48
3	0.100	5.23	5.22	5.20	5.18	5.17	5.17	5.16	5.15	5.15
3	0.050	8.79	8.74	8.70	8.66	8.63	8.62	8.59	8.58	8.57
3	0.025	14.42	14.34	14.25	14.17	14.12	14.08	14.04	14.01	13.99
3	0.010	27.23	27.05	26.87	26.69	26.58	26.50	26.41	26.35	26.32
3	0.001	129.25	128.32	127.37	126.42	125.84	125.45	124.96	124.66	124.47
4	0.100	3.92	3.90	3.87	3.84	3.83	3.82	3.80	3.80	3.79
4	0.050	5.96	5.91	5.86	5.80	5.77	5.75	5.72	5.70	5.69
4	0.025	8.84	8.75	8.66	8.56	8.50	8.46	8.41	8.38	8.36
4	0.010	14.55	14.37	14.20	14.02	13.91	13.84	13.75	13.69	13.65
4	0.001	48.05	47.41	46.76	46.10	45.70	45.43	45.09	44.88	44.75
5	0.100	3.30	3.27	3.24	3.21	3.19	3.17	3.16	3.15	3.14
5	0.050	4.74	4.68	4.62	4.56	4.52	4.50	4.46	4.44	4.43
5	0.025	6.62	6.52	6.43	6.33	6.27	6.23	6.18	6.14	6.12
5	0.010	10.05	9.89	9.72	9.55	9.45	9.38	9.29	9.24	9.20
5	0.001	26.92	26.42	25.91	25.39	25.08	24.87	24.60	24.44	24.33
6	0.100	2.94	2.90	2.87	2.84	2.81	2.80	2.78	2.77	2.76
6	0.050	4.06	4.00	3.94	3.87	3.83	3.81	3.77	3.75	3.74
6	0.025	5.46	5.37	5.27	5.17	5.11	5.07	5.01	4.98	4.96
6	0.010	7.87	7.72	7.56	7.40	7.30	7.23	7.14	7.09	7.06
6	0.001	18.41	17.99	17.56	17.12	16.85	16.67	16.44	16.31	16.21
7	0.100	2.70	2.67	2.63	2.59	2.57	2.56	2.54	2.52	2.51
7	0.050	3.64	3.57	3.51	3.44	3.40	3.38	3.34	3.32	3.30
7	0.025	4.76	4.67	4.57	4.47	4.40	4.36	4.31	4.28	4.25
7	0.010	6.62	6.47	6.31	6.16	6.06	5.99	5.91	5.86	5.82
7	0.001	14.08	13.71	13.32	12.93	12.69	12.53	12.33	12.20	12.12
8	0.100	2.54	2.50	2.46	2.42	2.40	2.38	2.36	2.35	2.34
8	0.050	3.35	3.28	3.22	3.15	3.11	3.08	3.04	3.02	3.01
8	0.025	4.30	4.20	4.10	4.00	3.94	3.89	3.84	3.81	3.78
8	0.010	5.81	5.67	5.52	5.36	5.26	5.20	5.12	5.07	5.03
8	0.001	11.54	11.19	10.84	10.48	10.26	10.11	9.92	9.80	9.73
9	0.100	2.42	2.38	2.34	2.30	2.27	2.25	2.23	2.22	2.21
9	0.050	3.14	3.07	3.01	2.94	2.89	2.86	2.83	2.80	2.79
9	0.025	3.96	3.87	3.77	3.67	3.60	3.56	3.51	3.47	3.45
9	0.010	5.26	5.11	4.96	4.81	4.71	4.65	4.57	4.52	4.48
9	0.001	9.89	9.57	9.24	8.90	8.69	8.55	8.37	8.26	8.19

Table A.5 Critical Values for the *F* Distribution (continued)

Denominator Degrees of Freedom	Area	Numerator Degrees of Freedom								
		1	**2**	**3**	**4**	**5**	**6**	**7**	**8**	**9**
10	0.100	3.29	2.92	2.73	2.61	2.52	2.46	2.41	2.38	2.35
10	0.050	4.96	4.10	3.71	3.48	3.33	3.22	3.14	3.07	3.02
10	0.025	6.94	5.46	4.83	4.47	4.24	4.07	3.95	3.85	3.78
10	0.010	10.04	7.56	6.55	5.99	5.64	5.39	5.20	5.06	4.94
10	0.001	21.04	14.91	12.55	11.28	10.48	9.93	9.52	9.20	8.96
11	0.100	3.23	2.86	2.66	2.54	2.45	2.39	2.34	2.30	2.27
11	0.050	4.84	3.98	3.59	3.36	3.20	3.09	3.01	2.95	2.90
11	0.025	6.72	5.26	4.63	4.28	4.04	3.88	3.76	3.66	3.59
11	0.010	9.65	7.21	6.22	5.67	5.32	5.07	4.89	4.74	4.63
11	0.001	19.69	13.81	11.56	10.35	9.58	9.05	8.66	8.35	8.12
12	0.100	3.18	2.81	2.61	2.48	2.39	2.33	2.28	2.24	2.21
12	0.050	4.75	3.89	3.49	3.26	3.11	3.00	2.91	2.85	2.80
12	0.025	6.55	5.10	4.47	4.12	3.89	3.73	3.61	3.51	3.44
12	0.010	9.33	6.93	5.95	5.41	5.06	4.82	4.64	4.50	4.39
12	0.001	18.64	12.97	10.80	9.63	8.89	8.38	8.00	7.71	7.48
13	0.100	3.14	2.76	2.56	2.43	2.35	2.28	2.23	2.20	2.16
13	0.050	4.67	3.81	3.41	3.18	3.03	2.92	2.83	2.77	2.71
13	0.025	6.41	4.97	4.35	4.00	3.77	3.60	3.48	3.39	3.31
13	0.010	9.07	6.70	5.74	5.21	4.86	4.62	4.44	4.30	4.19
13	0.001	17.82	12.31	10.21	9.07	8.35	7.86	7.49	7.21	6.98
14	0.100	3.10	2.73	2.52	2.39	2.31	2.24	2.19	2.15	2.12
14	0.050	4.60	3.74	3.34	3.11	2.96	2.85	2.76	2.70	2.65
14	0.025	6.30	4.86	4.24	3.89	3.66	3.50	3.38	3.29	3.21
14	0.010	8.86	6.51	5.56	5.04	4.69	4.46	4.28	4.14	4.03
14	0.001	17.14	11.78	9.73	8.62	7.92	7.44	7.08	6.80	6.58
15	0.100	3.07	2.70	2.49	2.36	2.27	2.21	2.16	2.12	2.09
15	0.050	4.54	3.68	3.29	3.06	2.90	2.79	2.71	2.64	2.59
15	0.025	6.20	4.77	4.15	3.80	3.58	3.41	3.29	3.20	3.12
15	0.010	8.68	6.36	5.42	4.89	4.56	4.32	4.14	4.00	3.89
15	0.001	16.59	11.34	9.34	8.25	7.57	7.09	6.74	6.47	6.26
16	0.100	3.05	2.67	2.46	2.33	2.24	2.18	2.13	2.09	2.06
16	0.050	4.49	3.63	3.24	3.01	2.85	2.74	2.66	2.59	2.54
16	0.025	6.12	4.69	4.08	3.73	3.50	3.34	3.22	3.12	3.05
16	0.010	8.53	6.23	5.29	4.77	4.44	4.20	4.03	3.89	3.78
16	0.001	16.12	10.97	9.01	7.94	7.27	6.80	6.46	6.19	5.98
17	0.100	3.03	2.64	2.44	2.31	2.22	2.15	2.10	2.06	2.03
17	0.050	4.45	3.59	3.20	2.96	2.81	2.70	2.61	2.55	2.49
17	0.025	6.04	4.62	4.01	3.66	3.44	3.28	3.16	3.06	2.98
17	0.010	8.40	6.11	5.18	4.67	4.34	4.10	3.93	3.79	3.68
17	0.001	15.72	10.66	8.73	7.68	7.02	6.56	6.22	5.96	5.75
18	0.100	3.01	2.62	2.42	2.29	2.20	2.13	2.08	2.04	2.00
18	0.050	4.41	3.55	3.16	2.93	2.77	2.66	2.58	2.51	2.46
18	0.025	5.98	4.56	3.95	3.61	3.38	3.22	3.10	3.01	2.93
18	0.010	8.29	6.01	5.09	4.58	4.25	4.01	3.84	3.71	3.60
18	0.001	15.38	10.39	8.49	7.46	6.81	6.35	6.02	5.76	5.56
19	0.100	2.99	2.61	2.40	2.27	2.18	2.11	2.06	2.02	1.98
19	0.050	4.38	3.52	3.13	2.90	2.74	2.63	2.54	2.48	2.42
19	0.025	5.92	4.51	3.90	3.56	3.33	3.17	3.05	2.96	2.88
19	0.010	8.18	5.93	5.01	4.50	4.17	3.94	3.77	3.63	3.52
19	0.001	15.08	10.16	8.28	7.27	6.62	6.18	5.85	5.59	5.39

Table A.5 Critical Values for the *F* Distribution (continued)

Denominator Degrees of Freedom	Area	Numerator Degrees of Freedom								
		1	2	3	4	5	6	7	8	9
20	0.100	2.97	2.59	2.38	2.25	2.16	2.09	2.04	2.00	1.96
20	0.050	4.35	3.49	3.10	2.87	2.71	2.60	2.51	2.45	2.39
20	0.025	5.87	4.46	3.86	3.51	3.29	3.13	3.01	2.91	2.84
20	0.010	8.10	5.85	4.94	4.43	4.10	3.87	3.70	3.56	3.46
20	0.001	14.82	9.95	8.10	7.10	6.46	6.02	5.69	5.44	5.24
21	0.100	2.96	2.57	2.36	2.23	2.14	2.08	2.02	1.98	1.95
21	0.050	4.32	3.47	3.07	2.84	2.68	2.57	2.49	2.42	2.37
21	0.025	5.83	4.42	3.82	3.48	3.25	3.09	2.97	2.87	2.80
21	0.010	8.02	5.78	4.87	4.37	4.04	3.81	3.64	3.51	3.40
21	0.001	14.59	9.77	7.94	6.95	6.32	5.88	5.56	5.31	5.11
22	0.100	2.95	2.56	2.35	2.22	2.13	2.06	2.01	1.97	1.93
22	0.050	4.30	3.44	3.05	2.82	2.66	2.55	2.46	2.40	2.34
22	0.025	5.79	4.38	3.78	3.44	3.22	3.05	2.93	2.84	2.76
22	0.010	7.95	5.72	4.82	4.31	3.99	3.76	3.59	3.45	3.35
22	0.001	14.38	9.61	7.80	6.81	6.19	5.76	5.44	5.19	4.99
23	0.100	2.94	2.55	2.34	2.21	2.11	2.05	1.99	1.95	1.92
23	0.050	4.28	3.42	3.03	2.80	2.64	2.53	2.44	2.37	2.32
23	0.025	5.75	4.35	3.75	3.41	3.18	3.02	2.90	2.81	2.73
23	0.010	7.88	5.66	4.76	4.26	3.94	3.71	3.54	3.41	3.30
23	0.001	14.20	9.47	7.67	6.70	6.08	5.65	5.33	5.09	4.89
24	0.100	2.93	2.54	2.33	2.19	2.10	2.04	1.98	1.94	1.91
24	0.050	4.26	3.40	3.01	2.78	2.62	2.51	2.42	2.36	2.30
24	0.025	5.72	4.32	3.72	3.38	3.15	2.99	2.87	2.78	2.70
24	0.010	7.82	5.61	4.72	4.22	3.90	3.67	3.50	3.36	3.26
24	0.001	14.03	9.34	7.55	6.59	5.98	5.55	5.23	4.99	4.80
25	0.100	2.92	2.53	2.32	2.18	2.09	2.02	1.97	1.93	1.89
25	0.050	4.24	3.39	2.99	2.76	2.60	2.49	2.40	2.34	2.28
25	0.025	5.69	4.29	3.69	3.35	3.13	2.97	2.85	2.75	2.68
25	0.010	7.77	5.57	4.68	4.18	3.85	3.63	3.46	3.32	3.22
25	0.001	13.88	9.22	7.45	6.49	5.89	5.46	5.15	4.91	4.71
30	0.100	2.88	2.49	2.28	2.14	2.05	1.98	1.93	1.88	1.85
30	0.050	4.17	3.32	2.92	2.69	2.53	2.42	2.33	2.27	2.21
30	0.025	5.57	4.18	3.59	3.25	3.03	2.87	2.75	2.65	2.57
30	0.010	7.56	5.39	4.51	4.02	3.70	3.47	3.30	3.17	3.07
30	0.001	13.29	8.77	7.05	6.12	5.53	5.12	4.82	4.58	4.39
40	0.100	2.84	2.44	2.23	2.09	2.00	1.93	1.87	1.83	1.79
40	0.050	4.08	3.23	2.84	2.61	2.45	2.34	2.25	2.18	2.12
40	0.025	5.42	4.05	3.46	3.13	2.90	2.74	2.62	2.53	2.45
40	0.010	7.31	5.18	4.31	3.83	3.51	3.29	3.12	2.99	2.89
40	0.001	12.61	8.25	6.59	5.70	5.13	4.73	4.44	4.21	4.02
60	0.100	2.79	2.39	2.18	2.04	1.95	1.87	1.82	1.77	1.74
60	0.050	4.00	3.15	2.76	2.53	2.37	2.25	2.17	2.10	2.04
60	0.025	5.29	3.93	3.34	3.01	2.79	2.63	2.51	2.41	2.33
60	0.010	7.08	4.98	4.13	3.65	3.34	3.12	2.95	2.82	2.72
60	0.001	11.97	7.77	6.17	5.31	4.76	4.37	4.09	3.86	3.69
120	0.100	2.75	2.35	2.13	1.99	1.90	1.82	1.77	1.72	1.68
120	0.050	3.92	3.07	2.68	2.45	2.29	2.18	2.09	2.02	1.96
120	0.025	5.15	3.80	3.23	2.89	2.67	2.52	2.39	2.30	2.22
120	0.010	6.85	4.79	3.95	3.48	3.17	2.96	2.79	2.66	2.56
120	0.001	11.38	7.32	5.78	4.95	4.42	4.04	3.77	3.55	3.38

Table A.5 Critical Values for the *F* Distribution (continued)

Denominator Degrees of Freedom	Area	Numerator Degrees of Freedom								
		10	**12**	**15**	**20**	**25**	**30**	**40**	**50**	**60**
10	0.100	2.32	2.28	2.24	2.20	2.17	2.16	2.13	2.12	2.11
10	0.050	2.98	2.91	2.85	2.77	2.73	2.70	2.66	2.64	2.62
10	0.025	3.72	3.62	3.52	3.42	3.35	3.31	3.26	3.22	3.20
10	0.010	4.85	4.71	4.56	4.41	4.31	4.25	4.17	4.12	4.08
10	0.001	8.75	8.45	8.13	7.80	7.60	7.47	7.30	7.19	7.12
11	0.100	2.25	2.21	2.17	2.12	2.10	2.08	2.05	2.04	2.03
11	0.050	2.85	2.79	2.72	2.65	2.60	2.57	2.53	2.51	2.49
11	0.025	3.53	3.43	3.33	3.23	3.16	3.12	3.06	3.03	3.00
11	0.010	4.54	4.40	4.25	4.10	4.01	3.94	3.86	3.81	3.78
11	0.001	7.92	7.63	7.32	7.01	6.81	6.68	6.52	6.42	6.35
12	0.100	2.19	2.15	2.10	2.06	2.03	2.01	1.99	1.97	1.96
12	0.050	2.75	2.69	2.62	2.54	2.50	2.47	2.43	2.40	2.38
12	0.025	3.37	3.28	3.18	3.07	3.01	2.96	2.91	2.87	2.85
12	0.010	4.30	4.16	4.01	3.86	3.76	3.70	3.62	3.57	3.54
12	0.001	7.29	7.00	6.71	6.40	6.22	6.09	5.93	5.83	5.76
13	0.100	2.14	2.10	2.05	2.01	1.98	1.96	1.93	1.92	1.90
13	0.050	2.67	2.60	2.53	2.46	2.41	2.38	2.34	2.31	2.30
13	0.025	3.25	3.15	3.05	2.95	2.88	2.84	2.78	2.74	2.72
13	0.010	4.10	3.96	3.82	3.66	3.57	3.51	3.43	3.38	3.34
13	0.001	6.80	6.52	6.23	5.93	5.75	5.63	5.47	5.37	5.30
14	0.100	2.10	2.05	2.01	1.96	1.93	1.91	1.89	1.87	1.86
14	0.050	2.60	2.53	2.46	2.39	2.34	2.31	2.27	2.24	2.22
14	0.025	3.15	3.05	2.95	2.84	2.78	2.73	2.67	2.64	2.61
14	0.010	3.94	3.80	3.66	3.51	3.41	3.35	3.27	3.22	3.18
14	0.001	6.40	6.13	5.85	5.56	5.38	5.25	5.10	5.00	4.94
15	0.100	2.06	2.02	1.97	1.92	1.89	1.87	1.85	1.83	1.82
15	0.050	2.54	2.48	2.40	2.33	2.28	2.25	2.20	2.18	2.16
15	0.025	3.06	2.96	2.86	2.76	2.69	2.64	2.59	2.55	2.52
15	0.010	3.80	3.67	3.52	3.37	3.28	3.21	3.13	3.08	3.05
15	0.001	6.08	5.81	5.54	5.25	5.07	4.95	4.80	4.70	4.64
16	0.100	2.03	1.99	1.94	1.89	1.86	1.84	1.81	1.79	1.78
16	0.050	2.49	2.42	2.35	2.28	2.23	2.19	2.15	2.12	2.11
16	0.025	2.99	2.89	2.79	2.68	2.61	2.57	2.51	2.47	2.45
16	0.010	3.69	3.55	3.41	3.26	3.16	3.10	3.02	2.97	2.93
16	0.001	5.81	5.55	5.27	4.99	4.82	4.70	4.54	4.45	4.39
17	0.100	2.00	1.96	1.91	1.86	1.83	1.81	1.78	1.76	1.75
17	0.050	2.45	2.38	2.31	2.23	2.18	2.15	2.10	2.08	2.06
17	0.025	2.92	2.82	2.72	2.62	2.55	2.50	2.44	2.41	2.38
17	0.010	3.59	3.46	3.31	3.16	3.07	3.00	2.92	2.87	2.83
17	0.001	5.58	5.32	5.05	4.78	4.60	4.48	4.33	4.24	4.18
18	0.100	1.98	1.93	1.89	1.84	1.80	1.78	1.75	1.74	1.72
18	0.050	2.41	2.34	2.27	2.19	2.14	2.11	2.06	2.04	2.02
18	0.025	2.87	2.77	2.67	2.56	2.49	2.44	2.38	2.35	2.32
18	0.010	3.51	3.37	3.23	3.08	2.98	2.92	2.84	2.78	2.75
18	0.001	5.39	5.13	4.87	4.59	4.42	4.30	4.15	4.06	4.00
19	0.100	1.96	1.91	1.86	1.81	1.78	1.76	1.73	1.71	1.70
19	0.050	2.38	2.31	2.23	2.16	2.11	2.07	2.03	2.00	1.98
19	0.025	2.82	2.72	2.62	2.51	2.44	2.39	2.33	2.30	2.27
19	0.010	3.43	3.30	3.15	3.00	2.91	2.84	2.76	2.71	2.67
19	0.001	5.22	4.97	4.70	4.43	4.26	4.14	3.99	3.90	3.84

Table A.5 Critical Values for the *F* Distribution (continued)

Denominator Degrees of Freedom	Area	Numerator Degrees of Freedom								
		10	12	15	20	25	30	40	50	60
20	0.100	1.94	1.89	1.84	1.79	1.76	1.74	1.71	1.69	1.68
20	0.050	2.35	2.28	2.20	2.12	2.07	2.04	1.99	1.97	1.95
20	0.025	2.77	2.68	2.57	2.46	2.40	2.35	2.29	2.25	2.22
20	0.010	3.37	3.23	3.09	2.94	2.84	2.78	2.69	2.64	2.61
20	0.001	5.08	4.82	4.56	4.29	4.12	4.00	3.86	3.77	3.70
21	0.100	2.96	2.57	2.36	2.23	2.14	2.08	2.02	1.98	1.95
21	0.050	4.32	3.47	3.07	2.84	2.68	2.57	2.49	2.42	2.37
21	0.025	2.73	2.64	2.53	2.42	2.36	2.31	2.25	2.21	2.18
21	0.010	8.02	5.78	4.87	4.37	4.04	3.81	3.64	3.51	3.40
21	0.001	14.59	9.77	7.94	6.95	6.32	5.88	5.56	5.31	5.11
22	0.100	2.95	2.56	2.35	2.22	2.13	2.06	2.01	1.97	1.93
22	0.050	4.30	3.44	3.05	2.82	2.66	2.55	2.46	2.40	2.34
22	0.025	2.70	2.60	2.50	2.39	2.32	2.27	2.21	2.17	2.14
22	0.010	7.95	5.72	4.82	4.31	3.99	3.76	3.59	3.45	3.35
22	0.001	14.38	9.61	7.80	6.81	6.19	5.76	5.44	5.19	4.99
23	0.100	2.94	2.55	2.34	2.21	2.11	2.05	1.99	1.95	1.92
23	0.050	4.28	3.42	3.03	2.80	2.64	2.53	2.44	2.37	2.32
23	0.025	2.67	2.57	2.47	2.36	2.29	2.24	2.18	2.14	2.11
23	0.010	7.88	5.66	4.76	4.26	3.94	3.71	3.54	3.41	3.30
23	0.001	14.20	9.47	7.67	6.70	6.08	5.65	5.33	5.09	4.89
24	0.100	2.93	2.54	2.33	2.19	2.10	2.04	1.98	1.94	1.91
24	0.050	4.26	3.40	3.01	2.78	2.62	2.51	2.42	2.36	2.30
24	0.025	2.64	2.54	2.44	2.33	2.26	2.21	2.15	2.11	2.08
24	0.010	7.82	5.61	4.72	4.22	3.90	3.67	3.50	3.36	3.26
24	0.001	14.03	9.34	7.55	6.59	5.98	5.55	5.23	4.99	4.80
25	0.100	2.92	2.53	2.32	2.18	2.09	2.02	1.97	1.93	1.89
25	0.050	4.24	3.39	2.99	2.76	2.60	2.49	2.40	2.34	2.28
25	0.025	2.61	2.51	2.41	2.30	2.23	2.18	2.12	2.08	2.05
25	0.010	7.77	5.57	4.68	4.18	3.85	3.63	3.46	3.32	3.22
25	0.001	13.88	9.22	7.45	6.49	5.89	5.46	5.15	4.91	4.71
30	0.100	1.82	1.77	1.72	1.67	1.63	1.61	1.57	1.55	1.54
30	0.050	2.16	2.09	2.01	1.93	1.88	1.84	1.79	1.76	1.74
30	0.025	2.51	2.41	2.31	2.20	2.12	2.07	2.01	1.97	1.94
30	0.010	2.98	2.84	2.70	2.55	2.45	2.39	2.30	2.25	2.21
30	0.001	4.24	4.00	3.75	3.49	3.33	3.22	3.07	2.98	2.92
40	0.100	1.76	1.71	1.66	1.61	1.57	1.54	1.51	1.48	1.47
40	0.050	2.08	2.00	1.92	1.84	1.78	1.74	1.69	1.66	1.64
40	0.025	2.39	2.29	2.18	2.07	1.99	1.94	1.88	1.83	1.80
40	0.010	2.80	2.66	2.52	2.37	2.27	2.20	2.11	2.06	2.02
40	0.001	3.87	3.64	3.40	3.14	2.98	2.87	2.73	2.64	2.57
60	0.100	1.71	1.66	1.60	1.54	1.50	1.48	1.44	1.41	1.40
60	0.050	1.99	1.92	1.84	1.75	1.69	1.65	1.59	1.56	1.53
60	0.025	2.27	2.17	2.06	1.94	1.87	1.82	1.74	1.70	1.67
60	0.010	2.63	2.50	2.35	2.20	2.10	2.03	1.94	1.88	1.84
60	0.001	3.54	3.32	3.08	2.83	2.67	2.55	2.41	2.32	2.25
120	0.100	1.65	1.60	1.55	1.48	1.44	1.41	1.37	1.34	1.32
120	0.050	1.91	1.83	1.75	1.66	1.60	1.55	1.50	1.46	1.43
120	0.025	2.16	2.05	1.94	1.82	1.75	1.69	1.61	1.56	1.53
120	0.010	2.47	2.34	2.19	2.03	1.93	1.86	1.76	1.70	1.66
120	0.001	3.24	3.02	2.78	2.53	2.37	2.26	2.11	2.02	1.95

Table A.6 Critical Values for the Studentized Range

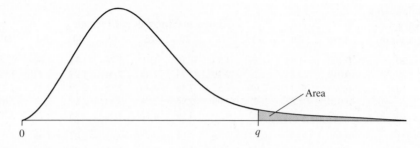

		I												
N − *I*	Area	2	3	4	5	6	7	8	9	10	11	12	13	14	15
1	0.10	8.93	13.44	16.36	18.49	20.15	21.51	22.64	23.62	24.48	25.24	25.92	26.54	27.10	27.62
1	0.05	17.97	26.98	32.82	37.08	40.41	43.12	45.40	47.36	49.07	50.59	51.96	53.20	54.33	55.36
1	0.01	90.03	135.0	164.3	185.6	202.2	215.8	227.2	237.0	245.6	253.2	260.0	266.2	271.8	277.0
2	0.10	4.13	5.73	6.77	7.54	8.14	8.63	9.05	9.41	9.72	10.01	10.26	10.49	10.70	10.89
2	0.05	6.08	8.33	9.80	10.88	11.74	12.44	13.03	13.54	13.99	14.39	14.75	15.08	15.38	15.65
2	0.01	14.04	19.02	22.29	24.72	26.63	28.20	29.53	30.68	31.69	32.59	33.40	34.13	34.81	35.43
3	0.10	3.33	4.47	5.20	5.74	6.16	6.51	6.81	7.06	7.29	7.49	7.67	7.83	7.98	8.12
3	0.05	4.50	5.91	6.82	7.50	8.04	8.48	8.85	9.18	9.46	9.72	9.95	10.15	10.35	10.52
3	0.01	8.26	10.62	12.17	13.33	14.24	15.00	15.64	16.20	16.69	17.13	17.53	17.89	18.22	18.52
4	0.10	3.01	3.98	4.59	5.03	5.39	5.68	5.93	6.14	6.33	6.49	6.65	6.78	6.91	7.02
4	0.05	3.93	5.04	5.76	6.29	6.71	7.05	7.35	7.60	7.83	8.03	8.21	8.37	8.52	8.66
4	0.01	6.51	8.12	9.17	9.96	10.58	11.10	11.55	11.93	12.27	12.57	12.84	13.09	13.32	13.53
5	0.10	2.85	3.72	4.26	4.66	4.98	5.24	5.46	5.65	5.82	5.97	6.10	6.22	6.34	6.44
5	0.05	3.64	4.60	5.22	5.67	6.03	6.33	6.58	6.80	6.99	7.17	7.32	7.47	7.60	7.72
5	0.01	5.70	6.98	7.80	8.42	8.91	9.32	9.67	9.97	10.24	10.48	10.70	10.89	11.08	11.24
6	0.10	2.75	3.56	4.07	4.44	4.73	4.97	5.17	5.34	5.50	5.64	5.76	5.87	5.98	6.07
6	0.05	3.46	4.34	4.90	5.30	5.63	5.90	6.12	6.32	6.49	6.65	6.79	6.92	7.03	7.14
6	0.01	5.24	6.33	7.03	7.56	7.97	8.32	8.61	8.87	9.10	9.30	9.48	9.65	9.81	9.95
7	0.10	2.68	3.45	3.93	4.28	4.55	4.78	4.97	5.14	5.28	5.41	5.53	5.64	5.74	5.83
7	0.05	3.34	4.16	4.68	5.06	5.36	5.61	5.82	6.00	6.16	6.30	6.43	6.55	6.66	6.76
7	0.01	4.95	5.92	6.54	7.01	7.37	7.68	7.94	8.17	8.37	8.55	8.71	8.86	9.00	9.12
8	0.10	2.63	3.37	3.83	4.17	4.43	4.65	4.83	4.99	5.13	5.25	5.36	5.46	5.56	5.64
8	0.05	3.26	4.04	4.53	4.89	5.17	5.40	5.60	5.77	5.92	6.05	6.18	6.29	6.39	6.48
8	0.01	4.75	5.64	6.20	6.62	6.96	7.24	7.47	7.68	7.86	8.03	8.18	8.31	8.44	8.55
9	0.10	2.59	3.32	3.76	4.08	4.34	4.54	4.72	4.87	5.01	5.13	5.23	5.33	5.42	5.51
9	0.05	3.20	3.95	4.41	4.76	5.02	5.24	5.43	5.59	5.74	5.87	5.98	6.09	6.19	6.28
9	0.01	4.60	5.43	5.96	6.35	6.66	6.91	7.13	7.33	7.49	7.65	7.78	7.91	8.03	8.13
10	0.10	2.56	3.27	3.70	4.02	4.26	4.47	4.64	4.78	4.91	5.03	5.13	5.23	5.32	5.40
10	0.05	3.15	3.88	4.33	4.65	4.91	5.12	5.30	5.46	5.60	5.72	5.83	5.93	6.03	6.11
10	0.01	4.48	5.27	5.77	6.14	6.43	6.67	6.87	7.05	7.21	7.36	7.49	7.60	7.71	7.81
11	0.10	2.54	3.23	3.66	3.96	4.20	4.40	4.57	4.71	4.84	4.95	5.05	5.15	5.23	5.31
11	0.05	3.11	3.82	4.26	4.57	4.82	5.03	5.20	5.35	5.49	5.61	5.71	5.81	5.90	5.98
11	0.01	4.39	5.15	5.62	5.97	6.25	6.48	6.67	6.84	6.99	7.13	7.25	7.36	7.46	7.56
12	0.10	2.52	3.20	3.62	3.92	4.16	4.35	4.51	4.65	4.78	4.89	4.99	5.08	5.16	5.24
12	0.05	3.08	3.77	4.20	4.51	4.75	4.95	5.12	5.27	5.39	5.51	5.61	5.71	5.80	5.88
12	0.01	4.32	5.05	5.50	5.84	6.10	6.32	6.51	6.67	6.81	6.94	7.06	7.17	7.26	7.36

Table A.6 Critical Values for the Studentized Range (continued)

$N - I$	Area	2	3	4	5	6	7	8	9	10	11	12	13	14	15
13	0.10	2.50	3.18	3.59	3.88	4.12	4.30	4.46	4.60	4.72	4.83	4.93	5.02	5.10	5.18
13	0.05	3.06	3.73	4.15	4.45	4.69	4.88	5.05	5.19	5.32	5.43	5.53	5.63	5.71	5.79
13	0.01	4.26	4.96	5.40	5.73	5.98	6.19	6.37	6.53	6.67	6.79	6.90	7.01	7.10	7.19
14	0.10	2.49	3.16	3.56	3.85	4.08	4.27	4.42	4.56	4.68	4.79	4.88	4.97	5.05	5.12
14	0.05	3.03	3.70	4.11	4.41	4.64	4.83	4.99	5.13	5.25	5.36	5.46	5.55	5.64	5.71
14	0.01	4.21	4.89	5.32	5.63	5.88	6.08	6.26	6.41	6.54	6.66	6.77	6.87	6.96	7.05
15	0.10	2.48	3.14	3.54	3.83	4.05	4.23	4.39	4.52	4.64	4.75	4.84	4.93	5.01	5.08
15	0.05	3.01	3.67	4.08	4.37	4.59	4.78	4.94	5.08	5.20	5.31	5.40	5.49	5.57	5.65
15	0.01	4.17	4.84	5.25	5.56	5.80	5.99	6.16	6.31	6.44	6.55	6.66	6.76	6.84	6.93
16	0.10	2.47	3.12	3.52	3.80	4.03	4.21	4.36	4.49	4.61	4.71	4.81	4.89	4.97	5.04
16	0.05	3.00	3.65	4.05	4.33	4.56	4.74	4.90	5.03	5.15	5.26	5.35	5.44	5.52	5.59
16	0.01	4.13	4.79	5.19	5.49	5.72	5.92	6.08	6.22	6.35	6.46	6.56	6.66	6.74	6.82
17	0.10	2.46	3.11	3.50	3.78	4.00	4.18	4.33	4.46	4.58	4.68	4.77	4.86	4.93	5.01
17	0.05	2.98	3.63	4.02	4.30	4.52	4.70	4.86	4.99	5.11	5.21	5.31	5.39	5.47	5.54
17	0.01	4.10	4.74	5.14	5.43	5.66	5.85	6.01	6.15	6.27	6.38	6.48	6.57	6.66	6.73
18	0.10	2.45	3.10	3.49	3.77	3.98	4.16	4.31	4.44	4.55	4.65	4.75	4.83	4.90	4.98
18	0.05	2.97	3.61	4.00	4.28	4.49	4.67	4.82	4.96	5.07	5.17	5.27	5.35	5.43	5.50
18	0.01	4.07	4.70	5.09	5.38	5.60	5.79	5.94	6.08	6.20	6.31	6.41	6.50	6.58	6.65
19	0.10	2.45	3.09	3.47	3.75	3.97	4.14	4.29	4.42	4.53	4.63	4.72	4.80	4.88	4.95
19	0.05	2.96	3.59	3.98	4.25	4.47	4.65	4.79	4.92	5.04	5.14	5.23	5.31	5.39	5.46
19	0.01	4.05	4.67	5.05	5.33	5.55	5.73	5.89	6.02	6.14	6.25	6.34	6.43	6.51	6.58
20	0.10	2.44	3.08	3.46	3.74	3.95	4.12	4.27	4.40	4.51	4.61	4.70	4.78	4.85	4.92
20	0.05	2.95	3.58	3.96	4.23	4.45	4.62	4.77	4.90	5.01	5.11	5.20	5.28	5.36	5.43
20	0.01	4.02	4.64	5.02	5.29	5.51	5.69	5.84	5.97	6.09	6.19	6.28	6.37	6.45	6.52
24	0.10	2.42	3.05	3.42	3.69	3.90	4.07	4.21	4.34	4.44	4.54	4.63	4.71	4.78	4.85
24	0.05	2.92	3.53	3.90	4.17	4.37	4.54	4.68	4.81	4.92	5.01	5.10	5.18	5.25	5.32
24	0.01	3.96	4.55	4.91	5.17	5.37	5.54	5.69	5.81	5.92	6.02	6.11	6.19	6.26	6.33
30	0.10	2.40	3.02	3.39	3.65	3.85	4.02	4.16	4.28	4.38	4.47	4.56	4.64	4.71	4.77
30	0.05	2.89	3.49	3.85	4.10	4.30	4.46	4.60	4.72	4.82	4.92	5.00	5.08	5.15	5.21
30	0.01	3.89	4.45	4.80	5.05	5.24	5.40	5.54	5.65	5.76	5.85	5.93	6.01	6.08	6.14
40	0.10	2.38	2.99	3.35	3.60	3.80	3.96	4.10	4.21	4.32	4.41	4.49	4.56	4.63	4.69
40	0.05	2.86	3.44	3.79	4.04	4.23	4.39	4.52	4.63	4.73	4.82	4.90	4.98	5.04	5.11
40	0.01	3.82	4.37	4.70	4.93	5.11	5.26	5.39	5.50	5.60	5.69	5.76	5.83	5.90	5.96
60	0.10	2.36	2.96	3.31	3.56	3.75	3.91	4.04	4.16	4.25	4.34	4.42	4.49	4.56	4.62
60	0.05	2.83	3.40	3.74	3.98	4.16	4.31	4.44	4.55	4.65	4.73	4.81	4.88	4.94	5.00
60	0.01	3.76	4.28	4.59	4.82	4.99	5.13	5.25	5.36	5.45	5.53	5.60	5.67	5.73	5.78
120	0.10	2.34	2.93	3.28	3.52	3.71	3.86	3.99	4.10	4.19	4.28	4.35	4.42	4.48	4.54
120	0.05	2.80	3.36	3.68	3.92	4.10	4.24	4.36	4.47	4.56	4.64	4.71	4.78	4.84	4.90
120	0.01	3.70	4.20	4.50	4.71	4.87	5.01	5.12	5.21	5.30	5.37	5.44	5.50	5.56	5.61
∞	0.10	2.33	2.90	3.24	3.48	3.66	3.81	3.93	4.04	4.13	4.21	4.28	4.35	4.41	4.47
∞	0.05	2.77	3.31	3.63	3.86	4.03	4.17	4.29	4.39	4.47	4.55	4.62	4.68	4.74	4.80
∞	0.01	3.64	4.12	4.40	4.60	4.76	4.88	4.99	5.08	5.16	5.23	5.29	5.35	5.40	5.45

I (column spanning header over columns 2–15)

Table A.7 Critical Values for the Sign Test

	One-tailed			
	$\alpha = 0.005$	$\alpha = 0.01$	$\alpha = 0.025$	$\alpha = 0.05$
	Two-tailed			
n	$\alpha = 0.01$	$\alpha = 0.02$	$\alpha = 0.05$	$\alpha = 0.10$
5	*	*	*	0
6	*	*	0	0
7	*	0	0	0
8	0	0	0	1
9	0	0	1	1
10	0	0	1	1
11	0	1	1	2
12	1	1	2	2
13	1	1	2	3
14	1	2	3	3
15	2	2	3	3
16	2	2	3	4
17	2	3	4	4
18	3	3	4	5
19	3	4	4	5
20	3	4	5	5
21	4	4	5	6
22	4	5	5	6
23	4	5	6	7
24	5	5	6	7
25	5	6	6	7

*It is impossible for the test statistic to be in the critical region.

Table A.8 Critical Values for the Signed-Rank Test

n	$\alpha = 0.10$	$\alpha = 0.05$	$\alpha = 0.02$	$\alpha = 0.01$
5	1	*	*	*
6	2	1	*	*
7	4	2	0	*
8	6	4	2	0
9	8	6	3	2
10	11	8	5	3
11	14	11	7	5
12	17	14	10	7
13	21	17	13	10
14	26	21	16	13
15	30	25	20	16
16	36	30	24	19
17	41	35	28	23
18	47	40	33	28
19	54	46	38	32
20	60	52	43	37
21	68	59	49	43
22	75	66	56	49
23	83	73	62	55
24	92	81	69	61
25	101	90	77	68
26	110	98	85	76
27	120	107	93	84
28	130	117	102	92
29	141	127	111	100
30	152	137	120	109

*It is impossible for the test statistic to be in the critical region.

TI-84 PLUS Stat Wizards

Stat Wizards is a feature on TI-84 PLUS calculators running operating system OS 2.55 MP or higher. This feature provides a wizard interface for selected commands and functions. The latest operating system may be downloaded from http://education.ti.com/calculators/downloads/.

The Stat Wizards may be turned on or off through an option on the **MODE** screen. The **MATHPRINT/CLASSIC** options affect the appearance of the calculator output. These options also appear on this screen. The screenshots in this text are generated using the **CLASSIC** setting.

Wizards are available for all commands in the **[STAT]CALC** menu and the **DISTR** menu.

1-Var Stats

The **1-Var Stats** wizard is accessed by selecting the **1-Var Stats** option under the [STAT] CALC menu. **L1** is the default setting for the **List** field. **FreqList** is an optional argument. **FreqList** accepts list names only. To run the command, select **Calculate** and press ENTER.

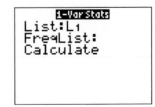

LinReg(a + bx)

The **LinReg(a + bx)** wizard is accessed by selecting the **LinReg(a + bx)** option under the [STAT] CALC menu. **L1** and **L2** are the default settings for the **Xlist** and **Ylist** fields. **FreqList** is an optional argument. A function name such as **Y1** may be entered in the **Store RegEQ** field as a location to store the regression equation. To run the command, select **Calculate** and press ENTER.

binompdf/binomcdf

The wizards for **binompdf** or **binomcdf** are accessed by selecting either the **binompdf** or **binomcdf** option under the **DISTR** menu. After entering the values for n, p, and x in the **trials**, **p**, and **x value** fields, select **Paste** and press [ENTER] to paste the command on the home screen.

poissonpdf/poissoncdf

The wizards for **poissonpdf** or **poissoncdf** are accessed by selecting either the **poissonpdf** or **poissoncdf** option under the **DISTR** menu. After entering the values for λ and x in the λ and **x value** fields, select **Paste** and press [ENTER] to paste the command on the home screen.

normalcdf

The **normalcdf** wizard is accessed by selecting the **normalcdf** option under the **DISTR** menu. The lower and upper endpoints are entered in the **lower** and **upper** fields. The default for the **lower** field is -1E99. The mean and standard deviation are entered in the μ and σ fields. The default for the μ field is 0 and the default for the σ field is 1. After entering all values, select **Paste** and press [ENTER] to paste the command on the home screen.

invNorm

The **invNorm** wizard is accessed by selecting the **invNorm** option under the **DISTR** menu. The area to the left of the desired value is entered in the **area** field. The mean and standard deviation are entered in the μ and σ fields. The default for the μ field is 0 and the default for the σ field is 1. After entering all values, select **Paste** and press to paste the command on the home screen.

```
DISTR DRAW
1:normalpdf(
2:normalcdf(
3:invNorm(
4:invT(
5:tpdf(
6:tcdf(
7↓Χ²pdf(
```

```
      invNorm
area:
μ:0
σ:1
Paste
```

Answers to Odd-Numbered Exercises

CHAPTER 1

Section 1.1

Exercises 1–6 are the Check Your Understanding exercises for this section. Answers to these exercises are on page 12.

7. population

9. simple random sample

11. cluster

13. False

15. True

17. Statistic

19. Parameter

21. Answers will vary.

23. Answers will vary. Cluster sample

25. Stratified sample

27. Cluster sample

29. Voluntary response sample

31. Sample of convenience

33. Stratified sample

35. Simple random sample

37. Systematic sample

39. It will be necessary to draw a sample of convenience. There is no list of all headache sufferers from which to draw a simple random sample.

41. Answers will vary. A simple random sample could be drawn from a list of all registered voters in the town.

43. Answers will vary. A stratified sample, consisting of simple random samples of 100 men and 100 women, could be drawn.

45. Answers will vary.

Section 1.2

Exercises 1–4 are the Check Your Understanding exercises for this section. Answers to these exercises are on page 18.

5. variables

7. Quantitative

9. discrete

11. False

13. True

15. Qualitative

17. Quantitative

19. Quantitative

21. Qualitative

23. Qualitative

25. Ordinal

27. Ordinal

29. Nominal

31. Nominal

33. Continuous

35. Discrete

37. Continuous

39. Discrete

41. Ordinal

43. Ordinal

45. Nominal

47. **a.** Game title, Publisher
    b. Percentage of gaming audience, Average minutes played per week
    c. Publisher
    d. Game title

49. **a.** Ordinal
    b. Yes, it reflects a more favorable opinion of the construction of a new shopping mall.
    c. No, we cannot say that Jason's opinion is twice as favorable.
    d. Quantitative
    e. Yes, Brenda's answer reflects the ownership of more cars, and specifically, the ownership of twice as many cars.
    f. Nominal
    g. No, Brenda's answer reflects neither more of something nor twice as much of something.

Section 1.3

Exercises 1–4 are the Check Your Understanding exercises for this section. Answers to these exercises are on page 26.

5. randomized

7. observational

9. prospective

11. True

13. False

15. True

17. **a.** Randomized experiment
    b. Yes, because the assignment to treatments is made at random, there is no systematic difference between the groups other than the drug taken that can explain the difference in pain relief.

19. **a.** Randomized experiment
    b. Yes, because the assignment to treatments is made at random, there is no systematic difference between the groups other than the amount of exercise that can explain the difference in blood pressure.

21. An observational study will be necessary, because one can't assign people to live in areas with high pollution levels.

23. The result may be due to confounding. Areas with denser populations may have both more crime and more taxicabs.

25. a. False   **b.** True

27. a. Heart rate   **b.** Maternal smoking
   c. Cohort   **d.** Prospective
   e. Yes. The level of prenatal care may differ between smoking and nonsmoking mothers.

29. a. Yes, because the subjects were randomly assigned to treatment
   b. If a doctor knew whether a child had received the vaccine, it might influence the diagnosis.
   c. It could be due to confounding. The children who received the placebo were more likely to be middle- or upper-income than those who did not participate, and this may be the reason that the rate of polio was higher.

Section 1.4
Exercises 1 and 2 are the Check Your Understanding exercises for this section. Answers to these exercises are on page 29.

3. Voluntary response surveys

5. population

7. True

9. Nonresponse bias

11. Self-interest bias

13. Voluntary response bias

15. Nonresponse bias

17. a. No
   b. No. Both questions are leading. The first question leads to a "yes" response, and the second leads to a "no" response.

19. Yes. People who do not have landline phones may tend to have different opinions on some issues than people who do have landline phones.

21. a. The poll oversampled higher-income people.
   b. Nonresponse bias. The response rate was low — only 23%. This results in nonresponse bias.
   c. A sample that is not drawn by a valid method can produce misleading results, even when it is large.

CHAPTER 1 Quiz
1. Answers will vary.

2. Qualitative

3. True

4. Continuous

5. False

6. Stratified sample

7. acceptable

8. Sample of convenience

9. True

10. Observational study

11. Randomized experiment

12. differences in treatment

13. Seniors may be more likely to have better preparation for the class than sophomores.

14. True

15. Not reliable. This is a voluntary response survey, so the people who respond tend to hold stronger opinions than others.

CHAPTER 1 Review Exercises
1. Quantitative

2. Nominal

3. Continuous

4. a. True   **b.** True   **c.** False

5. Stratified sample

6. Voluntary response sample

7. Cluster sample

8. Simple random sample

9. a. Observational study
   b. Yes. People who live in areas with fluoridated water may have different dental habits than those who live in areas without fluoridated water.

10. a. Randomized experiment
   b. Because this is a randomized experiment, the results are unlikely to be due to confounding.

11. a. Observational study
   b. Yes. People who talk on cell phones while driving may be more careless in general than those who do not.

12. a. Randomized experiment
   b. Because this is a randomized experiment, the results are unlikely to be due to confounding.

13. The sample is a voluntary response sample.

14. Nonresponse bias; living people not included. People who are still alive are not included in the sample.

15. There is a considerable level of nonresponse bias.

CHAPTER 1 Case Study
1. 450

2. 41

3. 9.1%

4. 43

5. 2

6. 4.7%

7. Yes

8. The high-exposure people and the school-return people are the same people.

9. The low-exposure people and the mail-return people are the same people.

10. People who respond by mail will be responding during a period of lower PM.

11. People with symptoms may tend to respond earlier; therefore, people with symptoms are more likely to be school-return people.

12. There would be no tendency for people with symptoms to respond earlier.

CHAPTER 2

Section 2.1

Exercises 1–4 are the Check Your Understanding exercises for this section. Answers to these exercises are on pages 48–49.

5. frequency

7. Pareto chart

9. False

11. True

13. a. Meat, poultry, fish, and eggs   **b.** False   **c.** True

15. a.

b.

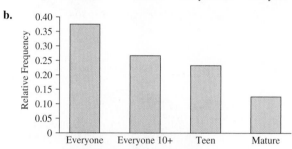

c. Everyone (E)   **d.** False   **e.** True

17. a. Families and individuals, Governments, Businesses
b. Produced at home
c. No
d. Yes

19. a.

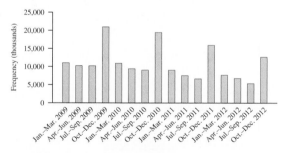

b.

Quarter	Relative Frequency
Jan.–Mar. 2009	0.064
Apr.–Jun. 2009	0.059
Jul.–Sep. 2009	0.059
Oct.–Dec. 2009	0.122
Jan.–Mar. 2010	0.063
Apr.–Jun. 2010	0.055
Jul.–Sep. 2010	0.053
Oct.–Dec. 2010	0.113
Jan.–Mar. 2011	0.052
Apr.–Jun. 2011	0.044
Jul.–Sep. 2011	0.038
Oct.–Dec. 2011	0.089
Jan.–Mar. 2012	0.045
Apr.–Jun. 2012	0.039
Jul.–Sep. 2012	0.031
Oct.–Dec. 2012	0.074

c.

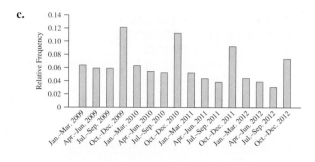

d. True

21. a.

Quarter	Frequency (thousands)
Jan.–Mar.	38,591
Apr.–Jun.	33,916
Jul.–Sep.	31,183
Oct.–Dec.	68,513

b.

c.

Quarter	Relative Frequency
Jan.–Mar.	0.224
Apr.–Jun.	0.197
Jul.–Sep.	0.181
Oct.–Dec.	0.398

d.

e.

f. False

23. a.

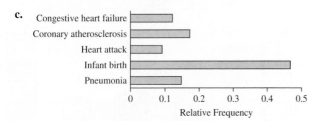

b.

Reason	Relative Frequency
Congestive heart failure	0.122
Coronary atherosclerosis	0.172
Heart attack	0.091
Infant birth	0.467
Pneumonia	0.148

c.

d.

e.

f. True

25. a.

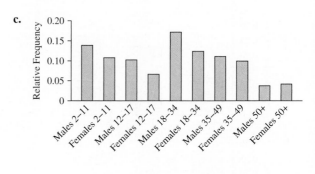

b.

Gender and Age Group	Relative Frequency
Males 2–11	0.139
Females 2–11	0.108
Males 12–17	0.102
Females 12–17	0.066
Males 18–34	0.172
Females 18–34	0.124
Males 35–49	0.111
Females 35–49	0.099
Males 50+	0.037
Females 50+	0.042

c.

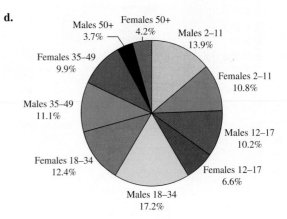

d.

Males 50+ 3.7% Females 50+ 4.2% Males 2–11 13.9%
Females 35–49 9.9%
Females 2–11 10.8%
Males 35–49 11.1%
Males 12–17 10.2%
Females 18–34 12.4%
Females 12–17 6.6%
Males 18–34 17.2%

e. True **f.** True **g.** 0.289

27. a.

Response	Relative Frequency
More than once per month	0.166
Once every 1–3 months	0.209
Once every 4–6 months	0.104
Once every 7–11 months	0.063
Once per year or less	0.153
Never	0.305

b.

c.

d.

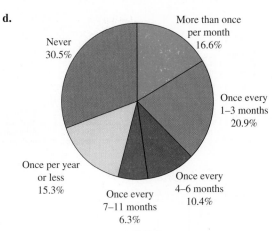

e. True **f.** False

29. a.

Type of Music	Relative Frequency
CD	0.148
Download single	0.687
Mobile	0.110
Other	0.056

b.

Type of Music	Relative Frequency
CD	0.136
Download single	0.735
Mobile	0.065
Other	0.064

c.

d. True

31. a.

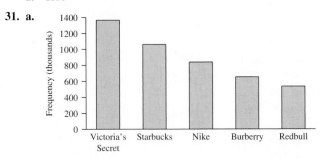

b.

Brand	Relative Frequency
Victoria's Secret	0.306
Starbucks	0.239
Nike	0.188
Burberry	0.146
Redbull	0.120

c.

d.

e. 0.239

33. a.

b.

Company	Relative Frequency
Samsung	0.308
Apple	0.182
LG Electronics	0.048
Huawei Technologies	0.044
ZTE	0.038
Others	0.379

c.

d.

e. False

35. a.

b.

c.

d.

e. The total frequency is equal to the sum of the frequencies for the two cities.

f. The total relative frequency is the total frequency divided by the sum of all total frequencies. The relative frequency for each city is the frequency for that city divided by the sum of the frequencies for that city. Since the sum of the frequencies for each city is not the same as the sum of the total frequencies, the total relative frequency is not the sum of the relative frequencies for the two cities.

Section 2.2

Exercises 1–4 are the Check Your Understanding exercises for this section. Answers to these exercises are on page 67.

5. symmetric

7. bimodal

9. False

11. True

13. Skewed to the left

15. Approximately symmetric

17. Bimodal

19. a. 11   **b.** 1   **c.** 70–71
　　d. 9%   **e.** Approximately symmetric

21. a. 30%
　　b. 240–260

23. a. Right
　　b. Left
　　c. Left

25. a. 9

b. 0.020

c. Lower limits: 0.180, 0.190, 0.200, 0.220, 0.240, 0.260, 0.280. 0.300, 0.320, 0.340. Upper limits: 0.199, 0.219, 0.239, 0.259, 0.279, 0.299, 0.319, 0.339, 0.359.

d.

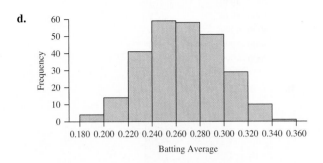

e.

Batting Average	Relative Frequency
0.180–0.199	0.015
0.200–0.219	0.052
0.220–0.239	0.154
0.240–0.259	0.221
0.260–0.279	0.217
0.280–0.299	0.191
0.300–0.319	0.109
0.320–0.339	0.037
0.340–0.359	0.004

f.

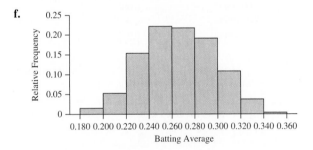

g. 15.0%

h. 6.7%

27. a. 10 **b.** 3.0

c. Lower limits: 1.0, 4.0, 7.0, 10.0, 13.0, 16.0, 19.0, 22.0, 25.0, 28.0. Upper limits: 3.9, 6.9, 9.9, 12.9, 15.9, 18.9, 21.9, 24.9, 27.9, 30.9.

d.

e.

Number of Hours	Relative Frequency
1.0–3.9	0.125
4.0–6.9	0.170
7.0–9.9	0.240
10.0–12.9	0.145
13.0–15.9	0.115
16.0–18.9	0.085
19.0–21.9	0.065
22.0–24.9	0.035
25.0–27.9	0.015
28.0–30.9	0.005

f.

g. 53.5%

h. 12.0%

29. a.

Price ($1000s)	Frequency
30–39.9	7
40–49.9	7
50–59.9	8
60–69.9	6
70–79.9	4
80–89.9	6
90–99.9	5
100–109.9	1
110–119.9	1
120–129.9	0
130–139.9	0
140–149.9	1

b.

c.

Price ($1000s)	Relative Frequency
30–39.9	0.152
40–49.9	0.152
50–59.9	0.174
60–69.9	0.130
70–79.9	0.087
80–89.9	0.130
90–99.9	0.109
100–109.9	0.022
110–119.9	0.022
120–129.9	0.000
130–139.9	0.000
140–149.9	0.022

d.

e.

Price ($1000s)	Frequency
30–49.9	14
50–69.9	14
70–89.9	10
90–109.9	6
110–129.9	1
130–149.9	1

Price ($1000s)	Relative Frequency
30–49.9	0.304
50–69.9	0.304
70–89.9	0.217
90–109.9	0.130
110–129.9	0.022
130–149.9	0.022

f. Answers will vary. Both choices seem reasonably good.

31. a.

Number of Words	Frequency
0–999	4
1000–1999	22
2000–2999	18
3090–3999	7
4000–4999	4
5000–5999	1
6000–6999	0
7000–7999	0
8000–8999	1

b.

c.

Number of Words	Relative Frequency
0–999	0.070
1000–1999	0.386
2000–2999	0.316
3090–3999	0.123
4000–4999	0.070
5000–5999	0.018
6000–6999	0.000
7000–7999	0.000
8000–8999	0.018

d.

e. Skewed right

f.

Number of Words	Frequency
0–1999	26
2000–3999	25
4000–5999	5
6000–7999	0
8000–9999	1

g.

Number of Words	Relative Frequency
0–1999	0.456
2000–3999	0.439
4000–5999	0.088
6000–7999	0.000
8000–9999	0.018

h. Answers will vary.

33. a.

b.

c. Skewed to the right

35. a.

b.

37. a.

b.

39. a.

Batting Average	Cumulative Frequency
0.180–0.199	4
0.200–0.219	18
0.220–0.239	59
0.240–0.259	118
0.260–0.279	176
0.280–0.299	227
0.300–0.319	256
0.320–0.339	266
0.340–0.359	267

b.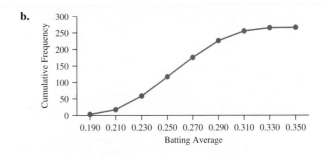

c.

Batting Average	Cumulative Relative Frequency
0.180–0.199	0.015
0.200–0.219	0.067
0.220–0.239	0.221
0.240–0.259	0.442
0.260–0.279	0.659
0.280–0.299	0.850
0.300–0.319	0.959
0.320–0.339	0.996
0.340–0.359	1.000

d.

d.

41. a.

Number of Hours	Cumulative Frequency
3.9	25
6.9	59
9.9	107
12.9	136
15.9	159
18.9	176
21.9	189
24.9	196
27.9	199
30.9	200

b.

c.

Number of Hours	Cumulative Relative Frequency
3.9	0.125
6.9	0.295
9.9	0.535
12.9	0.680
15.9	0.795
18.9	0.880
21.9	0.945
24.9	0.980
27.9	0.995
30.9	1.000

43. It is not possible to construct a histogram for this data set because the last class is open-ended.

45. 0.15

47. (i) is skewed, (ii) is approximately symmetric

Section 2.3
Exercises 1 and 2 are the Check Your Understanding exercises for this section. Answers to these exercises are on page 78.

3. leaf

5. time-series plot

7. True

9. True

11.

1	1225566
2	0012779
3	19
4	556
5	02578

13. The list is: 30, 30, 31, 32, 35, 36, 37, 37, 39, 42, 43, 44, 45, 46, 47, 47, 47, 47, 48, 48, 49, 50, 51, 51, 51, 52, 52, 52, 52, 54, 56, 57, 58, 58, 59, 61, 63

15.

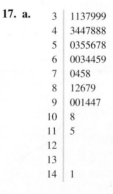

17. a.

3	1137999
4	3447888
5	0355678
6	0034459
7	0458
8	12679
9	001447
10	8
11	5
12	
13	
14	1

b.

3	113
3	7999
4	344
4	7888
5	03
5	55678
6	00344
6	59
7	04
7	58
8	12
8	679
9	00144
9	7
10	
10	8
11	
11	5
12	
12	
13	
13	
14	1
14	

c. Answers will vary. The plot with the split stems has more detail than is needed for stems larger than 9.

19. a.

0	3
0	55669999
1	01111112222333344
1	555666889
2	11124
2	556777
3	0111334
3	555678
4	02
4	6
5	
5	9
6	
6	66

b. Answers will vary. The split stem-and-leaf plot provides more detail in the range 0–39, where there is a lot of data. It is more spread out than necessary for values greater than 40, where the data are sparse.

21. a.

Wimbledon		Master's
	1	
87	1	
4444443332222222221111110	2	33
998777766665555	2	5666777888899
1100	3	01111222222333333
	3	5567888999
	4	123
	4	6

b. The Wimbledon champions are generally younger than the Master's champions.

23.

Yes

25. a.

b. Increasing 1989–1992, 2000-2003, 2007–2010. Decreasing 1992–2000, 2003–2007, 2010–2012.

27. a.

b. Increased: 1950s, 1960s, 1980s, 2000s. Decreased: 1970s, 1990s.

c. Decreased

d. Increased 1965–1969, then decreased

29. a. Approximately $800 billion
   b. Approximately $300 billion
   c. True   **d.** False

31. a. Approximately 115 inches
   b. 1910
   c. Less than
   d. True
   e. False

33. a. False
   b. True
   c. False
   d. True

35. a.

0	3333333344444
0	555566666677788999
1	00001111234
1	5668
2	00
2	99
3	
3	8
4	
4	
5	
5	5

b.

c. Each line of the stem-and-leaf plot corresponds to a class in the histogram.

Section 2.4

Exercises 1 and 2 are the Check Your Understanding exercises for this section. Answers to these exercises are on page 85.

3. zero

5. (i). Graph (A) presents an accurate picture, because the baseline is at zero. Graph (B) exaggerates the decline, because the baseline is above zero.

7. Graph (B) presents the more accurate picture. The baseline is at zero, and the bars are of equal width. The dollar bill graphic does not follow the area principle. The length and width of the smaller image are about 40% less than the length and width of the larger image, so the area of the smaller image is about 64% less than that of the larger image. This exaggerates the difference.

9. It presents an accurate picture of the increase, because the baseline is at zero.

11. a. The bars appear shorter than they really are.

b.

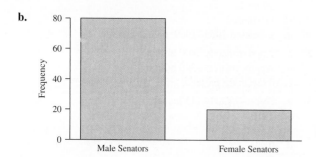

13. (ii) is more accurate. The plot on the left has its baseline at zero, and presents an accurate picture. The plot on the right exaggerates the increase.

15. a.

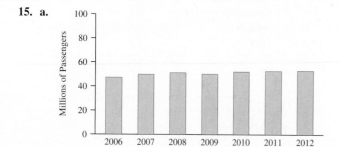

b. It makes the differences look smaller, because the scale on the y-axis extends much farther than the largest bar height.

c. Answers will vary. Figure 2.23 has the baseline at zero, and the scale on the y-axis is appropriate for the bar height.

CHAPTER 2 Quiz

1.

Grade	Frequency
A	9
B	5
C	6
D	3
F	4

2.

Grade	Relative Frequency
A	0.333
B	0.185
C	0.222
D	0.111
F	0.148

3.

4.

Grades in Algebra Class

5. 5.0–7.9, 8.0–10.9, 11.0–13.9, 14.0–16.9, 17.0–19.9. The class width is 3.

6. True

7. a.

b.

8.

9.

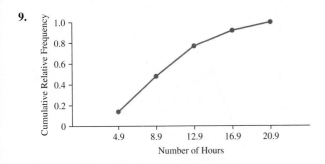

10. The list is: 11, 11, 15, 15, 19, 19, 19, 22, 22, 23, 25, 27, 28, 30, 30, 38, 44, 45, 47, 48, 50, 51, 53, 53, 55, 56, 58

11.

1	9
2	22889
3	579
4	1
5	
6	8

12.

Espresso Makers		Coffee Makers
	1	9
5	2	22889
10	3	579
0	4	1
600	5	
5	6	8
70	7	
	8	
99	9	

13.

14.

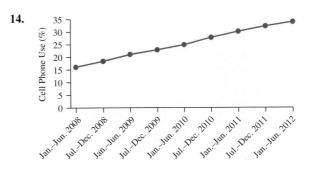

15. Twice

CHAPTER 2 Review Exercises

1. a. Somewhat   **b.** True   **c.** False   **d.** True

2. a.

b.

c. False

3. a.

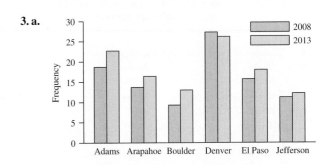

b. False   **d.** Adams

4. a.

b.

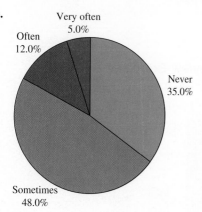

c. False

5. a. 7
b. 10
c. 10%
d. Unimodal

6. a. 8
b. 20
c. Lower limits: 20, 40, 60, 80, 100, 120, 140, 160.
Upper limits: 39, 59, 79, 99, 119, 139, 159, 179.

d.

e.

Number of Freshmen	Relative Frequency
20–39	0.039
40–59	0.255
60–79	0.196
80–99	0.275
100–119	0.137
120–139	0.059
140–159	0.020
160–179	0.020

f.

g. 23.5%   **h.** 29.4%

7. a.

b.

c.

d.

8. a.

Age	Frequency
10–19	2
20–29	1
30–39	3
40–49	10
50–59	9
60–69	9
70–79	4
80–89	2

b.

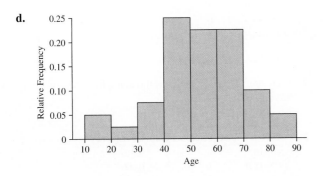

c.

Age	Relative Frequency
10–19	0.050
20–29	0.025
30–39	0.075
40–49	0.250
50–59	0.225
60–69	0.225
70–79	0.100
80–89	0.050

d.

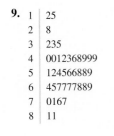

9.

```
1 | 25
2 | 8
3 | 235
4 | 0012368999
5 | 124566889
6 | 457777889
7 | 0167
8 | 11
```

10. a.

Age	Frequency
45–49	2
50–54	1
55–59	4
60–64	6
65–69	6
70–74	6
75–79	4
80–84	3
85–89	2
90–94	4

b.

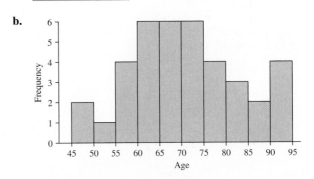

c.

Age	Relative Frequency
45–49	0.053
50–54	0.026
55–59	0.105
60–64	0.158
65–69	0.158
70–74	0.158
75–79	0.105
80–84	0.079
85–89	0.053
90–94	0.105

d.

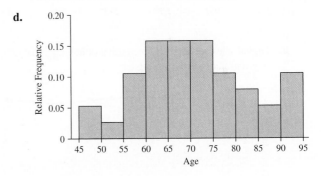

11. a.

```
Presidents        | Monarchs
                 1 | 25
                 2 | 8
                 3 | 235
              96 4 | 0012368999
           87763 5 | 124566889
    877765443300 6 | 457777889
      9887432110 7 | 0167
           85310 8 | 11
            3300 9 |
```

b.

Presidents		Monarchs
	1	2
	1	5
	2	
	2	8
	3	23
	3	5
	4	00123
96	4	68999
3	5	124
8776	5	566889
443300	6	4
877765	6	57777889
432110	7	01
9887	7	67
310	8	11
85	8	
3300	9	
	9	

c. Answers will vary. The split stem-and-leaf plot provides a more appropriate level of detail.

12. a.

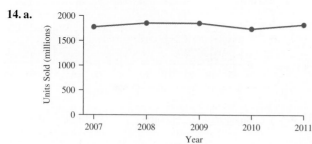

13. a.

b. Digital sales are generally increasing or holding steady; physical sales are decreasing.

14. a.

b.

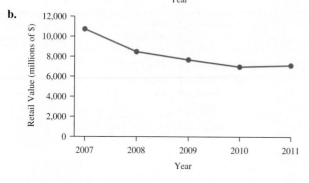

c. The number of inexpensive download singles has been increasing, while the number of expensive CDs has been decreasing. The total number of units has been increasing or holding steady, but because the number of CDs has been decreasing, the total retail value has been decreasing.

15. (i). The birth rate decreased somewhat between 1975 and 2010. The plot on the right is misleading, because the baseline is not at zero.

CHAPTER 2 Case Study

1.

Mileage	Frequency
16.0–16.9	1
17.0–17.9	0
18.0–18.9	0
19.0–19.9	0
20.0–20.9	0
21.0–21.9	3
22.0–22.9	0
23.0–23.9	3
24.0–24.9	3
25.0–25.9	0
26.0–26.9	3
27.0–27.9	9
28.0–28.9	8
29.0–29.9	3
30.0–30.9	6
31.0–31.9	6
32.0–32.9	3
33.0–33.9	4
34.0–34.9	3
35.0–35.9	1
36.0–36.9	1
37.0–37.9	1
38.0–38.9	3
39.0–39.9	0
40.0–40.9	1

2. There are 24 classes, which is too many for a data set with only 62 values.

3.

Mileage	Frequency	Relative Frequency
15.0–16.9	1	0.016
17.0–18.9	0	0.000
19.0–20.9	0	0.000
21.0–22.9	3	0.048
23.0–24.9	6	0.097
25.0–26.9	3	0.048
27.0–28.9	17	0.274
29.0–30.9	9	0.145
31.0–32.9	9	0.145
33.0–34.9	7	0.113
35.0–36.9	2	0.032
37.0–38.9	4	0.065
39.0–40.9	1	0.016

4.

The histogram is unimodal. There is little skewness.

5. Answers will vary. Here is a frequency distribution with a class width of 2.

Mileage	Frequency
22–23.9	1
24–25.9	1
26–27.9	1
28–29.9	6
30–31.9	8
32–33.9	10
34–35.9	7
36–37.9	12
38–39.9	6
40–41.9	4

6. Answers will vary. Here is a relative frequency distribution with a class width of 2.

Mileage	Relative Frequency
22–23.9	0.018
24–25.9	0.018
26–27.9	0.018
28–29.9	0.107
30–31.9	0.143
32–33.9	0.179
34–35.9	0.125
36–37.9	0.214
38–39.9	0.107
40–41.9	0.071

7. Answers will vary. Here is a histogram with a class width of 2.

Unimodal, slightly skewed left

8. 2013 cars tend to have higher mileage.

9.

2000 Cars		2013 Cars
	1	
6	1	
444333111	2	2
999888888888777777777666	2	57999999
444333332222111111000000	3	00000111222222333333444
88765	3	555666666777777778889996
0	4	0000

CHAPTER 3

Section 3.1

Exercises 1–6 are the Check Your Understanding exercises for this section. Answers to these exercises are on page 109.

7. mean

9. extreme values

11. False

13. False

15. Mean: 23.4; median: 26; mode: 27

17. Mean: 5.5; median: 14; mode: 28

19. 24.6

21. 145.0

23. (ii)

25. (i)

27. Mean: 30.4; median: 29; mode: 27

29. Skewed to the right, because the mean is greater than the median.

31. a. 290
  b. 300

33. a. 201.3
  b. 200
  c. 170, 200, 210, 220
  d. Approximately symmetric

35. a. Mean: 10.76; median: 10.05
  b. Mean: 4.35; median: 4.05   **c.** Yes

37. a. Mean: 1.241; median: 1.22
  b Mean 3.36; median: 3.38   **c.** Median

39. a. Mean: 296; median: 302.5
  b. Mean: 285.5; median: 285
  c. Offensive linemen are somewhat heavier.

41. a. Mean: 2145; median: 999
  b. Mean 528.5; median: 359   **c.** Yes, the data support this claim.

43. a. 373.49   **b.** 335.65   **c.** The mean would increase to 1023.7; the median would be unchanged.

45. a. 37.2   **b.** Too small, because the average age within the first class is greater than 16, but the midpoint is only 15.

47. a. 81.1   **b.** Too large, because the mean would be 0, while the midpoint would be 10.

49. a. 37.5   **b.** 36   **c.** Skewed to the right
  d. Answers will vary. Here is one possibility:

The results agree with the expectation.

51. Fiction

53. 160 pounds. The mean would be greater than the median, because the histogram is skewed to the right.

55. iii. Between 510 and 520

57. a. 161,000   **b.** 139,000   **c.** 66.8
   d. There are more women than men.

59. a. 13   **b.** 12

61. 208

63. a. 220,600   **b.** 20,000
   c. The mean is more appropriate, because it better reflects the amount of money the family now has.

65. Answers will vary.

67. Answers will vary.

69. No. If the largest or smallest value is an outlier, the value of the midrange will be strongly affected.

71. a. 68.4   **b.** 68   **c.** 5.417, 6.000, 5.667, 5.583, 5.833
   d. 5.7; yes   **e.** 5.667; yes

73. a. They are both equal to 5.
   b. The median. The mean increases to 6, and the median increases to 8.
   c. The mean. The mean increases to 9.5, and the median increases to 8.
   d. As the value becomes more extreme, the mean steadily increases, but the median stays the same. At some point, the mean becomes greater than the median.

Section 3.2
Exercises 1–8 are the Check Your Understanding exercises for this section. Answers to these exercises are on page 128.

9. zero

11. 68%

13. False

15. False

17. Variance is 100; standard deviation is 10.

19. Variance is 49; standard deviation is 7.

21. Variance is 289; standard deviation is 17.

23. Variance is approximately 228.41; standard deviation is approximately 15.11.

25. Variance is approximately 5680; standard deviation is approximately 75.37.

27. a. 68%   **b.** 20 and 44   **c.** 75%

29. a. 95%   **b.** 140 and 212   **c.** 88.9%

31. a. 13.46   **b.** 18.84   **c.** Yes

33. a. 26.8   **b.** 23.2   **c.** Offensive

35. a. 2.36   **b.** 1.05   **c.** 2007–2008: 7.9; 2012–2013: 4.5
   d. Decrease   **e.** Decrease

37. a. 18.528   **b.** 18.482   **c.** June

39. a. 7.58   **b.** 2.75

41. a. Almost all   **b.** 68%   **c.** $38.98 and $50.22

43. a. 422   **b.** 590

45. a. 16%   **b.** 2.5%   **c.** 34%

47. a. 521 or 522   **b.** 605 or 606   **c.** 295

49. Not appropriate; histogram is skewed.

51. Approximately 68%

53. At least 75% of the days had temperatures between 56.2°F and 68.6°F.

55. 2.5. We expect that almost all of the data will be within 3 standard deviations of the mean. The largest and smallest values in the data set are between 7 and 8 inches from the mean. We therefore expect the standard deviation to be closest to 2.5.

57. a. Impossible   **b.** Possible   **c.** Impossible   **d.** Possible

59. a. Yes. Answers will vary.

61. a. 0.045   **b.** 0.351   **c.** Weight

63. a. 4.8
   b.

x	$x - \bar{x}$	$(x - \bar{x})^2$	$|x - \bar{x}|$
1	−3.8	14.44	3.8
3	−1.8	3.24	1.8
4	−0.8	0.64	0.8
7	2.2	4.84	2.2
9	4.2	17.64	4.2

   c. SD = 3.1937; MAD = 2.56
   d. SD = 10.677; MAD = 7
   e. The MAD is more resistant. Its value changed less when the outlier was added to the data set.

Section 3.3
Exercises 1–4 are the Check Your Understanding exercises for this section. Answers to these exercises are on page 145.

5. Quartiles

7. interquartile range

9. False

11. False

13. a. −1   **b.** 1.5   **c.** 11

15. The outlier is 4.91. It seems certain to be an error.

17. a. $Q_1 = 20$; $Q_3 = 44$ **b.** 24
   c. Lower: −16; upper: 80   **d.** 82 is the only outlier.

19. a. 34   **b.** 14   **c.** 40   **d.** 8

21. a. 1.13   **b.** 1.16   **c.** SAT   **d.** 25   **e.** 280

23. a. 19   **b.** Lower: 79.5; upper: 155.5   **c.** No

25. a. $Q_1 = 11$; $Q_3 = 32$ **b.** 15
   c. Lower: −20.5; upper: 63.5   **d.** 67, 86, 97, 97, 116
   e.

   f. Skewed to the right   **g.** 12   **h.** 49   **i.** 55th percentile

27. a. $Q_1 = 19$; $Q_3 = 22$ **b.** 21
   c. Lower: 14.5; upper: 26.5   **d.** 31, 36, 38, 39

e.

f. Skewed to the right   g. 17   h. 26   i. 90th percentile

29. a. $Q_1 = 14$; $Q_3 = 41$ b. 27
   c. Lower: -26.5; upper: 81.5   d. No
   e. No, neither is an outlier.

f.

g. Approximately symmetric   h. 24   i. 39
   j. 66th percentile

31. a. $Q_1 = 3$; $Q_3 = 21.5$.   b. 7.5
   c. Lower: -24.75; Upper: 49.25   d. 70, 114

e.

f. Skewed right   g. 4   h. 31   i. 63rd

33. No, only 25% of the class scored lower than Ed.

35. a. Median: 1.3; $Q_1 = 0.48$; $Q_3 = 5.0$
   b. Median: 1.1; $Q_1 = 0.5$; $Q_3 = 5.5$
   c. Lower: -6.3; Upper: 11.78
   d. Lower: -7.0; Upper: 13.0

e.

Both sets of salaries are skewed to the right. The Giants have a few more high salaries than the Tigers.

37. a. Denver   b. 25%   c. Denver
   d. Skewed right   e. Skewed right

39. a. Yes

b. Skewed right

41. a. More than half the data values are equal to the minimum value of 0.

b.

c. The first quartile is the same as the minimum value.

43. a. $Q_1 = 20$; $Q_3 = 60$; IQR $= 40$
   b. Upper outlier boundary is $60 + 1.5 \cdot 40 = 120$. Since $150 > 120$, 150 is an outlier.
   c. $Q_1 = 25$; $Q_3 = 105$; IQR $= 80$ d. No
   e. Both the third quartile and the IQR increased.

45. a. Computation   b. $z = \dfrac{0 - 3.759}{3.489} = -1.08$
   c. $z = \dfrac{7.52 - 3.759}{3.489} = 1.08$ d. 2.2%   e. 17.8%
   f. Less extreme
   g. The right tail extends out farther than the left tail, so there are more data to the right of $z = 1.08$ than to the left of $z = -1.08$.

CHAPTER 3 Quiz

1. The mode
2. Mean: 520; median: 550; mode: 550
3. The mean
4. mean, median
5. a. 9.76   b. 4.51
6. 4
7. 3.96, 2.28
8. 95%
9. 75
10. 0.5
11. 0.037
12. False
13. 44
14. a. $Q_1 = 22$; $Q_3 = 39$
   b. Outlier boundaries are -3.5 and 64.5.
   c. 68 is the only outlier.

15.

CHAPTER 3 Review Exercises

1. a. 145.27   b. 146
2. a. 10.08   b. 9.2   c. Approximately symmetric
3. a. Mean of process 1 is 92.87; mean of process 2 is 91.46.
   b. Median of process 1 is 92.2; median of process 2 is 93.3.
   c. They are about the same.
4. a. Variance of process 1 is 9.40; variance of process 2 is 53.36.
   b. Standard deviation of process 1 is 3.07; standard deviation of process 2 is 7.30.
   c. Process 1 produces a more uniform thickness.
5. a. Mean in May is 34.479; median in May is 34.67.
   b. Mean in June is 34.052; median in June is 34.115.
   c. They are about the same.
6. a. 0.791   b. 0.814   c. Greater in June
7. a. A: 2.87; B: 0.75
   b. Method A. Estimating by eye is less precise than measuring with a ruler.
   c. It is better to have a smaller standard deviation. With a small standard deviation, there is less need to remeasure, since all measurements will be reasonably close to the first one.
8. 153 and 172.6
9. 25%
10. 68%
11. At least 8/9 of the rents are between $350 and $1250.
12. a. 8.752   b. 7.365   c. 14.616   d. 3.82   e. 6.34
   f. 8.98   g. 6.705   h. 8.66
13. a. (3)   b. (1)   c. (4)   d. (2)
14. a. $Q_1 = 14.6$; $Q_3 = 17.4$ b. 16.2
   c. Lower: 10.4; upper: 21.6   d. No outliers

e.

15. a. $Q_1 = 2.1$; $Q_3 = 15.7$ **b.** 9.2
   c. Lower: -18.3; upper: 36.1   **d.** $-38.7, 41.1$

e.

CHAPTER 3 Case Study

1.

2. For the recycled wafers, 77.3, 77.5, and 97.4 are outliers. There are no outliers for the new wafers.

3. The outliers 77.3 and 77.5 should be deleted, because they resulted from an error.

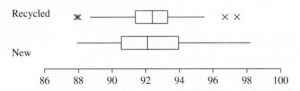

4. For the recycled wafers, 87.9, 88.0, 96.7, and 97.4 are outliers. There are no outliers for the new wafers. No outliers should be deleted, because they do not result from errors.

5. Approximately symmetric

6. New: 92.33; recycled: 92.31

7. New: 92.1; recycled: 92.4

8. New: 2.36; recycled: 2.21

9. Answers will vary. The interquartile range is more useful, because the standard deviation is not resistant to outliers.

CHAPTER 4

Section 4.1

Exercises 1–8 are the Check Your Understanding exercises for this section. Answers to these exercises are on page 164.

9. pairs

11. linear

13. False

15. True

17. 0.824

19. 0.515

21. Appropriate. The variables have a weak linear relationship.

23. Not appropriate. The variables have a nonlinear relationship.

25. positive

27. positive

29. positive

31. a.

b. 0.845   **c.** Above average; r is positive.   **d.** iii

33. a.

b. 0.332   **c.** Below average; r is positive.   **d.** ii

35. a.

b. 0.812   **c.** Cooler; r is positive.   **d.** i

37. a.

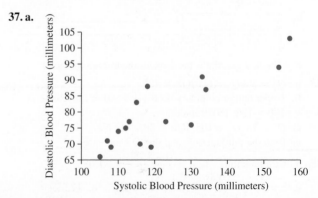

b. 0.857   **c.** Above average; r is positive.

39. No. Larger cities have more police officers and also tend to have higher crime rates than smaller cities.

41. Close to −1. The difference between the ages of two people is in most cases very close to the difference between their graduation years, with the older person graduating in the earlier year. Therefore, the two variables have a nearly perfect negative linear relationship.

43. **a.** $\bar{x} = 1.75$; $\bar{y} = 10.828$; $s_x = 1.031$; $s_y = 1.981$
   b. 0.906
   c. $\bar{y} = 11.828$; $s_y = 1.981$
   d. $\bar{y}$ increased by 1; s_y was unchanged.
   e. 0.906. The quantities $y - \bar{y}$ are unchanged.
   f. $\bar{x} = 21$; $s_x = 12.369$
   g. Each was multiplied by 12.
   h. 0.906. The quantities $(x - \bar{x})/s_x$ are unchanged.
   i. unchanged
   j. unchanged

Section 4.2

Exercises 1–4 are the Check Your Understanding exercises for this section. Answers to these exercises are on page 176.

5. explanatory, outcome

7. 15

9. True

11. False

13. $\hat{y} = 5.2 + 0.6x$

15. $\hat{y} = 0.8951 + 1.7674x$

17. $\hat{y} = 1175 + 35x$

19. $\hat{y} = 86.5 + 0.0003x$

21. **a.** $\hat{y} = 2.7592 + 0.3991x$ **b.** 0.10 **c.** 3.54

23. **a.** $\hat{y} = 107.2685 + 1.588$
   b. No. The x-values are all positive.
   c. 3.176 **d.** 225.6 **e.** Less

25. **a.** $\hat{y} = 33.7754 + 0.5930x$

   b.

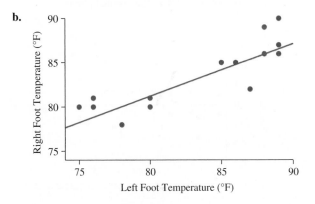

   c. 1.186 **d.** 81.8

27. **a.** $\hat{y} = 9.1828 + 0.5748x$
   b. No. The x-values are all positive.
   c. 5.748 **d.** 81.0

29. **a.** $\hat{y} = 49.7124 + 4.288759685x$
   b. 0.9186464394
   c. 92.6

31. **a.** $\hat{y} = 33.8127 + 1.21025x$ **b.** 59.7

33. **a.** $\hat{y} = 55.91275257 + 2.58289361x$ **b.** 68.827

35. **a.** 0.750
   b. $\bar{x} = 424$; $s_x = 123.26$
   c. $\bar{y} = 499.33$; $s_y = 143.93$
   d. $\hat{y} = 128.1812 + 0.8754x$
   e. 0.0325, −0.3083, 1.8579, −0.8762, 0.1136, −0.8194
   f. −0.8778, 0.4979, 1.297, −1.253, 0.7481, −0.4122
   g. 0.750
   h. $\hat{z}_y = 0.750z_x$. The means of z_x and z_y are both 0. Since the least-squares regression line goes through the point of averages, the y-intercept is 0. The slope of the least-squares regression line is $r\dfrac{s_{z_y}}{s_{z_x}}$. Since $s_{z_y} = s_{z_x} = 1$, the slope is equal to the correlation coefficient r.

Section 4.3

Exercises 1–4 are the Check Your Understanding exercises for this section. Answers to these exercises are on page 189.

5. extrapolation

7. sum of squared residuals

9. correlation coefficient

11. True

13. False

15. **a.** 0.36 **b.** 36%

17. **a.** $\hat{y} = 19.3734 + 0.5101x$
   b. (3, 21)
   c. $\hat{y} = 19.2354 + 0.5123x$
   d. No. Removing the outlier produced only a small change in the least-squares regression line.

19. No. The residual plot has a curved pattern.

21. Yes. The residual plot does not have a distinct pattern.

23. **a.** $y = 740.3765 - 21.8708x$ **b.** 0.3225

   c.

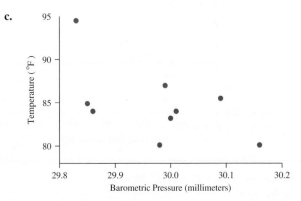

   d. (29.83, 94.5) **e.** $y = 320.54 - 7.8999x$
   f. Yes. Removing the outlier strongly affects the position of the least-squares line.
   g. 0.113. The proportion of variance explained is less without the outlier.

25. a. $\hat{y} = -76.2258 + 3.2499x$

b.

c. (82, 449)   **d.** $\hat{y} = 31.8632 + 1.2282x$

e. Yes. Removing the outlier strongly affects the position of the least-squares regression line.

27. a. $\hat{y} = 1062.4048 + 0.3600x$ **b.** 0.9603

c.

The relationship does not appear to be linear.   **d.** False

29. a. $\hat{y} = 757.2920 + 3.3009x$ **b.** 0.9043

c.

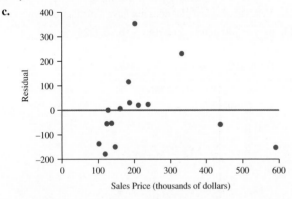

The relationship does not appear to be linear.   **d.** False

31. a. $-47, -21, -19, -12, 22, 28, 41, 8$

b. 6168

c. $-10.2721, 12.3552, -3.5004, -7.5114, 8.4346, 12.3514, 10.5709, -22.4291$

d. 1165.35

e. 5002.65

f. 0.811

g. Compute the correlation coefficient to be $r = 0.9005918$. The coefficient of determination is $r^2 = 0.811$.

33. a. $\hat{y} = 0.4636 + 0.5711x$

b.

The linear model does not seem appropriate.

c. For $R_1 < 4$: $\hat{y} = 1.2330 + 0.2636x$. The residual plot is

The linear model does not seem appropriate. For $R_1 \geq 4$: $\hat{y} = -0.1900 + 0.7100x$. The residual plot is

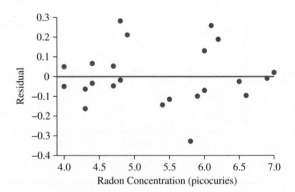

The line fits the data with $R_1 \geq 4$, but not the data with $R_1 < 4$.

d. A line describes the relationship well when $R_1 \geq 4$. A nonlinear model would be a better fit when $R_1 < 4$.

CHAPTER 4 Quiz

1. −0.959

2.

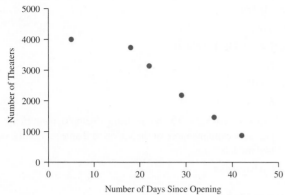

3. −0.965; strong negative

4. The points lie along a line with negative slope.

5. No. Larger cities have more paved streets and more cars.

6. $\hat{y} = -6.195 + 1.2474x$

7. 66.78

8. 0.747

9. 2.495

10. 0

11. $\hat{y} = 6.0667 - 0.6x$

12. iii

13. No, 10 hours is outside the range of the data.

14. i

15. 36%

CHAPTER 4 Review Exercises

1. a. $\hat{y} = 44.534 + 2.5313x$
   b. 2.5313
   c. 68.581

2. a. $\hat{y} = -1.0295 + 0.0523x$
   b. 0.4187
   c. 1.2735

3. a. $\hat{y} = 8.5593 - 0.1551x$
   b.

   c. 0.7755 mile per gallon   **d.** 6.2328

4. a. −0.945   **b.** 0.892   **c.** 89.2%

5. a. $\hat{y} = 2.8827 + 0.8895x$

b.

c. 10.674
d. 47.358

6. a. 0.960
   b. 0.921
   c. 92.1%

7. a. $\hat{y} = 7.5659 + 0.2015x$
   b.

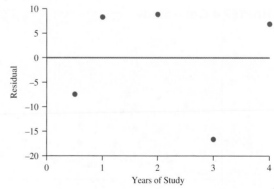

   c. Florida, Alabama, Georgia, Ole Miss, LSU, and South Carolina

8. a. 0.643
   b. 0.413
   c. 41.3%

9. a. $\hat{y} = 417.2012 + 18.4756x$
   b.

   c. 463

10. a. 0.917
    b. 0.8416
    c. 84.16%

11. a. $\hat{y} = -3.9357 + 0.2857x$ **b.** 0.937

c.

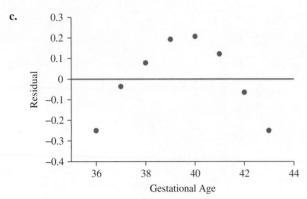

The relationship does not appear to be linear.   **d.** False

12. a.

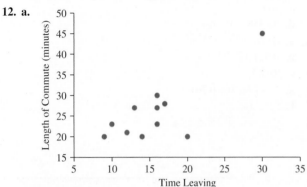

b. $\hat{y} = 9.7885 + 1.0192x$ **c.** 0.635   **d.** (30, 45)
e. $\hat{y} = 19.5162 + 0.3066x$
f. Yes, the least-squares regression line changes considerably when the outlier is removed.
g. 0.0744; the relationship is weaker without the outlier.

13. a. $\hat{y} = 0.3946 + 3.643061791x$ **b.** 36.825   **c.** 0.064
d. The linear relationship is weak, because the correlation is close to 0.

14. a. $\hat{y} = -1.04349 + 3.045141227x$ **b.** 151.21   **c.** 0.994
d. The linear relationship is strong, because the correlation is close to 1.

15. a. $\hat{y} = 4.99971 + 0.20462x$ **b.** 10.1

CHAPTER 4 Case Study

1.

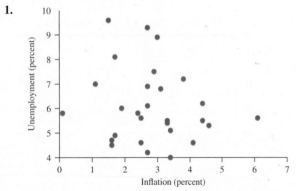

There is no strong nonlinearity.

2. $\hat{y} = 6.6594 - 0.1990x$

3. 6.06

4. −0.164

5. 2.7%

6. $\hat{y} = 6.7017 - 0.2251x$

7. 6.03

8. −0.184

9. 3.4%

10. $\hat{y} = 1.0548 + 0.8304x$

11. 5.21

12. 0.810

13. 65.6%

14. Answers will vary. The model using unemployment in the previous explains more of the variation than either of the other two models.

15. Underpredict

CHAPTER 5
Section 5.1
Exercises 1–4 are the Check Your Understanding exercises for this section. Answers to these exercises are on page 209.

5. 0

7. sample space

9. True

11. False

13. 1/6

15. 1/3

17. 0

19. a. No   **b.** No   **c.** Yes

21. No. The probabilities do not add up to 1.

23. Yes. The probabilities are all between 0 and 1, and they add up to 1.

25. a. ii   **b.** vi   **c.** iv   **d.** i
e. iii   **f.** vii   **g.** v   **h.** vii

27. a. $275/500 = 0.55$
b. We estimate that 55% of all voters plan to vote to reelect the mayor.

29. a. {TTTT, TTTF, TTFT, TTFF, TFTT, TFTF, TFFT, TFFF, FTTT, FTTF, FTFT, FTFF, FFTT, FFTF, FFFT, FFFF}
b. 1/8   **c.** 1/4   **d.** 3/8

31. $180/400 = 0.45$

33. a. $1912/2825 = 0.6768$ **b.** $685/2825 = 0.2425$

35. a. $7792/11{,}217 = 0.6947$ **b.** $10{,}270/11{,}217 = 0.9156$

37. a. $18/38 = 0.4737$
b. The law of large numbers says that in the long run, the percentage of the time you win will approach 47.37%.

39. a. 0.8674   **b.** 0.3118   **c.** No

41. a. 0.1944   **b.** 0.2378   **c.** Yes

43. Answers will vary.

45.

(1, 1)	(1, 2)	(1, 3)	(1, 4)	(1, 5)	(1, 6)
(2, 1)	(2, 2)	(2, 3)	(2, 4)	(2, 5)	(2, 6)
(3, 1)	(3, 2)	(3, 3)	(3, 4)	(3, 5)	(3, 6)
(4, 1)	(4, 2)	(4, 3)	(4, 4)	(4, 5)	(4, 6)
(5, 1)	(5, 2)	(5, 3)	(5, 4)	(5, 5)	(5, 6)
(6, 1)	(6, 2)	(6, 3)	(6, 4)	(6, 5)	(6, 6)

47. 1/6

49. {(3, 1), (3, 2), (3, 3), (3, 4), (3, 5), (3, 6)}

51. 1/6; yes

Section 5.2
Exercises 1–4 are the Check Your Understanding exercises for this section. Answers to these exercises are on page 218.

5. $P(A \text{ and } B)$

7. complement

9. True

11. False

13. 0.9

15. 0.7

17. Yes

19. 0.65

21. 0.73

23. 1

25. Not mutually exclusive

27. Mutually exclusive

29. Not mutually exclusive.

31. a. 200 or fewer of them use Google as their primary search engine.
   b. Fewer than 200 of them use Google as their primary search engine.
   c. At least 200 of them use Google as their primary search engine.
   d. The number that use Google as their primary search engine is not equal to 200.

33. a. {RR, RY, RG, YR, YY, YG, GR, GY, GG}
   b. {RR, YY, GG}
   c. {RY, RG, YR, YG, GR, GY}
   d. {RG, YG, GR, GY}
   e. Yes; they have no outcomes in common.
   f. No; they both contain the event GG.

35. a. 0.11   **b.** 0.90

37. a. 336/800 = 0.42   **b.** 734/800 = 0.9175

39. a. 0.226   **b.** 0.854

41. a. 18/25 = 0.72   **b.** 10/25 = 0.4   **c.** 0.88
   d. 0.12

43. a. 0.42   **b.** 0.56   **c.** 0.46   **d.** 0.54
   e. 0.52   **f.** 0.02   **g.** 0.54

45. No. The events of having a fireplace and having a garage are not mutually exclusive.

47. Answers will vary.

Section 5.3
Exercises 1–6 are the Check Your Understanding exercises for this section. Answers to these exercises are on page 231.

7. conditional

9. 5

11. False

13. False

15. 0.12

17. 0.18

19. 0.16

21. 0.28

23. 1/16 = 0.0625

25. 1/216 = 0.0046

27. Mutually exclusive

29. Neither

31. a. Yes; $P(A \text{ and } B) = P(A)P(B)$ **b.** 0.55
   c. No; $P(A \text{ and } B) \neq 0$

33. a. 0.3   **b.** No; $P(A \text{ or } B) \neq P(A) + P(B)$
   c. No; $P(A \text{ and } B) \neq P(A)P(B)$

35. 91/216 = 0.4213

37. 1/42 = 0.0238

39. a. 0.5660   **b.** 0.2333   **c.** 0.7391
   d. 0.2537   **e.** 0.8113

41. a. 0.8   **b.** 0.7   **c.** 0.7   **d.** Yes

43. a. 0.5123   **b.** 0.0555   **c.** 0.0316
   d. 0.0617   **e.** 0.5691

45. 0.0125

47. 0.05

49. a. 0.15   **b.** 0.05

51. a. 3/10 = 0.3   **b.** 2/9 = 0.2222   **c.** 1/15 = 0.0667
   d. No; if the first component is defective, the second component is less likely to be defective.

53. No. If someone owns two vehicles, the types of vehicles are not independent.

55. 0.7903

57. 0.743

59. 0.98

61. 1/3 = 0.3333

63. Since $P(A) = P(B) = 0$, $P(A \text{ and } B) = 0$. Therefore, A and B are independent because $P(A \text{ and } B) = P(A)P(B)$. Also, A and B are mutually exclusive, because $P(A \text{ and } B) = 0$.

65. No. If A and B are mutually exclusive, then $P(A \text{ and } B) = 0$. But since $P(A) > 0$ and $P(B) > 0$, $P(A)P(B) > 0$. Therefore, $P(A \text{ and } B) \neq P(A)P(B)$, so A and B are not independent.

Section 5.4
Exercises 1–6 are the Check Your Understanding exercises for this section. Answers to these exercises are on page 239.

7. mn

9. False

11. 362,880

13. 1

15. 1

17. 210

19. 1190

21. 1

23. 126

25. 2300

27. 1

29. 48

31. a. 336   **b.** 56

33. a. $10^4 \cdot 26^3 = 175{,}760{,}000$ **b.** $10^4 = 10{,}000$
c. $1/26^3 = 0.0000569$

35. a. $_{12}C_8 = 495$ **b.** 0.0303 **c.** 0.9697

37. a. $4^3 = 64$ **b.** $_4P_3 = 24$ **c.** 8
d. $24/64 = 0.375$ **e.** $1/8 = 0.125$

39. a. 1326 **b.** $_4C_2 = 6$ **c.** $6/1326 = 0.00452$

41. $\dfrac{1}{_{39}C_5} = 0.00000174$

43. a. 8 **b.** $2/8 = 0.25$

CHAPTER 5 Quiz

1. (i)

2. a. The sample space is the population of one million voters.
b. 0.56

3. a. $P(A \text{ or } B) = P(A) + P(B) - P(A \text{ and } B)$
b. $P(A \text{ or } B) = P(A) + P(B)$
c. $P(A^c) = 1 - P(A)$
d. $P(A \text{ and } B) = P(A)P(B \mid A)$ or $P(A \text{ and } B) = P(B)P(A \mid B)$
e. $P(A \text{ and } B) = P(A)P(B)$

4. a. $132/400 = 0.33$ **b.** 0.865

5. (i)

6. $79/100 = 0.79$

7. $2/3 = 0.6667$

8. 0.48

9. $4/11 = 0.3636$

10. (i)

11. $0.38^3 = 0.0549$

12. 0.9084

13. a. $21^3 \cdot 5^3 \cdot 10 = 11{,}576{,}250$
b. $1/11{,}576{,}250 = 0.0000000864$

14. $_{24}C_5 = 42{,}504$

15. a. $15! = 1.308 \times 10^{12}$ **b.** $_{15}P_3 = 2730$

CHAPTER 5 Review Problems

1. a. $\{R, W, W, B, B, B\}$ **b.** $1/2$

2. $18/30 = 0.6$

3. a. 0.5 **b.** 0.6667

4. a. $570/1200 = 0.475$ **b.** $1010/1200 = 0.8417$

5. a. 0.03 **b.** 0.32

6. a. 0.9998 **b.** 0.0098

7. a. $\{DDD, DDG, DGD, DGG, GDD, GDG, GGD, GGG\}$
b. 0.6815
c. 0.9603
d. Yes

8. a. $5/10 = 0.5$ **b.** $4/9 = 0.4444$ **c.** 0.2222

9. a. 0.6 **b.** 0.75 **c.** 0.2727
d. No; $P(\text{Female and Business major}) \neq$
$P(\text{Female})P(\text{Business major})$
e. No; $P(\text{Female and Business major}) \neq 0$

10. a. 0.4035 **b.** 0.3647 **c.** 0.1066 **d.** 0.6616
e. 0.2922 **f.** 0.2641

11. a. That the two days are independent with regard to rain
b. If it rains on Saturday, it is more likely that it will rain on Sunday.
c. Too low. In the equation, $P(\text{Rain Sunday})$ should be $P(\text{Rain Sunday} \mid \text{Rain Saturday})$, which is greater than 0.1.

12. $_5C_3 = 10$

13. $1/10 = 0.1$

14. a. $_6P_3 = 120$ **b.** $_5P_3 = 60$

15. a. $1/120 = 0.00833$ **b.** $1/20 = 0.05$

CHAPTER 5 Case Study

1. 0.99613

2. 0.98739

3. 0.97801

4. 0.96533

5. 50: 0.93782; 60: 0.88164; 70: 0.76666; 80: 0.54385; 90: 0.22931; 100: 0.02809

6. 0.028444

7. 0.81749

8. The probability that a person aged 20 is still alive at age 50 is 0.94979. The probability that a person aged 50 is still alive at age 60 is 0.94010. It is more probable that a person aged 20 will still be alive at age 50.

9. 0.79419

CHAPTER 6

Section 6.1

Exercises 1–8 are the Check Your Understanding exercises for this section. Answers to these exercises are on page 260.

9. random variable

11. Continuous

13. True

15. False

17. Discrete

19. Continuous

21. Discrete

23. Discrete

25. Continuous

27. Represents a probability distribution

29. Does not represent a probability distribution; probabilities do not add up to 1.

31. Does not represent a probability distribution; probabilities do not add up to 1.

33. Mean: 2.9; standard deviation: 2.042

35. Mean: 6.94; standard deviation: 2.087

37. Mean: 19.45; standard deviation: 3.232

39. 0.2

41. a. 0.2 **b.** 0.3 **c.** 0.1 **d.** 0.1 **e.** 2
f. 1.095

43. a. 0.1 **b.** 0.5 **c.** 0.2 **d.** 0.9 **e.** 0.8
f. 0.980 **g.** 0.8

45. a. 0.38   **b.** 0.96   **c.** 0.67   **d.** 0.10   **e.** 1.1
   f. 1.054

47. a.

x	P(x)
0	0.0680
1	0.1110
2	0.2005
3	0.1498
4	0.1885
5	0.1378
6	0.0889
7	0.0197
8	0.0358

   b. 0.282   **c.** 0.068   **d.** 3.36   **e.** 1.97

49. a.

x	P(x)
1	0.1278
2	0.1241
3	0.1236
4	0.1222
5	0.1227
6	0.1247
7	0.1266
8	0.1283

   b. 0.122   **c.** 0.255   **d.** 4.51   **e.** 2.31

51. −$0.50, an expected loss

53. $-1/6 = -\$0.17$, an expected loss

55. a. 0

   b. Answers will vary. If you don't answer a question, your score is the same as the expected value of a random guess.

57. $4500. It would be wise to make the investment.

59. a.

x	P(x)
0	0.5
1	0.25
2	0.125
3	0.125

   b. 0.875   **c.** 1.0533

61. a. 0, 1, 2, 3   **b.** 0.512   **c.** 0.128
   d. $P(\text{SFS}) = P(\text{SSF}) = 0.128$ **e.** 0.384   **f.** 0.096
   g. 0.008   **h.** 2.4   **i.** 0.6928

Section 6.2
Exercises 1–4 are the Check Your Understanding exercises for this section. Answers to these exercises are on page 271.

5. two

7. $\sqrt{np(1-p)}$

9. False

11. Does not have a binomial distribution because the sample is more than 5% of the population.

13. Has a binomial distribution with 7 trials.

15. Does not have a binomial distribution because it is not the number of successes in independent trials.

17. 0.3087, mean = 3.5, variance = 1.05, SD = 1.025

19. 0.0355, mean = 12, variance = 4.8, SD = 2.191

21. 0.2160, mean = 1.2, variance = 0.72, SD = 0.849

23. 0.7969, mean = 1.6, variance = 1.28, SD = 1.131

25. 0.8108, mean = 1.5, variance = 1.455, SD = 1.206

27. a. 0.2051   **b.** 0.0547
   c. No, P(7 or more) = 0.1719.

29. a. 0.0798   **b.** 0.2897   **c.** 0.5940
   d. No. P(11 or more) = 0.3043.

31. a. 0.0691   **b.** 0.2919   **c.** 0.0496
   d. Yes, P(Fewer than 12) = 0.0149.

33. a. 0.4131   **b.** 0.9529   **c.** No, P(0) = 0.2342.
   d. 1.4   **e.** 1.1411

35. a. 0.1166   **b.** 0.9486   **c.** 0.0010   **d.** Yes.
   P(More than 25) = 0.0101.   **e.** 19.8   **f.** 2.595

37. a. 0.1472   **b.** 0.3231   **c.** 0.0332
   d. No, P(More than 10) = 0.0978.   **e.** 7.5   **f.** 2.2913

39. a. 0.000135   **b.** Yes
   c. Yes, because if the shipment were good it would be unusual for 7 or more of 10 items to be defective.
   d. 0.456   **e.** No
   f. No, because if the shipment were good it would not be unusual for 2 of 10 items to be defective.

41. a. 0.16116   **b.**

x	P(x)
0	0.0134627
1	0.0724915
2	0.1756524
3	0.2522188
4	0.2376677
5	0.1535699
6	0.0689096
7	0.0212029
8	0.0042814
9	0.0005123
10	0.0000276

Section 6.3
Exercises 1–4 are the Check Your Understanding exercises for this section. Answers to these exercises are on page 277.

5. time, space

7. False

9. 0.0378

11. 0.1839

13. 0.8647

15. 0.7851

17. 0.2834

19. a. 0.1339   **b.** 0.9826   **c.** 0.4589   **d.** 6   **e.** 2.4495

21. a. 0.2231   **b.** 0.8647

23. a. 0.0916   **b.** 0.1048   **c.** 0.2381

25. a. 0.0844   **b.** 20   **c.** 4.4721

27. a. 0.1221   **b.** 0.0138   **c.** 0.1841   **d.** 8   **e.** 2.8284

29. a. $P(X = 0) = e^{-\lambda}\dfrac{\lambda^0}{0!} = e^{-\lambda}$
   b. $X = 0$ means that no events occurred in 1 second. Therefore, the amount of time that elapses until the next event will be greater than 1.
   c. $P(T > 1) = P(X = 0) = e^{-\lambda}$

d. $P(X = 0) = e^{-2\lambda}\dfrac{(2\lambda)^0}{0!} = e^{-2\lambda}$

e. $X = 0$ means that no events occurred in 2 seconds. Therefore, the amount of time that elapses until the next event will be greater than 2.

f. $P(X = 0) = e^{-\lambda t}\dfrac{(\lambda t)^0}{0!} = e^{-\lambda t}$

g. $X = 0$ means that no events occurred in t seconds. Therefore, the amount of time that elapses until the next event will be greater than t.

h. $P(T > t) = P(X = 0) = e^{-\lambda t}$

CHAPTER 6 Quiz

1. The probabilities do not add up to 1.

2. 2

3. a. 9   **b.** 3

4. 0.19

5.

6. a. 0.4   **b.** 0.2   **c.** 0.9   **d.** 0.6

7. 1.3

8. 1.3454

9. a. 0.0769   **b.** 0.9231   **c.** 0.7604

10. Mean: 2.5; standard deviation: 1.5411

11.

x	$P(x)$
0	0.24
1	0.52
2	0.24

12. a. 0.1755   **b.** 0.0671   **c.** 0.9179

13. 80

14. Variance: 80; standard deviation: 8.9443

15. 0.1755

CHAPTER 6 Review Exercises

1. a. Yes   **b.** No   **c.** No   **d.** Yes

2. a. 8.37   **b.** 3.3131   **c.** 1.8202

3. 0.41

4. a.

x	$P(x)$
1	0.5632
2	0.2500
3	0.1147
4	0.0473
5	0.0171
6	0.0053
7	0.0018
8	0.0006

b. 0.5632   **c.** 1.731   **d.** 1.049

5. a. 0.6630   **b.** 0.5760

6. a. 0.6943   **b.** 0.1353   **c.** Yes, $P(0) = 0.00564$.

7. −$0.11, an expected loss

8. a. 0.0480   **b.** 0.9854   **c.** Yes, $P(10) = 0.00511$.

9. a. 0.2756   **b.** 0.1844   **c.** Yes, $P(8) = 0.0360$.

10. No, the trials are not independent. If it rains on one day, it is more likely to rain the next day.

11. Yes; because the sample size is less than 5% of the population, X may be considered to have a binomial distribution.

12. a. 0.1606   **b.** 0.0174   **c.** 0.9826

13. 6

14. 2.4495

15. 0.0873

CHAPTER 6 Case Study

List (ii) is the fraud.

CHAPTER 7

Section 7.1

Exercises 1–8 are the Check Your Understanding exercises for this section. Answers to these exercises are on page 299.

9. density

11. 1

13. standard

15. True

17. False

19. a. 0.25   **b.** 0.6   **c.** 0.4

21. a. 0.8944   **b.** 0.3594   **c.** 0.7085   **d.** 0.5633

23. a. 0.7704   **b.** 0.0154   **c.** 0.8461   **d.** 0.4404

25. a. 0.9394   **b.** 0.3745   **c.** 0.0007   **d.** 0.9916

27. a. 0.7288   **b.** 0.8492   **c.** 0.0744   **d.** 0.0936

29. a. 0.5059   **b.** 0.1153   **c.** 0.5078   **d.** 0.6402

31. 0.10

33. 1.48

35. −0.99

37. 0.39

39. −0.67 and 0.67

41. −1.28 and 1.28

43. 0.0401

45. 0.6956

47. 0.3

Section 7.2

Exercises 1–8 are the Check Your Understanding exercises for this section. Answers to these exercises are on page 310.

9. standardization

11. True

13. False

15. False

17. a. 0.3085 **b.** 0.1056

19. a. 0.2061 [Tech: 0.2066] **b.** 0.0869 [Tech: 0.0863]

21. a. 0.8314 [Tech: 0.8315] **b.** 0.1549 [Tech: 0.1548]

23. 11.25 [Tech: 11.24]

25. 38.08 [Tech: 38.10]

27. a. 0.1446 [Tech: 0.1444] **b.** 0.5438 [Tech: 0.5421]
c. 0.1685 [Tech: 0.1686]
d. 0.0582 [Tech: 0.0587]; not unusual

29. a. 75.35 [Tech: 75.31] **b.** 84.86 **c.** 87.13 [Tech: 87.18]

31. a. 0.2236 [Tech: 0.2248] **b.** 0.6533 [Tech: 0.6544]
c. Yes, P(Less than 400) = 0.0239 [Tech: 0.0241]

33. a. 489.20 [Tech: 489.28] **b.** 510.11 [Tech: 509.96]
c. 453.53 [Tech: 453.35] **d.** 438.36 [Tech: 438.51]

35. a. 0.3085 **b.** No **c.** 0.0062 **d.** Yes

37. a. 4.022 [Tech: 4.023] **b.** 4.382 [Tech: 4.381] **c.** 4.1

39. a. 0.0548 **b.** 0.3811 [Tech: 0.3812] **c.** 0.8849

41. a. 34.80 [Tech: 34.82] **b.** 42.35 [Tech: 42.34]
c. 36.65 [Tech: 36.63] **d.** 29.75 [Tech: 29.73]

43. a. 0.3085 **b.** 0.1525 [Tech: 0.1524]
c. Yes, P(Less than 12) = 0.0062.

45. a. 12.055 **b.** 12.015 **c.** 12.011 and 12.089

47. a. 0.1056 **b.** 0.0001 **c.** 0.8882 [Tech: 0.8881]

49. a. 25.1200 [Tech: 25.1203] **b.** 25.0624 [Tech: 25.0626]
c. 24.9136 [Tech: 24.9138]
d. 25.0464 and 25.1536 [Tech: 25.0460 and 25.1540]

51. 34%

53. They are not approximately normal. If they were normally distributed, approximately 26% of the wells would have negative concentrations, which is impossible.

55. Mean: 74.26; SD: 8.02 [Tech: 8.01]

Section 7.3
Exercises 1–4 are the Check Your Understanding exercises for this section. Answers to these exercises are on page 318.

5. sampling

7. True

9. a. 0.9232 [Tech: 0.9236] **b.** 8.56

11. a. 0.7549 [Tech: 0.7540] **b.** 95.73

13. a. 0.0465 [Tech: 0.0461] **b.** 31.81 [Tech: 31.80]

15. a. No, $n \leq 30$. **b.** Not appropriate
c. Not appropriate

17. a. Yes, the population is normal.
b. 0.9372 [Tech: 0.9376] **c.** 64.51 [Tech: 64.52]

19. a. $\mu = 75.4, \sigma = 5.1225$
b. (69, 69), (69, 75), (69, 79), (69, 83), (69, 71), (75, 69), (75, 75), (75, 79), (75, 83), (75, 71), (79, 69), (79, 75), (79, 79), (79, 83), (79, 71), (83, 69), (83, 75), (83, 79), (83, 83), (83, 71), (71, 69), (71, 75), (71, 79), (71, 83), (71, 71)
c. $\mu_{\bar{x}} = 75.4, \sigma_{\bar{x}} = 3.62215$
d. $\mu_{\bar{x}} = 75.4 = \mu; \sigma_{\bar{x}} = 3.62215 = \sigma/\sqrt{2}$

21. a. 0.9015 [Tech: 0.9016] **b.** 0.2966 [Tech: 0.2979]
c. Yes, P(Less than 26) = 0.0049.

23. a. 0.9671 [Tech: 0.9674] **b.** 0.0228 [Tech: 0.0229]
c. 2.17 **d.** Yes, P(Less than 2) = 0.0329 [Tech: 0.0326]
e. No, the Central Limit Theorem applies only to the sample mean, not to individual values.

25. a. 0.4013 [Tech: 0.4001] **b.** 0.1840 [Tech: 0.1853]
c. 8000.5 [Tech: 7999.9]
d. Yes, P(Less than 7500) = 0.0003.
e. No, the Central Limit Theorem applies only to the sample mean, not to individual values.

27. a. 200 pounds **b.** 0.2483 [Tech: 0.2472] **c.** 0.0001

29. a. 0.1038 [Tech: 0.1030] **b.** 0.5646 [Tech: 0.5651]
c. 42.79 [Tech: 42.80]
d. Yes, P(Less than 35) = 0.0136 [Tech: 0.0134].
e. No, the Central Limit Theorem applies only to the sample mean, not to individual values.

31. a. 0.1922 [Tech: 0.1932] **b.** No
c. No, because this result would not be unusual if the claim were true.
d. 0.0047 **e.** Yes
f. Yes, because this result would be unusual if the claim were true.

33. a. 2.01 **b.** smaller
c. The correction factor is 0, so the standard deviation is 0. If the whole population is sampled, the sample mean will always be equal to the population mean, and therefore constant.

Section 7.4
Exercises 1–4 are the Check Your Understanding exercises for this section. Answers to these exercises are on page 324.

5. proportion

7. True

9. 0.2358 [Tech: 0.2354]

11. 0.0918 [Tech: 0.0922]

13. Not appropriate; $np < 10$

15. a. 0.63 **b.** 0.0305 **c.** 0.951
d. 0.8255 [Tech: 0.8261] **e.** 0.3707 [Tech: 0.3716]
f. Yes, P(Less than 0.57) = 0.0250 [Tech: 0.0247]

17. a. 0.67 **b.** 0.05100 **c.** 0.0853 [Tech: 0.0850]
d. 0.6463 [Tech: 0.6471] **e.** 0.0582 [Tech: 0.0584]
f. No, P(Less than 65%) = 0.3483 [Tech: 0.3475].

19. a. No, $np < 10$. **b.** 0.1492 [Tech: 0.1503]
c. 0.3011 [Tech: 0.2997] **d.** 0.9901
e. Yes, P(Greater than 0.25) = 0.0099.

21. a. Yes, $np \geq 10$ and $n(1 - p) \geq 10$.
b. 0.7088 [Tech: 0.7073] **c.** 0.2631 [Tech: 0.2646]
d. 0.1379 [Tech: 0.1376]
e. No, P(Less than 0.25) = 0.2676 [Tech: 0.2689].

23. a. 0.2709 [Tech: 0.2717] **b.** 0.2459 [Tech: 0.2456]
c. 0.2877 [Tech: 0.2874]
d. Yes, P(Less than 0.25) = 0.0375 [Tech: 0.0379].

25. 200

27. a. 0.0001 **b.** Yes
c. No, because a sample proportion of 0.075 would be unusual if the goal had been reached.

d. 0.3300 [Tech: 0.3317]   **e.** No

f. Yes, because a sample proportion of 0.053 would not be unusual if the goal had been reached.

Section 7.5

Exercises 1–4 are the Check Your Understanding exercises for this section. Answers to these exercises are on page 331.

5. np, $\sqrt{np(1-p)}$

7. True

9. 0.0559 [Tech using normal approximation: 0.0557; tech using binomial: 0.0563]

11. 0.6517 [Tech using normal approximation: 0.6524; tech using binomial: 0.6538]

13. 0.3102 [Tech using normal approximation: 0.3097; tech using binomial: 0.3008]

15. a. 0.2514 [Tech using normal approximation: 0.2529; tech using binomial: 0.2583]

b. 0.0228 [Tech using normal approximation: 0.0229; tech using binomial: 0.0271]

c. 0.2512 [Tech using normal approximation: 0.2527; tech using binomial: 0.2473]

17. a. 0.7549 [Tech using normal approximation: 0.7558; tech using binomial: 0.7544]

b. 0.0934 [Tech using normal approximation: 0.0929; tech using binomial: 0.0918]   **c.** 0.2115 [Tech using normal approximation: 0.2103; tech using binomial: 0.2133]

19. a. 0.0885 [Tech using binomial: 0.0887]

b. 0.7389 [Tech using normal approximation: 0.7388; tech using binomial: 0.7392]

c. 0.4598 [Tech using normal approximation: 0.4605; tech using binomial: 0.4593]

21. a. 0.9192 [Tech using normal approximation: 0.9190; tech using binomial: 0.9163]

b. 0.7389 [Tech using normal approximation: 0.7387; tech using binomial: 0.7341]

c. 0.3232 [Tech using normal approximation: 0.3230; tech using binomial: 0.3340]

23. 0.9744 [Tech using normal approximation: 0.9745; tech using binomial: 0.9833]

25. 0.0796 [Tech using normal approximation: 0.0797]

Section 7.6

Exercises 1–6 are the Check Your Understanding exercises for this section. Answers to these exercises are on page 341.

7. outlier, skewness, mode

9. No. The sample is strongly skewed to the right.

11. Yes. There are no outliers, no strong skewness, and no evidence of multiple modes.

13. Yes. There are no outliers, no strong skewness, and no evidence of multiple modes.

15. No. The sample is strongly skewed to the right.

17. Yes. The points approximately follow a straight line.

19.

No. The data contain an outlier.

21.

Yes. There are no outliers, no strong skewness, and no evidence of multiple modes.

23. Answers will vary. Here is one possibility.

```
0 |
0 | 99
1 | 0000111112234
1 | 69
2 | 1
2 | 79
```

The data are skewed to the right. It is not appropriate to treat this sample as coming from an approximately normal population.

25. Answers will vary. Here is one possibility.

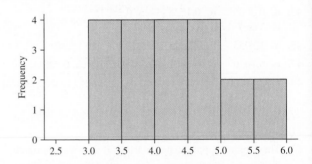

There are no outliers, no strong skewness, and no evidence of multiple modes. It is appropriate to treat this sample as coming from an approximately normal population.

27.

The points do not follow a straight line, so it is not appropriate to treat this sample as coming from an approximately normal population.

29. a. The following boxplot shows that the data do not come from an approximately normal population.

b. The following boxplot shows that the square-root-transformed data come from an approximately normal population.

31. a. The following boxplot shows that the data do not come from an approximately normal population.

b. The following boxplot shows that the square-root-transformed data do not come from an approximately normal population.

c. The following boxplot shows that the reciprocal-transformed data come from an approximately normal population.

CHAPTER 7 Quiz

1. a. 0.32   **b.** 0.41
2. a. 0.9616   **b.** 0.3409   **c.** 0.8977
3. a. −0.44   **b.** 0.81
4. $z = -1.28$ and $z = 1.28$
5. 1.04
6. $40,104 and $45,196 [Tech: $40,087 and $45,213]
7. a. 0.6141 [Tech: 0.6142]   **b.** 0.5910 [Tech: 0.5893]
8. 208.32 [Tech: 208.33]; because bowling scores are whole numbers, this can be rounded up to 209.
9. Let $\bar{x}$ be the mean of a large ($n > 30$) simple random sample from a population with mean μ and standard deviation σ. Then $\bar{x}$ has an approximately normal distribution, with mean $\mu_{\bar{x}} = \mu$ and standard deviation $\sigma_{\bar{x}} = \dfrac{\sigma}{\sqrt{n}}$.
10. $\mu_{\bar{x}} = 193$, $\sigma_{\bar{x}} = 5.25$
11. a. 0.5793 [Tech: 0.5803]   **b.** 0.8997 [Tech: 0.9001]
12. a. $\mu_{\hat{p}} = 0.34$, $\sigma_{\hat{p}} = 0.063875$ **b.** 0.9793 [Tech: 0.9791]
    c. 0.8264 [Tech: 0.8262]
13. a. 0.2843 [Tech: 0.2842]
    b. Yes, $P[\text{More than } 0.25] = 0.0228$ [Tech: 0.0230].
14. a. 0.8461 [Tech: 0.8471]   **b.** 0.7025 [Tech: 0.7031]
    c. 0.7157 [Tech: 0.7163]
15. The following boxplot shows that the data come from an approximately normal population.

CHAPTER 7 Review Exercises

1. a. 0.6368   **b.** 0.9406   **c.** 0.3564
2. −1.23
3. 0.0122
4. a. 0.5621 [Tech: 0.5598]   **b.** 13.280 [Tech: 13.286]
    c. 0.0009
5. a. 0.0793 [Tech: 0.0786]   **b.** 0.9977
    c. 0.7609 [Tech: 0.7600]
6. a. 0.0013   **b.** Yes
7. Yes, $P(\text{Greater than } \$2150) = 0.0037$ [Tech: 0.0036].
8. a. 85 pounds   **b.** 0.9429 [Tech: 0.9431]
9. a. 0.0985 [Tech: 0.0984]
    b. Yes, $P[\text{More than } 70\%] = 0.0019$ [Tech: 0.0018].
10. a. 0.1112 [Tech: 0.1115]   **b.** 0.6879 [Tech: 0.6870]
    c. 0.2009 [Tech: 0.2015]
11. a. 0.0119 [Tech: 0.0118]   **b.** Yes
    c. No, $P[\text{Less than half}] = 0.2843$ [Tech: 0.2856].
12. 0.9495
13. 0.8554 [Tech: 0.8547]
14. The following boxplot shows that the data do not come from an approximately normal population.

15. The following boxplot shows that the data come from an approximately normal population.

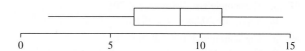

CHAPTER 7 Case Study

1. $\bar{x} = 99.6$, $s = 2.8363$
2. 0.0004
3. Yes
4. If the cans in each sample are arranged in increasing order of strength, each can in the second sample is stronger than the corresponding can in the first sample.
5. $\bar{x} = 101.9$, $s = 6.7404$
6. 0.0384 [Tech: 0.0387]
7. No
8. The following boxplot shows that the data do not come from an approximately normal population. For this reason, the method is not appropriate for the second shipment.

CHAPTER 8
Section 8.1
Exercises 1–16 are the Check Your Understanding exercises for this section. Answers to these exercises are on page 366.

17. point

19. margin of error

21. True

23. True

25. 1.96

27. 2.05

29. 98%

31. 99.5%

33. **a.** 1.344   **b.** Smaller, because the sample size is larger.

35. **a.** 5.510   **b.** Smaller, because the confidence level is lower.

37. **a.** $6.37 < \mu < 9.47$
    b. No, because the sample size is less than 30.

39. **a.** 482   **b.** Smaller, because the confidence level is lower.

41. **a.** 109   **b.** Larger, because the margin of error is smaller.

43. **a.** $428 < \mu < 488$
    b. Larger, because the sample size is smaller.
    c. Smaller, because the confidence level is lower.
    d. No, because the values in the confidence interval are all less than 500.

45. **a.** $25.0 < \mu < 26.0$
    b. No, because the sample consisted entirely of boys.
    c. Yes, because all the values in the confidence interval are less than 28.

47. **a.** The sample size is small ($n \le 30$)
    b. Yes   **c.** $202.6 < \mu < 303.1$

49. **a.** The sample size is small ($n \le 30$)
    b. No   **c.** Not appropriate

51. **a.** $120 < \mu < 130$ **b.** 295

53. **a.** Sally's confidence interval will have the larger margin of error because the confidence level is higher.
    b. Sally's confidence interval is more likely to cover the population mean because the confidence level is higher.

55. **a.** Bob's confidence interval will have the larger margin of error because the standard error will be larger.
    b. Both are equally likely to cover the population mean because they both have the same confidence level.

57. The 90% confidence interval is $7.2 < \mu < 12.8$, the 95% confidence interval is $6.6 < \mu < 13.4$, and the 99% confidence interval is $5.6 < \mu < 14.4$.

59. The students in the class are not a simple random sample of the students in the college.

61. **a.** True; this is the appropriate interpretation of a confidence interval.
    b. False; the confidence interval is for the population mean, not the sample mean.
    c. False; the confidence level is not the probability that the interval contains the true value.

d. False; the confidence level is about the population mean, not the proportion of the population contained in the interval.

63. **a.** 95%, 56.019, 60.881
    b. Yes, because the sample size is large ($n > 30$).

65. **a.** 95%, 9.6956, 15.0084
    b. $8.861 < \mu < 15.843$ **c.** 73   **d.** 125

67. 75.4

Section 8.2
Exercises 1–6 are the Check Your Understanding exercises for this section. Answers to these exercises are on page 378.

7. 11

9. False

11. **a.** 2.074   **b.** 2.920   **c.** 2.567   **d.** 2.763

13. **a.** 2.110   **b.** Smaller

15. **a.** 2.718   **b.** No

17. **a.** $1.2 < \mu < 3.0$
    b. Narrower

19. **a.** $86.1 < \mu < 88.3$ **b.** Wider

21. **a.** $5.42 < \mu < 5.64$
    b. No, because 5.55 is contained in the confidence interval.

23. **a.** $11.7 < \mu < 16.5$
    b. No, because all of the values in the confidence interval are greater than 10.

25. **a.** $132.9 < \mu < 140.9$
    b. Wider, because the confidence level is higher.

27. **a.** $65.1 < \mu < 95.9$
    b. Narrower, because the sample size is greater.

29. **a.** $9.986 < \mu < 12.808$
    b. Yes, because 11.5 is contained in the confidence interval.

31. **a.** $19.50 < \mu < 21.26$
    b. Yes, it is reasonable to believe that the mean mineral content may be as high as 21.3%.

33. **a.** Yes; there are no outliers and no strong skewness.
    b. $856.9 < \mu < 1168.9$

35. **a.** Yes; there are no outliers, and no strong skewness.
    b. $2.91 < \mu < 4.26$
    c. It does not contradict the claim, because the value 3.51 is contained in the confidence interval.

37. **a.** $13.27 < \mu < 17.61$
    b. $12.36 < \mu < 22.07$. The results are noticeably different. It is important to check for outliers in order to avoid misleading results.

39. 85.9%

41. We have data on the whole population of presidents, not just a sample. We know that the population mean is 70.8, so we don't need to construct a confidence interval.

43. **a.** 98%, 178.08, 181.58
    b. Yes, because the sample is large ($n > 30$).

45. a. 14 **b.** No, because the sample is small ($n \leq 30$).

   c. 2.624

   d. $4.5561 < \mu < 7.3185$ [Tech: $4.5558 < \mu < 7.3188$]

47. a. $69.3 < \mu < 100.7$

   b. No, because we can't compute a sample standard deviation from a sample of size 1.

Section 8.3

Exercises 1–6 are the Check Your Understanding exercises for this section. Answers to these exercises are on page 390.

7. standard error

9. True

11. Point estimate: 0.1916; standard error: 0.01426; margin of error = 0.02794

13. Point estimate: 0.4979; standard error: 0.02297; margin of error = 0.0378

15. $0.325 < p < 0.550$

17. $0.341 < p < 0.448$

19. a. 0.429

   b. $0.366 < p < 0.491$

   c. No, because 0.45 is contained in the confidence interval.

21. a. 0.244

   b. $0.190 < p < 0.297$

   c. Yes, because all the values in the confidence interval are greater than 0.09.

23. a. 0.400 **b.** $0.373 < p < 0.426$

   c. Yes, because all the values in the confidence interval are less than 0.50.

25. a. $0.367 < p < 0.446$

   b. $0.360 < p < 0.454$

   c. Increasing the confidence level makes the interval wider.

27. a. 2090 **b.** 2401

   c. About the same. The necessary sample size does not depend on the population.

29. a. 306 **b.** 1037

31. a. $0.154 < p < 0.266$

   b. Larger, because the sample size is smaller.

33. $0.657 < p < 0.979$

35. a. 99%, 0.41911, 0.73714 **b.** $0.457 < p < 0.699$

37. a. 98%, 0.732082, 0.870128 **b.** $0.752 < p < 0.850$

39. This is not a sample; it is the whole population of senators.

41. a. $0.357 < p < 0.802$ **b.** $0.357 < p < 0.801$

   c. $0.352 < p < 0.848$

43. a. 90%: $0.392 < p < 0.777$; 95%: $0.357 < p < 0.802$; 99%: $0.296 < p < 0.842$

   b. 90%: $0.393 < p < 0.765$; 95%: $0.357 < p < 0.801$; 99%: $0.287 < p < 0.871$

   c. The 95% confidence interval. The reason is that $z_{\alpha/2} = 1.96$ for the 95% confidence interval, which is very close to 2.

Section 8.4

Exercises 1–4 are the Check Your Understanding exercises for this section. Answers to these exercises are on page 395.

5. 14

7. True

9. 6.262, 27.488

11. $11.71 < \sigma < 20.87$

13. a. $63.04 < \sigma < 137.26$

   b. No, because the value 100 is contained in the confidence interval.

15. a. 1.798 **b.** $1.27 < \sigma < 3.05$

   c. Yes, because the value 2.7 is contained in the confidence interval.

17. a. 0.838 **b.** $0.46 < \sigma < 2.92$

   c. Yes, because some of the values in the confidence interval are less than 1.5.

19. $\chi^2_{0.025} = 74.22$, $\chi^2_{0.975} = 129.56$; 95% confidence interval is $35.14 < \sigma < 46.43$

21. Lower bound differs by 0.25, upper bound differs by 0.62.

23. Lower bound differs by 0.07, upper bound differs by 0.14.

Section 8.5

Exercises 1–4 are the Check Your Understanding exercises for this section. Answers to these exercises are on page 398.

5. Population mean; $63.47 < \mu < 79.17$

7. Population standard deviation, $0.99 < \sigma < 4.50$

9. Population proportion, $0.686 < p < 0.794$

11. Population mean, $7.32 < \mu < 9.58$

13. $235.38 < \mu < 261.38$

15. $6.26 < \sigma < 16.62$

17. $37.91 < \mu < 52.11$

19. $0.077 < p < 0.116$

CHAPTER 8 Quiz

1. a. A single number that is used to estimate the value of a parameter

   b. An interval that is used to estimate the value of a parameter

   c. The percentage of confidence intervals that will cover the true value of the parameter in the long run

2. 1.706

3. $21.7 < \mu < 24.7$

4. $7.60 < \sigma < 14.61$

5. $46.1 < \mu < 63.5$

6. $11.70 < \mu < 13.14$

7. $14.79 < \sigma < 45.97$ [Tech: $14.79 < \sigma < 45.96$]

8. 1.75 [Tech: 1.751]

9. 7.015, 34.805

10. a. 141 **b.** 2.53

11. a. 1.96 **b.** 4.958

   c. $136 < \mu < 146$

12. 984

13. $0.288 < p < 0.373$

14. 230

15. 260

CHAPTER 8 Review Exercises

1. 271

2. a. $10.79 < \mu < 14.57$

 b. No, because the value 13 is contained in the confidence interval.

 c. 180

3. a. $0.235 < p < 0.410$ **b.** 1612   **c.** 1844

4. a. $56.3 < \mu < 98.4$

 b. Yes, the value 27 is an outlier.

5. No, because the data do not come from a normal population.

6. a. $1.85 < \mu < 2.07$

 b. No, because all the values in the confidence interval are greater than 0.5.

 c. 737

7. a. $0.034 < p < 0.136$ **b.** 323   **c.** 1037

8. a.

Yes. There are no outliers, no strong skewness, and no evidence of multiple modes.

 b. $18.3 < \mu < 31.2$

9. $4.03 < \sigma < 12.39$

10. a. $0.453 < p < 0.618$

 b. No, because the value 0.481 is contained in the confidence interval.

 c. 1529

11. a. $10.3 < \mu < 14.1$

 b. Yes, because all the values in the confidence interval are less than 15.

12. a. $6.75 < \mu < 7.37$

 b. No, because all the values in the confidence interval are less than 8.

 c. 818

13. a. 0.168   **b.** $0.097 < p < 0.239$ **c.** 598

14. a. $36.3 < \mu < 37.5$

 b. No, because the value 38.7 is not contained in the confidence interval.

15. The days are not independent trials. If it rains on one day, it is more likely to rain the next day.

CHAPTER 8 Case Study

1.

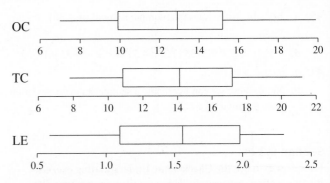

2. PM: $17.94 < \mu < 22.48$;   OC: $11.08 < \mu < 14.19$; TC: $12.428 < \mu < 15.847$;   LE: $1.291 < \mu < 1.784$

3. It is reasonable to conclude that the mean levels were lower than the baseline for all the pollutants, because all the values in the confidence intervals are less than the corresponding baselines.

4.

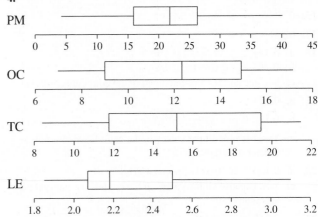

95% Confidence intervals:
PM: $17.33 < \mu < 25.00$;   OC: $10.493 < \mu < 13.615$; TC: $13.149 < \mu < 17.169$;   LE: $2.143 < \mu < 2.496$

5. It is reasonable to conclude that the mean levels were lower than the baseline for all the pollutants, because all the values in the confidence intervals are less than the corresponding baselines.

CHAPTER 9
Section 9.1
Exercises 1–6 are the Check Your Understanding exercises for this section. Answers to these exercises are on page 409.

 7. null, alternate

 9. False

 11. False

 13. Left-tailed

 15. Two-tailed

 17. Correct decision

 19. Type II error

 21. $H_0: \mu = 400$, $H_1: \mu > 400$

 23. The mean amount spent by diners is greater than $30.

 25. There is not enough evidence to conclude that the mean weight of adult German shepherd dogs differs from 75 pounds.

27. a. The conclusion will be that the mean breaking strength is greater than 50.

b. The conclusion will be that the mean breaking strength is equal to 50.

c. With a one-tailed hypothesis, we can conclude that the mean breaking strength is greater than 50. With a two-tailed hypothesis, we will not know whether the mean breaking strength is less than 50 or greater than 50.

29. a. (ii) **b.** (iii) **c.** Type I error

d. No, a Type I error occurs if H_0 is rejected when it is true. Therefore, a Type I error cannot occur when H_0 is false.

e. Yes, if H_0 is not rejected, a Type II error will occur.

Section 9.2

Exercises 1–22 are the Check Your Understanding exercises for this section. Answers to these exercises are on pages 432–433.

23. P-value

25. critical

27. increase

29. True

31. False

33. False

35. a. $z = 2.60$

b. Critical value: 1.645, P-value: 0.0047; H_0 is rejected.

c. Critical value: 2.326, P-value: 0.0047; H_0 is rejected.

37. a. $z = 0.99$

b. Critical values: $-1.96, 1.96$, P-value: 0.3222 [Tech: 0.3211]; H_0 is not rejected.

c. Critical values: $-2.576, 2.576$, P-value: 0.3222 [Tech: 0.3211]; H_0 is not rejected.

39. 0.035

41. a. True **b.** False **c.** True **d.** False

43. a. Yes **b.** Type I error **c.** Correct decision

45. a. No **b.** Type II error **c.** Correct decision

47. iii

49. i

51. a. $H_0: \mu = 20.8, H_1: \mu > 20.8$ **b.** $z = 2.75$

c. Critical value: 1.645, P-value: 0.0030. Reject H_0. We conclude that the mean time of Facebook visits has increased.

53. a. $H_0: \mu = 69.4, H_1: \mu < 69.4$ **b.** $z = -2.44$

c. Critical value: -2.326, P-value: 0.0073 [Tech: 0.0074]. Reject H_0. We conclude that the mean height of men aged 60–69 is less than the mean height of all U.S. men.

55. $H_0: \mu = 2.1, H_1: \mu < 2.1; z = -5.00$; Critical value: -1.645, P-value: 0.0001 [Tech: 0.000000287]. Reject H_0. We conclude that the mean FEV_1 in the high-pollution community is less than 2.1 liters.

57. a. $H_0: \mu = 260.7, H_1: \mu \neq 260.7$; Test statistic: $z = 1.40$. Critical values: $-1.96, 1.96$; P-value: 0.1616 [Tech: 0.1601]. Do not reject H_0. There is not enough evidence to conclude that the mean price in 2013 differs from the mean price in April through June of 2008.

b. There are two outliers.

c. The sample size is large ($n > 30$).

59. a. The sample is small ($n \leq 30$).

b. Yes, there are no outliers and no evidence of strong skewness.

c. $H_0: \mu = 15, H_1: \mu < 15$; Test statistic: $z = -3.02$; Critical value: -2.326, P-value: 0.0013. Reject H_0. We conclude that the mean concentration meets the EPA standard.

61. a. $H_0: \mu = 45, H_1: \mu \neq 45$ **b.** $z = 3.094063348$

c. 0.0019744896 **d.** Yes **e.** Yes

63. a. $H_0: \mu = 225, H_1: \mu \neq 225$ **b.** $z = 2.085$ **c.** 0.037

d. Yes **e.** No **f.** $z = 1.07$

g. 0.1423 [Tech: 0.1412]

h. No

65. a. True **b.** False

67. a. $z = 2.50$ **b.** 0.0062 **c.** Yes

d. No, an increase of 2.5 points on the SAT is small.

69. Yes, the actual P-value should have been given, rather than just saying "$P < 0.05$."

71. If the value of 100 specified by the null hypothesis is not contained in the 95% confidence interval, then the P-value must be less than 0.05.

Section 9.3

Exercises 1–6 are the Check Your Understanding exercises for this section. Answers to these exercises are on page 447.

7. skewness, outliers

9. True

11. a. P-value is between 0.025 and 0.05 [Tech: 0.0399]

b. P-value is between 0.20 and 0.50 [Tech: 0.2086]

c. P-value is between 0.0025 and 0.005 [Tech: 0.0030]

d. P-value is between 0.10 and 0.25 [Tech: 0.1793]

13. a. $-2.056, 2.056$ **b.** 2.390 **c.** $-1.753, 1.753$

d. -1.812

15. a. $H_0: \mu = 178,258, H_1: \mu > 178,528$

b. $t = 2.464$; 54 degrees of freedom

c. Critical value: 1.676; P-value is between 0.005 and 0.01 [Tech: 0.00848]. Reject H_0. We conclude that that the mean salary for family practitioners in Los Angeles is greater than the national average.

17. $H_0: \mu = 25, H_1: \mu > 25$, Test statistic: $t = 1.790$, Critical value: 2.326, P-value is between 0.025 and 0.05 [Tech: 0.0372]. Do not reject H_0. There is not enough evidence to conclude that the mean weight of one-year-old baby boys is greater than 25 pounds.

19. $H_0: \mu = 25, H_1: \mu < 25$. Test statistic: $t = -1.687$, Critical value: -1.645; P-value is between 0.025 and 0.05 [Tech: 0.0459]. Reject H_0. We conclude that the mean commute time is less than 25 minutes.

21. a. Yes, there are no outliers and no evidence of strong skewness.

b. $H_0: \mu = 10, H_1: \mu > 10$, Test statistic: $t = 3.669$, Critical value: 1.771, P-value is between 0.001 and 0.0025 [Tech: 0.0014]. Reject H_0. We conclude that the mean weight loss is greater than 10 pounds.

23. a. No, there is an outlier.

b. Not appropriate

25. a. Yes, there are no outliers and no evidence of strong skewness.

   b. $H_0: \mu = 300$, $H_1: \mu > 300$, Test statistic: $t = 1.927$, Critical value: 2.821, P-value is between 0.025 and 0.05 [Tech: 0.0431]. Do not reject H_0. There is not enough evidence to conclude that the mean price is greater than \$300.

27. a.

   There are no outliers and no evidence of strong skewness. It is appropriate to perform a hypothesis test.

   b. $H_0: \mu = 2$, $H_1: \mu > 2$, Test statistic: $t = 1.638$, Critical value: 1.895, P-value is between 0.05 and 0.10 [Tech: 0.0727]. Do not reject H_0. There is not enough evidence to conclude that the mean amount absorbed is greater than 2 micrograms.

29. a. $H_0: \mu = 2$, $H_1: \mu < 2$ **b.** 1.88   **c.** 0.21   **d.** 17

   e. Yes. The P-value is 0.0133859404, which is less than 0.05.

   f. $t = 1.616$

   g. P-value is between 0.05 and 0.10 [Tech: 0.0622].

   h. No, because the P-value is greater than 0.05.

31. a. $H_0: \mu = 5.5$, $H_1: \mu > 5.5$ **b.** 5.92563   **c.** 0.15755

   d. 4   **e.** Yes. The P-value is 0.002, which is less than 0.05.

   f. $t = -8.152$

   g. P-value is between 0.0005 and 0.001 [Tech: 0.00062].

   h. Yes, because the P-value is less than 0.05.

33. a. $t = 1.25$

   b. P-value is between 0.10 and 0.25 [Tech: 0.1071].   **c.** No

35. No, the people on a certain block may not be representative of the population of the city.

37. The P-value will be less, because the sample mean is farther from the mean specified by the null hypothesis.

39. a. No, we can only conclude the mean is greater than 1500. The P-value does not tell us how much greater it is.

   b. Yes, this is the correct interpretation of the P-value.

41. a. 1.645   **b.** 99   **c.** 0.0516

Section 9.4

Exercises 1–4 are the Check Your Understanding exercises for this section. Answers to these exercises are on page 458.

5. 10

7. True

9. a. 0.675

   b. Yes, the number of individuals in each category is greater than 10.

   c. $z = -2.80$

   d. Yes. Critical value: -1.645, P-value: 0.0026.

11. a. 0.560

   b. Yes, the number of individuals in each category is greater than 10.

   c. $z = -0.71$

   d. No. Critical values: $-1.96, 1.96$, P-value: 0.4778 [Tech: 0.4795].

13. a. $H_0: p = 0.8$, $H_1: p \neq 0.8$

   b. $z = -2.01$

   c. Critical values: $-1.96, 1.96$, P-value: 0.0444 [Tech: 0.0442]. Reject H_0. We conclude that the percentage of emails that are spam differs from 80%.

   d. Critical values: $-2.576, 2.576$, P-value: 0.0444 [Tech: 0.0442]. Do not reject H_0. There is not enough evidence to conclude that the percentage of emails that are spam differs from 80%.

15. $H_0: p = 0.55$, $H_1: p > 0.55$, Test statistic: $z = 0.56$, Critical value: 2.326, P-value: 0.2877 [Tech: 0.2875]. Do not reject H_0. There is not enough evidence to conclude that more than 55% of children aged 8–12 have cell phones.

17. $H_0: p = 0.5$, $H_1: p < 0.5$, Test statistic: $z = -4.94$, Critical value: -1.645, P-value: 0.0001 [Tech: 0.00000040]. Reject H_0. We conclude that less than half of HIV-positive smokers have used a nicotine patch.

19. $H_0: p = 0.25$, $H_1: p < 0.25$, Test statistic: $z = -4.17$, Critical value: -2.326, P-value: 0.0001 [Tech: 0.0000155]. Reject H_0. We conclude that more than 25% of video gamers prefer consoles.

21. $H_0: p = 0.73$, $H_1: p > 0.73$, Test statistic: $z = 1.05$, Critical value: 1.96, P-value: 0.1469. Do not reject H_0. There is not enough evidence to conclude that more than 73% of technology companies have Twitter accounts.

23. $H_0: p = 0.5$, $H_1: p < 0.5$, Test statistic: $z = -4.13$, Critical value: -2.326, P-value: 0.0001 [Tech: 0.000018]. Reject H_0. We conclude that less than half of patients prefer a doctor with high interpersonal skills to one with high technical skills.

25. a. $H_0: p = 0.5$, $H_1: p < 0.5$ **b.** $\hat{p} = 0.3382352941$

   c. Yes. The P-value is 0.0038164856, which is less than 0.05.

   d. $z = 1.68$, P-value: 0.0930 [Tech: 0.0929]. Do not reject H_0.

27. a. $H_0: p = 0.6$, $H_1: p > 0.6$ **b.** $\hat{p} = 0.618829$

   c. No. The P-value is 0.129, which is greater than 0.05.

   d. $z = -1.93$, P-value: 0.0268 [Tech: 0.0269]. Reject H_0.

29. a. $z = 2.68$, $P = 0.004$ **b.** No   **c.** Yes

31. This is a voluntary response sample, not a simple random sample.

33. a. $n = 10$, $X = 9$ **b.** $n = 10$, $p = 0.5$ **c.** $P = 0.0107$

   d. Yes

Section 9.5

Exercises 1–4 are the Check Your Understanding exercises for this section. Answers to these exercises are on page 461.

5. 14

7. False

9. Test statistic: $\chi^2 = 19.6$, Critical value: 18.307. Reject H_0.

11. Test statistic: $\chi^2 = 15.790$, Critical value: 10.196. Do not reject H_0.

13. Test statistic: $\chi^2 = 69.360$, Critical values: 12.401, 39.364. Reject H_0.

15. $H_0: \sigma = 5$, $H_1: \sigma < 5$, Test statistic: $\chi^2 = 17.75$, Critical value: 13.848. Do not reject H_0. There is not enough evidence to conclude that the pediatrician's claim is true.

17. H_0: $\sigma = 15$, H_1: $\sigma \neq 15$, Test statistic: $\chi^2 = 6.827$, Critical values: 12.401, 39.364. Reject H_0. We conclude that the standard deviation differs from 15.

19. H_0: $\sigma = 0.1$, H_1: $\sigma \neq 0.1$, Test statistic: $\chi^2 = 32.656$, Critical values: 2.700, 19.023. Reject H_0. We conclude that the standard deviation changed, and now differs from 0.1.

21. Test statistic: $\chi^2 = 177.778$, Critical values: 74.222, 129.561. Reject H_0.

23. Test statistic: $\chi^2 = 177.778$, Critical values: 73.772, 129.069. Reject H_0.

Section 9.6

Exercises 1–4 are the Check Your Understanding exercises for this section. Answers to these exercises are on page 464.

5. The parameter is the population mean. H_0: $\mu = 9$, H_1: $\mu \neq 9$, Test statistic: $t = -2.5607$, Critical values: -3.055, 3.055, P-value is between 0.02 and 0.05 [Tech: 0.0250]. Do not reject H_0.

7. The parameter is the population standard deviation. H_0: $\sigma = 5$, H_1: $\sigma > 5$, Test statistic: $\chi = 23.064$, Critical value: 24.996. Do not reject H_0.

9. The parameter is the population mean. H_0: $\mu = 40$, H_1: $\mu \neq 40$, Test statistic: $z = 2.32$, Critical values: -1.96, 1.96, P-value: 0.0204 [Tech: 0.0203]. Reject H_0.

11. The parameter is the population proportion. H_0: $p = 0.6$, H_1: $p \neq 0.6$, Test statistic: $z = 2.72$, Critical values: -2.576, 2.576, P-value: 0.0066 [Tech: 0.0065]. Reject H_0. We conclude that the proportion of families with one or more pets differs from 0.6.

13. H_0: $p = 0.3$, H_1: $p > 0.3$, Test statistic: $z = 2.63$, Critical value: 1.645, P-value: 0.0043. Reject H_0. We conclude that more than 30% of of U.S. adults with children have saved money for college.

15. H_0: $\sigma = 0.1$, H_1: $\sigma < 0.1$, Test statistic: $\chi = 4.195$, Critical value: 1.239. Do not reject H_0. There is not enough evidence to conclude that the standard deviation of weights of cookie boxes is less than 0.1.

17. H_0: $\mu = 50,000$, H_1: $\mu \neq 50,000$, Test statistic: $z = -1.50$, Critical values: -2.576, 2.576, P-value: 0.1336 [Tech: 0.1329]. Do not reject H_0. There is not enough evidence to conclude that the mean salary of public school teachers in Georgia differs from $50,000.

19. H_0: $\mu = 1$, H_1: $\mu > 1$, Test statistic: $t = 2.377$, Critical value: 1.943, P-value is between 0.025 and 0.05 [Tech: 0.0275]. Reject H_0. We conclude that the mean concentration is greater than 1 milligram per cubic meter.

Section 9.7

Exercises 1 and 2 are the Check Your Understanding exercises for this section. Answers to these exercises are on page 468.

3. Type II
5. increase
7. False
9. **a.** True **b.** True **c.** False **d.** False

11. **a.** 0.2578 [Tech: 0.2595]
    b. 0.6406 [Tech: 0.6388]
    c. 0.0918 [Tech: 0.0924]
    d. 0.3707 [Tech: 0.3721]

13. **a.** 0.5319 [Tech: 0.5316] **b.** 0.9962
    c. 0.2743 [Tech: 0.2735] **d.** 0.9761 [Tech: 0.9764]

15. 0.5160

17. 0.9582 [Tech: 0.9586]

19. **a.** 1.645 **b.** 0.3342 **c.** 0.8186 [Tech: 0.8197]

CHAPTER 9 Quiz

1. 0.035
2. iv
3. **a.** True **b.** False **c.** True **d.** False
4. **a.** False **b.** True **c.** True **d.** True
5. **a.** No, because the population standard deviation is unknown.
   b. Yes, because the population is normal.
6. 1.711
7. True
8. **a.** $z = 2.50$ **b.** Yes
   c. We conclude that more than 85% of people pass their driver's test.
9. **a.** Yes, because the P-value is less than 0.05.
   b. No, because the P-value is greater than 0.01.
10. False
11. Either the P-value or the value of the test statistic should be reported.
12. Yes
13. True
14. Test statistic: $\chi^2 = 8.444$, Critical value: 10.117, reject H_0. We conclude that the population standard deviation is less than 15.
15. 0.9871

CHAPTER 9 Review Exercises

1. i.
2. **a.**

   Yes

   b. H_0: $\mu = 8.62$, H_1: $\mu < 8.62$. Test statistic: $t = -0.387$, Critical value: -1.714; P-value is between 0.25 and 0.40 [Tech: 0.3511]. Do not reject H_0. There is not enough evidence to conclude that the mean number of runs in 2013 is less than the mean number of runs in 2012.

3. **a.** H_0: $p = 0.6$, H_1: $p \neq 0.6$
   b. $z = 2.02$
   c. Critical values: -1.96, 1.96, P-value: 0.0434 [Tech: 0.0433]. Reject H_0.
   d. We conclude that the proportion of students who log in to Facebook daily differs from 0.6.

4. **a.** H_0: $\mu = 25$, H_1: $\mu \neq 25$
   b. t-test, because the population standard deviation is unknown.

c. $t = 0.478$

d. Critical values: $-2.032, 2.032$; P-value is between 0.50 and 0.80 [Tech: 0.6360]. Do not reject H_0.

e. There is not enough evidence to conclude that the mean price differs from \$25.00.

5. 0.2776 [Tech: 0.2762]

6. The power would be greater. When the significance level is increased, the probability of rejecting H_0 increases as well.

7. a. $H_0: \mu = 2.5, H_1: \mu < 2.5$

b. We should perform a t-test, because the population standard deviation is unknown.

c. $t = -1.633$

d. Critical value: -2.364, P-value is between 0.05 and 0.10 [Tech: 0.0523]. Do not reject H_0.

e. There is not enough evidence to conclude that the mean number of people per household is less than 2.5.

8. a. $H_0: p = 0.45, H_1: p > 0.45$,

b. Test statistic: $z = 3.14$

c. Critical value: 2.326, P-value: 0.0008. Reject H_0.

d. We conclude that more than 45% of people have considered a career in science.

9. $H_0: \sigma = 2, H_1: \sigma > 2$, Test statistic: $\chi^2 = 15.75$, Critical value: 18.475. Do not reject H_0. There is not enough evidence to conclude that the population standard deviation is greater than 2.

10. a. $H_0: \mu = 4.7, H_1: \mu \neq 4.7$ **b.** $z = 2.074$
c. P-value is 0.038.   **d.** Yes   **e.** No

11. a. Proportion   **b.** $H_0: p = 0.2, H_1: p < 0.2$
c. $z = -2.25$ **d.** P-value is 0.0122244334   **e.** Yes
f. No

12. a. $H_0: \mu = 3, H_1: \mu > 3$

b. We should perform a z-test, because the population standard deviation is known.

c. $z = 0.52$

d. Critical value: 2.326, P-value: 0.3015 [Tech: 0.3028]. Do not reject H_0.

e. There is not enough evidence to conclude that the mean number of TV sets per household is greater than 3.

13. $H_0: \sigma = 10, H_1: \sigma < 10$, Test statistic: $\chi^2 = 6.00$, Critical value: 13.848. Reject H_0. We conclude that the population standard deviation is less than 10.

14. a. $H_0: \mu = 1000, H_1: \mu > 1000$

b. We should perform a t-test, because the population standard deviation is unknown.

c. $t = 2.108$

d. Critical value: 1.685, P-value is between 0.01 and 0.025 [Tech: 0.0207]. Reject H_0.

e. We conclude that the mean monthly rent is greater than \$1000.

15. a. $H_0: p = 0.23, H_1: p > 0.23$,

b. Test statistic: $z = 3.36$

c. Critical value: 2.326, P-value: 0.0004. Reject H_0.

d. We conclude that more than 23% of students at the university watch cable news regularly.

CHAPTER 9 Case Study

1. Test statistic: $z = 3.61$, P-value: 0.0002, H_0 is rejected at any reasonable level, including $\alpha = 0.05$ and $\alpha = 0.01$.

We conclude that the probability is greater than 0.5 that a record high occurred more recently than a record low.

2. Answers will vary. The days are not independent and thus do not constitute a simple random sample.

3. Test statistic: $t = 3.606$, P-value is less than 0.0005 [Tech: 0.00016]. H_0 is rejected at any reasonable level, including $\alpha = 0.05$ and $\alpha = 0.01$. We conclude that the probability is greater than 0.5 that the mean year of record high is greater than 1942.

4. Answers will vary. When two years are both record highs, only the later one counts.

5. Test statistic: $t = 3.351$, P-value is between 0.001 and 0.0005 [Tech: 0.00084]. H_0 is rejected at any reasonable level, including $\alpha = 0.05$ and $\alpha = 0.01$. The conclusion does not change. We conclude that the probability is greater than 0.5 that the mean year of record high is greater than 1942.

6. Test statistic: $t = -6.446$, P-value is less than 0.0005 [Tech: 0.000000050]. H_0 is rejected at any reasonable level, including $\alpha = 0.05$ and $\alpha = 0.01$. We conclude that the probability is greater than 0.5 that the mean year of record low is less than 1942.

7. Answers will vary.

CHAPTER 10

Section 10.1

Exercises 1–4 are the Check Your Understanding exercises for this section. Answers to these exercises are on page 487.

5. independent

7. False

9. $8.5 < \mu_1 - \mu_2 < 14.9$ [Tech: $8.6 < \mu_1 - \mu_2 < 14.8$]

11. $36.70 < \mu_1 - \mu_2 < 137.14$
[TI-84 Plus: $39.99 < \mu_1 - \mu_2 < 133.85$]
[MINITAB: $39.90 < \mu_1 - \mu_2 < 133.94$]

13. $-17.2 < \mu_1 - \mu_2 < 27.4$
[TI-84 Plus: $-16.5 < \mu_1 - \mu_2 < 26.7$]
[MINITAB: $-16.6 < \mu_1 - \mu_2 < 26.8$]

15. $8.0 < \mu_1 - \mu_2 < 14.0$ [Tech: $8.1 < \mu_1 - \mu_2 < 13.9$]

17. a. The sample sizes are small ($n \leq 30$).

b. Yes; there are no outliers and no evidence of strong skewness.

c. $31.4 < \mu_1 - \mu_2 < 94.9$
[TI-84 Plus: $34.1 < \mu_1 - \mu_2 < 92.2$]
[MINITAB: $34.0 < \mu_1 - \mu_2 < 92.3$]

19. a. $-12.6 < \mu_1 - \mu_2 < 10.8$
[TI-84 Plus: $-11.9 < \mu_1 - \mu_2 < 10.1$]
[MINITAB: $-12.0 < \mu_1 - \mu_2 < 10.2$]

b. Because the confidence interval contains 0, it is reasonable to believe that the mean IQ may be the same for both firstborn and secondborn sons.

21. a. $0.3 < \mu_1 - \mu_2 < 1.5$

b. Because the confidence interval does not contain 0, it contradicts the claim that the mean weight is the same for both boys and girls.

23. a. $0.52 < \mu_1 - \mu_2 < 2.02$

    b. Because the confidence interval does not contain 0, it is not reasonable to believe that the mean number of energy drinks consumed is the same for both men and women.

25. a. $-8.0 < \mu_1 - \mu_2 < 2.2$ [Tech: $-7.9 < \mu < 2.1$]

    b. Because the confidence interval contains 0, it does not contradict the claim that the mean scores are the same.

27. a. 47.519   **b.** 12.28537157   **c.** 20.904, 74.134

29. a. 20.7429   **b.** 30   **c.** 98%, 12.9408, 28.5450

31. a. 9 degrees of freedom   **b.** 2.262

    c. $39.241 < \mu_1 - \mu_2 < 85.359$

Section 10.2

Exercises 1 and 2 are the Check Your Understanding exercises for this section. Answers to these exercises are on page 495.

3. independent

5. True

7. $0.066 < p_1 - p_2 < 0.384$

9. $-0.489 < p_1 - p_2 < -0.193$

11. $0.099 < p_1 - p_2 < 0.415$

13. a. $0.038 < p_1 - p_2 < 0.079$

    b. Because the confidence interval does not contain 0, it contradicts the claim that the proportion of rear-end accidents is the same at the two intersections.

15. a. $-0.027 < p_1 - p_2 < 0.272$

    b. Because the confidence interval contains 0, it does not contradict the claim that the proportion of patients requiring pain medication is the same for both procedures.

17. a. $0.120 < p_1 - p_2 < 0.178$

    b. No, it does not contradict the claim. It is reasonable to believe that the proportion of defective parts may have decreased by as much as 17.8%.

19. a. $0.022 < p_1 - p_2 < 0.034$

    b. Because the confidence interval does not contain 0, it contradicts the claim that the proportion of patients who have had colonoscopies is the same for those with and without colorectal cancer.

21. a. 0.086666667   **b.** −0.0815, 0.25484

23. a. 0.177294   **b.** 99%, −0.051451, 0.406038

25. 301

Section 10.3

Exercises 1 and 2 are the Check Your Understanding exercises for this section. Answers to these exercises are on page 502.

3. paired

5. True

7. a. $-5.33 < \mu_d < 2.36$

    b. Because the confidence interval contains 0, it does not contradict the claim that the mean speeds of the processors are the same.

9. a. $14.0 < \mu_d < 52.4$

b. Because the confidence interval does not contain 0, it contradicts the claim that the mean strengths are the same.

11. a. $70.4 < \mu_d < 88.3$

    b. No; it is reasonable to believe that the difference may be as large as 88.3.

13. a. $5.18 < \mu_1 - \mu_2 < 5.55$

    b. Because the confidence interval contains 5.5, it does not contradict the claim that the mean increase is 5.5.

15. a. 8.7385   **b.** 13   **c.** 6.5788, 10.898

17. a. 2.5324   **b.** 5   **c.** 99%, −3.411394, 8.476194

19. a. $-0.48 < \mu_1 - \mu_2 < 4.72$
    [TI-84 Plus: $-0.27 < \mu_1 - \mu_2 < 4.51$]
    [MINITAB: $-0.28 < \mu_1 - \mu_2 < 4.53$]

    b. The confidence interval using matched pairs is narrower, because the standard error is smaller.

CHAPTER 10 Quiz

1. Paired

2. Independent

3. 0.1183

4. 1.96

5. 0.0806

6. $0.038 < p_1 - p_2 < 0.199$

7. 8

8. 3.499

9. 8.0172 [Tech: 8.0183]

10. $-0.02 < \mu_d < 16.02$

11. 2543

12. 1.671 [Tech: 1.655]

13. 717.161 [TI-84 Plus: 710.262] [MINITAB: 710.269]

14. $1825.8 < \mu_1 - \mu_2 < 3260.2$
    [Tech: $1832.7 < \mu_1 - \mu_2 < 3253.3$]

15. These are not independent samples.

CHAPTER 10 Review Exercises

1. $0.041 < p_1 - p_2 < 0.063$

2. The samples are not independent, because they are based on the same set of days.

3. $6.5 < \mu_d < 8.3$

4. $(-7153.4 < \mu_1 - \mu_2 < 5323.6)$
    [TI-84 Plus: $(-6710.1 < \mu_1 - \mu_2 < 4880.3)$]
    [MINITAB: $(-6733.7 < \mu_1 - \mu_2 < 4903.9)$]

5. $0.7 < \mu_1 - \mu_2 < 21.2$ [TI-84 Plus: $2.0 < \mu_1 - \mu_2 < 20.0$]
    [MINITAB: $1.8 < \mu_1 - \mu_2 < 20.2$]

6. $-0.020 < p_1 - p_2 < 0.156$

7. $-2.44 < \mu_1 - \mu_2 < -2.16$

8. $-0.043 < \mu_d < 0.003$

9. $-0.241 < p_1 - p_2 < 0.018$

10. a. 11.402
    b. 113.2701584
    c. 9.8059, 12.998

11. a. 2.3515 **b.** 7 **c.** 95%, −2.02947, 6.73247

12. a. 0.31614 **b.** 0.12307, 0.5092

13. a. 0.249243 **b.** 95%, 0.135735, 0.362751

14. a. 8.533 **b.** 17 **c.** 7.7632, 9.3028

15. a. 9.8612
 b. 14
 c. 95%, 7.803957, 11.918443

CHAPTER 10 Case Study

1.

Characteristic	95% Confidence Interval
Age	$-2.03 < \mu_1 - \mu_2 < 0.03$
Systolic BP	$-4.65 < \mu_1 - \mu_2 < -1.35$
Diastolic BP	$-1.94 < \mu_1 - \mu_2 < -0.06$
Treatment for hypertension	$-0.050 < p_1 - p_2 < 0.040$
Atrial fibrillation	$-0.015 < p_1 - p_2 < 0.045$
Diabetes	$-0.039 < p_1 - p_2 < 0.047$
Cigarette smoking	$-0.017 < p_1 - p_2 < 0.045$
Coronary bypass surgery	$-0.048 < p_1 - p_2 < 0.036$

2. Systolic blood pressure and Diastolic blood pressure

3. No; the differences are too small to be of practical significance.

CHAPTER 11
Section 11.1
Exercises 1 and 2 are the Check Your Understanding exercises for this section. Answers to these exercises are on page 521.

3. skewness, outliers

5. False

7. a. 9 **b.** $t = -3.41$
 c. Critical value: −1.833, P-value is between 0.0025 and 0.005 [Tech: 0.0039] [TI-84 Plus: 0.0010] [MINITAB: 0.0011]. Reject H_0.

9. a. $H_0: \mu_1 = \mu_2, H_1: \mu_1 < \mu_2$
 b. $t = -1.105$
 c. 1017
 d. Critical value: −1.645, P-value is between 0.10 and 0.25 [Tech: 0.1347] [TI-84 Plus: $P = 0.1346$] [MINITAB: $P = 0.1346$]. Do not reject H_0. There is not enough evidence to conclude that the mean number of hours on the internet increased between 2010 and 2012.

11. $H_0: \mu_1 = \mu_2, H_1: \mu_1 \neq \mu_2$, Test statistic: $t = -0.173$, Critical values: −3.250, 3.250, P-value is greater than 0.80 [Tech: 0.8662] [TI-84 Plus: $P = 0.8646$] [MINITAB: $P = 0.8647$]. Do not reject H_0. There is not enough evidence to conclude that there is a difference in mean IQ between firstborn and secondborn sons.

13. $H_0: \mu_1 = \mu_2, H_1: \mu_1 > \mu_2$, Test statistic: $t = 5.127$, Critical value: 2.132, P-value is between 0.0025 and 0.005 [Tech: 0.0034] [TI-84 Plus: $P = 0.00025$] [MINITAB: $P = 0.00031$]. Reject H_0. We conclude that the mean benzene concentration is less in treated water than in untreated water.

15. $H_0: \mu_1 = \mu_2, H_1: \mu_1 < \mu_2$, Test statistic: $t = -1.043$, Critical value: −1.812, P-value is between 0.10 and 0.25 [Tech: 0.1607] [TI-84 Plus: $P = 0.1554$] [MINITAB: $P = 0.1557$]. Do not reject H_0. There is not enough evidence to conclude that the mean life span of those exposed to the mummy's curse is less than the mean of those not exposed.

17. $H_0: \mu_1 = \mu_2, H_1: \mu_1 \neq \mu_2$, Test statistic: $t = -2.559$, Critical values: −2.131, 2.131, P-value is between 0.02 and 0.05 [Tech: 0.0218] [TI-84 Plus: $P = 0.0164$] [MINITAB: $P = 0.0164$]. Reject H_0. We conclude that the mean score differs between men and women.

19. $H_0: \mu_1 = \mu_2, H_1: \mu_1 \neq \mu_2$, Test statistic: $t = -1.166$, Critical values: −2.032, 2.032, P-value is between 0.20 and 0.50 [Tech: 0.2519] [TI-84 Plus: $P = 0.2479$] [MINITAB: $P = 0.2479$]. Do not reject H_0. There is not enough evidence to conclude that the mean score differs between paper and online tests.

21. a. Right-tailed
 b. 24.99965945
 c. 0.101223442
 d. No

23. a. $H_1: \mu_1 \neq \mu_2$ **b.** Yes, the P-value is 0.003.
 c. 54 **d.** 35
 e. The P-value is between 0.002 and 0.005 [Tech: 0.0036].

25. a. 1.833 **b.** 26.727 **c.** 0.039

Section 11.2
Exercises 1 and 2 are the Check Your Understanding exercises for this section. Answers to these exercises are on page 531.

3. 20

5. False

7. a. $z = -1.00$ **b.** No **c.** No

9. a. $H_0: p_1 = p_2, H_1: p_1 \neq p_2$
 b. $z = 0.62$
 c. Critical values: −1.96, 1.96, P-value is 0.5352 [Tech: 0.5376]. Do not reject H_0. There is not enough evidence to conclude that the proportion of boys who are overweight differs from the proportion of girls who are overweight.

11. $H_0: p_1 = p_2, H_1: p_1 > p_2$, Test statistic: $z = 3.11$, Critical value: 2.326, P-value is 0.0009. Reject H_0. We conclude that the proportion of patients suffering a heart attack or stroke is less for ticagrelor.

13. $H_0: p_1 = p_2, H_1: p_1 > p_2$, Test statistic: $z = 3.26$, Critical value: 1.645, P-value is 0.0006. Reject H_0. We conclude that the proportion of patients who needed retreatment is less for those who received drug-coated stents.

15. These are not independent samples.

17. a. Right-tailed **b.** 0.0037512809 **c.** Yes

19. a. Left-tailed **b.** 0.192
 c. No, because the P-value is greater than 0.05.

21. a. $H_0: p_1 - p_2 = 0.05, H_1: p_1 - p_2 > 0.05$
 b. $z = 2.11$
 c. Critical value: 1.645, P-value is 0.0174 [Tech: 0.0175]. Reject H_0.
 d. The more expensive machine

Section 11.3

Exercises 1 and 2 are the Check Your Understanding exercises for this section. Answers to these exercises are on page 540.

3. skewness, outliers

5. True

7. a. $1, -4, 6, 9, 7$ **b.** $t = 1.614$

   c. Critical value: 2.132, P-value is between 0.05 and 0.10 [Tech: 0.0909]. Do not reject H_0.

   d. Critical value: 3.747, P-value is between 0.05 and 0.10 [Tech: 0.0909]. Do not reject H_0.

9. a. $H_0: \mu_d = 0, H_1: \mu_d > 0$ **b.** $t = 1.369$

   c. Critical value: 1.943, P-value is between 0.10 and 0.25 [Tech: 0.1100]. Do not reject H_0. There is not enough evidence to conclude that the mean pain level is less with drug B.

11. a. $H_0: \mu_d = 0, H_1: \mu_d \neq 0$ **b.** $t = -4.790$

   c. Critical values: $-2.776, 2.776$, P-value is between 0.005 and 0.01 [Tech: 0.0087]. Reject H_0. We conclude that the mean strength after three days differs from the mean strength after six days.

13. a. $H_0: \mu_d = 5, H_1: \mu_d > 5$

   b. $t = 4.009$

   c. Critical value: -1.685, P-value is less than 0.0005 [Tech: 0.000133]. Reject H_0. We conclude that the mean increase is greater than 5 inches.

15. a. Two-tailed **b.** 15 **c.** 0.0296591111

   d. Yes, because the P-value is less than 0.05.

17. a. Right-tailed

   b. Yes, because the P-value is less than 0.05.

   c. No, because the P-value is greater than 0.01.

19. a. P-value is between 0.05 and 0.10 [Tech: 0.0726] [TI-84 Plus: 0.0653] [MINITAB: 0.0658].

   b. The P-value is greater because the standard error is larger.

Section 11.4

Exercises 1–4 are the Check Your Understanding exercises for this section. Answers to these exercises are on page 547.

5. normal

7. False

9. 2.51

11. a. The critical value is 4.00. Do not reject H_0.

   b. The critical value is 7.72. Do not reject H_0.

13. a. $H_0: \sigma_1 = \sigma_2, H_1: \sigma_1 < \sigma_2$, sample standard deviation for day 1 is 0.13868, sample standard deviation for day 2 is 0.38481. $F = 7.7$. There are 12 and 12 degrees of freedom. Critical value is $f_{0.05} = 2.69$. Reject H_0. We conclude that the variability of the process is greater on the second day than on the first day.

   b. $H_0: \sigma_2 = \sigma_3, H_1: \sigma_2 < \sigma_3$, sample standard deviation for day 2 is 0.38481, sample standard deviation for day 3 is 0.51826. $F = 1.8139$. There are 12 and 12 degrees of freedom. Critical value is $f_{0.01} = 4.16$. Do not reject H_0. There is not enough evidence to conclude that the

variability of the process is greater on the third day than on the second day.

15. $H_0: \sigma_1 = \sigma_2, H_1: \sigma_1 < \sigma_2$, sample standard deviation for month 1 is 14.169, sample standard deviation for month 7 is 59.757. $F = 17.787$. There are 8 and 6 degrees of freedom. Critical value is $f_{0.01} = 8.10$. Reject H_0. We conclude that the time to freeze-up is more variable in the seventh month than in the first month.

17. 0.177

Section 11.5

Exercises 1 and 2 are the Check Your Understanding exercises for this section. Answers to these exercises are on page 550.

3. less

5. True

7. a. Hypothesis 4 can be rejected at the 0.05 level.

   b. None of the hypotheses can be rejected at the 0.01 level.

9. The P-value needs to be adjusted to $20(0.01) = 0.2$ because 20 tests were performed. The adjusted P-value is greater than 0.05, so the conclusion is not justified.

11. 0.0025

13. a. 0.05

   b. Binomial with 500 trials and success probability 0.05

   c. 0.002701 (exact), 0.0014 using normal approximation

   d. Yes, because the P-value for the hypothesis that all the null hypotheses are true is very small.

CHAPTER 11 Quiz

1. ii

2. iv

3. i

4. iii

5. Do not reject H_0. The difference $p_1 - p_2$ may be equal to 0.

6. Yes, the test statistic is $F = 3.199$ and the critical value is 2.65.

7. 14

8. Test statistic is 2.598. Critical value is 1.796, P-value is between 0.01 and 0.025 [Tech: 0.0124]. Reject H_0.

9. $H_0: p_1 = p_2, H_1: p_1 < p_2$

10. $z = -2.72$

11. Critical value: -2.326, P-value is 0.0033. Reject H_0.

12. $H_0: \mu_1 = \mu_2, H_1: \mu_1 \neq \mu_2$

13. $t = -1.971$

14. Critical values: $-2.120, 2.120$, P-value is between 0.05 and 0.10 [Tech: 0.0662] [TI-84 Plus: 0.0576] [MINITAB: 0.0577]. Do not reject H_0.

15. One

CHAPTER 11 Review Exercises

1. $H_0: \mu_1 = \mu_2, H_1: \mu_1 < \mu_2$. Test statistic: $t = -2.409$, Critical value: -2.374, P-value is between 0.005 and 0.01 [Tech: 0.0089] [TI-84 Plus: 0.0085] [MINITAB: 0.0085]. Reject H_0.

We conclude that the mean number of days missed is less with flextime.

2. $H_0: p_1 = p_2, H_1: p_1 > p_2$. Test statistic: $z = 2.04$, Critical value: 1.645, P-value is 0.0207 [Tech: 0.0206]. Reject H_0. We conclude that the proportion of voters who favor the proposal is greater in county A than in county B.

3. $H_0: \sigma_1 = \sigma_2, H_1: \sigma_1 \neq \sigma_2$, Test statistic: $F = 1.4745$, Critical value: 9.20. Do not reject H_0. There is not enough evidence to conclude that the standard deviation of benzene concentration differs between treated water and untreated water.

4. $H_0: \mu_d = 0, H_1: \mu_d \neq 0$. Test statistic: $t = -2.945$, Critical values: $-2.306, 2.306$, P-value is between 0.01 and 0.02 [Tech: 0.0186]. Reject H_0. We conclude that the mean sales differ between the two programs.

5. $H_0: \sigma_1 = \sigma_2, H_1: \sigma_1 \neq \sigma_2$, Test statistic: $F = 1.3514$, Critical value: 5.12. Do not reject H_0. There is not enough evidence to conclude that the standard deviation of distances walked differs between treatment and control.

6. $H_0: \mu_1 = \mu_2, H_1: \mu_1 \neq \mu_2$, Test statistic: $t = -0.306$, Critical values: $-2.000, 2.000$, P-value is between 0.50 and 0.80 [Tech: 0.7606] [TI-84 Plus: $P = 0.7601$] [MINITAB: $P = 0.7601$]. Do not reject H_0. There is not enough evidence to conclude that the mean number of hours differs between men and women.

7. $H_0: \mu_d = 0, H_1: \mu_d \neq 0$. Test statistic: $t = -3.145$, Critical values: $-2.201, 2.201$, P-value is between 0.005 and 0.01 [Tech: 0.0093]. Reject H_0. We conclude that the mean time differs between the two applications.

8. **a.** $H_0: p_1 = p_2, H_1: p_1 < p_2$

   b. Test statistic: $z = -1.04$, Critical value: -1.645, P-value is 0.1492 [Tech: 0.1501]. There is not enough evidence to conclude that machine B produces a larger proportion of rods meeting specifications. Machine A should be used.

9. **a.** Right-tailed   **b.** 21.18819537   **c.** 0.0012913382
   d. Yes, because the P-value is less than 0.05.

10. **a.** Left-tailed   **b.** 99   **c.** 0.161
    d. No, because the P-value is greater than 0.05.

11. **a.** Two-tailed   **b.** 0.2041558545
    c. No, because the P-value is greater than 0.05.

12. **a.** Right-tailed   **b.** 0.003
    c. Yes, because the P-value is less than 0.05.

13. **a.** Right-tailed   **b.** 14   **c.** 0.0021327513
    d. Yes, because the P-value is less than 0.05.

14. **a.** Two-tailed   **b.** 11   **c.** 0.012
    d. Yes, because the P-value is less than 0.05.

15. iii

CHAPTER 11 Case Study

1. The P-values are 0.1164 [Tech: 0.1174], 0.2628 [Tech: 0.2640], 0.0052, 0.0802 [Tech: 0.0795], 0.0258.

2. Proportion with underground wiring and proportion with father Hispanic are significant at the $\alpha = 0.05$ level.

3. For proportion with underground wiring: 0.0260 [Tech: 0.0258]; for proportion with father Hispanic: 0.129.

4. Answers will vary. The evidence is strong only for underground wiring.

CHAPTER 12
Section 12.1
Exercises 1–8 are the Check Your Understanding exercises for this section. Answers to these exercises are on page 567.

9. expected

11. False

13. 23.685

15. 0.01

17. **a.** 10.125   **b.** 4   **c.** Critical value is 9.488. Reject H_0.

19. **a.**

Category	1	2	3	4
Expected	100	60	30	10

   b. 3.427   **c.** 3
   d. Critical value is 11.345. Do not reject H_0.

21. **a.**

Category	1–7	8–14	15–21	22–28	29–35
Expected	30.2	30.2	30.2	30.2	30.2

   b. 2.874   **c.** 4
   d. Critical value is 9.488. Do not reject H_0. There is not enough evidence to conclude that the lottery is unfair.

23. Test statistic: 27.792; 11 degrees of freedom; Critical value: 24.725. Reject H_0. We conclude that fire alarms are more likely in some months than in others.

25. Test statistic: 11.782; 3 degrees of freedom; Critical value: 7.815. Reject H_0. We conclude that the proportions of people living in the various regions changed between 2000 and 2011.

27. Test statistic: 15.891; 3 degrees of freedom; Critical value: 11.345. Reject H_0. We conclude that the proportions of people giving the various responses changed between 2011 and 2012.

29. **a.** Test statistic: 7.120; 5 degrees of freedom; Critical value: 11.070. Do not reject H_0. There is not enough evidence to conclude that the die is not fair.
   b. 1: 0.1556 [Tech: 0.1544]; 2: 0.9124 [Tech: 0.9128]; 3: 0.5092 [Tech: 0.5110]; 4: 0.0376 [Tech: 0.0374]; 5: 0.3788 [Tech: 0.3808]; 6: 0.3222 [Tech: 0.3242].
   c. The P-value is less than 0.05, so we reject H_0 at the $\alpha = 0.05$ level.
   d. The P-value must be adjusted for the fact that multiple tests have been performed.
   e. The adjusted P-value is $6(0.0376) = 0.2256$ [Tech: $6(0.0374) = 0.2244$]. The adjusted P-value is greater than 0.05, so we do not reject H_0 at the $\alpha = 0.05$ level.

Section 12.2
Exercises 1 and 2 are the Check Your Understanding exercises for this section. Answers to these exercises are on page 577.

3. grand

5. homogeneity

7. False

9. **a.** Row totals: 37, 25, 35; Column totals: 27, 35, 35; Grand total: 97

b.

	1	2	3
A	10.299	13.351	13.351
B	6.959	9.021	9.021
C	9.742	12.629	12.629

c. 6.481   **d.** 4

e. Critical value: 9.488. Do not reject H_0.

11. a.

	Shift		
	Morning	**Evening**	**Night**
Influenza	16.323	19.982	10.695
Headache	21.880	26.784	14.335
Weakness	11.114	13.605	7.281
Shortness of Breath	8.683	10.629	5.689

b. 17.572   **c.** 6

d. Critical value: 12.592. Reject H_0. We conclude that occurrence of symptoms and shift are not independent. The frequencies of the symptoms vary among the shifts.

13. a.

	Household Size				
	1	**2**	**3**	**4**	**5**
Agree	66.443	105.581	52.335	44.144	35.497
No Opinion	32.784	52.096	25.823	21.781	17.515
Disagree	46.772	74.323	36.841	31.075	24.988

b. 6.377   **c.** 8

d. Critical value: 20.090. Do not reject H_0. There is not enough evidence to conclude that household size and opinion are not independent.

15. Test statistic: $\chi^2 = 1.485$; Critical value: 7.815. Do not reject H_0. There is not enough evidence to conclude that people in some age groups are more likely to be promoted than those in other age groups.

17. a. 22   **b.** 39.86   **c.** C2   **d.** C3   **e.** 0.301

f. No, because the P-value is greater than 0.05.

g. No, we cannot conclude that the null hypothesis is true.

19.

	1	2	3	4	Row Total
A	10	52	29	7	98
B	25	38	10	19	92
C	30	27	44	46	147
Column Total	65	117	83	72	

CHAPTER 12 Quiz

1. Yes. There are 12 degrees of freedom, and the critical value is 21.026.

2. False

3. True

4. H_0: The failure probabilities are the same for each line. H_1: The failure probabilities are not the same for all the lines.

5.

	Line			
	1	**2**	**3**	**4**
Pass	485.395	473.687	445.771	406.147
Fail	53.605	52.312	49.229	44.853

6. 4.676

7. 3

8. 7.815

9. Do not reject H_0. There is not enough evidence to conclude that the failure probabilities are not all the same.

10. H_0: The age distribution is the same at each site. H_1: The age distributions are not the same at all sites.

11.

	Ages of Skeletons		
Site	**0–4 Years**	**5–19 Years**	**20 Years or More**
Casa da Moura	32.349	60.219	121.433
Wandersleben	32.651	60.781	122.567

12. 2.123

13. 2

14. 9.210

15. Do not reject H_0. There is not enough evidence to conclude that the age distributions are not all the same.

CHAPTER 12 Review Exercises

1. H_0: $p_0 = 0.0625, p_1 = 0.25, p_2 = 0.375, p_3 = 0.25,$ $p_4 = 0.0625$

2.

Number of Boys	0	1	2	3	4
Expected	12.5	50	75	50	12.5

3. Test statistic: $\chi^2 = 5.193$; Critical value: 9.488. Do not reject H_0. There is not enough evidence to conclude that the number of boys does not follow a binomial distribution.

4.

	Educational Level				
	No High School Diploma	**High School Diploma**	**Associate's Degree**	**Bachelor's Degree**	**Graduate Degree**
Men	165.749	653.433	101.999	230.864	127.954
Women	198.251	781.567	122.001	276.136	153.046

5. 17.419

6. Critical value is 13.277. Reject H_0. We conclude that education level and gender are not independent.

7.

	Number of Welds		
	High Quality	**Moderate Quality**	**Low Quality**
Day Shift	466.400	196.400	37.200
Evening Shift	433.086	182.371	34.543
Night Shift	266.514	112.229	21.257

8. 5.760

9. Critical value is 13.277. Do not reject H_0. There is not enough evidence to conclude that the quality varies among shifts.

10.

	Outcome		
Hospital	**Substantial Improvement**	**Some Improvement**	**No Improvement**
A	118	59	23
B	118	59	23
C	118	59	23
D	118	59	23

11. 17.524

12. Critical value is 12.592. Reject H_0. We conclude that the distribution of outcomes varies among the hospitals.

13.

	1	2	3	Total
A	8.333	3.333	13.333	25
B	3.333	1.333	5.333	10
C	13.333	5.333	21.333	40
D	25.000	10.000	40.000	75
Total	50	20	80	150

14. Test statistic: $\chi^2 = 49.979$; Critical value: 83.298. Do not reject H_0. There is not enough evidence to conclude that the numbers are not equally likely to come up.

15. Test statistic: $\chi^2 = 12.075$; Critical value: 9.488. Reject H_0. We conclude that absences are not equally likely on each day of the week.

CHAPTER 12 Case Study

1. Test statistic: 92.205; Critical value: 6.635. Reject H_0. We conclude that acceptance rates differ between men and women.

2. Test statistic: 778.907; Critical value: 15.086. Reject H_0. We conclude that acceptance rates differ among departments.

3. Test statistic: 1068.372; Critical value: 15.086. Reject H_0. We conclude that gender and department are not independent.

4. The critical value in each case is 6.635. A: Test statistic is 17.248. Reject H_0. B: Test statistic is 0.254. Do not reject H_0. C: Test statistic is 0.754. Do not reject H_0. D: Test statistic is 0.298. Do not reject H_0. E: Test statistic is 1.001. Do not reject H_0. F: Test statistic is 0.384. Do not reject H_0. In department A, 82.4% of the women were accepted, but only 62.1% of the men were accepted.

5. A: 65.4%; B: 63.2%; C: 35.1%; D: 34.0%; E: 25.2%; F: 6.4%

6. A: 88.4%; B: 95.7%; C: 35.4%; D: 52.7%; E: 32.7%; F: 52.2%

7. Yes

8. ii

CHAPTER 13
Section 13.1

Exercises 1–6 are the Check Your Understanding exercises for this section. Answers to these exercises are on page 599.

7. 18

9. False

11. a. 28   **b.** 2.048   **c.** 2.073   **d.** $-0.861 < \beta_1 < 3.285$

13. Test statistic: $t = 1.1975$, Critical values: $-2.763, 2.763$, P-value is between 0.20 and 0.50 [Tech: 0.2411]. Do not reject H_0.

15. a. 3.9724   **b.** 1.1315   **c.** 277.33   **d.** 0.067946
   e. 2.776   **f.** 0.1886   **g.** $3.7837 < \beta_1 < 4.161$
   h. Test statistic: $t = 58.464$, Critical values: $-2.776, 2.776$, P-value is less than 0.001 [Tech: 0.00000051]. Reject H_0.

17. a. 3.1236   **b.** 0.19353   **c.** 71.2   **d.** 0.022935
   e. 3.182   **f.** 0.0730   **g.** $3.0506 < \beta_1 < 3.1966$
   h. Test statistic: $t = 136.19$, Critical values: $-3.182, 3.182$, P-value is less than 0.001 [Tech: 0.00000087]. Reject H_0.

19. a. $\hat{y} = 302.5212 - 0.5082x$
   b. $-7.8848 < \beta_1 < 6.8684$ [Tech: $-7.8845 < \beta_1 < 6.8680$]
   c. Test statistic: $t = -0.1461$, Critical values: $-2.120, 2.120$, P-value is greater than 0.80 [Tech: 0.8857]. Do not reject H_0. There is not enough evidence to conclude that the amount of protein is useful in predicting the number of calories.

21. a. $\hat{y} = 52.0434 - 0.8445x$
   b. $-2.0322 < \beta_1 < 0.3432$ [Tech: $-2.0324 < \beta_1 < 0.3434$]

c. Test statistic: $t = -2.0229$, Critical value: -1.725, P-value is between 0.025 and 0.05 [Tech: 0.0283]. Reject H_0. We conclude that wingspan is useful in predicting lifespan.

d. Shorter, because we conclude that $\beta_1 < 0$.

23. a. $\hat{y} = 71.4363 + 0.2392x$
   b. $0.1669 < \beta_1 < 0.3116$
   c. Test statistic: $t = 8.0886$, Critical values: $-3.707, 3.707$, P-value is less than 0.001 [Tech: 0.00019]. Reject H_0. We conclude that speed is useful in predicting noise level.

25. a. $\hat{y} = 73.2662 - 0.4968x$
   b. $-0.9521 < \beta_1 < -0.0415$
   [Tech: $-0.9520 < \beta_1 < -0.0416$]
   c. Test statistic: $t = -2.5806$, Critical values: $-2.365, 2.365$, P-value is between 0.02 and 0.05 [Tech: 0.0364]. Reject H_0. We conclude that horizontal expansion is useful in predicting vertical expansion.

27. a. $\beta_1 \neq 0$ **b.** 3.143104276   **c.** 6   **d.** 0.0199889322
   e. Yes, because the P-value is less than 0.05.

29. a. Slope: -0.7524; Intercept: 88.761.   **b.** Yes

31. $24.091 < \mu_{y|20} - \mu_{y|15} < 32.3025$

Section 13.2

Exercises 1 and 2 are the Check Your Understanding exercises for this section. Answers to these exercises are on page 605.

3. confidence

5. False

7. a. $6.35 < $ Mean response $ < 9.43$
   b. $0.43 < $ Individual response $ < 15.35$

9. a. $b_0 = -12.2183$, $b_1 = 2.7919$ **b.** 21.284   **c.** 3.7885
   d. 32.833   **e.** 2.776   **f.** $15.43 < $ Mean response $ < 27.14$
   g. $9.24 < $ Individual response $ < 33.32$

11. a. 294.9   **b.** $245.69 < $ Mean response $ < 344.1$ **c.** 294.9
   d. $108.17 < $ Individual response $ < 481.63$

13. a. 26.708   **b.** $23.367 < $ Mean response $ < 30.05$
   c. 26.708   **d.** $12.247 < $ Individual response $ < 41.17$

15. a. 79.81   **b.** $79.29 < $ Mean response $ < 80.33$
   c. 79.81   **d.** $78.588 < $ Individual response $ < 81.032$

17. a. 60.847   **b.** $53.103 < $ Mean response $ < 68.59$
   c. 60.847   **d.** $37.813 < $ Individual response $ < 83.88$

19. a. 43.62   **b.** $41.23 < $ Mean response $ < 46.00$ **c.** 43.62
   d. No, because the prediction interval is
   $20.86 < $ Individual response $ < 66.37$

21. a. 9   **b.** 14   **c.** 10

Section 13.3

Exercises 1–6 are the Check Your Understanding exercises for this section. Answers to these exercises are on page 622.

7. residual

9. True

11. a. $\hat{y} = 10.372 + 0.5215x_1 + 1.4426x_2 + 2.8834x_3$
   b. 35.26   **c.** 94.1%   **d.** Yes, $F = 31.98$, $P = 0.0004$.

e. P-value for β_1 is 0.3531; do not reject H_0. P-value for β_2 is 0.0298; reject H_0. P-value for β_3 is 0.0004; reject H_0.

13. a. $\hat{y} = -1.9446 + 1.1325x_1 + 3.0963x_2 + 0.4224x_3 + 1.2384x_4$
   b. 39.263
   c. 72.0%
   d. Yes, $F = 6.42$, $P = 0.0080$.
   e. P-value for β_1 is 0.0286; reject H_0. P-value for β_2 is 0.0007; reject H_0. P-value for β_3 is 0.4382; do not reject H_0. P-value for β_4 is 0.0162; reject H_0.

15. a. 166.475   **b.** 113.46   **c.** Yes, $F = 15.649$, $P = 0.000$.
   d. 51.6%   **e.** $P = 0.000$, reject H_0.
   f. $P = 0.013$, do not reject H_0.

17. ii

19. a. $\hat{y} = 0.1210 + 0.5408x_1 + 0.002146x_2 + 0.0002816x_3$
   b. 3.01
   c. (2.8720, 3.1483)
   d. (2.4147, 3.6056)
   e. 72.5%   **f.** Yes, $F = 14.04$, $P = 0.000095$.
   g. P-value for β_1 is 0.0075; reject H_0. P-value for β_2 is 0.1445; do not reject H_0. P-value for β_3 is 0.7475; do not reject H_0.

21. a. $\hat{y} = 8.8721 - 0.006795x_1 + 0.71632x_2 + 2.030x_3$
   b. 42.202   **c.** (36.934, 47.470)   **d.** (32.888, 51.517)
   e. 70.2%   **f.** Yes, $F = 34.63$, $P = 0.00000000012$.
   g. P-value for β_1 is 0.9536; do not reject H_0. P-value for β_2 is 0.0078; reject H_0. P-value for β_3 is 0.000000026; reject H_0.

23. $\hat{y} = 524.7948 + 0.0022154x_1 - 113.3986x_2 + 2.8074x_3 + 11.8733x_4$
   b. 759.55   **c.** (750.368, 768.751)   **d.** (741.385, 777.734)
   e. 90.0%   **f.** Yes, $F = 45.16$, $P = 0.00000000097$.
   g. P-value for β_1 is 0.2202; do not reject H_0. P-value for β_2 is 0.00099; reject H_0. P-value for β_3 is 0.0056; reject H_0. P-value for β_4 is 0.3489; do not reject H_0.

CHAPTER 13 Quiz

1. 18

2. True

3. False

4. False

5. (65.3, 68.2) must be the confidence interval, because it is narrower.

6. $b_0 = 13.0508$, $b_1 = 1.0574$

7. $0.2247 < \beta_1 < 1.8902$

8. Test statistic: 3.107, Critical values: -3.707, 3.707, P-value is between 0.02 and 0.05 [Tech: 0.0209]. Do not reject H_0.

9. $29.195 < $ Mean response $ < 39.205$

10. $19.327 < $ Individual response $ < 49.073$

11. $\hat{y} = -13.452 + 0.204x_1 + 2.056x_2 + 0.181x_3$

12. 37.09

13. 78.2%

14. Yes, $F = 9.59$, $P = 0.0050$.

15. P-value for β_1 is 0.6397; do not reject H_0. P-value for β_2 is 0.00083; reject H_0. P-value for β_3 is 0.5681; do not reject H_0.

CHAPTER 13 Review Exercises

1. a. $\hat{y} = 8.5593 - 0.1551x$
   b. $-0.217 < \beta_1 < -0.093$
   c. Test statistic: $t = -6.437$, Critical value: -2.015, P-value is between 0.0005 and 0.001 [Tech: 0.00067]. Reject H_0. We conclude that weight is useful in predicting mileage.

2. a. 5.4567
   b. $5.0064 < $ Mean response $ < 5.907$
   c. 5.4567
   d. $4.1857 < $ Individual response $ < 6.7278$

3. a. $\hat{y} = -1.0123 + 0.0519x$ **b.** $0.0389 < \beta_1 < 0.0650$
   c. Test statistic: $t = 8.8546$, Critical value: 2.764, P-value is less than 0.0005 [Tech: 0.0000024]. Reject H_0. We conclude that diameter is useful in predicting volume.

4. a. 1.2733
   b. $1.1224 < $ Mean response $ < 1.4241$
   c. 1.2733
   d. $0.8631 < $ Individual response $ < 1.6834$

5. a. $\hat{y} = 12.1933 - 0.8333x$ **b.** $-1.078 < \beta_1 < -0.589$
   c. Test statistic: $t = -7.8523$, Critical value: -1.860, P-value is less than 0.0005 [Tech: 0.000025]. Reject H_0. We conclude that concentration is useful in predicting drying time.

6. a. 7.9433
   b. $7.7945 < $ Mean response $ < 8.0922$
   c. 7.9433
   d. $7.4745 < $ Individual response $ < 8.4122$

7. a. $\hat{y} = 2.9073 + 0.8882x$ **b.** $0.677 < \beta_1 < 1.100$
   c. Test statistic: $t = 9.6778$, Critical value: 2.896, P-value is less than 0.0005 [Tech: 0.0000054]. Reject H_0. We conclude that income is useful in predicting energy consumption.

8. a. 47.319   **b.** $34.48 < $ Mean response $ < 60.158$
   c. 47.319   **d.** $10.954 < $ Individual response $ < 83.685$

9. a. $\beta_1 \neq 0$ **b.** 4.953149177   **c.** 8
   d. 0.0011164587   **e.** Yes, $P < 0.01$.

10. a. Slope is 5.5582; Intercept is 1.9167.
   b. Yes, $P = 0.006$.

11. a. $\hat{y} = -42.354 + 0.65394x_1 + 3.662x_2 + 0.57857x_3$
   b. 19.47   **c.** (13.44, 25.50)   **d.** (-4.55, 43.50)
   e. No, the particulate concentration cannot be negative.
   f. 53.8%   **g.** Yes, $F = 13.20$, $P = 0.0000071$.
   h. P-value for β_1 is 0.0246; reject H_0. P-value for β_2 is 0.00000074; reject H_0. P-value for β_3 is 0.0322; reject H_0.

12. a. $\hat{y} = -372.98 + 3.5368x_1 + 3.7345x_2 - 2.1661x_3$
   b. 701.76   **c.** (656.34, 747.18)   **d.** (602.97, 800.55)
   e. 72.9%   **f.** Yes, $F = 8.06$, $P = 0.0064$.
   g. P-value for β_1 is 0.0160; reject H_0. P-value for β_2 is 0.0082; reject H_0. P-value for β_3 is 0.0316; reject H_0.

13. a. 40.0%   **b.** 31.0%   **c.** Yes, $F = 4.44$, $P = 0.015$

14. a. 10.933   **b.** 6.805, 15.061   **c.** 0.6874, 21.178

15. ii

CHAPTER 13 Case Study

1. $\hat{y} = -26.373 + 9.2915x$

2.

The plot exhibits a clear pattern, which indicates that the assumptions of the linear model are violated.

3. $\hat{y} = -0.01502 + 0.110265x^3$

4.

The assumptions of the linear model appear to be satisfied.

5. 13.768

6. 12.023 < Mean response < 15.513

7. 3.363 < Individual response < 24.173

CHAPTER 14
Section 14.1

Exercises 1–6 are the Check Your Understanding exercises for this section. Answers to these exercises are on page 647.

7. equal

9. True

11. True

13. a. 4   **b.** 3 for $SSTr$, 30 for SSE

    c. $MSTr = 0.08333$, $MSE = 0.0703$ **d.** 1.1848

    e. Critical value is 2.92, P-value is greater than 0.10 [Tech: 0.3321]. Do not reject H_0. There is not enough evidence to conclude that two or more of the population means are different.

15. a. $SSTr = 3.1511$, $SSE = 2.2868$

    b. 2 for $SSTr$, 20 for SSE

    c. $MSTr = 1.5755$, $MSE = 0.11434$ **d.** 13.779

    e. Critical value is 5.85, P-value is less than 0.001 [Tech: 0.00017]. Reject H_0. We conclude that two or more of the population means are different.

17. a.

Source	DF	SS	MS	F	P
Duration	4	1.328	0.332000	8.5862	0.00082
Error	15	0.580	0.038667		
Total	19	1.908			

    b. Yes. $F = 8.5682$, $P = 0.00082$.

19. $\mu_1 \neq \mu_4$, $\mu_1 \neq \mu_{24}$, $\mu_2 \neq \mu_{24}$

21. a. The three standard deviations are $s_{16} = 1.2309$, $s_{28} = 0.7031$, and $s_{36} = 0.9585$. The largest is less than twice as large as the smallest.

    b.

Source	DF	SS	MS	F	P
Day	2	24.001	12.000	10.803	0.0053
Error	8	8.8867	1.1108		
Total	10	32.887			

    c. Yes. $F = 10.803$, $P = 0.0053$.

23. $\mu_{16} \neq \mu_{36}$

25. a. $n_1 = n_2 = n_3 = n_4 = 30$

    b. $s_1 = 24.85$, $s_2 = 35.26$, $s_3 = 33.53$, $s_4 = 23.09$

    c. 12712.7   **d.** 102027.8   **e.** 3 for $SSTr$, 116 for SSE

    f. $MSTr = 4237.57$, $MSE = 879.55$ **g.** $F = 4.818$

    h. Yes, $P = 0.003375$.

27. a. H_0: The mean volume is the same at all restaurants.

    b. 3   **c.** 4

    d. $SSTr = 0.7845$, $SSE = 0.278$, $MSTr = 0.39225$, $MSE = 0.030888889$

    e. 12.69874101   **f.** 0.0023972943

    g. Reject H_0. We conclude that the mean volume differs among restaurants.

29. a. $SSTr = 50946.6$, $SSE = 120550.9$

    b. $MSTr = 25473.3$, $MSE = 5023.0$

    c. 5.071

    d. No, $P = 0.015$, so $P > 0.01$. Do not reject H_0. There is not enough evidence to conclude that the mean energy absorbed differs among pillars.

31.

Pair of Means	a. Test Statistic	b. P-value
$\mu_A - \mu_B$	1.8582	0.0816
$\mu_A - \mu_C$	3.4001	0.0037
$\mu_A - \mu_D$	1.6210	0.1246
$\mu_B - \mu_C$	1.5419	0.1426
$\mu_B - \mu_D$	0.2372	0.8155
$\mu_C - \mu_D$	1.7791	0.0942

    c. A pair of means differs if the P-value is less than $0.05/12 = 0.004167$. The only pair of means that we can conclude to be different is $\mu_A \neq \mu_C$.

    d. In this example, the results of the Bonferroni correction are the same as the results of the Tukey–Kramer test.

Section 14.2

Exercises 1–4 are the Check Your Understanding exercises for this section. Answers to these exercises are on page 660.

5. interaction

7. False

9. **a.** No, the *P*-value for interactions is greater than 0.10.

   b. Yes. We conclude that there are differences in the mean responses among levels of the row factor; $P = 0.023$.

   c. Yes. There is not enough evidence to conclude that there are differences in the mean responses among levels of the column factor; $P = 0.34$.

11. No large interactions

13. Some large interactions

15. **a.** No, $P = 0.82946$.

   b. Yes. We conclude that there are differences in the mean exam scores between freshmen and sophomores; $P = 0.010673$.

   c. Yes. There is not enough evidence to conclude that there are differences in the mean exam scores among subjects; $P = 0.82946$.

17.

There are no strong interactions.

19. **a.** No, $P = 0.61534$.

   b. Yes. We conclude that there are differences in the mean amounts of foam produced between distilled water and tap water; $P = 0.0017558$.

   c. Yes. There is not enough evidence to conclude that the presence or absence of glycerol makes a difference in the mean amount of foam produced; $P = 0.12282$.

21.

There are no strong interactions.

23. **a.** Yes, $P = 0.000039$.

   b. No. Because we reject the hypothesis of no interactions, it is not appropriate to interpret the main effect of adhesive.

   c. No. Because we reject the hypothesis of no interactions, it is not appropriate to interpret the main effect of pressure.

25.

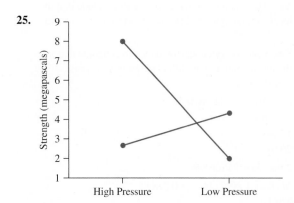

There is evidence of a strong interaction.

27. **a.** No, $P = 0.321$.

   b. Yes. We conclude that the mean yield differs among pesticides; $P = 0.000$.

   c. Yes. There is not enough evidence to conclude that the mean yield differs among varieties; $P = 0.103$.

29. **a.** These have the largest cell means.

   b. The main effect of design is
   $(51 + 49 + 43 + 41)/4 - (51 + 49 + 43 + 41)/4 = 0$.

   c. The main effect of material is
   $(51 + 49 + 43 + 41)/4 - (51 + 49 + 43 + 41)/4 = 0$.

   d. No. There is an interaction between design and material; $P = 0.00132$.

CHAPTER 14 Quiz

1. 3 for *SSTr*, 194 for *SSE*

2. $SSTr = 1.0518$, $SSE = 43.017$, $MSTr = 0.35059$, $MSE = 0.22174$

3. 1.5811

4. No, $P = 0.19527$.

5. No, $P = 0.74823$.

6. Yes. There is not enough evidence to conclude that the mean number of parts differs among operators; $P = 0.9376$.

7. Yes. We conclude that the mean number of parts differs among machines; $P = 0.0033212$.

8. Yes, $P = 0.001$.

9. No. Because we reject the hypothesis of no interactions, it is not appropriate to interpret the main effect of gender.

10. No. Because we reject the hypothesis of no interactions, it is not appropriate to interpret the main effect of diet.

CHAPTER 14 Review Exercises

1.

Source	DF	SS	MS	F	P
Gypsum	3	0.013092	0.0043639	0.28916	0.83215
Error	8	0.12073	0.015092		
Total	11	0.13383			

2. No, $P = 0.83215$.

3. a. No, $P = 0.72162$.

    b. Yes. There is not enough evidence to conclude that the mean conductivity differs with the amount of gypsum; $P = 0.43865$.

    c. Yes. There is not enough evidence to conclude that the mean conductivity differs between soil types; $P = 0.2057$.

4. a. $n_1 = 16, n_2 = 31, n_3 = 30$

    b. $s_1 = 0.6, s_2 = 0.4, s_3 = 0.6$

    c. 3.7995

    d. 20.64

5. a. 2 for $SSTr$, 74 for SSE

    b. $MSTr = 1.8997, MSE = 0.27892$

    c. 6.8111

    d. We conclude that there are differences among the mean pH levels, $P = 0.0019269$.

6. $\mu_1 \neq \mu_2, \mu_1 \neq \mu_3$

7. The largest standard deviation is more than twice as large as the smallest standard deviation. We cannot use ANOVA to determine whether conductivity varies with location.

8. a. No, $P = 0.6387$.

    b. Yes. We conclude that the mean yield differs between fertilizers; $P = 0.011843$.

    c. Yes. We conclude that the mean yield differs among soil types; $P = 0.024787$.

9.

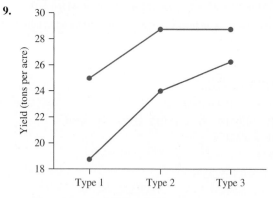

There are no strong interactions.

10. a. The standard deviations are $s_1 = 122.87, s_2 = 122.71, s_3 = 185.31, s_4 = 138.78$. The largest standard deviation is less than twice as large as the smallest standard deviation.

    b.

Source	DF	SS	MS	F	P
Gypsum	3	378610.4	126203.5	6.210	0.0059
Error	15	304838.1	20322.5		
Total	18	683448.5			

11. $\mu_A \neq \mu_B, \mu_A \neq \mu_C$

12. a. No, $P = 0.19474$.

    b. Yes. We conclude that the mean number of miles traveled differs among car types; $P = 0.00005992$.

    c. Yes. We conclude that the mean number of miles traveled differs among tires; $P = 0.00000001$.

13.

There is some evidence of interaction, but not enough to be statistically significant.

14. a. No, $P = 0.452$.

    b. Yes. We conclude that the mean number of satisfactory parts differs among materials; $P = 0.010$.

    c. Yes. We conclude that the mean number of satisfactory parts differs among machines; $P = 0.000$.

15.

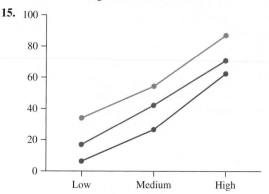

There are no strong interactions.

CHAPTER 14 Case Study

1.

Source	DF	SS	MS	F	P
Wafer	2	114661.4	57330.722	11340.14	0.0000
Operator	2	136.7778	68.388889	13.52747	0.0019
Interaction	4	6.555556	1.6388889	0.324176	0.8549
Error	9	45.50000	5.0555556		
Total	17	114850.3			

2. Yes, we conclude that there are differences in the measured weights among operators: $F = 13.52747, P = 0.00193977$.

3.

Source	DF	SS	MS	F	P
Wafer	2	262627.1	131313.6	24118.8	0.0000
Operator	2	24.77778	12.38889	2.27551	0.1586
Interaction	4	16.22222	4.055556	0.74490	0.5852
Error	9	49.00000	5.444444		
Total	17	262717.1			

4. The effect of operator is no longer statistically significant: $P = 0.15856864$. The balance should be powered up continuously so that there are no discernible differences in the measured weights among operators.

CHAPTER 15

Section 15.1

Exercises 1–2 are the Check Your Understanding exercises for this section. Answers to these exercises are on page 673.

3. less than

5. True

7. $n = 10 \leq 25, x = 4$

9. a. $n = 20 \leq 25, x = 4$ **b.** Critical value is 5. Reject H_0.
c. Critical value is 3. Do not reject H_0.

11. a. $H_0: m = 15, H_1: m < 15$ **b.** $n = 19 \leq 25, x = 2$
c. Critical value is 5. **d.** Reject H_0. We conclude that the median weight loss is less than 15 pounds.

13. $H_0: m = 60, H_1: m > 60; n = 16 \leq 25$; Test statistic: $x = 3$. Critical value is 4. Reject H_0. We conclude that the median height of oak trees in this forest is greater than 60 feet.

15. $H_0: m = 1000, H_1: m < 1000; n = 35 > 25$; Test statistic: $x = 3, z = -1.69$. Critical value is -1.645. Reject H_0. We conclude that the median rent is less than \$1000 per month.

17. $H_0: m = 480, H_1: m < 480; n = 20 \leq 25$; Test statistic: $x = 7$. Critical value is 5. Do not reject H_0. There is not enough evidence to conclude that the median fraud complaint rate is less than 480.

19. $H_0: m = 1, H_1: m < 1; n = 70 > 25$; Test statistic: $x = 25$, $z = -2.27$. Critical value is -1.645. We reject H_0 at the $\alpha = 0.05$ level. However, H_0 is true. The population median is 1, and the sample median is also 1. This shows that the sign test is unreliable for discrete populations.

Section 15.2

Exercises 1–2 are the Check Your Understanding exercises for this section. Answers to these exercises are on page 679.

3. average

5. False

7. $\mu_S = 255, \sigma_S = 27.659, z = -0.40, P = 0.3446$ [Tech: 0.3454]

9. $\mu_S = 348.5, \sigma_S = 36.550, z = -2.09, P = 0.0366$ [Tech: 0.0363]

11. a. $H_0: m_1 = m_2, H_1: m_1 < m_2$
b. $S = 71.5, z = -3.24$
c. $P = 0.0006$
d. Reject H_0. We conclude that the mileage is less for compact cars than for midsize cars.

13. $H_0: m_1 = m_2, H_1: m_1 \neq m_2; \mu_S = 232.5, \sigma_S = 24.109$. If Cured 2 days is taken to be sample 1, then $S = 194.5$ and $z = -1.58$. If Cured 6 days is taken to be sample 1, then $S = 270.5$ and $z = 1.58$. $P = 0.1142$ [Tech: 0.1150]. Do not reject H_0. There is not enough evidence to conclude that the strengths differ between the two curing times.

15. $H_0: m_1 = m_2, H_1: m_1 > m_2; \mu_S = 174, \sigma_S = 21.541$. $S = 141$, $z = -1.53, P = 0.0630$ [Tech: 0.0628]. Do not reject H_0. There is not enough evidence to conclude that the median load time

for 3-D Blu-ray players is greater than for standard Blu-ray players.

17. a. The sample median is 30 for both groups.
b. $S = 183.5$ (or 281.5), $\mu_S = 232.5, \sigma_S = 24.109, z = -2.03$ (or 2.03)
c. $P = 0.0424$ [Tech: 0.0421]. Reject H_0.

Section 15.3

Exercises 1–2 are the Check Your Understanding exercises for this section. Answers to these exercises are on page 684.

3. less than or equal to

5. True

7. Test statistic: $S = 1$, Critical value: 14. Reject H_0.

9. Test statistic: $S = 19$, Critical value: 11. Do not reject H_0.

11. Test statistic: $S = 8.5$, Critical value: 14. Reject H_0. We conclude that the median scores differ between pretest and posttest.

13. Test statistic: $S = 17.5$, Critical value: 2. Do not reject H_0. There is not enough evidence to conclude that the median traffic times differ.

15. Test statistic: $S = 25.5$, Critical value: 17. Do not reject H_0. There is not enough evidence to conclude that the median score before the seminar differ from the median score after.

17. a. The median is the middle value, which is 0.
b. $S = 78$
c. The critical value is 81. Reject H_0.

CHAPTER 15 Quiz

1. a. $n = 10 \leq 25, x = 4$ **b.** Do not reject H_0
c. Do not reject H_0

2. a. $n = 16 \leq 25, x = 0$ **b.** Reject H_0 **c.** Reject H_0

3. a. $n = 24 \leq 25, x = 5$ **b.** Reject H_0 **c.** Reject H_0

4. a. $n = 16 \leq 25, x = 15, z = -2.80$
b. Critical value is -1.645. Reject H_0.
c. Critical value is -2.33. Reject H_0.

5. $\mu_S = 510, \sigma_S = 50.498, z = -2.18, P = 0.0146$ [Tech: 0.0147]

6. $\mu_S = 725, \sigma_S = 62.183, z = 2.01, P = 0.0444$

7. $\mu_S = 255, \sigma_S = 27.659, z = 0.90, P = 0.1841$ [Tech: 0.1830]

8. $S = 37$, Critical value: 30. Do not reject H_0.

9. $S = 0$, Critical value: 21. Reject H_0.

10. $S = 0$, Critical value: 11. Reject H_0.

CHAPTER 15 Review Exercises

1. $S = 29.5$, Critical value: 25. Do not reject H_0. There is not enough evidence to conclude that the median weights differ.

2. $n = 20 \leq 25, x = 5$, Critical value: 3. Do not reject H_0. There is not enough evidence to conclude that the median price differs from \$300.

3. $S = 23$, Critical Value: 7. Do not reject H_0. There is not enough evidence to conclude that the median number of customers differs between Monday and Tuesday.

4. $S = 210.5$, $z = 2.07$, $P = 0.0192$ [Tech: 0.0190]. Reject H_0. We conclude that the median systolic blood pressure is less for people who exercise than for people who are sedentary.

5. $n = 17 \leq 25$, $x = 7$, Critical value: 5. Do not reject H_0. There is not enough evidence to conclude the median is different from 8.

6. $n = 11 \leq 25$, $x = 3$, Critical value: 1. Do not reject H_0. There is not enough evidence to conclude that the median is less than 5 days.

7. $S = 202$, $\mu_S = 217$, $\sigma_S = 24.055$, $z = -0.62$, $P = 0.5352$ [Tech: 0.5329]. Do not reject H_0. There is not enough evidence to conclude that the median weight differs between boys and girls.

8. $S = 157$, $z = 2.77$, $P = 0.0056$. Reject H_0. We conclude that the median times differ between the two routes.

9. $S = 24.5$, Critical value: 14. Do not reject H_0. There is not enough evidence to conclude the measurements differ between gauges.

10. $n = 13 \leq 25$, $x = 5$, Critical value: 1. Do not reject H_0. There is not enough evidence to conclude that the median differs from 12 ounces.

CHAPTER 15 Case Study

1. $S = 53.5$, $\mu_S = 99$, $\sigma_S = 14.071$, $z = -3.23$, $P = 0.0006$. Reject H_0. We conclude that the median number of microorganisms is greater for a retention time of 12 days than for 3 days.

2. $S = 66.5$, $\mu_S = 126$, $\sigma_S = 19.442$, $z = -3.06$, $P = 0.0011$. Reject H_0. We conclude that the median number of microorganisms is greater for a retention time of 30 days than for 3 days.

3. $S = 206$, $\mu_S = 186$, $\sigma_S = 23.622$, $z = 0.85$, $P = 0.3954$ [Tech 0.3972]. Do not reject H_0. There is not enough evidence to conclude that the median number of microorganisms differs between a retention time of 30 days and a retention time of 12 days.

4. iii.

Credits

PHOTOGRAPHS

Chapter 1

Opener: US Fish and Wildlife Services; p. 5: © Creatas/Jupiter RF; p. 7: © Getty RF; p. 11: © Photoalto/Picturequest RF; p.12: © Comstock/Jupiter RF; p. 19: © Corbis RF; p. 25: © Corbis RF. Line Art p. 7 Figure 1.1 © Sarah Navidi; 1.2 © Sarah Navidi; 1.3 © Sarah Navidi; 1.4 © Sarah Navidi.

Chapter 2

Opener: © Getty RF; p. 46, p. 77: © Corbis RF; p. 79: © David R. Frazier Photolibrary RF.

Chapter 3

Opener: © Getty RF; p. 97: © Corbis RF; p. 99, p. 120, p. 123: © Getty RF; p. 132: © Corbis RF.

Chapter 4

Opener: © Stockbyte/Punchstock RF; p. 158: © Getty RF; p. 163: © Creatas/Punchstock RF; p. 174: © Getty RF; p. 180: © Digital Vision/Punchstock RF; p. 188: © Getty RF.

Chapter 5

Opener: © Getty RF; p. 217: © Veer RF; p. 230, p. 232: © Getty RF.

Chapter 6

Opener: © Getty RF; p. 255: © Ingram Publishing/age fotostock RF; p. 258: © Getty RF; p. 262: © Alamy RF; p. 269: © Getty RF; p. 277: © Comstock/Jupiter RF.

Chapter 7

Opener: © Getty RF; p. 309: © Getty RF; P. 310: © Mark Steinmetz RF; p. 317: © Getty RF; p. 330: © Pixtal /age fotostock RF.

Chapter 8

Opener: © Corbis RF; p. 363: © Blend Images RF; p. 370: © Getty RF; p. 375: © Alamy RF; p. 380: © Comstock/Getty RF; p. 387: © McGraw-Hill Education.

Chapter 9

Opener: © Corbis RF; p. 409: © Corbis/Picture Quest RF; p. 430: © Getty RF; p. 444: © Comstock RF; p. 456: © Stockbyte/Getty RF; p. 467: © Ingram Publishing/SuperStock RF.

Chapter 10

Opener: © Comstock/JupiterImages RF; p. 494: © Getty RF.

Chapter 11

Opener: © Corbis RF; p. 512: © Alamy RF; p. 530: Photo by Jocelyn Augustino/FEMA; p. 534: © Comstock/PictureQuest RF; p. 550: © Banana Stock/Jupiter RF.

Chapter 12

Opener: © Getty RF; p. 560: © Brand X/Punchstock RF; p. 567: © Banana Stock/Jupiter RF.

Chapter 13

Opener: USDA; p. 597: © Getty RF.

Chapter 14

Opener: © Banana Stock/Punchstock RF; p. 646: USGS/photo by Don Becker. p. 648: © Getty RF.

Chapter 15

Opener: © Brand X RF.

INDEX